D1174780

Geological and Geophysical Investigations of Continental Margins

AAPG Memoir 29

Geological and Geophysical Investigations of Continental Margins

Edited by
Joel S. Watkins, Lucien Montadert, and Patricia Wood Dickerson

Published by
The American Association of Petroleum Geologists
Tulsa, Oklahoma, U.S.A., 1979

Published April 1979

Library of Congress Catalog Card No. 79—50442
ISBN: 0-89181-305-5

Those responsible:

John W. Shelton, AAPG Editor
Ronald Hart, Project Coordinator
Jack Kleinecke, Production
Nancy G. Wise, Production

Printed by
Edwards Brothers, Inc.
Ann Arbor, Michigan

Foreword

Knowledge of continental margins has advanced rapidly during the 1970s. Multichannel seismic reflection whose cost formerly restricted its use largely to the immediate vicinity of shallow-water prospects has become more common in deeper waters. The use of the technique by government and academic groups has helped solve basic structural and evolutionary problems of rocks of the deeper offshore. Better sources and more sophisticated processing yield better and deeper resolution of the data.

The Deep Sea Drilling Project holes, while for safety reasons generally restricted to lower slopes, rises, and abyssal plains, provided valuable insight to lithologies, depositional environments, and ages of the sediments. Interpolation and extrapolation of DSDP data using seismic reflection have made it possible to project information from DSDP holes into undrilled areas of the slopes and rises. Similarly, data from petroleum exploration holes on shelves have been projected into deeper water, and in places, the geological-knowledge gap between shelf and deeper sea rocks has been closed with some confidence.

Economic concerns rising from diminishing hydrocarbon reserves have given continental-margin research a strong push. The realization that continental slopes, rises, and plateaus represent the largest remaining frontier for petroleum exploration has resulted in a greater flow of funding, manpower, and overall effort into studies of continental borderlands. Studies such as (for example) the detailed aeromagnetic surveys of the U.S. East Coast, might yet be 10 or more years in the future without a pervasive feeling of economic concern.

Concurrent with the better, more readily available instrumentation and increased effort has been the development of new theories. Problems of attenuation and subsidence of continental crust along rifted margins, the evolution of the subduction complexes and forearc basins located between deep-sea trenches and volcanic chains, and the origin and emplacement of hydrocarbon deposits (though not solved) appear to be yielding to well-designed and informed investigations. It is clear that the level of investigation has risen from pre-paradigm reconnaissance in the early to mid-sixties to well-focused paradigm testing in the mid-seventies.

To better disseminate new knowledge of continental margins, the American Association of Petroleum Geologists Research Activities Subcommittee initiated three meetings in 1977, to review the current status of our knowledge. The first was a Research Conference convened at Galveston, Texas, in January by Joel Watkins and Lucien Montadert on "Geophysical investigations of continental slopes and rises;" the second and third, respectively, were held at the Annual Meeting of AAPG. These were a short course on the "Geology of Continental Margins" organized by Edward McFarlan, Jr., and Charles L. Drake, and a Research Symposium on "Petroleum Potential of Slopes, Rises, and Plateaus" chaired by McFarlan, Drake, and C. A. Burk. Because of the overlapping and complementary nature of papers presented at the three sessions, the papers were combined into a single volume.

The editors gratefully acknowledge the assistance and contributions of Ted McFarlan, Chuck Drake, and Creighton Burk, in the solicitation of papers, organization of sessions, and firm support throughout preparation of this volume. Georges Pardo, chairman of the AAPG Research Committee, Dick Sheldon, past chairman, and members of the committee had the foresight to initiate the sessions, guided the integration of the papers from the various sessions, and otherwise provided firm support.

The Geophysics Laboratory, University of Texas Marine Science Institute at Galveston, provided substantial administrative and logistical support for the January conference. Leone Barnes served unfailingly as a contact point for the authors and editors throughout, and handled many of the administrative arrangements for the January conference.

Editor John W. Shelton, Science Director, Gary Howell, and the AAPG editorial staff coordinated manuscript review and publication.

On behalf of authors and editors, we thank the captains and crews, the core describers and seismic observers, the draftsmen and secretaries, and all the rest on whose dedicated efforts this work relies so heavily.

Joel Watkins
Lucien Montadert
Pat Dickerson

Table of Contents

Rifted Margins

Subsidence Mechanisms at Passive Continental Margins[1]

M. H. P. BOTT[2]

Abstract Some mechanisms in favor of later regional and earlier graben type subsidence at passive continental margins are reviewed. Gravity loading can explain thick sediment piles beneath some deltas but fails to account for most shelf subsidence except as a contributory factor. Heating of the lithosphere resulting in crustal thinning by uplift and erosion and/or increased lower crustal density by metamorphism or intrusion can account for shelf subsidence of up to about 3 to 5 km after sediment loading. Seaward creep of continental crustal material may cause substantial subsidence owing to thinning of the crust beneath the shelf. Thinning of the crust by "necking" prior to continental splitting appears only viable in a narrow graben setting where igneous intrusion can extend the upper crust. Graben subsidence prior to splitting can best be explained by the wedge subsidence mechanism applied to the upper 10 km of the crust rather than to the whole crust. Most of these mechanisms probably contribute to observed subsidence although their relative importance is not yet clear.

INTRODUCTION

Passive continental margins undergo a history of tectonic development of predominantly vertical type despite their location within plates. The most obvious aspect is the strong subsidence which affects particularly the shelf and slope at rifted margins. Less obviously, there appears to be a progressive widening of the crustal[3] transition with maintenance of approximate isostatic equilibrium. This paper reviews some of the mechanisms which have been suggested recently to explain subsidence at rifted margins.

Four main stages can be recognized in the tectonic development of a typical passive margin of rifted type. (1) The rift valley stage which may not be ubiquitous, involves early graben formation prior to continental splitting and is possibly exemplified by the present East African rift system. This stage may be associated with domal uplift caused by hot underlying upper mantle material, but it is not clear whether such doming is typical of passive margins or is mainly restricted to hot spot regions. (2) The youthful stage, lasting about 50 m.y. after the onset of spreading while thermal effects of the split are dominant, is exemplified by the Red Sea margins. This stage is characterized by rapid regional subsidence of the outer shelf and slope, but some graben subsidence may locally persist. (3) The mature stage, during which more subdued regional subsidence may continue after the initial thermal event ceases to be an important influence, represents the present stage of development of most of the Atlantic and Indian Ocean margins. (4) The fracture stage, when subduction starts, terminates the history of a passive margin.

Although passive margins differ greatly in their style of tectonic development, two contrasting types of subsidence can be recognized (Fig. 1): (1) graben-type subsidence occurs during the rift valley stage and may possibly persist later. It is attributable to normal faulting in the basement in response to crustal stretching, possibly in association with doming; (2) subsidence by regional downwarping of the margin without obvious fault control is characteristic of the youthful and mature stages, with most rapid subsidence characteristically occurring at the start of spreading. Theories to explain the mechanism of subsidence fall into three main groups depending upon whether gravity loading, temperature, or extensional stress is the primary cause. There is much controversy concerning the primary mechanism of regional downwarping, with all three types of mechanisms probably contributing. A stress mechanism associated with a thermal event causing doming can best explain the graben formation.

Distinguishing among the several hypotheses of

[1]Manuscript received, April 5, 1977; accepted, September 1, 1977.

[2]Department of Geological Sciences, University of Durham, Durham, DH1 3LE, England.

[3]Author's note: Throughout the paper, *crust* refers to the region above the Mohorovičić discontinuity and *lithosphere* refers to the relatively strong upper layer of the Earth comprising the crust and the top part of the upper mantle.

Article Identification Number:
0065-731X/78/MO29-0001/$03.00/0.
(see copyright notice, front of book)

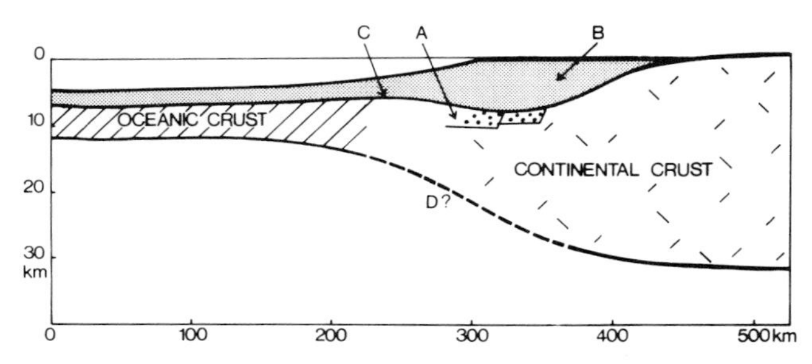

FIG. 1. Section across a passive margin showing some of the characteristic features of sediment and crustal structure: **A**, pre-split graben sediments; **B**, post-split sediments associated with flexural subsidence; **C**, the problematic position of the continent-ocean crustal contact beneath the sediments; **D**, the apparent gradational contact between deep continental and oceanic crust.

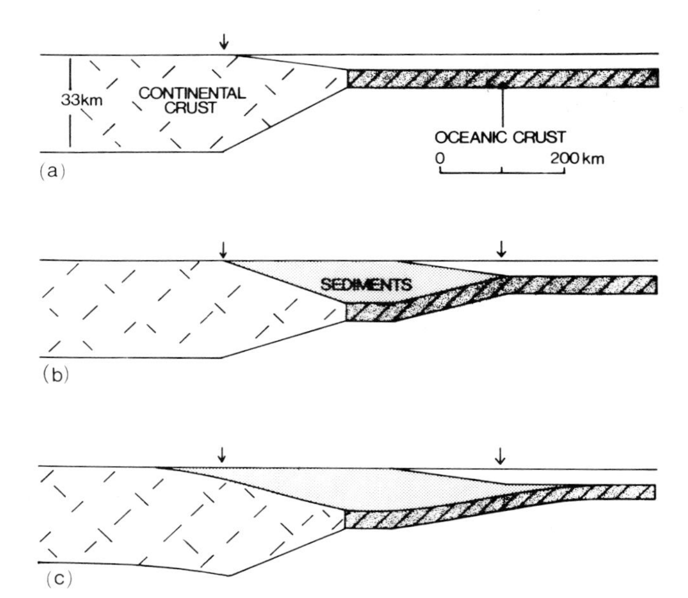

FIG. 2. The gravity loading hypothesis: (a) the initial situation prior to loading, assuming a pre-existing 200-km-wide transition between oceanic and continental crust following Walcott's (1972) model; (b) the result of local Airy sediment loading, assuming density of the sediments of 2450 kg/m^3 and of the upper mantle of 3,300 kg/m^3; (c) the result of flexural loading, assuming that the lithosphere has a flexural rigidity of 2×10^{22} Newton meters and that densities are as in (b), adapted from Walcott (1972) with change of sediment density.

subsidence is mainly a task for the future. One aspect is the determination of the detailed history of subsidence as revealed by the stratigraphy and structure of the sediment pile. The other main aspect is location of the continent-ocean crustal contact and investigation of the nature of the crustal and upper mantle transition.

GRAVITY LOADING HYPOTHESIS

The gravity loading hypothesis (Dietz, 1963; Walcott, 1972) attributes subsidence to sediment load. In its simplest form, it is based on local Airy isostasy. Suppose that the initial water depth is d and that densities of water, sediment, and upper mantle are p_w, p_s, and p_m respectively. If the sea is filled by sediments, then the total thickness of marine sediments which may form with local isostatic adjustment is given by $t = d \cdot (p_m - p_w)/(p_m - p_s)$. Putting $p_w = 1{,}030\ kg/m^3$ and $p_m = 3{,}300\ kg/m^3$, we see that a sediment thickness of about twice the initial depth can develop for sediment with mean density of 2,150 kg/m^3, and of nearly three times the initial depth for mean density of 2,550 kg/m^3. This shows that thick shelf successions with initial water depths of less than 200 m cannot form by this mechanism, but that substantial subsidence of the slope and rise may occur where thick piles of sediments are deposited, such as at deltas. Figure 2b shows that a total thickness of sediment of about 14 km can form near the base of the initial slope.

A more sophisticated approach is to treat the lithosphere as a thin elastic plate and investigate its flexure in response to the sediment loading by elastic beam theory. Walcott (1972) used this technique to show the growth of a sedimentary lens at a continental margin, with particular relevance to deltas such as that of the Niger. Figure 2c shows a modification of Walcott's result using a sediment density of 2,450 kg/m^3 (Walcott himself used the rather high density of 2,800 kg/m^3). Over most of the original slope and adjacent oceanic crust, the resulting pile is closely similar to that predicted for classical isostasy (Fig. 2b), but the main distinction is that the downwarping extends about 150 km beyond the local sediment load. Thus some significant subsidence of the shelf can be produced.

The gravity loading hypothesis thus gives a viable explanation of thick delta wedges, with progressive seaward migration of slope and shelf over the continent-ocean crustal contact. However, Watts and Ryan (1976) show that neither local nor flexural loading can explain the substantial thicknesses of shallow water sediments typically observed on the shelves of passive margins. A driving force other than sediment loading is apparent. Here the importance of the sediment loading effect is to increase the amount of subsidence caused by other mechanisms by a factor of between about two and three, depending upon the mean sediment density. Thus sediment loading appears to be a contributory factor in most subsidence, although it appears to be the primary cause only where great sediment volumes are deposited in initially deep water.

Another variation of the gravity loading hypothesis, suggested by Collette (1968) for North Sea subsidence, is that the sediment load causes a phase transition, such as basalt to eclogite, to occur within the crust. This produces a density increase in the lower crust, or adds material from crust to mantle, thus leading to further subsidence. This hypothesis is difficult to test but should be considered as a possibility.

THERMAL HYPOTHESES

The basic thermal hypothesis of Hsu (1965) was developed theoretically in an elegant way by Sleep

(1971, 1973). This hypothesis (Fig. 3) assumes that the continental lithosphere near the embryo margin is heated at the time of continental splitting. This causes reduction of the density of the lithosphere by thermal expansion and phase transitions with consequent isostatic uplift. After the initial split, as the ocean widens the lithosphere will cool and recover towards its initial elevation with a time constant of about 50 m.y. If, however, the crust has been thinned by surficial erosion at the time of uplift or by some other process, then the recovery on cooling will involve isostatic subsidence of the shelf which may be amplified by sediment load. This hypothesis would be expected to give rise to a smooth exponential decay type of subsidence reflected in sediment thicknesses, with time constant of about 50 m.y., but Sleep (1976) showed that apparently jerky subsidence can be produced by superimposed eustatic sea level changes.

There are some factors which suggest that Sleep's hypothesis based on surficial erosion alone cannot be the sole explanation of regional subsidence at passive margins. The amount of crustal thinning (h) caused by erosion depends on the initial uplift and the time constants of cooling and erosion. Taking a maximum likely initial uplift of 2 km and an erosional time constant of 50 m.y., then h is about 4 km. The maximum possible sediment thickness to sea level which this will allow is given by $h \cdot (p_m - p_c)/(p_m - p_s)$ which is always less than h and typically about $\frac{1}{2}h$. Thus sediment thicknesses of up to about 2 km on the outer shelf can be explained, but unacceptably large amounts of supracrustal erosion would be needed to account for sediment thicknesses of 5 to 10 km which are commonly observed. Another difficulty is to explain local regions of differential subsidence such as the Hatton Basin on the Rockall microcontinent (Matthews and Smith, 1971). These difficulties are avoided if some of the crustal thinning occurs through seaward creep of continental crustal material (Bott, 1971) as described below.

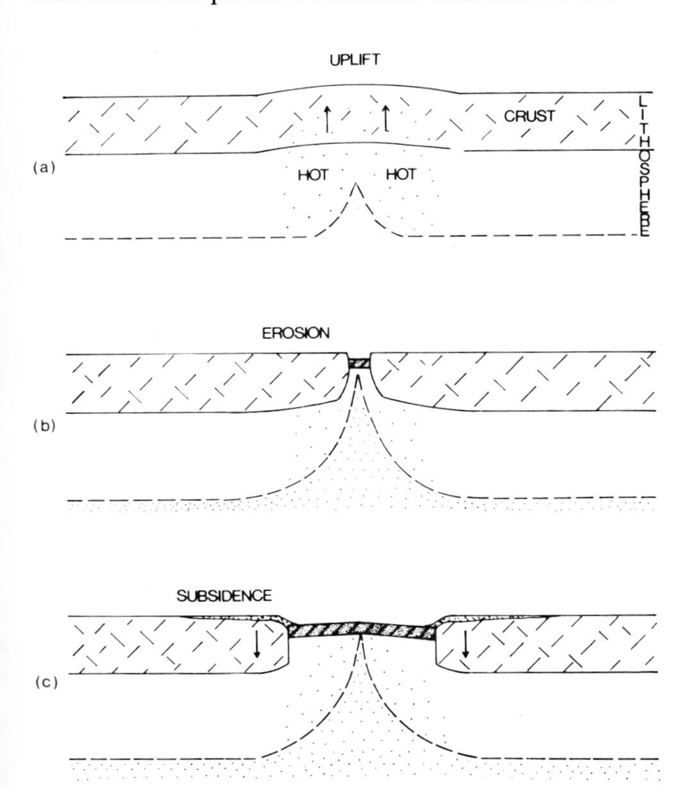

FIG. 3. The basic thermal hypothesis of Sleep (1971): (a) uplift following heating of the lithosphere; (b) initiation of new ocean accompanied by erosion of continental uplifted region, causing crustal thinning; (c) subsidence of continental margins as underlying lithosphere cools. Note the nearly vertical edges of continental crust predicted by this model and that of Figure 4.

A modification of the simple thermal hypothesis is that the thermal event causes increase in the density of the lower continental crust by metamorphism (Falvey, 1974; Haxby, et al, 1976) or by igneous intrusion (Beloussov, 1960; Sheridan, 1969). According to Falvey (Fig. 4), rise in lithospheric temperature at the time of splitting causes greenschist facies rocks in the lower continental crust to be metamorphosed to amphibolite facies, with an increase in density of between 150 and 200 kg/m^3. This causes a slight thinning of the crust which to some extent counters the thermal uplift. As the continental lithosphere cools after spreading starts, subsidence below the initial level occurs because of the more dense and slightly thinner crust. Taking the crustal thinning to be h, then the maximum sediment thickness to sea level which can result is given by $h \cdot p_m/(p_m - p_s)$. Let us take an extreme example of an increase in density of 200 kg/m^3 affecting 15 km of lower crust, given h of about 1 km. The maximum sediment thickness is then between about 3 and 4 km. Thus, this version of the thermal hypothesis also fails to account for the great sediment thicknesses which occur on some margins. A similar difficulty faces the other versions dependent on metamorphism or intrusion.

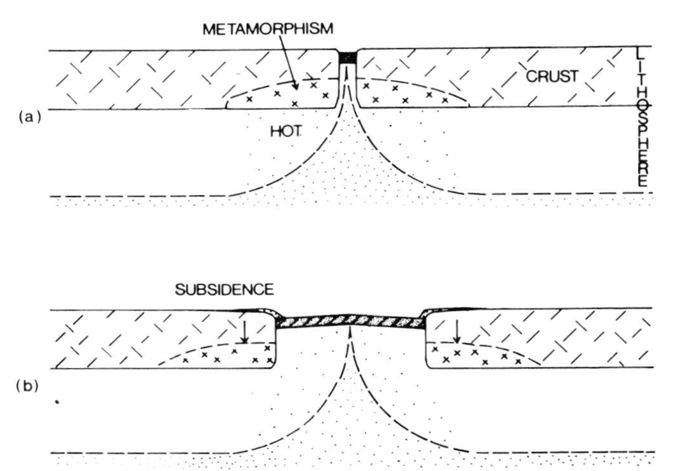

FIG. 4. The thermal hypothesis of Falvey (1974), omitting the rift stage: (a) heating of the continental lithosphere prior to and during split causes metamorphic transition of lower crust from greenschist to amphibolite facies, raising the mean crustal density; (b) subsidence of continental margins as underlying lithosphere cools.

Thermal hypotheses thus provide an explanation of marginal subsidence producing up to about 4 km of sediments, but fail to explain some observed sediment thicknesses of 5 to 15 km without crustal thinning by processes other than surface erosion. They predict a possible gap of up to 50 m.y. between onset of spreading and first marine sedimentation, which should be observable. Evidence for thermal uplift also needs to be sought. Exponential lessening of subsidence rate with time is predicted, but this may be indistinguishable from the pattern of subsidence associated with the crustal creep hypothesis. Thermal hypotheses leave the continental crustal boundary (Figs. 3 and 4) unmodified except for slight thinning of crust, but the versions depending on metamorphism or intrusion should yield a distinctive and detectable lower crustal structure adjacent to the margin.

CRUSTAL THINNING HYPOTHESES

Regional subsidence of more than about 4 km at margins requires a more radical mechanism of crustal thinning than the thermal hypotheses can provide. Such mechanisms are possible provided that crustal material can undergo significant flow on a sufficiently short time scale.

There is now evidence that the continental crust and the lithosphere can be subdivided into a relatively strong upper elastic layer about 10 to 20 km thick which yields by brittle fracture, overlying a much weaker lower layer which deforms by ductile flow (Artemjev and Artyushkov, 1971; Bott, 1971; Fuchs, 1974). This concept is supported by experimental and theoretical work starting with Griggs et al (1960) and by lithospheric flexural studies (Walcott, 1970). The boundary between brittle and ductile layers is likely to be gradational and its depth will depend upon the local geothermal gradient. It should be emphasized that the ductile layer is likely to be much stiffer than the asthenosphere and it may possess finite strength.

Crustal Thinning by Creep at Margins

A stress-based hypothesis which may account for major subsidence at passive margins appeals to thinning of the continental crust near the margin by progressive creep of middle and lower crustal material towards the suboceanic upper mantle (Bott, 1971, 1973). Such flow will cause the continent-ocean crustal transition to become progressively more gradational, and will release gravitational energy. The hypothesis depends upon the ability of the lower and middle continental crust to flow significantly by steady-state creep while the overlying elastic layer subsides by elastic flexure or by normal faulting.

Bott and Dean (1972) showed by finite element analysis that a passive margin is associated with a differential stress system as a result of unequal topographic loading across the margin and associated upthrust of the low density continental crust in isostatic equilibrium with the oceanic region. This can be understood as follows. Down to about 5 km depth, seawater of density 1,030 kg/m^3 on the oceanic side lies opposite rocks of density greater than 2,000 kg/m^3 on the continental side, so that the continental crust is more heavily loaded than the oceanic crust. On the other hand, the root of thick, low-density continental crust causes an upthrust which is not present on the oceanic side. If the margin is in isostatic equilibrium, loading and upthrust will be equal. This will effectively squeeze the continental crust, causing a horizontal deviatoric tension in the continental crust of up to 20 MPa (200 bar; 1 MPa = 10 bar), or a stress difference of 40 MPa. The tension peaks in the middle of the crust, decreasing to zero at the surface and at the Moho. Kusznir and Bott (1977) have recently shown that if the lower crust is treated as visco-elastic, then stress differences in the upper elastic layer will be increased. The main point is that a stress system does exist in the continental crust adjacent to a passive margin due to the density distribution, and that this will tend to drive lower continental crustal material towards the sub-oceanic upper mantle.

According to this hypothesis (Fig. 5), the crustal thinning will cause isostatic subsidence of the shelf and possibly slope, which will be accentuated by the sediment load. At the same time there will be a progressive broadening of the transition between oceanic and continental crust, in contrast to the predictions of the thermal hypotheses. The addition of low-density continental crustal material at the base of the crust or into the upper mantle beneath lower slope and rise should cause some complementary

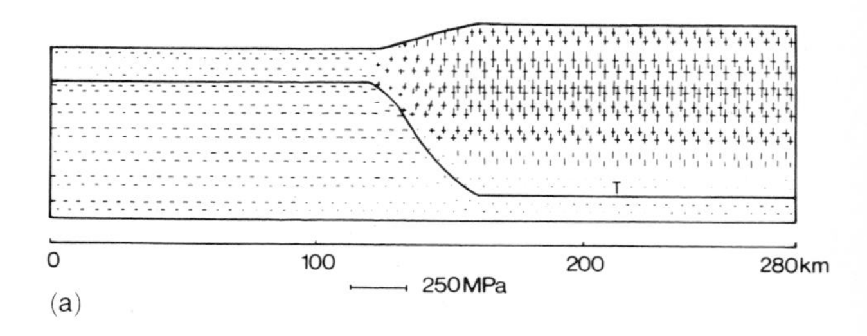

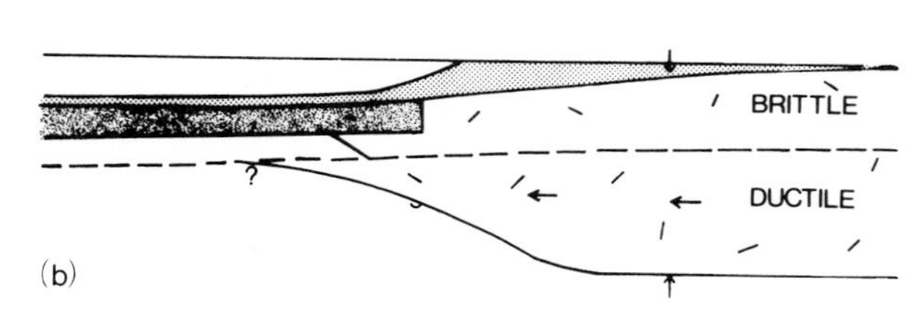

FIG. 5. The crustal flow hypothesis of Bott (1971): (a) magnitude and direction of the principal compressive stresses arising from differential gravitational body forces across a passive margin, the intermediate principal stress being everywhere perpendicular to the model (after Bott and Dean, 1972). (b) progressive gradation of the transition between continental and oceanic crust caused by seaward flow of the lower and middle continental crust in response to the stress system. This causes thinning of the crust beneath the continent with consequent isostatic subsidence, and thickening of crustal material beneath the rise with complementary uplift.

uplift there. Rona (1974) presented evidence which might be interpreted this way.

The rate of subsidence is likely to be controlled by both temperature and superimposed stresses. Solid-state creep is a thermally activated process, so that the most rapid subsidence would be expected to occur at and shortly after the initiation of spreading when lithospheric temperatures are high. Thus an approximately exponential decay of subsidence with time might be expected for this hypothesis, making distinction from the thermal hypotheses difficult on these grounds. The other controlling factor is likely to be the external stress system associated with the plate driving mechanisms or other sources such as membrane stresses (Turcotte, 1974). This will be superimposed on the nearly steady stress system associated with the unequal loading effect. Rapid subsidence would be encouraged by a superimposed tension but inhibited by a superimposed compression. Thus, subsidence according to this mechanism is likely to be controlled partly by thermal means and partly by stress. The creep hypothesis would thus be expected to give rise to more jerky subsidence than would the thermal hypotheses.

Can Necking of the Crust Occur?

It has been suggested that extreme thinning of the continental crust within a rift valley setting can occur by plastic necking (Artemjev and Artyushkov, 1971; Kinsman, 1975). If the rift zone subsequently splits to form complementary passive margins, then the region of thinned crust would give rise to a zone of crustal transition beneath the slope (Fig. 6).

Because of the brittle nature of the upper crust, the writer is skeptical as to whether this process can cause such great thinning of the crust unless there is accompanying igneous activity. An extreme criss-cross pattern of normal faulting would be needed to thin the upper crust comparably. The brittle upper crust could more easily be disrupted and extended by extensive magma invasion. If sediments were deposited during necking, evidence of extension of increasing intensity with depth within the sediment pile should be detectable (Bott, 1972).

A typical rift zone is about 50 km wide. Thus, a transition zone about 25 to 30 km wide at the continent-ocean crustal contact might be caused by this process.

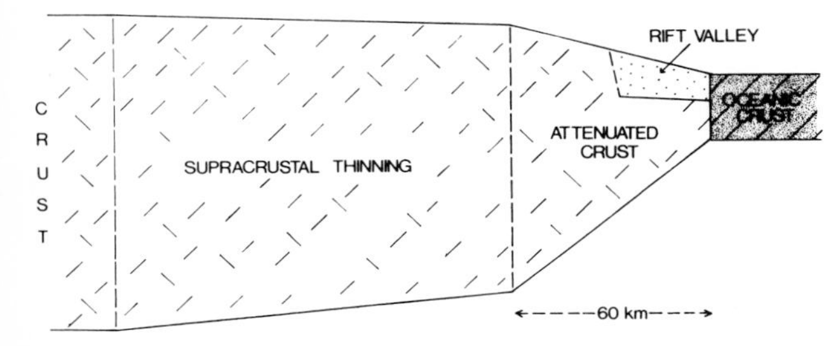

FIG. 6. "Necking" of the continental crust according to Kinsman (1975) with attenuated margin produced by stretching during 40-my-long rift phase prior to spreading.

NORMAL-FAULT BASED MECHANISMS

Graben Formation

Modern concepts of graben formation stem from the hypothesis that a downward-narrowing wedge of continental crust about 65 km wide can form by normal faulting in response to crustal tension and that this wedge can subside isostatically to form a rift valley between flanking uplifts (Vening Meinesz, 1950). This hypothesis fails in practice because the predicted crustal root is not observed and because grabens tend to be narrower than 65 km. A viable mechanism of graben formation has recently been suggested in which the Vening Meinesz wedge subsidence concept is applied to the brittle upper 10 to 20 km of the crust rather than to the crust as a whole (Bott, 1971, 1976; Fuchs, 1974). This mechanism depends critically upon subdivision of the continental crust into brittle and ductile layers.

The current graben formation hypothesis is such that (Fig. 7) a horizontal deviatoric tension is applied to the lithosphere and the extensional stress is enhanced in the brittle layer as a result of viscoelastic response in the ductile layer beneath (Kusznir and Bott, 1977). The brittle layer yields by normal faulting and a second converging normal fault may then develop where curvature is greatest or along a pre-existing line of weakness. This produces a downward-narrowing wedge of brittle upper crust which can subside provided that the incremental loss of gravitational plus elastic strain energy is sufficient to overcome friction and other dissipation of energy. Bott (1976) showed that this mechanism can cause substantial subsidence provided that water pressure reduces friction on the faults. As the wedge sub-

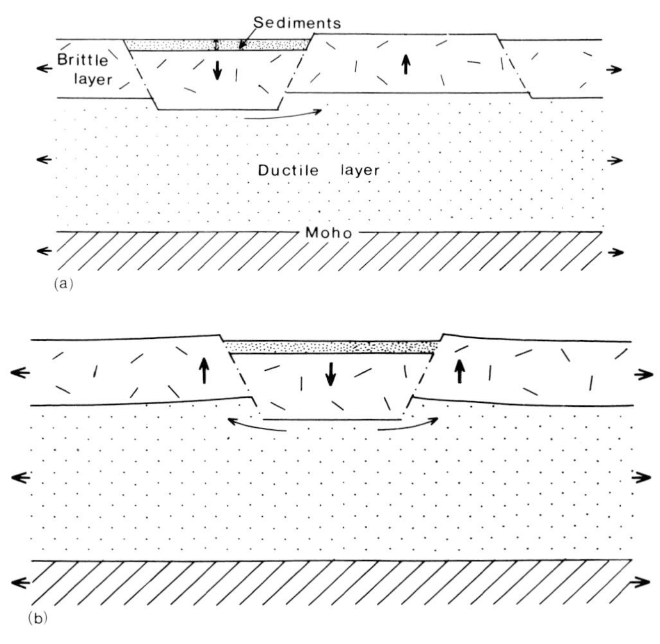

FIG. 7. The mechanism of graben formation by wedge subsidence affecting the upper continental crust with outflow in the lower crust: (a) subsidence compensated by horst uplift; (b) subsidence compensated by elastic upbending. After Bott (1976).

sides, outflow of material in the underlying ductile part of the crust accompanies uplift of the adjacent regions either by horst formation (Fig. 7a) or by elastic upbending (Fig. 7b). Complementary overall thinning of the ductile part of the crust must accompany subsidence so that the Moho may rise slightly beneath the graben vicinity, although this would not be by more than one or two kilometers in general. Within the setting of major rift systems, the above process may be complicated by associated igneous activity.

Grabens of between 25 and 50 km width are predicted for realistic crustal parameters, with wider or narrower troughs being possible if there are basement weaknesses. A subsidence of about 5 km can occur for an initially 20-km-wide trough under a maintained deviatoric tension of 500 bar, provided that sediments load the trough to its initial depth. In general, greater subsidence can occur for narrower troughs and for larger applied deviatoric crustal tension.

This hypothesis can readily be applied to early graben formation along passive continental margins, either prior to continental splitting or continuing after it. The deviatoric tension causing graben formation may possibly arise from a domal uplift such as that of East Africa according to a mechanism suggested by Kusznir and Bott (1977).

Faulting Near the Continent-Ocean Crustal Contact

Another mechanism which may cause limited subsidence of the continental crust adjacent to a passive margin is normal faulting accompanying downdrag of the cooling oceanic lithosphere. After initial formation at an ocean ridge, the oceanic lithosphere subsides on cooling with a time constant of about 50 m.y. (Sleep, 1969). Subsidence may be accentuated by sediment loading. At a typical passive margin, the continental side probably subsides at about the same rate as the oceanic side, obviating the need for a faulted contact. However, if the continental shelf is relatively unaffected by subsidence, as along parts of the margins of Rockall Trough, then one would expect the oceanic side to be progressively downfaulted relative to the continental side.

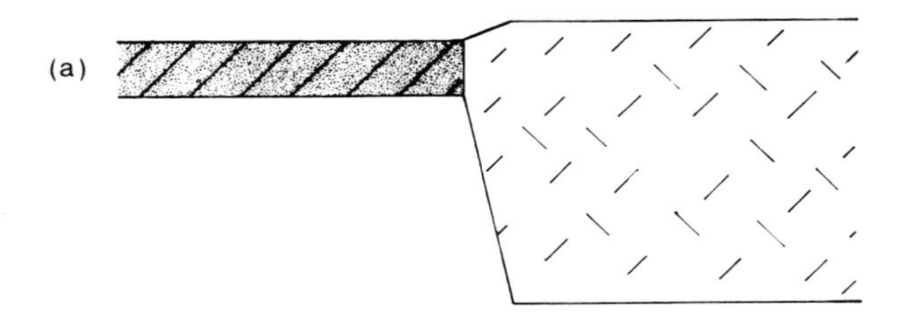

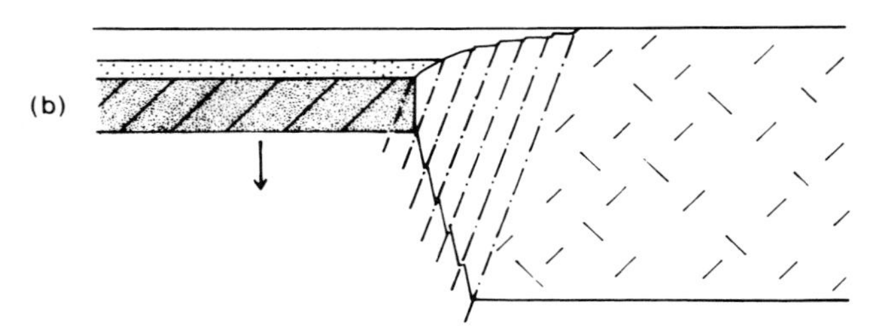

FIG. 8. Downdrag of the continental slope by normal faulting associated with the cooling and subsiding oceanic lithosphere. (a) Initial situation; (b) After subsidence of the oceanic lithosphere, the differential vertical movement being taken up by normal faulting beneath the slope.

A model showing the structure such faulting might produce is shown in Figure 8. The normal faults are shown penetrating the whole crust, but more realistically they may only penetrate to about 10 km depth with adjustment occurring by flow at greater depths. By taking up the relative vertical movement on a series of spaced faults, a relatively steep initial slope can readily be extended over a much wider horizontal distance. On the other hand, the faulting can only produce a relatively small effect on the slope of the Moho, unless other processes such as creep are also operative.

CONCLUSIONS AND DISCUSSION

In this paper are reviewed a series of recently proposed mechanisms for subsidence at passive margins. Each mechanism has been discussed in isolation, but it is probable that most of them contribute to the observed subsidence. The writer's broad assessment is as follows:

1. Gravity loading mechanisms based on local or flexural isostasy may account for sediment piles at such deltas as the Niger, but this mechanism can be shown not to be the primary cause of most shelf subsidence. However, the gravity loading mechanism is a contributory factor in most subsidence, increasing the possible sediment thickness by a factor of between two and three.
2. Thermal hypotheses give an elegant explanation of the observed decay of shelf sediment thicknesses with time, but they appear to be inadequate to account for shelf sediment piles of more than about 4 km thickness.
3. The crustal creep hypothesis provides a complementary mechanism to the thermal ones in explanation of regional subsidence of the shelf, and may, account for accumulation of major sediment thicknesses by seaward creep of the lower continental crust.
4. Necking of the crust at incipient margins is not favored, although it may occur within the early rift valley setting provided that there is extensive igneous invasion of the upper crust. The necked zone would not normally be wider than about 30 km, unless asymmetrically split.
5. A mechanism for early graben formation prior to splitting is provided by applying the Vening Meinesz wedge subsidence hypothesis to the upper brittle part of the crust.

The main outstanding problem seems to be to establish the relative contributions of gravity- , thermal- , and stress-based hypotheses to the regional downwarping of margins. Some critical questions are: What is the detailed history of subsidence? Is there evidence for uplift and erosion just prior to splitting? Where is the continent-ocean crustal contact beneath the sediments? Is the early rift stage universal, and what is the type of crust underlying early rifts? What is the nature of the overall crustal transition? Does continental crustal material now reside in the upper mantle beneath the rise? Is the structure of the lower crust anomalous?

The answers to some of these questions may come from deep drilling, but many of them will require carefully planned deep geophysical investigations of the crust and topmost mantle across marginal zones.

REFERENCES CITED

Artemjev, M. E., and E. V. Artyushkov, 1971, Structure and isostasy of the Baikal rift and the mechanism of rifting: Jour. Geophys. Research, v. 76, p. 1197–1211.

Beloussov, V. V., 1960, Development of the earth and tectogenesis: Jour. Geophys. Research, v. 65, p. 4127–4146.

Bott, M. H. P., 1971, Evolution of young continental margins and formation of shelf basins: Tectonophysics, v. 11, p. 319–327.

——— 1972, Subsidence of Rockall Plateau and of the continental shelf: Geophys. Jour. Royal Astron. Soc., v. 27, p. 235–236.

——— 1973, Shelf subsidence in relation to the evolution of young continental margins, *in* D. H. Tarling and S. K. Runcorn, eds., Implication of continental drift to the earth sciences: New York, Academic Press, v. 2, p. 675–683.

——— 1976, Formation of sedimentary basins of graben type by extension of the continental crust: Tectonophysics, v. 36, p. 77–86.

——— and D. S. Dean, 1972, Stress systems at young continental margins: Nature (Phys. Sci.), v. 235, p. 23–25.

Collette, B. J., 1968, On the subsidence of the North Sea area, *in* D. T. Donovan, ed., Geology of shelf seas: London, Oliver and Boyd, p. 15–30.

Dietz, R. S., 1963, Collapsing continental rises—an actualistic concept of geosynclines and mountain building: Jour. Geology, v. 71, p. 314–333.

Falvey, D. A., 1974, The development of continental margins in plate tectonic theory: Jour. Australian Petroleum Exploration Assoc., v. 14, p. 95–106.

Fuchs, K., 1974, Geophysical contributions to taphrogenesis, *in* J. H. Illies and K. Fuchs, eds., Approaches to taphrogenesis: Stuttgart, Schweitzerbart, p. 420–432.

Griggs, D. T., F. J. Turner, and H. C. Heard, 1960, Deformation of rocks at 500° to 800°C: Geol. Soc. America, Mem. 79, p. 39–104.

Haxby, W. F., D. L. Turcotte, and J. M. Bird, 1976, Thermal and mechanical evolution of the Michigan Basin: Tectonophysics, v. 36, p. 57–75.

Hsu, K. J., 1965, Isostasy, crustal thinning, mantle changes, and the disappearance of ancient land masses: Amer. Jour. Science, v. 263, p. 97–109.

Kinsman, D. J. J., 1975, Rift valley basins and sedimentary history of trailing continental margins, *in* A. G. Fischer and S. Judson, eds., Petroleum and global tectonics: Princeton Univ. Press, p. 83–126.

Kusznir, N. J. and M. H. P. Bott, 1977, Stress concentration in the upper lithosphere caused by underlying visco-elastic creep: Tectonophysics, v. 43, p. 247–256.

Matthews, D. H., and S. G. Smith, 1971, The sinking of Rockall Plateau: Geophys. Jour. Royal Astron. Soc., v. 23, p. 491–498.

Rona, P. A., 1974, Subsidence of Atlantic continental margins: Tectonophysics, v. 22, p. 283–299.

Sheridan, R. E., 1969, Subsidence of continental margins: Tectonophysics, v. 7, p. 219–229.

Sleep, N. H., 1969, Sensitivity of heat flow and gravity to the mechanism of sea-floor spreading: Jour. Geophys. Research, v. 74, p. 542–549.

——— 1971, Thermal effects of the formation of Atlantic continental margins by continental break-up: Geophys. Jour. Royal Astron. Soc., v. 24, p. 325–350.

——— 1973, Crustal thinning on Atlantic continental margins: evidence from older margins, *in* D. H. Tarling and S. K. Runcorn, eds., Implications of continental drift to the Earth sciences: New York, Academic Press, v. 2, p. 685–692.

——— 1976, Platform subsidence mechanism and "eustatic" sea-level changes: Tectonophysics, v. 36, p. 45–56.

Turcotte, D. L., 1974,embrane tectonics: Geophys. Jour. Royal Astron. Soc. v. 36, p. 33–42.

Vening Meinesz, F. A., 1950, Les grabens africains, résultat de compression ou de tension dans la croûte terrestre?: Bull. Inst. Recherche Colon. Belge, v. 21, p. 539–552.

Walcott, R. I., 1970, Flexural rigidity, thickness, and viscosity of the lithosphere: Jour. Geophys. Research, v. 75, p. 3941–3954.

——— 1972, Gravity, flexure, and the growth of sedimentary basins at a continental edge: Geol. Soc. America Bull., v. 83, p. 1845–1848.

Watts, A. B., and W. B. F. Ryan, 1976, Flexure of the lithosphere and continental margin basins: Tectonophysics, v. 36, p. 25–44.

Geology of the Offshore Southeast Georgia Embayment, U.S. Atlantic Continental Margin, Based on Multichannel Seismic Reflection Profiles[1]

RICHARD T. BUFFLER,[2] JOEL S. WATKINS[2,3]
AND WILLIAM P. DILLON[4]

Abstract A geologic interpretation of the offshore Southeast Georgia Embayment is based on an 1,100-km multichannel seismic reflection survey conducted jointly by the University of Texas Marine Science Institute and the U.S. Geological Survey. The Southeast Georgia Embayment consists of a wedge of Cretaceous and Cenozoic sedimentary rocks that thins from 5 to 8 km beneath the Blake Plateau to about 1 km over the Cape Fear Arch. North of about 31°N lat. the sedimentary section onlaps a regional unconformity characterized by a strong, smooth reflector/high-velocity refractor (5.8 to 6.3 km/sec). Rocks below the unconformity may consist of Lower Jurassic volcanic(?) rocks, crystalline basement, or a high-velocity sedimentary layer. Acoustic basement south of 31°N lat. is an irregular unconformable surface that probably truncates a varied lithology consisting of Paleozoic (and Precambrian?) igneous, metamorphic, and sedimentary rocks or Triassic to Early Jurassic igneous and sedimentary rocks in fault basins, all of which crop out or have been drilled onshore.

The sedimentary section is divided into three major seismic intervals. The intervals are separated by unconformities and can be mapped regionally. The oldest interval ranges in age from Early Cretaceous through middle Late Cretaceous, although it may contain Jurassic rocks where it thickens beneath the Blake Plateau. It probably consists of continental to nearshore clastic rocks where it onlaps basement and grades seaward to a restricted carbonate platform facies (dolomite-evaporite). The middle interval (Upper Cretaceous) is characterized by prograding clinoforms interpreted as open marine slope deposits. This interval represents a Late Cretaceous shift of the carbonate shelf margin from the Blake Escarpment shoreward to about its present location, probably due to a combination of continued subsidence, an overall Late Cretaceous rise in sea level, and strong currents across the Blake Plateau. The youngest (Cenozoic) interval represents a continued seaward progradation of the continental shelf and slope. Cenozoic sedimentation on the Blake Plateau was much abbreviated owing mainly to strong currents.

INTRODUCTION

This paper presents a geologic interpretation of the offshore Southeast Georgia Embayment located along the southeastern United States Atlantic continental margin (Fig. 1). The interpretation is based primarily on an 1,100-km multichannel seismic reflection survey conducted jointly by the University of Texas Marine Science Institute at Galveston (UTMSI) and the U.S. Geological Survey (USGS). The survey consists of one long north-south line along the continental shelf (Fig. 2; SE-1) and five east-west cross lines extending from the shelf out onto the adjacent Blake Plateau (Fig. 2; SE-2 to SE-6). The lines supplement other USGS multichannel seismic lines collected in the area as part of a study of the geology and resources of the Southeast Georgia Embayment-Blake Plateau area (Dillon et al, 1978, this volume).

The multichannel seismic data presented here are 24-fold. They were collected aboard the UTMSI R/V *Ida Green* in May, 1975, and were processed by UTMSI at Galveston.

Discussion of the geology of the Southeast Georgia Embayment is facilitated through a series of generalized interpretive cross-sections along each of the seismic lines (Figs. 3 to 8). The sections show seis-

[1]Manuscript received, July 28, 1977; accepted, April 6, 1978.

[2]Geophysics Laboratory, University of Texas Marine Science Institute, Galveston, Texas 77550.

[3]Current address: Gulf Research and Development Co., Houston, Texas 77036.

[4]Office of Marine Geology, U.S. Geological Survey, Woods Hole, Massachusetts 02543.

The writers wish to acknowledge the assistance of Cecil and Ida Green, Exxon Production Research Co., Texas Instruments, Inc., Western Geophysical Company, Texaco Inc., Continental Oil Company, Mobil Oil Corp., Shell Oil Co., University of Texas Medical Branch at Galveston Service Computation Center, Chevron Oil Co., the captain and crew of the R/V *Ida Green*, Rainer Dehm, and Gerry T. Schuster.

Critical review was provided by M. M. Ball, K. J. McMillen, T. H. Shipley, and R. J. Wold.

This project was sponsored by U.S. Geological Survey Contract No. 14-08-0001-14942.

This paper is The University of Texas Marine Science Institute Contribution No. 302, Galveston Geophysics Laboratory.

Article Identification Number:
0065-731X/78/MO29-0002/$03.00/0.
(see copyright notice, front of book)

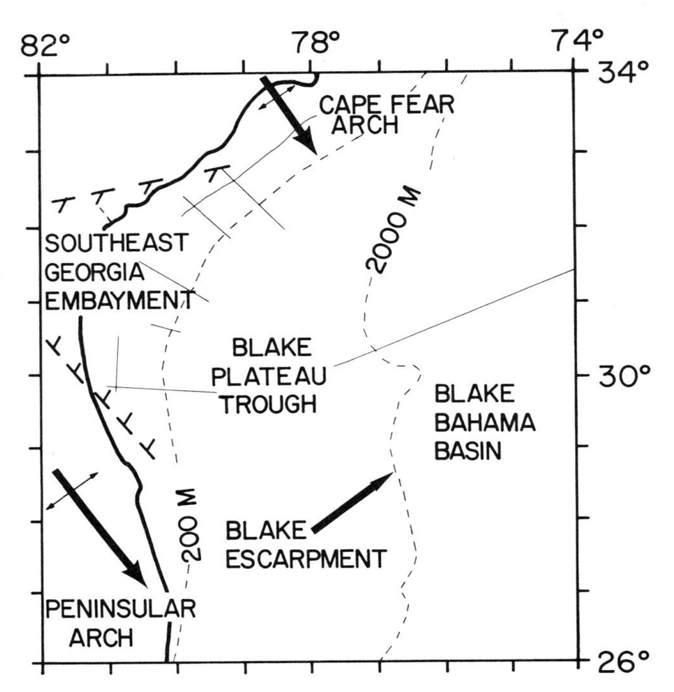

FIG. 1. Generalized geologic setting of Southeast Georgia embayment, U.S. Atlantic continental margin. Contours in meters.

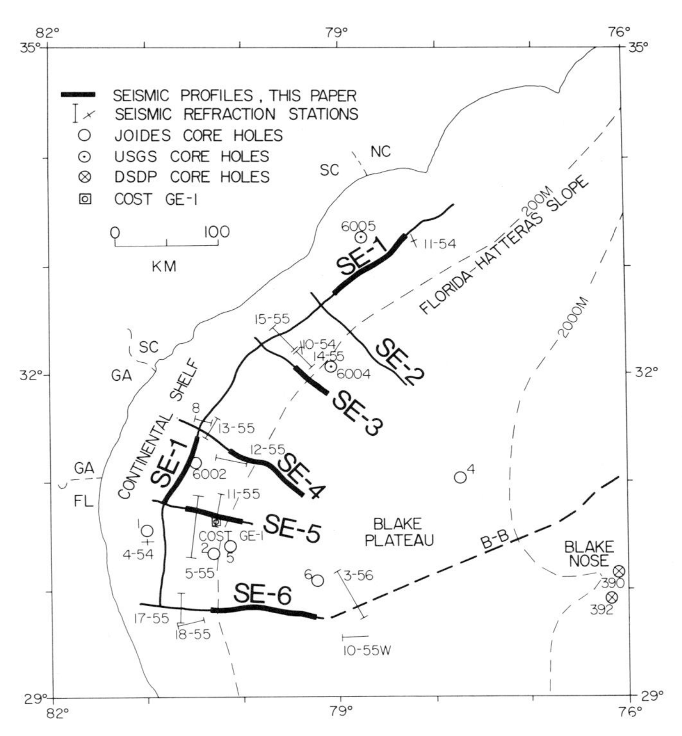

FIG. 2. Location of UTMS/USGS multichannel seismic lines SE-1 through SE-6 (solid), Southeast Georgia embayment, U.S. Atlantic continental margin. Line designated B-B (dashed) collected by UTMSI (Shipley et al, 1978; Buffler et al, 1978). Also shown is location of pertinent refraction stations (Antoine and Henry, 1965; Hersey et al, 1959; Sheridan et al, 1966); early JOIDES core holes (JOIDES, 1965); DSDP core holes (Benson et al, 1976); and USGS core holes (Hathaway et al, 1976). Contours in meters.

mic intervals mapped, inferred ages, general environments of deposition, overall structural configuration, and other pertinent geologic features. Time-depth conversions were made using interval velocities obtained from common-depth point (CDP) reflection velocity analyses as well as published refraction data from areas nearby. Velocity analyses were determined at approximately 18-km intervals along each seismic line, and interval velocities were calculated using methods as discussed by Tanner and Koehler (1969) and Montalbetti (1971). Average interval velocities from one or more analyses are shown on the cross-sections (Figs. 3 to 8). The published refraction velocity data used are also shown on the sections (in parentheses), and the locations of the stations are shown on Figure 2. Examples of seismic sections from selected areas are included to demonstrate the seismic character of the units mapped and other geologic features (Figs. 9 to 15). Figure captions explain pertinent details of each section and give appropriate references.

GEOLOGIC SETTING

The Southeast Georgia Embayment is a broad sedimentary and structural embayment along the southeastern U.S. Atlantic continental margin bounded on the north by the Cape Fear Arch and on the southwest by the Peninsular Arch (Fig. 1). The embayment is filled with a wedge of mainly Cretaceous and Cenozoic sedimentary rocks that thickens southeastward into a deep sedimentary basin beneath the Blake Plateau known as the Blake Plateau trough (Fig. 1). The Cretaceous rocks pass beneath the present continental shelf-slope without any apparent structural break and thin and onlap shoreward onto a gently seaward-dipping unconformity surface. The present continental shelf-slope is a Late Cretaceous-Cenozoic constructional feature controlled in part by the north-flowing Gulf Stream. The entire sedimentary sequence represents a transition zone between dominantly carbonate rocks to the south in Florida and dominantly siliceous clastic rocks north of Cape Hatteras. This basic structural and sedimentary configuration is evident on all the cross-sections and seismic sections (Figs. 3 to 15).

Prior to the multichannel seismic surveys sponsored by the USGS, not much was known about the details of the deeper structure and geologic history of the offshore Southeast Georgia Embayment–Blake Plateau area. Seismic-refraction surveys had established the gross configuration and velocity structure of the basin (Woollard et al, 1957; Hersey et al, 1959; Antoine and Henry, 1965; Sheridan et al, 1966), and the shallow structure and sedimentary history had been delineated using single-channel, shallow penetration seismic profiles (Ewing et al, 1966; Emery and Zarudski, 1967; Uchupi, 1967, 1970; Uchupi and Emery, 1967; Zarudski and Uchupi, 1968). A few shallow core holes have been drilled on the shelf, slope, and adjacent Blake Plateau as part of the early JOIDES program (JOIDES, 1965) and a recent USGS drilling program (Hathaway et al, 1976), but pre-Tertiary (Upper Creta-

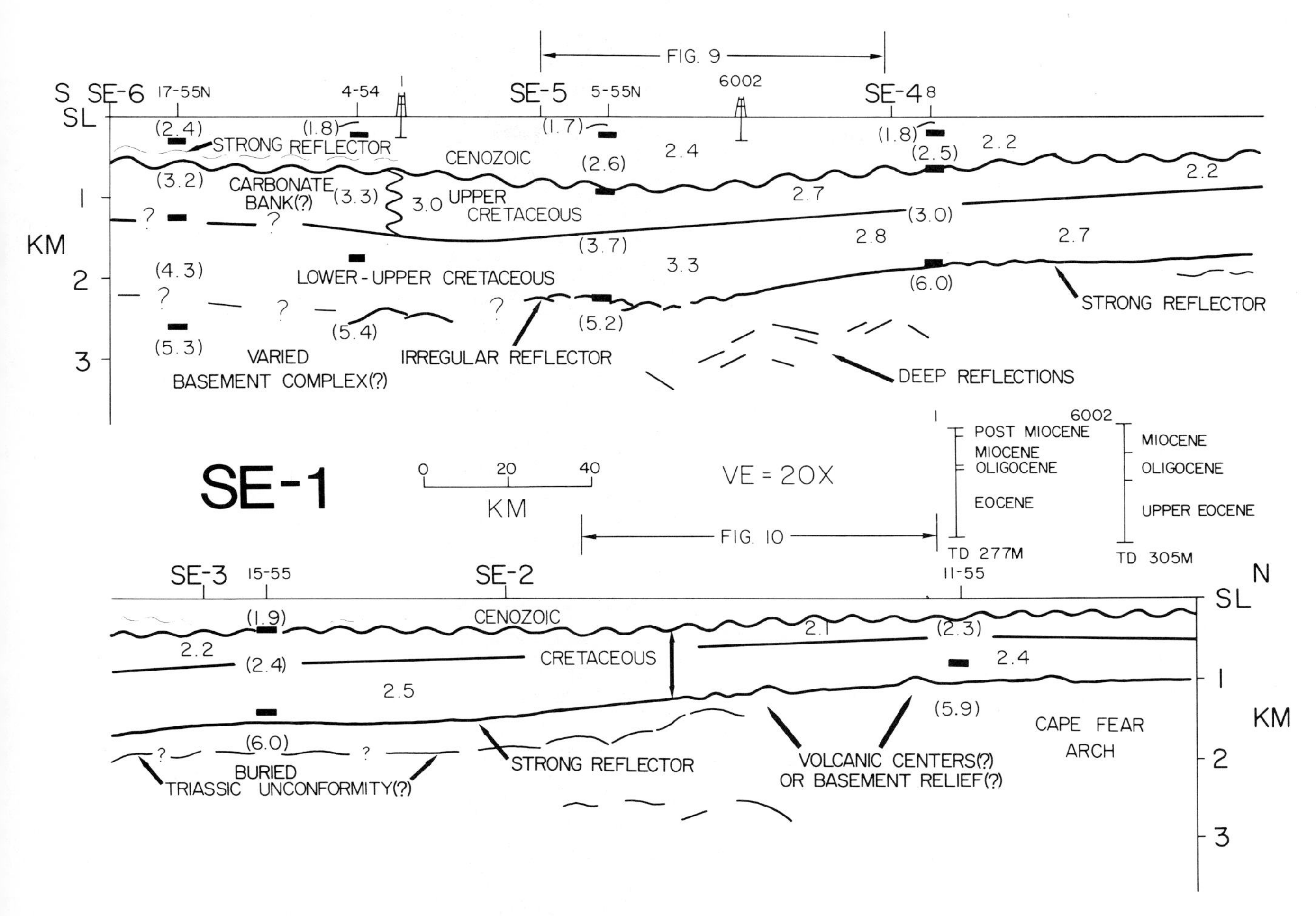

FIG. 3. Generalized interpretive geologic cross-section along seismic line SE-1 showing intervals mapped, inferred ages, core hole locations, and pertinent geologic features. Interval boundaries are heavy solid lines. Numbers are average interval velocities determined from CDP velocity analyses located at 18-km intervals along line. Numbers in parentheses are interval velocities projected from nearby refraction stations (Antoine and Henry, 1965; Hersey et al, 1959; Sheridan et al, 1966). Short solid dashes are refraction interval boundaries. Increase in interval velocities from north to south may reflect facies change from terrigenous clastics to carbonates and/or increase in depth of burial. Zigzag line designates lateral change in seismic character; deep reflections south of this line masked by strong, irregular reflector (early Tertiary unconformity?), possibly marking top of high-velocity carbonate bank deposits. Sedimentary section thins north onto Cape Fear arch and onlaps a smooth, strong reflector/high-velocity refractor (±6.0 km/sec; acoustic basement). This unconformity possibly represents either the top of a Lower Jurassic volcanic layer, the top of crystalline basement, or the top of high-velocity sedimentary rocks. Relief on strong reflector near Cape Fear arch may represent relief on basement surface or volcanic center. Discontinuous reflectors beneath strong reflector may represent buried Triassic-Early Jurassic unconformity. Irregular reflector to the south possibly represents Triassic-Early Jurassic breakup unconformity underlain by varied basement complex with lower refraction velocities (5.3 to 5.4 km/sec).

ceous) rocks were penetrated at only one site, #6004 (Fig. 2). An analysis of the structure and geologic development of the Southeast Georgia Embayment–northern Blake Plateau area as interpreted from all the USGS multichannel seismic data is presented in a companion paper in this volume (see Dillon et al, 1978). Other published multichannel data in the area consist of one regional line shot by UTMSI that extends northeast from the end of Line SE-6 across the Blake Plateau and into the deep ocean (Shipley et al, 1978; Buffler et al, 1978; Fig. 2).

GEOLOGIC INTERPRETATION

"Basement"

The seaward-dipping unconformity surface onto which the Cretaceous sediments lap has two salient seismic characteristics. North of about 31°N lat., the unconformity is a strong, smooth reflector, while to the south it is a weak, discontinuous, and irregular reflector. This change in character is seen on the cross-sections and the seismic sections along the southern end of line SE-1 (Figs. 3 and 9) and also on line SE-5 (Figs. 7 and 14).

The strong, smooth reflector north of 31°N lat. forms an effective acoustic basement over a widespread area. It is seen along most of line SE-1 (Figs. 3, 9, 10) as well as on lines SE-2 through SE-5 (Figs. 4–7, 13, 14) and is remarkably smooth over most of the area. A small fault interrupts the reflector along line SE-2 (Fig. 4). Along the north end of SE-1 near

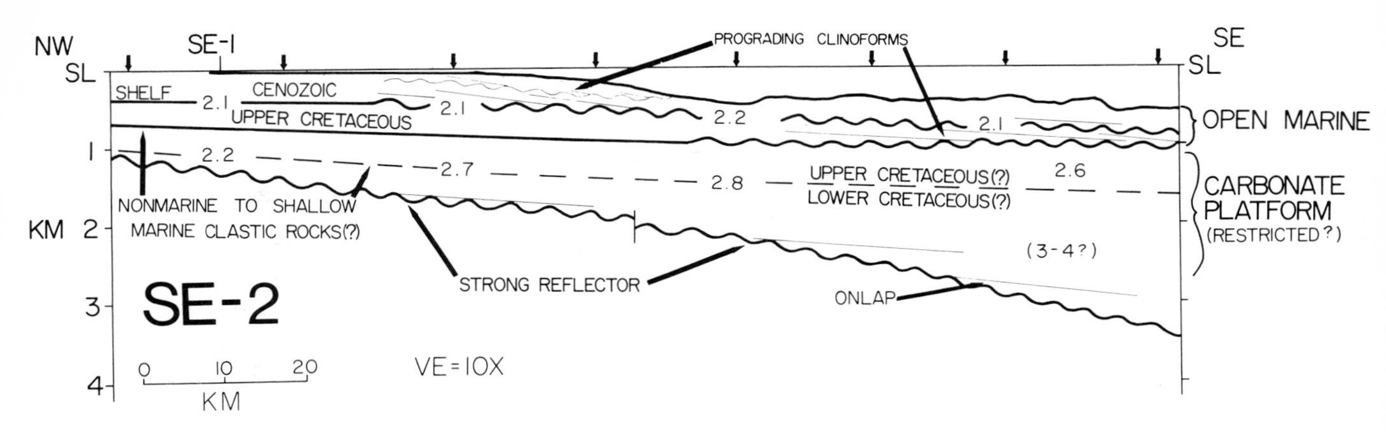

FIG. 4. Generalized interpretive geologic cross-section along seismic line SE-2 showing intervals mapped, inferred ages, and inferred depositional environments. Interval boundaries are heavy solid lines. Numbers are average interval velocities determined from CDP velocity analyses made at locations shown by arrows. Numbers in parentheses are interval velocities estimated from nearby refraction data. Lower Cretaceous strata onlap strong reflector (acoustic basement). This unconformity possibly represents the top of Jurassic volcanic rocks, the top of crystalline basement, or the top of high-velocity sedimentary rocks. Prograding clinoforms are interpreted to represent slope deposits.

the Cape Fear Arch the reflector has several hundred meters of relief and forms local topographic highs (Figs. 3, 10). This reflector correlates with the top of a widespread refraction layer characterized by velocities ranging from 5.8 to 6.3 km/sec.

The strong reflector/high-velocity refractor has not been drilled offshore. Dillon et al (1978, this volume) suggest that it may represent the top of a wide-

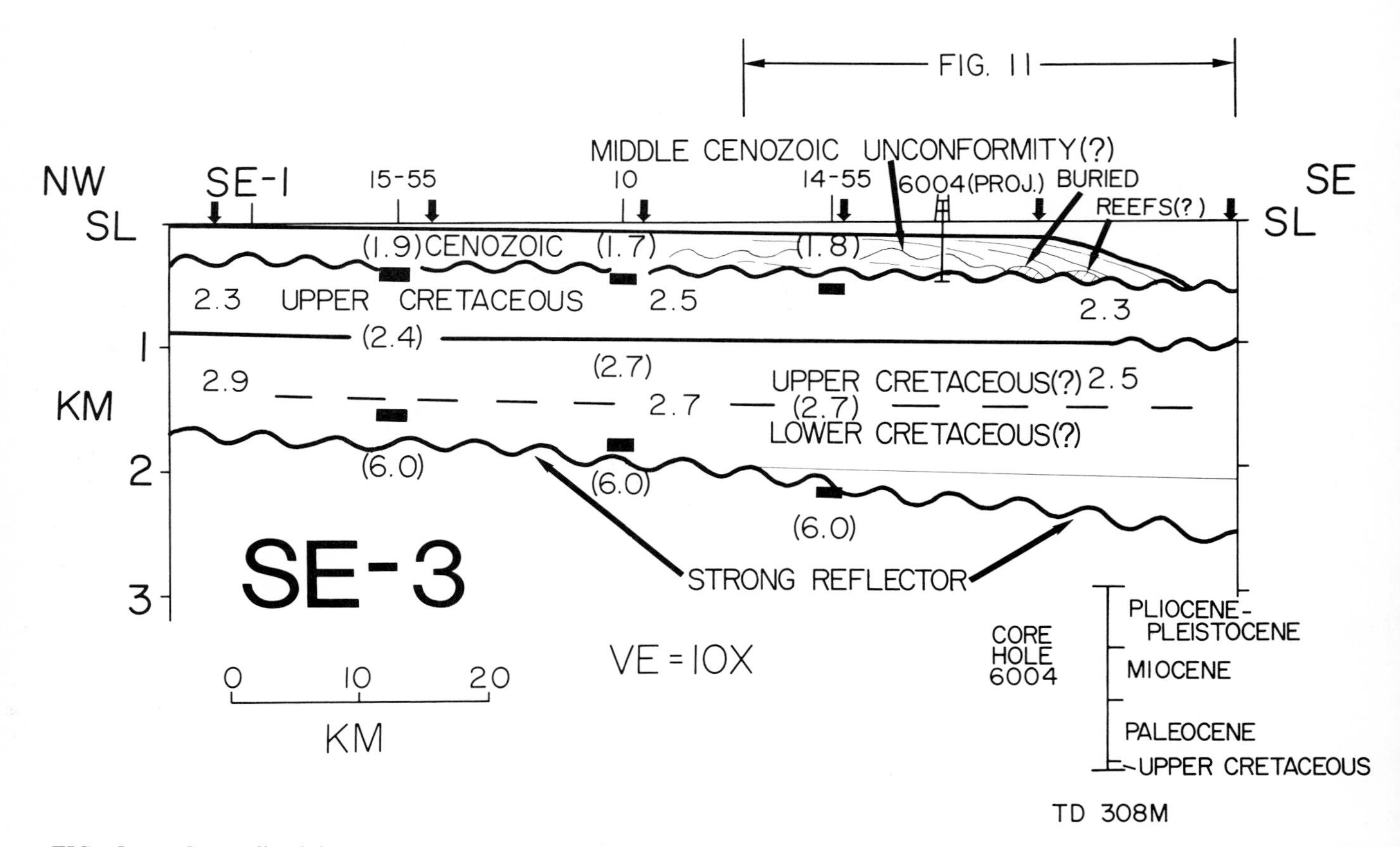

FIG. 5. Generalized interpretive geologic cross-section along seismic line SE-3 showing intervals mapped, inferred ages, and pertinent geologic features. Interval boundaries are heavy solid lines. Numbers are average interval velocities determined from CDP velocity analyses made at locations shown by arrows. Numbers in parentheses are interval velocities projected from nearby refraction stations (Hersey et al, 1959). Short solid dashes are refraction interval boundaries. Lower Cretaceous strata onlap strong reflector/high-velocity refractor (acoustic basement). This unconformity possibly represents the top of a Lower Jurassic volcanic layer, the top of crystalline basement, or the top of high-velocity sedimentary rocks. USGS core hole 6004 (Hathaway et al, 1976) projected approximately 20 km southwest onto section.

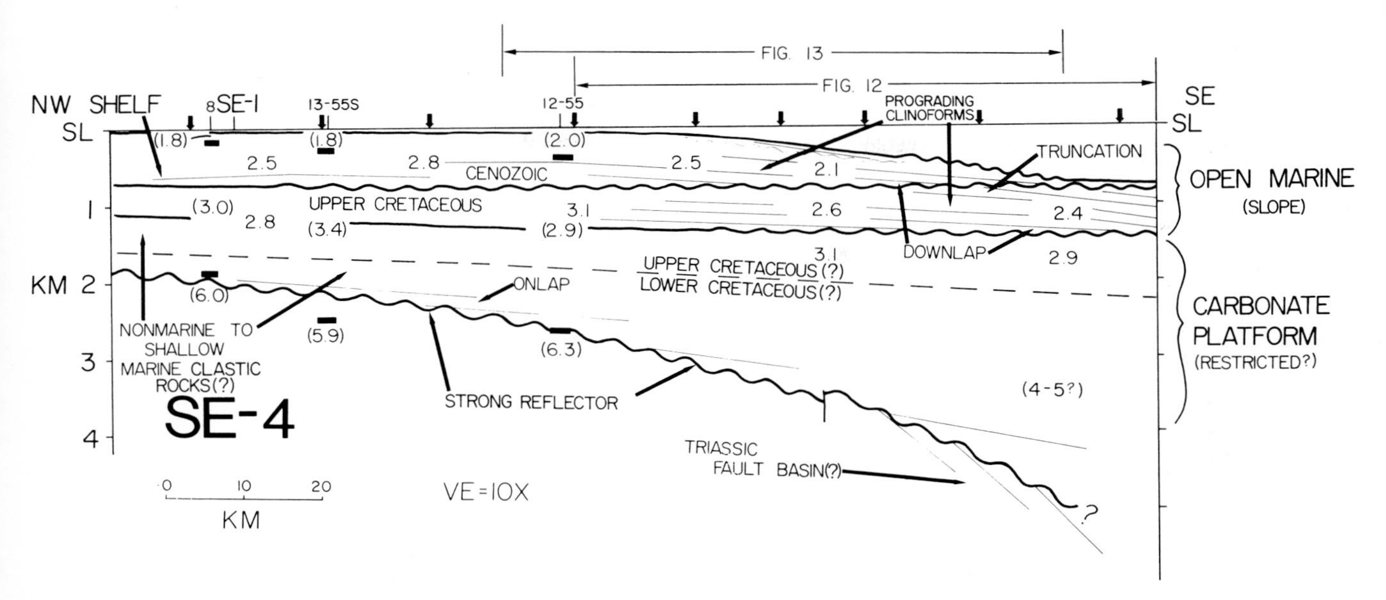

FIG. 6. Generalized interpretive geologic cross-section along seismic line SE-4 showing intervals mapped, inferred ages, inferred depositional environments, and various unconformable relationships. Interval boundaries are heavy solid lines. Numbers are average interval velocities determined from CDP velocity analyses made at locations shown by arrows. Numbers in parentheses are interval velocities projected from nearby refraction stations (Hersey et al, 1959; Antoine and Henry, 1965). Short solid dashes are refraction interval boundaries. Lower Cretaceous strata onlap strong reflector/high-velocity refractor. This unconformity possibly represents the top of Lower Jurassic volcanic rocks, the top of crystalline basement, or the top of high-velocity sedimentary rocks. To the southeast, this unconformity may be underlain by tilted Triassic fault basin.

spread volcanic unit that was deposited in Early Jurassic time during the early rifting and breaking apart of the continent. The volcanic lithology is based on a tentative correlation with volcanic rocks drilled in onshore wells throughout northern Florida, Georgia, and South Carolina. An Early Jurassic age is inferred by Dillon et al (1978) because the unit appears to cover an older unconformity truncating possible Triassic-Early Jurassic tilted fault basins. A similar relationship may occur in the area of line SE-4 (Figs. 6, 13), but here it is not clear whether the strong reflector is a surface separate from the unconformity truncating the tilted beds or is the same surface.

It is also not clear on our seismic data whether or not the inferred volcanic unit has any considerable thickness. As mentioned above, the strong reflector acts as an acoustic basement over most of the area, but in places irregular, discontinuous reflections can be seen below it. In some cases these reflections may be internal multiples, but in other cases they appear to have no relationship with the overlying strong reflector and may be real reflections (e.g., SE-1, Figs. 3, 10). In places these deeper reflections may represent the older Triassic–Early Jurassic unconformity truncating older basement or the tilted fault basins.

Assuming a volcanic origin for the rocks below the strong reflector, the topographic highs on the flank of the Cape Fear Arch may represent large volcanic centers (SE-1, Figs. 3, 10). The presence of large gravity highs in the vicinity of these features supports this hypothesis (Krivoy and Eppert, 1977).

Alternatively, this smooth reflector may simply be a peneplaned erosional surface overlying crystalline basement rocks. Such rocks occur in the subsurface and in outcrops onshore to the northwest. The relief on the unconformity, therefore, may be erosional, and the possible reflections beneath the reflector may represent changes within the basement complex.

The refraction velocities for the rocks beneath the strong reflector are somewhat higher (5.8 to 6.3 km/sec) than might be expected for crystalline rocks or even volcanic rocks. This suggests the possibility of even a third origin for the strong, smooth reflector (i.e., it may be the top of a well-lithified sedimentary layer consisting of dolomites or evaporites).

The weak, discontinuous reflector characterizing the pre-Cretaceous unconformity south of about 31° N lat., is best seen along the southern part of line SE-1 and on line SE-5 (Figs. 3, 7, 9, 14). This unconformity may represent the widespread Triassic-Early Jurassic unconformity discussed above, formed during the initial breakup of the continents (*breakup unconformity* of Falvey, 1974). The nature of the rocks underlying this unconformity offshore is unknown, but it probably consists of a varied basement complex. Refraction data indicate that the rocks have a significantly lower velocity (5.2 to 5.3 km/sec) than those to the north. Based on projections from onshore well data and outcrops, the basement here probably consists of Paleozoic volcanic rocks, Paleozoic sedimentary rocks, Precambrian and Paleozoic igneous and metamorphic rocks, and

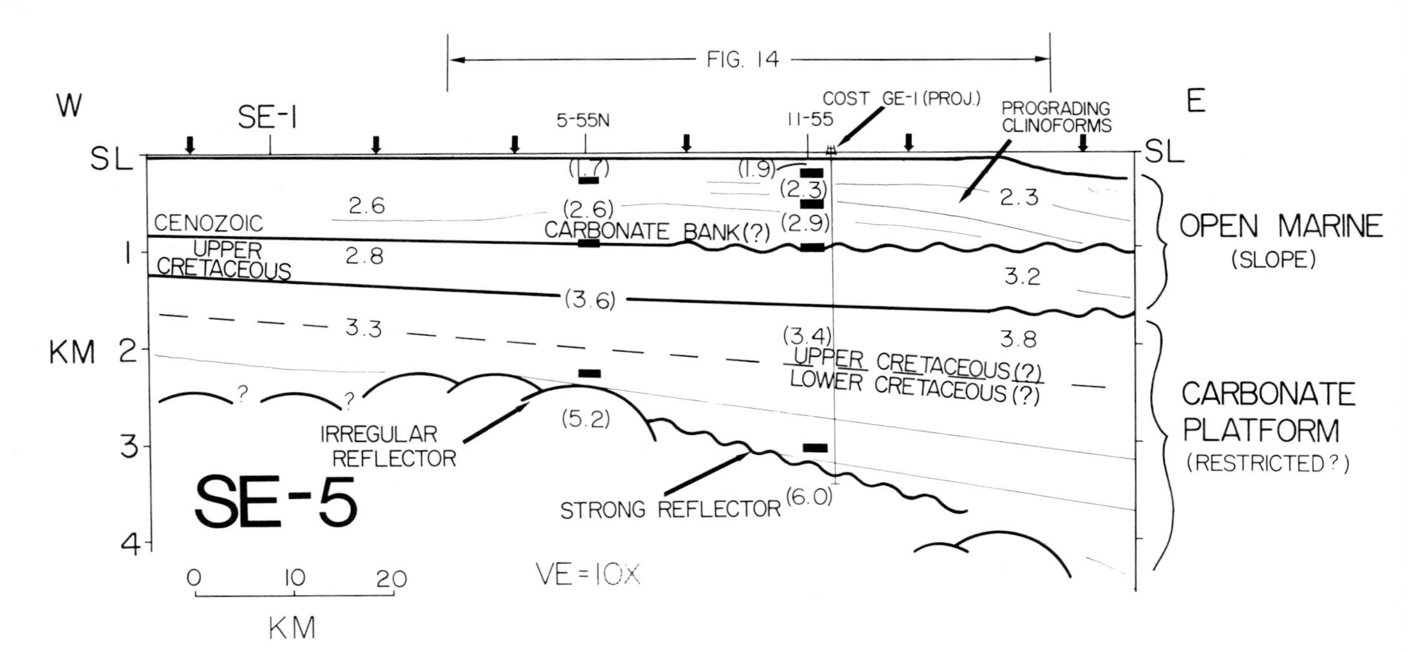

FIG. 7. Generalized interpretive geologic cross-section along seismic line SE-5 showing intervals mapped, inferred ages, inferred depositional environments, and other geologic features. Interval boundaries are heavy solid lines. Numbers are average interval velocities determined from CDP velocity analyses made at locations shown by arrows. Numbers in parentheses are interval velocities projected from nearby refraction stations (Hersey et al, 1959). Short solid dashes are refraction interval boundaries. Lower Cretaceous strata onlap both irregular discontinuous reflector (west and east) and strong reflector (center). Irregular reflector may represent Triassic unconformity underlain by varied basement complex. Strong reflector may represent top of Lower Jurassic volcanic rocks, top of crystalline basement, or top of high-velocity sedimentary rocks. Expected section penetrated by nearby COST GE-1 well is shown on section.

Triassic to Upper Jurassic sedimentary and igneous rocks in fault basins.

Some clues as to the nature of the "basement" should be forthcoming when the results of the recently drilled (1977) Continental Offshore Stratigraphic Test (COST) well No. GE-1 are released. This well is located near line SE-5 and should have drilled through the pre-Cretaceous unconformity and bottomed in "basement" (Figs. 7 and 14).

Seismic Stratigraphy

The sedimentary section in the Southeast Georgia Embayment is divided into three major seismic intervals that can be mapped throughout the study area. The intervals are separated by prominent unconformities and correspond to major depositional sequences. Depositional sequences are defined by Vail et al (1977) as "stratigraphic units composed of relatively conformable successions of genetically related strata bounded at their tops and bases by unconformities or their correlative conformities." The unconformities are recognized on the seismic sections by reflection relationships such as truncation, downlap, onlap, or irregular surfaces. Examples of these relationships are noted on the cross-sections and seismic sections. Interval boundaries are outlined by heavy solid lines on the cross-sections. They are wavy where unconformable relationships are obvious, and straight where conformable. The intervals are discussed below and are summarized in Table 1.

Lower to Upper Cretaceous interval—The lowermost and oldest interval mapped is characterized generally by discontinuous, flat-lying, variable-amplitude reflections. Beneath the inner Blake Plateau the upper part of the interval contains several prominent groups of strong reflections (Figs. 13 and 15). Along the seaward end of lines SE-4 and SE-6, the pre-Cretaceous unconformity plunges into the Blake Plateau trough, and the overlying sedimentary section thickens to more than 4 to 7 km (Figs. 6, 8, 13, 15). Shoreward the unit thins to <1 km as it onlaps basement. This unconformable relationship forms the lower boundary of the interval.

The age of strata in this interval is estimated to range from Early Cretaceous through middle Late Cretaceous, although where the section thickens into the Blake Plateau trough, it may contain Upper Jurassic strata. This age estimation is made mainly by extrapolating inferred data from the Blake Escarpment across the Blake Plateau using a regional UTMSI seismic line (Shipley et al, 1978; Buffler et al, 1978). The Blake Escarpment age data are from dredge hauls (Heezen and Sheridan, 1966), from DSDP core holes (Benson et al, 1976), and from tentative correlations of prominent unconformities at the escarpment with major global unconformities proposed by Vail et al (1977). The Lower-Upper Cretaceous boundary, shown on the cross-sections (Figs. 3 to 8) as a bold dashed line within the interval, is extrapolated across the Plateau from a prominent unconformity at the escarpment that

Table 1. Major Seismic Intervals, Southeast Georgia Embayment

Seismic Interval	Seismic Characteristics	Velocity (km/sec)	Thickness (km)	Inferred Rock Types — Depositional Setting
Cenozoic	Strong and weak, semidiscontinuous reflections. Flat-lying beneath inner shelf. Gently seaward-dipping, prograding clinoforms beneath outer shelf and slope. Lowermost beds downlap unconformably onto underlying unit. Composed of several depositional sequences separated by unconformities. Prominent unconformity/buried erosion surface beneath slope.	1.7-2.8	0-1.0	Generally fine-grained, low-energy, open marine, outer shelf-slope deposits. Lithology variable; mainly carbonates. Becomes more clastic and coarser-grained to north and in upper part of section.
Upper Cretaceous	Generally weak, semidiscontinuous reflections. Flat-lying beneath shelf. Gently seaward-dipping sigmoidal prograding clinoforms beneath slope that downlap unconformably onto underlying unit and are truncated above by prominent unconformity. Change in seismic character in south due to dispersement of seismic energy by irregular Cenozoic unconformity above.	2.1-3.2	0.3-0.8	Fine-grained, low energy, open marine, outer shelf-slope deposits. Mainly carbonates.
Lower Cretaceous to middle Upper Cretaceous. Possibly includes Upper Jurassic(?)	Strong and weak, discontinuous, flat-lying reflections. Onlaps basement. Prominent strong, high amplitude reflections beneath outer slope-inner Blake Plateau.	2.2-(5.0)	<1.0->7.0	Terrigenous clastics intertonguing seaward with restricted carbonate platform facies, including dolomites and evaporites.

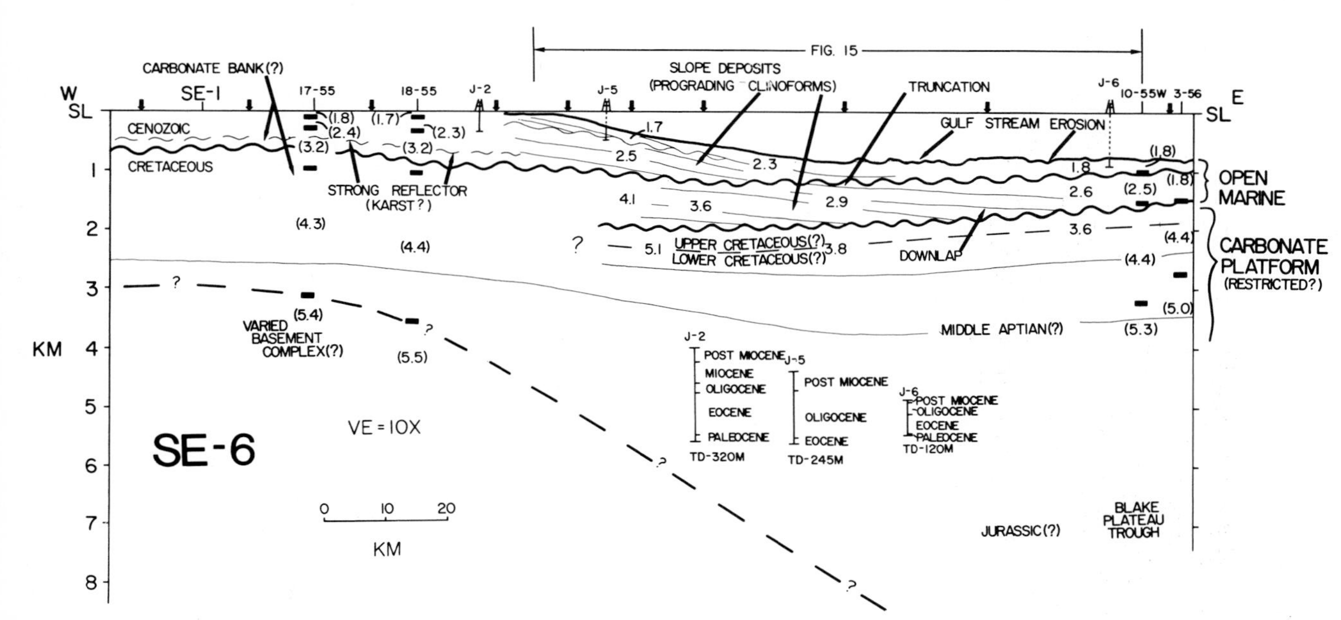

FIG. 8. Generalized interpretive geologic cross-section along seismic line SE-6 showing intervals mapped, inferred ages, inferred depositional environments, and other geologic features. Interval boundaries are heavy solid lines. Numbers are average interval velocities determined from CDP velocity analyses made at locations shown by arrows. Numbers in parentheses are interval velocities projected from nearby refraction stations (Hersey et al, 1959; Sheridan et al, 1966). Short solid dashes are refraction interval boundaries. Depth to inferred basement complex increases to east beneath thicker sedimentary section of Blake Plateau Trough. Middle Aptian(?) and Upper-Lower Cretaceous(?) time lines are projected from Blake Escarpment along line B-B (Buffler et al, 1978; Shipley et al, 1978). A strong irregular reflector, possibly a lower Cenozoic unconformity at the top of carbonate bank, masks deeper seismic reflections beneath shelf. JOIDES core holes projected onto line of section (see Fig. 2 for location of core holes).

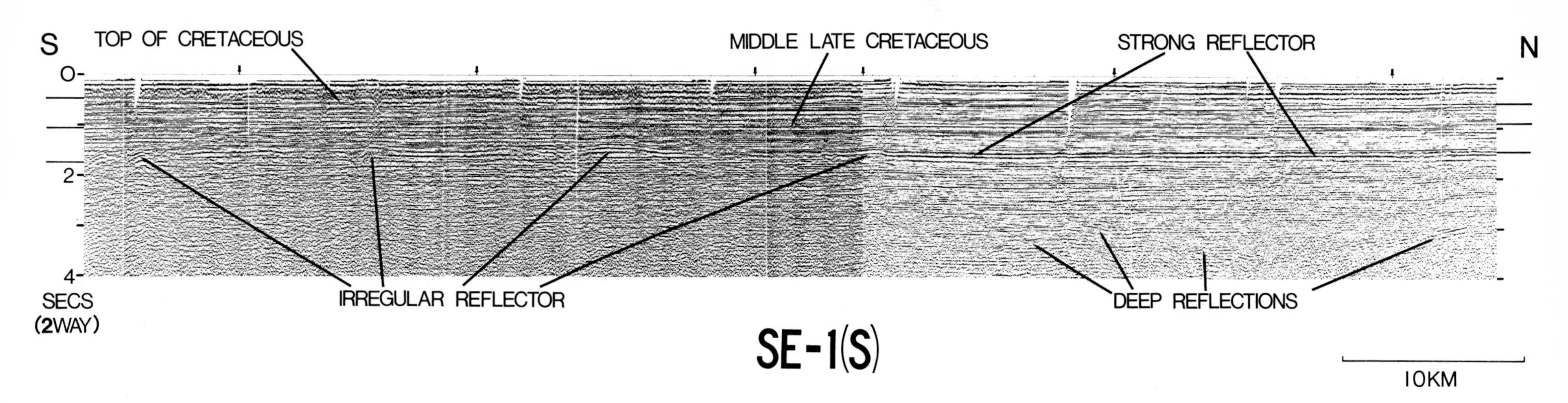

FIG. 9. Multichannel seismic profile from southern part of line SE-1 showing irregular reflector (possibly Triassic-Lower Jurassic unconformity underlain by a varied basement complex) changing northward into strong, smooth reflector/top of high-velocity refractor (possibly represents top of Lower Jurassic volcanic unit, top of crystalline basement, or top of high-velocity sedimentary layer). Deep reflections occur within inferred basement. Lines in margin are interval boundaries.

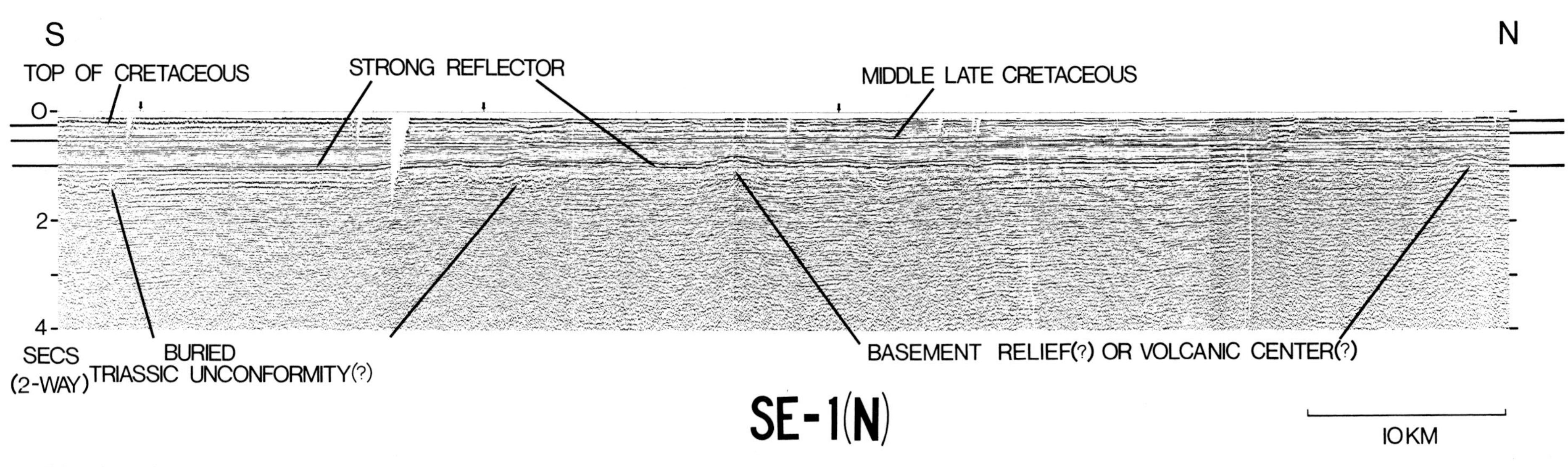

FIG. 10. Multichannel seismic profile from northern part of Line SE-1 showing smooth, strong reflector/top of high-velocity refractor (possibly represents top of Lower Jurassic volcanic layer, top of crystalline basement, or top of high-velocity sedimentary layer). Relief on reflector may represent volcanic centers or basement relief. Deeper irregular reflectors may represent buried Triassic unconformity. Lines in margin are interval boundaries.

possibly correlates with the major middle Cenomanian unconformity defined by Vail et al (1977). The unconformity forming the upper boundary of the interval, therefore, is thought to be approximately middle Late Cretaceous in age, as it lies between the inferred middle Cenomanian horizon and a younger unconformity believed to represent the Cretaceous-Tertiary boundary (discussed below).

This Lower to Upper Cretaceous interval has not yet been drilled in the Southeast Georgia Embayment, and the rock types and depositional setting are not known. A general depositional model for the area is proposed, however, based on extrapolation from adjacent areas. As stated above, the embayment lies between a dominantly carbonate province to the south and a dominantly siliceous clastic province to the north. During Early and most of Late Cretaceous time the Blake Plateau apparently was a broad carbonate platform with its outer margin at the Blake Escarpment (Shipley et al, 1978; Buffler et al, 1978). Onshore wells in northern Florida, Georgia, and South Carolina indicate that similar age rocks contain abundant clastics (e.g., Herrick and Vorhis, 1963). The offshore Southeast Georgia Embayment, therefore, probably is transitional between continental to nearshore clastic facies to the north and west and carbonate platform facies to the south and east. A schematic model of this relationship is presented in Figure 16. The carbonate platform facies along the inner Blake Plateau probably reflect more restricted platform environments and may consist, in part, of interbedded dolomite and evaporite. The strong reflections in the upper part of the interval beneath the inner plateau may, in fact, represent the interbedded relationship of high-velocity dolomites and evaporites with lower velocity carbonates and clastics. The COST well drilled near line SE-5 in 1977 will provide the first data on the lithology and depositional history of this interval.

Upper Cretaceous interval—The middle interval mapped in the Southeast Georgia Embayment is characterized by generally low-amplitude, semidiscontinuous reflectors. Beneath the shelf the strata are generally flat-lying, but beneath the slope and inner plateau they dip seaward and downlap onto the underlying interval. This unconformable relationship forms the lower boundary of the interval. The upper boundary is another regional unconformity characterized by truncation of the seaward-dipping strata. These relationships occur along all the crosslines. They are best observed on lines SE-4 and SE-6 (Figs. 6, 8, 13, 15) and are particularly evident on the profiler record of line SE-4 shown in Figure 12.

The age of this interval is estimated to be Late Cretaceous. As discussed above, the lower boundary unconformity is considered middle Late Cretaceous

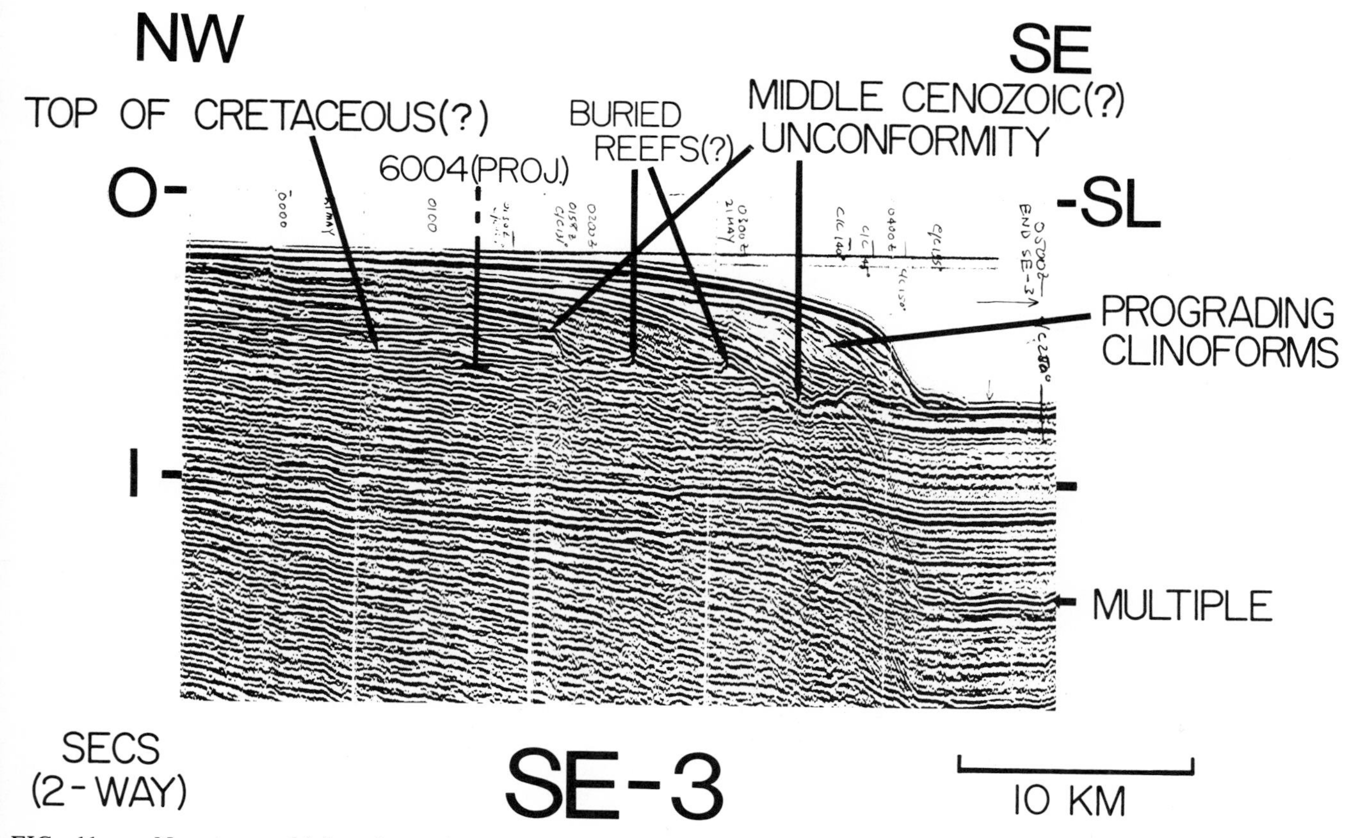

FIG. 11. Near-trace shipboard monitor record of line SE-3 across outer continental shelf-slope showing details of middle Cenozoic(?) erosional surface and possible reefs buried by prograding slope deposits. Core hole 6004 (projected approximately 20 km south onto section) bottomed in Upper Cretaceous shales (Hathaway et al, 1976). Upper Cretaceous strata exposed at or near seafloor along eastern end of section (Uchupi, 1967, 1970).

on the basis of extrapolations across the Blake Plateau and its occurrence between a possible middle Cenomanian unconformity and an unconformity thought to represent the Cretaceous-Tertiary boundary. This latter unconformity forms the upper boundary of the interval. The Cretaceous-Tertiary boundary age is assigned with some confidence, as JOIDES holes J-4 and J-6 on the Blake Plateau drilled into Paleocene rocks not far above this unconformity (Figs. 8 and 15). Unless there was a rapid change in sedimentation rate and the Paleocene is uncommonly thick on the plateau, this age would be a reasonable extrapolation. This unconformity is tentatively correlated with an important global unconformity at the Cretaceous-Tertiary boundary (Vail et al, 1977).

A Late Cretaceous age for this interval is based on other data as well. Upper Cretaceous strata have been dredged from the seafloor near the ends of lines SE-3 and SE-4 (Uchupi, 1967, 1970), an area where the upper part of the interval is exposed at or near the seafloor as shown by the seismic sections (Figs. 5, 6, 11, 12, 13). In addition, the Upper Cretaceous rocks reached in core hole No. 6004 tentatively are correlated with the upper part of the interval (Figs. 5 and 11).

A distinct change in the seismic record is seen in the southeastern part of the study area along the southern end of line SE-1 between lines SE-5 and SE-6 (Fig. 3), and also along line SE-6 beneath the slope (Figs. 8 and 15). It apparently is caused by a strong irregular reflector that effectively disperses and prevents most seismic energy from penetrating deeper. This surface appears to correlate with an unconformity slightly younger than the Cretaceous-Tertiary boundary unconformity discussed above. It probably marks a change to hard, dense, high-velocity (4.3 km/sec) Cretaceous-early Tertiary shallow-water carbonate rocks. The irregular nature of the surface may reflect reef growth on a carbonate bank, or it may represent karst topography developed during subaerial exposure.

The gently seaward-dipping reflectors within the Upper Cretaceous interval beneath the slope and inner plateau probably represent prograding clinoforms; i.e., gently sloping depositional surfaces (Vail et al, 1977). These clinoforms have a sigmoidal configuration, which is generally thought to be typical of relatively low-energy depositional regimes (Vail et al, 1977). These strata, therefore, probably were deposited in a relatively deep-water, open marine, prograding slope environment, not too unlike the present slope environment. The generally low-amplitude reflection characteristic of the unit (transparent) suggests a relatively uniform sedimentation regime.

A dominantly carbonate section is inferred for most of the interval based upon the presence of a possible carbonate shelf in the southwest part of the area (discussed above) and a carbonate section characterizing most of the rocks overlying the interval (discussed below). The rocks probably contain progressively more terrigenous clastic components to the northwest. These relationships are shown schematically in Figure 16.

Deposition of this interval followed a major Late Cretaceous shift of the continental shelf margin from the Blake Escarpment to about its present location and establishment of deep water over the Blake Plateau. Evidently shallow-water carbonate

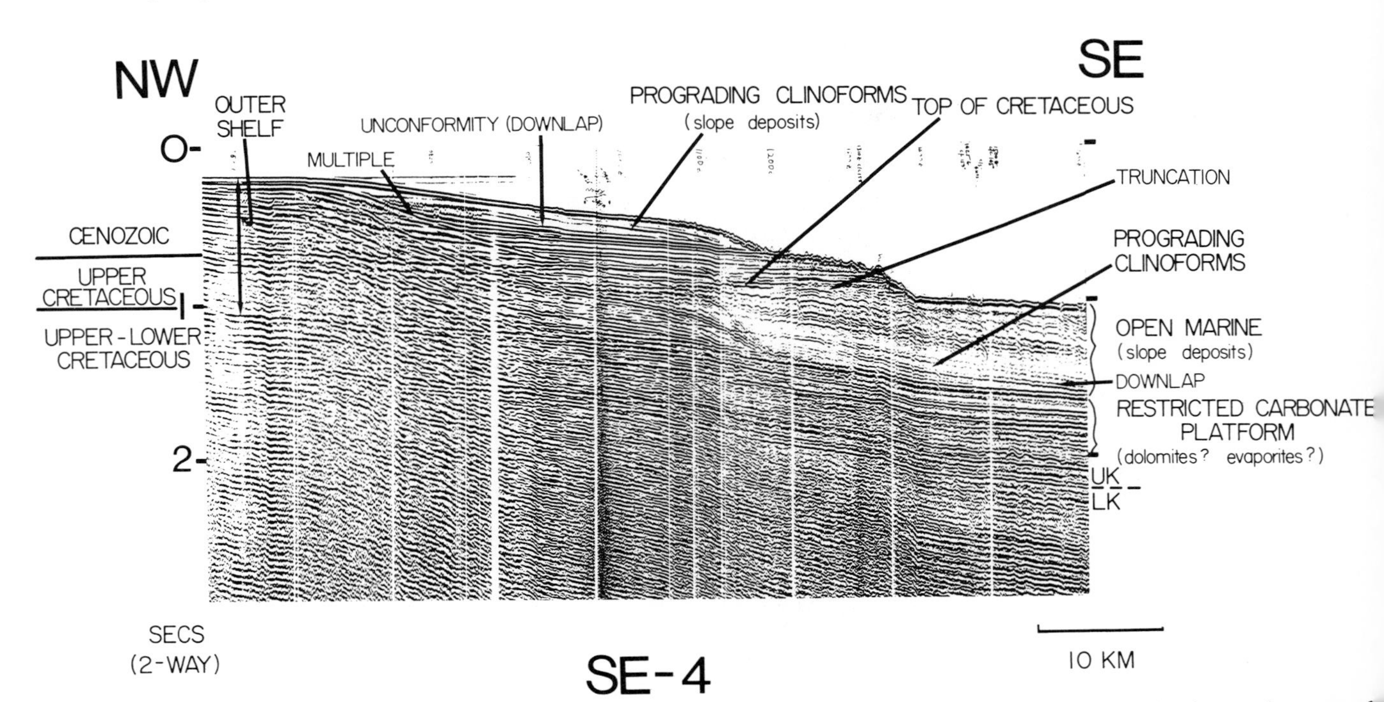

FIG. 12. Near-trace shipboard monitor record of line SE-4 across outer continental shelf-slope showing intervals mapped and unconformable relationships defining unit boundaries. Gently seaward-dipping reflections interpreted to be prograding clinoforms and probably represent continental slope deposits. Shown also are other inferred depositional environments.

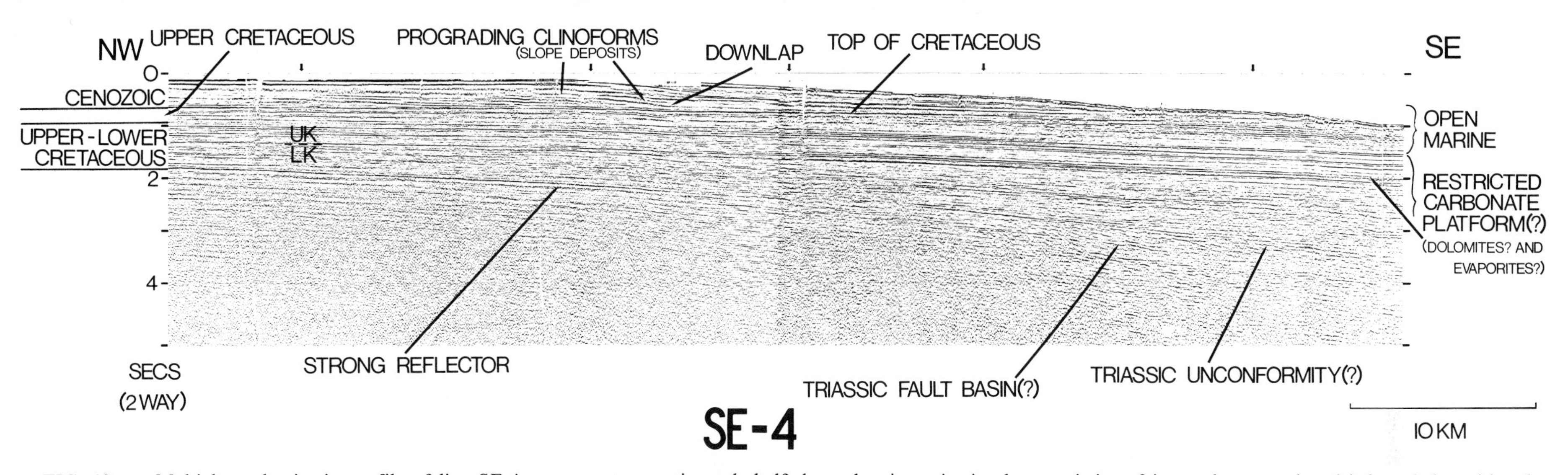

FIG. 13. Multichannel seismic profile of line SE-4 across outer continental shelf-slope showing seismic characteristics of intervals mapped and inferred depositional environments. Lower Cretaceous strata onlap strong reflector (possibly top of Jurassic volcanic layer, top of crystalline basement, or top of high-velocity sedimentary layer). Possible truncated fault basin beneath unconformity along eastern end of line. Strong Cretaceous reflector sequences may be high-velocity dolomites and evaporites interbedded with lower velocity carbonates and clastics.

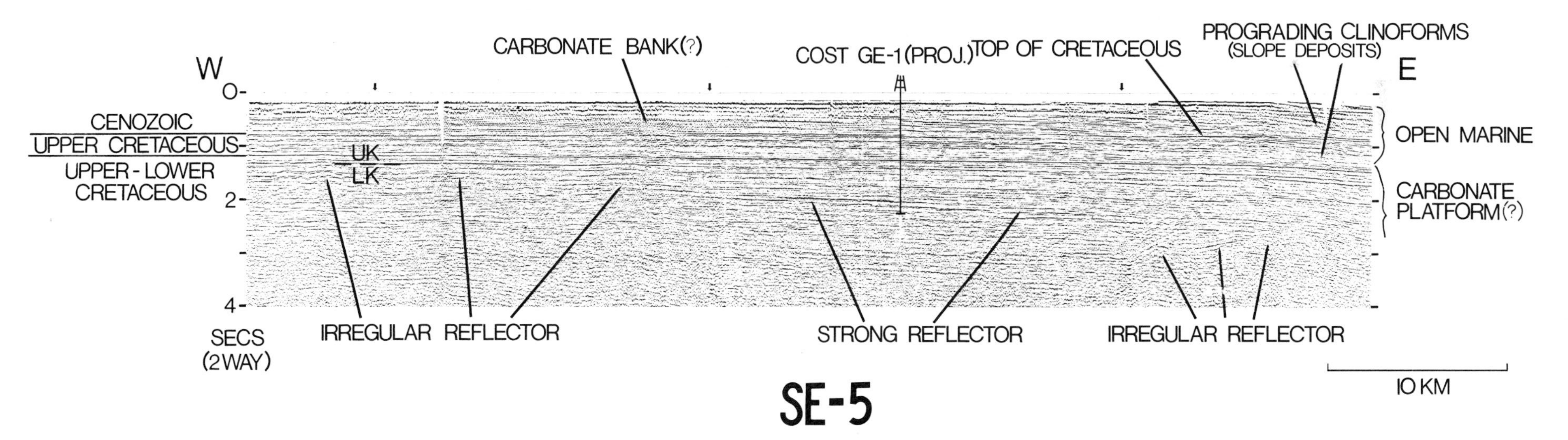

FIG. 14. Multichannel seismic profile of line SE-5 across outer continental shelf-slope showing seismic intervals mapped and inferred depositional environments. Irregular reflector at east and west end of line (possibly top of varied basement complex) changes to strong, smooth reflector in middle (possibly top of Lower Jurassic volcanic layer, top of crystalline basement, or top of high-velocity sedimentary layer). Expected section penetrated by COST well GE-1 is shown.

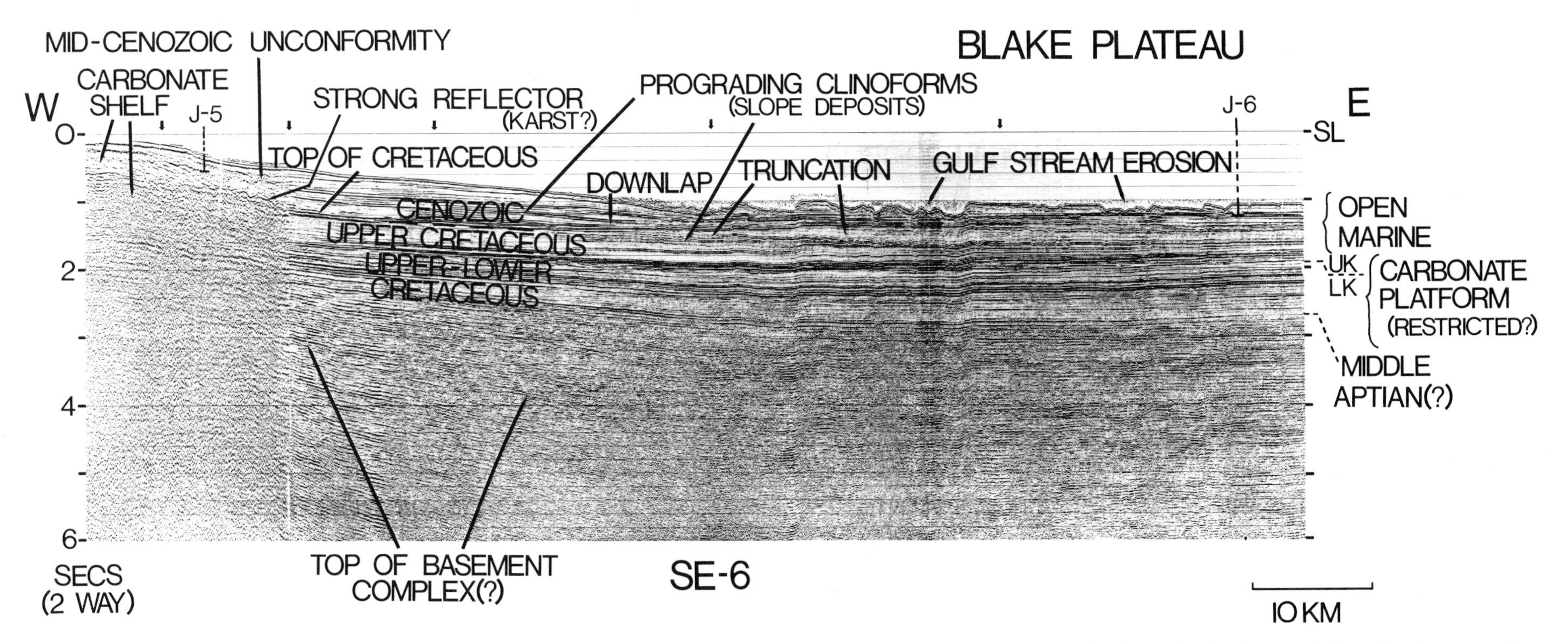

FIG. 15. Multichannel seismic profile of line SE-6 across outer continental shelf-slope and inner Blake Plateau showing seismic characteristics of units mapped, unconformable relationships of interval boundaries, and inferred depositional environments. Gently seaward-dipping reflections are interpreted to be prograding clinoforms and probably represent continental slope deposits. Change in seismic character beneath slope apparently due to strong, irregular reflector that masks deeper reflectors (possibly Cenozoic unconformity on top of dense carbonate shelf deposits—karst topography?).

sedimentation was not able to keep pace with the continued subsidence of the margin and the overall Late Cretaceous rise in sea level. Perhaps this was caused, in part, by the establishment of vigorous bottom currents across the plateau.

Cenozoic interval—The youngest interval mapped in the study area consists internally of several depositional sequences, all separated by unconformities. No attempt is made to map individual sequences in this preliminary study, although one prominent buried unconformity within the interval seen on most of the crosslines is shown on the sections. This buried erosion surface is strikingly displayed on the monitor record of line SE-3 (Fig. 11). Beneath the outer shelf and slope the interval is characterized by gently seaward-dipping, semidiscontinuous, variable-amplitude reflectors. The lowermost strata downlap onto the Cretaceous-Tertiary unconformity that forms the lower boundary. The upper boundary is the seafloor, which itself is a prominent erosion surface along the inner Blake Plateau. The interval is best displayed on lines SE-4 and SE-6 (Figs. 6, 8, 12, 13, 15).

This interval consists of Cenozoic rocks. As stated above, the lower boundary unconformity probably represents approximately the Cretaceous-Tertiary boundary. The Cenozoic age is confirmed by the early JOIDES core holes drilled in the area as well as the recent USGS core holes (JOIDES, 1965; Hathaway et al, 1976), all of which penetrated the interval to varying depths (Figs. 2, 3, 5, 8, 11, 15). Only one core hole (site 6004) actually penetrated the entire unit and reached Upper Cretaceous strata (Figs. 5 and 11). The prominent unconformity within the unit on line SE-3 is tentatively correlated with an Oligocene unconformity drilled in core hole no. 6004 (Figs. 5 and 11).

The gently seaward-dipping reflectors have a sigmoidal configuration and are interpreted to be prograding clinoforms and to represent a prograding slope environment similar to the present slope. Although the sigmoidal configuration suggests relatively low-energy depositional regimes, the variable nature of the reflections within the interval indicates changing or nonuniform depositional conditions. This interpretation is confirmed by the varied lithologies found in the core holes, which included calcareous silt, calcareous clay, calcareous oozes, calcilutites, chert, dolomite, calcarenitic sand, and quartzose sand (JOIDES, 1965; Hathaway et al, 1976). The rocks are dominantly carbonates, although more terrigenous clastics are found in the younger part of the section and to the north.

The internal geometry of this Cenozoic interval indicates a history consisting of overall progradation of the shelf and slope interrupted by periods of erosion or nondeposition. The present shelf-slope, therefore, is a constructional feature and has no apparent structural origin. The unconformities within the section probably reflect the many lowerings of sea level during the Cenozoic discussed by Vail et al (1977), but no attempt has been made here to correlate the unconformities with the Cenozoic cycle chart of Vail et al (1977). Strong currents similar to the present-day Gulf Stream undoubtedly played an important role in the Cenozoic sedimentation patterns and probably were responsible for forming the buried erosion surfaces as well as the erosional topography seen on the seafloor today. The currents apparently also influenced the location of the prograding Cenozoic shelf-slope margin and were responsible for the much-abbreviated Cenozoic section on the Blake Plateau. The detailed history of the Cenozoic section in the Southeast Georgia Embay-

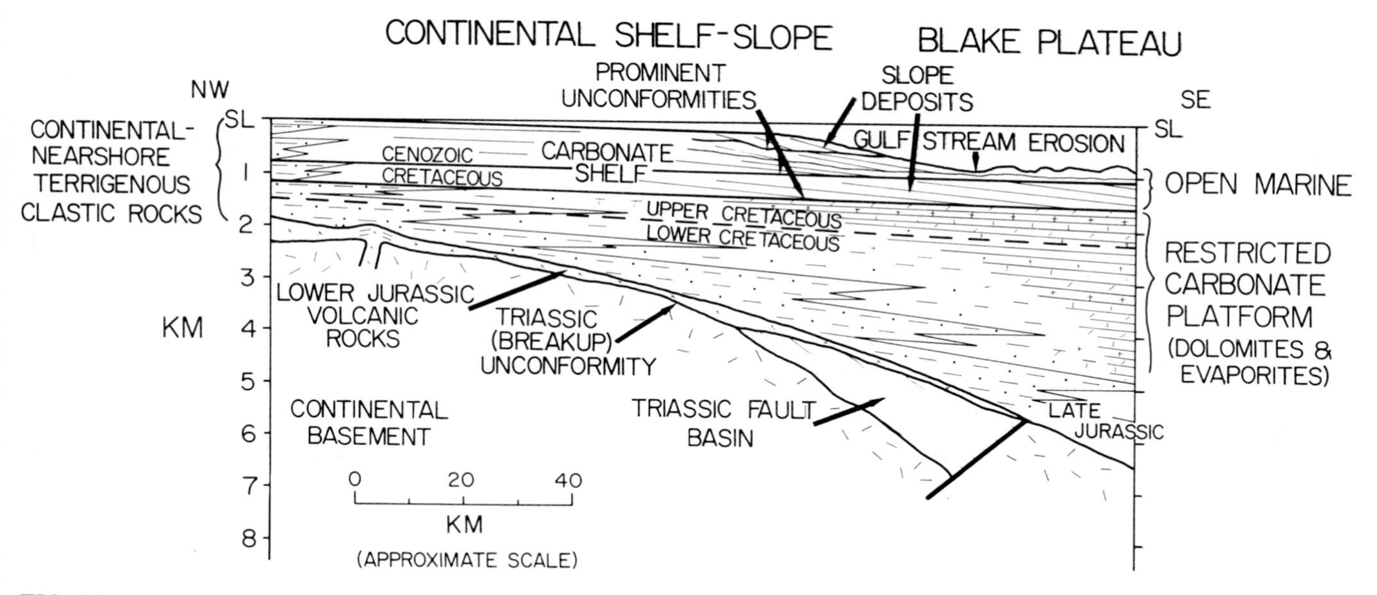

FIG. 16. Generalized schematic northwest-southeast cross-section across Southeast Georgia embayment showing inferred geologic and structural configuration of basement and inferred facies relationships and depositional settings of seismic intervals mapped.

Table 2. Summary of Geologic History, Southeast Georgia Embayment

Age	Approximate Time (m.y.)	Major Events
Cenozoic	65–0	Continued progradation of continental shelf-slope. Variable sedimentation. Deposition of mainly fine-grained carbonates. Numerous periods of erosion or nondeposition (unconformities). Sedimentation influenced by changes in sea level and vigorous bottom currents.
Late Cretaceous	80–65	Continued subsidence, overall rise in sea level and establishment of Gulf Stream-like currents cause major shift of shelf margin from Blake Escarpment to approximately present location and establishment of deeper water over Blake Plateau. Progradation of uniform, fine-grained (mainly carbonate) sediments from shelf onto inner Blake Plateau.
Early Cretaceous to middle Late Cretaceous	135–80	Broad carbonate platform established across Blake Plateau. Overall transgression and progressive onlap of sediments onto tilted unconformity. Sediments probably transitional between siliceous clastics (onshore and north) and restricted carbonate platform facies--evaporites and dolomites (offshore and south).
Middle Jurassic to Late Jurassic	170–135	Subsidence of continental margin. Transgression of Jurassic seas, but generally erosion or nondeposition as far west as Southeast Georgia embayment.
Early Jurassic	180–170	Possible widespread deposition of volcanics north of approximately 31°N filling in erosion surface and forming smooth, high-velocity acoustic basement.
Late Triassic to Early Jurassic	210–180	Uplift and rifting. Deposition of continental sediments and volcanics in fault basins. Widespread erosion (*breakup unconformity*). Initial separation of continents along outer edge of Blake Plateau.

ment area is currently being studied by the U.S. Geological Survey (Edsall and Dillon, 1977).

Geologic History

The geologic history of the Southeast Georgia Embayment is based on the geologic interpretations discussed above and is summarized in the schematic depositional model (Fig. 16) and also in Table 2. The model shows the interpreted transition from carbonate rocks in the south and east to terrigenous clastic rocks in the north and west. It also reflects the major shift from shallow water to deeper water deposition on the Blake Plateau in Late Cretaceous time and the establishment of a prograding shelf margin at about the present location.

REFERENCES CITED

Antoine, J. W., and V. J. Henry, Jr., 1965, Seismic refraction study of shallow part of continental shelf off Georgia coast: AAPG Bull., v. 49, p. 601–609.

Benson, W. E., et al, 1976, In the North Atlantic deep-sea drilling: Geotimes, v. 21, p. 23–26.

Buffler, R. T., T. H. Shipley, and J. S. Watkins, 1978, Blake continental margin seismic section: AAPG Seismic Section No. 2.

Dillon, W. P., et al, 1978, Structure and development of the Southeast Georgia embayment and northern Blake Plateau: preliminary analysis: this volume.

Edsall, D. W., and W. P. Dillon, 1977, Geologic development of Cenozoic continental margin of Southeast Georgia embayment (abs.): AAPG Bull., v. 61, p. 782.

Emery, K. O., and E. F. K. Zarudski, 1967, Seismic reflection profiles along the drill holes on the continental margin off Florida: U.S. Geol. Survey Prof. Paper 581-A, 8 p.

Ewing, J. I., M. Ewing, and R. Leyden, 1966, Seismic profiler survey of Blake Plateau: AAPG Bull., v. 50, p. 1948–1971.

Falvey, D. A., 1974, The development of continental margins in plate tectonic theory: Australian Petrol. Explor. Assoc. Jour., v. 14, p. 95–106.

Hathaway, J. C., et al, 1976, Preliminary summary of the 1976 Atlantic margin coring project of the U.S. Geological Survey: U.S. Geol. Surv., Open File Report 76-844, 217 p.

Heezen, B. C., and R. E. Sheridan, 1966, Lower Cretaceous rocks (Neocomian-Albian) dredged from Blake Escarpment: Science, v. 154, p. 1644–1647.

Herrick, S. M., and R. C. Vorhis, 1963, Subsurface geology of the Georgia Coastal Plain: Georgia Geol. Survey, Inf. Cir. 25, 78 p.

Hersey, J. B., et al, 1959, Geophysical investigation of the conti-

nental margin between Cape May, Virginia, and Jacksonville, Florida: Geol. Soc. America Bull., v. 70, p. 437–466.
JOIDES, 1965, Ocean drilling on the continental margin: Science, v. 150, p. 709–716.
Krivoy, H. L., and H. C. Eppert, 1977, Simple Bouguer anomaly representation over a part of the Atlantic continental shelf and adjacent land areas of Georgia, the Carolinas, and northern Florida: U.S. Geol. Survey, Open File Report 77-316.
Montalbetti, J. F., 1971, Computer determinations of seismic velocities—a review: Jour. Canadian Soc. Explor. Geophys., v. 7, p. 32–35.
Sheridan, R. E., et al, 1966, Seismic refraction study of continental margin east of Florida: AAPG Bull., v. 50, p. 1972–1991.
Shipley, T. H., R. T. Buffler, and J. S. Watkins, 1978, Seismic stratigraphy and geologic history of the Blake Plateau and adjacent western Atlantic continental margin: AAPG Bull., v. 62, p. 792–812.
Tanner, M. T., and F. Koehler, 1969, Velocity spectra—digital computer derivation and application of velocity functions: Geophysics, v. 34, p. 859–881.
Uchupi, E., 1967, The continental margin south of Cape Hatteras, North Carolina—shallow structure: Southeastern Geol. v. 8, p. 155–177.
——— 1970, Atlantic continental shelf and slope of the United States—shallow structure: U.S. Geol. Survey, Prof. Paper 529-I, 44 p.
——— and K. O. Emery, 1967, Structure of continental margin off Atlantic coast of United States: AAPG Bull., v. 51, p. 223–234.
Vail, P. R., et al, 1977, Seismic stratigraphy and global changes in sea level: AAPG Memoir 26, p. 49–212.
Woollard, G. P., W. E. Bonini, and R. P. Meyer, 1957, A seismic refraction study of the subsurface geology of the Atlantic coastal plain and continental shelf between Virginia and Florida: Univ. of Wisconsin Tech. Rept., Contract N7onr-28512, 128 p.
Zarudski, E. F. K., and E. Uchupi, 1968, Organic reef alignments on the continental margin south of Cape Hatteras: Geol. Soc. America Bull., v. 79, p. 1867–1870.

Structure and Development of the Southeast Georgia Embayment and Northern Blake Plateau: Preliminary Analysis[1]

WILLIAM P. DILLON,[2] CHARLES K. PAULL,[3]
RICHARD T. BUFFLER,[4] AND JEAN-PIERRE FAIL[5]

Abstract Multichannel seismic reflection profiles from the Southeast Georgia Embayment and northern Blake Plateau show reflectors that have been correlated tentatively with horizons of known age. The top of the Cretaceous extends smoothly seaward beneath the continental shelf and Blake Plateau, unaffected at the present shelf edge. A reflector inferred to correspond approximately to the top of the Jurassic section onlaps and pinches out against rocks below. A widespread smooth reflector probably represents a volcanic layer of Early Jurassic age that underlies only the northwestern part of the research area. A major unconformity beneath the inferred volcanic layer is probably of Late Triassic or Early Jurassic age. This unconformity dips rather smoothly seaward beneath the northern Blake Plateau, but south of a geological boundary near 31°N, it has subsided much more rapidly, and reaches depths of more than 12 km. Development of the continental margin north of the boundary began with rifting and subsidence of continental basement in the Triassic. An episode of volcanism may have been due to stresses associated with a spreading center jump at about 175 million years ago. Jurassic and Cretaceous deposits form an onlapping wedge above the inferred early Jurassic volcanics and Triassic sedimentary rocks. During Cenozoic times, development of Gulf Stream flow caused a radical decrease in sedimentation rates so that a shelf that was much narrower than the Mesozoic shelf was formed by progradation against the inner edge of the stream. South of the 31°N geological boundary, the basement probably is semi-oceanic and reef growth, unlike that in the area to the north, has been very active at the outer edge of the plateau.

INTRODUCTION

The post-Triassic development of the continental margin off the southeastern United States is characterized by major subsidence and deposition of thick sedimentary sections. The region, comprising the Southeast Georgia Embayment and Blake Plateau Trough (Fig. 1) (Maher, 1971) is one of the three areas on the U.S. Atlantic continental margin that are considered good prospects for petroleum exploration. As part of its resource studies, the U.S. Geological Survey, Office of Marine Geology, has obtained 4680 km of multichannel seismic reflection profiles south of Cape Fear. High-accuracy gravity profiles and a closely spaced aeromagnetic survey have also been made, but this paper will present only a preliminary analysis of the seismic profiles for the northern part of the area (Fig. 2). Many single-channel, shallow-penetration profiles have been presented by Ewing, Ewing and Leyden (1966), Uchupi (1967, 1970), Uchupi and Emery (1967), Emery and Zarudzki (1967), Zarudzki and Uchupi (1968), Pilkey, MacIntyre and Uchupi (1971). Some multichannel data have been presented for the deep sea (Dillon et al, 1976). Due to the very thick sedimentary section, modern, deep penetration, multichannel seismic profiling techniques are necessary to understand the geological structure and history of this region. This paper represents the first publication of a grid of such profiles for the southeastern U.S. margin which is adequate to convey a generalized understanding of the area.

GEOLOGICAL SETTING

Submarine topography in the U.S. South Atlantic region is somewhat unusual (Fig. 2). A continental shelf is connected by the Florida-Hatteras slope to a broad flat plateau, the Blake Plateau, at about 800 m below sea level. The seafloor drops very abruptly into the deep ocean at the seaward edge of the plateau to form the Blake Escarpment south of lat. 30°N, whereas to the north, the transition to abyssal depths is less abrupt.

The Southeast Georgia Embayment (Fig. 1) is an east-plunging depression recessed into the Atlantic

[1]Manuscript received, July 20, 1977; accepted, January 5, 1978.

[2]U.S. Geological Survey, Woods Hole, Massachusetts 02543.
[3]U.S. Geological Survey, Woods Hole, Massachusetts 02543.
[4]University of Texas, Marine Science Institute, Galveston, Texas.
[5]Institut Français du Pétrole, Rueil Malmaison, FRANCE.

We wish to thank John Schlee and Kim Klitgord who reviewed the manuscript. Discussions with Mahlon Ball, John Grow, Robert Sheridan, and Kim Klitgord were invaluable in developing this paper.
Article Identification Number:
0065-731X/78/MO29-0003/$03.00/0.
(see copyright notice, front of book)

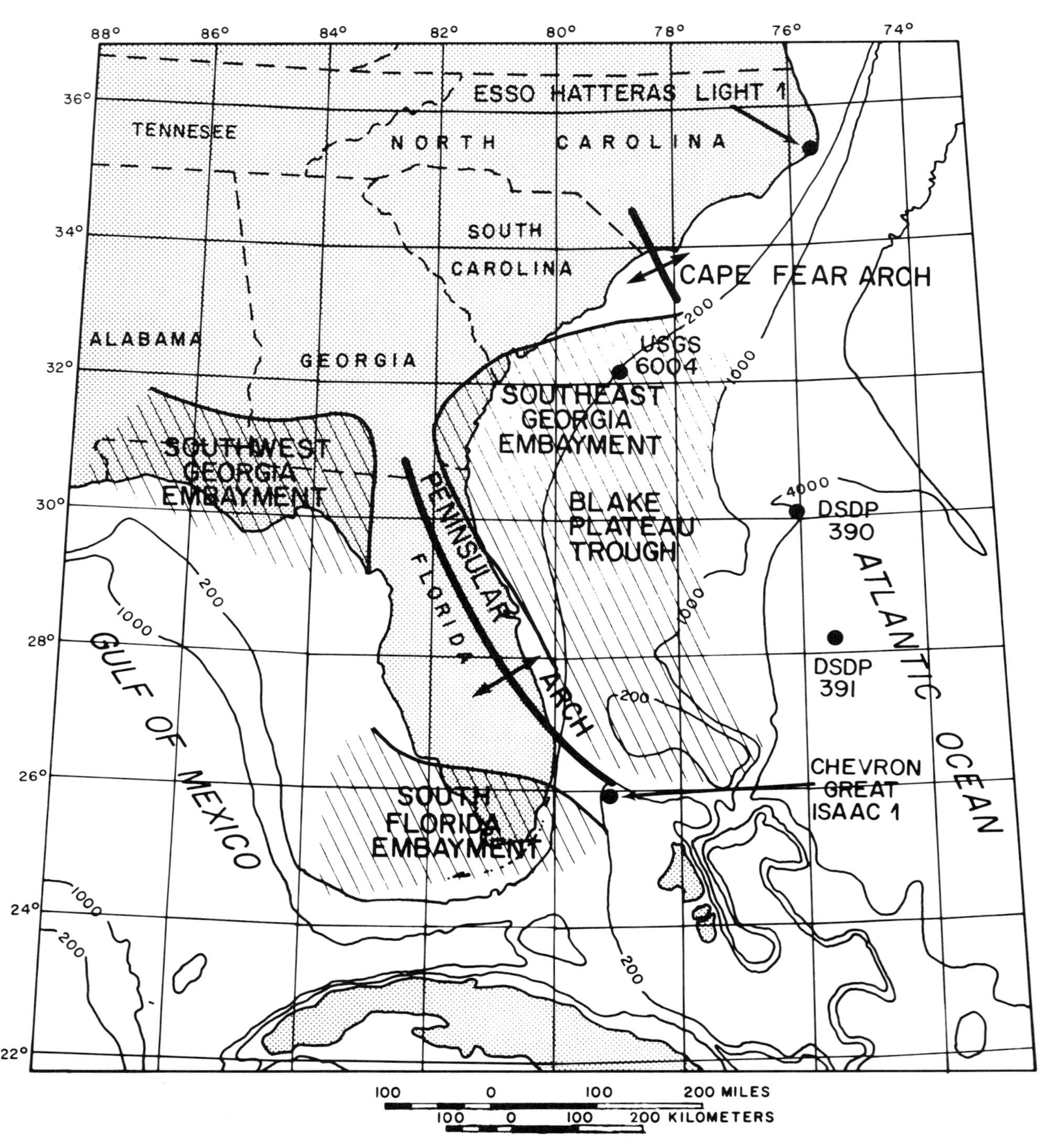

FIG. 1. U.S. South Atlantic region: sedimentary basins of the continental margin and selected drill sites. Basins generalized from King (1969), Meyerhoff and Hatten (1974), Barnett (1975), and Peter Popenoe and Isidore Zietz (unpub. ms.). Contours in meters below sea level.

Coastal Plain between Cape Fear, North Carolina and Jacksonville, Florida. The Cape Fear Arch bounds the embayment on the north, and the Peninsular Arch forms the southwestern limit. The embayment opens into the Blake Plateau Trough with no structural demarcation between them.

The Paleozoic and Precambrian basement, as exposed in the Piedmont, is formed of igneous and metamorphic rocks (Overstreet and Bell, 1965). The pre-Cretaceous "basement" in wells in Georgia and northern Florida is Ordovician to Devonian sandstone and shale (Applin, 1951; Rainwater, 1971). However, much of the onshore part of the Southeast Georgia Embayment apparently is underlain by undeformed tuffaceous clastic deposits intermixed with basaltic and rhyolitic flows (Milton and Hurst, 1965; Milton, 1972; Barnett, 1975; Peter Popenoe and Isidore Zietz, unpub. ms.; Gohn et al, in

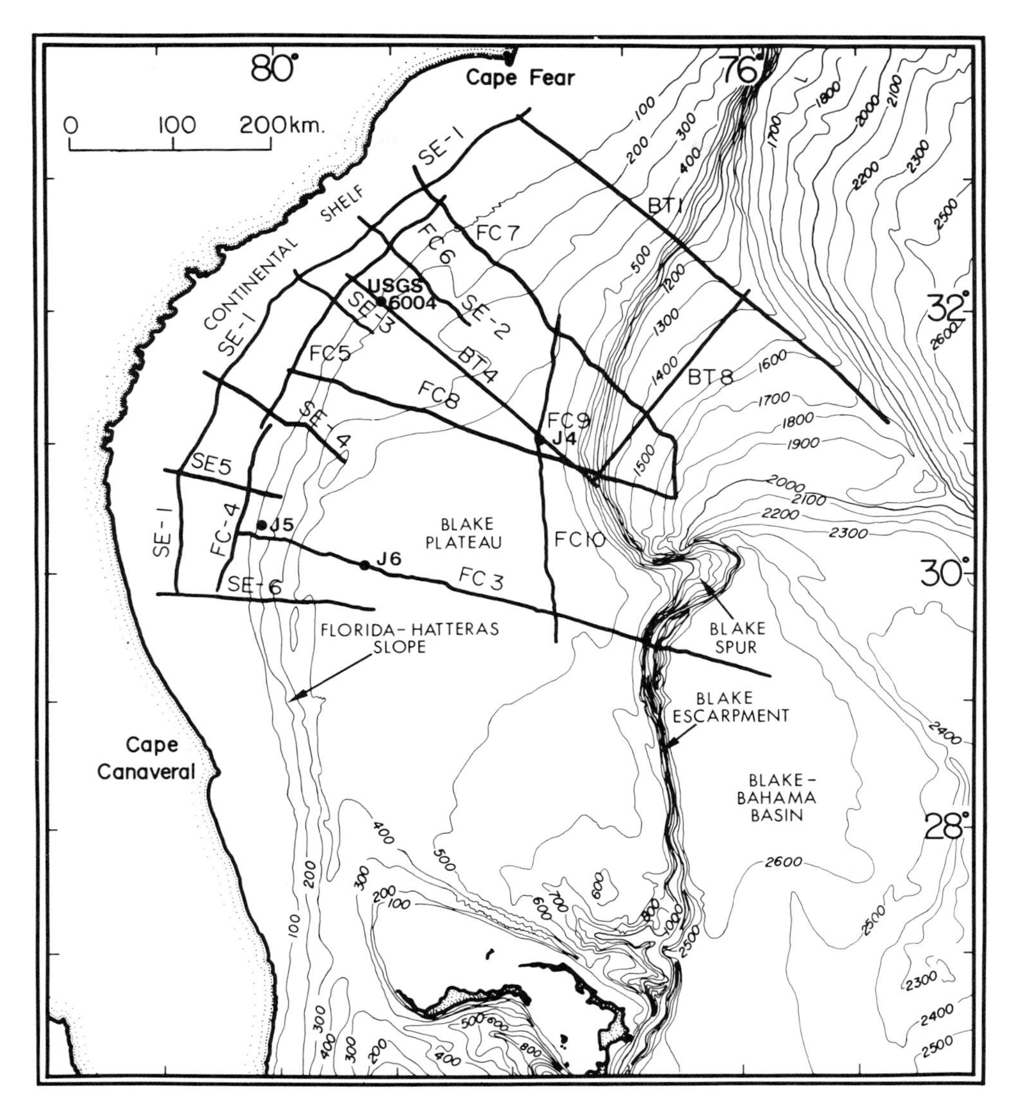

FIG. 2. Multichannel seismic reflection profile tracks in the northern part of the U.S. South Atlantic region and drill sites crossed by the profiles. Contours in fathoms below sealevel. The profiles were collected by three organizations. Those designated FC were collected during two cooperative cruises with the Institut Français du Pétrole. Profiles identified by SE were collected by the University of Texas, Marine Science Institute under a contract with the U.S. Geological Survey. The BT lines were purchased from Geophysical Service, Inc.

press; T. M. Chowns, personal communication, 1976). Some of these volcanic rocks are associated with continental terrigenous clastic deposits, which may have been deposited in troughs of a failed rift arm during early opening of the Atlantic, presumably in Triassic time (D. W. Rankin, unpub. ms.).

Rocks of Jurassic age do not crop out in the southeastern United States. Possibly, Upper Jurassic or Lower Cretaceous deposits may be present as finely crystalline, partly oolitic limestone and gray shale beds grading downward to conglomerate and sandstone, possibly continental, which were drilled at Cape Hatteras (ESSO-Hatteras Light 1, see Fig. 1) (Spangler, 1950; Maher, 1971; Brown, Miller and Swain, 1972). The Lower Cretaceous-Jurassic section drilled in Florida and the Bahamas (for example, in the Chevron Great Isaac 1 well, Fig. 1) consists of shallow marine, highly organic carbonate rocks, and has increasing amounts of evaporite deposits at depth in the Bahamas (Tator and Hatfield, 1975; Jacobs, 1977).

The Southeast Georgia Embayment occurs in a transitional zone between a Cretaceous and Cenozoic, predominantly clastic depositional province north of Cape Hatteras and a carbonate province which includes Florida and the Bahamas. More than 400 wells have been drilled in Georgia and Florida, some of which reached basement rocks, and a few wells have been drilled in the Bahamas area. In the offshore area, several test holes have been drilled by the USGS (U.S. Geological Survey) and the JOIDES (Joint Oceanographic Institutions Deep Earth Sampling) programs several hundred meters into the continental shelf off Georgia and Florida, but only one of these has penetrated pre-Tertiary strata. At this site (USGS, 6004, Fig. 1) Upper Cretaceous silty calcareous clay was sampled, which was characteristic of an outer shelf or upper slope environment (Hathaway et al, 1976; C. W. Poag, personal communication, 1976). Cretaceous deposits were also sampled at DSDP site 390 at the outer edge of the Blake Plateau (Fig. 1) (Benson, et al, 1976).

ESTIMATES OF AGES OF REFLECTORS

During deposition, many horizons were produced which act as reflectors and may be traced through our grid of profiles. We have chosen some of these as representing stratigraphically significant interfaces (Figs. 3, 4, 5) and have mapped them to determine regional structure. Before considering the structure, however, we will discuss the reasons for these choices and the degree of confidence to be placed in them.

Top of the Cretaceous

Our most dependable stratigraphic correlation to a reflector probably occurs at the top of Cretaceous rocks, because that surface was penetrated at USGS 6004 (Fig. 1) (Hathaway et al, 1976). This site was chosen on our profile BT4 (Fig. 4) and thus can be tied to our grid of profiles. This horizon also was drilled in the offshore part of the region at the DSDP (Deep Sea Drilling Project) site 390 (Fig. 1) but this site is not easily tied to our grid. Six other JOIDES drill sites sampled Eocene, and two encountered Paleocene rocks (JOIDES, 1965; Charm et al, 1969). Estimates of depth to the Cretaceous top on profiles taken in conjunction with this drilling have been made by Emery and Zarudzki (1967). As noted on the profiles, our tracks passed near JOIDES site 4 (J4 on profile BT4, Fig. 4) and sites 5 and 6 (J5 and J6 on profile FC3, Fig. 5). Thus, we have some basis for choosing the position of the Cretaceous top. Locations of dredged Cretaceous rocks as indicated by Uchupi (1967, 1970) also have been taken into account.

In the deep sea, in the Blake-Bahama Basin and Blake Ridge areas (Fig. 2), our choice of Cretaceous top is firm because it is associated in all drill holes with a major hiatus in deposition. At DSDP site 391 (Fig. 1) Miocene material overlays Albian (Benson, et al, 1976) and such breaks were also noted on Leg XI of DSDP (Ewing and Hollister, 1972). The hiatus is visible clearly in the profiles as a major, channeled unconformity, as for example, at the seaward end of profile FC7 (Fig. 3).

Top of the Lower Cretaceous

On line FC3 (Fig. 5) an attempt has been made to pick the top of the Lower Cretaceous by comparison to DSDP site 390 (Fig. 1). As the line does not pass through the drillsite and the depositional environment of the site is not typical of that for the plateau (Benson et al, 1976), this pick is not considered dependable and has not been carried to the rest of our net.

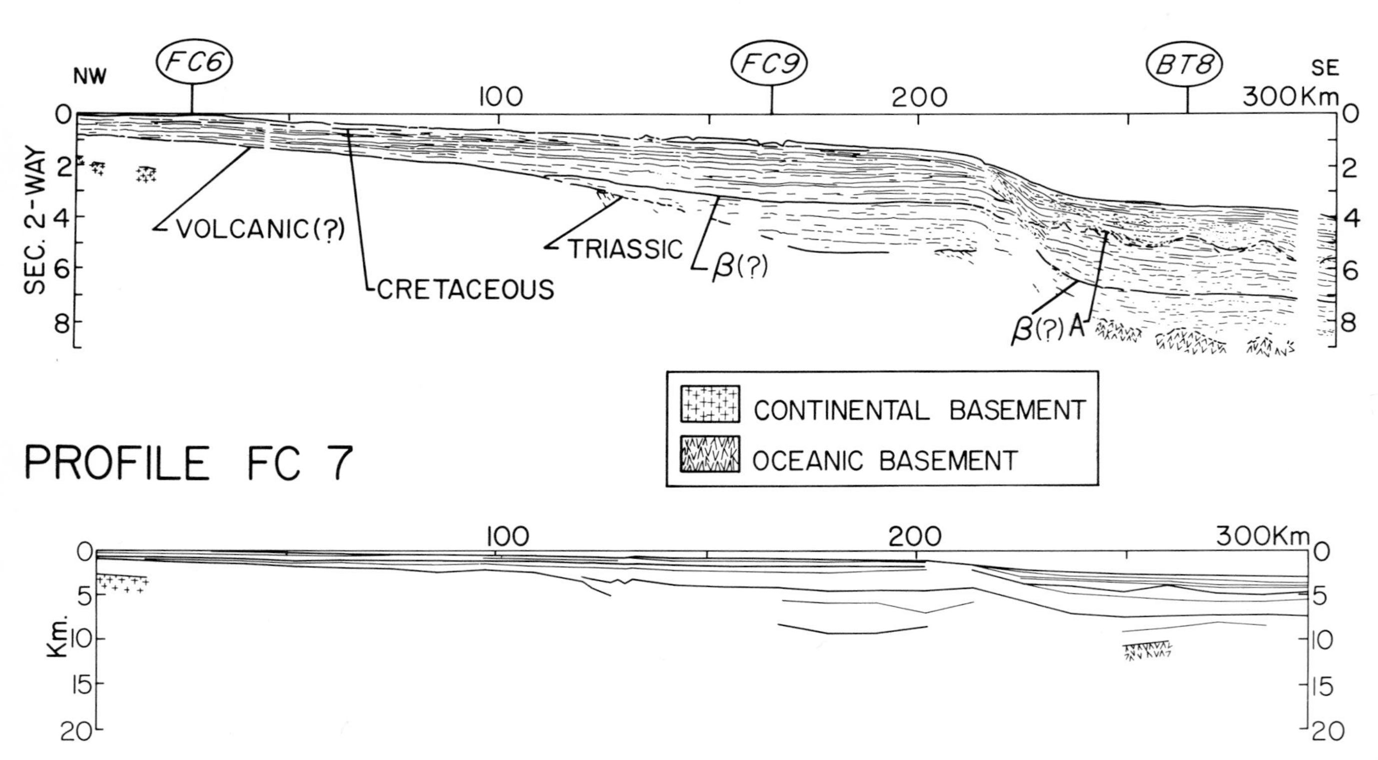

FIG. 3. Interpretation of multichannel seismic reflection profile FC7. Location shown on Figure 2. The time section (upper profile) has a vertical exaggeration of about 8:1 on the basis of water velocity. The depth section (lower profile) has a vertical exaggeration of about 2:1. Horizons identified as β (?) have subsequently been adjudged older as discussed in the text. Profile collected by the Institut Français du Pétrole.

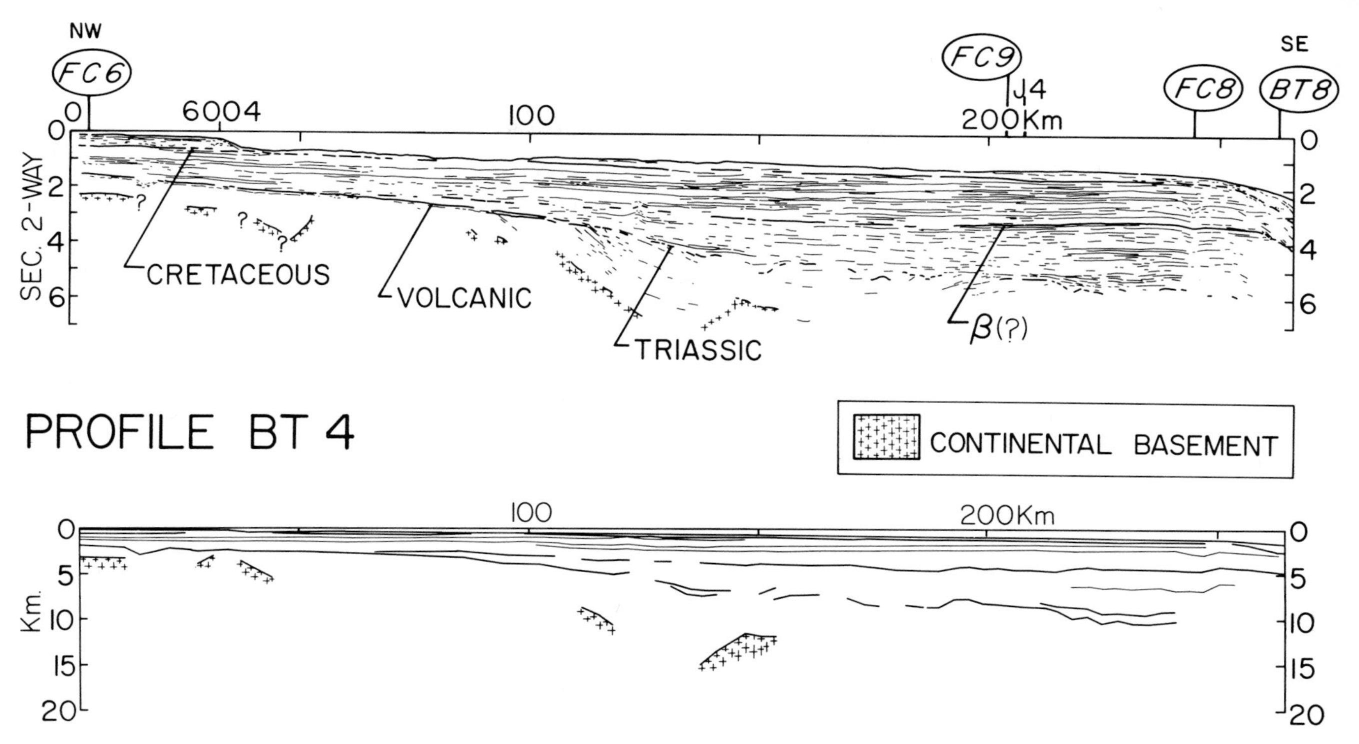

FIG. 4. Interpretation of multichannel seismic reflection profile BT7. Location shown on Figure 2. The time section (upper profile) has a vertical exaggeration of about 8:1 on the basis of water velocity. The depth section (lower profile) has a vertical exaggeration of about 2:1. Horizons identified as β (?) have subsequently been adjudged older, as discussed in the text. J4 is JOIDES drill site 4. Profile collected by Geophysical Service Inc.

Top of the Neocomian (Horizon Beta) or Top of the Tithonian

An attempt was made to pick Horizon Beta in the deep sea where it is commonly a recognizable reflector, and then to trace it beneath the Blake Plateau. This horizon has been drilled in the deep sea at DSDP site 391 (Fig. 1) where its reflection characteristics result from a change from Albian-Aptian carbonaceous clays to Barremian limestone (Benson et al, 1976). DSDP site 391 was drilled on one of our profiles (Fig. 6) that extends southward from the seaward end of profile FC3 (Dillon et al, 1976). Thus, we have great confidence in our stratigraphic pick on the seaward end of profile FC3. This profile crosses the steep Blake Escarpment, however, and the reflectors cannot be carried up onto the Blake Plateau. In order to estimate the age of an equivalent reflector beneath the plateau, we used profiles FC7 and FC8, which cross a much gentler slope between the plateau and deep sea. As these profiles are not tied to any drillsite, we picked reflectors on the deep sea sections which appeared similar in reflection characteristics to Horizon Beta in the region. This reflecting horizon can be traced from abyssal depths to the plateau. The reflector chosen as Horizon Beta may be older, however, as it appears to correlate with a reflector estimated to be Late Jurassic (Oxfordian) by T. H. Shipley, R. T. Buffler and J. S. Watkins (unpub. ms.). This older age is based on estimation of age of the ocean crust, based on assumed oceanic spreading rates, where the reflector appears to pinch out on a regional seismic line. Of course, sufficient sediment to produce a discrete reflector on a low-frequency multichannel profile probably does not accrete on a spreading ridge until well after the time that the basement was formed. We now believe that the reflector identified as β (?) or Barremian (?) (Figs. 3, 4, 5), therefore, may be the continuation of strong reflectors occurring within the Tithonian section at DSDP 391 (Fig. 6), except at the seaward end of line FC3 (Fig. 5) where it has been tied directly to DSDP 391 and does represent Horizon Beta.

Jurassic Volcanic (?) Layer

A very strong smooth reflector is commonly observed beneath the inner Blake Plateau and the continental shelf north of about 30.5°N (Fig. 7, inferred volcanic layer). It correlates with a high-velocity refractor of refraction velocity about 5.8 to 6.2 km/sec, a velocity previously correlated with basement (Woolard et al, 1957; Hersey et al, 1959; Antoine and Henry, 1965; Sheridan et al, 1966; Dowling, 1968; Emery et al, 1970). We believe that this reflector represents basaltic flows and pyroclastic deposits because it can be projected landward to volcanic rocks (basalts) of the U.S. Geological Survey test wells at Clubhouse Crossroads, west of Charleston (Fig. 11) (Gohn et al, in press). Reflection profiles carried out on land near these wells (H. D. Akermann, personal communication, 1977) produce reflections from the volcanic layer having

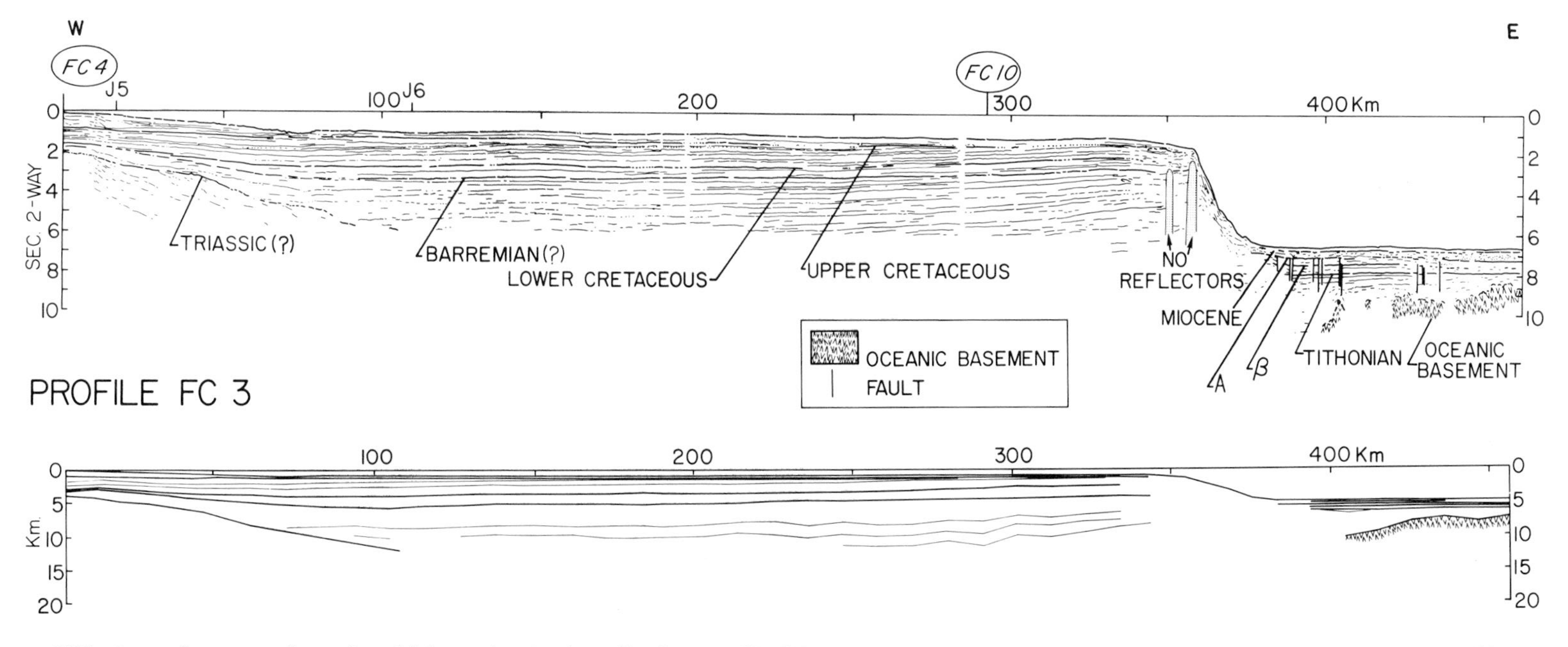

FIG. 5. Interpretation of multichannel seismic reflection profile FC3. Location shown on Figure 2. The time section (upper profile) has a vertical exaggeration of about 8:1 on the basis of water velocity. The depth section (lower profile) has a vertical exaggeration of about 2:1. Horizons identified as β (?) or Barremian(?) have subsequently been adjudged older, as discussed in the text. J5 and J6 are JOIDES drill sites 5 and 6. Profile collected by the Institut Français du Pétrole.

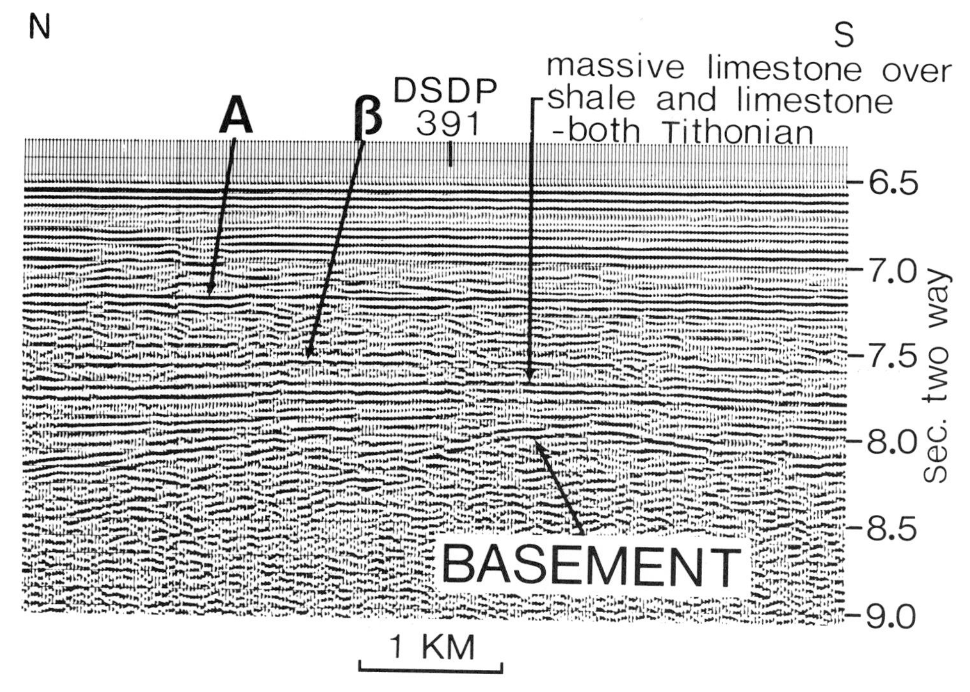

FIG. 6. Part of a seismic profile through DSDP drill site 391 showing some of the reflectors correlated with drill data. The continuation of this profile connects to the seaward end of profile FC3 (Fig. 5).

a very similar acoustic return to those collected at sea, as shown in Fig. 7. Relation of this reflector to rhyolite tuff and porphyry drilled onshore in Georgia (Maher, 1971) is also shown in Fig. 11. The rhyolites may be older, however—perhaps Paleozoic, according to T. M. Chowns (personal communication, 1977). The high-velocity layer might be carbonate, but this seems less likely, given its relation to volcanic rocks drilled on land (admittedly 80 km away from our closest data). Furthermore, its presence only in the northern part of the area and its disappearance to the south seem inconsistent with the

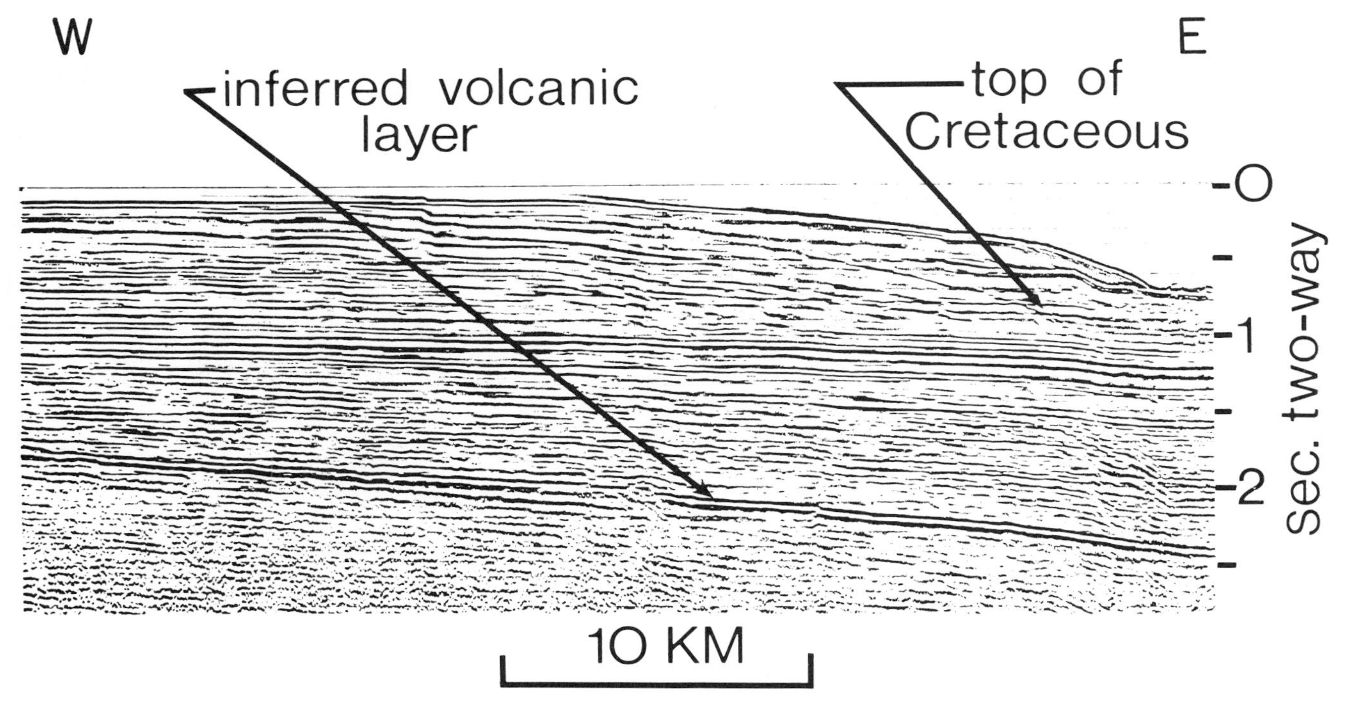

FIG. 7. Part of seismic reflection profile FC8 showing the outer continental shelf off the central Southeast Georgia Embayment. A strong reflector is inferred to be a volcanic layer.

expected distribution of carbonates, because one finds in the more recent sediment, and would expect in the ancient ones, an increase in carbonate southward toward warmer water and away from the terrigenous clastic source formed by the Appalachian orogen (Milliman, 1972).

Both Precambrian-Paleozoic and Triassic-Jurassic volcanic rocks are present in Georgia and Florida (Milton and Hurst, 1965; Milton, 1972; Barnett, 1975; T. M. Chowns, personal communication, 1977). The strong reflector described here is assumed to be of Mesozoic age, however, as it covers an inferred Triassic-Early Jurassic unconformity (see discussion below) and is smooth and unfractured over a broad area as shown in our profiles. Two samples of basalt from the USGS Clubhouse Crossroads well, west of Charleston, discussed above, produced minimum K/Ar ages of Middle Cretaceous. However, these rocks are highly altered (Gohn et al, in press), and we feel that the minimum values are not reliable ages. The volcanic layer may be equivalent in age to the dike swarms of the Appalachians; according to de Boer (1967), the dike swarms resulted from a tensional episode in the Jurassic. Such tensional stresses might have been caused by the reorganization of plate movements associated with the spreading center jump 175 m.y. ago in the early drift phase of the Atlantic opening as discussed by Vogt (1973) and Klitgord and Behrendt (this volume).

Triassic Unconformity

An angular unconformity is inferred to cut Triassic deposits (Figs. 4, 5 and 8) and to have formed at the end of the rifting phase, prior to the opening of the Atlantic Ocean. The strata beneath the unconformity probably are not Paleozoic. At a location where we believe, on the basis of correlation with wells onshore, that we are receiving reflections from Paleozoic sedimentary rocks, the reflections are weaker and less continuous than those observed beneath this "Triassic" unconformity.

REGIONAL STRUCTURE

Structure on Top of the Cretaceous

The top of the Cretaceous rocks (Fig. 9) dips gently seaward beneath the continental shelf and Blake Plateau, with no structural effect at the present shelf

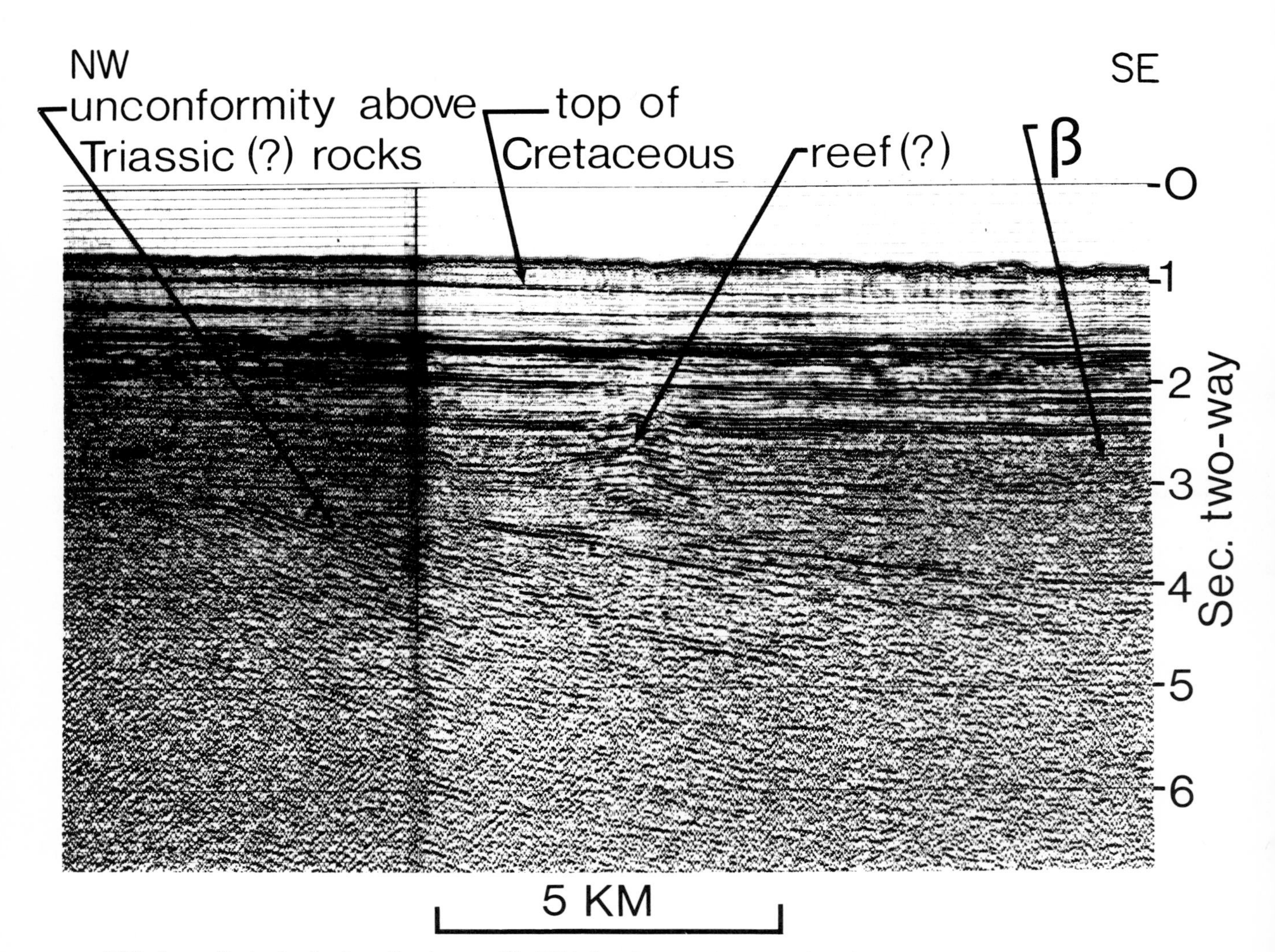

FIG. 8. Part of seismic reflection profile BT4 showing angular unconformity of inferred Late Triassic age.

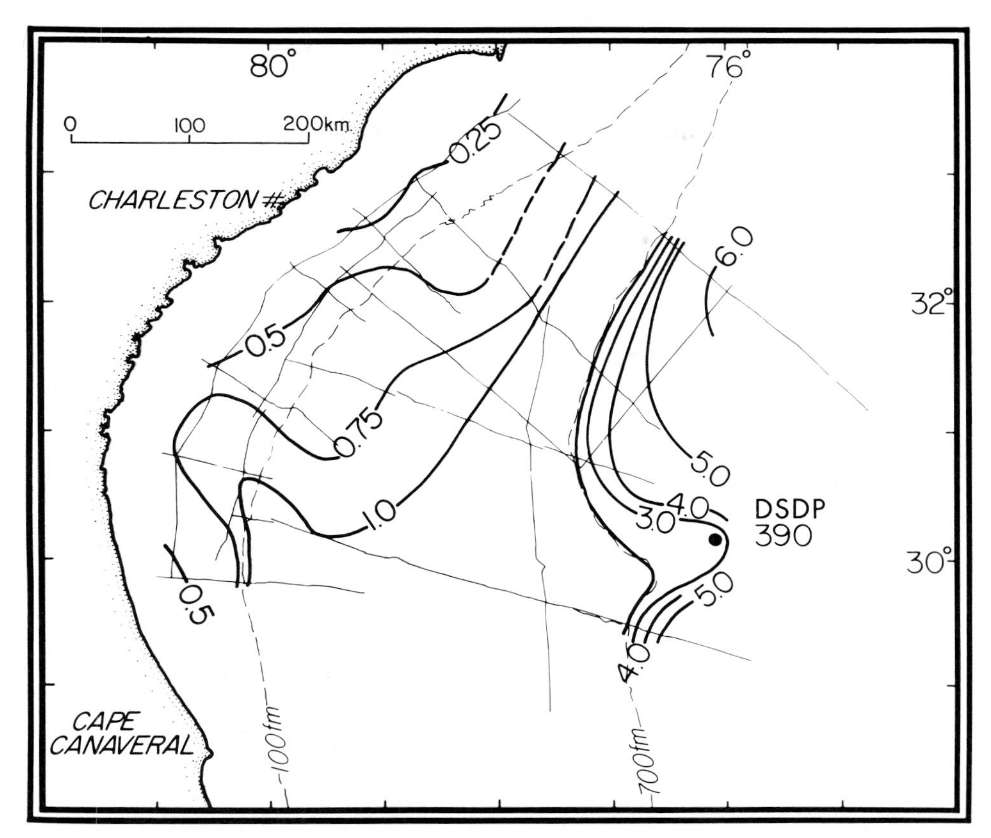

FIG. 9. Depth in kilometers below sea level to top of Cretaceous rocks. Light lines show profiling coverage. The 100- and 700-fathom curves (dashed) represent the approximate shelf edge and approximate seaward limit of the Blake Plateau. Contour interval 0.25 km from 0 to 1 km and 1 km below.

edge. The shelf, as shown in the profiles (Figs. 3, 4 and 5), is a sedimentary constructional feature of Cenozoic age atop the broad Cretaceous platform which forms the Blake Plateau. The Cape Fear Arch appears as a very weak warping of the Cretaceous top. A depression appears at 30° to 31°N beneath the shelf in the southwest corner of the area contoured in Figure 9. The southwest side of this re-entrant was supported by a reef near the end of Cretaceous time, and the re-entrant itself may have been formed in part by currents controlled by the position of the reef.

Structure on the Tithonian(?) Reflector

Configuration of the horizon that we have correlated with reflectors within the Tithonian is shown in Figure 10. This horizon terminates by depositional onlap against deeper reflectors as seen in Figs. 3 and 4. Therefore, as the pinchout presumably was horizontal when formed, its present configuration may be assumed to have been caused by subsequent warping. The shallow depth of the reflector at the northern end of the area, less than 2 km, probably is the result of subsequent differential subsidence of the Southeast Georgia Embayment. The horizon has subsided generally toward the south, but a main axis of subsidence extends northeastward from the inner Blake Plateau at about 30°N toward the outer plateau at about 31°N. This trough formed a Cretaceous depocenter.

Structure on Volcanic(?) Layer

Structure contours on the strong reflector and high-velocity refractor, which is inferred to be a volcanic layer of Early Jurassic age, show a generally smooth, seaward dip. Although the reflector does not have sharply defined boundaries, its approximate southern and southeastern limits are shown in Figure 11. The southern limit, if extended westward, is found to extend into the boundary between Paleozoic sedimentary "basement" to the south and volcanic rocks (dominantly rhyolites and tuffs) to the north, as found in drill samples (T. M. Chowns, personal communication, 1976; Peter Popenoe, personal communication, 1977). This reinforces the concept that the reflector is produced by volcanic material, although the possibility that some of the volcanic rocks in Georgia are much older than the maximum possible age for this reflector, as discussed above, remains a concern.

The limit of the "volcanic" reflector (Fig. 11) corresponds closely to the boundary between an area of many, short-wavelength magnetic anomalies and an area to seaward of relatively much broader ones (Klitgord and Behrendt, this volume). This boundary seems to form the seaward limit of a type of "base-

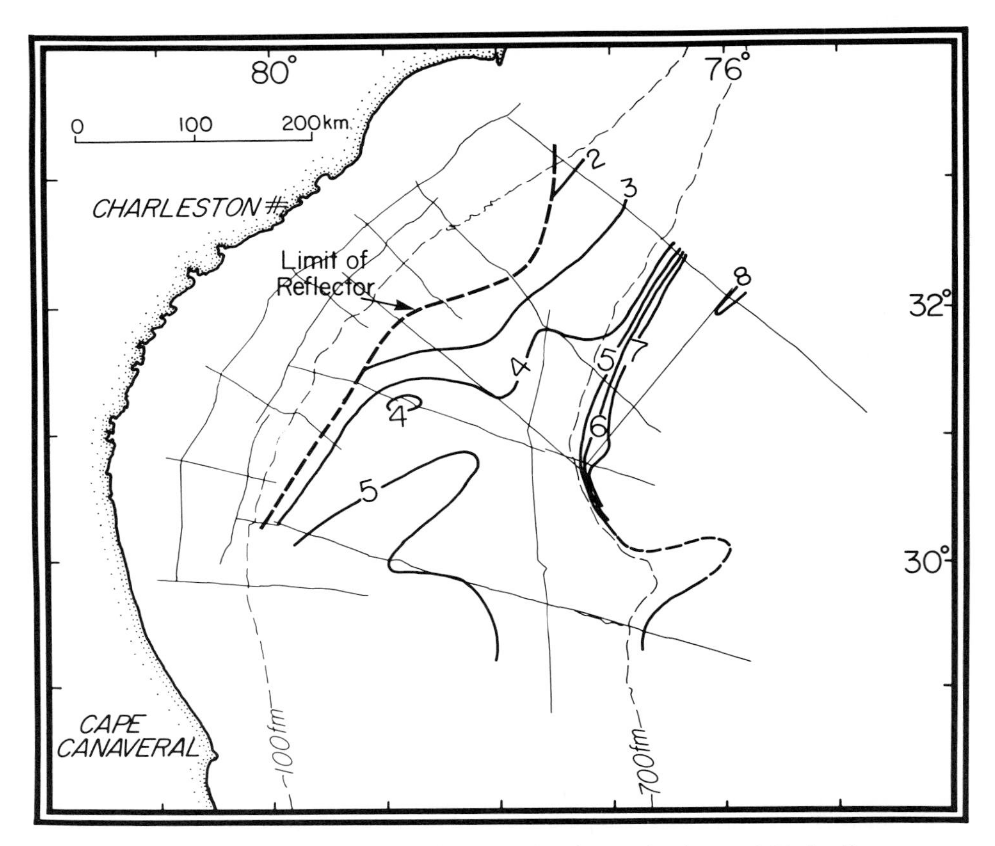

FIG. 10. Depth in kilometers below sea level to a horizon of Early Cretaceous or Late Jurassic age (probably Tithonian). Light lines show profiling coverage. The 100- and 700-fathom curves (dashed) represent the approximate shelf edge and approximate seaward limit of the Blake Plateau. Contour interval 1 km.

ment'' found beneath the southeastern part of the coastal plain of Georgia and South Carolina, as defined by Peter Popenoe and Isidore Zietz (unpub. ms.); these authors believe that the basement consists of undeformed volcaniclastic deposits with basaltic and rhyolitic flows and intrusions and pyroclastic deposits.

Structure on Angular Unconformity of Late Triassic(?) or Early Jurassic(?) Age

A deep angular unconformity is observed beneath the northern Blake Plateau (Fig. 12). This can be traced landward beneath the ''volcanic'' reflector, although that reflector tends to obscure returns from below because of its high reflectivity (Fig. 4). East of Charleston (Fig. 12) the unconformity is gently warped on the extension of the Cape Fear Arch, then deepens to a platform at about 9 km depth. A trough more than 11 km deep appears to be located seaward of the edge of the Blake Plateau. We have shown the depth contours both on the unconformity and on oceanic basement in Figure 12, but do not mean to imply that this is necessarily a continuous surface. Although the depth information can be contoured as a single surface, the reflector cannot be traced from one to the other. If our estimate is correct, however, the oceanic basement and unconformity probably did form the top of the crust in Early Jurassic time. The oceanic basement appears to dip smoothly landward (Fig. 12) and we observe no basement ridge which might have been associated with the formation of the Blake Ridge as was suggested by Le Pichon and Fox (1971).

At about 31°N the unconformity drops abruptly toward the south, reaching more than 12 km below sea level before exceeding the maximum penetration of our profiles. This inflection is coincident with the extension of the Blake Spur Fracture Zone (Fig. 12) recognized by offsets of seafloor spreading anomalies (H. Schouten and K. Klitgord, unpub. data), and with the southern terminus of the U.S. East Coast magnetic anomaly (Taylor et al, 1968; Vogt, 1973; Klitgord and Behrendt, this volume). South of the inflection the magnetic field is composed of much longer wavelength components (Taylor et al, 1968; Klitgord and Behrendt, this volume) and basement probably is formed of mantle-derived volcanic and plutonic rocks (Vogt, 1973; Sheridan, 1976), highly contaminated by continental sialic material. This would be a transitional basement between typical oceanic and continental types.

This conclusion regarding the nature of basement requires that the zone of the inflection must have acted as a transform fault during the early rifting stage. After migration of the present Blake Plateau region away from the spreading center, the major

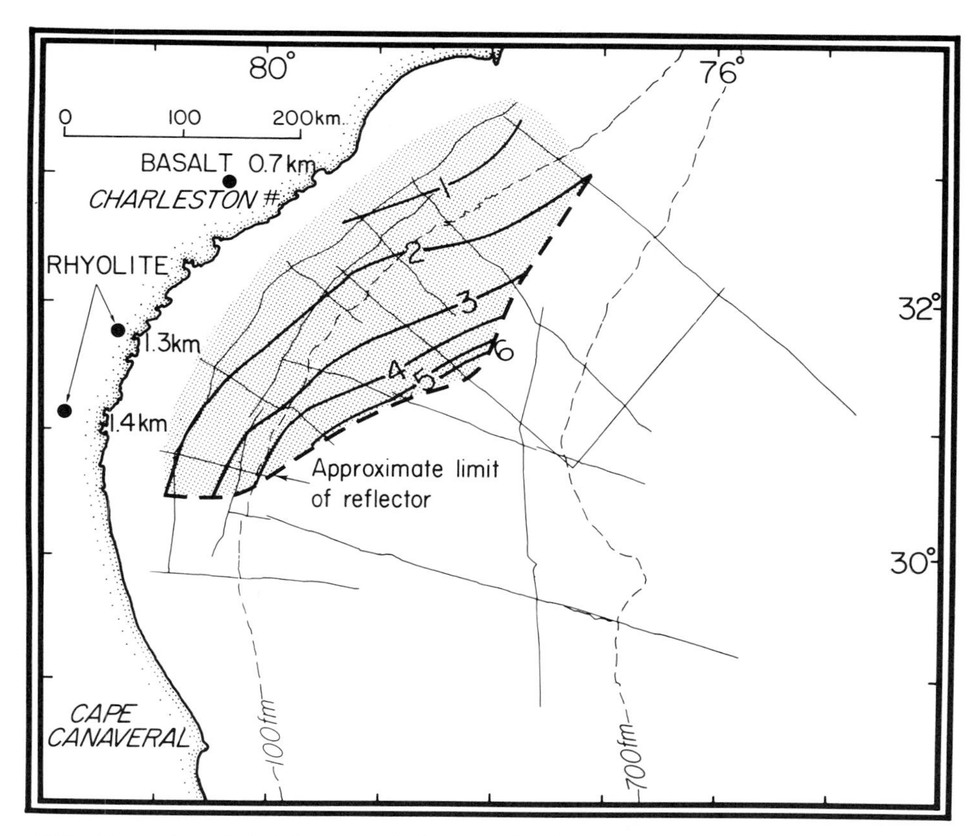

FIG. 11. Depth in kilometers below sea level to a strong reflector—high-velocity refractor, probably representing volcanic layer of Early Jurassic age. Dots indicate locations of USGS drill hole near Charleston, South Carolina (at Clubhouse Crossroads) and of two drillholes to the south in Georgia (Maher, 1971). Light lines show profiling coverage. The 100- and 700-fathom curves (dashed) represent the approximate shelf edge and approximate seaward limit of the Blake Plateau. Contour interval 1 km.

transcurrent movement on the fault would have ceased, of course. However, this patch of transitional crust south of the transform fault, being thinner and denser, subsided much farther than crust to the north.

The Blake Spur Fracture Zone and the inflection may mark a long-active zone of lithospheric weakness because its extension onto land is marked by a well-defined zone of seismic activity (Bollinger, 1972; Hadley and Devine, 1974), which includes the site of the major Charleston earthquake of 1886 (Dutton, 1889; Bollinger, in press; D. W. Rankin, unpub. ms.)

DEVELOPMENT OF THE NORTHERN BLAKE PLATEAU AND ADJACENT CONTINENTAL SHELF

The present continental margin began to develop with the initiation of stretching, normal faulting and volcanism in a rift phase, perhaps during the Triassic. Initially, erosion of the block faulted terrain was rapid and the grabens quickly were filled by terrigenous continental deposits along with volcanic materials (Fig. 13, middle of Triassic). Perhaps the material derived from the horsts, as well as that transported from the Appalachians, which were much more rugged at the time, created a relatively smooth surface above sea level, while the area slowly sank as a result of heat loss. Alternatively, a series of uplifts and subsidences caused by tectonic adjustments in this rift stage, as well as resultant shore line shifts, may have produced much more complicated geologic history. In any case, an eroded subaerial surface atop Triassic deposits eventually resulted.

The phase of drift began and oceanic crust was formed until, about 175 m.y. ago, a spreading center jump apparently took place (Pitman and Talwani, 1972; Vogt, 1973; Klitgord and Behrendt, this volume). We infer that changes in the stresses associated with this jump may have produced a major episode of volcanism, covering parts of the Triassic deposits with basalt flows and pyroclastic deposits (Fig. 13, 175 m.y. ago).

Salt is commonly deposited in the early stages of ocean development, but the only possible evidence for salt in the northern Blake Plateau area is a small diapir discovered by Grow near line BT1 (Fig. 2) in about 2000 m of water (J. Grow, personal communication, 1976). Evaporite minerals have been noted in the Esso-Hatteras Light 1 well (Fig. 1) (Spangler, 1950) and bedded salt in the Chevron Great Isaac well (Fig. 1) (Tator and Hatfield, 1975). Other evaporite deposits in southern Florida and the Baha-

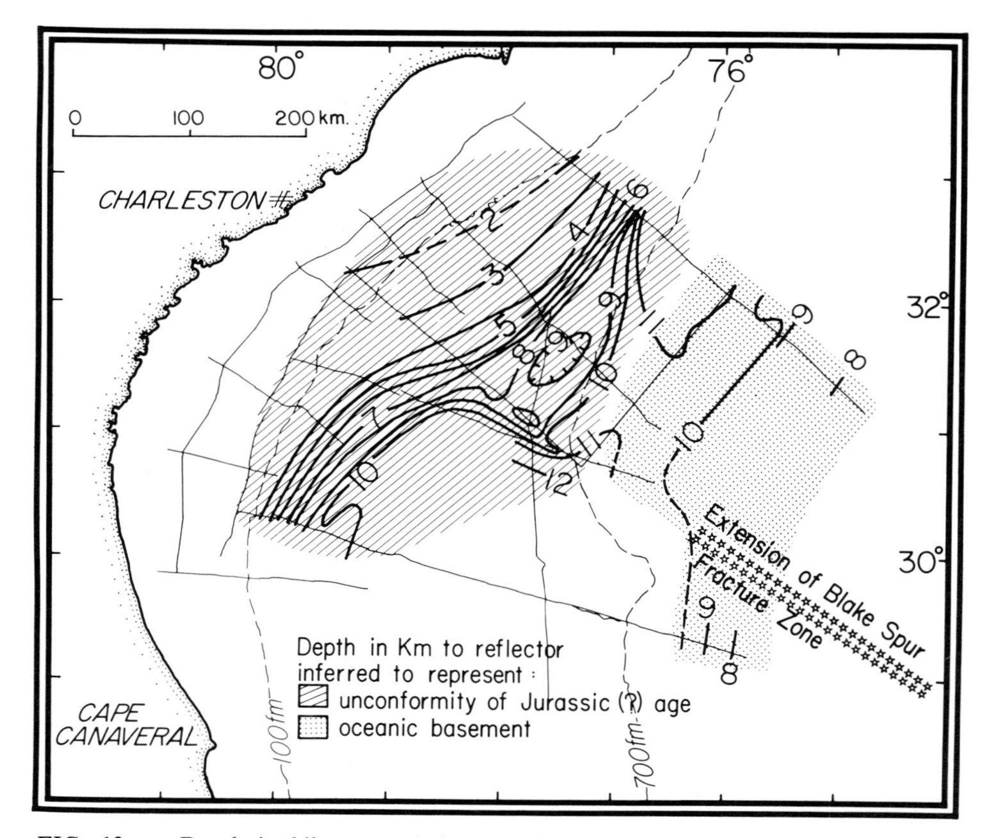

FIG. 12. Depth in kilometers below sea level to an unconformity assumed to be of Late Triassic or Early Jurassic age and to oceanic basement. Light lines show profiling coverage. The 100- and 700-fathom curves (dashed) represent the approximate shelf edge and approximate seaward limit of the Blake Plateau. Contour interval 1 km.

mas were discussed by Meyerhoff and Hatten (1974), and diapirs have been reported in the Bahamas by Ball and others (1968) and Lidz (1973). We conclude that little salt was deposited beneath the northern Blake Plateau.

As the continental margin subsided, Jurassic marine sediments onlapped across an unconformity cut on Triassic deposits and then across the Lower Jurassic volcanic layer that partially covered the unconformity (Fig. 13, end of Jurassic). Subsidence continued through Cretaceous time and a broad continental shelf was formed on which terrigenous clastic sediments deposited on the northwest graded to carbonate materials toward the southeast (Herrick and Vorhis, 1963). The shelf surface was maintained near sea level by accretion as it subsided, as shown by disconformities scattered through the section. In the northern Blake Plateau area, the structure and position of the shelf break resulted from a balance between subsidence, accretion and erosion. The maximum subsidence took place well back from the platform edge, near the middle of the Blake Plateau.

The present sedimentary-structural framework of shallow, narrow shelf and broad, intermediate-depth plateau results from a disruption of the prior balance of subsidence and deposition by the Gulf Stream near the beginning of the Tertiary. This high-speed current swept the sediment from the plateau during the Cenozoic (Schlee, 1977), so that even though the subsidence rate decreased, sedimentation was not able to keep up with it except near the coast where the much narrower Cenozoic continental shelf was constructed (Fig. 13, present)(Edsall and Dillon, 1977).

In early Tertiary time, extensive erosion of the slope and rise deposits resulted in a major deep sea unconformity having channels as much as a kilometer deep. Above the unconformity in the deep sea are upper Tertiary turbidites appearing as discontinuous reflectors on seismic profiles (Sheridan et al, 1974; Benson, et al, 1976; Dillon et al, 1976). The deep sea sequence is completed by upper Tertiary and Quaternary hemipelagic muds.

CONTRASTS BETWEEN THE DEVELOPMENT OF THE NORTHERN AND THE SOUTHERN BLAKE PLATEAU

The southern Blake Plateau developed quite differently in some aspects from the northern part, the geologic history of which is described above and illustrated in Figure 13. South of about 30°N, the structure of the seaward edge of the plateau is dominated by Cretaceous reef development rather than by a depositional-erosional balance. Furthermore, south of 31°N the pre-Jurassic "basement" is different, and the amount of subsidence—especially in

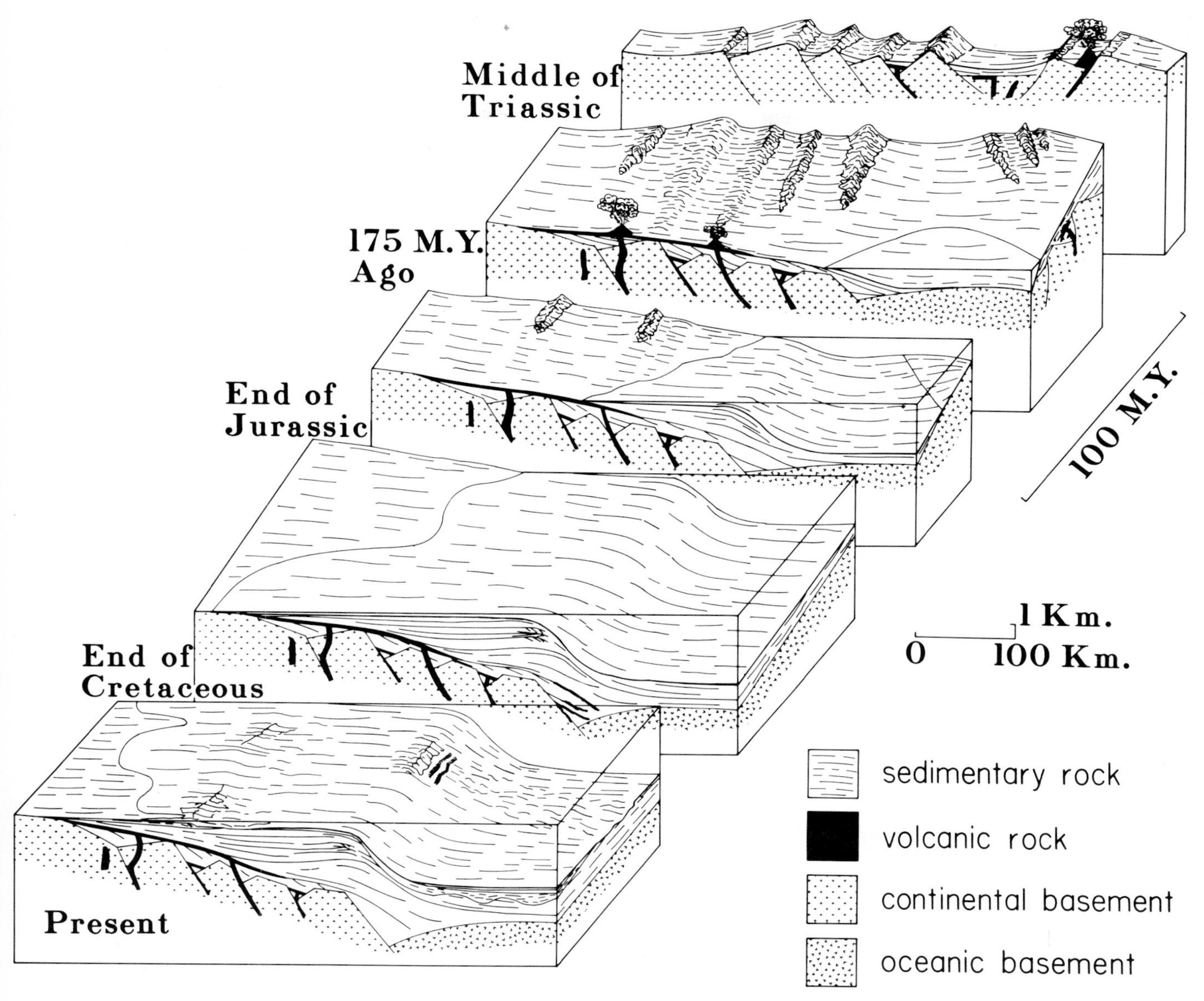

FIG. 13. Model of northern Blake Plateau geologic history. Section crosses the margin near Charleston, South Carolina.

the Jurassic—was much greater. The nonlineated, broad-wavelength character of the magnetic field over the southern Blake Plateau suggests the presence of a noncontinental basement—that is, a rift stage crust, in which the basement consists of mafic, mantle-derived rocks, highly contaminated by sialic continental material covered by a mixture of basalt and sedimentary rocks (Klitgord and Behrendt, this volume). Vogt (1973) noted that the southern Blake Plateau overlaps Africa on his reconstruction, thus he also concluded that this area is underlain by oceanic material and referred to the area (indelicately) as an excrescence on North America. In the Triassic, this process of basement formation probably resulted in development of a thinner, denser crust beneath the southern Blake Plateau and therefore led to much more rapid subsidence during early development of the continental margin. The unconformity at the top of inferred Triassic rocks is more than 12 km deep in this area (Fig. 12) and certainly seems to be significantly deeper than the oceanic basement (layer 2) seaward of the plateau (Fig. 5, depth section).

Apparently, as the continental margin subsided, it dragged down the adjacent seafloor. This took place mainly prior to the end of the Jurassic, because Tithonian and younger horizons are essentially horizontal. The margin subsided smoothly in the north (Figs. 3, 4, 13), but to the south (FC3, Fig. 5, and another profile at 28°N, not shown), where the amount of margin subsidence was much greater, a series of faults having small offsets appear just seaward of the escarpment. These faults probably result from a partial decoupling of the subsiding Blake Plateau block from the normal oceanic crust to the east. The faults appear to extend with minimal offsets to the top of Cretaceous units (Horizon A), but they do not continue into the Miocene to Quaternary deposits above.

SUMMARY AND CONCLUSIONS

Regional Structure—Northern Blake Plateau and Southeast Georgia Embayment

In general, reflectors that we infer to represent sedimentary strata of Middle Jurassic through Cretaceous age form an onlapping wedge above an older surface. Atop this major wedge, a minor buildup of Cenozoic deposits near the shore forms the present continental shelf. The top of Cretaceous rocks dips gently and evenly seaward from the inner shelf to the edge of the plateau; some minor irregularities beneath the shelf at about 30° to 31°N may be related to reef development. All reflectors appear to be slightly warped because of differential subsidence in the Southeast Georgia Embayment with respect to the Cape Fear Arch.

A reflector inferred to be near the top of the Jurassic section has subsided more than 5 km beneath the central Blake Plateau at the southern end of our study area and pinches out beneath the inner Blake Plateau. At greater depth a strong reflector/high-velocity refractor probably represents a Lower Jurassic(?) volcanic(?) layer that is present only in the northwestern part of the area. Below it, an angular unconformity, inferred to have been cut into Triassic deposits, shows a fairly abrupt differential subsidence toward the south at about 31°N and reaches 12 km below sea level before exceeding the maximum penetration of our profiles.

Geologic History of the Northern Blake Plateau

Formation of the continental margin of the northern Blake Plateau probably began with fracturing, extension of the continental lithosphere, and subsidence in Triassic. The rough basement surface was covered by Triassic deposits and eventually a rather smooth subaerial surface was produced. Rapid subsidence of the newly forming margin caused beds to develop a seaward dip. The landward part of the unconformity above these Triassic(?) deposits was covered by a volcanic layer, probably produced during an episode of minor tectonism 175 million years ago when the newly formed Atlantic basin experienced a spreading-center jump. During the remainder of Jurassic and Cretaceous time the margin subsided smoothly as a very broad shallow platform on which carbonate sediments were deposited toward the southeast but with major terrigenous clastic input toward the Appalachian source. Position of the shelf break was controlled by a balance of deposition-erosion and subsidence. The sedimentary units formed a wedge, onlapping toward the northwest. Maximum transgression occurred in the early Tertiary followed by general regression. The dominant factor in Tertiary sedimentation has been the Gulf Stream which has swept the Blake Plateau and allowed only a very thin accumulation of Tertiary deposits. The present narrow continental shelf is the remainder of a complexly prograded and eroded sediment pile built out during the Cenozoic against the edge of this rapidly flowing current.

Differences in Development toward the South

South of about 30° or 31°N some significant differences in continental margin development are apparent. A sharp geological boundary appears to cross the seaward edge of the Blake Plateau at about 31°N and form a northwest-trending extension of the Blake Spur Fracture Zone of the deep sea. To the south of the boundary, the inferred Triassic unconformity has subsided much deeper and the magnetic field is much flatter than to the north. We believe that these differences are due to a change in basement type from that formed of fractured continental sialic rocks, intruded by basaltic material on the north, to a basement formed of mantle-derived simatic material, contaminated by remnants of sialic continental crust and terrigenous clastic sediments on the south. The anticipated difference in thickness and density between two such crustal types might account for the increased subsidence to the south.

A second major difference between the northern and southern Blake Plateau lies in the factors controlling its seaward limit during its Cretaceous (and Late Jurassic?) period of development. Whereas the northern part of the plateau, north of the Blake Spur (30°N), has built up with its seaward limit controlled mainly by depositional-erosional processes, the southern part was limited by reef development, which produced the steep Blake Escarpment.

Structural Relationship between the Crust beneath the Blake Plateau and That beneath the Adjacent Deep Sea

Seaward of the northern Blake Plateau the oceanic crust appears to be gently bent down against the crust of the subsiding continental margin. This presumably is due to affects of sediment loading, heat loss and coupling with the adjacent margin. However, south of the geological boundary defined above, the amount of subsidence of oceanic basement (layer 2) is greater. A series of faults is observed extending about 80 km seaward of the Blake Escarpment. Little throw is apparent on near-surface layers and little subsidence has occurred since the Tithonian, but this trough subsided at least 4 km before then. We believe that the subsidence here was caused dominantly by dragging down of the oceanic crust by the sinking Blake Plateau. The abyssal sea floor subsided much less than did the part of the plateau, south of 31°N, and the faulting was the mechanism that partially decoupled the deep sea basement from the plateau basement.

REFERENCES CITED

Antoine, J. W., and V. J. Henry, Jr., 1965, Seismic refraction study of shallow part of continental shelf off Georgia coast: AAPG Bull., v. 49, p. 601–609.

Applin, P. L., 1951, Preliminary report on buried pre-Mesozoic rocks in Florida and adjacent states: U.S. Geol. Survey Circ. 91, 28 p.

Ball, M. M., R. M. Gaudet, and G. Leist, 1968, Sparker reflection seismic measurements in Exuma Sound, Bahamas (abstract): Am. Geophys. Union, Trans. v. 49, p. 196–197.

Barnett, R. S., 1975, Basement structure of Florida and its tectonic implications: Gulf Coast Assoc. Geol. Socs. Trans., v. 25, p. 122–142.

Benson, W. E., et al, 1976, Deep-sea drilling in the North Atlantic: Geotimes, v. 21, no. 2, p. 23–26.

Bollinger, G. A., 1972, Historical and recent seismic activity in South Carolina: Seismol. Soc. America Bull., v. 62, p. 851–864.

——— in press, Reinterpretation of the intensity effects of the 1886 Charleston, South Carolina earthquake, *in* Rankin, D. W., ed., Geological, geophysical and seismological studies related to the Charleston, South Carolina earthquake of 1886: A preliminary report: U.S. Geol. Survey Prof. Pap. 1028.

Brown, P. M., J. A. Miller, and F. M. Swain, 1972, Structural and stratigraphic framework and spatial distribution of permeability of the Atlantic Coastal Plain, North Carolina to New York: U.S. Geol. Survey Prof. Pap. 796, 79 p., 59 plates.

Charm, W. B., W. D. Nesteroff, and Sylvia Valdes, 1969, Detailed stratigraphic description of the JOIDES cores on the continental margin off Florida: U.S. Geol. Survey Prof. Pap. 581-D, 13 p.

de Boer, J., 1967, Paleomagnetic-tectonic study of Mesozoic dike swarms in the Appalachians: Jour. Geophys. Research, v. 72, p. 2237-2250.

Dillon, W. P., R. E. Sheridan, and J. -P. Fail, 1976, Structure of the western Blake-Bahama basin as shown by 24-channel CDP profiling: Geology, v. 4, p. 459–462.

Dowling, J. J., 1968, The East Coast onshore-offshore experiment, II. Seismic refraction measurements on the continental shelf between Cape Hatteras and Cape Fear: Seismol. Soc. America, Bull. v. 58, p. 821-834.

Dutton, C. E., 1889, The Charleston earthquake of August 31, 1886: U.S. Geol. Survey Ninth Annual Rept., p. 203–528.

Edsall, D. W., and W. P. Dillon, 1977, Geologic development of the Cenozoic continental margin of the Southeast Georgia Embayment (abstract): AAPG Bull., v. 61, no. 5, p. 782.

Emery, K. O., and E. F. K. Zarudzki, 1967, Seismic reflection profiles along the drill holes on the continental margin off Florida: U.S. Geol. Survey Prof. Pap. 581–A, 8 p.

——— et al, 1970, Continental rise off eastern North America: AAPG Bull., v. 54, p. 44–108.

Ewing, J. I., and C. H. Hollister, 1972, Regional aspects of deep sea drilling in the western North Atlantic, *in* C. D, Hollister, J. I. Ewing, et al, eds., Initial reports of the Deep Sea Drilling Project: Washington D. C., U.S. Govt., v. 11, p. 951–973.

——— Maurice Ewing, and Robert Leyden, 1966, Seismic profiler survey of Blake Plateau: AAPG Bull., v. 50, p. 1948–1971.

Gohn, G. S., B. B. Higgins, C. C. Smith and J. P. Owens, in press, Preliminary report on the lithostratigraphy of the deep core hole (Clubhouse Crossroads corehole 1) near Charleston, South Carolina, *in* D. W. Rankin, ed., Geological, geophysical and seismological studies related to the Charleston, South Carolina earthquake of 1886: A preliminary report: U.S. Geol. Survey Prof. Pap. 1028.

Hadley, J. D., and J. F. Devine, 1974, Seismotectonic map of the eastern United States: U.S. Geol. Survey Misc. Field Studies Map 620, 8 p., 3 sheets, scale 1:500,000.

Hathaway, J. C., et al, 1976, Preliminary summary of the 1976 Atlantic margin coring project of the U.S. Geological Survey: U.S. Geol. Survey open file rept. No. 76–844, 217 p.

Herrick, S. M., and R. C. Vorhis, 1963, Subsurface geology of the Georgia Coastal Plain: Georgia Geol. Survey Inf. Circ. 25, 78 p.

Hersey, J. B., et al, 1959, Geophysical investigation of the continental margin between Cape Henry, Virginia and Jacksonville, Florida: Geol. Soc. America Bull., v. 70, p. 437–466.

Jacobs, C., 1977, Jurassic lithology in Great Isaac 1 well, Bahamas: Discussion: AAPG Bull., v. 61, p. 443.

JOIDES (Joint Oceanographic Institutions Deep Earth Sampling Program), 1965, Ocean drilling on the continental margin: Science, v. 150, no. 3697, p. 709–716.

King, P. B., compiler, 1969, Tectonic map of North America: Washington, D. C., U.S. Geol. Survey, 2 sheets, scale 1:500,000.

Le Pichon, X., and P. J. Fox, 1971, Marginal offsets, fracture zones and the early opening of the North Atlantic: Jour. Geophys. Research, v. 76, p. 6294–6308.

Lidz, B., 1973, Biostratigraphy of Neogene cores from Exuma Sound diapirs, Bahama Islands: AAPG Bull., v. 57, p. 841–857.

Maher, J. C., 1971, Geologic framework and petroleum potential of the Atlantic coastal plain and continental shelf: U.S. Geological Survey Prof. Pap. 659, 98 p.

Meyerhoff, A. A., and C. W. Hatten, 1974, Bahamas salient of North America: Tectonic framework, stratigraphy and petroleum potential: AAPG Bull., v. 58, p. 1201–1239.

Milliman, J. D., 1972, Atlantic continental shelf and slope of the United States—Petrology of the sand fraction of sediments, northern New Jersey to southern Florida: U.S. Geol. Survey Prof. Pap. 529–J, 40 p.

Milton, C. R., 1972, Igneous and metamorphic basement rocks of Florida: State of Florida Dept. of Natural Resources, Bull. No. 55, 125 p.

——— and V. J. Hurst, 1965, Subsurface "basement" rocks of Georgia: Georgia Geol. Survey Bull. No. 76, 56 p.

Overstreet, W. C., and Henry Bell, III, 1965, The crystalline rocks of South Carolina: U.S. Geol. Survey Bull. 1183, 126 p.

Pilkey, O. H., I. G. MacIntyre, and E. Uchupi, 1971, Shallow structures; shelf edge of continental margin between Cape Hatteras and Cape Fear, North Carolina: AAPG Bull., v. 55, p. 110–115.

Pitman, W. C., III, and M. Talwani, 1972, Sea-floor spreading in the North Atlantic: Geol. Soc. America Bull., v. 83, p. 619–649.

Rainwater, E. H., 1971, Possible future petroleum potential of peninsular Florida and adjacent continental shelves; *in* I. H. Cram, ed., Future petroleum provinces of the United States—their geology and potential: AAPG Mem. 15, v. 2, p. 1211–1341.

Schlee, J., 1977, Stratigraphy and Tertiary development of the continental margin east of Florida: U.S. Geol. Survey Prof. Pap. 581–F, 25 p.

Sheridan, R. E., 1976, Sedimentary basins of the Atlantic margin of North America: Tectonophysics, v. 36, p. 113–132.

——— X. Golovchenko and J. I. Ewing, 1974, Late Miocene turbidite horizon in Blake-Bahama Basin: AAPG Bull., v. 58, p. 1797–1805.

——— et al, 1966, Seismic refraction study of continental margin east of Florida: AAPG Bull., v. 50, p. 1972–1991.

Spangler, W. B., 1950, Subsurface geology of Atlantic coastal plain of North Carolina: AAPG Bull., v. 34, 100–132.

Tator, B. A., and L. E. Hatfield, 1975, Bahamas present complex geology: Oil and Gas Jour., Part 1, v. 73, no. 43, p. 172–176; Part 2, v. 73, no. 44, p. 120–122.

Taylor, P. T., I. Zietz, and L. S. Dennis, 1968, Geologic implications of aeromagnetic data for the eastern continental margin of the United States: Geophysics, v. 33, p. 755–780.

Uchupi, Elazar, 1967, The continental margin south of Cape Hatteras, North Carolina: Shallow structure: Southeastern Geology, v. 8, p. 155–177.

——— 1970, Atlantic continental shelf and slope of the United States—shallow structure: U.S. Geol. Survey Prof. Pap. 529-I, 44 p.

——— and K. O. Emery, 1967, Structure of continental margin off Atlantic coast of United States: AAPG Bull., v. 51, p. 223–234.

Vogt, P. R., 1973, Early events in the opening of the North Atlantic; *in* D. H. Tarling, and S. K. Runcorn, eds., Implications of continental drift to the earth sciences: London, Academic Press, p. 693–712.

Woolard, G. P., W. E. Bonini, and R. P. Mayer, 1957, A seismic refraction study of the subsurface geology of the Atlantic coastal plain and continental shelf between Virginia and Florida: Univ. of Wisconsin Tech. Rept., Contract N7onr-28512, 128 p.

Zarudzki, E. F. K., and E. Uchupi, 1968, Organic reef alignments on the continental margin south of Cape Hatteras: Geol. Soc. America Bull., v. 79, p. 1867–1870.

Erosional and Depositional Structures of the Southwest Iceland Insular Margin: Thirteen Geophysical Profiles[1]

JULIUS EGLOFF[2] AND G. LEONARD JOHNSON[3]

Abstract The insular margin of Iceland has been surficially formed by glacial action which beveled existing strata and then deposited morainal debris on the erosional surface. The Outer Insular Shelf (OIS) has prograded out to 30 km beyond the previous paleo-shelf level. In an area 30 to 50 km from the mid-Atlantic Ridge axis through the Iceland margin, the shelf has been upwarped since the Late Pleistocene. Snaefellsnes Arch comprises the coalesced bases of former volcanoes southwest of Snaefellsnes Peninsula. Oblique to the mid-oceanic ridge axis are basement scarps beneath the Insular Margin that have trapped sediment. Basement seems to be highly reflective volcanogenic material over sediment in many areas. Marginal plateaus on the lower slope have entrained sediment-laden bottom water currents on the rise. Contourite deposits west-southwest of Iceland are delineated as Snorri Mid-Ocean Drift.

INTRODUCTION

Thirteen seismic reflection profiles collected over the Iceland insular margin in the past seven years are presented and discussed in this paper (Figs. 1, 2). Several profiles were collected in an orthogonal survey pattern; most ship tracks were completed in transit to and from Reykjavik, Iceland. Detailed descriptions of each profile accompany Plates 1 through 16.

The OIS, Insular Slope, Insular Rise, and the Reykjanes Ridge (mid-ocean ridge) are the major physiographic provinces of the study area. Marginal plateaus, mid-ocean drifts, and faults, slumps and prograded shelf structures accent the main physiography (Figs. 2, 3). Basement is defined as the deepest observed reflector (DOR) in seismic profiles wherever there is question about the presence of crustal basement. Johnson and Palmason (in preparation) present a profile across the shelf west of Reykjavik, Iceland, showing velocity data averaging 3.2 km/sec. References to thickness of sediment in meters are minima.

Data Control

The locations of the following thirteen seismic profiles are shown in Figure 2, numbered counter-clockwise from the northwest. Major physiographic provinces are identified. The position for the shelf break was selected from the seismic profile, and is utilized as a geographic reference point throughout the discussion of each profile (Plates 1–16). Satellite navigation and commonly LORAN C were employed on all cruises, relying upon the high frequency of satellite passes to give the best confidence in the smoothed tracklines.

Copies of the data are available through the National Geophysical and Solar-Terrestrial Data Center (NGSDC), Boulder, CO 80302.

Sediment core data were collected by The International Program of Ocean Drilling at sites 407, 408, and 409; these data provide stratigraphic control for further interpretation of the seismic reflection data.

Instrumentation

Equipment used to record sections 1 through 13

[1]Manuscript received, December 12, 1977; accepted, June 15, 1978.

[2]Sea Floor Division, Naval Ocean Research and Development Activity, NSTL Station, MS 39529.

[3]Arctic Programs, Office of Naval Research, Arlington, VA 22217.

The cooperation of the Icelandic people who made this study possible by providing the use of their support facilities is worthy of the highest regard and thanks. The patience and understanding of the half-dozen or more different crews of the ships, of the funding agencies of the U.S. Navy, and of members of the scientific groups are greatly appreciated. The participants in the six separate expeditions involved U.S. Navy civilian oceanographers, geophysicists, and technicians, as well as representatives of several academic organizations and nations—from Iceland, Denmark, Canada, and West Germany. Logistic support was liberally provided by the personnel employed and stationed at the NATO Air Base, Keflavik, Iceland. W. F. Ruddiman, G. L. Johnson, and J. Egloff were chief scientists. This manuscript benefited by comments made after its original presentation at the AAPG Research Conference (1977), and from internal reviews by T. L. Holcombe and F. A. Bowles. Personal communications with P. R. Vogt, A. S. Laughton, and C. Neumann contributed beneficial suggestions. This research was supported by the Chief of Naval Research through the auspices of the Naval Oceanographic Office and the Naval Ocean Research and Development Activity.

Article Identification Number:
0065-731X/78/MO29-0004/$03.00/0.
(see copyright notice, front of book)

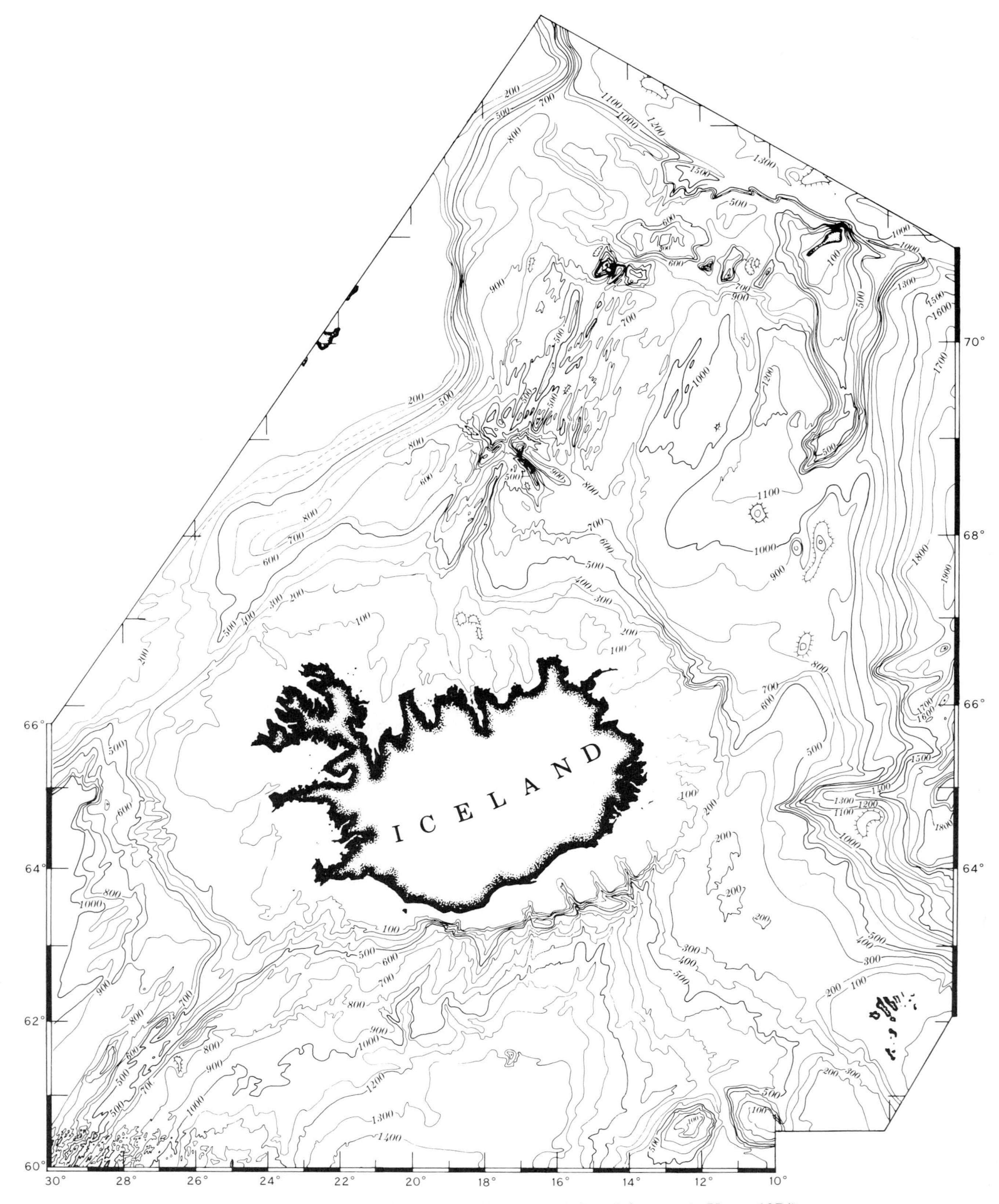

FIG. 1. Bathymetric contours in meters (after Johnson, *in* Voyt, 1974).

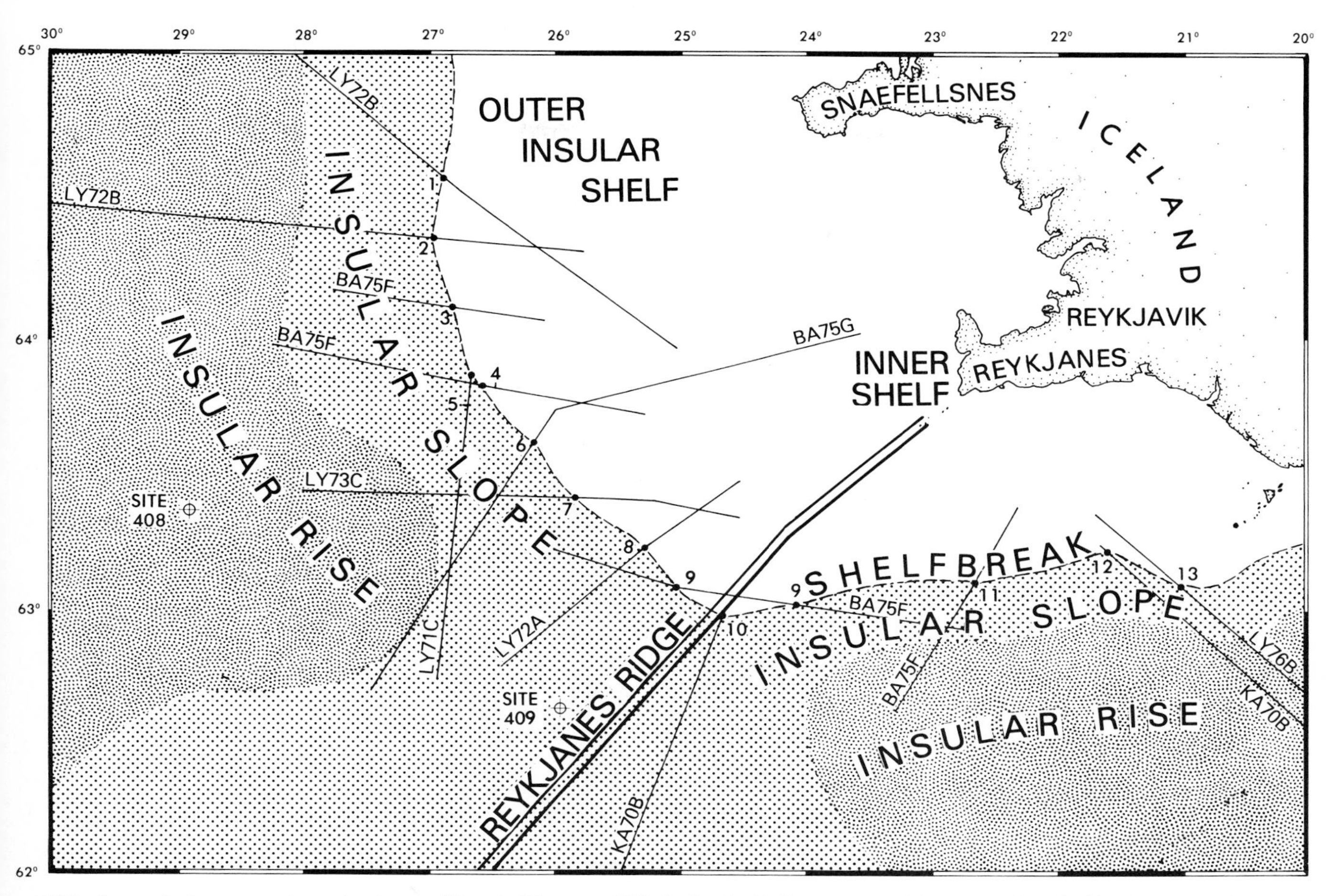

FIG. 2. Index map to seismic profiles 1–13 near SW Iceland. Reference positions at the shelf edge are given on each profile (Plates 1 to 16), and all distances in km to geologic features are given from the shelf edge. DSDP Sites 408 and 409 are shown; Site 407 is located west of Site 408 at 63° 56′N lat., 30° 35′W long. Major physiographic provinces are shaded.

included 33,000- and 66,000-joule seismic sparkers and hydrophone arrays, with Raytheon Precision Seismic Recorders. Exceptions occurred in making profiles 10 and 12 when an airgun sound source was used, and profile 13 when a Raytheon Universal Graphic Recorder was used.

Magnetic and 3.5-kHz bathymetric survey data were collected on all lines, with the addition of gravity data on USNS *Bartlett* sections utilizing an experimental, 3-gyro Lacoste & Romberg meter (Fig. 2; profiles 3, 4, 6, 9, 11).

Navigation equipment consisted of a Magnavox 702 and ITT SRN-9 satellite receiver/computer systems.

OUTER INSULAR SHELF (OIS)

This province, developed during the Pliocene-Pleistocene ice ages (Johnson and Palmason, in preparation), is largely a level plain with a few surface irregularities. A 100- to 200-m rise (incline) is adjacent and subparallel to the active mid-ocean ridge where it crosses the shelf (Fig. 3). A mostly basaltic subbottom arch west and southwest of Snaefellsnes Peninsula is bordered on the south by a syncline with more than a half-kilometer of sediment near the synclinal axis (64°N, 25°W). Glacial moraines can be identified along the rim of the outer shelf (Olafsdottir, 1975). Iceberg plough marks are widespread.

Beneath the Shelf and Paleo-Shelf

Within the uppermost 500 m of sediments underlying the OIS, probable volcanogenic and glacial debris cause zones of high reflectivity, hence rapid attenuation of seismic energy. An acoustically transparent (100- to 350 m thick) upper layer is present to the OIS edge, the shelf break. This layer, overlying what is probably an erosional surface of a paleo-shelf is interpreted to represent morainal deposits of late Pleistocene-Holocene age (Fig. 2; profiles 1, 4, 6, 7, 8, 11). A highly reflective layer just below the first paleo-shelf could be an even older paleo-shelf and erosional surface. Ash layers are another possibility. Variation in thickness of the 100- to 350-m OIS layer may be a function of positioning of moraines. Moraines partially fill glacial troughs on the south OIS edge (data in preparation). Reported typical morainal sediments were sampled along a 100-km-long feature near the shelf break west of Iceland at 65°N (Fig. 3).

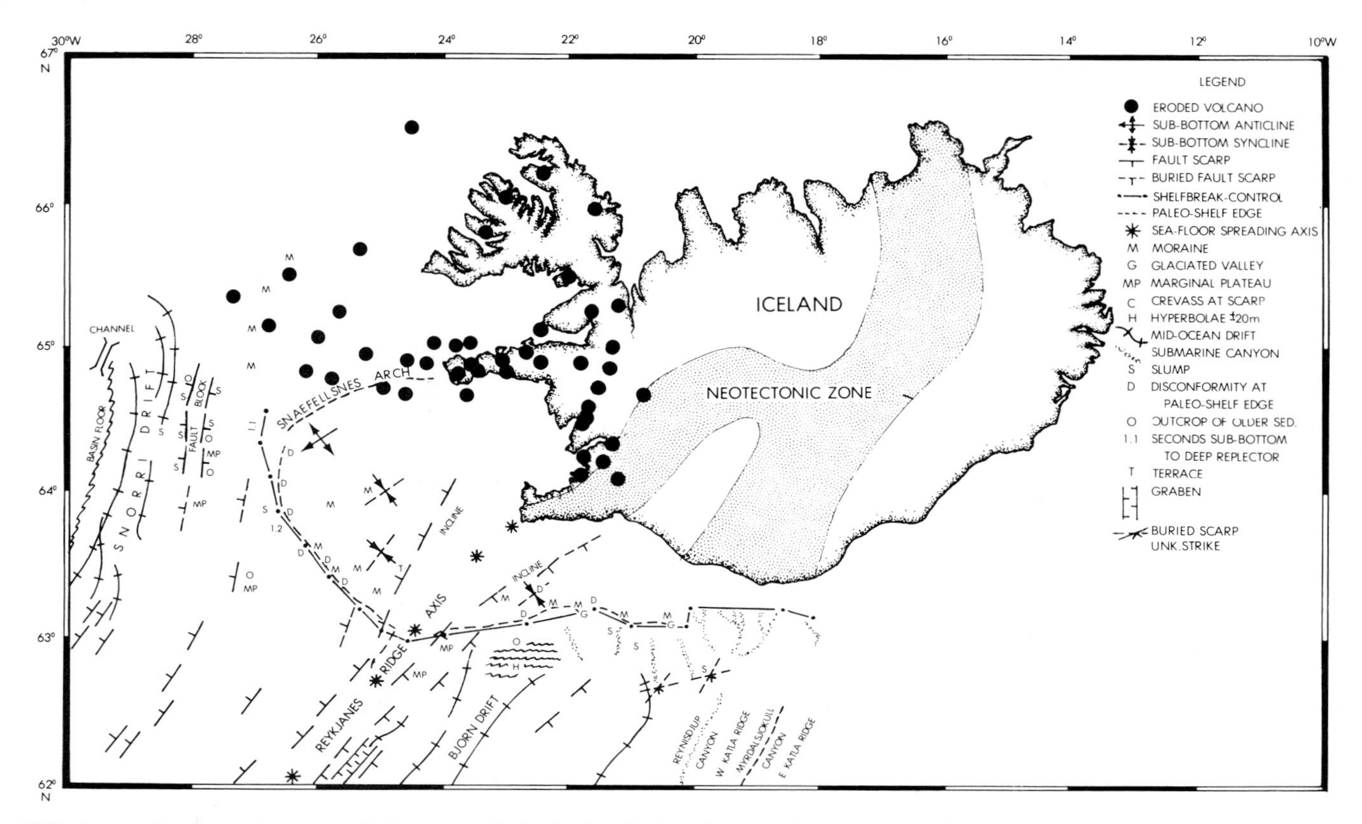

FIG. 3. Geological structural diagram of the Icelandic insular margin. Major geologic features are indicated as *Snorri Drift, Snaefellsnes Arch, Reykjanes Ridge, Bjorn Drift,* and the *Katla Ridges.*

Within 30 to 50 km northwest of, and converging with, the mid-ocean ridge to the southwest across the insular shelf, the sea floor has been uplifted 100 to 200 m (Profiles 6, 7, 8). Uplift is localized near the spreading center, where sedimentary strata which occur elsewhere at 500 m subbottom are differentially upwarped and now conform with the seafloor incline. If we assume that the shelf was leveled during the latest glaciation, then the uplift (100 to 200 m) along the mid-ocean ridge crest has occurred in the Holocene. Faulting of the sediment sequence under the incline is not seen; therefore, the uplift has been spread out in time and space. Two mechanisms would fit our data: 1) thermal uplift in the proximity of the mid-ocean ridge crest, and 2) a difference in rate of isostatic rebound of about 1 m per 100 years. The arguments for either, and the combined effect, should be considered for detailed geological investigation, including by means of geothermal borings. Palmason (1974) mapped a 20-mgal free-air gravity anomalq on the Iceland Shelf in the vicinity of the incline converging with the mid-ocean ridge spreading center.

Snaefellsnes Arch

An acoustically opaque rise in the basement (DOR) beneath the insular shelf is probably related to the eroded remains of many central volcanoes (Fig. 3 and Profile 1). These were located by Thors and Kristjansson (1974) and Kristjansson et al (1976). The zone of volcanism extended westward of the Snaefellsnes Peninsula. After repeated glaciations and beveling of the shelf, the insular slope prograded west of the arch. We propose to describe it further as the Snaefellsnes Arch (Profile 1). Prominent reflectors down to 1.0 sec (two-way travel time) subbottom are obscured in the vicinity of the arch at its seaward flank. The west flank of the arch crest once formed a shelf-edge prominence. The prograded paleo-slope can be seen up to its flank (Profile 2, Plate 4). Southeastward, a sedimentary sequence of reflective and transparent layers lap onto the flank of the Snaefellsnes Arch. Beneath the shelf, traversing the axis of the basin (64°N, 25°W), sediment spanning more than 0.6 sec tapers to an interval covering less than 0.15 sec on the arch crest (near 64°20′N, 26°W). Subsidence rates complicate history of the shelf (Tryggvason, 1974).

The Shelf Break

The OIS edge has prograded 10 to 25 km seaward from the first paleo-shelf, or more reflective surface found at 0.15 to 0.35 sec subbottom (profiles 2, 3, 4, 6, 7, 8, 11). The present shelf break north of 64°20′N off the west coast is not well defined due to the aggradation of sediments on the slope and shelf. The former shelf edge once lapped onto the Snaefellsnes Arch between 64°20′N and 65°00′N, east of 27°W (Fig. 3). The slope change is more distinct to the east of the Reykjanes Ridge intersection with the shelf, where the upper insular slope is steeper. A smooth, almost level reflector at 0.8 to 1.1 sec in most sections, under the OIS edge and landward, may represent a DOR type of sediment horizon—either an erosion surface, or more likely, a widespread volcanic event that scattered ash throughout southwestern Iceland (Profiles 1, 2, 4, 5, 11). This reflector shoals seaward and thus defines the sediment wedge of the slope province. Submarine canyons begin on the upper insular slope (Profiles 12, 13).

INSULAR SLOPE

Hemipelagic sediment interbedded with volcanogenic and glacial debris are the materials believed to be represented in the insular slope sections. These wedge out within 50 km seaward of the shelf break. The slope and upper rise near the flanks of the Reykjanes Ridge are more reflective than elsewhere, denoting a larger fraction of volcanogenic debris. Locally, zones of nondeposition are present due to bottom current activity (Lonsdale and Hollister, 1976). The slope south of Reykjavik has steepened, possibly due to ocean currents and nondeposition (Fig. 1, Profile 11). Canyons and sedimentary processes affecting the southern insular slope will be considered by Johnson and Palmason (in preparation).

Paleo-Slopes and Deepest Observed Reflectors (DOR)

Older strata may crop out on the lower slope west of Reykjavik, Iceland, 25 to 30 km west of the shelf edge, where there is a possible slump scar on the mid-slope. A transparent layer up to 350 m in thickness overlying the 0.8- to 1.1-sec-deep reflector (63°30′N to 64°40′N, 26°00′W to 27°40′W), probably represents a quiescent period in the volcanic development of southwestern Iceland, with continued pelagic deposition (Profiles 1, 2, 4, 5). Along the flank of the Reykjanes Ridge, the 0.8- to 1.1-sec-deep reflector extends southwestward from beneath the shelf break, becoming a mid-sediment reflector. Deeper, more irregular basementlike reflectors are visible below the former shelf break DOR.

Basement Ridges

There is a possible basement ridge beneath the lower slope/upper rise 40 to 60 km west of the shelf break, west-northwest of Reykjavik (Profiles 1, 2). Landward of this block, beneath the adjacent strong mid-sediment reflector (elsewhere described as the shelf break 0.8- to 1.1-sec DOR), eastward-dipping layers are much older than any previously mentioned. On the western side of the basement block, a scarp of nearly one second in relief is found. Some rotation of the sediment column has occurred above the east side of the basement block, 40 to 60 km west of the shelf break, near 64 to 65°N. A Recent chasm 100 m deep and over 200 m wide (east to west), is located at the base of the fault scarp (Profile 1). It is not yet filled with sediment from the insular slope and shows a steep face on the downthrown side. This scarp and other features nearby may be traced southward. They tend to converge with the Reykjanes Ridge trend (southwest), and are similar to oblique trends reported by Vogt and Johnson (1973). The scarps do not appear to be radial structures surrounding southwestern Iceland (Vogt and Johnson, 1975).

INSULAR RISE

West of Snaefellsnes and Reykjanes Peninsula, on the insular rise of Iceland, the sediment section is highly faulted and shaped by bottom-current activity (Profiles 1, 2). No reflector continues more than 20 km. South of the Reykjanes Peninsula, a portion of the lower insular rise is accented by hyperbolic swales. The prominent 0.8- to 1.1-sec-deep shelf break DOR becomes a mid-sediment reflector, and the sediment above this nearly level horizon wedges to an interval spanning less than 0.3 sec toward the foot of the rise. West of Reykjanes Peninsula, on the rise, a 0.2- to 0.3-sec subbottom reflector, which may be an ash layer, seems to conform with adjacent underlying structures. Marginal plateaus formed by sediment infilling exist between or beside fault scarps (Fig. 3, Profile 2). Faults within the sediment of the slope and upper rise (63°30′N, 27°W) are numerous, possibly indicating more rapid subsidence landward (Profiles 5, 7).

Examples of canyons eroding down to, but not through, more reflective mid-sediment layers were noted by Johnson, Sommerhoff and Egloff (1975). Just as in the case off southeastern Greenland, canyons expose an older sediment wedge off southwestern Iceland on the slope and upper continental rise. The highly reflective layer a few tenths of a second below the remaining inter-canyon divides is believed to be an erosion surface or top of a former insular rise (Profile 12, Plate 15). The proposed situation off southern Iceland west of 20°W is as follows: the more reflective 0.15- to 0.25-sec-deep stratum is probably the erosional surface of a paleo-rise; a new rise was later built upon this and thence starved; at present, the new rise is an area of relative nondeposition and is being eroded by the submarine canyons. This schema of a vertically subdivided insular rise, in both southeastern Greenland and southern-southwestern Iceland, is a product of Pliocene-Pleistocene glacial/interglacial sedimentary regimes.

MID-OCEAN DRIFTS

On the insular rise off southwestern Iceland, there is an acoustically transparent sediment drift. The north-south trending sediment body ranges in width from 40 km near 65°N, to 100 km at 64°20′N, and extends from 63° to 65°N (Fig. 3). On the west, the turbidite-filled Irminger Sea basin floor has buried the toe with flat-lying basin sediments (Profile 2). The drift stretches eastward at this latitude to the fault scarp near the base of the insular slope. South-southwest of 64°N, between 29° and 30°W, the transparent drift deposits are scattered and disoriented, overlying irregular basement structures.

The mid-ocean drift west of Iceland (Fig. 3) is composed of 200 to 500 m of late Miocene-Pliocene nannofossil ooze redistributed by bottom currents, with minor amounts of ash-related glauconite, similar to a contourite deposit (Luyendyk et al, 1977). Sediment waves or elongate crests modify the morphology of the drift (Egloff and Johnson, 1975). The absence of terrigenous sediments throughout the transparent drift section suggests that turbidites have bypassed hummocky relief of the insular rise. We propose to name this drift the Snorri Drift, in honor of the Icelander, Snorri Sturluson, who wrote the sagas of Iceland in the thirteenth century.

South of Iceland on the lower insular rise and Reykjanes Ridge flanks, several mid-ocean drifts occur. These are separated by basement scarps, varying in relief from 400 m to over 1 km, that are buried northward beneath the insular margin. The scarps, which divide the Reykjanes Ridge into three distinct steps, including the crest, have entrained different streams of Norwegian Sea overflow water moving southwestward (Lonsdale and Hollister, 1976; Talwani et al, 1971; Ruddiman, 1972). Adjacent to the scarps and currents, beneath zones of relatively reduced current velocity, deposition of transparent silty clay and nannofossil ooze has resulted (Davies and Laughton, 1972). The drift sediments formed elongate crests in the direction of bottom currents (Johnson and Schneider, 1969; Egloff and Johnson, 1975).

The eastern steps of the Reykjanes Ridge flank bear another thick section of acoustically transparent sediments. These fine silts and clays were drilled at site DSDP 114 (Laughton et al, 1972). This mid-ocean drift has formed entirely separately from the Gardar Drift of the deeper ocean basin. East of this separate drift crest, and west of the bulk of Gardar Drift, a submarine canyon/mid-ocean channel system has divided the sedimentary provinces (Johnson and Palmason, in preparation). We therefore propose that this section of drift on the middle step be named Bjorn Drift after the first man to map all of Iceland, Bjorn Gunnlaugsson, a mathematician and geographer of the mid-nineteenth century (Fig. 3). Currents are influencing the insular slope (Shor et al, 1977) on the northern end of the drift.

These sedimentary and basement structural features will be the subject of a comprehensive paper presenting seismic reflection and other geophysical data over the northern Reykjanes Ridge.

SUMMARY

The insular margin of southwestern Iceland has been extensively modified by glacial erosion and morainal deposition with 10 to 30 km of shelf-edge progradation. A subbottom structural arch and adjoining sediment-filled syncline west-northwest of Reykjavik are boundary features to the extension of the Snaefellsnes volcanic zone. The mid-ocean ridge junction with the Iceland margin has locally upwarped the shelf strata in post-glacial times. Recently there has opened a 100-m-deep chasm near the prominent fault scarp west of Snaefellsnes on the slope/rise boundary. The fault block has trapped lower slope sediments and formed a marginal plateau. Dynamic bottom-water conditions on the insular rise in the late Miocene-Pliocene are indicated by sediment redistribution and accumulation in elongated drifts. These have formed along basement scarps on the steplike flanks of the Reykjanes Ridge.

APPENDIX: PROFILES 1 THROUGH 13

Profile 1

Snaefellsnes Arch spans the seismic reflection profile, 10 to 80 km southeast of the shelf-edge reference position (Plate 1). The crest of this arch, 0.2 sec subbottom, is 30 to 50 km southeast of the shelf break. The arch also borders on Profile 2 (Plate 4) at the east end where profiles cross. Positions along profiles are measured from the shelf break.

At 60 km southeast, there is a sedimentary basin. Due to strong interference by the second reflection, the minimum interval apparent in the basin is 0.3 sec at 92 km southeast. This layer is dipping southward beneath flat-lying strong reflectors. The basin is topped by 0.25 sec of probable morainal sediment.

At the shelf edge, a landward thickening of sediment above a 0.4- to 1.0-sec reflector is an excellent example of the typical southwestern Iceland section. Slight landward subsidence due to isostatic loading by shelf deposits would explain the eastward deepening of the DOR. A moraine (M) is spread over the shelf edge, with disturbed or slumped structures, at 5 to 25 km NE. The insular slope is buried and the shelf break is not well defined (Plates 2, 3). Ten km landward of the shelf-edge reference position, steeply dipping strata at 0.2 sec subbottom are truncated by the 0.2-sec-depth discontinuity (D). Below 0.3 sec is a paleo-shelf edge and dipping paleo-slope at the base of strong reflecting layers. This former shelf break can be correlated with the flank of the Snaefellsnes Arch. Dip of the arch flank and paleo-slope are visible at 0.5 sec through the second reflection. A slump (S) must have occurred 7 to 20 km NW of the shelf break reference position.

A fault displacing all mid-sediment reflectors and the sea floor is observed at 40 km NW (Plate 2). At 50 km, the DOR may have been rotated, with faulting of the basement. Further discussion follows under Profile 2.

On the upper portion of the insular rise, 40 to 62 km NW, turbidite deposits become more distinct (Plate 2). An especially transparent layer, spanning 0.1 sec at 60 km NW, increasing to 0.35 sec at 15 km NW, may increase even further under the shelf near the Snaefellsnes Arch (10 km SE). Another reflector uniformly at 1.1 sec below sea level can be traced 0 to 33 km NW and through the second reflection.

The fault scarp at 62 km NW with the 100-m-deep chasm at the base of the scarp, 200 m wide (east to west), has been active recently; otherwise, the normal downslope influx of sediment would have filled the chasm. Note the unusual undercut face on the downthrown side, rarely detected by sounding systems. Investigation by submersible at this location would be rewarding.

Northwestward of the fault scarp on the rise, sedi-

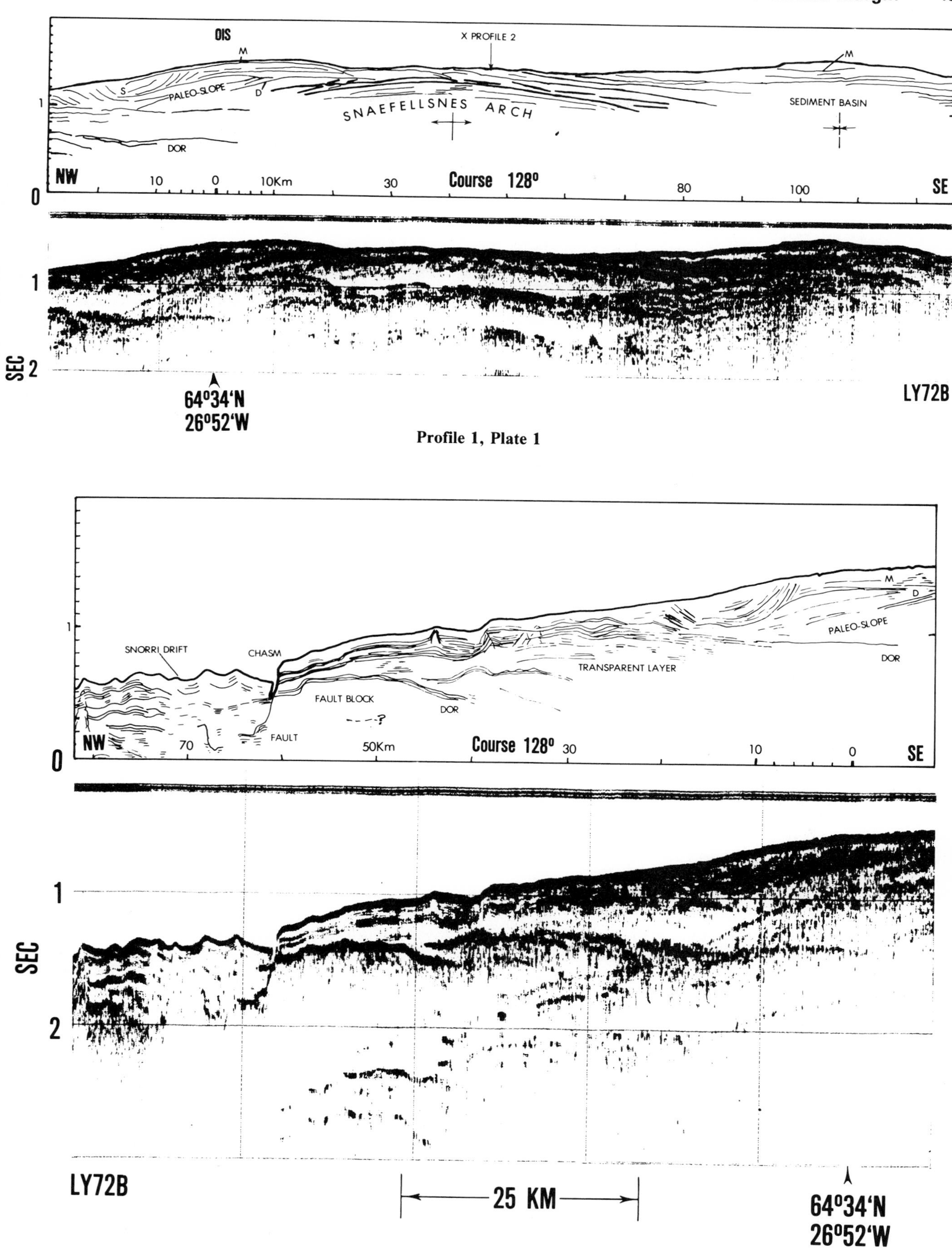

Profile 1, Plate 1

Profile 1, Plate 2

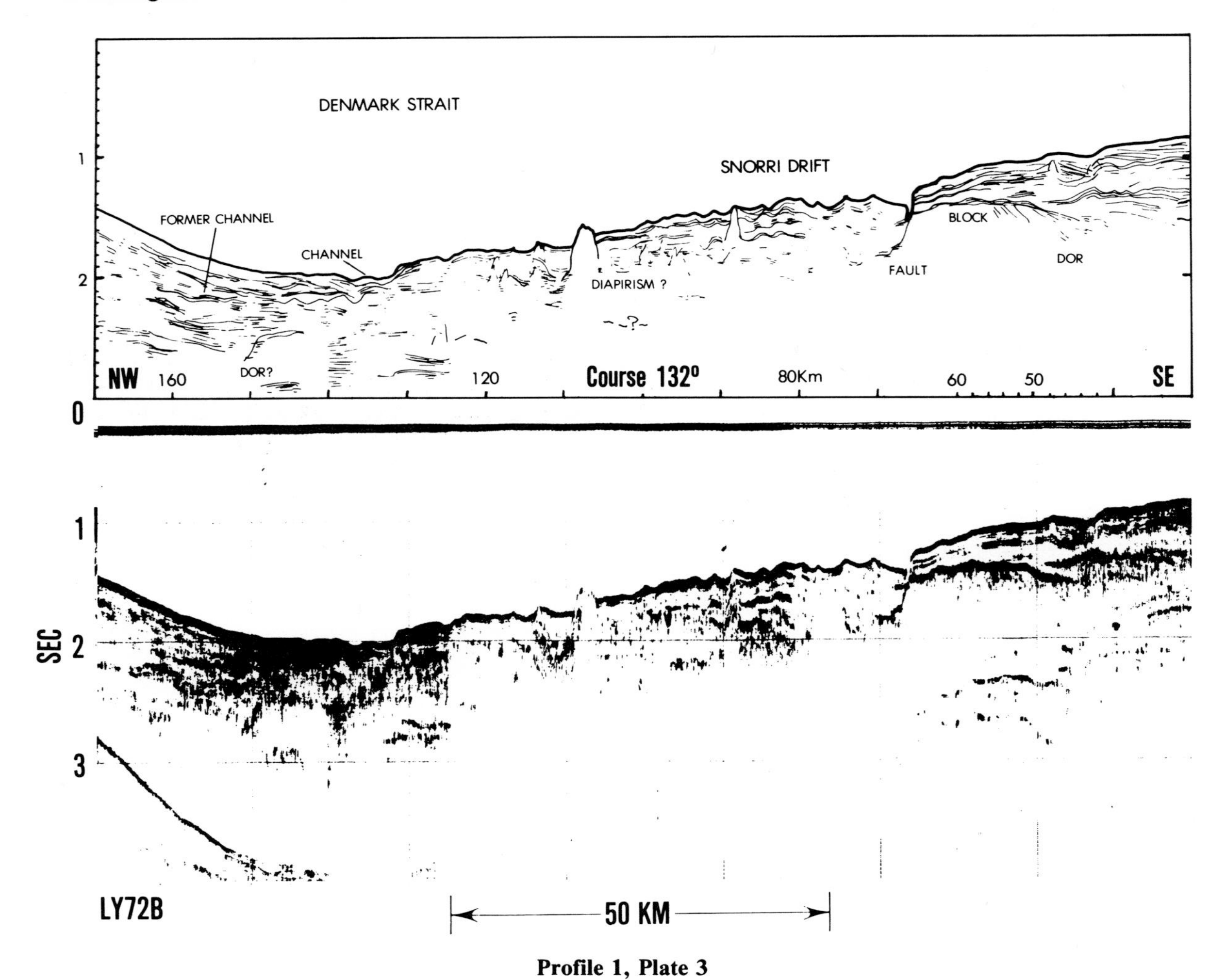

Profile 1, Plate 3

ments of this section (69 to 129 km NW) have been additionally deformed by processes other than the obvious faulting and differential compaction (Plate 3). A 0.2- to 0.5-sec layer in mid-rise is typically more transparent than either adjacent turbidite section, the ocean basin (beyond 130 km NW), or the slope (0 to 60 km NW).

The axis of Denmark Strait is in 1500 m of water, with the deepest observed subbottom reflection nearly 3 seconds below sea level. This is 67 km NW of the basement scarp described above. Across the section 67 to 130 km NW, basement has relief of about one second, topped by 0.6 to 1.0 sec of redistributed hemipelagic sediments with some turbidites or ash layers. Near the axis of greatest depth, higher reflectivity indicates a zone of nondeposition except by coarse material.

The sections in Plates 1 and 3 are at the same scale and can be matched. The northwest end is at 65°31.1′N, 29°25.4′W.

Profile 2

Intersecting the OIS edge at 64°20′N, 26°55′W, only 23 km S of the previous seismic reflection line, this profile is markedly different in morphology and internal structure (Plate 4). The shelf-edge progradation is obvious to an earlier shelf edge 25 km E under 0.3 sec of horizontal reflectors. The oldest paleo-slope seems to have lapped upon the west flank of the Snaefellsnes Arch crest as if it had once been the shelf edge of Iceland (at 20 to 40 km E; Plate 2). Beneath the slope are seen the most recent transparent sediments wedging to a thinness beyond the resolution of the equipment, 25 to 30 km W of the shelf break. Highly reflective, older outcropping layers may be related to the paleo-shelf edge.

The slope has a marginal plateau between 30 and 55 km that widens southward. Eastward-dipping structures of the DOR are 35 to 45 km W of the shelf edge. The dip of this structure corresponds to the east flank of a faulted (basalt?) block beneath the lower insular slope. The deepest reflector dips steeply landward 0.3 sec under another extended reflector, which is, itself, buried at least a km at the shelf break (Johnson and Palmason, in preparation, report velocity data near here in excess of 2.7 km/sec). The oldest sediments within 0.6 sec that are found more than one km beneath the insular shelf could be found 40 km W of 64°20′N, 26°55′W.

A basement fault scarp is evident on the mid-rise. On both Profiles 1 and 2, the scarp is approximately 60 to 85 km W of the shelf break. It is indeterminable if there has been a recent slump, noted only

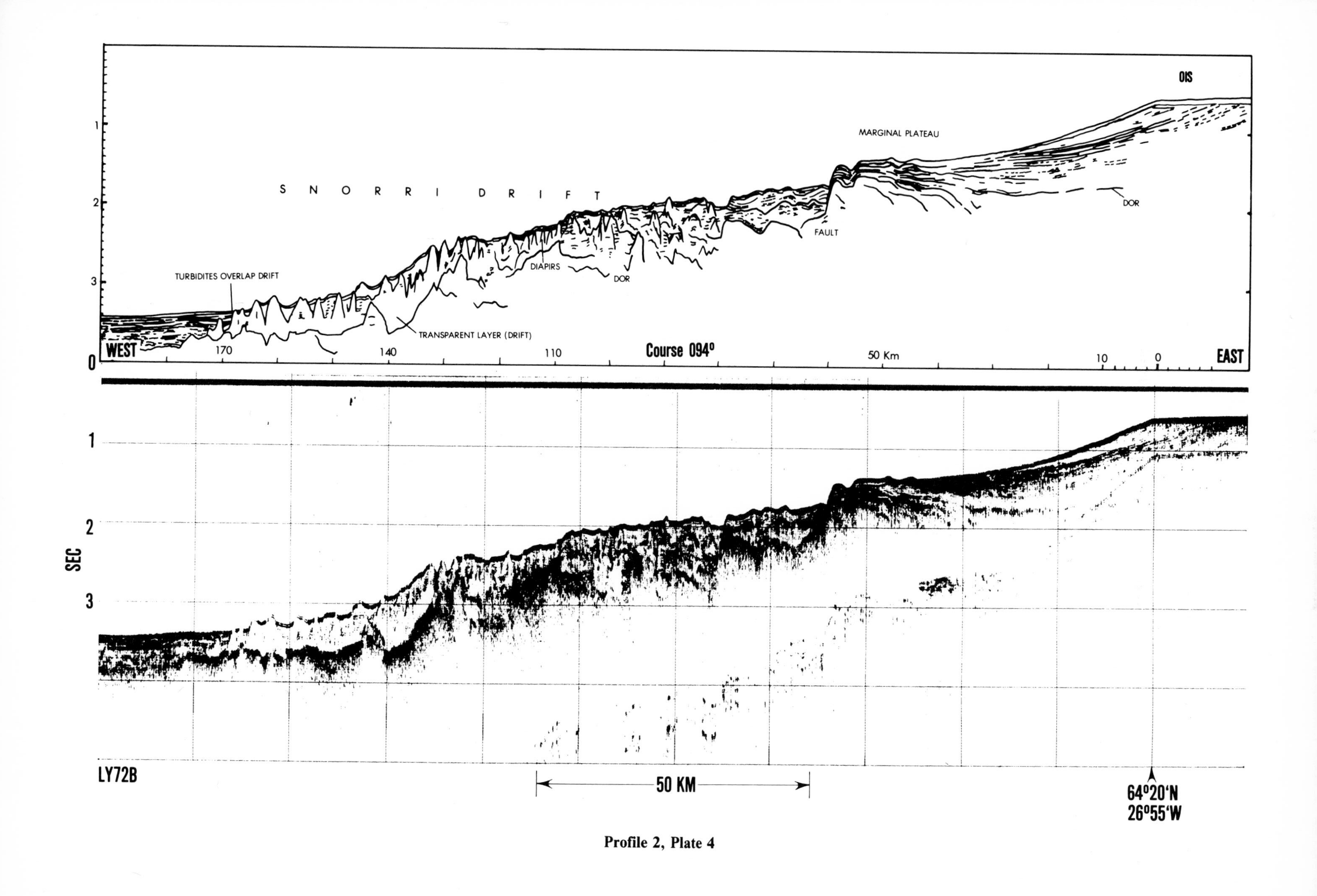

Profile 2, Plate 4

50 km to the north. Three km upslope (east) of the top of the scarp, a grabenlike depression 70 m deep could be the initiation scar of a slump that has caused mid-sediment reflectors to be so disturbed.

West of the basement scarp (65 km from the shelf break), next to the lowest step of the highly disturbed DOR (130 km W), there is another scarp. The DOR does not seem to represent crustal basement, and it lies beneath about 0.4 sec of sediment for 100 km. Between the two described scarps intermittent reflectors are impossible to trace more than 20 km.

Overlying the mid-rise and foot of the insular rise of Iceland is an acoustically transparent, 0.2- to 0.3-sec wedge of sediment. DSDP Site 408 at 63°22.63′N, 28°54.71′W, cored 256.5 m of late Miocene-Pliocene nannofossil oozes. The absence of terrigenous turbidites and large grain-size variations in the sequence corresponds with the acoustic characteristics of this layer. The thickness is 200 to 300 m (at 2 km/sec), out to 160 km west of the shelf break. Turbidites of the Irminger Sea basin overlap the foot of the drift. At 150 to 180 km W, the DOR is not crustal basement, but appears to be a highly reflective volcanogenic layer overlying sediment.

The physiographic traits of a mid-ocean drift are present where the transparent sediment layer lies on the insular rise 60 to 170 km W (Plate 4), and NW of the Reykjanes Ridge. Scientists onboard *Glomar Challenger* cruise 49 interpreted the thick late Miocene-Pliocene nannofossil oozes and a few, thin, ash-related, glauconite laminations as evidence of contourite sedimentation. The sediment body described probably was deposited in the counterclockwise turn of northward-flowing bottom waters deflected at the northernmost end of the Reykjanes Ridge by the Iceland margin. In the turn, decreasing current velocities would result in deposition of suspended sediment, with subsequent development of the mid-ocean drift. Additional characteristics of this drift similar to those previously described for mid-ocean drift morphology are present, such as superficial dunes and elongated crests (Egloff and Johnson,

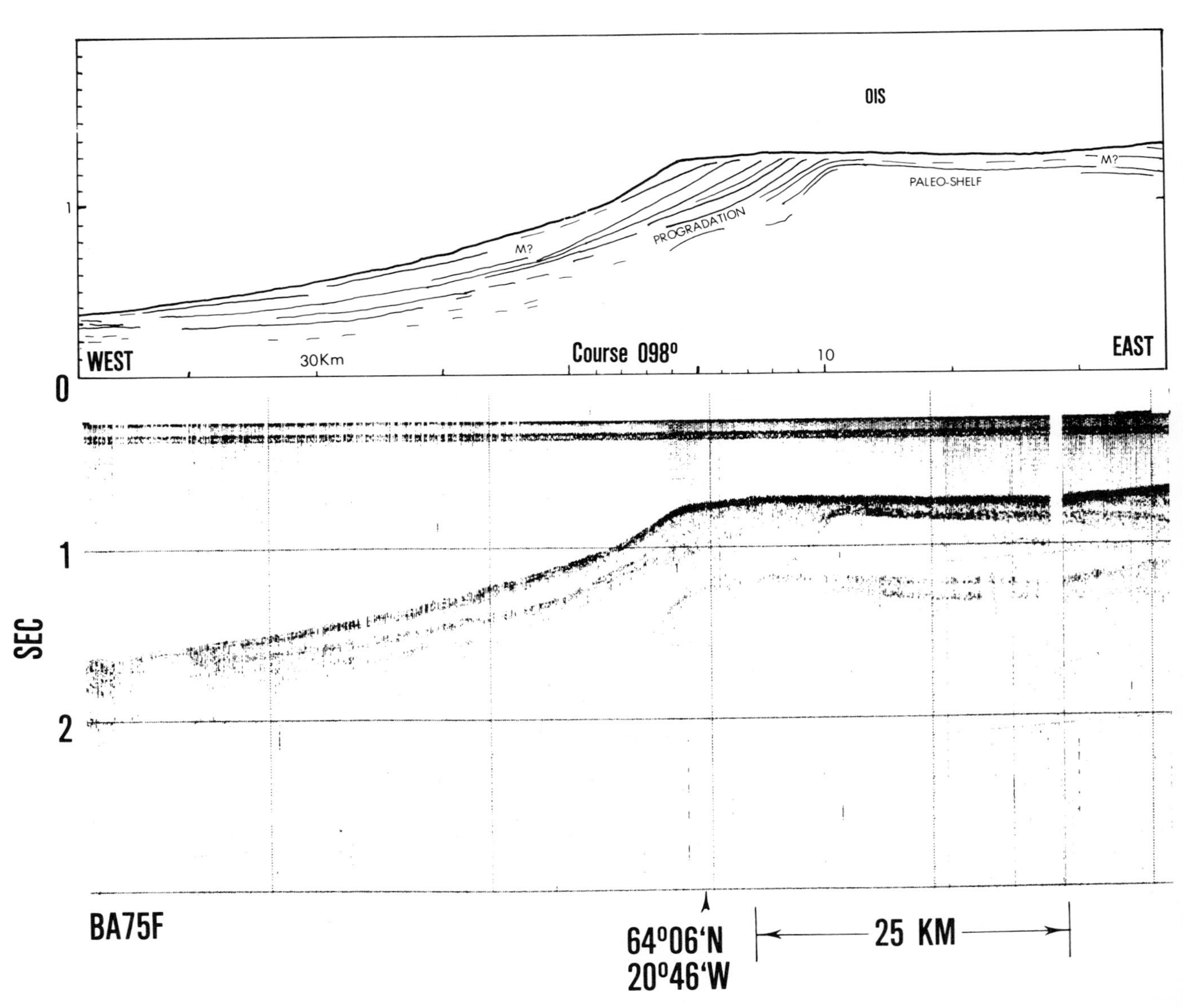

Profile 3, Plate 5

1975). Diapirism may have resulted due to loading by later turbidite sequences on top of the nannofosil ooze.

Profile 3

The OIS less than 30 km S of Profile 2 has prograded 12 to 13 km from its former location at 0.1 sec beneath horizontal reflectors (Plate 5). The sequence of reflectors at 0.25 sec beneath the shelf edge can be correlated with those at 0.3 sec under the shelf edge in Profile 2. These wedge out 40 km W of the shelf break. Also evident in Profile 3 is the more steeply dipping upper slope reflector beneath the shelf 20 km E and landward of the present upper slope.

Profile 4

Parallel to, but 15 km S of Profile 3, this seismic section shows a most intriguing difference in the progradation of the shelf edge (Plate 6). A continuous reflector at 0.2 to 0.35 sec is present below the outer shelf. This discontinuity may represent the paleo-shelf surface at its most deeply ice-eroded level. Note the shape and depth of the OIS in Profiles 3 and 4: In Profile 3 there has been a zone of nondeposition since the retreat of the ice; however, Profile 4 on the OIS has been buried under a 0.35-sec transparent lens of deposits. This is probably a large body of terminal moraine material which was preferentially deposited south of section 3 (Profile 3 has only a 0.1-sec surficial morainal deposit). The deepest paleo-shelf edge detected is 15 to 17 km E, marked by a disconformity and delineated by the buried paleo-slope bedding.

On the upper insular slope, 0 to 15 km W, wedging occurs at 0.1 sec depth. None of the uppermost transparent material, visible in a lens beyond 25 km westward, is present between 5 and 25 km W. The niche just below the shelf edge, 2 km W of the position given, may be a slump scar. At 25 to 30 km W, there is a discordant boundary diagonally across the mid-sediment reflector just above a basement scarp. The extended DOR, seen through the second reflection 1.1 km below the OIS edge, can be followed westward to the top of the basement scarp. Between

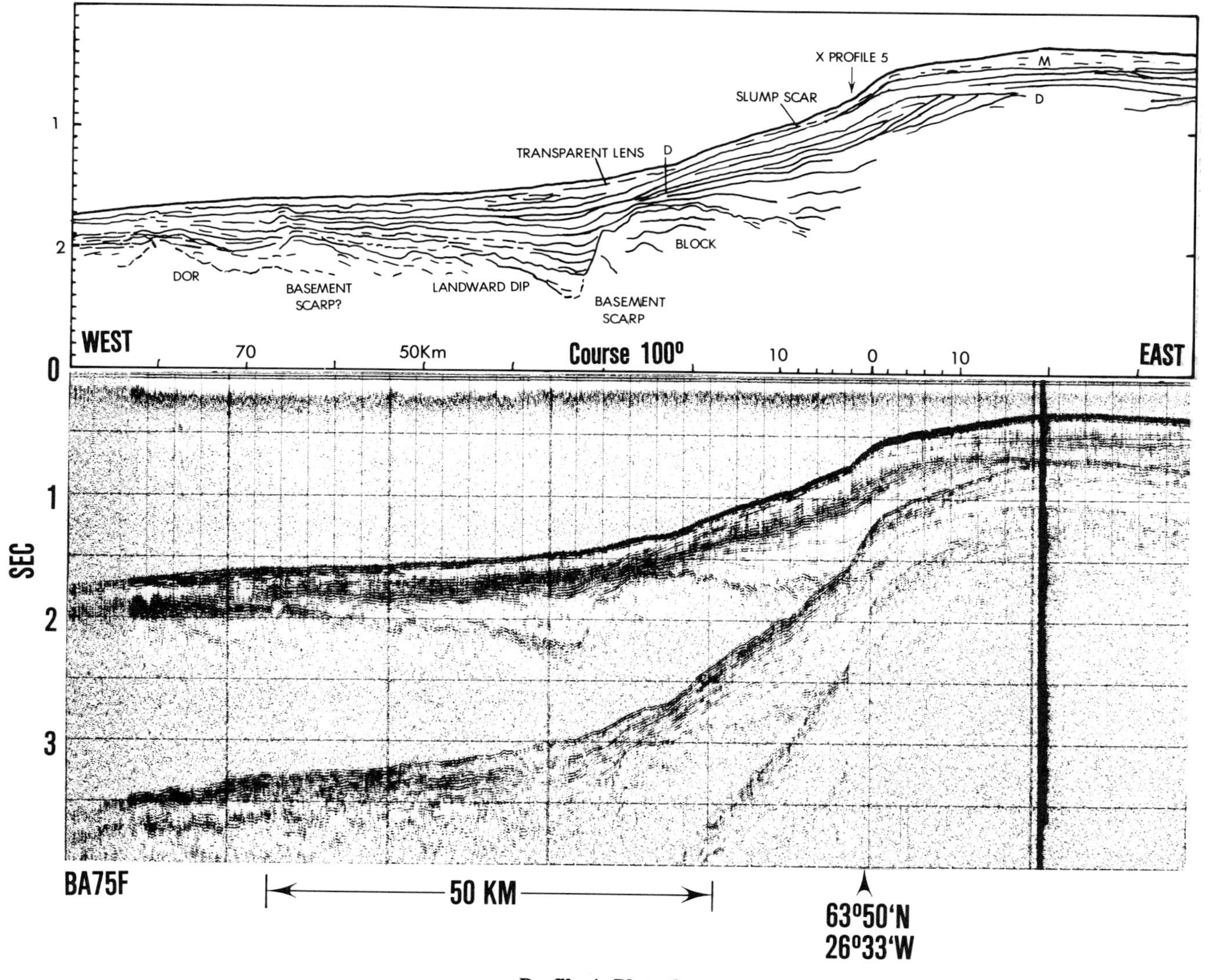

Profile 4, Plate 6

5 and 15 km W, this deepest reflector is irregular and dips landward as if it is a discontinuity similar to that of Profile 2. The basement feature landward of the basement scarp at 33 km W is similar to the fault block described landward of prominent scarps of Profiles 1 and 2. Another step in the DOR of at least 0.5 sec is at 65 km W.

One additional 0.3-sec lens of acoustically unstratified (disturbed?) sediment is at 25 to 55 km W (Plate 6). This is near the beginning of deposits of the Snorri Drift sediment but in an area of some downslope turbidite flows. The rise contains a regional eastward-dipping sediment wedge within the top second. The landward dip may be due to subsidence of Iceland beneath the increasing basalt crustal pile.

Profile 5

Under the shelf edge, the 0.2-sec transparent layer ("M") is present on Profiles 4 and 5 (Plates 6, 7). High recording gain useful for detecting the DOR at 1.1 sec obscured definition in the top 0.5 sec of the sediment. The 1.1-sec reflector is too smooth to be volcanic basement. On the slope and rise there is mid-layer faulting. Profile 5 intersects Profile 7 where both indicate a basement rise of 0.2 sec just under a prominent ash or turbidite mid-layer sequence. (This mid-layer is the "DOR" under the shelf break already noted.)

The basement scarp on Profile 7 and 70 km W is *not* an equidistant radial structure surrounding Iceland, because it is absent on this profile at a projected distance of 35 to 40 km S. At no point along Profile 5, out to 85 km S, does the basement or deepest reflector exceed 2 sec beneath the sea surface; however, the minimum bottom or basement depth at the foot of the escarpment is at 2.3 sec in Profiles 2, 4, 7. South of this crossing, as well as just west of it, neither profile gives satisfactory evidence of the prominent escarpmant described above, nor does the DOR obscure the basement trace. Basement is definitely indicated beyond 75 km S by more normal mid-ocean ridge structures.

Profile 6

Profile 6 nears the Reykjanes Peninsula, within 45 km of Reykjanes, Iceland, the active spreading center of the Mid-Atlantic Ridge. At 110 km NE (Plate 8) there is a change in the general depth of the shelf: At the base of a 200 m bathymetric incline, only 50 km from the spreading ridge axis, there is a 0.55 sec interval of stratified deposits. Across the 200 m incline, stratification conforms with the rise. Except for a few beds near the sea floor, all strata bend in-

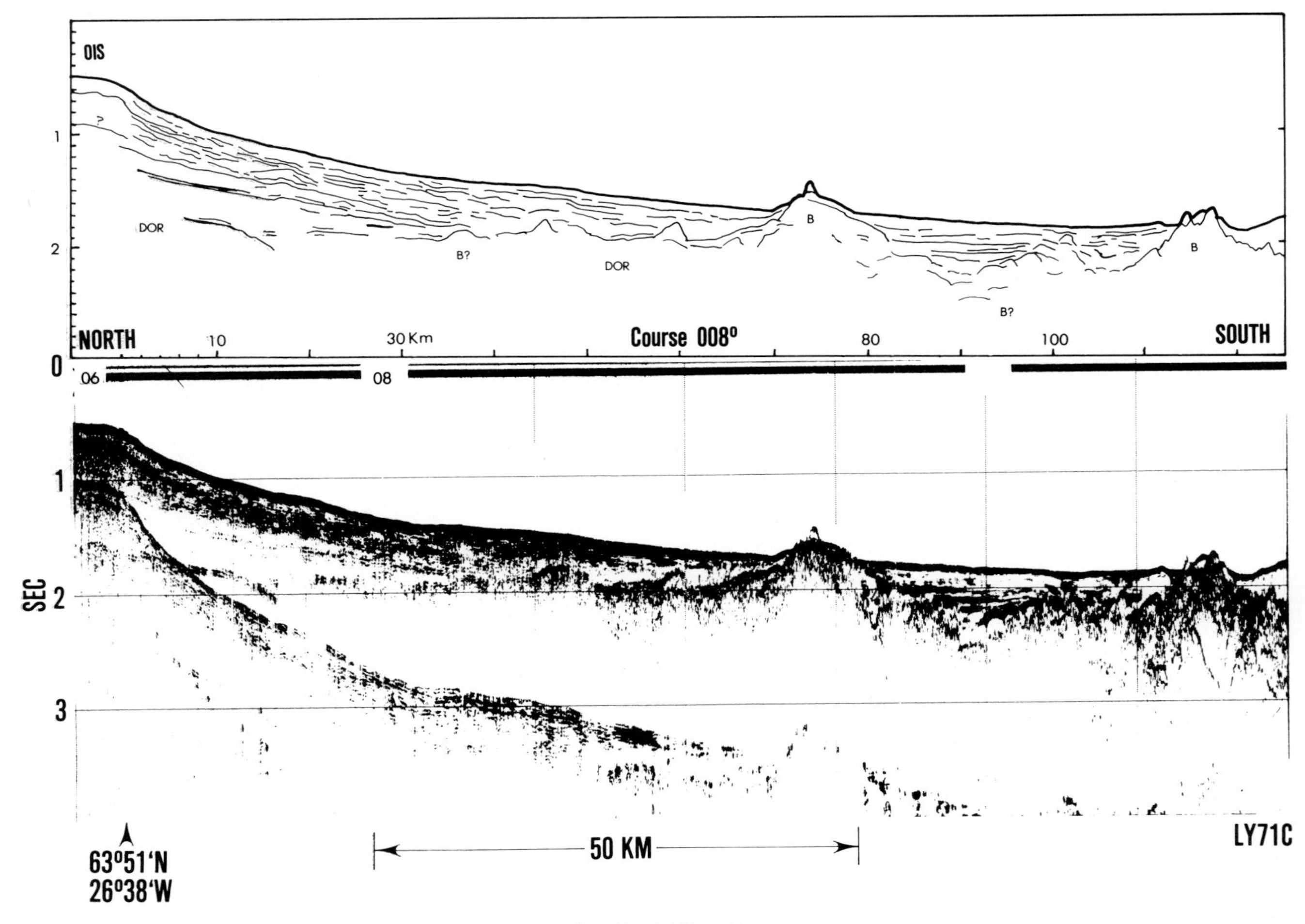

Profile 5, Plate 7

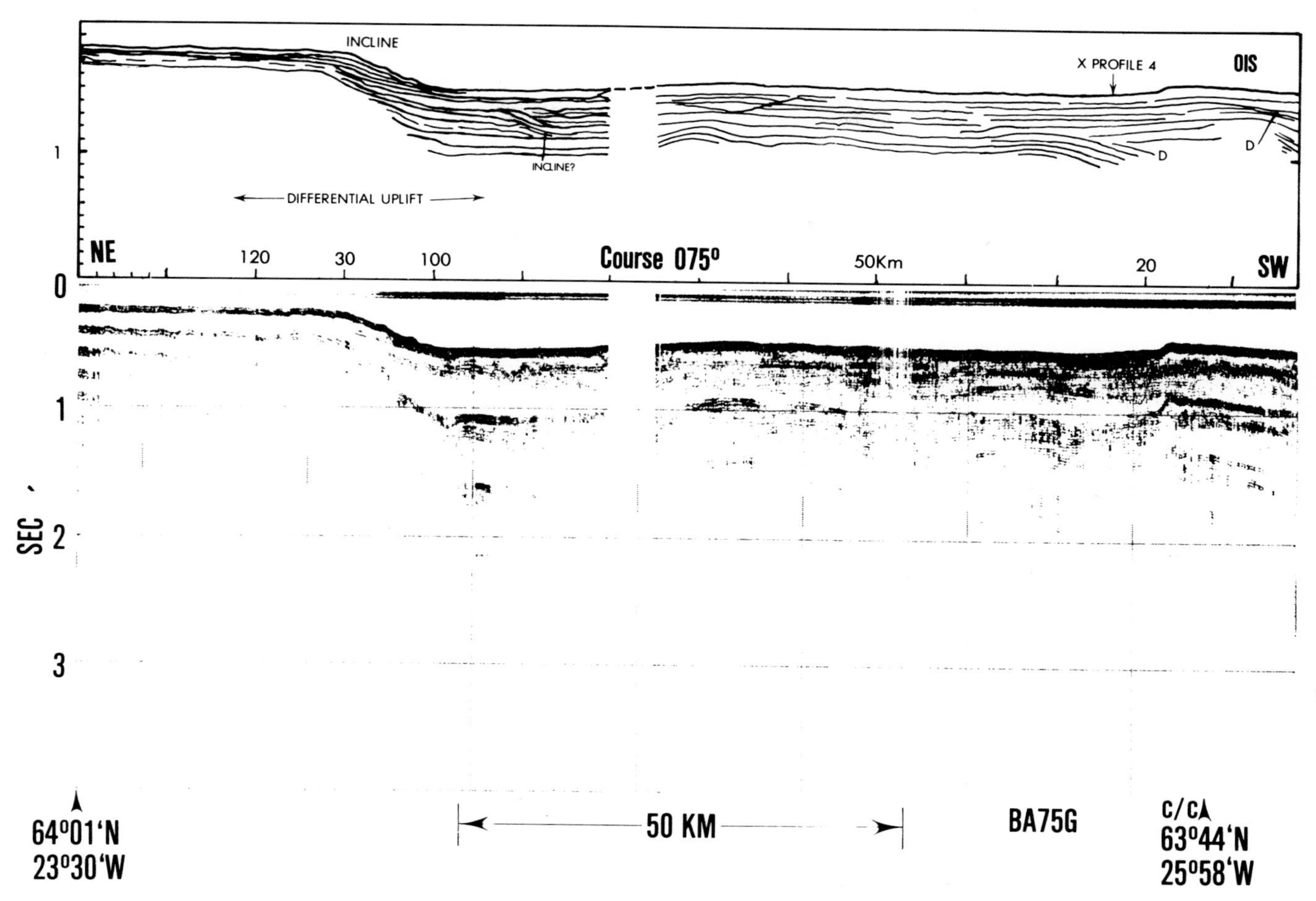

Profile 6, Plate 8

tact and are not faulted or otherwise disturbed. This is not an erosional feature: The shelf has been differentially uplifted since the latest glacial period.

The 50-m rise of the sea floor at 20 km NE is quite likely a product of the movement of retreating ice as the moraine ("M") was deposited (Plate 9). There is no deformation of the paleo-shelf at 0.15 sec depth below the 50 m sea-floor rise. Other than the latter rise, all reflectors are conformal on the northeast of the shelf edge near 80 km, to 0.55 sec in the subbottom. Another incline of 0.2 sec is possible 0.3 sec below the sea floor near 90 km NE.

At 8 and 25 km NE (landward) of the shelf break, two paleo-shelf levels are noted by disconformities, with associated adjacent former slopes dipping more steeply at 0.15 and 0.3 sec, respectively. The intersection along Profile 4 further substantiates the paleo-shelf-break morphology.

Deeper under the shelf break is a regular section of lower reflectivity, which denotes a less dramatic phase of progradation of the paleo-slope down to at least 1 sec subbottom. Wedging (thinning) of all layering above a very reflective sequence ("R") at 0.35 sec is seen toward the south. The latest wedge of transparent prograded sediment on the OIS and upper insular rise rests conformably on the highly reflective sequence. Beneath the shelf break, this highly reflective interface probably also was the sea floor prior to the last glaciation. The sections previously described and Profile 7 show this same pattern.

The mid-insular rise of section 6 is highly reflective down to 0.15 sec, at 20 to 50 km S and in 900 to 1300 m of water. It may be a zone of nondeposition due to increased current activity. On the lower reaches of the insular rise, normal turbidites are recognizable above and below the highly reflective mid-layer. It is this same mid-layer, very likely a sequence of volcanic ash, which may be traced to the DOR described near 1.0 sec under the shelf in Profiles 1, 2, 4, and 5. South of a turning point at 63°44′N, 25°58′W, the profile is almost coterminous with magnetic anomaly 5. It is only beyond 60 km S that the highly reflective mid-layer permits deeper acoustic detection of the irregular oceanic basement at 2.4 sec below the ship.

Profile 7

The uplifted inner shelf (55 km E) rising 100 m, the acoustically transparent moraine spanning 0.2 sec overlying the OIS and slope, prograded paleo-shelf edge under the moraine, and the prominent fault scarp at 88 km W, are found on profiles 5, 6, and 7.

The 100-m differential uplift, at the east end of Profile 7 (Plate 10), occurs 28 km from the mid-

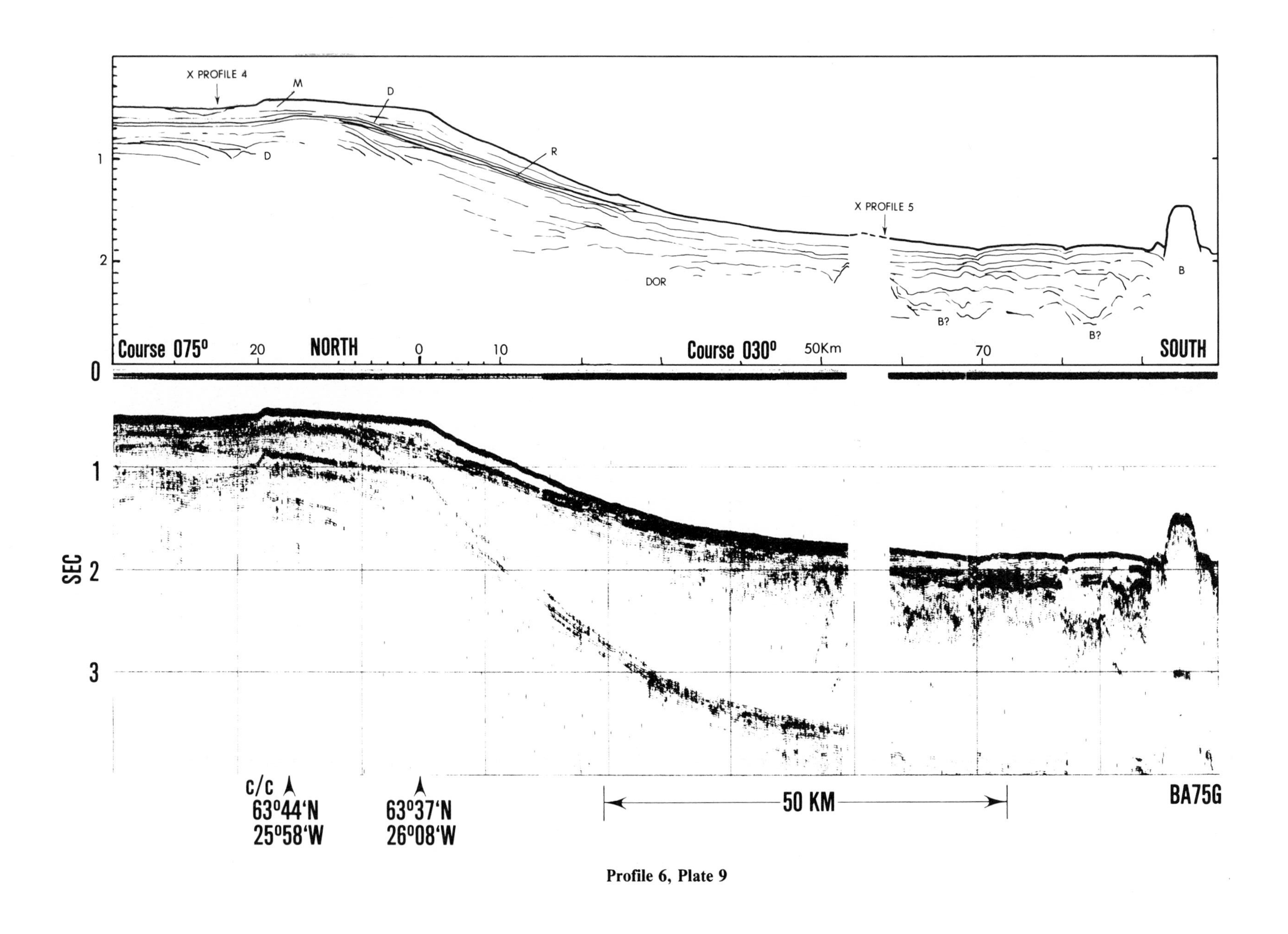

Profile 6, Plate 9

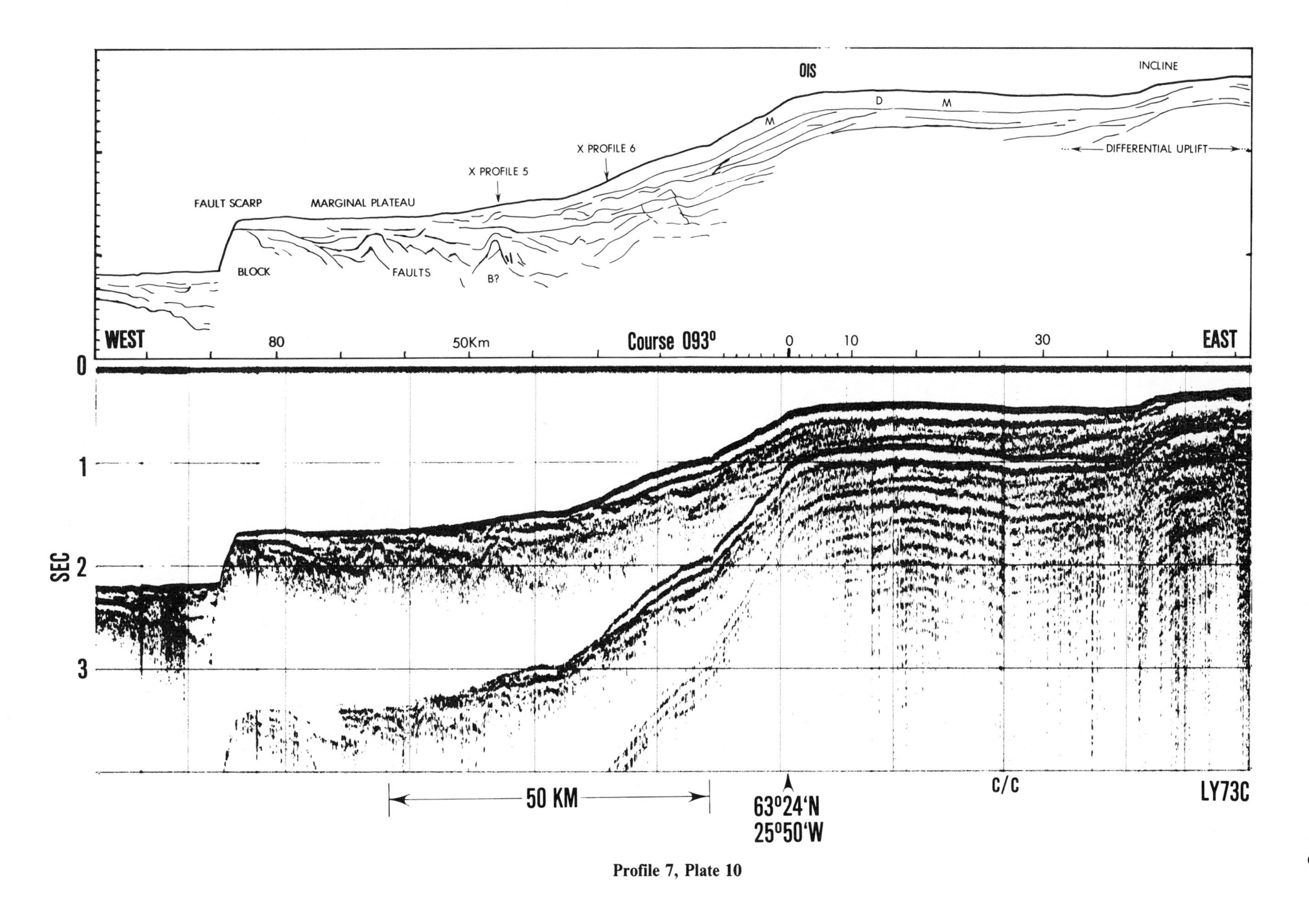

Profile 7, Plate 10

ocean ridge spreading axis. Similar to Profile 6 (at 50 km from the ridge axis), the extra 100 m uplift above the outer shelf level in such a narrow (5 km) distance may be due to isostatic rebound of the ridge crest, and a difference on either side in rates of 1 m per 100 years.

The acoustically transparent, 0.2-sec layer on the OIS and slope is believed to be a moraine over a paleo-shelf level. It may be correlated between adjacent profiles to beyond 30 km W. An acoustically transparent feature is at 15 km E and 0.25 sec, and rests upon a distinct reflector at 0.35 sec. Between 33 and 41 km E, another arched reflector at 0.3 sec overlies more transparent material to 0.5 sec. Within 15 km on either side of the OIS edge, prograded upper slope bedding can be traced to 0.8 sec. Thinning of the beds occurs on the insular slope about 750 m below sea level.

The base of the insular slope is on a marginal plateau crossed obliquely by Profiles 5 and 6. Under the marginal plateau, at the insular slope base, there is a 1.0 sec basement scarp (88 km W), and the sedimentary reflectors from 0.1 to 0.6 sec show evidence of faulting. Faults may be located at 10, 18, 45, 67, and 88 km W in the basement, which has been rotated down on the east and up on the west of each block between the faults.

Profile 8

Morphological and subbottom features correlate with those found in the two adjacent profiles. The profile crossings tie together the traces of the uplifted incline (Plate 11). The incline due to differential uplift is 30 km W of the mid-ocean ridge spreading axis. At this point also, there is a terrace 2 km wide (NE-SW) and 375 m below present sea level.

The strongly reflecting mid-sediment layer ("R") at 0.15 ($\pm$0.05) sec over the OIS is believed to be a paleo-shelf. The acoustically transparent layer called "M," superimposed on the paleo-shelf, is a moraine which wedges out on the slope at 10 km SW. A paleo-shelf edge noted at 10 to 12 km NE is near the discontinuity ("D") below the moraine. Beneath and just seaward of the shelf break, the deepest reflectors are at 0.7 to 1.0 sec, 2 to 10 km SW. On the mid-insular rise, turbidite stratification predominates to 0.3 sec depth. Basement is not well defined except beyond 50 km SW on the MAR.

Profile 9

The Reykjanes Ridge crest merges with the Iceland Shelf very close to this profile, making it a unique section (Plate 12). Recent topographic expression of the axial uplift, and over 0.7 sec of sediment less than 30 km from the spreading axis (roughly equat-

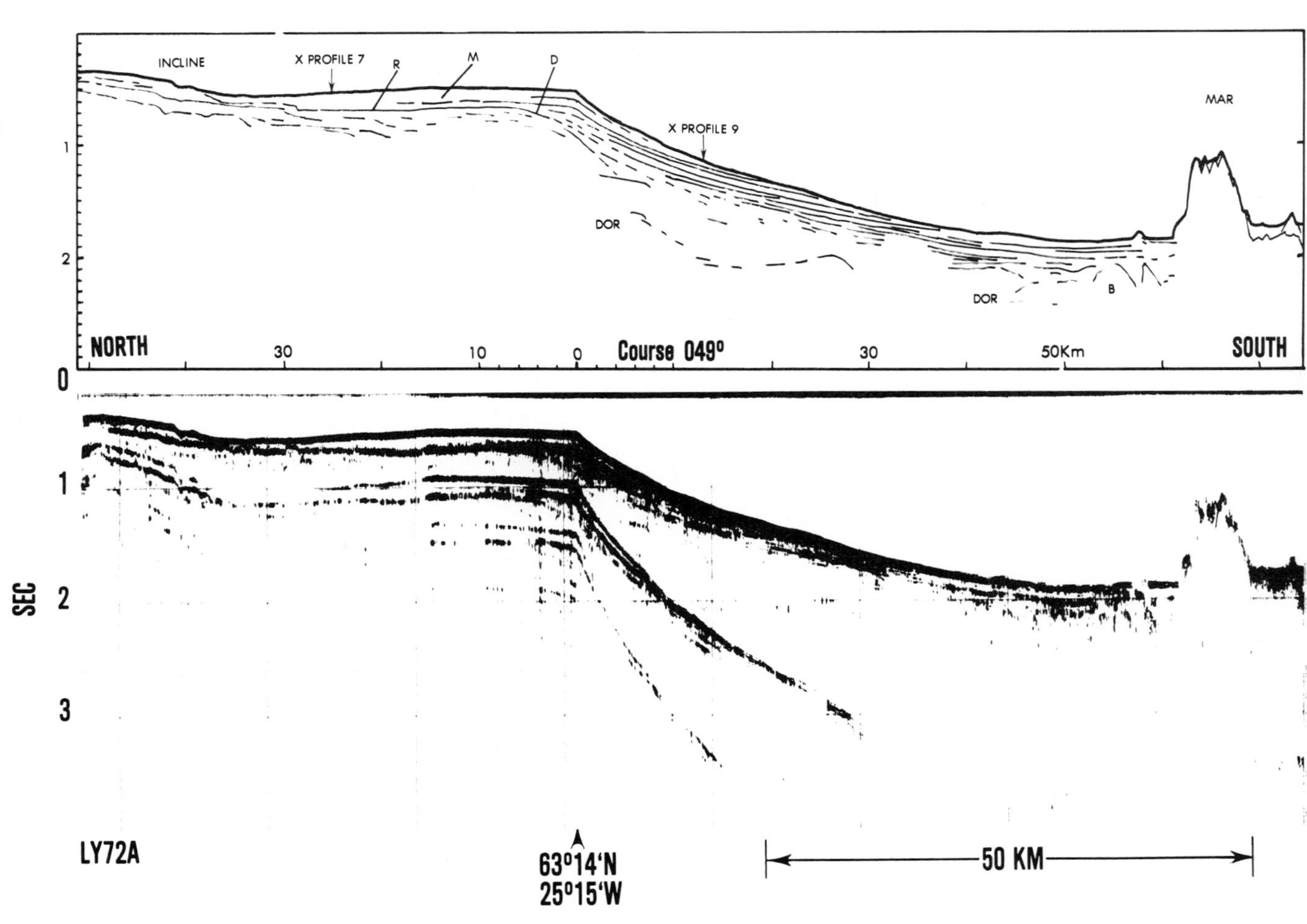

Profile 8, Plate 11

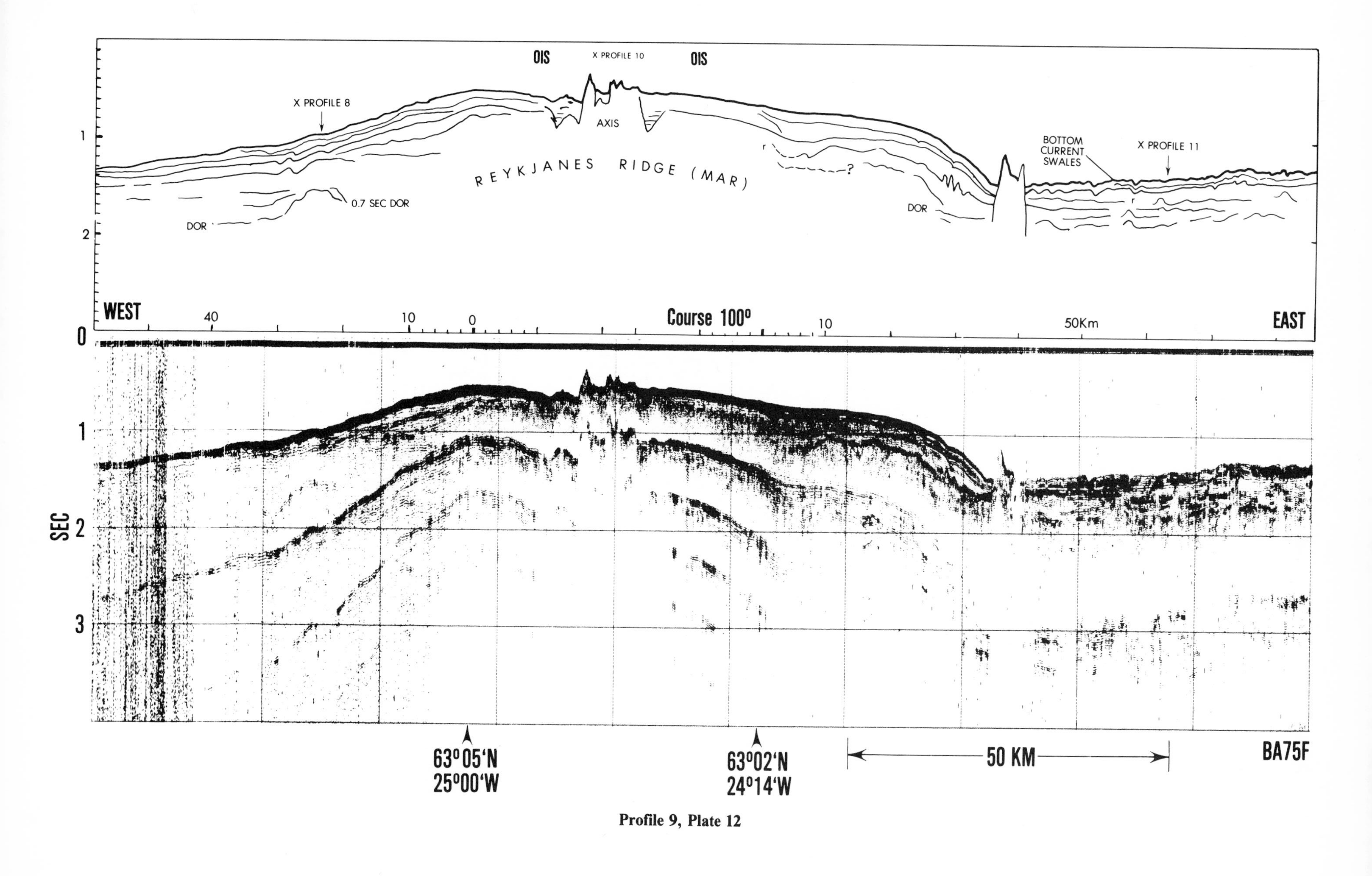

Profile 9, Plate 12

ing to 16 cm per 1,000 years), accentuate sedimentary and tectonic processes.

The OIS is indicated on the geologic map (Fig. 3) as "MP" (marginal plateau). The two chosen shelf-break positions given on the profile are not the OIS edge, but rather, are estimated paleo-shelf edges of the Reykjanes Ridge central block. Nonetheless, typical subslope bedding, thinning of layers, and discontinuities prevail, just as on the adjacent insular slopes near the OIS edge.

To the east of "X Profile 11," beyond 40 km E, is a zone of hummocky redistributed sediments spanning more than 0.8 sec (probably more than 1.0 sec). The more reflective beds are possibly volcanogenic and glacial debris. This layering may be traced where the prominent 0.2-sec mid-sediment reflector ("r"), at 45 km E (near the small basement peak), is extended up onto the ridge flank where it becomes the DOR within 20 km of the ridge axis.

Profile 10

Longitudinally, from an oblique crossing of the rift valley, Profile 10 begins only 5 km S of Profile 9. The generally low relief of the eastern half of the ridge crest, resembling a marginal plateau, was at one time probably leveled by glacial or wave action, as was the subsided small guyot at 75 km S (Plate 13). A sediment pond ("P") perched high on the central block is filled with about 0.3 sec of highly reflective strata at 55 to 70 km S. Another perched sediment pond may be present only 30 to 40 km S, up to the edge of the maximum height of the ridge.

The scarp off the central block, 80 to 115 km S, has nearly 1.5 sec of basement relief and is highly irregular, noticed here in an oblique section. There is little sediment on this slope, possibly due to active mid-water currents. The basin, 115 to 125 km S at the slope base, has at least 0.5 sec of redistributed sediment.

Profile 11

An isochronous crustal section along the eastern magnetic anomaly 5 may be contrasted with those west of the Reykjanes Ridge/OIS intersection. The southern Iceland shelf break is sharp, the slope steepened (Plate 14). Prograded paleo-slope bedding is clear as far landward as 23 km NE, where a discontinuity at 0.3 sec may occur at a buried analog of the differential uplift seen in Profiles 6 and 7. The intermittent paleo-shelf was tilted or subsequently

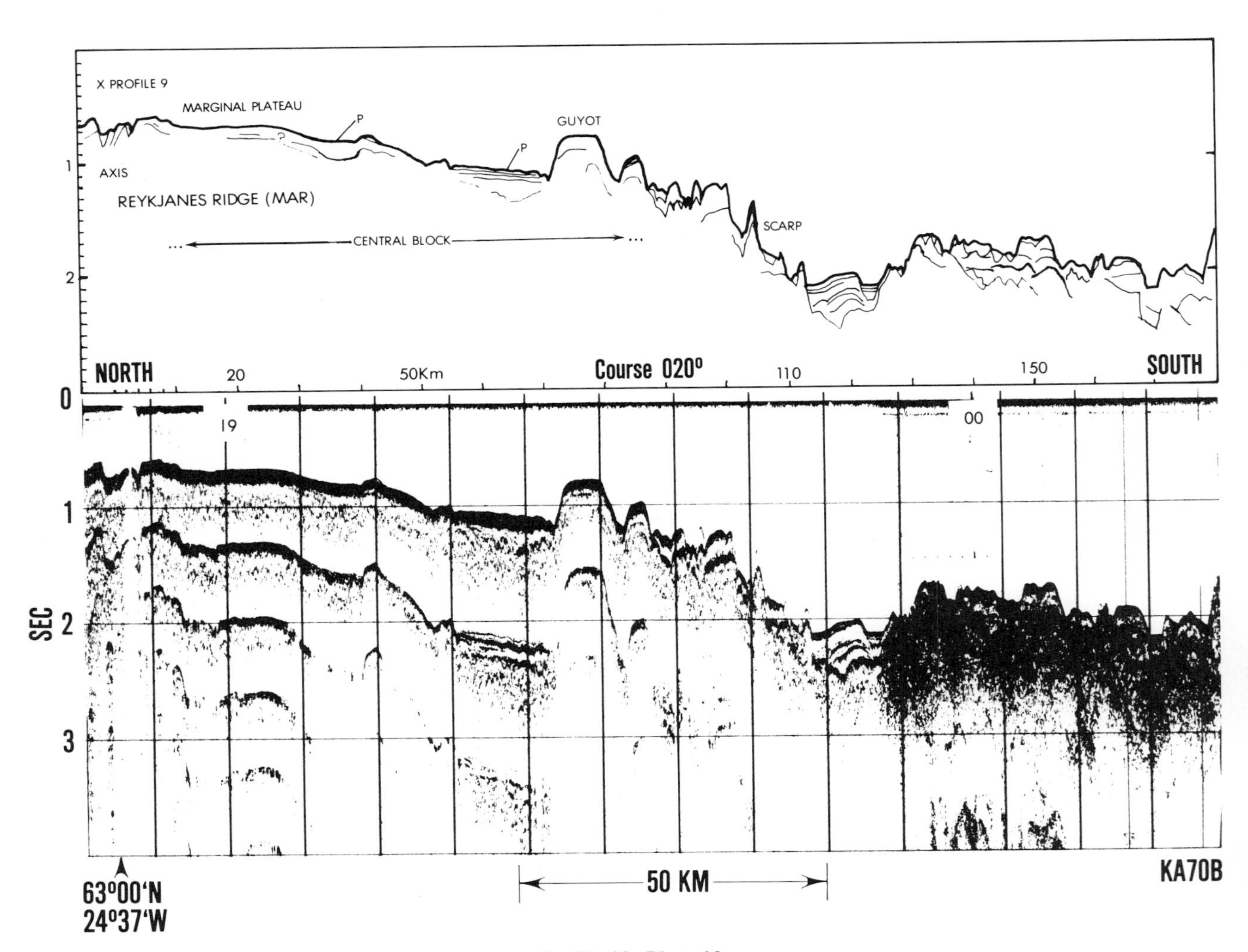

Profile 10, Plate 13

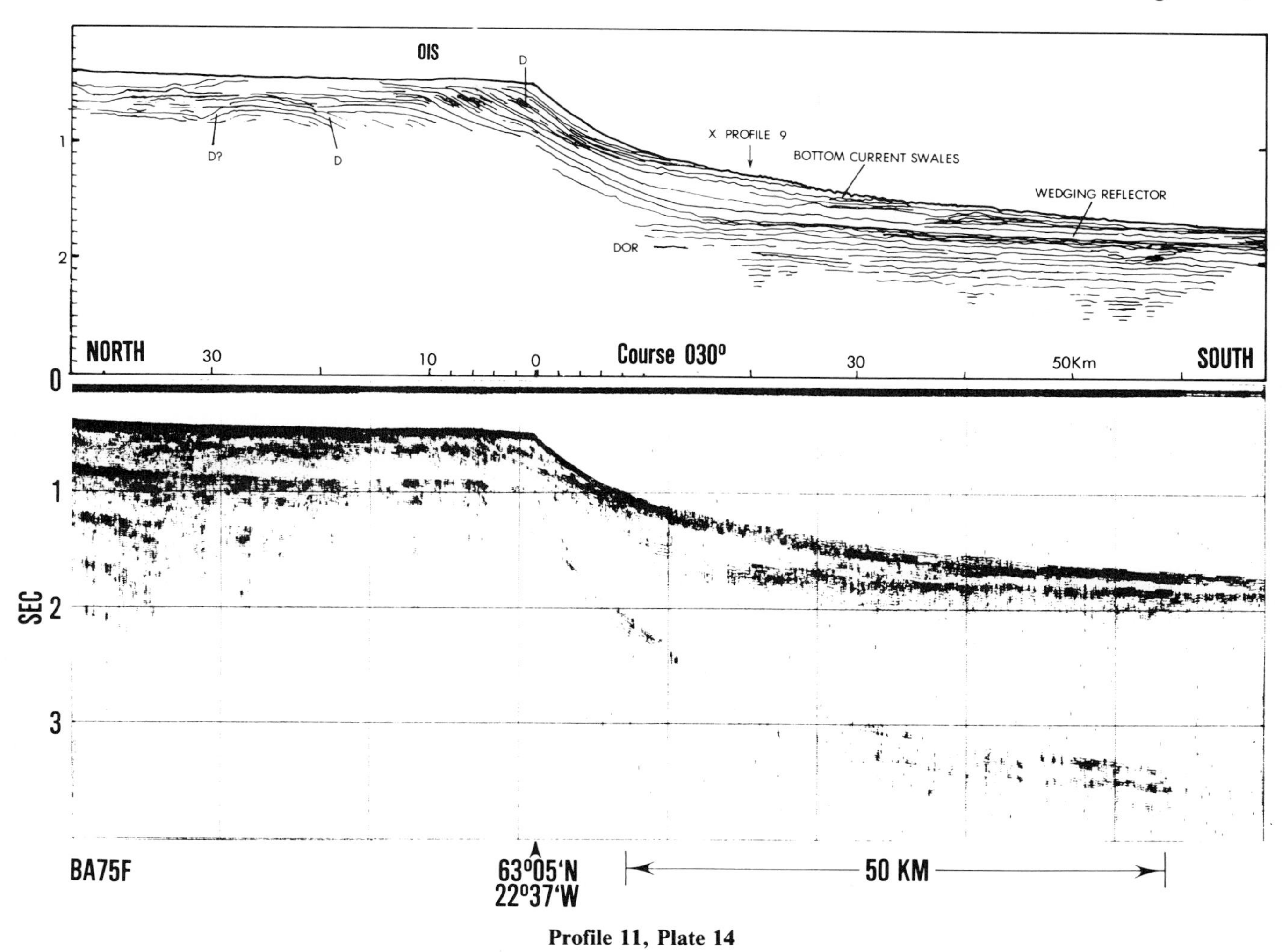

Profile 11, Plate 14

slumped in a saw-tooth, irregular pattern, along slope-parallel bedding planes 50 to 150 m between 0 and 10 km NE. Later glacial grading and filling of the OIS covered these irregularities. Under the sea floor 30 km N (northeast) at 0.3 to 0.45 sec is a filled syncline and discontinuity.

Lower on the slope, highly reflective sediments are concentrated between 700 and 850 m water depth. This evidence accentuates the argument for nondeposition on the insular slope due to ocean currents. Beneath this, the strata are subparallel with the sea floor to 0.8 sec. On the insular rise, 12 to 32 km S, and to a lesser extent, to 45 km S, the sea floor has been shaped into hyperbolic swales by bottom-water currents. They were active as long as the sedimentation rate allowed the deposition of the top 0.15 sec of disturbed strata, from 12 to 50 km S. These features, bottom-current-derived swales, are also shown on Profile 9 (Plate 12). On the lower insular slope and rise to the south, other layers very gradually thin or wedge to one-third their depth at the slope base. A prominent reflecting series at 0.7 sec, at 20 km S, tapers and rises to only 0.25 sec at 68 km S.

Profile 12

The 265 m of water at the shelf break (position given) overlies a 0.2-sec-depth reflective series of parallel layers which could indicate the thickness of the moraine (Plate 15). The slope and upper rise is gullied and traversed by canyon heads about 100 m deep. A discontinuity ("D?"), or just a very reflective layer ("R"), at 0.1 to 0.15 sec on the upper rise (20 km SE) limits the canyon erosion of more deeply buried rise materials ("R" and "D," like "M," are not meant to imply a connection everywhere they are found). The top transparent sediment pile on the upper rise thickens on the mid-rise at 30 to 70 km SE, to cause the very reflective layer to reach 0.2 to 0.25 sec below the pile. At 42 km SE, the 0.25-sec layer and those around it have slumped (S?) or are faulted more deeply than elsewhere. Coring "R" would determine if it is a former erosion surface or a zone of nondeposition in the southern insular rise.

Beneath the mid-rise the DOR is below 1.0 sec. The rise sediment wedges (thins) away from the insular slope in the upper 0.5 sec; however, the DOR remains at 1.0 sec from 15 to 100 km SE of the shelf break.

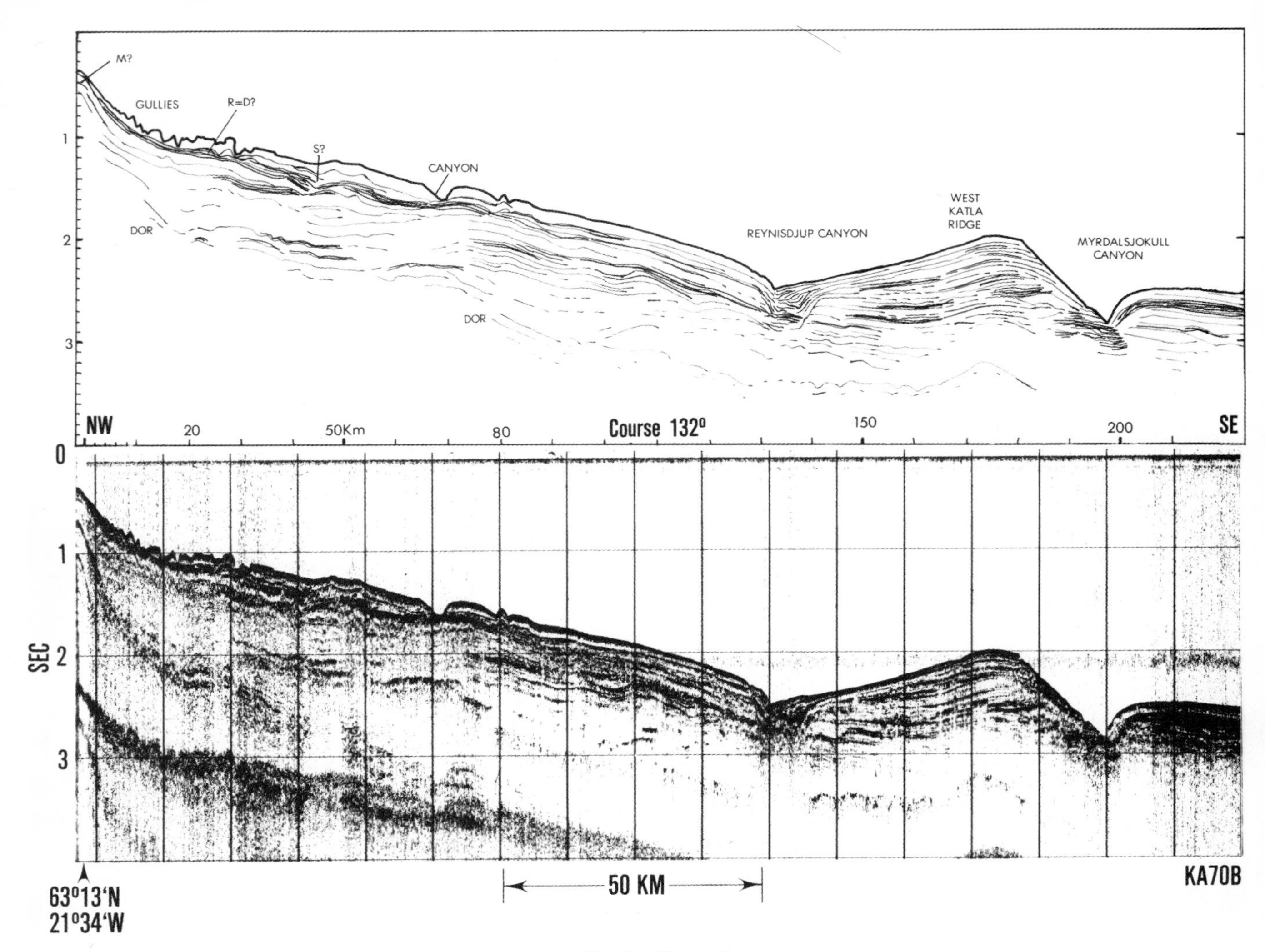

Profile 12, Plate 15

A submarine canyon head may be studied at 69 km SE. At reflector R, it widens its transverse profile due to the higher density of the canyon floor. It has eroded down to the reflective 0.2-sec "R" layer and has begun to broaden its channel rather than deepen the canyon.

Reynisdjup and Myrdalsjokull Canyons (Fig. 3) isolate the West Katla Ridge, perhaps once a more substantial portion of the insular rise. Excellent definition of strata on either side of Reynisdjup Canyon may be alternating ash and/or glacial debris. Highly reflective turbidite materials have been deposited in the canyons.

Profile 13

Parallel, and less than 10 km from section 12, Profile 13 lies just upslope (Plate 16), but depth of penetration as compared with Profile 12 is unchanged. Note the DOR below the mid-rise at more than 1 sec depth (25 to 85 km SE). Sea-floor morphology is unchanged in either case. The canyon head at 50 km SE corresponds with features 65 to 80 km SE on Profile 12. Wedging and discontinuities abound as at 40 to 50 and 70 to 80 km SE, and are equally visible on both sections 12 and 13. However, the muting by the recording system of the variations in sediment reflectivity interferes with the interpretation that may be given or implied for section 13.

REFERENCES CITED

Davies, T. A. and A. S. Laughton, 1972, Sedimentary processes in the North Atlantic: Wash., D.C., U.S. Government Printing Office, Initial Reports of the Deep Sea Drilling Project, v. XII, p. 905–930.

Egloff, J. and G. L. Johnson, 1975, Morphology and structure of the southern Labrador Sea: Canadian Jour. Earth Sciences, v. 12, p. 2111–2133.

Herron, E. M. and M. Talwani, 1972, Magnetic anomalies on the Reykjanes Ridge: Nature, v. 238, p. 390–392.

Johnson, G. L. and E. D. Schneider, 1969, Depositional ridges in the North Atlantic: Earth and Planet. Sci. Letters, v. 6, p. 416–422.

——— G. Sommerhoff, and J. Egloff, 1975, Structure and morphology of the West Reykjanes Basin and the southeast Greenland continental margin: Mar. Geology, v. 18, p. 175–196.

——— and G. Palmason, in preparation, Morphology and structure off Southern Iceland.

Jones, E. J. W., M. Ewing, J. I. Ewing, and S. L. Eittreim, 1970, Influences of Norwegian Sea overflow water on sedimentation in the northern North Atlantic and Labrador Sea: Jour.

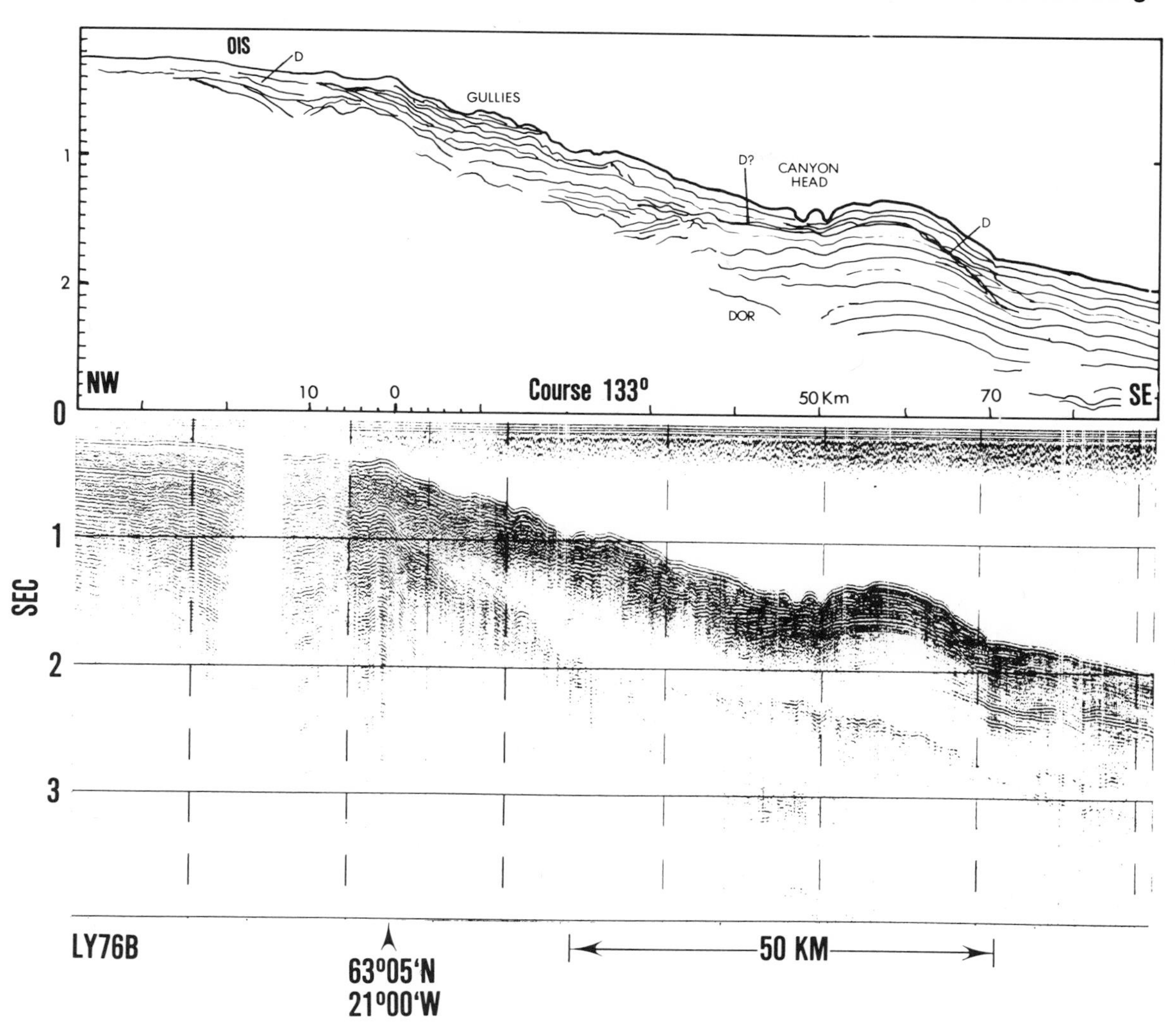

Profile 13, Plate 16

Geophys. Research, v. 75, p. 1655–1680.

Kristjansson, L., H. Karlsson, and K. Thors, 1976, In search of central volcanoes on the Iceland Shelf: Natturufraedingurinn, v. 46, p. 209–216.

Laughton, A. S., et al., 1972, Initial Reports of the Deep Sea Drilling Project: Wash., D.C., U.S. Government Printing Office, v. 12.

Lonsdale, P. and C. D. Hollister, 1976, Cut-off of an abyssal meander on the Icelandic insular rise (Abs.): Trans. Amer. Geophys. Union, EOS, v. 57, p. 269.

Luyendyk, B., et al, 1977, Young and hot drilling in the North Atlantic: Geotimes, v. 22, p. 25–28.

Olafsdottir, Thordis, 1975, A moraine ridge on the Iceland Shelf, west of Breidafjördur: Natturufraedingurinn, v. 45, p. 31–36.

Palmason, G., 1974, Insular margins of Iceland, *in* C. A. Burk and C. L. Drake, eds., Geology of continental margins: New York, Springer-Verlag, p. 375–379.

Ruddiman, W. F., 1972, Sediment redistribution on the Reykjanes Ridge: Seismic evidence: Geol. Soc. of America Bull., v. 83, p. 2039–2062.

Shor, A., D. Muller, and D. Johnson, 1977, Transport of Norwegian Sea overflow—preliminary results: Presented at International Council for the Exploration of the Sea, Reykjavik.

Talwani, M., C. C. Windisch, and M. G. Langseth, Jr., 1971, Reykjanes Ridge crest: A detailed geophysical study: Jour. Geophys. Research, v. 76, p. 473–517.

Thors, K. and L. Kristjansson, 1974, Westward extension of the Snaefellsnes volcanic zone of Iceland: Jour. Geophys. Research, v. 79, p. 413–415.

Tryggvason, E., 1974, Vertical crustal movement in Iceland, *in* L. Kristjansson, ed., Geodynamics of Iceland and the North Atlantic area: p. 241–262.

Vogt, P. R., 1974, The Iceland phenomenon: imprints of a hot spot on the ocean crust, and implications for flow below the plates, *in* L. Kristjansson, ed., Geodynamics of Iceland and the North Atlantic Area: p. 105–126.

——— and G. L. Johnson, 1973, A longitudinal seismic reflection profile of the Reykjanes Ridge, Part II: Implications for mantle hot spot hypothesis: Earth and Planet. Sci. Letters, v. 18, p. 49–58.

——— ——— 1975, Transform faults and longitudinal flow below the Mid-Oceanic Ridge: Jour. Geophys. Research, v. 80, p. 1399–1428.

——— and J. Egloff, in preparation, Geophysical profiles of the northern Reykjanes Ridge.

Worthington, L. V. and G. H. Volkmann, 1965, The volume transport of the Norwegian Sea overflow water in the North Atlantic: Deep-Sea Research, v. 12, p. 667–676.

Multichannel Seismic Depth Sections and Interval Velocities over Outer Continental Shelf and Upper Continental Slope between Cape Hatteras and Cape Cod[1]

JOHN A. GROW,[2] ROBERT E. MATTICK,[3] JOHN S. SCHLEE[2]

Abstract Six computer-generated seismic depth sections over the outer continental shelf and upper slope reveal that subhorizontal Lower Cretaceous reflectors continue 20 to 30 km seaward of the present shelf edge. Extensive erosion on the continental slope has occurred primarily during the Tertiary, causing major unconformities and retreat of the shelf edge to its present position. The precise age and number of erosional events is not established, but at least one major erosional event is thought to be Oligocene and related to a marine regression in response to a worldwide eustatic lowering of sea level.

Velocities derived from the multichannel data reveal distinctive ranges and lateral trends as functions of sediment age, depth of burial, and distance from the coastline. Seismic units beneath the shelf and slope of inferred Tertiary age range from 1.7 to 2.7 km/sec, increasing with age and depth of burial. Units interpreted as Upper Cretaceous rocks beneath the shelf range from 2.3 to 3.6 km/sec and show a distinct lateral increase across the shelf followed by a decrease beneath the present continental slope. Inferred Lower Cretaceous and Upper Jurassic rocks beneath the shelf increase from 3.7 to 4.8 km/sec from nearshore to offshore and indicate a change in facies from clastic units below the inner shelf to carbonate units beneath the outer shelf and upper continental slope. Both reflection and refraction data suggest that thin, high-velocity limestone units (5.0 km/sec) are present within the Lower Cretaceous and Upper Jurassic units beneath the outermost shelf edge, but that these change lithology or pinch out before reaching the middle shelf. Although lateral changes in velocity across the shelf and local velocity inversions appear, the interval velocities along the length of the margin show excellent continuity between Cape Hatteras and Cape Cod. The high-velocity horizons within the Lower Cretaceous and Upper Jurassic shelf-edge complex indicate the presence of a carbonate bank or reef.

The continental shelf off New Jersey is underlain by a trough approximately 150 km wide with up to 12 km of sedimentary fill. The oceanic basement beneath the upper continental rise is usually at a depth of 10 to 11 km and is overlain by 6 to 8 km of sediments. The rise sediment trough is separated from the shelf trough by an acoustic basement ridge 25 to 75 km wide where penetration never exceeds 6 km beneath sea level, although faulting and carbonate bank or reef complexes frequently limit penetration to 3 to 4 km in this zone. The acoustic basement ridge coincides with the East Coast Magnetic Anomaly and is interpreted as thick oceanic crust formed during the initial phase of seafloor spreading between North America and Africa. Rapid differential subsidence occurred on opposite sides of the basement ridge during the Jurassic and Early Cretaceous. Differential subsidence beneath the shelf also occurred along the margins, with narrower and shallower shelf troughs occurring off platform areas such as Cape Cod and Cape Hatteras.

INTRODUCTION

Marine seismic reflection systems record the echos of acoustic energy from geologic horizons beneath the sea floor. These are recorded as a function of travel time between an acoustic source (such as an airgun) and a sensor system (such as a hydrophone). In recent years, streamers as long as 3,600 meters and with as many as 48 separate hydrophone groups have been used. Longer streamers with 96 or more hydrophone groups may be used in the near future. Grouping of channels about a common-depth point (CDP) and analysis of the increase in travel time due to an increase in source to detector distance can then yield interval velocity determinations for the upper horizons (see Taner and Koehler, 1969). The velocity data can then be used to properly sum or stack all the reflections from a common depth point into a single trace in a manner that improves signal-to-noise ratios on primary reflections and discriminates against undesired multiples (Mayne, 1962). Time varying filters, deconvolution, and scaling may also be used to enhance the signal-

[1]Manuscript received, July 14, 1977; accepted, May 12, 1978.

[2]U.S. Geological Survey, Woods Hole, Massachusetts 02543.

[3]U.S. Geological Survey, Reston, Virginia 22092.

The seismic data processing for lines 5, 6, 9, 10, and IPOD was performed by John Anderson and William McBride of Geophysical Services, Inc., whereas line 2 was processed by Richard Wise of Digicon Geophysical, Inc. Discussions with Kim Klitgord, Bob Sheridan, Bill Dillon, John Behrendt, Wylie Poag, Mahlon Ball, and Frank Manheim stimulated many of the interpretations presented in this paper. Kim Klitgord and Mahlon Ball critically reviewed the manuscript. Linda Sylwester drafted all figures; Janet Burke and Janet Gelinas typed the manuscript.

to-noise ratio. Finally, the reflection time to a seismic horizon can be multiplied by the average velocity to that horizon, and visual displays can be obtained with reflections plotted as a function of depth (i.e., depth sections).

Many single channel reflection profiles over the Atlantic continental margin (e.g., Emery et al, 1970) and a few multichannel profiles (Grow and Schlee, 1976; Dillon et al, 1976; Schlee et al, 1976) have been published which display reflections as a function of travel time. In this paper, we will refer to such profiles as time sections, and sections that display reflections as a function of depth will be referred to as depth sections. Sections that summarize interval velocity as a function of depth also will be presented in this paper and will be referred to as velocity sections.

Six recently processed depth sections over the Atlantic outer shelf and continental slope between Cape Hatteras and Cape Cod (Fig. 1) will be discussed here, along with some geologic implications inferred from interval velocity analysis. Although the exploration industry has had the capability to prepare depth sections for at least five years, these are the first publically available depth sections on any U.S. continental margin, and they provide a much improved picture of the subsurface. Line 2 was recorded in 1973 using a 24-channel streamer (2,400 m long) and a tuned airgun array totaling 22.9 cubic liters. Line 2 was recorded and processed by

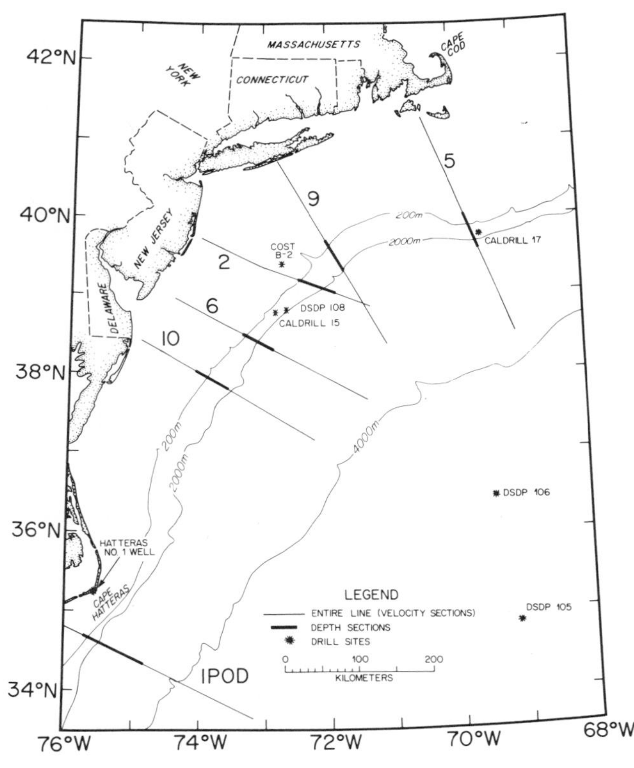

FIG. 1. Location map of six seismic profiles plus several drill holes referred to in text. Interval velocity summaries for the entire length of each profile are shown in Figure 2. Portions converted to depth sections (heavy line) are shown in Figures 4–9.

Digicon Geophysical Corporation. Lines 5, 6, 9, 10, and IPOD were recorded during 1974 and 1975 by Digicon Geophysical Corporation using a 48-channel streamer (3,600 m long) and a 27.9 cubic liter airgun array; they were processed by Geophysical Services, Inc.

Previous interpretations of the time sections for lines 2, 5 and 6 have been made by Grow and Schlee (1976), Schlee et al, (1976), and Schlee et al, (1977). A detailed interpretation of the depth sections for the IPOD line and the velocity structure near the Cape Hatteras area has been made by Grow and Markl (1977). A detailed discussion of the petroleum potential along Line 2 using both the time and depth sections has been presented by Mattick et al, (1978).

INTERVAL VELOCITY ANALYSIS

Compressional velocities within sedimentary rocks vary as functions of lithology, age, depth of burial, and diagenetic history. Analysis of velocity data derived from multichannel seismic profiles permits the continuous mapping of vertical and horizontal velocity variations. In cases where age and depth of burial are known to be constant, velocity variations can be restricted to differences in either lithology or diagenetic history. Although our analysis of these variations on the U.S. Atlantic margin has just started, certain major trends are apparent. For example, Schlee et al, (1976) observed that velocity sections for lines 1, 2, and 3 showed that sedimentary horizons between 2 and 4 km deep increased in velocity from nearshore to offshore. This is interpreted as indicating a trend from nearshore clastic facies to more limestone-rich facies offshore such as a carbonate bank or reef complex.

Velocity analyses were made every 3 km along the entire length of the six profiles, except for the outer shelf and upper slope portions of lines 2, 6, and IPOD, where either 1-km or continuous (every shotpoint) velocity analyses were made. Interval velocities were derived from root-mean-square or stacking velocities according to the methods of Dix (1955) and Taner and Koehler (1969). The interval velocities along key reflecting horizons were averaged over 20- to 50-km lengths and were plotted by depth and distance along the profiles (Fig. 2).

Standard deviations of interval velocities in the upper 3 km of shelf sediments and the upper 5 km (beneath sea level) of continental rise sediments generally are less than 10% of the mean velocities, and these are generally considered very reliable. Between approximately 3 to 5 km depth on the shelf and 5 to 7 km on the continental rise, the accuracy of the interval velocities degrades somewhat but is still considered good. At depths greater than 5 to 7 km, the interval velocities frequently have standard deviations of 15 percent or more and are considered fair to poor.

The correlation of seismic horizons with geologic age has been inferred from the following drill hole data: Deep Sea Drilling Project sites (Sites 105–108: Hollister, Ewing et al, 1972), COST B-2 well (Smith et al., 1976, Scholle, 1977), Caldrill Sites 15 and 17

(Manheim and Hall, 1976) and the Hatteras Light well (Maher, 1971), all of which are shown in Figure 1. The geologic ages shown on the velocity sections are preliminary and will probably be modified as more holes are drilled, as more tie lines are run, and as reanalysis of the present lines proceeds.

The interval velocities of sedimentary rocks between Cape Hatteras and Cape Cod have very characteristic ranges for specific ages and specific locations on the shelf, slope, and rise. Below the shelf, velocities of Tertiary rocks are typically between 1.7 and 2.3 km/sec (lines 5, 9, 10, and IPOD). On lines 2 and 6, where the Tertiary section below the shelf was divided into the upper and lower Tertiary units, velocity values of 1.7 to 2.0 and 2.1 to 2.7 km/sec were calculated, respectively. Beneath the continental slope and rise, the velocity of Tertiary units also ranges between 1.7 to 2.7 km/sec.

Velocities in Upper Cretaceous rocks over the continental shelf and slope on line 9 and those lines to the south have a very distinctive trend. The velocities range from 2.3 to 3.0 km/sec nearshore, and gradually increase to 2.9 to 3.6 km/sec on the outer shelf. Beneath the continental slope, the velocities of Upper Cretaceous rocks systematically decrease to 2.7 to 2.9 km/sec.

The COST B-2 well was drilled to a depth of 4.8 km and was originally thought to have terminated in Lower Cretaceous shallow-water sedimentary rocks (Scholle, 1977). However, recent identification of Upper Jurassic (Kimmeridgian) dinoflagellates has moved the top of the Jurassic up to at least 4.1 km (E. Robbins, personal commun., 1978). The section between 4.1 and 4.8 km is an alternating sequence of sandstone, shale, limestone, and coal (Scholle, 1977). Velocities in inferred Lower Cretaceous rocks range from 3.0 to 3.5 km/sec nearshore to 4.2 km/sec below the outer shelf (Fig. 2c). Interval velocities in the Upper Jurassic strata range from 3.2 km/sec nearshore to 4.8 km/sec beneath the outer shelf. Interval velocity values of 5 to 6 km/sec are observed between depths of 5 and 12 km. However, since the depth to the base of this section can be over three times the length of the hydrophone array, the velocities at this depth are less accurate and could be off by 15% or more.

Early refraction studies of the Atlantic shelf revealed four mappable velocity units (Drake et al, 1959). Velocities less than 2 km/sec were generally found at depths less than 1 km, and these were assumed to represent "unconsolidated sediments." Velocities between 2.0 to 3.0 km/sec generally were found at depths between 1 and 2 km and were assumed to represent "semiconsolidated sediments." Velocities between 3.0 to 4.5 km/sec were found at depths of 2 to 5 km in the middle shelf and at depths of only 2 to 3 km along the outer shelf; these velocities were assumed to represent "consolidated sediments." Material with velocities of over 4.5 km/sec underlie the "consolidated sediments," and as such velocities correlated with basement velocities onshore in the coastal plain, they were also assumed to represent basement on the outer continental shelf.

More recent interpretations have assumed that the high-velocity refracting horizons at depths of 2 to 5 km on the middle and outer shelf are limestone units (Emery and Uchupi, 1972; Behrendt et al, 1974; Mayhew, 1974; Sheridan, 1974; Schlee et al, 1976).

Thin, high-velocity refracting horizons (5.0 km/sec) at a depth of 2.6 km beneath the shelf edge off New Jersey were shown by Grow and Sheridan (1976) and Sheridan (1977) to correlate with a sedimentary horizon which is at the top of the Lower Cretaceous. A similar high-velocity sedimentary horizon is present at a depth of 1.8 km beneath the shelf edge under Georges Bank (Jaworski, et al, 1976), which is also probably Lower Cretaceous. These thin, high-velocity horizons appear to change laterally toward shore into lower velocity units over distances of 20 km or less. We think that these represent limestone beds of a carbonate or reef facies which grade shoreward into clastic units. Recent sampling of Lower Cretaceous limestones including reefal debris has been accomplished by Miller and Ryan (1978) using the *Alvin* submarine in Heezen Canyon, which cuts into the continental slope of northeastern Georges Bank.

A general correlation of refraction and interval velocities indicates that virtually all of the high-velocity horizons (>4.5 km/sec) found beneath the middle and outer continental shelf in the early refraction studies are sedimentary units of Early Cretaceous age and older as previously pointed out by Schlee et al (1976). Velocity inversions beneath some of these beds and lateral changes in facies remain a problem to be delineated.

The velocity measurements from sonic logs in the COST B-2 well (Scholle, 1977) show a more gradational increase with depth than interval velocities derived from line 2 (Fig. 3). The velocities measured from a Schlumberger sonic log were digitized from the original log every 30 meters down to the bottom of the hole at a depth of 4.8 km. Velocity inversions of 0.3 to 0.5 km/sec on the sonic log are numerous. Although refraction velocities of 5.0 km/sec were measured at the top of the Lower Cretaceous farther out on the shelf (Grow and Sheridan, 1976; Sheridan, 1977), a maximum of only 3.7 km/sec is seen on the sonic log for this horizon at the COST B-2 well. A 4.3 km/sec maximum, at a depth of 2.8 km, is also significantly lower (Fig. 3) than the measured refraction velocities. We attribute this difference to a lateral lithologic change in facies in the Lower Cretaceous refracting horizons. The velocity inversions may also be more significant farther out on the shelf.

Figure 3 shows that interval velocity measurements are single-value averages over depth intervals of 500 to 1,000 meters, whereas sonic velocities can vary nearly 1 km/sec above or below the interval value. In spite of the general problems of velocity inversions and lateral velocity variations, the interval velocities determined for line 2 near the COST B-2 well show very good overall agreement with the sonic data (Fig. 3). Because of the good correlation, we feel that the velocity sections in Figure 2 repre-

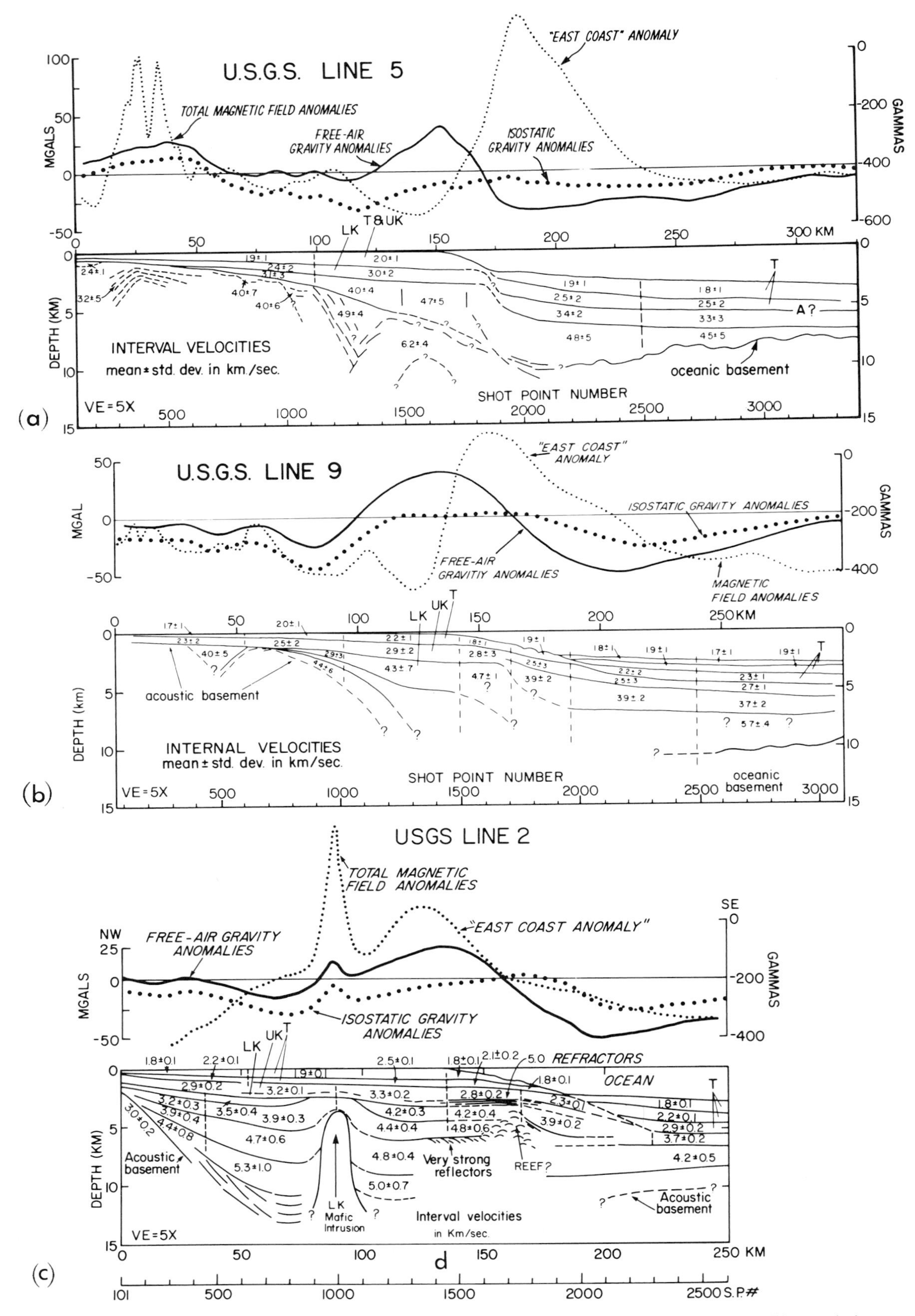

FIG. 2. Velocity sections summarize interval velocity data sampled every 3 km along profiles and then averaged over 20- to 50-km-wide bands. Approximate series assignment for the layers are Tertiary (T), Upper Cretaceous (UK), and Lower Cretaceous (LK) where correlations are presently possible. Velocity values at depths greater than 5 to 7 km are less reliable and may be biased slightly on the high side (especially deep rise sediments on lines

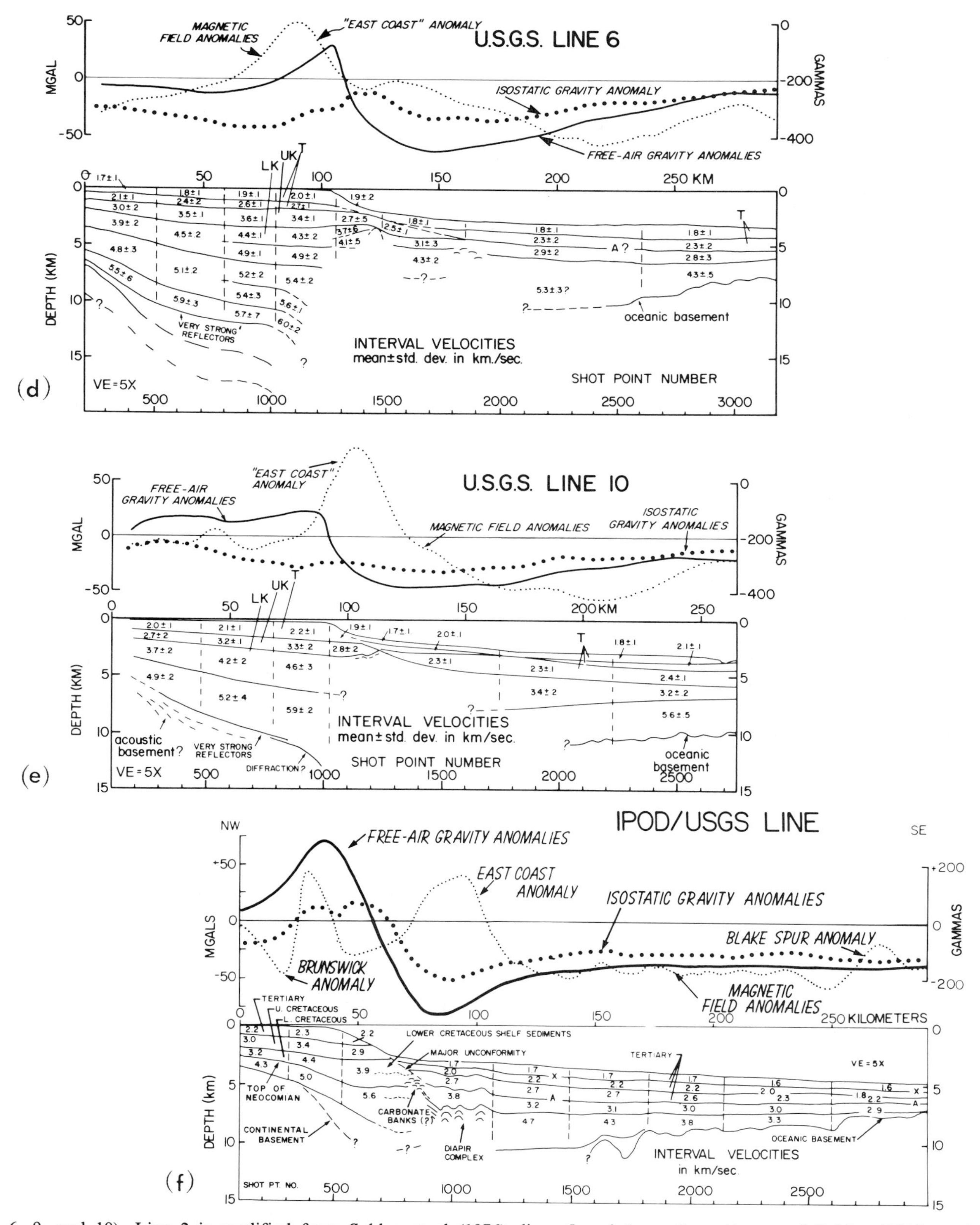

6, 9, and 10). Line 2 is modified from Schlee et al (1976); lines 5 and 6 are from Grow and Schlee (1976); and IPOD is from Grow and Markl (1977). Magnetic field and free-air anomaly data collected during 1975–76 cruises are plotted for reference above the velocity sections in Figure 2, but will not be discussed in detail in this paper. Shotpoint numbers will be referred to as SPN.

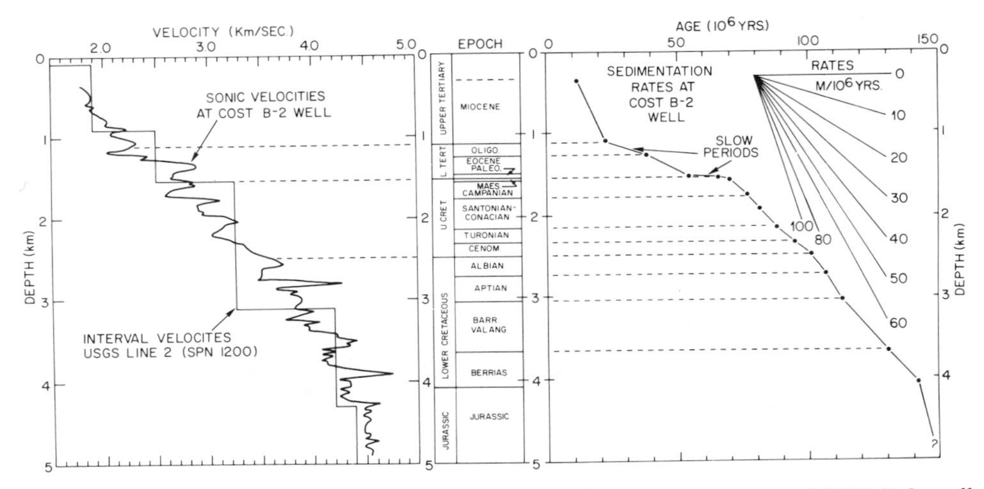

FIG. 3. Ages, depths, sonic velocities, and sedimentation rates at COST B-2 well locality (Smith et al, 1976; Scholle, 1977). Interval velocities from SPN 1200 of line 2 (Fig. 2c) compare well with the general trend of the sonic velocities. Sonic velocities were digitized directly from Schlumberger sonic logs every 30 m. Although several high-velocity lenses and numerous velocity inversions are obvious, the general correlation between sonic velocities and interval velocities indicates that the interval velocities are generally reliable down to a depth of 5 km. The top of Jurassic at 4,100 m has recently been identified on the basis of dinoflagellates (E. Robbins, personal commun., 1978).

sent an adequate preliminary summary of the interval velocities over the Cape Hatteras to Cape Cod section of the Atlantic continental margin. Furthermore, the interval velocity patterns over the outer shelf and upper slope are now well enough known to allow calculation of depth sections.

DEPTH SECTIONS

The depth sections, in contrast to the time sections, permit the sedimentary wedge to be presented in a less distorted perspective. A comparison of the time and depth sections for line 6 (Fig. 4) shows the changes that occur across the critical area of the continental slope off Delaware. Although the time section was stacked using velocity analyses every 3 km, a continuous velocity analysis was made prior to conversion to depth. The continuous velocity analyses use every airgun firing (50-m horizontal spacing) and average them at 1-km intervals. The velocity values displayed at the base of figure 4 are averaged over 5-km horizontal bands. Marine geologists and geophysicists are aware that the low velocity of sound in water (1.5 km/sec) delays reflections; however, this delay is of such a magnitude that misinterpretations of structure are likely in areas of steep or rough bottom topography such as continental slopes, unless the section is converted to depth.

The inferred top of Lower Cretaceous sediments on line 6 continues as a strong subhorizontal reflector about 20 km seaward of the present continental shelf edge, thus indicating the former seaward extension of the shelf in earlier time. A major unconformity appears to exist between shotpoint numbers (SPN) 1390 and 1480 where Upper Cretaceous and lower Tertiary units have been truncated. The diffraction pattern, at a depth of 3.5 km near shotpoint 1480, is interpreted as indicating the top of a large rotated fault block of Lower Cretaceous strata.

The same seaward extension of the Lower Cretaceous shelf edge can be seen on line 2 (Fig. 5). A very strong reflector at a depth of 2.6 km, correlated with the top of the Lower Cretaceous in the COST B-2 well, appears to extend to shotpoint 1750, approximately 25 km seaward of the present shelf edge. The very strong reflectors between shotpoints 1500 to 1750 and depths 2.6 to 3.2 km correlate with refraction velocities of about 5.0 km/sec (Grow and Sheridan, 1976). Since the interval velocities above and below the refracting horizons average 2.8 and 4.2 km/sec, respectively (Fig. 2), the 5.0 km/sec refractors must be from thin, high-velocity beds within a section having a lower overall velocity. Also, note that the amplitude of the reflectors off the top of Lower Cretaceous rocks dramatically decreases toward the shore (Fig. 5). This lateral decrease in amplitude suggests a change in sedimentary facies that may account for the velocities of only about 3.7 km/sec at the top of the Lower Cretaceous at the B-2 well, compared with velocities of 5.0 km/sec recorded by refraction surveys on the outer shelf and slope. A number of authors (Behrendt et al, 1974; Mayhew, 1974; Sheridan, 1974; Schlee et al, 1976) have proposed that carbonate banks or reef complexes underlie the faster refracting surfaces along the shelf edge. The irregular reflectors and diffractions at 3 to 6 km depth between SPN 1550 and 1750 on line 2 are consistent with this hypothesis. Beneath the continental rise, gently landward-dipping horizons are present at depths of 9 to 10 km between SPN 1800 to 1900 (the strong reflector 11 km deep at SPN 1850 could be either basement or a multiple). The strong reflector observed at a depth of 13 km at SPN 2000 is probably a multiple, and no

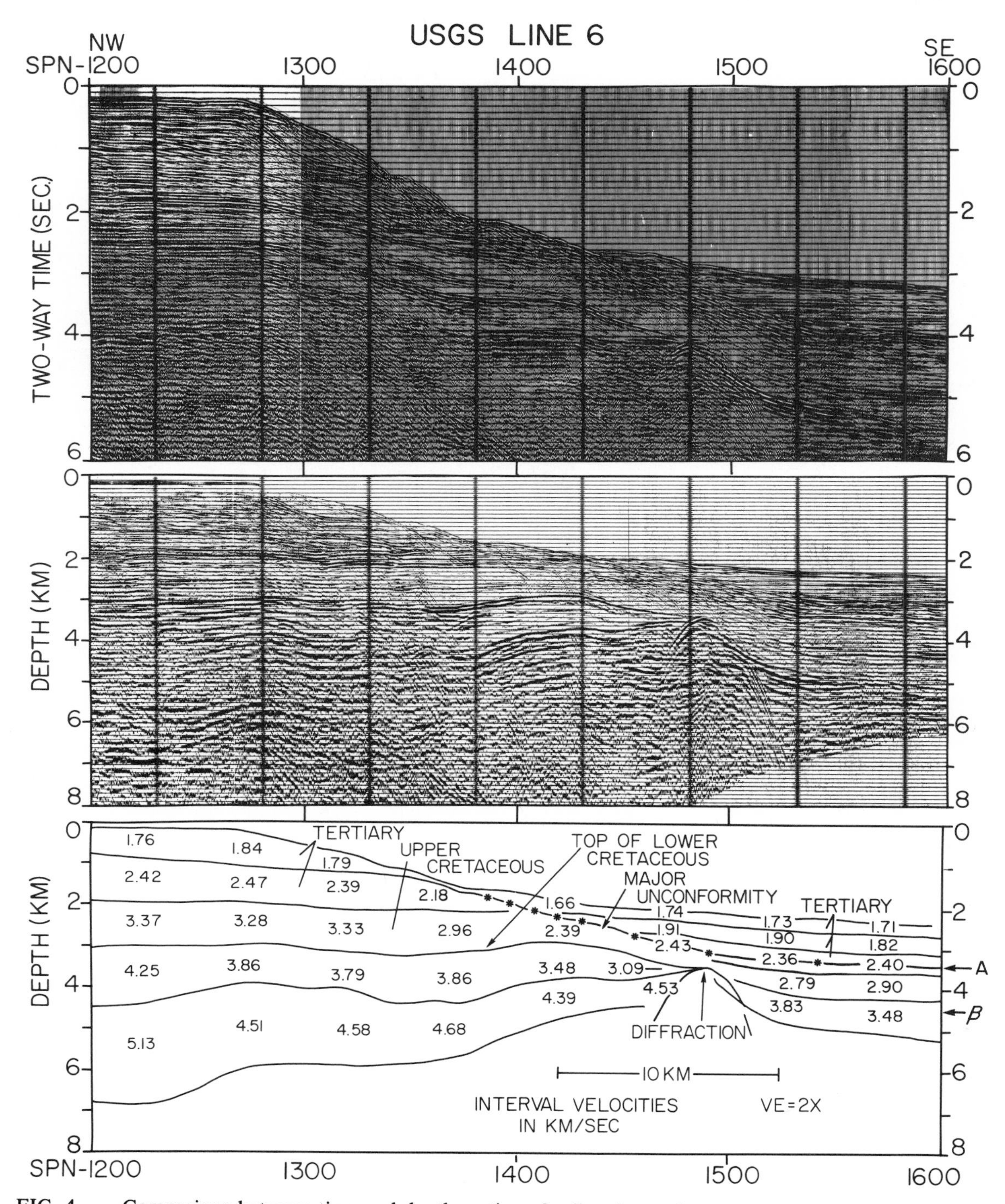

FIG. 4. Comparison between time and depth sections for line 6 reveals remarkable change in structural relationships. Interval velocity measurements were made continuously (every depth point; i.e., 50-m spacing), summarized every 1 km, and finally averaged over 5-km blocks to yield values at base of figure. Note that top of Lower Cretaceous shelf horizons continue over 20 km seaward of present shelf break and that a major unconformity exists between Tertiary and Upper Cretaceous horizons on the slope.

clear oceanic basement reflection is observed. The depth to basement in this zone appears to be at least 10 km.

The inferred Lower Cretaceous shelf sediments on the IPOD line continue out to shotpoint 800, approximately 30 km seaward of the shelf break (Fig. 6). Again a major unconformity shows deep erosion of inferred Upper and Lower Cretaceous shelf horizons between SPN 720 and 820. Note also the similarity of velocities for Upper Cretaceous and Tertiary units between this line and line 6 (Fig. 4).

On line 10, very high-amplitude reflections were recorded at a depth of 12 km that are interpreted as interbedded evaporites and carbonates over base-

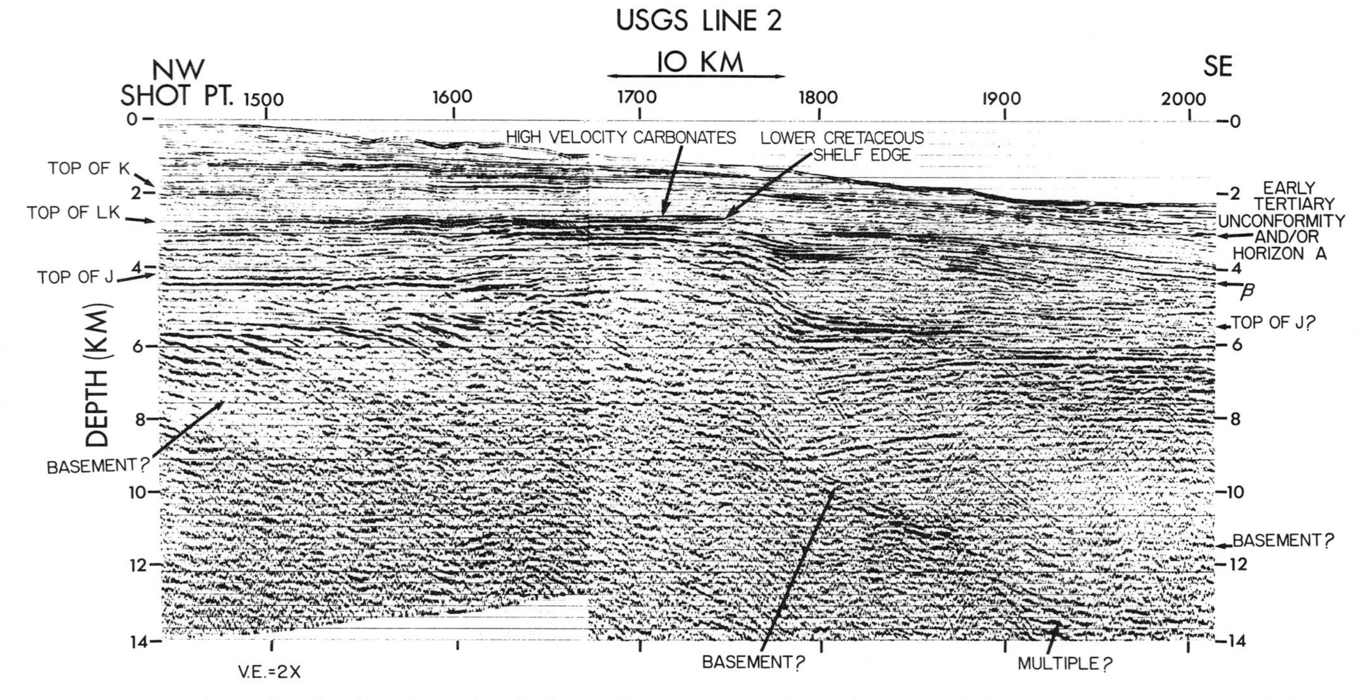

FIG. 5. Depth section for line 2 used velocity analyses every 1 km. The tops of the Upper Cretaceous and Lower Cretaceous at SPN 1700 are at 1.8 and 2.6 km, respectively. The strong reflectors off the upper portion of the Lower Cretaceous between SPN 1600 to 1750 have refraction velocities of about 5.0 km/sec (Grow and Sheridan, 1976) whereas interval velocities of above and below are 2.8 and 4.2 km/sec (Fig. 2c). Note that the edge of the Lower Cretaceous shelf is 25 km seaward of present shelf edge. The zone between 3 and 6 km deep between SPN 1500 to 1750 is thought to be a carbonate bank or reef complex.

ment (Fig. 7). The top of inferred Upper and Lower Cretaceous units beneath the shelf edge are at 1.8 and 3.0 km, respectively. The top of the Lower Cretaceous continues horizontally to shotpoint 1140. A rotated fault block of Lower Cretaceous material appears between SPN 1140 and 1200. At SPN 1300, a major unconformity, similar to the one found on lines previously discussed, occurs at 3.6 km depth where middle or upper Tertiary units lie directly on top of Lower Cretaceous units.

Off Long Island and southern New England (lines 9 and 5, Figs. 8 and 9), the sedimentary section is thinner under the outer shelf and upper slope than those shown in profiles farther south. Velocities at depths greater than 6 km on both these lines were difficult to calculate. Undulations at depths of 6 to 9 km may be due to poor velocity selection rather than to true structure. On line 9, few coherent reflectors were observed beneath 4 km depths on the shelf. However, the seaward end of the Lower Cretaceous shelf edge on line 9 probably is represented by the diffraction pattern at 3.0 km depth near SPN 1700. On line 5, the seaward end of the Lower Cretaceous shelf edge is probably near 3.0 km depth at SPN 1860, which is about 18 km seaward of the present shelf edge at 1680. The base of the Tertiary section at SPN 2100 on line 5 is probably at 3.4 km. Although our age correlations for lines 9 and 5 are not as reliable as for lines 2, 6, and 10 and IPOD, it is still evident that during the Early Cretaceous, the edge of the continental shelf south of Cape Cod extended approximately 20 km seaward of its present position.

LOWER CRETACEOUS SHELF EDGE

One of the most important geologic conclusions derived from these six depth sections is that the location of the continental shelf edge during the Early Cretaceous was 20 to 30 km seaward of its present position. Although a few published profiles could be reinterpreted to suggest this possibility (Rona and Clay, 1967; Emery et al, 1970; Scott and Cole, 1975), the continuation of Lower Cretaceous shelf horizons beneath the present continental slope as a systematic phenomenon has not been previously noted in the northwest Atlantic. However, Seibold and Hinz (1974, p. 194) have presented evidence off northwest Africa for a Lower Cretaceous ". . . shelf-edge anticline located 40 to 75 km west of the present shelf edge."

The presence of an older buried shelf edge seaward of the present shelf edge has also been described by Lehner and De Ruiter (1977) off Senegal and Guinea. The ancestral slope appears to be cut into a sequence of probable limestones (Jurassic-Cenomanian); they are buried by 1 to 2 km of post-Cenomanian sediments.

Deep-sea drilling at DSDP site 369 on the continental slope off northwest Africa, above the Lower Cretaceous "shelf-edge anticline" described by Sei-

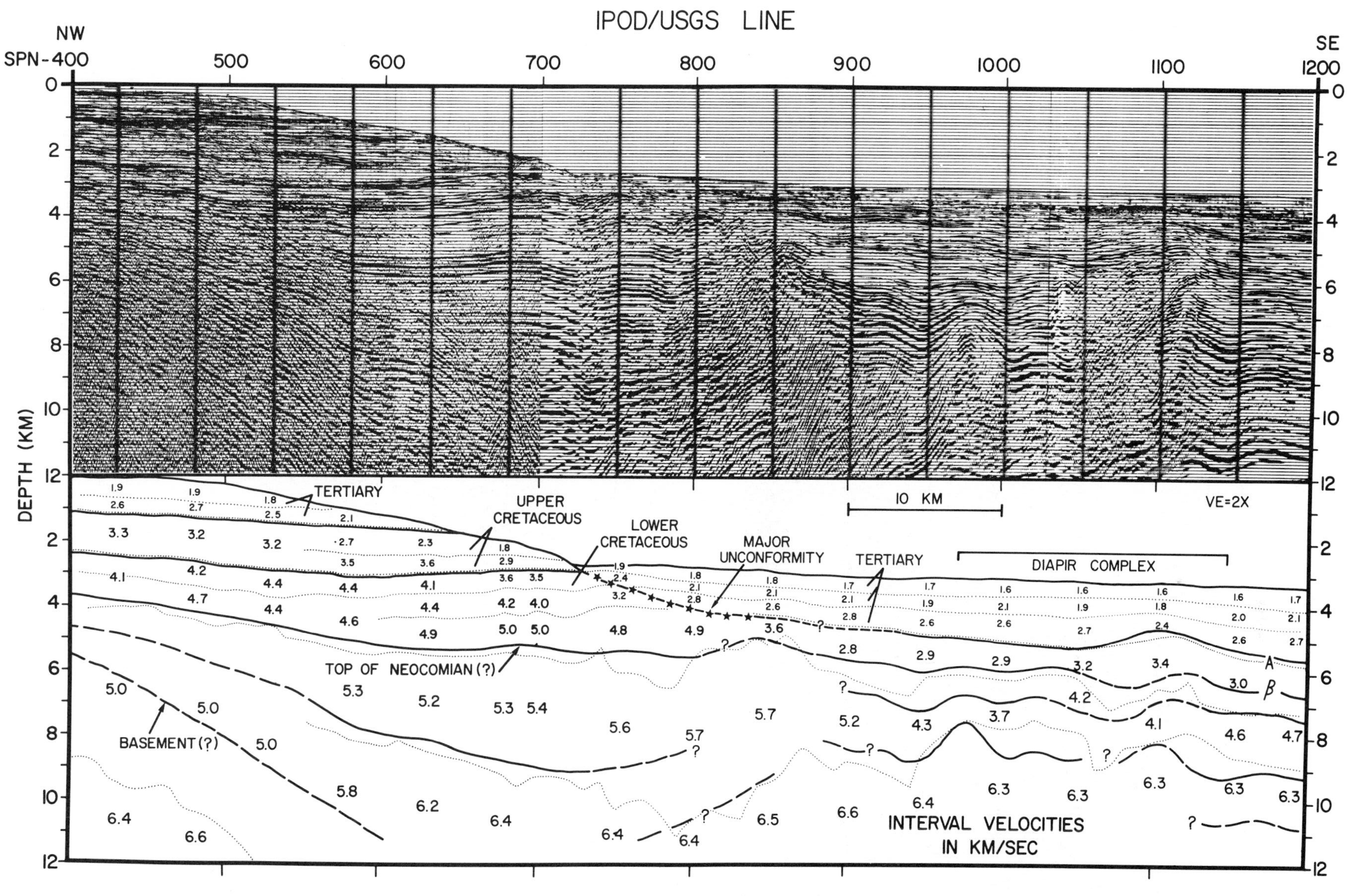

FIG. 6. Depth section for IPOD/USGS line from Grow and Markl (1977) is based on continuous velocity scans. Portion between SPN 800 and 1200 has been migrated. Jurassic and Lower Cretaceous shelf horizons appear to continue out to SPN 850, which is over 30 km seaward of present shelf edge.

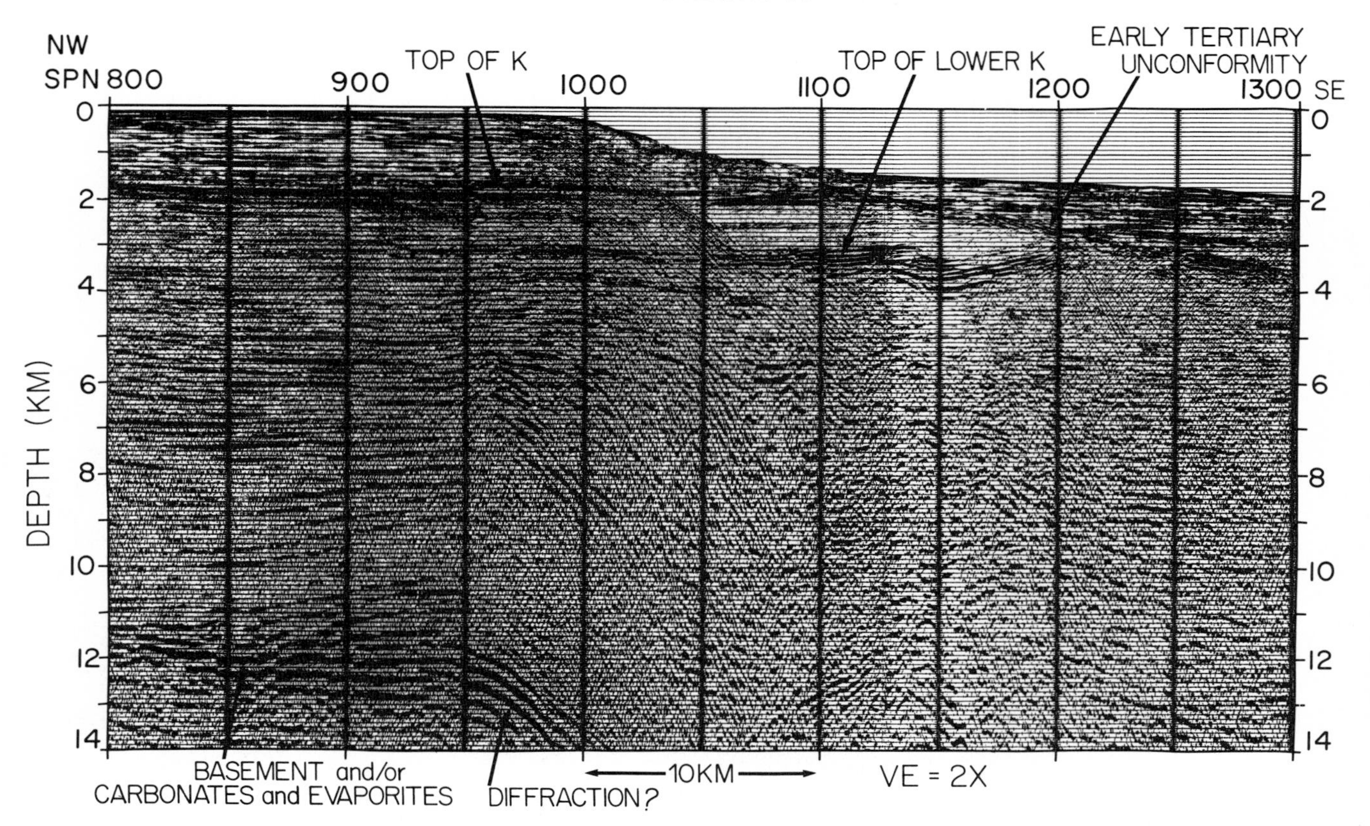

FIG. 7. Depth section for line 10 used velocity scans every 3 km. The very strong reflectors at 12 km are assumed to be carbonates and evaporites above the basement (the segment crossing the turnover at SPN 950 to 1000 has not been migrated, so that interpretation will not be discussed). The tops of the Upper Cretaceous and Lower Cretaceous at SPN 1000 are thought to be at 1.9 and 3.1 km, respectively. Note the continuation of the Lower Cretaceous shelf edge to SPN 1200 and the backtilted sedimentary block between SPN 1140 and 1200. The base of the Tertiary unconformity on the slope at SPN 1300 is at 3.6 km depth. The disturbed zone between SPN 1000 to 1300 and between 4 to 14 km depth with no coherent reflections is very typical of the "disturbed zone" between shelf sediment trough and rise sediment trough.

bold and Hinz, sampled Lower Cretaceous (Albian-Aptian) hemipelagic, highly organic, nannofossil marlstones which were inferred to have been deposited above carbonate compensation depth (CCD) in an upper rise or lower slope environment (Lancelot, Seibold et al, 1975; Von Rad and Eiselle, in press). Von Rad and Eiselle inferred that a thick deltaic deposit of Early Cretaceous age prograded seaward from a shallow-water carbonate platform of Jurassic age prior to the retreat of the shelf edge to its present position. A structural section through site 369 presented by Von Rad and Eiselle also suggests numerous rotated fault blocks within the Lower Cretaceous, similar to those on lines 6 and 10 (Figs. 4 and 7), with differential subsidence between a thinner section beneath the upper rise and a faster subsiding basin beneath the shelf, where deposits are much thicker.

Our results, combined with those of the West African studies, suggest that the continuation of Lower Cretaceous shelf horizons beneath the present continental slope may be a common feature of older Atlantic margins. The failure to note this as a systematic phenomenon in most previously published studies of continental margins is probably due to the failure to convert the time sections into depth sections.

TECTONIC DEFORMATION DURING THE JURASSIC AND EARLY CRETACEOUS

Two of the six depth sections (Fig. 4 and 7) show block faulting in Lower Cretaceous shelf units. Little or no deformation is observed in the younger overlying sediments. Line 2 also crossed a major mafic intrusion near shotpoint 1000 (Fig. 2c) which was emplaced during the Early Cretaceous (Schlee et al, 1976).

Except for differential subsidence over a broad region, the Early Cretaceous deformation observed in lines 2, 6, and 10 appears to be the last major tectonic adjustment of the U.S. Atlantic margin. Small faults of 10 to 20-m displacement or less do occur within the Upper Cretaceous and Tertiary on line 6 (Grow and Schlee, 1976) and line 2 (Sheridan and Knebel, 1976), but these can be considered as very minor adjustments of the underlying crustal blocks (Sheridan, 1976).

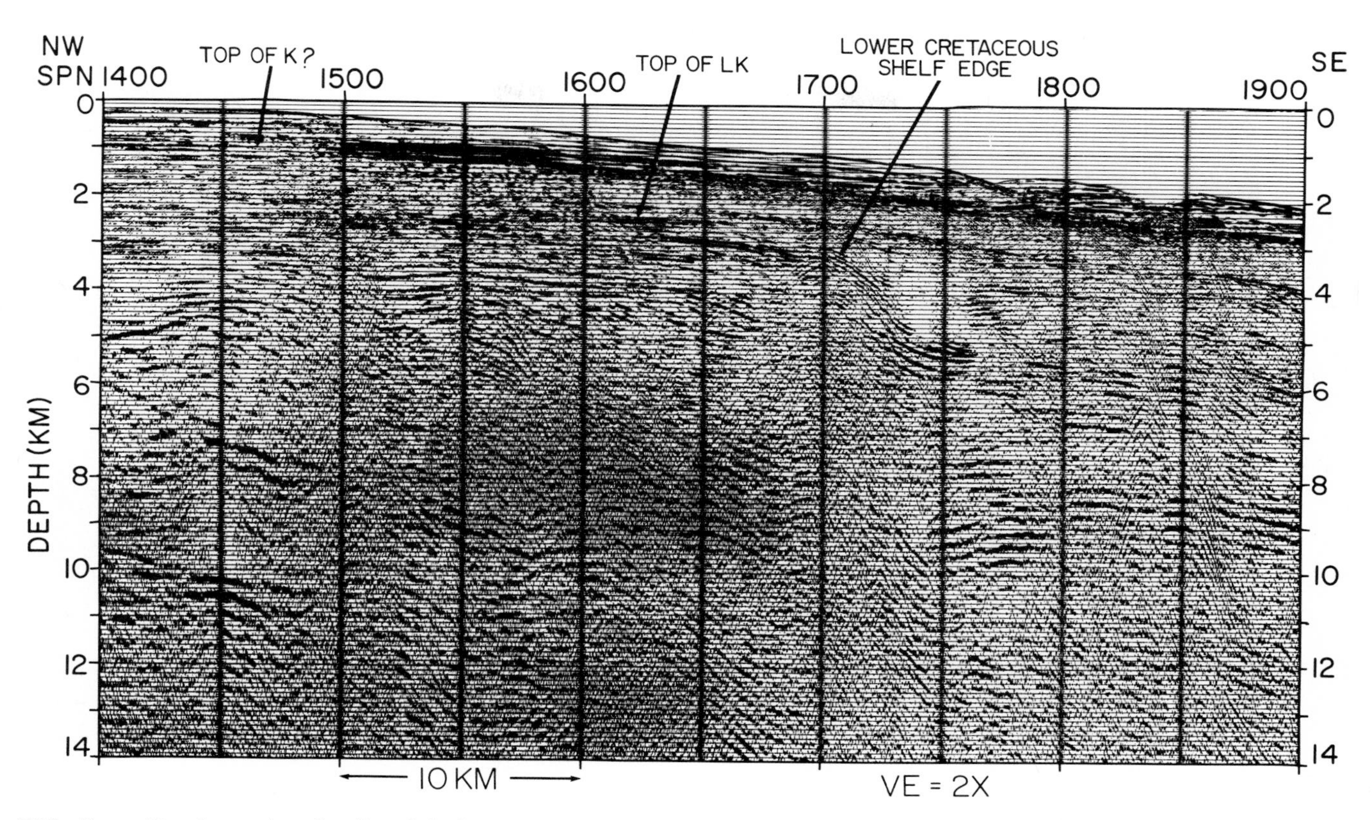

FIG. 8. Depth section for line 9 is based on velocity scan every 3 km. Undulations at depths more than 6 km on this line are somewhat influenced by noisy data and scatter in the velocity values. The Lower Cretaceous shelf edge is probably about 3 km depth at SPN 1700.

UNCONFORMITY ON THE CONTINENTAL SLOPE

The erosional unconformity beneath the continental slope, which truncates lower Tertiary, Upper Cretaceous, and in a few places, Lower Cretaceous strata (Figs. 5 and 7) is inferred to be the result of submarine erosion of the slope during a major regression of the sea during Oligocene time. This conclusion is based primarily on seismic evidence from line 2 (Fig. 2 and 5). On line 2, the unconformity truncates beds inferred to be Eocene and older, and according to Schlee et al (1976), inferred early Miocene and younger units form a separate prograding foreset sequence on the continental shelf. In addition, Oligocene sedimentary rocks, although penetrated in the COST B-2 well (Scholle, 1977), are not present in any onshore wells in the vicinity of New Jersey (Perry et al, 1975). On seismic line 2, it is possible to trace Oligocene strata landward from the COST B-2 well to about the middle of the shelf, where they appear to wedge out. Apparently during Oligocene time, a period of lowstand of sea level, most of the river sediments were carried across the continental shelf and funneled through submarine canyons (perhaps as turbidite flows) directly onto the continental rise and abyssal plain; Oligocene sediments thus bypassed the continental slope and erosion occurred. Eustatic lowering of sea level by perhaps 100 m during Oligocene time would have also allowed direct wave erosion along what is now the upper continental slope.

Deep sea drilling results off eastern North America (Hollister, Ewing et al, 1972; Benson, et al, 1976) have revealed major unconformities at sites 105 (between Upper Cretaceous or early Tertiary and Miocene), at site 106 (between Eocene and middle Miocene), and at site 391 in the Blake Basin (between Upper Cretaceous and Miocene). All of these unconformities are consistent with a major erosional event on both the slope and rise during the Oligocene.

Other evidence, including the COST B-2 well data (Fig. 3), indicates that depositional rates were extremely slow during Paleocene and Oligocene time. Studies of eustatic sea level changes inferred from seismic records by Vail et al (1977) indicate a moderate regression during the middle Paleocene, a major regression during the late Oligocene, a moderate regression during the late Miocene, and finally a series of regressions during the Pliocene and Pleistocene, which reflect the onset of glaciation. A theoretical analysis of eustatic sea level changes, seafloor spreading rates, margin subsidence patterns, and stratigraphic sequences during the Late Cretaceous and early Tertiary by Pitman (in press) also

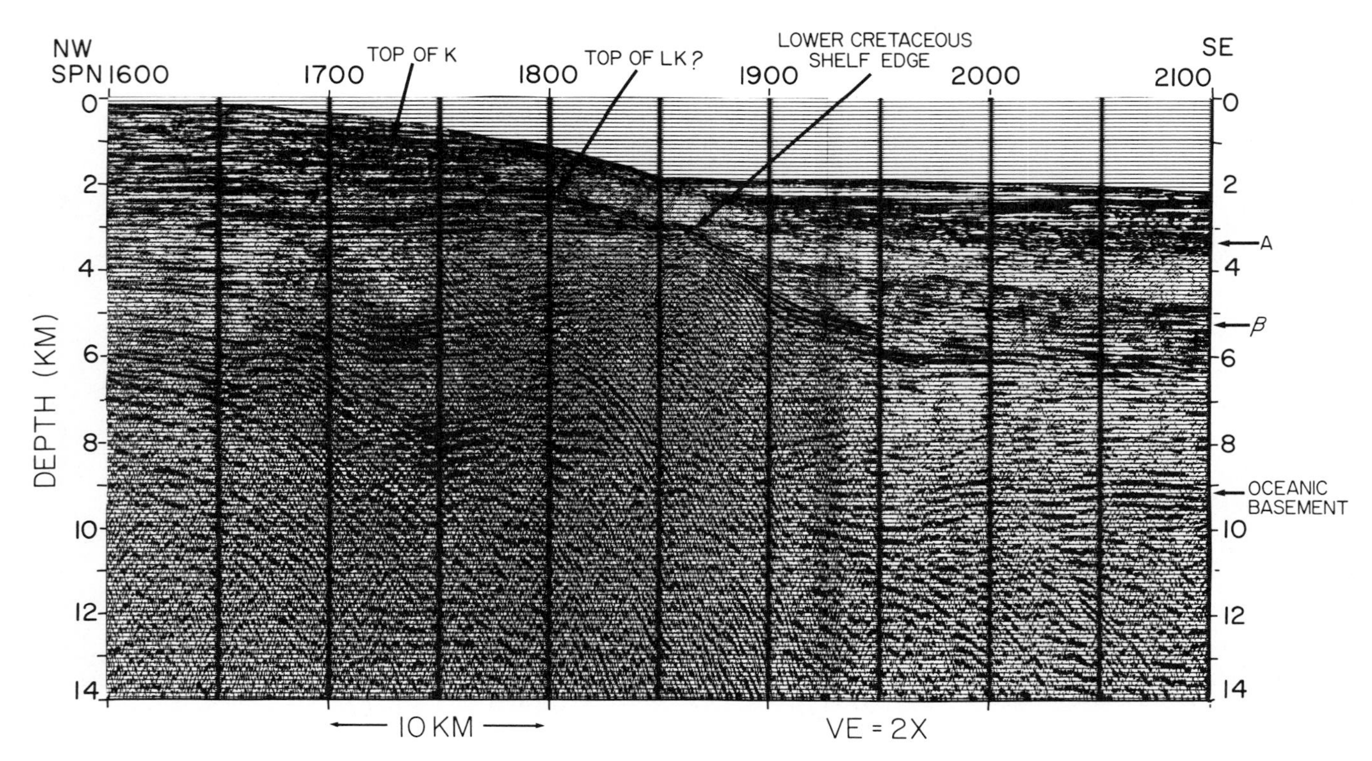

FIG. 9. Line 5 used velocity scans every 3 km. The top of the Lower Cretaceous is probably between 2 and 3 km deep at SPN 1700. Continuation of Lower Cretaceous shelf sedimentary units extend out to SPN 1860.

shows large regressions during the Paleocene and Oligocene.

Seibold and Hinz (1974) described two major unconformities below the continental slope and rise off northwest Africa, which they inferred to be Middle Cretaceous (Cenomanian) and Oligocene to early Miocene in age, respectively. They concluded that during the latter regression and erosion, "large parts of the pre-Oligocene continental margin were removed." Subsequent drilling of Deep Sea Drilling Sites 369 and 397 (Lancelot, Seibold, et al, 1975; Ryan, Sibuet et al, 1976; respectively) revealed an unconformity between the uppermost Cretaceous and middle Eocene at the slope site (369) and an unconformity between the Lower Cretaceous (Hauterivian) and lower Miocene at an upper rise site (397). Although the major unconformity beneath the upper rise could have been caused by increased turbidite activity during a eustatic lowering of sea level during the Oligocene or early Miocene, as proposed by Seibold and Hinz, the apparent absence of a major Oligocene-early Miocene unconformity at site 369 has not been explained to date. The age of such unconformities on the continental slope and rise remains an important matter for continuing investigation (Vail et al, 1977).

INFERRED BASIN STRUCTURE AND THE OCEAN-CONTINENT TRANSITION ZONE

Although the widths and depths of the sedimentary troughs beneath the continental shelf and beneath the continental rise may vary by a factor of two (Fig. 2), the general pattern suggests a trough beneath the shelf with up to 12 km of Mesozoic and Tertiary sediments and another trough beneath the rise with up to 8 km of sediments overlying basement at a depth of 10 to 11 kilometers. These troughs are separated by a disturbed zone where the Lower Cretaceous and Jurassic outer shelf horizons between 3 and 6 km deep are capped by high-velocity carbonates and sometimes block faulted or intruded by diapirs. No coherent seismic returns are detected between depths of 6 and 12 km in this disturbed zone. The axis of the East Coast Magnetic Anomaly generally occurs between the center and the landward edge of the disturbed zone.

Numerous hypotheses to account for the disturbed zone and the East Coast Anomaly have been proposed. To explain the East Coast Magnetic Anomaly, Taylor et al (1968) proposed an igneous intrusion rising to depths of 3 to 7 km, whereas Keen and Keen (1974) preferred to model the anomaly off Nova Scotia as an edge effect between continental and oceanic crust. In addition, positive local isostatic gravity anomalies over some parts of the East Coast Anomaly off eastern North America were interpreted as indications of a zone of high-density crust by Rabinowitz (1974). Mattick et al (1974) proposed basement horst blocks for the disturbed zone, while Schlee et al (1976) suggested a thick carbonate bank or reef complex above a deeply buried block-faulted transitional crust.

Klitgord and Behrendt (this volume) present preliminary depth estimates for magnetic sources over the East Coast Anomaly from a recent aeromagnetic survey. Their results indicate moderate-strength sources at depths of 7 to 8 km in the middle of the disturbed zone with stronger sources on the landward side of the anomaly at depths of 10 or more km. Their results suggest that a major fault in the basement between the oceanic crust and the deep sediment troughs landward of the East Coast Anomaly is the major source of the East Coast Anomaly (Klitgord and Behrendt, this volume).

Although very little is known about the velocity structure of the crust beneath the shelf, seismic refraction data do indicate that a 7.1- to 7.2-km/sec refracting horizon dips landward beneath the continental rise, slope, and shelf edge off New Jersey from a depth of 13 km beneath the rise to a depth of 15 km directly beneath the East Coast Magnetic Anomaly (Sheridan, 1977; Sheridan, Grow, Behrendt, and Bayer, unpubl. data). This 7.1- to 7.2-km/sec layer is interpreted as oceanic layer 3B and is inferred to mark the landward edge of oceanic crust. Following Keen and Keen's (1974) hypothesis and the magnetic depth-to-basement evidence of Klitgord and Behrendt, we interpret the East Coast Anomaly as marking the landward edge of thick oceanic crust and the seaward edge of thin continental crust. However, the continental crust beneath the shelf sediments is probably less than 15 km thick beneath the major sediment basins and probably was fragmented and attenuated in thickness during the initial rifting stage of continental separation between Africa and North America.

Schematic cross-sections along multichannel seismic lines 5, IPOD, 2 and 6 are shown in Figures 10 to 13, respectively, with our inferred basin and crustal structure in the ocean-continent transition zone. As systematic inflections of the magnetic and isostatic anomalies have been noted off Norway (Talwani and Eldholm, 1973), eastern North America (Rabinowitz, 1974), and southwest Africa (Rabinowitz and La Breque, 1976), we have displayed these above profiles for comparison with the seismic data. The isostatic anomalies shown here assume a water layer with a density of 1.03 g/cc, a crust at 2.7, mantle at 3.3 g/cc, and a depth of compensation of 30 kilometers at the shoreline; these imply density contrasts of -1.67 and $+0.6$ g/cc at the water/crust and crust/mantle boundaries. As Rabinowitz (1974) suggested that sediment corrections are small and do not change the isostatic anomaly greatly, we have tentatively accepted that assumption and have ignored sediment corrections for the isostatic anomalies used in this paper. Although a more detailed analysis of the effect of sediment corrections and the crustal model used in the isostatic anomaly calculations is needed, that subject is beyond the scope of this paper.

The transition zone between oceanic and continen-

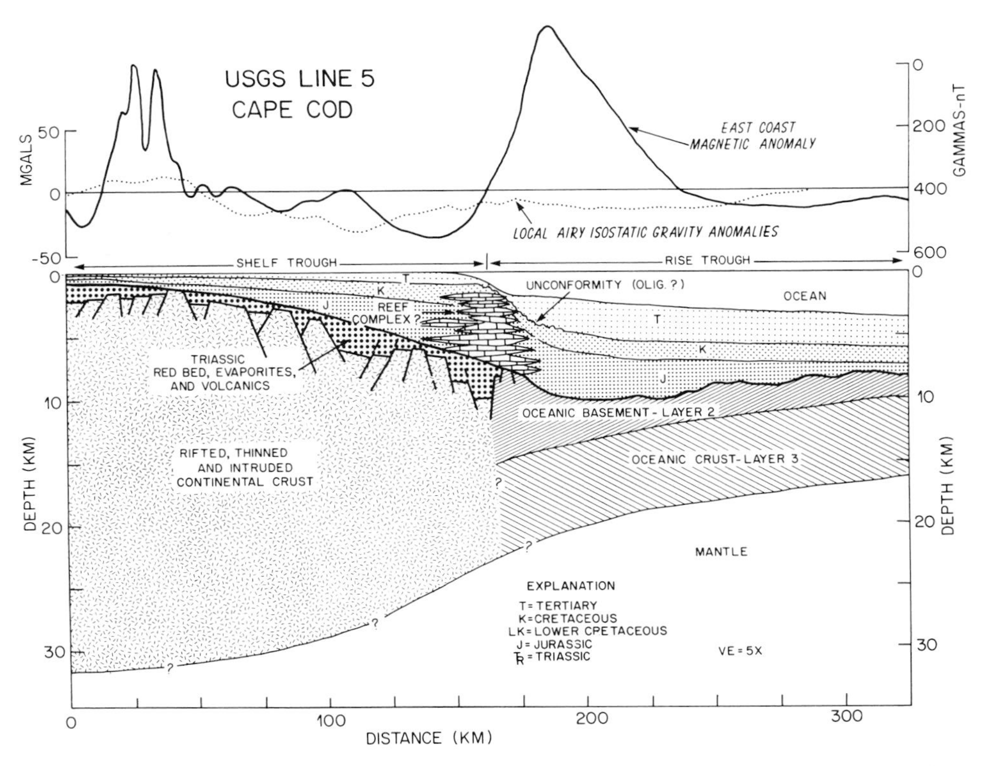

FIG. 10. Schematic cross-section of CDP line 5 summarizing reflection data along with magnetic and isostatic gravity anomalies. Line 5 is unique in that oceanic basement can be traced into SPN 1950 (Fig. 9), which is directly beneath the axis of the East Coast Magnetic Anomaly (Fig. 2a). Line 5 crosses the eastern end of the Long Island Platform (Klitgord and Behrendt, this volume), and is typical of a narrow transition zone between continental and oceanic crusts.

tal crust is most clearly defined in our multichannel line 5 (Fig. 10), where the top of oceanic crust can be traced beneath the continental rise to a depth of 10 km directly beneath the axis of the East Coast Magnetic Anomaly (Figs. 2A and 9). The rise of oceanic basement to a depth of 8 km (Fig. 10) beneath the Lower Cretaceous and Jurassic shelf edge is inferred from the magnetic depth-to-source studies of Klitgord and Behrendt (this volume). Although the East Coast Anomaly has a signature of over 600 gammas on line 5, the isostatic anomalies along this profile are very small and imply near-isostatic equilibrium.

The schematic cross-section for the IPOD line (Fig. 11) shows a narrow transition zone similar to that on line 5, but with the added complication of diapirs. The diapirs evident on the IPOD line (Grow and Markl, 1977) have been found to extend 50 km north and 200 km south of the IPOD line along the axis of the East Coast Anomaly in a more recent survey (Grow et al, 1977). None of the diapirs have had a detectable magnetic or gravity signature and are therefore thought to be composed of either shale or salt, probably the latter.

Schematic cross-sections across a wide ocean-continent transition zone off New Jersey, in the center of a major basin known as the Baltimore Canyon trough, are shown for seismic lines 2 and 6 in Figures 12 and 13, respectively. The diapir shown in Figure 12 has been projected from 10 km south of the seismic line and is located over the northwest flank of the East Coast Magnetic Anomaly in a similar position to those on the IPOD line and along the slope off Cape Hatteras (Grow et al, 1977). In our discussion of the depth section for line 10 (Fig. 7), we suggested that the strong reflectors at 12 km represent interbedded evaporites and carbonates which might serve as a source bed for the diapir. The fact that only one diapir is known in Baltimore Canyon trough suggests that either the salt beds are thin or that the carbonates are too massive for the salt to penetrate. Additional evidence for evaporites at depth has recently been given by Manheim and Hall (1976) who reported hypersaline brines of 55 parts per thousand at a depth of 1,700 m in the Caldrill 15 hole off New Jersey (Fig. 1). They project that saturated brines should exist at 3 to 4 km depth (i.e., 1 to 2 km beneath the Lower Cretaceous shelf edge (see line 2, 6, and 10). Although these concentrations could be explained by presence of local evaporite basins, they could also be explained by salt diapirs.

Deep-sea drilling at site 369 off northwest Africa on the landward flank of Seibold and Hinz's Lower Cretaceous "shelf-edge anticline" found Aptian-Albian marls with barite, which suggested ". . . upward or lateral migration of solutions from Jurassic evaporites underlying the continental shelf and slope" (Lancelot, Seibold, et al, 1975). Therefore, the importance of evaporites and possibly salt struc-

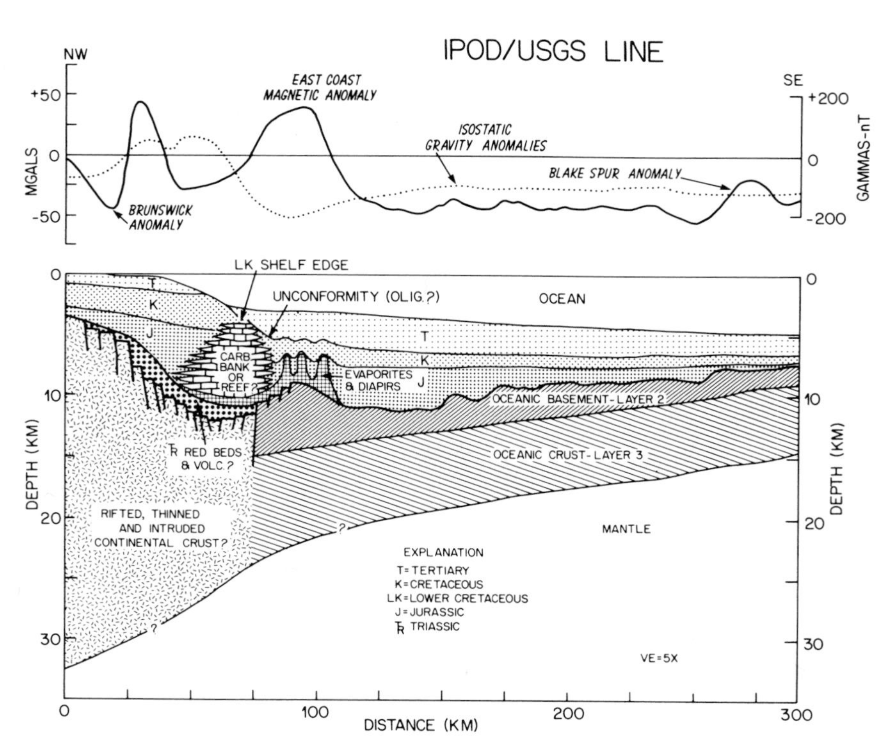

FIG. 11. Schematic cross-section of IPOD line which crosses the Carolina Platform and northern tip of the Carolina trough (Klitgord and Behrendt, this volume). The features displayed on the IPOD line is similar to the narrow transition zone between ocean and continent observed on line 5, except for the presence of diapirs near the ocean-continent boundary along the continental slope off Cape Hatteras.

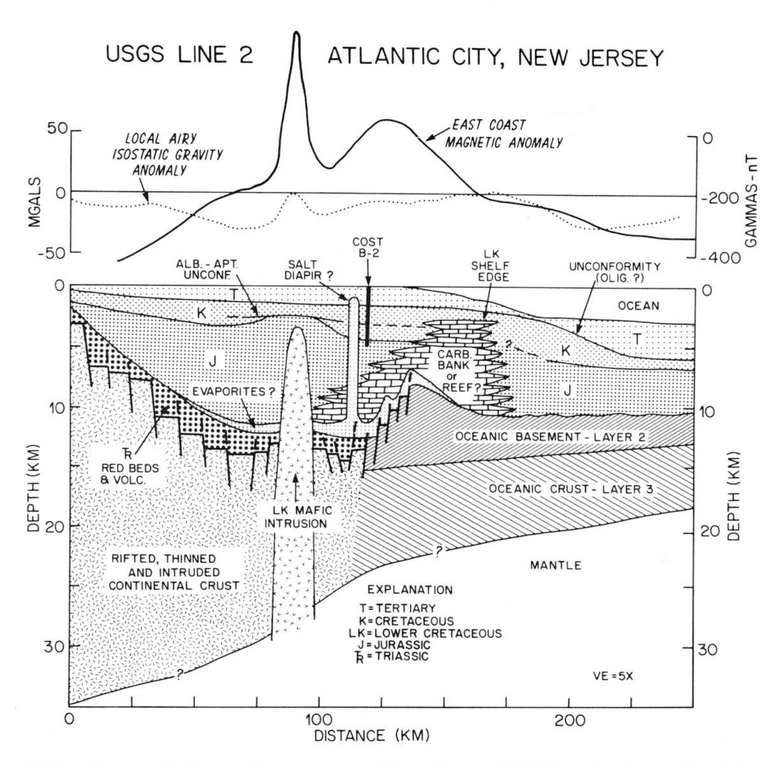

FIG. 12. Schematic cross-section along CDP line 2 through wide ocean-continent transition zone associated with the Baltimore Canyon trough off New Jersey. The evidence for high-velocity carbonates or reef complexes in the Jurassic and Lower Cretaceous is discussed within the text. The diapir is projected from 10 km south of line 2 and the COST B-2 well is projected from 10 km north of the line.

tures in the disturbed zone is suggested by recent results on both the U.S. Atlantic and northwest Africa margins.

Evans (1978) reviewed additional evidence off the Scotian slope, Morocco, Angola, and Brazil to suggest that formation of evaporites may be a common occurrence during the initial rifting and sea floor spreading stages of continental drift. This is presumably due to the existence of local shallow seas which had not yet been connected by a continuous proto-Atlantic seaway.

The ocean-continent transition zone for the lines off Cape Cod and Cape Hatteras is much narrower compared to that shown on the two lines through the Baltimore Canyon trough. The section for Line 6 is compatible with gravity modeling (Grow, Bowin, and Hutchinson, unpubl. data) and displays the widest transition zone, where thick oceanic crust extends nearly 100 km seaward of the East Coast Anomaly. Rabinowitz and La Breque (1976) suggested that abnormally thick oceanic crust may be typical during the initial stages of rifting and sea floor spreading.

PROGRADATION AND RETREAT OF THE SHELF EDGE

In the Baltimore Canyon trough, the shelf edge appears to have prograded 40 to 50 km seaward of the initial ocean-continent boundary during the Jurassic and Early Cretaceous (Figs. 12 and 13). During the same period, the shelf edges off Cape Cod and Cape Hatteras (Figs. 10 and 11) appeared to have had little net movement, although many small advances (progradation) and retreats (erosion) could have achieved this balance. All four areas indicate systematic retreat by 20 to 30 km during the Tertiary. The Oligocene erosional event appears the most significant in terms of retreat of the shelf break. Vail et al (1977) suggested that, although the Oligocene erosion is due to the most dramatic worldwide lowering of sea level during Mesozoic and Cenozoic time, many lesser worldwide transgressions and regressions have left important unconformities along continental margins

RIFTING AND SUBSIDENCE HISTORY

Rifting between North America and northwest Africa has generally been inferred to have initiated between Late Triassic and Early Jurassic. Upper Jurassic sediments have been drilled at DSDP Site 105 (Hollister, Ewing, et al, 1972), approximately 400 km seaward of the shelf edge off New Jersey (Fig. 1). Late Triassic continental redbeds and basalts in the Newark and Hartford basins of New Jersey and Connecticut have generally been taken as evidence for the initial rifting and pre-drift stage of continental separation. Recent biostratigraphic correlations suggest that although the redbeds straddle

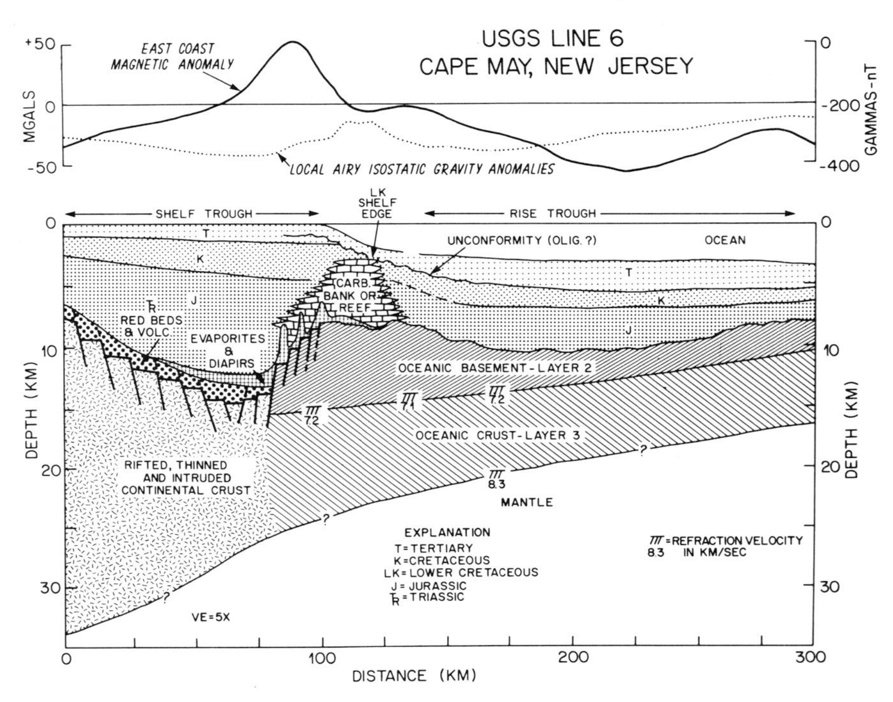

FIG. 13. Schematic cross-section along CDP line 6 through wide ocean-continent transition zone associated with the Baltimore Canyon trough off southern New Jersey. The 7.1-km/sec refraction data is from Ewing and Ewing (1959) and the 7.2 and 8.3-km/sec refractors are from Sheridan, Grow, Behrendt, and Bayer (unpub. data) and Sheridan (1977). The 7.1- to 7.2-km/sec layer is assumed to represent oceanic layer 3 B. The evaporites assumed at depth are inferred on the basis of diapirs near the East Coast Anomaly along the slope off Cape Hatteras and the diapir near line 2 (Fig. 12). The sediment and crustal thicknesses suggested are compatible with a two-dimensional gravity model for line 6 (Grow, Bowin, and Huchinson, unpub. data). The wide zone of thick oceanic crust is some what influenced by a NW to SE fracture zone (see second magnetic positive feature southeast of East Coast Magnetic Anomaly).

the Triassic-Jurassic boundary the basalt flows, sills, and dikes were emplaced during the Early Jurassic (Cornet, 1977; Hubert et al, 1978). Potassium-argon whole-rock dates for lava-flow units in the Connecticut Valley average 184 million years (m.y.) ago with a standard deviation of 8 m.y. (Reesman et al, 1973), whereas the Palisades sill of New Jersey was intruded about 190 m.y. ago (Dallmeyer, 1975). The Triassic-Jurassic boundary is dated at about 192 m.y. (Van Hinte, 1976).

The age-versus-depth relationship within the COST B-2 well is summarized in Figure 3. The Upper Jurassic through Tertiary sediments encountered in the COST B-2 well have inferred depositional environments ranging from nonmarine (e.g., Upper Jurassic coal) to shallow-water marine and upper slope. If our assumption is correct that evaporites exist at depths of 10 to 12 km below sea level beneath the New Jersey shelf and off Cape Hatteras, then a shallow-water environment of deposition may be inferred for the entire sedimentary section beneath the COST B-2 well. A generalized subsidence curve for the deepest part of Baltimore Canyon Trough off New Jersey can be constructed (Fig. 14) based on the COST B-2 well and on the assumption that all sedimentary rocks between 4.8 and 12.0 km were also deposited at or near sea level. The deepest horizons at 12 km are assumed to be at the Triassic-Jurassic boundary. The subsidence curve illustrates the fact that nearly 70% of the subsidence beneath the New Jersey shelf took place during the Jurassic, followed by slower subsidence during the Cretaceous and Tertiary.

Subsidence of a shelf edge after continental separation can be explained in part in terms of sediment loading and (in part) in terms of cooling of the lithosphere (Watts and Ryan, 1976). Where the effect of sediment loading is removed from the subsidence curve, the subsidence rate appears to decay exponentially after the initial rifting in a manner similar to the age-versus-depth curves of mid-ocean ridges, which have been explained in terms of cooling and contraction of the lithosphere (Sclater et al, 1971). As mid-ocean ridge cooling models only account for 3 to 4 km of subsidence (sea floor subsidence from 2.5 to 6.0 km depth over 150 m.y.), at least 8 of the 12 km of subsidence beneath the New Jersey shelf may be due to sediment loading and other effects. The subsidence curve of the shelf off New Jersey (Fig. 14) suggests that the rate of sediment loading may also follow an exponentially decreasing curve, which is reasonable if highlands adjacent to the

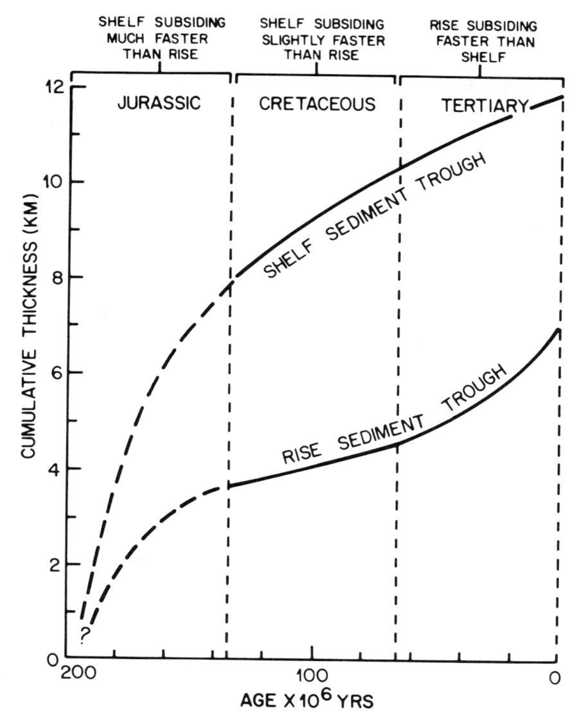

FIG. 14. The subsidence history of the shelf trough off New Jersey can be inferred from the COST B-2 well, and from an assumption that the deepest sedimentary units at 12 km depth were shallow-water deposits of earliest Jurassic age. The rise sediment trough contains 6 to 8 km of sediment, of which the last 2.5 km are Tertiary units. Accumulation on the rise during the Middle and Late Cretaceous appears to have been much slower than during the Jurassic and earliest Cretaceous, i.e. pre-Beta Horizon (Grow and Markl, 1977). Rapid differential subsidence between the shelf trough and the adjacent rise trough appears to have occurred during the Lower Cretaceous. Rapid differential subsidence during the Jurassic also has been important along the margin as well as across the margin.

basin were rapidly uplifted and eroded during the rift and early drift stages of continental breakup.

No deep drill-hole data are available beneath the continental slope or rise to quantitatively document the paleobathymetry and subsidence rates in that environment. However, as the sediment thickness beneath the upper rise appears to have a maximum value of only 6 to 8 km, it appears reasonable to assume that the subsidence may only be 50 to 70% of that observed on the New Jersey shelf (Fig. 14). Furthermore, the differential subsidence between the basins underlying the shelf and rise may have been localized along the landward edge of oceanic crust (i.e., the East Coast Magnetic Anomaly), especially during the Jurassic when the subsidence rates were greatest. Differential subsidence probably continued into the Early Cretaceous; local faulting and arching have been observed in rocks of that age (Figs. 4 and 7). Although the thicknesses of Upper Cretaceous strata are similar on both the shelf and rise, the Tertiary sedimentary units are twice as thick as on the rise because of lowered sea levels and sediments bypassing the shelf.

Although the subsidence history of the U.S. Atlantic continental margin has only been related to one deep offshore well so far, our analysis of that well and our geophysical profiles suggest that differential subsidence both *across* the ocean-continent transition zone and *along* the length of the margin was important during the Jurassic and Early Cretaceous. The localization of this differential subsidence near the landward edge of the oceanic crust appears to have been probable during this early phase of subsidence. A structural section across the Aaiun Basin off northwest Africa by Von Rad and Eiselle (in press) suggests that differential subsidence between thick shelf strata and thinner rise strata was also important in that region.

CONCLUSIONS

Interval velocities in the upper 5 km below the continental shelf show consistent patterns between Cape Hatteras and Cape Cod. Although sonic velocities from the COST B-2 well increase gradually with depth and display numerous small and large velocity inversions, they still show good general agreement with the interval velocities at USGS line 2 (Fig. 3). Lateral variations in velocities across the shelf are consistently observed, especially in the Mesozoic section where high-velocity units occur near the Lower Cretaceous shelf edge. These high-velocity trends indicate the presence of carbonate bank or reef complexes near the shelf edge during Late Jurassic and Early Cretaceous time.

Finally, conversion of seismic sections to depth sections has provided a valuable way to look at the geologic evolution of the outer shelf and upper continental slope. Careful velocity studies are required before this can be attempted, as artificial structures can be created if the velocities are not accurate and properly smoothed. The velocities and depth sections presented here are fairly reliable down to a depth of 5 to 7 km, but we are continuing to reprocess these data to improve our picture of the deeper structure.

The depth sections over the outer continental shelf and upper slope reveal that Lower Cretaceous shelf horizons continue as flat-lying units beneath the upper slope. Although portions of these horizons are tilted and arched on two of our six profiles, the edge of the continental shelf during the Early Cretaceous can clearly be seen to occur 20 to 30 km seaward of the present shelf edge.

At least one major unconformity is recorded in the sedimentary rocks beneath the continental slope, which marks a retreat of the continental shelf edge to its present position. The age of this erosional unconformity is not well established, but our present data suggest that it correlates with a major marine regression during the Oligocene. Although later erosion on the outer shelf and upper continental slope during the Miocene to Pleistocene took place in response to rapid glacial-eustatic fluctuations of sea level, it is thought to be less significant than the Oligocene erosion.

The major sediment troughs beneath the continental shelf and continental rise have accumulated upwards of 12 and 8 km of sedimentary strata, respectively. They are separated by a 25- to 75-km-wide disturbed zone beneath the Lower Cretaceous shelf edge where no coherent reflections can be seen deeper than 6 km. The landward edge of the disturbed zone coincides with the East Coast Magnetic Anomaly. Although many hypotheses have been proposed for what lies between 6 and 12 km depth in this disturbed zone, no unique solution is obvious from the present data. Whether it is composed of carbonate banks or reefs, broken sedimentary blocks, diapirs, dikes, sills, volcanic ridges, basement ridges, or some combination of the above remains unresolved. Despite these uncertainties, our present interpretation of the seismic and potential field data indicates that both the disturbed seismic zone and the East Coast Magnetic Anomaly can be explained as representing the landward edge of thick oceanic crust, which is overlain by Jurassic and Lower Cretaceous carbonate bank or reef complexes. Furthermore, faulting and diapiric activity may be important locally above or adjacent to the landward edge of oceanic crust, where differential subsidence has occurred between the shelf basin and the adjacent oceanic crust.

REFERENCES CITED

Behrendt, J. C., J. Schlee, and R. Foote, 1974, Seismic evidence for a thick section of sedimentary rock on the Atlantic outer continental shelf and slope of the United States (abs.): EOS, v. 55, no. 4, p. 278.

Benson, W. E., et al, 1976, Deep sea drilling in the North Atlantic—leg 44: Geotimes, v. 21, no. 2, p. 23–26.

Cornet, B., 1977, Palynostratigraphy and age of the Newark Supergroup: Pennsylvania State Univ., Ph.D. dissertation, 505 p.

Dallmeyer, R. D., 1975, The Palisades sill: a Jurassic intrusion: Geology, v. 3, p. 243–245.

Dillon, W. P., R. E. Sheridan, and J. P. Fail, 1976, Structure of the western Blake-Bahama basin as shown by 24-channel CDP profiling: Geology, v. 4, p. 459–462.

Dix, H. C., 1955, Seismic velocities from surface measurements: Geophysics, v. 20, no. 1, p. 68–86.

Drake, C. L., M. Ewing, and G. H. Sutton, 1959, Continental margins and geosynclines: The east coast of North America north of Cape Hatteras, *in* L. H. Ahrens et al, eds., Physics and chemistry of the earth: London, Pergamon Press, v. 3, p. 110–198.

Emery, K. O., and E. Uchupi, 1972, Western North Atlantic Ocean: Topography, rocks, structure, water life and sediments: AAPG Memoir 17, 532 p.

Emery, K. O., et al, 1970, Continental rise off eastern North America: AAPG Bull., v. 54, no. 1, p. 44–108.

Evans, R., 1978, Origin and significance of evaporites in basins around the Atlantic margin: AAPG Bull., v. 62, no. 2, p. 223–234.

Grow, J. A., and J. Schlee, 1976, Interpretation and velocity analysis of U.S. Geological Survey multichannel reflection profiles 4, 5, and 6, Atlantic continental margin: U.S. Geological Survey Misc. Field Studies Map, MF-808.

——— and R. E. Sheridan, 1976, High-velocity sedimentary horizons beneath the outer continental shelf off New Jersey (abs.): EOS, v. 57, no. 4, p. 265.

——— and R. G. Markl, 1977, IPOD-USGS multichannel seismic reflection profile from Cape Hatteras to the Mid-Atlantic Ridge: Geology, v. 5, p. 625–630.

——— W. P. Dillon, and R. E. Sheridan, 1977, Diapirs along continental slope off Cape Hatteras (abs.): Soc. Explor. Geophysicists, 47th An. Mtg. (1977) Calgary, p. 51.

Hollister, C. D., J. I. Ewing et al., 1972, Initial reports of the Deep Sea Drilling Project: Washington, D.C. (U.S. Govt., v. 11).

Hubert, J. F., et al, 1978, Guide to the redbeds of central Connecticut: Eastern Section, SEPM, field trip guidebook; also Univ. Massachusetts, Amherst, Dept. Geol. and Geography Contr. No. 32, 129 p.

Jaworski, B. L., J. A. Grow, and C. A. Meeder, 1976, Airgun-sonobuoy refraction measurements on Georges Bank (abs.): EOS, v. 57, no. 4, p. 265.

Keen, C. E., and M. J. Keen, 1974, Continental margins of eastern Canada and Baffin Bay, *in* C. A. Burk and C. L. Drake, eds., The Geology of continental margins: New York, Springer-Verlag, p. 381–390.

Klitgord, K. D., and J. C. Behrendt, 1978, Basin structure of the U.S. Atlantic margin: this volume.

Lancelot, Y., E. Seibold et al, 1975, The eastern North Atlantic—Deep Sea Drilling Project, Leg 41: Geotimes, v. 20, no. 7, p. 18–21.

Lehner, P., and P. A. D. De Ruiter, 1977, Structural history of Atlantic margin off Africa: AAPG., v. 67, no. 7, p. 961–981.

Maher, J. C., 1971, Geologic framework and petroleum potential of the Atlantic Coastal Plain and continental shelf: U.S. Geol. Survey Prof. Paper 659, 98 p.

Manheim, F. T., and R. E. Hall, 1976, Deep evaporitic strata off New York and New Jersey—Evidence from interstitial water chemistry of drill holes: U.S. Geol. Survey, Jour. Research, v. 4, no. 6, p. 697–702.

Mattick, R. E., et al, 1974, Structural framework of United States Atlantic outer continental shelf north of Cape Hatteras: AAPG Bull., v. 58, p. 1179–1190.

——— et al, 1978, Petroleum potential of U.S. Atlantic slope, rise, and abyssal plain: AAPG Bull., v. 62, no 4, p. 592–607.

Mayhew, M. A., 1974, Geophysics of the Atlantic North America, *in* C. A. Burk and C. L. Drake, eds., Geology of continental margins: New York, Springer-Verlag, p. 409–427.

Mayne, W. H., 1962, Common reflection point horizontal data stacking techniques: Geophysics, v. 27, no. 6, part II, p. 927–938.

Miller, E. L., W. B. F. Ryan, et al 1978, Geologic sampling with submersible in Georges Bank canyons (abs.): Geol. Soc. America, Northeastern Sect. Mtg., Abs. with Programs, v. 10, no. 2, p. 75.

Perry, W. J., et al, 1975, Stratigraphy of the Atlantic continental margin of the United States north of Cape Hatteras—brief survey: AAPG Bull., v. 59, p. 1529–1548.

Pitman, W., 1978, Relationship between eustacy and stratigraphic sequences of passive margins: Geol. Soc. America Bull., v. 89, no. 9, p. 1389–1403.

Rabinowitz, P. D., 1974, The boundary between oceanic and continental crust in the western North Atlantic: *in* C. A. Burk and C. L. Drake, eds., The geology of continental margins: New York, Springer-Verlag, p. 67–83.

Rabinowitz, P. D., and J. L. LaBreque, 1977, The isostatic gravity anomaly: Key to the evolution of the ocean-continent boundary at passive continental margins: Earth and Planetary Sci. Letters, v. 35, p. 145–150.

Reesman, R. H., C. R. Filbert, and H. W. Krueger, 1973, Potassium-argon dating of the Upper Triassic lavas of the Connecticut Valley, New England (abs.): Geol. Soc. America, Northeastern Sect. Mtg., Abs. with Programs, v. 5, no. 2, p. 211.

Rona, P. A., and C. S. Clay, 1967, Stratigraphy and structure along a continuous seismic reflection profile from Cape Hatteras, North Carolina, to the Bermuda Rise: Jour. Geophys. Research, v. 72, no. 8, p. 2107–2130.

Ryan, W. B. F., J. C. Sibuet, et al, 1976, Passive contiental margin—Deep Sea Drilling Project, leg 47: Geotimes, v. 21, no. 10, p. 21–24.

Schlee, J., et al, 1976, Regional geologic framework off northeastern United States: AAPG Bull., v. 60, no. 6, p. 21–24.

——— et al, 1977, Petroleum geology on the United States Atlantic/Gulf of Mexico margins, *in* V. S. Cameron, ed., Proceedings of the Southwest Legal Foundation—Exploration and economics of the petroleum industry: New York, Matthew Bender and Company, Inc., . 15, p. 47–93.

Scholle, P. A. (ed.), 1977, Geological studies on the COST No. 2 B-2 well, U.S. Mid-Atlantic outer continental shelf area: U.S. Geol. Survey Circ. 750, 71 p.

Sclater, J. G., R. N. Anderson, and M. L. Bell, 1971, Elevation of ridges and the evolution of the central eastern Pacific: Jour. Geophys. Research, v. 76, no. 32, p. 7888–7915.

Scott, K. R., and J. M. Cole, 1975, U.S. Atlantic margin looks favorable: Oil and Gas Jour. v. 73, no. 1, p. 95–99.

Seibold, E., and K. Hinz, 1974, Continental slope—construction and destruction; *in* C. A. Burk and C. L. Drake, eds., The geology of continental margins: New York, Springer-Verlag, p. 179–196.

Sheridan, E., 1974, Atlantic continental margin off North America, *in* C. A. Burk and C. L. Drake, eds., The geology of continental margins: New York, Springer-Verlag, p. 391–407.

——— 1976, Sedimentary basins of the Atlantic margin of North America: Tectonophysics, v. 36, p. 113–132.

——— 1977, Older Atlantic crust and the continental edge based on seismic data (abs.): Geol. Soc. America, Northeastern Sect. Mtg., Abs. with Programs, v. 9, no. 3, p. 317.

——— and J. J. Knebel, 1976, Evidence of post-Pleistocene faults on New Jersey Atlantic outer continental shelf: AAPG Bull. v. 60, no. 7, p. 1112–1117.

Smith, M. A., et al, 1976, Geological and operational summary, COST No. B-2 well, Baltimore Canyon trough area, mid-Atlantic OCS: U.S. Geol. Survey, Open File Rep. 76–774, 79 p.

Talwani, M., and O. Eldholm, 1973, Boundary between continent and oceanic crust at the margin of rifted continents: Nature, v. 241, p. 325–330.

Taner, M. T., and F. Koehler, 1969, Velocity spectra-digital computer derivation and applications of velocity functions: Geophysics, v. 34, p. 859–881.

Taylor, P. T., I. Zietz, and L. S. Dennis, 1968, Geologic implications of aeromagnetic data for the eastern continental margin of the United States: Geophysics, v. 33, p. 755–780.

Vail, P. R., R. M. Mitchum, Jr., and S. Thompson, III, 1977, Global cycles of relative changes of sea level, *in* C. E. Payton, ed., Seismic stratigraphy—applications to hydrocarbon exploration: AAPG Memoir 26, p. 83–97.

Van Hinte, J. E., 1976, A Jurassic time scale: AAPG Bull. v. 60, p. 489–497.

Von Rad, U., and G. Eiselle, in press, Mesozic-Cenozic subsidence history and paleobathymetry of the northwest African continental margin (Aaiun Basin to DSDP Site 397): Royal Soc. London Philos. Trans.

Watts, A., and W. B. F. Ryan, 1976, Flexure of the lithosphere and continental margin basins: Tectonophysics, v. 35, p. 25–44.

Basin Structure of the U.S. Atlantic Margin[1]

KIM D. KLITGORD[2] AND JOHN C. BEHRENDT[3]

Abstract A detailed magnetic study of the U.S. Atlantic continental margin north of Cape Hatteras delineates the pattern of basins and platforms that form the basement structure. A 185,000-km, high-sensitivity aeromagnetic survey acquired in 1975 over the entire U.S. Atlantic continental margin forms the basis of this study. Magnetic depth-to-source estimates were calculated for the entire survey using a Werner "deconvolution" type method. These depth-to-basement estimates are integrated with multichannel seismic reflection profiles to interpolate basement structures between seismic profiles.

The deep sediment-filled basins along the margin are bounded on their landward sides by blockfaulted continental crust; their seaward sides are marked by the East Coast magnetic anomaly. The trends of the landward sides of these basins vary from 030° in the south to 040° in the north, consistent with a common pole of opening for all of the basins. The ends of these basins are controlled by sharp offsets in the continental crust that underlie the various platforms. These offsets are the result of the initial breakup of North America and Africa and are preserved as fracture zones under the continental rise.

The regions west of the various basins are comprised of platforms of Paleozoic and older crust and embayments of Triassic-Jurassic age. The Long Island platform is a series of ridges and troughs. These troughs are oriented northeastward, parallel with the Baltimore Canyon trough and the Georges Bank trough. The Connecticut Valley Triassic basin has a broad magnetic low associated with it that can be traced across Long Island. A similar magnetic signature is associated with the trough between Martha's Vineyard and Nantucket Island, suggesting that it also may be a Triassic basin. The Salsbury Embayment with its Triassic-Jurassic age sediments lies just west of the Baltimore Canyon trough while the Carolina platform, which has a few smaller Triassic basins within predominantly Paleozoic and older crust, lies landward of the Carolina trough. The area around Charleston is another major embayment of Triassic-Jurassic age, and west of the Blake Plateau is the Florida platform with Paleozoic and older crust.

A magnetic basement high associated with the East Coast magnetic anomaly separates oceanic crust from the deep sediment-filled troughs. The minimum depth of this high ranges from 6 to 8 km and the susceptibility contrast suggests that it is more likely an uptilted block of oceanic crust than a massive intrusive body. The magnetic anomaly probably is produced by a combination of a basement high and an "edge effect," where the edge is between the uptilted block and flat-lying, nonmagnetic sediments to the west.

INTRODUCTION

The U.S. Atlantic continental margin is a Mesozoic and Tertiary sedimentary wedge overlying a basement composed of continental crust, transitional crust, and oceanic crust. The general shape of the margin, as reflected by the 1,000-m depth contour, bears only slight resemblance to the major basement structure, and completely hides the pattern of basins and platforms along the margin (Fig. 1). Multichannel seismic reflection profiles across the margin (Behrendt et al, 1974; Grow and Schlee, 1976; Schlee et al, 1976) show a great thickness of sediments in the basins along the margin. The existence of these basins had been inferred from earlier geophysical studies (Drake et al, 1959; Drake et al, 1968; Emery et al, 1970; Emery and Uchupi, 1972; Mayhew, 1974; Sheridan, 1974).

The structural boundaries and depths of these basins have been estimated by various authors (Drake et al, 1959; Emery et al, 1970; Maher and Applin; 1971; Sheridan, 1974; Schlee et al, 1976), but because of the limited seismic coverage, only a rough determination of their shape has been possible. Additionally, the use of complementary geo-

[1]Manuscript received, November 16, 1977; accepted, July 27, 1978.

[2]U.S. Geological Survey, Woods Hole, Massachusetts 02543.

[3]U.S. Geological Survey, Federal Center, Denver, Colorado 80225.

Discussions with W. Dillon, J. Grow, P. Popenoe, H. Schouten, and J. Schlee are gratefully acknowledged. The multichannel seismic reflection interpretations were provided by W. Dillon, J. Grow, and J. Schlee. The writers thank L. Morse (U.S. Geological Survey, Woods Hole, Mass.) for her aid in compiling the aeromagnetic data; F. Navazio and D. Boggs (LKB Resources, Inc., Huntingdon, Pa.) for their cooperation in collecting and compiling the aeromagnetic data; and R. Hartman and C. Curtis (INTernational EXploration, Ivyland, Pa.) for their processing of the depth-to-source estimates. The figures were drafted by P. Forrestal and L. Sylwester.

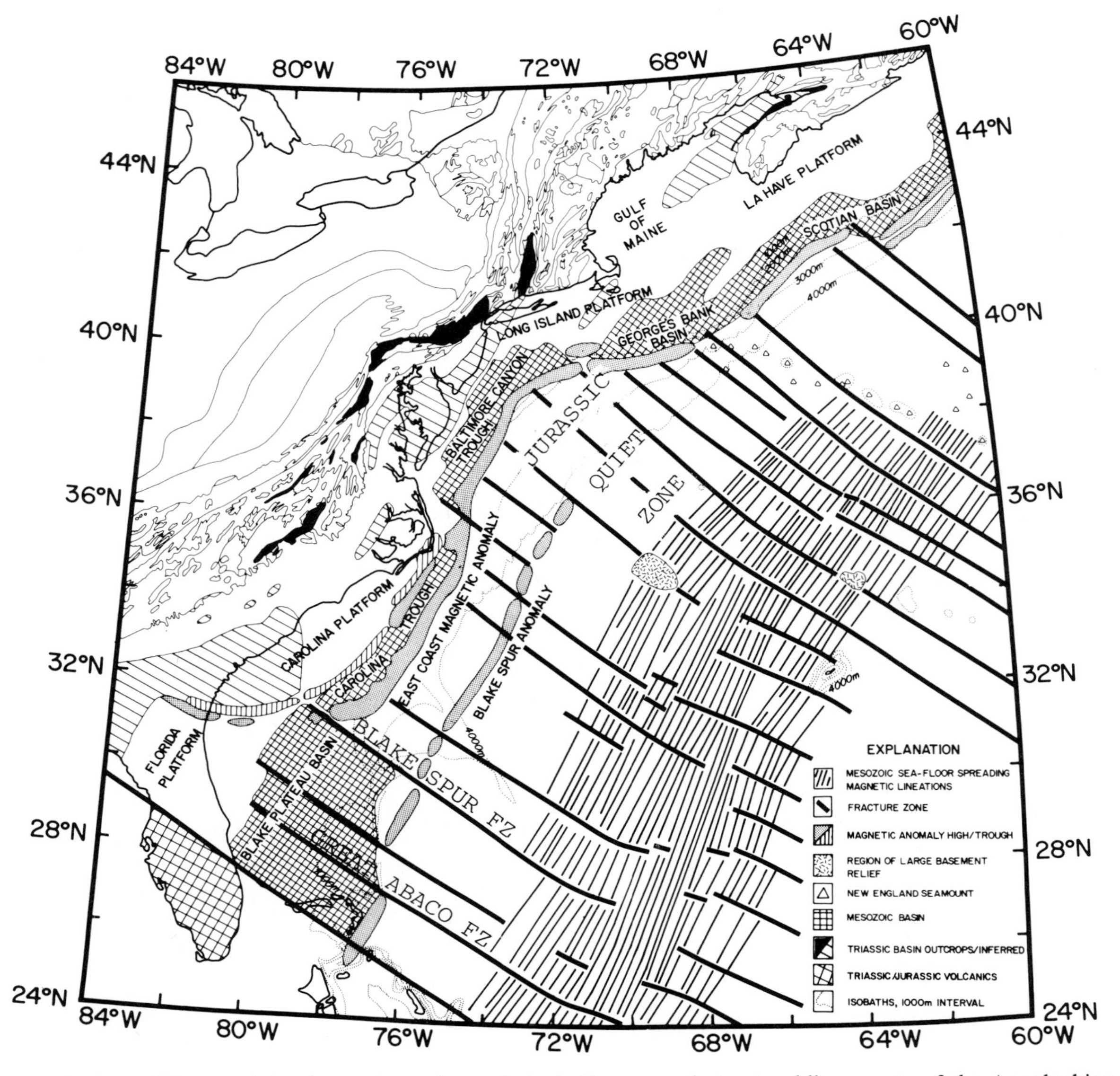

FIG. 1. Western Atlantic continental margin including general structural lineaments of the Appalachian Orogenic System (after King, 1969), Mesozoic sea-floor spreading magnetic lineations and fracture zones (after Schouten and Klitgord, 1977), prominent magnetic lineaments within the Jurassic Quiet Zone, the major sediment-filled troughs of Mesozoic age, and the major platforms which are underlain by continental crust. E.B. = Essaouira Basin.

physical information (magnetic and gravity) has been limited (Emery et al, 1970; Mayhew, 1974; Sheridan, 1974). This paper uses magnetic source depth analyses from a new high-sensitivity aeromagnetic survey over the U.S. Atlantic continental margin, in conjunction with multichannel seismic profiles, to determine the detailed basement structure of the margin. The earlier detailed magnetic study of the Atlantic continental margin by Taylor et al (1968) concentrated on the continental foldbelt region and the continental slope and rise but contained almost no discussion of the continental shelf area and the buried basins. The writers compared magnetic depth estimates and multichannel seismic depth profiles to test the reliability of the magnetic depth analyses. Trends of prominent magnetic sources on adjacent closely spaced magnetic profiles are used to interpolate the locations of the associated basement features between seismic profiles.

DATA SOURCES

1975 Aeromagnetic Survey

As part of the U.S. Geological Survey program to study the U.S. Atlantic continental margin, a 185,000-km high-sensitivity, nonexclusive aeromagnetic survey was flown, compiled and contoured by LKB Resources, Inc. The track coverage (Fig. 2) ranged from a 2.5 × 16-km (1.5 × 10-mi) grid to a 5 × 32-km (3 × 20 mi) grid between the coastline and the 2,000-m depth contour and from a 9.5 × 32-km (6 × 20-mi) grid to a 32.5 × 64.5-km (20 × 40-mi) grid between the 2,000 m and 4,000-m depth contours. The 2.5 × 16-km grid over Georges Bank was flown in

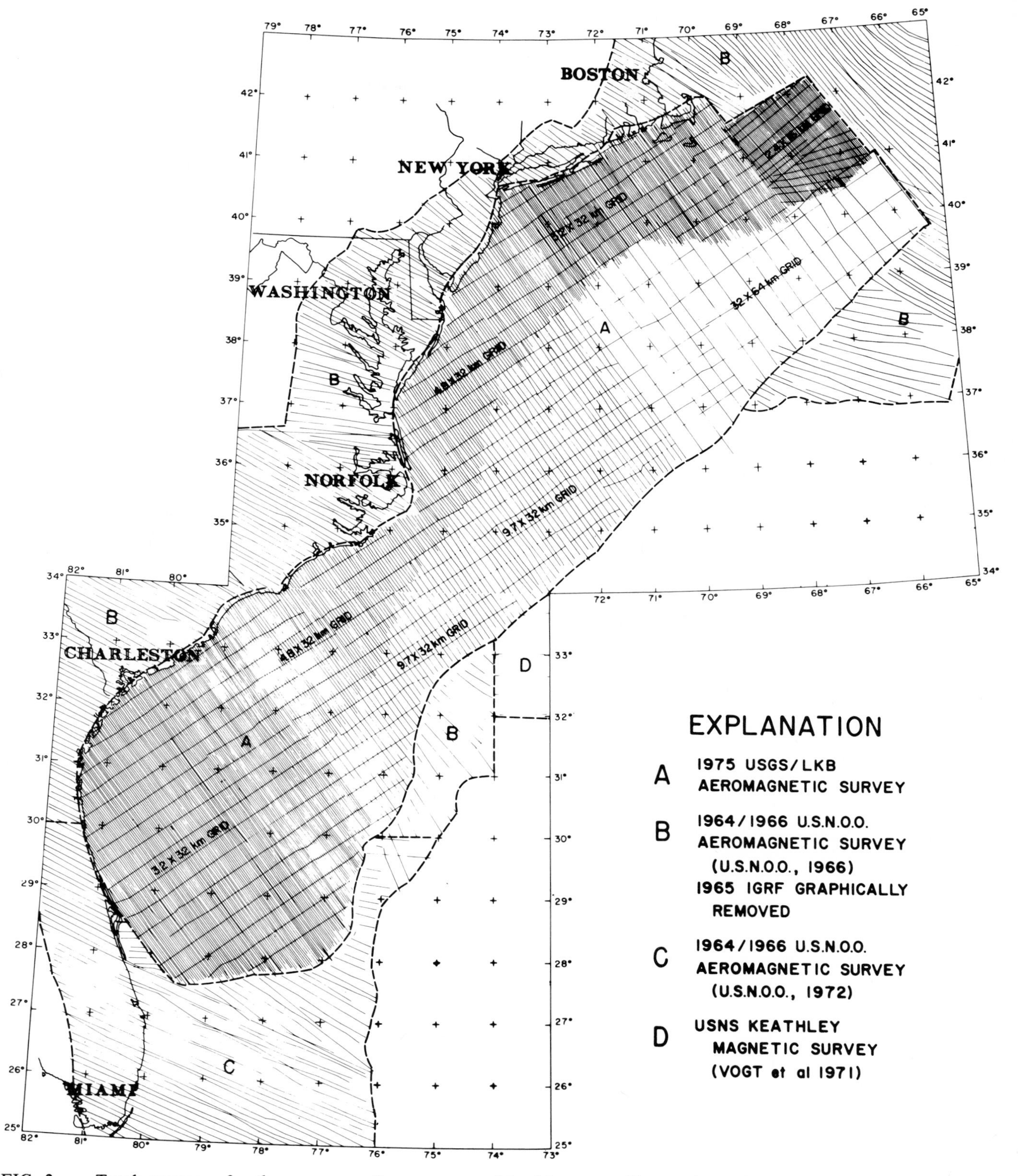

FIG. 2. Track coverage for the aeromagnetic surveys used in this report. The track spacings are indicated in kilometers for Area A.

1966 by Aero Service and the data were recompiled by LKB Resources, Inc. The flight elevation for the survey was about 450 m except for the 2.5 × 16-km grid over Georges Bank, which was flown at about 300 m. A *LORAN* C-doppler radar-VLF (very low frequency) integrated navigation system was used for the 1975 survey. *LORAN* A was used for the navigation of the 1966 survey. A high-sensitivity

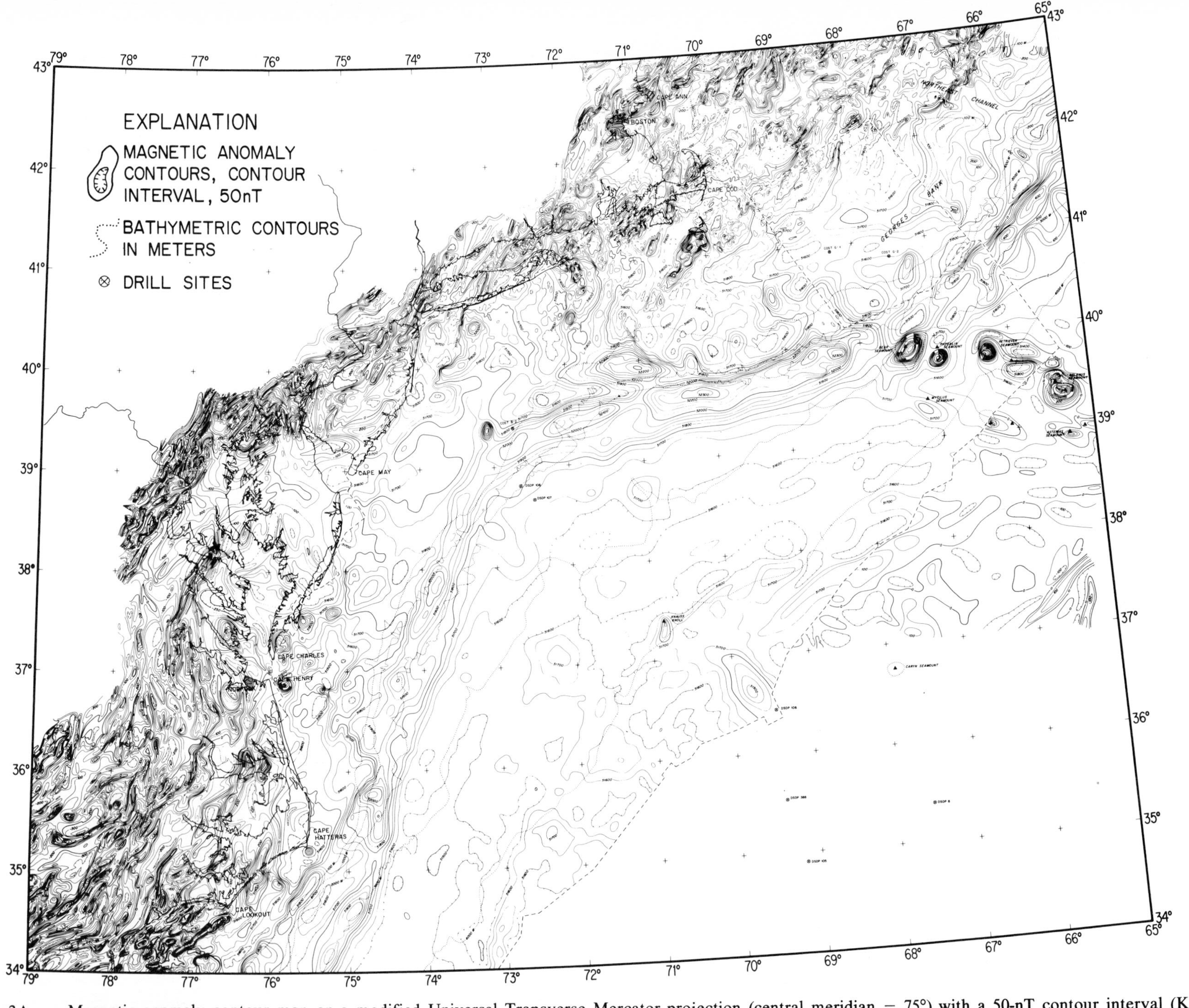

FIG. 3A. Magnetic anomaly contour map on a modified Universal Transverse Mercator projection (central meridian = 75°) with a 50-nT contour interval (Klitgord and Behrendt, 1977). **3A shows area north of 34°N lat. The magnetic contours are based on the track coverage shown in Figure 2.**

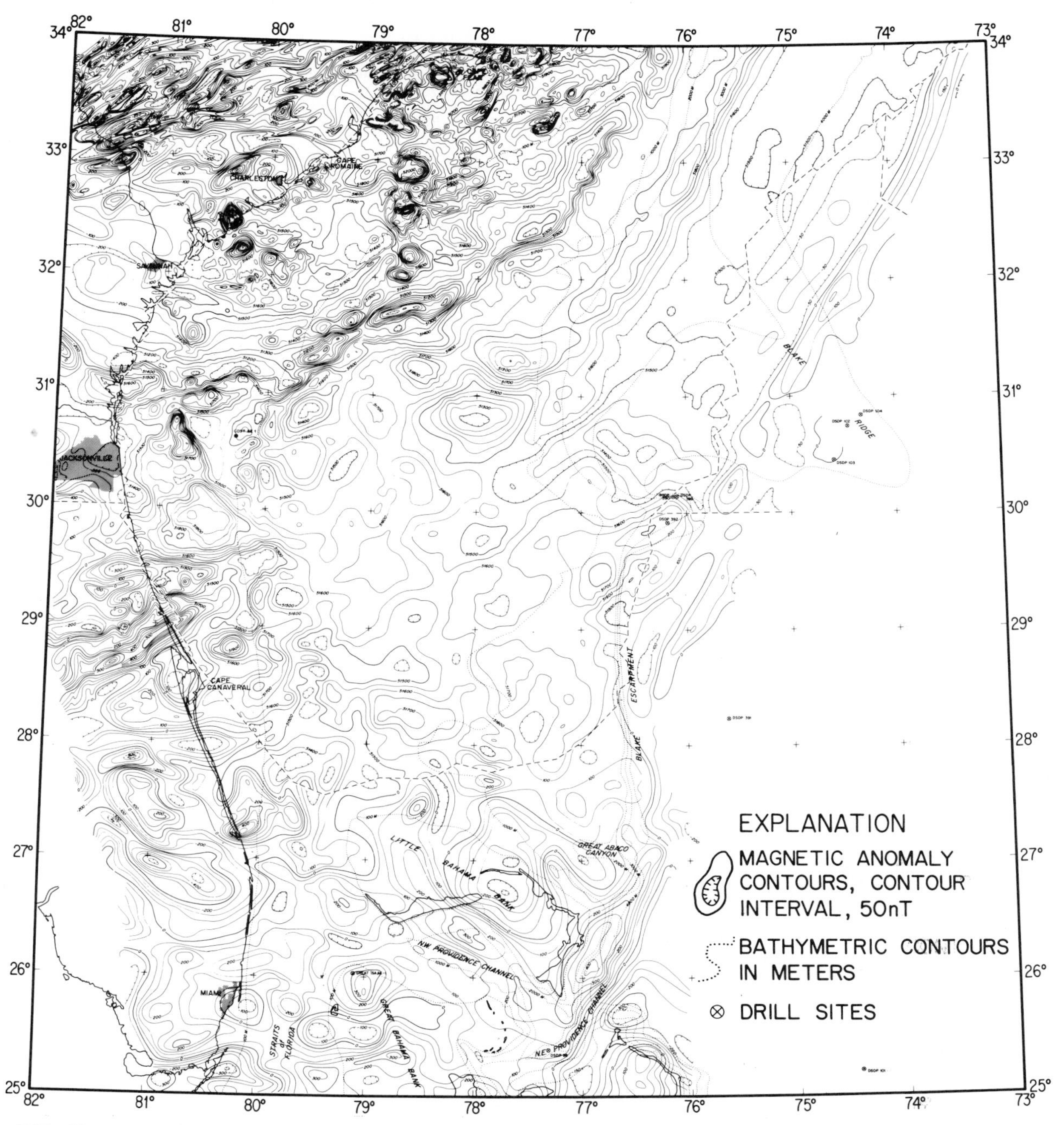

FIG. 3B. Magnetic anomaly contour map on a modified Universal Transverse Mercator projection (central meridian = 75°) with a 50-nT contour interval (Klitgord and Behrendt, 1977). **3B** shows area south of 34°N lat. The magnetic contours are based on the track coverage shown in Figure 2.

magnetometer utilizing optical absorption of energized helium vapor (rubidium vapor for the 1966 survey) was used in the airplane for the survey and at a base station to monitor the diurnal field. Flight lines were reflown when the magnetic noise level was determined to be too high by the base station monitor. The aeromagnetic data, navigational data, and altitude were digitally recorded.

The compilation of the magnetic data included cross-line leveling and regional field removal. The total magnetic field readings were adjusted by utilizing a modified control-line leveling system. This system adjusts the magnetic field values at each flight-line intersection to minimize the crossover errors caused by diurnal variations and ground-positioning errors. The resulting adjusted total field data then had the regional field removed using the IGRF 1965 tables (IAGA, 1969) updated to mid-1975 (mid-1966 for the 1966 survey). A datum of 52,000 nanotesla (nT, 1 nT = 1 gamma; 43,850 nT for the 1966 survey) was then added to the magnetic anomalies.

Magnetic anomaly maps were compiled at a contour interval of 2 nT by LKB Resources, Inc. and recompiled by the authors (Klitgord and Behrendt,

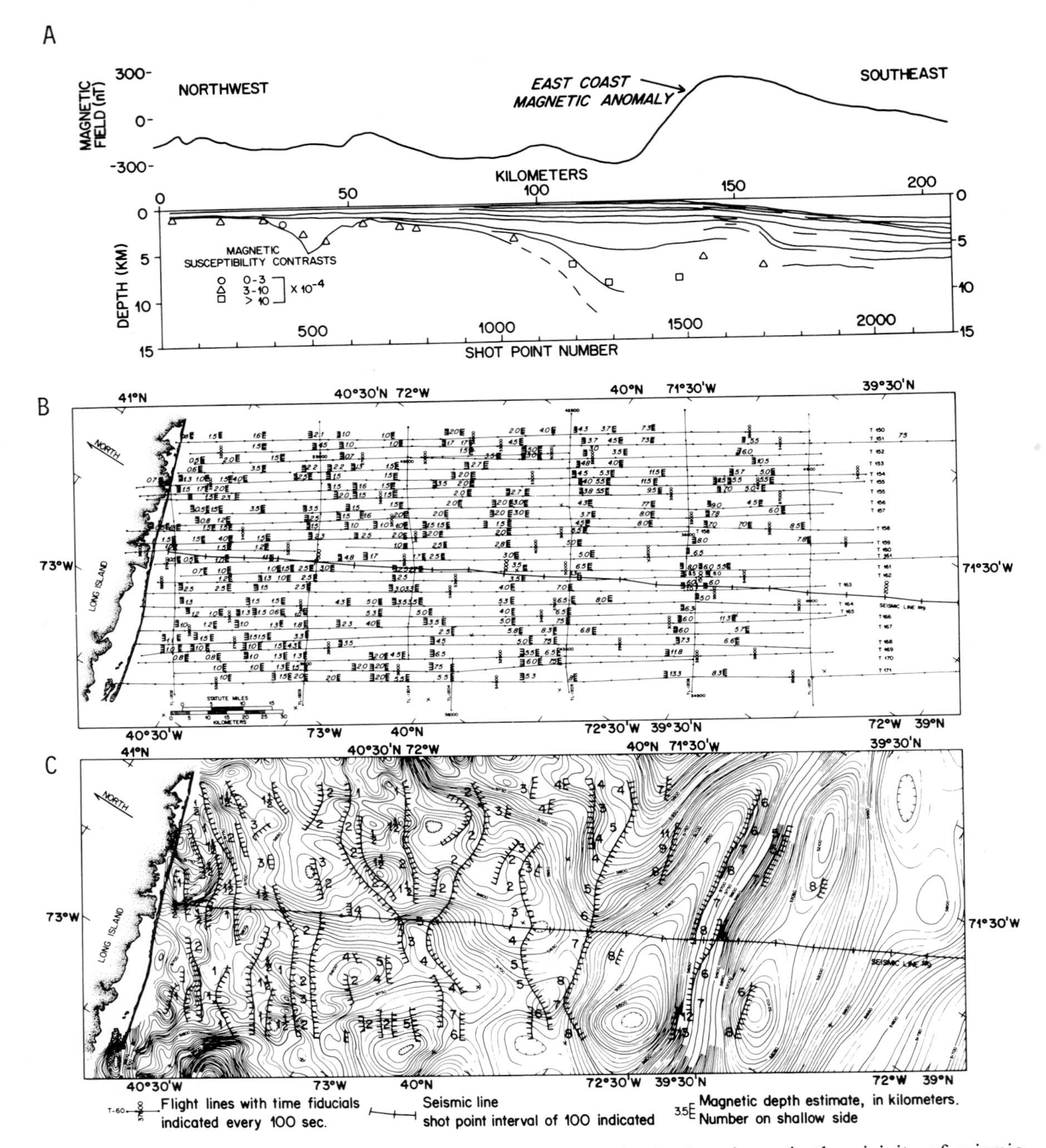

FIG. 4. Example showing the general use of the magnetic depth estimate in the vicinity of seismic line 9: **A**—comparison between magnetic depth estimates and seismic reflection profile; **B**—comparison between magnetic depth estimates on adjacent flight lines; **C**—smoothed version of magnetic depth estimates from **B** superimposed on the magnetic anomalies, contour interval is 10 nT.

1977) at a contour interval of 50 nT (Fig. 3) and a scale of 1:1,000,000. The reliability of these contours is extremely good except in the northeastern deep-water area, where the grid spacing is only 32.5 × 64.5 km (Fig. 2). In this latter area, the contours agree reasonably well with previous aeromagnetic surveys (Taylor et al, 1968). The magnetic contours have been modified in the vicinity of the New England Seamounts, taking into consideration more detailed magnetic surveys by the Naval Oceanographic Office (Walczak, 1963).

1964/1966 Aeromagnetic Survey

Aeromagnetic data from the 1964–1966 U.S. Naval Oceanographic Office survey (U.S. Naval Oceanographic Office, 1966) were used to augment to LKB Resources, Inc. survey (see Fig. 2). This older survey was flown between 1964 and 1966 at flight elevations of about 150 m over water, 450 m over land south of Washington D.C., and 750 m over land north of Washington D.C. The flight lines (Fig. 2) were spaced at about 8-km intervals, with no tie-lines.

The 50 nT magnetic contours south of 30°N lat. were taken from a map previously published by the U.S. Naval Oceanographic Office (1972). North of 30°N lat., the regional field based upon the IGRF 1965 tables was graphically removed by the writers from the total-intensity magnetic maps (U.S. Naval Oceanographic Office, 1966).

U.S.N.S. *Keathley* Magnetic Survey

A small portion of the marine magnetic data collected by the U.S.N.S. *Keathley* (Vogt et al, 1971) is included in figure 3. The magnetic data were collected on a track spacing of 35 km and an average regional field was removed from the total field data.

METHOD OF ANALYSIS OF MAGNETIC DATA

Magnetic Depth-to-Source Estimates

The determination of the depth-to-magnetic basement can be a valuable tool for investigating the structure of sedimentary basins (for example, Vacquier et al, 1951; Dobrin, 1960). In general, the sediments within a basin have very weak susceptibilities compared with those of crystalline/volcanic basement at the bottom of the sediment pile. Therefore, it could be expected that the major source of magnetic anomalies over these basins would be basement structure, susceptibility variations within the basement, and intrusive bodies or extrusive volcanic bodies within the basin.

All the methods for estimating magnetic source depths require assumptions about these sources. Usually the reliability of the depth estimates is closely tied to the validity of these assumptions. When other geophysical data, such as seismic reflection or refraction profiles, or drill-hole data can be used to help limit these assumptions, one can often increase the reliability of these depth estimates. The best method for carrying out the depth estimates will usually be determined by these assumptions and the desired results.

Estimates of the depth to magnetic sources were calculated for the U.S. Geological Survey by International Exploration, Inc., using a Werner "deconvolution" type method (Hartman et al, 1971; Jain, 1976). This method assumes that the sources are either two-dimensional dikes or edges of bodies. The mathematical formula for the magnetic anomaly and horizontal magnetic gradient associated with these two types of sources can be expressed as a truncated expansion (Werner, 1953) having a fixed number of unknowns. These unknowns include the horizontal location of the source, it's depth, susceptibility contrast, and dip.

The magnetic field measurements at a small number of points (e.g., 7), called a sampling window, are used to solve a set of simultaneous equations for the desired unknowns. This window is moved, point by point, across the entire data set; consistency criteria are used for selecting the most reliable source-location estimates. Several passes are made of the data set, each pass having wider sampling intervals and appropriate anti-alias filters, to look at the deeper sources. The number of data

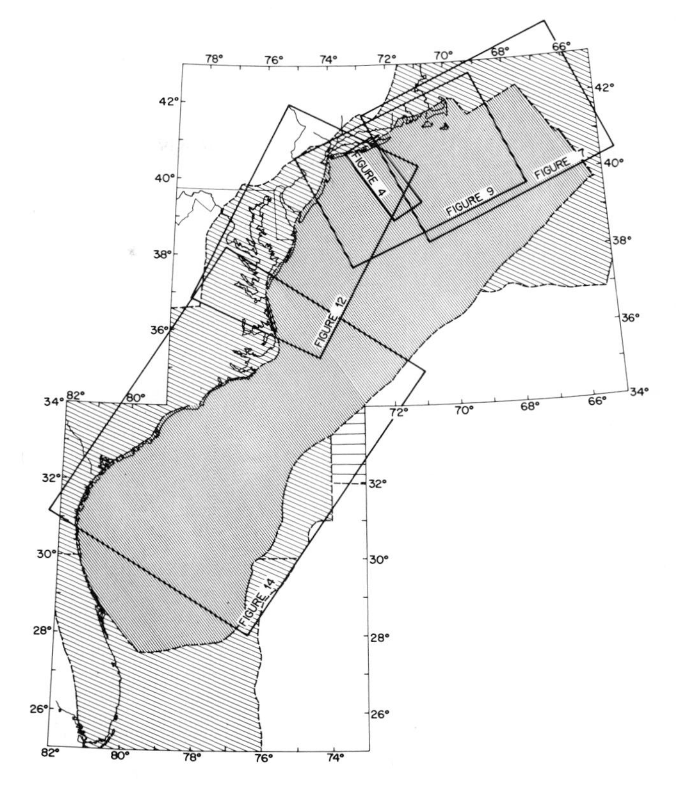

FIG. 5. Location map for areas shown in more detail on other figures. The magnetic surveys shown in Figure 2 are indicated.

points is kept the same for each of the passes and the window is moved with the same distance increment for each pass. As might be expected, the shallower sources producing the shortest wavelength anomalies will be seen by only a few sampling windows because only a few shifts of the window will rapidly bring it beyond a particular magnetic anomaly.

In contrast, the deeper sourced depth estimates, generally associated with the broader wavelength anomalies, will be seen by more sampling windows, but the resolution will be poorer than for the shallow-sourced depth estimates. Each pass of the window produces a set of solutions based on the block edge (interface) model and a set of solutions based on the dike model. Because both sets of solutions are determined from the same data, one or the other must be chosen at any given location; however, the set of solutions chosen along any given profile can be a combination of the two types. In general, for the continental shelf area, the edge solutions were chosen in this study because most of the source bodies appear to be wider than the depth to the source for the basement features and because of the agreement of the solutions with seismic data. Over oceanic crust with its thin source layer, the dike solutions provided a better estimate of the basement depth.

Susceptibility contrasts within the basement can affect the depth analyses. Large-scale sharp variation in the susceptibility, such as at contacts be-

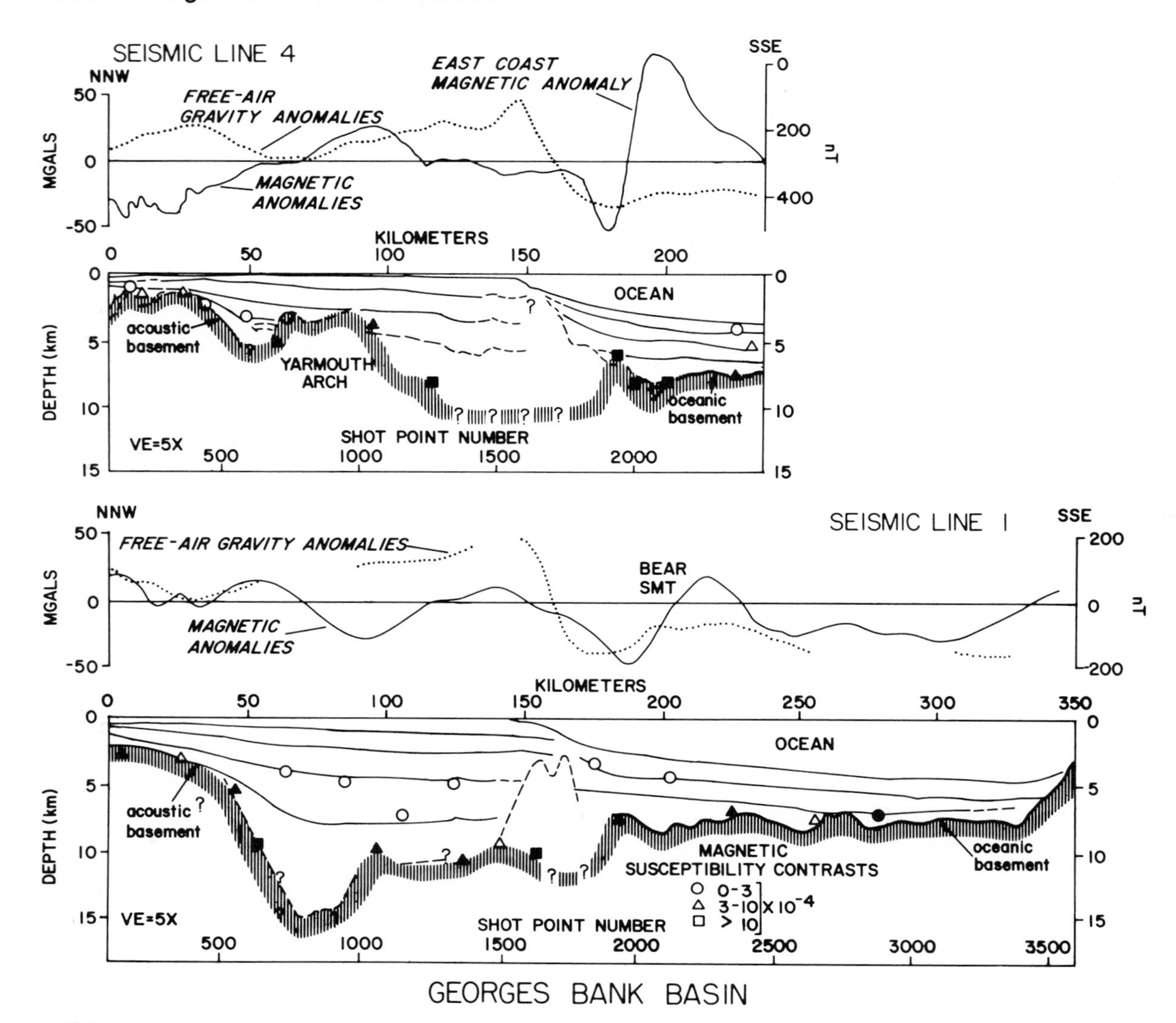

FIG. 6. The comparison between magnetic depth-to-basement estimates and multichannel seismic lines 4 and 1 (Schlee et al, 1976; Grow and Schlee, 1976) over the Georges Bank basin. See Figure 7b for their locations. The heavy lines are prominent seismic horizons; these lines are dashed where the horizon is indefinite. The magnetic basement is shaded. Where the depth to basement is unknown, a set of question marks (?) is used. The reliable magnetic depth estimates are shown. Depth estimates that have more than 50 solutions have been filled in. Seismic line 4 did not cross the area for which magnetic depth estimates were calculated. The magnetic depth estimates have been projected onto the seismic line in the appropriate positions relative to the magnetic anomalies.

tween rock types, will look like edges or dikes and reliable depth estimates can be calculated by the above method. Moderately long wavelength variations (i.e., those that take place gradually over a long distance—for example, the susceptibility variations associated with a broad metamorphic contact zone) could produce erroneously deep depth estimates. Fortunately, comparison with seismic reflection data and with the many adjacent magnetic profiles for consistent solutions, plus the existence of reliable shallow depth estimates, will usually result in the discarding of most of these erroneous deep estimates. Depth estimates greater than approximately 20 km can be immediately discarded because this depth is probably below the Curie point isotherm (Vacquier and Affleck, 1941).

Depth estimates have been categorized by susceptibility contrasts and the number of solutions that determined the estimate. Susceptibility contrasts have been designated as weak for 0-3 $\times$ 10^{-4} (C.G.S. units), moderate for 3-10 $\times$ 10^{-4}, and strong for $>10 \times 10^{-4}$. This division tends to separate the sources within the sediments ($<3 \times 10^{-4}$) from the basement sources and to separate the sources at the edges of the basins ($>10 \times 10^{-4}$) from the other sources. The depth estimates with greater than 50 solutions have been indicated by filled symbols (Figs. 6, 8, 10, 11, 13) to earmark the sources of the most prominent anomalies.

The depth analyses just described were carried out using the digital data on the entire 185,000 km track coverage of the 1975 LKB Resources, Inc.

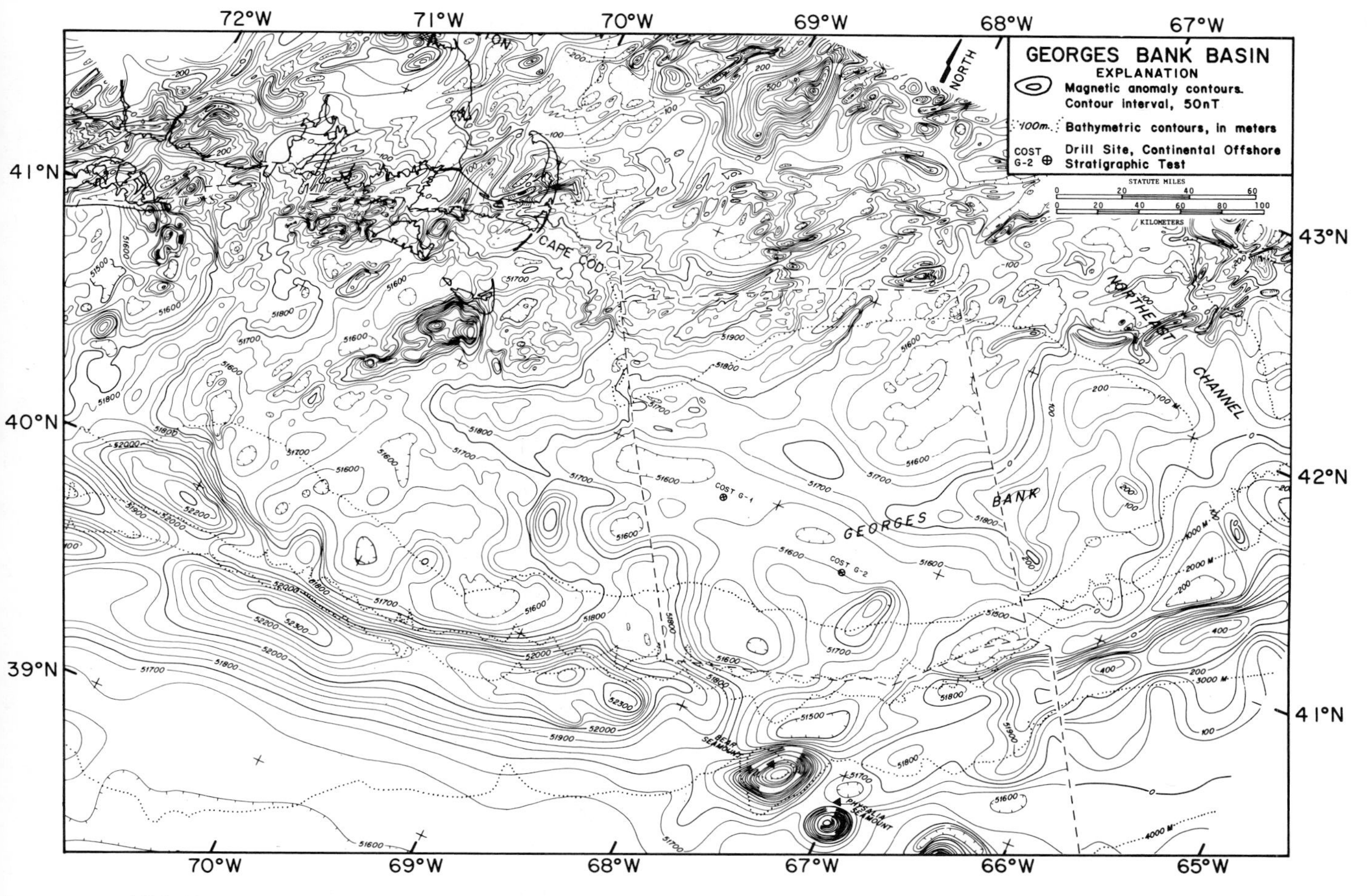

FIG 7A. Magnetic anomaly contour map for the Georges Bank basin region. Contour interval is 50 nT.

aeromagnetic survey and the 1966 Aero Services aeromagnetic survey. Interpretation of these depth analyses included the comparison of adjacent profiles for compatibility. The validity of the above method and its assumptions for this area were checked by examination of the results from adjacent magnetic profiles (Figs. 7b, 9b, 12b, 14b), the linearity of the magnetic anomalies on the contour map (Figs. 3, 7a, 9a, 12a, 14a), and a comparison between the magnetic data and the seismic reflection profiles (Figs. 6, 8, 10, 11, 13).

Example: Seismic Line 9

The results of the magnetic depth estimate calculations in the vicinity of seismic line 9 (Figs. 4, 9) provide an example of the general method utilized for all of the magnetic profile interpretations and for the integration with the seismic profiles. Seismic line 9 crosses the Long Island platform and the northern end of the Baltimore Canyon trough, traversing both shallow and deep basement structures.

The magnetic depth estimates along the flight lines adjacent to seismic line 9 are shown on Figure 4b. These depth estimates are assumed to be associated with basement (in this region, basement probably corresponds to the base of the Mesozoic sediments). They represent the set of depth estimates which are compatible with the seismic reflection data (Fig. 4a) and which are consistent on adjacent magnetic profiles (Fig. 4b). The agreement in the areal location of depth estimates on adjacent profiles is typical for the whole margin. The scatter in the estimated depths is commonly found to be about 20%, probably reflecting noise in the data, influence from adjacent magnetic anomalies, non two-dimensionality of the sources, and other variations in the sources from our assumed dike or interface models. The depth estimates (Fig. 4b) were averaged and rounded off to the nearest kilometer to produce the depth-to-basement location maps presented here (Figs. 4c, 7b, 9b, 12b, 14b). The quality of the depth estimates and desired results did not warrant a sophisticated smoothing of the results.

The comparison between the magnetic depth estimates and the seismic reflection data (Fig. 4a) was produced by projecting the depth estimates onto the seismic line, using the trends in the depth locations from Figure 4b. The troughs centered at shotpoints 500 and 1500 are clearly distinguished by both the magnetic depth estimates and the seismic reflection data. The peak in the magnetic basement near shotpoint 1600 is characteristic of the indications of

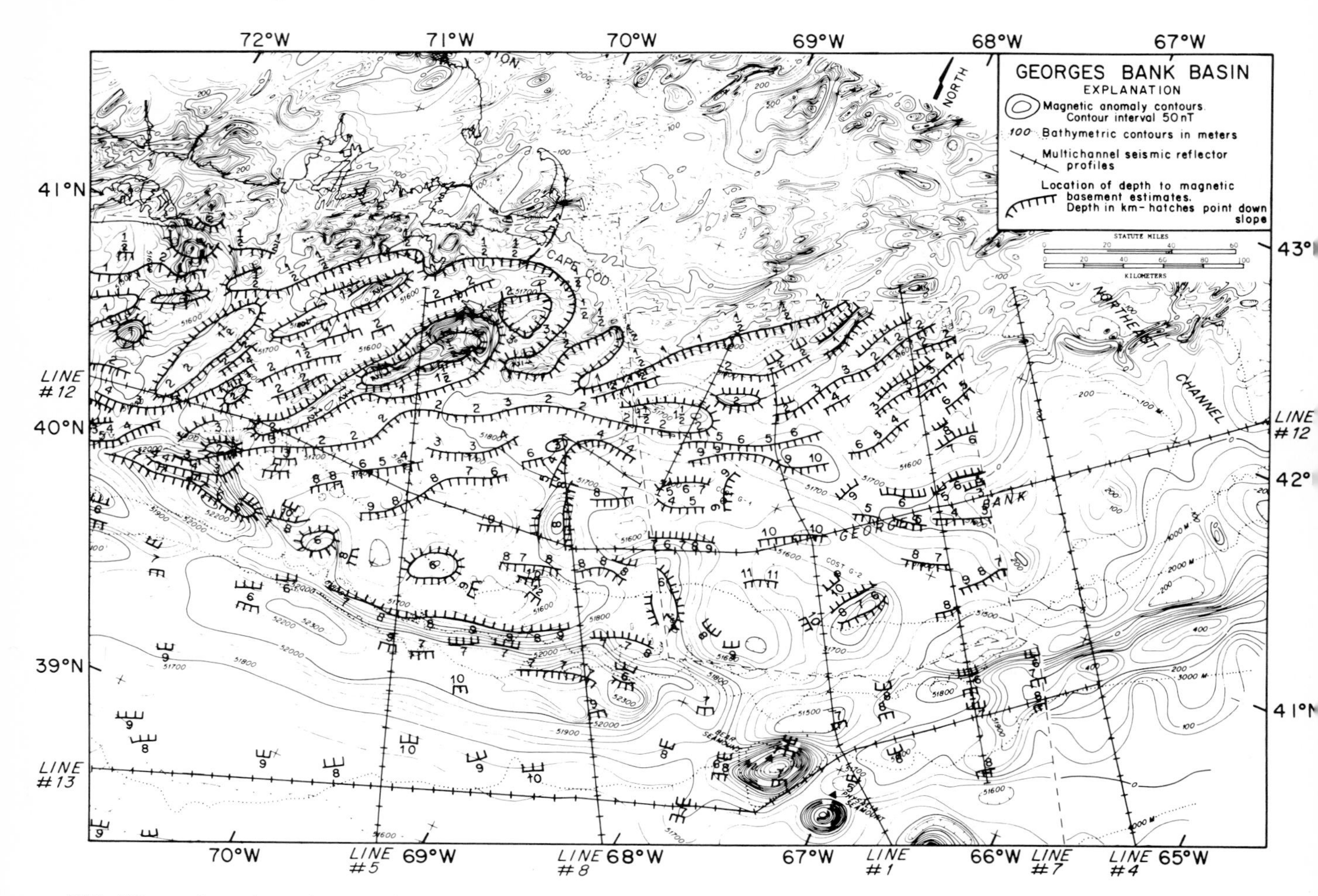

FIG. 7B. Location of magnetic depth estimates for the Georges Bank basin. The background contours are the 50-nT magnetic anomaly contours. Multichannel seismic lines are identified by numbers along the map margins.

a basement high beneath the East Coast magnetic anomaly which we found along most of the margin. The seismic structures directly above this peak have been inferred to be caused by a carbonate bank or reefal deposits (Schlee et al, 1976), and probably represent acoustic basement in this region on most of the multichannel seismic profiles.

The magnetic anomalies which are associated with the magnetic depth estimates (Fig. 4c) are very linear seaward of shotpoint 1100 on seismic line 9, while the anomalies landward of this point (on the Long Island platform) are more three dimensional. The more chaotic magnetic anomaly pattern on the platform is typical of the other platforms, and only the broader basement features such as the trough at shotpoint 500 can be identified with reasonable confidence. Within the major basins, the longer wavelength magnetic anomalies have fewer associated depth estimates, and it is easier to interpret them in terms of the source structure.

RESULTS

Our study of the U.S. Atlantic continental margin has been sub-divided into a series of areas, focusing on the individual basins and platforms which are found along the margin (Figs. 1, 5). These regions have been identified as the Georges Bank Basin (Fig. 7), the Long Island platform (Fig. 9), the Baltimore Canyon trough (Fig. 12), and the Carolina platform (Fig. 14). Our analyses of the Blake Plateau Basin are not complete, but they will be included in our discussion using preliminary results with some of the area included in Figure 14. For each area, we will discuss the magnetic anomaly lineations, compare the seismic reflection data with the magnetic depth estimates, compare the magnetic depth estimates on adjacent profiles, and, finally, discuss the depth to basement contour maps.

Comparison Between Magnetic Profiles

The agreement among depth estimates on adjacent magnetic profiles (Figs. 7b, 9b, 12b, 14b) is very good. This consistency of the depth estimates adds to the reliability of the results. The linearity of these sources and the linearity of the magnetic anomalies (Figs. 3, 7a, 9a, 12a, 14a) show that most of the sources are longer in extent than their depth, so the two-dimensionality assumption is reasonably accurate. The depth estimates near the ends of the anomalies commonly have to be discarded and a rough correction has to be made for the anomalies that were crossed obliquely by the flight lines. The variation in depth between an estimate for the same source on adjacent profiles may reflect the limits of the resolution of the method, the influence of noise on the estimates, the fact that the anomalies are not two dimensional, and other deviations in the actual source from our assumed dike or interface models. The depth estimates also indicate that many of the "edges" are not horizontal. Because these depth estimates are only for the top edges of the source blocks, we have little information concerning the depth to basement between the edges. The depths on these maps are given to the nearest 1/2 kilometer for the shallower estimates and to the nearest kilometer for the deeper estimates, reflecting the lack of resolution for these deeper estimates.

Comparison Between Magnetic and Seismic Data

The magnetic depth-to-basement estimates along the seismic lines were compared with the seismic depth sections (Figs. 6, 8, 10, 11, 13) to minimize the ambiguities in the magnetic depth estimates. On the landward side of the deep, sediment-filled basins, multichannel seismic reflection profiles across the margin (Grow and Schlee, 1976; Schlee et al, 1976) show an acoustic basement that appears to be composed of large blocks of continental(?) crust rather than many intrusive bodies. This interpretation would be compatible with the assumption that major magnetic anomaly sources are the edges of bodies. On the seaward side of the East Coast magnetic anomaly, the hyperbolic acoustic signature of the acoustic basement (Schlee et al, 1976) and the seismic refraction data from the region (Emery et al, 1970; Mayhew, 1974; Sheridan, 1974) indicate that the acoustic basement is oceanic crust. In general, the dike solution depth estimates are more compatible with the seismic data for the oceanic crust because the source layer is fairly thin. In the areas where the acoustic basement was thought to be continental crust, the "edge" solutions of the magnetic depth estimates having reasonably high susceptibility contrasts were consistent with the location of acoustic basement.

There are some features that are found on the magnetic/seismic depth comparisons for all the seismic profiles. A consistent set of weak susceptibility solutions within the sediment column is associated with prominent seismic horizons (e.g., the long shallow horizon seen on line 9 [Fig. 8]). In a previous

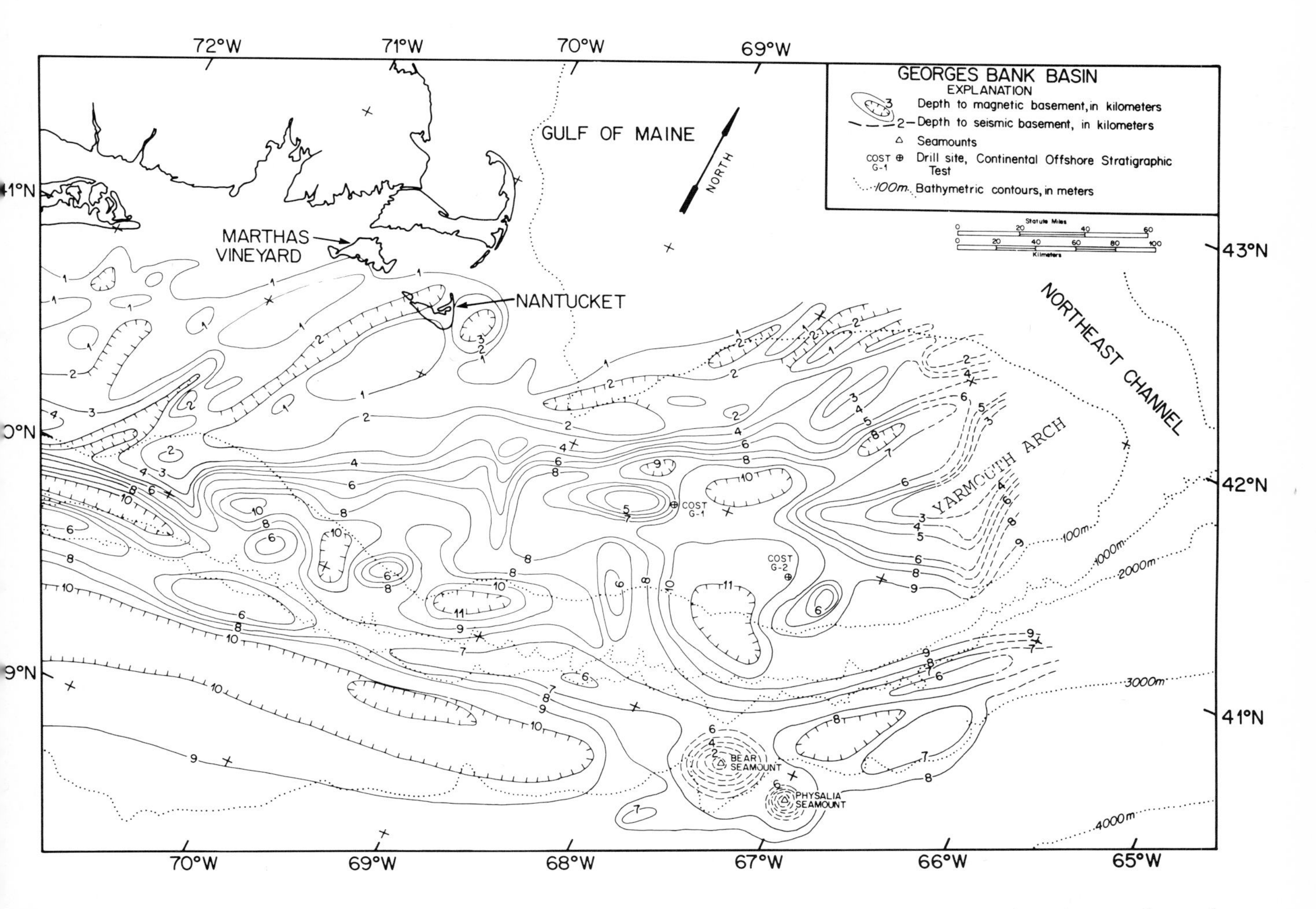

FIG. 7C. Depth-to-basement contours for the Georges Bank basin based on the magnetic depth estimates shown in Figure 7B. The depth-to-basement contours are in kilometers; the contour interval is 1 km. Hachures point down slope.

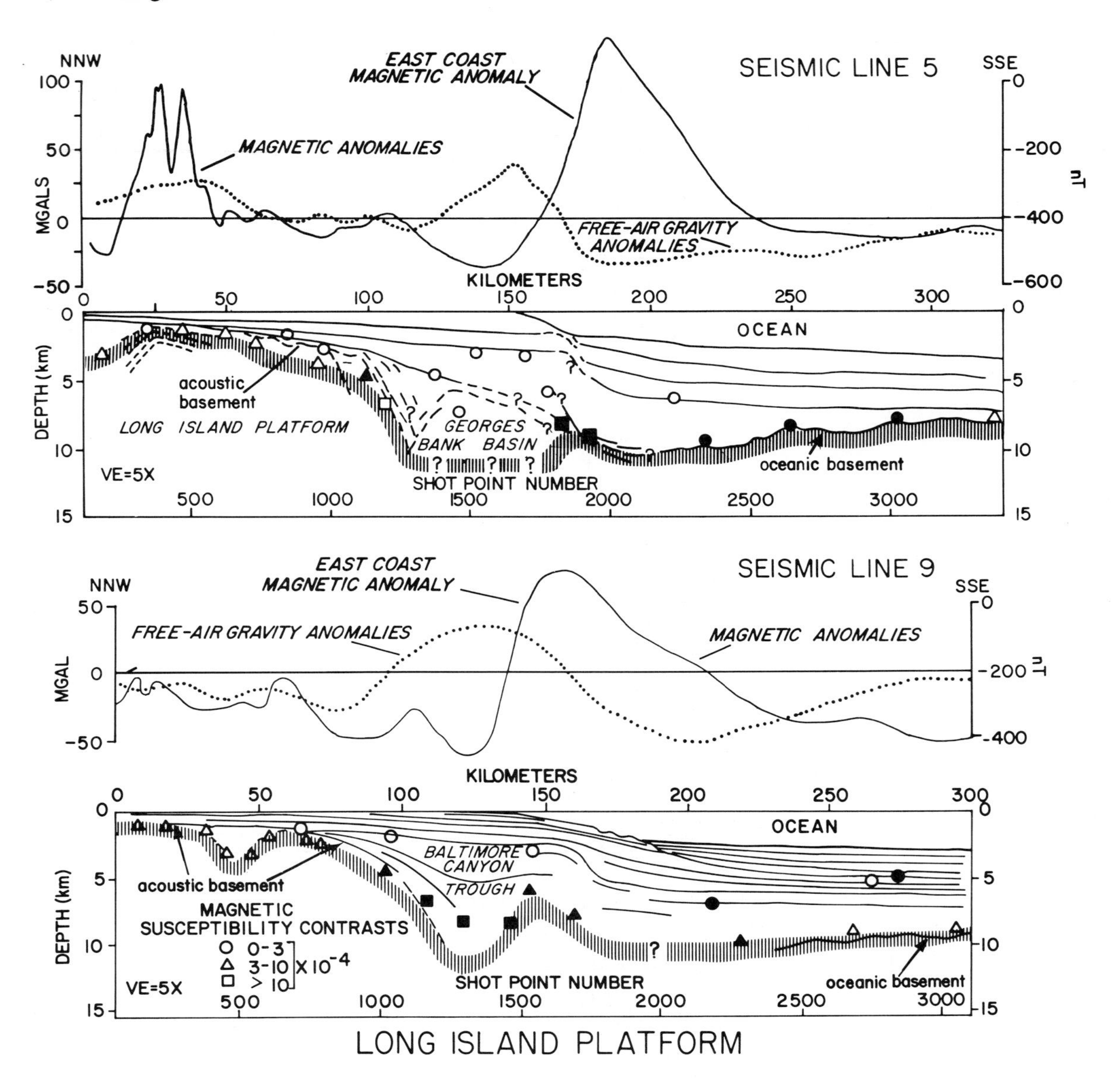

FIG. 8. Comparison between magnetic depth-to-basement estimates and multichannel seismic lines 5 and 9 (Grow and Schlee, 1976) over the Long Island platform. See Figure 9B for locations. Symbols are the same as in Figure 6.

study of high-resolution aeromagnetic data on the U.S. Atlantic continental margin, Steenland (1977) found a very shallow magnetic basement. This basement may have been the same layer as the one associated with the weak susceptibility contrasts that are found within the sediment column. The small susceptibility contrasts of these shallow depth estimates suggest that they are not produced by crystalline/volcanic rock. The consistent appearance of these solutions on adjacent magnetic profiles and their association with prominent seismic horizons suggest that they are real, but further discussion of them would be extraneous to the subject of basement structures.

Associated with the landward edges of the deep, sediment-filled troughs are intermediate to strong susceptibility contrast solutions that have depths which are progressively deeper towards the center of the basins. On the seaward side of the East Coast magnetic anomaly, where seismic reflection data (Grow and Schlee, 1976; Schlee et al, 1976) and seismic refraction data (Mayhew, 1974; Sheridan, 1977) indicate that the acoustic basement is probably oceanic crust, the associated magnetic depth estimates show weak to moderate susceptibility contrasts. Between the two above-mentioned zones are the deep, sediment-filled basins and the East Coast anomaly. Many deep magnetic depth estimates (>10 km) are found in this region, and a narrow set of shallow depth estimates (6 to 8 km) is obtained from beneath the East Coast magnetic anomaly. These facts indicate a peak in the magnetic basement between the oceanic crust and the deep troughs (Figs. 6, 8, 10, 11, 13). A high susceptibility contrast ($>10 \times 10^{-4}$) is usually associated with the landward edge of this magnetic high. The seismic profiles do not show any identifiable structures below this magnetic basement high, but in most places, above this high is a disturbed seismic zone that has been interpreted as carbonate rocks (Schlee et al, 1976).

Depth-to-Basement Maps

Depth-to-basement maps for the Atlantic margin were compiled for each of the basins and platforms (Figs. 7c, 9c, 12c, 14c). Although these maps are based on the depth-to-magnetic basement estimates, the control provided by the magnetic/seismic comparisons in the selection of reliable magnetic depth estimates results in a map of basement depths that is compatible with the available multichannel seismic reflection data. The controls for these contours are the depth-to-magnetic basement locations and multichannel seismic reflection profiles shown on Figures 7b, 9b, 12b, and 14b. For each of the areas to be discussed, the depth-to-basement maps (Figs. 7c, 9c, 12c, 14c) are used in conjunction with the magnetic/seismic comparisons (Figs. 6, 8, 10, 11, 13), providing an aerial and cross-sectional view of each region.

Georges Bank Basin

The Georges Bank sediment pile masks a large variation in basement depth. The sediment pile overlaps the shallow basement of the Long Island platform (Garrison, 1970) on the northwest, and that of the Gulf of Maine (Ballard and Uchupi, 1972) on the north. The main part of the bank is underlain by a deep trough (Schultz and Grover, 1974) with possible sediment thicknesses of more than 8 km (Schlee et al, 1976). The Yarmouth arch protrudes into the trough from the northeast, splitting it into two basins. On the seaward side of Georges Bank, oceanic crust at a depth of 7 to 8 km extends from the deep ocean basin almost to the East Coast magnetic anomaly (Schlee et al, 1976). Two published multichannel seismic reflection profiles (Fig. 6) cross Georges Bank (USGS seismic line 1, Schlee et al, 1976; USGS seismic line 4, Grow and Schlee, 1976). Seismic line 4 crosses the northeast end of the bank over the Yarmouth arch, and seismic line 1 crosses the central part of Georges Bank over the deepest part of the Georges Bank Basin.

The magnetic anomalies over Georges Bank (Figs. 3, 7a) reflect the variations in basement depth (Kane et al, 1972) and demonstrate the linearity of many of the basement features. Shallow magnetic depth estimates (<4 km) are obtained over the landward edges of the Georges Bank Basin. Intermediate magnetic depth estimates (4 to 6 km) are associated with the Yarmouth arch and several smaller intrusive(?) bodies within the basin. Magnetic depth estimates of 7 to 8 km are generally found seaward of the East Coast magnetic anomaly, over oceanic crust. The

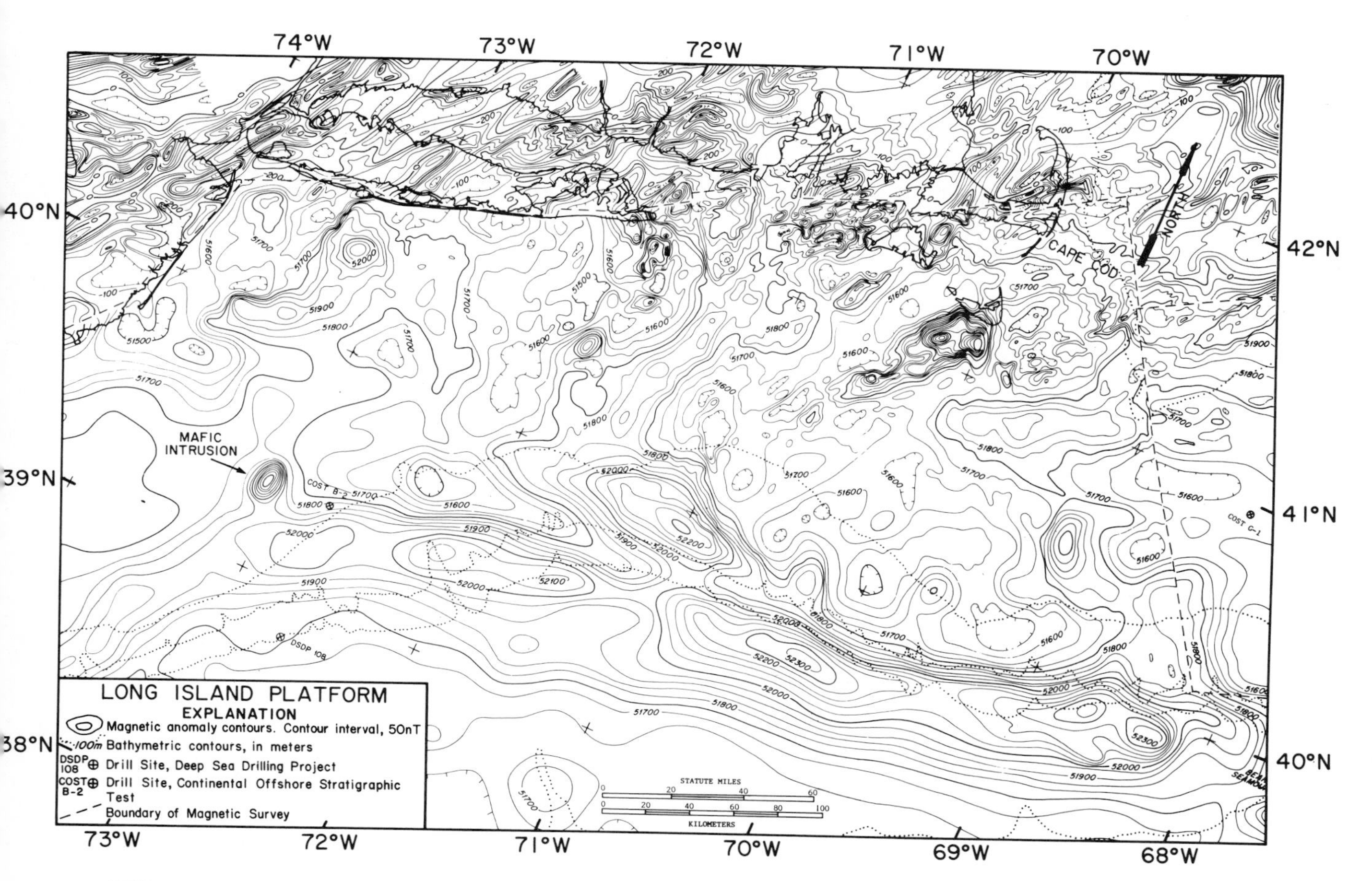

FIG. 9A. Magnetic anomaly contour map for the Long Island platform region. Contour interval is 50 nT.

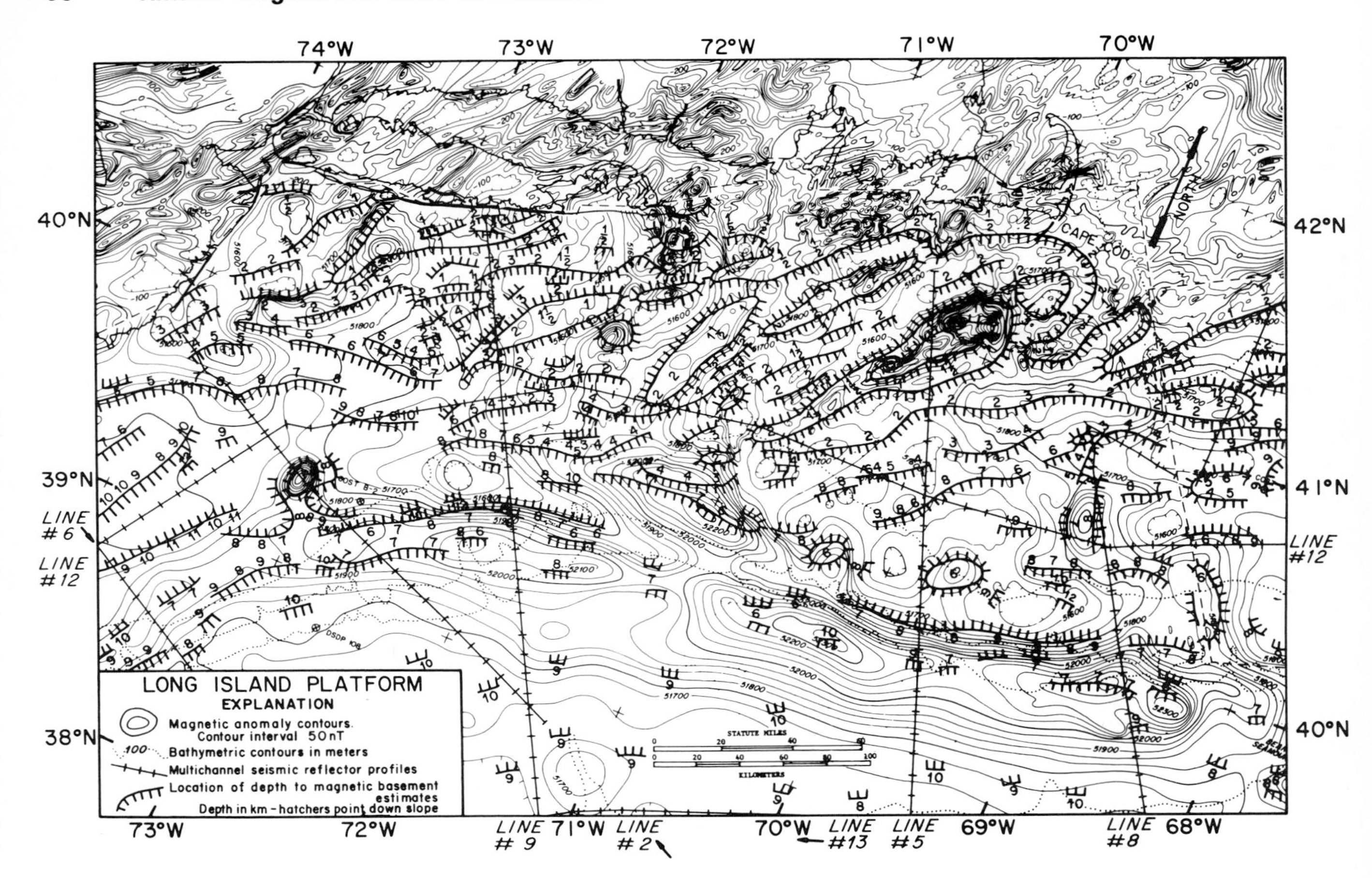

FIG. 9B. Location of magnetic depth estimates for the Long Island platform. Symbols are the same as in Figure 7b.

deepest magnetic depth estimates (>10 km) are located just landward of the East Coast magnetic anomaly.

The structure of the Georges Bank Basin (Figs. 1, 7) trends 030° to 040°, roughly parallel with the Baltimore Canyon trough and the general trend of the western Atlantic continental margin. The northwest side of the trough appears to be composed of block-faulted basement that rapidly deepens between the Long Island platform (<3 km) and the axis of the trough. The southwestern end of the trough terminates near 40°N lat., 70°W long., at the East Coast magnetic anomaly. Two magnetic features at this point may be caused by intrusive bodies (buried seamounts?) that reach a minimum depth of about 6 km. Northeast of these intrusive bodies, the trough deepens abruptly at about 40.5°N lat., 69°W long. The trough then branches—a shallow part going on the north side of the Yarmouth arch, and the deeper part continuing south of the arch towards the Scotian Basin. Near 40.8°N lat., 67.3°W long., there appears to be another buried seamount having a minumum depth of about 6 km.

The seaward edge of the Georges Bank Basin is marked by the East Coast magnetic anomaly. This anomaly is discontinuous at the southwestern end of the Georges Bank Basin and where Bear Seamount intersects the anomaly. Northeast of Bear Seamount, the East Coast magnetic anomaly is fairly narrow (~35 km) and separates the shallow (7 to 8 km) oceanic crust from the deep basin (8 to 12 km). The anomaly southwest of Bear Seamount is much broader (~50 km) and trends approximately east to west. A magnetic basement high that has a large susceptibility contrast ($>20 \times 10^{-4}$) is associated with the east to west part of the East Coast magnetic anomaly. This high has a minimum depth of 6 to 7 km and a deeper (>10 km) basement to the north and south.

Long Island Platform

The Long Island platform has a fairly shallow acoustic basement, which is broken into a series of ridges and troughs. Single-channel seismic reflection profiles show that the top of the basement dips seaward (Garrison, 1970; McMaster, 1971; Emery and Uchupi, 1972). The existence of a few small basins was suggested by Ballard and Uchupi (1972); these basins can be seen on multichannel seismic reflection profiles (Grow and Schlee, 1976; Schlee et al, 1976). The platform is crossed by only two published multichannel seismic profiles (Fig. 8); therefore, a reasonable determination of the orientation of the troughs or of the extent of the platform by this means is precluded.

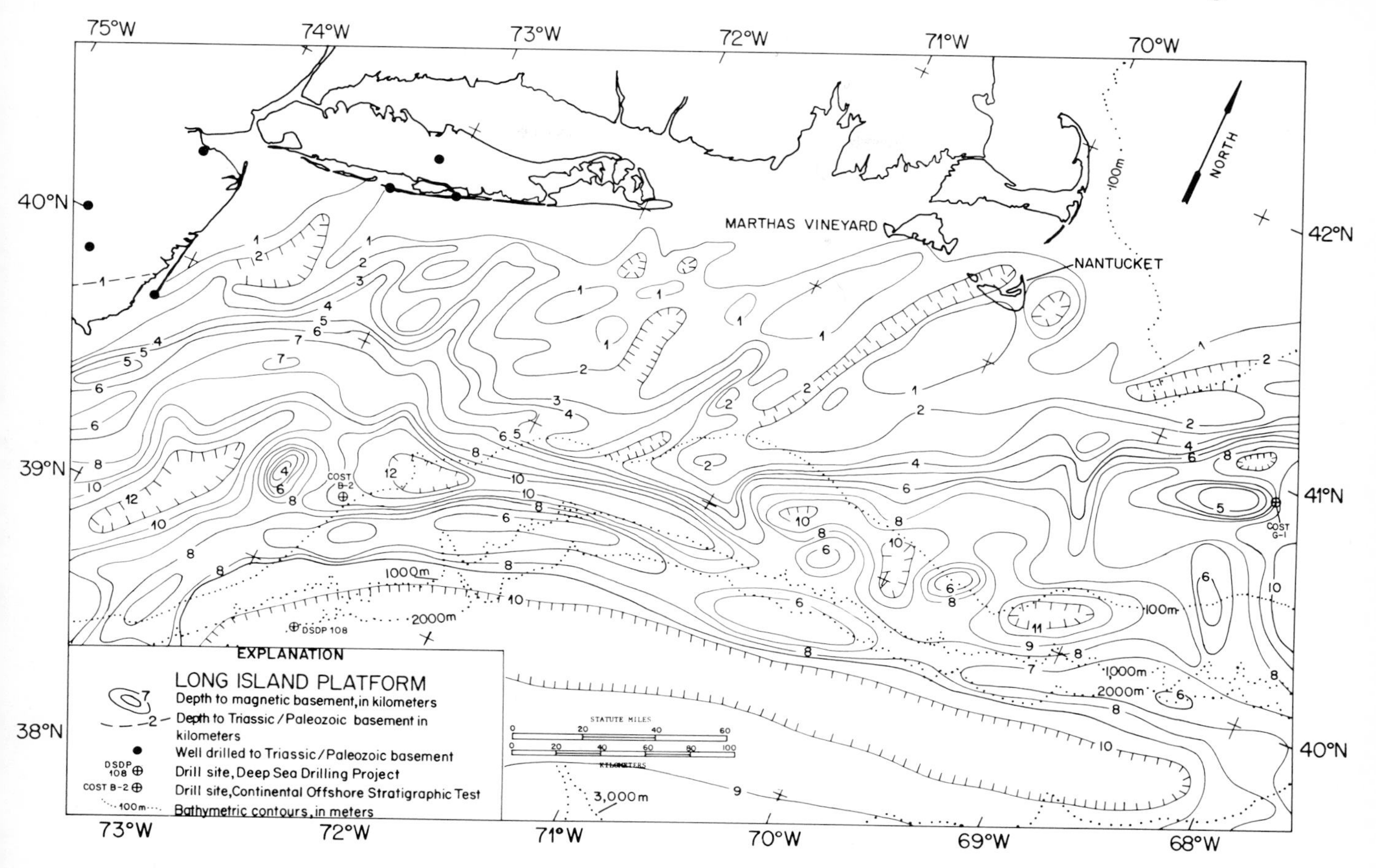

FIG. 9C. Depth-to-basement contours for the Long Island platform based on the magnetic depth estimates shown in Figure 9B. The symbols are the same as in Figure 7C.

This region of shallow basement is clearly distinguished by the magnetic field (Figs. 3a, 9a) which has many more short wavelength anomalies than does that over the deep sediment-filled basins. There are at least four small basins which trend parallel with the Baltimore Canyon trough and the Georges Bank Basin (Fig. 9). Seismic line 5 (Fig. 8) crossed two of these troughs, and seismic line 9 (Fig. 8) crossed one of them. The generally high-amplitude, short-wavelength character of the magnetic anomalies (Figs. 3, 9a) is fairly typical of the continental crust under New England. The platform probably is composed of blocks of continental crust that were compressed by the Appalachian orogeny and then faulted apart by the Triassic and Jurassic rifting.

At least two of the troughs in the Long Island platform are probably Triassic and Jurassic basins. The broad magnetic low at about 40.5°N lat., 73.5°W long. (Fig. 9a) is a continuation of the low over the Connecticut Valley Triassic and Jurassic basin, indicating that this basin continues beneath Long Island. Well data also show the location of this basin beneath Long Island (Wheeler, 1938). A similar magnetic signature is associated with the basin between Martha's Vineyard and Nantucket; Minard et al (1974) suggested that this basin was Triassic. The other small troughs also may be Triassic and/or Jurassic in age.

A rapid increase in the depth-to-magnetic basement (Fig. 9) marks the southern and southeastern edges of the Long Island platform. South of Long Island, this edge is at about 40°N lat. where the depth increases from about 4 km to more than 8 km into the Baltimore Canyon trough. The change in depth along this zone is quite variable because this is where the basement highs of the Long Island platform are truncated and the small basins broaden and deepen into the Baltimore Canyon trough. Farther east, at about 40°N lat., 71°W long., a large magnetic high is associated with the southwestern end of the Nantucket basement high. This magnetic anomaly (Fig. 9a), which appears to be an offset part of the East Coast magnetic anomaly, separates the Long Island platform (<3 km depth) from the northeastern end of the Baltimore Canyon trough (>8 km depth). This region has the most rapid drop in basement depth from the Long Island platform to an adjacent basin. Along the southeastern edge of the Nantucket basement high, the deepening of the basement is more gradual into the Georges Bank Basin. The drop takes place in four steps (or blocks). The two upper steps closest to the Nantucket basement high have large linear magnetic anomalies associated with them.

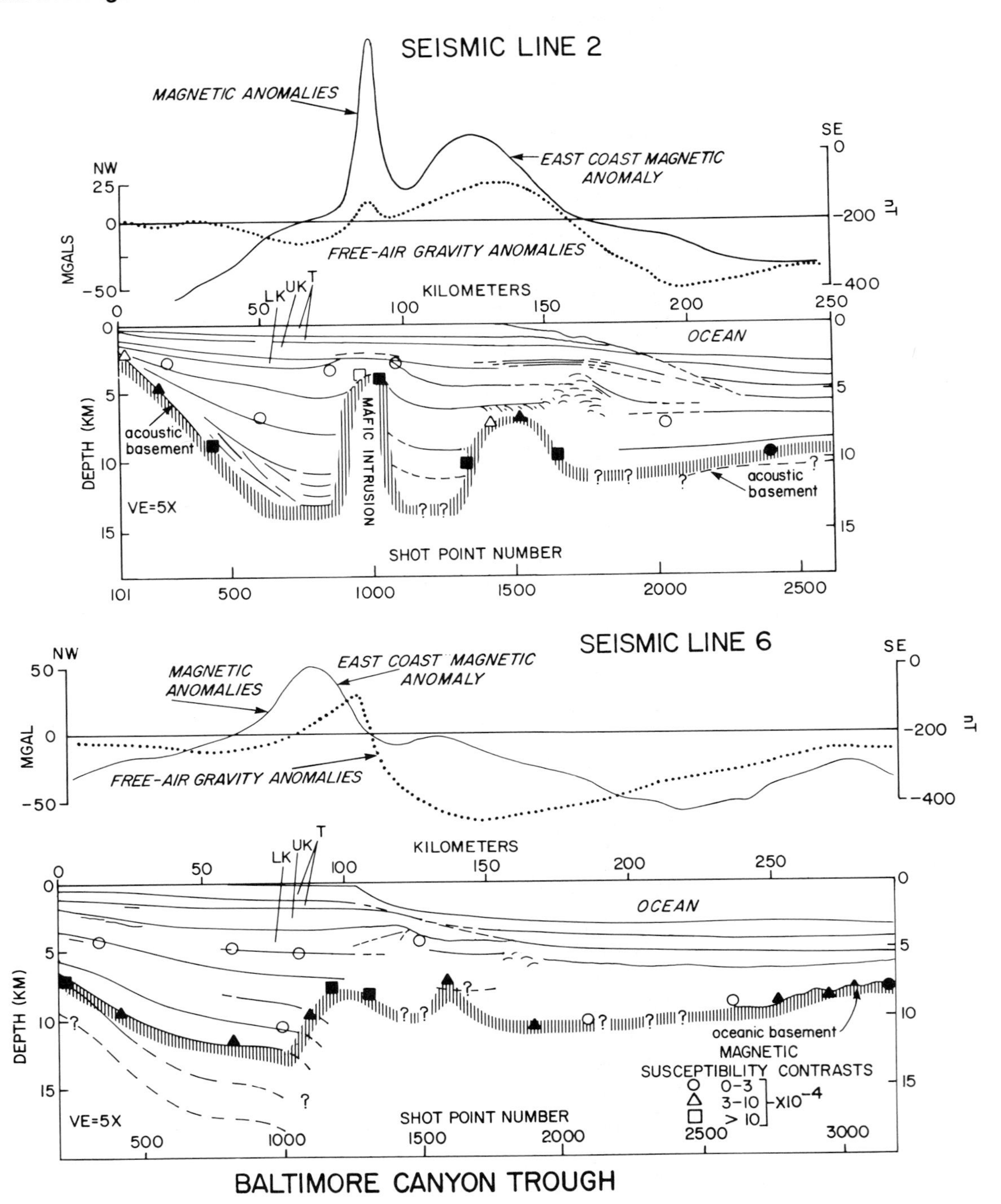

FIG. 10. The comparison between magnetic depth-to-basement estimates and multichannel seismic lines 2 and 6 (Schlee et al, 1976; Grow and Schlee, 1976) over the Baltimore Canyon trough. See Figure 12B for their locations. The symbols are the same as in Figure 6.

Baltimore Canyon Trough

The continental margin between Long Island and Cape Hatteras contains two major offsets in the continental crust that form a large embayment into continental North America. Along the coastline of New Jersey and Delaware, drill holes reached Triassic or Paleozoic basement at a depth of 1.5 to 2 km (Brown et al, 1972). Beneath the continental shelf is a deep, sediment-filled trough (Drake et al, 1968; Emery et al, 1970) called the Baltimore Canyon trough. Multichannel seismic reflection profiles (Figs. 10, 11) indicate that the basement deepens into the Balitmore Canyon trough in a series of faulted blocks (Schlee et al, 1976). This trough contains more than 12 km of sediments and is oriented along 030°.

Magnetic anomalies over the Baltimore Canyon Trough (Figs. 3, 12a) in general have very broad wavelengths, suggesting a source depth of more than 8 to 10 km. The exceptions are a few circular highs

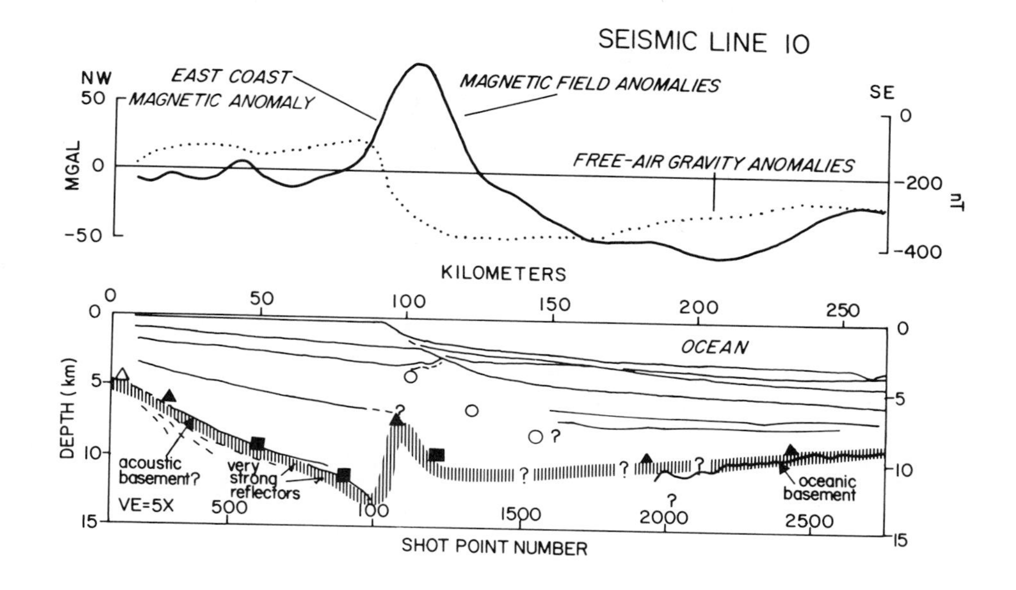

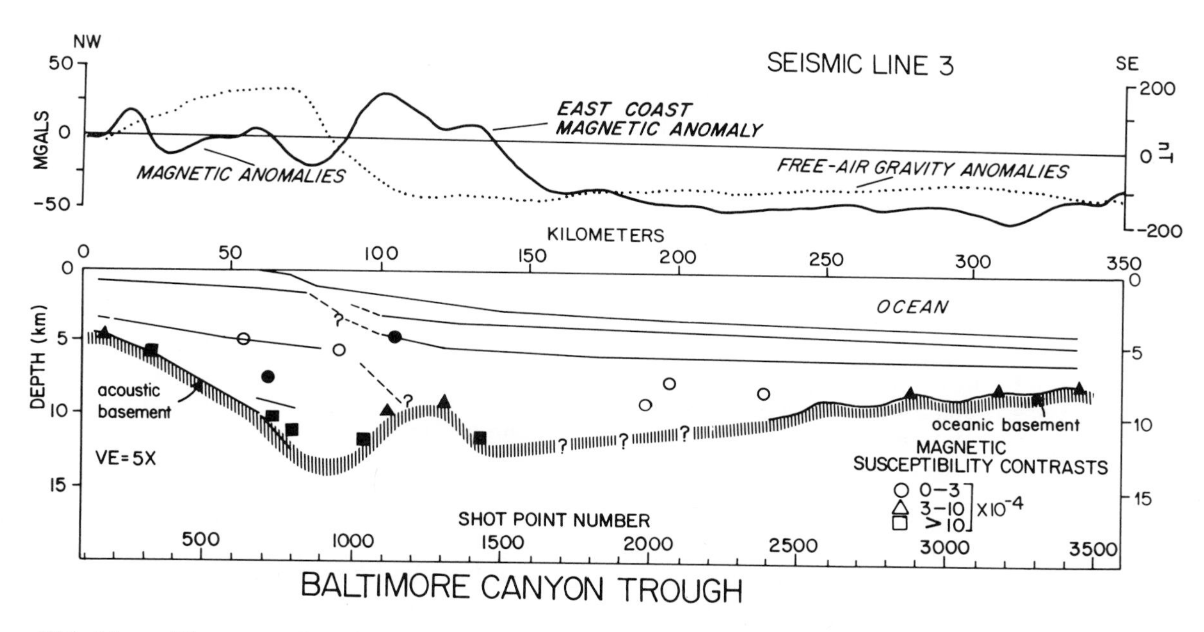

FIG. 11. The comparison between magnetic depth-to-basement estimates and multichannel seismic lines 10 and 3 (Schlee et al, 1976) over the Baltimore Canyon trough. See Figure 12B for their locations. Symbols are the same as in Figure 6.

which are probably intrusive bodies. The magnetic depth estimates (Figs. 10, 11, 12b) get progressively deeper in the seaward direction until reaching the East Coast magnetic anomaly. Nearshore, the magnetic basement drops off rapidly from about 2 km (in agreement with drill-hole information) to about 6 km near the beginning of the multichannel lines. These estimates reach a maximum depth of greater than 10 km just landward of the East Coast magnetic anomaly.

A few intrusive bodies are associated with the Baltimore Canyon trough. The most conspicuous of these bodies is the mafic intrusion at about 39.3°N lat., 73°W long. (Fig. 12a; Schlee et al, 1976). Minimum magnetic depth estimates are about 3.5 km for this body, which has a susceptibility contrast of about 25×10^{-4} with the surrounding sediments. The other major intrusive bodies indicated by the magnetic data are near Cape Charles and Cape Henry. They are near the edge of the Baltimore Canyon trough and may be intruded into older continental crust.

The magnetic depth estimates are of particular interest near the shelf break and the East Coast magnetic anomaly because the seismic technique has not been able to penetrate more than a few kilometers in this region. Several depth estimates have been obtained in this region, ranging from about 7 km to

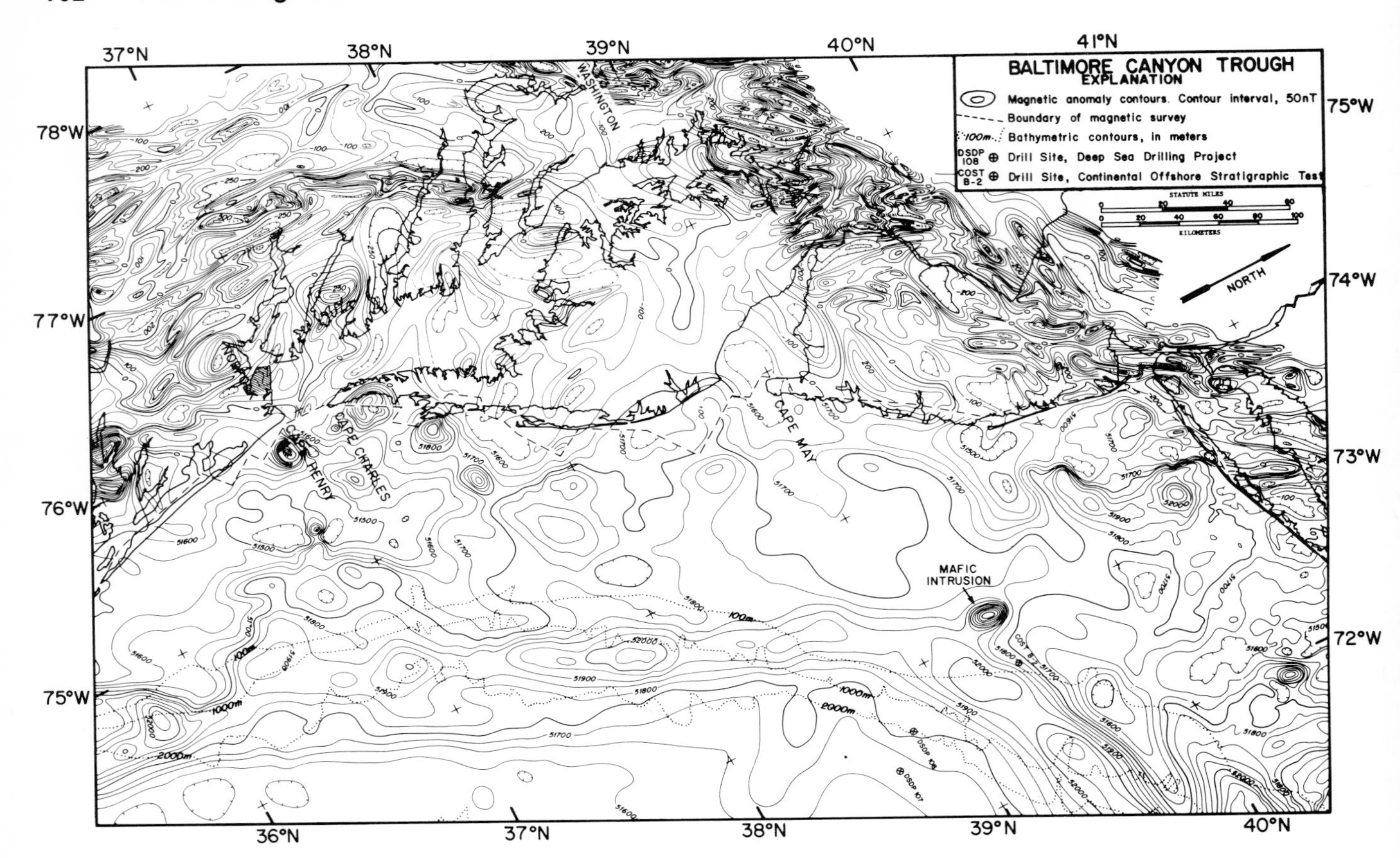

FIG. 12A. Magnetic anomaly contour map for the Baltimore Canyon trough region. Contour interval is 50 nT.

more than 12 km; the deeper depth estimates are well landward or seaward of the East Coast magnetic anomaly. Many of the shallower depth estimates are at about the same location as the truncation of prominent seismic reflectors (Figs. 10, 11). The susceptibility contrasts are about 10×10^{-4}, lower than the contrast for the other intrusive bodies ($>25 \times 10^{-4}$). Basement is at a depth of about 10 to 12 km on the seaward side of the East Coast magnetic anomaly, and the magnetic depth estimates suggest that the basement shallows to an average depth of 8 km beneath this anomaly.

The Baltimore Canyon trough is composed of three basins (Fig. 12c) of different widths and depths. The broadest and deepest basin is on the north, and the narrowest and shallowest basin is on the south. The landward edges of these basins have the same trend and have fairly abrupt offsets to the west. The most southern part of the Balitmore Canyon trough terminates at the Carolina platform, and the most northern part terminates at the Long Island platform. Each of these offsets probably marks an initial offset in the continental crust as Africa and North America separated about 185 m.y.B.P. These offsets were propagated as transform faults in the adjacent oceanic crust (Fig. 1).

Carolina Platform and Trough

The Carolina platform is the region of shallow, Triassic and older crust which lies between the Baltimore Canyon trough and the Blake Plateau Basin (Fig. 1). The Carolina trough is just seaward of the platform and is the narrowest of the deep sediment-filled basins which are found along the entire U.S. Atlantic continental margin. The East Coast magnetic anomaly marks the seaward edge of the Carolina trough, and the Brunswick magnetic anomaly (Fig. 13), described by Taylor et al (1968) as the inner branch of the East Coast magnetic anomaly, marks the seaward edge of the platform.

Magnetic anomalies (Fig. 14a) and magnetic depth estimates (Fig. 14b) suggest that the crust of the Carolina platform is similar to that of the Long Island platform. A number of Triassic basins have been noted on the platform (King, 1969; Brown et al, 1972; Marine and Siple, 1974), and the whole southern part of the platform appears to be a region of Triassic and Jurassic material similar to the Salsbury Embayment (Popenoe and Zietz, 1977). The shallow depth estimates are confined to the region landward of the Brunswick anomaly (Figs. 13, 14b), a magnetic anomaly which is a prominent peak north of 33.5°N lat., a prominent trough with a small peak between 33.5°N lat., 36.5°W long., and 31°N lat., 80°W long., and a peak and trough west of 80°W long.

At the boundary between the Carolina trough and platform, there is a small basement trough and high which are associated with the Brunswick magnetic anomaly. These two features can be seen in both the

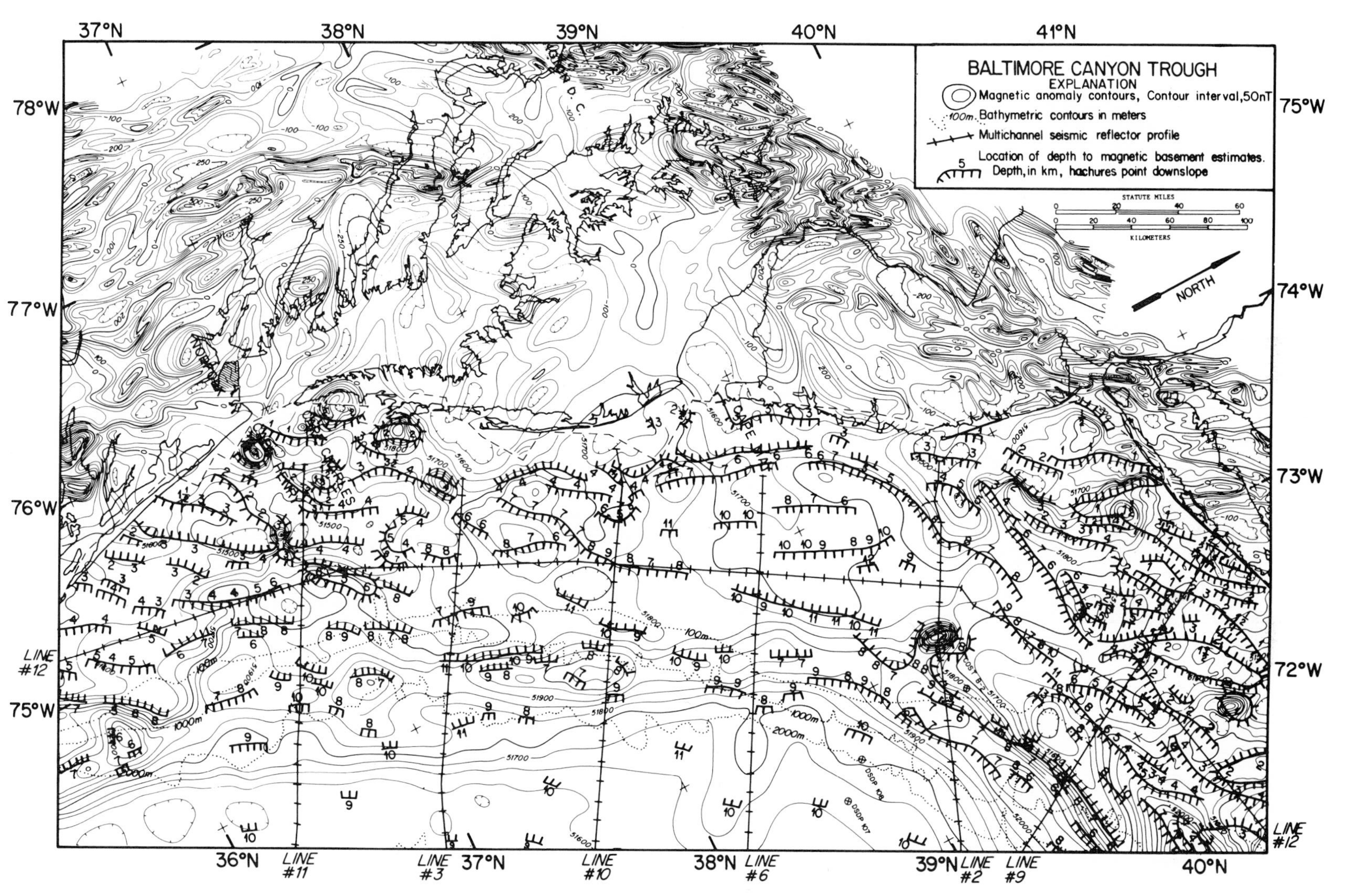

FIG. 12B. Location of magnetic depth estimates for the Baltimore Canyon trough. Symbols are the same as in Figure 7A.

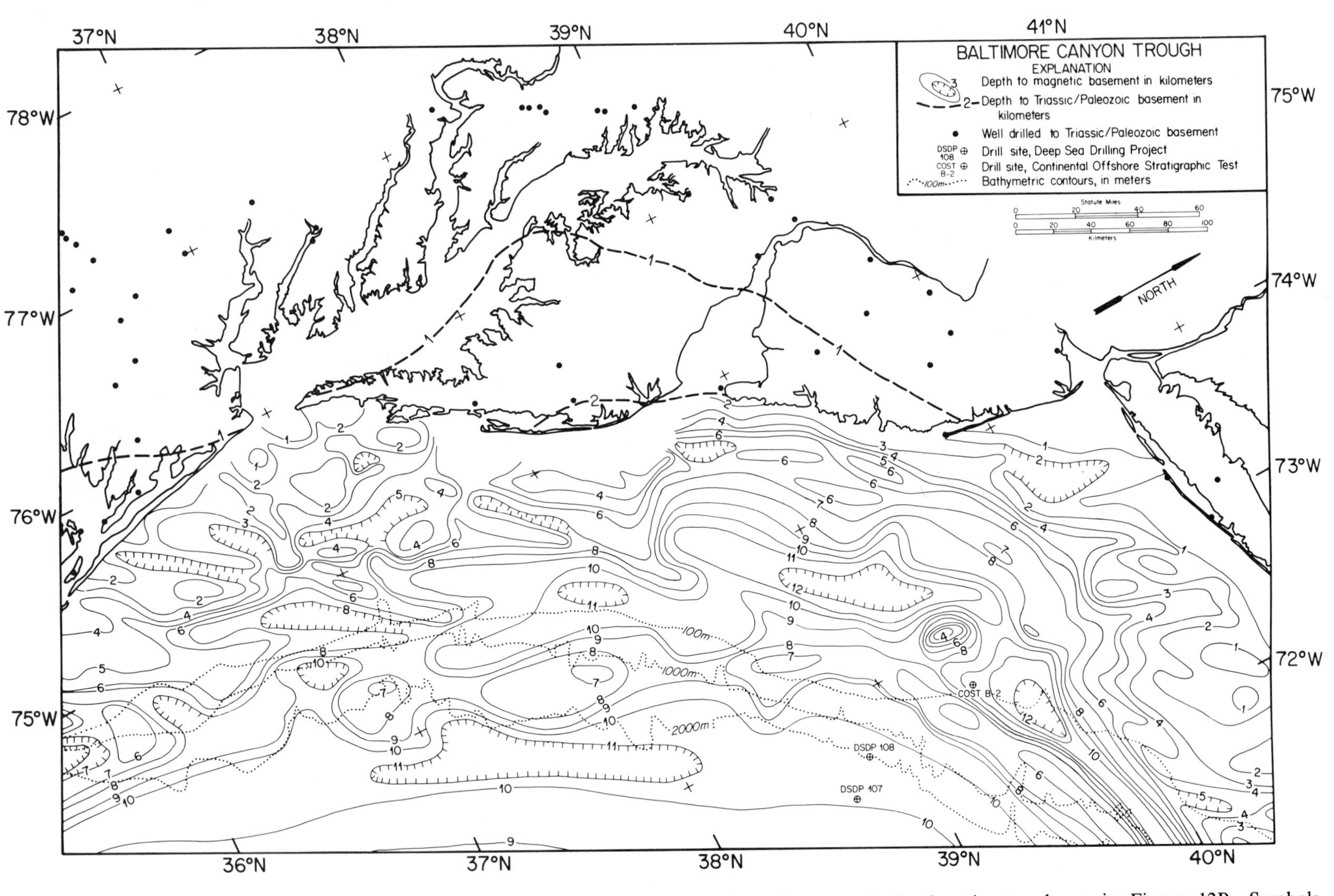

FIG. 12C. Depth-to-basement contours for the Baltimore Canyon trough based on the magnetic depth estimates shown in Figure 12B. Symbols are the same as in Figure 7B.

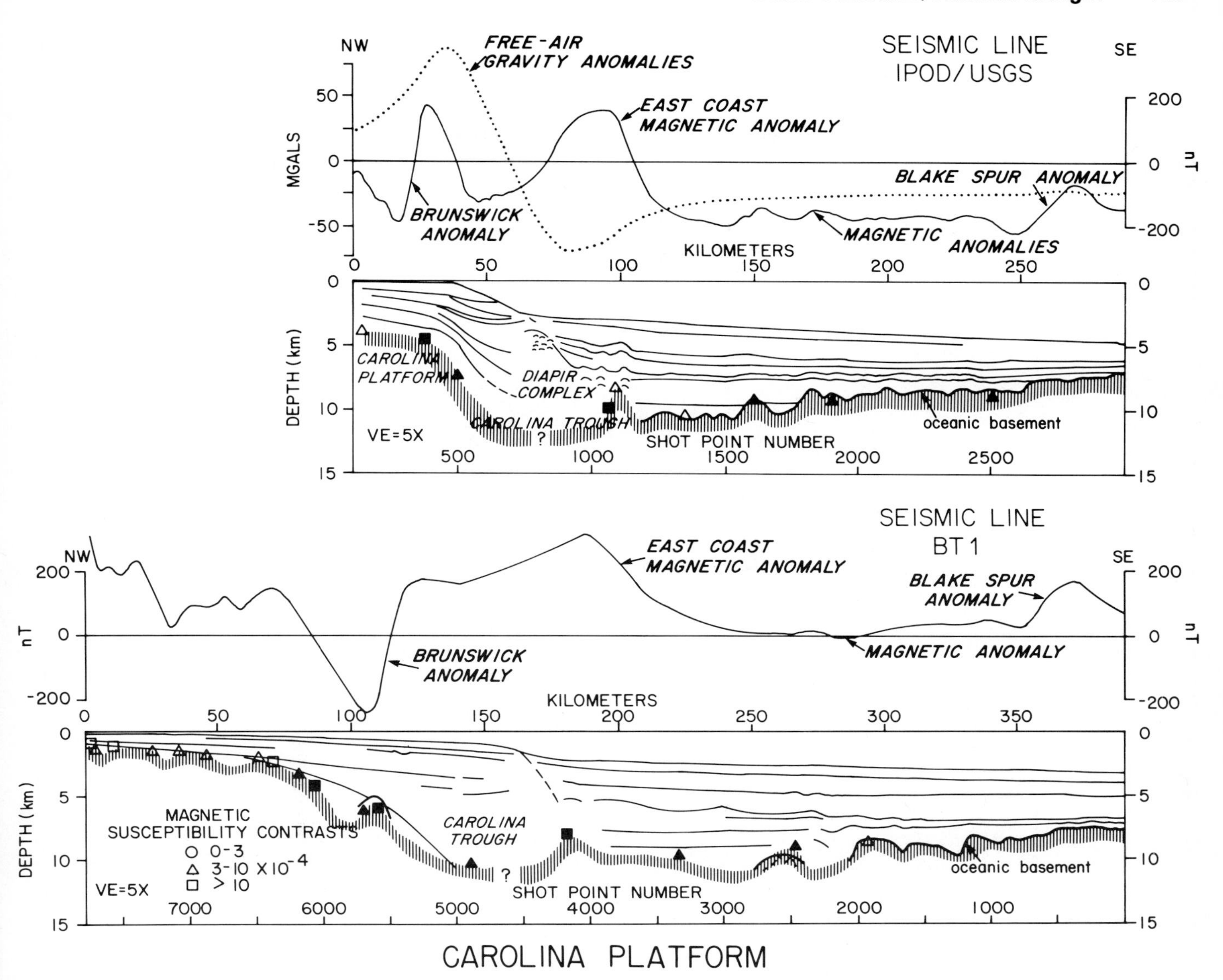

FIG. 13. Comparison between magnetic depth-to-basement estimates and multichannel seismic lines IPOD/USGS and BT-1 (Grow and Markl, 1977; Dillon et al, 1978) over the Carolina platform. See Figure 14B for their locations. Symbols are the same as in Figure 6.

magnetic data and seismic data for line BT1 (shot point 5800 and 5600, Fig. 13). This basement high and trough is found along the entire edge of the Carolina platform south of 34°N lat. (Figs. 14b, 14c) and is similar to the source of the Brunswick magnetic anomaly within Georgia suggested by Popenoe and Zietz (1977) to be Triassic or Early Jurassic in age. Additional multichannel seismic profiles across this region (Fig. 14b) also indicate the presence of this ridge and trough (Dillon and Paull, 1978). Just south of the Brunswick anomaly and west of the COST GE-1 well (Fig. 14a), there is another small basin (Fig. 14c) centered at 30.5°N lat., 80.5°W long. inferred from magnetic depth estimates. It is a triangular-shaped basin about 50 km across, probably of Triassic age, and the main target area for the OCS lease sale 43 (Dillon et al, 1975).

The Carolina trough is the narrowest of the marginal basins and forms one of the most lineated features on the entire U.S. Atlantic margin. As with the basins to the north, its landward edge is probably block-faulted continental(?) crust, while its seaward edge is marked by the East Coast magnetic anomaly.

Blake Plateau Basin

The Blake Plateau Basin is the largest of the sediment-filled basins on the U.S. Atlantic continental margin. The existence of the thick sequence of sediments beneath the plateau has been known for a long time (Hersey et al, 1959; Sheridan et al, 1966; Emery and Uchupi, 1972; Sheridan, 1974), but the base of the sediment pile has been hard to define with either seismic refraction data. This basin was formed during the Jurassic rifting of Africa from

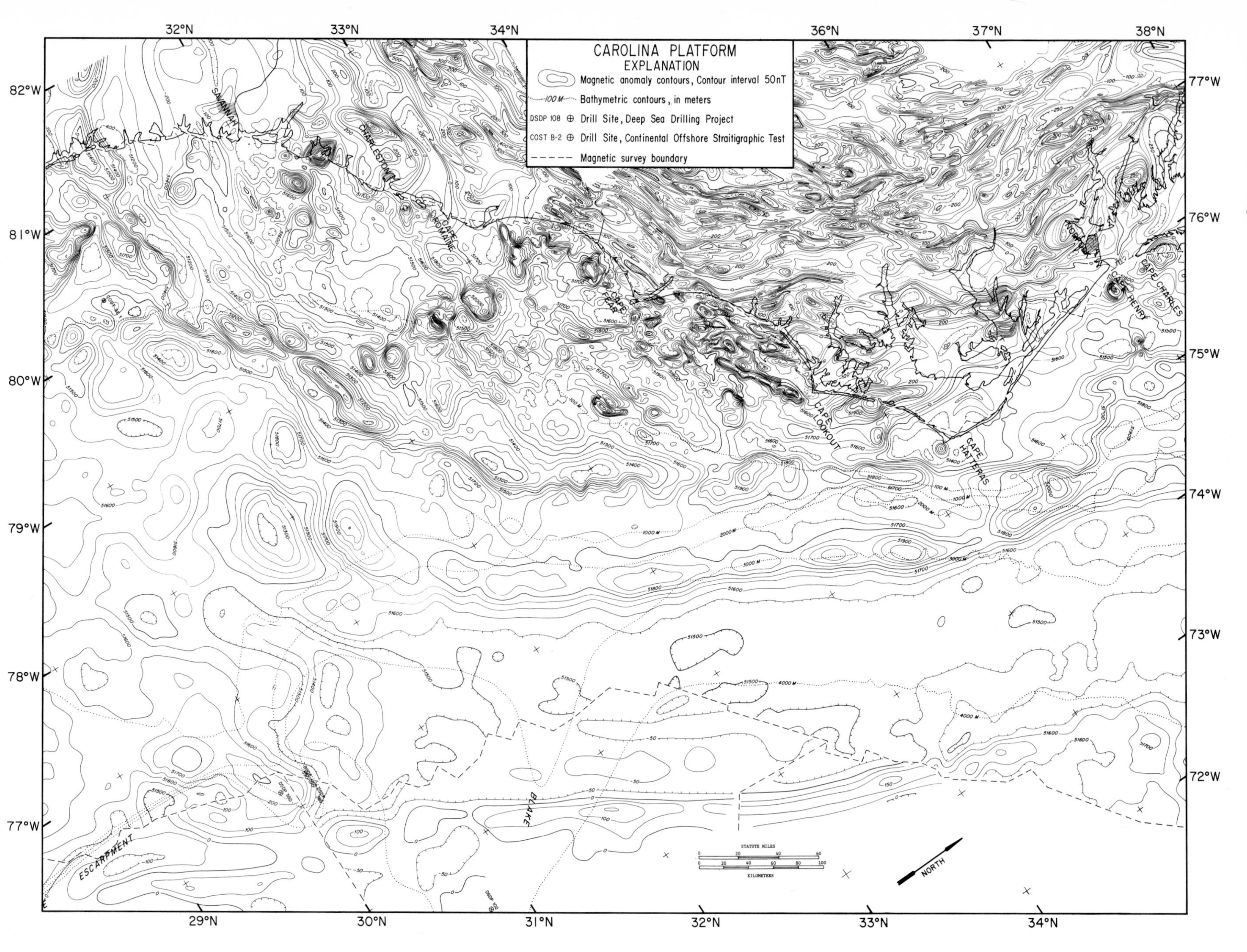

Figure. 14A. Magnetic anomaly contour map for the Carolina platform region. Contour interval is 50 nT.

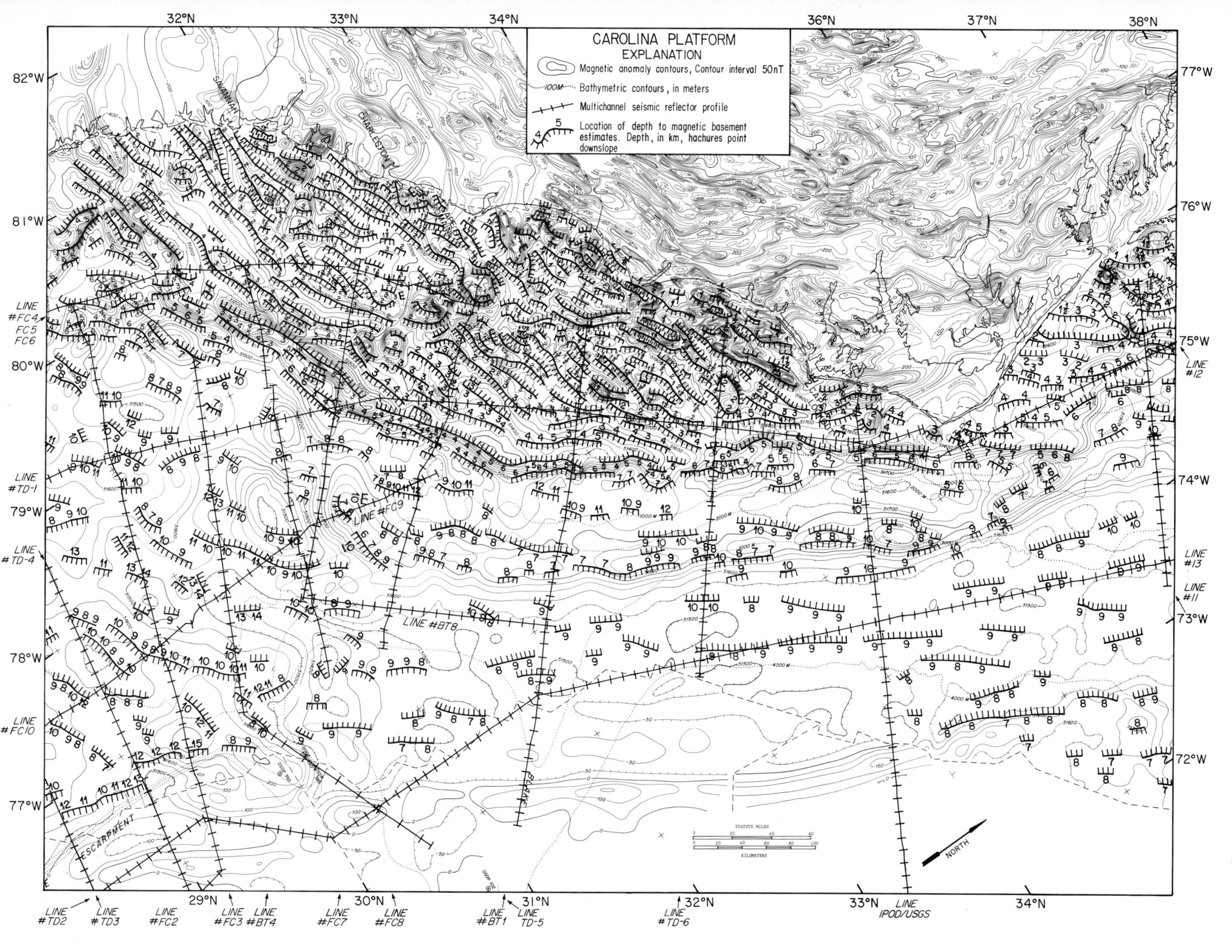

FIG. 14B. Location of magnetic depth estimates for the Carolina platform. The symbols are the same as in Figure 7B.

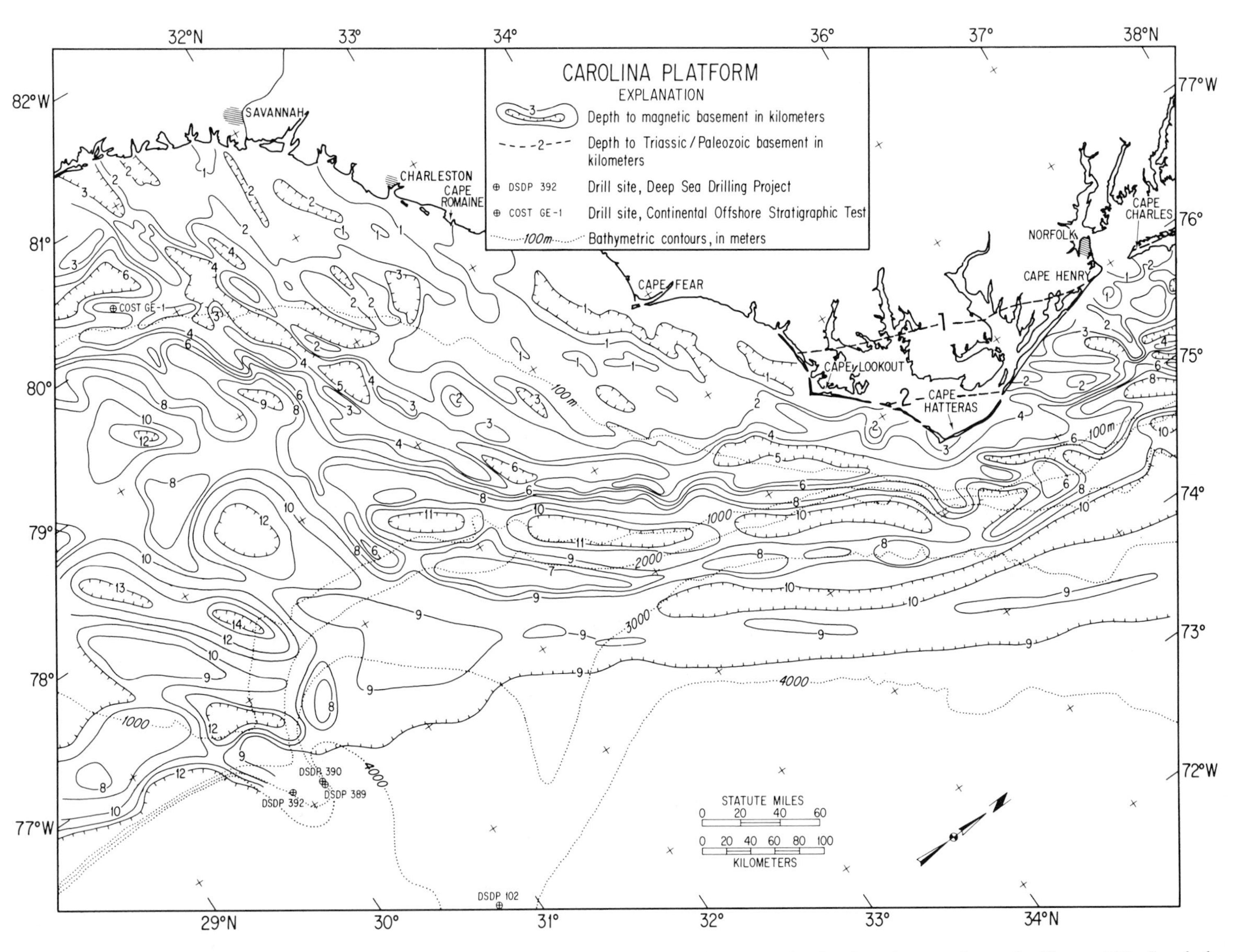

FIG. 14C. Depth-to-basement contours for the Carolina platform based on the magnetic depth estimates shown in Figure 14B. Symbols are the same as in Figure 7C.

North America and is bounded by the Blake Spur fracture zone on the north, the Blake Escarpment on the east, the Florida platform and the Triassic embayment on the west, and the Bahama platform on the south (Fig. 1).

The Blake Plateau Basin is segmented by two fracture zones which can be traced from the offsets in the landward edge of the basin into the deep sea and offsets in the seafloor spreading magnetic anomalies (Fig. 1). The shallowest part of the basin is found between the Bahamas and the southernmost fracture zone within the basin (called the Great Abaco fracture zone by Sheridan, 1974). This is the portion of the basin which has the Florida platform (Barnett, 1975) at its landward edge. The northern two thirds of the basin is the deepest, with more than 10 to 12 km of sediment (Sheridan, 1974), and it has the Triassic basin at 40.5°N lat., 80.5°W long. at its landward edge.

The magnetic depth estimates for the Blake Plateau Basin clearly define the landward edge (Fig. 14b, c) but are very hard to interpret within the basin. There is a scatter in the depth estimates for the deep portion of the basin from 9 km to 14 km (Fig. 14b). The reconstructions of the North Atlantic (Le Pichon et al, 1977; Klitgord and Schouten, 1977) indicate that crust under the Blake Plateau was formed at the same time as oceanic crust was formed to the north. The region may be similar to the Gulf of California where the magma does not extrude onto the seafloor but is intruded within the overlying sediment column, with the magnetic depth estimates picking out these various sills within the sediments.

DISCUSSION

Comparisons of multichannel seismic reflection profiles and magnetic depth estimates (Figs. 6, 8, 10, 11, 13) show that the use of magnetic depth estimates for interpolating between seismic lines is reasonable. The linear block-faulted nature of the landward edges of the various basins is ideal for the Werner deconvolution method of determining magnetic depth estimates. Although some errors exist in the nonunique depth estimates determined by such a method, the results do provide a good regional picture of the relative depths in various areas. The primary reason for this success is the relatively nonmagnetic property of the flat-lying sediments above the basement.

The structure of the Atlantic continental margin consists of a series of platforms and basins (Fig. 1; Maher and Applin, 1971; Emery and Uchupi, 1972; Jansa and Wade, 1975) stretching from the Bahamas to Nova Scotia. From south to north, these features are the Blake Plateau Basin, the Carolina trough, the Carolina platform, the Baltimore Canyon trough, the Long Island platform, the Georges Bank Basin, the LaHave platform, and the Scotian Basin. The platforms have a Paleozoic or older continental basement while the basins probably have a mixture of continental and rift stage crust, covered by Triassic or younger sediments.

The platforms have a distinctive high-amplitude, short-wavelength magnetic pattern which reflects the shallow basement. They also contain smaller basins which have been interpreted as Triassic in age (see Ballard and Uchupi, 1972; Minard et al, 1974; Popenoe and Zietz, 1977) and which have broad magnetic lows associated with them. These platforms tend to be terminated sharply, and the ends probably mark the location of continental to continental transform faults which appeared when Africa and North America split up about 185 m.y.B.P. The seaward edges of these platforms have long and linear magnetic anomalies associated with them. Along the Carolina platform, this large magnetic anomaly is the inner anomaly described by Taylor et al (1968), where their East Coast magnetic anomaly bifurcates near Cape Hatteras (Figs. 3, 14a). Small offsets of the Long Island platform/Gulf of Maine result in a series of en echelon linear magnetic anomalies. The magnetic data along the LaHave platform are fairly sparse but do suggest the presence of a conspicuous magnetic anomaly along the edge of the platform (e.g. Rabinowitz, 1974).

Deep, sediment-filled troughs are found along the entire continental margin from the Bahamas to Newfoundland. The orientation of the axes of all of these troughs suggest that they were all formed by a rifted system having the same pole of rotation. This axial orientation and the orientation of the landward edges of the troughs vary from about 030° in the south to about 050° in the north off Nova Scotia, but the orientation of the seaward edges departs from this pattern. The lengths of the troughs are determined by the locations of the prominent offsets in the continental crust (i.e., the ends of the platforms).

The prominent East Coast magnetic anomaly (Drake et al, 1968; Taylor et al, 1968) is found at the seaward edge of all of the deep, sediment-filled troughs north of the Blake Spur fracture zone. The two major gaps in the East Coast magnetic anomaly are at the southeastern edge of the Long Island platform (where the deep basins pinch out) and at the intersection of the New England Seamounts where the large intrusive body of Bear Seamount drastically alters the magnetic field and the basement shape. The major offsets in the anomaly (Fig. 1) are at offsets in the continental crust along the intersections of associated fracture zones with the margin (Klitgord and Schouten, 1977). A magnetic basement high is indicated by the magnetic depth estimates for the area around the East Coast magnetic anomaly. Oceanic basement can be traced on multichannel seismic profiles almost to the seaward edge of this anomaly (Grow and Schlee, 1976; Schlee et al, 1976), whereas a thick sediment layer (>10 km) is often found just landward of the anomaly.

Seismic reflection profiles over the region of the East Coast magnetic anomaly are characterized by a disturbed zone. An acoustic basement that could be attributed to crystalline/volcanic basement has not been found (Schlee et al, 1976). A ridge feature has been reported at the location of the East Coast mag-

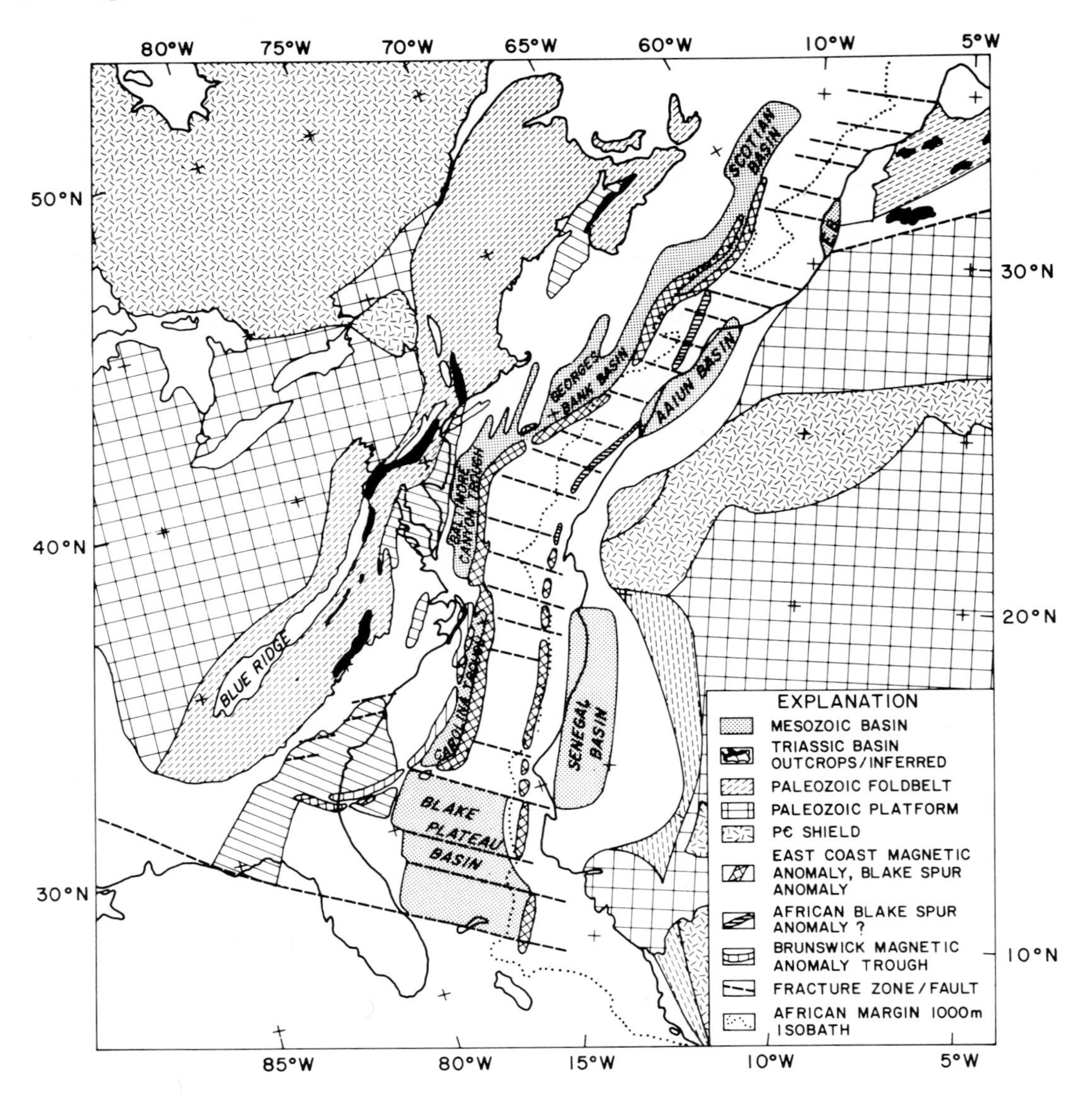

FIG. 15. Reconstruction of the North Atlantic at the time of the Blake Spur anomaly (~175 m.y.B.P.; after Klitgord and Schouten, 1977). Locations of the major sediment filled basins of Mesozoic age are indicated for the African and North American margins.

netic anomaly (Drake et al, 1959; Taylor et al, 1968; Emery et al, 1970; Sheridan, 1974; Schlee et al, 1976; Uchupi et al, 1977). This feature has been interpreted as a basement ridge at a depth of 2–4 km (Drake et al, 1959; Emery et al, 1970; Uchupi et al, 1977), but multichannel seismic reflection data indicate that it is probably composed of carbonate rocks (Sheridan, 1974; Schlee et al, 1976). Magnetic data suggest a basement ridge at about 6 to 8 km, which could be the feature upon which the carbonate rocks accumulated. This interpretation is similar to that suggested by Sheridan (Fig. 8, *in* Sheridan, 1974). The susceptibility contrast of this basement high is much lower than that found for intrusive bodies along the margin (e.g., $>25 \times 10^{-4}$ for the mafic intrusion in the Baltimore Canyon trough). We believe that it is more likely an uptilted block of oceanic(?) crust, the steep landward edge of the block forming the seaward edge of the deep, sediment-filled troughs. The prominent East Coast magnetic anomaly would then reflect a combination of a basement high and an edge effect (Keen, 1969), where the edge is between an oceanic(?) type crust and flat-lying sediments.

The widths of the major basins along the margin range from about 50 km (the Carolina trough) to about 350 km (Blake Plateau Basin), the other three troughs (Baltimore Canyon trough, Georges Bank Basin, and the Scotian Basin southwest of Sable Island) each having a width of about 100 km. This large variation in widths may be a result of the initial rifting process or the later sediment-loading patterns. If the initial rifting stage is important, then the basin pattern on the African side should be related to that of the western Atlantic margin. The broadest basins on one side should be matched by the narrowest basins on the other, with the offsets in the edge of continental crust at the time of rifting controlling the locations.

Reconstructions of the north-central Atlantic (Klitgord and Schouten, 1977; LePichon et al, 1977) show that at least 200 km of oceanic crust still existed between Africa and North America at the time (~175 m.y.B.P.) that the eastern edge of the Blake Plateau was against the continental(?) crust of the Guinea marginal plateau (Fig. 15). This reconstruction places the Blake Plateau adjacent to the part of west Africa that has almost no marginal basin, whereas the small Carolina trough runs parallel with the Senegal Basin (Aymé, 1965), the largest Mesozoic basin on the west African margin. North of Cape Blanc (Africa)/Cape Hatteras (North America), the basin pattern on the African side is not as well known, but basins do exist—the small Aaiun Basin and the Essaouira Basin (Dillon and Sougy, 1974; Van Houten, 1977). The Cape Blanc Paleozoic platform is adjacent to the Baltimore Canyon trough, and the Aaiun Basin is adjacent to the Georges Bank Basin. This relation between the basins on the two margins suggests that there may be a direct correlation between the basin widths on the two sides, but the information concerning the basin sizes on the African margin is too vague.

The locations of the basins and oceanic crust at about 175 m.y.B.P. (Fig. 15) provides evidence for the type of crust beneath some of the basins. The Blake Plateau crust must have been formed by means of at least 200 km of extension at the time that oceanic crust was being generated to the north. This crust could have been rift-stage-type crust or oceanic crust. The nonlinear magnetic anomaly pattern associated with the Blake Plateau is very different from the linear magnetic anomalies associated with oceanic crust just north of the plateau and east of the East Coast magnetic anomaly (Figs. 3, 14a). This nonlinear pattern suggests more of a rift-stage-type crust for the Blake Plateau and a more random location of the zone of intrusion/extrusion than is found for oceanic crust.

CONCLUSIONS

Detailed magnetic studies of the U.S. Atlantic continental margin have yielded valuable information concerning the structure of the margin. The pattern of platforms and deep, sediment-filled troughs is brought out by the magnetic anomaly contour maps and magnetic depth analyses. The integration of magnetic data with multichannel seismic reflection profiles provides a limit on the number of interpretations of the magnetic data and allows the interpolation of basement structures between seismic profiles.

The general basement structure is controlled by sharp horizontal offsets in the continental crust, which were formed during the initial rifting stage. The edges of the platforms are marked by large linear magnetic anomalies. The landward edges of the troughs are of a block-faulted construction; they trend parallel with the basin axes and are formed by an opening about the same pole of rotation. The seaward edges of the basins are identified by the prominent East Coast magnetic anomaly which may be caused by an uptilted block of oceanic(?) crust adjacent to the flat-lying, relatively nonmagnetic sediments within the deep, sediment-filled trough. The major basins are found to be segmented into smaller basins whose locations are controlled by the initial offsets in the continental crust.

REFERENCES CITED

Aymé, J. M., 1965, The Senegal salt basin, *in* Salt basins around Africa: Amsterdam, Elsevier Publishing Co., p. 83–90.

Ballard, R. D., and E. Uchupi, 1972, Carboniferous and Triassic rifting: A preliminary outline of the tectonic history of the Gulf of Maine: Geol. Soc. America Bull., v. 83, p. 2285–2302.

Barnett, R. S., 1975, Basement structure of Florida and its tectonic implications: Gulf Coast Assoc. Geol. Socs. Trans., v. 25, p. 122–142.

Behrendt, J. C., J. Schlee, and R. Q. Foote, 1974, Seismic evidence for a thick section of sedimentary rock on the Atlantic outer continental shelf and slope of the United States (abs.): EOS, v. 55, p. 278.

Brown, P. M., J. A. Miller, and F. M. Swain, 1972, Structural and stratigraphic framework, and spatial distribution of permeability of the Atlantic coastal plain, North Carolina to New York: U.S. Geol. Survey Prof. Paper 796, 79 p.

Dillon, W. P., and J. M. A. Sougy, 1974, Geology of West Africa and Canary and Cape Verde Islands: *in* A. E. M. Nairn and F. G. Stehli, eds., The ocean basins and margins: New York, Plenum Press, p. 315–390.

——— and C. K. Paull, 1978, Interpretation of multichannel seismic reflection profiles of the Atlantic continental margin off the coasts of South Carolina and Georgia: U.S. Geol. Survey Misc. Field Studies Map MF-936.

——— et al, 1975, Sediments, structural framework, petroleum potential, environmental conditions, and operations considerations of the United States South Atlantic outer continental shelf: U.S. Geol. Survey Open File Report 75-411, 202 p.

Dobrin, M. B., 1960, Introduction to geophysical prospecting: New York, McGraw Hill, 2d ed., 446 p.

Drake, C. L., M. Ewing, and G. H. Sutton, 1959, Continental margins and geosynclines: The east coast of North America north of Cape Hatteras, *in* L. H. Ahrens et al, eds, Physics and chemistry of the earth, v. 3.: London, Pergamon Press, p. 110–198.

——— J. I. Ewing, and H. Stockard, 1968, The continental margin of the eastern United States: Canadian Jour. Earth Sci., v. 5, p. 993–1010.

Emery, K. O., and E. Uchupi, 1972, Western North Atlantic Ocean: topography, rocks, structure, water, life and sediments: AAPG Memoir 17, 532 p.

——— et al, 1970, Continental rise off eastern North America: AAPG Bull., v. 54, p. 44–108.

Garrison, L. E., 1970, Development of continental shelf south of New England: AAPG Bull., v. 54, p. 109–124.

Grow, J. A., and J. Schlee, 1976, Interpretation and velocity analysis of U.S. Geological Survey multichannel reflection profiles 4, 5, and 6, Atlantic continental margin: U.S. Geol. Survey Misc. Field Studies Map MF-808.

——— and R. G. Markl, 1977, IPOD-USGS multichannel seismic reflection profile from Cape Hatteras to the mid-Atlantic ridge: Geology, v. 5, p. 625–630.

Hersey, J. B., et al, 1959, Geophysical investigation of the continental margin between Cape May, Virginia, and Jacksonville, Florida: Geol. Soc. America Bull., v. 70, p. 437–466.

Hartman, R. R., D. J. Teskey, and J. L. Friedberg, 1971, A system for rapid digital aeromagnetic interpretation: Geophysics, v. 36, p. 891–918.

IAGA Commission Two Working Group 4, 1969, Analysis of the geomagnetic field, International Geomagnetic Reference Field, 1965: Jour. Geophys. Research, v. 74, p. 4407–4408.

Jansa, L. F., and J. A. Wade, 1975, Geology of the continental margin off Nova Scotia and Newfoundland, *in* W. J. M. Van der Linden and J. A. Wade, eds., Offshore geology of eastern Canada: Canada Geol. Survey Paper 74–30, v. 2, p. 51–105.

Jain, S., 1976, An automatic method of direct interpretation of magnetic profiles: Geophysics, v. 41, p. 531–541.

Kane, M. F., et al, 1972, Gravity and magnetic evidence of lithology and structure in the Gulf of Maine region: U.S. Geol. Survey Prof. Paper 726-B, p. B1–B22.

Keen, M. J., 1969, Possible edge effect to explain magnetic anomalies off the eastern seaboard of the U.S.: Nature, v. 222, p. 72–74.

King, P. B., 1969, Tectonic map of North America: Washington, D.C., U.S. Geol. Survey, 2 sheets, scale 1:5,000,000.

Klitgord, K. D., and J. C. Behrendt, 1977, Aeromagnetic anomaly map of the U.S. Atlantic continental margin: U.S. Geol. Survey Misc. Field Studies Map MF-913, 2 sheets, scale 1:1,000,000.

——— and H. Schouten, 1977, The onset of sea-floor spreading from magnetic anomalies, *in* Symposium on the geological development of the New York Bight: Palisades N.Y., Lamont-Doherty Geol. Obsv., p. 12–13.

Le Pichon, X., J. C. Sibuet, and J. Francheteau, 1977, The fit of the continents around the North Atlantic Ocean: Tectonophysics, v. 38, p. 169–209.

Maher, J. C., and E. R. Applin, 1971, Geologic framework and petroleum potential of the Atlantic coastal plain and continental shelf: U.S. Geol. Survey Prof. Paper 659, 98 p.

Marine, I. W., and G. E. Siple, 1974, Buried Triassic basin in the central Savannah River area, South Carolina and Georgia: Geol. Soc. America Bull., v. 85, p. 311–320.

Mayhew, M. A., 1974, "Basement" to east coast continental margin of North America: AAPG Bull., v. 58, p. 1069–1088.

McMaster, R. L., 1971, A transverse fault on the continental shelf off Rhode Island: Geol. Soc. America Bull., v. 82, p. 2001–2004.

Minard, J. P., et al, 1974, Preliminary report of geology along Atlantic continental margin of northeastern United States: AAPG Bull., v. 58, p. 1169–1178.

Popenoe, P., and I. Zietz, 1977, The nature of the geophysical basement beneath the coastal plain of South Carolina and northeastern Georgia: U.S. Geol. Survey Prof. Paper 1028-I, p. 119–137.

Rabinowitz, P. D., 1974, The boundary between oceanic and continental crust in western North Atlantic, *in* C. A. Burk and C. L. Drake, eds., Geology of continental margins: New York, Springer-Verlag, p. 67–84.

Schlee, J., et al, 1976, Regional geologic framework off northeastern United States: AAPG Bull., v. 60, p. 926–951.

Schouten, H., and K. D. Klitgord, 1977, Map showing Mesozoic magnetic anomalies; western North Atlantic: U.S. Geol. Survey Misc. Field Studies Map MF-915, scale 1:2,000,000.

Schultz, L. K., and R. L. Grover, 1974, Geology of Georges Bank basin: AAPG Bull., v. 58, p. 1159–1168.

Sheridan, R. E., 1974, Atlantic continental margin of North America, *in* C. A. Burk and C. L. Drake, eds., Geology of continental margins: New York, Springer-Verlag, p. 391–407.

——— 1977, Older Atlantic crust and the continental edge based on seismic data (abs.): Abstracts with Programs, Northeast Sect., Geol. Soc. America, v. 9, no. 3, p. 285.

Steenland, N. C., 1977, Regional geologic framework off northeastern United States: discussion: AAPG Bull., v. 61, p. 741–743.

Taylor, P. T., I. Zietz, and L. S. Dennis, 1968, Geologic implications of aeromagnetic data for the eastern continental margin of the United States: Geophysics, v. 33, p. 755–780.

Uchupi, E., R. D. Ballard, and J. P. Ellis, 1977, Continental slope and upper rise off western Nova Scotia and Georges Bank: AAPG Bull., v. 61, p. 1483–1492.

U.S. Naval Oceanographic Office, 1966, Total magnetic intensity aeromagnetic survey 1964–1966—United States Atlantic coastal region: Washington, D.C., U.S. Naval Oceanog. Office, 15 sheets, scale 1:500,000.

——— 1972, Residual magnetic intensity contour chart: Gulf of Mexico–Caribbean Sea–North American Basin, *in* Environmental and acoustic atlas of the Caribbean Sea and Gulf of Mexico, v. II, Marine environment: Washington, D.C., U.S. Naval Oceanog. Office, scale 1 inch per degree longitude.

Vacquier, V., and J. Affleck, 1941, A computation of the average depth to the bottom of the earth's magnetic crust based on a statistical study of local magnetic anomalies: Am. Geophys. Union, Trans. 1941, p. 446.

——— et al, 1963, Interpretation of aeromagnetic maps: Geol. Soc. America Memoir 47, 151 p.

Van Houten, F. B., 1977, Triassic-Liassic deposits of Morocco and eastern North America: AAPG Bull., v. 61, p. 79–99.

Vogt, P. R., C. N. Anderson, and D. R. Bracey, 1971, Mesozoic magnetic anomalies, sea-floor spreading and geomagnetic reversals in the southwestern North Atlantic: Jour. Geophys. Research, v. 76, p. 4796–4823.

Walczak, J. E., 1963, A marine magnetic survey of the New England seamount chain, project M-9: U.S. Naval Oceanog. Office technical report TR-159, 37 p.

Werner, S., 1953, Interpretation of magnetic anomalies at sheet-like bodies: Sveriges Geologiska Undersokning, Arsbok 43, no. 6, Ser. c, No. 508, 130 p.

Wheeler, G., 1938, Further evidence of broad-terrane Triassic: Jour. Geomorphology, v. 1, p. 140–142.

Structure of Colorado Basin and Continent-Ocean Crust Boundary off Bahia Blanca, Argentina[1]

W. J. LUDWIG[2], J. I. EWING[2], C. C. WINDISCH[2], A. G. LONARDI[3], AND F. F. RIOS[3]

Abstract New seismic refraction measurements made with radio-sonobuoys, ocean bottom seismographs, and ship-deployed hydrophones are used to improve the knowledge of structure of the Colorado Basin and the nature of the boundary between continental crust and oceanic crust. Included are the tabulated results of 120 airgun-sonobuoy stations on the continental shelf, slope, and rise off Bahia Blanca and a schematic (seismic) structure section of these features. The seismic data and information from boreholes on the shelf are interpreted to show the presence of an outer shelf-slope basement ridge, perhaps capped by a volcanic or reefal complex, that may have barred marine deposition from the Colorado Basin until the close of the Cretaceous. The ridge of continental basement rock is located beneath sediments of the upper continental slope, about 80 km beyond the edge of the shelf, at the boundary between continental and oceanic crust. Closely spaced refraction profiles made parallel to isobaths in the vicinity of the continent-oceanic crust boundary suggest stretching and thinning of lower crustal rock, perhaps accompanied by subsidence of continental basement rock to form the continental slope. Alternatively, as is suggested by others to explain a coincident belt of magnetic and gravity anomalies in the vicinity of the outer shelf-slope, the basement of the continental slope is elevated material of composition similar to oceanic crust.

INTRODUCTION

The continental margin off Bahia Blanca, Argentina, was rifted from Africa during the fragmentation of Gondwanaland and the initial opening of the South Atlantic Ocean. It is characterized by a 400-km-wide continental shelf and a gentle continental slope that dips toward the Argentine Basin (Fig. 1). M. Ewing et al (1963), Ludwig et al (1968), and M. Ewing and A. Lonardi (1971) showed from seismic

[1]Manuscript received, August 19, 1977; accepted, March 2, 1978.

A complete set of time-distance graphs has been deposited with the American Society of Information Science, National Auxiliary Publications Service.

See NAPS document no. 03361 for 8 pages of supplementary material. Order from: NAPS, c/o Microfiche Publications, P.O. Box 3513, Grand Central Station, New York, New York 10017. Remit in advance for each NAPS accession number. Institutions and organizations may use purchase orders when ordering, however there is an additional $5.00 charge for this service. Make checks payable to Microfiche Publications.

Photocopies are $5.00; microfiche are $3.00. Outside the U.S. and Canada, add extra postage: $3.00 for photocopies and $1.00 for fiche.

[2]Lamont-Doherty Geological Observatory of Columbia University, Palisades, New York 10964. Ewing's new address: Woods Hole Oceanographic Institution, Woods Hole, Massachusetts 02543.

[3]Instituto Argentino de Oceanografía, Bahia Blanca, Argentina. Lonardi's new address: Consejo Nacional de Investigaciones, Científicas y Técnicas and I.T.B.A., Buenos Aires, Argentina.

The airgun-sonobuoy data used in this report were collected on board R/V *Conrad* (1972,1973) and R/V *Vema* (1974) of Lamont-Doherty Geological Observatory. The two-ship seismic refraction measurements were made during 1973 with R/V *Conrad* serving as the shooting vessel and A.R.A. *El Austral* (formerly *Atlantis I* of WHOI) of Instituto Argentino de Oceanografía serving as the receiving vessel. W. J. Ludwig and L. Garcia installed and operated the seismic refraction equipment aboard *El Austral;* F. F. Rios, E. Fernandez, R. Ercoli, D. Galfon, A. Ferrante, and R. Zibecchi assisted in the collection of the data. The ocean-bottom seismograph measurements were made on board *Vema,* under the supervision of G. Carpenter.

J. Ewing, C. Windisch, G. Carpenter, and G. Bryan served on one or more occasions as Chief Scientist of *Conrad* and *Vema;* also participating in the work were A. Yung, F. Mouzo, P. Margalot, R. Adamini, J. Isquierdo, and E. Schnaak. H. P. Casal, Comite Nacional de Oceanografía, and A. Lonardi and R. Olivera of Instituto Argentino de Oceanografía planned and coordinated the cooperative work in Argentina.

The work at sea would not have been possible without the cooperation of the officers and crews of R/V *Conrad,* R/V *Vema,* and A.R.A. *El Austral.* Explosives used for seismic refraction studies were supplied by the U.S. Navy, and were stored and loaded aboard *Conrad* and *Vema* by the Argentine Navy; this support is gratefully acknowledged.

This study was supported by various government agencies of Argentina and by the U.S. National Science Foundation (Grants GX 34410 and IDO/OCE 72-06426) as part of the International Decade of Ocean Exploration. Additional support was provided by the Oceanography Section of NSF (Grants GA 27281) and by the U.S. Office of Naval Research contracts N00014-67-A-0108-0004 and N00014-75-C-0210.

We thank R. E. Houtz for assistance with the analyses of the airgun-sonobuoy data. P.D. Rabinowitz and R.E. Houtz critically read the manuscript.

Lamont-Doherty Geological Observatory contribution 2741.

Article Identification Number:
0065-731X/78/MO29-0007/$03.00/0.
(see copyright notice, front of book)

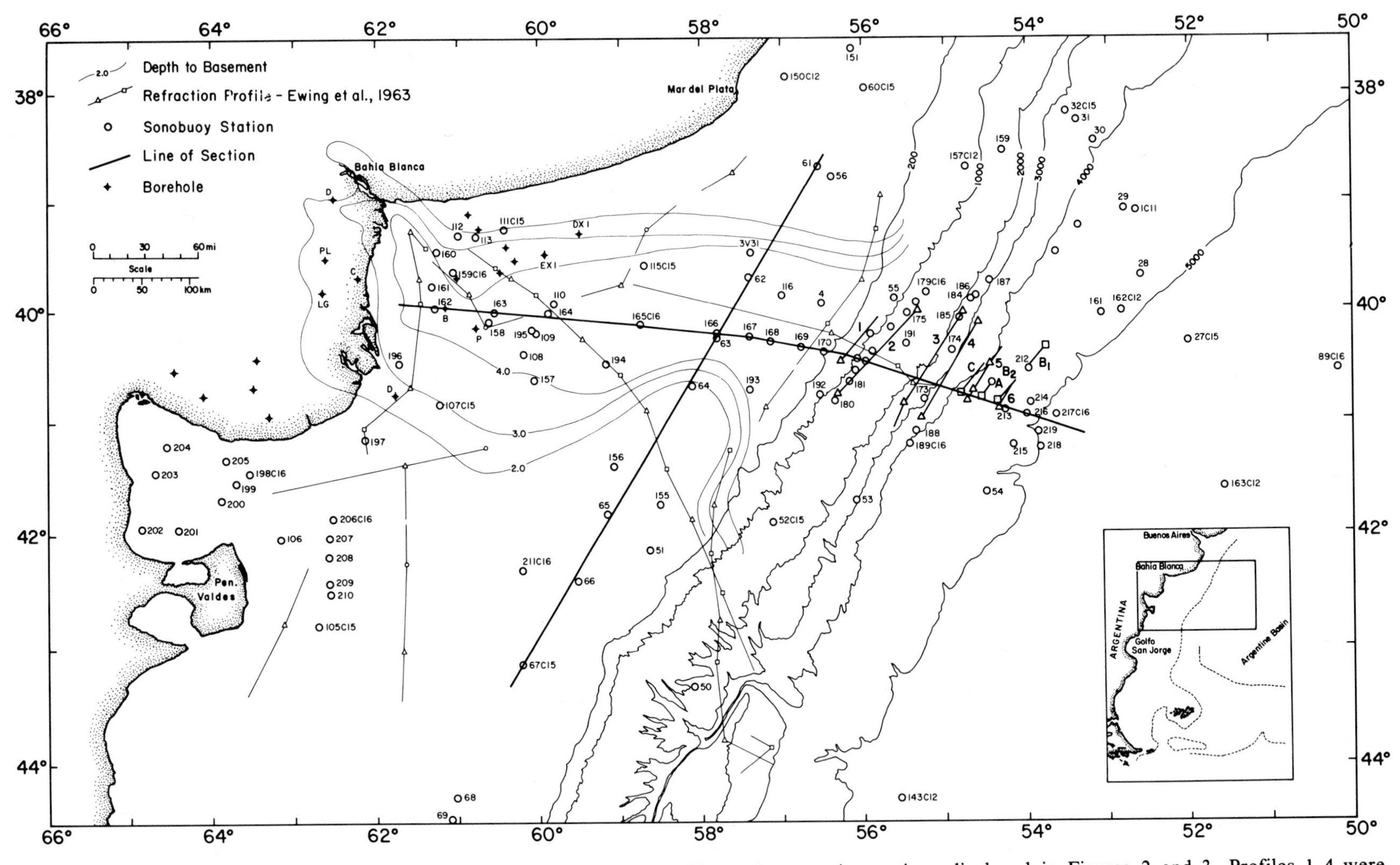

FIG. 1. Bathymetric map showing locations of seismic refraction profiles and composite sections displayed in Figures 2 and 3. Profiles 1–4 were recorded with ship-deployed hydrophones; 5 and 6 with long-range sonobuoys; and A, B, C with an ocean-bottom seismograph. Bathymetry in meters. Onshore boreholes: O, Ombucta; PL, Pedro Luro; LG, Los Gauchos; C, Colorado. Offshore boreholes: B, La Ballena; D, El Delfin; P, El Penguino. Contours on basement from Ludwig et al, (1977).

refraction profiles that the shelf is underlain by thick sediments in an embayment-like depression which extends well out under the continental slope. The Colorado Basin thus developed is a true composite graben, bounded both to the north and south by a series of normal faults (see Zambrano, 1972, 1974). It was formed during Late Jurassic-Early Cretaceous time by basement fracturing and continued to subside intermittently during the Tertiary (Shell Production Company of Argentina, 1962).

The Colorado Basin, covering some 126,000 km^2 and containing 425,000 km^3 of sediment, has 71 percent of its area and 87 percent of its sediment volume located beneath the continental shelf (Zambrano, 1974; Yrigoyen, 1975). The western onshore end of the basin is bordered to the north and northeast by the southern hills of Buenos Aires; to the west, by the scattered exposures of Precambrian granites and Permian-Triassic igneous and metamorphic rocks; and to the southwest, by the Sierra Grande area of the Rio Negro province west of Golfo San Matias. The results from one of several boreholes in the basin on land (Fig. 1) permitted Ludwig et al (1968) tentatively to correlate seismic velocities with geologic ages. Zambrano (1972) further correlated seismic velocities with stratigraphic units and constructed isopach maps of major velocity layers. In a series of articles, Zambrano and Urien (1970, 1974), Urien and Zambrano (1973), and Urien et al (1976) describe the structural framework and formative history of the offshore basins of Argentina, based on Lamont-Doherty refraction data and the results of exploratory wells drilled on the shelf.

Between June 1969 and March 1972, approximately 50,000 km of multichannel reflection measurements were acquired by the petroleum industry and 32 wells were drilled in the Salado, Colorado, and San Jorge Basins (Urien and Zambrano, 1973). The basins are filled with Cretaceous, Tertiary, and Quaternary sediments. In the Salado and Colorado Basins, the Cretaceous units are largely continental red beds; the oldest marine sediments found by drilling to date are of latest Cretaceous (Maestrichtian or ~70 m.y.b.p.) age (Urien and Zambrano, 1973).

Unlike the gently dipping continental and shallow marine sediments of the inner continental shelf, the outer shelf-slope-rise region is sedimentologically and structurally complex. It is the site of the boundary between continental and oceanic crust, which may have complex differences in basement morphology and petrology as well as complex sedimentological contrasts on either side. Rabinowitz and LaBrecque (in press) and LaBrecque and Rabinowitz (1977) mapped the location of a prominent linear magnetic anomaly near the shelf break, called Anomaly G, which is at the conjugate location of Anomaly G observed off southwest Africa (Rabinowitz, 1976). The location of the anomaly is independent of the location of the shelf edge. Coincident with the magnetic anomaly are steep gradients in the isostatic gravity anomaly. The magnetic anomaly may be modeled as an edge-effect anomaly separating continental from oceanic crust; the gravity anomaly may result from elevated oceanic basement near the continental-oceanic crust boundary (Rabinowitz and LaBrecque, 1977, in press).

In order to further define velocity structure of sedimentary strata of the Argentine continental margin and to determine the gross morphology of crustal layers and mantle of the continent-ocean crust boundary, additional seismic refraction measurements were made from Lamont-Doherty research vessels during 1972–74 (Fig. 1). The project was done in cooperation with Consejo Nacional de Investigaciones Cientificas y Técnicas, Instituto Argentino de Oceanografía, and Servicio Hidrografía Naval, as part of the International Decade of Ocean Exploration. In this report, we present velocity solutions from 120 airgun-sonobuoy refraction profiles of the shelf-slope-rise (Table 1) and from nine long-range refraction profiles made with two ships, an ocean-bottom seismograph, and long-range sonobuoys on the continental slope and rise (Table 2).

THE COLORADO BASIN

In Figure 1 we have plotted with open circles the locations of refraction stations made with radio-sonobuoys and an airgun sound source. These data were obtained while engaged in single-channel vertical reflection profiling, often with a large airgun sound source (approximately 1 ft^3 volume at 500 psi) developed by the Mobil Oil Company and a 400-ft-long hydrophone streamer designed at Lamont-Doherty. This combination of sound source and receivers provided a good signal/noise ratio in the low-frequency part of the spectrum and resulted in deep reflection/refraction penetration of the sediments in some areas. The velocities and layer thicknesses from the sonobuoy data (Table 1) verify the velocity structure and basinal forms mapped by M. Ewing et al (1963) and Ludwig et al (1968), but show the presence of additional layering. On a revised map Ludwig et al (1978a) show the general distribution of sediments on the Argentine continental margin based on the results of the new sonobuoy data. Maps of the magnetic anomalies and free-air gravity anomalies of the margin are presented by LaBrecque and Rabinowitz (1977) and Rabinowitz (1977), respectively.

The seismic profiles on the shelf reveal a strongly reflective interface, generally more than 1 km below sea level, whose morphology and tendency to truncate older layers suggest that it is an erosional unconformity developed near former sea level. From correlations with boreholes (Fig. 1), deposits above this reflector appear to correspond to the base of the earliest marine beds (of latest Cretaceous-earliest Tertiary age) to be deposited in the offshore Buenos Aires province. Deposits below the reflector are nonmarine. In a subsequent paper, we shall be concerned with mapping the reflector from examination of all available data.

The significant results of the seismic refraction measurements are readily apparent in the schematic structure sections of Figures 2 and 3. Correlation of geologic age and tentative identification of stratigraphic units with velocity layers in Figure 2 are

TABLE 1

Results of Airgun-Sonobuoy Stations, Argentine Continental Margin off Bahia Blanca

Sonobuoy	Water	Thickness in Km							Velocity in km/sec								Lat. (S)	Long. (W)
		h_2	h_3	h_4	h_5	h_6	h_7	h_8	V_2	V_3	V_4	V_5	V_6	V_7	V_8	V_9		
1C11	4.72	1.00	1.41						1.87*	2.85*							39°06.4'	52°38.6'
149C12	.81	.69	.50	.45					2.00*	2.10*	2.78*	4.26					39°57.4'	55°19.4'
150C12	.03	.22	.74						1.72	2.09	5.85						37°50.9'	56°57.2'
151C12	.08	.52	.87	1.10					1.92	2.27	3.06	5.04					37°34.9'	56°07.8'
152C12	.08	.96	.63	1.23					1.92	2.45	3.24	4.26					37°31.0'	55°54.3'
153C12	.08	.37	.90	1.00	1.50				1.91	2.22	2.85	3.85	5.00				37°25.4'	55°37.2'
154C12	.09	.35	.39	.27	.86	.50			1.86	2.13	2.35	2.66	3.41	4.08			37°20.8'	55°22.5'
155C12	.11	.33	.49	1.37					1.91	2.15	2.53	4.11					37°16.1'	55°10.1'
156C12	.14	.51	1.48	2.05					1.87	2.30	3.67	5.25					37°11.2'	54°35.4'
157C12	1.05	.94	.55	1.58					(1.90)	2.64	3.13	7.50					38°42.1'	54°45.3'
159C12	1.18	.74	.27	.41	.42	1.27			1.86*	2.06*	2.49*	3.25	5.17	6.16			38°32.9'	54°17.9'
160C12	4.32	.40	1.07	.67					1.96*	3.18*	4.18*						39°29.0'	53°39.1'
161C12	5.08	.82	.45	.53	.70				1.62*	2.33*	3.62*	3.98*					40°02.8'	53°03.5'
162C12	5.48	1.51							2.05*								40°01.4'	52°49.3'
50C15	2.14	.61	.66	.75	1.50				1.61*	2.36*	3.00	4.50	6.45				43°20.9'	58°05.1'
51C15	.10	.30	.85						1.90	2.25	5.40						42°09.4'	58°38.6'
52C15	1.42	.53	.84	.53					1.68*	2.55*	3.53*	4.80					41°54.7'	57°06.6'
53C15	3.42	.42	.78						1.77*	2.33*	6.07						41°42.2'	56°03.8'
54C15	4.95	.69	.92	1.28					2.03*	2.07*	3.74*						41°38.3'	54°28.0'
55C15	1.11	.58	.91	.82	2.73				1.55*	2.54*	3.18*	4.50	5.80				39°53.9'	55°35.2'
56C15	.09	1.04							2.20	5.60							38°47.7'	56°22.0'
57C15	.04	1.05							1.95	2.70							36°15.4'	54°58.8'
58C15	.04	1.08	.57						2.02	2.39	2.86						36°14.8'	55°00.6'
59C15	.05	1.31							1.92	2.89							36°41.5'	55°08.2'
60C15	.08	1.43	.37						2.16	2.35	2.82						37°57.7'	55°58.4'
61C15	.08	1.15							2.05	5.16							38°41.6'	56°31.5'
62C15	.10	.19	.41	.97	.50	2.68			1.97	2.00	2.30	3.05	4.10	4.80			39°42.4'	57°23.9'
63C15	.10	1.59	.83	1.02	.40				2.27	3.05	3.72	4.08	4.66				40°16.3'	57°48.2'
64C15	.09	.86	1.27						2.21	2.60	6.38						40°41.9'	58°06.1'
65C15	.08	1.41							2.46	5.30							41°50.2'	59°09.2'
66C15	.10	1.33							2.40	5.60							42°25.1'	59°30.2'
67C15	.10	1.00							2.00	6.35							43°08.5'	60°11.9'
68C15	.10	1.06	.64	.94					2.00	5.00	5.65	6.08					44°17.3'	61°02.6'
69C15	.11	.51	.68	.68					2.10	2.50	4.95	5.70					44°28.3'	61°06.4'
104C15	.08	.62	.47						2.20	4.35	5.80						44°34.3'	63°55.9'
105C15	.07	.06	1.45	.54					1.89	2.32	4.90	5.65					42°47.5'	62°43.4'
106C15	.04	.21	.75						1.85	2.13	5.50						42°02.6'	63°10.5'
107C15	.04	.50	.74						2.03	2.24	5.44						40°50.9'	61°13.6'
108C15	.05	1.40							2.12	3.07							40°24.2'	60°11.2'
109C15	.06	1.01	.45	1.06					2.10	2.63	3.14	3.77					40°11.9'	60°07.2'
110C15	.05	1.33	.78	1.14	1.63				2.13	2.97	3.71	4.58	5.94				39°56.5'	59°48.4'
111C15	.03	.69	.59	.49					2.17	3.57	4.45	5.92					39°15.6'	60°24.9'
112C15	.03	1.12							2.00	4.50							39°19.1'	60°59.6'
113C15	.03	.97	.10						2.20	2.60	4.90						39°20.2'	60°46.4'
115C15	.08	1.31	.50	.56	.55				2.30	2.65	3.10	3.80	4.30				39°35.6'	58°41.3'
116C15	.10	.76	1.10	1.11	.77	1.45			2.15	2.65	3.50	4.05	5.10	6.00			39°52.5'	56°59.6'
27C15	5.41	.99	.99						2.09*	2.95*	6.55						40°18.1'	52°00.4'
28C15	4.97	2.58	2.05						2.30	5.50	6.65						39°42.1'	52°35.6'
29C15	4.64	2.29	2.30	2.32	.31	1.51			2.26	4.40	5.30	6.50	7.35	8.10			39°05.8'	52°48.1'
30C15	3.99	.44	.39	1.94	.99	3.20			1.42*	1.99*	2.50*	3.71*	5.01*				38°27.2'	53°09.8'
31C15	3.77	.36	.58						1.48*	2.46*							38°16.4'	53°22.6'
32C15	3.68	1.40	1.31						2.22*	4.13*	5.20*						38°11.5'	53°30.2'
89C16	5.36	1.05	1.16						1.98*	3.01*							40°33.4'	50°11.2'
155C16	.08	.33	.79						1.95	2.25	5.75						41°45.0'	58°30.4'
156C16	.09	.42	.73	1.09					(1.80)	2.40	5.00	5.60					41°24.2'	59°04.3'
157C16	.07	1.29	.92	.56	1.13	1.85			2.15	3.10	3.65	4.95	5.65	6.00			40°38.0'	60°03.1'
158C16	.05	.28	1.05	1.01	1.17	.87			1.97	2.17	3.02	3.63	4.31	5.71			40°06.6'	60°36.4'

160C16	.03	.42	1.10	.90	1.39			1.90	2.20	3.60	5.80	6.60			39°27.7'	61°15.1'
161C16	.03	1.14	1.32	1.23	1.14	1.08		2.15	3.00	3.80	5.00	5.70	6.10		39°46.6'	61°19.1'
162C16	.03	1.01	1.23	.82	.95			2.10	2.75	3.75	4.40	4.70			39°59.2'	61°16.5'
163C16	.05	.87	.87					2.05*	2.60	3.30					40°01.3'	60°32.9'
165C16	.09	1.08	.84	.50	.97	1.03		2.10*	2.80	3.50	3.90	4.50	5.00		40°08.1'	58°44.6'
166C16	.09	1.16	1.02	.89	1.21			2.15*	2.85	3.65	4.15	5.10			40°13.0'	57°47.6'
167C16	.10	.64	.67	1.02	.96	.86	1.93	2.10	2.40	2.90	3.60	4.05	4.80	5.40	40°15.1'	57°23.6'
168C16	.10	1.14	.95	1.43	1.39			(2.20)	2.65	3.60	4.50	5.20			40°17.6'	57°08.0'
169C16	.10	1.24	.86	1.35	.57	.85		2.00	3.00	3.55	4.25	4.70	5.25		40°20.7'	56°45.0'
170C16	.16	.62	.76	.96	1.05	1.27		1.99	2.35	2.66	3.46	4.34	5.51		40°23.3'	56°28.0'
171C16	1.02	.87	.65	1.24				1.94*	2.46	3.25	4.31	5.02	5.52		40°26.8'	56°02.3'
172C16	1.27	.43	.62	.17	.49	.70		1.80*	2.00	2.95	4.30	4.70	5.15		40°28.4'	55°56.5'
173C16	3.34	.82	.37	.86	.54	.93		2.09*	2.50	3.30	4.60	5.46	6.52		40°48.7'	55°13.4'
174C16	2.88	1.33	.69	.89				2.21*	4.00	5.65	6.25				40°22.4'	54°53.2'
175C16	1.31	1.51	.70	.68	1.46	2.71		2.34*	3.25	4.00	4.75	5.30	6.15		40°03.7'	55°27.2'
176C16	1.26	1.43	1.02	.30	2.12	1.73		2.33*	3.40	4.23	4.90	5.55	6.05		40°11.5'	55°38.8'
177C16	1.40	1.63	.20	1.03	3.59			2.32*	3.54	4.15	5.15	6.60			40°23.6'	55°53.3'
178C16	1.38	.74	.30	1.13	.73			(1.80)	2.25	3.40	4.40	5.20			39°56.5'	55°19.4'
179C16	1.36	.79	.52	.66	.90	.86		1.80*	2.52	3.80	4.40	4.85	5.45		39°50.9'	55°11.9'
180C16	1.31	1.96	.34	.76				2.46	3.11	4.20	5.00				40°49.6'	56°19.9'
181C16	1.31	.89	.05	1.14	1.07	1.23		1.81	2.05	2.90	4.30	5.25	5.60		40°39.1'	56°08.5'
182C16	1.32	1.87	.59	.73	3.17			2.60	3.10	4.15	4.90	5.85			40°22.9'	55°51.4'
184C16	2.62	1.35	.43	.54	1.61			2.22	2.75	3.40	5.05	6.50			39°55.4'	54°40.0'
185C16	2.70	.54	1.04	.25				2.04	3.10	4.10	5.35				40°05.1'	54°46.7'
186C16	2.70	2.23	.69	.88	1.67	1.58		2.50	3.65	4.60	5.60	6.71	8.70		39°53.4'	54°36.5'
187C16	3.30	1.95	.52	1.10	.93	2.11		(2.50)	3.34	4.50	5.03	7.10	8.80		39°44.3'	54°25.6'
188C16	3.82	1.25	.70	.90	1.53			2.26*	3.21*	3.61*	3.65*	4.75			41°06.2'	55°19.5'
189C16	4.03	1.30	1.14	1.96				(2.26)	3.23	5.10	6.20				41°12.8'	55°24.4'
190C16	1.15	1.24	.90	.51				(2.00)	3.40	5.10	5.55				40°13.7'	55°53.6'
191C16	1.50	.56	1.03	.87	1.60			1.71*	2.92*	3.71*	4.35	5.30			40°18.8'	55°27.1'
192C16	1.15	.59	1.35	1.91				1.78*	2.55*	3.71*	5.55				40°46.4'	56°30.4'
193C16	.09	1.86	1.69	3.75				2.46*	3.40*	5.20	6.00				40°43.5'	57°22.7'
194C16	.09	1.35	1.24	.79				(2.10)	3.20	3.70	4.50				40°29.8'	59°08.9'
195C16	.06	1.24	1.04	1.04	.86	.62		(2.10)	3.10	3.75	4.55	5.00	5.45		40°11.4'	60°04.1'
196C16	.03	1.10	1.19	1.09				(2.05)	2.90	3.95	5.90				40°29.2'	61°43.7'
197C16	.03	.59	.68	1.57	1.23			2.05	2.30	5.75	6.60	7.30			41°09.7'	62°09.3'
198C16	.05	.85	1.15					1.95	3.10	6.70					41°27.6'	63°33.5'
199C16	.07	1.19	.60					2.10	4.30	5.45					41°33.7'	63°43.4'
200C16	.08	.69	.41	1.24				2.00	4.10	5.35	6.20				41°42.3'	63°53.9'
201C16	.15	.28	.35					(2.00)	3.40	4.10					41°57.5'	64°25.7'
202C16	.11	.63						5.00	5.70						41°56.1'	64°52.2'
203C16	.17	.60	.56	2.10				3.15	4.70	5.25	7.35				41°26.9'	64°42.5'
204C16	.16	.32	.27					3.00	4.30	5.80					41°13.0'	64°33.8'
205C16	.07	.56	.25					3.00	4.15	5.80					41°20.3'	63°50.7'
206C16	.06	.91	.57					1.90	3.30	5.95					41°51.8'	62°32.9'
207C16	.06	1.11	.43					2.15*	5.25	5.85					42°02.4'	62°35.1'
208C16	.07	.86						2.22*	5.25						42°12.7'	62°35.8'
209C16	.08	1.33	.59					2.20	3.70	4.45					42°26.2'	62°35.0'
210C16	.08	1.29						2.10	5.30						42°31.7'	62°34.4'
211C16	.09	.74	.67					2.10	2.50	5.45					42°19.4'	60°12.5'
212C16	4.85	1.16	1.86					1.88*	3.85*	4.50	5.45	7.30			40°32.6'	53°56.9'
213C16	4.88	.80	.51	.85	.73	.91	2.26	1.71*	2.48*	3.29*	4.10*	5.40	6.45	8.00	40°54.8'	54°14.6'
214C16	4.99	.49	.70	.93	.72	2.14		1.70*	2.18*	3.15*	3.37*	6.40	7.35		40°50.6'	53°55.9'
215C16	4.97	.47	.83	1.65				1.63*	2.14*	3.68*					41°13.3'	54°08.5'
216C16	5.04	1.36	1.54	1.09	3.55			2.13*	3.48*	5.70	6.60	7.85			40°56.9'	53°58.7'
217C16	5.20	1.64	.83					2.16*	3.65*						40°57.8'	53°36.8'
218C16	5.19	.45	.43	.72	.64			1.57*	1.88*	2.69*	3.95*	5.45	6.70		41°15.0'	53°48.8'
219C16	5.14	3.45						3.00	5.60						41°06.5'	53°49.7'

NOTES:

Asterisks denote interval velocity from wide-angle reflection data.
Parenthesis indicate assumed velocity.
All other velocities are unreversed refraction velocities.
Sonobuoy solutions outside the area of the Colorado Basin and its adjacent continental rise and slope are taken directly from preliminary listings.

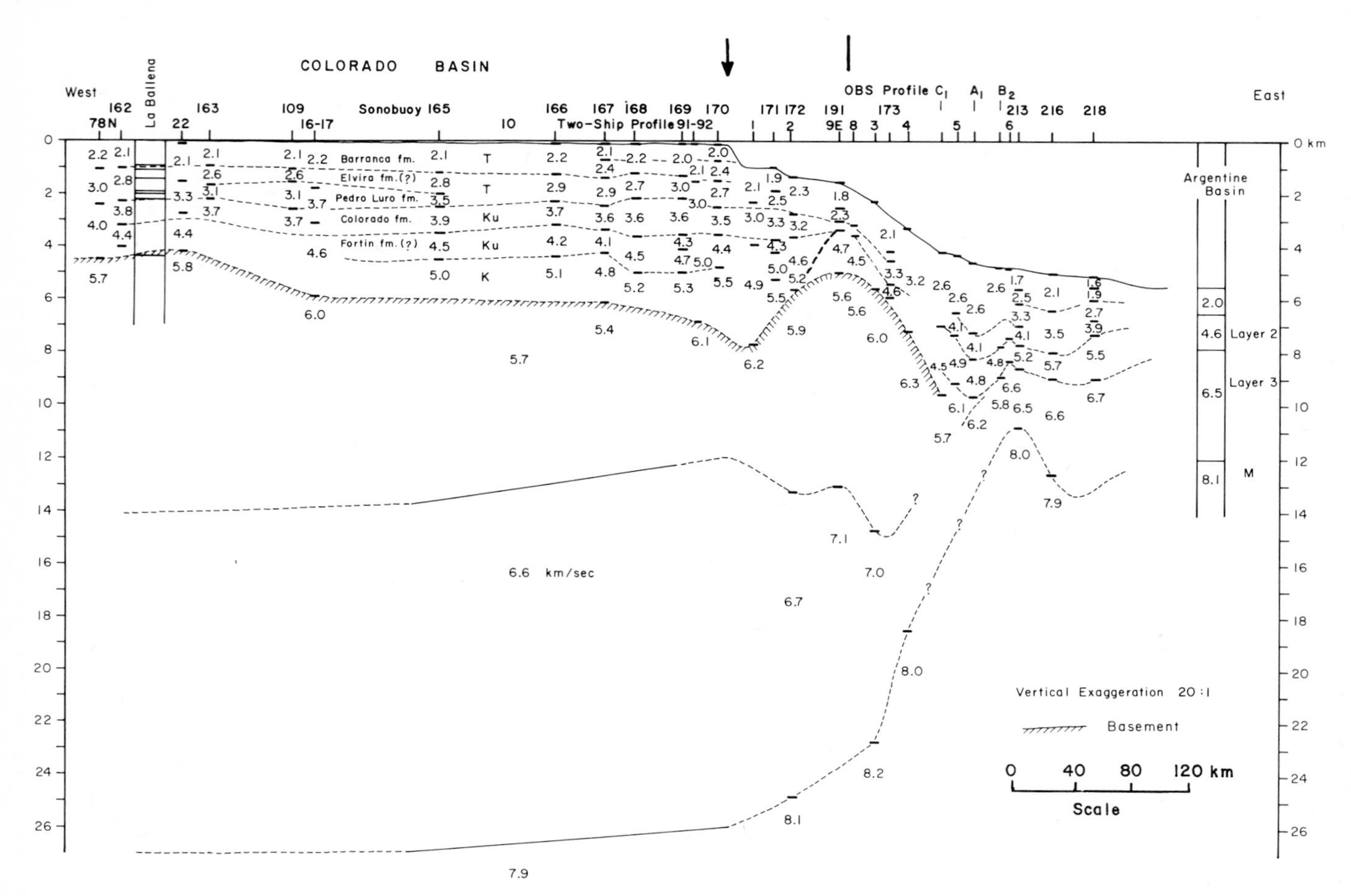

FIG. 2. Schematic west-east structure section of the Argentine continental margin off Bahia Blanca. Seismic boundaries do not necessarily represent time-stratigraphic boundaries. The interface between the continental and marine beds corresponds approximately to the top of a layer with velocities of 3.2 to 3.7 km/sec. Volcanic or reefal material of velocity near 4.5 km/sec may cap the outer basement ridge beneath the continental slope. There are no measurements of gravity and magnetics along the line of section. The area of maximum declination of the basement towards the ocean basin corresponds in position with the extrapolated position of a prominent magnetic anomaly, called Anomaly G, modeled as an edge-effect anomaly separating continental from oceanic crust (see Rabinowitz and LaBrecque, 1978; cf., Figs. 5 and 6). Arrow denotes the shelf break. Vertical line marks the location of Anomaly G. We emphasize that the marked thinning of layer 3 beneath the upper continental rise is based on the results of only one sonobuoy station. Reflection (CDP) data are needed for positive correlation of velocity layers, particularly those beneath the continental slope and rise.

based on comparison of the seismic section with a generalized section of La Ballena well (Fig. 4) and correlations of velocity with time-stratigraphic units as given by Zambrano (1972, 1974) and Yrigoyen (1975) for the onshore portion of the Colorado Basin. In the western half of the Colorado Basin the identification is as follows:

	Age	*Formation*	*Velocity (km/sec)*
Marine	Quaternary and Tertiary	Belen (Rio Negro) Barranca Final Elvira/Ombucta	1.8–2.7
	Early Tertiary (Paleocene-Maestrichtian)	Pedro Luro	2.1–3.2
Continental	Late Cretaceous (Maestrichtian)	Colorado	2.8–4.0
	Late Cretaceous	Fortin	4.0–4.5
	Early Cretaceous (with Jurassic?)	Basement(?)	4.7–5.4
	Pre-Cretaceous	Basement	5.9

Cretaceous (Valanginian-Berriasian or ~130 m.y.b.p.) time (Larson and Ladd, 1973; Rabinowitz, 1976) did not receive marine sediments until almost the end of the Cretaceous. A possible explanation is that a carbonate bank or volcanic ridge was built up on the outer basement ridge, which kept pace with subsidence and sea level rise; ridge development re-

TABLE 2

Results of Long-range Seismic Refraction Profiles, Argentine Continental Margin off Bahia Blanca

Profile		Thickness in kms						Velocity in km/sec						Location	
		Water	h_2	h_3	h_4	h_5	h_6	V_2	V_3	V_4	V_5	V_6	V_7	Lat. (S)	Long. (W)
1	N	1.00	1.29	1.64	3.83			2.13*	2.98	4.89	6.21			40°04.3'	55°48'
	S	1.00	1.29	1.64	3.83									40°27.5'	56°14.6'
2	N	1.33	1.53	0.67	2.14	9.13	10.41	2.33*	3.20	4.61	5.90	6.70	8.05	40°00.2'	55°17.6'
	S	1.33	1.31	1.10	1.84	6.13	12.74							40°45.2'	56°18'
3	N	2.29	3.12	10.94	6.75			3.19*	5.97	6.96	8.16			40°02.6'	54°46'
	S	2.29	3.49	7.23	9.42									40°51'	55°29'
4	N	3.29	3.71	15.23				3.19*	6.27	8.00				40°06.8'	54°34'
	S	3.29	4.13	7.33										40°59.8'	55°16.7'
C_1		4.20	2.82	2.57				2.60*	4.53	5.73				40°46.1'	54°48.3'
5	N	4.37	2.12	0.91	1.79			2.60*	(4.10)	4.91	6.14			40°43.8'	54°38'
	S	4.37	2.12	0.91	1.79										
A_1		4.58	2.70	0.97	1.41			2.60*	4.11	4.83	6.25			40°49.4'	54°32.6'
B_1		4.92	3.47	0.94				2.85*	4.71	6.00				40°20'	53°44'
B_2		4.80	2.99	1.17				2.60*	4.79	5.84				40°51.7'	54°21.0'
6	N	4.83	2.62	0.90				2.60*	(4.80)	6.61				40°53.5'	54°19'
	S	4.83	2.62	0.90											

NOTES:

Asterisks denote mean velocity in the sediments determined from sonobuoy from nearby sonobuoy stations (Table 1).

Parenthesis indicate assumed velocity.

All other velocities, except those of profiles 2-4, are unreversed refraction velocities.

Water thickness refers to the depth of the base line used for topographic corrections. The topography has not been restored into the calculations.

Profiles 1-4 were recorded by ship-deployed hydrophones; A, B, C by a ocean bottom seismograph; and 5 and 6 by long-range (low frequency) sonobuoys. The profiles are listed in west-east order.

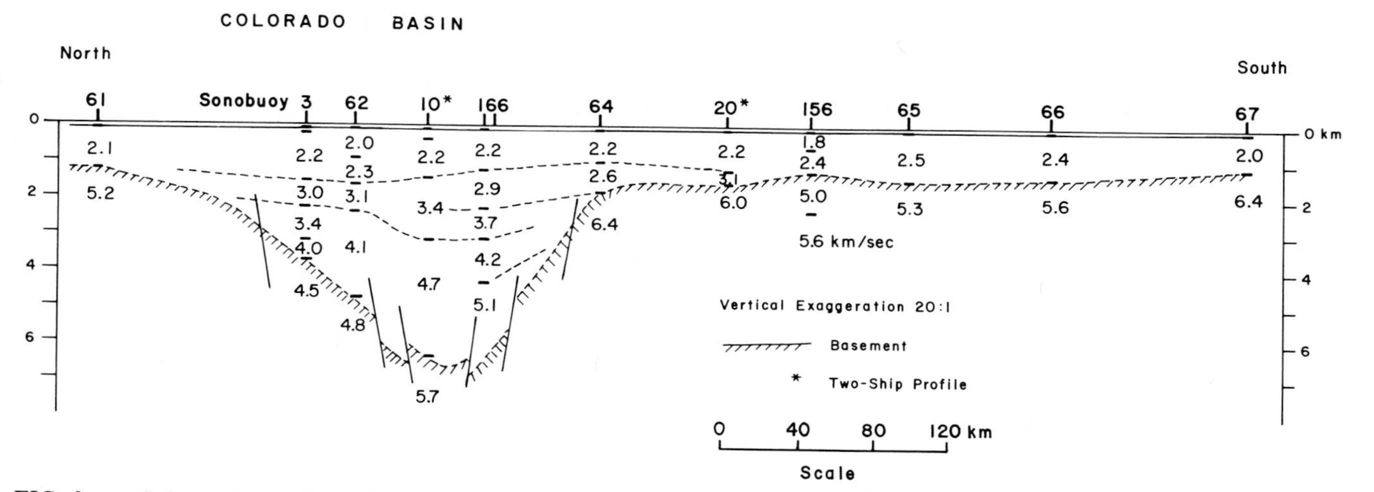

FIG. 3. Schematic north-south structure section of the Colorado Basin. The taphric nature of the basin is discussed by Zambrano (1974).

At this writing, no wells have been drilled in the eastern, deeper portion of the basin where appreciable changes in lithology and/or structure may occur. Note that the continental crust continues seaward at least 80 km beyond the edge of the shelf, an observation that will be emphasized later in the discussion of the slope profiles. The important point here is that there appears to be seaward closure of the Colorado Basin in the deeper high-velocity layers; i.e., there is a considerable thickness of Early Cretaceous continental sediment behind an outer basement ridge (cf. Ludwig et al, 1978a). This has great significance because it remains to be explained why the Colorado Basin (and Salado Basin), which was supposedly open to the South Atlantic since Early sulted in entrapment of the Early Cretaceous continental section and the feature was an effective barrier to marine transgression. The barrier ridge now lies buried beneath the continental slope on continental crust, near the junction of continental and oceanic crust. Alternately, the continental margin remained above sea level until the end of the Cretaceous.

CONTINENT-OCEAN CRUST BOUNDARY

From examination of Figure 2, it is apparent that layers of velocity and thickness indicative of continental crust extend 110 km beyond the shelf edge to the vicinity of two-ship refraction profile 3; velocity structures typical of oceanic crust are found

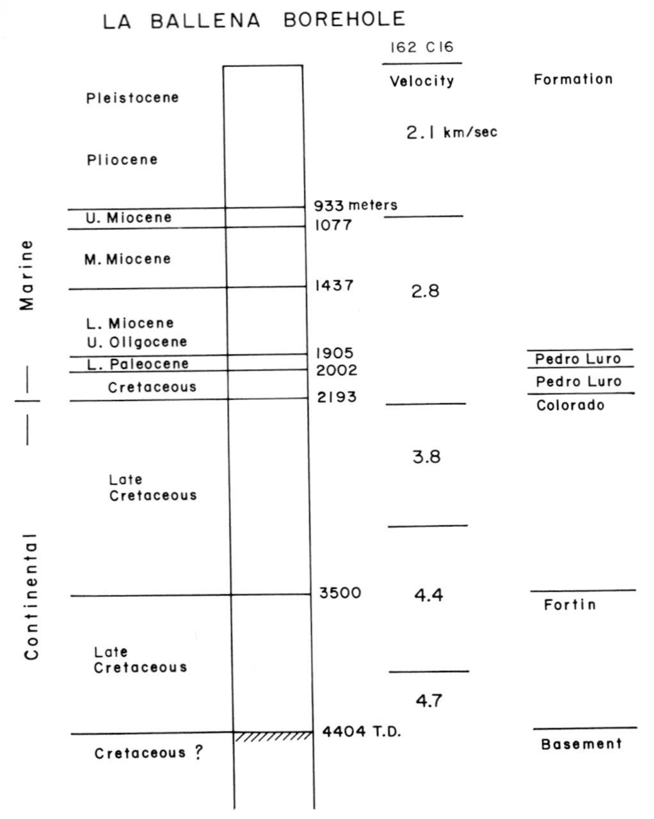

FIG. 4. Generalized log of La Ballena well drilled by Hunt International Petroleum Co. (courtesy of Yacimientos Petrolíferos Fiscales), showing correlation with velocity structure.

seaward of profile 5. Hence, the change from continental crust to oceanic crust takes place between profiles 3 and 5, within a distance of 60 km. Furthermore, the transition is marked by appreciable thinning of the lower crustal rocks.

As was mentioned previously, the outer shelf–upper slope region of the Argentine margin is characterized by coincident linear magnetic and isostatic gravity anomalies that denote a change in character of the basement rock. The plotted position of magnetic anomaly G from the map of Rabinowitz and LaBrecque (in press, Fig. 2) onto Figure 2 places it near the location of profile 3, the area at which the basement is observed to dip steeply beneath the continental slope. In Figures 5 and 6, we have plotted the magnetic and gravity anomalies in relation to observed seismic structure at other locations. Note from comparison of Figures 2, 5, and 6 that the presence of the anomalies is independent of the location of the shelf edge and independent of whether thick sediments are located landward or seaward of them (cf. Rabinowitz and LaBrecque, in press). The anomalies are associated with the hingeline of a steeply dipping basement surface.

According to Rabinowitz and LaBrecque (1978), the anomalies off eastern Argentina mark the contact between continental and oceanic crust. They rotate the location of the anomalies to fit the anomalies of southwest Africa and produce a remarkably good paleofit of the two continents. Some support for their proposed uncompensated oceanic basement high lying adjacent to continental crust, to account for the isostatic gravity anomaly, may come from

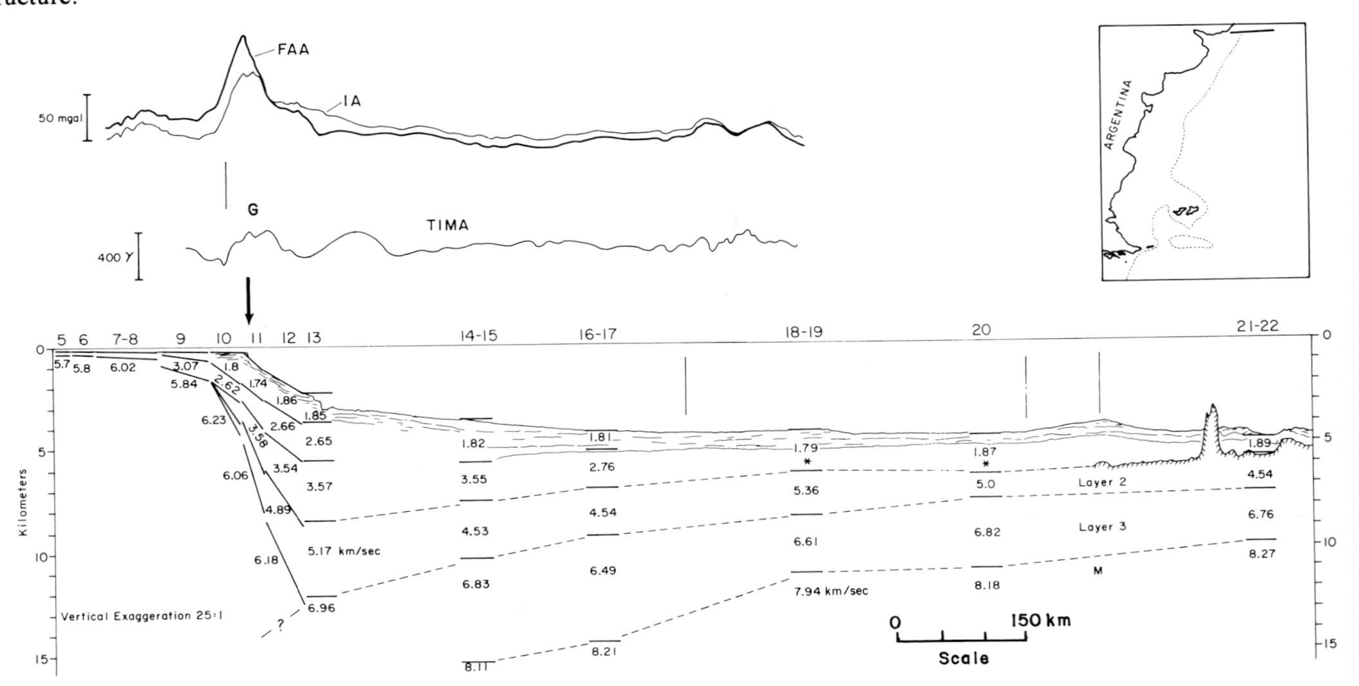

FIG. 5. Schematic structure section of the margin off Punta del Este, Uruguay, from Leyden et al, (1971). Arrow denotes the shelf break. Plotted above is the free-air gravity anomaly (FAA), isostatic gravity anomaly (IA), and total intensity magnetic anomaly (TIMA). The area of maximum declination of the basement towards the ocean basin corresponds in position with the coincident anomalies. The magnetic anomaly, called Anomaly G, is modeled as an edge-effect anomaly separating continental and oceanic crust; the steep landward gradient of the isostatic anaomaly may result from an uncompensated oceanic basement high adjacent to continental crust (Rabinowitz and LaBrecque, 1977, 1978).

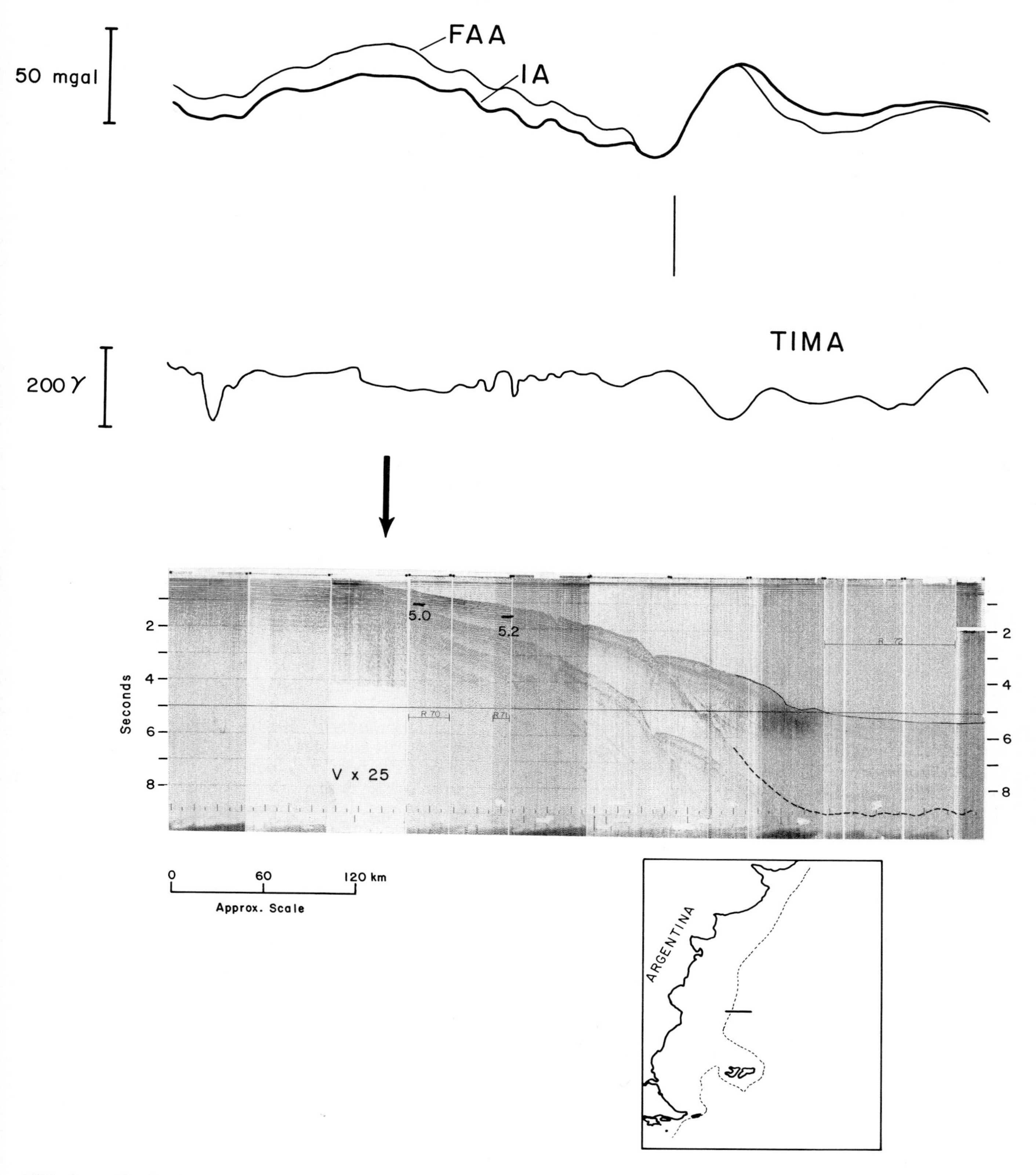

FIG. 6. Single-channel seismic reflection profile of the margin off Golfo San Jorge. The vertical scale represents two-way reflection time. Weak reflections from the basement are retraced for clarity. Explanation same as for Figure 5. The velocity in the sediments overlying basement beneath the continental rise (72C15) is 4.60 km/sec (Table 1 of Ludwig et al, 1978).

the refraction work of Keen and Keen (1974) and Keen et al (1974) on the continental margin of eastern Canada.

Like the margin of Argentina, the outer shelf–upper slope region off eastern Canada and the United States has a prominent magnetic anomaly, called the East Coast magnetic anomaly, that lies on the landward side of the magnetic quiet zone. Off the eastern United States there is a free-air gravity high over the outer shelf edge, but it is not always coincident with the magnetic slope anomaly (e.g., Schlee et al, 1976). The isostatic gravity anomaly has not yet been compared with the magnetic anomaly. Various hypotheses incorporating the presence of basement ridges, volcanic ridges, and reefs have been advanced to explain the magnetic and free-air gravity anomalies. Keen et al (1974) made a series of refraction profiles of the continental slope off Nova Scotia. The results of one profile located near the magnetic anomaly showed sediments resting directly on rocks intermediate in velocity between that of oceanic layer 3 and mantle (7.4 km/sec). Although it has not yet been determined if a gravity high is coincident with the magnetic anomaly, this lends some support to the idea of Keen et al (1974) and Rabinowitz and LaBrecque (1977, in press) that, during early rifting of the continents, material of composition similar to oceanic crust is injected between the separating blocks.

Our refraction profiles, on the contrary, do not indicate anomalous high-velocity (high-density) crust or oceanic crust lying immediately adjacent to continental crust. Velocities near 7 km/sec are observed beneath the continental slope anomaly. These are associated with a lower continental crustal layer or a seismic discontinuity (Conrad discontinuity?) which may, like the crust/mantle boundary, be continuous across the continent-ocean crust boundary. Velocities in the basement beneath the lower continental slope are representative of either oceanic basement (layer 2C of Houtz and Ewing, 1976) or continental basement. However, the thickness of the layer is of present-day continental proportions, indicating perhaps that continental basement has subsided to form the continental slope and, possibly therefore, that the seaward side of the anomalies might better define the boundary between continental and oceanic crust.

CONCLUSION

The Argentine continental margin north of 46°S is an old, well-developed rifted margin. From all indications, separation from Africa took place in the Early Cretaceous (Rabinowitz and LaBrecque, in press). Thinning of the edge of the continental crust has occurred either early in the development of the margin or was due to subsidence of the margin after rifting. It seems probable that, in places, the pre-rifting or rifting stage involved the formation of an outer shelf ridge which received a capping of reefal and/or volcanic material that prevented deposition of marine sediments in the developing shelf basins until the end of the Cretaceous. The cause of the breaching and drowning of the proposed barrier ridge by South Atlantic waters in Maestrichtian-Paleocene time might be related to a relative rise of sea level due to increased subsidence of the shelf platform in response to sediment loading (cf. Sleep, 1976; Pitman, 1978). Schlee et al (1976) indicated the presence of a deeply buried outer shelf ridge, which they interpreted as possible carbonate rocks, in their reflection profiles of the margin of the eastern United States.

The question of whether the basement of the continental slope is subsided continental crust or elevated oceanic crust must await the result of deep sea drilling. The continental margin off Golfo San Jorge (Fig. 6) appears to have comparatively little sediment over the supposed contact between continental and oceanic crust; core holes could be drilled to basement on either side. The possible presence of an outer basement ridge at the continent-ocean crust boundary capped with a thick reefal complex is of significance with respect to the hydrocarbon potential of the outer shelf, as structural-stratigraphic traps might localize accumulations of oil and gas on either side of the basement ridge or over the top.

In this report we have offered one possible explanation for the nondeposition of marine sediments in the Colorado (and Salado) Basin until the close of the Cretaceous—because of the presence of an outer barrier ridge. We wish to also point out that, exclusive of the basins, the sediment thickness over the Argentine continental shelf is about 1500 m; over the Deseado high area of the shelf, between the San Jorge and Malvinas basins, the sediments are only about 500 m thick (Ludwig et al, 1978a). Hence, these thicknesses of sediment represent the total accumulation over a 130+ m.y. period. Either very little post-rift subsidence of the shelf has occurred or there have been long periods of lowstand of eustatic sea level and erosion of sediments. In the latter instance, it is difficult to account for the present-day location of the eroded sediments.

APPENDIX

The seismic refraction measurements were made using short- and long-range sonobuoys, ocean bottom seismograph, and ship for recording. Short-range sonobuoys (type SSQ supplied by the U.S. Navy) were launched while reflection profiling with a 25-cu. in. or 1-cu. ft airgun, and refraction data transmitted by radio from the buoy were recorded (variable density) at two frequency bands on a shipboard two-channel drum recorder (e.g., Ewing et al, 1969; Ludwig et al, 1975). Aquatronics, Inc. (now Fairfield Industries) SLF–73 sonobuoys launched from *Conrad* and hydrophones suspended from *Austral* while on listening station were the detectors used in the long-range refraction experiments. In these instances, explosions set off by *Conrad* generated the acoustic signals. Incoming signals from the buoys and hydrophones were recorded as oscillograms on board *Conrad* and *Austral*, respectively. Both vessels were equipped with a 12-channel Dresser S.I.E. portable analog reflection-refraction

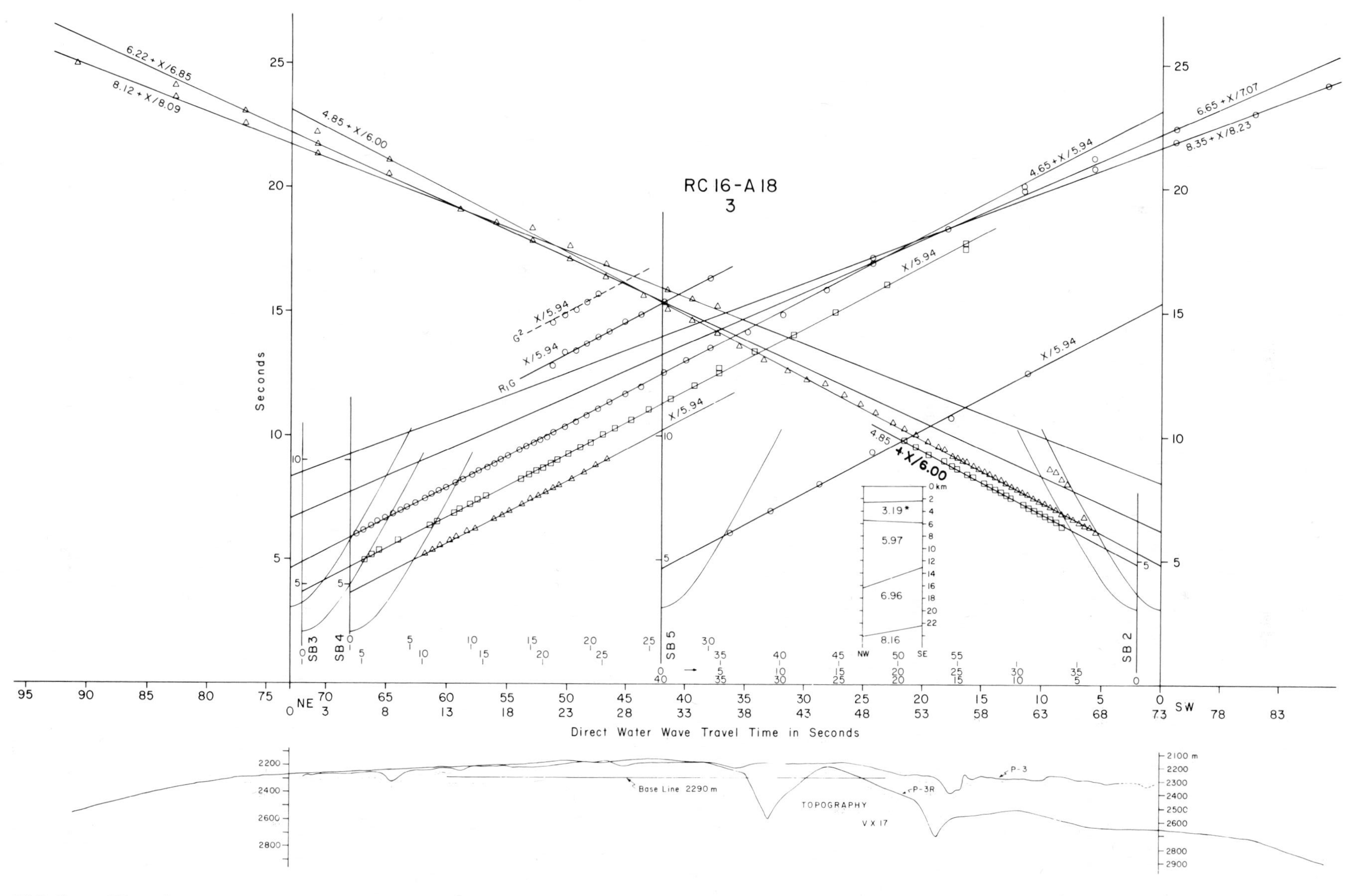

FIG. 7. Time-distance graph of profile 3. The profile was recorded near magnetic anomaly G and gives velocities and layer thicknesses similar to those measured beneath the shelf.

system (modified to receive the input of hydrophones and sonobuoys). Low-frequency amplifiers for the detection of refracted waves were operated without low-cut filtering, hence the low-frequency response was set by the sonobuoys or hydrophones. High-cut filters of 16 Hz, 35 Hz, and 75 Hz were used, depending upon range and energy propagation. High-frequency-rectified (water break) amplifiers operated in the frequency passband 600 Hz–3.5 Hz were used to distinguish direct water waves and bottom-reflected waves.

The analysis of the long-range refraction data was done in conventional manner: picking the seismograms, reducing the travel-time data to surfaces of reference, constructing a travel-time graph for each profile, fitting the data points by eye to straight-line segments, and computing velocity structure sections on the assumption of constant-velocity layers separated by plane dipping boundaries (e.g., Ludwig et al, 1968). The mean velocity in the sediments at each long-range refraction station was determined from nearby airgun-sonobuoy stations.

The interpretations of the travel-time graphs are fairly straightforward.[4] The forward and reverse halves of profiles 1–4 were recorded on board *Austral; Conrad* launched sonobuoys during the shooting runs and recorded the data transmitted by the buoys. By this method, up to four refraction profiles were recorded along a single shooting line for the determination of velocity layers (e.g., Fig. 7). Profiles 5 and 6 are unreversed long-range sonobuoy profiles.

Profiles A, B, and C were recorded with an ocean-bottom seismograph. The Lamont-Doherty OBS is a 3-unit (vertical and horizontal seismometers and a hydrophone) self-recording pop-up system that is contained in a buoyant sphere with a time-release mechanism (Carmichael et al, 1973). The experimental procedure of making seismic refraction measurements with a detector on the ocean floor and method of data reduction are described by Ewing and Ewing (1961) and Davis et al (1976), among others.

REFERENCES CITED

Carmichael, D., et al, 1973, A recording ocean-bottom seismograph: Jour. Geophys. Research, v. 78, p. 8748–8750.

Davis, E. E., C. R. B. Lister, and B. T. R. Lewis, 1976, Seismic structure of the Juan de Fuca ridge: ocean-bottom seismometer results from the median valley: Jour. Geophys. Research, v. 81, p. 3541–3555.

Ewing, J., and M. Ewing, 1961, A telemetering ocean-bottom seismograph: Jour. Geophys. Research, v. 66, p. 3863–3878.

——— R. Leyden, and M. Ewing, 1969, Refraction shooting with expendable sonobuoys: AAPG Bull., v. 53, p. 174–181.

Ewing, M., and A. Lonardi, 1971, Sediment transport and distribution in the Argentine basin; 5, Sedimentary structure of the Argentine margin, basin, and related provinces, *in* L. H. Ahrens and others, eds., Physics and Chemistry of the Earth: Pergamon Press, New York, v. 8, p. 125–249.

——— W. J. Ludwig, and J. I. Ewing, 1963, Geophysical investigations in the submerged Argentine coastal plain: Part I, Buenos Aires to Peninsula Valdez: Geol. Soc. America Bull., v. 74, p. 275–292.

Houtz, R., and J. Ewing, 1976, Upper crustal structure as a function of plate age: Jour. Geophys. Research, v. 81, p. 2490–2498.

Keen, C. E., and M. J. Keen, 1974, The continental margins of eastern Canada and Baffin Bay, *in* C. A. Burk and C. L. Drake, eds., The geology of continental margins: New York, Springer-Verlag, p. 381–389.

——— et al, 1974, Some aspects of the ocean-continent transition at the continental margin of eastern North America: Geol. Survey Canada Paper 74-30, v. 3, p. 188–197.

LaBrecque, J. L., and P. D. Rabinowitz, 1977, Magnetic anomalies bordering the continental margin of Argentina: AAPG Argentine Map Series.

Larson, R. L., and J. W. Ladd, 1973, Evidence for the opening of the South Atlantic in the Early Cretaceous: Nature, v. 246, p. 209–212.

Leyden, R., W. J. Ludwig and M. Ewing, 1971, Structure of continental margin off Punta del Este, Uruguay, and Rio de Janeiro, Brazil: AAPG Bull., v. 55, p. 2161–2173.

Ludwig, W. J., J. I. Ewing, and M. Ewing, 1968, Structure of Argentine margin: AAPG Bull., v. 52, p. 2337–2368.

——— R. E. Houtz, and J. I. Ewing, 1975, Profiler-sonobuoy measurements in the Colombia and Venezuela basins, Caribbean Sea: AAPG Bull., v. 59, p. 115–123.

——— et al, 1978a, Sediment isopach map of the Argentine continental margin: AAPG, Argentine Map Series.

——— et al, 1978b, Structure of Falkland plateau and offshore Tierra del Fuego, Argentina: this volume.

Pitman, W. C., 1978, The relationship between eustasy and stratigraphic sequences of passive margins: Geol. Soc. America Bull., v. 89, no. 9, p. 1389–1401.

Rabinowitz, P. D., 1976, Geophysical study of the continental margin of southern Africa: Geol. Soc. America Bull., v. 87, p. 1643–1653.

——— 1977, (Map of) Free-air gravity anomalies bordering the continental margin of Argentina: AAPG Argentine Map Series.

——— and J. L. LaBrecque, 1977, The isostatic gravity anomaly: key to the evolution of the ocean-continental boundary at passive margins: Earth and Planetary Sci. Letters, v. 35, p. 145–150.

——— ——— in press, The Mesozoic South Atlantic Ocean and evolution of its continental margin: Jour. Geophys. Research.

Schlee, J., et al, 1976, Regional geologic framework off northeastern United States: AAPG Bull., v. 60, p. 926–951.

Shell Production Company of Argentina, Ltd., 1962, Algunas observaciones geológicas a lo largo del borde septentrional del escudo Patagonico: Buenos Aires, Anales Primeras Jornadas Geológicas Argentinas, v. 2, p. 323–336.

Sleep, N. H., 1976, Platform subsidence mechanisms and eustatic sea-level change: Tectonophysics, v. 36, p. 45–56.

Urien, C. M., and J. J. Zambrano, 1973, The geology of the basins of the Argentine continental margin and Malvinas Plateau, *in* A.E.M. Nairn, and F. G. Stehli, eds., The ocean basins and margins, 1. The South Atlantic: New York, Plenum Press, p. 135–170.

——— L. R. Martins, and J. J. Zambrano, 1976, The geology and tectonic framework of southern Brazil, Uruguay and north Argentina continental margin: their behavior during the southern Atlantic opening, *in* Continental margins of Atlantic type: Anais da Academia Brasileria de Ciencias, v. 48, p. 365–376.

Yrigoyen, M., 1975, Geología del subsuelo y platforma continental, *in* Relatorio sobre la geología de la provincia de Buenos Aires, p. 139–435.

Zambrano, J. J., 1972, La cuenca del Colorado, *in* Geología regional Argentina; Acad. Nac. Ciencias, p. 419–438.

——— 1974, Cuencas sedimentarias en el subsuelo de la provincia de Buenos Aires y zonas adyacentes: Revista de la Asociación Geológica Argentina, v. 29, p. 443–438.

——— and C. M. Urien, 1970, Geological outline of the basins in southern Argentina and their continuation off the Atlantic shore: Jour. Geophys. Research, v. 75, p. 1363–1396.

——— ——— 1974, Pre-Cretaceous basins in the Argentine continental shelf, *in* C. A. Burk and C. L. Drake, eds., The geology of continental margins: New York, Springer-Verlag, p. 463-473.

[4]For the complete set of travel-time graphs, order from the National Auxiliary Publications Service. See footnote #1 on the first page of this article for information.

Structure of Falkland Plateau and Offshore Tierra del Fuego, Argentina[1]

W. J. LUDWIG,[2] C. C. WINDISCH,[2]
R. E. HOUTZ,[2] AND J. I. EWING[2,3]

Abstract Velocity solutions from 165 airgun-sonobuoy reflection/refraction stations on the continental margin of southernmost South America are presented in 10 seismic structure sections. Also included are an isopach map of Cretaceous and Tertiary sediments (seismic velocities generally less than 4.2 km/sec) in the Magellan, Malvinas, Falkland Plateau, and Falkland Trough basins, and three north-south single-channel seismic reflection traverses of the Falkland Plateau, Falkland Trough, and North Scotia Ridge. The Falkland Trough is a sliver of oceanic crust between the Falkland Plateau and the North Scotia Ridge. The acoustic basement and deep sedimentary layers of the plateau extend beneath the trough and then are subducted beneath the northern flank of the ridge. Movement of the ridge and the plateau toward each other has apparently resulted in the deformation and uplift of the overlying sediments to form the northern flank of the ridge.

INTRODUCTION

Lithic and structural mapping on South Georgia Island at the eastern end of the North Scotia Ridge, on Isla de los Estados (Staten Island) at the southeastern tip of Tierra del Fuego, in Cordillera Darwin of Tierra del Fuego, and in the Patagonian Andes (Fig. 1), has resulted in several papers that attempt to establish the interrelationships of the southernmost Andes, Scotia Arc, and Falkland[4] (Malvinas) Plateau and their relation to Gondwanaland and the early opening of the South Atlantic Ocean (Dalziel and Elliot, 1973; Dalziel, 1974; Dalziel et al, 1974a,b, 1975; Barker et al, 1976; de Wit, 1977). These authors view the western margin of southernmost South America as being the site of an andesitic volcanic arc initiated in Late Jurassic time, prior to the separation of the continents. The Falkland Plateau region of South America was attached to South Africa along the present southern margin of Agulhas Bank (Rabinowitz and La Brecque, in press). South Georgia was attached to South America along the present southern margin of Burdwood Bank. During Early Cretaceous time the arc broke away from the continent to form a small marginal ocean basin between the arc and the continental interior (Fig. 1, upper inset). During early Late Cretaceous (Cenomanian) time, the arc moved back toward the continent, resulting in deformation and uplift of the marginal basin floor and the adjacent arc to produce parts of the Andean cordillera. Later fragmentation of the southernmost Andes carried South Georgia 2,000 km eastward to its present location. In this scheme, the Burdwood Bank and North Scotia Ridge would represent a Late Jurassic–Early Cretaceous remnant arc. It also has been proposed that the Falkland Trough may represent an even older marginal basin (de Wit, 1977).

According to Dalziel et al (1974, 1975), Natland et al (1974), and H. J. Harrington (*in* Ludwig et al, 1965), the Magellan Basin of Patagonia was initiated in Early Cretaceous time, and received sediments from an eastern to northeastern source to fill depressions developed on the Serie Tobifera, a largely

[4]Followed here is the usage common in English scientific literature of Falkland Islands, Falkland Plateau, etc. In Argentine scientific literature, the name Malvinas is used.

[1]Manuscript received, August 19, 1977; accepted, January 27, 1978.

[2]Lamont-Doherty Geological Observatory of Columbia University, Palisades, New York 10964.

[3]New address: Woods Hole Oceanographic Institution, Woods Hole, Massachusetts 02543.

This survey work was undertaken as part of the International Decade of Ocean Exploration, through support from the U.S. National Science Foundation (Grants GX 34410 and IDO/OCE 72-06426). Additional support was provided by the Oceanography Section of NSF (Grant GA 27281) and by the U.S. Office of Naval Research contracts N00014-67-A-0108-0004 and N00014-75-C-0210.

The seismic data were collected on board R/V *Conrad* (1972–73) and R/V *Vema* (1974) under the supervision of J. Ewing, C. Windisch, G. Carpenter, and G. Bryan. Argentine participants on one or more legs of the survey vessels were A. Yung, F. Mouzo, P. Margalot, R. Adamini, J. Isquierdo, and E. Schnaak. The planning of the survey work was done by J. Ewing in cooperation with H. P. Casal, Comite Nacional de Oceanografia, and A. Lonardi of Instituto Argentino de Oceanografia.

B. Tucholke and G. Bryan critically read the manuscript and provided helpful suggestions for improvement.

Lamont-Doherty Contribution number 2740.

Article Identification Number:
0065-731X/78/MO29-0008/$03.00/0.
(see copyright notice, front of book)

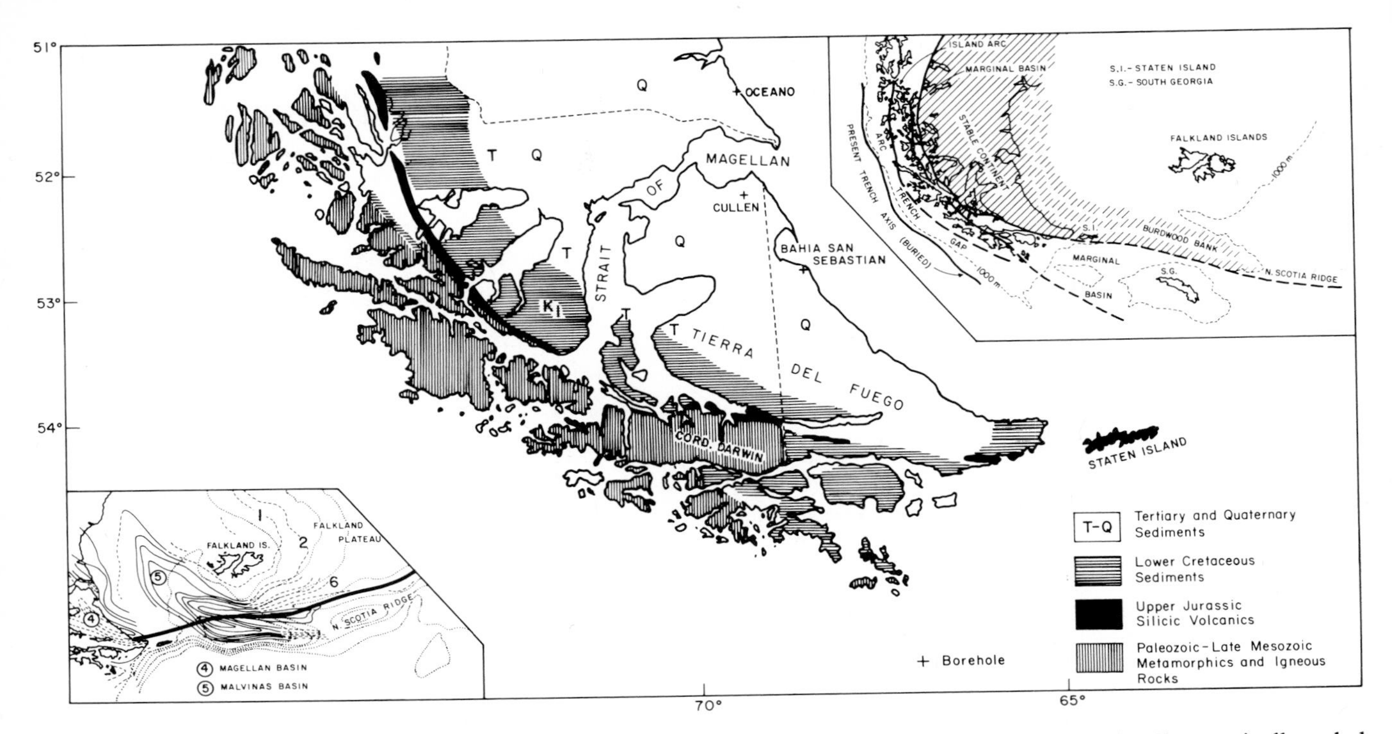

FIG. 1. Simplified map of the distribution of rocks in the Magellan region, after Tezón (1964). The vertically ruled areas (metamorphic and igneous rocks) constitute the Andean Cordillera. The Patagonian Cordillera is approximately defined by exposures of Cretaceous and Jurassic rocks represented by the horizontal ruled areas and black areas, respectively. Upper right corner inset shows tectonic setting of the Magellan region in Early Cretaceous time (modified from Dalziel et al, 1975). Lower left corner inset shows relations between the Magellan Basin and the Malvinas Basin. Contours in kilometers. Contour datum is top of 4.5 to 6.4 km/sec basement layers (after Ludwig et al, 1968; cf. Fig. 8). Heavy line represents axis of gravity minimum from Rabinowitz (1977).

Jurassic tuffaceous formation. The basal sediments of the Springhill Formation were succeeded by marine sediments which onlap Tobifera topographic highs. As a result of the initial uplift of the Andean cordillera, the Magellan Basin subsided rapidly, and a thick wedge of marine Upper Cretaceous-lower Tertiary flysch and molasselike sediments was deposited. During Oligocene time, the Patagonian cordillera was uplifted for the first time and a new marine basin came into existence and began to receive sediments from the cordillera in the west.

Petroleum and gas production in the Magellan Basin comes from the Springhill Formation, which has been exploited extensively in the Chilean part of the basin by Empresa Nacional del Petroleo (Olea and Davis, 1977). Exploration by Yacimientos Petroliferos Fiscales of Argentina (YPF) has involved drilling near the coast north and south of the Magellan Straits and (CDP) reflection seismic surveys between Tierra del Fuego and the Falkland Islands. It is generally agreed by explorationists that the offshore part of the Magellan Basin has potential as a future petroleum province.

Our purpose is to present the results of 165 airgun-sonobuoy reflection/refraction profiles offshore from Tierra del Fuego and on the Falkland Plateau that bear on the structural framework of this part of the continental margin (Table 1). Included are three north-south-oriented reflection record sections of the Falkland Plateau. A subsequent paper will deal with the details of these and other reflection measurements of the plateau. In previous papers (Ludwig et al, 1965, 1968; J. Ewing et al, 1971), we described the velocity structure and outline of the Magellan Basin and the Malvinas Basin between Tierra del Fuego and the Falkland Islands (Fig. 1). We showed that the Falkland Plateau is a southward-tilted continental block, capped by a thick wedge of sediments which extend southward into the Falkland Trough. Ocean crust velocities underlie the Falkland Trough. Left unanswered was the structural-stratigraphic relationship of the Magellan Basin, Malvinas Basin, and Falkland Plateau.

The new seismic data were obtained during 1972–1974 by Lamont-Doherty Geological Observatory personnel aboard R/V *Conrad* and R/V *Vema*, as part of the International Decade of Ocean Exploration (Fig. 2). The survey work was carried out in cooperation with Consejo Nacional de Investigaciones Científicas y Técnicas, Instituto Argentino de Oceanografía, and Servicio Hidrografía Naval.

OFFSHORE TIERRA DEL FUEGO

Figures 3 and 4 show schematic structure sections of the continental shelf off Tierra del Fuego, based on the results of the airgun-sonobuoy measurements.

TABLE 1. Results of Airgun-Sonobuoy Stations, Argentine Continental Margin South of 45°S

		Thicknesses, km							Velocity, km/sec									
Sonobuoy	Water	h_2	h_3	h_4	h_5	h_6	h_7	h_8	V_2	V_3	V_4	V_5	V_6	V_7	V_8	V_9	Lat. (S)	Long. (W)
4C11	2.30	1.36	1.41	.97					2.00	2.80	3.90	4.50					50°14.1'	51°52.3'
143C12	.82	.47							(1.8)	5.10							51°25.2'	56°10.1'
144C12	1.59	.72	.56	1.63					1.87*	2.21*	3.15*						50°28.4'	54°02.8'
145C12	.62	.53	1.15						2.00	2.80	6.22						49°07.5'	56°50.7'
146C12	5.72	.61							2.18*								46°23.9'	51°26.0'
147C12	5.73	.34	.23						2.04*	2.14*							46°21.4'	51°20.5'
148C12	5.28	.57	1.34						1.70*	2.55*	6.25						44°18.2'	55°34.3'
15C15	.48	1.07	1.78	1.98					(1.8)	2.15	3.80	5.30					53°14.9'	63°15.3'
16C15	.37	.88	1.14	1.72					(1.8)	2.55	3.15	4.85					53°04.0'	63°18.9'
18C15	.27	.62	1.24	1.64					(1.8)	2.30	3.40	4.90					52°41.5'	64°01.5'
19C15	.13	.65							1.90	2.35							54°14.8'	64°32.9'
20C15	.07	.11	1.08	2.46					3.35	4.20	5.00	5.60					54°36.6'	64°00.8'
21C15	3.14	.64	1.10	3.36	1.68				1.76*	2.34*	3.44*	5.46*					52°50.6'	52°59.7'
22C15	2.32	1.16	2.57						2.45	3.70	5.65						49°56.8'	43°16.4'
23C15	3.99	.99	1.03	2.29					1.69*	2.25*	5.10	6.40					51°12.9'	39°10.5'
24C15	4.03	1.59	.77						1.84*	(2.8)							51°20.2'	38°52.6'
37C15	2.48	.24	.39						2.50	4.00	4.90						49°44.5'	47°35.3'
38C15	2.68	.47	.71	1.07	.64				1.65*	2.48*	4.00	4.70	5.45				50°07.7'	47°55.0'
39C15	2.78	.41	.86	.89	.87	1.47			1.78*	2.19*	2.75*	3.09*	(4.0)	5.90			51°01.4'	48°51.0'
41C15	2.44	.96	.53	.74	.53	.47			1.72*	2.51*	2.79*	3.12*	(4.0)	4.80			51°39.5'	49°30.1'
42C15	.14	.70							2.40	3.40							54°28.6'	58°29.2'
43C15	.10	1.04							2.00	2.85							54°07.3'	58°19.3'
44C15	.71	.44	.48	1.28	.76				(1.8)	2.40	3.25	5.40	6.95				53°01.9'	57°55.6'
45C15	.33	1.84							5.01*								52°18.1'	57°40.9'
46C15	.47	.98	.42						2.00	2.90	5.40						50°24.3'	56°34.2'
47C15	.79	.54	.48						(2.0)	3.70	4.30						49°30.4'	55°48.7'
48C15	.90	.59	.96						(2.0)	3.85	5.50						49°22.5'	55°43.7'
49C15	5.27	.49	2.02	1.13	.90	1.98			1.79*	2.85*	5.60	6.80	7.45	8.25			44°56.6'	56°23.1'
68C15	.10	1.06	.64	.94					2.00	5.00	5.65	6.10					44°17.3'	61°02.6'
69C15	.11	.52	.68	.68					2.10	2.50	4.95	5.70					44°28.3'	61°06.4'
70C15	.34	.57	1.12						2.00	4.95	6.50						46°51.7'	60°32.4'
71C15	.70	.49							1.83*	5.15							46°52.4'	59°59.4'
72C15	3.66	1.00	1.37	.54	1.17				2.00*	2.20*	2.61*	4.30*	4.60				47°05.0'	57°51.4'
73C15	5.97	.48	.48	.88	2.41				1.60*	2.36*	2.59*	5.10	7.10				48°21.3'	45°57.4'
74C15	2.48	.52	.49	.68	.69	.64			1.62*	1.87*	2.66*	2.91*	4.40	5.40			52°04.8'	48°06.9'
75C15	2.14	.41	.39	.92	.93	1.25	.63	.64	1.59*	1.84*	2.26*	2.83*	3.45*	4.00*	4.23*	5.10	51°36.6'	52°17.3'
76C15	1.76	.58	.89	.64	1.00	1.49	1.29		1.71*	2.23*	2.76*	3.45*	3.90*	4.67*			51°28.3'	53°52.3'
77C15	1.29	.94	1.08						1.87*	2.71*	3.80						51°20.2'	55°01.4'
78C15	.59	.36	1.12						(2.0)	5.30	6.05						51°11.5'	56°17.9'
79C15	.14	.96							4.35	5.40							50°58.8'	58°10.7'
80C15	.13	.11	.41						(1.8)	4.70	6.10						50°45.3'	61°13.5'
81C15	.18	.37							4.85	5.80							51°07.6'	61°47.0'
82C15	.24	.20							(1.8)	6.25							51°49.1'	62°24.4'
83C15	.23	.69							1.80	5.80							52°03.7'	62°48.5'
84C15	.25	1.09	.24						1.90	2.50	5.30						52°19.0'	63°07.9'
85C15	.26	1.28	.97	.69	.26				2.15	2.50	4.05	4.40	5.45				52°31.3'	63°21.6'
86C15	.28	.99	1.55	1.80					1.90	2.45	4.30	5.70					52°43.0'	63°38.0'
87C15	.29	.29	.42	.81	1.06	.57	.64	1.90	1.73*	1.94*	2.74*	3.17*	3.90	4.80	6.15		52°52.2'	63°59.1'
88C15	.20	.93	1.82	1.33					2.05	2.55	4.80	6.50					53°09.4'	64°26.9'
89C15	.13	.85	1.76	.89					1.95	2.50	4.70	5.95					53°26.1'	64°51.2'
90C15	.10	.41	.47	.94					1.90	2.45	2.90	3.95					53°52.3'	65°32.3'
91C15	.09	.19	.84						2.30	2.60	3.40						54°07.6'	65°45.9'
92C15	.07	1.04							2.65	3.70							54°12.9'	65°57.2'
93C15	.05	.52							2.60	3.30							53°50.7'	67°07.4'
94C15	.12								2.10								51°57.7'	66°01.1'
95C15	.12								2.15								51°51.4'	65°53.3'
96C15	.14	.81							2.05	2.55							51°43.0'	65°41.1'
97C15	.13	.74	.69	.55					1.90	2.70	3.05	5.30					51°07.4'	64°50.0'
98C15	.14	.42	.65						2.00	4.15	5.30						50°10.4'	63°26.0'
99C15	.14	.52	.86						1.95	4.45	6.50						49°12.2'	61°55.6'
100C15	.14	.54							2.05	6.05							47°35.2'	61°58.5'
101C15	.10	.53							2.20	5.05							46°49.7'	64°04.7'
102C15	.07	.45	.75						2.00	3.60	4.55						46°28.6'	65°21.4'
103C15	.09	.43	.61	.39	2.32				2.10	2.80	4.30	5.10	6.15				45°38.9'	66°25.6'
104C15	.08	.62	.47						2.20	4.35	5.80						44°34.3'	63°55.9'
90C16	.15	.53	.38	.81					1.95	2.80	3.40	4.75					47°45.8'	61°39.5'
92C16	.12	.40							(1.8)	5.25							49°22.5'	64°33.9'
93C16	.11	.32	.50						(1.8)	2.30	4.75						50°21.5'	65°55.5'
94C16	.11	.58	1.67						1.80	4.95	6.25						50°41.3'	66°28.4'
95C16	.11	.48	.49	.98					(1.8)	4.20	5.05	6.10					51°00.8'	67°00.7'
96C16	.10	.22	.32	1.19					(1.8)	2.20	4.15	6.55					51°24.2'	67°32.2'
97C16	.07	.53							1.95	2.65							51°42.7'	68°07.6'
98C16	.08	.25	1.21	1.83					1.90	2.25	4.95	6.50					52°16.9'	67°47.0'
99C16	.08	1.33	.40	1.10					2.25*	3.75	5.25	6.55					52°49.4'	67°34.1'
100C16	.07	.33	1.28	.89					2.10	2.60	4.95	6.25					53°24.5'	67°08.3'
101C16	.11	.56	.96	1.12					2.00	2.50	4.70	5.45					53°18.5'	66°30.2'
102C16	.12	1.45	2.57						2.05	5.00	6.55						52°50.7'	65°43.6'
103C16	.16	.28	.42	.96	1.00	.90			(1.6)	1.95	2.35	3.35	4.65	5.60			52°21.6'	64°55.4'
104C16	.16	.83	.37	1.37					1.90	2.50	5.10	6.45					51°37.1'	64°00.3'
105C16	.16	.55	2.35						1.90	5.00	6.40						51°01.2'	63°06.7'
106C16	.18	.18							(1.8)	5.30							50°39.5'	62°36.9'

TABLE 1 (continued)

Sonobuoy	Water	Thicknesses, km							Velocity, km/sec								Lat. (S)	Long. (W)
		h_2	h_3	h_4	h_5	h_6	h_7	h_8	V_2	V_3	V_4	V_5	V_6	V_7	V_8	V_9		
107C16	.17	.39	.48	.79					(1.8)	3.55	4.40	5.25					49°40.7'	61°22.5'
108C16	.17	.75							(1.8)	5.80							49°18.4'	60°56.6'
109C16	.17	.63	1.35						(1.8)	4.40	5.60						49°24.6'	60°50.9'
110C16	.15								5.05								51°13.0'	61°33.4'
111C16	.13	.82							4.00	5.00							51°40.1'	61°34.7'
112C16	.46	.50	.44	1.46	1.05	1.51	2.61		1.57*	1.68*	2.26*	3.00*	4.10*	4.68	5.60		53°19.6'	62°58.3'
113C16	.51	2.71	1.60	1.07					2.23*	3.14*	3.48*						53°28.7'	63°12.9'
114C16	.56	2.40	2.00	.91					2.20*	3.47*	4.73*	5.45					53°35.5'	63°26.0'
115C16	.50	.57	1.56	2.03	1.53	.60			1.62*	2.45*	3.40*	4.35*	6.10	6.80			53°41.8'	63°37.2'
116C16	.15	2.23	2.85	.51					2.38*	3.59*	(4.0)	6.00					53°52.3'	63°55.0'
117C16	.14	.68	2.10	2.04					1.80	2.50	4.10	5.50					54°01.1'	64°04.6'
118C16	.09	.86							2.00	2.80							54°07.0'	64°16.9'
119C16	.12	.69	.55	1.04					1.80	2.50	3.30	3.85					54°15.1'	64°40.6'
120C16	.11								2.15								54°21.5'	64°53.0'
121C16	.11	1.25	1.86						2.15	2.80	4.75						54°15.3'	65°22.7'
122C16	.13	2.25	1.41						2.49*	4.15*							53°52.0'	64°33.5'
123C16	.10	.26	.87	1.93					(1.6)	1.90	2.50	4.20					53°25.4'	64°09.5'
124C16	.35	.77	1.89	.93	1.59				1.80	2.30	3.65	4.10	5.00				53°01.4'	63°11.4'
125C16	.38	.83	1.03	1.35	1.72				1.80*	2.19*	2.65*	4.01*					53°10.7'	63°04.5'
126C16	.45	3.68	1.50						2.26*	3.56*							53°20.2'	62°51.2'
127C16	.50	1.28	.71	2.62	2.30				(1.8)	2.35	2.85	3.70	5.40				53°37.5'	62°41.2'
128C16	.47	.87	1.13	1.15	1.61				(1.8)	2.05	2.40	3.75	5.80				53°50.2'	62°29.0'
129C16	.35	.79	1.23						1.80	2.85	3.90						54°00.9'	62°19.7'
130C16	.49	.44	.77						(1.8)	2.30	3.15						54°02.5'	62°33.2'
131C16	.41	1.75	.94						2.45	3.25	4.55						54°05.3'	62°47.9'
132C16	.35	2.02							2.70	4.15							54°11.0'	63°02.4'
133C16	No Data																54°27.5'	63°35.8'
134C16	.11	.23	2.13	1.35					2.45	3.00	4.90	6.20					54°27.7'	64°04.2'
135C16	.20	.59	.98						4.35	5.10	5.95						54°57.8'	65°18.4'
141C16	.14	.54							4.05	5.10							52°48.5'	59°25.3'
142C16	2.93	1.37	2.85	1.87					2.03*	3.51*	4.20						53°31.1'	56°36.4'
143C16	2.04	.88	1.37	2.18	1.67	1.05			1.86*	2.34*	3.66*	5.25*	(6.2)				51°06.4'	52°11.8'
144C16	2.48	.98	1.03	1.17	.65				2.01*	2.27*	3.50	4.15	4.60				49°59.2'	51°44.6'
145C16	1.94	.24	.57						(1.8)	4.35	5.20						49°27.4'	53°23.3'
146C16	.98	.71	.55						1.79*	2.36*	5.35						49°55.1'	54°57.6'
147C16	1.39	1.05							2.05*	5.10							50°03.1'	54°32.4'
148C16	1.87	1.33	1.48	.48					2.02*	3.16*	3.95	5.25					50°08.2'	53°12.6'
149C16	2.35	1.30	.59	1.23	1.23				2.17*	2.22*	3.35*	3.86*	5.00				50°12.3'	51°43.1'
150C16	2.81	2.04	.45	1.03	1.70	1.72			2.26*	2.75	4.20	4.65	5.35	6.10			50°20.1'	49°36.4'
151C16	2.70	1.09	.90	1.05	.17	1.04	.99		1.76*	2.72*	3.06*	(4.0)	5.55	5.95	6.60		51°25.9'	47°29.2'
152C16	1.58	3.24							2.70*								51°08.4'	43°19.2'
153C16	3.75	.96	.72	1.91	1.11	1.45			1.63*	2.51*	3.15*	(4.0)	5.35	6.90			52°07.1'	40°09.4'
154C16	3.70	1.06	1.88	1.47	1.34				1.65*	2.55*	4.85	6.00	6.95				51°49.4'	39°31.3'
4V22	5.07	.75	1.85						1.82*	2.54*							44°23.2'	41°22.3'
15V31	.80	.61	.63						1.62*	4.75	5.40						48°26.9'	59°06.7'
16V31	.43	.70	.58						1.83*	3.90	4.90						48°58.3'	59°42.7'
17V31	.18	.42	.32	.60					(1.8)	1.90	2.50	4.60					49°30.4'	60°21.1'
18V31	.19	.90							4.15	5.80							50°54.0'	62°02.9'
19V31	.20	.28	.67	1.25					(1.9)	4.75	5.20	6.15					51°28.4'	62°36.0'
20V31	.23	.14	.86						(1.7)	1.95	5.00						52°02.8'	63°14.2'
21V31	.23	1.28	1.51	2.75					1.98*	2.28*	4.16*						52°28.9'	64°00.1'
22V31	.21	1.26	.96	1.05	2.17				1.92*	2.53*	3.48*	5.40*					52°54.8'	64°36.6'
23V31	.13	.41	.91	1.22					2.06*	2.58*	(3.7)						53°08.6'	65°06.8'
24V31	.12	1.06	1.02	.66					2.18*	2.84*	3.32*						53°16.9'	65°26.9'
25V31	.11	.54	1.14	.73					2.31*	2.79*	3.35*						53°25.3'	65°39.4'
26V31	.10	.22	.94	1.45	1.34				2.00	2.35	3.20	4.95	5.90				53°36.1'	65°53.9'
27V31	.09	.44	.66	1.22					2.30	2.75	3.30	4.60					53°44.5'	66°08.9'
28V31	.09	.42	1.42						2.45	3.35	4.90						53°50.5'	66°25.3'
29V31	.07	.74	1.23	1.36					2.75	3.30	5.05	5.50					53°43.9'	67°07.5'
30V31	.07	.46	.78						(1.8)	3.00	4.25						53°30.8'	67°22.3'
31V31	.06	.31	1.35	.74					2.10	2.45	4.70	5.85					53°22.7'	67°30.7'
32V31	.08	.28	.88						2.10	2.35	5.10						53°15.7'	67°19.3'
33V31	.08	.88							2.10	5.45							53°07.3'	67°07.5'
34V31	.08								2.20								52°57.7'	66°53.2'
35V31	.10	1.21	.69						2.15	4.15	4.90						52°51.9'	66°44.3'
36V31	.11	.98	1.08						2.10	3.85	5.05						52°44.6'	66°27.0'
37V31	.11	.13	1.05	.97					(1.9)	2.25	3.70	5.40					52°38.0'	66°10.6'
38V31	.12	.72	.55	1.20					1.95	2.40	4.30	5.40					52°30.4'	65°53.6'
39V31	.15	.61	.70	1.17					1.74*	2.35*	2.79*						52°02.9'	65°11.2'
40V31	.16	.27	1.09	.77	.71				1.90	2.25	3.00	4.10	5.35				52°07.1'	65°11.1'
41V31	.14	.65	1.29	1.08					1.95	2.60	3.70	4.70					52°45.2'	65°04.0'
42V31	.14	.44	1.38	.99	1.08				1.85	2.25	3.60	4.60	5.80				52°57.0'	65°00.9'
43V31	.13	.44	1.10						(1.9)	2.35	3.15						53°34.7'	65°02.3'
44V31	.11	.20	.44						1.80	2.20	2.80						54°07.9'	64°55.3'
45V31	.11	.33	1.69						1.90	2.25	3.50						54°15.7'	64°53.7'
46V31	.45	.25	.74	.61					1.58*	4.05	5.15	5.95					52°52.9'	58°30.7'
47V31	1.03	.47							1.82*								52°59.5'	57°21.9'
48V31	1.06	.66	1.16	2.17					1.76*	2.49*	3.81*						52°58.2'	57°18.4'
49V31	1.22	.76	.38	.48	.52	1.20			(2.0)	2.35	2.80	3.30	3.75	4.30			52°54.7'	57°08.8'
50V31	.62	1.53	.40	1.03	1.09	.71			2.30	3.25	3.55	5.05	5.90	6.80			52°23.2'	57°00.3'
51V31	.46	.47	.40	.92					(2.0)	2.35	4.60	5.50					51°33.1'	57°46.8'
52V31	.57	.26	.26	.92					2.00	4.00	4.50	5.60					51°11.1'	56°21.0'

TABLE 1 (continued)

Sonobuoy	Water	Thicknesses, km							Velocity, km/sec								Lat. (S)	Long. (W)
		h_2	h_3	h_4	h_5	h_6	h_7	h_8	V_2	V_3	V_4	V_5	V_6	V_7	V_8	V_9		
53V31	1.06	.39	.44	.27	.57	.86	2.30		1.56*	2.28*	2.47*	3.71*	3.76*	6.11*			51°07.8'	55°42.0'
54V31	1.31	.85	.27	.76	1.46	2.21	1.73		1.81*	2.00*	2.95*	3.30*	4.37*	6.32*			51°06.2'	55°00.4'
55V31	1.53	.64	.35	.43	.55	.63	1.08	.85	1.75*	2.04*	2.41*	3.09*	3.18*	3.45*	4.05*		50°50.4'	54°06.2'

Notes:
Asterisks denote interval velocity from wide-angle reflection data.
Parentheses indicate assumed velocity.
All other velocities are unreversed refraction velocities.

The sections show layers that are correlated on the basis of seismic velocity. Presently no wells have been drilled on the shelf to correlate velocity layers with lithology and by geologic age. Drilling in the onshore portion of the Magellan Basin has shown it to be filled with Cretaceous and Tertiary sediment (Olea and Davis, 1977), but the age of sediments in the Malvinas Basin is not known.

The potential pitfalls in correlating layers solely on the basis of similar velocity may be particularly great in the Magellan Basin because the onshore portion of the basin rests on the Tobifera Formation which contains an upper sedimentary section with thin interspersed lava flows and a lower volcanic section with velocities in the same range. The correlations are more convincing when interval velocities are plotted on the sections because an identifiable reflection event has been used, which can be correlated from station to station. Ludwig et al (1965) identified four major velocity layers above a 5.7-km/sec continental crustal layer in the Magellan Basin. By comparison with a generalized log of a well drilled in Manantiales Province near the Atlantic entrance to the Magellan Straits they assigned velocities of 1.8 to 3.0 km/sec to two layers of Tertiary sediments; 3.3 to 3.5 km/sec to Upper Cretaceous sediments; and 4.0 to 4.6 km/sec to sediments (and volcanics?) of Jurassic and Early Cretaceous age. Sonic logs of the Oceano well north of the Magellan Straits and San Sebastian (Fig. 1) gave, respectively, interval velocities of 2.2 and 2.4 km/sec for the Tertiary section and 2.5 and 3.0 km/sec for the Cretaceous sediments; the top of the Jurassic Serie Tobifera drilled in the Oceano well has a velocity of 2.9 km/sec (YPF staff, personal commun., 1975), but we believe this value to be representative of a weathering layer. In the Cullen well, just south of the Atlantic entrance to the Magellan Straits, 2,200 m of the Serie Tobifera was drilled without reaching basement; 20 km to the northeast the formation was missing (Olea and Davis, 1977; p. 562).

In Figure 8, we show the distribution of sediment with velocities less than 4.2 km/sec. In the deep onshore portion of the Magellan Basin, a mean velocity of 4.2 km/sec is generally regarded as representing the Serie Tobifera of Jurassic age, but may include Lower Cretaceous sediments as well (Ludwig et al, 1965). Therefore, the contours on land are probably nearly correct in outlining the Magellan Basin of Upper Cretaceous–Quaternary sediments; i.e., the basin that formed contemporaneously with the uplift of the southernmost Andean (Coast) cordillera in early Late Cretaceous time.

The Magellan and Malvinas basins are separated by a basement high that trends southward from the Deseado massif, which lies south of the Golfo San Jorge (Ludwig et al, 1965, 1968, 1977). The east and west margins of the Malvinas Basin between Tierra del Fuego and the Falkland Islands are characterized by short-wavelength magnetic anomalies, indicating the presence of shallow basement or igneous intrusions or sediments with a large susceptibility contrast (LaBrecque and Rabinowitz, 1977).

A contour map of free-air gravity anomalies shows belts of positive free-air anomalies greater than 150 mgal over the North Scotia Ridge, negative anomalies less than −150 mgal associated with the northern flank of the North Scotia Ridge and Falkland Trough that trend into southern Tierra del Fuego, and steep gradients between the gravity high and low (Rabinowitz, 1977). The axis of the positive anomaly is located over the southern edge of Burdwood Bank. Ludwig et al (1968) showed that the northern edge of the bank is underlain by low-velocity sediments about 7 km thick, an interpretation seemingly confirmed by Davey (1972) who reconciled the thick sediments with the gravity low. However, it must be remembered that westward of Burdwood Bank the gravity low trends directly into southern Tierra del Fuego (Fig. 1, lower inset); it does not bend northwestward and follow the axis of the Malvinas Basin. There is no pronounced free-air gravity low associated with the northwesternmost end of the Malvinas Basin, where the sedimentary thickness is nearly the same as that north of Burdwood Bank (sections D and E of Figure 4). Obviously, the gravity low must be associated with deeper structure.

The sonobuoy stations made immediately east of southernmost Tierra del Fuego (in the vicinity of the gravity low) consistently fail to reveal layers of velocity greater than 4.2 km/sec. Refracted waves do not always align in straight-line segments, which makes interpretation difficult and open to question. This phenomenon suggests to us that the deep-rooted gravity low is manifested by an upper zone of highly deformed sediments which do not return coherent reflections. The northern flank of the North Scotia Ridge, including Burdwood Bank, appears to be a zone of deformed sediments (Figure 7).

Off southernmost Tierra del Fuego, there appears

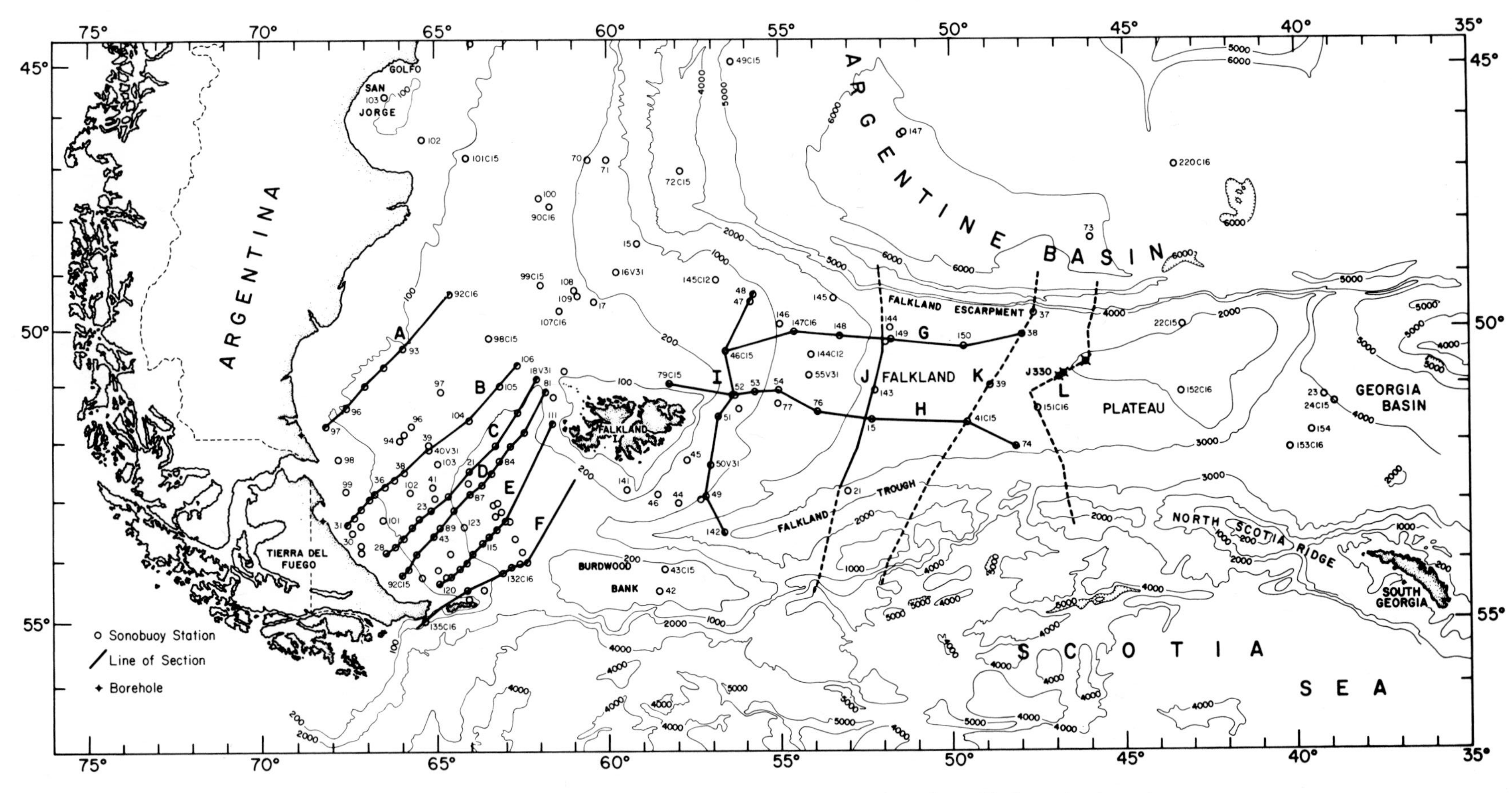

FIG. 2. Location map of sonobuoy stations and lines of sections. Bathymetry in meters.

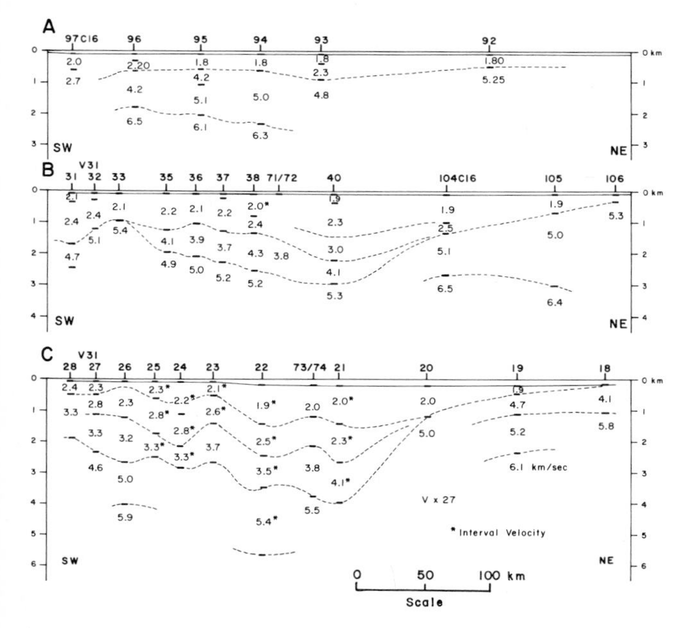

FIG. 3. Seismic structure sections of the Malvinas Basin.

to be a connection between the Magellan and the Malvinas Basins, but only by way of the upper 2 km or less of lower velocity sediments which are observed to thicken seaward (Ludwig et al, 1968; see also Fig. 8). There is undoubtedly a 5.8 km/sec basement high separating the two basins just east of

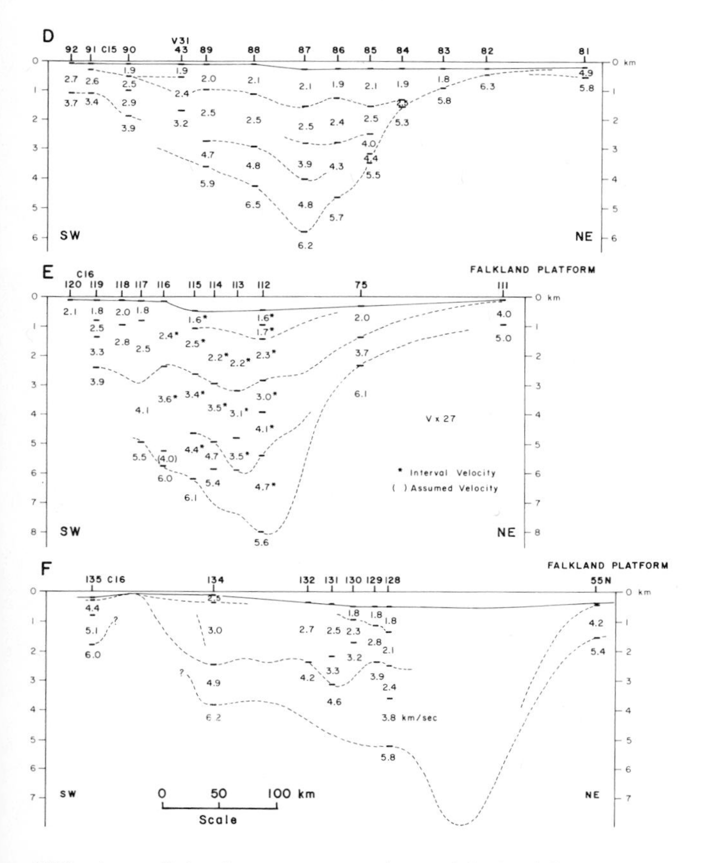

FIG. 4. Seismic structure sections of the Malvinas Basin.

Tierra del Fuego (Fig. 1). In the absence of borehole data, the age of the sediments filling the Malvinas Basin remains unknown. However, we speculate that the basin was formed at about the same time as the Magellan Basin, i.e., that it contains sediments of Jurassic and younger age, but that the major subsidence took place in the latest Eocene-earliest Oligocene at the same time as the uplift of the Patagonian cordillera.

FALKLAND PLATEAU-FALKLAND TROUGH-NORTH SCOTIA RIDGE

The Falkland Plateau is bounded on the north by a ridge associated with the Falkland Escarpment and on the south by the Falkland Trough and North Scotia Ridge (Barker and Griffiths, 1972). Between the ridges is a sediment-filled basin that is partially closed on the eastern side by a domelike basement feature called Maurice Ewing Bank by Barker et al (1976; outlined approximately by the 2,000-m isobath in Fig. 2). The sediments in the basins are 4 to 5 km thick and generally dip to the south (Ewing et al, 1971; Ludwig et al, 1977; see also Fig. 8).

The base of the Falkland Escarpment is characterized by a linear belt of negative magnetic anomalies and by negative free-air gravity anomalies with values less than –75 mgals. Steep gradients in the isostatic gravity anomalies are coincident with the linear magnetic anomalies. The Falkland Plateau has positive free-air anomalies greater than 75 mgals and relatively smooth magnetic character except over the eastern domelike feature where fairly high-amplitude anomalies may be associated with intrusive volcanics or shallow basement structure (Rabinowitz et al, 1976; Rabinowitz, 1977; La Brecque and Rabinowitz, 1977).

Rabinowitz et al (1976) and La Brecque and Rabinowitz (1977) modeled the Falkland Escarpment magnetic anomaly as an edge-effect anomaly separating continental and oceanic crust. There are no diagnostic anomalies associated with the boundary between the continental crust of the Falkland Plateau and oceanic crust of the Falkland Trough. Linear belts of gravity and/or magnetic anomalies are rarely found at the continent-ocean crust boundary of marginal basins unless an apparent trench (subduction zone) is present.

In Figures 5 and 6 we have plotted the velocity solutions of sonobuoy stations along two west-east and two north-south section lines of the western part of the Falkland Plateau. A dramatic five-fold increase in sediment thickness occurs over a short distance about 150 km east of the Falkland Islands. The velocity regression equation (least-squares fit) of the interval velocities of seismic layers to one-way travel time on the Falkland Plateau (Houtz, 1977) is based on velocity values which smoothly increase to 4.7 km/sec. This suggests that the 4.2 to 4.7 km/sec velocities in the Falkland Plateau–Malvinas Basin(s) represent a continuous body of sedimentary material the velocity of which is dependent upon the overburden thickness.

Deep sedimentary horizons on the plateau are

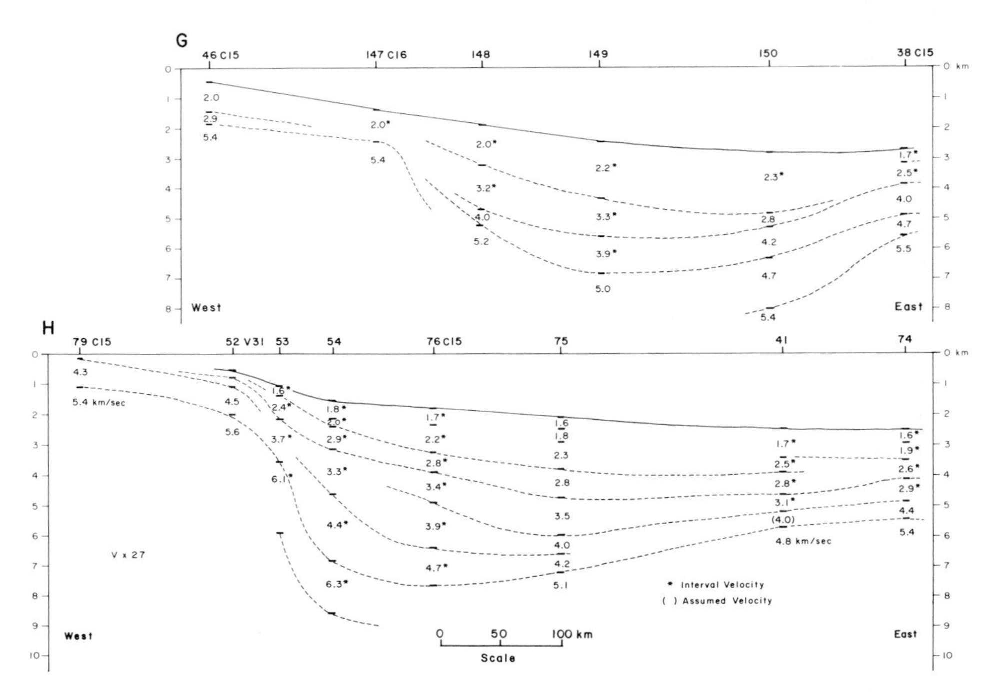

FIG. 5. West to east structure sections of the western part of the Falkland plateau.

truncated near the edge of the Falkland Escarpment (Fig. 7). The layers appear to crop out in some places, but in others are covered by a few hundred meters of younger sediments. According to Burckle and Hays (1974), Eocene sediments containing a deep-water fauna crop out in a southwest-northeast direction across the Falkland Plateau; Miocene and Pliocene sediments in piston cores are entirely of pelagic composition. They believe that Miocene erosion exposed the Eocene sediments; seismic structure and outcrops of the sediments indicate that the Falkland Plateau was subjected to intense erosion. Subsequent deposition must have been extremely slow to account for the thin capping of Neogene sediments and for the accumulation of manganese (piston-cored and dredged) on the plateau.

DSDP drilling on the eastern part of the Falkland Plateau (section L; Site 330) recovered 576 m of Mesozoic and Cenozoic sediments above a basement of metasedimentary gneiss and granite that represents continental crust (Scientific Staff, 1974; Barker et al, 1976). The sediments directly above basement are continental fluviatile deposits laid down during or before the Middle Jurassic. During Middle to Late Jurassic to late Early Cretaceous (Albian) time shallow marine sedimentation under euxinic conditions prevailed. Pelagic deposition (open ocean circulation) was established on the eastern part of the plateau during Aptian-Albian time, when the plateau subsided (Scientific Staff, 1974).

It seems likely that subsidence of the Falkland Plateau occurred during periods of major uplift of the southern Andes cordillera and subsidence of the Magellan Basin. As mentioned previously, the Andean cordillera was uplifted during Cenomanian time, accompanied by subsidence that formed the Magellan Basin. At the close of the Eocene, or sometime during the Oligocene, the Patagonian cordillera was first uplifted. Although there is no evidence for post-Cenomanian subsidence of the plateau in the DSDP cores, subsidence could also have occurred during the Eocene-Oligocene and thus may be *indirectly* related to the Andean orogenic events.

The acoustic basement (deepest reflector observed on single-channel seismic records and generally correlative with velocities near 4.2 km/sec) extends beneath the Falkland Trough and then dips beneath the northern flank of the North Scotia Ridge (Fig. 7). The Falkland Trough is proposed by de Wit (1977) to be a Mesozoic extensional marginal basin between the plateau and the North Scotia Ridge. Oceanic crust underlies the trough and is covered by a 3 to 4-km-thick sequence of sediments containing reflectors which are continuous with those of the pre-Eocene sequence of sediments on the plateau, indicating that the trough may indeed be very old. In any event, the reflection data suggest subduction of the deeper sedimentary layers and basement beneath the northern flank of the ridge. Overlying sediments appear to be deformed and uplifted to form the northern flank of the ridge. Sediments in the eastern part of the Falkland Trough appear to be deformed by thrusting against the North Scotia Ridge (section L). A seismic refraction profile of Ewing et al (1971) shows the northern flank of the North Scotia Ridge to be composed of thick low-velocity material. These observations indicate that the northern flank of the North Scotia Ridge is composed of deformed sediments perhaps related to relative movement of the ridge and the plateau to close the intervening section of oceanic crust.

The axis of sea floor spreading in the western

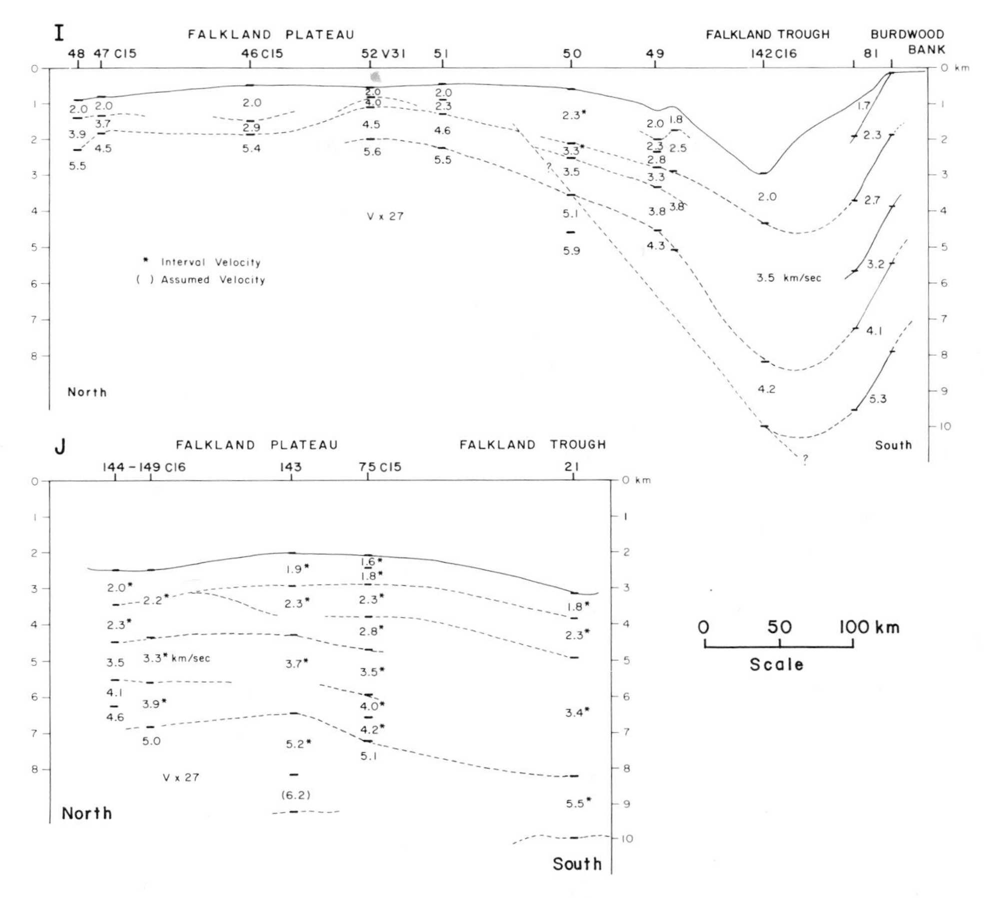

FIG. 6. North-south structure sections of the western part of the Falkland plateau. Profile 81 over the northern flank of Burdwood Bank is from Ludwig et al (1968).

Scotia Sea Basin, as indicated by magnetic anomalies (Barker, 1970, 1972) and by the pattern of sediment distribution (Ewing et al, 1971) trends northeast and intersects the North Scotia Ridge at about 50° W long. Because of orientation of ships' tracks there is some uncertainty about the correlation of the anomalies, which have provisionally been dated as late Tertiary to Recent. It seems probable that during this time the North Scotia Ridge–Falkland Trough region was the site of crustal subduction resulting from movement of the ridge towards the Falkland Trough. We view the deformed sediments of the northern flank of the North Scotia Ridge as a manifestation of this event.

CONCLUSION

Airgun-sonobuoy reflection/refraction profiles, the results of deep-sea drilling, and land geological mapping indicate that the Magellanes–Falkland Plateau region of southern South America contains thick sediments in at least two (Magellan and Malvinas) basins and possibly a third (Falkland Plateau) basin, the development of which may be related to orogenic cycles of development of the southernmost Andes. The onshore basins are filled with sediments of Jurassic, Cretaceous, and Tertiary age (Urien et al, 1976). The Malvinas Basin contains the largest thickness of low-velocity (1.8 to 2.6 km/sec) sediment, presumably of Tertiary age.

Deformation of the sediments on the northern flank of the North Scotia Ridge indicates that the ridge and the plateau moved toward each other, deforming the section between. However, we do not know when the deformation occurred. Large-scale folding of the post-Jurassic rocks of Isla de los Estados (Staten Island) occurred in Late Cretaceous time (Dalziel et al, 1974b), but, as was mentioned previously, it seems likely that deformation occurred during late Tertiary to Recent times as a result of northwest-southeast sea floor spreading in the western Scotia Sea Basin. If the deformation of the ridge sediments is largely the result of compression in a north-south direction, it is difficult to reconcile the postulated eastward migration of South Georgia from its position off Burdwood Bank.

The writers speculate that the southern edge of the North Scotia Ridge represents an east-west-trending shear zone along which South Georgia moved to its present position. The western Strait of Magellan represents a transcurrent fault(s) that follows the bend of the Andes into southernmost Tierra del Fuego as a major left-lateral transcurrent fault system (the Magellan shear zone of Katz, 1964) which aligns with the North Scotia Ridge. Earth-

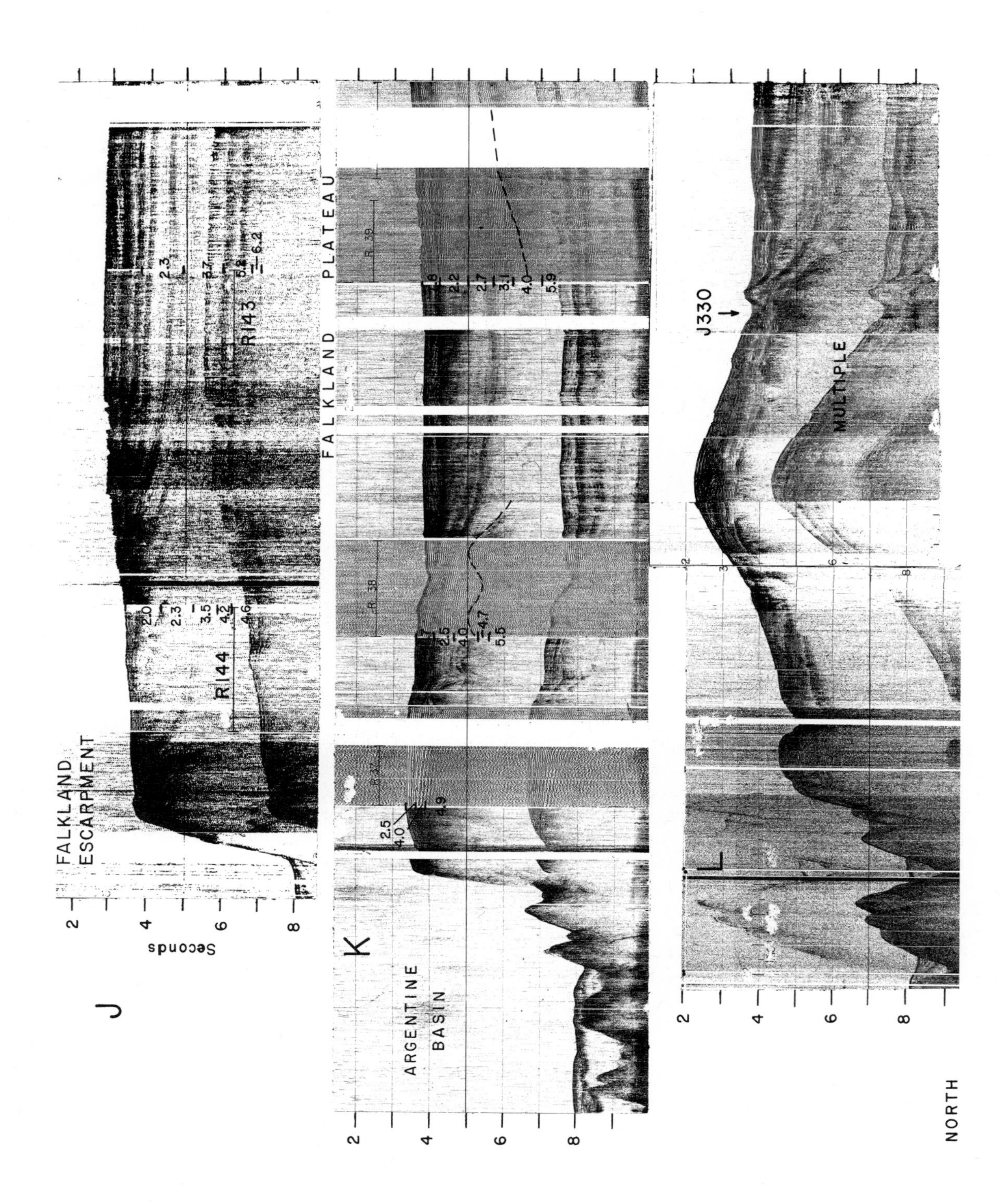

J
K
L
FALKLAND ESCARPMENT
R144
R143
FALKLAND PLATEAU
ARGENTINE BASIN
R.37
R.38
R.39
J330
MULTIPLE
Seconds
NORTH

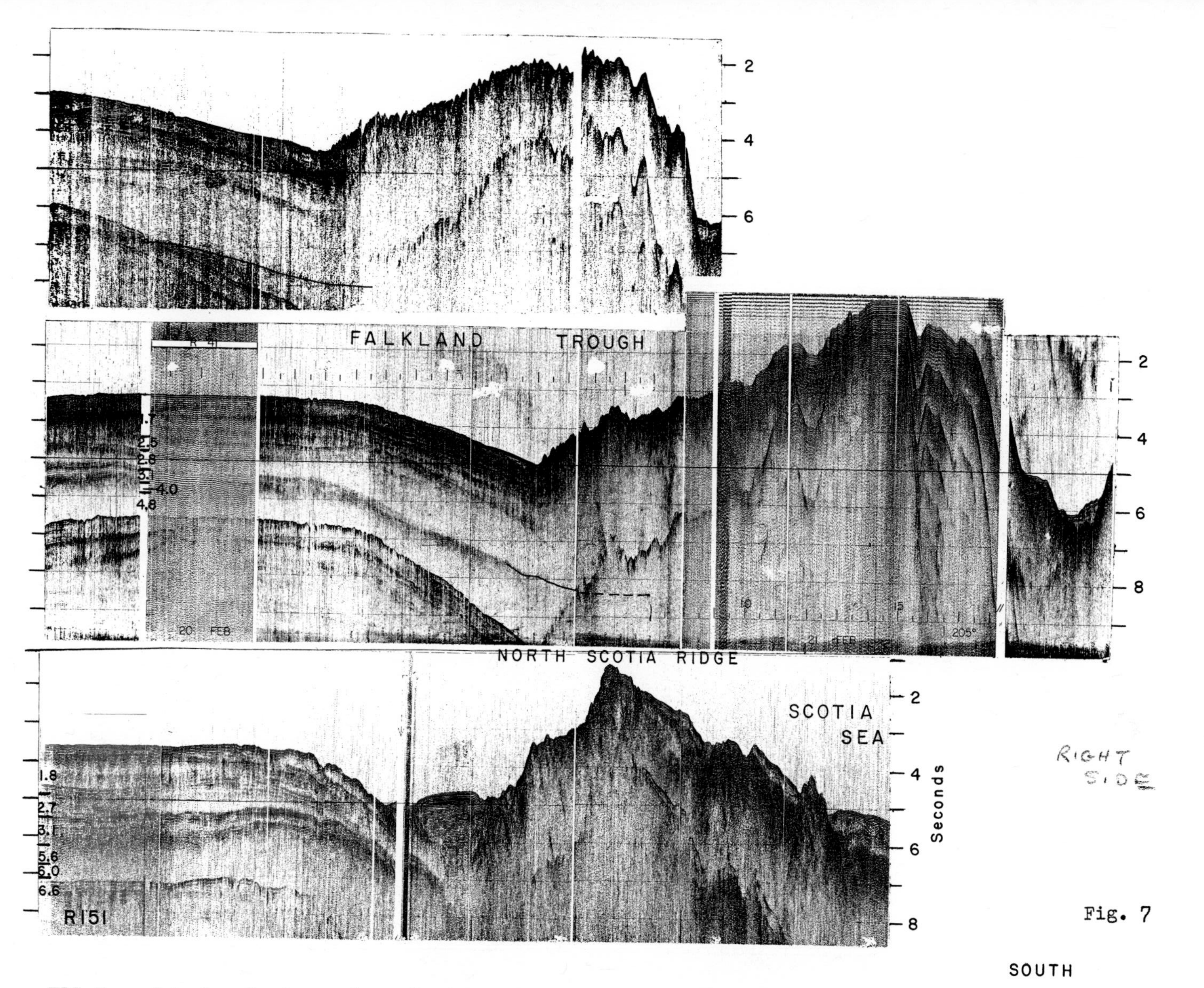

FIG. 7. Seismic reflection sections of Falkland Plateau, Falkland trough, and North Scotia Ridge (25 cu in. airgun source; passband 40–100 Hz). The vertical scale represents two-way reflection time. The horizontal scale varies with ship's speed, 7 to 10 knots during regular reflection profiling and 5 to 6 knots during the simultaneous recording of variable-angle (sonobuoy) data. The vertical exaggeration is about 25:1. Interpretation of record section L in the vicinity of drill site 330 is given by Barker (1977). The velocities and thicknesses from sonobuoy stations are plotted for comparison purposes. Note that there is not exact agreement between the layer depths in reflection time from sonobuoy data with prominent reflecting horizons on the record sections, largely because of a lower passband (5 to 20 Hz) used in recording the sonobuoy data.

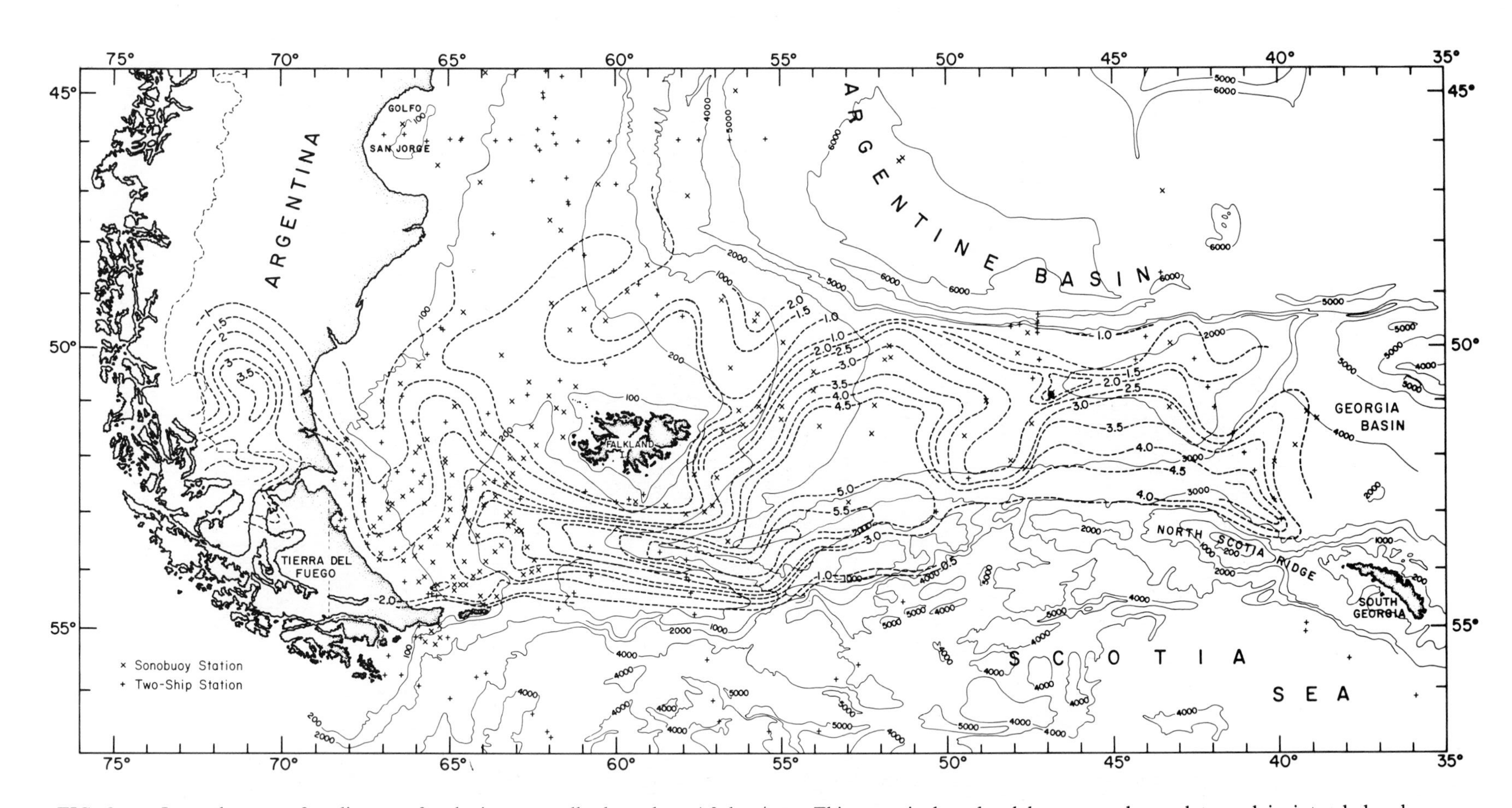

FIG. 8. Isopach map of sediment of velocity generally less than 4.2 km/sec. This map is based solely on sonobuoy data and is intended only as a guide to the relative thickness of sediments. Isopachs over Maurice Ewing Bank (outlined by 2,000-m isobath) are particularly uncertain because of the sparsity of data. Isopachs on land are drawn from Yacimientos Petroliferos Fiscales (Argentina) and Empresa Nacional del Petroleo (Chile) refraction profiles.

quake focal mechanisms (Forsyth, 1975) and geological mapping in Tierra del Fuego (Winslow, 1978; Bruhn et al, 1978) indicate that the South America plate is moving left-laterally with respect to the Scotia plate; the relative motion between the two plates is taken up on both the North and South Scotia Ridges. However, the plate boundary in the North Scotia Ridge area has not yet been well defined. Clearly, a careful study of the style of deformation of the North Scotia Ridge sediments and the pattern and age of sea floor spreading in the Scotia Sea basins is needed in order to understand the effects of relative plate motions in the region. For example, if underthrusting of the North Scotia Ridge has resulted from the proposed pattern of spreading in the West Scotia Sea Basin, we would expect to find evidence of compression in the North Scotia Ridge west of 50° W long. and transverse motion (or even extension) east of 50° W long.

REFERENCES CITED

Barker, P. F., 1970, Plate tectonics of the Scotia Sea region: Nature, v. 228, p. 1293–1297.

——— 1972, Magnetic lineations in the Scotia Sea, *in* Adie, R. J., ed., Antarctic Geology and Geophysics: IUGS, ser. B, no. 1, p. 17–26.

——— 1976, Correlations between sites on the eastern Falkland Plateau by means of seismic reflection profiles, leg 36, DSDP, *in* P. F. Barker et al, Initial Reports of the Deep Sea Drilling Project: Washington, D.C., U.S. Govt., v. 36, p. 971–990.

——— and D. H. Griffiths, 1972, The evolution of the Scotia Ridge and Scotia Sea: Royal Soc. London Philos. Trans., A. 271, p. 151–183.

——— et al, 1976, Initial reports of the Deep Sea Drilling Project: Washington, D.C., U.S. Govt., v. 36, 1080 p.

Bruhn, R. L., M. A. Winslow, and I. W. D. Dalziel, 1978, Late Tertiary to recent structural evolution of southernmost South America: in preparation.

Burckle, L. H., and J. D. Hays, 1974, Pre-Pleistocene sediment distribution and evolution of the Argentine continental margin and Falkland plateau (ab.): Geol. Soc. America Abstracts with Programs, v. 6, p. 673.

Dalziel, I. W. D., 1974, Evolution of the margins of the Scotia Sea, *in* C. A. Burk and C. L. Drake, eds., The geology of continental margins: New York, Springer-Verlag, p. 567–579.

——— and D. H. Elliot, 1973, The Scotia Arc and Antarctic margin, *in* A. E. M. Nairn and F. G. Stehli, eds., The ocean basins and margins, 1., The South Atlantic: New York, Plenum Press, p. 171–246.

——— M. J. deWit, and K. F. Palmer, 1974a, A fossil marginal basin in the southern Andes: Nature, v. 250, p. 291-294.

——— et al, 1974b, South extremity of Andes: Geology of Isla de los Estados, Argentine Tierra del Fuego: AAPG Bull., v. 58, p. 2502–2512.

——— et al, 1975, Tectonic relations of South Georgia Island to the southernmost Andes: Geol. Soc. America Bull., v. 86, p. 1034–1040.

Davey, F. J., 1972, Gravity measurements over Burdwood bank: Marine Geophys. Researches, v. 1, p. 428–435.

de Wit, M. J., 1977, The evolution of the Scotia Arc as a key to the reconstruction of southwestern Gondwanaland: Tectonophysics, v. 37, p. 53–81.

Ewing, J., et al, 1971, Structure of the Scotia sea and Falkland plateau: Jour. Geophys. Research, v. 76, p. 7118–7137.

Forsyth, D. W., 1975, Fault plane solutions and tectonics of the south Atlantic and Scotia sea: Jour. Geophys. Research, v. 80, p. 1429–1443.

Houtz, R. E., 1977, Sound velocity characteristics of sediment from the eastern South American margin; Geol. Soc. America Bull., v. 88, p. 720–722.

Katz, H. R., 1964, Some new concepts on geosynclinal development and mountain building at the southern end of South America: Proc. 22nd Internat. Geol. Cong. (New Delhi), v. 4, p. 241–255.

La Brecque, J. L., and P. D. Rabinowitz, 1977, Magnetic anomalies bordering the continental margin of Argentina: AAPG Argentine Map Series.

Ludwig, W. J., J. I. Ewing, and M. Ewing, 1965, Seismic-refraction measurements in the Magellan Straits: Jour. Geophys. Research, v. 70, p. 1855–1876.

——— ——— ——— 1968, Structure of Argentine margin: AAPG Bull., v. 52, p. 2337–2368.

——— et al, 1977, Sediment isopach map of the Argentine continental margin: AAPG Argentine Map Series.

Natland, M. L., et al, 1974, A system of stages for correlation of Magellanes basin sediments: Geol. Soc. America Memoir 139, 126 p.

Olea, R. C., and J. C. Davis, 1977, Regionalized variables for evaluation of petroleum accumulation in Magellan basin: AAPG Bull., v. 61, p. 558–572.

Rabinowitz, P. D., 1976, Geophysical study of the continental margin of southern Africa: Geol. Soc. America Bull., v. 87, p. 1643–1653.

——— 1977, Free-air gravity anomalies bordering the continental margin of Argentina: AAPG Argentine Map Series.

——— and J. L. LaBrecque, in press, The Mesozoic south Atlantic ocean and evolution of its continental margins: Jour. Geophys. Research.

——— S. C. Cande, and J. L. LaBrecque, 1976, The Falkland escarpment and Agulhas fracture zone: the boundary between oceanic and continental basement at conjugate continental margins, *in* Continental margins of Atlantic type: Anais da Acad. Brasileira de Ciencias, v. 48, p. 241–251.

Scientific Staff, 1974, Southwestern Atlantic, leg 36: Geotimes, v. 19, p. 16–18.

Tezón, R. V., 1964, Mapa geológica de la República Argentina: Ministerio Economía de la Nación, Buenos Aires, Secretaría Estado Industria y Minera.

Urien, C. M., and J. J. Zambrano, 1973, The geology of the basins of the Argentine continental margin and Malvinas Plateau, *in* A. E. M. Nairn and F. G. Stehli, eds., The ocean basins and margins, 1. The South Atlantic: New York, Plenum Press, p. 135–170.

——— L. R. Martins, and J. J. Zambrano, 1976, The geology and tectonic framework of southern Brazil, Uruguay and northern Argentina continental margin: their behavior during the southern Atlantic opening, *in* Continental margins of Atlantic type: Anais da Acad. Brasileira de Ciencias, v. 48, p. 365–376.

Winslow, M. A., 1978, Neotectonics and Cenozoic sedimentation in southernmost South America: late Tertiary transcurrent faulting and the foreland fold and thrust belt of southern Chile: in preparation.

Variations in Bottom Processes Along the U.S. Atlantic Continental Margin[1]

BONNIE A. MCGREGOR[2]

Abstract Seismic reflection data on strike lines along the slope and rise of the U.S. East Coast continental margin show variations in morphology and sedimentation pattern parallel to the margin. From these data, the regional versus localized extent of erosional or depositional features and events within each physiographic province of the margin can be determined. Two localized areas of extensive slope deposition were identified, one seaward of Albemarle Sound and the other seaward of the Delmarva Peninsula. A regional unconformity on the rise extends from Georges Bank to Wilmington Canyon and is believed to be Pliocene in age. The upper slope between Cape Hatteras and Hudson Canyon is dissected by many small valleys cut during the late Pleistocene, whereas northeast of Hudson Canyon the upper slope is only cut by major submarine canyons. The Florida-Hatteras slope and Blake Plateau have many unconformities with one in late Tertiary truncating an extensive sequence of foreset beds on the north end of the Blake Plateau.

INTRODUCTION

The sea floor adjacent to the continents is an important source of energy and mineral resources, in addition to being a potential area for waste disposal. The National Oceanic and Atmospheric Administration (NOAA) program RUSEF (Rational Use of the Sea Floor), through geological and geophysical research, is attempting to understand the environment of the continental margin and the dynamic bottom and near-bottom processes that affect the margin through time. Data have been collected over the past three years on the U.S. Atlantic continental margin (outer shelf, slope, and rise) using the NOAA ship *Researcher*. These data consist of single-channel seismic reflection profiles collected on strike lines parallel to the continental margin approximately 15 km apart, extending from Hydrographer Canyon on Georges Bank to Cape Canaveral, Florida (Fig. 1). Most seismic reflection data have been collected on lines perpendicular to the margin (e.g., Uchupi and Emery, 1967), with strike lines serving only as tie lines. The objective of using a suite of strike lines was to look at the variation in the sedimentation record as a continuum within each physiographic province of the margin (outer shelf, slope, and rise) as defined by Heezen et al (1959). From these data one can locate areas with different depositional histories for subsequent detailed study.

In order to assess recent sedimentation patterns, bottom sediment samples were collected coincidently with the seismic work. Transects of four evenly spaced 6-m piston cores with 1.5-m pilot cores were taken approximately 80 km apart along the margin, sampling the outer shelf, slope, and rise (Fig. 1). Sediment analyses of the 6-m piston cores are being done by Larry J. Doyle of the University of South Florida and Orrin H. Pilkey of Duke University. Preliminary results indicate that rapid deposition is occurring on the continental slope at a high sedimentation rate and/or by mass movement of sediment downslope (Doyle et al, 1975, 1976). The 1.5-m cores have been analyzed at AOML-NOAA to assess variations in slope stability and variations in the depositional environment (Keller et al, 1978).

Detailed studies are being carried out on three submarine canyons (labeled A, B, C, Fig. 1) to determine active processes and to assess the role of the canyons as conduits to the deep sea. These canyons were selected for their varying physiographic setting: Veatch Canyon (A) on Georges Bank is in a glaciated area, and Washington (B) and Norfolk (C) are in a nonglaciated area with a major river drainage system from the Chesapeake Bay region. High-resolution narrow-beam echo sounding surveys were run to determine maximum depths and to define the morphology; a suite of grab samples, hydroplastic cores, and seismic reflection profiles were also

[1]Manuscript received, April 18, 1977; accepted, December 22, 1977.

[2]Atlantic Oceanographic and Meteorological Laboratories, National Oceanic and Atmospheric Administration, Miami, Florida 33149. Present address: Department of Oceanography, Texas A&M University, College Station, Texas 77843.

The author would like to thank George H. Keller, Richard H. Bennett, Douglas Lambert, William Sawyer, George Merrill, and Evan Forde for assisting in the data collection. Appreciation is given to William L. Stubblefield and John W. Kofoed for critically reviewing the manuscript. Thanks are also given to the officers and crew of the NOAA ship *Researcher* for diligence and excellence in executing this study. Funds for this work were provided by the National Oceanic and Atmospheric Administration.

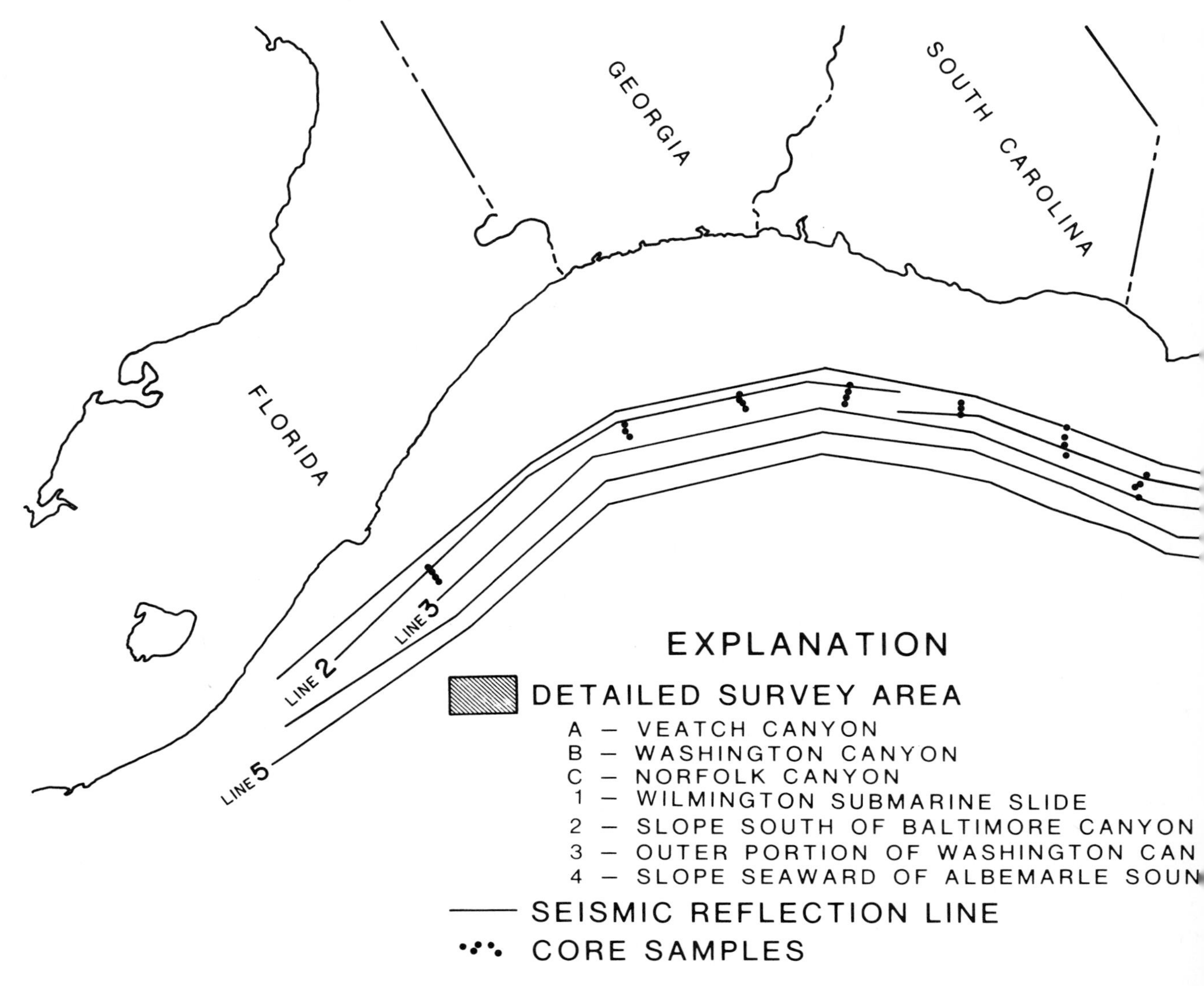

FIG. 1. Index map from Uchupi (1968)

taken in each canyon in order to understand their respective erosional and depositional histories and evaluate their present activity.

Four other portions of the margin have been studied in detail (Fig. 1): (1) a submarine slide on the slope north of Wilmington Canyon, (2) a portion of the slope where ridges of stratified sediment have undergone extensive erosion, (3) outer part of Washington Canyon to understand levee formation and to determine relative activity of two channels, and (4) the slope seaward of Albemarle Sound where several periods of deposition and erosion have occurred.

From the general seismic reflection data and the bottom samples, variations in the sedimentary history of the margin can be determined. Detailed studies may now focus on specific areas to understand the processes that have been and still remain active.

REGIONAL VARIATIONS IN SEDIMENTATION PATTERN

The morphology of the continental margin changes as one moves offshore from the outer shelf to the rise. The outer shelf is flat, except where cut by the heads of the major submarine canyons. The continental slope has irregular topography and is dissected by major submarine canyons as well as by numerous small valleys. The rise has irregular but subdued topography with many submarine canyons present. Noteworthy are the variations, not only among the margin provinces but also within each physiographic province.

Profile 2

Profile 2 (Fig. 2) is along the upper slope from Hydrographer Canyon on Georges Bank to Cape

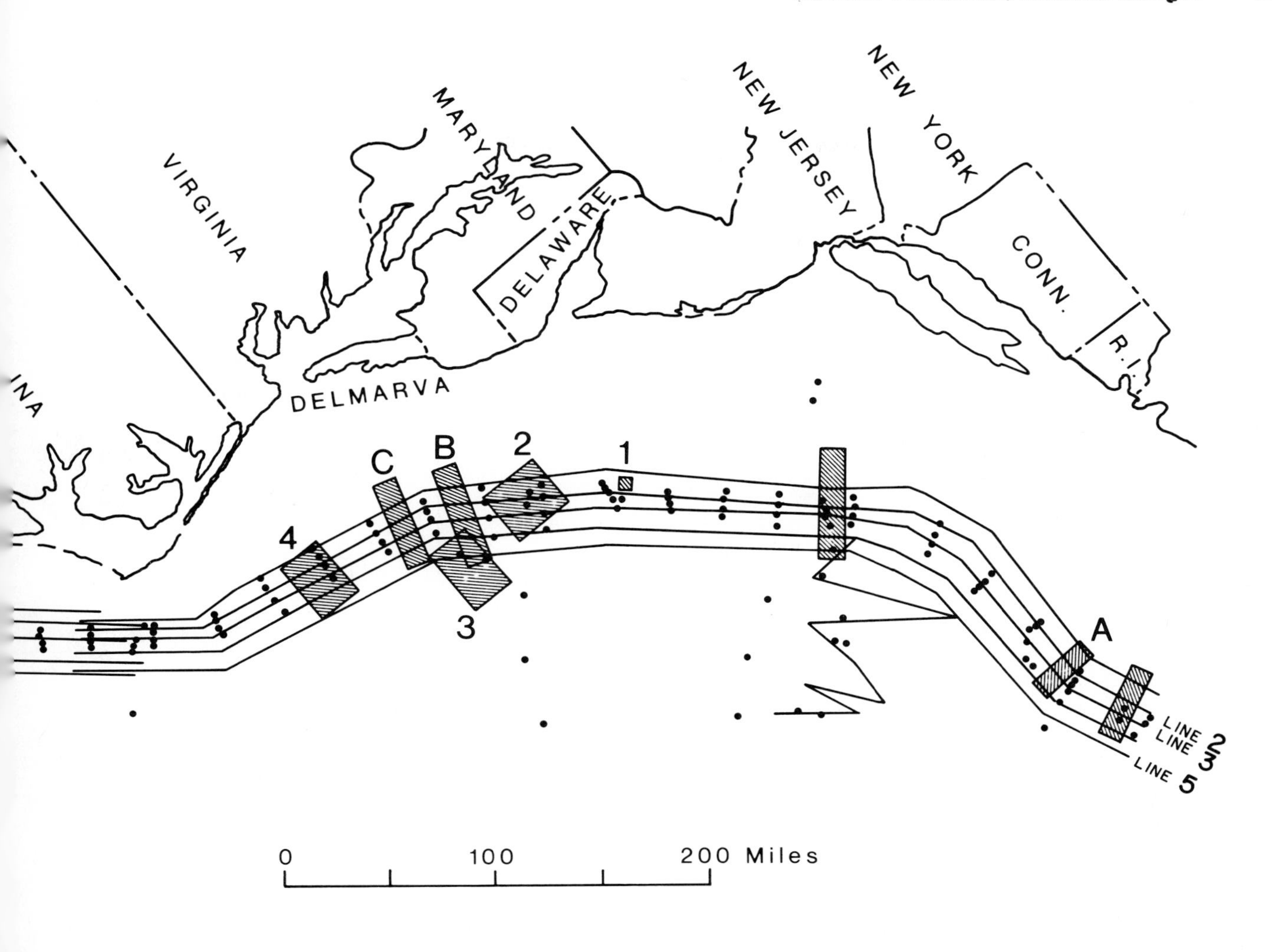

showing track lines and core locations.

Canaveral, Florida (Fig. 1). The total seismic record running north to south is shown in Figure 2 with the individual profiles reading from right to left, top to bottom. The profiles around Cape Hatteras (latitude 35°N) do not overlay exactly, due to ship's navigational problems in the Gulf Stream. In each profile the major canyons are labeled for reference. The numbers and x's refer to the location of the crossing profiles of Uchupi and Emery (1967).

In general, reflecting horizons can be traced over tens of kilometers, indicating that processes governing sedimentation were active over a wide area. A large wedge of sediment believed to be Tertiary, based on correlation with units of Hoskins (1967) and Garrison (1970), thickens toward Hudson Canyon (profiles 2a and 2b) and also slopes seaward with conformable reflectors on the slope (Uchupi and Emery, 1967). The sediment gives the slope a smooth appearance between Hudson and Veatch Canyons except for a few large canyons. Between Veatch and Hydrographer Canyons the slope is very dissected, and Uchupi and Emery (1967), on crossing profiles 55 and 56, show reflecting horizons cropping out on the slope, indicating extensive erosion. Upper Cretaceous material was cored 100 m below the sea floor at Caldrill Site 17 which is east of Veatch (Manheim and Hall, 1976). This region is seaward of Great South Channel, an erosional feature on the shelf that in the past probably was a drainage system around Georges Bank connecting with Hydrographer Canyon (Uchupi, 1968). The large, stratified block in the twin canyon west of crossing profile 62 appears to have been rotated or to have slid, because the reflectors within the block

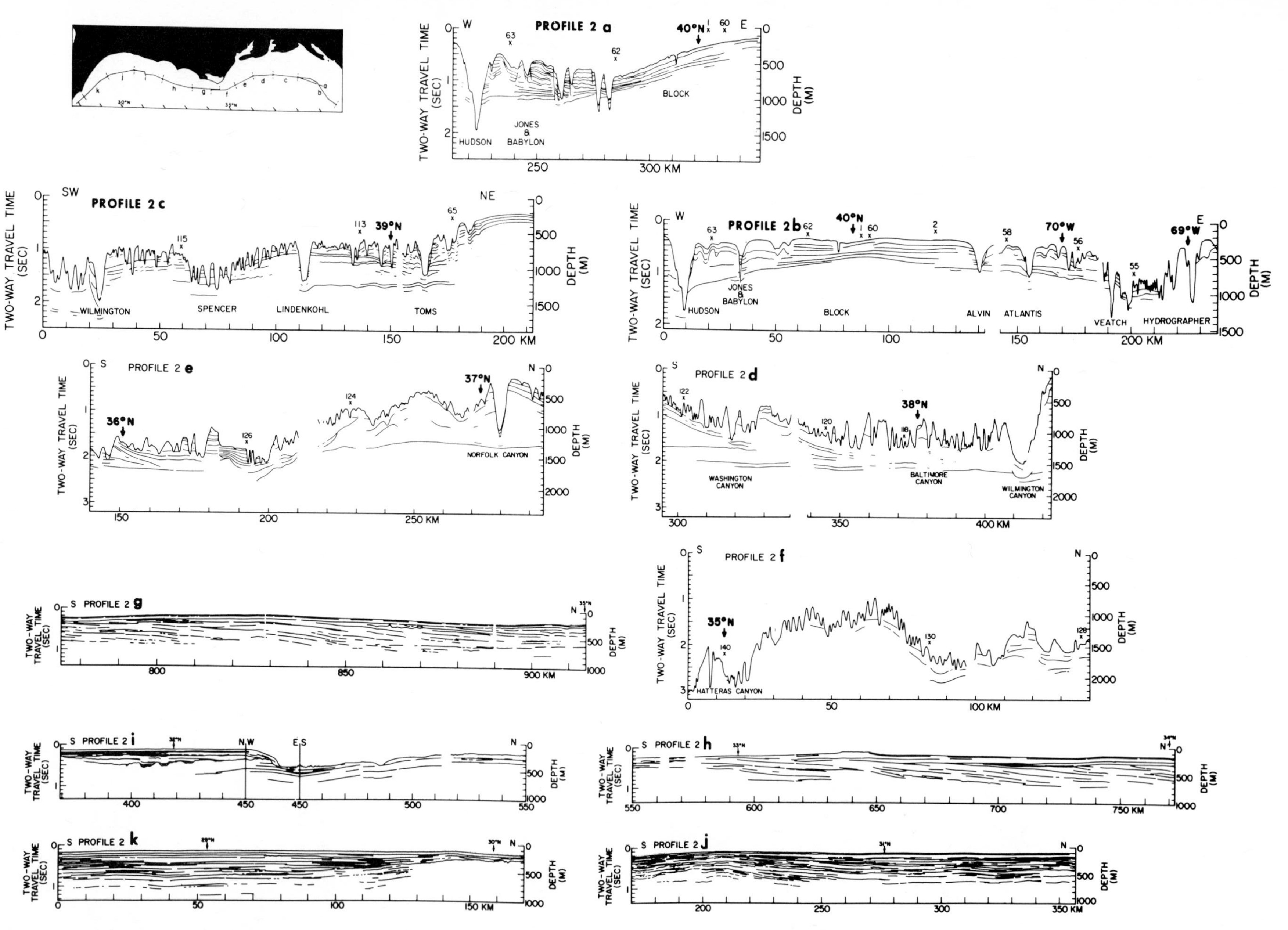

FIG. 2. Line drawing of original seismic reflection profile record for profile 2 (see Fig. 1 for location). Air guns with 160×10^{-6} m^3 (10 cu in.) and 1920×10^{-6} m^3 (120 cu in.) chambers were used as the sound source north of Cape Hatteras and a 640×10^{-6} m^3 (40 cu in.) chamber south of Cape Hatteras. A 50-element hydrophone was used to receive the returning signal which was filtered at 100 to 430 Hz. Vertical exaggeration = 30 × (assuming 1500 m/sec velocity). "X"'s and numbers refer to crossing profiles of Uchupi and Emery (1967).

are not continuous with those of either wall as is the case with other canyons. Approximately 100 m of fill is present in each of the valleys flanking this rotated central block. Seismic penetration to reflecting horizons beneath the valleys was not achieved to demonstrate conclusively fault control in this area. Buried valleys are present northeast of Hudson Canyon and may be part of an old Hudson drainage system. Profile 2a shows that slumping has occurred in the walls of several canyons.

South of Hudson Canyon and north of Cape Hatteras (profile 2c-2f) the slope is dissected by many small valleys as well as by major canyons. Slumping is evident in many of the canyon walls. Just south of crossing profile 126 (profile 2e), which is seaward of Albemarle Sound, the sedimentation pattern has been different from that of the adjacent areas. Profile 126 (Uchupi and Emery, 1967) shows that the zone of stratified sediments is part of a wedge of sediments that tapers seaward, but landward it laps onto truncated strata of the continental slope. The stratified sediments represent renewed deposition after a major period of erosion on the slope. This deposition appears to have been confined to a narrow region by a physical barrier on the north which confined the multilayered sequence and since has been eroded. The small channel in the surface of this sequence suggests recent erosion. This region seaward of Albemarle Sound will be studied in detail (Fig. 1, area 4). Deposition in this area may be related to the Roanoke River or the James River, as shelf valleys and a shelf delta are present on the adjacent shelf (Swift, 1976). The reflecting horizons shown on profiles 2c-2e are believed to be Tertiary. In some cases the reflecting horizons appear to control the depth to which valleys are cut; for example, the four valleys southwest of Lindenkohl Canyon all stop at the horizon 200 m below the sea floor. The age of this horizon is believed to be Miocene, based on correlation with the profiles of Sheridan (1975).

South of Cape Hatteras on the Florida-Hatteras slope, a maximum of 800 m of sediment was penetrated seismically (profiles 2g-2k). Many buried unconformities are present with a major one north of 33°N (Fig. 2h). The unconformity truncates a 200 to 400-m-thick sequence of reflecting horizons that dip to the north; both this unconformity and the sequence of dipping horizons can be traced offshore on other profiles. On the short east-west segment at kilometer 450 (profile 2i), an erosional unconformity is present and the Florida-Hatteras slope can be seen to have prograded seaward about 25 km. The unconformity 300 m below the sea floor near 32°N (profile 2i) is Late Cretaceous based on drilled material from U.S.G.S. drill hole 6004 (Hathaway et al, 1976). Drill holes 2 and 5 in the vicinity of kilometer 180 (profile 2j) north of 30°N indicate that the deepest reflecting horizons are middle Eocene with less than 100 m of post-Miocene sediment (JOIDES, 1965). Bottom samples (Fig. 1) were primarily sand or sandy silt all along the slope. The location of the Gulf Stream appears to have controlled the erosional and depositional pattern on the Florida-Hatteras slope.

Profile 3.

Profile 3 (Fig. 3) is along the lower slope, which throughout much of the margin is well dissected. The portion of the slope between Block and Hudson Canyons is not shown. The irregular surface topography (profile 3a) from west of Veatch to Hydrographer Canyon is an eroded portion of the slope with the crossing profiles 56 and 55 (Uchupi and Emery, 1967) showing strata truncated by the slope. Hoskins (1967) believes this region has been eroded by current scour from meandering of the Gulf Stream. Some of the erosion may be related to the Great South Channel drainage system mentioned on profile 2. On profile 3b some of the valleys appear to have cut down to the same horizon as that in the vicinity of profile 113, suggesting a resistant stratum influencing the topography. DSDP Site 108 is located between Lindenkohl and Toms Canyons, and the ages of the reflecting horizons (Fig. 3, profile 3b) are determined on the basis of the drill information. The deepest reflector is believed to be the Eocene-Cretaceous boundary and the intermediate reflector Eocene, correlated with Horizon A (Hollister et al, 1972).

Three major canyons are present in the area of profile 3c (Fig. 3). The slope is not dissected by as many small valleys south of Wilmington Canyon as it is to the north. Norfolk and Washington Canyons appear to have associated levees with the levee on the south side of Washington Canyon being the largest. Baltimore Canyon is flanked on the south by a large ridge of stratified sediments, which is similar in appearance to the Nyckel Ridge described by Stanley and Kelling (1970). Because of the large volume of sediment composing the Nyckel Ridge and its internal structure Stanley and Kelling (1970) believe it is pre-Quaternary and not formed by overbank deposition. The slope between Washington and Baltimore Canyons has a different morphology and sedimentation pattern compared with the slope between Wilmington and Toms Canyon. Ridges of stratified sediments with many unconformities are present transverse to the slope. The ridges do not have the same spacing and frequency as the topography up-slope on profile 2 (Fig. 2). These ridges are similar in internal structure, although not as large as the ridge on the south side of Baltimore Canyon. From the volume and distribution of sediment, this region has been a major depositional site with subsequent modification by bottom currents and by slumping on the flanks of the ridges. This portion of the slope is between the major drainage systems of Chesapeake and Delaware Bays; the ancestral Delaware River drainage system can be traced across the shelf to the head of Wilmington Canyon (Twichell et al, 1977). However, the morphology of the slope south of Baltimore Canyon suggests a major drainage system existed in the area in the past and has since become buried on the shelf. The deposi-

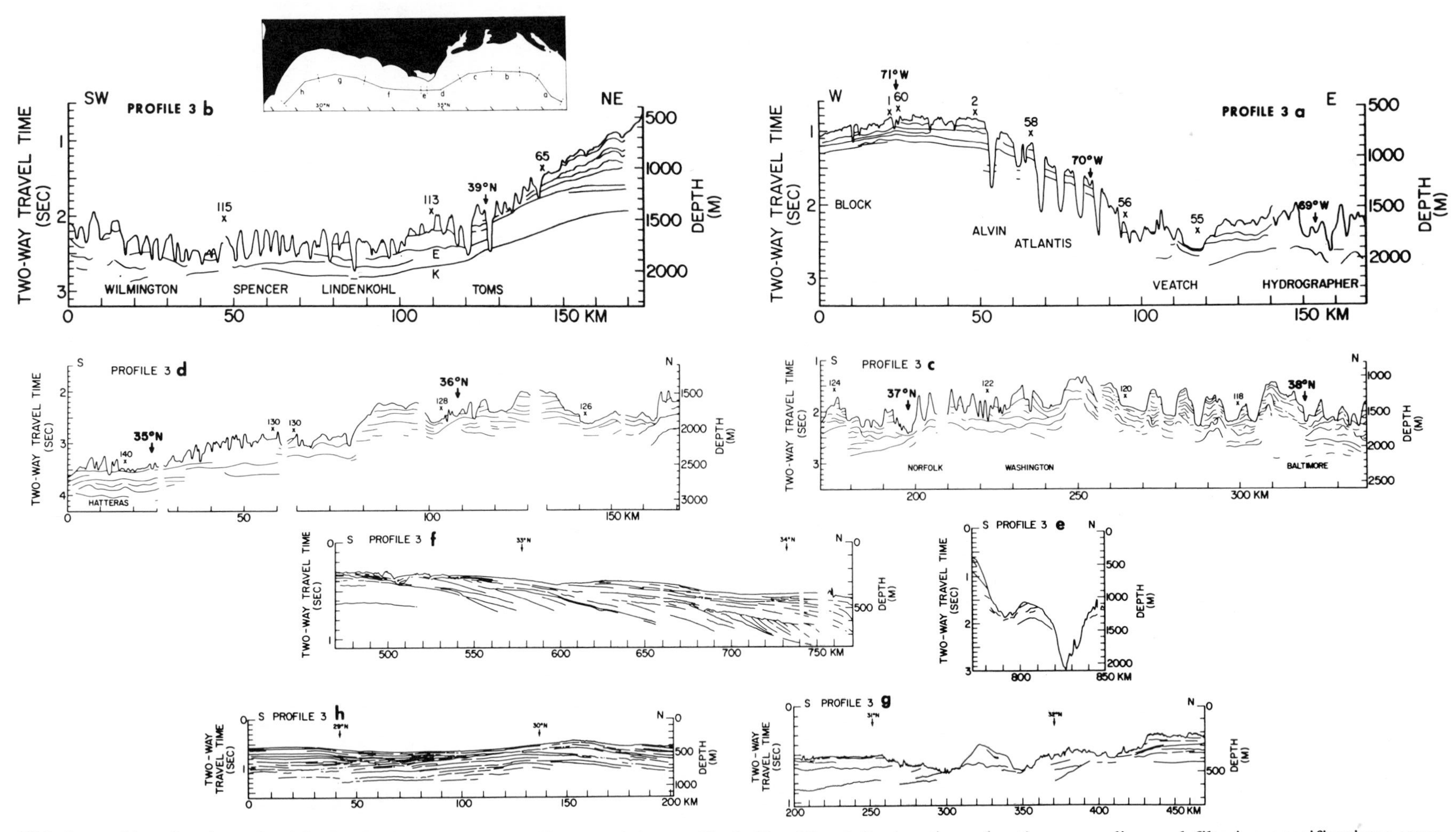

FIG. 3. Line drawing of original seismic reflection profile record for profile 3 (See Fig. 1 for location; shooting, recording and filtering specifications same as for Fig. 2). Vertical exaggeration = 20 to 30 × (assuming 1500 m/sec velocity). "X"s and numbers refer to crossing profiles of Uchupi and Emery (1967). Index map shows profile segment locations.

tional history of this area is being studied in detail (Fig. 1, area 2).

The stratified mass of sediments just south of crossing profile 126 (profile 3d, Fig. 3) is a continuation of the stratified sediments noted on profile 2. This area seaward of Albemarle Sound has been the site of slope deposition of a large volume of sediments. Hatteras Canyon and adjacent areas of the lower slope are different from previously discussed East Coast canyons, as Hatteras has no apparent shoaler wall on the south and both levees are very small. Pratt (1967) suggested that the origin of this canyon is due to slumping, as the slope is very steep. The absence of levees is perhaps due to the strong current regime in the vicinity of Cape Hatteras composed of the Gulf Stream and the Western Boundary Undercurrent, both of which transport large quantities of suspended matter (Betzer et al, 1974). These currents with opposite flow directions and fluctuations in their position should be responsible for the lack of preferential deposition as well as for erosion in the vicinity of Hatteras Canyon.

At the northern end of the Blake Plateau, reflecting horizons crop out. The buried unconformity noted on profiles 2h and 2g (Fig. 2) extending from south of 33°N to the edge of the plateau can be seen on profile 3f (Fig. 3). Beneath the unconformity, reflecting horizons dip to the north. The surface of the unconformity is irregular, indicating varying resistance of the truncated horizons. The age of this unconformity is believed to be late Miocene or Pliocene and will be discussed in the section on profile 5. Between 30°30′N and 32°45′N (profiles 3f-3h) the surface topography is very irregular, suggesting erosion and nondeposition attributable to the Gulf Stream. In contrast is the smooth surface topography south of 30°30′N (profile 3h) which indicates that deposition has dominated over erosion. Although unconformities are present in the sedimentary sequence (profile 3h), they are not as extensive as farther north on the Blake Plateau (profile 3f).

Profile 5.

Profile 5 was run along the continental rise where the topography is subdued (Figs. 4 and 5); only the major canyons have extensions at this depth. The sedimentary sequence has many reflecting horizons and unconformities. A very prominent reflecting horizon about 200 m below the surface can be traced from near Hydrographer Canyon to Wilmington Canyon, a distance of 440 km (profiles 5a and 5b). This implies a uniform environment on the rise for a time during the past. An age for this horizon can be estimated from the area south of Hudson Canyon. A buried trough 400 m deep is present between Hudson and Wilmington Canyons which is partially filled with acoustically transparent sediment, topped with a zone of stratified sediment, in turn truncated by an erosional unconformity. DSDP Site 106 is seaward of profile 5b in this area (Hollister et al, 1972). Correlation with lithology and seismic stratigraphy of the site suggests the bottom of the trough may be in material of Eocene age, with the acoustically transparent material which fills the bottom of the trough being Miocene clay. The stratified sediments would be Pliocene with the unconformity Pliocene, and the sediments burying the unconformity would be Pleistocene to Recent. Vail et al (1977) have identified several global unconformities in the late Tertiary; in the mid-Oligocene, late Miocene, and mid-Pliocene. Correlating these unconformities with the DSDP Site 106 data and with seismic reflection profile 5b between kilometer 70 and 120, erosion of the trough into Eocene material would have occurred during the mid-Oligocene: the late Miocene unconformity would represent the transition from the transparent material filling the trough to the overlying stratified sediments; and the erosional surface on the stratified sediments would correlate with the mid-Pliocene unconformity. The continuous reflecting horizon traced between Hydrographer and Wilmington Canyons would be, therefore, mid-Pliocene and represent the mid-Pliocene unconformity. If the age correlation is correct, the large volume of sediment in the levees around Hudson Canyon would be late Pliocene and Pleistocene in age, as the continuous horizon can be traced beneath them. This portion of Hudson Canyon has formed by both depositional and erosional processes, as described by Nelson and Kulm (1973).

A large levee is present on the south side of Baltimore Canyon (profile 5c) and a smaller levee on the south side of Wilmington Canyon (profile 5b). Reflecting horizons are truncated in the walls of all the canyons indicating that erosion has occurred at this depth on the rise. A 14-cm-thick graded sand layer was cored in Norfolk Canyon just seaward of profile 5, indicating transport of coarse sand this far seaward on the rise.

Just north of Hatteras Canyon a diapiric structure is present, with approximately 100 m of overlying sediment (profiles 5d and 5e). The level or horizon from which the structure originates cannot be identified on the basis of one crossing. It has no gravity or magnetic expression, indicating that it probably is composed of mud. More deeply buried diapirs have been identified in this area on CDP data by Grow Hatteras slope, one encounters the Blake Plateau (profile 5e). Near 34°N at 2,000 m depth on the edge of the Plateau, very friable middle Eocene material was cored (Florentine Maurrasse, personal communication, 1976). This cored sample is from the stratified sequence of horizons dipping to the north that extends from 32°30′N and crops out on the northern edge of the Blake Plateau. Erosional unconformities are present within this sequence and a major unconformity truncates the sequence. The surface of this unconformity is very irregular, with approximately 100 m of material eroded north of 33°N at kilometer 730 (profile 5e). Approximately 150 m of sediment blankets the unconformity. Because Eocene material was cored within a sequence of horizons that are truncated by the unconformity, the age of the unconformity is post-Eocene. Vail et al (1977) have identified three post-Eocene unconformities, as previously mentioned. As the Eocene horizon is buried

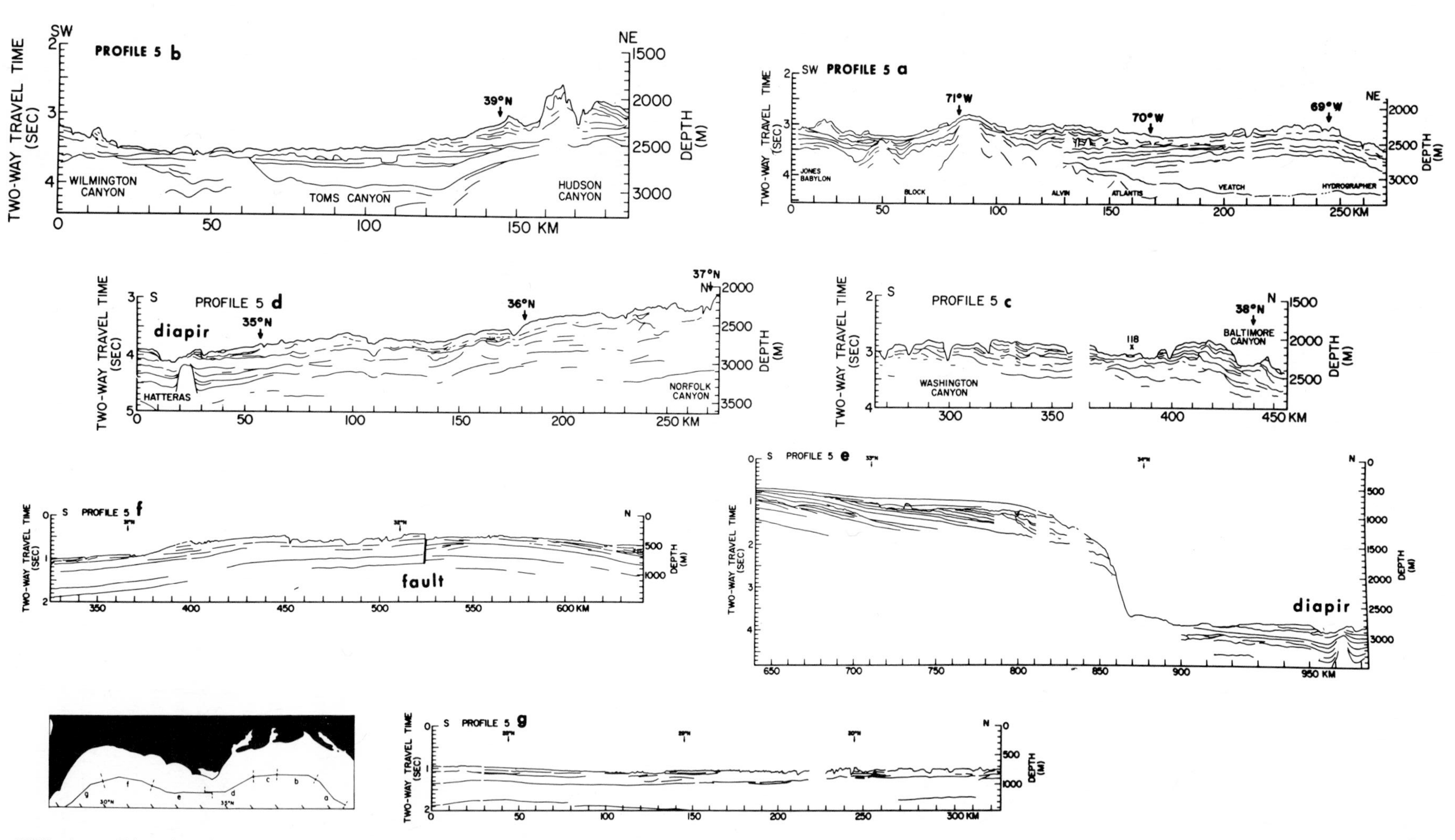

FIG. 4. Line drawing of original seismic reflection profile record for profile 5 (see Fig. 1 for location). Air guns with 160×10^{-6} m^3 (10 cu in.) and 1920×10^{-6} m^3 (120 cu in.) chambers were used as a sound source. A 50-element hydrophone was used to receive the returning signal which was filtered at 100 to 430 Hz. Vertical exaggeration = 20 to 30 × (assuming 1500 m/sec velocity). Index map shows profile segment locations.

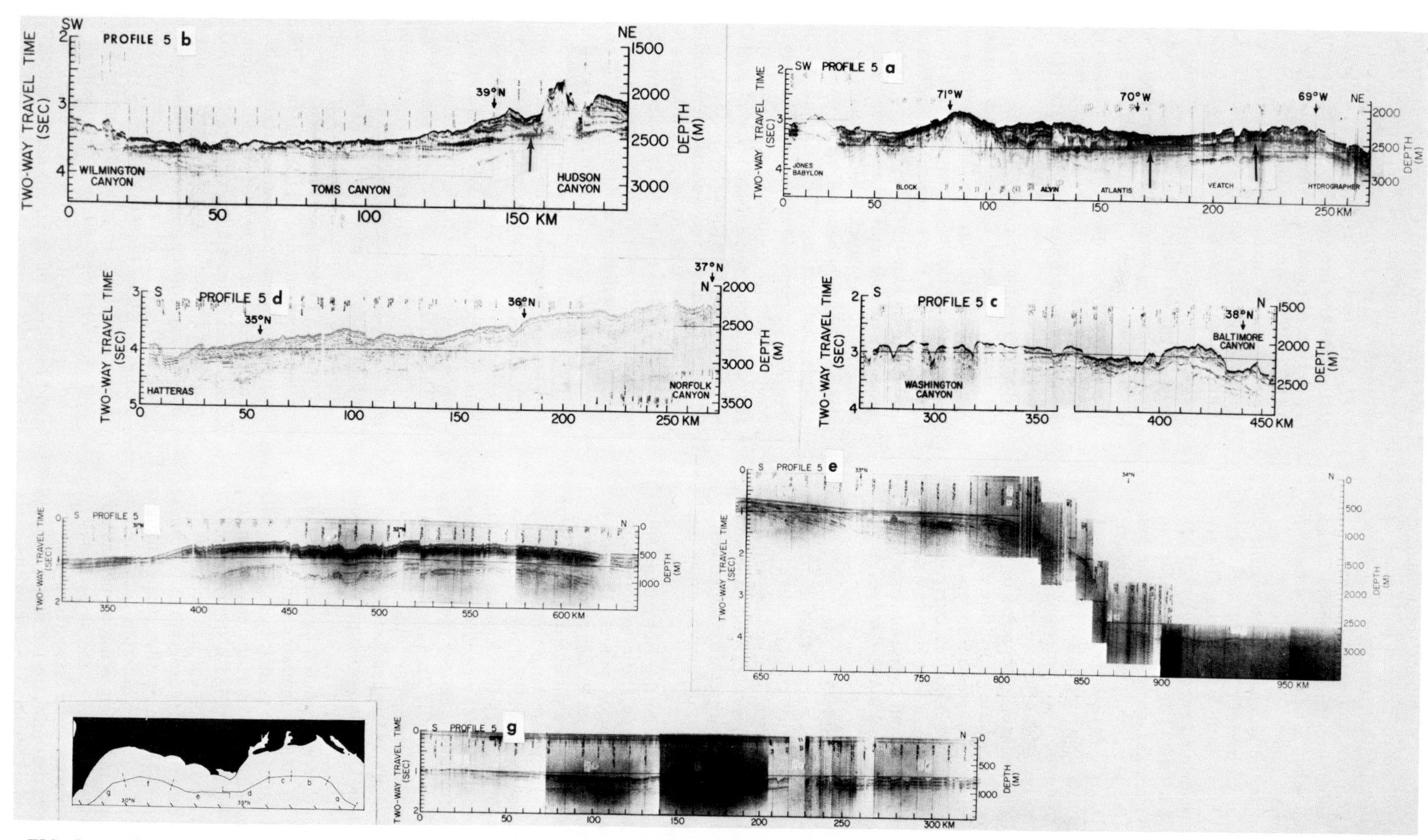

FIG. 5. Photograph of original seismic reflection profile record for profile 5. Index map shows profile segment locations. Arrow denotes continuous reflecting horizon.

by approximately 500 m of sediment and an unconformity near kilometer 800 is present within the sequence, the unconformity that truncates this sequence of dipping horizons and extends from 32°30′N off Cape Romain to the edge of the Blake Plateau off Cape Lookout is believed to be late Miocene or mid-Pliocene.

A fault is present in the area shown on profile 5f just north of 32°N at kilometer 522; it offsets buried horizons as well as the sea floor. Sheridan (1974) shows several faults in this general area based on magnetic anomaly patterns. The general surface of the Plateau is very irregular along much of the profile resulting from the activity of the Gulf Stream. Drill hole J-6 (JOIDES, 1965) which is just east of profile 5 near kilometer 280 penetrated lower Eocene material 100 m down. The continuous reflecting horizon about 100 m below the sea floor on profile 5g north of 30°N is, therefore, interpreted to be middle to lower Eocene. Very little deposition and/or a great deal of erosion has occurred in the area from middle Eocene to the present. Bottom currents, part of the Gulf Stream, have been and are very active in controlling the topography and the sedimentation pattern on this portion of the Blake Plateau.

DISCUSSION

Seismic reflection profiles parallel to the margin provide a different perspective from profiles oriented perpendicular to the margin. Variability within each province—slope and rise—is readily apparent, and as most morphologic features are perpendicular to the margin, they are better defined by track lines parallel to the margin.

The continental margin is cut by many large submarine canyons between Cape Hatteras and Cape Cod. Where seismic penetration beneath the canyons was achieved, there is no evidence of fault control for the canyons, even though alignment with some of the North Atlantic fracture zones suggests a correlation (Sheridan, 1974) and Kelling and Stanley (1970) believe a fault underlies the head of Wilmington Canyon. Extensive infilling of sediments in the canyons has not occurred except for the two valleys near crossing profile 62 (profile 2a, Fig. 2). The lack of infilling suggests the canyons are actively channelling material offshore. A graded sand bed found on the rise in Norfolk Canyon also supports the activity of this canyon.

The continental slope is dissected by many small valleys in addition to major submarine canyons (Figs. 2, 3). Small valleys are primarily located north of Veatch Canyon and south of Hudson Canyon to Hatteras Canyon. Profile 2b (Fig. 2) shows that much of the late Tertiary sediment has been removed between Veatch and Hydrographer Canyons. Uchupi and Emery (1967), on profiles perpendicular to the margin, show reflecting horizons truncated by the continental slope, indicating erosion such as on profile 55. On profiles parallel to the slope these eroded areas can be seen to be dissected by numerous small valleys that indicate the major process by which the erosion occurred. The age of this erosion is inferred to be Pleistocene, correlative with the valleys northeast of Wilmington Canyon shown on profile 2c of Figure 2 (McGregor and Bennett, 1977). The lack of infilling in the valleys also suggests that they are not very old, as the upper slope is a region of rapid deposition according to Doyle et al (1975, 1976), on the basis of the cores shown in Figure 1. Grow and Markl (1977) have found that the present shelf edge off New Jersey and Cape Hatteras was eroded 30 km landward of its Early Cretaceous location. This major erosion, they believe, occurred during the Tertiary. Several of the late Tertiary unconformities noted by Vail et al (1977) were possibly identified on the rise (profile 5, Figs. 4 and 5) and on the Blake Plateau (profiles 3 and 5, Figs. 3 and 4); however, the major erosion of the slope found by Grow and Markl (1977) was not identified.

Regions of major deposition are also present on the slope. Between Baltimore and Washington Canyons (profile 3c, Fig. 3) a large volume of sediment in the form of ridges is present on the slope. The age of some of this sedimentary sequence is believed to be older than Pleistocene, because of the thickness of sediment (over 600 m) and because of similarity to the Nyckel Ridge south of Wilmington Canyon, which is believed to be pre-Quaternary (Stanley and Kelling, 1970). Drainage from Delaware Bay was responsible in the late Tertiary and Pleistocene for a large volume of sediment deposited on the continental slope (McGregor et al, 1977). The slope seaward of Albemarle Sound (profiles 2e and 3d, Figs. 2 and 3) was another depocenter. Drainage from the Albemarle River or the James River, both shown by Swift (1976) to have been present on the adjacent shelf, may have been responsible for transporting the sediment and depositing it on the slope near the shelf edge. Since deposition, these sediments have undergone erosion. A detailed study of this area is presently underway using bottom samples and geophysical techniques. Deposition seaward of Albemarle took place on the upper slope, whereas between Baltimore and Washington Canyons deposition was primarily on the middle and lower slope.

The northern portion of the Blake Plateau, north of Cape Romain, can be seen to have built upward and to the northeast by a series of foreset beds (profiles 2h, 3f, 5e; Figs. 2–4). This depositional pattern lasted for most of the Tertiary on the basis of the middle Eocene age of material cored at 2,000 m on the northern edge of the Plateau (Florentine Maurrasse, personal communication, 1976). This Eocene horizon is within the dipping sequence of horizons, with older material beneath it, and it is truncated by a possible late Miocene or mid-Pliocene unconformity correlating with unconformities noted by Vail et al (1977). Many unconformities are present along the Blake Plateau, and even the surface topography is variable (profiles 3g and 3h, Fig. 3). Pratt and Heezen (1964) believe that the Gulf Stream is the controlling factor for determining the topography and locations of sediment deposition. The position of the Gulf Stream in the vicinity of 32°N has permitted the progradation of the Florida-Hatteras Slope 25 km seaward (profile 2i, Fig. 2), while on the adjacent Blake Plateau erosion is occurring (profiles 3g and 5f, Figs. 3 and 4), producing very irregular topography with less than 100 m of sediment overlying lower Eocene material.

SUMMARY

The physiographic provinces of the slope and rise are very different in morphology and erosional and depositional history. The slope has been well dissected as far south as Cape Hatteras. On the basis of the detailed study of the slope north of Wilmington Canyon, the general dissection of the upper slope by small valleys appears to have occurred in the Pleistocene. The rise is smoother in morphology with only the major canyons having extensions at this depth.

Within each physiographic province parallel to the margin, there are regional variations in the erosional and depositional history which show up on strike lines. The Florida-Hatteras slope and adjacent portions of the Blake Plateau have been and still are actively influenced by the Gulf Stream. The interaction of the Gulf Stream and the Western Boundary Undercurrent with the sea floor has exposed or inhibited deposition atop Eocene material in the north end of the Blake Plateau near Cape Hatteras. This same interaction has influenced deposition in the vicinity of Hatteras Canyon, as it has very small levees compared with those of other East Coast canyons. A graded sand layer with coarse to fine sand on the rise in Norfolk Canyon suggests that this canyon is serving as a conduit in transporting sediment to the rise. In addition to the vicinity of canyons, several other portions of the slope have been deposition sites for large volumes of sediment in the past; these areas are seaward of the Delmarva Peninsula and Albemarle Sound.

During the mid-Pliocene, the rise between Hydrographer and Wilmington Canyons appears to have been subjected to very uniform erosion followed by deposition. This unconformity can be traced beneath the levees of Hudson Canyon, suggesting that they are Pleistocene to Recent features.

On the basis of strike-line profiles, variations in morphology and sedimentation history along the continental margin can be readily assessed with identification of regional versus localized events and processes.

REFERENCES CITED

Betzer, P. R., P. L. Richardson, and H. B. Zimmerman, 1974, Bottom currents, nepheloid layers, and sedimentary features under the Gulf Stream near Cape Hatteras: Marine Geology, v. 16, p. 21–29.

Doyle, L. J., C. C. Woo, and O. H. Pilkey, 1976, Sediment flux through inter-canyon slope area: U.S. Atlantic continental margin (abstract): Geol. Soc. America, Abstracts with Programs, v. 8, p. 843.

——— et al, 1975, Sedimentation on the northeastern continental slope of the United States: Proc., IX Cong. Internat. de Sedimentologie (Nice), v. 6, p. 51–56.

Garrison, L. E., 1970, Development of continental shelf south of New England: AAPG Bull., v. 54, p. 109–124.

Grow, J. A., and R. G. Markl, 1977, IPOD-USGS multichannel seismic reflection profile from Cape Hatteras to the Mid-Atlantic Ridge: Geology, v. 5, p. 625–630.

Hathaway, J. C., et al, 1976, Preliminary summary of the 1976 Atlantic margin coring project of the U.S. Geological Survey: U.S. Geol. Survey Open File Report 76-844, 217 p.

Heezen, B. C., M. Tharp, and M. Ewing, 1959, The floors of the oceans, 1. The North Atlantic: Geol. Soc. America Spec. Paper 65, 122 p.

Hollister, C. D., et al, 1972, Site 108—continental slope, *in* Initial reports of the Deep Sea Drilling Project: Washington, D.C., U.S. Govt., v. 11, p. 357–363.

Hoskins, H., 1967, Seismic reflection observations on the Atlantic continental shelf, slope and rise southeast of New England: Jour. Geology, v. 75, p. 598–611.

JOIDES, 1965, Ocean drilling on the continental margin: Science, v. 150, p. 709–716.

Keller, G. H., D. N. Lambert, and R. H. Bennett, 1978, Geotechnical properties of continental slope deposits, Cape Hatteras to Hydrographer Canyon (abstract): AAPG Bull., v. 62, p. 529.

Kelling, G., and D. J. Stanley, 1970, Morphology and structure of Wilmington and Baltimore submarine canyons, eastern United States: Jour. Geology, v. 78, p. 637–660.

McGregor, B. A., and R. H. Bennett, 1977, Continental slope sediment instability northeast of Wilmington Canyon: AAPG Bull., v. 61, p. 918–928.

——— ——— and G. F. Merrill, 1977, Continental slope south of Baltimore Canyon, U.S. East Coast (abstract): Geol. Soc. America, Abstracts with Program, v. 9, p. 1089.

Manheim, F. T., and R. E. Hall, 1976, Deep evaporitic strata off New York and New Jersey—evidence from interstitial water chemistry of drill cores: Jour. of Research U.S. Geol. Survey, v. 4, p. 697–702.

Nelson, C. H., and L. D. Kulm, 1973, Submarine fans and deep-sea channels, *in* Turbidites and Deep-Water Sedimentation: Pacific Section S.E.P.M. Short Course, Anaheim, p. 39–78.

Pratt, R. M., 1967, The seaward extension of submarine canyons off the northeast coast of the United States: Deep-Sea Research, v. 14, p. 409–420.

——— and B. C. Heezen, 1964, Topography of the Blake Plateau: Deep-Sea Research, v. 11, p. 721–728.

Sheridan, R. E., 1974, Atlantic continental margin of North America, *in* C. A. Burke and C. L. Drake, *eds.*, Geology of Continental Margins: Berlin, Springer-Verlag, p. 391–407.

——— 1975, Dome structure, Atlantic continental shelf east of Delaware: Preliminary geophysical report: AAPG Bull., v. 59, p. 1203–1211.

Stanley, D. J., and G. Kelling, 1970, Interpretation of a levee-like ridge and associated features, Wilmington Submarine Canyon, eastern U.S.: Geol. Soc. America Bull., v. 81, p. 3747–3752.

Swift, D. J. P., 1976, Continental shelf sedimentation, *in* D. J. Stanley and D. J. P. Swift, eds., Marine Sediment Transport and Environmental Management: New York, John Wiley and Sons, Inc., p. 311–350.

Twichell, D. C., H. J. Knebel, and D. W. Folger, 1977, Delaware River: evidence for its former extension to Wilmington Submarine Canyon: Science, v. 195, p. 483–485.

Uchupi, E., 1968, Atlantic continental shelf and slope of the United States—physiography: U.S. Geol. Survey Prof. Paper 529, p. C1–C30.

——— and K. O. Emery, 1967, Structure of the continental margin off Atlantic coast of the United States: AAPG Bull., v. 51, p. 223–234.

Vail, P. R., R. M. Mitchum, Jr., and S. Thompson III, 1977, Seismic stratigraphy and global changes in sea level, *in* C. E. Payton, ed., Seismic stratigraphy—applications to hydrocarbon exploration: AAPG Memoir 26, p. 49–212.

The Crustal Structure and Evolution of the Area Underlying The Magnetic Quiet Zone on the Margin South of Australia[1]

MANIK TALWANI,[2] JOHN MUTTER,[3]
ROBERT HOUTZ,[4] AND MICHAEL KÖNIG[2]

Abstract Recently acquired deep crustal refraction measurements made aboard R/V *Vema* in January and February, 1976, are combined with new and existing seismic reflection, shallow-refraction, gravity, magnetic and other data, to present a picture of the structure of the Magnetic Quiet Zone south of Australia from its surface morphology down to the mantle.

We deduce that the quiet zone is bounded by primary discontinuities in the earth's crust that extend to great depths beneath the surface. The inner (landward) boundary is marked by a prominent magnetic trough, a steplike change in crustal thickness and velocity structure, and commonly an isostatic gravity high. Its outer (seaward) boundary is marked by anomalously shallow basement topography, and represents the sharp boundary with normal oceanic crust produced at the Southeast Indian Ridge.

The crust within the quiet zone lies at a relatively uniform depth, but has a variable velocity structure. Computed seismic results indicate "continental" and "oceanic" sections and still others that are neither "oceanic" nor "continental." No systematic gradation from one kind to the others was observed. Rather, the quiet zone appears to be an inhomogeneous amalgamation of different crustal types. We propose that this type of crust may be unique, being neither continental nor oceanic.

We suggest that the evolution of this margin took place in two stages. In the first stage, a continental rift valley about 150 to 200 km wide was formed. Its width stayed nearly constant while predominantly vertical motions took place within it and the unique rift crust evolved during a period of several tens of millions of years. During the second stage, on the other hand, horizontal motions dominated. The rift valley broke apart at its axis and seafloor spreading started. Thus, in contrast to a continuous evolution of the margin as many previous authors have suggested, we propose a two-stage evolution: the first in which the width of the rift valley remained constant but the crustal composition evolved, and the second in which the geometry changed (that is, drift took place) but the composition of the new oceanic crust remained relatively constant.

INTRODUCTION

Many rifted continental margins include large areas where the magnetic field is relatively smooth, known as Magnetic Quiet Zones. Where the Magnetic Quiet Zone cannot be readily associated with creation by seafloor spreading during a period of uniform polarity of the Earth's magnetic field, its origin is not certain. A closely related question concerns the continental rifting that precedes seafloor spreading. Is there always a long interval of rifting before drift starts? How is the break related to the position of the rift? Are the rifts which preceded spreading now located under the continental shelves or do they also underlie the continental slope and rise?

An ideal place to examine these questions appeared to be the southern margin of Australia, where there is a well-developed Magnetic Quiet Zone. Furthermore, the continental margin south of Australia has been regarded as a classic example of a rifted margin (Sproll and Dietz, 1969; Smith and Hallam, 1970; von der Borch et al, 1970; Griffiths, 1971). In addition, recent studies of deep-penetration multichannel seismic reflection data by Boeuf and Doust (1975) and Willcox (in press), and a synthesis of structural and drilling information by Deighton and others (1976) have led to proposals of detailed evolutionary models for the margin. The relationship between these models and the Magnetic Quiet Zone and the description of the results of the experiment to determine the deep structure of the margin, par-

[1]Manuscript received, October 24, 1977; accepted, June 8, 1978.

[2]Lamont-Doherty Geological Observatory of Columbia University and Department of Geological Sciences, Columbia University, Palisades, New York 10964

[3]Bureau of Mineral Resources, Geology and Geophysics, Canberra City, Australia

[4]Lamont-Doherty Geological Observatory of Columbia University, Palisades, New York 10964

We thank Captain Kohler and the scientists and crew aboard *Vema* for help in conducting the *Vema* Cruise 33 project, which was carried out cooperatively by Lamont-Doherty and the Bureau of Mineral Resources, Geology and Geophysics, Australia. National Science Foundation grant OCE 75-20438 was the major source of funds for this study.

L-DGO Contribution No. 2736.

Article Identification Number:
0065-731X/78/MO29-0010/$03.00/0.
(see copyright notice, front of book)

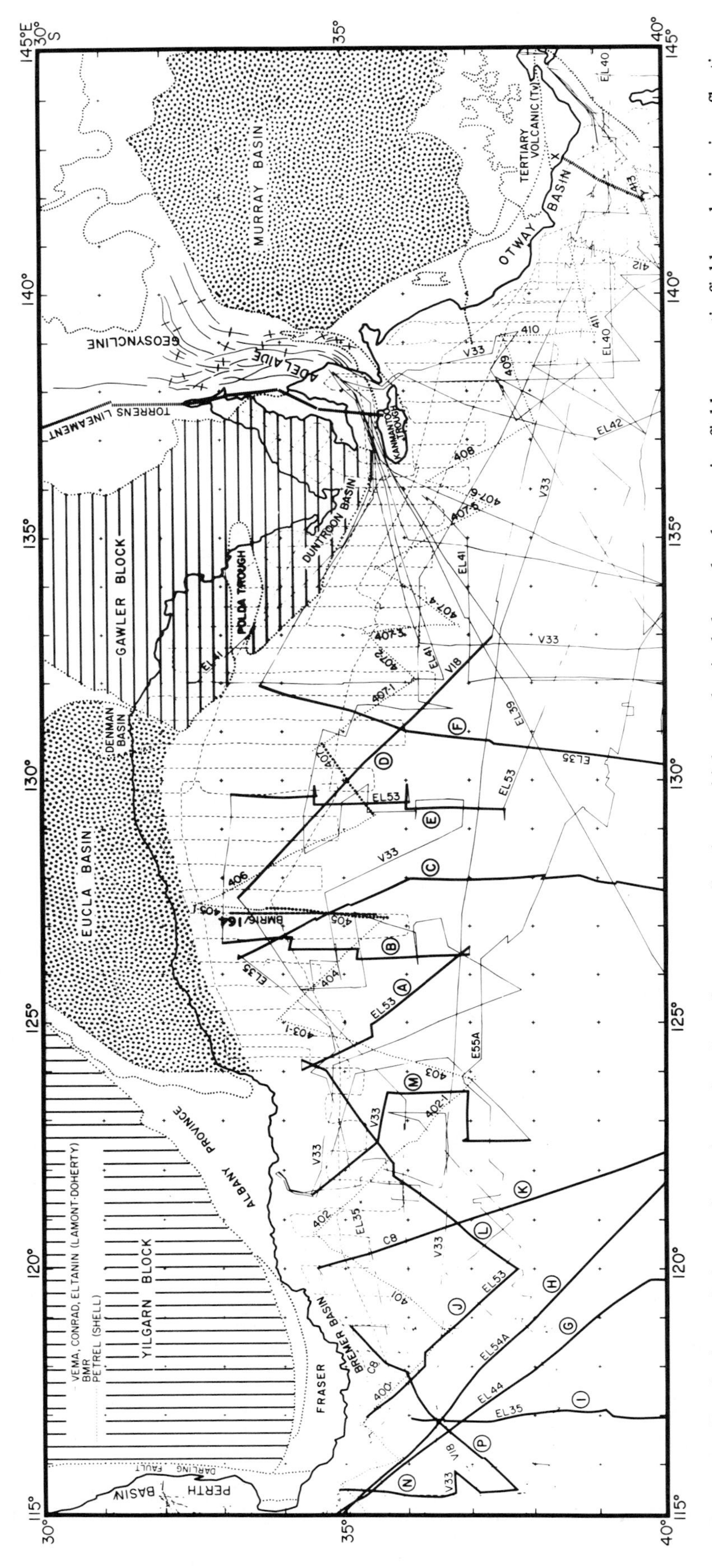

FIG. 1. Track chart of the continental margin of southern Australia along which geophysical data: depth, gravity-field, magnetic-field, and seismic-reflection profiles were obtained.

ticularly the area underlying the Magnetic Quiet Zone, is the subject of this study.

The large number of ship tracks on the southern margin of Australia on which geophysical data were obtained is shown in Figure 1. These include Lamont-Doherty tracks (vessels *Vema, Conrad* and *Eltanin*), Australian Bureau of Mineral Resources, Geology and Geophysics (BMR) surveys, and a Shell Oil Company survey (R/V *Petrel*). Measurements of depth, gravity and magnetics as well as seismic reflection profiling were carried out on all of these surveys. *Vema* and *Eltanin* employed single-channel seismic equipment; BMR and Shell used multichannel seismic reflection equipment. The geophysical profiles along some of the tracks in Figure 1 are used as illustrations in this paper. These tracks are emphasized by heavy lines.

The principal new results which form the core of the present study were obtained in a one-month period during *Vema* Cruise 33 along the Australian margin. Gravity, magnetic, and seismic reflection results help in establishing the geological framework of the area. The seismic refraction results enable us to examine in detail the crustal structure of the area underlying the Magnetic Quiet Zone.

The Diamantina Fracture Zone was named before the recognition of fracture zones as transform faults. The Diamantina Fracture Zone is not a transform fault and for this reason, we shall hereafter refer to it as Diamantina Zone.

GEOLOGICAL AND GEOPHYSICAL FRAMEWORK OF THE MAGNETIC QUIET ZONE

Physiography

The physiographic provinces of the margin south of Australia are sketched in Figure 2. Bathymetric profiles in the western and central part of the margin (between longitudes 115°E and 132°E) are shown in Figure 3. The continental shelf is widest in the area of the Great Australian Bight, and narrows to the east and west. In the west the continental slope is steep and narrow. The continental slope south of the Eyre Plateau is fairly steep, whereas the Ceduna Plateau is constructional and seems to be built up at the expense of the rise, and its slope is gentler. The rise is very narrow at the foot of the Ceduna Plateau. With the exception of this narrow part of the rise, the continental margin is remarkably uniform from about 115°E to 130°E. South of the slope, the Diamantina Zone extends to about 125°E.

The bathymetric contours in Figure 4 are taken principally from the Hayes and Conolly (1972) bathymetric chart of the Southeast Indian Ocean. New bathymetric soundings made following the completion of this chart have not altered the physiographic interpretation sufficiently to justify a complete recontouring of the map. Some new observations do alter the picture somewhat and these are described in the following.

The Diamantina Zone on Hayes and Conolly's bathymetric chart is shown to extend no farther east than about 121°30′E. Heezen and Tharp (1973), working on results from *Eltanin* Cruise 55, indicated that they were able to trace key features of the Diamantina Zone from 125°E and 300 miles westward to 115°E. On *Vema* Cruise 33, we made several crossings of the fracture zone at its eastern extremity, which showed that its characteristic topography was present to at least 124°E. Prominent basement ridges with deep intervening valleys are present to this longitude. To the east, the valleys are occupied by narrow, elongate extensions of the abyssal plain, and east of about 124°E, the topography breaks into isolated seamounts surrounded by a large expanse of abyssal plain. With the track spacing available, it is

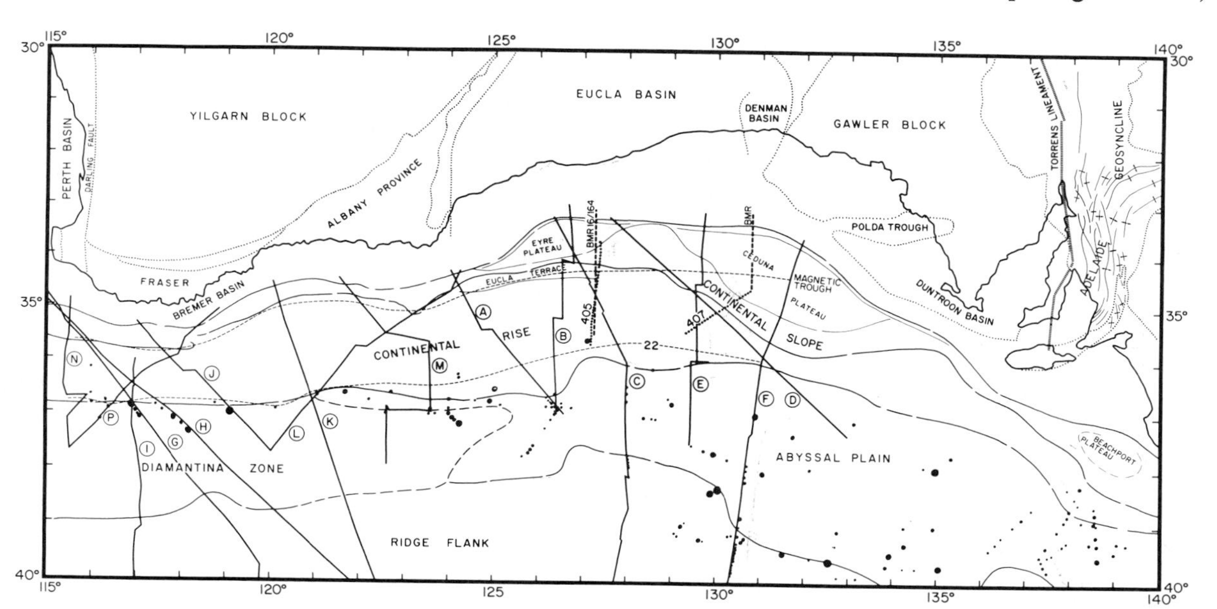

FIG. 2. Physiographic provinces of the continental margin of southern Australia. Profiles used in this paper are shown. The location of the Magnetic Trough and the location of magnetic lineation 22 are also shown. Black dots indicate seamounts or basement peaks.

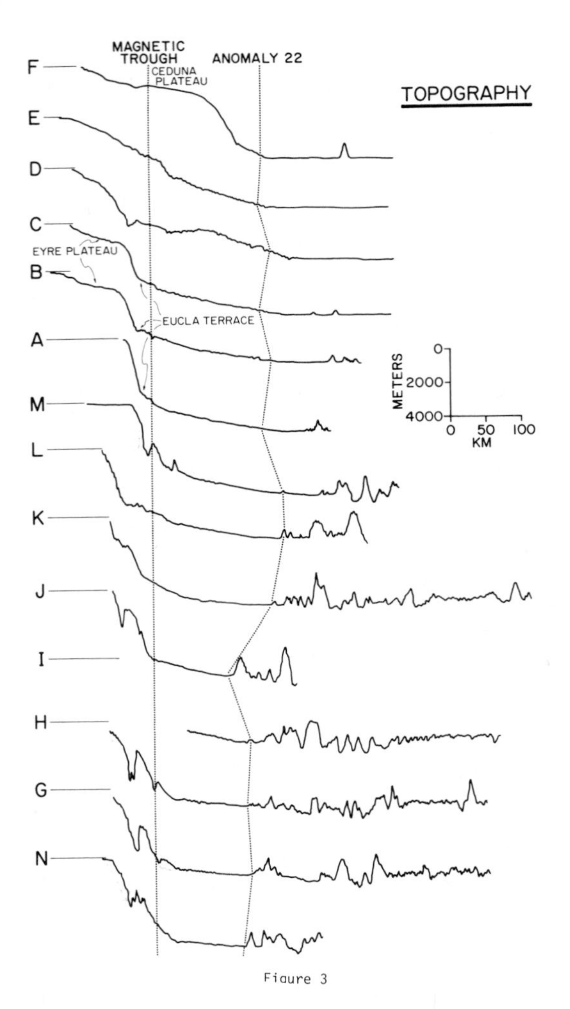

FIG. 3. Selected depth profiles on the continental margin of southern Australia. Locations are shown in Figures 1 and 2.

frequently not possible to determine whether seamounts observed on a single crossing are isolated structures or form part of an elongate ridge. Near the northern edge of the abyssal plain just east of 125°E, the depth to seafloor increases in three steps from 7.2 sec reflection time to 7.32 sec and then to 7.41 sec. Seamounts are located at the changes in sill depth, suggesting that they may have some east-west linearity. However, at 126°E the east-west linearity is gone and does not reappear to the east.

The contours of Hayes and Conolly (1972) show the northern extension of the Australian/Antarctic discordance occupying the region east of 121°E. More recent coverage of this area, however, suggests that the contours run east-west rather than north-south so that the north-south linearity associated with the discordance farther south does not extend as far north as 38°S. The continental rise and slope in most areas west of about the central Great Australian Bight show extensive excavation in the form of major canyons and broad erosional depressions.

Willcox (in press) describes a narrow terrace at the head of the continental rise south of Eyre Plateau from BMR profiles in the region. The name Eucla Terrace is suggested for this structure and will be used throughout this paper. At 127°E we indicate a sharp northward step in the edge of the abyssal plain which is not present on the Hayes and Conolly (1972) chart.

Figure 2 shows 175 individual seamounts; considering the distribution of ship tracks and the frequency with which they occur along these tracks, it is likely that we have mapped less than one-third of those present. Most occur north and east of the Diamantina Zone and near the southern edge of the abyssal plain, where they are topographic expressions of fracture zones and ridge flank basement

FIG. 4. Bathymetric map modified slightly from Hayes and Conolly (1972). Contours are in fathoms.

features. A significant number also occur in the northern and central parts of the abyssal plain and the lower continental rise.

Magnetic Pattern

Weissel and Hayes (1972) first outlined the sea-floor spreading history of the Southeast Indian Ridge from studies of the magnetic anomaly pattern. Their identification of older anomalies near the margin was limited by the spacing between ship tracks and by the disturbing influence of the great topographic relief in the Diamantina Zone. Weissel and Hayes identified older anomalies between the longitudes of 126°E and 135°E. They were able to identify anomaly 20 with certainty. There was, however, some doubt about the locations of anomalies 21 and 22. Weissel and Hayes made an important contribution by recognizing the Magnetic Quiet Zone north of the identifiable anomaly sequence in the continental margin south of Australia.

König and Talwani (1977) extended the anomaly identifications made by Weissel and Hayes. They identified anomaly lineations 21 and 22 in the area of the Bight and extended lineation 19 west to 115°E. They also noted the presence of the Magnetic Trough. This is the most consistent feature of the magnetic-anomaly field south of Australia. Closely spaced tracks made by BMR further established the persistence of the Magnetic Trough.

Figure 5 shows selected projected magnetic profiles and Figure 6 shows the revised picture of magnetic lineations including the Magnetic Trough, superimposed on the physiographic chart. (Locations of seismic-refraction stations are also given in Fig. 6.) Anomaly lineations 19, 20, 21, and 22 have been identified through most of the Diamantina Zone, although the influence of steep basement topography

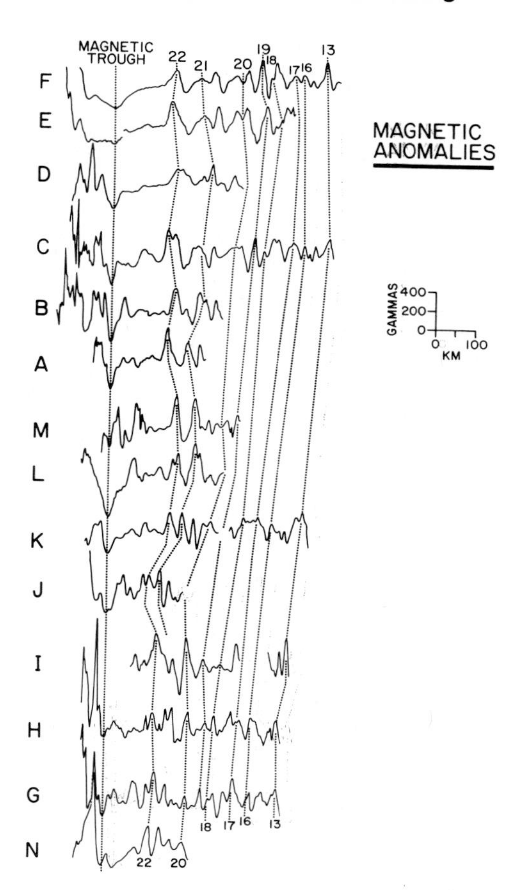

FIG. 5. Magnetic profiles on the continental margin of southern Australia. Locations are shown in Figures 1 and 2.

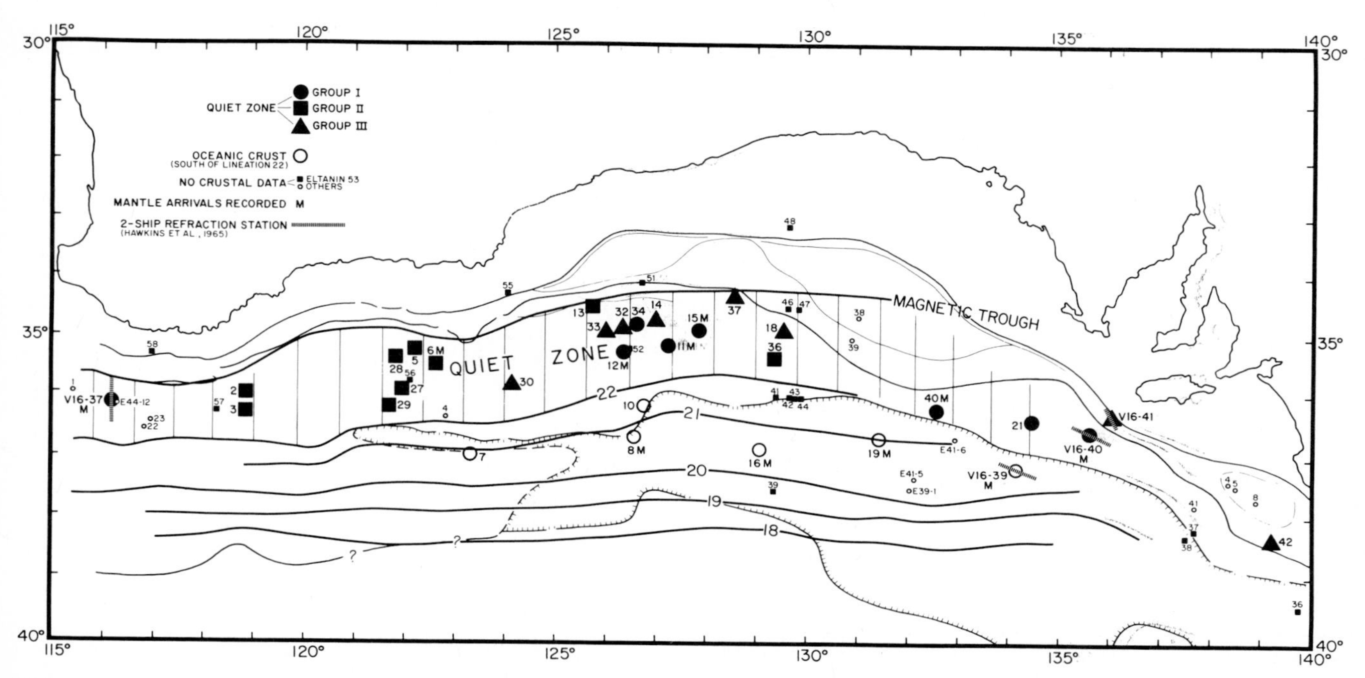

FIG. 6. Magnetic lineations on the continental margin of southern Australia. Physiographic provinces are demarcated (see names in Fig. 2). The locations of seismic refraction stations are also given.

makes the identification of spreading anomalies less reliable there than in other areas.

The new pattern shows the important results that lineation 22, the first in the spreading sequence, lies along the northern flank of the Diamantina Zone. Lineation trends 22 to 19 form a nearly east-west pattern within the zone. East of the zone, lineations 22, 21 and 20 swing northward with the oldest anomalies showing the maximum deflection, whereas lineation 19 remains at nearly constant latitude. This amounts to a compression of the anomaly pattern in the Diamantina Zone.

New tracks in the eastern part of the area confirm the absence of lineation 22 east of 131°30′E. Lineation 21 forms the first in the seafloor spreading sequence east of this longitude. Loss of lineation 22 occurs fairly rapidly.

The position of the Magnetic Trough is now known in considerable detail. Figure 7 shows a series of profiles crossing the trough obtained on BMR surveys. The anomaly minima have been aligned and the profiles are spaced so that they do not seriously overlap. This emphasizes the continuity of the trough across more than 20° of longitude. The trough is narrower and deeper in the west and broadens and shallows somewhat in the east. It has been mapped as far west as the limit of the Australian continent. It is, therefore, possible to extend the trough to the edge of the map in Figure 6.

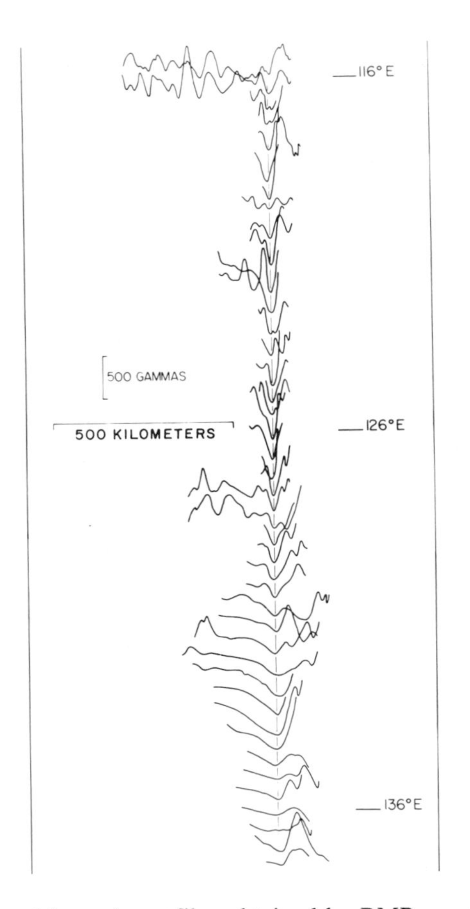

FIG. 7. Magnetic profiles obtained by BMR are aligned to show the continuity of the Magnetic Trough. See Figure 1 for the location of BMR profiles. (These profiles are plotted with the westernmost profiles on top, in contrast with Figure 5 in which the eastern profiles are plotted on top.)

East of 136°E, at the western tip of Kangaroo Island, the trough loses clear definition. If it exists east of this longitude, it is severely reduced in amplitude. Although a Magnetic Quiet Zone may exist off southwest Victoria and west of Bass Strait, its inner or landward boundary is not marked by a prominent anomaly but is created by a fairly sharp change in anomaly wave length and amplitude. There is some indication that the main magnetic trough bifurcates and rapidly loses amplitude east of 136°E but this is not well established.

The relation of the trough to physiographic features can be seen in Figures 3 and 6. In the west, the trough lies near the base of the continental slope but this changes at 130°E. South of the Eyre Plateau the trough lies immediately over the outer edge of the Eucla Terrace. The trough crosses the Ceduna Plateau about mid-way down its length, and terminates near 136°E. Lineation 22 lies along the northern flank of the Diamantina Zone as demonstrated very well in Figure 3. East of the Diamantina Zone lineation 22 continues to be associated with rugged basement topography (König and Talwani, 1977). North of lineation 22, the basement is much deeper throughout the margin.

In the eastern part of the margin the magnetic field is very smooth between anomaly 22 and the Magnetic Trough, and the designation Magnetic Quiet Zone has been applied to the area (Fig. 5). The magnetic field becomes progressively more disturbed toward the west in the area lying between anomaly 22 and the Magnetic Trough. Strictly speaking the western area does not constitute a Magnetic Quiet Zone. Detailed examination of the existing magnetic data suggests that the anomalies are not very strongly aligned in this area and are therefore not seafloor-spreading-type anomalies. We also note that the gentle southward increase in the magnetic field from the trough to the crest of anomaly 22, as perhaps typified in profile F, exists in all or most of the profiles in the area even when the field is disturbed. Hereafter, when we refer to the Magnetic Quiet Zone, we will include the area north of the Diamantina Zone where the field is considerably rougher but is irregular and unlike the field associated with the linear magnetic pattern generated by seafloor spreading. Portions of other Magnetic Quiet Zones, as in the Norwegian margin, display irregular magnetic anomalies and, therefore, the Australia situation is not unique.

Gravity Pattern

König and Talwani (1977) have described and interpreted the gravity field variations in the region. The new gravity data collected on *Vema* Cruise 33 have not significantly altered the picture of the gravity field they presented.

The profiles of isostatic gravity anomalies given by König and Talwani (1977) show two major belts of relative highs; one along the inner boundary of the quiet zone, the other along the outer boundary with a pronounced relative gravity low of about 40 mgal between. Figure 8 shows the relation of isostatic gravity anomalies to key magnetic features. This relation is maintained with some consistency on Profiles A, B, C, D and E. Profile F, which crosses the Ceduna Plateau, does not show a high that is associated with the Magnetic Trough. The absence of a relative low over the Magnetic Quiet Zone in the Ceduna Plateau can be attributed to the constructional nature of the Plateau as clearly seen in the topographic profile.

On the isostatic gravity profiles in the western part of the margin, a gravity high is generally associated with the Magnetic Trough, although in some profiles it is somewhat displaced—seaward in profiles M and L and landward in profiles G and H.

The overall isostatic gravity field is depressed with very few positive values. Negative values are

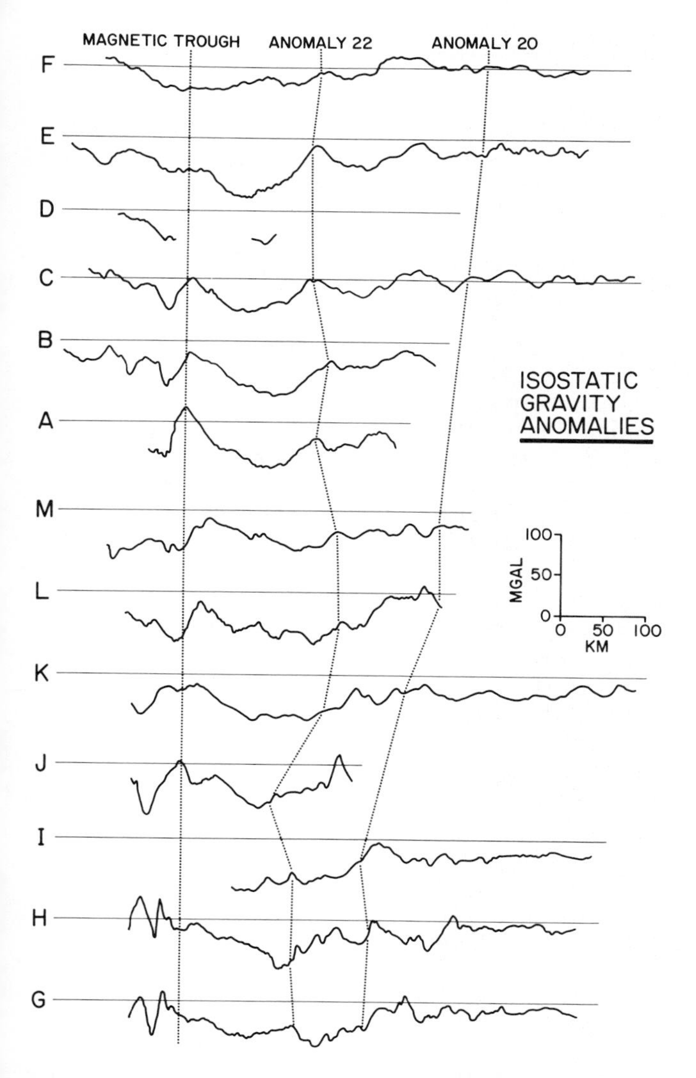

FIG. 8. Isostatic gravity profiles on the continental margin of southern Australia. The Airy hypothesis and simplifying two-dimensional assumptions were used in computing isostatic anomalies. Locations are shown in Figures 1 and 2.

recorded from the edge of the shelf to beyond anomaly 20. The minimum of the isostatic gravity field lies in the Magnetic Quiet Zone, although it is displaced toward the southern end of the quiet zone in the western profiles.

Seismic Reflection Data

DATA SOURCES

Seismic reflection data are available along most of the tracks shown in Figure 1. The data collected by Lamont-Doherty aboard R/V *Vema*, R/V *Conrad* and USNS *Eltanin* (which was operated by the U.S. National Science Foundation) involved a relatively small airgun sound source (15 cu in at 2200 psi pressure) and a single-channel streamer. Penetration with this 18-Hz source is usually restricted to a few hundred meters. Some very early *Vema* records (*Vema* Cruise 18) were made using half-pound explosive charges as the sound source.

BMR recordings were made with a relatively large sparker sound source (120 kilojoule) and a six-channel receiving streamer. The recording shown in Figure 12 is the monitor record of the second channel of the six-channel streamer. More than 1000 m of penetration was often achieved. The source frequency was nearly 55 Hz.

Probably the best data recorded in the region is the Shell data obtained aboard *Petrel* and presented by Boeuf and Doust (1975). The *Petrel* employed a 1090 cu in airgun array, 24-channel streamer and digital recording. Boeuf and Doust (1975) showed interpreted geological cross-sections in which they indicated structures as much as 9.5 km below the seabed.

DEPTH TO BASEMENT

The depth-to-basement map shown in Figure 9 was produced from the sonobuoy refraction data discussed later in this paper, and was supplemented by seismic reflection profiles in areas away from sonobuoy control. Reflection profile intervals were converted to thickness on the basis of a statistical analysis of sonobuoy data, which is also discussed later.

The positions of the prominent Magnetic Trough and lineation 22 are shown. There is some uncertainty as to basement depths beneath the Ceduna Plateau and the continental slope in the western part of the margin. Basement appears very close to the seabed on reflection profiles made on the continental shelf, and plunges steeply beneath the slope. It is sometimes difficult to establish basement depths beneath the slope due to its dip and to the interference of seabed multiple reflections.

The contour map shows that the Magnetic Trough lies either within or immediately seaward of the steepest descent of the basement. The anomaly is clearly related to an edge effect caused by the step in basement level which occurs from shelf to rise (König and Talwani, 1977).

The basement highs at the base of the rise have been contoured as a set of continuous ridges that strike east-northeast, trending into the Diamantina Zone. There is a consistent relationship of lineation

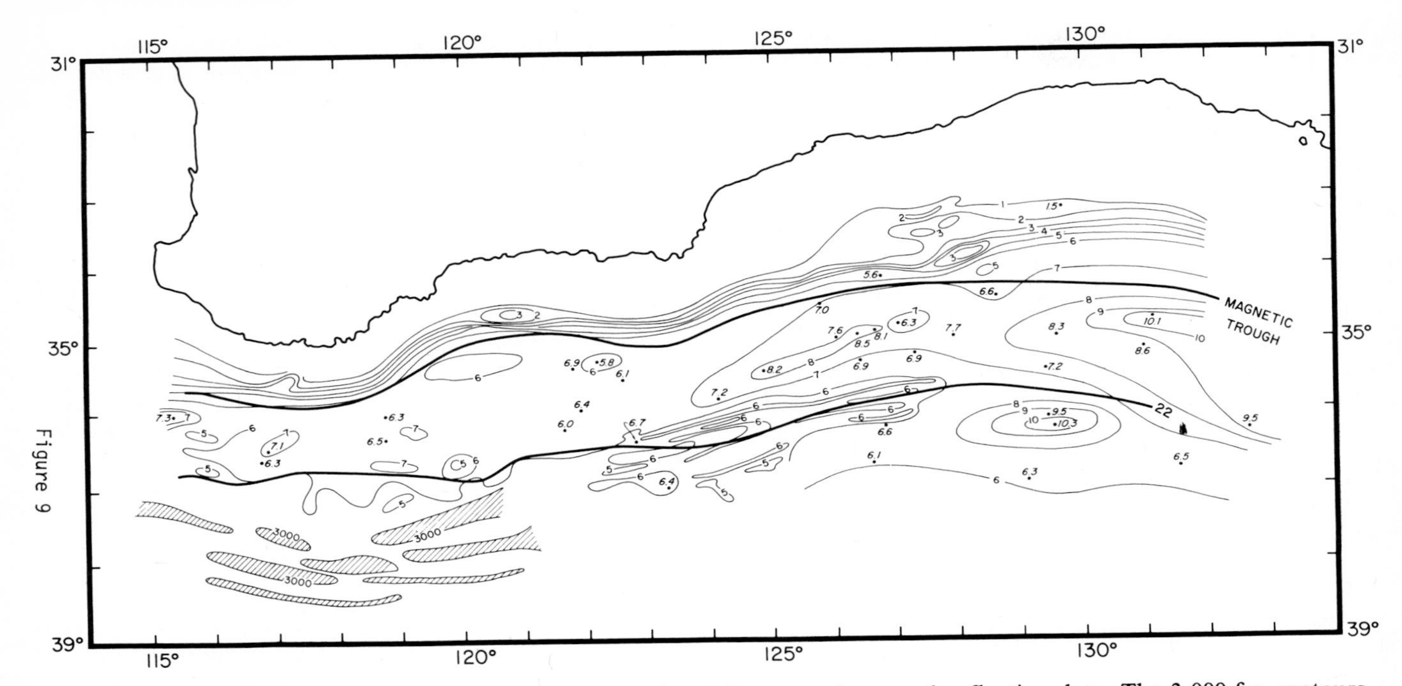

FIG. 9. Depth-to-basement map (contours in km) produced from sonobuoy and reflection data. The 3,000-fm contours in the southwestern part of the map are bathymetric contours which indicate ridges that form the Diamantina Zone. Note the association of the magnetic trough with an escarpment in the basement, and of lineation 22 with ridges in the basement beneath the lower part of the continental rise, and with the northern edge of the Diamantina Zone.

22 with this system of ridges. The ridge system probably extends farther east but is covered by thick sediments. Isolated seamounts or basement highs immediately north of the Diamantina Zone have not been contoured as ridges although such an interpretation would be possible.

On mainland southern Australia, Precambrian rocks outcrop in two major regions: the Yilgarn Block (see Figs. 1 and 2) which lies west of 124°E and the Gawler Block which lies east of 131°E. Between these two blocks lies the Eucla Basin which contains Tertiary and Mesozoic sediments throughout most of its area. The total thickness of Mesozoic and Tertiary sediments does not exceed 1000 meters and is usually less than half that. Sediments are either mildly deformed, or undeformed.

Precambrian basement rocks and Eucla Basin sediments extend seaward to form the substructure of the continental shelf, and at least some parts of the continental slope. There is ample evidence (Conolly et al, 1970; Willcox, in press) that the gross structures known onshore extend uninterrupted to at least the shelf edge, and that south of the Yilgarn Block (that is, west of 124°) and south of the Gawler Block (that is, east of 132°) the thickness of sediments on the shelf is very small; between, the Eucla Basin extends southward onto the shelf with a thickness of sediments somewhat larger and more variable.

The sediment thickness increases on the slope, especially on the Eyre and Ceduna Plateaus where it exceeds 10 km. The reflection profiles that are described later give some details of the sedimentary sections and illustrate their variability.

The sediment thickness under the continental rise which roughly coincides with the Magnetic Quiet Zone, varies from about 2 to 4 km, generally thickening east of 128°E.

There is a sharp decrease in sediment thickness farther south as anomaly 22 is approached. This decrease corresponds to the appearance of the Diamantina Zone in the west or of buried basement ridges just east of the Diamantina Zone.

In summary, sediments over the Precambrian basement in the shelf area are thin. The basement goes down sharply under the shelf edge giving rise to an increased sediment thickness on the slope. The sediment thickness is particularly large under the Eyre and Ceduna Plateaus. It varies from about 2 to 4 km on the rise which roughly coincides with the Magnetic Quiet Zone. There is a sharp decrease of sediments over basement ridges farther south. These ridges which are exposed in the west, as the Diamantina Zone, but are buried to the east are associated with the earliest seafloor-spreading-type crust.

REFLECTION PROFILES ACROSS THE MARGIN

The seismic reflection profiles, together with available well-log information on the shelf, have been used to develop geologic sections across the continental margin. The interpretation of four of these sections is summarized with their description.

Profile Across Otway Basin.—The Otway Basin is the best known of the continental margin basins as it has been the subject of detailed seismic exploration and exploratory drilling by oil companies for some years.

Figure 10 shows an interpreted geological cross-

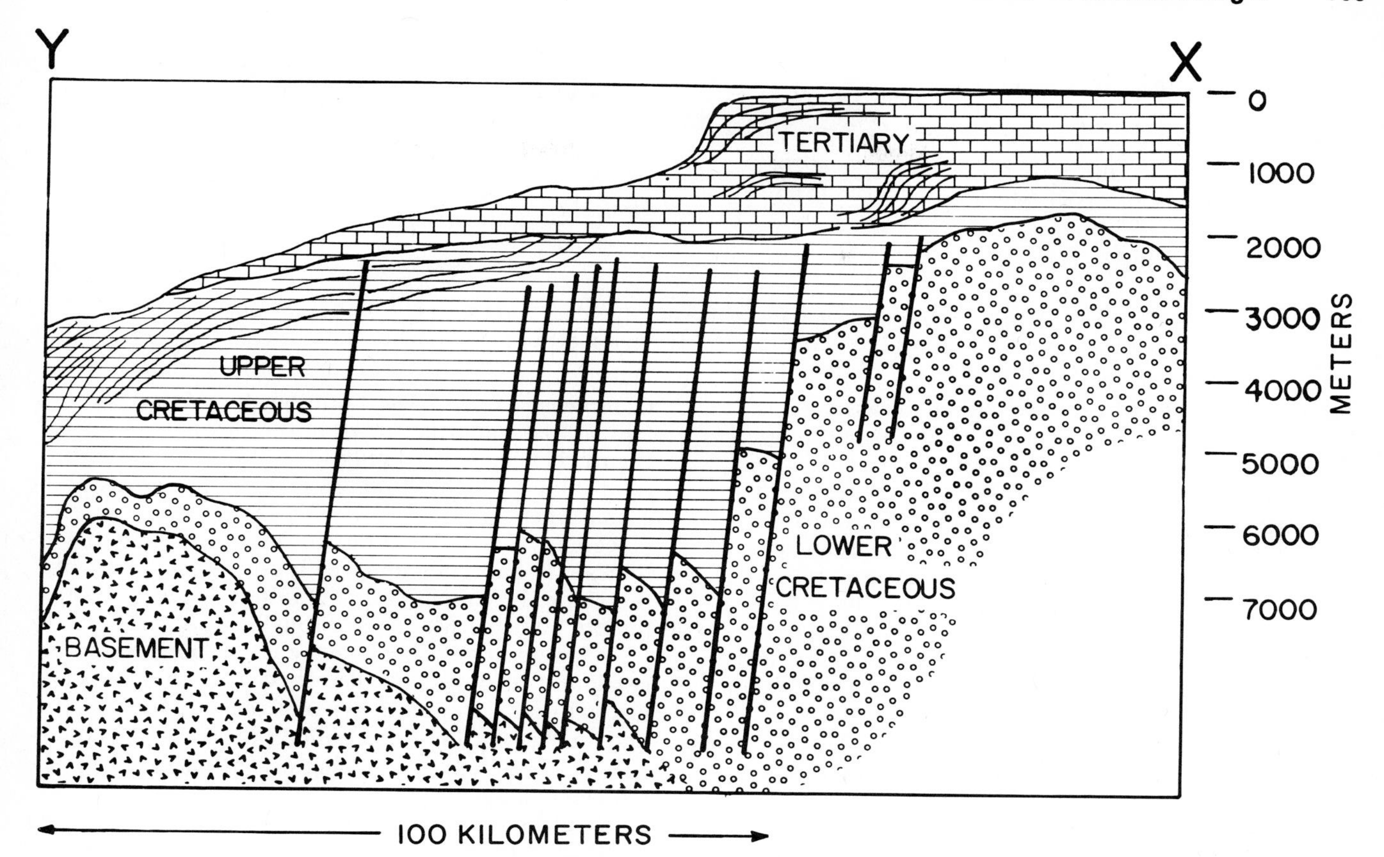

FIG. 10. Cross-section of the Otway Basin (location shown in Fig. 1). The section is simplifed after Denham and Brown (1976) and is based on exploration drilling on the continental shelf and seismic profiling in deeper water. The steeply inclined heavy lines mark normal faults. The structural style represented in this figure is generally applicable from the Otway Basin to the western margin of the Great Australian Bight.

section of the Otway Basin reproduced with some simplifications from Denham and Brown (1976). It is based on data from exploration drilling on the continental shelf and from seismic exploration in deeper water. The location of the section is indicated by *XY* in Figure 1.

The Otway Basin began its history as a broad basement downwarp in earliest Cretaceous or possibly Jurassic times. The trough trended nearly east-west across the dominant north-south structures of the Paleozoic Tasman Geosyncline. During Jurassic(?) and Early Cretaceous times up to 2500 m of fluviatile-lacustrine deposits were laid down in a single large trough which extended east beneath Bass Strait and the Gippsland Basin (Denham and Brown, 1976). No major fault tectonics accompanied basin development in the Early Cretaceous. Source areas for sediments lay both to the north and south.

During the Late Cretaceous extensive normal faulting divided the Otway Basin from the eastern basins. The northern margin of the basin was uplifted, and subsequent erosion of the Paleozoic basement rocks led to the deposition of a dominantly argillaceous suite. Deposition in the early part of the Late Cretaceous was rapid and according to Denham and Brown (1976), large quantities of underlying Lower Cretaceous units were removed and redeposited. Hence, the Lower/Upper Cretaceous boundary is marked by an erosional unconformity except in the deepest troughs of the basin, and by the commencement of fault tectonics throughout.

Denham and Brown (1976) favored marine conditions of sedimentation in the latter part of the Late Cretaceous, whereas Deighton and others (1976) suggested that the depositional environment was marginal marine with weak marine incursions in Late Cretaceous time.

Fault tectonics diminished throughout the Late Cretaceous and had generally ceased by the close of the Cretaceous. A major unconformity at the base of the Tertiary represents a change to a sedimentary regime in a rapidly deepening ocean, in accordance with the time of continental dispersal deduced from magnetic anomalies. Tertiary deposits formed in a seaward-building foreset sequence and consist of marls, clays and, in the upper part, predominantly limestones.

The scheme outlined above for the Otway Basin is generally considered to be applicable to the major areas of deposition on the continental margin west through the Great Australian Bight. For instance, Boeuf and Doust (1975) presented a schematic geological development of the southern margin of Australia that is very similar to the Otway Basin

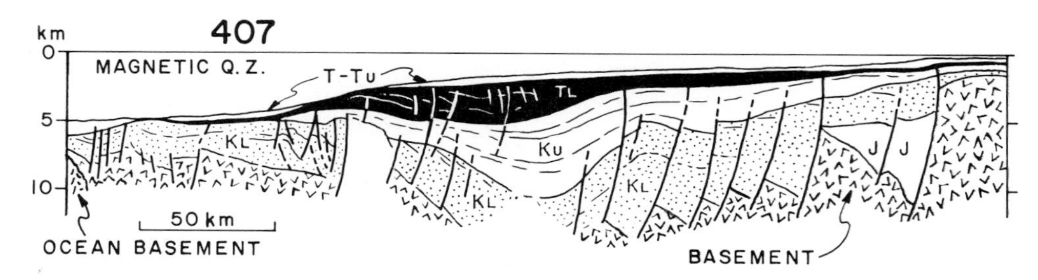

FIG. 11. Interpretation of Shell *Petrel* line 407 by Boeuf and Doust (1975). For location see Figures 1 and 2. This line runs from the Ceduna Plateau to the continental rise.

scheme. They, however, considered Jurassic—Lower Cretaceous sediments to have formed in a rift valley rather than a downwarp. According to their interpretation, relatively larger thicknesses of Tertiary sediments occur in the Great Australian Bight Basin with a reduced Upper Cretaceous section, and they suggest that marine transgressive conditions prevailed in the Late Cretaceous.

Boeuf and Doust's Interpretation of Shell Petrel Profile 407 from the Ceduna Plateau to the Continental Rise.—Figure 11 shows an interpretive sketch by Boeuf and Doust. They make an important distinction between basement that appears to be block-faulted in the reflection records and basement characterized by a much more prominent envelope of diffractions. They interpret the former as block-faulted continental basement and the latter as oceanic basement created by seafloor spreading. In Figure 11, the block-faulted continental basement underlies the Magnetic Quiet Zone including the Ceduna Plateau.

The basement faults as interpreted in Figure 11 extend upward through the Lower and Upper Cretaceous sequences. Faulting commenced in the Lower Cretaceous and died in the Upper Cretaceous. Beneath the Lower Cretaceous unit, a nonmarine sequence of Permian and Jurassic sediments with associated basalts may be locally preserved in faulted inliers.

From an extrapolation of well-log results on land and in shallow water, the Lower Cretaceous unit is believed to consist of fluviatile-lacustrine deposits. There is generally an unconformity between the Lower Cretaceous and Upper Cretaceous units, although none is seen in Figure 11 for the deepest part of the basin. The Upper Cretaceous beds are deltaic. Marine conditions had not been fully developed in the east though they may have progressed further in the west. The Upper Cretaceous units thin considerably seaward (Fig. 11); they are completely absent in the south where Tertiary beds directly overlie the Lower Cretaceous beds. The unconformity at the base of the Tertiary is generally very prominent, and the Tertiary beds are believed to represent well-developed marine conditions for the first time.

Generally there is little faulting in the Tertiary beds although faults in the lower Tertiary section are seen in some areas, particularly the Ceduna Plateau as shown in Figure 11.

BMR Profile 16/164 from the Eyre Plateau to the Continental Rise.—The fault pattern of the Eyre Plateau basement is shown in the profile given in Figure 12. The shelf break is underlain by a major fault, along which basement is downthrown by about 2 km. Another large fault with about 1.5 km of displacement is located about 16 km south of that. Farther south and coinciding with the steepest part of the slope, basement is downfaulted by several kilometers near the Magnetic Trough.

Sediments overlying basement can be grouped into two units: a lower unit disturbed by numerous small normal faults and an upper unit that is essentially undisturbed by faulting. Willcox (in press) has interpreted the two units as being of Tertiary and Upper Cretaceous ages.

Boeuf and Doust's profile 405, located quite close to BMR Profile 16/164 shows very much the same structural relation. Profile 405 reveals faulted basement under the Magnetic Quiet Zone, whereas in the vicinity of anomaly 22 the envelope of diffractions identifies oceanic basement. The fault with the large throw below the slope is clearly seen as the major fault that displaces basement downward in the Magnetic Quiet Zone. The faulting extends upward through Cretaceous beds but is absent in the overlying Tertiary units.

Shallow Structures from Eltanin and Vema Profiles.—Shallow structures are shown well in *Eltanin* and *Vema* profiles. Tracings of five selected reflection profiles are shown in Figure 13. Sonobuoy solutions which accompany the profiles have been projected from nearby stations to the closest points in the profiles. Although some of these lines are close to the Shell and BMR profiles, the recording of the *Eltanin* and *Vema* profiles show shallow features in somewhat better detail. The sonobuoy stations give velocities for the various layers and hence aid in their identification. The locations of these profiles are shown in Figures 1 and 2.

Profile F crosses near the center of the Ceduna Plateau and extends south to the flank of the Southeast Indian Ridge. A prominent reflector is located 300 to 400 msec beneath the Plateau surface. The indicated refraction solutions show an increase in acoustic velocity from 1.75 to 2.4 km/sec at about 500 msec below seabed and the prominent reflector probably represents the Tertiary/Cretaceous boundary.

Cretaceous sediments are probably exposed on the outer slope of the Ceduna Plateau. The continental slope here is convex and smoothly joins the Tertiary/Cretaceous boundary reflector on the Plateau. The continental rise is formed of a wedge of largely transparent sediments. The uppermost sedimentary layers show evidence of slumping, but erosional structures also seem to be present, particularly near the foot of the rise.

A buried basement peak is located near the foot of the rise at about the rise/abyssal plain junction. It immediately underlies magnetic lineation 22. A similar feature was located by nearly every margin crossing indicating a ridge that is at least partially continuous across the whole margin.

The larger basement peaks are more pronounced than the basement relief of the adjacent ridge flank. The peaks are clearly anomalous features and may represent the equivalent of the Diamantina Zone in the eastern part of the area.

The slope and rise in Profile E from 2.5 sec reflection time to the abyssal plain are dominated by erosional features. A reflector corresponding to the top of the Cretaceous sequence is present and erosion of the slope sediments appears to have exposed it on the slope. South of the Cretaceous outcrop, it appears that all Tertiary strata have been removed by erosion. Extrapolation of the dip of the Creta-

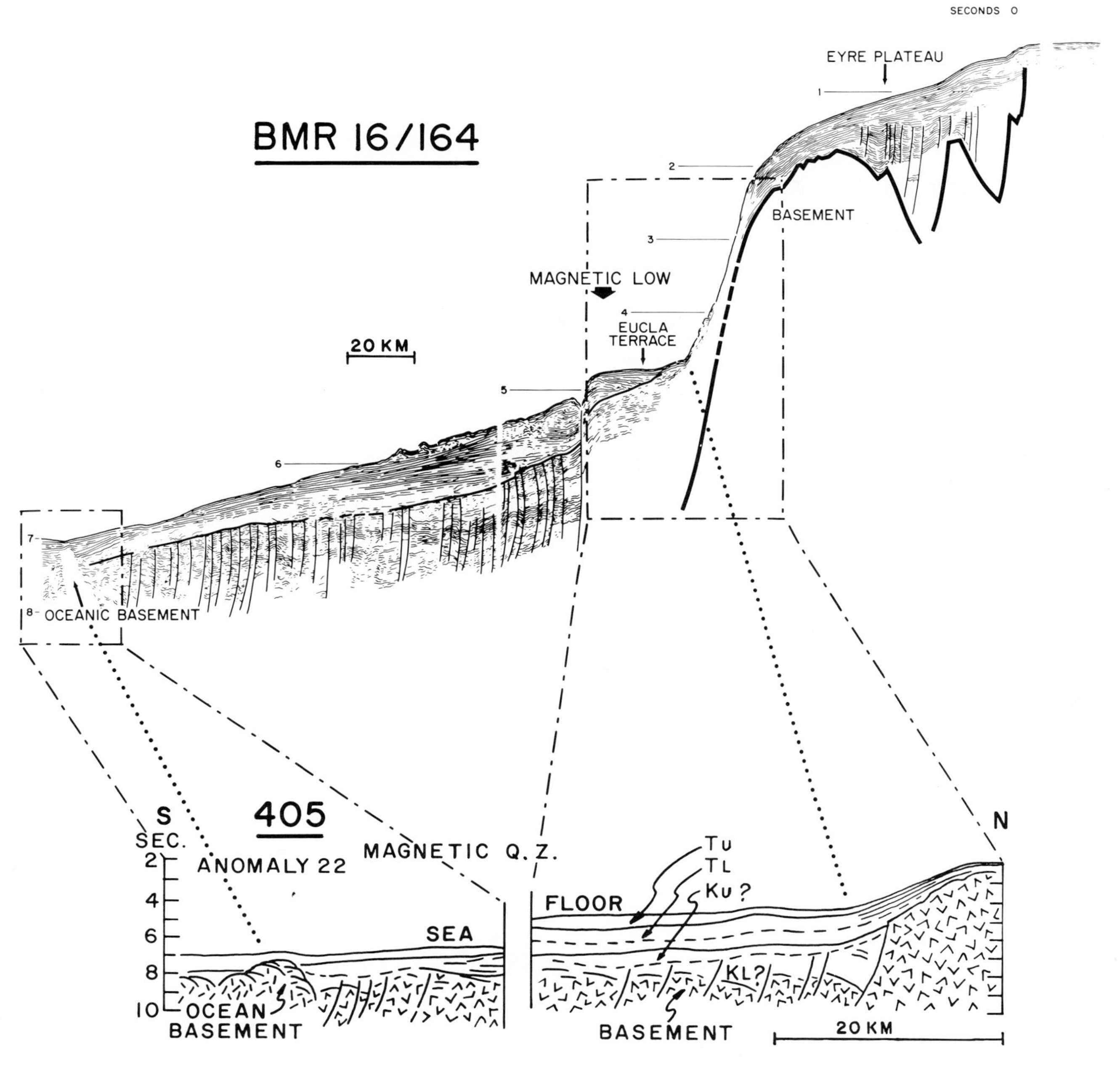

FIG. 12. Tracing of BMR profile 16/164 from the Eyre Plateau to the continental rise. Also shown are portions of Shell *Petrel* line 405 (interpretation by Boeuf and Doust, 1975). Locations of both lines are shown in Figures 1 and 2.

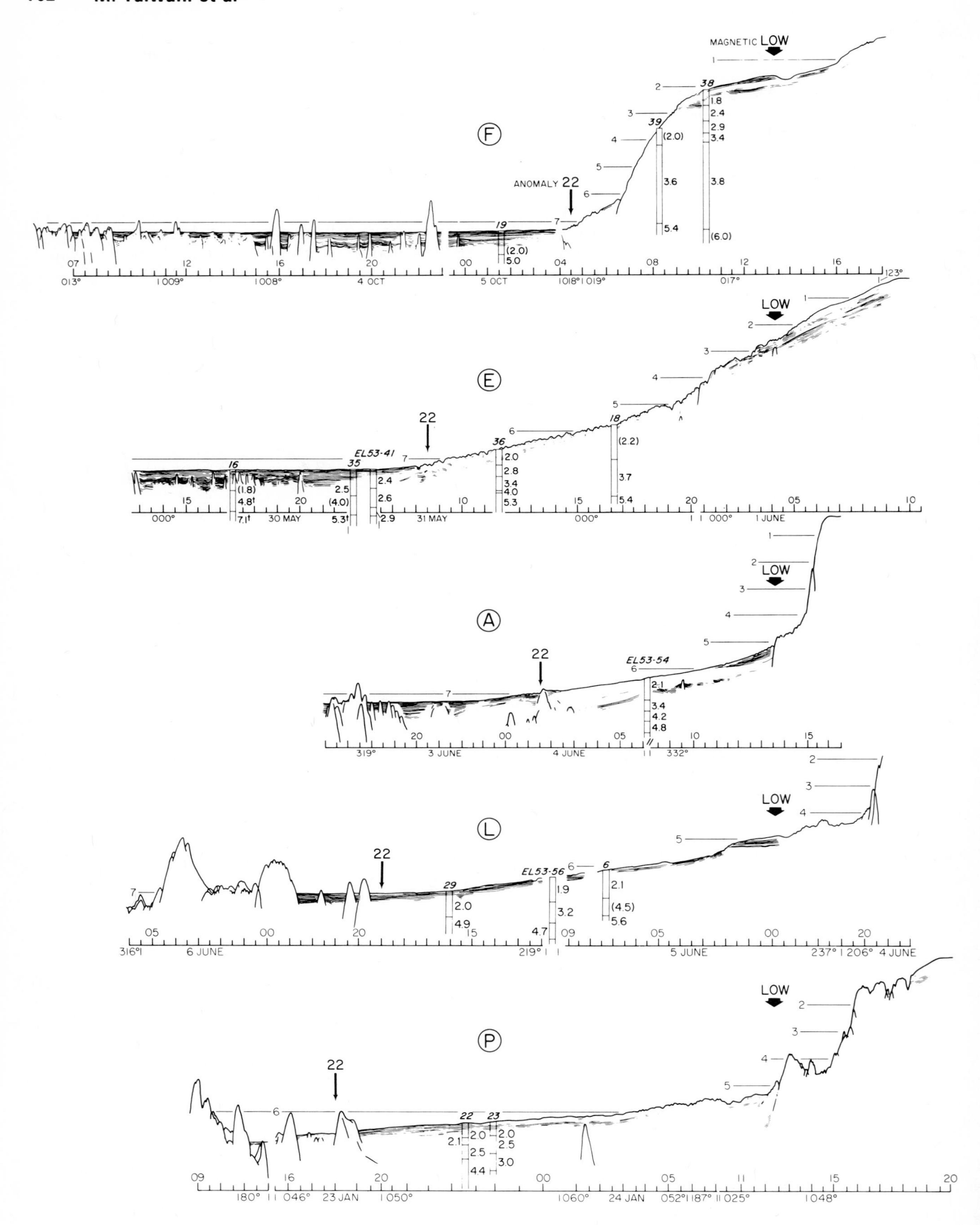

FIG. 13. Tracing of selected *Eltanin* and *Vema* profiles on the margin. For locations see Figures 1 and 2. Sonobuoy results are projected onto reflection lines. Vertical exaggeration is approximately 30:1.

ceous reflector places it above the seafloor to a point near the base of the rise, but the solution of values from the two sonobuoys on the rise show a substantial low-velocity layer at the surface. The Cretaceous sequence is apparently downthrown at about the location of the base of the slope. The material at depth beneath the rise with higher refraction velocity may represent a thinned equivalent of the Cretaceous Ceduna Plateau sequence.

A prominent subbottom reflector appears 700 to 800 msec below seabed in Profile A. Sonobuoy measurements show that the reflector lies at the top of a relatively high-velocity layer of 3.4 km/sec, again suggesting that this reflector represents the Tertiary/Cretaceous boundary. In Profile A, prominent basement peaks are very close to magnetic anomaly 22, and mark the seaward (south) boundary of the quiet zone.

In Profile A as in Profile P, a basement peak is located about half-way up the continental rise that coincides with a relative high in the magnetic field. Other magnetic highs are located in the quiet zone but do not coincide with obvious basement peaks on reflection profiles; it is possible, however, that undetected buried basement peaks cause the highs in this region.

Profiles L and P show an absence of deep subbottom reflection horizons beneath the rise. The association of magnetic lineation 22 with the edge of the Diamantina Zone is clear in these profiles. In Profile P, at the indicated location of the Magnetic Trough, we note a sharp, high-angle contact between sediments that abut, or wedge out against the steeply inclined wall of the continental slope. The structure is similar to the rise/slope boundary on Profile F. Throughout the margin, the contact between slope and rise appears to be formed of the same sort of abutment of rise sediments against steep slope escarpment.

SEISMIC REFRACTION MEASUREMENTS

Descriptions of the Experiment

During *Vema* Cruise 33 thirty-six new refraction stations were established on the margin south of Australia and in the adjacent deep ocean basin. Their positions are plotted in Figure 6, together with those of the earlier *Eltanin* refraction stations. Recordings were made in the present work with short- and long-range sonobuoys shot with explosives and/or airgun. Values of sediment thickness, and basement and sub-basement velocities were obtained at nearly all stations, and mantle arrivals were recorded at nine locations.

Prior to a mid-cruise port call at Esperance, W.A., the only sonobuoys available to us were relatively short-range models, which provided data up to ranges of about 24 sec of direct wave (D) travel time. Some of these sonobuoys (numbers 22 to 29) were used with an airgun sound source to a range of about 10 sec, most of which were extended with 1-lb charges to the maximum range. Sonobuoy numbers 1 to 5 were shot with large explosive charges.

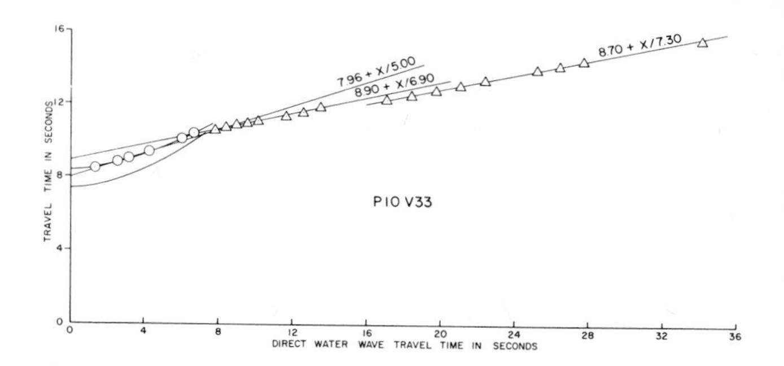

FIG. 14. Travel-time graph recorded at station 10 showing incompatability of short- and long-range data. Circles denote reflected arrivals; triangles denote refracted arrivals.

After our port call, we used long-range commercial sonobuoys and nearly doubled our range. We used small charges in the beginning of the profiles (numbers 6 to 21), but then charge sizes were increased up to 67 lbs for the maximum ranges. However, we found that much better reflection curves could be obtained with the airgun at the near ranges. As a result, the latter half of our sonobuoy data (numbers 30 to 43) were recorded with an airgun source out to the cross-over range for the first refractions, then explosives were used. The cross-over range, when necessary, was computed from *Eltanin* sonobuoys that had previously been deployed in the study area.

Results

Results listed in Table 1 include published and unpublished solutions from the earlier airgun/sonobuoy data obtained aboard *Eltanin*. The seismic refraction profile at station 10 is shown in Figure 14, where the short-range data are incompatible with the long-range data, implying an important change within the profile. The short-range solution appears as 10A in Table 1 and the long-range one as 10B. An example of a variable-density recording, extended to maximum range with 1-lb explosives, is shown in Figure 15. The variable nature of the refraction line in Figure 15 is typical of our data from the Magnetic Quiet Zone. The seemingly spurious events seen in Figure 15 (shortly after explosives were used) are possibly similar to those that caused the contradictory data obtained at Profile 10 (Fig. 14). This kind of data may reflect the broken-up or heterogeneous nature of the crustal layers within the Magnetic Quiet Zone. Although the heterogeneous crust degrades the data quality, it emphasizes an important difference between the Magnetic Quiet Zone crust and the oceanic crust south of anomaly 22, where the refraction data are much more orderly.

The designation "MT" for minimum thickness in Table 1 results from assuming a velocity line that passes through the final (most distant) refraction arrival. Note that we have slightly modified two published sonobuoy solutions, as indicated in the table.

Interval velocities from the present work and published airgun/sonobuoy data are plotted in Figure

Table 1. Sonobuoy velocity solutions from the South Australian margin.

Station (shots only)	Water Km	H_2	H_3	H_4	H_5	H_6	H_7	H_8	V_2 km/sec	σ_2	V_3	σ_3	V_4	σ_4	V_5	σ_5	V_6	σ_6	V_7	V_8	V_9	South Lat.	East Long.
1 V33	5.12	2.16							(2.2)		4.50											-35.94	115.36
2	4.56	.80	.93	1.36	2.41				(1.8)		2.30		4.50*		(5.5)		7.20*					-35.98	118.77
3	4.79	1.71	1.07	1.90					(2.2)		4.60*		5.30		7.20*							-36.30	118.78
4	5.35	.33	1.88						(1.8)		4.50		5.50									-36.34	122.78
5	4.35	1.40	1.86	1.50					(2.2)		4.40		(5.5)		7.30							-35.27	122.17
6	4.69	1.36	.95	3.41	3.11				2.10		(4.5)		5.60		7.20*		8.00					-35.52	122.57
7	5.47	.97	1.35						(2.0)		(5.0)		7.30*									-36.95	123.27
8	5.59	.46	1.87	4.14					(1.8)		(4.8)		7.00		8.30							-36.65	126.59
10A	5.56	1.03	2.66						(2.0)		(5.0)		6.90									-36.13	126.78
10B	5.56	1.03	1.83						(2.0)		(5.0)		7.30									-36.13	126.78
11	5.00	1.08	.84	2.97	4.57				(2.0)		3.15		4.70		6.70		8.50					-35.16	127.25
12	5.05	1.89	.42	3.11	2.55				(2.2)		3.25*		5.40		6.70*		8.10					-35.27	126.37
13	4.03	.63	2.34	2.11	4.76MT				(1.8)		3.00		5.30*		7.50		(8.3)					-34.50	125.72
14	4.63	1.67	1.75	2.16	4.82MT				(2.2)		4.15		5.05		6.15		(7.3)					-34.78	127.00
15	4.57	1.13	2.01	3.27	5.78				(2.0)		3.10*		5.30		6.70		8.30					-34.95	126.86
16	5.61	.70	1.96	2.99					(1.8)		4.75		7.10*		8.20*							-36.89	129.05
18	4.40	1.43	2.48	2.99	4.71MT				(2.2)		3.70		5.40		5.95		(7.2)					-34.95	129.53
19	5.61	.86	3.02	2.23					(2.0)		5.00		6.60		8.50							-36.71	131.48
21	5.00	.67	2.49	2.13	5.31MT				(1.8)		3.10		5.50		6.75		(8.5)					-36.39	134.52
(shots with airgun)																							
22 V33	4.84	.53	.32	.65	2.41				1.99	.04	2.13	.30	2.51	.09	4.40	.12						-36.57	116.76
23	4.80	.53	.82	.97					1.95	.05	2.52	.05	3.01	.04								-36.44	116.90
27	5.05	1.37	1.38	2.26	3.02MT				(2.2)		4.50		5.20		7.20		(8.3)					-35.91	121.91
28	4.46	.65	1.78	1.22	1.58				1.85	.05	(3.0)		4.66		5.70*		7.42					-35.38	121.78
29	5.26	.76	1.05	2.41					(2.0)		4.85		5.55		7.30							-36.19	121.66
30	5.02	.65	1.49	1.30					1.82	.04	2.95		4.85		6.00							-35.80	124.11
32	4.72	.35	.62	.53	.83	.38	1.08	1.65	1.64	.08	2.57	.05	2.74	.14	3.87	.21	4.30		4.75	5.35	6.00	-34.86	126.32
33	4.59	.55	.49	.44	.68	.87	1.26		1.93	.07	2.06	.07	2.70	.09	3.66	.13	3.71	.12	5.00	6.15		-34.92	125.99
34	4.68	.53	.39	.87	1.00	.61	1.81	1.78	2.00	.18	2.07	.30	3.34	.47	(3.7)		4.45		5.10	5.45	6.70	-34.81	126.60
35	5.60	2.13	2.60						2.53	.05	(4.0)		5.30*									-36.15	129.51
36	5.10	.53	.63	.79	.17	6.01	2.68MT		2.03	.07	2.81	.47	3.35		4.00		5.25		7.40	(8.3)		-35.37	129.36
37	3.86	.96	.98	.77	4.94	7.28MT			(1.8)		2.85		3.45		5.00		6.00		(7.2)			-34.34	128.56
38	1.59	.57	.94	.60	.62	5.76MT			1.75	.08	2.43	.03	2.89	.20	3.41	.53	3.77		(6.0)			-34.67	131.04
39	2.67	.62	5.30	8.67					(2.0)		3.60		5.40		(7.2)							-35.03	130.90
40	5.03	1.32	3.18	2.67	3.67				(2.1)		3.45		5.20*		7.00		8.50					-36.20	132.60
41	5.20	1.00							2.36	.05												-37.79	137.76
42	2.98	1.01	2.82	9.28MT					(2.1)		2.95		5.80		(7.3)							-38.32	139.29
43	2.54	.50	.69	.52					(1.8)		2.85		3.25		3.65							-35.25	141.90

Station (airgun only)	Water Km	H_2	H_3	H_4	H_5	V_2	σ_2	V_3	σ_3	V_4	σ_4	V_5	South Latitude	East Longitude	References
13E37	4.41	.48	1.26			2.37	.10	2.47	.04				-39.89	140.71	Houtz & Markl (1972)
14	3.76	1.03	.99	2.45		2.26	.10	2.70	.03	3.50			-39.68	141.26	Houtz & Markl (1972)
15	2.93	1.72	4.05			(2.5)		3.65					-39.48	141.73	Houtz & Markl (1972)
16	2.04	.41	.37	5.40		1.79	.05	1.77	.12	3.20		5.25	-39.40	141.93	Houtz & Markl (1972)
13E44	.14	.36				(2.0)		5.65					-34.71	114.93	Houtz & Markl (1972)
14	.05	.28	.82			(2.0)		4.41		6.48			-33.78	114.76	Houtz & Markl (1972)
38E53	5.58	1.04	1.41			(2.0)		5.50		6.78			-38.29	137.61	
41	5.56	.84	.98	.52	1.59	2.44	.13	2.56	.12	2.90	.23		-35.99	129.43	
48	.13	.30	1.04			(2.0)		2.90		6.60			-33.14	129.68	König & Talwani (1977)
51	3.46	.99	1.18	.63		2.58	.10	3.00		4.90		5.80	-34.12	126.74	König & Talwani (1977)
54	4.86	.88	.66	.78	1.00	2.10		3.40		4.20		4.80	-35.42	124.93	König & Talwani (1977)
55	.09	.22				(2.0)		5.50					-34.29	124.05	König & Talwani (1977)
56	4.87	.96	1.22	1.59		1.87	.11	3.18	.13	4.70			-35.76	122.10	König & Talwani (1977)
57	4.63	3.10				(3.4)							-37.27	118.25	
58	.09	.46	1.54			2.40		5.28		6.70			-35.32	116.94	König & Talwani (1977)

MT = minimum thickness in last layer
() = assumed velocity
* = poorly determined velocity

16. The regression coefficients relating interval velocity and one-way vertical travel time are shown in the figure. A published regression (Houtz, 1975) based on only 22 points gave results that are very similar to those in Figure 16. The regression V = f(time) representing all our data from the Australian margin has been integrated to v = g(depth) and compared with observational data from a borehole on the Ceduna Plateau. The comparison appears in Figure 17 where the agreement is surprisingly close and suggests a margin-wide uniformity in the sediment-velocity function.

LAYER WITH VELOCITY OF ABOUT 4.6 KM/SEC

West of about 124°E, layers with velocity in the range of 2.0 to 2.3 km/sec (presumably consisting of Tertiary sediments) directly overlie a layer with velocity ranging from 4.4 to 4.8 km/sec, but which sometimes has a basal reflector, suggestive of sedimentary material. However, on the velocity-versus-depth plot in Figure 16 the high velocities (shown as solid circles) are grouped off the main trend, occurring at shallower depths than expected. Hence, the high velocities are unlikely to result from the normal

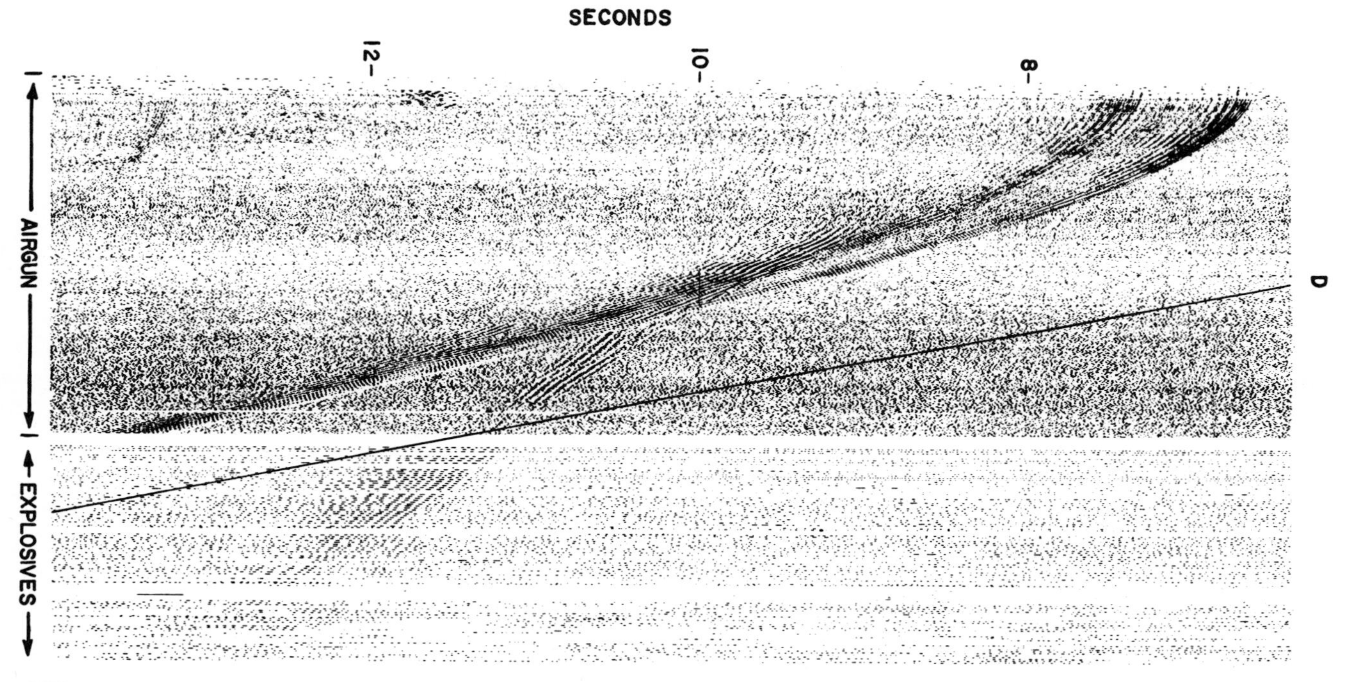

FIG. 15. Example of sonobuoy recording made by extending an airgun profile using rapidly fired explosive charges.

velocity increase in sediments due to age and overburden pressure. We have compared data from stations 3, 28 and 4 with appropriate Shell lines, 400/401, 402 and 402(1) respectively (refraction stations are located on Figure 6, the Shell lines on Figure 1), and it seems that the layer with a velocity of about 4.6 km/sec is not a sedimentary layer, as the corresponding reflector in the Shell lines has the appearance of basement with many diffractions. There is a suggestion on 402(1) that the 4.5 km/sec layer (of station 4) can be distinguished by the presence of a basal reflector from the lower 5.5 km/sec layer, perhaps suggesting that the upper layer is a volcanic/sedimentary mix. On the other hand, on the basis of a comparison of station 32 data with Shell profile 404 in the central part of the Magnetic Quiet Zone, the layers with velocity 4.3 and 4.75 km/sec appear to be sedimentary layers. We consider the 4.6 km/sec layer to be igneous in the western part of the quiet zone where it occurs most frequently (although some doubt must remain about its nature) but sedimentary elsewhere.

SEDIMENTARY ROCKS

Inspection of the results in Table 1 shows that the greatest thickness of sedimentary rocks is beneath the Ceduna Plateau in the eastern part of the area.

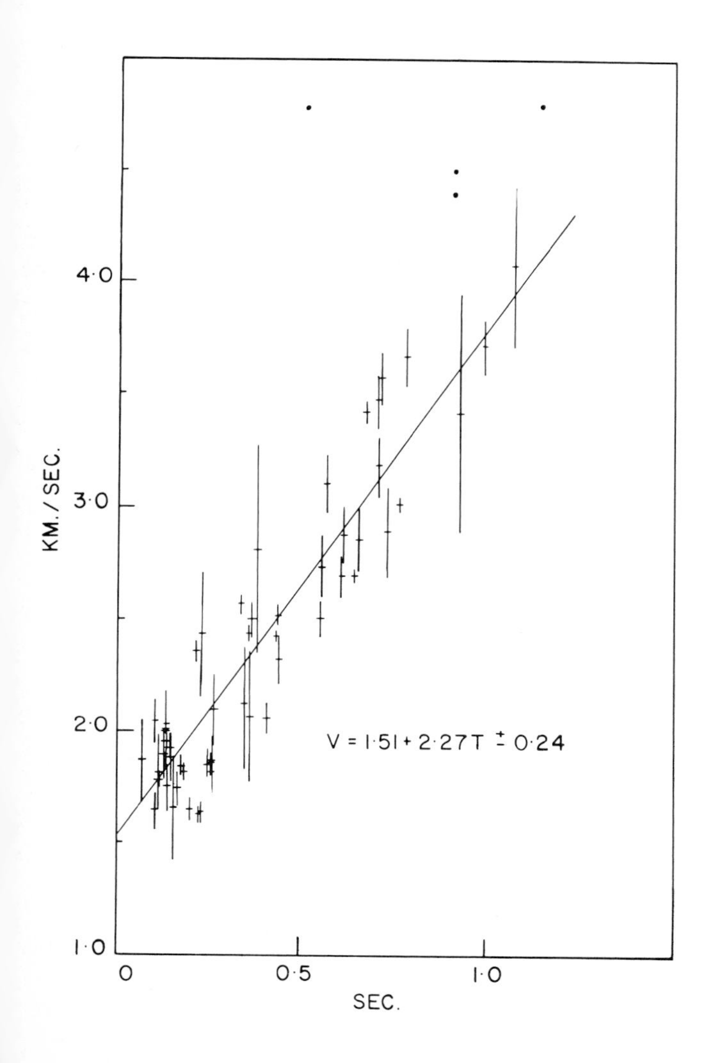

FIG. 16. Plot of interval velocities against one-way travel time with regression line shown. Note that dots have been plotted for the refraction velocities near 4.6 km/sec. They lie well off the regression line. The error bars on the interval velocities are one standard deviation in length (plus and minus) as given in Table 1.

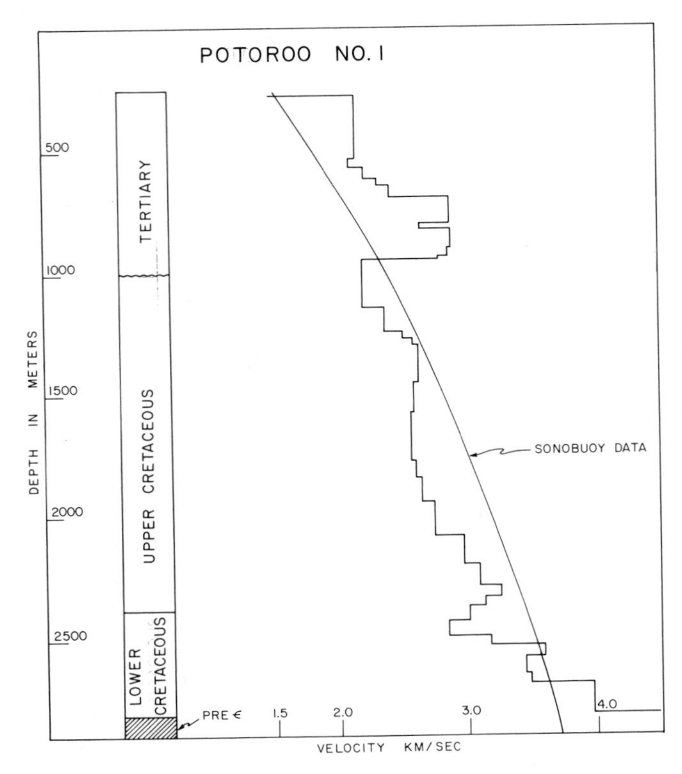

FIG. 17. Comparison of interval velocity-versus-depth relation from Potoroo 1, with velocity-versus-depth relation determined from the statistical analysis of sonobuoy data in the region. Age information is from the Potoroo 1 well.

At station 38, 8.5 km of sedimentary rock was measured without reaching basement, the thickest section we recorded. West of the Ceduna Plateau, the thickness reduces fairly rapidly to a nearly constant value of about 3 km.

Boeuf and Doust (1975) have suggested that sedimentary rock as much as 10 km thick underlies the Ceduna Plateau, with the axis of maximum deposition somewhat north of station 38. At station 39, downslope of station 38, we measured 5.9 km of sediment overlying a basement with a 5.4 km/sec velocity. These two stations establish an overall landward dip to the basement, suggesting that thicker sedimentary sections may lie north of station 38, in agreement with Boeuf and Doust.

Willcox (in press) has interpreted a north-south reflection profile about 16 km west of station 38, on which he identified Tertiary, and Upper and Lower Cretaceous rock units. A comparison of station 38 data with Willcox's interpreted profile suggests that the interval velocity of 1.75 km/sec probably represents Tertiary rocks, the velocities 2.43 and 2.89 km/sec represent Upper Cretaceous rocks, and that the Lower Cretaceous rocks have velocities of 3.41 to 3.77 km/sec. This velocity-to-age correlation is in broad agreement with the velocity log from the Potoroo No. 1 well drilled at the head of the Ceduna Plateau in 255 m of water. The velocity log is shown in Figure 17 together with the curve describing the velocity-versus-depth relation determined from the sonobuoy data of this report, and the sediment age information from Potoroo No. 1.

The thinning of the sedimentary section west of the Ceduna Plateau is achieved largely at the expense of layers with velocities in the 3.1 to 3.81 km/sec range. If the above age identification is assumed, our sections suggest much more restricted deposition in the Early Cretaceous in this region. Denham and Brown (1976) have accounted for locally thick Upper Cretaceous sections in the Otway Basin by erosion and redeposition of Lower Cretaceous strata. Boeuf and Doust (1975) and Deighton and others (1976) also indicate a period of basement uplift and consequent erosion of exposed beds at around the Lower/Upper Cretaceous boundary.

West of 124°E, Lower Cretaceous sediments are generally absent or very restricted in occurrence, unless they form part of the layer with a velocity of about 4.6 km/sec, a possibility that we feel is fairly remote.

Whether pre-Tertiary rocks are extremely thin in the western part of the margin, or whether they are represented by very high velocity rocks, the east-to-west variation in sediments implies a corresponding variation in Mesozoic depositional environments. An extreme interpretation would be that Mesozoic rocks are completely missing in the western part and that Tertiary sediments overlie oceanic basement.

CRUSTAL ROCKS

Most of the seismic refraction stations were shot within the Magnetic Quiet Zone. A few were shot south of magnetic lineation 22 in the area of oceanic crust. Sections obtained at the latter stations are shown in Figure 18. Although two of the stations show anomalously high crustal velocities of 7.1 and 7.3 km/sec, these velocities are poorly determined. A sedimentary layer of variable thickness lies over what is presumably layer 2 with a velocity close to 5 km/sec, which in turn overlies the main crustal layer with a velocity ranging from 6.6 to 7.0 km/sec. Mantle velocities were determined below the main crustal layer.

Because of the large water depth and an absence of any very large gravity anomalies over the Magnetic Quiet Zone, isostatic considerations suggest that even if the crust had originated as continental crust it would have to be substantially different at the present time. Such differences could result from massive intrusions that would make the crustal column heavier, or attenuation of the crust could have taken place, with most of the continental crust having been eroded from below and replaced by the mantle. (Beloussov, 1968; Bott, 1971; Falvey, 1974; have considered various mechanisms of modification of continental crust.)

Our seismic results in the quiet zone vary greatly and do not easily fall into any typical category. For the purpose of examination of the data we divided those refraction stations in the quiet zone at which crustal velocities had been determined into three groups: Group I, in which the velocity of the main crustal layer ranges from 6.5 to 7.1 km/sec; Group

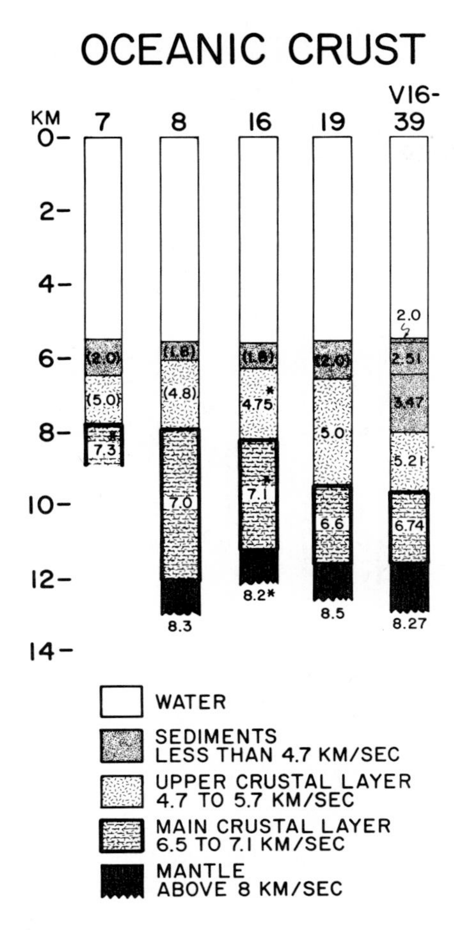

FIG. 18. Seismic sections obtained using sonobuoys in the area of oceanic crust south of magnetic lineation 22. Station V16-39 was obtained by two-ship refraction surveying (Hawkins et al, 1965). Velocities typical of oceanic crust are obtained. For locations see Figure 6.

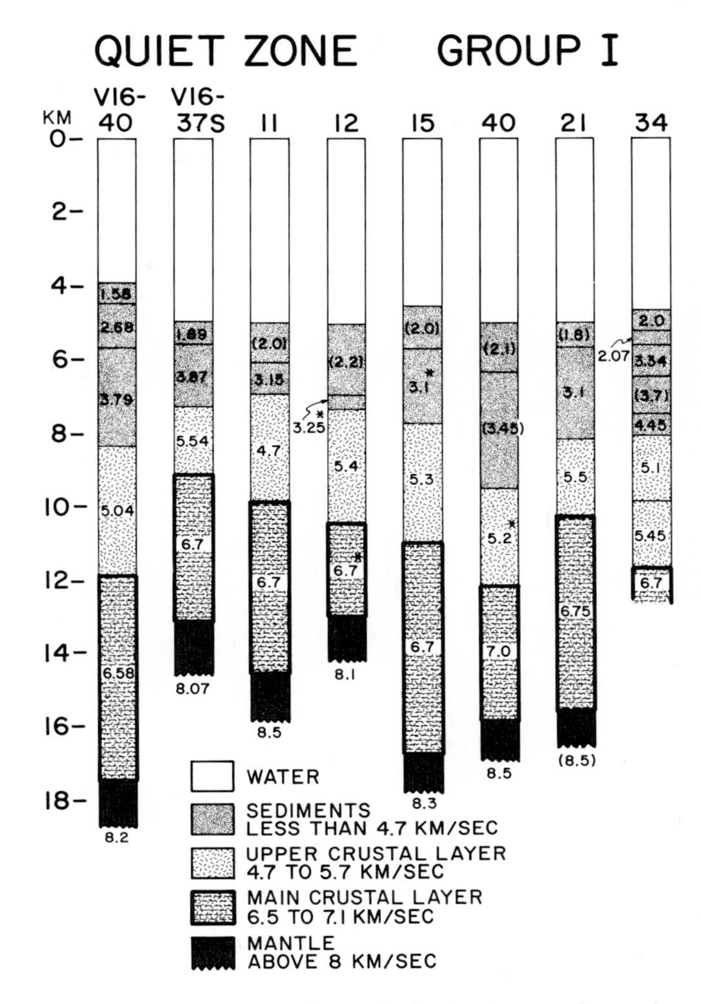

FIG. 19. Seismic sections obtained using sonobuoys in the area of the Magnetic Quiet Zone. Stations V16-37 and V16-40 were obtained by two-ship refraction surveying (Hawkins et al, 1965). Velocities typical of oceanic crust were obtained but the crustal sections range in thickness from 13 to 18 km and are thus abnormally thick when compared to typical oceanic crustal sections. For locations see Figure 6.

II, in which the velocity of the main crustal layer ranges from 7.2 to 7.5 km/sec; and Group III, where the velocity ranges from 5.8 to 6.2 km/sec. When the stations are divided into groups in this fashion, a geographical pattern does become apparent. Group II stations are mainly concentrated in the western part of the Magnetic Quiet Zone and Group III stations in the central part. Group I stations are found in both the eastern and the central part. However, there are overlaps in all cases.

The velocities of Group I stations are not atypical of oceanic crust. A sedimentary sequence of variable thickness overlies a layer of velocity close to 5 km/sec, which in turn overlies the main crustal layer with typical oceanic crustal velocities. More or less normal mantle velocities are found beneath. The values obtained at Group I stations are therefore not too different from velocities found in the oceanic crust. However, the important difference is that the crustal thickness ranges from about 13 to about 18 km which is 2 to 7 km greater than that of normal oceanic crust (about 11 km). This increase in thickness is not simply due to thicker sediments. This can be judged best by comparing the sections in Figure 19 with the oceanic crustal sections in Figure 18.

The sections in Group II (Fig. 20) represent a fairly unusual crustal structure. The main crustal layer has a velocity greater than 7.2 km/sec. Such a layer has been found in other places, particularly in the Pacific (Sutton and others, 1971). However, in the Pacific and in other places where it has been noted, this 7+ km/sec layer underlies a more or less normal oceanic crustal layer with velocity of about 6.7 km/sec. This is not true for the Group II sections where this layer underlies an upper crustal layer of velocity close to 5 km/sec. We have discussed the layer with a velocity of about 4.6 km/sec that is seen in most of the Group II stations. If this layer is igneous basement, then the upper crustal layer is divided into two parts—one with a velocity of around 4.6 and the other with a velocity slightly higher than 5 km/sec.

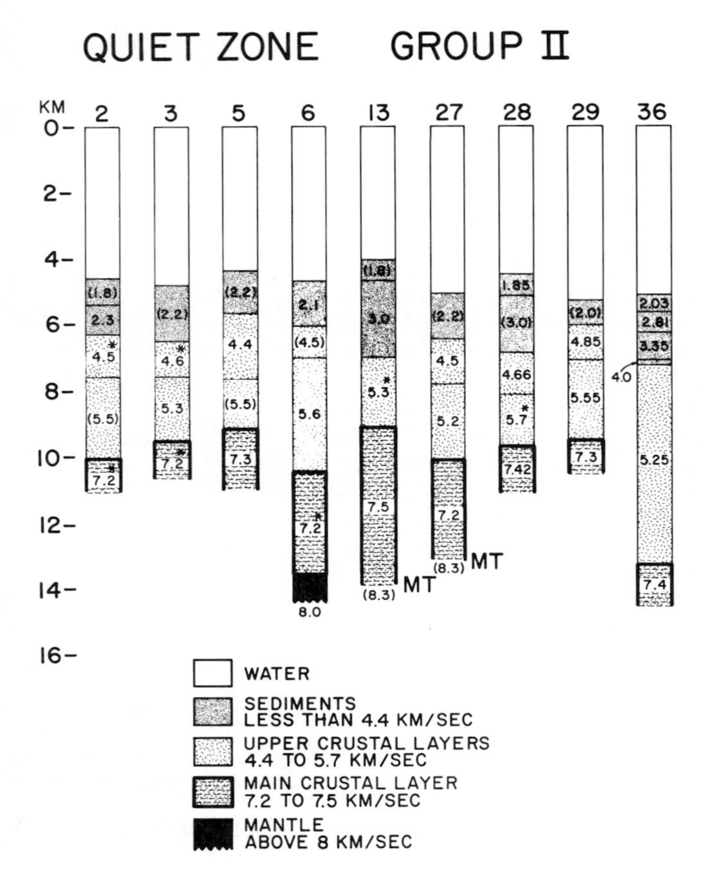

FIG. 20. Seismic sections obtained using sonobuoys in the area of the Magnetic Quiet Zone (Group II). On comparison with seismic reflection records, the layer with a velocity of about 4.6 km/sec is believed to represent basement. Note that neither typical continental crustal velocity of 6.0 km/sec nor typical oceanic crustal velocity of 6.7 km/sec is present. Rather, a layer with a velocity of 7.2 km/sec (or greater) is present.

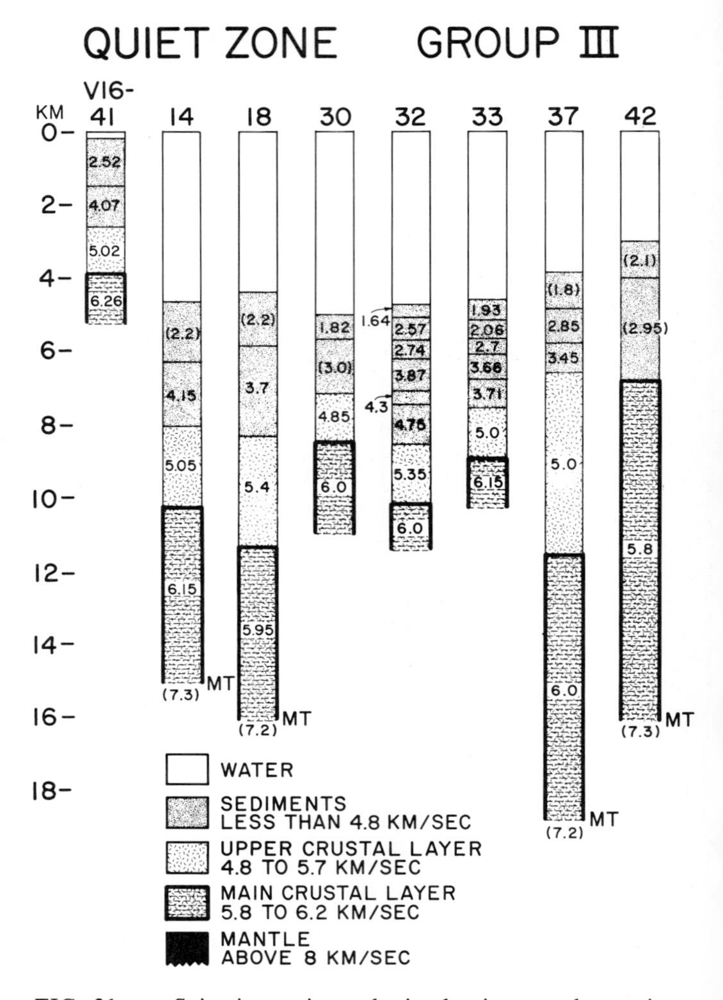

FIG. 21. Seismic sections obtained using sonobuoys in the area of the Magnetic Quiet Zone (Group III). The crustal layer with velocity close to 6 km/sec is typical of continental crust.

Group III stations are the only ones that indicate a main crustal layer with the velocity range of 5.8 to 6.2 km/sec that is generally associated with continental crust. Mantle velocities were not obtained at any of the Group III stations. The minimum crustal thickness indicated for these stations (Fig. 21) was determined on the assumption of an underlying layer with a velocity of about 7.3 km/sec. Group III stations are on the average slightly shallower than the stations in Group I and Group II, but the difference in crustal structure is so large and the differences in water depth generally so small that one cannot simply relate the differences in the crustal structure of the three groups to changes in depth. The water depth over the oceanic crust stations (south of lineation 22) is greater than that over the stations within the quiet zone, which of course reflects the difference in depth between the abyssal plain and continental rise.

The seismic refraction results in the area where the magnetic lineations point to seafloor spreading show a structure that is normal for oceanic crust. Lineation 22, which lies at the boundary between the seafloor-spreading-type anomalies and the Magnetic Quiet Zone, thus also separates regions of different crustal structure. In other words, a major east-west structural boundary exists near magnetic lineation 22.

A more unexpected result is the east-west variation of crustal structure within the Magnetic Quiet Zone. We notice that Group II-type stations are generally concentrated in the western part of the quiet zone, whereas Group I and II stations are located in the central and eastern part of the zone. This east-west difference in the quiet zone is also reflected in the magnetic profiles. The field is really smooth only in the eastern part, whereas in the western part, north of the Diamantina Zone, the magnetic field is quite disturbed.

In spite of these large east-west differences within the Magnetic Quiet Zone, we are reluctant to invoke a completely different origin for the western part and for the eastern part. The two main structural discontinuities in the area are the east-west discontinuities, one of them at the Magnetic Trough and the other at anomaly 22, which represent the most consistent geophysical lineations in the area. These discontinuities are expected to separate areas which have had different histories of evolution.

Representative stations from Group I, II and III and from the oceanic crust are presented for comparison in Figure 22 together with a station obtained by Hinz (1972) on the Inner Vøring Plateau. The Inner Vøring Plateau lies off Norway within a magnetic quiet zone in an area which may have similarities to the quiet zone south of Australia.

INTERPRETATION OF THE SEISMIC REFRACTION RESULTS IN TERMS OF THE NATURE AND EVOLUTION OF THE MAGNETIC QUIET ZONE

Questions about Interpretation of the Magnetic Quiet Zone Crust as Rifted Continental Crust

The interpretation of the Magnetic Quiet Zone crust as continental crust, which appears to be the prevailing view in the literature (Falvey, 1974; Boeuf and Doust, 1975; Deighton et al, 1976), is directly supported by sonobouy refraction measurements only in the eastern part of the quiet zone. Here we recorded several crustal sections with a thick basement layer having an average velocity of 6.0 km/sec typical of continental crustal structure. To apply this solution to the quiet zone as a whole is difficult. Clearly the crustal structure in the central and western regions bears little resemblance to that of normal continental crust. Although the crust is at present totally atypical of a continental crust, it might be argued that it was formerly continental, and that it has been altered to its present structure by some process of attenuation or "oceanization."

Deighton and others (1976), among many authors, have proposed a method of crustal thinning. Theirs is based on an earlier model of Falvey (1974), and they suggest that it has application to the southern margin of Australia. In their scheme, arching of the continental crust leads to its erosion, and hence thinning. The amount of erosional reduction in crustal thickness is related to the elevation of the initial arched uplift, and in their example crustal thickness is reduced from 32 to 20 km, following an initial uplift of 2.6 km. To produce subsidence of the crust they propose that lithospheric heating induces metamorphism of the lower crustal layers from a low- to a high-density phase, which also alters the lower crustal velocity structure. This hypothesis is used to account for the subsidence of crust in rift valley systems following the initial arched uplift. Subsequent to continental dispersal, lithospheric cooling results in further subsidence of the continental margin.

It is difficult to apply these schemes to the seismic results which yield sections of the Group II and Group I types. Group I sections show a crustal

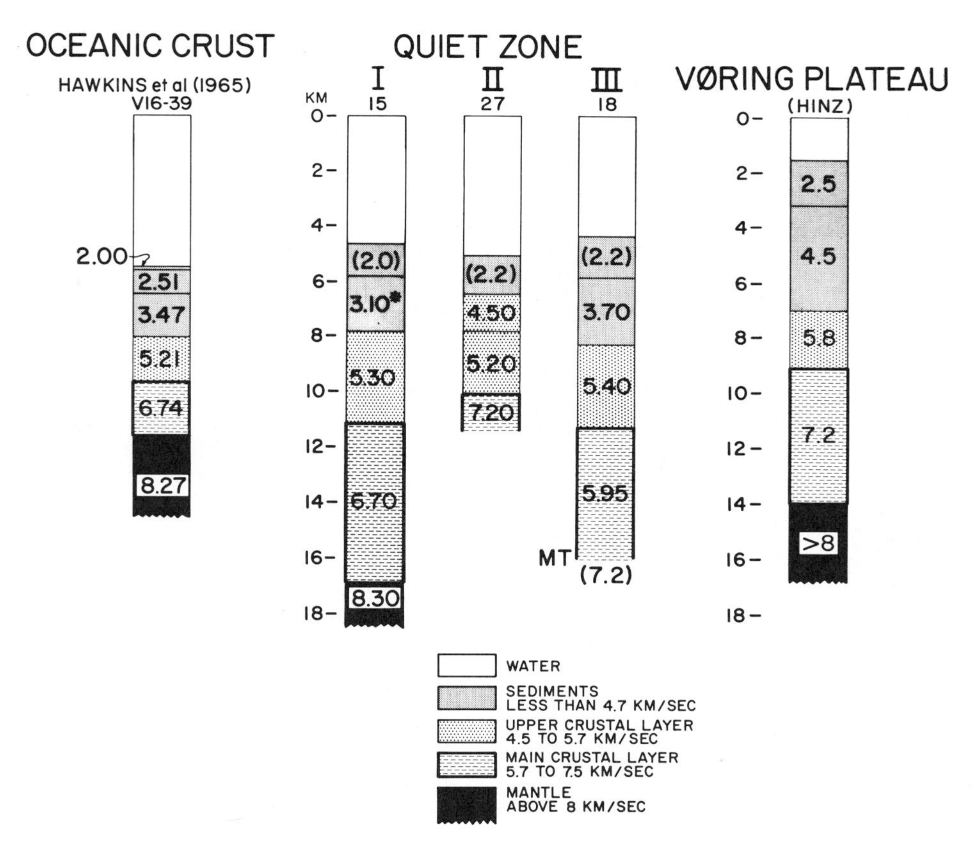

FIG. 22. Representative sections from Groups I, II, and III in the Magnetic Quiet Zone and from the area of oceanic crust. Also included for comparison is a section obtained by Hinz (1972) in the Magnetic Quiet Zone underlying the Inner Vøring Plateau off Norway.

thickness of about 6 to 13 km (excluding the water layer) and the proposed modes of crustal attenuation would be inadequate by a factor of two or three to achieve the required thinning. Group II sections show a velocity structure so different from that of typical continental sections that, again, the proposed attenuation scheme would be difficult to apply. Another objection to the thinning hypothesis is the absence of large amounts of Mesozoic sediments produced by erosion of the upwarped continental crust. No vast Mesozoic basins are located onshore in Australia in the region, and none has been discovered in Antarctica. The notion that crustal thinning has been achieved solely by erosion is considered wholly untenable.

Thus, while we do not deny the possibility that the Magnetic Quiet Zone crust was originally continental, the explanations that have been advanced for its modification do not explain the magnitude and nature of the alterations or the variability of the quiet zone crust in an east-west direction.

Interpretation of the Magnetic Quiet Zone Crust as Oceanic Crust

It is possible to postulate a Mesozoic seafloor-spreading type of origin for the crust in the Magnetic Quiet Zone. The lack of magnetic anomalies could be attributed to one of the following hypotheses: (1) generation of Mesozoic oceanic crust during an extended uniform-polarity interval; (2) thermal blanketing and/or metamorphism of the crust with attendant reduction of magnetic properties; (3) very slow spreading. Boeuf and Doust's (1975) assessment of the continental edge is based on a change in basement character from a highly faulted margin to one defined by an envelope of diffractions. It could be argued that instead this reflects a boundary between Mesozoic seafloor and Tertiary seafloor. The former could well have been faulted and subsided to great depth following cessation of Mesozoic spreading. The thick sedimentary sequence beneath the Ceduna Plateau could then have developed in a narrow ocean basin rather than in a wide rift valley. The depositional environment might be very similar to that of a rift valley, especially if seafloor was generated near sea level and sediment infilling was rapid.

There are, however, two major problems with this thesis. The seismic stations in Group II and Group III yield sections so different from normal oceanic crustal sections that it seems difficult to invoke schemes which can be used to obtain them by conversion from normal oceanic crust. Secondly, there does not appear to be any ready explanation for the east-west variation within the quiet zone crust if it is oceanic.

Interpretation of the Magnetic Quiet Zone Crust as a Hybrid Crust

If we consider the Group III stations as indicative of continental crust and Group II as indicative of oceanic crust (although with very unusual velocities) we can conceive a solution in which the western part of the Magnetic Quiet Zone is underlain by oceanic crust and the eastern part by continental crust. Some support of this thesis comes from the reflection records of *Petrel.* The basement in the east appears to be block-faulted whereas diffraction patterns indicative of oceanic basement seem to be more prevalent in the west (as seen on unpublished records).

The hybrid interpretation has two principal difficulties. Firstly, the magnetic trough, which forms the inner boundary of the quiet zone, is an extremely consistent and continuous feature across the whole margin. If the crust varies in its composition along the margin, the trough requires a different source structure at different places along the margin. In the east the trough would result from the contact between normal continental crust and subsided (and perhaps altered) continental crust, whereas in the west the same trough would be generated by an ocean/continent boundary. It would then require unreasonable coincidence to justify the extremely prominent continuity of the trough along the whole margin.

Secondly, the quiet zone has a nearly constant width across the entire margin. If the quiet zone crust were generated by plate separation about a pole in Bass Strait we would expect it to increase in width to the west, away from the pole. Dietz et al (1970) recognized this requirement and described the separation in their initial rifting as resulting in a progressively widening "lunate gore." The Magnetic Quiet Zone, in fact, narrows somewhat to the west, has a nearly constant width over most of the margin, but apparently does narrow to the east past the Ceduna Plateau. The observed geometry of the quiet zone could be explained by postulating large transform offsets between the continental and oceanic sections of the margin with attendant relative movements of eastern and western Antarctica or Australia. No such features are obvious on these continents.

EVOLUTION OF THE MARGIN

The Uniqueness of the Quiet Zone Crust

We propose that the quiet zone crust can best be treated as something unique, being neither continental nor oceanic. The inner boundary of the quiet zone, as defined by the magnetic trough anomaly, is far from being a superficial or shallow crustal boundary. It is associated with a primary discontinuity in the deep structure of the earth's crust. The outer boundary of the quiet zone is created by the junction of the quiet zone crust with normal oceanic crust of Tertiary age accreted at the Southeast Indian Ridge. It too forms a primary structural discontinuity. These two crustal discontinuities isolate the quiet zone crust from continental crust which adjoins it on the landward side, and oceanic crust which adjoins it on the seaward side.

Within the quiet zone the velocity layering of the crust is variable. The velocity structure is typically continental in some places; in others it is oceanic, although the layer thicknesses are much larger than those for typically oceanic crust; and in still others it is neither oceanic nor continental. The distribution of these crustal types is not systematic but rather, they form a geographically scattered set. The quiet zone is seismically passive, and lack of recent faulting suggests that it is stable at present. It therefore cannot be transitional—at least not in a temporal sense.

Complementary Margin off Antarctica

The geophysical profiles off the Antarctic margin are very similar to the profiles off the southern margin of Australia. The profile designated *Eltanin* 35 in Figure 24 is an extension of profile C of this paper. The magnetic, gravity, and topographic data in this profile are almost identical to the data of profile *Eltanin* 37. The only major difference is that the sea floor is about 1 km shallower near the Antarctic margin (Houtz and Markl, 1972). In Figure 25 we have combined the two profiles after eliminating the portions of these profiles which are over Tertiary ocean floor. Figure 25 therefore represents the situation just prior to Tertiary opening (except for later sedimentation).

Mesozoic Evolution

Mesozoic sedimentation took place within a rift valley, but that rift valley must have had much more sharply defined borders than those predicted by generalized models of rift-valley development. The borders are marked at present by the magnetic trough anomalies found in the margins of Australia and Antarctica and form the landward boundaries of the conjugate Magnetic Quiet Zones (Figs. 24, 25). The rift valley is thus about 150 km wide, considerably wider than the continental rift valleys observed at the present time.

However, it is unlikely that the rift valley widened from a simple, small graben equivalent to the Gulf of Suez to a wider structure equivalent to the Red Sea.

Although there is ample evidence that crustal subsidence has taken place beneath the southern margin of Australia, there is little to suggest that subsidence took place on a widening front. The vast majority of Mesozoic sediments are confined within steep fault blocks in the basement which underlie the continental slope and Ceduna Plateau. We suggest that the rift valley was initiated along boundary faults now represented by buried basement scarps, and that the rift valley subsided within the area bounded by these faults, but did not progress outward. Horizontal extension did take place but was small. The limiting of the zone of deformation between the major boundary faults is a significant point of divergence from the generalized models.

We propose that the Magnetic Quiet Zone which is sharply defined by magnetic trough anomalies was formerly a continental rift valley. The boundary faults formed in the Mesozoic and continued to develop during the period of rift tectonics. The fall in basement elevation (across the landward fault) is approximately the same throughout the margin at around 7 km, and in the western part it forms the structural fabric of the continental slope.

Beneath the Ceduna Plateau the basement scarp forms the boundary between thick Mesozoic basin fill and very thin shelf cover. Tertiary faulting is minor and restricted to the outer part of the Plateau. The basement escarpment clearly controlled Mesozoic sedimentation and was little affected by Tertiary marginal subsidence.

West of the Ceduna Plateau the buried basement scarp underlies the continental slope, a morphologic feature which is usually considered to have formed as a result of subsidence following continental dispersal. However, it is clear that the scarp had a con-

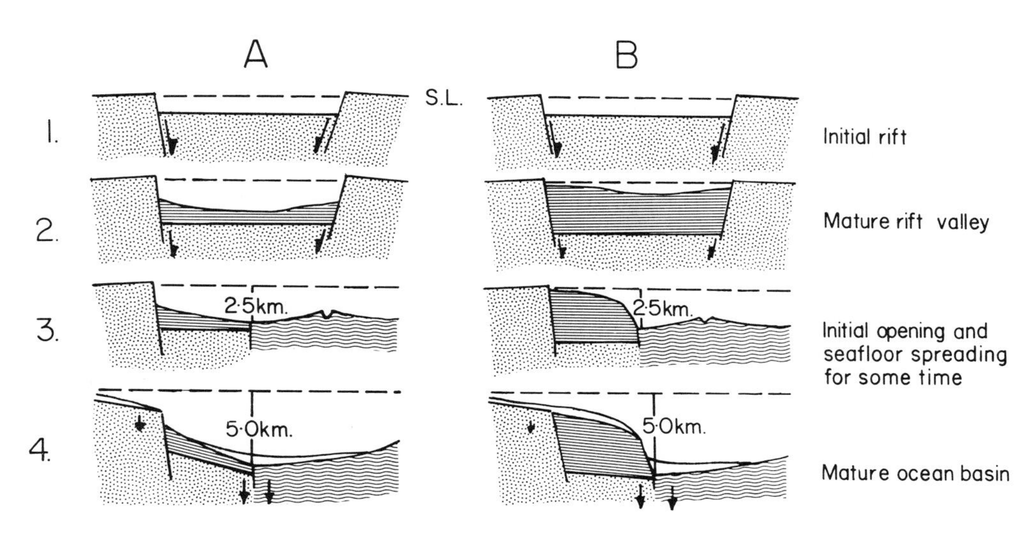

FIG. 23. Simplified model for the evolution of the margin: **A**, The Eyre Plateau or western areas; and **B**, the Ceduna Plateau area. Stippled area represents original continental crust with wavy lines representing Tertiary oceanic crust. Mesozoic sediments are represented by horizontal lines; water and Tertiary sediments are left blank. This model outlines only the (most) major events. The base of the crust is not shown.

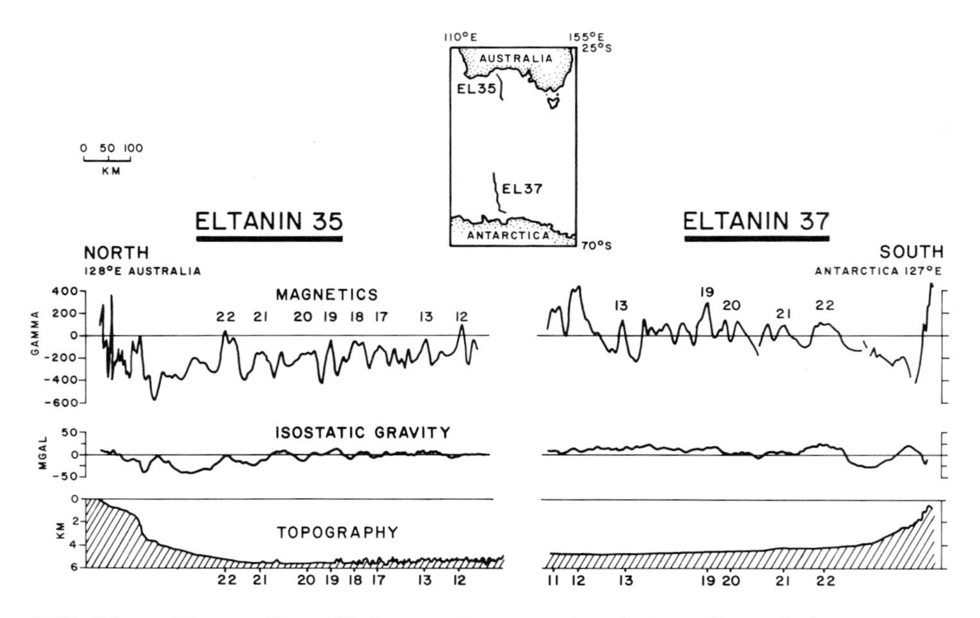

FIG. 24. The profiles off the southern margin of Australia and the conjugate Antarctic margin show great similarity in their geophysical characteristics.

siderable influence on Cretaceous sediment distribution, as thicknesses change by almost an order of magnitude across the scarp, in a similar manner to the Ceduna Plateau situation. We must therefore accept the presence of the scarp in Cretaceous times. That is, the structure and physiography of the feature which is now the continental slope must have existed in essentially the same form as we see it today for a considerable period (perhaps 30 m.y.) prior to the dispersal of the continents. This also represents a major deviation from the published generalized models of margin evolution.

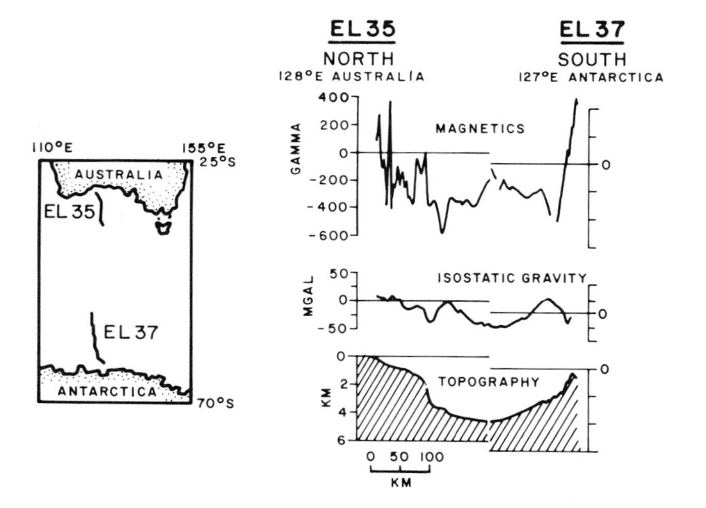

FIG. 25. The profiles in Figure 24 have been joined together and the portion over the Tertiary seafloor has been eliminated. Except for later sedimentation (opposite from which, incidentally, are about 1 km thicker on the Antarctic side), this figure should be representative of the morphology and the gravity and magnetic fields over the Australian-Antarctic rift in the later Mesozoic before seafloor spreading was initiated.

Post-breakup subsidence does not appear to have added substantially to the displacement on the scarp. Vertical faults with minor extension are confined primarily to Cretaceous—that is, pre-breakup—formations. We believe that post-breakup subsidence was in the form of tilting subsidence of shelf and shallow regions, and of tilting subsidence of the continental rise, and was not achieved by vertical movement at the continental slope.

In Figure 23 we show a diagrammatic representation of the subsidence history of two places along the margin, Figure 23B representing the Ceduna Plateau situation, and Figure 23A the Eyre Plateau/continental rise. Note that fault movements ceased by the end of the rift-valley stage and all subsequent subsidence was accomodated by large-scale tilting. Note also that in Figure 23A, the "continental slope" is formed during continental rifting, and that sediments are deposited below sea level from an early stage of rifting. Boeuf and Doust (1975) noted depositional patterns in Upper Cretaceous sediments in the Great Australian Bight which suggested a marine influence at that time, and Denham and Brown (1976) have postulated the same conditions for the Upper Cretaceous in the Otway Basin.

Although sediments were deposited below sea level in the western part of the margin, open-marine conditions had not fully developed during continental rifting. Deighton et al (1976) have concluded that "there must have been some inhibiting mechanism which prevented the penetration of oceanic currents east of the Naturaliste Plateau." In the Ceduna Plateau region very high rates of sediment influx essentially maintained shallow-water conditions there.

Thus, we envisage a Mesozoic tectonic regime marked by subsidence of the basement within a region defined by major boundary faults. Within the major graben taphrogenic basement deformation took place, resulting in the block-faulting described by Boeuf and Doust (1975) and Willcox (in press). The basement elevation probably fell below sea level fairly early in the rifting history but marine conditions did not fully develop due to high sedimentation rates in the east and the presence of a western barrier. The structural and morphologic feature which is now the continental slope west of about 128°E was present during the middle to late stages of Mesozoic graben tectonics.

While subsidence and block-faulting took place, the continental crust evolved for a period of several tens of millions of years prior to breakup. This was accomplished without significant extensional faulting. Although Boeuf and Doust (1975) indicated large-scale vertical offsets in their schematic sections, there is no evidence in our (or their) profile sections for large faults in the Cretaceous sediments. Igneous activity which is commonly associated with rift valleys occurred during this period; this activity must have varied in different sections of the rift valley giving rise to east-west, that is, along-strike variations in the crustal structure and also to the close juxtaposition of areas with different crustal structures. An important difference between the seismic results in the Magnetic Quiet Zone and in areas of normal oceanic crust is the presence of apparently spurious seismic arrivals in the former area, which probably result from the broken-up nature of the crust. Hence, the crust in this Mesozoic rift valley is nearly continental in some areas, but in others it has been subjected to a very large amount of igneous activity that appreciably altered the crustal velocity, reflected in the Group I or II crustal sections. Greater igneous activity in the western part of the quiet zone presumably resulted not only in higher crustal velocities but also in a more disturbed and irregular magnetic field there.

An alternative explanation for the irregular magnetic field in the west is that it has been inherited from the original continental crust. The Fraser-Albany Province to the north (Figs. 1 and 2) is a middle Proterozoic orogenic zone consisting of basic pyroxene granulites and garnetiferous granulites. This granulite complex also contains basic lenses of iron-rich minerals which give rise to magnetic anomalies. The belt of granulites is known to extend into Antarctica, and this set of "paired metamorphic belts" is a criterion used from the Precambrian rocks to back up the usual fit of Australia and Antarctica. If the western part of the magnetic quiet zone is indeed underlain by a granulite basement, this might also explain the unusually high crustal velocities (corresponding to Group II) in this area. An argument against this alternative explanation is that the seismic velocity structure on land in Australia does not clearly reflect the east-west variation found in the Magnetic Quiet Zone.

Tertiary Evolution of Shallow Structures

Seafloor spreading on the Southeast Indian Ridge began in the late Paleocene. It is apparent, however, that spreading was not entirely systematic at first and produced several anomalous features.

First, the breakup of the continents was not isochronous across the whole of the margin. Lineation 22 does not exist east of 131°30′E. The first recognizable anomaly east of this longitude is lineation 21. Second, anomalous basement topography is associated with the early stages of seafloor generation as shown by the ridges contoured in Figure 9 and by the Diamantina Zone.

Diachronous breakup of the continents may be explained by a short period of marginal stretching in the areas east of 131°30′E. The amount of stretching required is not large—being about the width of anomaly 22. We are therefore not postulating that there was large-scale stretching of the margin, nor that stretching was the agent of crustal thinning.

Subsidence of the margin following continental dispersal marked the onset of fully marine sedimentation and gave rise to the present physiographic features. This subsidence was largely achieved by oceanward tilting of the basement and rift-valley sedimentary rocks, and not by slip along the major boundary faults formed during the Mesozoic. There was nearly complete coupling of continental, quiet zone, and oceanic crusts following the generation of the latter. The fault on the outer edge of the Eucla Terrace appears to be the only evidence of decoupling.

Marine carbonate sediments began to form a capping layer on the Ceduna Plateau shortly after continental dispersal. Water depths gradually increased from near zero to bathyal due to marginal subsidence. In regions west of the Ceduna Plateau the floor of the rift valley probably lay below sea level before dispersal and deep marine conditions were established shortly after. Tertiary pelagic sediments now blanket the rift-valley sequence on the continental rise and are capped by a thin turbidite layer equivalent to the abyssal plain turbidite sequence. Extensive erosion of the shelf edge, slope and upper rises, the evidence for which lies in the presence of the major canyons in these regions, contributed large volumes of terrigeneous material across the continental rise to the abyssal plain.

At present continental shelf carbonate sediments are building out the shelf edge and blanket pelagic sedimentation is covering the deeper areas. The margin is tectonically passive, experiencing slow subsidence not accompanied by faulting.

Comparison with Continental Rifts

Because we believe that the Magnetic Quiet Zone evolved from a continental rift, a comparison of it with present-day continental rifts is in order.

Recent studies of the Rhine Graben and the African Rift system emphasize the confinement of the rift zones' anomalous deep structure to those areas immediately beneath the rift valley proper (Prodehl et al, 1976, and Edel et al, 1975, on the Rhine Graben; Maguire and Long, 1976, and Long and Backhouse, 1976, on the African Rift). These studies showed earlier ideas that the rifts were underlain by a broad region of arched mantle to be incorrect. In the Long and Backhouse (1976) study it was reported that essentially unaltered crust of the African Shield continues virtually to the rim of the rift, and "this steep boundary probably extends to a depth of at least 55 km." In the Rhine Graben study (Edel et al, 1975) the formerly held idea of a "rift cushion" was dismissed in the light of new data. The new model shows a zone of anomalous crustal structure formed by a sharp decrease in the depth of high-velocity layers, occupying the region immediately beneath the rift proper. Virtually no crustal modification occurred in regions outside the rift.

Prodehl et al (1976) also commented on peculiar aspects of the crust in the Rhine Graben. In discussing the crustal refraction data they stated "in 25 record sections prepared from the data . . . it is difficult to find two that look alike. These differences suggest a fine structure that is beyond the resolving power of the methods which can be used and which require certain assumptions not being true in all cases." A very similar statement could well have been made with reference to the extremely variable crustal structure of the Magnetic Quiet Zone on Australia's southern continental margin.

Although authors of previous models of rift evolution have commented on the similarity of rift-graben structure to the basement structure of rifted continental margins, the results from the Australian margin taken with the newer results from the African and German rifts provide the first real link between the deep structures of both systems. It seems very reasonable to us to suppose that the anomalous quiet zone crust is the relic of the anomalous rift-zone crust. The Magnetic Trough marks the location of the northern rift-zone bounding escarpment. It constitutes a sharp discontinuity in the deep crust across which normal continental crust abruptly changes to the anomalous rift zone/quiet zone crust. The gross crustal structure of the Australian margin shows that the continent-to-ocean transition is achieved by a step-wise change from normal continental to quiet zone to oceanic crust and that this is the clear outcome of rupturing a rift-zone with its anomalous crust and prominent boundary discontinuities.

CONCLUSIONS

1. The Magnetic Trough is a ubiquitous feature of the southern margin of Australia, extending between the longitudes of about 115°E and 135°E. It is generally located in the vicinity of the continental slope. The oldest seafloor-spreading magnetic lineation 22 has now been identified westward to 115°E. In the east it disappears rather suddenly near 131°30′E.

2. Seafloor spreading started near anomaly-22 time and was associated with the generation of rugged topography. The topography was especially rugged and shallow in the western area where it now stands as the Diamantina Zone. Eastward the rugged topography is buried under a progressively increasing thickness of sediments.

3. The Magnetic Trough and magnetic anomaly lineation 22 are the expressions of fundamental boundaries in this area. North of the Magnetic Trough lies continental crust; at anomaly 22 and south of it lies oceanic crust. In between lies an area of smooth magnetic field in the east and disturbed but irregular magnetic field in the west. We designate the entire area as the Magnetic Quiet Zone.

4. Seismic reflection data indicate that the greatest thickness of sediments and the greatest depth to basement occurs in the Magnetic Quiet Zone which roughly coincides with the continental rise.

5. Seismic refraction data show great variability of the crustal structure in the Magnetic Quiet Zone. Crustal sections have been divided into three groups—Group I with typical oceanic crustal velocities but with abnormally great crustal thicknesses; Group II with a velocity of greater than 7.2 km/sec for the main crustal layer; and Group III with the main crustal layer with a velocity close to 6.0 km/sec. The entire Magnetic Quiet Zone cannot be said to be underlain either by continental or by oceanic crust.

6. In the first stage of the evolution of the margin south of Australia, starting in the Early Cretaceous, two prominent boundary faults defined a 150- to 200-km-wide continental rift. Sedimentation during the Early Cretaceous was fluviatile-lacustrine. Vertical movements continued into Late Cretaceous times but without further widening of the rift valley. The Upper Cretaceous sediments are less extensive than the Lower Cretaceous ones, especially in what was then the axial region of the rift valley. The Upper Cretaceous sediments are deltaic and marine incursive, conditions being progressively more marine to the west. There is a possibility that Cretaceous sediments are either very thin in the western part of the rift valley or mixed with volcanics. During the Cretaceous the crustal structure evolved in the rift valley to yield the extremely variable crustal structure now represented by Groups I, II, and III. Igneous activity may have been especially high in the west.

7. In the early Tertiary the second stage began in the evolution of the margin with breakup and initiation of seafloor spreading. Normal oceanic crust was generated during the Tertiary. The arching prior to breakup is probably responsible for some removal of Upper Cretaceous rocks and the "breakup" unconformity. Early Tertiary rocks are clastic, late Tertiary rocks are carbonates; both represent open-marine conditions.

8. Thus, the Australian-Antarctic margin can be conceived to have evolved in two stages: a Mesozoic stage during which the rift geometry remained constant but the composition of the crust continuously evolved through igneous activity, and a Tertiary stage during which the geometry continuously evolved (that is, drift took place) but the composition of the newly created crust remained nearly constant.

REFERENCES CITED

Beloussov, V. V., 1968, Some problems of development of the earth's crust and upper mantle of oceans, *in* L. Knopoff,

C. L. Drake, and P. J. Hart, eds., The crust and upper mantle of the Pacific area: Wash., D.C., Amer. Geophys. Union, Geophys. Monograph 12, p. 449–459.

Boeuf, M. G. and H. Doust, 1975, Structure and development of the southern margin of Australia: Austral. Petroleum Explor. Assoc. Jour., v. 15, p. 33–43.

Bott, M. H. P., 1971, Evolution of young continental margins and formation of shelf basins: Tectonophysics, v. 11, p. 319–327.

Conolly, J. R., A. Flavelle and R. S. Dietz, 1970, Continental margin of the Great Australian Bight: Marine Geology, v. 8, p. 31–58.

Deighton, I., D. A. Falvey and D. J. Taylor, 1976, Depositional environments and geotectonic framework: Southern Australian continental margin: Austral. Petroleum Explor. Assoc. Jour., v. 16, p. 25–36.

Denham, J. I. and B. R. Brown, 1976, A new look at the Otway Basin: Austral. Petroleum Explor. Assoc. Jour., v. 16, p. 91–98.

Dietz, R. S., J. C. Holden and W. P. Sproll, 1970, Geotectonic evolution and subsidence of Bahama Platform: Geol. Soc. America Bull., v. 81, p. 1915–1927.

Edel, J. B., et al, 1975, Deep structure of the southern Rhine Graben area from seismic refraction investigations: Jour. Geophys. Res., v. 41, p. 333–356.

Falvey, D. A., 1974, The development of continental margins in plate-tectonic theory: Austral. Petroleum Explor. Assoc. Jour., v. 14, p. 95–106.

Griffiths, J. R., 1971, Continental margin tectonics and the evolution of southeast Australia: Austral. Petroleum Explor. Assoc. Jour., v. 11, p. 75–79.

Hawkins, L. V., et al, 1965, Marine seismic refraction studies on the continental margin to the south of Australia: Deep-Sea Res., v. 12, p. 479–495.

Hayes, D. E. and J. R. Conolly, 1972, Morphology of the Southeast Indian Ocean, *in* D. E. Hayes, ed., Antarctic oceanology II: The Australian-New Zealand sector: Wash., D.C., Amer. Geophys. Union. Antarctic Res. Series, v. 19, p. 125–145.

Heezen, B. C. and M. Tharp, 1973, USNS *Eltanin* cruise 55: Antarctic Jour. U.S., v. 8, p. 137–140.

Hinz, K., 1972, The seismic crustal structure of the Norwegian margin in the Vøring Plateau, in the Norwegian deep sea, and on the eastern flank of the Jan Mayen Ridge between 66° and 68°N: 24th Int. Geol. Congr., Section 8, p. 28–36.

Houtz, R. E., 1975, Comparison of sonobuoy and sonic-probe measurements with drilling results, *in* J. P. Kennet, R. E. Houtz, and others, Initial Reports of the Deep-Sea Drilling Project: Wash., D.C., U.S. Govt., vol. 29, p. 1123–1131.

——— and R. Markl, 1972, Scismic profiler data between Antarctica and Australia, *in* D. Hayes, ed., Antarctic oceanology II—The Australian-New Zealand sector: Antarctic Res. Ser., v. 19, p. 147–164 (Am. Geophys. Union).

König, M. and M. Talwani, 1977, A geophysical study of the southern continental margin of Australia: Great Australian Bight and western sections: Geol. Soc. America Bull., v. 88, p. 1000–1014.

Long, R. E. and R. W. Backhouse, 1976, The structure on the western flank of the Gregory Rift, Part II: The mantle: Geophys. Jour. Roy. Astron. Soc., v. 44, p. 676–688.

Maguire, P. K. H. and R. E. Long, 1976, Structure on the western flank of the Gregory Rift, Part I: The crust: Geophys. Jour. Roy. Astron. Soc., v. 44, p. 661–675.

Prodehl, C. J., et al, 1976, Explosion-seismology research in the central and southern Rhine Graben—a case history, *in* P. Giese, C. Prodehl, and A. Stein, eds., Explosion seismology in central Europe: data and results: Springer-Verlag, New York, Monograph 1, European Seismological Commission.

Smith, A. G. and A. Hallam, 1970, The fit of the southern continents: Nature, v. 225, p. 139–144.

Sproll, W. P. and R. S. Dietz, 1969, Morphological continental-drift fit of Australia and Antarctica: Nature, v. 222, p. 345–348.

Sutton, G. H., G. L. Maynard and D. M. Hussong, 1971, Widespread occurrence of a high-velocity basal crustal layer in the Pacific crust found with repetitive sources sonobuoys, *in* Heacock, J. G., ed., The structure and physical properties of the Earth's crust: Wash., D.C., Amer. Geophys. Union, Geophys. Monograph 14, p. 193–209.

von der Borch, C. C., J. R. Conolly, and R. S. Dietz, 1970, Sedimentation and structure of the continental margin in the vicinity of the Otway Basin, southern Australia: Marine Geology, v. 8, p. 59–83.

Weissel, J. K. and D. E. Hayes, 1972, Magnetic anomalies in the Southeast Indian Ocean, *in* Hayes, D. E., ed., Antarctic oceanology II: The Australian-New Zealand sector: Wash., D.C., Amer. Geophys. Union, Antarctic Res. Series, v. 19, p. 165–196.

Willcox, J. B., in press, The Great Australian Bight: A regional interpretation of gravity, magnetic, and seismic data.

Structure and Stratigraphy of the Blake Escarpment Based on Seismic Reflection Profiles[1]

R. E. SHERIDAN,[2] C. C. WINDISCH,[3]
J. I. EWING,[3,4] AND P. L. STOFFA[3]

Abstract Recently obtained high-energy multichannel seismic reflection profiles across the Blake Escarpment show that the oceanic basement, identified as a hyperbolic reflector, forms a deep sediment-filled trough at the base of the escarpment. The western boundary of the trough is formed by a steep rise (45°–60°) of acoustic basement producing an apparent structural relief of more than 2.5 sec for the trough. Seismic velocities were determined as 3.23 to 3.95 km/sec and indicate 4.03 to 4.94 km of sediments in the trough, which deepens to approximately 11 to 12 km. The steep west side suggests fault control and deposition in a fault-bounded half-graben trough. Undisturbed sediments lie over Horizon β which passes across the trough without deflection on one profile, and which is the shallowest disturbed and upturned reflector on another profile.

Correlation of reflectors involved in the structure of the trough with the nearby DSDP Site 391 indicates that the deeper seismic layers include sediments older than Late Jurassic. These old sediments filling the fault-bounded trough at the base of the Blake Escarpment should include facies deposited during the earliest opening of the Atlantic, perhaps in the Early Jurassic.

The seismic reflection profiles also show reflectors at the edge of the Blake Plateau which can be correlated with DSDP Drill Site 390. Important reflectors of earliest Eocene, Campanian/Albian and Barremian age can be traced westward to depths of 3.4 km (2.8 sec) under the Blake Plateau. Hyperbolic reflectors, not associated with igneous basement, occurring over acoustically opaque zones are interpreted to be well-cemented, high-velocity carbonate bank-margin complexes as drilled at DSDP Site 392.

INTRODUCTION

As part of the first multichannel-recording cruise of Lamont-Doherty Geological Observatory vessel R/V *Conrad,* two seismic reflection profiles across the Blake Escarpment were made on Cruise 19. These lines also were to provide acoustic-stratigraphic ties between recently completed Sites 390 and 391 of Leg 44 of the Deep Sea Drilling Project (Benson et al, 1976) (Fig. 1) and to extend the core-controlled seismic stratigraphy landward under the Blake Plateau into the existing multichannel data network of the cooperative U.S. Geological Survey—Institut Francais du Petrole survey. The U.S.G.S.-I.F.P. cross lines FCI-II-III were recorded by the R/V *Florence* in 1974 (Dillon et al, 1976) (Fig. 1) and these data provided the basis for the selection of Site 391 for drilling by the *Glomar Challenger.*

The *Conrad* multichannel lines MC1 and MC2 were shot using a Texas Instruments DFS IV acquisition system with a 2400-m-long 24-channel streamer. Shots at 2,000 psig, fired on a 20-sec interval simultaneously from 4 Bolt Associates 460-cu-in air guns, provided the high-energy sound source. Completed processing of line MC1 in the Blake-Bahama Basin at the base of the escarpment includes demultiplexing, common depth point gather, 24-channel velocity analysis, normal move-out correction, 24-fold stacking, trace equalization, deconvolution, and time variable gain. Velocity analyses were done using the semblance technique of Taner and Koehler (1969).

The U.S.G.S.-I.F.P. profiles across the base of the escarpment have been interpreted by Dillon et al (1976), but in these data the basement is not distinquishable as a strong reflection event within about

[1]Manuscript received, July 15, 1977; accepted, November 23, 1977.

[2]Department of Geology, University of Delaware.
[3]Lamont-Doherty Geological Observatory of Columbia University.
[4]Present address—Woods Hole Oceanographic Institution.

The seismic reflection profiles reported here were made from the R/V *Conrad* during its first multichannel-data-gathering Cruise 19-01 in 1975. This work was supported by National Science Foundation grants DES-75-21594 and DES-75-15865. Additional support was obtained from the Office of Naval Research grant N00014-75-C-0210. Much appreciation must be given to the scientists and crew on this cruise for their conscientious efforts in carrying out this work. We are grateful for the review of this manuscript by B. Tucholke and R. Houtz.
This is Lamont-Doherty contribution 2735.
Article Identification Number:
0065-731X/78/MO29-0011/$03.00/0.
(see copyright notice, front of book)

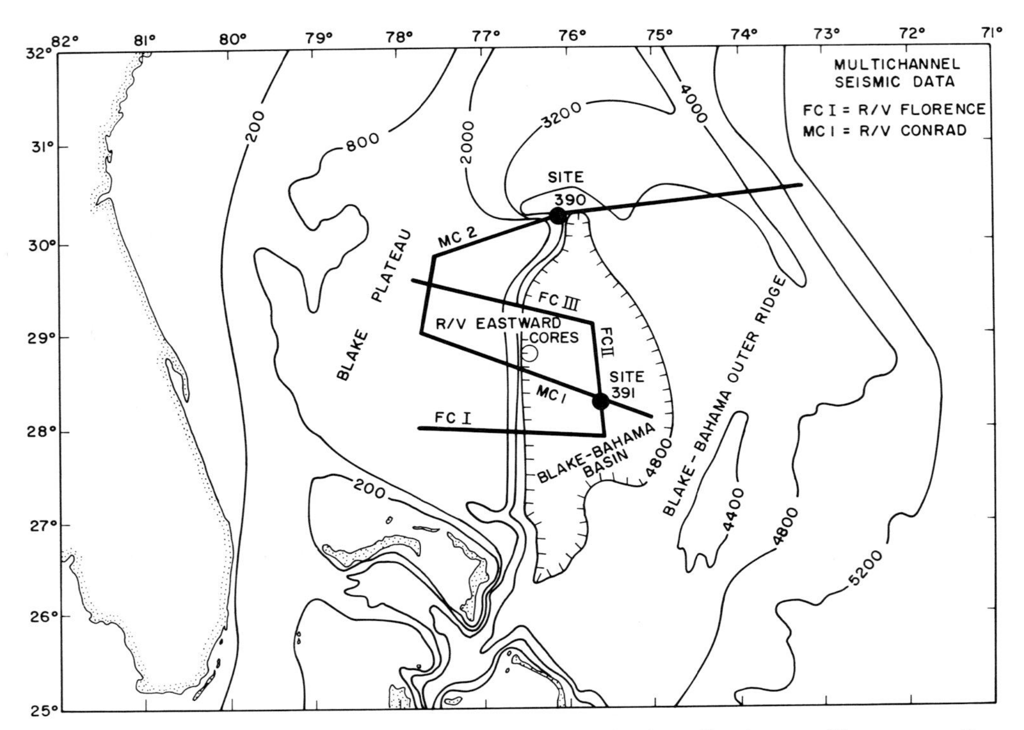

FIG. 1. Location map of recent multichannel seismic reflection profiles across the Blake Escarpment. FC I, II, and III are U.S. Geological Survey–Institut Français du Petrol profiles published by Dillon et al (1976). MC 1 and 2 are new data recorded by Lamont-Doherty Geological Observatory. Also shown are the location of piston cores which recovered Miocene turbiditic intraclastic chalks from an outcrop. Water depths are in meters. Note that the prominent topographic spur in the Blake Escarpment on which DSDP Site 390 was drilled is the Blake Nose.

50 km of the escarpment. Consequently, little was learned about the basement structure in the critical area of the base of the Blake Escarpment from these seismic reflection data. Older two-ship refraction profiles at the base of the escarpment indicated a thick layer of sediments (more than 2.5 km) with seismic velocities of 3.64 km/sec (profile 102, Sheridan et al, 1966), and this suggested a deepening of the oceanic basement toward the escarpment with the accumulation of older, more consolidated, higher velocity sediments in a trough-like feature, as depicted by Sheridan (1974).

Other evidence, including gravity and magnetic data, indicates that the structure of the Blake Escarpment is apparently controlled by deep basement structure. Strong Bouguer gradients, from more than 300 mgal on the east to less than 100 mgal near the edge of the plateau (Fig. 2), suggest a very steep boundary between rocks with density differences beneath the Blake Escarpment. This fundamental boundary could be modeled as representing the density contrast between oceanic and continental crustal rocks, although other models are possible (Sheridan and Osburn, 1975). The aeromagnetic data of Taylor et al (1968) show that a positive magnetic anomaly parallels the escarpment and terminates the northeast-striking Blake Spur anomaly, which extends into the Blake Escarpment from the deeper part of the Atlantic basin (Fig. 3). Apparently the ig-

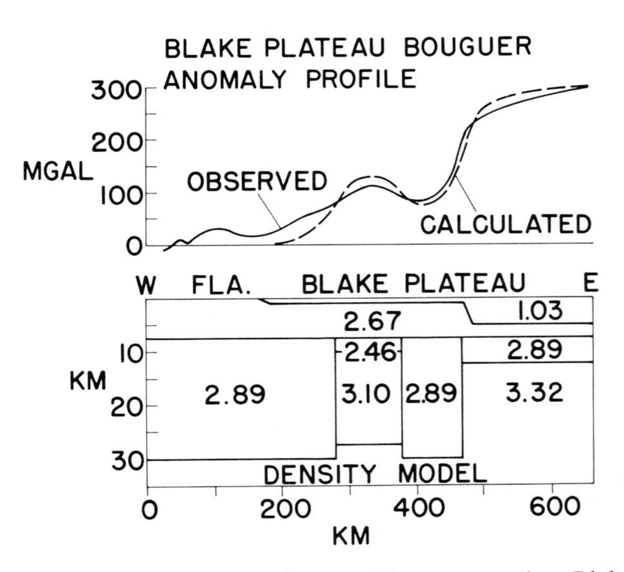

FIG. 2. Bouguer gravity profile across the Blake Escarpment with simple subsurface density models (after Sheridan and Osburn, 1975). Densities are in g/cm^3. The 2.67 g/cm^3 layer = carbonates; 2.46 g/cm^3 = salt, evaporitic sediments; 2.89 = granitic crust; 3.10 = intermediate "rift" crust; and 3.32 = mantle. The location of the profile is east-west off Cape Canaveral (Fig. 3).

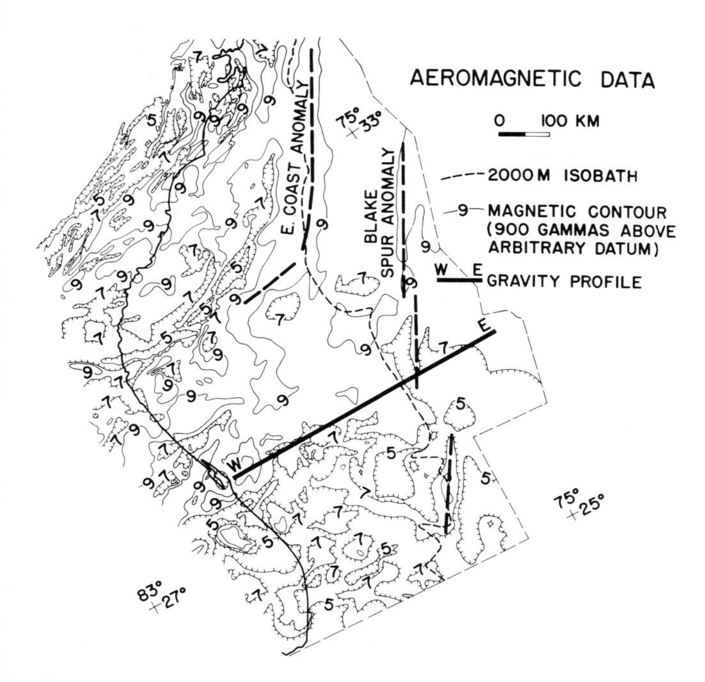

FIG. 3. Aeromagnetic map of the Blake Escarpment area (after Taylor et al, 1968). The prominent East Coast and Blake Spur anomalies are indicated by dashed lines. Note the apparent termination of the Blake Spur anomaly at the Blake Escarpment and the non-linear positive magnetic anomaly near the Blake Nose. Also, noted by the solid line is the east-west gravity profile discussed in Figure 2.

neous body forming the source for the Blake Spur anomaly is truncated along the basement structure causing the magnetic anomaly coincident with the escarpment.

In hopes of better defining the structural relationships along the base of the Blake Escarpment, multichannel lines MC1 and 2 were recorded using the full energy of the 4 large Bolt air guns. Based on a comparison of these results with the USGS-IFP profiles FCI-II-III, the higher source-energy of the air guns appears to be required in this region for deeper penetration. We believe that this is because the shallower sediments in the basin are Miocene intraclastic chalks, of relatively high seismic velocity and impedance and, therefore, highly reflective, whereas the Cretaceous black shales and claystones below the Miocene are of lower velocity and slightly gaseous (Benson et al, 1976). Apparently, above the velocity inversion much of the energy is reflected back, and below the inversion the less consolidated sediments absorb much of that which is transmitted, making the Blake-Bahama Basin a difficult area for seismic profiling.

MULTICHANNEL SEISMIC PROFILES

Profile MC1

Profile MC1 runs generally northwest-southeast across DSDP Site 391 and across the edge of the Blake Plateau (Fig. 4). This profile also passes close to an area along the base of the Blake Escarpment (Fig. 1) where piston cores taken from R/V *Eastward* revealed the outcrop of Miocene chalks, very

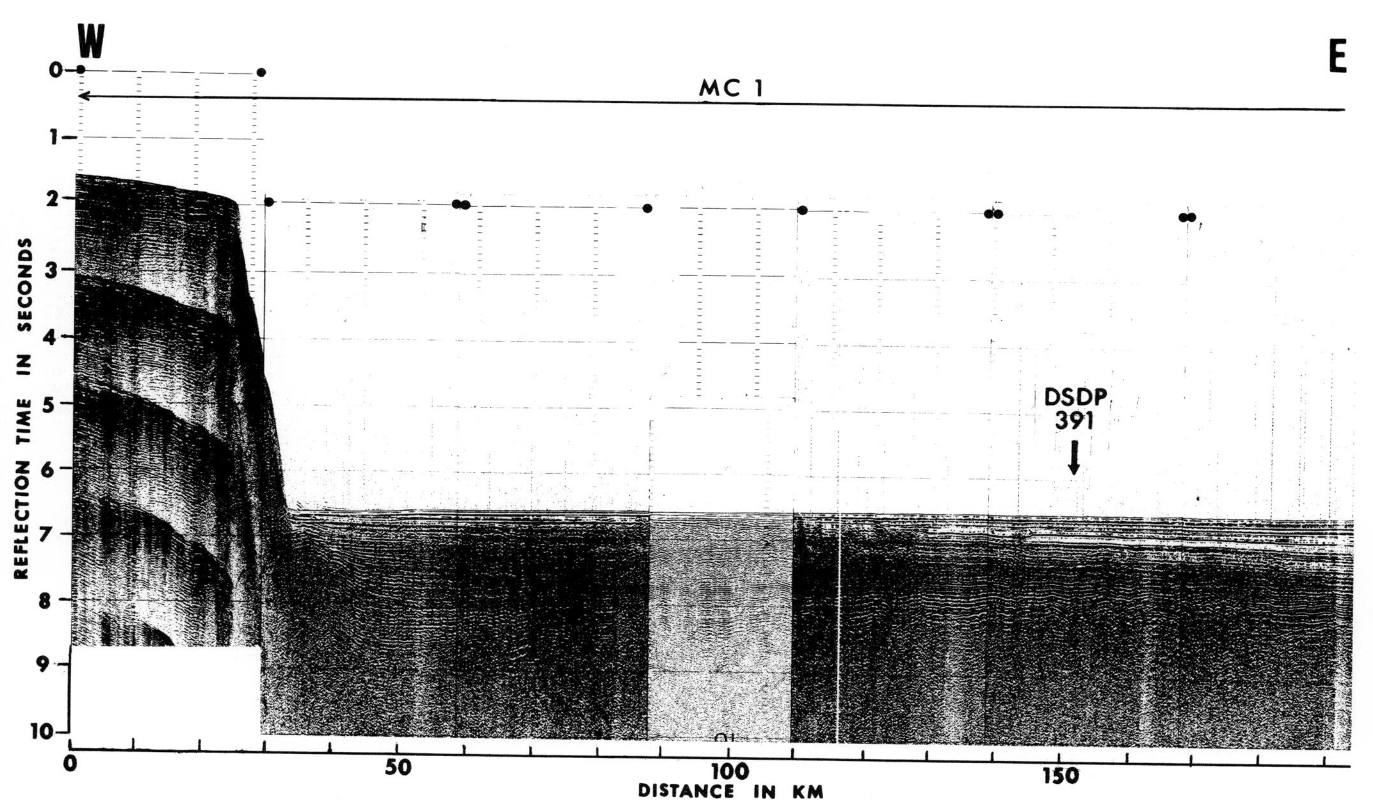

FIG. 4. Photograph of analog record of single channel monitor for profile MC1.

similar in lithology to those drilled at Site 391 (Sheridan et al, 1974; Benson et al, 1976). The widespread extent of these chalks across the Blake-Bahama Basin is well documented and can be seen on profile MC1 by the tracing of reflector M, first correlated by Dillon et al (1976) to the Miocene turbiditic chalks.

An interesting feature associated with the shallow Miocene reflectors in the Blake-Bahama Basin is the generation of diffractions and hyperbolic reflectors in a zone about 50 km east of the escarpment (Fig. 5). This same phenomenon was observed on the USGS-IFP line FCI to the south where the diffractions are so abundant and closely spaced that no coherent reflectors could be seen deeper than reflector M. It was interpreted that oceanic basement terminated in a zone of very closely spaced faults of insignificant throw, probably along an older fracture zone reactivated by later Tertiary subsidence of the Blake Plateau (Dillon et al, 1976).

On the *Conrad* lines the deeper reflectors overlying basement and the basement can be seen to pass beneath the hyperbolic reflector zone (Fig. 5). These deeper reflectors can be traced to within 20 km of the base of the escarpment (Fig. 6). The hyperbolic diffractions apparently originate from steep-walled rills once formed as erosional notches in the Miocene sediments. Relief of the order of 20 m on a 200-m spacing could be the cause of the observed disruption in the M reflectors. This type of relief

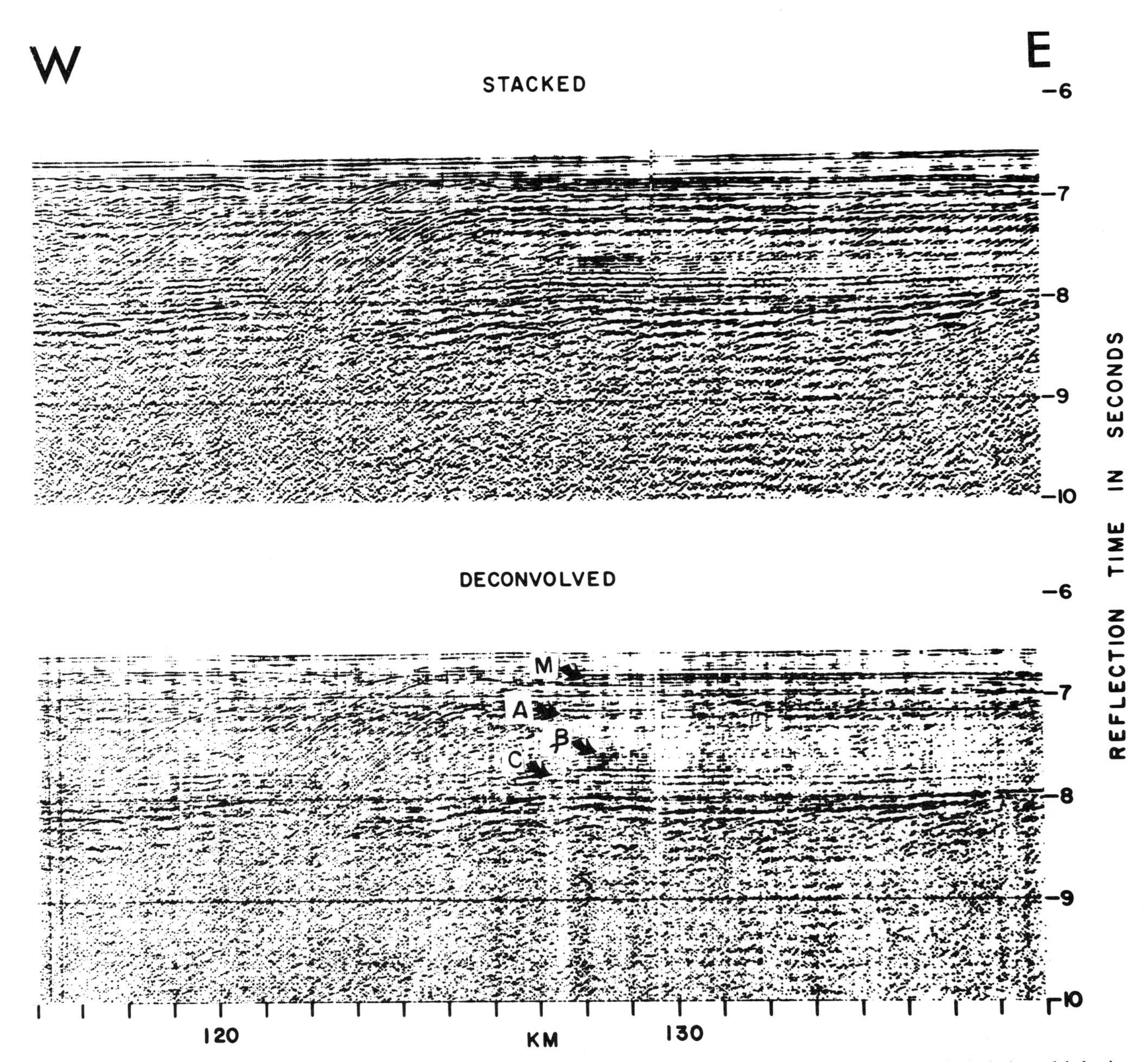

FIG. 5. Example of processed multichannel data from R/V *Conrad* profile MC1. Processing included demultiplexing, common-depth point gather, velocity analysis, normal-move-out correction, 24-fold stacking, time variable gain, and deconvolution. Note shallow hyperbolic reflectors at 125 km. Distance scale relates to Figure 6.

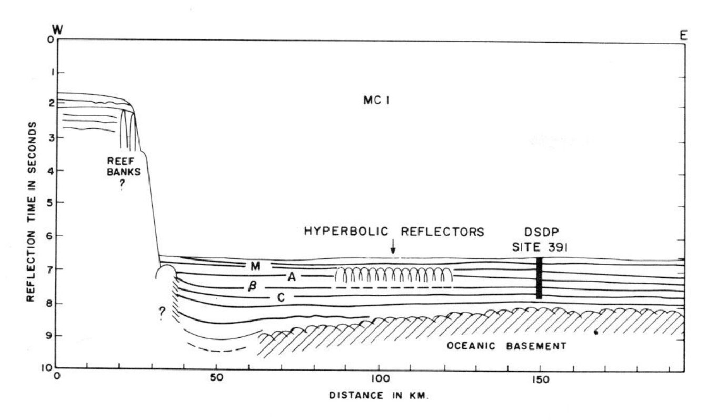

FIG. 6. Line tracing of R/V *Conrad* profile MC1 across the Blake Escarpment tied to DSDP Drill Site 391. Note major reflectors identified as M, A, β, C and oceanic basement, and the shallow hyperbolic reflectors. Also indicated by a question mark is an apparent salient of acoustic basement(?) at the base of the Blake Escarpment, and the thick sediments in the trough bounded by this basement feature.

and diffraction effect is well documented for the recent sea floor by deep-tow side-scan sonar and photographic studies in the Blake-Bahama Basin (Flood and Hollister, 1974). A relatively restricted area on the present sea floor on the western flank of the Bahama Outer Ridge is marked by many of these parallel rills. Deep ocean currents apparently have scoured the sea floor muds into rows of deep gullies and channels, and the eroded sediment has been transported to be deposited elsewhere.

The presence of these Miocene rills on profiles FCI and MC1 (Figs. 1, 6) suggests that the Miocene rilled zone trends north-south in the central part of the Blake-Bahama Basin. There is noticeable, although gentle, relief on the reflector involved with the diffractions, and it appears that on both profiles FCI and MCl the Miocene rills were on a slight ridge now buried in the abyssal Blake-Bahama Basin, much like the situation in which they are found today. As the reflectors below M and the oceanic basement can now be traced through the hyperbolic zone without offset, the suggestion that the diffractions are caused by closely spaced minor faults, as put forward by Dillon et al, (1976), does not seem to be correct.

Below the Miocene reflector M, Horizon A can be traced as a nearly horizontal reflector across the Blake-Bahama Basin to the base of the Blake Escarpment (Fig. 6). Horizon A is the interface between the Miocene units, which appear as both hyperbolic and acoustically laminated reflectors, and the Cretaceous black shales and claystones, which are seismically non-laminated (Benson et al, 1976; Dillon et al, 1976). A weaker reflection event at the top of a laminated and reverberant group of reflectors is identified as Horizon β. This reflector can be traced to the base of Blake Escarpment, although in places it is discontinuous. Where it was drilled at nearby Site 391, β is the downward transition from Aptian-Albian black shales to Neocomian white-gray limestones. Similar correlations of Horizon β hold at other DSDP sites in the western North Atlantic basin (Tucholke et al, 1975).

Below Horizon β in the Blake-Bahama Basin is a more prominent reflector, designated reflector C in this paper (Fig. 6). At Site 391 it corresponds to the transition from white limestones to red argillaceous limestones of Tithonian-Kimmeridgian age. This same lithologic and age boundary was correlated to a similar widespread reflector, reflector C, in the eastern North Atlantic basin by Lancelot et al, (1975), and we are following their use of this designation. Reflector C appears to be very widespread in the western North Atlantic, being identified from Cape Hatteras to the Cat Gap area.

Below reflector C another sedimentary reflector is seen just above basement at DSDP site 391 (Fig. 6) and an even deeper sedimentary reflector is found at 8.7 sec reflection time lapping onto the oceanic basement more than 50 km west of the drill site. These two reflectors involve sediments older than the Kimmeridgian-Oxfordian red argillaceous limestones drilled at several places in the western North Atlantic basin, perhaps involving sediments of mid-Early Jurassic age.

On line MC1 basement deepens to at least 11 km (9.5 sec) within about 20 km of the base of the Blake Escarpment. At the base of the escarpment the acoustic basement shoals to 6 km (6.8 sec) defining a steep-sided diffracting mass. This basement rise lies along the west side of what appears to be a basement trough. The slope of the basement rise is steeper than 45°. It is also evident that the reflectors of Horizon β and the older sedimentary units are flexed upward or differentially compacted against the steep western wall of this trough (Fig. 6). This indicates that the basement rise is a real structure rather than an acoustic blanking.

Velocity Analyses—To observe better the lateral facies changes in the sediments, especially in the area of the shallow hyperbolic reflectors, and to determine the sedimentary velocities of the layers in the trough at the base of the Blake Escarpment, velocity analyses were made along profile MC1 using the semblance technique of Taner and Koehler (1969).

Velocity analyses were made for approximately every five kilometers in three groups: east of the hyperbolic zone, in the hyperbolic zone, and west of the hyperbolic zone. These data were then averaged in the three areas for the distinguishable acoustic intervals between the major reflectors (Fig. 7). The resulting velocities compare very favorably with those determined by drilling at DSDP Site 391. Based on drilling depths to distinct lithologic contacts correlated to key reflectors, the interval velocities were calculated at Site 391 (Benson et al, 1976). Six acoustic units (I–VI) were identified (Fig. 7).

The shallowest acoustic unit I (1.93–2.05 km/sec) correlates with Quaternary and late Miocene muds and oozes; Unit II (2.21–2.92 km/sec) correlates with more consolidated intraclastic chalks of middle Miocene age; Unit III (2.25–3.04 km/sec) correlates

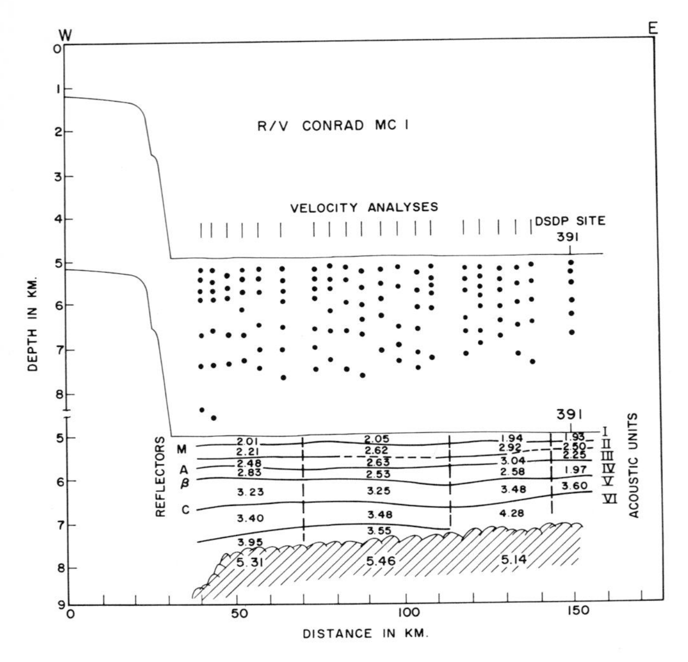

FIG. 7. Velocity analyses along profile MC1 correlated with the seismic reflection horizons and the acoustic velocity units determined by drilling at DSDP Site 391. The individual velocity scans determined by the semblance technique of Taner and Koehler (1969) were averaged in the three distinct acoustic zones: west of the hyperbolic reflectors, within the hyperbolic reflectors, and east of the hyperbolic reflectors. The dots are amplitude picks for RMS velocity curve for each analysis.

with the mixed early Miocene intraclastic chalks and cemented chalky muds; Unit IV (1.97–2.83 km/sec) correlates with Cretaceous carbonaceous claystones and shales; Unit V (3.23–3.60 km/sec) correlates with Tithonian-Neocomian white limestones; and Unit VI (3.40–4.28 km/sec) correlates with Kimmeridgian-Oxfordian red argillaceous limestones and possibly older Jurassic sediments.

It is interesting to note that the velocity inversion noticed in drilling at Site 391 between acoustic Unit III, the Miocene intraclastic chalk, and Unit IV, the carbonaceous black shale of Cretaceous age, is also indicated on the CDP profiles away from the site. The velocity inversion persists over much of the basin west of 391, implying that the same facies relation occurs throughout the Blake-Bahama Basin.

In the western part of the Blake-Bahama Basin at the base of the Blake Escarpment the CDP interval velocities for the sediments are very similar to velocities measured on a long, deep crustal refraction profile, 102 (Sheridan et al, 1966). On profile 102 velocities of 2.84 km/sec were measured for acoustic Units III and IV, and 3.64 km/sec for Units V and VI. This agreement of the CDP interval velocities with those of refraction profiles and those of the drilling calculations provides some confidence in the values determined by CDP method in deep water.

The sediments along the base of the Blake Escarpment in the sedimentary trough have seismic velocities of 3.23–3.95 km/sec. Using this range of values, the thickness of sediments in the trough below Horizon β is from 4.03 to 4.94 km. The age and lithology of the deeper section of these sediments are unknown, but they must be at least pre-Oxfordian in age, older than any sediments yet drilled in the oceans. Conceivably they could have been deposited very rapidly and thus may not be of much greater age than the overlying unit. On the other hand, the half-graben structure of the trough is compatible with rift features thought to have formed with the initial opening of the Atlantic, and thus these sediments might be quite old.

Profile MC2

Profile MC2 crosses the escarpment at the Blake Nose, a prominent topographic spur upon which DSDP Site 390 has recently been drilled (Fig. 1 and Fig. 8). Profile MC2 extends eastward across the Blake Outer Ridge, but only the section of the profile across the Blake Escarpment at the nose will be discussed here. At the base of the Escarpment the acoustic stratigraphy is markedly different from that of profile MC1 to the south (Fig. 9). In the Tertiary sediments the shallowest reflector is identified as Reflector X of Markl et al (1970). Thus far this reflector and the sediments below it have not been drilled or sampled by coring, and therefore, the lithology of this horizon is unknown.

No evidence is found on this profile of the prominent M Reflector identified on Profile MC1. Presumably the Miocene intraclastic chalks, formed of turbiditic deposits and debris flows did not extend as far north as the Blake Nose and were confined to the Blake-Bahama Basin by the Outer Ridge as it existed in Miocene time. Note that Reflector X on MC2 is at 5.1 km (6.4 sec) depth, shallower than the 5.3 km (6.6 sec) depth of the M Reflector on profile MC1. The acoustically non-laminated appearance of the sediments below X suggest that these are deposits of hemipelagic mud similar to those forming the Blake Outer Ridge today. The hemipelagic nature of these sediments agrees with the interpretation of Ewing and Hollister (1972) that the X Reflector defined the nucleus of the Outer Ridge structure in the early Miocene.

Below Reflector X there is a prominent horizontal and strongly reverberant reflector sequence, the top of which identified as Horizon β by correlation with other seismic lines (Fig. 9). Farther east on profile MC2, Horizon A can be traced in from the deep basin and is seen to pinch out against Horizon β under the northeast flank of the Blake Outer Ridge. If this seismic interpretation is correct, the early Miocene-Albian hiatus drilled at DSDP Site 391 and representing at least 800 m of erosion (Benson et al, 1976) must be an extensive erosional event observable southward from the margin at the Blake Nose possibly to the Horizon β outcrop area near Cat Island. On profile MC2 this erosional hiatus might represent the early Miocene-Neocomian interval, in as much as erosion has stripped the section completely down to Horizon β.

It is suggested that the lithology across this hiatus

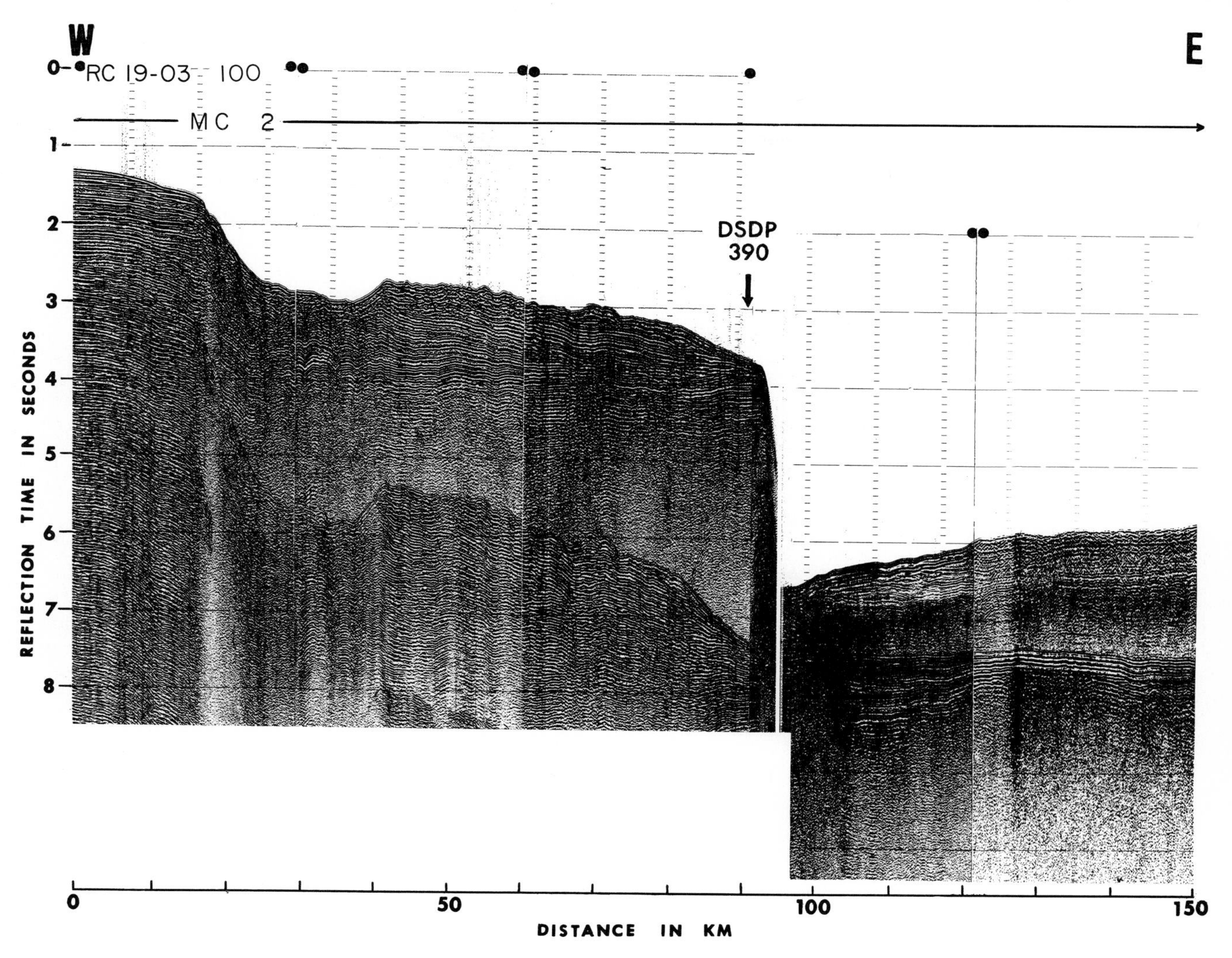

FIG. 8. Photograph of analog record of single-channel monitor for profile MC2.

could change from Miocene hemipelagic chalks to Neocomian white and gray limestones, the Middle Cretaceous carbonaceous shales being omitted. This is important because the carbonaceous shales are possibly a source of petroleum for the Atlantic margin, provided that they exist under the continental rise to the north. Burial there might have resulted in adequate and timely hydrocarbon maturation, which has not taken place in the Blake-Bahama Basin. However, the evidence on profile MC2 suggests that the carbonaceous shales are eroded to the north and west as well and might not be continuous under the margin.

Between Horizon β and the hyperbolic reflector identified as oceanic basement, a prominent sedimentary reflector is correlated by its depth and reflectivity with Horizon C of profile MC1 (Fig. 6 and Fig. 9). This reflector is conformable to the overlying Horizon β along most of the profile except in the area near the base of the Blake Nose, where on profile MC2 evidence exists of downflexing causing the addition of a growth-wedge of overlying sediments. On profile MC2 there is also some distinct upturning of both Reflector C and the prominent reflector below C against a steep-sided (>60°) salient of opaque acoustic basement at the base of the escarpment. This steep-sided acoustic basement can be seen to depths of 12 km (10.5 sec) beneath the trough. The steep west side of the trough with the upturned reflectors can be interpreted as a normal fault bordering a graben. The eastern trough edge is rather a downwarp or a series of normal faults which give the impression of a flexure. Structural relief is again more than 4 km (2.5 sec), as on MC1 where the seismic velocities for the sediments suggested more than 4 km of sediments in the trough.

The origin and nature of the acoustic basement shoulder of the trough is still uncertain, as suggested by the question mark in Figure 9. The magnetic and gravity anomalies associated with the base of the escarpment indicate that its form is controlled by the underlying basement, possibly associated with the transition from oceanic to continental rocks (Figs. 2 and 3). Erosion has apparently affected the acoustic basement of this shoulder, as it appears smoother and flatter than the typical oceanic basement reflec-

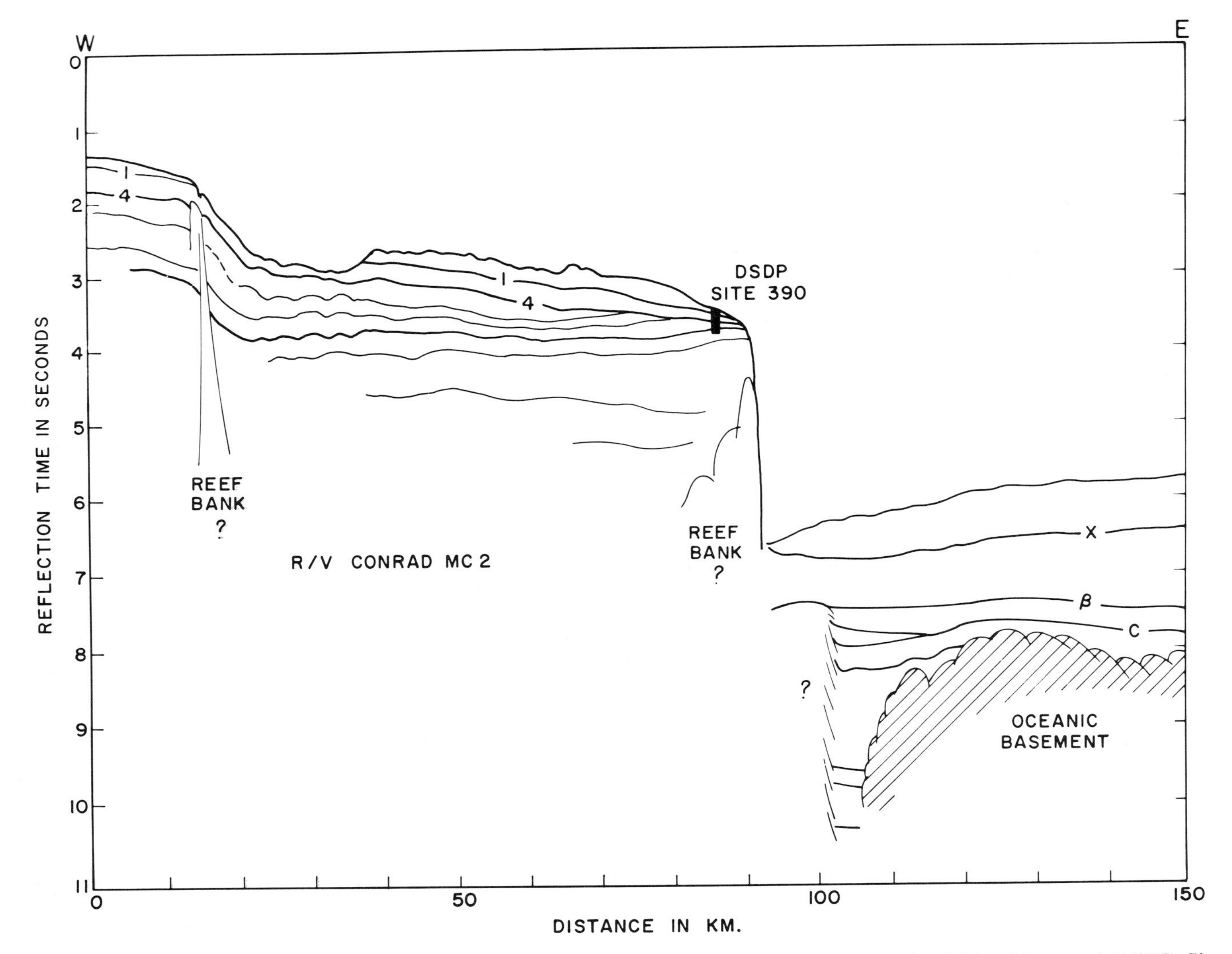

FIG. 9. Line tracing of R/V *Conrad* profile MC2 across the Blake Escarpment at the Blake Nose and DSDP Site 390. Note the identification of the prominent reflectors of the Blake-Bahama Basin as X, β, C and oceanic basement, and of the Blake Nose and Blake Plateau as 1 and 4. Possible reef-bank margin cemented limestones are interpreted from their acoustic character. Note the apparent salient of acoustic basement (?) at the base of the Blake Nose and the deep narrow sedimentary trough at the base of the Blake Escarpment.

tor. Presumably the erosional stripping of Horizon A and the Cretaceous black shale unit took place in a submarine environment and this same erosion flattened the acoustic basement shoulder on the west of the trough at the base of the Blake Escarpment.

On the Blake Nose, profile MC2 passes through DSDP Site 390 and ties the drilled reflectors to the west on the Blake Plateau (Fig. 9). Reflector 1, identified by Ewing et al (1966) on the Blake Plateau, extends to Site 390 where it corresponds to a chert horizon in basal Eocene-topmost Paleocene nannofossil oozes. This is similar in age and lithology to those rocks drilled and sampled from reflector 1 on the Plateau to the west (JOIDES, 1965). Widespread chert deposition apparently occurred over the plateau at that time. Eocene cherts are also commonly found in the deep western North Atlantic basin (Peterson et al, 1970) indicating even more widespread siliceous deposition. At Site 390 the Eocene is below only a thin veneer of Quaternary lag deposits indicating strong and possibly recent, submarine erosion which has sculpted the surface of the Blake Nose on profile MC2.

Reflector 4 (Ewing et al, 1966) can be traced from the Blake Plateau to Site 390 where it corresponds to an angular unconformity between Campanian and Albian nannofossil oozes (Fig. 9). Two well-developed reflectors are truncated by the unconformity, suggesting that all or part of the missing Cretaceous units are present farther west on the Blake Nose and on the Blake Plateau. Apparently strong submarine currents eroded the Blake Nose sometime after Albian but prior to Campanian deposition (Benson et al, 1976). The possible presence of the missing Albian-Campanian sediments to the west suggests that the erosional event was closer in time to the Campanian than to Albian. The Campanian is approximately the time of the rifting in the Labrador Sea region (Laughton et al, 1972), and this occurrence might have permitted access of colder Arctic waters to increase the circulation of the western North Atlantic. Such a change in bottom circulation after the Cenomanian is recorded in the important facies change in the deep western North Atlantic between the Aptian-Cenomanian black carbonaceous shales and the Late Cretaceous variegated clays (Tucholke et al, 1975). Apparently the anoxic or stagnating conditions caused by lack of circulation in the deep basin during deposition of the black shale ended, and some circulation later developed the more oxygenated conditions which resulted in deposition of multicolored clays.

The deepest reflector drilled at Site 390 on the Blake Escarpment corresponds to the contact between Aptian-Albian nannofossil ooze deposits and a lithified shallow-water limestone of Barremian or older Cretaceous age. Twenty-five kilometers to the south on the edge of the Blake Nose this reflector tops a hummocky hyperbolic reflector, which appears acoustically opaque, with the well-stratified deeper reflectors terminating along its flank in what appears to be a reef-bank mass (Fig. 9). At site 392 the drill penetrated this opaque mass and an Early Cretaceous highly cemented limestone of variable carbonate bank-margin facies was recovered. Seismic velocities of 5.0–5.5 km/sec were measured for these rocks onboard ship (Benson et al, 1976). Apparently these bank-margin rocks, probably formed as oolite shoal ridges that were shallower than the adjacent bank sediments, were the first features to emerge during regressive cycles and, therefore, the first rocks to be diagenetically cemented. Thus a mass of high-velocity rock was generated along the bank margin as the rim complex was cemented progressively through geologic time. With the control provided by Site 392, similar hyperbolic reflective features are interpreted as reef-bank margin complexes cemented into high-seismic-velocity masses.

In tracing the deep prominent reflector, correlated with the Barremian and older limestone at the bottom of the hole at Site 390, there appears to be a westward dip under Blake Nose and an apparent shallowing under the edge of the Blake Plateau. Velocity analysis has not been done on profile MC2, but older refraction data on the plateau (Hersey et al, 1959) near this profile indicate a relatively high seismic velocity (~4.1 km/sec) for the sediments below reflector 4 and above the prominent Barremian reflector at 2.8 sec (Fig. 9). Also, cemented reef-bank margin limestones are interpreted as being present on the Blake Plateau edge on profile MC2. Allowing for the velocities in these limestones, the apparent shallowing of the prominent Barremian and older reflector can be interpreted as a velocity effect, which appears only on the time section. In depth section the deep reflector would continue to dip westward to a depth of approximately 3.4 km under the Blake Plateau from the shallower depth of 2.8 km where drilled at Site 390 (Fig. 9).

DISCUSSION

Although lines MC1 and MC2 were intended to expand upon DSDP drilling results, the discovery of a sediment-filled trough at the base of the Blake Bahama escarpment raises some old arguments about the origin of the escarpment itself. The massive basement salients at the western edge of the trough are suggestive of old fault blocks or slump features whereas the eastern flank is less abrupt, suggesting downwarping or progressive normal faulting. The trough is most clearly defined on MC2 east of the Blake Nose probably due to the absence of acoustically absorptive or strongly reflective sedimentary horizons. Clearly the trough contains a number of deep sedimentary reflectors below the deepest and oldest sampled units in the western North Atlantic.

Questions remain as to the extent of the trough and whether it is really a primitive feature formed at the time of early opening of the Atlantic. If so, the deep reflectors are probably very old. An alternative hypothesis is that it is a drag feature formed by subsidence of the Blake-Bahama block relative to the normal oceanic crust. If so, the subsidence probably ceased prior to the deposition of Horizon β in the area east of the Blake Nose, because β and younger reflectors are relatively undisturbed (Fig. 9).

The picture is not so clear south of the Nose on MC1 (Fig. 6) where the trough is a broader feature and the attitude of the pre-β section is more indicative of differential compaction of sediments in the trough rather than of differential movement between the Blake Escarpment and the adjoining oceanic crust. The depth to β is approximately the same on profiles MC1 and 2. Assuming that all pre-β layers were deposited as near-horizontal strata, the gentle westerly dip of the deeper strata argues for some regional subsidence prior to Neocomian time, modified by differential subsidence in the vicinity of the Blake Nose but not of the order of the several hundred meters which would be required to drown the shallow reefs along the Blake Escarpment. We conclude that if differential subsidence occurred, it took place in a very narrow zone, possibly between the basement salient along the western edge of the trough and the escarpment itself.

Describing the overall morphology and extent of the trough requires more crossings in other areas using at least equally powerful geophysical tools and/or drawing upon existing commercial offshore seismic lines that extend across the escarpment, if such data exist and are of useful quality. It is quite possible that the strongly reflecting Miocene intraclastic deposits lie quite close to the escarpment in many places and will, to some extent, complicate these studies. However, the possibility of being able to pin down the structural/morphologic boundary of the early continent seems worth the effort. The deeper reflectors in the trough appear to be accessible to existing DSDP drilling capability and should not be overlooked as desirable targets for continental margin geologic studies in the near future.

REFERENCES CITED

Benson, W. E., et al, 1976, Deep-sea drilling in the North Atlantic: Geotimes, v. 21, N. 2., p. 23-26.

Dillon, W. P., R. E. Sheridan, and J. P. Fail, 1976, Structures of the western Blake-Bahama Basin as shown by 24-channel CDP profiling: Geology, v. 4, p. 459–462.

Ewing, J. I., M. Ewing, and R. Leyden, 1966, Seismic profiler survey of Blake Plateau: AAPG Bull., v. 50, p. 1948–1971.

——— and C. D. Hollister, 1972, Regional aspects of deep sea drilling in the western North Atlantic, *in* Initial Reports of the Deep Sea Drilling Project: U.S. Govt. Printing Office, v. XI, p. 951–976.

Flood, R. D., and C. D. Hollister, 1974, Current-controlled topography on the continental margin off the eastern United States: *in* C. A. Burk and C. L. Drake, eds., Geology of continental margins: New York, Springer-Verlag, p. 197–205.

Hersey, J. B., et al, 1959, Geophysical investigation of the continental margin between Cape Henry, Virginia, and Jacksonville, Florida: Geol. Soc. Amer. Bull., v. 70, p. 437–466.

JOIDES, 1965, Ocean drilling on the continental margin: Science, v. 150, p. 709–716.

Lancelot, Y., et al, 1975, The eastern North Atlantic: Geotimes, v. 20, No. 7, p. 18–21.

Laughton, A. S., 1972, The southern Labrador Sea—a key to the Mesozoic and early Tertiary evolution of the North Atlantic: *in* Initial Reports of the Deep Sea Drilling Project: U.S. Govt. Printing Office, v. XII, p. 33–82.

Markl, R. G., G. M. Bryan, and J. I. Ewing, 1970, Structure of the Blake-Bahama Outer Ridge: Jour. Geophys. Res., v. 75, p. 4539–4555.

Peterson, M. N. A., et al, 1970, Initial Reports of Deep Sea Drilling Project: U.S. Govt. Printing Office, v. II, 493 p.

Sheridan, R. E., 1974, Atlantic margin of North America, *in* C. A. Burk and C. L. Drake, eds., Geology of continental margins: New York, Springer-Verlag, p. 391–407.

——— and W. L. Osburn, 1975, Marine geological and geophysical studies of the Florida-Blake Plateau-Bahamas area: Can. Soc. Petroleum Geologists, Mem. 4, p. 9–32.

——— X. Golovchenko, and J. I. Ewing, 1974, Late Miocene turbidite horizon from the Blake-Bahama Basin: AAPG Bull., v. 58, p. 1797–1805.

——— et al, 1966, Seismic refraction study of the continental margin east of Florida: AAPG Bull., v. 50, p. 1972–1991.

Taner, M. T., and F. Koehler, 1969, Velocity spectra digital computer derivation and applications of velocity functions: Geophysics, v. 34, p. 859–881.

Taylor, P. T., I. Zietz, and L. S. Dennis, 1968, Geologic implications of aeromagnetic data for the eastern continental margin of the United States: Geophysics, v. 33, p. 765–780.

Tucholke, B., et al, 1975, Glomar Challenger drills in the North Atlantic: Geotimes, v. 20, N. 12, p. 18–21.

Convergent Margins

Tectonics of the Andaman Sea and Burma[1]

J. R. CURRAY, D. G. MOORE, L. A. LAWVER, F. J. EMMEL,
R. W. RAITT, M. HENRY, AND R. KIECKHEFER[2]

Abstract The Andaman-Nicobar Ridge and the Indo-Burman Range are composed of sediments of the Bengal and Nicobar Fans scraped off the underthrusting Indian plate at the Sunda subduction zone. The Andaman Sea and the eastern part of the central valley of Burma represent a Neogene-Quaternary extensional basin underlain by oceanic crust formed by northwestward rifting of this orogenic ridge away from continental crust of the eastern Burma highlands and the Malay Peninsula. The plate edge is defined by a north-south transform in Burma, by the Sumatran fault system longitudinally bisecting Sumatra, and by a complex system of short spreading rifts and transforms in the central basin of the Andaman Sea. Where sedimentation rates are high over the newly formed oceanic crust, no magnetic anomalies are identifiable; where sedimentation rates are low or negligible, clear magnetic anomalies have permitted dating this phase of opening at approximately 11 m.y. B.P.

INTRODUCTION

Marginal seas are one type of small ocean basin that constitute an important element in plate tectonic concepts of geosynclines; they have recently been of considerable interest as zones of possible crustal extension lying above and behind active subduction zones. However few marginal basins believed to have formed by crustal extension have been studied in enough detail for understanding of either present three-dimensional tectonics or geological history. For this reason and because of the close relationship to our studies of the history and orogeny of the geosyncline underlying the Bay of Bengal and Bengal Basin (Curray and Moore, 1971, 1974; Moore et al, 1974, 1976), we have made the surveys (Fig. 1) and tectonic analysis of the Andaman Sea which we report here.

In terms of regional tectonics, the Andaman Sea lies along the Indian and China (Eurasian) plate boundary, which extends from the Sunda Arc and Trench off Java and Sumatra, along the west side of the Andaman-Nicobar Ridge and northward through the Indo-Burman Ranges (Fig. 2).

Previous modern published studies of the Andaman Sea resulted from two reconnaissance cruises of the U.S. Coast and Geodetic Survey (now NOAA; Burns, 1964; Peter et al, 1966; Weeks et al, 1967; Rodolfo, 1969; and Frerichs, 1971). We have drawn heavily on data and ideas from those publications but now add our own new data and conclusions from geological and geophysical surveys on research vessels of the Scripps Institution of Oceanography in 1968, 1971, 1973, 1975, and most recently from work on R/V *Thomas Washington* during March 1977 (Fig. 1).

For this report, marine seismic reflection data constitute a primary source of new structural information, supplemented by both free-drifting and moored sonobuoy wide-angle reflection and refraction measurements, and preliminary analyses of gravity, magnetics, heat flow, and bottom samples. This paper presents our favored working hypothesis, which may be subject to change as we continue our

[1]Manuscript received, October 24, 1977; accepted, February 15, 1978.

This preliminary review of the tectonics and geological history of Burma and the Andaman Sea was first completed and submitted for consideration for publication in January and February 1977, before our departure for the March 1977 work at sea. Excessive delays in the review process for the journal to which it was submitted made us decide to withdraw it, revise it slightly in the light of the new data, and submit it for publication in this volume. All major conclusions are, however, the same as in the earlier version.

[2]Scripps Institution of Oceanography, La Jolla, California 92093.

We wish to thank our many colleagues who have assisted us in collection of data and samples at sea and in the processing and interpretation of those data, including LeRoy Dorman, Carrel Ramsey, and Perry Crampton. The manuscript was reviewed and significantly modified by suggestions from Chris von der Borch, Seiya Uyeda, Aung Tin U, Chiramit Rasrikriengkrai, Suvit Sampattavanija, M. J. Terman, and especially Doyle Paul, with whom we enjoyed lengthy discussions on the interpretation of the tectonics of Burma and the Andaman Sea. In addition, we have most recently had the pleasure of long discussions on the geology of Burma and the Andaman Sea with Hla Tin and Aung Min. This study has been entirely supported by the Office of Naval Research.

Article Identification Number:
0065–731X/78/MO29–0012/$03.00/0.
(see copyright notice, front of book)

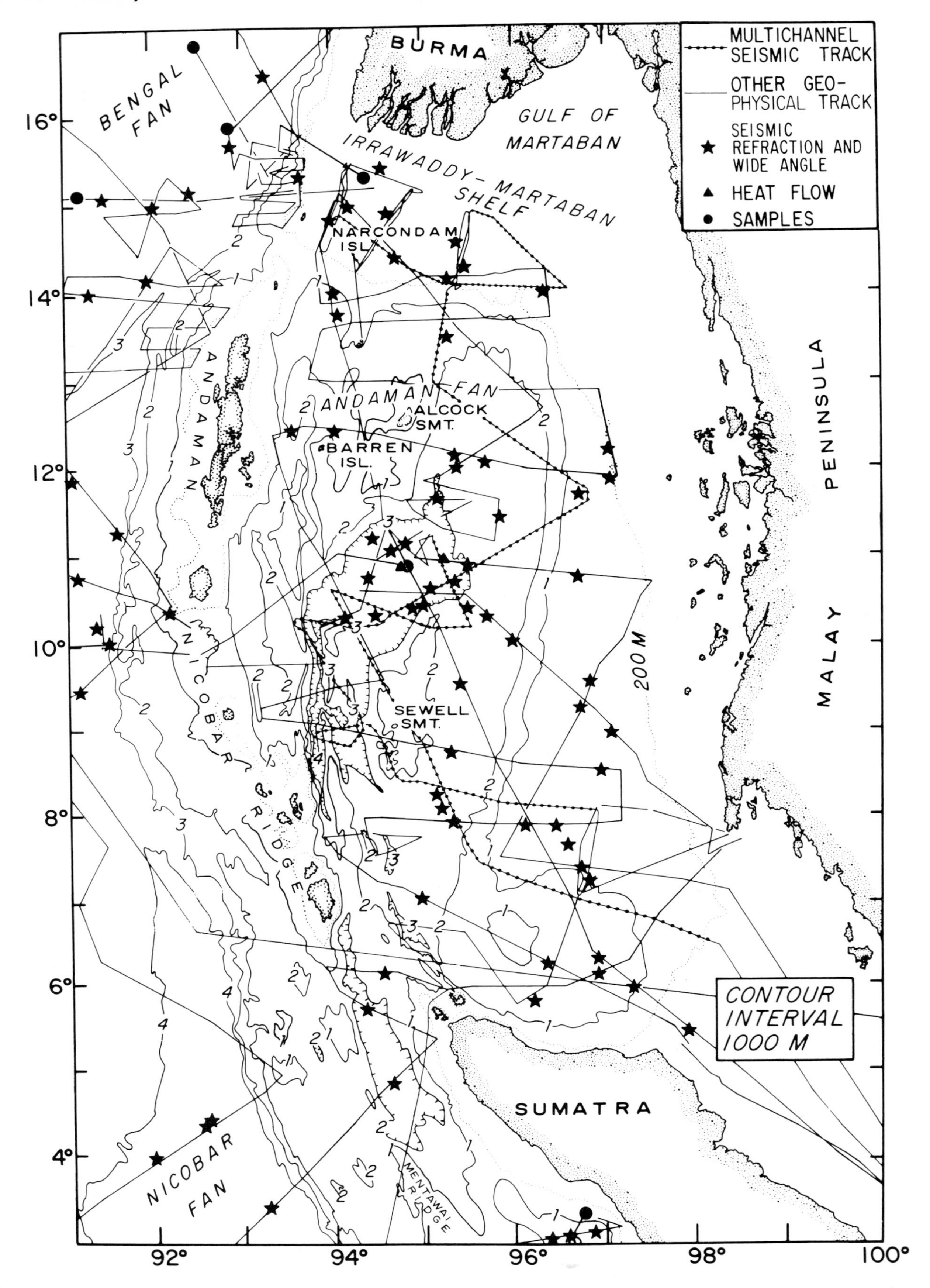

FIG. 1. Bathymetry and geophysical tracks in the Andaman Sea. All soundings are in uncorrected meters at 1,500 m/sec.

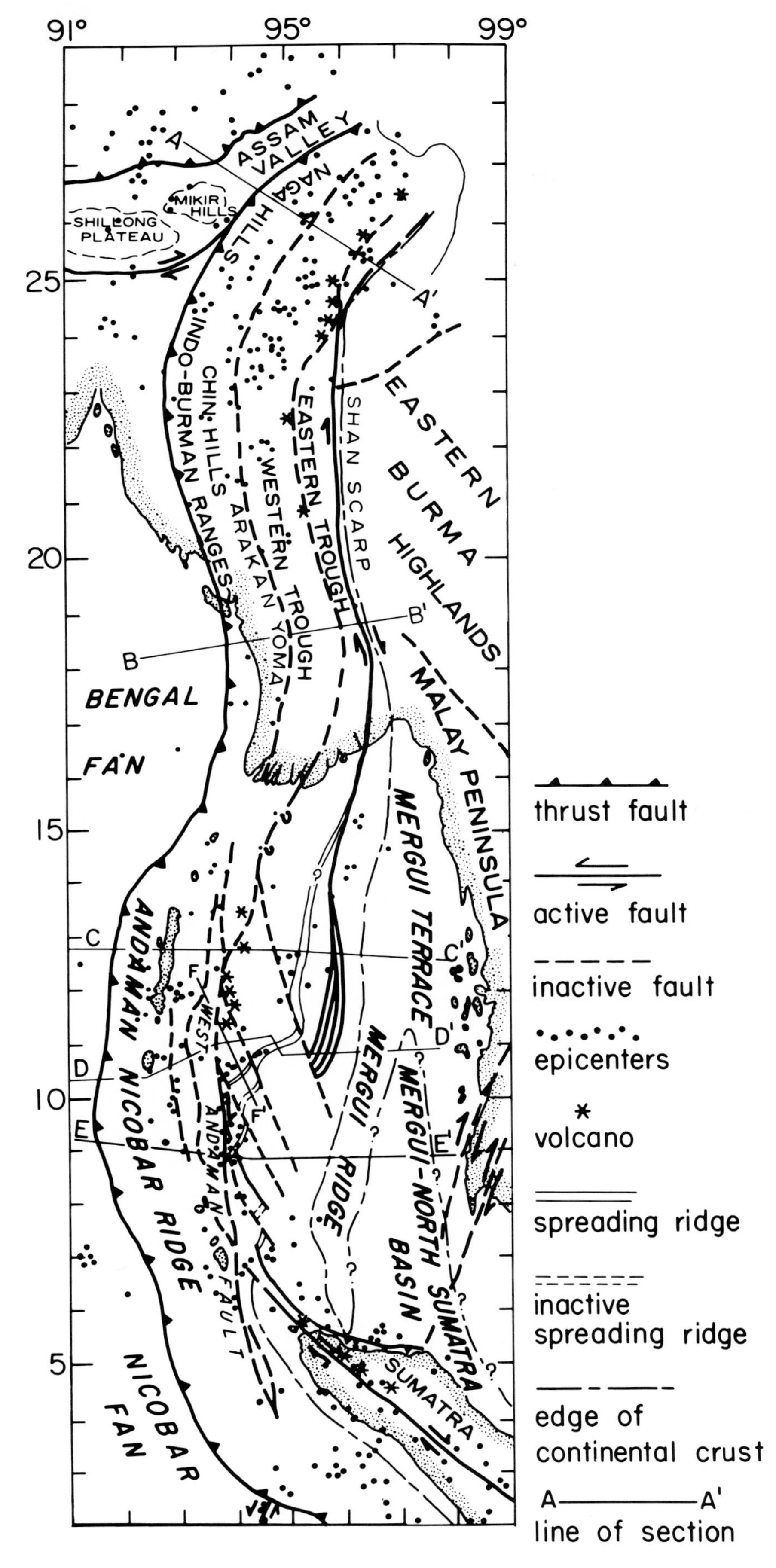

FIG. 2. Tectonic map of Burma and the Andaman Sea, with locations of the sections of Figures 3 and 5.

work and complete the analysis of our data and observations.

MAJOR PHYSIOGRAPHIC AND TECTONIC PROVINCES

Bathymetry of the Andaman Sea is exceedingly complex (Fig. 1), and we show only a simplified chart. Physiographic provinces coincide in general with tectonic provinces (Fig. 2) except in the northern filled portion of the structural depression, where sediments derived from south-flowing rivers from the Tibetan highlands have prograded the shoreline and shelf edge southward parallel to the structural trends.

The Andaman-Nicobar Ridge (Figs. 2, 3) lies between and is contiguous with the Indo-Burman Ranges to the north, the sedimentary arc of the Mentawai Island Ridge (Fig. 1) off Sumatra, and the "nonvolcanic" outer ridge off Java. The surface trace of the subduction zone which forms the Sunda Trench lies at the western base of the ridge where the trench is nearly filled with sediment of the Bengal and Nicobar Fans. The section in the Andaman and Nicobar Islands consists of Cretaceous ophiolites (Hutchison, 1975) with radiolarian cherts, pelagic limestones, and quartzites of unknown origin; Cretaceous to Eocene cherty pelagic limestones, grits, and conglomerates with clasts of primarily ophiolitic and sedimentary origin; and a thick section of the Eocene and Oligocene Andaman Flysch, overlain by Neogene shallow-water sediments of the ridge top and islands (Karunakaran et al, 1968, 1975). An analogous section occurs in the Indo-Burman Ranges (Brunnschweiler, 1966, 1974; Mitchell and McKerrow, 1975). We interpret these published descriptions as indicating that the rocks are sea floor ophiolites and sediments scraped off the underthrusting Indian plate, overlain by autochthonous sediments of shallow-water forearc environment, and we believe that the Andaman and Indo-Burman flysch units are from the middle sedimentary unit underlying the modern Bengal and Nicobar Fans (Curray and Moore, 1971, 1974).

Subduction probably started along the western Sunda Arc following breakup of Gondwanaland in the Early Cretaceous (about 130 m.y. BP), and the Andaman-Nicobar Ridge and Indo-Burman Ranges were uplifted as a sedimentary arc during the Oligocene or late Eocene, probably contemporaneously with uplift of the nonvolcanic ridge to the southeast along the Sunda Arc. The structure of all of this orogen is predominantly east-dipping nappes, with folded offscraped sediments deformed to varying degrees toward mélange (Fig. 3). The sediments are gently folded within the nappes in the northern part of the arc where the Bengal Fan section is thicker, and are more tightly folded and intensely deformed within the nappes off Sumatra and Java, where the Nicobar Fan and pelagic sections are thinner. This subduction and offscraping process continues today. Even the youngest Pleistocene sediments of the Bengal Fan are being deformed and folded at the base of the landward wall of the filled trench (Figs. 3C and D).

The Malay Peninsula, its continental shelf, and the Mergui Terrace form the opposite, eastern boundary of the Andaman Sea (Figs. 2, 3). This province consists of continental crust overlain locally by Paleozoic and Mesozoic sedimentary rocks and Tertiary sedimentary cover of variable thickness, reflecting a history of subsidence. We suggest that the western margin of this continental crust, a former continental slope, extends southward from the Shan scarp at the western edge of the eastern Burma highlands, under the Martaban Shelf, and the slope of the Mergui Terrace, to where it is apparently offset by the Sumatran fault system (Fig. 2). An alternative explanation is that the Mergui Ridge is an old magmatic arc.

The Mergui-North Sumatra Basin is apparently continuous with the onshore portion of the North Sumatra Basin. Our seismic data do not delineate this section well, but we do measure locally up to 5 to 6 km of sediment cut by north- or north-northeast-trending block faulting. We can locally interpret deltaic structure in reflection records of the Neogene section and suggest that some of the section represents shore zone and continental shelf sediments deposited during a period of rapid subsidence. Preliminary interpretation of our newer refraction data suggests that thinned continental crust underlies this basin.

The Irrawaddy-Martaban shelf represents the southern limit of filling of the structural depression of the combined Andaman Sea and central lowlands of Burma (the Burma Tertiary geosyncline) from the huge Tibetan sediment supplies. The Andaman Sea tectonic provinces extend generally under this shelf and into the tectonic provinces of Burma (Chhibber, 1934; Brunnschweiler, 1966, 1974; Aung Khin and Kyaw Win, 1969; and Mitchell and McKerrow, 1975). Mitchell and McKerrow (1975, p. 305) referred to ". . . up to 17 km of Tertiary marine and fluviatile sediment" in the Burma central lowlands. Our seismic reflection information is inadequate to confirm this thickness of sediment underlying the northern shelf, but preliminary interpretation of our refraction data does suggest such thicknesses under the eastern trough part of the shelf. Isostatic considerations demand that such a sediment thickness, especially if it is in a vertical succession rather than stratigraphic succession, must be underlain by oceanic crust.

The Andaman Sea Basin includes the deep terrace, Sewell Seamount, and Alcock Seamount of Rodolfo (1969). Our seismic refraction results show up to 4 km of sediment overlying typical oceanic second layer and crust within this province. The complex bathymetry controls deposition, and thick sediment fill is generally confined to the foot of the Mergui Terrace and in the Andaman Fan to the south of the Irrawaddy-Martaban shelf.

The boundary between the Andaman-Nicobar Ridge and the Andaman Sea Basin is a complex of north-south-trending faults, dividing the sea floor

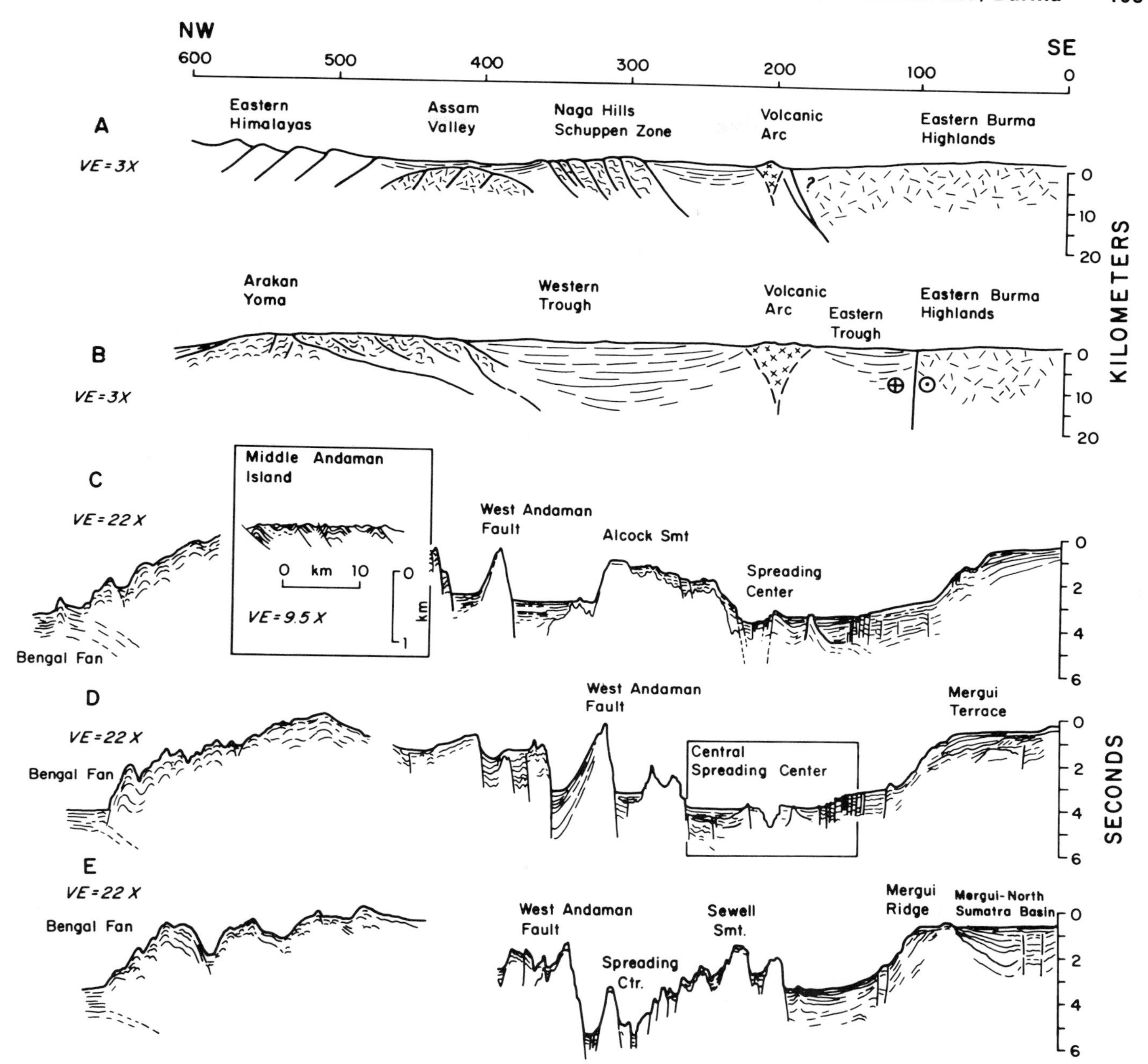

FIG. 3. Structure sections across Burma and the Andaman Sea. See Figure 2 for locations.
A-A′ Naga Hills to Eastern Burma Highlands, a composite diagrammatic section adapted from Evans (1964), Raju (1968), Brunnschweiler (1974), and Mitchell and McKerrow (1975).
B-B′ Arakan Yoma to Eastern Burma Highlands, a composite diagrammatic section adapted from Brunnschweiler (1974), and Mitchell and McKerrow (1975).
C-C′ Northern Andaman Sea, prepared from line drawings of seismic reflection records and refraction data. Middle Andaman Island section adapted from Geological Survey of India (1944).
D-D′ Central Andaman Sea, prepared from line drawings seismic reflection records and refraction data. For photograph of record in inset, see Figure 4.
E-E′ Southern Andaman Sea, prepared from line drawings of seismic reflection records and refraction data.

into a series of ridges. The most prominent of these faults, the West Andaman Fault (Fig. 2), appears to be continuous from west of Sumatra to where it is lost beneath the fill of the Irrawaddy-Martaban Shelf. It generally is marked by a west-dipping cuesta of sedimentary rock determined by Frerichs (1971) and Rodolfo (1969) to be pelagic Miocene calcarenites and calcilutites, indicating deposition in water depths in excess of those from which they were dredged. As indicated in Figure 2, this fault appears in our reflection records to be inactive except for the central section, where focal mechanism studies (Fitch, 1972) indicate a north-south right-lateral sense of motion.

The line of this fault is also marked by the proximity of the volcanic islands Barren and Narcondam and other probable volcanic seamounts. We correlate and extrapolate this boundary northeastward into the line of inactive volcanoes of the central lowlands of Burma, the boundary between the eastern and western troughs.

EXTENSION IN THE ANDAMAN SEA

Through combined study of all data available to us, we have interpreted a complex system of spreading centers and transform faults in the Andaman Sea Basin (Fig. 2). In general, the pattern is rather closely constrained by our survey control.

The spreading rift is best illustrated in the central part of the basin with seismic reflection records (Figs. 3D, 4). The rift here is typically 5 to 8 km wide, 500 m deep, and trends about 070°–250°, separating formerly flat-lying ponded sediments. Symmetrically arranged on either side are upturned edges of these ponded sediments, now rifted, separated and upturned as if by doming. A third, older, more deeply buried pair of upturned edges is evident farther from the present rift. A heat flow measurement (Fig. 1) in the flat ponded sediments several km away from the rift is 3.3 hfu, and one within the rift valley but not in the bottom has a value of 5.9 hfu. The writers previously interpreted the presence of these upturned edges as indicative of episodic spreading (Curray and Moore, 1974, p. 626), but we now suggest that the spreading rate may have been uniform, and that sediment deposition and ponding were episodic, coinciding with periods of Pleistocene lowered sea level.

Segments of the spreading centers and transforms in the southern part of the Andaman Sea are generally deep, narrow valleys where sediment fill is negligible, and no ponding and upturned edge effect are visible. The northern spreading center is particularly well constrained by our close-spaced survey lines, and the trend is sinuous in contrast with the rectilinear pattern of the southern Andaman Sea. The spreading center here lies along the eastern flank of Alcock Seamount, which is thinly sediment-veneered basement rock, described from one dredge haul as augite basalt (Rodolfo, 1969). In contrast, the eastern flank of the spreading center is made up of flat-ponded sediment as in the central Andaman Sea, again with upturned ridges which we can correlate between our survey lines. The fact that the seamount lies on one flank and ponded sediment with upturned edges lies on the other flank may be interpreted as indicating either asymmetric spreading, i.e., accretion of new crust on the southeast side only, or as evidence of uplift on one flank of the spreading center as Alcock Seamount and subsidence on the other flank. Data from our magnetic measurements favor the latter interpretation.

The sediment of the central basin of the Andaman Sea has been derived primarily from the prograding shelf at the north and appears to be the distal deposits of the Andaman deep-sea fan. These distal fan deposits have filled and buried the spreading center

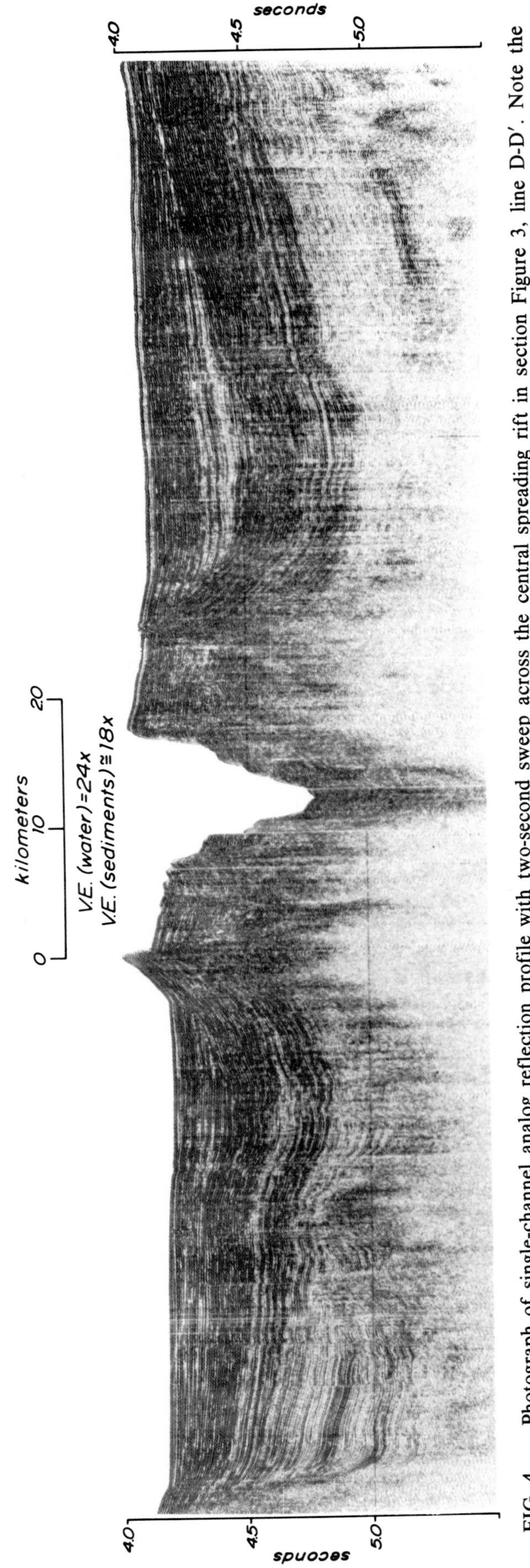

FIG. 4. Photograph of single-channel analog reflection profile with two-second sweep across the central spreading rift in section Figure 3, line D-D′. Note the older buried, upturned, and rifted sections of sediment.

in the entire northern Andaman Sea and down to approximately the western edge of the prominent central segment of the spreading rift at about 94°20′E long. The Andaman rift system is, therefore, a most uncommon combination of a fan valley and an active spreading rift. It has been alternately filled with sediments of the fan and rifted apart with accompanying tectonic disturbance of the sediments. Our seismic studies show that sediments near this rift system range in thickness from about 2.3 km in the wide northeast-trending central trough to at least 5.9 km beneath the filled northern part of the trough underlying the southern Gulf of Martaban. Typical oceanic second layer and crust underlie these sediments.

Where sedimentation has been rapid, the magnetic pattern is broad, of low-amplitude, and nondefinitive. This suggests that magnetic anomalies must be near-surface effects which can be easily destroyed by contemporaneous burial with sediments. Where sedimentation has been slow or negligible, clearly identifiable magnetic anomalies exist. The pattern along one line which crosses the central anomaly and extends northwest to the back edge of the Andaman-Nicobar Ridge (Section F, Fig. 2) is shown in Figure 5. Segments of other lines, including some over both Alcock and Sewell Seamounts, also show magnetic anomalies which may ultimately be identifiable. From this one line, we obtain a half spreading rate of 1.86 cm/yr, with the oldest identifiable magnetic evidence in anomaly 5 at about 10.8 m.y. From this, we may conclude that the entire central Andaman Basin has opened in slightly more than 10.8 m.y.

In the central segment of the spreading center (Figs. 3D, 4), the open central part of the rift valley is 5 km wide; the edges of the rift valley are 10 km apart; the most recent paired upturned edges are 24 km apart, and the oldest are approximately 65 km apart. At a total opening rate of 3.72 cm/yr, these would correspond to ages of 150,000 years, 270,000 years, 650,000 years, and 1.75 m.y. and would indicate the ending times of periods of rapid ponding of sediments. These correlate approximately with the ends of cool periods or low Pleistocene sea level stands (Mesolella et al, 1969; Hays et al, 1976). When deposition ceased with rise of sea level, spreading continued, and the process of intrusion of volcanics beneath the sediment pile resulted in doming, rifting, and separation of the edges. Thus, spreading may have been continuous, as suggested by the magnetic pattern, but ponding of sediments may have been rapid and episodic. If rapid deposition occurred in this particular locality during the last lowering of sea level, between 15,000 and 20,000 years B.P., it would have been terminated at approximately 10,000 B.P., corresponding to a rift width of only 375 m. The period of low sea level was short enough that the rift system or incised fan valley may have served only as a conduit through which the sediment load passed, to have been deposited at the distal end. Our survey control does not define this youngest low sea level deposit.

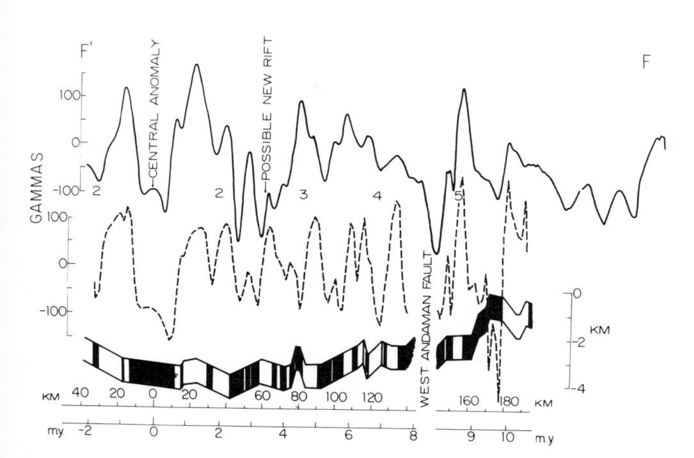

FIG. 5. Magnetic profile and synthetic magnetics from line F-F′ in the Andaman Sea. See Figure 2 for location.

This rift and transform system is strikingly similar to that in the Gulf of California (Moore, 1973), in its oblique opening of the seaway, and with high heat flow in the spreading rifts (Lawver et al, 1975). It appears to be similar in yet another respect in that short spreading segments are in the process of realigning, with an apparent tendency toward lengthening into fewer, longer spreading rifts and a simpler pattern, as is occurring in the Gulf of California (Sharman, 1976). Such a realignment is apparently occurring even in the ocean crust slice through which we have our identifiable magnetics. An incipient rift is now opening between anomalies 2 and 3 in line with the longer rift segment lying to the northeast. This is shown both in our reflection records and in seismicity. Similar shifts are apparently occurring farther to the south. In all of this central area, the otherwise dormant West Andaman Fault appears to be reactivated, with first motion studies indicating north-south, right lateral offset (Fitch, 1972).

The southernmost segment of the spreading-transform system immediately northwest of Sumatra is not well located by our seismic lines, and we are not certain how it relates to the Sumatran fault system, or whether it is an en echelon branch, curving around the coastline of Sumatra (Fig. 2). First-motion studies on shallow-focus earthquakes near the north coast of Sumatra suggest northwest-trending right-lateral motion, combined in some focal mechanisms with a component of underthrusting toward the south (Fitch, 1972). The possibility does exist, then, that some decoupling and underthrusting occurs, with the Mergui-North Sumatra Basin underthrusting northern Sumatra.

TECTONICS OF THE ANDAMAN SEA AND BURMA

The Andaman rift system appears to mark a plate edge between the larger China (Eurasian) plate on the east and a small elongate plate which we shall here call the Burma plate lying on the west (Fig. 6).

Separation is occuring in a northwest-southeast direction and is transformed southward in a complex way into the Sumatran right-lateral fault system and northward into a major north-south transcurrent fault trending through Burma. This fault in Burma is clearly shown by epicenters of major historic earthquakes (Chhibber, 1934), although it has been rather quiet in recent years (Tarr, 1974). This fault has been described in its central segment, where it controls the drainage pattern of the Irrawaddy River (Win Swe, 1972); and it can be traced clearly on ERTS-LANDSAT imagery from the eastern part of the Irrawaddy delta to at least 25°N lat., a distance of almost 1,000 km. At that latitude it may splay out near the northern end of the part of the lowlands that we postulate to be underlain by oceanic crust (Fig. 6).

Various interpretations have been made of the fault or faults lying between the eastern Burma highlands and the central lowlands. It was suggested as early as 1898 by Middlemiss that a major normal fault lay along this prominant scarp, downthrown to the west. Dey (1968) delineated the trace of a fault lying west of the scarp on aerial photos and suggested that this lineament was a major left-lateral strike-slip fault. Aung Khin et al (1970) presented the results of gravity surveys which also delineated the fault west of the scarp, and they interpreted the Shan scarp as a fault-line scarp. Win Swe (1972) recognized right-lateral offset features along the central segment of this fault near Mandalay and suggested the possibility of the existence of two separate faults, the normal fault of the Shan scarp and the strike-slip fault. Mitchell (1976) referred to the general fault zone as the Hninzee-Sagaing fault and attributed in excess of 300 km right-lateral displacement to it. It appears to us in examining ERTS-LANDSAT imagery that the strike-slip fault may be unrelated to the fault which formed the Shan scarp. Our reconstructions may require as much as 460 km right-lateral displacement along this major transform fault, which we will refer to as the Sagaing fault.

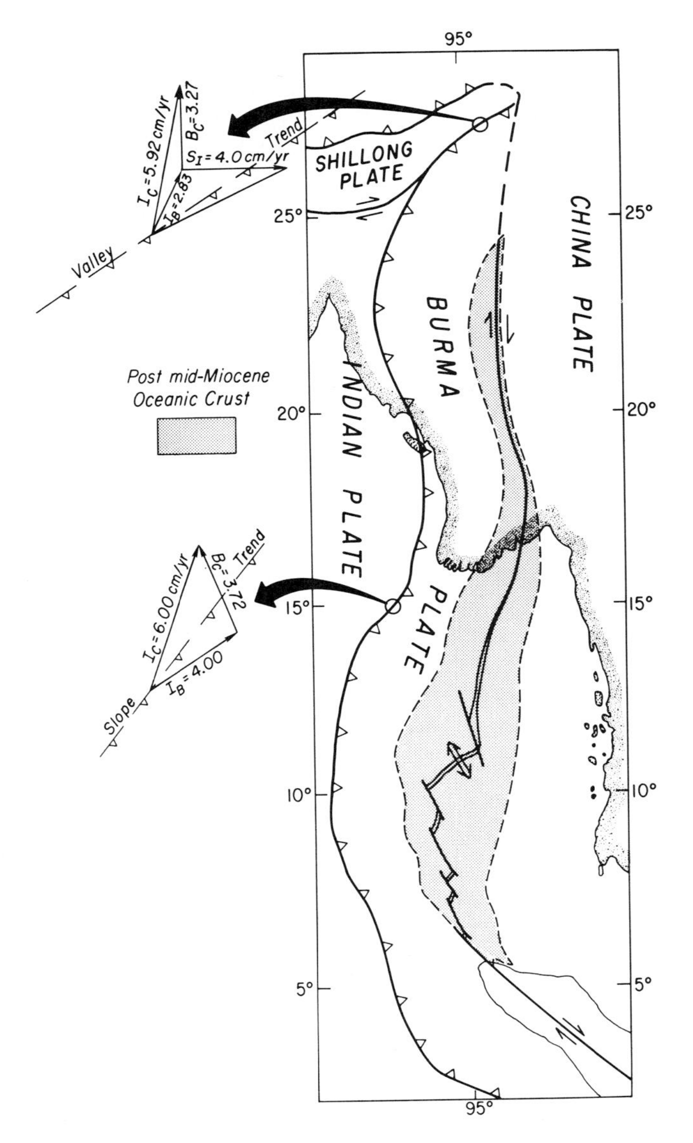

FIG. 6. Plates, plate edges, relative vectors, and post-middle Miocene oceanic crust in Burma and the Andaman Sea.

Seismicity of the Sunda Arc shows a well defined Benioff zone with shallow- , intermediate- , and deep-focus earthquakes beneath Java. Under Sumatra, there are shallow- and intermediate-focus but no deep-focus earthquakes. North from Sumatra, the seismicity pattern becomes ragged and appears to diverge into two lines (Fig. 2). Shallow- , and occasional intermediate-focus earthquakes delineate the subduction zone lying under the Andaman-Nicobar Islands and passing up through the Indo-Burman ranges. A distinct, separate lineation of shallow-focus earthquakes passes under the central basin of the Andaman Sea and indistinctly follows the line of our proposed transform fault in Burma toward the eastern Himalayan syntaxis. The southern termination of the small Burma plate could be in the vicinity of Krakatoa and the Sunda Strait between Sumatra and Java, and the northern termination lies between the Naga Hills, the Assam Valley, and the eastern Himalayan syntaxis.

A pole of rotation has been calculated on the assumption that the right lateral fault in Burma is a transform, that the spreading direction is indicated by the alignment of the central segment of the rift, and that the Sumatran fault system is a transform. This pole (George Sharman, personal commun., 1976) lies at 23.86°N lat., 125.12°E long., with a rate of rotation of 0.32°/m.y. Opening of the Andaman Sea, apparently commencing in middle Miocene time, sometime prior to 10.8 m.y. B.P., separated the Andaman-Nicobar Ridge northwestward from the edge of continental crust. The area of post-middle Miocene sea floor created (Figs. 2, 6) is therefore bounded by the foot of the continental slope and the Shan scarp at the edge of the eastern Burma highlands on the east, and the topographic slope at the back of the Andaman-Nicobar Ridge and the line of volcanoes separating the eastern from the western troughs, of the central Burma lowlands. This interpretation would then imply that the western trough of the central Burma lowlands is underlain by old offscraped sedimentary "mélange" of Cretaceous and Paleogene subduction now subsided and covered with molasse. This interpretation would also imply that the present magmatic ridge at the back of the Andaman-Nicobar Ridge and its northern continuation formerly lay adjacent to the continental slope of the Mergui Terrace and Shan scarp.

An alternative interpetation is that the Mergui Ridge represents a former magmatic arc of Cretaceous and Paleogene subduction. Mitchell and McKerrow (1975) have called attention to Jurassic and Lower Cretaceous deformed turbidites in the Kalaw Syncline on the west edge of the eastern Burma highlands, between about 20° and 22°N lat. These lie adjacent to Upper Cretaceous and lower Tertiary granitic rocks in a belt as much as 50 km wide. Hutchison (1975) classified some of the ultramafic rocks along this line to the north as true ophiolites. This interpretation would be compatible with first formation of the Cretaceous subduction zone adjacent to the edge of the eastern Burma highlands and continental crust of the Malay Peninsula, and formation of the Cretaceous and lower Tertiary magmatic arc along the line including the Mergui Ridge before Neogene-Quaternary opening of the Andaman Sea and the eastern trough of the Central Lowlands of Burma.

Our refraction data suggest that the Mergui-North Sumatra Basin is underlain by thinned continental crust. A north-south-trending block faulting pattern furthermore suggests a similarity to tensional faulting characteristic of rifting young continental margins. This basin may be an Oligocene or early Miocene extensional basin or a proto-Andaman Sea which aborted before drifting and formation of oceanic crust started. It was succeeded by the middle Miocene oblique extension of the Andaman Sea on the west side of Mergui Ridge.

Previously, two regions along the western Sunda Arc showed apparent contradiction between geological and geophysical observations. The first, along the arc at approximately 15°N lat. (Fig. 2), is where the vector of convergence between the India and Eurasian plates would have required pure transform motion, or even a component of separation. Our records from a rather detailed reflection survey of the trench and its landward slope in this locality (Fig. 1) instead show clear evidence of compression requiring a component of convergence. A vector solution for this locality at 15°N lat., using the Indian-Eurasian pole from Minster et al (1974), and the Burma-China (or Burma-Eurasian) relative motion derived from this study is shown in Figure 6. This indeed shows a component of convergence of at least 1.25 cm/yr where the trend of the subduction zone is inclined farthest to the northeast. Thus consideration of the proposed Burma plate and Andaman Sea spreading apparently resolves this contradiction, if we accept the Minster et al pole as being representative of Indian-China plate motion. This assumption will be discussed later.

The other area of apparent contradiction lies on the southeastern flank of the Assam Valley north of the Naga Hills. Geological field studies in this area (Evans, 1964; Brunnschweiler, 1966, 1974; Raju, 1968) have shown that the structure of the Naga Hills is a series of southeast-dipping thrusts, the Schuppen zone of Evans (Fig. 3A). Thus continental crust of the Assam Valley spur has apparently, since sometime in the Miocene, had some component of underthrusting or subduction beneath the Naga Hills. However, a vector solution (Fig. 6) does not resolve this contradiction, and suggests instead that a component of divergence should exist, rather than a component of convergence. Two possible explanations for this contradiction can be offered. First is that perhaps the Dauki Fault south of the Shillong plateau has had a sufficient right-lateral component to provide a component of convergence between the Assam Valley and the Naga Hills. A minimum of approximately 4 cm/yr right-lateral offset would be required, as indicated in Fig. 6.

The second possible explanation is that the Indian-Eurasian pole of rotation of Minster et al (1974) does not apply very well to this region. Minster and Jordan (in press) have subsequently slightly revised this pole position and rotation vectors, but not sufficiently to have much effect. The probability that Europe and Asia do not behave perfectly as a single plate is recognized by referring here to the China rather than Eurasian plate. Furthermore, Minster and Jordan (in press) and Stein and Okal (1978) have suggested that the Indian plate is probably not a single plate, but is instead divided into two plates approximately along the Ninetyeast Ridge.

Finally, despite the obvious and significant differences, we would like to point out some of the striking tectonic similarities between the Andaman Sea and the Gulf of California:

1. Both were formed as marginal extensional basins behind an elongate ridge lying adjacent to an offshore subduction zone.
2. Both show oblique shearing of the outer ridge away from the mainland as a result of interaction between two major lithospheric plates.
3. In contrast to other small ocean basins which are probably formed by extension, both the Andaman Sea and the Gulf of California lie on the east sides of oceans and the west sides of continents.
4. They have complex systems of rifts and transforms, with high heat flow and a tendency to simplify themselves into fewer and longer spreading rifts.
5. Where sedimentation rates are high, magnetic anomalies of the sea floor are subdued, but anomalies are identifiable where sedimentation is slower.
6. Both have mainland margins which are behaving like youthful passive continental margins, with block faulting and high rates of subsidence. In contrast with some passive margins created from rifting of normal continental crust, neither showed a prerift stage of broad regional doming or uplift.
7. Both are excellent examples of geosynclines filling longitudinally (Kuenen, 1957).

REFERENCES CITED

Aung Khin and Kyaw Win, 1969, Geology and hydrocarbon prospects of the Burma Tertiary geosyncline: Union Burma Jour. Sci. Technol. v. 2, p. 53–81.

——— et al, 1970, A study on the gravity indication of the Shan scarp fault: Union Burma Jour. Sci. and Technol., v. 3, p. 91–113.

Brunnschweiler, R. O., 1966, On the geology of the Indoburman ranges: Geol. Soc. Australia Jour., v. 13, p. 127–194.

——— 1974, Indoburman ranges, from Mesozoic-Cenozoic orogenic belts, *in* A. M. Spencer, ed.: Geol. Soc. London Spec. Pub. No. 4, p. 279–299.

Burns, R. E., 1964, Sea bottom heat-flow measurements in the Andaman Sea: Jour. Geophys. Research, v. 69, p. 4918–4919.

Chhibber, H. L., 1934, Geology of Burma: London, Macmillan, 538 p.

Curray, J. R., and D. G. Moore, 1971, Growth of the Bengal deep-sea fan and denudation in the Himalayas: Geol. Soc. America Bull., v. 82, p. 563–572.

——— ——— 1974, Sedimentary and tectonic processes in the Bengal deep sea fan and geosyncline, *in* C. A. Burk and C. L. Drake, eds., The geology of continental margins: New York, Springer-Verlag, p. 617–627.

Dey, B. P., 1968, Aerial photo interpretation of a major lineament in the Tamethin–Pyawbwe quadrangle: Union Burma Jour. Sci. and Technol., v. 1, p. 431–443.

Evans, P., 1964, The tectonic framework of Assam: Geol. Soc. India Jour., v. 5, p. 80–96.

Fitch, T. J., 1972, Plate convergence, transcurrent faults, and internal deformation adjacent to Southeast Asia and the western Pacific: Jour. Geophys. Research, v. 77, p. 4432–4462.

Frerichs, W. E., 1971, Paleobathymetric trends of Neogene foraminiferal assemblages and sea floor tectonism in the Andaman Sea area: Marine Geology, v. 11, p. 159–173.

Geological Survey of India, 1944, The Andaman and Nicobar archipelagoes: Strategic Branch, Geol. Survey of India, Tech. Note No. 16.

Hays, J. D., J. Imbrie, and N. J. Shackleton, 1976, Variations in the earth's orbit—pacemaker of the ice ages: Science, v. 194, no. 4270, p. 1121–1132.

Hutchison, C. S., 1975, Ophiolite in Southeast Asia: Geol. Soc. America Bull., v. 86, p. 797–806.

Karunakaran, C., K. K. Ray, and S. S. Saha, 1968, Tertiary sedimentation in the Andaman-Nicobar geosyncline: Jour. Geol. Soc. India, v. 9, p. 32–39.

——— et al, 1975, Geology of Great Nicobar Island: Jour. Geol. Soc. India, v. 16, no. 2, p. 135–142.

Kuenen, P. H., 1957, Longitudinal filling of oblong sedimentary basins: Kon. Ned. Mijnbouwk Gen. Verhand, v. 18, p. 189–196.

Lawver, L. A., D. L. Williams, and R. P. von Herzen, 1975, A major geothermal anomaly in the Gulf of California: Nature, v. 257, no. 5521, p. 23–28.

Mesolella, K. J., et al, 1969, The astronomical theory of climatic change—Barbados data: Jour. Geology, v. 77, p. 250–274.

Middlemiss, C. S., 1898, Report on the Bengal earthquake of 14th July, 1885: Rec. Geol. Survey India, v. 4.

Minster, J. B. and T. H. Jordan, in press, Present-day plate motions: Jour. Geophys. Research

——— et al, 1974, Numerical modelling of instantaneous plate tectonics: Royal Astron. Soc. Geophys. Jour., v. 36, p. 541–576.

Mitchell, A. H. G., 1976, Southeast Asian tin granites: magmatism and mineralization in subduction- and collision-related setting: U.N., CCOP Newsletter, v. 3, nos. 1 and 2, p. 10–19.

——— and W. S. McKerrow, 1975, Analogous evolution of the Burma orogen and the Scottish Caledonides: Geol. Soc. America Bull., v. 86, p. 305–315.

Moore, D. G., 1973, Plate-edge deformation and crustal growth, Gulf of California structural province: Geol. Soc. America Bull., v. 84, p. 1883–1906.

——— J. R. Curray, and F. J. Emmel, 1976, Large submarine slide (olistostrome) associated with Sunda Arc subduction zone, northeast Indian Ocean: Marine Geology, v. 21, p. 211–226.

——— et al, 1974, Stratigraphic-seismic section correlations and implications to Bengal Fan history, *in* C. C. von der Borch, J. G. Sclater, et al, eds., Initial reports of the Deep Sea Drilling Project: Washington, D. C., U. S. Govt., v. 22, p. 403–412.

Peter, G., L. A. Weeks, and R. E. Burns, 1966, A reconnaissance geophysical survey in the Andaman Sea and across the Andaman-Nicobar Island area: Jour. Geophys. Research, v. 71, p. 495–509.

Raju, A. T. R., 1968, Geological evolution of Assam and Cambay Tertiary basins of India: AAPG Bull., v. 52, p. 2422–2437.

Rodolfo, K. S., 1969, Bathymetry and marine geology of the Andaman Basin and tectonic implications for Southeast Asia: Geol. Soc. America Bull., v. 80, p. 1203–1230.

Sharman, G. F., 1976, The plate tectonic evolution of the Gulf of California: Scripps Inst. Oceanography, Univ. California, San Diego, unpubl. Ph.D. thesis, 160 p.

Stein, S., and E. A. Okal, 1978, Seismicity and tectonics of the Nintyeast Ridge area: evidence for internal deformation of the Indian Plate: Jour. Geophys. Research, v. 83, p. 2233–2245.

Tarr, A. C., 1974, World seismicity map: Washington, D. C., U.S. Geol. Survey.

Weeks, L. A., R. N. Harbison, and G. Peter, 1967, Island arc system in the Andaman Sea: AAPG Bull., v. 51. p. 1803–1815.

Win Swe, 1972, Strike-slip faulting in Central belt of Burma (abs.), *in* N. S. Haile, ed., Regional conference on the geology of Southeast Asia: Kuala Lumpur, Geol. Soc. Malaysia, Annex to Newsletter No. 34, p. 59.

Extensional Tectonics in the Okinawa Trough[1]

BRUCE M. HERMAN, ROGER N. ANDERSON AND MAREK TRUCHAN[2]

Abstract The Okinawa Trough is a marginal sea that opened when the Ryukyu Arc was rifted away from the Asian mainland. Geophysical surveys in the southwestern portion of the Okinawa Trough have delineated two grabens formed in the basement and overlying 2.5 seconds of sediment. One graben extends along the trend of the basin from 124°E to 125°E longitude, to the east and west of which the well-developed graben structure is replaced by a more dispersed pattern of normal faults. About 40 km north of this broadening of the fault pattern, a second graben is found which continues to the northeast where it, too, is replaced by basin-wide normal faulting.

The sediments in the basin can be divided into two formations. Widespread deformation by folding and faulting is found in the lower unit, whereas the unfolded upper unit is primarily deformed by the faulting associated with the graben. A tentative scheme for the evolution of the graben would have the lower unit deposited in the basin while extension was active, with the upper unit deposited after the main phase of spreading had ceased and as a result, subject only to limited deformation. Faulting in the grabens is presently active although extension is occurring at a very slow rate. These faults have offsets which in many cases increase with depth to at least the bottom of the upper unit. The dip of the bedding beneath the graben also increases with depth, indicating that incipient spreading has been occurring since the beginning of the deposition of the upper unit (at rates on the order of a few mm/yr).

An east-west trending ridge between the two grabens was dredge-sampled and found to be composed of biotite-rich quartz diorite and metamorphosed pillow basalts. It is thought to be a sliver of the Ryukyu Arc stranded in the basin during the rifting event which formed the Okinawa Trough.

In spite of the extremely slow spreading rate in the Okinawa Trough, two new heat flow measurements of 126 and 443 mW/sq m, in addition to previously reported values, show the heat flow to be high in the basin.

INTRODUCTION

Crustal extension has been proposed to explain the morphology and sediment distribution in the marginal basins (Karig, 1970; 1971), as well as the similarity of the prevolcanic lithologies found on either side of the marginal basins (e.g. Dietz, 1954; Murauchi and Den, 1966). However, the mechanism by which extension occurs is poorly defined. Data required to study the history of spreading in the back-arc basin include (a) magnetic anomalies which could date the oceanic basement, and/or (b) seismic profiler records showing the history of deposition and deformation in the sediment above the spreading basin. Few marginal basins have a well developed marine magnetic anomaly pattern associated with back-arc spreading, possibly because the spreading does not originate at a simple mid-ocean ridge spreading center, or because the plates involved are non-rigid. The Okinawa Trough magnetic anomaly pattern appears to be dominated by shallow igneous intrusions (Ishihara and Murakami, 1976; Miyazaki et al, 1976), and therefore does not have an identifiable seafloor spreading magnetic anomaly pattern. It does have a sufficient thickness of sediments to record the evolution of spreading within the basin. We use these data to interpret the history of extension in the Okinawa Trough.

The Okinawa Trough is the marginal basin behind

[1]Manuscript received, May 26, 1977; accepted, February 7, 1978.

[2]Lamont-Doherty Geological Observatory of Columbia University, Palisades, New York 10964.

We would like to thank the officers and crews of the R/V *Vema* and R/V *Conrad* who collected the magnificent profiles shown in this paper, and in particular Wm. Ludwig, C. Windisch, and D. E. Hayes, who were the chief scientists in the earlier L-DGO cruises over the Okinawa Trough.

We are grateful to J. I. Ewing, D. E. Hayes, and J. E. Nafe for critically reviewing this manuscript. E. Honza generously provided a prepublication copy of the excellent Geological Society of Japan publication, 20–22, *Ryukyu Island (Nansei-Shoto) Arc, GH 75-1 and GH75-5 Cruises.*

This work was supported by the Office of Naval Research under contract N000-14-75-C-0210 and by the National Science Foundation through grant DES 74-24112.

Lamont-Doherty Geological Observatory Contribution No. 2731.

Article Identification Number:
0065-731X/78/MO29-0013/$03.00/0.
(see copyright notice, front of book)

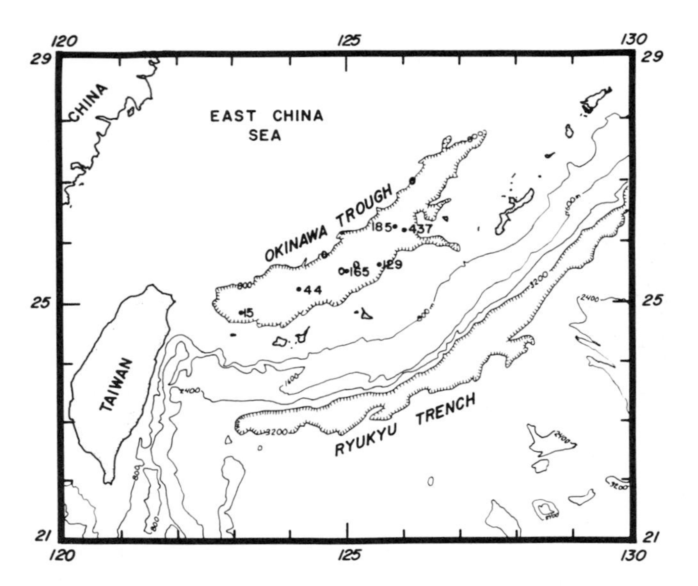

FIG. 1. Index map of the area around the Okinawa Trough. Solid dots are heat flow values in mWm^{-2}. Circles are dredge-sample locations.

the Ryukyu Island Arc; it is bounded by the arc to the south and east and by the continental margin beneath the East China Sea to the north and west (Fig. 1). Compared to other marginal seas, the Okinawa Trough is shallow—less than 2200 m deep. It also has a very thick deposit of sediments (up to 2.5 sec, two-way travel time).

Wageman et al, (1970) provided the first verification of a tectonic origin of the Okinawa Trough. They found evidence in seismic data of recent vertical offsets in the walls of the basin and concluded that the Okinawa Trough had a fault origin. High heat flow (124.4 ± 63.2 mW/sq m) in the basin was reported in the same year by Yasui et al (1970). Because the best seismic refraction work available at that time reported seismic velocities similar to that of continental crust beneath the Okinawa Trough (Murauchi, et al., 1968), Yasui et al (1970) interpreted the high but scattered heat flow to result from shallow igneous intrusions into continental crust. Additional seismic data showing recent extensional faulting in the Okinawa Trough were presented by Tamaki et al (1976). Shallow seismicity in the basin (Katsumata and Sykes, 1969) further substantiated the recent tectonic activity. Murakami (1976) concluded that the gravity data over the basin indicated a thin crust. Ishihara and Murakami (1976) extended the interpretation by stating that the high Bouguer gravity anomaly over the southwestern Okinawa Trough was not inconsistent with the formation of new oceanic crust in this portion of the Okinawa Trough.

The above data were generally regarded as evidence for crustal extension in the Okinawa Trough. Seismic reflection and high-resolution 3.5-kHz Precision Depth Recorder (PDR) data obtained by Lamont-Doherty Geological Observatory (L-DGO) ships during the fall and winter of 1976 in the southwestern portion of the basin provide graphic evidence for extensional tectonics in the Okinawa Trough: the presence of a well-developed graben, of classic form, extending along the axis of the basin floor. We interpret this, along with the other available data, as evidence that the Okinawa Trough opened by back-arc spreading.

SEISMIC REFLECTION DATA

Selected seismic profiler data from a series of tracks over the Okinawa Trough (Fig. 2) are shown in Figure 3; the individual sections are oriented with the side toward the island arc to the right, and are presented from west to east. Because the records obtained on the R/V *Vema* (profiles with prefix "V") were made while the ship was traveling at a speed of 8 to 10 knots and those of the R/V *Robert D. Conrad* (profiles with the prefix "C") were made while traveling at 5 to 6 knots, the horizontal scales on the respective profiler records is not the same. The vertical scale, in two-way travel time, is the same, so that the vertical exaggeration for the two is different.

Below each profiler record is an interpretive line drawing, based not only on the profiler record, but also on the PDR data. The acoustic layering of the sediment is obscured to varying degrees by the complex bubble pulse of the airgun sound source of the seismic profiler system and also by internal reflections within the sediment layers. In order to determine which acoustic horizons were most likely to represent actual sedimentary layers, the data were recorded on four channels, each with a different frequency bandwidth, and these were then compared before deciding which horizons were significant. (This method does not completely eliminate subjectivity, but one hopes that it limits it to some

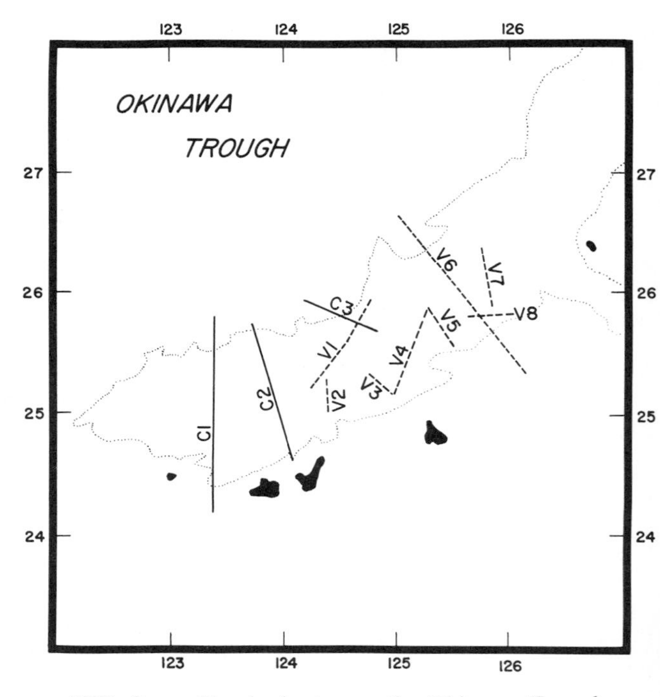

FIG. 2. Track chart over the Okinawa Trough.

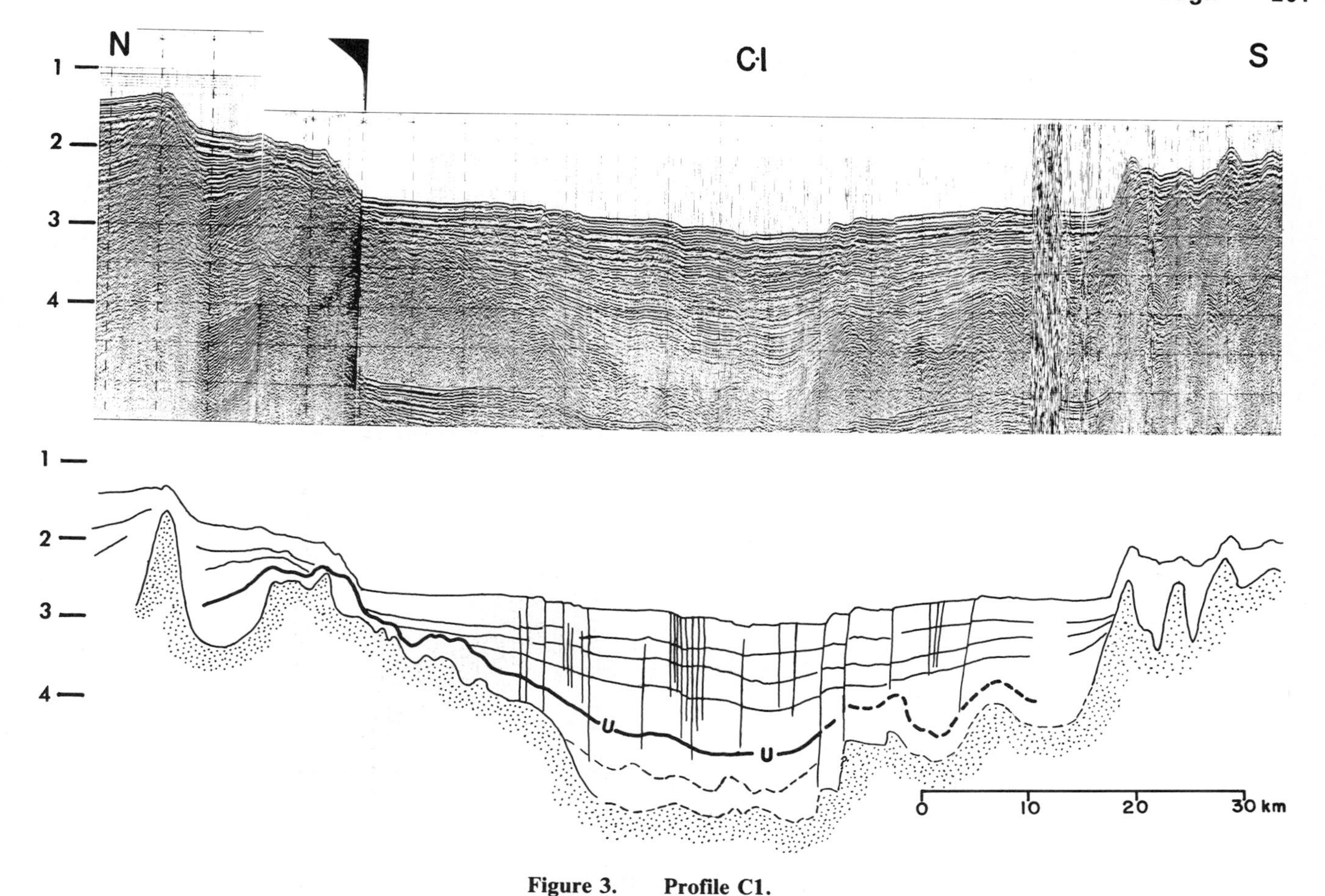

Figure 3. Profile C1.

FIG. 3. Seismic profiler records from the Okinawa Trough, with the associated line drawings below each record. Profiles obtained on the R/V *Conrad* have a prefix "C," and those from the R/V *Vema* have a "V" before the line number. Because the speed of the ships varied during the cruises, horizontal scale is indicated on the drawing for each line. Vertical scale is in seconds of two-way travel time. The tracks are shown in order from west to east, and the lines are oriented so that the continental margin is to the left and the Ryukyu Arc is to the right. Line V8 runs west to east.

extent). Faults were drawn on the basis of offsets in the acoustic layering and, where the offsets were small, on the basis of the diffraction pattern resulting from the interaction of the sonic energy and the edge of the layering at the fault scarp (e.g., Dobrin, 1976, p. 224).

The sediments of the Okinawa Trough are acoustically well-stratified, but the degree of layering changes with depth (Fig. 3). In a number of the cross-sections, two formations can be distinguished on the basis of differing degrees of deformation. The upper unit, which we call Unit A, is more highly stratified and less deformed than the underlying Unit B. An unconformity separates the two units.

Unit A

Unit A appears to be deformed only by faulting, and in large areas there is essentially no deformation (e.g., VI and C3). Cores taken by us and by Honza et al (1976) in the southwestern Okinawa Trough yielded intercalated turbidites and clays. As the PDR data show the uppermost sediments of Unit A to be acoustically highly stratified and the seismic records show that the entire section of Unit A is highly stratified, it is probable that this unit is composed entirely of turbidites and clays. The thickness of Unit A increases by more than 50 percent from the edges of the basin toward the well-defined graben in the central portion of the trough (C1, C2, V1 and V8). This increase is essentially uniform, indicating that the depositional axis of the basin has remained fixed during the period these sediments were deposited.

Unconformity

An unconformity separates Unit A from Unit B and is represented in the line drawings by a heavy solid line marked "U" (C3, V1, V6 and V8). As the lower unit appears to have been structurally deformed prior to the deposition of the upper unit, the contact is an angular unconformity, although it is possible for the bedding in the two units to be locally parallel. Not all sections show the unconformity well (e.g., C1 and C2). In these, the unconformity was drawn along the upper surface of beds which appear to be truncated. Perhaps the relatively low frequency sound source used to obtain these records could not resolve the deformation in the lower unit,

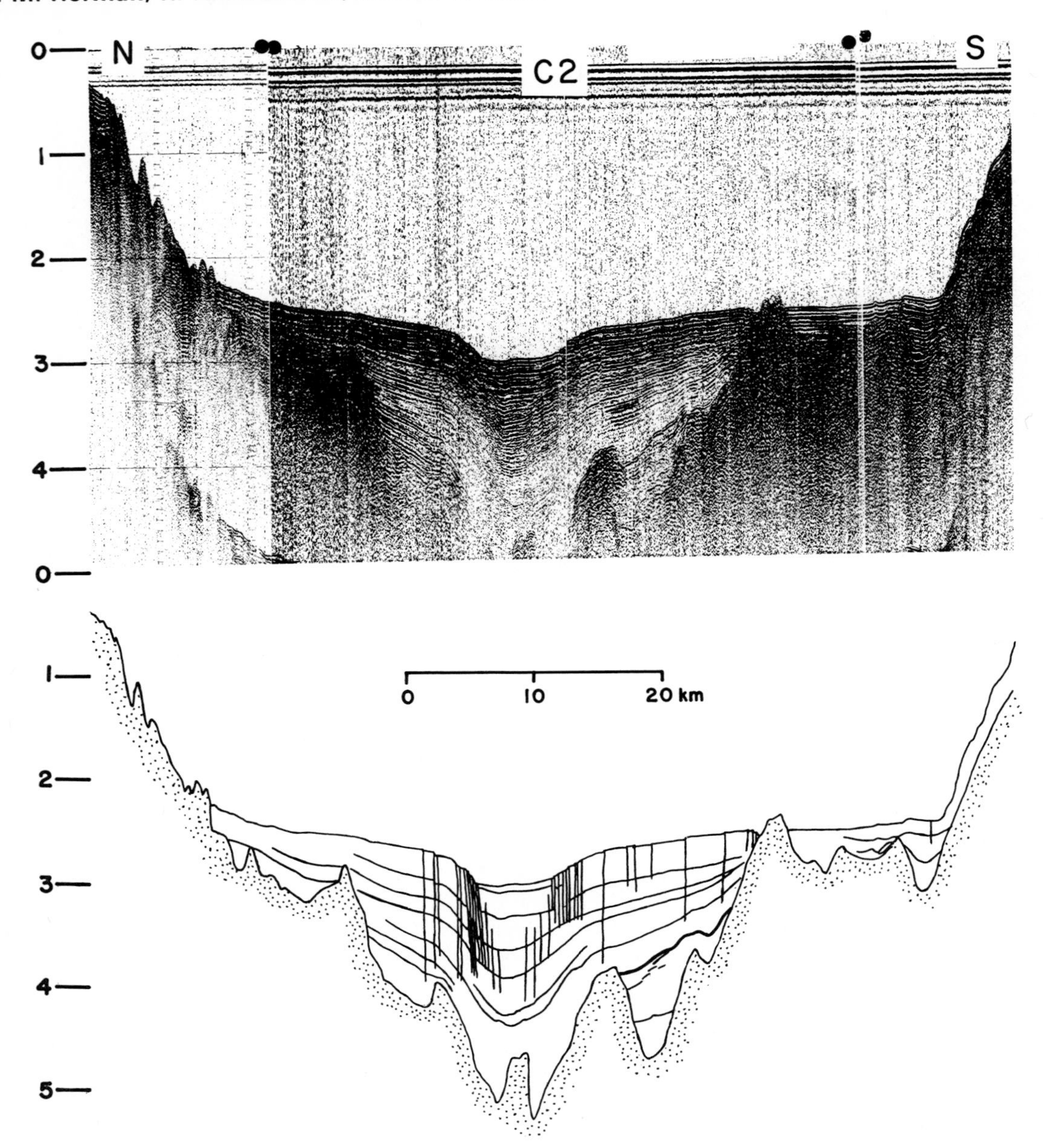

Figure 3. Profile C2.

but the possibility exists that the nature of the unconformity changes in the westernmost part of the basin (C2 is about 60 km from the nearest positive identification of the unconformity). An angular unconformity has also been reported on the continental shelf by Wageman et al (1970) and Leyden et al (1973). This unconformity has been dated as Miocene-Pliocene (unpublished well information from the East China Sea). Although this unconformity is between a lower, structurally deformed unit, and an upper, undeformed unit, the relationship between it and the one in the Okinawa Trough remains obscure, as we cannot trace the Okinawa Trough unconformity across the northern margin of the basin.

Unit B

Below the unconformity, the degree of stratification seen in the higher frequency records generally decreases (V1, V6 and V8). The greater intensity of deformation in the lower unit may sufficiently disrupt the layering in the sediments so as to obscure the minor reflections. Alternatively, Unit B may have more massive bedding than Unit A.

The lowermost section of Unit B, where it can be seen, is often so deformed that it is difficult to pick the contact between the sediment and the underlying crystalline rock. Increased compaction of the sediments with depth may increase their seismic velocity to nearly that of basalt, but the lack of a discernible sediment-basalt contact may also be due to metamorphism of the sediments. The mean heat flow in the Okinawa Trough is ~170 mW/sq m (see Appendix) and if a typical conductivity for turbidite sediments is used, 1.3 to 1.7 W/m·°K, then the temperature gradient is ~100 to 130°C/km. Preliminary sonobuoy refraction results show the average sedi-

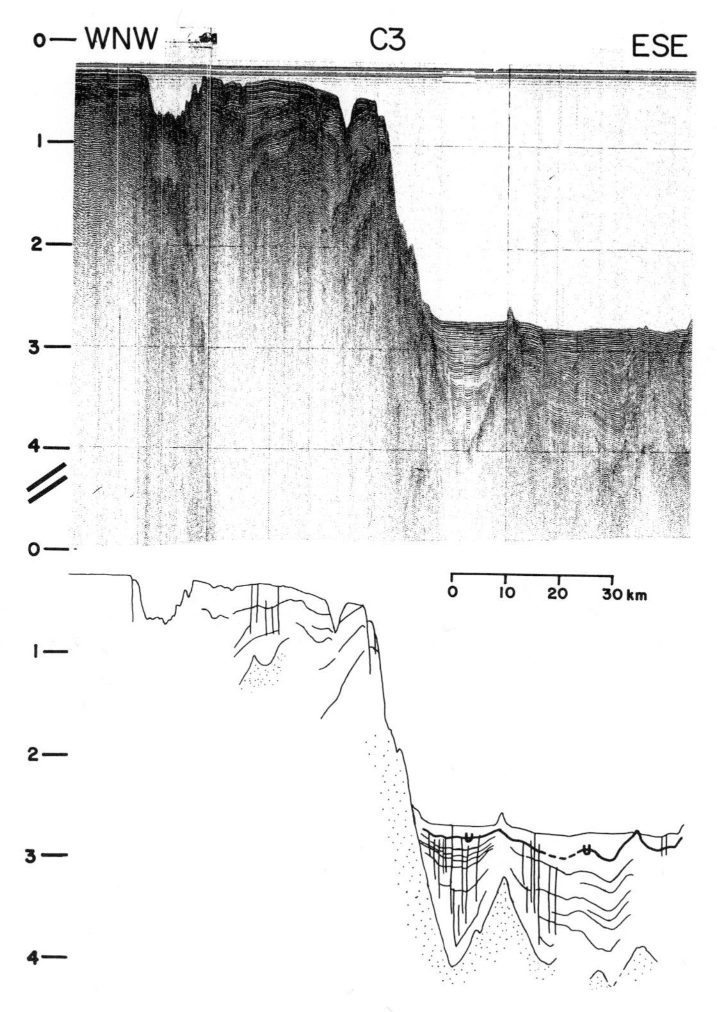

Figure 3. Profile C3.

ment thickness in the basin to be approximately 3 km, so the temperature at the sediment-basalt interface is 300 to 400°C. As the pressure at this depth is ~1 kb, metamorphism would most likely be of the albite-epidote hornfels facies (Turner, 1968). Contact metamorphism, directly from intrusion of basalt into the sediment, is also likely. The piercement structure in the graben floor shown in profile V3 was dredge-sampled and found to be composed of relatively fresh basalt, suggesting that many of the piercement structures seen below the graben floor (C2, V1, V3, and V8) are basalt intrusions. The high scatter in the heat-flow data also might indicate shallow igneous intrusions (Yasui et al, 1970).

FAULTING

Recent normal faults subparallel to the trend of the basin have been found to occur along nearly the entire length of the Okinawa Trough (Wageman et al, 1970; Honza et al, 1976). In the southwestern portion of the Okinawa Trough, the faulting is associated with an east-west trending graben. This graben extends less than 300 km and is quite irregular in its size and shape, as can be seen by comparing the various cross-sections in Figure 3. Previous workers in this area reported only a narrow valley on the floor of the basin, which they thought to be the result of erosion by turbidity flows (Wageman et al, 1970; Honza et al, 1976). Possibly the sound sources used by these workers operated at too low a frequency to resolve the faults in the graben walls.

The form of the graben is that of a valley, the walls of which are step-faulted as a result of offsets along near-vertical normal faults. In the central portion of the L-DGO survey area (about 125.5°E) the graben is well developed (V1), but 70 km to the east (V4) or west (C1) the basin contains only depressed areas with widespread normal faulting. A second graben develops about 40 km north of the depressed area at the eastern end of the first graben, just north of a prominent east-west-trending basement ridge (V4 and V5). The northern graben continues eastward to about 125°45′ where it is replaced by a more irregular, basinwide complex of normal faults (V7). In such areas, it is not possible to determine if a specific axis of extension exists. However, along most of the southwestern Okinawa Trough, the pattern of faulting indicates that recent extension is tak-

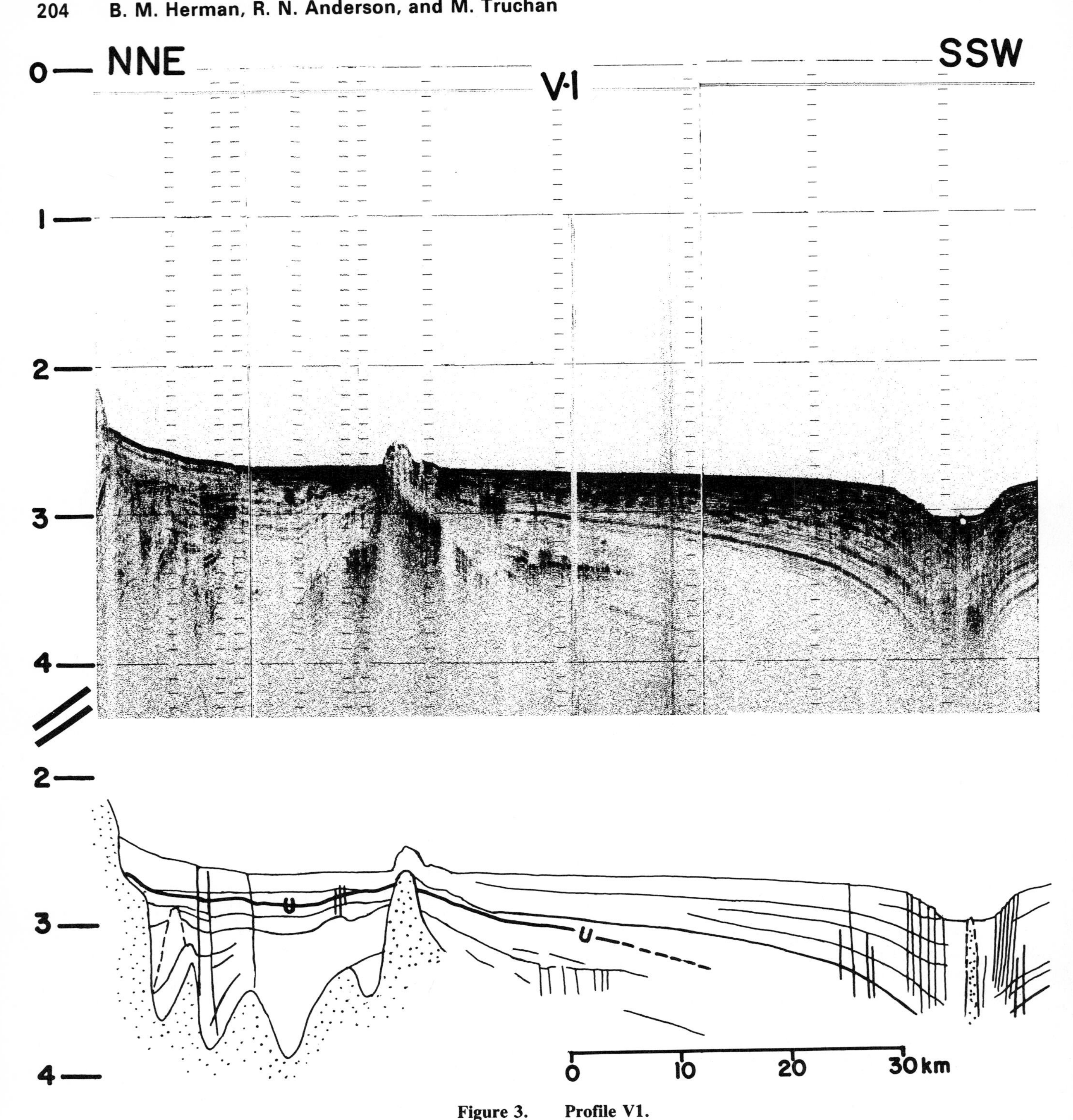

Figure 3. Profile V1.

ing place along a fairly well-defined line 15 km wide or less.

Most vertical offsets along the faults in the Okinawa Trough extend from the surface sediments through the deepest observable reflections. The vertical displacement averages about 20 m, and reaches as much as 40 m. On faults where the beds are visibly offset, the amount of offset does not generally appear to increase with depth (C2, C3, V4 and V5). Because of compaction in the sediment column, the speed of sound will increase with depth in the sediments, and as a consequence of this, the true offset along the faults must increase with depth. Based upon preliminary sonobuoy reductions, Unit A has a velocity of about 1.9 km/sec and a velocity gradient of about 1 km/sec per second of one-way travel time. Over the upper one second of sediment (two-way travel time), the seismic velocity increases about 25 percent, so that a fault with a 20-m offset at the sea floor would have a 25-m offset one sec-

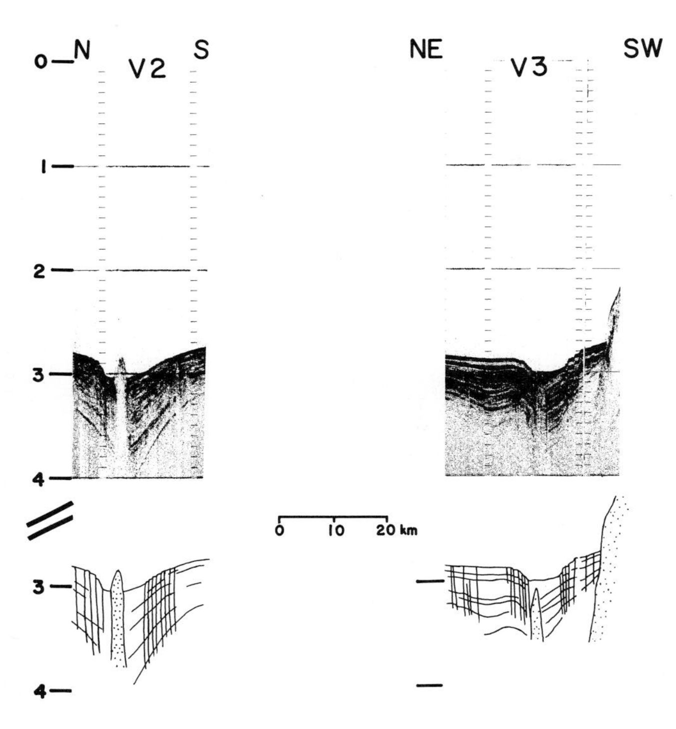

Figure 3. Profiles V2 and V3.

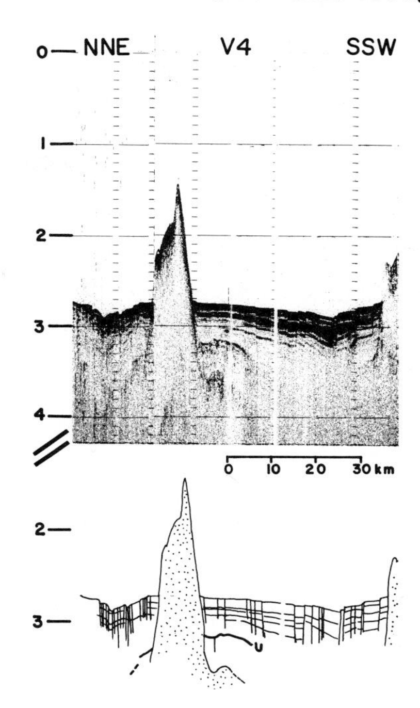

Figure 3. Profile V4.

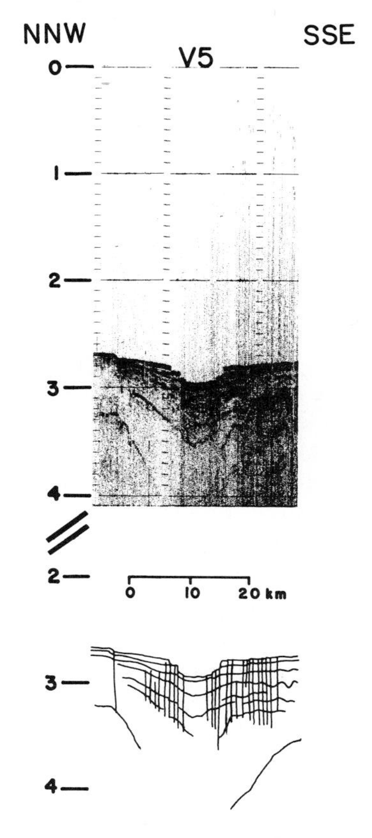

Figure 3. Profile V5.

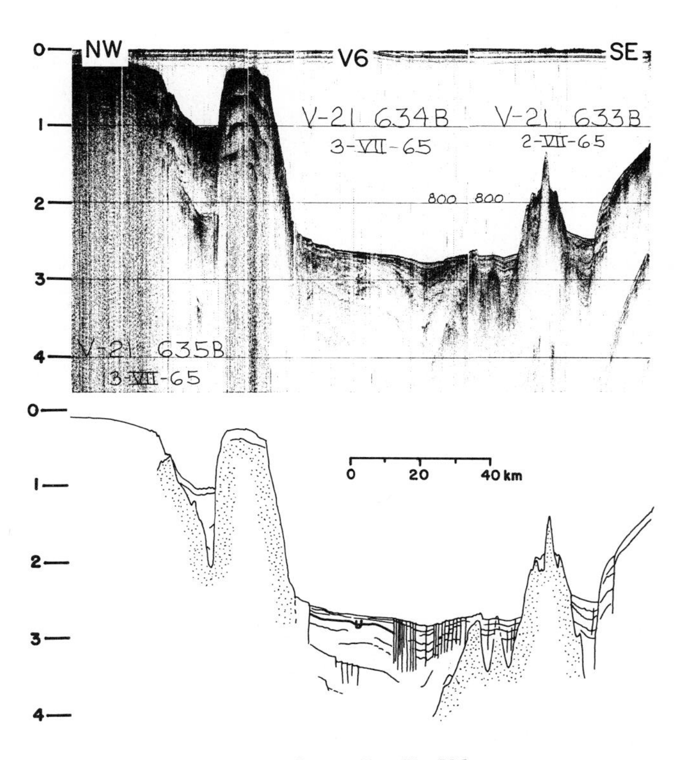

Figure 3. Profile V6.

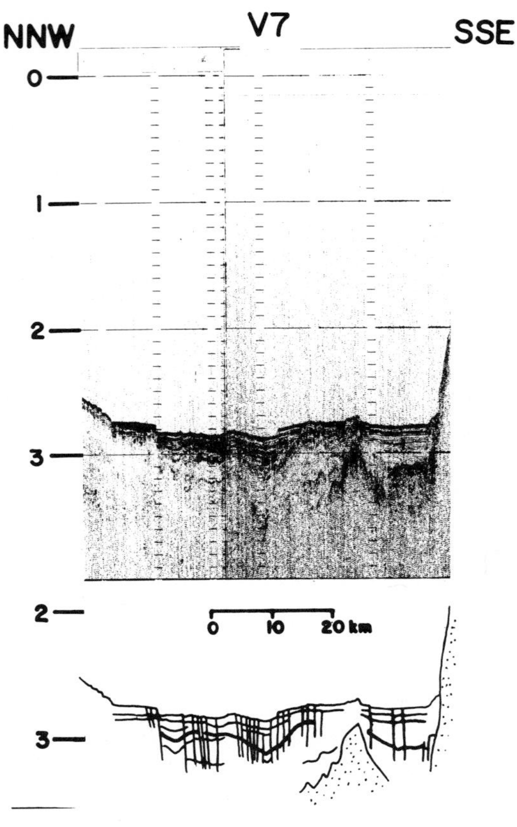

Figure 3. Profile V7.

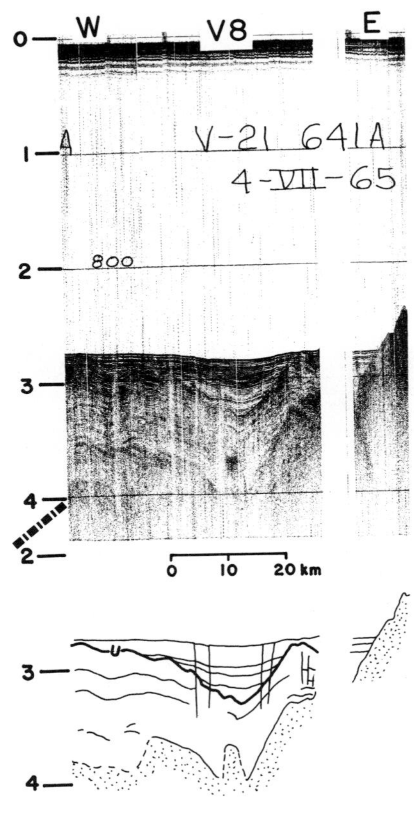

Figure 3. Profile V8.

ond (two-way travel time) below the sea floor. Because of the small offsets on many of the faults and the complex reflection pattern from intensely faulted areas such as beneath the graben walls, it is not possible to observe if most of the faults associated with the graben are growth faults. Indirect evidence for slowly ongoing deformation during the deposition of Unit A may be provided by the increase of the bedding dip with depth (C2, and V1), which would indicate that the deeper sediment has been down-faulted more than the shallow sediment.

The growth faults in the graben walls and the increasing dip of the bedding beneath the graben indicate that in the southwestern Okinawa Trough, the graben has been in existence at least since early in the period of deposition of Unit A. Unit B is rarely seen beneath the graben, so it is not possible to determine if a graben had been present during the time it was deposited.

During the deposition of Unit A very little spreading has taken place, which can be documented in two ways. First, direct measurement on the seismic records of the dip and displacements on the faults associated with the graben indicates the total extension since the deposition of Unit A to be no more than about 300 m. Also, if there had been significant extension, Unit A would have been deformed at depths beyond the graben walls as the spreading process transported that part of Unit A away from the spreading center. Although there is a spreading center in the southwestern Okinawa Trough, it is probably best described as incipient: its axis is well defined and the spreading is ongoing, but it is very slow (a few mm/yr).

EARLY SPREADING IN THE OKINAWA TROUGH

Data are sparse concerning the initial opening of the Okinawa Trough. Wageman and others (1970) related the Neogene faulting found in Kyushu to the opening of the Okinawa Trough. Citing seismic evidence, Takahashi (1975) stated that the Okinawa Trough opened during the Pliocene. There are no published deep drill hole data in the basin, from which a reliable age of the oldest sediments could be obtained; thus, the discussion in this section, as that of the previous one, will be primarily within a seismic stratigraphic framework.

In the previous section, evidence was presented which indicated that the basin had opened to almost its present-day width before the deposition of Unit A began. Unit B is moderately deformed and was probably deposited in the basin before the major period of spreading ceased, which would account for its more severe deformation. All profiles in which Unit B can be observed show the lower formation to be deformed, and it appears that this occurred over most of the width of the basin. The unconformity between Units A and B may correlate with a change in the spreading process from one that was basin-wide to one primarily confined to the grabens which are somewhat south of the center of the basin. We postulate that Unit A was deposited over the irregular surface of Unit B, and that only minor faulting of the upper unit has occurred in areas away from the graben (V1, C3 and V8).

Further evidence of a more disorganized pattern of spreading during the initial opening of the Okinawa Trough may be inferred from the presence of a large ridge in the center of the basin from which biotite-rich quartz diorite and metamorphosed pillow basalts were dredged (V4)(L-DGO and Honza et al, 1976). Quartz diorite and pillow basalts are typical of the crustal section of a magmatic arc (Hamilton and Meyers, 1967) and we propose that the ridge is a stranded piece of the batholith and overlying volcanics which were associated with the Ryukyu Arc prior to the time it was rifted from the Asian continent. Honza et al (1976) identified sediments dredged from this ridge and also from the walls of the Okinawa Trough as upper Miocene-lower Pliocene. Similar pieces of stranded continental crust have been reported between Sardinia and Italy in the Tyrrhenian Sea by Boccaletti and Guazzone (1974) and Heezen et al (1971). Possible examples of slivers of crust only partially rifted from the continental margin can be seen on lines C3 and V6.

CONCLUSIONS

The geologic section of the Okinawa Trough has been discussed from the most recent to the oldest events, because resolution of the data, which constrain our interpretation of these events, decreases with depth in the basin. Here the evolution of the Okinawa Trough is summarized chronologically.

Subduction of oceanic lithosphere beneath the eastern Asian continental margin began in the early Tertiary (Bowin and Reynolds, 1975). Back-arc spreading was initiated some time after late Miocene-early Pliocene time and the Ryukyu Arc was rifted from the continental margin. At least two slivers of crust were only partially rifted from the continental margin (C3 and V6), and a third was completely rifted away but then stranded in the middle of the basin (V4).

Turbidites deposited in the basin were deformed by folding and faulting as the basin continued to open. A change in the style and rate of spreading occurred at the time corresponding to the unconformity between Units B and A. Extension in the southwestern Okinawa Trough slowed to only a few mm/yr and was confined to a fairly well defined line source. The resulting graben has been maintained by growth faulting from that time through the present. Unit A is less deformed than Unit B because it was deposited at a time of incipient extension. Intrusive and extrusive events have accompanied the incipient extension.

There are several unsolved problems related to the Okinawa Trough. Particularly important is the timing of the various episodes outlined above, which might be resolved if the unconformity could be sampled by drilling. The graben is only well developed in the southwestern Okinawa Trough. Perhaps this is related to the direction of subduction of the Philippine plate beneath the island arc. Katsumata and Sykes (1969) hypothesized that the relative motion between the Ryukyu Arc and the Philippine Plate is north-south, on the basis of strike-slip faulting and first-motion studies along the Taiwan-Luzon region. A related problem concerns the fate of the back-arc spreading system as it approaches Taiwan, where it can be traced ashore on the Ilan Plain (Lee and Lu, 1977). In any event, a great deal promises to be learned about the early stages of opening of a marginal basin from further investigations of the Okinawa Trough. What can be conclusively stated is that the mode of extension varies from southwest to northeast along the trough, from extension within a well-defined zone less than 15 km wide to basinwide normal faulting.

REFERENCES CITED

Boccaletti, M., and G. Guazzone, 1974, Remnant arcs and marginal basins in the Cainozoic development of the Mediterranean: Nature, v. 252, p. 18–21.

Bowin, C., and P. H. Reynolds, 1975, Radiometric ages from Ryukyu Arc region and an $^{40}Ar/^{39}Ar$ age from biotite dacite on Okinawa: Earth and Planetary Sci. Letters., v. 27, p. 363–370.

Dietz, R. S., 1954, Marine geology of northwestern Pacific: description of Japanese bathymetric chart 6901: Geol. Soc. America Bull., v. 65, p. 1199–1224.

Dobrin, M. B., 1976, Introduction to geophysical prospecting: New York, McGraw Hill, 630 p.

Hamilton, W., and W. B. Meyers, 1967, The nature of batholiths: U.S. Geol. Survey Prof. Paper 554-C, 30 p.

Heezen, B. C., et al, 1971, Evidence of foundered continental crust beneath the central Tyrrhenian Sea: Nature, v. 229, p. 327–329.

Honza, E., M. Arita, and K. Onodera, 1976, Dredged material, *in* E. Honza, ed., Ryukyu Island (Nansei-Shoto) Arc, GH 75-1 and GH 75-5 Cruises: Geol. Survey of Japan, p. 25–26.

——— et al, 1976, Continuous seismic reflection profiling survey, *in* E. Honza, ed., Ryukyu Island (Nansei-Shoto) Arc, GH 75-1 and GH 75-5 Cruises: Geol. Survey of Japan, p. 20–22.

Ishihara, T., and F. Murakami, 1976; Gravity and geomagnetic survey, *in* E. Honza, ed., Ryukyu Island (Nansei-Shoto) Arc, GH 75-1 and GH 75-5 Cruises: Geol. Survey of Japan, p. 13–19.

Karig, D. E., 1970, Ridges and basins of the Tonga-Kermadec island arc system: Jour. Geophys. Research, v. 75, p. 239–254.

——— 1971, Structural history of the Mariana Island arc system: Geol. Soc. America Bull., v. 82, p. 323–344.

Katsumata, M., and L. Sykes, 1969, Seismicity and tectonics of the Western Pacific: Izu-Mariana-Caroline and Ryukyu-Taiwan regions: Jour. Geophys. Research, v. 74, p. 5923–5948.

Langseth, M. G., X. Le Pichon, and M. Ewing, 1966, Crustal structure of the mid-ocean ridges, 5. Heat flow through the Atlantic Ocean floor and convection currents: Jour. Geophys. Research, v. 71, p. 5321–5355.

Lee, C. S., and R. S. Lu, 1977, Significance of the southwestern section of the Ryukyu Inner Ridge in the exploration of geothermal resources in the Ilan area: Mining Technology, v. 14, p. 114–120.

Leyden, R., M. Ewing, and S. Murauchi, 1973, Sonobuoy refraction measurements in the East China Sea: A.A.P.G. Bull., v. 57, p. 2396–2403.

Lister, C. R. B., 1970, Heat flow west of the Juan de Fuca Ridge: Jour. Geophys. Research, v. 75, p. 2648–2654.

——— 1972, On the thermal balance of a mid-ocean ridge: Roy. Astron. Soc. Geophys. Jour., v. 26, p. 515–535.

Miyazaki, T., K. Tamaki, and F. Murakami, 1976, Geomagnetic survey, *in* E. Honza, ed., Ryukyu Island (Nansei-Shoto) Arc, GH 75-1 and GH 75-5 Cruises: Geol. Survey of Japan, p. 52–54.

Murauchi, S., and N. Den, 1966, Origin of the Sea of Japan: Paper presented at monthly colloquium of the Earthquake

Institute, Tokyo University, Tokyo.
——— et al, 1968, Crustal structure of the Philippine Sea: Jour. Geophys. Research, v. 73, p. 3143–3171.
Takahashi, R., 1975, The Okinawa Trench area, *in* H. Kagami, ed., Preliminary report of the Hakuho Maru Cruise KH-72-2, The Southwest Japan Arc and Ryukyu Arc Areas, 14–15: Ocean Research Inst., Univ. of Tokyo, 144 p.
Tamaki, K., T. Miyazaki, and E. Honza, 1976, Continuous seismic reflection profiling survey, *in* E. Honza, ed., Ryukyu Island (Nansei-Shoto) Arc, Gh 75-1 and GH 75-5 Cruises: Geol. Survey of Japan, p. 55–61.
Turner, F. J., 1968, Metamorphic Petrology: New York McGraw-Hill, 403 p.
Wageman, J. M., T. W. C. Hilde, and K. O. Emery, 1970, Structural framework of East China Sea and Yellow Sea: A.A.P.G. Bull., v. 54, p. 1611–1643.
Yasui, M., et al, 1970, Terrestrial heat flow in the seas around the Nansei Shoto (Ryukyu Islands): Tectonophysics, v. 10, p. 225–234.

APPENDIX

Two successful heat-flow measurements have been made by L-DGO in the Okinawa Trough. The main purpose of these measurements was to provide a check on those made with a short (two-meter) probe by Yasui et al (1970). We had hoped to achieve deeper penetration in the sediment in order to determine whether the high scatter in their data (125 ± 63 mW/sq m) was due to environmental effects, water circulation in the mud, or if it was indeed the result of a highly variable geothermal heat flux. The heat-flow data from L-DGO and Yasui et al in the L-DGO survey area are plotted in Figure 1, and the L-DGO data are listed in Table 1.

Both measurements were made in flat-lying sediments, and well away from known basement exposures. However, V33-071 may be within 5 to 10 km of the edge of the basin, so the observed heat flow may be affected by both the edge effect found near the edges of sedimentary basins and also the effect of topography.

Crossings over the positions of the heat-flow measurements made by Yasui et al (1970) revealed that two of the measurements (23 and 24) were made either on, or next to, structures which pierced the seafloor. Both measurements were low (15 and 44 mW/sq m) compared to the mean heat flow. As dredging revealed that at least one piercement structure in the basin is composed of basalt, and as all these structures have similar acoustic reflection character, it is probable that they are all composed of basalt. Lister (1970, 1972) and many others have shown that heat-flow measurements made near basalt outcrops in areas of high heat flow are anomalously low, possibly as a result of seawater circulating through and cooling the basalt. This hypothesis could account for these anomalously low heat-flow values.

The average heat flow for the southwestern Okinawa Trough, including only stations V33-072 and 25 and 26 of Yasui et al (165.0 and 185.1 mW/sq m, respectively) is 262.3 mW/sq m. The average heat flow is higher than that predicted by Yasui et al (1970). Because of the sparseness of reliable data and the large scatter, the computed average is not considered to be well determined. A value of about 170 mW/sq m has been chosen as a conservative estimate of the near-surface heat flow for this area. We believe that the large scatter in the data is, as assessed by Yasui et al (1970), the result of local intrusion. Until further data are available, it will be difficult to provide a more meaningful estimate of the mean heat flow from the southwestern Okinawa Trough.

Table 1. New Heat Flow Measurements in the Okinawa Trough.

T'G#	*Latitude (North)*	*Longitude (East)*	*Depth corr. meters*	*Gradient (°C/m)*	*Number of probes in mud*	*K*[1]	*Heat Flow*[2]	*Penetration (meters)*	*Eval.*[3]
V33-071	25°38.3′	125°36.9′	2,054	0.103	2	1.25	128.9	3.7	6
V33-072	26°12.5′	126°01.6′	2,050	0.426	3	1.03	437.0	4.4	9

[1] W/m °K
[2] mW/m²
[3] Langseth et al (1966)

Seismic Refraction and Reflection Studies in the Timor-Aru Trough System and Australian Continental Shelf[1]

R. S. JACOBSON,[2] G. G. SHOR, JR.,[2]
R. M. KIECKHEFER[2] AND G. M. PURDY[3]

Abstract Seismic refraction and reflection profiles were recorded on the continental shelf and slope north of Australia and in the Timor-Tanimbar-Aru Trough system of the Banda Sea. This trough system, not deeper than 3.6 km, is the eastern extension of the Java Trench. Morphologically, the area is similar to other circum-Pacific subduction zones, although continental crust, rather than oceanic crust, is being thrust under a series of emergent islands consisting of nonvolcanic imbricated crustal blocks. The refraction results reveal a close similarity between the crust underlying the continental shelf and that under the trough system. Typical continental crustal thicknesses (up to 40 km) and velocities were observed. Reflection profiles reveal that the continental slope was formed by predominantly normal faulting and active subsidence, presumably related to the downwarping of the continental shelf into the subduction zone. A tectonic front at the landward (northern) wall of the trough system compressionally deforms unlithified sediments. Uplifting of small crustal blocks into the imbricated island trend is also apparent. The data strongly support the idea that the Timor-Tanimbar-Aru Trough system is the surface trace of a subduction zone.

INTRODUCTION

In southeast Asia, the Eurasian, Pacific and Australian-Indian plates converge into a complex system of small platelets, many boundaries of which are poorly defined. The Sunda (or Java) Trench marks the southern boundary of the Eurasian plate where the Australian-Indian plate is being subducted. The Sunda Trench appears to terminate east of longitude 118°. A curvilinear system of troughs (Fig. 1) consisting of the Timor Trough, Tanimbar Trough (or Saddle) and the Aru Trough (or Basin), may well be the eastern extension of the Sunda Trench, with a marked southerly offset south of the island of Sumba. This trough system, nowhere greater than 3.6 km deep, separates the Australian continental platform from the Banda Sea, the southeasternmost part of Indonesia. Between the Banda Sea and the trough system lie two arcuate island chains. Between these two island chains lies the Weber Deep, a broad, horseshoe-shaped depression up to 7 km in depth. The more southerly island chain is nonvolcanic, consisting of the islands of Timor, the Tanimbar group, and the Kai Islands. These islands are emergent sections of tectonic thrust sheets of mixed Asian and Australian origin (c.f., Audley-Charles, 1968; Audley-Charles and Milsom, 1974; Carter et al, 1976; Barber et al, 1977; Fitch and Hamilton, 1974; Hamilton, 1972, 1973). The other, more northerly, island chain is composed of active andesitic volcanoes, where recent activity is concentrated in the northeast.

Geophysical resemblances and differences between the Banda Arc and the simpler Sunda Arc to the west have long been recognized. Gutenberg and Richter (1954) note the extension of the trench, the linear negative gravity anomaly, the volcanic line

[1]Manuscript received, December 2, 1977; accepted, May 17, 1978.

[2]University of California, San Diego, Marine Physical Laboratory of the Scripps Institution of Oceanography, La Jolla, California 92093.

[3]Woods Hole Oceanographic Institution, Woods Hole, Massachusetts 02543.

The studies reported here were part of the Banda Sea transect of the SEATAR (Studies of the East Asia Tectonics and Resources) program, an international program of the International Decade of Ocean Exploration, the CCOP (Committee for Coordination of Joint Prospecting for Mineral Resources in East Asian Offshore Areas) and the International Oceanographic Commission. The Banda Sea program was organized and led by Carl Bowin of WHOI, to whom a great deal of credit is due for his planning, organization, and diplomacy in putting together a successful international multi-institutional operation.

We also thank the staff of the CCOP and the Geological Survey of Indonesia for their assistance, as well as the many volunteer participants in the work, and the masters, crews, and scientific parties of the R/V *Thomas Washington* and R/V *Atlantis II.* C. Von der Borch, L. A. Lawver and J. R. Curray provided many helpful suggestions and discussions.

The work was supported by NSF grant OCE 75-19150 to the Woods Hole Oceanographic Institution, and by NSF grant OCE 75-19387 and ONR contract N000-14-75-C-0749 to the Scripps Institution of Oceanography. Contribution of the Scripps Institution of Oceanography, new series.

Woods Hole Oceanographic Institution contribution 4145.

Article Identification Number:
0065-731X/78/MO29-0014/$03.00/0.
(see copyright notice, front of book)

with intermediate depth earthquakes, and the zone of deep-focus earthquakes from the Sunda Arc around the Banda Arc. They also point out that few shallow shocks have been observed from the southern area near Timor. Benioff (1954) presented the hypocentral locations in the area as an illustration of the dipping zones of earthquakes, now known as "Benioff zones." Cardwell and Isacks (in press) have recently analyzed a much larger and more accurate body of earthquake data, showing the continuity of the dipping zone of intermediate and deep earthquakes from the Sunda Arc around the Banda Arc to the Aru Trough, but again noting the curious gap in shallow seismicity over a segment from Timor to Tanimbar. Ian Reid (pers. comm.) has used field seismographs to study microseismic activity in the area, and again finds a near-continuous Benioff Zone lacking only the shallow portion of the zone in the same area.

In September of 1976, Woods Hole Oceanographic Institution, Scripps Institution of Oceanography and the Geological Survey of Indonesia conducted a cooperative geophysical survey of the Banda Sea as part of the SEATAR study of east Asia tectonics and resources as outlined by CCOP (1972). This paper presents results of seismic refraction and reflection studies of the Timor-Tanimbar-Aru Trough system. The purpose of the refraction program was to understand the interrelationship of the crust underlying the Australian continental shelf and that under the trough system, and to provide further constraints for proposed models of the origin and evolution of the trough system.

Previous refraction studies, reported by Curray, Shor, Raitt and Henry (1977), include three stations in the Arafura Sea, between Australia and Irian Jaya. These stations reveal a thin layer of Recent sediments overlying higher (4.8 to 5.9 km/s) velocity material. Velocities (6.1 to 6.4 km/s) indicative of crystalline basement at depths of 3 to 9 km are also present. A split profile in the Timor Trough was also discussed. A thick crustal section, comparable to continental crustal thicknesses, was suspected, although not adequately proven. This station has been reexamined for this present study.

The geology of the northern Australian platform has been studied primarily by use of multichannel seismic reflection and well data (Mollan et al, 1970; Martison et al, 1973; Robertson et al, 1976). These studies indicate that the continental shelf has been relatively stable since the Permian, although large-scale faulting during the Mesozoic has separated depositional basins. The sequence of Permian to Tertiary sediments is composed mainly of carbonates, shales, sandstones, and dolomitic limestones. Veevers et al (1974), have correlated subbottom reflectors over most of the northwest shelf. Beck and Lehner (1974) and Montecchi (1976) presented multichannel reflection profiles across the trough system. Von der Borch (in prep.) reviews all available reflection profiles across the trough system and both island chains.

The Banda Basin itself, to the north and west of the volcanic ridge, has been the subject of few reported studies. Raitt (1967) and Curray et al (1977) report on one refraction station in the Banda Sea, and Purdy et al (1977) on others, which indicate that the crust beneath the Banda Sea is similar to normal oceanic crust if not identical; Heezen and Fornari (1975) suggest that the crust is of Miocene age. Further information on the structure and history of the Banda Basin should result from the studies carried out during the SEATAR work.

REFRACTION PROFILES

During the INDOPAC Expedition of September, 1976, four refraction profiles were carried out on the continental shelf (Figure 1), and nine profiles in the Timor-Tanimbar-Aru Trough system, using the research vessels *Thomas Washington* and *Atlantis II* of the Scripps Institution of Oceanography and Woods Hole Oceanographic Institution, respectively. The two-ship refraction lines were standard reverse and split profiles. Appropriate corrections were applied to the data (Shor, 1963), and plane layer solutions (Officer, 1958; Ewing, 1963) were calculated. Extensive use of second arrivals, where correlation between shots permitted, helped in the interpretation of the data. In many cases, high-amplitude mantle reflections were observed with travel-time curves that asymptotically approached an apparent velocity of approximately 6.8 km/s at large ranges; this is interpreted to be evidence of a lower crustal layer of similar velocity. Where a layer of this velocity was not observed, a calculation was made of the shallowest depth to a masked layer of this velocity. The effect of such an undetected masked layer would be to deepen the Mohorovicic discontinuity by two to five kilometers. Detailed discussions of the quality of data in individual stations and of the methods of data analysis used, along with travel-time plots, can be found in the appendix. Table 1 lists the layer solutions of the INDOPAC (INDP) stations along with a slightly revised determination of MONSOON (MN) station 11 in the Timor Trough, and a corrected version of stations MN9A-9B, previously discussed by Curray et al (1977).

Two reversed profiles on the continental shelf north of Australia were carried out during the INDOPAC Expedition: stations INDP8-1 and 8-2 north of Melville Island on the Sahul Shelf and stations INDP8-14 and 8-15 south of Irian Jaya (Figure 1). The refraction results are shown in Figures 2 and 3. The layer with velocities ranging from 5.2 to 5.9 km/s may be correlated with the shallow, high-velocity carbonates of Eocene Age reported by Mollan et al (1970); Martison et al (1973); Robertson et al (1976), from wells on the northwest shelf of Australia. Beneath these carbonates Martison et al (1973) report finding shales and sandstones which have a lower seismic velocity, in which case our reported thicknesses of this layer may be too great. Alternatively, at some stations this layer may represent these Mesozoic sandstones and shales. Arrivals with velocities 6.0 to 6.5 km/s are interpreted as refracting from Precambrian crystalline basement of the

TABLE 1. SOLUTIONS FOR REFRACTION STATIONS.

STATION	RECEIVING SHIP	TYPE	LATITUDE	LONGITUDE	AZIMUTH	LAYER VELOCITY (km/s) WATER	A	B	C	D	E	F	G	LAYER THICKNESS (km) WATER	A	B	D	E	F	MANTLE DEPTH (km)
Shelf Stations																				
MN 9A MN 9B	AR MA	Reverse	7°55'S 7°37'S	134°04'E 133°43'E	124°	1.54	1.74			5.35	6.14			.07 .07	.28 .42		4.0 3.1			
INDP 8-1 INDP 8-2	TW A2	Reverse	10°54.3'S 10°18.5'S	129°30.9'E 131°26.0'E	072°	1.53	2.17			5.88	6.32		7.91	.08 .07	1.47 1.95		7.8 4.6	22.1 27.6		31.4 34.2
INDP 8-14	A2	One-Way	7°01.0'S	135°13.2'E	009°	1.53	(1.61)	2.50*		(5.8)	6.24^{+}			.04	.45$^{\Delta}$		2.2$^{\Delta}$			
INDP 8-15	TW	Split	4°35.1'S	135°32.0'E	011°	1.53	1.55 + .88 Z			5.25	6.24^{+}	(6.91)	(7.63)	.06	2.60		5.0	11.4	7.6	26.6
Trough Stations																				
INDP 8-3	A2	One-Way	9°04.5'S	127°52.1'E	077°	1.50	2.80*			(5.19)	(5.71)		(8.46)	3.28	2.07		0.8	24.0		30.1
MN 11	AR	Split	9°01'S	128°43'E	088°	1.50	2.80*	3.64		4.84	5.98			3.08	2.05	.76	1.7			
INDP 8-4	TW	One-Way	8°53.4'S	129°28.9'E	081°	1.50	(2.83)	(3.39)		(5.19)	(5.81)	(7.21)	(8.88)	2.28	.88	3.20	3.9	10.4	19.6	40.0
INDP 8-12	TW	Split	8°33.5'S	130°59.8'E	045°	1.49	2.45	3.28	4.03					1.47	.93	1.94				
INDP 8-12 INDP 8-13	TW A2	Reverse	8°33.5'S 7°30.1'S	130°59.8'E 132°05.5'E	045° 047°	1.49	2.45	3.49		5.17	6.27			1.47 1.31	1.14 2.11	3.36 2.07	1.8 5.4			
INDP 8-13	A2	Split	7°30.1'S	132°05.5'E	047°	1.49	2.50*	3.75		4.96	6.56			1.31	2.46	1.92	3.2			
INDP 8-13 INDP 8-17	A2 TW	Reverse	7°30.1'S 5°58.8'S	132°05.5'E 133°24.9'E	047° 225°	1.49 1.52	2.50*	3.50		5.42	6.22			1.31 3.60	2.49 3.46	2.25 1.42	1.2 2.9			
INDP 8-17	TW	Split	5°58.8'S	133°24.9'E	225°	1.52	2.50*	3.23		5.04	5.69			3.60	3.60	1.29	2.4			
INDP 8-16 INDP 8-17	A2 TW	Reverse	4°53.8'S 5°58.8'S	133°59.7'E 133°24.9'E	203° 225°	1.49 1.53	2.50*	2.98		5.03				3.43 3.60	2.43 4.23	1.40 1.34				
INDP 8-12 INDP 8-17	TW TW	Reverse	8°33.5'S 5°58.8'S	130°59.8'E 133°24.9'E	045° 225°	1.49 1.52	(2.47)	3.21		5.49	6.27	6.89	7.85	1.47 3.60	.98 3.11	3.42 1.60	2.0 2.8	16.7 15.9	15.7 5.0	40.3 32.0

KEY:

* Assumed velocity.

( ) Values in parenthesis represent velocity from a one-way run.

\+ Calculated from reverse of INDP 8-14 and INDP 8-15.

Δ Thicknesses listed are those under the receiver (see Figure 3). Layers are apparently faulted.

SHIPS: A2 - R/V Atlantis II; TW - R/V Thomas Washington; AR - R/V Argo (Monsoon Expedition, 1960); MA - M/V Malita (Monsoon Expedition, 1960).

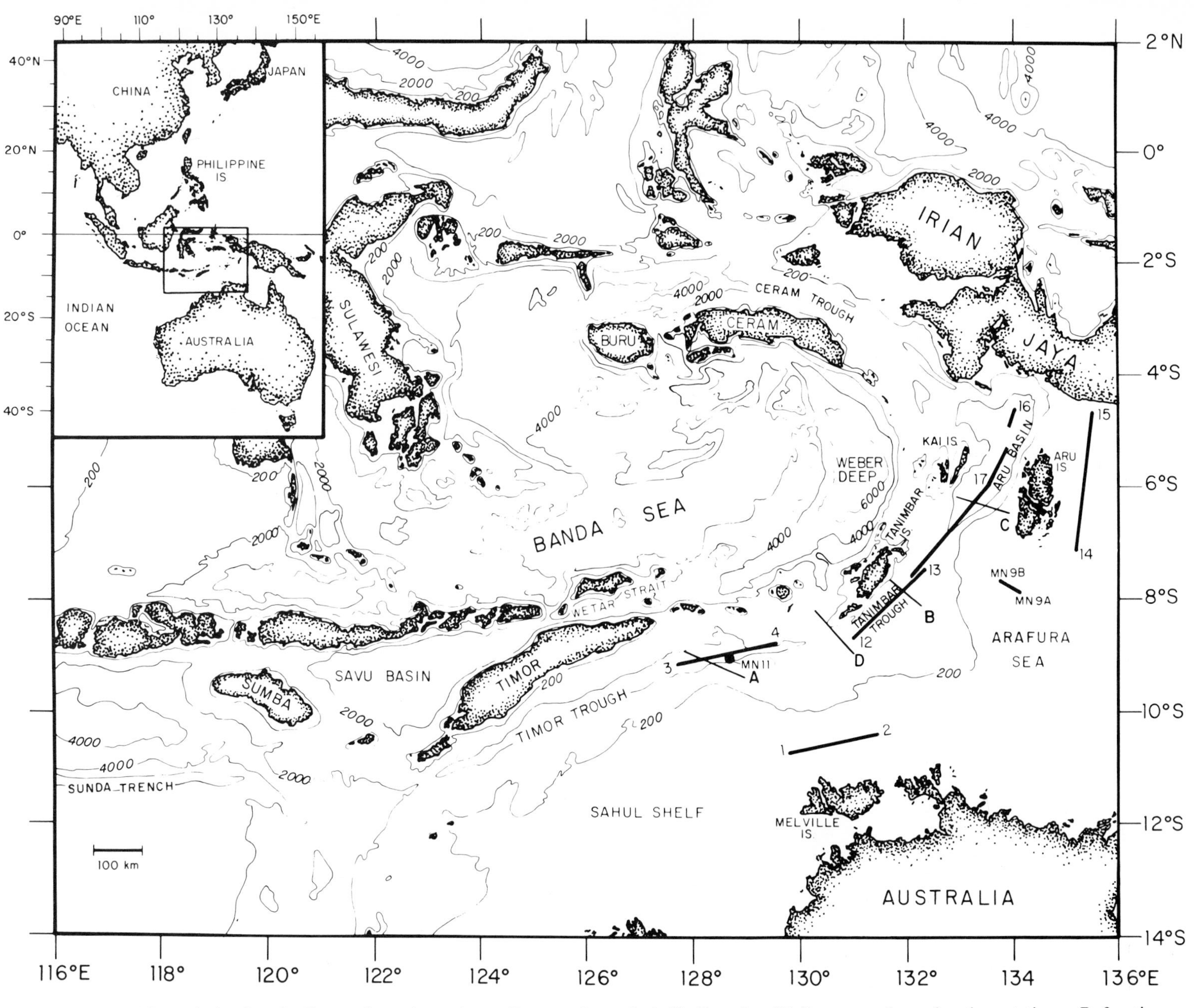

FIG. 1. Map of the Banda Sea and northern Australian continental shelf. Broad solid lines are the refraction stations. Refraction station locations are identified by the final digits of the station number (i.e., 17 indicates the receiver location of refraction station INDP8-17). Narrow lines marked A through D are reflection profiles corresponding to figures 7 through 10, respectively. Contours are 200 m, 2 km, 4 km, and 6 km.

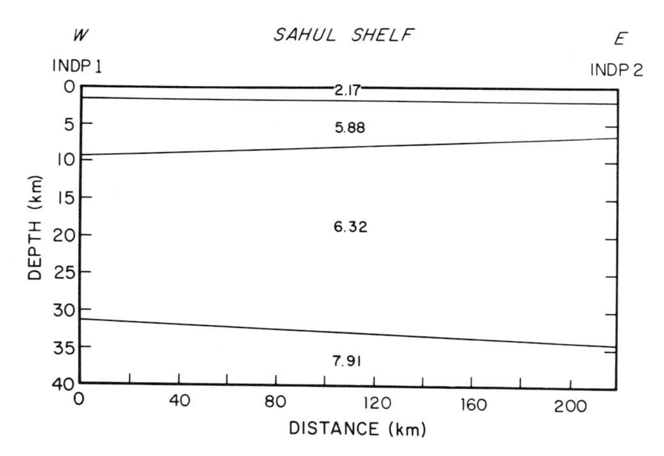

FIG. 2. Crustal cross-section of Australian continental shelf north of Melville Island derived from reversed solution of refraction stations INDP8-1 and 8-2.

continental shelf (Martison et al, 1973). Refracted arrivals from the lower crust were not observed at stations 1 and 2, although there is evidence for a higher velocity lower crust there. The probable depth to the lower crust at stations 1 and 2 would then be 20 to 22 km. Depth to the Mohorovicic discontinuity is typical of continental crust (26 to 35 km), although the observed mantle velocities are low, less than 8.0 km/s.

Refracted arrivals observed at station INDP8-14, south of Irian Jaya, reveal two major faults north of the receiving position. The uppermost sediments, velocity 1.6 km/s, overlie material of 5.8 km/s velocity near the receiving station; north of the station the material with velocity of 5.8 km/s has apparently been downfaulted (Fig. 3). The throw of the faults, calculated for an assumed velocity of 2.5 km/s for the deeper sediments, is 0.9 km for the fault 11.5 km north of the receiver, and an additional 1.4 km for the second fault 22.5 km from the receiving position. The 5.8 km/s refracting layer is probably the carbonate horizon found at the other continental shelf stations. Beneath this material lies crystalline basement with a seismic velocity of 6.2 km/s, calculated using both INDP8-14 and INDP8-15 refraction profiles. This subbottom morphology may be alternatively explained by presence of a reef which developed on a basement high during the early Tertiary. A large depositional basin would then lie north of the receiving position, consistent with the gravity profile along the shooting track (see Appendix) and the northward dipping crystalline basement (Figure 3). Refracted arrivals from the lower crystalline basement and mantle were not observed at INDP8-14. Refracted arrivals interpreted as coming from the lower crust were observed at station INDP8-15 having an apparent unreversed velocity of 6.9 km/s. Depth to the Mohorovicic discontinuity is 27 km, and the unreversed mantle seismic velocity is 7.6 km/s.

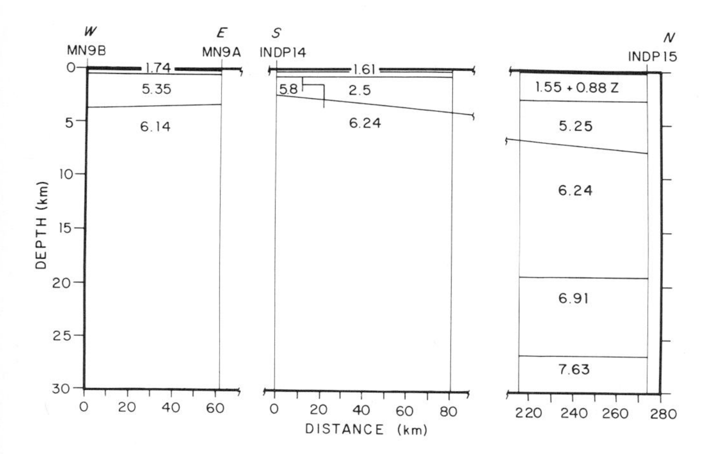

FIG. 3. Preferred crustal cross-section of Arafura Sea south of Irian Jaya as determined from refraction stations MN 9A, MN 9B, INDP8-14 and 8-15.

Nine refraction profiles were received at six locations in the Timor-Tanimbar-Aru Trougs system (Figure 1), including two profiles shot to complement split profile MN11 in the Timor Trough, previously discussed by Curray et al (1977). Seven of the refraction profiles were shot in a leapfrog fashion along the Tanimbar-Aru Trough system. Only two of these stations (INDP8-12 and INDP8-17) had observable mantle arrivals. Fortunately, these two stations could be treated as a nonoverlapping reverse pair, allowing good control of mantle velocity and depth.

The three refraction stations in the Timor Trough, INDP8-3 and 8-4 and MN11, all show similar, but not identical, crustal cross-sections (Figure 4). The sedimentary section exhibits the most variability. All stations, though, have a layer with velocities ranging from 4.8 to 5.2 km/s. These velocities are probably from a carbonate or sandstone layer, much like the layers with slightly higher velocities found on the shelf. Basement velocities show consistency between stations: 5.7 to 6.0 km/s. Only the INDOPAC profiles were of sufficient length to detect subcrustal and mantle arrivals. INDP8-4 has a travel-time branch corresponding to a 7.2 km/s velocity from the lower crust; INDP8-3 does not. However, there is evidence from mantle reflections that crustal material with higher velocity (approximately 6.8 km/s) is present at a depth of 22 km at station INDP8-3. Mantle velocities are very high (8.4 to 8.9 km/s); as there are very few observed refracted arrivals from this layer, however, the precision of the determinations of these velocities is poor. The Mohorovicic discontinuity deepens considerably from 31 km at station 8-3 to 40 km at station 8-4.

The multiple refraction profiles (INDP8-12, 8-13, 8-16, 8-17) in the Tanimbar-Aru Trough system allow slightly differing interpretations; the preferred crustal cross-section is depicted in Figure 4. Three distinct layers of sediment are present, the velocities showing consistency between stations. The sedimentary layer, with seismic velocities ranging from 5.0 to 5.5 km/s, compares well with the carbonate layer found on the shelf and in the Timor Trough. Upper and lower crystalline crust are detected, of velocities 6.2 to 6.6 km/s and 6.9 km/s, respectively. There is

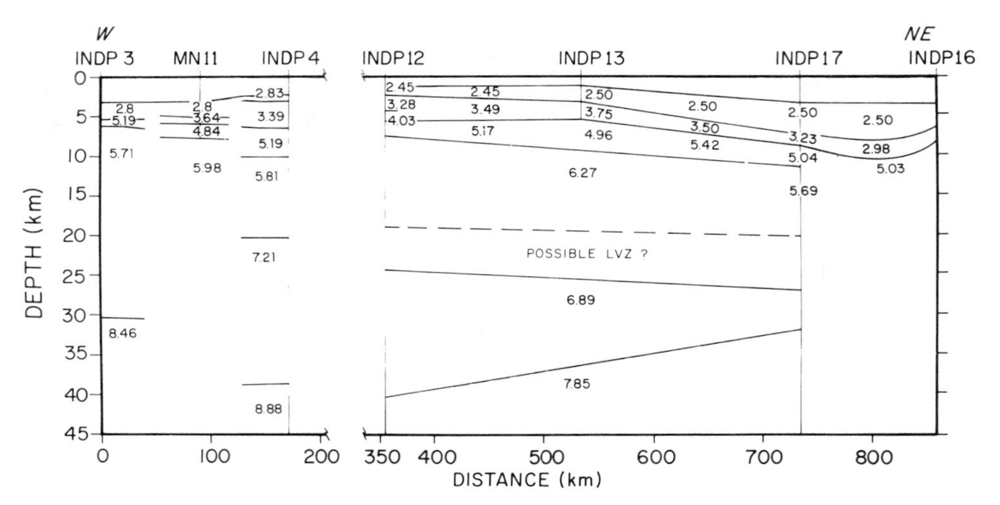

FIG. 4. Preferred crustal cross-section along the Timor-Tanimbar-Aru Trough system, including refraction stations INDP8-3, 8-4, 8-12, 8-13, 8-16, 8-17 and MN11. Determination of velocities and thicknesses of the lower crystalline crust and mantle were derived from the non-overlapping reverse of stations INDP8-12 and INDP8-17, assuming the low-velocity zone (LVZ) is not present. Bounds on the velocity and thickness of the LVZ are indeterminate.

some evidence for a low-velocity zone above the lower crust. The record section or travel-time plot of INDP8-12 (Figure 5) indicates that the lower crustal arrivals are delayed at 90 to 120 km range. On the record section of station INDP8-17 (Figure 6), however, the lower crustal arrivals appear as a masked layer. The mantle, velocity 7.9 km/s, dips steeply to the southwest.

The continental crust under the Timor-Tanimbar-Aru Trough system is remarkably similar to the continental shelf crust. The presence of thicker sediments and the lower apparent seismic velocity of the carbonate layer and the upper crystalline crust are the major differences. The crystalline crust under INDP-17 (Figure 4) in the Aru Basin has velocities and thicknesses similar to that at INDP8-15 (Figure 3) south of Irian Jaya, especially the comparatively thin lower crust. The crustal structure of the Timor Trough closely resembles that of the Sahul Shelf, although basement velocities are lower, and the carbonate layer has thinned in the trough.

REFLECTION PROFILES

Seismic reflection profiles across the trough and continental shelf confirm the idea that the crust in the trough system is continental shelf material. The locations of the reflection profiles are shown in Figure 1. An oblique crossing of the Timor Trough at 128°E (Figure 7) shows numerous erosional channels and faults on the continental slope. Deep subbottom

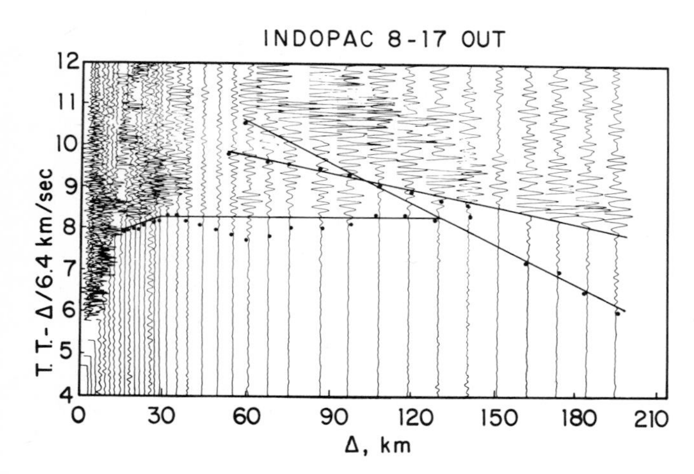

FIG. 5. Record section of the northeastward outgoing shooting run from of INDP8-12 near Tanimbar island, reduced at 6.4 km/s, with a horizontal datum.

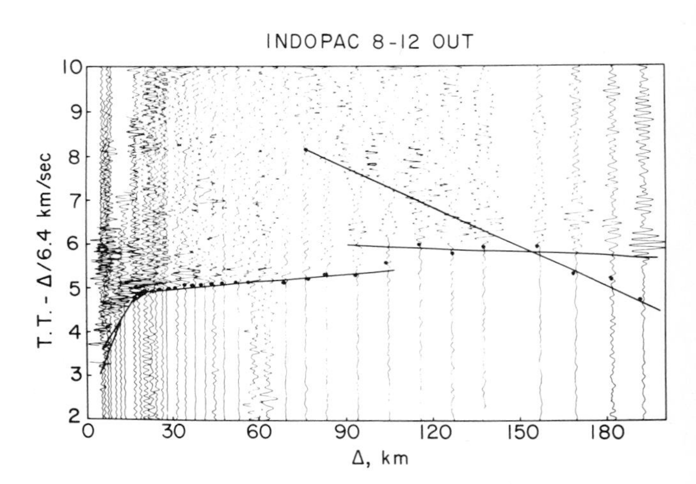

FIG. 6. Record section of the southwestward line from INDP8-17, in the southern Aru Basin; reduced at 6.4 km/s, with a sloping datum. Note the offset from the straight line in the travel-time plot near 60 km range. This is probably due to the steep scarp separating the Aru Basin from the Tanimbar Trough. The refracted arrivals which appear tangent to the mantle reflections at large range are probably from the lower crust, velocity 6.9 km/s.

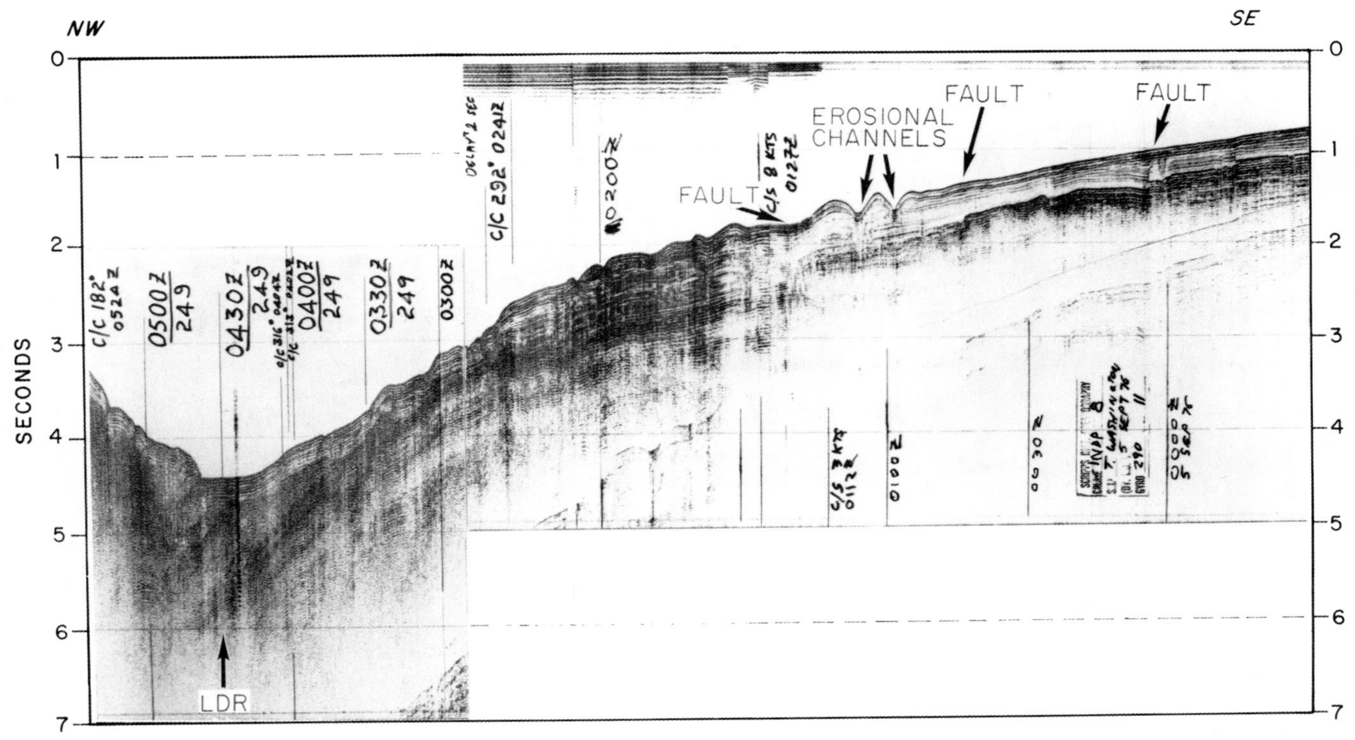

FIG. 7. Reflection profile A (Fig. 1) of the Timor Trough at 128°E long., showing two erosional valleys and numerous faults. LDR refers to landward dipping reflectors. Vertical exaggeration 12.7*x*.

reflectors dip under the Timor Trough smoothly until the northern wall of the trough is reached. Here, all subbottom reflectors become incoherent. This change, more clearly seen in Figure 8, marks a tectonic front north of which the upper unlithified sediments are highly disturbed. The point of disturbance, usually at the landward (northern) wall of the trough, has moved in Figure 8 to the seaward (southern) side of the trough. The deeper reflector, however, continues to dip beneath this disturbance. This smoothly reflecting bed is probably of lithified carbonates of Miocene age (Martison et al, 1973),

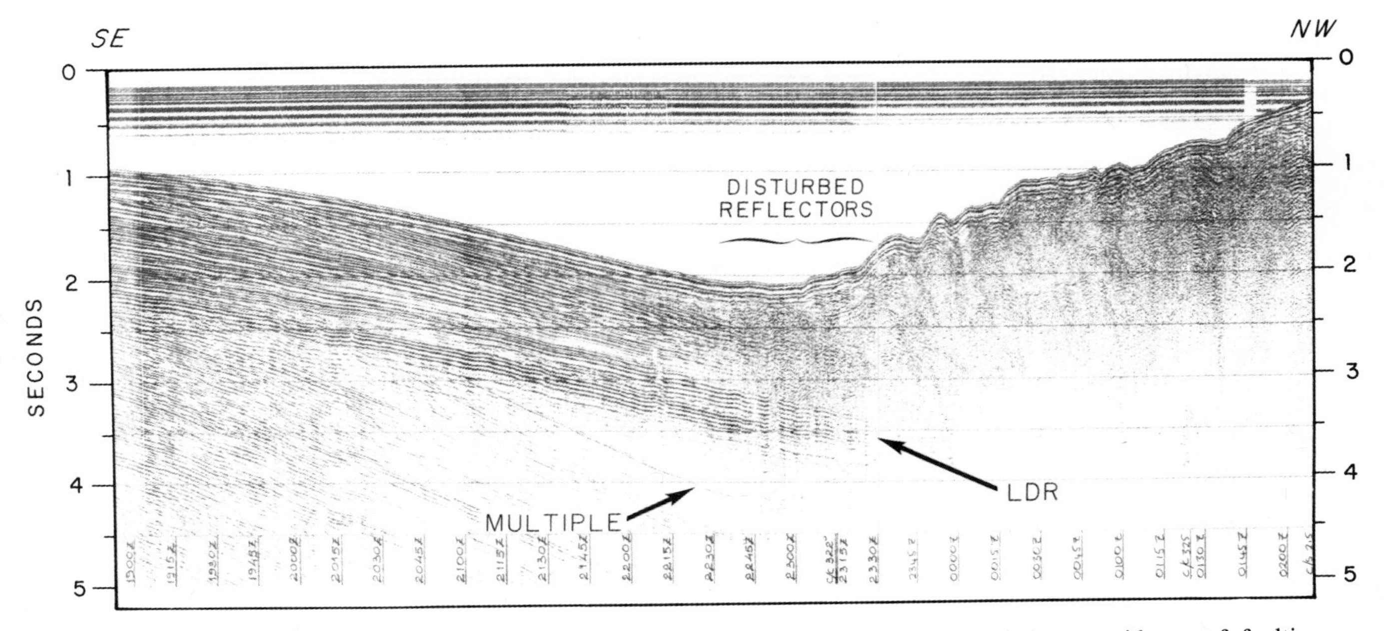

FIG. 8. Reflection profile B (Fig. 1) of the Tanimbar trough at 131.5°E long. There is no evidence of faulting on the continental slope. The tectonic front has advanced to the seaward wall of the trough, compressionally deforming the uppermost unlithified sediments. The strong reflector at 3 sec of 2-way travel time does not appear to be altered. Vertical exaggeration 3.7*x*.

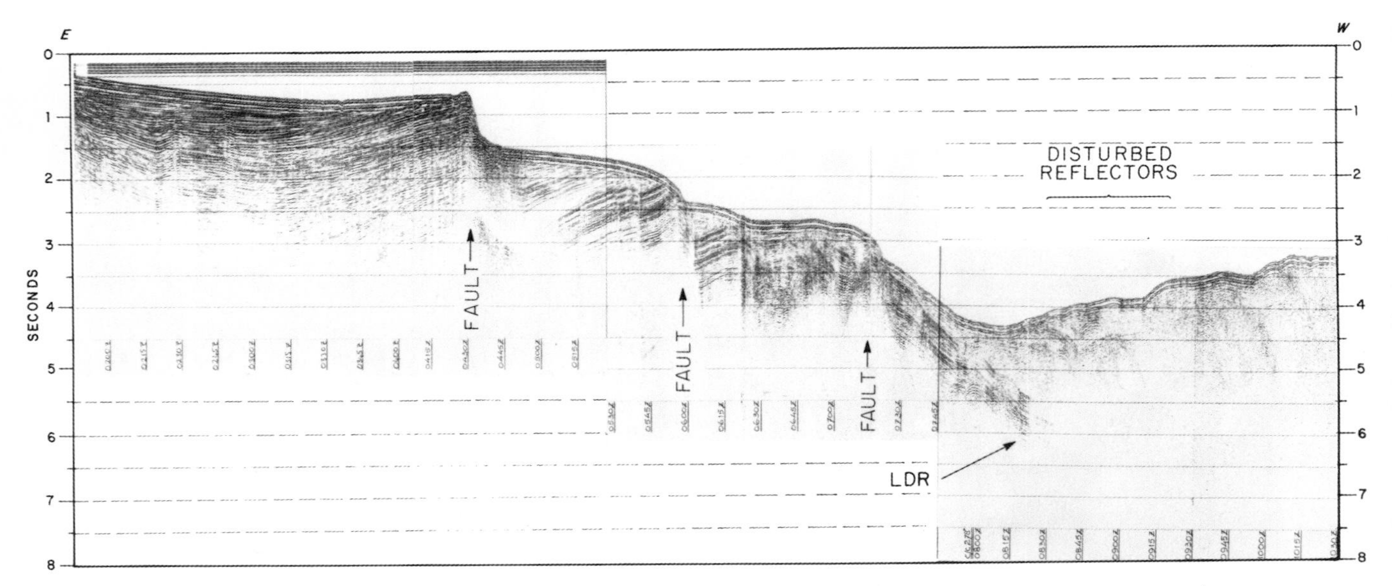

FIG. 9. Reflection profile C (Fig. 1) across the central Aru Basin at 6.5°S lat. Major normal faulting with offsets up to 700 m is clearly evident. Vertical exaggeration 4.7*x*.

corresponding to the 3.0 to 3.7 km/s layer found in the trough system. It appears that these carbonates are more competent than the overburden, and thus are more resistant to compressional deformation.

The reflection profile across the Aru Basin (Figure 9) shows block faulting with 0.5-km offsets on the east side of the basin. The offsets here are considerably larger than those south of the Timor Trough, causing the east side of the basin to bear a closer resemblance to a rifted margin than to a typical subducting slab. There is no equivalent rifting apparent on the west side of the basin, however, and the reflectors become "incoherent" at the foot of the west side in the same manner as in the sections farther to the southwest. On the basis of morphology alone it is hard to say that the evidence for subduction is indisputable here; the alternatives are no easier. It has been suggested that the Aru Basin is a rift zone; if so, it is not one in the sense of being a spreading center, as heat-flow measurements in the area are not exceptionally high (L. Lawver, pers. comm.).

On the shelf itself, an unconformity is apparent in the upper sedimentary section. Presumably, during the low-standing sea in the Pleistocene, the continental shelf was emergent, and the landward-dipping sedimentary section was eroded.

Figure 10, a reflection profile in the Tanimbar Trough, reveals a major fault at the landward edge of the trough, extending through a strong subbottom reflector. The landward (northern) block is uplifted, becoming part of the Timor-Tanimbar-Kai Island crustal material.

In each reflection profile, reflecting horizons, although in some places offset by faults, can be easily followed from the shelf to the trough. This faulting may be related to the bending and subsidence of the continental shelf. Where the gradient of the slope is about 1 to 40 (the Tanimbar Trough, Figure 8), no faults are observed. A slightly higher gradient correlates with minor faulting (the Timor Trough, Figure 7). Major faulting as seen in Figure 9 is associated with a gradient of 1 to 17, which is probably caused by the response of the crust to the sharp curvature (in plan as well as cross-section) of the crust at the east end of the Banda Arc. It appears that the continental crustal material cannot sustain a downwarp of much more than 1.5° (gradient of 1 to 40) without faulting. As the crust enters the trough system, the uppermost sedimentary section becomes highly disturbed. The front of disturbance is usually at the landward wall of the trough, but sometimes advances towards the seaward (or southern) wall (Figure 8).

DISCUSSION AND CONCLUSIONS

Various tectonic models have been proposed in the literature for the origin and evolution of the Timor-Tanimbar-Aru Trough system. One model, developed by Audley-Charles (1968), is based primarily on the observed geology of Timor; he proposes that the subduction zone is north of Timor, offset from the Java Trench by transform faults near Sumba (Audley-Charles, 1975). Permian sediments containing tropical fauna, suggesting Asian origin, have been overthrust upon high-pressure metamorphic rocks derived from Australia (Barber and Audley-Charles, 1976). Other strata show the same juxtaposition of rocks of differing origins, and Carter et al (1976) propose various thrust sheets from Asia overriding the Australian continental shelf south of the subduction zone (Figure 11a). The additional loading of these thrust sheets on the shelf would cause a broad elastic downwarping of the shelf, producing the Timor Trough.

Fitch (1972) and Hamilton (1972, 1973) envision subduction in the Timor Trough similar to that in other circum-Pacific subduction zones (Figure 11b).

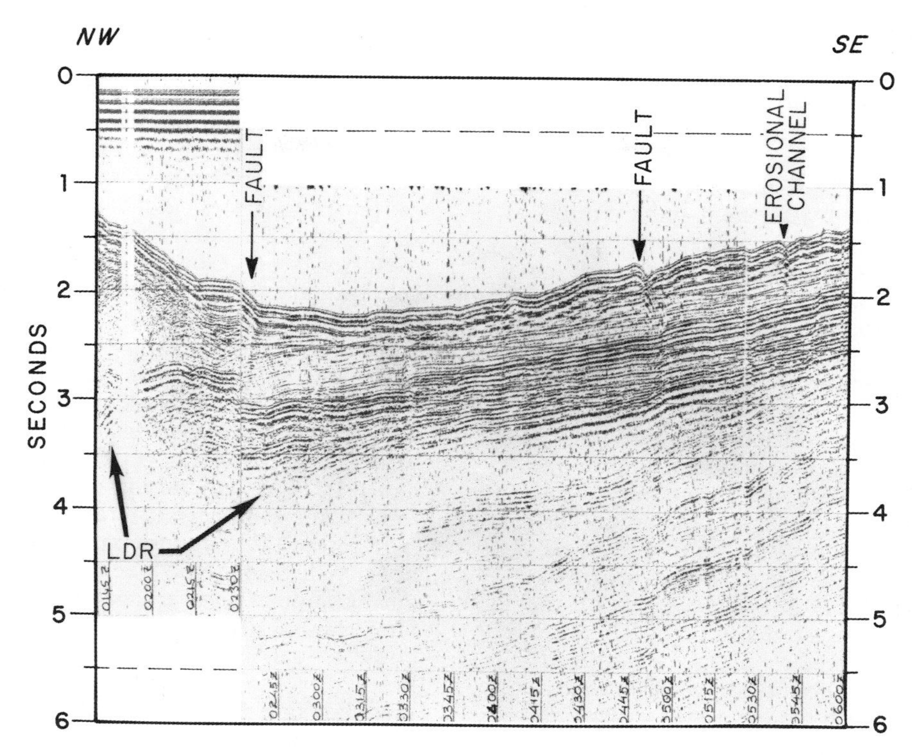

FIG. 10. Reflection profile D (Fig. 1) crossing the Tanimbar Trough at 130.5°E long. A major fault at the landward edge of the trough is evident. The uplifted block to the left of this fault represents the newly formed toe of the accretionary wedge. Vertical exaggeration 3.7*x*.

The chaotic assemblages on Timor are considered good indicators of a subduction zone (Seeley et al, 1974; Karig, 1974) lying south of Timor. Timor

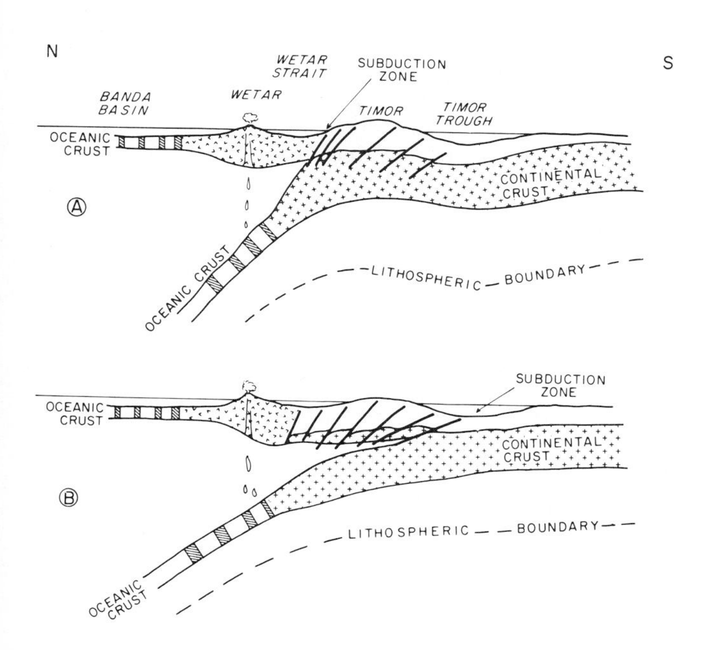

FIG. 11. **a**, Schematic crustal cross-section across Timor as envisioned in the model of Carter et al, (1976). **b**, Schematic crustal cross-section across Timor as proposed in this paper.

would then be an emergent accretionary wedge, similar to the Java ridge to the west. The apparent continuity of the Timor Trough with the Java Trench, and the shallow seismicity (Fitch, 1972; Cardwell and Isacks, 1978) under the Tanimbar-Kai Island chain point to a subduction zone south of the islands. Underthrusting of continental crust would result in a cessation of subduction, explaining the paucity of shallow earthquakes under Timor and the lack of active volcanism north of the island.

The difference between these two models is basically that of overthrusting versus underthrusting, which is not easy to resolve. The model of Audley-Charles and Carter (1977) places the true subduction zone north and under Timor, while Fitch and Hamilton (1974) maintain that the surface trace of the subduction zone lies in the Timor-Tanimbar-Aru Trough system.

The above-mentioned hypotheses for the evolution and origin of the Timor-Tanimbar-Aru Trough system are both supported by the results of the refraction data presented in this report. Both theories maintain that continental crust with strong affinities to the crust underlying the Sahul Shelf should exist in the trough system. The crystalline crust, in the trough system, shows continuity with that under the continental shelf. The reflection profiles confirm the idea that the crust underlying the trough is downwarped, subsided, and block-faulted continental shelf material. The landward dipping reflectors under the trough indicate at least partial underthrusting of continental crustal material under the nonvolcanic island chain.

Uplifting of small continental crustal blocks (Figure 10) is reasonable under either hypothesis. The uplift of these blocks could be explained by the "bulldozing" effect of the Asian blocks onto the continental shelf carrying with them small chunks of continental crustal material. Another explanation for this phenomenon, consistent with the model of Fitch and Hamilton (1974), would be that the crustal block seen in Figure 10 represents the newly-formed toe of an accretionary prism.

The diffuse reflectors on the nonvolcanic island chain may be a result of intense imbrication of the rocks, deformation of unlithified sediments, and/or olistostrome (gravity slide) formations. Figure 8, however, shows that the diffuse reflectors of the sea floor are due to active deformation of the unlithified sediments. The northern wall of the trough usually marks the leading edge of this deformational front.

Other relevant information by which one can determine the location of a subduction zone is provided by seismological studies reported by Cardwell and Isacks (1978). Their report shows a continuous, steeply dipping zone of hypocenters extending from the well-established subduction zone of the Sunda Trench, eastward along the north side of the Banda Arc, and curving northward beneath the Banda Sea parallel to the Timor-Tanimbar-Aru Trough lineation. Connected to this steeply dipping portion of the Benioff zone is a discontinuous band of shallow hypocenters, forming a zone of lesser dip, which serves to define the subduction zone in the critical area, allowing a distinction between the alternative theories. The zone of shallow dip is clear at the east end of the Java Trench, connecting from the trench axis south of the islands to the steep part of the Benioff zone north of the islands. A similar geometry prevails at the easternmost end of the trend, at the Aru Basin, where a gently dipping zone of hypocenters extends from the Aru Basin, dipping down to the west beneath the Banda Arc, and joining the steep Benioff zone beneath the Weber Deep. From Timor to Tanimbar, the shallow seismicity is missing. If, however, one assumes a moderate degree of continuity and similarity of shape of the Benioff zone between the areas in which it is well observed, one would draw its outcrop along the deepest part of the Timor-Tanimbar-Aru lineament, not north of the islands in the Wetar Strait and the Weber Deep.

It is clear that the trough system marks the southernmost edge of the zone of collision between two plates. It is also evident that the continental material of Australia extends to the bottom of the trough, and most probably well under the northern "wall" of the troughs. The overthrusting of the nonvolcanic island material onto the continental shelf as described by Audley-Charles may very well represent the latest stages of subduction, in which pelagic material entering the accretionary wedge was followed by large volumes of deposits from a continental rise, and finally by the edge of the continent itself. This latest stage of accretion would form a much larger accretionary prism than was formed during earlier periods when only normal sea floor was entering the subduction zone, as it is still doing in the Sunda Trench to the west, and would result in movement of the tectonic front and the intersection of the "plate boundary" with the sea floor farther to the south. The differences between the Timor-Tanimbar-Aru Trough lineation and the Sunda Trench to the west may well be only a matter of the size of the accretionary wedge. How far the continental crust itself may have been carried into the subduction zone is as yet unknown. Recent petrological studies of Holocene lavas in the andesitic chain northwest of the Timor-Tanimbar-Kai Island chain (Whitford et al, 1977) show high strontium isotope ($^{87}Sr/^{86}Sr$) ratios, suggesting magma contamination caused by deep (150 km) subduction of continental crust or thick sediments derived from a continental source.

If sinking of a subducting slab is caused by the density difference between cold oceanic lithosphere and the lighter, hotter aesthenosphere below, it is hard to understand how the lighter continental crust could be carried down a subduction zone. We do not pretend to have an answer to this. One possibility is that the Banda Arc is in a stage of transition: oceanic crust has been subducted down a zone extending east as far as Aru, and continental crust has only recently entered the zone of collision. The area near Timor, in which shallow seismicity has apparently ceased, would mark the point of greatest penetration, where if relative movement of the plate continues, future crustal shortening must either take place by folding or by formation of a new subduction zone farther to the north.

In conclusion, all available field evidence indicates that a subduction zone has existed in the recent past northwest of the Timor-Tanimbar-Aru trough system and that its leading edge, in the trough system, overlies continental crust. How far the continental crust has descended beneath the oceanic crust of the Banda Sea, or whether subduction still continues in this zone, cannot be decided on the basis of present evidence.

APPENDIX

All of the seismic refraction stations discussed here were carried out by the two-ship explosive method, in which each ship alternately shoots explosives charges to be received by the other (Ewing, 1963; Shor, 1963). The WHOI group on the R/V *Atlantis II* used a group of up to 4 Select International SLF-73 MHz sonobuoys, deployed near the receiving ship. Hydrophone suspensions were modified by the addition of wax floats and lead weights to decouple the hydrophone from the surface motion of the buoy. Data were recorded on analog paper records and on analog magnetic tape.

The Scripps group on the R/V *Thomas Washington* used a much older system consisting of AX-58 Rochelle salt hydrophones, balanced for neutral buoyancy and decoupled from surface wave noise by a system of floats and weights. Hydrophone cables were slacked for each shot. Preamplified signals were transmitted to the receiving ship by floating cables, where they both went through filtered and band-separated analog amplifiers to a paper oscillograph, and through broad-band amplifiers to a digital recording system. Most of the data discussed here was obtained from the analog records. On one station (INDP8-4) an 11-element linear array of the same type of hydrophones at 500-meter spacing was used. In this instance, two of the hydrophones were hardwired, and the other nine telemetered signals to the ship through modified transmitters from military sonobuoys.

The direct-cabled hydrophone system gave remarkably good signal/noise ratios. Station INDP8-1 was the quietest station, probably because shoal areas shielded the receiving location from water-borne noise, and the sea conditions were extremely calm. The signal/noise ratio (comparing the peak-to-peak amplitude of the second and third half-cycles of the refracted signal to the preceding background noise) was about 10 db for mantle arrivals from 55 kg charges at 208 to 220 km range.

Shot-receiver ranges were generally calculated from arrival times of sound traveling through the mixed layer in the water column, with velocities calculated from temperature measurements by expendable bathythermographs at each station and salinities from nearby hydrographic casts. Where direct water-wave transmissions were not received at the more distant shots, corrections were made to determine distance from travel times of bottom reflections. On stations INDP8-1, -2, -14, and -15, water-transmitted sound faded out before the most distant shots were reached, and distances were calculated from the combined navigation (satellite and dead-reckoning) of the two ships.

Corrections were made for distance from the shot to a shot-detector hydrophone streamer on each ship, for depth of shot and receiver, water depth, and topographic variations of the bottom. The effects of topographic variations are probably the largest source of scatter of the data. Reflection profiler records made by the *Washington* during shooting runs (except on the shallow-water stations 2 and 14) provided some information concerning the shallowest layers, to aid in the analysis of the refraction data. Because these data were not available on both runs from each station pair, however, it was not possible to make actual corrections for the variation of sediment thickness as revealed by the profiler. Instead, the usual assumption was made that the thickness of the first layer varied linearly between stations, and that all of the topographic variations from a sloping datum were in the second layer. Travel times were fitted to straight lines by a least-squares solution that requires that reverse points agree for reversed stations, and that intercept times agree on split stations. Layer solutions were made by the methods given by Ewing (1963) and Officer (1958) for reversed stations. The same method was used for split stations, with the reverse distance set to zero.

Stations INDP8-1 and -2 were a reversed pair, 220 km apart in shallow water north of Melville Island. Water-borne arrivals were detected to 55 km on station 1, to 116 km on station 2. To determine ranges beyond this, distances were calculated from the differences of satellite positions of the two ships, plotted against time, smoothed, checked against distances by water-wave travel time at the close-in ranges, and interpolated to determine shot range. The use of this method could introduce an error of about 1% in determination of velocity and depth of the mantle. High-amplitude mantle reflections were observed starting at about 90 km range. The travel time of the reflection at the critical point was used with the observed mantle refractions (which were received clearly from 8 shots of station 1, and 2 shots from station 2) to determine mantle velocity and depth.

Station INDP8-3 and -4 and the older station MN-11 were located in the Timor Trough. On the shooting run of station 3, the *Washington's* track led up onto the north side of the trough several times. When it became obvious that the ship was out of the trough, course changes were made to bring the track back into the trough. Reflection profiler records made during the run show numerous faults through the sediments, and irregular sediment thickness. The shooting run for station 4 was farther south, and stayed in the trough. The layering in the upper part of the section as observed on the two INDP runs differed so greatly that they could not be computed as a reversed pair. These differences may be due to absence of the second sedimentary layer at station 3, to topographic variations, or to the slight lateral offset of the shooting tracks, which placed the run for station 3 partly over the toe of the north slope of the trough. INDP8-3 and -4 were therefore calculated as separate one-way profiles, and MN-11 recalculated as a split.

Direct water waves were observed on INDP8-3 only to 33 km, and on INDP8-4 only to 18 km. Bottom-reflected arrivals of progressively higher order were observed as the range increased, with lower order reflections fading out with distance. Since, because of the high surface temperature, sound speed in the mixed layer was higher than in any deeper part of the water column, each successive reflection quickly came close to its limiting path in which most of the travel is in the mixed layer, at the same speed as the direct wave. The time difference between successive reflections thus approached a constant value. The equivalent direct time can be obtained in such cases, if the bottom depth is reasonably constant, by use of the reflection difference method, in which one determines the asymptotic value, K, of the difference in travel time between successive reflections, and then corrects back to the direct-wave time by subtracting $n \cdot K$ from the travel time of the nth reflection. This same method was used to obtain shot-receiver ranges for stations 3, 4, 12 and 17.

Refracted arrivals were not obtained from the sediments at the sea floor from stations INDP8-3, -4, or MN-11. However, strong reflected and refracted arrivals were obtained from the next deeper layer on

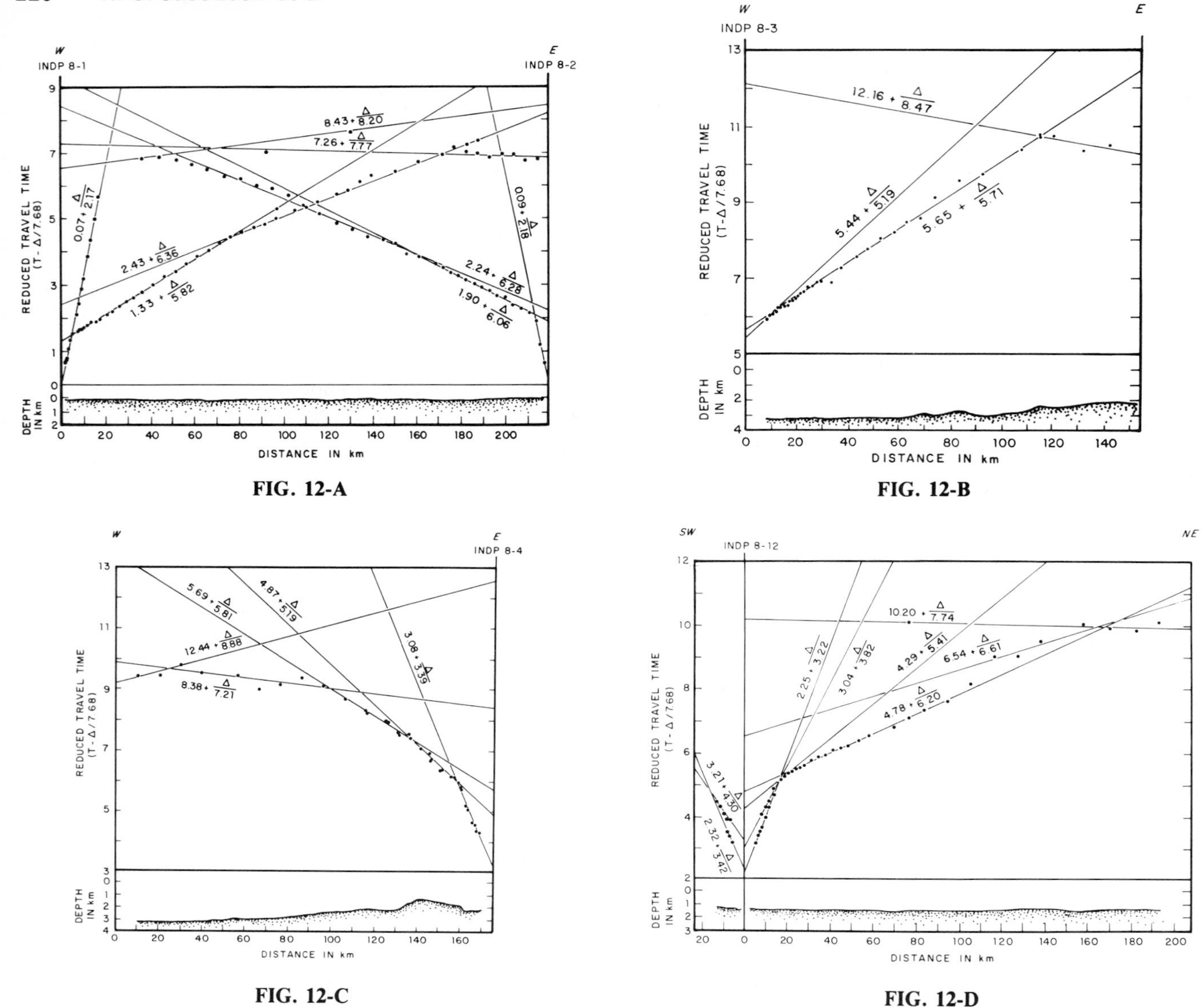

FIG. 12. Travel-time plots for INDOPAC stations. All times are reduced by a function of range as indicated on the vertical scale. The plots are corrected to horizontal datum in each case. The velocities on the travel-time plots are the apparent, unreversed velocities. Free-air gravity anomaly is also plotted along the outgoing run of station INDP8-14. Inflections of the gravity anomaly coincide with the offsets of the travel-time plot.

station 4, from which one could solve for the velocity of the first layer by the reflection/refraction method, using the equation $V_1 = V_2 (1 - k^2/T^2)^{1/2}$ where k is the intercept time of the refracted arrivals from the second layer, corrected to the sea floor, and T is the two-way vertical reflection time from the same layer. This method depends critically on identification of the same reflection at both vertical incidence and at the point of tangency with the refracted arrival. A velocity of 2.83 km/sec was obtained for layer 1 on INDP8-4 by this method. A velocity of 2.8 km/sec was assumed for the first layer on station INDP8-3 and for the recalculation of MN-11.

Stations INDP8-12 and -13 were two split profiles in the Tanimbar Trough, each of which had a long run forming a reverse between the two stations, and a short run at the other end. Direct water waves were recorded on the short run of station 12, out to 123 km on the long run of 12, and as far as refracted arrivals were received on both runs of station 13. The reflection difference method was used to compute ranges for the longest shots on 12. Again, refracted arrivals were not observed from the sea floor, but reflected and refracted arrivals from the second layer were good enough on station 12 to permit computation of the velocity of layer 1 by the reflection/refraction method, from which a velocity of 2.45 km/sec was obtained. A deeper layer, with velocity 4.0 km/sec was observed on station 12, but not on station 13; it was included in the calculation of the split profile for 12, but omitted from the reverse. Seismograms from station 12 are shown in figure 5. Weak mantle refracted arrivals and strong

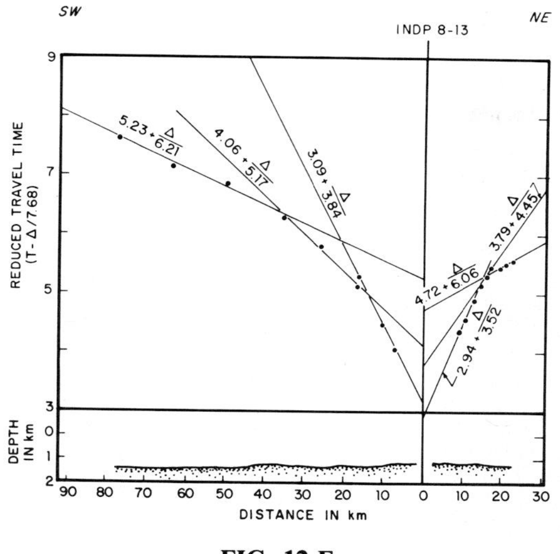

FIG. 12-E

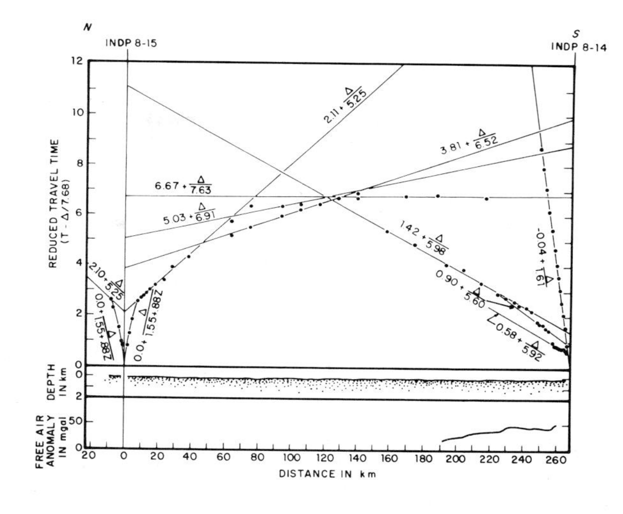

FIG. 12-F

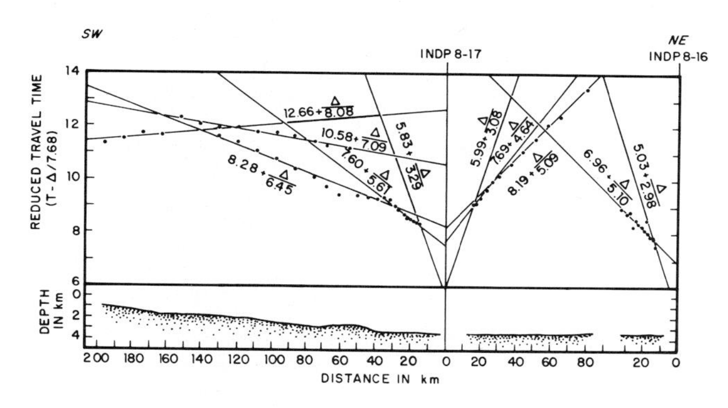

FIG. 12-G

crustal arrivals were received with 109 kg shots to 192 km range. There is evidence from fading of arrivals from the upper crust for a low velocity zone between it and the lower crust. Mantle arrivals are weak, but readable on the individual records where correlation between 3 hydrophones serves to determine first motion.

Stations INDP8-14 and 8-15 were intended as another reversed pair on the continental shelf, east of the Aru Islands and south of Irian Jaya (western New Guinea). INDP8-14 had a single run only; 8-15 had a long incoming run and a short outgoing run. INDP8-14 was close to earlier station MN9B, shot in 1960 and published by Curray et al (1977). As there is a typographical error in the position given for MN9B in that publication, the solution is repeated with the correct data in Table I. Direct water waves were observed for short distances from both receiving stations. As there were few satellite navigation fixes, distances were based almost entirely on dead-reckoning navigation for the longer shots on both stations. On station 14, refracted arrivals from the sea floor were observed out to 18 km with no indication of curvature, indicating nearly constant velocity in the upper sediments. The next deeper layer shows discontinuities in the travel time plot, assumed to be due to faulting. Basement arrivals were detected on 14, but not from deeper layers. On station 15, first layer arrivals form a smooth curve, calculated to represent a velocity-depth function $V = 1.55 + 0.88Z$ km/sec, where Z is depth in km. Because of the differences in structure between stations 14 and 15, each was computed as a one-way profile. The basement velocity however, was derived from the reversal of this layer from the two stations.

Stations INDP8-16 and 8-17 were located in the Aru Basin. Station 16, at the north edge of the basin, was a one-way profile. There were problems with the sonobuoy hydrophones, and good data were obtained only from the short-range shots. No refracted arrivals were detected from the sea floor, and an assumed velocity of 2.5 km/sec was used. Refracted arrivals were received from two deeper layers, which are assumed to correspond to the two deeper sedimentary layers seen on stations to the southwest along the trough system. Reflection profiler records taken along the shooting run and across the basin later show a thick sequence of sediments, dipping in toward the center of the basin. The profiler records did not reach basement. The incoming run of station 17 has arrivals from two sedimentary horizons, plus basement arrivals. Although the two runs did not actually overlap, a reversed solution was made for stations 16 and 17 for the first three layers. Velocities from the plane-layer solution are low in comparison with values for the same layers farther to the southwest; this is not surprising in view of the evidence from the reflection records that interfaces slope downwards from both receiving stations toward the center of the basin.

The outgoing run from station 17 led through the saddle between the Aru Basin and the Tanimbar Trough, and as far down the Tanimbar Trough as signals could be received, to a total range of 196 km. The farthest end of the run overlapped the outgoing run of INDP8-13, and barely reached the end of INDP8-12. On this run, ranges were computed by the reflection-difference method, and ranges on the last few shots were extrapolated assuming constant ship speed.

Refracted arrivals from the second layer were observed on the incoming run of INDP8-17 but not on the longer outgoing run. A velocity derived from the incoming run was used with the first refracted arrival on the outgoing run to define the second layer. There are several records with early first arrivals near the 60 km range; while these arrivals show an apparent velocity of about 8 km/sec and could have been interpreted as representing extremely shallow mantle if the line had been shot no farther, it is most probable that they are caused by thinning of the sediment in the saddle between the Aru Basin and the Tanimbar Trough. They could, alternatively, be the beginning of arrivals from a deep crustal layer that then deepens suddenly. Very strong second arrivals with a smooth travel-time plot were received that are interpreted as refracted arrivals from a deep high-velocity crustal layer. Mantle reflections are strong at critical range; mantle refractions are weak but detectable as first arrivals on the most distant shots.

Station INDP8-17 was computed as a split profile, and as a reverse with station 16, with 13, and with 12.

REFERENCES CITED

Audley-Charles, M. G., 1968, The geology of Portuguese Timor: Mem. Geol. Soc. London, v. 4, p. 1–76.

——— 1975, The Sumba Fracture: a major discontinuity between eastern and western Indonesia: Tectonophysics, v. 26, p. 213–288.

——— and J. S. Milsom, 1974, Comment on Plate convergence, transcurrent faults, and internal deformation adjacent to southeast Asia and western Pacific, by T. J. Fitch: Jour. Geophys. Research, v. 79, p. 4980–4981.

——— and D. J. Carter, 1977, Interpretation of a regional seismic line from Misool to Seram: implications for regional structure and petroleum exploration: Proc. Indonesian Petroleum Assoc., 6th Annual Convention, Jakarta.

Barber, A. J., and M. G. Audley-Charles, 1976, The significance of the metamorphic rocks of Timor in the development of the Banda Arc, Eastern Indonesia: Tectonophysics, v. 30, p. 119–128.

——— ——— D. J. Carter, 1977, Thrust tectonics in Timor: Jour. Geol. Soc. Australia, v. 24(1), p. 51–62.

Beck, R. H., and P. Lehner, 1974, Oceans, new frontier in exploration: AAPG Bull., v. 58, p. 376–395.

Benioff, H., 1954, Orogenesis and deep crustal structure: additional evidence from seismology: Geol. Soc. America Bull., v. 65, p. 385–400.

Cardwell, R. K., and B. L. Isacks, 1978, Geometry of the subducted lithosphere beneath the Banda Sea in eastern Indonesia from seismicity and fault-plane solutions: Jour. Geophys. Research, v. 83, p. 2825–2838.

Carter, D. J., M. G. Audley-Charles, and A. J. Barber, 1976, Stratigraphical analysis of island arc-continental margin collison in eastern Indonesia: Jour. Geol. Soc. London, v. 132, p. 179–198.

CCOP-IOC, 1974, Metallogenesis, hydrocarbons and tectonic patterns in eastern Asia: Bangkok, United Nations Development Programme (CCOP), 158 p.

Curray, J. R., et al, 1977, Seismic refraction and reflection studies of crustal structure of the eastern Sunda and western Banda Arcs: Jour. Geophys. Research, v. 82, p. 2479–2489.

Ewing, J. I., 1963, Elementary theory of seismic refraction and reflection measurements, *in* The sea: New York, Wiley Interscience, v. 3, p. 3–19.

Fitch, T. J., 1972, Plate convergence, transcurrent faults, and internal deformation adjacent to southeast Asia and western Pacific: Jour. Geophys. Research, v. 77, p. 4432–4460.

——— and W. J. Hamilton, 1974, Reply: Jour. Geophys. Research, v. 79, p. 4982–4985.

Gutenberg, G., and C. F. Richter, 1954, Seismicity of the Earth: Princeton, N.J., Princeton Univ. Press, 2nd Edition, 310 p.

Hamilton, W., 1972, Tectonics of the Indonesian region: U.S. Geol. Survey, Project Rept. (IR) IND-20.

——— 1973, Tectonics of the Indonesian region: Geol. Soc. Malaysia Bull., v. 6, p. 3–10.

Heezen, B. P., and D. J. Fornari, 1975, Geological map of the Pacific Ocean, *in* Initial reports of the Deep Sea Drilling Project: Washington, D.C., U.S. Govt., v. 30.

Karig, D. E., 1974, Evolution of arc systems in the Western Pacific: Ann. Rev. Earth and Planet. Sci., v. 2, p. 51–75.

Martison, N. W., D. R. McDonald, and P. Kay, 1973, Exploration on continental shelf off northwest Australia: AAPG Bull., v. 57, p. 972–989.

Mollan, R. G., R. W. Craig, M. J. W. Lofting, 1970, Geological framework of continental shelf off northwest Australia: AAPG Bull., v. 54, p. 583–600.

Montecchi, P. A., 1976, Some shallow tectonic consequences of subduction and their meaning to the hydrocarbon explorationist, *in* M. T. Halbouty, J. C. Maher, and H. M. Lian, eds., Circum-Pacific Energy and Mineral Resources: AAPG Mem., 25, p. 189–202.

Officer, C. B., 1958, Introduction to the Theory of Sound Transmission: New York, McGraw-Hill.

Purdy, G. M., R. Detrick, and G. G. Shor, Jr., 1977, Crustal structure of the Banda Sea and Weber Deep: EOS, Trans. Amer. Geophys. Union, v. 58, no. 6, p. 509.

Raitt, R. W., 1967, Marine seismic refraction studies of the Indonesian Island Arc (abst.): EOS, Trans. Amer. Geophys. Union, v. 48, p. 217.

Robertson, G. A., D. E. Powell, and G. M. Edmond, 1976, Australian northwest continental shelf—Results of ten years of exploration, *in* M. T. Halbouty, J. C. Maher and H. M. Lian, eds., Circum-Pacific Energy and Mineral Resources: AAPG Mem., 25, p. 231–238.

Seely, D. R., P. R. Vail, and G. G. Walton, 1974, Trench-slope model, *in* C. A. Burk and C. L. Drake, eds., The Geology of Continental Margins: New York, Springer-Verlag, p. 249–260.

Shor, G. G., Jr., 1963, Refraction and reflection techniques and procedures, *in* M. N. Hill, ed., The sea: New York, Wiley Interscience, v. 3, p. 20–38.

Veevers, J., et al, 1974, Seismic reflection measurements of northwest Australian margin and adjacent deeps: AAPG Bull., v. 58(9), p. 1731–1750.

Von der Borch, C. C., in press, Continent-island arc collision in the Banda Arc: Tectonophysics.

Whitford, D. J., et al, 1977, Geochemistry of late Cenozoic lavas from eastern Indonesia: Role of subducted sediments in petrogenesis: Geology, v. 5, p. 571–575.

Structure and Cenozoic Evolution of the Sunda Arc in the Central Sumatra Region[1]

D. E. KARIG,[2] SUPARKA S.,[3] G. F. MOORE,[2,4] AND P. E. HEHANUSSA[3]

Abstract The Cenozoic geologic history of west-central Sumatra is governed by the northward movement of the Indian plate with respect to Southeast Asia. The morphology and structure of the western Sumatran margin reflects the cumulative effects of the resulting subduction and right-lateral slip, especially of that since the late Oligocene. The characteristics of this margin are similar to those of other arc systems.

Nias, one of the islands of the trench slope break, consists of mid-Tertiary melange and younger, less deformed slope-basin strata. The trench slope break has been migrating westward since at least the mid-Miocene as shown by the Quaternary pattern of uplift and subsidence and by combined geological and geophysical data along the western flank of the forearc, or upper slope, basin. This forearc basin contains at least 4 km of Neogene sediments. On the western flank of the forearc basin, strata lie on a melange basement and are sharply flexed onto the trench slope break, but on the eastern flank the basin strata appear to lap onto an older continental slope and shelf. Seismic reflection profiles and drilling on the continental shelf delineate a marked unconformity that increases in depth from near zero at the coast to more than 2 km at the old shelf break. Subsidence and transgression of the old shelf by younger shelf sediments began in the early Miocene and may be continuing at present. Beneath the unconformity on the outer shelf there are Paleogene strata that may define an older forearc basin. The landward flank of this suspected basin, beneath the inner shelf, is probably underlain by Mesozoic and Paleozoic metamorphic and igneous rocks that are covered by littoral Paleogene strata. Scattered across the inner shelf and extending into the coastal mountains are numerous, probably Oligocene, andesitic intrusives and their associated extrusive debris. These igneous centers are anomalous in that they are closer to the trench than are the volcanic chains of younger and older ages. Both the uplift which cut the shelf unconformity and the andesites may have been related to the northward migration of a ridge-trench-trench triple junction along the Sunda arc.

The Paleogene littoral strata along the west coast show increasing intensity of folding toward the Barisan Range and at the mountain front are sharply flexed and sheared. The rate of deposition in the offshore basins and the lack of deformation of the shelf unconformity indicate that the major uplift of the Barisan block occurred during the late Miocene and Pliocene, although it may still be proceeding. There is no evidence of a mid-Miocene orogeny in central Sumatra, but instead, of continuous subduction since at least the late Oligocene which has been accompanied by westward migration of morphotectonic units.

INTRODUCTION

The Sunda arc system has served as a classic illustration of an island arc and as the basis for many concepts concerning the development of island-arc orogenic systems. Under Dutch leadership, intensive marine geological and geophysical investigations of arc systems were first undertaken in this general region. Unfortunately, few additional studies of the Sunda arc and its characteristics have been carried out in recent years (Hariadi and Soeparjadi, 1976). It is ironic that, at present, the Sunda arc is poorly known in comparison to other arc systems that have received more attention since the advent of newer oceanographic techniques. Partly as a result of this neglect, the regional kinematic plate evolution of Southeast Asia also remains poorly constrained.

Under the auspices of the IDOE/SEATAR Pro-

[1]Manuscript received, June 27, 1977; accepted, October 31, 1977.

[2]Department of Geological Sciences, Cornell University, Ithaca, New York 14853.

[3]National Institute of Geology and Mining, Bandung, Indonesia.

[4]Present Address: Geological Research Division, A-015, Scripps Institution of Oceanography, La Jolla, California 92093.

This paper is a partial result of cooperative United States-Indonesian studies funded primarily by the National Science Foundation through grants to the first author. We are particularly grateful to Dr. Fred Hehuwat of the National Institute of Geology and Mining (Indonesia) for his aid in facilitating the field studies. We are also pleased to acknowledge the continuous logistic assistance, the gracious permission for access to data, and the personal hospitality given us by the staff of the Union Oil Company. Mr. Harold Billman, formerly of Union Oil, has, throughout our program, provided us with paleontologic data of inestimable value.

Article Identification Number:
0065-731X/78/M029-0015/$03.00/0.
(see copyright notice, front of book)

gram, marine and related land-based geological and geophysical studies of the Southeast Asian arc complexes have been undertaken by geoscientists of Southeast Asian and cooperating countries in order to relate the basic processes occurring along consuming plate margins to the development of mineral and hydrocarbon resources. In particular, our group is studying the Sumatran section of the Sunda arc, principally in the area between Padang and Sibolga (Figure 1). Because the general characteristics of the arc system seem to vary gradually along trend, we have concentrated our efforts along a transect through the island of Nias, as proposed by the CCOP/IDOE Bangkok Workshop (CCOP-IOC, 1974). Our investigations to date include mapping on Nias, along the west coast of Sumatra, and on the offshore islands that lie between, as well as marine studies in the water-covered areas of the transect. These data have been augmented by the extensive seismic reflection and drilling data of the Union Oil Company of Indonesia, which, together with Pertamina, has made many of these data available to us. The primary goal of the transect study is to gain a clearer and more detailed picture of a subduction zone and of the suduction process itself. The results of the study are strictly applicable to the central Sunda arc system, but we feel that many of the conclusions can be extrapolated, in varying degrees, to other arc systems.

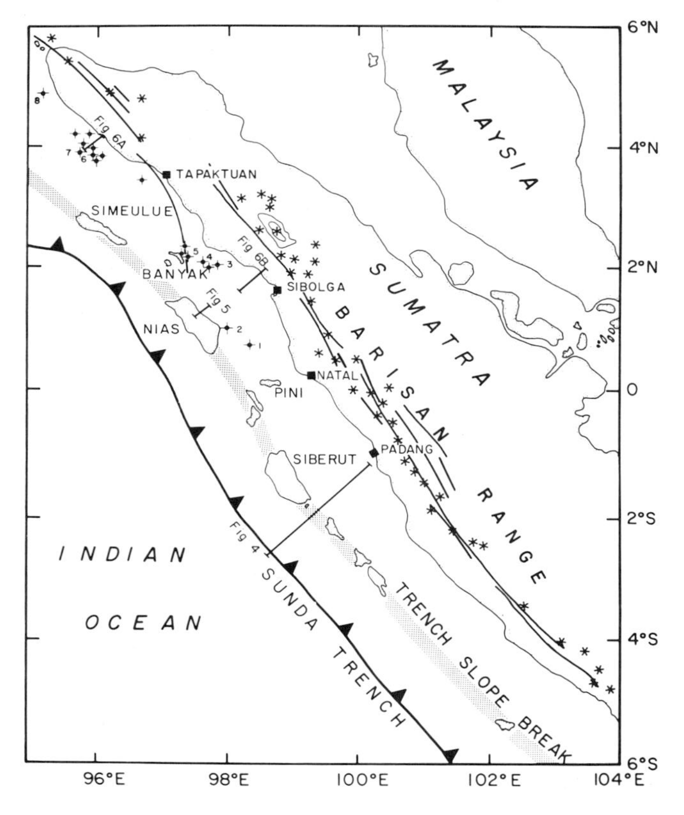

FIG. 1. Index map to Sumatran section of the Sunda arc. In addition to labelled tectonic elements the writers have noted Union Oil Company exploratory wells (filled circles) and the approximate locations of profiles shown in later figures. The numbers beside some wells are keyed to Table 1.

BACKGROUND

The kinematic framework of the Sumatra margin during the past 100 m.y. has been controlled by the northward motion of the Indian plate relative to the Asian plate (Sclater and Fisher, 1974; Molnar and Tapponier, 1975). This motion has given rise to a persistent but variable combination of subduction and right-lateral slip along the Sumatran sector of the plate margin. The known Cenozoic geology of Sumatra reflects this motion to a reasonable degree, although there are aspects still unresolved. Prior to 100 m.y. ago, Malaysia and, presumably, Sumatra lay well north of the Equator and far from the northern margin of Gondwanaland (McElhinney et al, 1974; Haile et al, 1977), and data with which to unravel this earlier history are scarce. Apparently Sumatra has formed a part of a convergent margin intermittently since at least the Permian, as implied by the granitic plutons and the arc-related lithologies along the length of the island (Hutchison, 1973; Katili, 1975).

The Cenozoic subduction history of Sumatra is constrained by the magnetic anomaly pattern of the Indian Ocean (Sclater and Fisher, 1974) and by finite-motion plate reconstructions using a global plate network (e.g., Molnar and Tapponier, 1975). The validity of this approach is weakened by the assumption that the southeastern part of the Asian plate has been rigidly attached to Eurasia throughout the Cenozoic. Not only has internal deformation within China been noted (e.g., Molnar and Tapponier, 1975), but Haile et al (1977) and McElhinney et al (1974) have represented paleomagnetic data indicating a 50-degree anticlockwise rotation of the Malay Peninsula since the mid-(?)Cretaceous. Unless there was extensive crustal deformation beneath the basins of eastern Sumatra, this rotation must also have affected Sumatra and the orientation of its subduction boundary. Unfortunately, the deformation history of the Mesozoic and older metamorphic and igneous rocks forming the basement of eastern Sumatra (de Coster, 1975), has not yet been unravelled. The geology of the strata within the basins indicates that since the late Eocene there has not been large-scale motion between Sumatra and Malaysia, with the likely exception of strike-slip faulting (de Coster, 1975). The late Cenozoic clockwise rotation of Sumatra and Malaysia postulated by Ninkovitch (1976) is not supported by data presented in this paper and has already been questioned by Tjia (1976).

If we assume, for the moment, that internal deformation within the Asian plate has not been large during the past 60 m.y., the following results emerge from the plate kinematic framework shown in Figure 2. These results should be viewed as semiquantitative and as primarily indicating periods of change in plate kinematics. Subduction along the Sumatran part of the Sunda arc from the Cretaceous to the late Paleocene was very rapid (greater than 15 cm

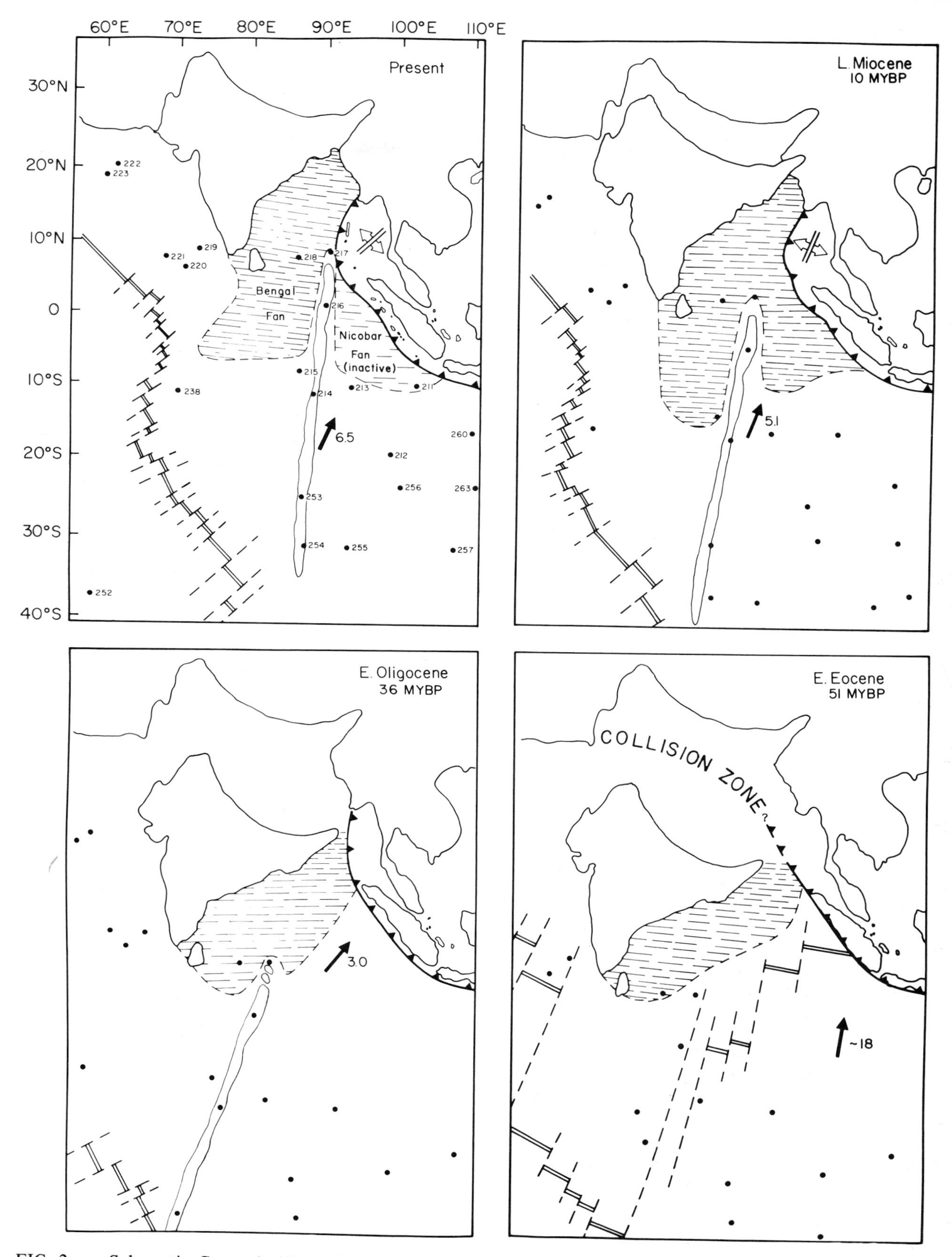

FIG. 2. Schematic Cenozoic kinematic history of the Sunda convergence zone, derived from a global set of relative finite plate movements (after Sclater and Fisher, 1974; Molnar and Tapponier, 1975). In this set of reconstructions, no attempt was made to model the India-Asian collision boundary. The extent of the sediment lobe south of India is largely taken from Curray and Moore (1974). The calculated subduction vector at Nias is given by the heavy arrow with the rate in cm/yr. DSDP sites are shown with site number on the "Present" panel.

per year; Molnar and Tapponier, 1975; Moore, 1978). Subduction during the Cretaceous is recorded in Sumatra by the igneous rocks of this age which are interpreted to represent a magmatic arc (Katili, 1973, 1975). Late Cretaceous and early Tertiary subduction is also inferred from ophiolite exposures on the Andaman Islands. Rocks of the ophiolite suite outcrop along the east coasts of the Andamans and include ultramafics, volcanics and chert and are overlain by Paleocene and Eocene slope sediments (e.g., Karunakaran et al, 1964; Chatterjee, 1967).

A change in relative plate motion in the Indian Ocean occurred in the late Paleocene (anomaly 21, Sclater and Fisher, 1974). Curray and Moore (1974) and others suggest that this time (55 m.y. BP) corresponds to the initial contact between the Indian continental margin and Eurasia. Graham et al (1975) and Hamilton (1977) suggest that at this time, subduction ceased and continental margin sediments began to prograde over the old subduction zone. We prefer to interpret this as a time of slowing of subduction rate from greater than 15 cm per year to approximately 3 cm per year, following Molnar and Tapponier (1975).

Some time after the time of anomaly 17 (41 m.y. BP), the spreading center east of the Ninety-East Ridge ceased spreading and has since been partially subducted (Sclater and Fisher, 1974). At the time of anomaly 13 (36 m.y. BP) renewed spreading in the Indian Ocean (Sclater and Fisher, 1974) caused an increase in the rate of subduction to approximately 5–6 cm per year at Sumatra. Oligocene uplift in the Himalayas due to this increased subduction rate has led to the influx of very thick fan sediments into the subduction zone (Graham et al, 1975).

DESCRIPTIVE ANALYSIS

The morphology and structure of the western Sumatran margin (Figure 3) reflects the cumulative effects of subduction during the Cenozoic, especially since the mid-Tertiary. The resulting morphotectonic belts or units of the Sunda orogen have long been recognized by van Bemmelen (1933), Vening Meinesz (1940), Umbgrove (1947) and others. What has changed since those earlier analyses is (1) a vastly increased supply of marine data, including oil well data and multichannel seismic reflection profiles, and (2) the new concept of plate tectonics into which to fit the observations. The continuing study of arc systems supports the early belief that these features show a fundamental similarity, but it has clarified their characteristics and has led to much reinterpretation of the behaviour of arc units. For this reason a new, more general arc terminology has been developed and is used in this paper.

From the Indian Ocean basin inboard, the principal morphotectonic units recognized in this study are trench, lower trench slope, trench slope break, upper slope or forearc basin, frontal arc and volcanic arc (Figure 3). These terms correspond with those used to describe arc systems in general (Dickinson, 1973; Karig, 1974). The trench slope break is equivalent to the nonvolcanic or outer arc (Umbgrove, 1947) and the upper slope basin was termed

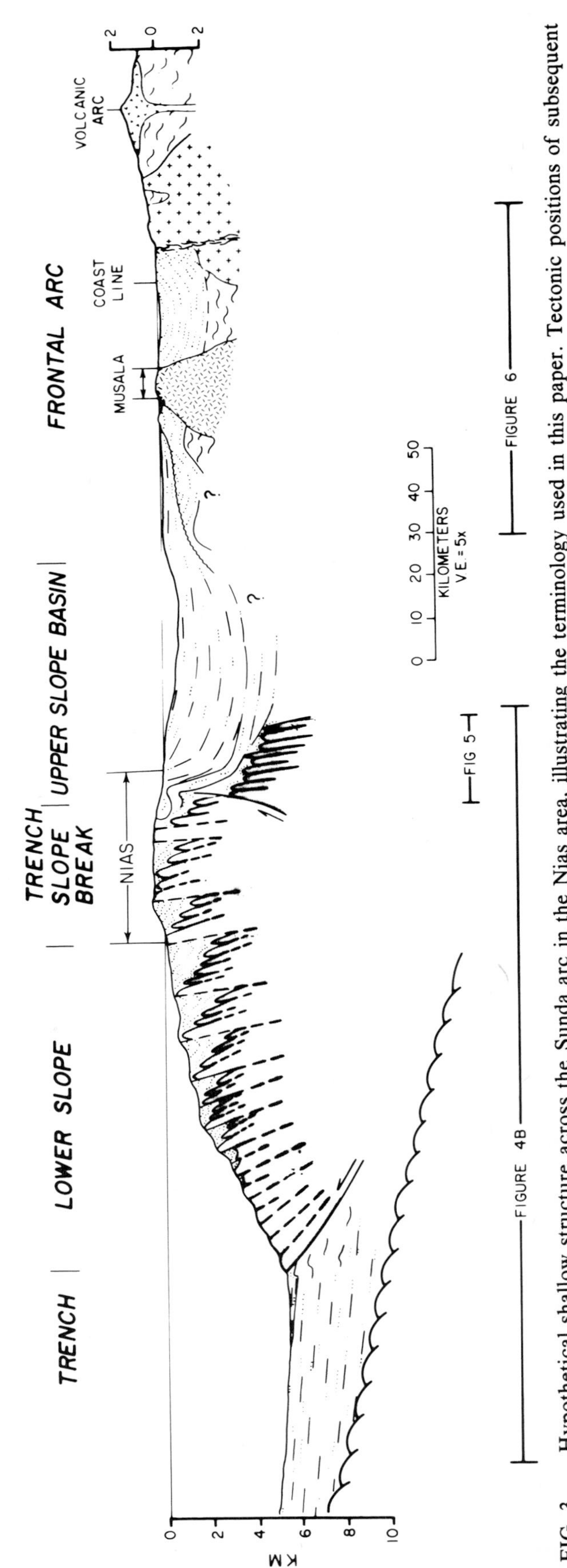

FIG. 3. Hypothetical shallow structure across the Sunda arc in the Nias area, illustrating the terminology used in this paper. Tectonic positions of subsequent profiles are indicated beneath the section. The geologic control for this section is discussed in the text.

the interdeep by van Bemmelen (e.g., 1970).

Although the Sunda Trench off Sumatra neither is as deep nor has as much relief as most other Pacific trenches, its characteristics are consistent with those of a typical trench. The characteristics noted above and the northward shallowing of the trench appear to be either directly or indirectly a result of the thick Bengal and Nicobar fan sediments on the downgoing plate. The relationship is not simple, however, because the thickness of fan sediments that has been subducted in the central Sumatran sector of the trench has varied greatly since the mid-Tertiary (Figure 2). Nevertheless, it appears likely that the accretion of thick fan sediments since the late Miocene (Curray and Moore, 1974) has caused a rapid outgrowth of the accretionary prism to its present 200-km width (Hamilton, 1973). The load of this accreted material has in turn caused the flattening of the downgoing plate and the moderation of the original flexure of the downgoing plate. The dip of the outer trench slope at the trench axis is thus only about 2 degrees, in contrast to 5- to 6-degree dips for the Mariana, Tonga, and Java Trenches (Karig et al, 1976).

Since about late Pliocene or Pleistocene time (Curray and Moore, 1974) the southward flow of Nicobar fan sediments on the east side of the Ninety-East Ridge has been stifled by the impingement of that ridge with the Sunda Trench in the Andaman region (Figure 2). At present, the surface distribution of fan sediments on the downgoing plate at the trench is restricted to the area north of about 10°N (Venkatarathnam and Biscaye, 1973; Curray and Moore, 1974), and the total thickness of fan sediments in the Sumatran trench sector increases northward from about 500 m at the Sunda Straits to about 750 m near Nias (McDonald, 1977). The wedge of turbidites that has been deposited in the trench itself along Sumatra is relatively thin (100–500 m) because the drainage and sediment flow in Sumatra is dominantly eastward and because most westward-flowing sediments are trapped in the basins on both the upper and lower trench slope.

The lower section of the inner trench slope is corrugated by ridges and troughs (Figure 4). Recent marine surveys in several arc systems and mapping on Nias indicate that this pattern reflects a system of active thrust faults which decrease in number and rate of displacement upslope (Karig et al, 1975; Moore and Karig, 1976). The thrusts are thought to surface primarily at the bases of the ridges and to deform sediments deposited by turbidite and hemipelagic processes in the slope basins. As a result of this thrust distribution, internal deformation in the slope basins decreases in intensity upward from the contact with the melange, but is generally much less than that of the melange. The slope basins in most arc systems increase in width upslope as thrusting diminishes. At Nias they average about 10 km in width and perhaps 3 to 4 km in depth by the point where they reach the trench slope break (Figure 4).

The island of Nias comprises both melange and slope sediments. The melange, informally termed the Oyo Complex, consists predominantly of siltstone, sandstone, and pebble conglomerate, all of which show clear evidence of having been transported by turbidity current mechanisms. Clast composition and the very coarse average grain size of the clastic rocks in the Oyo Complex indicate that these sediments were derived from Sumatran sources and were not parts of the Bengal or Nicobar fans (Moore, 1978). Shales make up only about 10 to 15 percent of the complex. Approximately 10 percent of the complex consists of basalts—commonly pillow-type, altered pelagic sediments, and a very small percentage of deeper crustal rocks. These lithologic units, which occur as blocks tectonically mixed into the clastics, are most easily interpreted as having been stripped off the oceanic crust. They also provide one piece of evidence that the Oyo clastics were deposited in a trench, rather than being highly deformed slope sediments.

No indigenous fossils have been found within the Oyo Complex, but apparently reworked Eocene foraminifera (Douville, 1912) imply that it is no older than Eocene. The probability that Oyo sediments were deposited in the trench, were deformed and were subsequently overlain by sediments deposited on the lower slope would indicate that the Oyo in the western half of Nias is earliest Miocene in age. However, the complex would be younger toward the trench (westward) and older eastward if there has been continuous accretion and westward migration of the trench. We have found no pre-Tertiary basement rocks on Nias. Outcrops on the southwest coast cited in van Bemmelen (1970, p. 165) as crystalline schist are actually Oyo melange, but with a greater than usual component of basaltic and other oceanic crustal rocks.

Our mapping study of the slope sediments on Nias and their structural relationship to melange has greatly clarified our understanding of the distribution of deformation across the inner slope of the Sunda Trench. In brief, the upward decrease in deformation in the slope basins is accompanied by an upward decrease in depth of sediment deposition and by an irregular but very distinct change in sediment type.

Near the base of the sediment column, calcareous lutite or siltite turbidites predominate, giving way upward to coarser turbidite units including coral rubble-bearing calcirudites. Shallow-water fossils and obvious littoral clast components, including large lignitic wood fragments, are included in the turbidites, leading van Bemmelen (1970, p. 166) and others to interpret these deposits as largely shallow-water to nonmarine. However, the mode of deposition and the benthic foraminifera that lived at bathyal to abyssal depths and were recovered from these slope sediments supply a minimum water depth (H. Billman, personal communication) and demonstrate that most clast types have been redeposited. The lutite turbidites probably represent slumps from adjacent ridges, whereas the coarser clastic sediments were transported down submarine canyons and

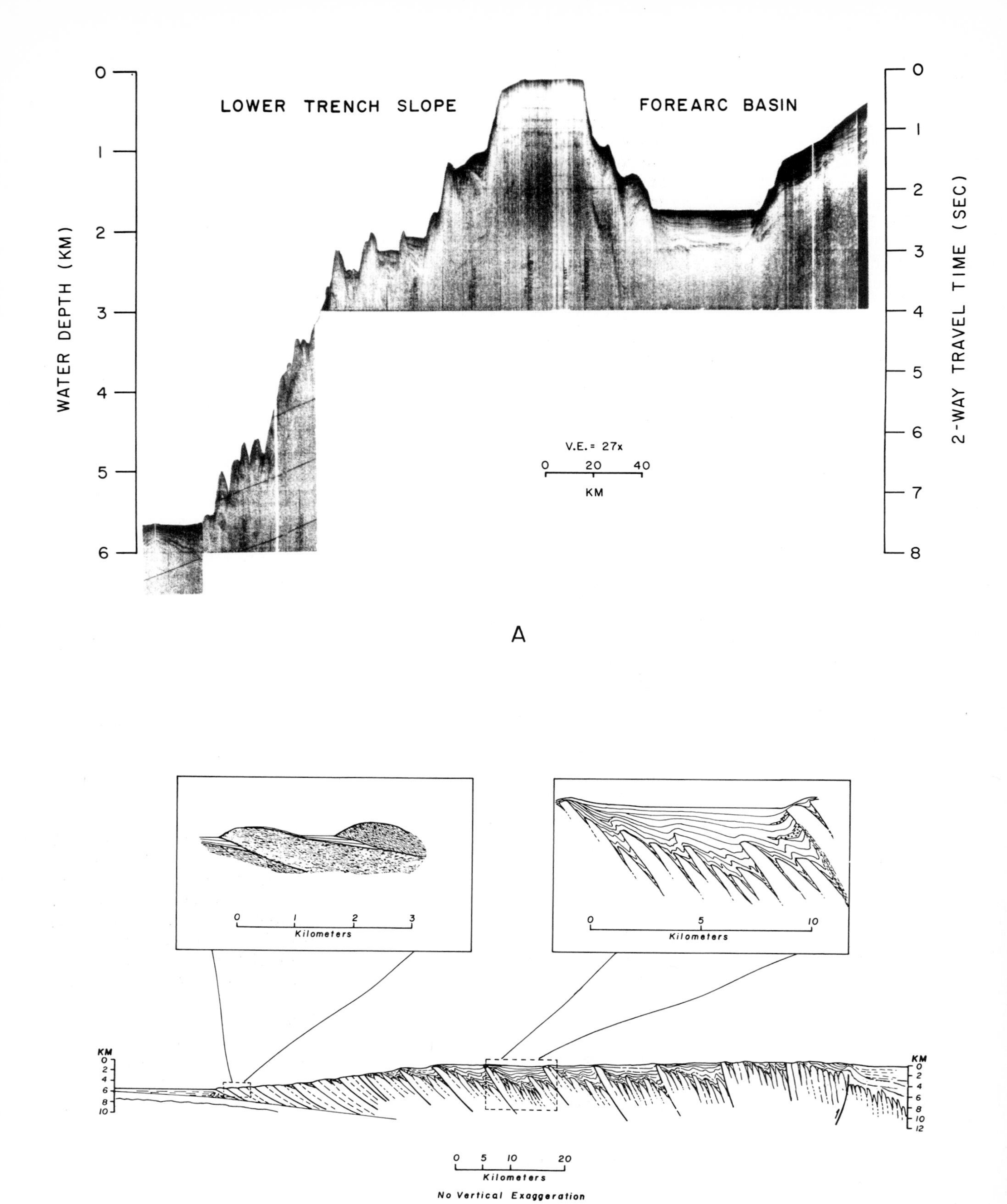

FIG. 4. **A, Seismic reflection profile** from Scripps Institution of Oceanography *Antipode* cruise across the Sunda arc near Siberut (Karig, 1977) illustrating ridges and basins on the lower trench slope and large, subsiding upper slope (forearc) basin east of Siberut. **B**, Deduced shallow structure across the lower trench slope (Moore and Karig, 1976) showing relation between accreted "melange" and slope sediments. Insets indicate the change in structural relations from the lowermost to upper sections of this slope.

channels, some of which could be identified in the field.

The age of the basal slope sediments that are interstructured or interthrust with the melange and those in quasi-depositional contact with it is consistent across an area the size of one slope basin (approximately 10 km). This consistency, and the paucity or lack of slope sediment slivers within more extensive and deeply eroded melange bodies is evidence that broad-scale churning and imbrication within the accretionary prism (Hamilton, 1973, 1977) has not occurred, at least on Nias. Our data suggest that the maximum depths of the slope basins, or the maximum throw on thrusts involving slope sediments is less than 5 km.

The trench slope break, from which the nonvolcanic islands rise, marks a balance point between uplift, caused by accretion, and subsidence, largely attributable to loading of the downgoing plate (Karig et al, 1976, and references therein). For example, the row of small uplifted coral islands west of Nias (Kep. Hinako, P. Wunga) marks a ridge which is rapidly rising in comparison to the main island. The eastern sides of Nias and of most of the trench slope break islands are either stable or are subsiding (Verstappen, 1973; and mapping, this study). These regimes of uplift and subsidence form bands parallel to the arc, but there are definite swales and culminations along trend. These appear on structural contour maps derived from reflection profiles and can be deduced from Quaternary shoreline data. On Nias, both ends of the island are presently submerged, whereas the center contains the highest elevations (886 m) suggesting that the topography closely reflects the relative uplift along the trench slope break.

The zone of relative subsidence along the eastern flank of the trench slope break is of major interest because it marks the seaward flank of the upper slope or forearc basin. Its shallow structural expression is a steep homocline or flexure, separating moderately deformed slope sediments from nearly flat-lying sediments of the forearc basin (Figure 5). Multichannel reflection profiles, nine paleontologically controlled geologic traverses, and data from the Union Oil Suma-1 exploratory well have here been used to decipher the shallow structural relationships and tectonic history of this feature in the central Sunda arc.

On Nias the flexure is a zone approximately 3 km wide in which strata dip steeply east or, in one area, are overturned with a steep westward dip. Strata within the flexure can be traced with reflection profiles and paleontologic data into the forearc basin but there are marked facies changes between the outcrop and the well section. In the flexure, strata of mid-Miocene (N 10 of Blow (1969) zonation) to latest Miocene (N 16) were deposited in bathyal water depths. In the Suma-1 well, strata of mid-Miocene (N 14) age are nonmarine whereas younger strata are marine and reflect subsidence in the forearc basin. Structural relief across the flexure is therefore on the order of 3 km, with both sides having moved in opposite directions with respect to sea level.

A northward plunge of the flexure, shown by geometries of secondary structures and by the strong northerly slope of reflecting horizons to the east of the flexure, results in the exposure over a vertical span of several kilometers within that structure. At shallow levels, a Pliocene and younger littoral sequence, capped by a flight of regressive reef terraces, unconformably overlies the eastern half of the flexure and is only slightly tilted. This unconformity is strongly angular within the flexure but rapidly dies out eastward. The age and environmental relationships associated with the unconformity point toward an inversion of structural relief primarily in late Miocene to early Pliocene time. The change in facies distribution was apparently contemporaneous with the development of the flexure.

To the south, the dips along the flexure steepen to near-vertical and, for several kilometers in south central Nias, beds are overturned and cut by reverse faults. Because of the high dips and degree of deformation a reflection profile that extends into the core of the flexure in this area (Figure 5) is incapable of defining the structure beneath the exposures. However, control is adequate to conclude that the basal strata of the forearc basin do outcrop within the flexure. Nowhere along the flexure are forearc basin strata converted to or imbricated into the melange as proposed by Hamilton (1973, 1977). Correlative strata west of the flexure on Nias are identified as lower slope sediments because they form part of upward-shallowing, westward-verging synclinal structures.

Over the range of structural depth exposed, the flexure is dominated by vertical displacement. This displacement may have resulted from underlying horizontal forces, as back-thrusting associated with accretion, or as strike-slip faulting associated with oblique subduction. The remarkably straight trace of the flexure over the entire length of Nias and into the offshore areas is more typical of strike-slip faulting than of back-thrusting, but the few secondary folds and other structures within the flexure indicate dip-slip displacements.

Seismic reflection profiles across the trench slope break from Siberut northward to and beyond Nias (Figs. 1, 4) are similar in showing a bench- or step-like descent of acoustic basement from the trench slope break into the forearc basin. On some of these profiles the strata on the treads have a slight westward tilt (e.g., Fig. 4). The risers between these treads are interpreted as flexures similar to that previously described on Nias. If such be the case, the back-tilted treads may have resulted from rotation associated with movement along underlying reverse faults that flatten downward.

The nature of the deeper part of the forearc basin and of the underlying basement is not yet known. The few industrual reflection profiles across the center of the basin show no evidence of acoustic basement at two-way travel times of 4 seconds. The deep basin strata might overlie a basement of either

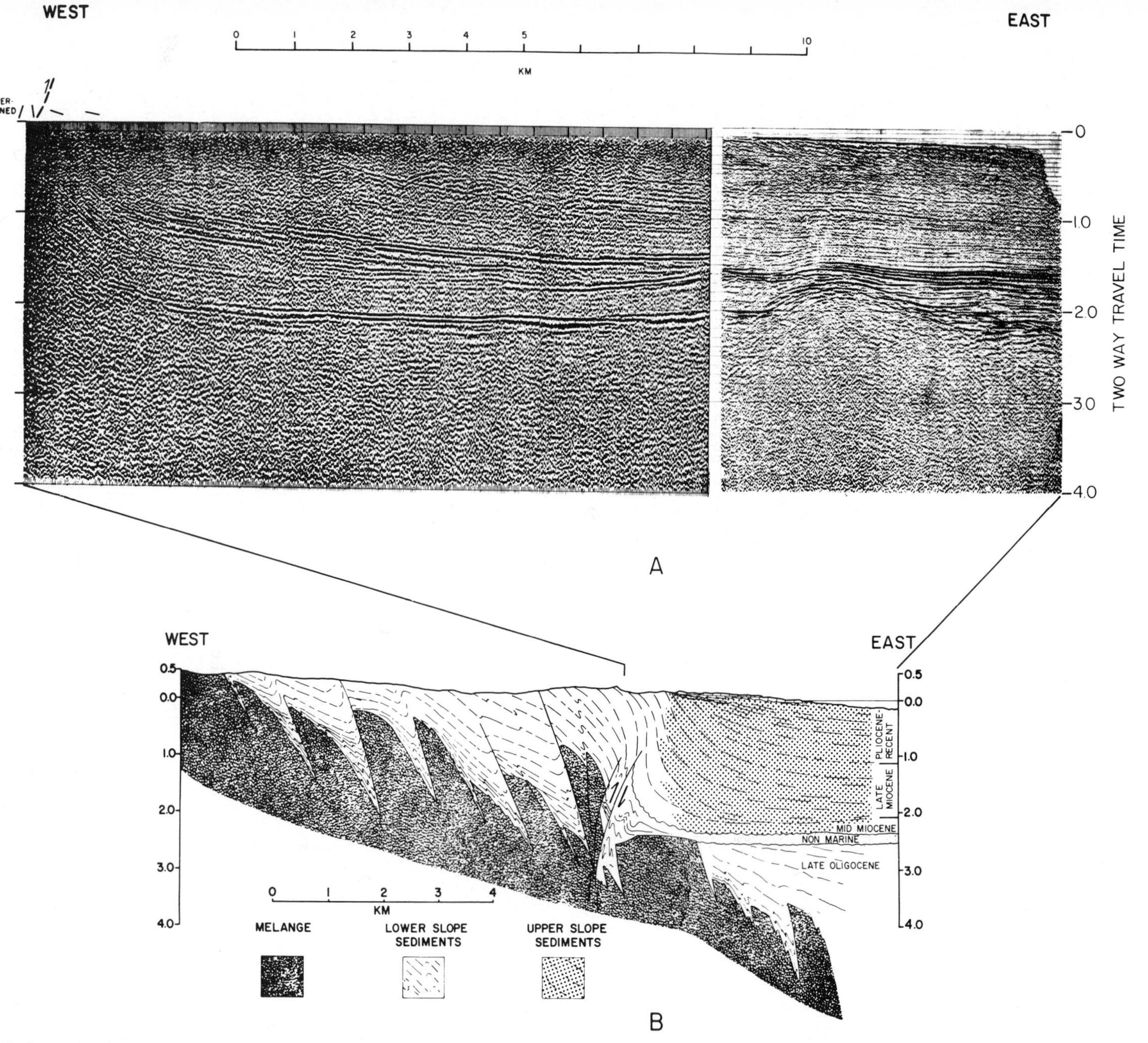

FIG. 5. Reflection profile and composite section across the flexure separating the trench slope break from the upper slope (forearc) basin at Nias. **A**, Seismic reflection profile across east Nias and offshore shelf area from Union Oil Company. This profile shows the degeneration of acoustic coherence in the flexure where steep dips have been mapped. Dips along a geologic traverse coincident with the seismic line are noted at the west end of the profile. Prominent basal reflector is the mid-Miocene calcarenite and reef horizon. **B**, Composite section based on numerous Union Oil reflection profiles, the Suma 1 well, and several geologic transects across eastern Nias

oceanic crust (Curray et al, 1977) or melange (Karig, 1977).

The interpretation that oceanic crust underlies forearc basins is supported by ophiolite occurrences on the seaward flank of some basins and by a seismic velocity structure similar to that of oceanic crust (Curray et al, 1977). However, ophiolite sheets can be accreted to the trench slope and thus would be structural entities within a larger subduction complex. Moreover, the velocity profiles beneath forearc basins (Curray et al, 1977; Shor and von Huene, 1972) show layer thicknesses quite different from that of oceanic crust and the velocities themselves are not close enough to oceanic velocities to be convincing. Oceanic crustal fragments, accreted in rough stratigraphic sequence to the base of the accretionary prism could also produce the observed velocity structure.

We presently favor a melange basement in the central Sunda arc for several reasons. If the subduction has been continuous since the Cretaceous, as claimed elsewhere in this paper, the material accreted during the Paleogene should occur along the Sumatran margin, although possibly disrupted by strike-slip faulting. These subduction complexes would have to have been emplaced west of the shelf edge because well data coupled with reflection profiles outline the existence of thick Paleogene basin strata in that area. A possible view of the Paleogene subduction complex beneath the forearc basin is afforded in the Banyak Islands (Fig. 1). These islands lie farther from the trench than Nias and occur on the west side of the prominent fault zone (Fig. 1). Undated melange on the westernmost of the islands apparently lies structurally beneath a sequence of pelagic calcareous mudstones on the next island to the east. These strata appear to be lower slope sediments with a latest Oligocene to earliest Miocene age (H. G. Billman, personal communication). They are thus slightly older than slope sediments mapped on Nias and, because of their moderate deformation and distance from the melange, there is very probably a significant thickness of older slope strata beneath.

During the Neogene the western flank of the forearc basin has migrated westward from the structural terrace on which the Suma-1 well (Fig. 1; Table 1) was drilled to its present position at the flexure on Nias. The mid-Miocene, or possibly slightly older, nonmarine strata in this well can be identified with a trench slope break because a deep structural basin to the east had already been in existence since at least the early Miocene. This mid-Miocene basin, which forms the eastern part of the present forearc basin, separated the Suma well site from the coeval Sumatran shelf edge (see next section and Fig. 6). The distribution of Quaternary vertical displacements around the present trench slope break demonstrate that the westward migration is continuing.

The structure of the east flank of the upper slope basin is dominated by a marked unconformity which underlies most of the Sumatran shelf at depths that increase westward from near zero at the coast and in the Natal area to as much as 2 km at the shelf edge north of the Banyak Group (Figs. 3 and 6). At the shelf edge the unconformity changes character and becomes a part of a series of westward-dipping reflectors. These are overlain by nearly flat-lying basin strata and together indicate the onlap of continental slope strata by sediments infilling the forearc basin (Fig. 6B), much as is presently occurring in the deeper subbasin to the south (Fig. 4A).

The sediments above the shelf unconformity are well explored by Union Oil Company by means of reflection profiles and drill holes, but the rocks and structures beneath the unconformity are nearly unknown. The strata immediately above the unconformity range from early Miocene (N 4) carbonate complexes at the shelf edge (e.g., Meulaboh-1 well, Fig. 1; Table 1) to very young clastics near shore (e.g., Lakota well, Fig. 1; Table 1), and record subsidence and transgression since the early Miocene which is possibly still proceeding. The present tectonism, however, is masked by the effects of Quaternary sea level oscillations resulting from glaciation.

The unconformity is relatively undeformed in most areas. In the Natal region (Fig. 1) it is warped into a broad arch that has developed transverse to the arc system. This late Tertiary arch has exposed Permian-Triassic metamorphic and Mesozoic igneous rocks along the coast and late Tertiary basinal sediments on the island of Pini (Fig. 1).

The most pronounced deformation of the unconformity is related to the fault zone of apparent large displacement that extends from the Banyak Islands north-northwestward to the Sumatran mainland near Tapaktuan (Fig. 1). This fault is interpreted by Union Oil Company geologists as having significant horizontal displacement because it throws the west side up in the Banyak Islands and has an apparent upthrown eastern side along the Sumatran coast. Although there has definitely been several kilometers of vertical offset of Miocene strata in the Banyak area, some of the offset at the north end of the fault appears to predate the unconformity. It is quite possible that this fault has been a major right-lateral shear associated with the oblique subduction in the Sunda arc.

There may be other faults of this nature cutting the accretionary prism off central Sumatra judging from the apparent offsets in the edge of continental crust and from zones of disturbance associated with north-south trending troughs cutting the trench slope break to the south and north of Nias. For example, the Suma and Panjang wells (Table 1) lie roughly along a line parallel to the trend of the arc (Fig. 1) but are separated by the trough south of Nias. However, the Suma well lies on the west flank of the forearc basin whereas Panjang appears to be on the Sumatran shelf edge. A right-lateral strike-slip fault along the trough could account for this situation but much more data analysis is needed to provide a reliable interpretation.

The sedimentary section beneath the unconformity seldom produces clear reflections, but local areas of

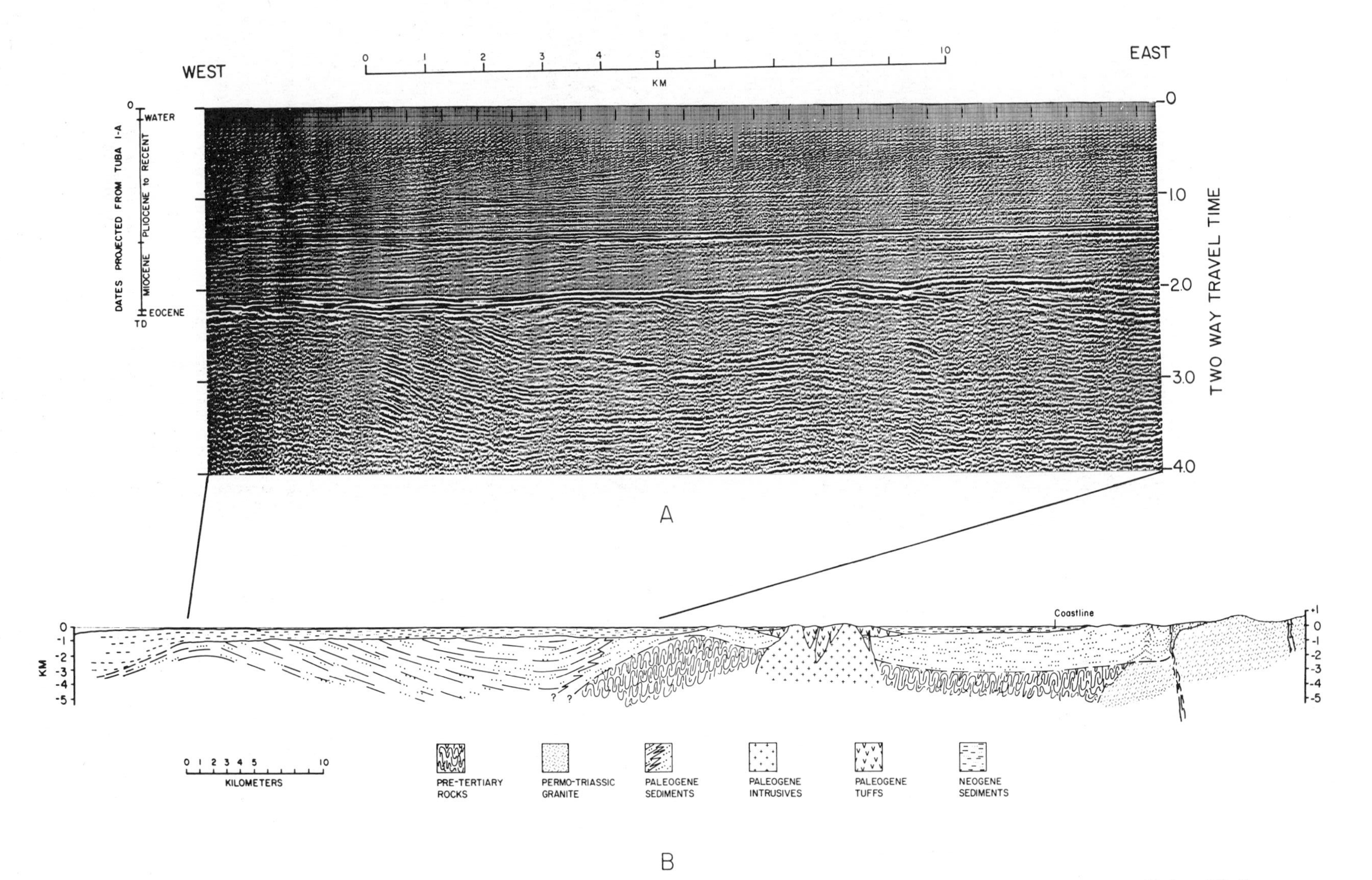

FIG. 6. Reflection profile and composite section across the shelf off northern Sumatra. **A,** Processed reflection profile from Union Oil Company across the outer shelf. Control above the marked unconformity is supplied by data from a number of nearby wells (see Fig. 1, Table 1). Age control is projected from nearby Tuba 1A well. Eocene and Oligocene units were penetrated beneath the unconformity (Table 1). **B,** Composite section across the Tapanuli district north of Sibolga, based on unpublished field work of Suparka and Karig, and Union Oil Company reconnaissance mapping. Seaward extent and depth of metamorphic basement is very approximate and is based on projection from the Natal district and from the Lakota well.

TABLE 1. CRITICAL WELLS OFF WESTERN SUMATRA

Well	Location	Total Depth		Synoptic Results
1. Panjang-1	00°45'N 98°18'E	7565	194-3794'	Recent to Pliocene, coral-algal detrital limestone over calc-sandstone, mud and silt
			3794-6190	Burdigalian-Aquitanian shoal detrital limestone
			6190-7125	Aquitanian sandstone, shale, silt; marginal marine
			7125-7250	Early Oligocene shale, silt, sandstone (turbidites?)
			7250-t.d.	Reportedly Jurassic but redetermined to be Oligocene and/or Eocene shale, silt, sandstone, bioclastic limestone.
2. Suma-1	01°01'N 98°00'E	8910	213-3400'	Recent to Pliocene bathyal to littoral clay and detrital limestone
			3400-6000	Late Miocene bathyal to littoral clays, silts, and reefal limestone
			6000-7434	Mid-Miocene shelf detritus over reef limestone
			7434-t.d.	Mid-Miocene or older non-marine sandstone, silt, coal
3. Lakota	02°01'N 97°54'E	4494	179-4220'	Recent-Pliocene sandstone, clay, silt, and detrital limestone, all of a shallow water nature
			4220-t.d.	Metatuff showing intense penetrative deformation: probably pre-Tertiary in age.
4. Singkel-1	02°02'N	6990	361-1890'	Recent Pleistocene reef limestone, clay, sandstone
	97°43'E	(2131)	1890-4060	Pliocene, sandstone and clay
			4060-t.d.	Late Miocene clay, reef limestone, bottoming in hard sandstone
5. Palembak-1	02°09'N	10,270	200-400'	Recent Pleistocene reef limestone
	97°28'E	3131	400-3150	Pliocene reef limestone over interbedded sandstone, clay, silt
			3150-6500	Late Miocene clay, silt, turbidites
			6500-t.d.	Mid-Miocene clay, silt, turbidites
6. Meulaboh-1	03°52'N 96°01'E	10,079	260-5440'	Recent-Pliocene conglomerate, sandstone, mud, with some reef or detrital limestone intervals
			5450-8262	Miocene mudstone over early Miocene shallow-water limestone
			8262-8445	Eocene mudstone, sandstone (turbidites?)
			8445-t.d.	Reportedly Jurassic but probably Eocene mudstone and shale
7. Tuba-1A	03°58'N	8615	275-5118'	Recent-Pliocene mudstone, sandstone, minor limestone
	95°53'E		5118-8245	Miocene mudstone over early Miocene shallow-water limestone
			8245-8528	Eocene limestone over shale and silt
			8528-t.d.	Shale, silt, reportedly Jurassic, but probably Eocene
8. Raja-1	04°55'N	4677	0-1136'	Recent-Pliocene reef limestone
	95°13'E	(1426)	1136-1356	Pliocene-Late Miocene sandstone, clay
			1356-2400	Mid-early Miocene clay
			2400-4640	Eocene-Oligocene clay, shale, calc-sandstone
			4640-t.d.	Unfossiliferous dolomitic limestone

steep dips and other complexities suggest that there was significant deformation of the shelf before the Miocene. The rocks beneath the unconformity include older Tertiary sediments, volcanics and intrusives and Paleozoic(?) metamorphics, but the extent and the geometric relationships among these units are virtually unknown. Along the coast in the Tapanuli district, the rocks beneath the unconformity are Paleogene quartzose to arkosic sandstones and minor noncalcareous shales and conglomerates. These strata are referred to as the Sibolga beds by Union Oil Company geologists and represent a littoral to lagoonal sequence. East of the coastal plain the Sibolga beds overlie granitic plutons of Permian-Triassic age (Katili, 1973b) and metasediments intruded by those granites. The Sibolga beds are cut and overlain by andesitic intrusives and their associated extrusives on islands and on headlands near Sibolga (Fig. 7). Clasts of this igneous sequence in early Miocene conglomerates on Nias provide a minimum age for volcanic activity. For this reason we feel that the Sibolga beds must be Paleogene, probably late Eocene and/or Oligocene.

The Sibolga beds around Tapanuli Bay are mildly and openly folded around axes trending 300° in the most seaward exposures but become increasingly more tightly folded toward the Barisan mountain front (Fig. 7). At or near the mountain front, the Sibolga beds are sharply flexed into a monocline with at least 2 km relief, but they resume a nearly horizontal attitude on top of the Barisan massif. In the few exposures of the flexure, north of Sibolga, there

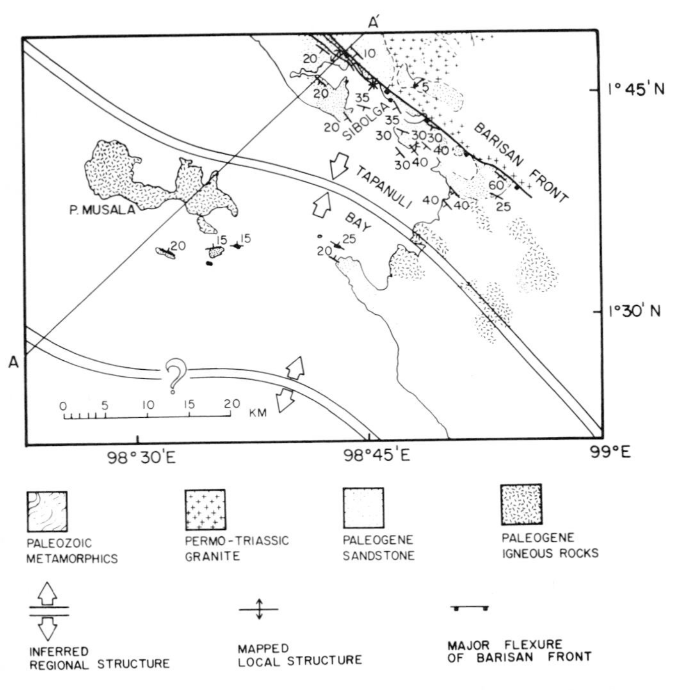

FIG. 7. Sketch geologic map of Tapanuli district from work in progress by Suparka and Karig. The structure shown is simplified and no attempt is made to show the complex deformation in and near the major flexure that bounds the Barisan Range. The profile in Figure 6B has been constructed along section A-A′.

are extremely tight folds and shear zones indicating that strong compression accompanied uplift. Although one might also suspect horizontal slip along this zone, we have as yet not been able to find any evidence supporting such displacement.

The age of deformation and uplift along the western Barisan front also remains uncertain. Sedimentation rates and total sediment volume in the upper slope basin, as determined using reflection profiles and well data, increased sharply during the late Miocene and remained at least as high through the Pliocene. Thus the data from the west flank of the Barisan uplift suggests a late Miocene to Pliocene, perhaps continuing, uplift that supports the conclusion of de Coster (1975) and others based on data from basins along the eastern flank of the Barisan Range. The folding of the Sibolga beds does not seem to have affected the units above the unconformity, which in the Sibolga area (Fig. 1) are of probable Pliocene age. It is most likely, therefore, that deformation and maximum uplift occurred together in the late Miocene–early Pliocene.

Because of their massive bedding, the Sibolga beds off the coast cannot be resolved on seismic profiles and thus cannot be physically correlated with early Tertiary marine units that were penetrated in several wells on the outer shelf (Table 1). Eocene and possibly Oligocene sands, shales, and limestones with sedimentary structures that indicate turbidite deposition underlie the unconformity and form a clearly delineated trough beneath the outer shelf that can be traced for over 100 km northwestward from the Banyak Islands-Tapaktuan fault zone. Strata within the trough dip up to 20° east and are at least 2 km thick. This Paleogene trough is bounded on the west side by an arch or complex zone (Fig. 6) beneath the shelf edge. Meulaboh-1 well, drilled into this feature, bottomed in Paleogene mudstones with dips near 50°, supplying further evidence of significant pre-Miocene deformation. Panjang-1 well penetrated the unconformity and bottomed in probable Oligocene turbidites with a 20° dip in this same setting. The landward connection of these marine Paleogene strata with the Sibolga and Babaluhung beds requires rapid facies changes and a possible synclinal axis (Fig. 6).

The outer or seaward boundary of pre-Tertiary basement is yet another unresolved problem. The only evidence of pre-Tertiary basement seaward of the Barisan front in the Tapanuli district is the metavolcanic rock that we identified in the basal core of the Lakota well (Fig. 1; Table 1). It is possible that this lithology is similar to the pre-Tertiary volcanics reported by Union Oil Company geologists on several small islands just offshore in the Natal area, and that there is a band of these basement rocks seaward of the late Paleozoic metaclastics and Permian-Triassic granites of the Barisan ranges (Zwierzycki, 1919; Katili, 1973a). How far this pre-Tertiary basement extends seaward remains to be determined. Using the admittedly poor criterion of the character of acoustic basement, we suggest that it underlies the inner part of the shelf area (Fig. 6B).

In the morphotectonic terminology of an active arc system, all the area out to the shelf edge, or to the structural east flank of the forearc basin, would act as a frontal arc—a basically rigid block that undergoes vertical and strike-slip displacement but very little compression. The outer half of the shelf, where a thick Paleogene and perhaps older section is suspected, may well represent a Late Mesozoic–Early Tertiary forearc basin and accretionary prism.

CENOZOIC HISTORY

As with most geologic records, that of Sumatra becomes increasingly confused with increasing age, but the Sumatran situation is exaggerated by the almost total lack of fossiliferous marine Paleogene strata. We are thus forced to treat over half the Cenozoic as virtually a single time unit.

Another serious problem is the ambiguity in the kinematic plate framework of Sumatra during the Paleogene. Paleomagnetic calculations demonstrate a general northward motion of India with respect to Eurasia (Stauffer, 1974), but there may have been discontinuities in rate, and there were certainly intraplate or small plate motions to modify the subduction kinematics off Sumatra. There is reasonably good evidence for Late Cretaceous subduction along south and central Sumatra (Katili, 1973a; de Coster, 1975), but the persistence of subduction through the Paleogene is debatable.

Hamilton (1977), Graham et al, (1975) and de Coster (1975) suggest a period from about late Paleocene (55 m.y. or anomaly 22) to late Eocene–early Oligocene (41 m.y. or anomaly 17) during which a stable shelf or a strike-slip boundary existed off Sumatra, Hamilton citing a lack of volcanic activity and the others relying on the postulated cessation of spreading along ridges in the eastern Indian Ocean (Sclater and Fisher, 1974).

Graham et al (1975) associate the Andaman flysch, that presently forms a major component of the accretionary prism in the Andaman-Nicobar sector of the Sunda arc, with deposition during a period of inactivity following an earlier period of ophiolite accretion. We would emphasize that this change in mode of accretion does not require a halt in the subduction, but more likely reflects a change in the type and thickness of sediment cover on the downgoing plate (Karig, 1974, 1977). In the Andaman sector, ophiolitic scraps could have been accreted when thinly sedimented oceanic crust was subducted during Late Cretaceous to early Tertiary time. Accretion of the Andaman flysch followed when the Bengal and Nicobar fan sediments entered the trench sometime during the early Tertiary.

Although a mid-Paleogene halt in subduction is certainly plausible, an equally good case can be made for continuous subduction off the Sumatra margin, together with an indeterminate amount of transcurrent motion. Volcanism was not at a high level during the Paleogene, but it certainly was not absent. The Paleogene clastics along the eastern flank of the Barisan Range, where penetrated by wells, occasionally contain significant amounts of tuff and/or volcanic debris, and imply a source to the southwest (de Coster, 1975). The better exposed Paleogene cover along the Barisan crest in northern Sumatra also is reported to have intercalations of andesitic flows and tuff (Zwierzycki, 1919). Paleogene volcanic activity is much more obvious in south and central Sumatra, where intrusives, extrusives and tuffs occur (Kastowo and Leo, 1973; Roezin, 1976; de Coster, 1975), than it is farther north. The timing of volcanic activity is based primarily on palynology and radiometric dating, which may not be very accurate, but the emerging picture shows that volcanic activity was spread through the Paleogene.

It might be argued that this volcanism was not related to subduction. The north-south-trending Paleogene troughs on the eastern side of Sumatra (de Coster, 1975) could be interpreted as rifts, possibly related to right-lateral slip along the plate margin or to basin-range type activity. Volcanism might then be related to extensional tectonics. In that case it is not clear why activity seemed restricted to a band lying within the present Barisan Range.

The lack or paucity of volcanic detritus in Paleogene sediments flanking the Barisan Range might be better understood by examining the Quaternary situation. Even though there is a well-defined volcanic chain, there is a remarkably minor volcanic component in Quaternary strata in wells off the west coast. This can be attributed to the dilution by other detritus, including degraded volcanic products, and by the strongly asymmetric drainage. Even in the eastern basins, the amount of recognizable volcanic debris noted in the Quaternary strata is low considering the massive eruptions that occurred.

The occurrence and characteristics of marine Paleogene strata off the west coast neither support nor detract from a continuous subduction model. If these sediments are turbidites, one might conclude that a Paleogene basin underlies the outer Sumatran shelf. Such a basin may have been an early Tertiary forearc basin (Fig. 6), but it might also have been a basin on a stable or strike-slip margin.

One late Paleogene event of particular interest in the study area was the Oligocene(?) emplacement of many andesitic intrusives that have been mapped along the western Sumatran coast from within the Barisan massifs to the shelf offshore (Fig. 7). The westernmost intrusives lie well trenchward of the later Miocene to Recent volcanic chain and with limited available data seem to also lie trenchward of the earlier Paleogene volcanic centers. It has been suggested (Marshak and Karig, 1977) that arc volcanism that lies anomalously close to the trench may be a result of migrating trench-trench-ridge triple junctions, which would lead to anatectic melting of crustal material. Although such a triple junction apparently did migrate northwestward along the Sunda arc sometime during the early Tertiary (Sclater and Fisher, 1974; Fig. 2), the temporal correlation of this migration with volcanism is not at all certain, and the andesitic nature of the anomalous igneous rocks cannot clearly be associated with anatectic melting of older crust.

This igneous event may also have played a role in the late Paleogene uplift which resulted in the cutting of a major unconformity beneath the shelf off central and northern Sumatra. Uplift may have resulted from thermal expansion of the upper plate or of the subducted ridge crest (Delong et al, 1977), followed by isostatic adjustments to erosional unloading. The geometry of the truncated Paleogene strata beneath the unconformity at the shelf edge (Fig. 6) suggests that the uplift totalled several kilometers. Cooling and loading due to sedimentation on the shelf would have caused the observed Neogene subsidence, which seems to be of the same order of magnitude as the uplift.

By the end of the Oligocene, subduction can be documented by the record of deformed rock. The position of the Oligocene-Miocene trench was near central Nias, where lowermost slope sediments have an earliest Miocene (N 4) age. The trench slope break apparently lay east of the Suma-1 and Panjang well sites (Table 1) in what is now a deeply subsided part of the upper slope basin. The dominance of clasts from Sumatra, and the component of pelagic sediments and basalt in the early Miocene melange indicates that the Bengal-Nicobar Fan had not yet impinged upon the trench at Nias and that the trench fill was not extremely thick. Comparison with contemporary trenches would suggest a fill thickness of less than 500 m. The amount of very coarse material suggests that there was reasonably free transport from the Sumatran shelf and thus a more subdued trench slope break. Very likely the geometry of the inner trench slope was much like that off Java at present.

Bands of tuff in early Miocene slope sediments on Nias demonstrate that the volcanic chain was active at that time. Marine transgression across the slowly subsiding Sumatran shelf was also occurring during the early Miocene and was recorded by deposition of a carpet of calcarenites and reefs. This transgression continued through the mid-Miocene with the deposition of a shoreline belt of carbonate and offshore detrital clastics. A general transgression, but with minor irregularities, continued through the Miocene and Pliocene, persisting even during the sharp uplift of the Barisan Range in late Miocene and Pliocene times.

There is little evidence for a mid- or intra-Miocene orogeny in western Sumatra (van Bemmelen, 1970). Recent subsurface data gathered during the search for petroleum along both flanks of the Barisan Range demonstrate that there was, instead, the beginning of uplift of the Barisan block and marine regression to the east in the late mid-Miocene with strong uplift through the late Miocene and Pliocene, and perhaps into the Quaternary. Published (Adinegoro and Hartoyo, 1975) and unpublished oil company data strongly indicate that the block-bounding structures are flexures and very high-angle faults rather than the low-angle, large-displacement thrusts postulated by van Bemmelen. Strike-slip faulting and deformation related to these displacements may have persisted throughout the entire Neogene and over a broad band along Sumatra, as is implied by structures in several oil fields (Mertosono, 1976; Roezin, 1976) and along the western flank of the Barisan block (Tjia and Posavec, 1972; Posavec et al, 1973).

Continuous subduction throughout the Neogene off Sumatra is indicated by the geological characteristics and by kinematic calculations. The presently available kinematic data (Molnar and Tapponier, 1975) suggest that the subduction rate increased from 5 to 6.5 cm per year sometime between 5 and 10 m.y. ago. This rate increase may be responsible for the Miocene change in geologic behavior along the Sumatran margin.

It is plausible that uplift of the Barisan block resulted from an input of heat to the upper mantle. Such a thermal doming would have been related to the subduction process and could have led to subsequent Quaternary ignimbritic activity and normal faulting along the Barisan in the Quaternary. The Barisan rift displays the same spatial relationship to the volcanic chain and the same sequence of uplift, ignimbrite eruptions and faulting that is commonly observed in volcanic-tectonic rifts associated with subduction zones (Karig, 1974). It differs in that normal faulting has not proceeded to the usual extent and strike-slip activity is now dominant.

Events that occurred closer to the trench from the mid-Miocene to present involve primarily the rapid migration of arc units westward following the contact with and accretion of the Bengal Fan (Curray and Moore, 1974). Although the record of this accretion still lies submerged beneath the slope west of Nias, it is probable that the accreted melange of these ages consists primarily of fan sediments. The enlargement of the trench slope break, which was at least in part a result of the rapid accretion (Hamilton, 1973) acted to trap most of the debris shed from Sumatra. The impingement of the Ninety-East Ridge on the Andaman section of the arc during the Quaternary has modified this scheme by cutting off the flow of turbidites to the Nicobar fan and by again reducing the amount of fan sediments being subducted in the trench.

REFERENCES CITED

Adinegoro, U., and Hartoyo, P., 1975, Paleogeography of northeast Sumatra: Proc. 3rd Ann. Convention (1974) Indonesian Petroleum Assoc., Jakarta.

Blow, W. H., 1969, Late middle Eocene to Recent planktonic foraminiferal biostratigraphy: Proc. First Int. Conf. Planktonic Microfossils, v. 1, p. 199–422.

CCOP-IOC, 1974, Metallogenesis, hydrocarbons, and tectonic patterns in eastern Asia: United Nations Development Program (CCOP), Bangkok, 158 p.

Chatterjee, P. K., 1967, Geology of the main islands of the Andaman area: Proc. Symposium on Upper Mantle Project, Geophys. Res. Board, N.G.R.I., Hyderabad, India, Session V, p. 348–360.

Curray, J. R., and Moore, D. G., 1974, Sedimentary and tectonic processes in the Bengal deep-sea fan and geosyncline, *in* The geology of continental margins, C. A. Burke and C. L. Drake, Eds.: New York, Springer-Verlag, p. 617–627.

——— et al, 1977, Seismic refraction and reflection studies of crustal structure of the eastern Sunda and western Banda arcs: Jour. Geophys. Research, v. 82, p. 2479–2489.
de Coster, G. L., 1975, The geology of the central and south Sumatran basins: Proc. 3rd Ann. Convention (1974) Indonesian Petroleum Assoc., Jakarta, p. 77–110.
Delong, S. E., Fox, P. J., and McDowell, F. W., 1977, Subduction of the Kula Ridge at the Aleutian Trench: Geol. Soc. America Bull., in press.
Dickinson, W. R., 1973, Widths of modern arc-trench gaps proportional to past duration of igneous activity in associated magmatic arcs: Jour. Geophys. Res. v. 78, p. 3376–3389.
Douville, H., 1912, Les foraminiferes de l'Ile de Nias: Samml. Geol. Reichsmus v. 8, p. 255–278.
Graham, S. A., Dickinson, W. R., and Ingersoll, R. V., 1975, Himalayan-Bengal model for flysch dispersal in the Appalachian-Ouachita system, Geol. Soc. America Bull., v. 86, p. 273–286.
Haile, N. S., McElhinny, M. W., and McDougall, I., 1977, Palaeomagnetic data and radiometric ages from the Cretaceous of West Kalimantan (Borneo), and their significance in interpreting regional structure: Jour. Geol. Soc. London, v. 133, p. 133–144.
Hamilton, W., 1973, Tectonics of the Indonesian region: Geol. Soc. Malaysia Bull., v. 6, p. 3–10.
——— 1977, Subduction in the Indonesian region: Amer. Geophys. Union Monogr., M. Ewing Series, v. 1, p. 15–31.
Hariadi, N., and Soeparjadi, R. A., 1976, Exploration of the Mentawai Block-West Sumatra: Proc. 4th Ann. Convention (1975) Indonesian Petroleum Assoc., v. 1, p. 55–66.
Hutchison, C. S., 1973, Tectonic evolution of Sunda land, a Phanerozoic synthesis: Geol. Soc. Malaysia Bull., v. 6, p. 61–86.
Karig, D. E., 1974, Evolution of arc systems in the Western Pacific: Ann. Rev. Earth Planet. Sci., v. 2, p. 51–75.
——— 1977, Growth patterns on the upper trench slope: Amer. Geophys. Union Monogr., M. Ewing Series, v. 1, p. 175–186.
——— Caldwell, J. G., and Parmentier, E. M., 1976, Effects of accretion on the geometry of the descending lithosphere: Jour. Geophys. Res., v. 81, p. 6281–6291.
——— et al, 1975, Initial Reports of the Deep Sea Drilling Project: Washington, D.C., U.S. Government Printing Office, v. 31, 927 p.
Karunakaran, C., Ray, K. K., and Saha, S. S., 1964, A new probe into the tectonic history of the Andaman and Nicobar Islands: Reports of the 22nd Intl. Geol. Cong. New Delhi, pt. 4, p. 507–515.
Kastowo and Leo, G. W., 1973, Geologic map of the Padang Quadrangle, Sumatra: Bandung, Direktorat Geologie Indonesia, scale 1:250,000.
Katili, J. A., 1973a, Geochronology of west Indonesia and its implication on plate tectonics: Tectonophysics, v. 19, p. 195–212.
——— 1973b, Plate tectonics and its significance in the search for mineral deposits in western Indonesia: ECAFE-CCOP Tech. Bull., v. 7, p. 23–37.
——— 1975, Volcanism and plate tectonics in the Indonesian island arcs: Tectonophysics, v. 26, p. 165–188.
McElhinney, M. W., Haile, N. S., and Crawford, A. R., 1974, Paleomagnetic evidence shows Malay Peninsula was not a part of Gondwanaland: Nature, v. 252, p. 641–645.
Marshak, R. S., and Karig, D. E., 1977, Triple junctions as a cause for anomalous igneous activity between the trench and volcanic arc: Geology, v. 5, p. 233–236.
McDonald, J. M., 1977, Sediments and structure of the Nicobar Fan, northeast Indian Ocean: Univ. of California, San Diego, unpublished Ph.D. dissert., 120 p.
Mertosono, S., 1976, Geology of Pungut and Tandun oil fields, central Sumatra: Proc. 4th Ann. Convention (1975) Indonesian Petroleum Assoc., v. 1, p. 165–179.
Molnar, P., and Tapponier, P., 1975, Cenozoic tectonics of Asia, effects of a continental collision: Science, v. 189, p. 419–426.
Moore, G. F., 1978, Structural geology and sedimentology of Nias Island, Indonesia, A study of subduction zone tectonics and sedimentation: Cornell Univ., unpublished Ph.D. dissert., 142 p.
——— and Karig, D. E., 1976, Development of sedimentary basins on the lower trench slope: Geology, v. 4, p. 693–697.
Ninkovich, D., 1976, Late Cenozoic clockwise rotation of Sumatra: Earth Planet. Science Letters, v. 29, p. 269–275.
Posavec, N., et al, 1973, Tectonic controls of volcanism and complex movements along the Sumatran fault system: Geol. Soc. Malaysia Bull., v. 6, p. 43–60.
Roezin, S., 1976, The discovery and development of the Petapahan oil field, central Sumatra: Proc. 4th Ann. Convention (1975), Indonesian Petroleum Assoc., v. 2.
Sclater, J. G., and Fisher, R. L., 1974, Evolution of the east-central Indian Ocean, with emphasis on the tectonic setting of the Ninety-East Ridge: Geol. Soc. America Bull., v. 85, p. 683–702.
Shor, G.C., Jr., and R. Von Huene, 1972, Marine seismic refraction studies near Kodiak, Alaska: Geophysics, v. 37, p. 697–700.
Stauffer, P. H., 1974, Malaya and southeast Asia in the pattern of continental drift: Geol. Soc. Malaysia Bull., v. 7, p. 89–138.
Tjia, H. D., 1976, Radiometric ages of ignimbrites of Toba, Sumatra: Warta Geologi (Geol. Soc. Malaysia Newsletter), v. 2, p. 33–34.
——— and Posavec, M. M., 1972, The Sumatra fault zone between Padang-Panjang and Muaralabuch: Sains Malaysiana, v. 1, p. 77–105.
Umbgrove, J. H. F., 1947, The pulse of the earth: The Hague, Martinus Nijhoff, 385 p.
van Bemmelen, R. W., 1933, On the geophysical foundations of the undation theory: Proc. Kon. Acad., v. Wetensch., Amsterdam, v. 36, p. 336–343.
——— 1970, The geology of Indonesia: The Hague, Government Printing Office, 2nd ed., 732 p.
Venkatarathnam, K., and Biscaye, P., 1973, Clay mineralogy and sedimentation in the eastern Indian Ocean: Deep-Sea Research v. 20, p. 727–738.
Vening Meinesz, F. A., 1940, The earth's crust and deformation in the East Indies: Proc. Kon. Acad. v. Wetensch., Amsterdam, v. 48, p. 278–293.
Verstappen, H. Th., 1973, A geomorphological reconnaissance of Sumatra and adjacent islands (Indonesia): Gröningen, Walters-Noordhoff Publ., Royal Dutch Geographical Society, Verhandelingen no. 1, 182 p.
Zwierzycki, J., 1919, Geologische overzichtskaart van den Nederlandsch-Oost-Indischen Archipel: Toelichting bij Blad I, Blad VII, Jaarboek van het Mijnwezen in Nederlandsch-Oost-Indie, v. 48, p. 11–108.

Subsidence of the Daito Ridge and Associated Basins, North Philippine Sea[1]

ATSUYUKI MIZUNO,[2] YOSHIHISA OKUDA,[2,5] SHOZABURO NIAGUMO,[3] HIDEO KAGAMI,[4] AND NORIYUKI NASU[4]

Abstract Cenozoic history of the Daito Ridge and two associated abyssal basins is preliminarily discussed, based on a shipboard near-trace monitor record section from a multichannel seismic reflection survey across the region and on dredge-sample data from the ridge area.

On the Daito Ridge, middle Eocene *Nummulites*-bearing deposits, shown by a dipping high-frequency pattern on the profile, accumulated in a shallow sea environment on a basement that included the Daito Metamorphic Rocks. Volcanic activity around the time of deposition is suggested by dredged rocks from the Daito north peak. The Daito Ridge has undergone a change in environmental conditions from neritic to pelagic and has subsided to a depth of 1.5 km below sea level from the middle Eocene through the Quaternary, resulting in the deposition of a transparent veneer on the main peak.

The acoustic sequence in the Minami-Daito Basin is divided into Units A through E, in descending order. Of them, Units C and D are partly correlative with the middle Eocene deposits on the ridge and include the earlier deposits. They represent archipelagic apron sedimentation on the slope and basin boundary area when the top of the ridge was extensively uplifted. Possibly related to the submergence of the Daito Ridge, a new sedimentary cycle characterized by turbiditic sedimentation (Unit B) started after the cease of apron sedimentation, and was followed by pelagic deposition (Unit A). A similar feature is observed in the Kita-Daito Basin and in the valley on the Daito Ridge.

Two types of post-opening intrusive activity are observed in the abyssal basins. One is indicated by Unit E, characterized by stratified layers and diffractions and having a smooth upper surface. It represents the anomalously smooth oceanic crust of the basin and resulted from repeated intrusion of sills after or during the deposition of Units C and D. Another is indicated by the acoustic basement with rugged upper surface and dense diffractions in the Kita-Daito Basin. It represents the rugged oceanic crust of the basin, which was likely formed by extensive intrusion of dikes after or during the deposition of Unit B.

INTRODUCTION

This article presents a preliminary geologic interpretation of the shipboard near-trace monitor record section of a multichannel seismic reflection survey across the Daito Ridges region of the North Philippine Sea (Nasu et al, 1976), combined with dredge-sample data from the topographic highs in the region.

The multichannel seismic profile was collected aboard the survey ship *Kaiyo-Maru* of the Japanese Petroleum Exploration Co., Ltd., contracted by the Ocean Research Institute, University of Tokyo, using a 48-channel 24-fold CDP (common-depth point) stack system with a 16-airgun array of 1,490 cu in. in total volume, which exploded at approximately 1,900 p.s.i.g. at a shot spacing of 50 m. A near-trace monitor signal was recorded after digital filtration between 8 and 64 Hz.

The Daito Ridges region consists of three major ridges (Amami Plateau, Daito Ridge, and Oki-Daito Ridge) and two associated deep basins (Kita-Daito Basin and Minami-Daito Basin). It is bounded by the Kyushu-Palau Ridge on the east and by the Ryukyu Trench on the west, facing the Philippine Basin on the south (Fig. 1). Generally, the ridge tops are at a depth of 1.5 to 2.0 km and are rather rugged, except some places where a very flat surface is developed as seen on the Daito Ridge. The associated basins are characterized by great water depths, in some places over 6 km.

[1]Manuscript received, May 31, 1977; accepted, July 17, 1978.

[2]Marine Geology Department, Geological Survey of Japan, Takatsu-ku, Kawasaki 213, Japan.

[3]Earthquake Research Institute, University of Tokyo, Bunkyo-ku, Tokyo 113, Japan.

[4]Ocean Research Institute, University of Tokyo, Nakano-ku, Tokyo 164, Japan.

[5]Present address: Technology Research Center, Japan Petroleum Development Corporation, Minato-ku, Tokyo 105, Japan.

The writers are grateful to Dr. J. Lyons of Technology Research Center, Japan Petroleum Development Corporation, for his critical review of the manuscript. Thanks are also due to Dr. Joel Watkins of the University of Texas for his invitation to contribute to the Galveston Symposium volume.

Article Identification Number:
OO65-731X/78/MO29-0016/$03.00/0.
(see copyright notice, front of book)

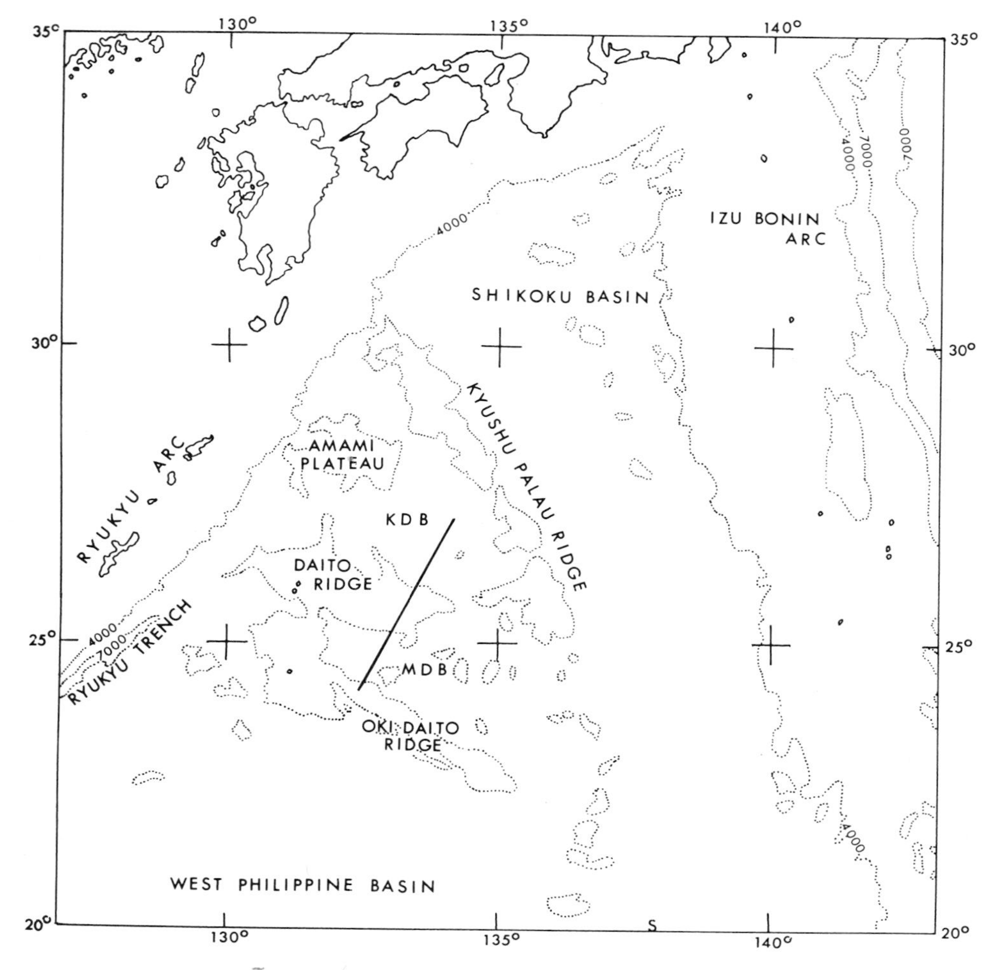

FIG. 1. Major topography of North Philippine Sea and the multichannel seismic reflection survey line across the Daito Ridge and associated basins. The survey line is shown by a straight solid line. Abbreviations: KDB (Kita-Daito Basin); MDB (Minami-Daito Basin).

After Karig (1975) discussed a possible Mesozoic-earliest Tertiary remnant arc origin of the Daito and Oki-Daito Ridges, many Japanese scientists contributed to the understanding of geology of the major ridges of the region particularly through the studies of dredged rocks and fossils from (about) 30 sites. On the basis of the dredge data and single-channel seismic reflection data, Mizuno et al (1975, 1976), Okuda et al (1976), and Shiki et al (1977) preliminarily discussed the geologic history of the region.

Mizuno et al (1977) presented a summary table of the dredged rock data and showed that the major ridges consist of four rock series: (1) pre-Eocene basement rocks; (2) middle Eocene *Nummulites*-bearing rocks presumably overlying the basement rocks; (3) early Pliocene pelagic calcareous mudstones; and (4) various kinds of igneous rocks of unknown age. In addition, Nishida et al (1978) recently reported late Oligocene nannofossil chalk from a site at the eastern part of the Daito Ridge, which may belong to the third rock series.

SEISMIC REFLECTION PROFILES AND GEOLOGIC INTERPRETATION

The monitor record section (Fig. 2) runs NNE-SSW with a length of 400 km and covers the southern half of the Kita-Daito Basin, the Daito Ridge, the Minami-Daito Basin, and the northern periphery of the Oki-Daito Ridge.

It is apparent from Figure 2 that both abyssal basins and small depressions on the ridges are filled with thick stratified sequences. The ridge top areas are characterized by different acoustic features. The Daito Ridge has three peaks on the profile, called the north, main, and south peaks for convenience.

Daito Ridge and Oki-Daito Ridge

The flat top of the Daito main peak is covered with horizontally layered younger sediments with a thickness of 0.12 to 0.15 sec. They are underlain by the strata which are characterized by several alignments of high-frequency events dipping southward (SP 4350-4400). The base of the dipping strata is uncertain and the style of their distribution along the whole ridge top also is vague.

Both the northern and southern slopes of the main peak appear to consist of continuations of the dipping strata and the related basement. Only the north peak consists of an opaque mass—acoustic basement—whereas a layered structure can be discriminated at the south peak.

There are some dredge-sampling data from the sites close to the seismic line (GDP-15-4, GDP-15-7; Mizuno et al, 1977; Fig. 3) and some single-channel seismic reflection data across the sites, in addition to the present seismic line (Misawa et al, 1976). They suggest that the dipping strata at the main peak represent a section of middle Eocene shallow sea deposits including carbonate rocks with *Nummulites boninensis, Asterocyclina penuria,* etc., which are extensively distributed on crestal parts of the main ridges of the Daito Ridges region.

Hornblende schist and serpentinite were collected from the solitary basement peak near the seismic line (GH 74-7-183), and greenschist, tuffaceous sandstone, and dolerite associated with *Nummulites boninensis,* were collected from the valley wall between the main and south peaks (GDP-15-7). These data indicate that the Daito Metamorphic Rocks of low-pressure type (Mizuno et al, 1975; Yuasa and Watanabe, 1977) probably underlie the middle Eocene strata and the volcanics, forming the basement.

From the data at three dredge sites on and very close to the seismic line (GDP-15-1, -2, and -3), the opaque mass of the north peak is thought to consist of the basement complex including schist, sandstone, arkosic wacke, etc., associated with andesite, dolerite, and tuff which likely overlie the basement rocks.

The volcanic and volcaniclastic rocks may represent either Eocene or younger volcanic activity. If the ridge top was largely within a shallow sea environment during the middle Eocene, there may have been volcanic activity which caused uplifting of the ridge around that time.

The layered reflectors of the south peak likely

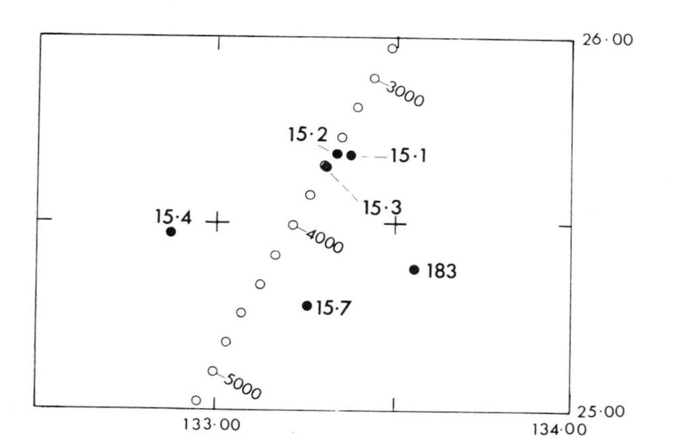

FIG. 3. Enlarged location map of the Daito Ridge showing shot points of the seismic line (open circles) and the dredging sites (solid circles).

represent a continuation of the dipping middle Eocene sequence from the main peak.

The sediment layer on the main peak shown by a transparent veneer is younger than middle Eocene. Recent discovery of late Oligocene pelagic nannofossil chalk from the eastern part of the Daito Ridge at a depth of 2.0 to 2.5 km (Nishida et al, 1978) shows that the deposition of the sediment layer had taken place in a pelagic environment since the late Oligocene. The Daito Ridge environmental conditions have changed from neritic to pelagic and the Ridge has subsided to a depth of 1.5 km below sea level between the middle Eocene and the Quaternary.

The subsidence might have been caused by isostatic sinking due to the thermal contraction following the Eocene volcanism. An amount of subsidence of guyots and atolls (Menard, 1964, p. 94, Fig. 4.23) suggests the possibility of isostatic sinking of about 1 km since the Eocene. Additional subsidence could be indicated by the negative free-air gravity anomaly under the Minami-Daito Basin (Segawa, 1976). On the other hand, Ingle et al (1975) considered that submergence of the Kyushu-Palau Ridge in excess of 150 m might have continued until the late Pleistocene. Nishimura et al (1977) suggested subsidence of the Amami Plateau of about 1 km during the latest Pliocene and early Pleistocene from paleontologic data. The subsidence of the plateau might have been partly related to subduction of the Ryukyu Trench and the Nankai Trough.

The Oki-Daito Ridge is composed of an opaque mass, except for a shallow depression at its crest. The dredge-sample data (Mizuno et al, 1977) show that basaltic rocks of unknown age are extensively distributed on the ridge, together with Eocene *Nummulites*-bearing rocks. The opaque mass probably represents the basaltic rocks.

Minami-Daito Basin

The stratigraphic sequence can be divided into Units A, B, C, D, and E in the Minami-Daito Basin. Units A through D are defined by seismic reflection characteristics at SP 5300 (Fig. 4).

Unit A, characterized by a transparent layer, blankets the entire basin with almost constant thickness (0.14 to 0.18 sec in two-way reflection time). Unit B, with parallel high-frequency reflections, covers almost the entire basin but gradually thins southward from a thickness of about 0.43 sec in the northern part to that of about 0.10 sec in the southern part. At the northern margin of the basin, it abuts against the upper surface of Unit C, which rises toward the Daito Ridge along its lower slope.

Unit C is largely characterized by relatively low-amplitude, low-frequency reflections, and Unit D consists of similar features with diffractions. They can apparently be traced from the deeper part of the basin to the Daito Ridge, almost half way to its south peak.

In the medial to southern parts of the basin, both

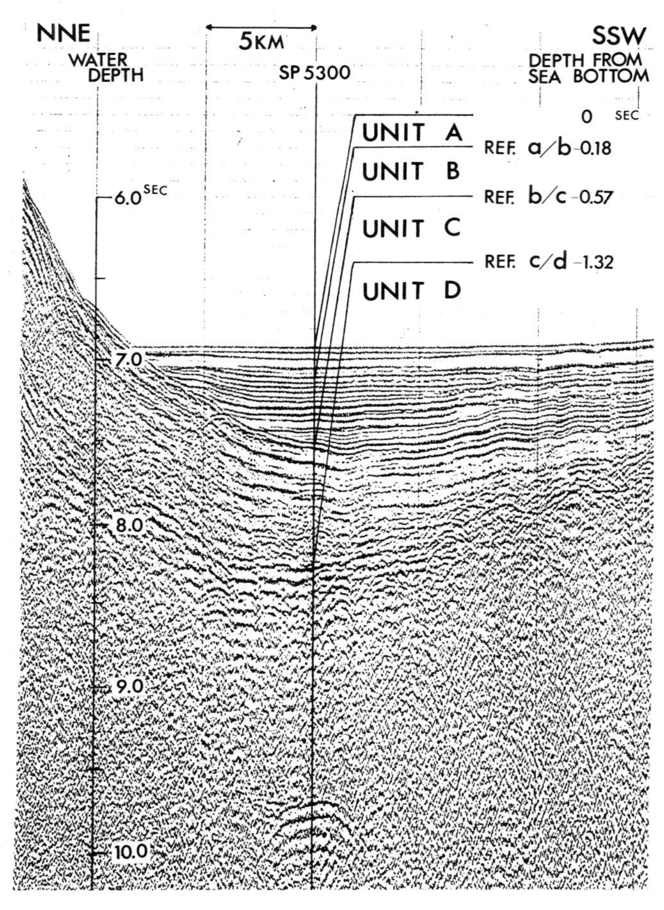

FIG. 4. Seismic stratigraphic division at SP 5300 in the Minami-Daito Basin. Vertical scale is in two-way reflection travel time.

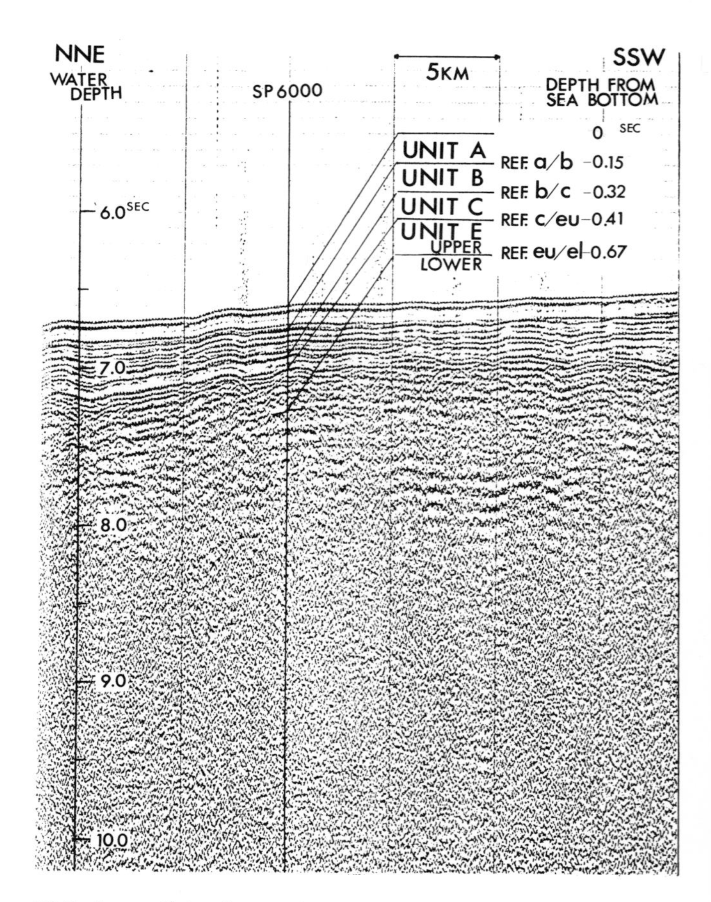

FIG. 5. Seismic stratigraphic division at SP 6000 in the Minami-Daito Basin. Vertical scale is in two-way reflection travel time.

units seem to be replaced for the most part by diffractive layers with several reflection horizons, suggesting a layered structure. Unit E comprises this interval beneath the uppermost part of Unit C at SP 6000 (Fig. 5). It is subdivided into upper and lower parts (called Eu and El, respectively) bounded by a reflector at a level of 0.67 sec from the sea bottom at SP 6000.

There is no direct stratigraphic connection between Units C and D and the Eocene sequence on the Daito Ridge. However, the general features of the profile at the southern slope of the Daito Ridge seem to suggest that Units C and D are partly correlative with the middle Eocene sequence on the ridge. This implies that some parts of the units are middle Eocene. The drilling results at DSDP Site 446 (24°42.04′N, 132°46.49′E)(Klein et al, 1978) suggest that they also include the earlier deposits.

Unit C thickens toward the upper portion of the slope. This indicates that Unit C represents archipelagic apron sediments on the slope and in the basin boundary area, when shallow sea sedimentation prevailed on the ridge. The sediments may have been supplied from the ridge top areas of Daito and Oki-Daito to the deeper part of the basin. The low-frequency reflection pattern of the unit suggests that they are dominantly fine materials.

Units A and B represent post-middle Eocene sediments. The transparent nature of Unit A indicates pelagic uniform sediments such as brown clay, as suggested by Karig (1975), and high-frequency reflections of Unit B may indicate turbidite sedimentation. Abutting of Unit B on Unit C at the northern periphery of the Minami-Daito Basin suggests that an abrupt change of direction of sediment supply occurred after the deposition of Unit C. The cease of the apron sedimentation of Unit C and the start of a new sedimentary cycle could have been related to submergence of the Daito Ridge.

Unit E is characterized acoustically by stratified layers with diffractions and a smooth upper surface. This suggests that it consists of sediments and basaltic sills, which was corroborated by the drilling at Site 446 which recovered samples of 16 tholeiitic sills intruding upper lower Eocene mudstones below a depth of about 400 m measured from the sea bottom (Klein et al, 1978). Unit E constitutes an anomalously smooth oceanic crust with sill-type basaltic intercalations, and it was formed by the post-opening volcanic activity in the Minami-Daito Basin during or after the deposition of Units C–D. Such a feature has been reported from the Venezuelan Basin, where a basaltic flow was sampled at Sites 146/149 and 150 (Talwani et al, 1977).

Kita-Daito Basin

The stratigraphic sequence in the Kita-Daito Basin is similar to that of the Minami-Daito Basin. It can be identified with Unit A (0.15 sec thick), Unit B (0.25 to 0.40 sec thick), and Unit C (0 to 0.90 sec thick), in descending order. Units A and B are characterized by transparent and high-frequency reflection intervals, respectively, and Unit C has intermediate acoustic character of Units C and B in the Minami-Daito Basin.

Densely diffractive basement underlies the sedimentary sequence in the deepest flat basin, and also is observed on the basement highs north and south of the basin. Different from Unit E in the Minami-Daito Basin, basement is characterized by a very rugged upper surface. The elevated surfaces of diffractions are consistent with the deformed sediments of Unit B, which are unconformably cut by Unit A. These features and the diffractive basement suggest extensive intrusion of basaltic dikes during or after deposition of Unit B. The interval reflects intense post-opening intrusive activity in the Kita-Daito Basin.

Thus, from the seismic reflection profiles, two types of post-opening intrusive activity can be recognized in the abyssal basins associated with the Daito Ridge. One is shown by the anomalously smooth oceanic crust in the Minami-Daito Basin, which was formed by repeated intrusion of sills after or during the deposition of Units C and D. Another is indicated by the rugged oceanic crust in the Kita-Daito Basin, which was formed by extensive intrusion of dikes after or during the deposition of Unit B. This activity might have contributed to basin subsidence, and also might have been related to the volcanism on the Daito Ridge.

Small Depression on the Daito Ridge

The seismic line crosses small valleys which sculpture the Daito Ridge. The sedimentary sequence in the valley between the north and main peaks (SP 3840-4030) consists of a transparent layer, a high-frequency reflection layer, and a low-frequency reflection layer, in descending order. The acoustic basement may be the continuation of the pre-Eocene rocks of the north peak and possibly of the main peak.

The low-frequency reflection layer appears to rise toward both the north and main peaks. It is supposedly the deeper facies of the middle Eocene sequence on the main crest, consisting of sedimentary materials derived from the Daito Ridge, as well as Unit C in the southern slope of the ridge. This is supported by the drilling results at DSDP Site 445 (25°31.36′N, 133°12.49′E)(Klein et al, 1978).

The high-frequency reflection layer and the transparent layer abut against the low-frequency layer and only occur in the valley. This shows a similar sedimentary history to that of the Minami-Daito Basin.

REFERENCES CITED

Ingle, J. C. Jr., et al, 1975, Site 296, *in* D. E. Karig, J. C. Ingle, Jr., et al, Initial reports of the Deep Sea Drilling Project: Washington, D.C., U.S. Govt., v. 31, p. 191–274.

Karig, D. E., 1975, Basin genesis in the Philippine Sea, *in* D. E. Karig, J. C. Ingle, Jr., et al, Initial reports of the Deep Sea Drilling Project: Washington, D.C., U.S. Govt., v. 31, p. 857–878.

Klein, G. deV., et al, 1978, Philippine Sea drilled: Geotimes, May 1978, p. 23–25.

Menard, H. W., 1964, Marine geology of the Pacific: New York, McGraw-Hill, 271 p.

Misawa, Y., et al, 1976, Geophysical results of the GDP-15th cruise in the Philippine Sea: Marine Sci. Monthly, v. 8, p. 702–707. (Japanese with English abstracts).

Mizuno, A., Y. Okuda, and K. Tamaki, 1976, Some problems on the geology of the Daito Ridges Region and its origin: Geol. Studies Ryukyu Islands, v. 1, p. 177–198. (Japanese with English abstracts).

——— T. Shiki, and H. Aoki, 1977, Dredged rock and piston and gravity core data from the Daito Ridges and the Kyushu-Palau Ridge in the North Philippine Sea: Geol. Studies Ryukyu Islands, v. 2, p. 107–119.

——— 1975, Marine geology and geologic history of the Daito Ridges area, northwestern Philippine Sea: Marine Sci. Monthly, v. 7, p. 484–491, p. 543–548. (Japanese with English abstract).

Nasu, N., et al, 1976, Multichannel seismic reflection data across the Shikoku basin and the Daito Ridges, 1976: IPOD-Japan Basin Data Series, n. 1, Ocean Research Inst., Univ. Tokyo, 17 p.

Nishida, S., et al, 1978, The Miocene and Oligocene deposits in the Daito Ridges area: Geol. Soc. Japan, The 85th Ann. Meeting, Abs., p. 187. (in Japanese)

Nishimura, A., et al, 1977, Microfossils of the core sample GDP-11-15 from the Amami Plateau, the northern margin of the Philippine Sea: Mem. Fac. Sci., Kyoto Univ., Ser. Geol. Mineral., v. 48, p. 111–130.

Okuda, Y., et al, 1976, Tectonic development of the Daito Ridges: Marine Sci. Monthly, v. 8, p. 414–422. (Japanese with English abstracts).

Segawa, J., 1976, Gravity in the Ryukyu Arc: Jour. Geodetic Soc. Japan, v. 22, p. 23–39.

Shiki, T., et al, 1977, Geology and geohistory of the northwestern Philippine Sea, with special reference to the results of the recent Japanese research cruises: Mem. Fac. Sci., Kyoto Univ., Ser. Geol. Mineral., v. 44, p. 67–78.

Talwani, M., et al, 1977, Multichannel seismic study in the Venezuelan Basin and the Curacao Ridge: Island arcs, deep-sea trenches and back-arc basins: Maurice Ewing Series, v. 1, p. 83–98.

Yuasa, M., and T. Watanabe, 1977, Pre-Cenozoic metamorphic rocks from the Daito Ridge in the northern Philippine Sea; Japanese Assoc. Mineralogists, Petrologists, and Econ. Geologists Jour., v. 72, p. 241–251.

The Evolution of Structural Highs Bordering Major Forearc Basins[1]

D. R. SEELY[2]

Abstract The evolution of structural highs bordering the seaward edges of major modern forearc basins is controlled by their location at inception and two primary processes—seaward accretion, and landward understuffing. Many highs probably initiate as crustal ruptures. Seaward accretion broadens the highs. Landward understuffing elevates the highs and tends to move them in a landward direction by tilting or by failure of their inner edges by compressional folding and faulting. Forearc basins behind the highs grow seaward if accretion is accompanied by submarine elevation of progressively younger accreted section added to the structural highs. Forearc basins become narrower, however, when the highs bordering their outer edges tilt landward or fail under compression in the absence of such contemporaneous accretionary growth. Accretionary broadening has previously been discussed and is exemplified by the postulated late Mesozoic history of the northern California forearc. Landward tilting, or compressional failure, is exemplified by the Cenozoic history of the northern California forearc and is suggested by a cross-section across the Guatemalan forearc, where small landward migration similar to that of highs bordering smaller Aleutian and Chilean forearc basins is recorded.

INTRODUCTION

Forearc terrace-edge, shelf-edge, and trench-inner-slope structural highs are commonly associated with subduction zones. Those highs bordering major forearc basins have been postulated to migrate progressively seaward to form the seaward edge of a steadily broadening basin (Seely et al, 1974; Karig, 1974; Moore and Karig, 1976; Dickinson, 1976). Karig in a recent article (1977) has discussed the postulate in depth. The concept has arisen as a corollary of subduction-related accretion. If the concept is correct, most forearc basins should have a basement consisting of materials scraped off the underthrust plate (subduction complex). Dickinson (1977), however, has pointed to several "thick forearc basins" underlain by a basement consisting of oceanic crust. Additionally, Grow (1973) considered that the central Aleutian terrace forearc basin could be underlain by either subduction complex or crustal remnant, and Curray et al (1977) concluded that the Java terrace forearc basin is underlain by trapped oceanic crust, as has Hamilton (1977b).

There has been a paucity of subsurface data from modern forearc areas that pertain to these two concepts. The purpose of this paper is to present such data pertaining to the evolution of one of these highs from an area on the Guatemalan shelf edge, to discuss it relative to the evolution of a forearc structural high off Chile recently discussed by Coulbourn and Moberly (1977) and of a high off Alaska documented by von Huene (1972), and to advance the hypothesis that structural highs bordering major forearc basins, such as those cited by Dickinson, Grow, Curray et al, and Hamilton (above), generally evolve from the ruptures that entrapped the crust, and develop under the control of two major processes—seaward accretion and landward understuffing. It will be seen that the Guatemalan shelf-edge high probably developed from an initial lithospheric rupture like that postulated for the northern California forearc by Hamilton (1977a), but which has retreated landward since its inception owing to understuffing. Before discussing it and the other highs, we need to place them in context by briefly reviewing forearc features and terminology discussed by Dickinson and Seely (1979).

The forearc is the area between an active volcanic arc and the outer edge of the oceanic trench associated with that arc (Fig. 1). Various bathymetric

[1]Manuscript received, June 16, 1977; accepted, March 22, 1978.

[2]Exxon Production Research Company, Houston, Texas 77001.

The writer is indebted to R. A. Dorsey and R. H. Jones of Esso Inter-America for the data supplied by their maps and cross-sections of offshore Guatemala; to R. L. Fleisher, J. Hardenbol, and J. L. Lamb of Exxon Production Research for discussion of the paleontology of the Esso Petrel well; to Esso Inter-America for encouraging the release of this paper; to R. von Huene for providing a copy of his seismic cross-section through the Aleutian slope basin; to K. H. Hadley and P. R. Vail for reviewing the manuscript; and to W. R. Dickinson and S. A. Graham for insight into northern California geology.

Article Identification Number:
0065–731X/78/MO29–0017/$03.00/0.
(see copyright notice, front of book)

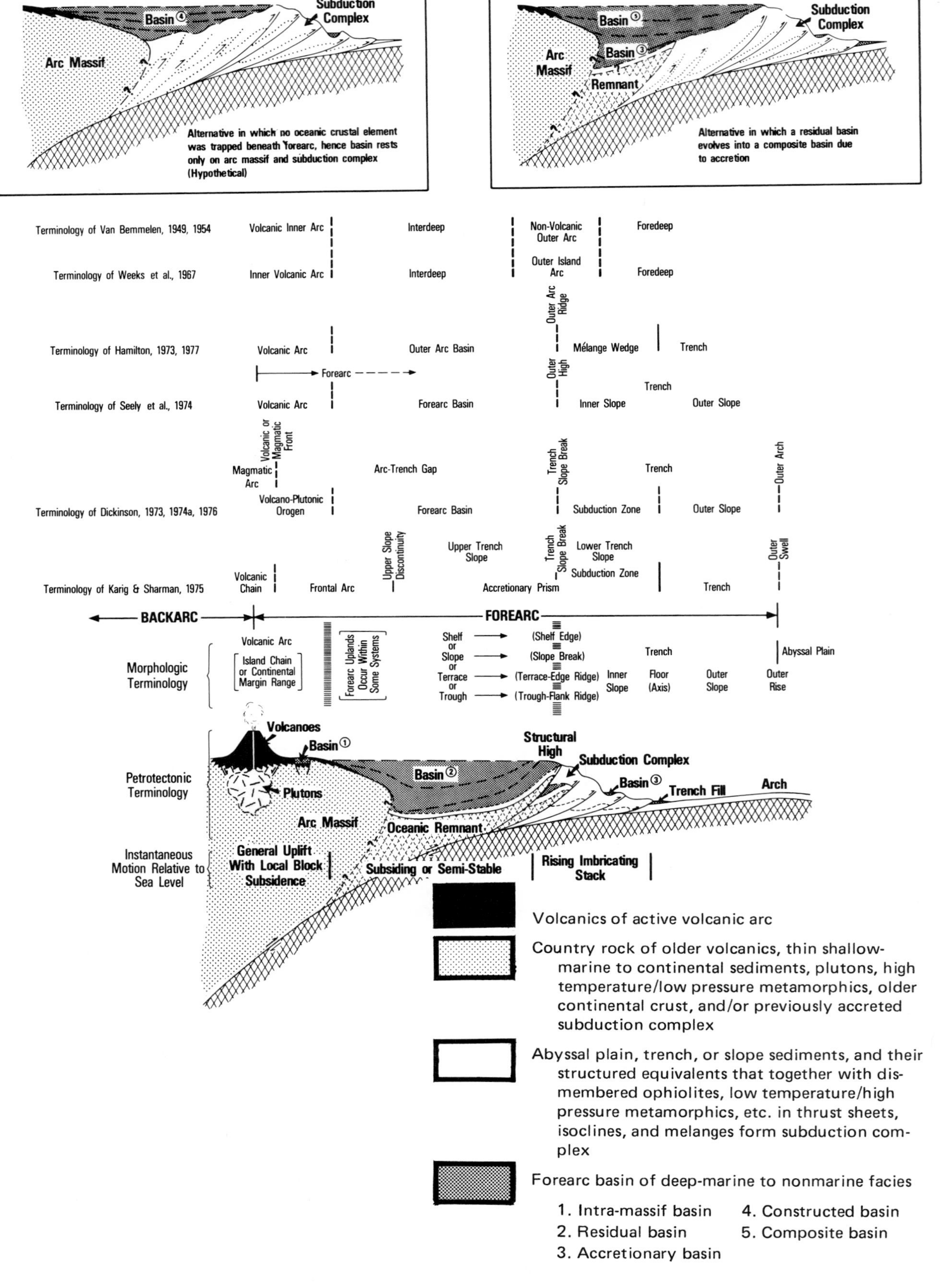

FIG. 1 Generalized forearc model for an instant of geologic time (Seely and Dickinson, 1977). The features are transient in time and space, and may become superposed. The terminology of the present paper begins at the "BACKARC-FOREARC" line.

and topographic features exist in this area; they differ from arc to arc and in different portions of the same arc. Common morphologic terms can be used to describe the features found. The volcanic arc itself is an oceanic island arc or a continental margin mountain range. The trench has an inner (or landward) slope, a floor (or an axis) and an outer (or seaward) slope. The trench may merge imperceptibly with the abyssal plain, or there may be a gentle positive bathymetric feature, the outer rise, at the outer edge of the trench that separates it from the abyssal plain.

Morphologic features between the volcanic arc and the trench are varied (Fig. 2). There may be a continuous, though irregular, slope from the volcanic arc to the trench. Forearcs having this configuration as their dominant characteristic can be roughly categorized as *sloped* forearcs. The slope may

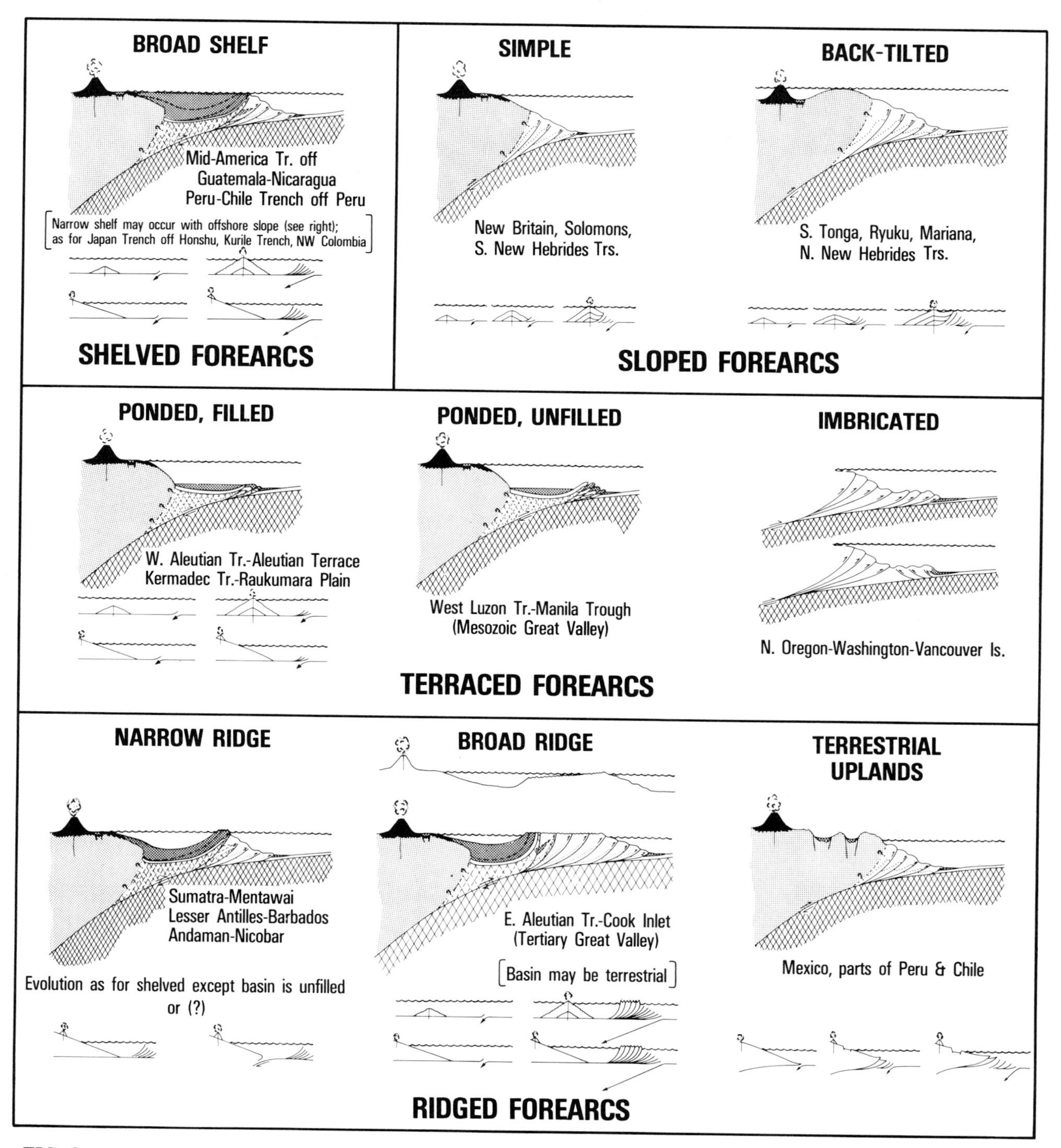

FIG. 2 Models of modern forearcs (Seely and Dickinson, 1977). Conceptual evolutionary cartoons shown at the bottoms of the diagrams. Question marks indicate the unknown nature and position of the massif-subduction complex boundary, which is a current subject of JOIDES research.

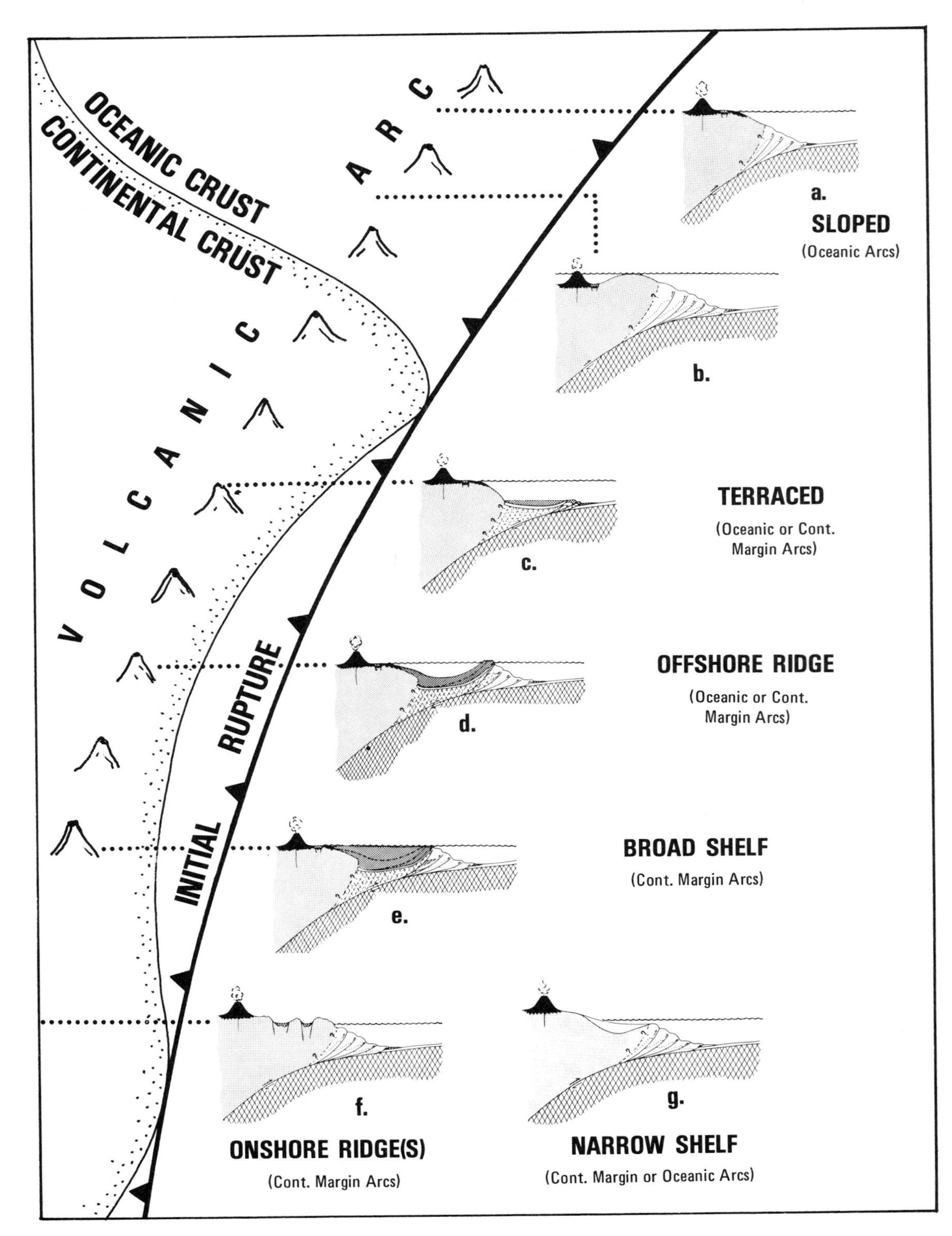

FIG. 3 Forearc morphologies relative to hypothetical genesis of arc-trench system (Idea for figure from P. R. Vail).

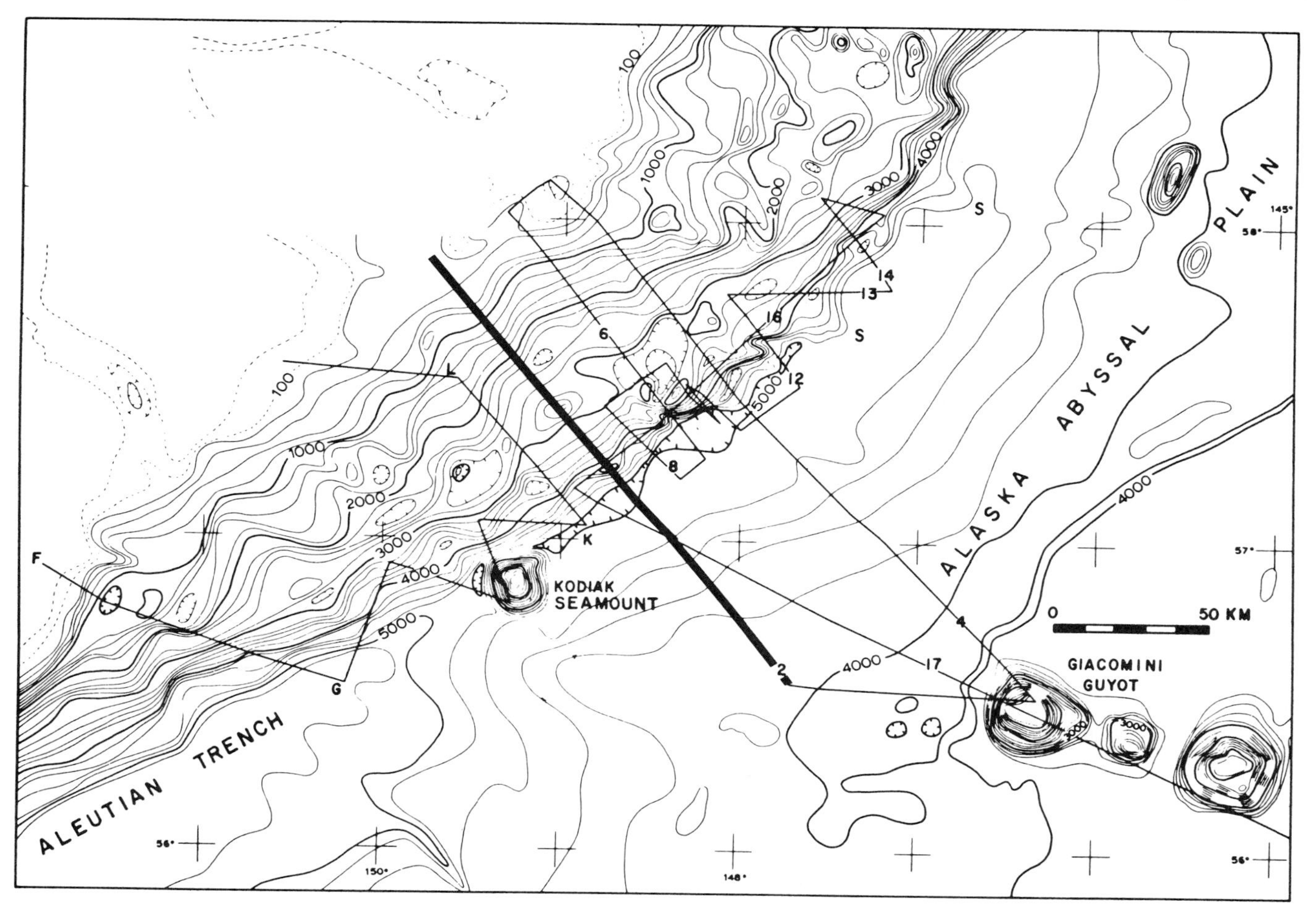

FIG. 4 Eastern Aleutian trench and accretionary basin (von Huene, 1972, and personal commun.). **4a** Map of von Huene's tracklines, with depths in meters.

FIG. 4B Line drawing of high-frequency seismic section along heavy black line, and enlarged segment of seismic section showing accretionary basin.

have a "knick point," or break, within it that is referred to as the trench-slope break. There may be a prominent terrace between the volcanic arc and trench, a characteristic categorizing these forearcs as *terraced* forearcs. Ponded terraces and structural terraces occur. The ponded, terraced forearcs are formed by sediment ponding behind an elevated terrace edge that is also a high. The structurally terraced forearcs have little, if any, sediment cover, the terrace being formed by a series of compressional structures, each of which culminates at about the same bathymetric depth.

A third morphology that is common between the volcanic arc and trench is characterized by a shelf. The *shelved* forearcs are like the ponded, terraced forearcs except that sediment has filled the basin behind the prominent structural high and has accumulated to shelf water depths. In the fourth and last general morphologic grouping of forearcs the structural high lies on the seaward edge of a submarine trough that is incompletely filled with sediment. The high produces a ridge that is commonly expressed as a chain of nonvolcanic islands. The ridge may be either broad or narrow. A special subcategory of *ridged* forearcs is recognized where the ridges are present on land, producing terrestrial uplands.

Forearc morphologies can be grouped according to the hypothetical position of the initial rupture that formed the subduction zone (Fig. 3). In this diagram, *a* and *b* have formed at an intraoceanic crustal rupture far from land, whereas *c, d,* and *e* formed at a similar rupture near the continent, thus trapping a remnant of oceanic crust beneath the forearc basin. The crustal rupture forming *f* and *g* occurred at the continent-ocean join. Note that *c, d,* and *g,* though shown on Figure 3 in association with continental margin arcs, are also common in association with intra-oceanic arcs, as indicated parenthetically on the figure.

A separate set of terms is used to describe the structured rock masses found in the forearc (Fig. 1). The *arc massif* is the terrain of volcanic sequences, underlying plutons, and metamorphic country rocks into which these igneous rocks were intruded. Structurally, the arc massif is characterized by block faulting and gentle warping.

The nature of the seaward edge of the arc massif is one of the major unknowns of convergent margins (Karig and Sharman, 1975, "upper slope discontinuity"). The seaward edge may be the contact between the massif and a remnant of oceanic or transitional crust, or it may be the contact between the massif and scrapings of oceanic materials that are collectively referred to as the *subduction complex.* Oceanic remnants should exist adjacent to intraoceanic island arcs between the point of initial crustal rupture and the point where oceanic lithosphere reaches sufficient depths for the melting necessary to form the arc. Oceanic crustal remnants should also exist adjacent to arc massifs in areas where the continental margin is so irregular that the smoothly bending oceanic lithosphere is incapable of following the margin irregularities when it breaks to form the subduction zone initially. On such margins the rem-

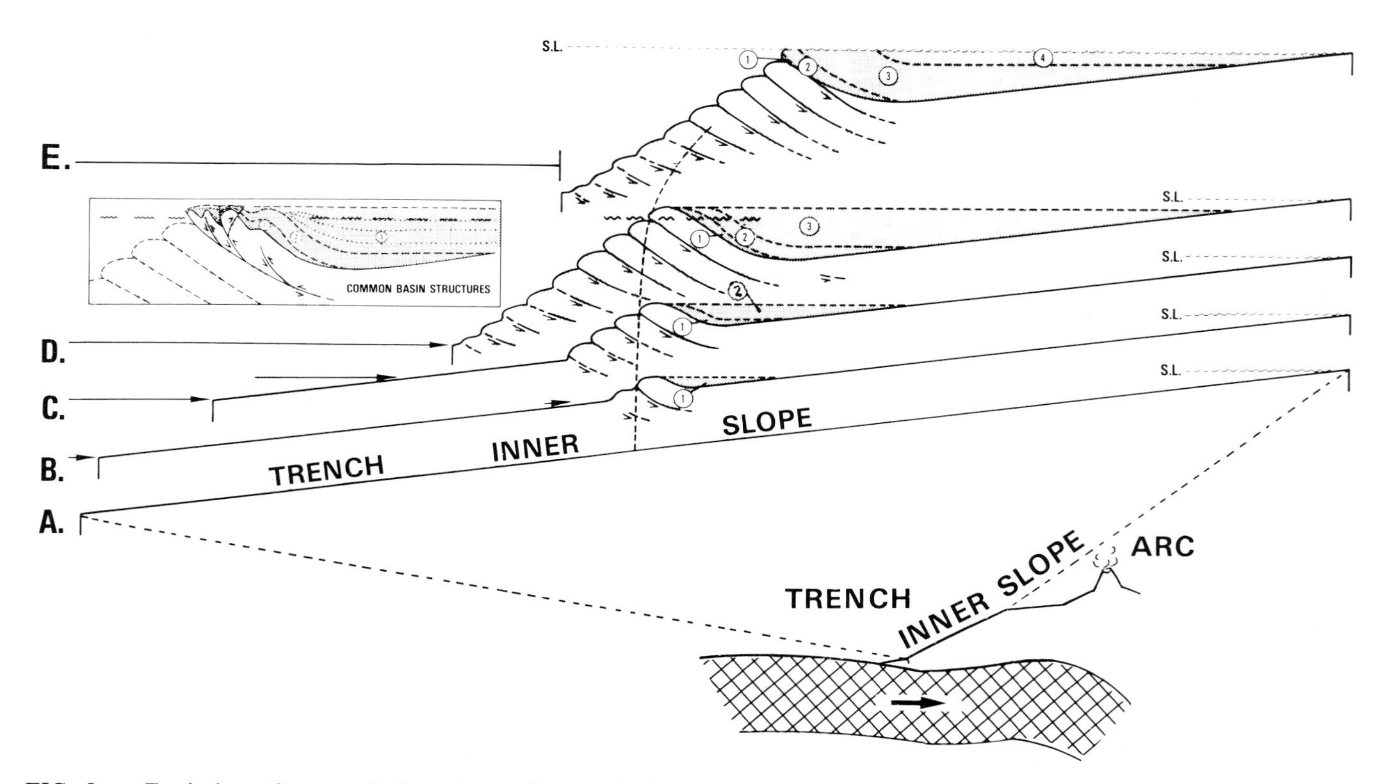

FIG. 5 Evolution of a trench-slope (accretionary) basin. The nearly vertical, gently curving dashed line connects the same point at various evolutionary stages and shows a slight component of motion in a landward direction of this point. See text for further explanation.

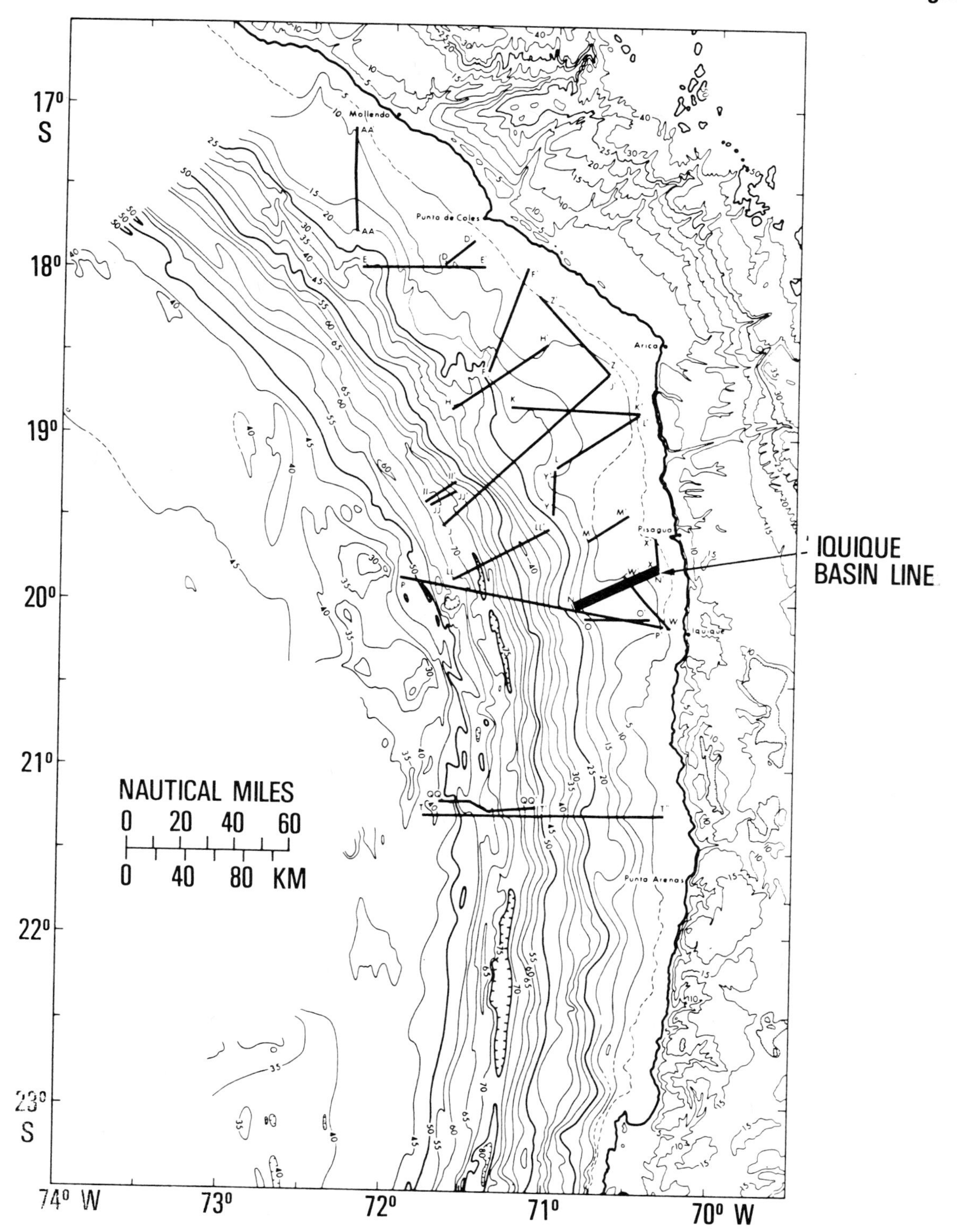

FIG. 6A Peru-Chile trench and Iquique constructed basin (Coulbourn and Moberly, 1977). Map of Coulbourn and Moberly's tracklines with depths in hundreds of meters.

nants would occur in marginal recesses (Fig. 3). Perhaps the preservation of transitional crust from an earlier cycle of margin evolution is also more likely in these recesses. If so, marginal salients may bring the massif into nearly direct contact with the subduction complex.

Two conceptual models for the evolution of forearcs result from the foregoing relationships: (1) the subduction complex extends from the inner edge of the trench to a contact with the arc massif, a contact which may be gradational, or (2) the subduction complex extends from the inner edge of the trench to the seaward edge of older transitional crust or an oceanic crustal remnant. Following these two con-

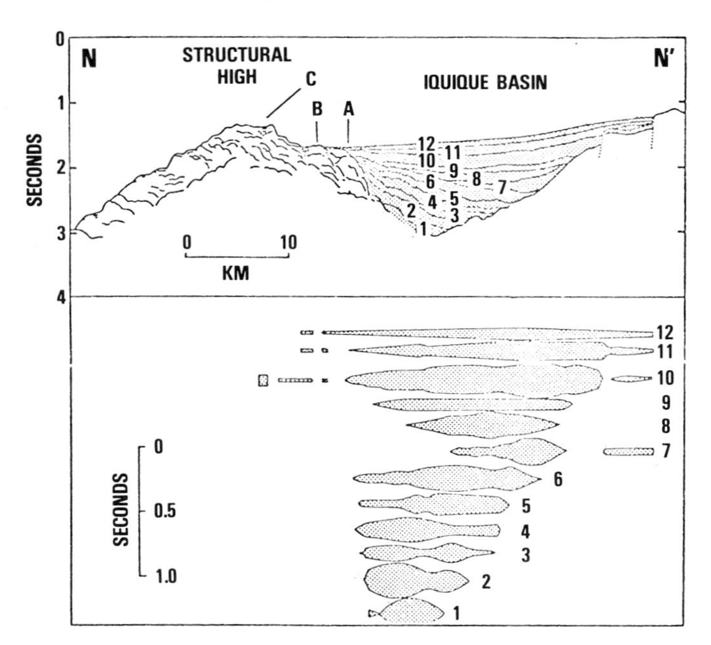

FIG. 6B Line drawing of high-frequency seismic section along heavy black line (above), and thickness distribution of basin-fill sediments (below). Note that the depocenter migrates landward as the structural high evolves.

cepts, the terrace-edge or shelf-edge structural high may be either the migrating crest of an accreting subduction complex or it may approximate the seaward edge of a crustal remnant.

We distinguish four basic types of forearc basins and a fifth, which is a composite of the basic types (bottom of Fig. 1). The first type is entirely included within the arc massif. We refer to that type of basin as an *intramassif basin.* It is commonly a fault trough of some type, but it may also be formed by warping within the arc massif. The second type of forearc basin lies primarily on an oceanic crustal remnant, or transitional crustal remnant. We refer to this type as a *residual basin.* A third type of forearc basin lies entirely on a subduction complex. This is an *accretionary basin.* Sediments deposited within accretionary basins are usually destined to become incorporated into the subduction complex. Most of them are located on the trench inner slope, but some occur on broad shelves typical of ridged forearcs where the ridge is formed predominantly of subduction complex material. The fourth type of forearc basin is the *constructed basin.* Sedimentary fill in this basin type rests unconformably on the arc massif on the inner side of the basin and on accreted strata of the subduction complex on the outer side of the basin. The fifth type of basin is a composite of two or more of the four basic types. For example, the basin may develop on a subsided arc massif, a crustal remnant, and subduction complex.

Three structural highs will be discussed. The first is located at the edge of a small Alaskan terrace that is underlain by an accretionary forearc basin. The second lies on the edge of a larger, Chilean terrace underlain by a basin that is probably a constructed forearc basin. The third lies beneath the edge of the Guatemalan shelf.

ACCRETIONARY BASIN STRUCTURAL HIGH

A structural high occurs at the edge of a terrace that is present at about midslope on the eastern Aleutian trench inner slope (Fig. 4a). The terrace has developed entirely within rocks of the subduction complex, as indicated by the continuation of the strong oceanic basement reflector from the trench outer slope across the trench and some distance back beneath the trench inner slope. The structural high, which is represented by a terrace-edge ridge (Fig. 4b), lies on the seaward edge of an accretionary basin. Within the basin, sediments on the flank of the structural high can be seen dipping landward at depth and gently seaward at shallower levels. Thus, the flank of the structural high was tilted landward after deposition of the older sequence of sediments and before deposition of the younger sequence that now fills the accretionary basin.

The landward tilting of the flank of the structural high can be explained as the result of understuffing of lower slope sediments beneath sediments higher on the slope that produced slight landward migration of the basin margin (Fig. 5 and Seely et al, 1974). Early ponded sediments deposited in the accretionary basin become part of the tilting flank of the structural high against which younger accretionary basin sediments onlap. Stages A, B, and C of Figure 5 show this process. Sediments of the lower slope are shown moving approximately horizontally beneath sediments of the upper slope to produce the structural high. Earliest accretionary basin sediments, sequence 1, are tilted back and onlapped by the younger sediments of sequence 2 as the underlying sediments of the thrust plate are wedged back by underthrusting. Stage C of the diagram is analogous to the Aleutian Slope accretionary basin described above. Stage D is a still later phase in the evolution of the accretionary basin, showing yet another sequence (3) onlapping the upturned beds of sequences 1 and 2, and also showing the possibility that the structural high may grow to such levels that it can be exposed on the edge of a shelf, at the surface, or at relatively shallow water depths, thus causing its erosion (dashed heavy wavy line). The inset of Stage D is included to show that generally the structural high is itself highly structured with smaller folds and faults, and also shows the surficial structural expression that would result from erosion of the high. Diagram E is one of the many possible variations that can develop during the evolution of the structural high in which the accretionary basin sediments develop a thrust directly at their contact with the underlying slope sediments.

CONSTRUCTED BASIN STRUCTURAL HIGH

A structural high has developed on the seaward edge of what appears to be a constructed basin near the top of the inner slope of the Peru-Chile trench (Fig. 6). The high forms a prominent trench-slope break. Coulbourn and Moberly (1977) detail develop-

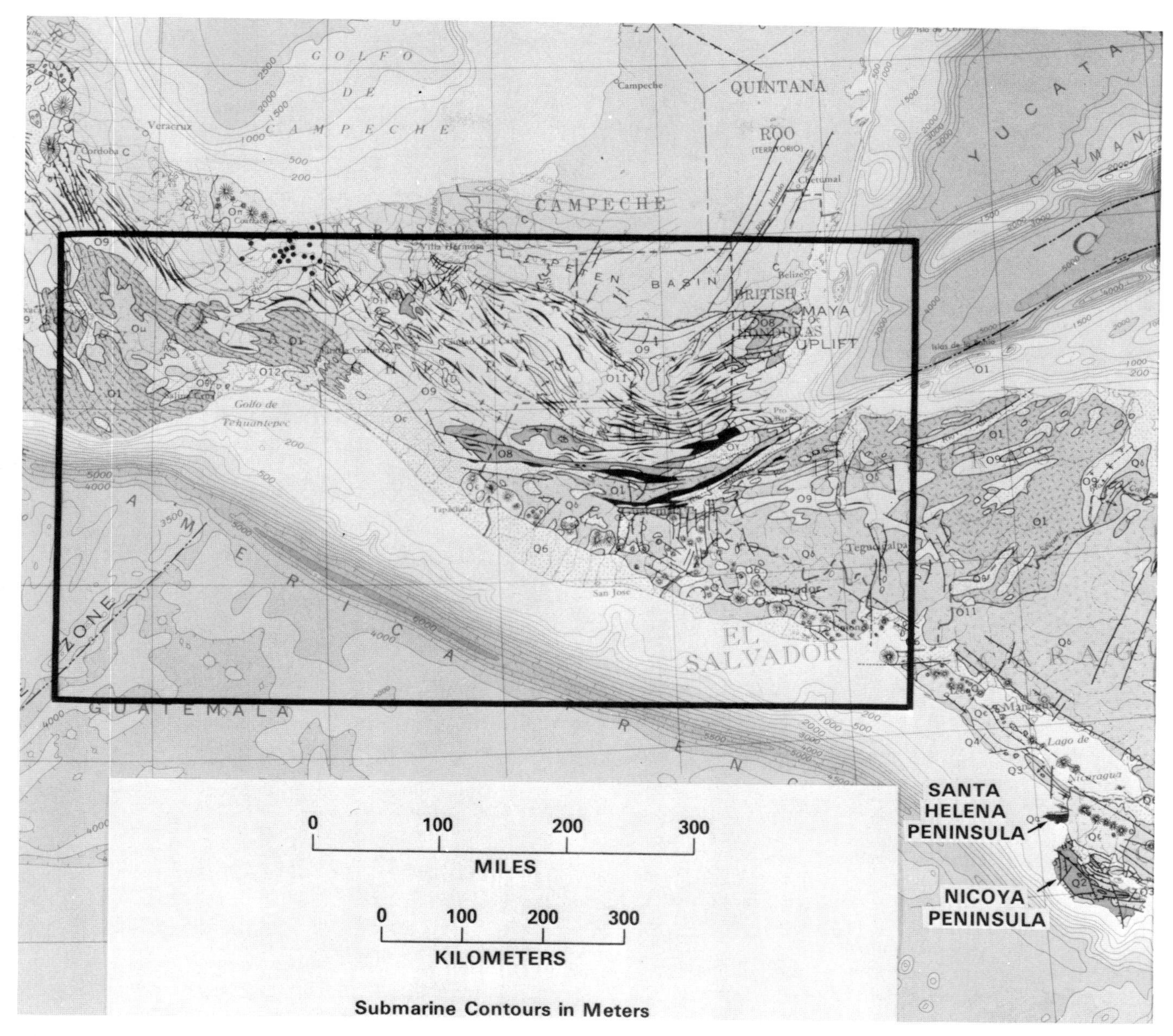

FIG. 7 Tectonic map of northern Central America (King, 1969). The area of Figure 8 is outlined. Note the location of the submarine canyon south of San Jose, Guatemala, for orientation of Figure 8.

ment of the basin and high, and show it to be similar to that described above for the accretionary basin and high on the inner slope of the eastern Aleutian Trench. The structuring on the landward side of the Iquique Basin, however, consists of normal faulted blocks and contrasts with the compressional structuring on the landward side of the Aleutian accretionary basin. The block faults indicate that the inner edge of the basin probably overlies the seaward edge of the arc massif, and the intricate structuring indicated by the reflectors in the structural high suggests that the outer edge of the Iquique Basin overlies the subduction complex. The basin would thus be a constructed basin. There has been considerable upward growth of the structural high relative to basin sediments, but in order for there to be significant seaward growth, large-scale accretion would have to occur on the slope below the high or else the slope would become oversteepened.

RESIDUAL FOREARC BASIN STRUCTURAL HIGH

The outer edge of what appears to be a residual forearc basin occurs beneath the outer shelf and upper slope of the Middle America Trench off Guatemala (Figs. 7 and 8). The Nicoya complex, which is designated by the symbol Q_2 to the south of Lake Nicaragua in the southeastern corner of Figure 7, was described as an ophiolite complex by Dengo (1962) and herein is presumed to consist of a thrust slice of oceanic crust (more studies are needed to confirm this). A magnetic basement high following

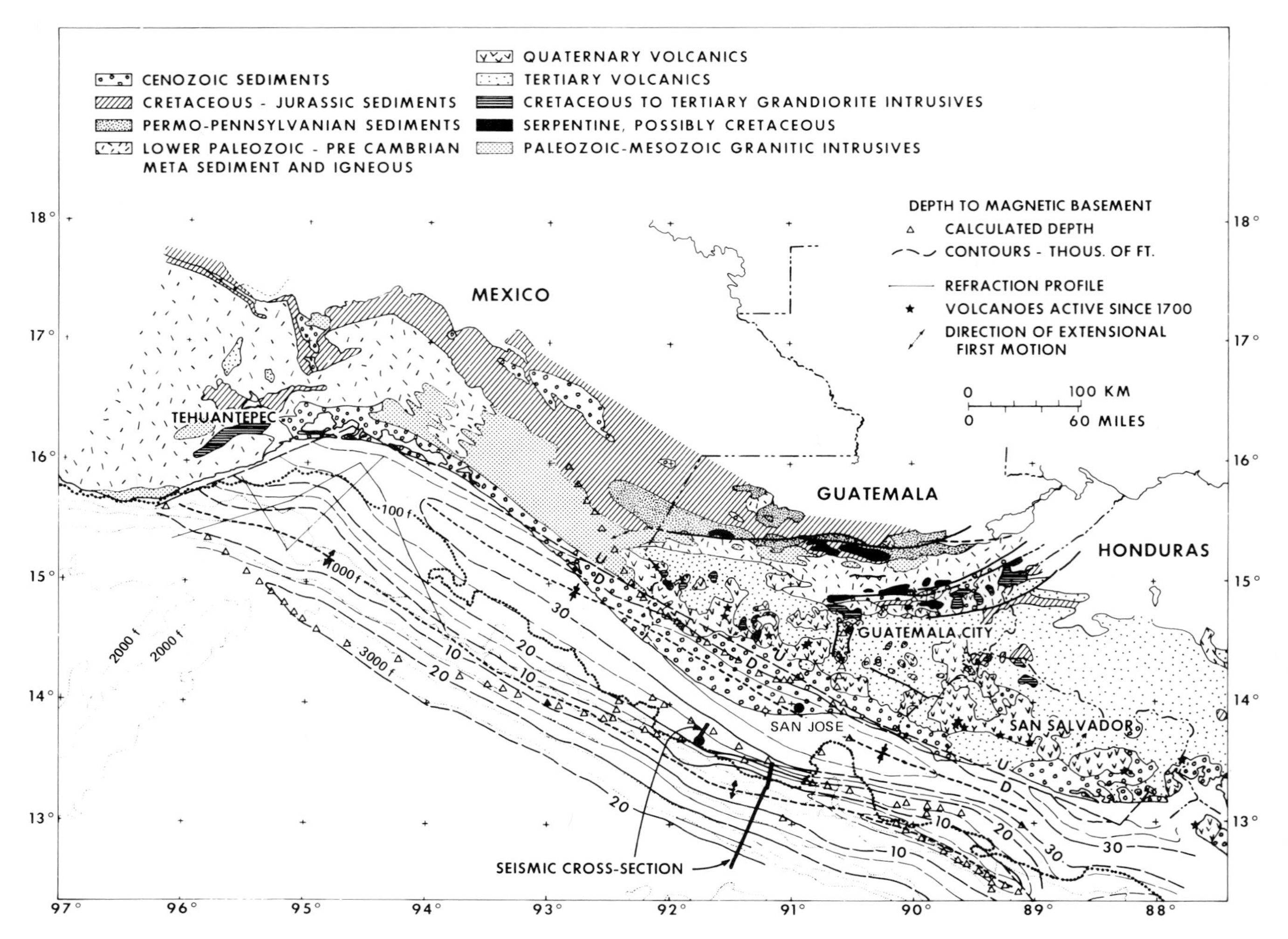

FIG. 8 Esso Petrel well, seismic lines, and magnetic basement off Guatemala (geology by R. A. Dorsey, personal commun.; refraction from Shor and Fisher, 1961; first motion from Isacks and Molnar, 1971). The well is located near the seaward end of the short seismic line (see Figs. 9, 13). The long seismic line is that of Figures 10 and 11.

the shelf edge to the northwest of the Nicoya Peninsula is suggested by Project Magnet reconnaissance magnetic profiles (Fig. 8) and may represent continuity of oceanic thrust slices from the peninsula into the subsurface. Likewise, a deep trough can be interpreted using the same reconnaissance data (Fig. 8) and may be floored by the northwestward extension in the subsurface of rocks that crop out on the Nicoya and Santa Helena Peninsulas (Fig. 7). R. W. Couch has described a similar high and trough based on his interpretation of gravity and magnetic data (personal communication, 1977). The high, however, is apparently discontinuous.

The development of the structural high is illustrated by a seismic line and well located near the shelf edge (Fig. 8). The seismic section (Fig. 9) shows that Cretaceous and Paleocene sediments were tilted landward on the flanks of the structural high prior to the deposition of the onlapping Eocene sequence. Paleontologic determination of the depositional environments of sediments penetrated by the well indicates that there was actual uplift relative to sea level at the wellsite between Paleocene and Eocene time (Fig. 10) as indicated by the abyssal depositional depths (below the lysocline) of the Paleocene sediments and the middle bathyal depositional depths of the Eocene sediments. The prominent angular unconformity at the base of the early Miocene sediments (Fig. 9), however, was not caused by similar uplift at the wellsite in post-Eocene pre-early Miocene time. Rather, the position of the structural high relative to present-day sea level probably stayed approximately fixed or the high subsided slightly while the basin to the landward of it subsided markedly to cause the pronounced tilting. Erosion of the tilted section may have occurred subaerially in mid-Oligocene time when sea level was low (P. R. Vail, personal communication). Tilting due to basin subsidence was renewed in post-middle Miocene time.

The basement on which the sediments were deposited was not penetrated by the well (Fig. 10), but several factors suggest that the basement is oceanic. First, the oldest sediments penetrated by the well were deposited in abyssal water depths below the lysocline and consist largely of montmorillonitic shales. The tectonic history recorded in sediments

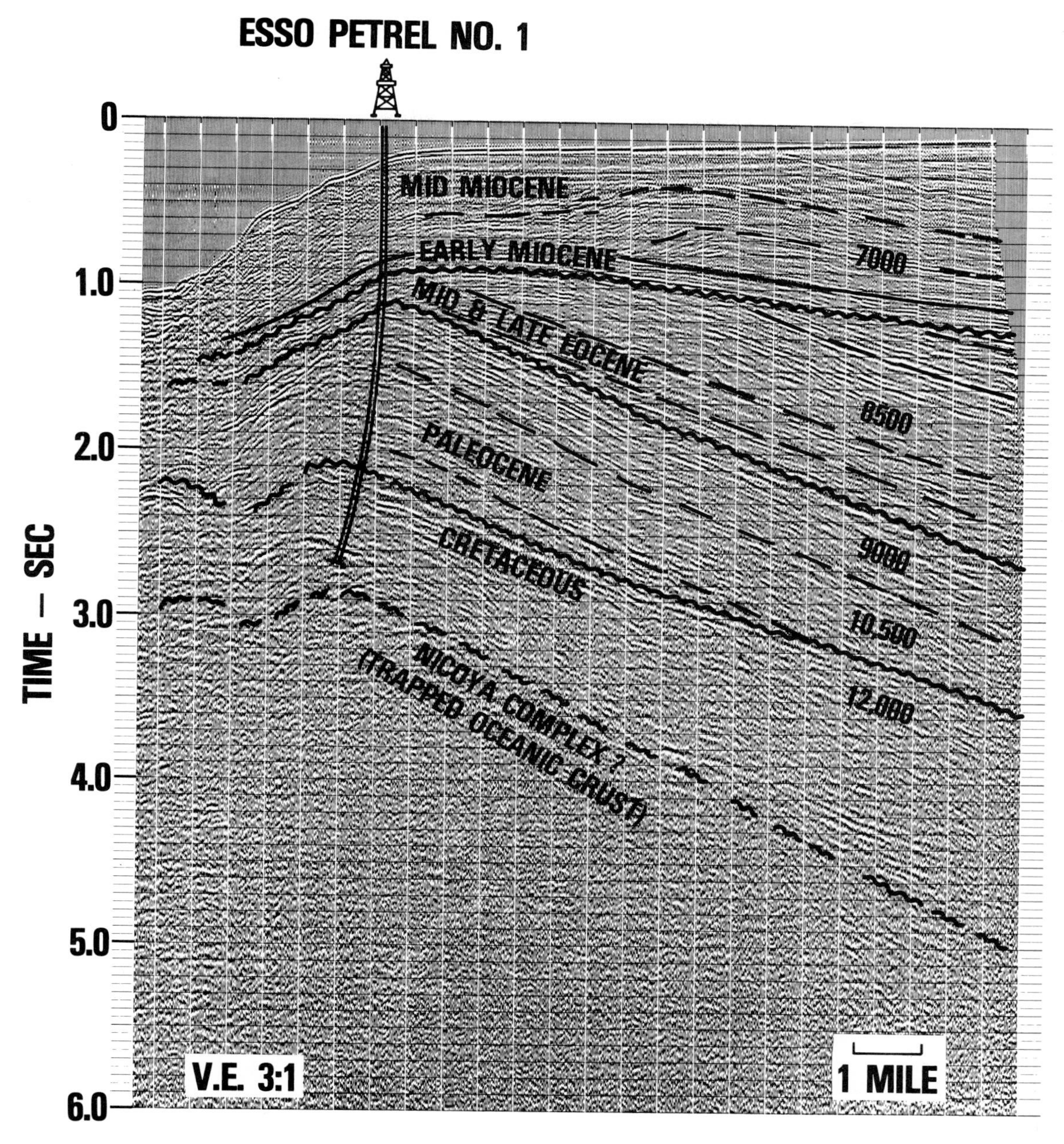

FIG. 9 Seismic time section across shelf edge near Esso Petrel well (see Fig. 8 for location).

penetrated by the well has been one of uplift, not of subsidence, since deposition of these oldest beds. The absence of coherent reflectors beginning at depths only slightly greater than the total depth of the well, and calculations of depths to magnetic sources, indicate that basement may be only a short distance deeper than the well. These factors appear to leave little room for a record of the great subsidence that would have had to occur had the Cretaceous sediments been deposited upon a subsided continental crust, but are consistent with their deposition on oceanic basement. Secondly, as previously mentioned, the reconnaissance magnetic and gravity data suggest that a basement high is relatively continuous from the wellsite to the Nicoya Peninsula outcrops, which probably consist of oceanic crust.

The uplifted magnetic basement is shown by the calculated times and depths to the top of magnetic basement, indicated by the black dots on Figures 11a and 11b. The dots are a short distance below the seafloor of the outer slope of the trench, confirming the seismic reflection evidence for the presence of only a thin sedimentary cover there. They follow the strong reflector back beneath the trench inner slope, the strong reflector probably being the continuation of oceanic igneous crust beneath the slope, as previously discussed (Seely et al, 1974). The five dots at the right end of the cross-section are displaced upward from the four in the left half of the section, and the displacement is more abrupt when the section is plotted with a depth vertical scale rather than a time vertical scale (Fig. 11b); that is, in depth the

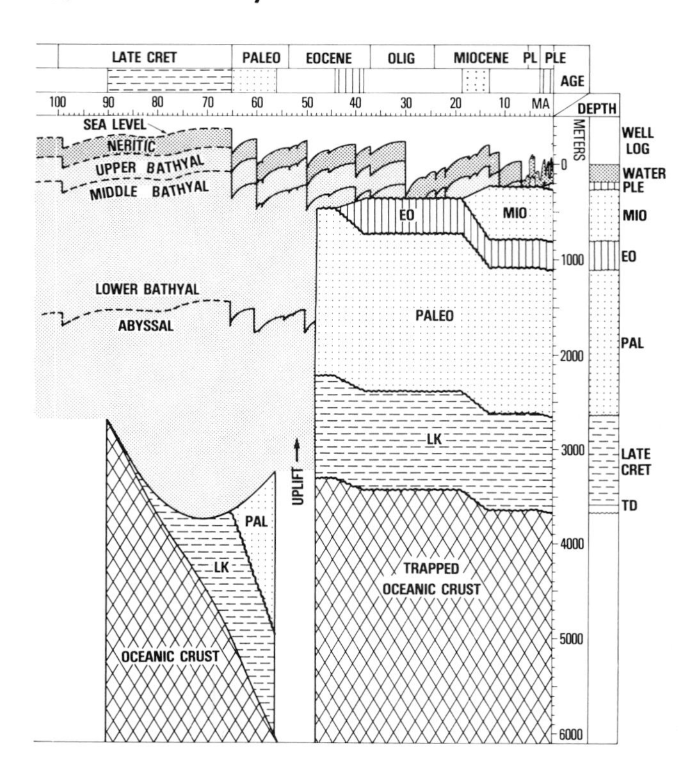

FIG. 10 Tectonic history diagram showing sediment accumulation and motion relative to present sea level (corrected for eustatic changes—shown by curve marked "SEA LEVEL") as a function of time at the Esso Petrel well site (P. R. Vail, personal commun.). The vertical scale at right marks depths and elevations relative to present sea level. The horizontal scale is in millions of years before the present (MA), with patterned intervals representing preserved sediments. The right edge of the diagram is a "stick section" of the well showing depths at which various time-rock units were penetrated. Similar "stick sections" can be constructed for past times by dropping "sticks" from the times selected and reading depths at right. Sediment thicknesses have been adjusted for the effects of compaction. From Eocene time onward, subsidence traces a curve approximating isostatic response to sediment loading. Small deviations from this curve due to the isostatic effects of eroded sediments and changing water depths should be expected, but the imprecision of geologic data prohibited their detection.

dots and the strong reflector beneath the trench inner slope are on an approximate straight-line extension of the dots and reflector beneath the trench outer slope to the left (see Seely et al, 1974). The dots beneath the trench inner slope on the strong reflector thus occur near the base of a depth seismic section, whereas dots farther landward still occur at the relatively shallow depths seen on the time section.

The shallow magnetic sources on Figure 12 appear to come from diapirs or intrusions into the sedimentary section (see also Figs. 6 and 13, Seely et al, 1974). The evidence for this is their very shallow depth—that is, a depth of about one second on the seismic section; the fact that dipping sediment reflectors, when migrated, would occur in the area below the shallow magnetic sources; and the seismic suggestion that sedimentary layers on the landward side of the shallow magnetic sources are actually dragged upward at the contact with the magnetic materials. The normal faults in the sediments above the magnetic sources also suggest extension above intrusions. One of several possible explanations for them is that they are serpentine plugs rising from the oceanic basement that probably underlies the sedimentary section along the shelf edge and under the forearc basin landward from it. If the interpretation based upon seismic evidence of intrusive relationships is incorrect, the magnetic sources could be shallow slices of oceanic crust.

The depth version (Fig. 13) of the time section across the shelf edge previously discussed (Fig. 9) shows the Paleocene section thinning seaward beneath the present upper part of the trench inner slope. Also at that location, the Cretaceous section appears to continue to rise and thin in a seaward direction. Interpretation of gentle structuring within the Cretaceous-Paleocene and the possible configuration of the basement surface are also shown on Figure 13. Note the absence of complex structuring in these sediments. Such structuring would be expected had the sediments been deposited on the abyssal plain or in the trench and passed through the subduction zone.

Sequential diagrams to show the possible evolution of the Guatemala shelf edge structural high are shown in Figure 14, which illustrates that initial rupture of the lithosphere reaches the seafloor some distance seaward from the seaward edge of the incipient arc massif. After early growth of this rupture (Fig. 14A–C) a deep terrace basin is formed. This basin would be a residual forearc basin, because it has formed above a remnant of oceanic lithosphere. With continued underthrusting, shown in Figure 14D–E, sediments of the abyssal plain and trench underthrust and uptilt the lip of the outer edge of the residual forearc basin to produce a prominent terrace-edge structural high. At a later stage of evolution, the terrace basin grows to become a shelf basin behind a structural high that has grown to where it underlies the shelf edge. Understuffing of abyssal plain and trench sediments beneath the structural high causes tilting and uplift in a fashion somewhat analogous to that described for the structural high bordering the Alaskan accretionary basin and the Chilean constructed basin, respectively. The present shelf edge was developed by progradation in Miocene time across a structural high, located under the shelf edge and upper slope, that was probably formed by compressional failure of the outer edge of the Cretaceous-Paleocene basin in middle Eocene time. The shelf edge is landward from the high interpreted to have formed at the seaward edge of the deep Cretaceous terrace.

Also diagrammed in Figure 14D–E is downwarp of the inner edge of the forearc basin relative to the structural high. This is included to explain the seeming marked subsidence of the Guatemalan forearc basin in Miocene and later time when the high acted as a "hinge," approximately fixed (or subsiding

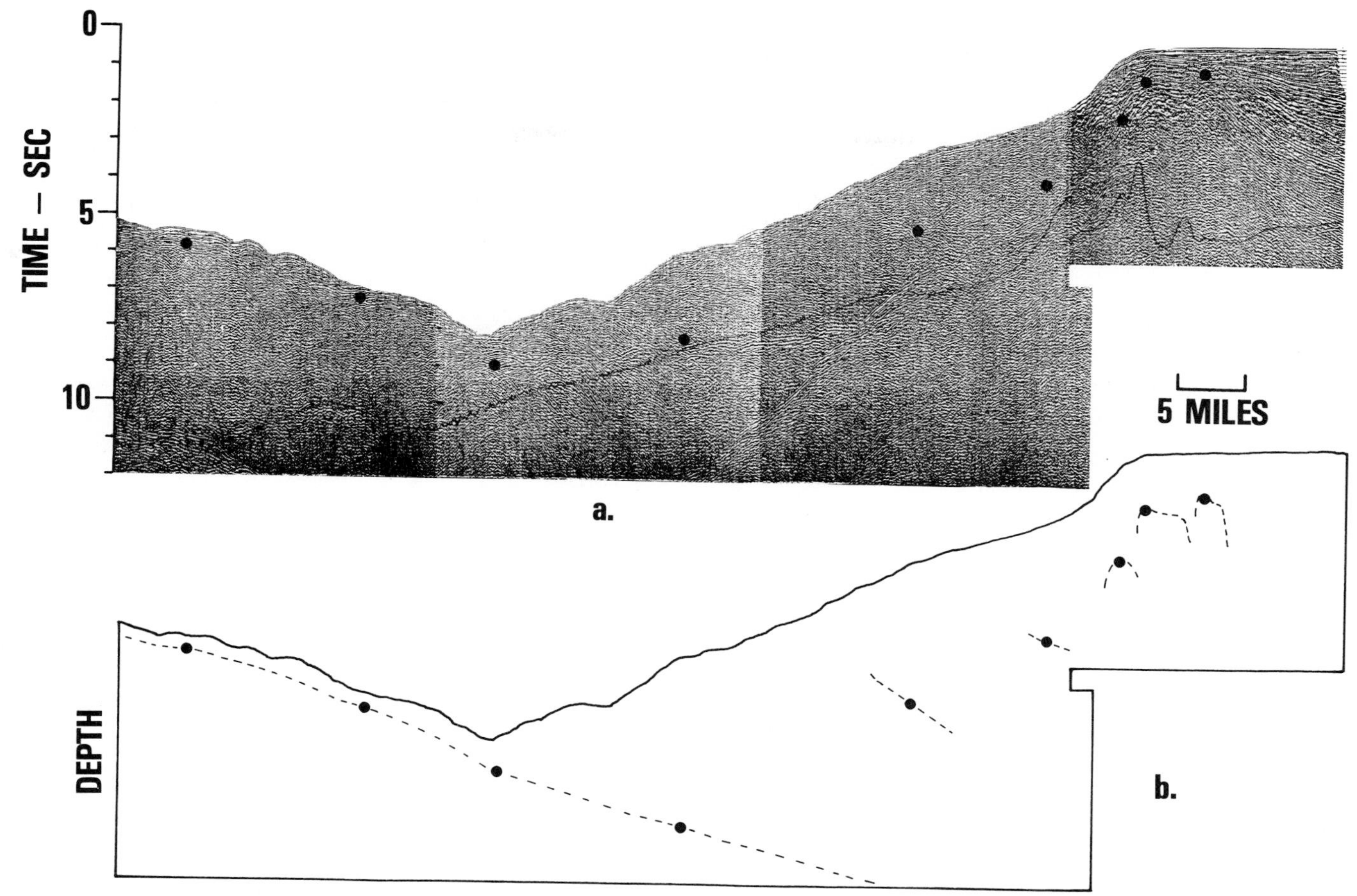

FIG. 11 Middle America trench seismic section with total intensity magnetic profile printed at base (see Fig. 8 for location). **a** Seismic time section showing calculated positions of magnetic basement (black dots). **b** Positions of black dots when velocities are used to convert times (11a) to depths. Generalized interpreted configuration of magnetic surfaces shown as light dashed line through dots (see Fig. 12 for detail of right end of section).

somewhat) relative to sea level (Fig. 10). Assuming that the eustatic cycles shown on Fig. 10 are correct, such behavior could be interpreted to mean that the high was being uplifted by understuffing coincidentally at nearly the same rate that the underthrusting oceanic plate was isostatically subsiding due to forearc loading or that, as the writer prefers, factors in addition to load subsidence (such as down-folding) are also important to the development of forearc basins. If the eustatic assumption is incorrect, uplift and erosion caused by understuffing in post-Eocene—pre-Miocene time, followed by subsidence to upper bathyal depths prior to Miocene deposition, would also explain the relationships observed.

DISCUSSION AND CONCLUSIONS

Structural highs on the seaward edges of three types of forearc basins have been discussed. These types of basins are the accretionary basin, the constructed basin, and the residual basin. In all three examples there has been slight landward migration of the inner limit of uplifting and tilting on the flanks of the structural highs that bordered the basins on their seaward sides. In these cases, understuffing is the probable cause of the tilting.

Landward migration of the edge of uplifting and tilting on the flank of the Guatemala shelf-edge high has occurred intermittently from Paleocene time to the present. The migration may have begun in Late Cretaceous time. Reconnaissance magnetic, well, and outcrop data suggest that the Guatemala structural high flanks an initial residual forearc basin that developed upon a fragment of oceanic lithosphere that was trapped between the incipient subduction zone and the evolving arc during Cretaceous time. Structural failure and uplift of the outer part of that fragment near the beginning of the Eocene are postulated to have caused a large part of the landward migration.

Although small forearc basins like the Iquique basin appear to grow by the progressive upward and outward accretion of a subduction complex that extends beneath the basin to near the arc massif (Fig. 15; Seely et al, 1974; Karig, 1974; Dickinson, 1976; Moore and Karig, 1976; Karig, 1977), similar development of major forearc basins is yet to be documented. The hypothesis proposed by Seely and Dickinson (1977) and elaborated here is that many, if not most, major forearc basins were initiated as re-

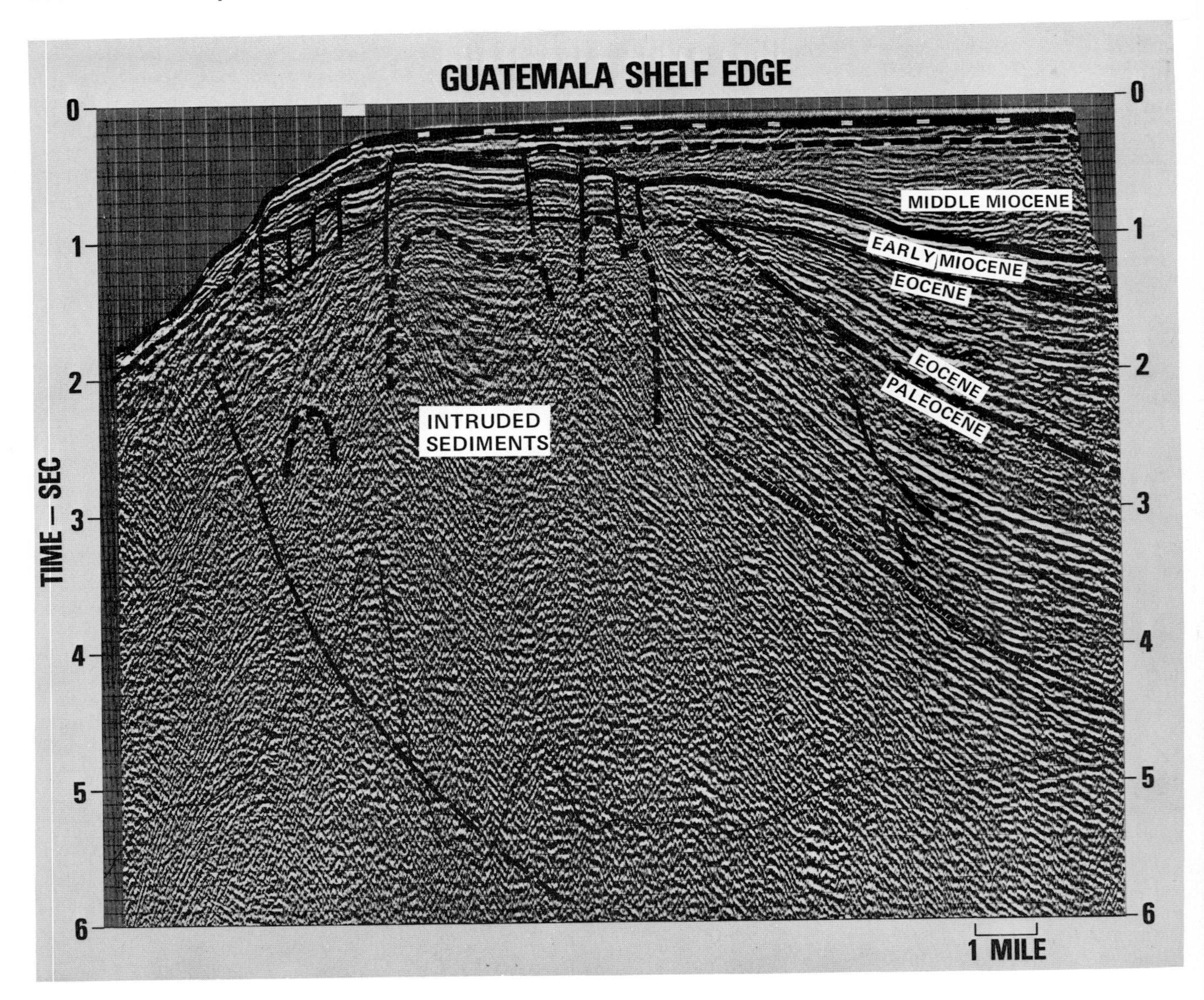

FIG. 12 Guatemala shelf edge southeast of Esso Petrel well (right end of Fig. 11). Black line at base of section is total intensity magnetic profile.

sidual forearc basins in which the structural high that underlies their outer flank began as a rupture within oceanic or transitional lithosphere which is now represented by a major lithologic boundary between overthrust lithosphere and subduction complex. This boundary is structurally unstable and may be reactivated or fail at anytime during subsequent subduction.

During periods when the boundary is relatively inactive, thrust highs within the subduction complex can migrate successively farther seaward from the boundary and form the outer margin of a broadening forearc basin (as in the Iquique basin) in arc-trench systems where large-scale accretion has occurred (Fig. 1, upper right inset). Such may have been the case in northern California where younger late Mesozoic forearc basin sediments occur westward from the boundary lying on Franciscan subduction complex above a tectonic contact that probably reflects original depositional relationships (Ingersoll et al, 1977).

When the boundary is active, however, the highs tilt or collapse in a landward direction, causing the forearc basin to narrow. In Guatemala, this appears to have resulted from understuffing of the relatively small volume of accreted section. In northern California, this probably resulted in Cenozoic time from rejuvenation of the boundary after large-scale accretion had occurred. The same may be true off Sumatra where forearc basin sediments on the landward edge of Nias Island apparently have been uplifted due to compressional tilting and folding (Karig, 1977, Fig. 5).

The evolution of the structural highs bordering

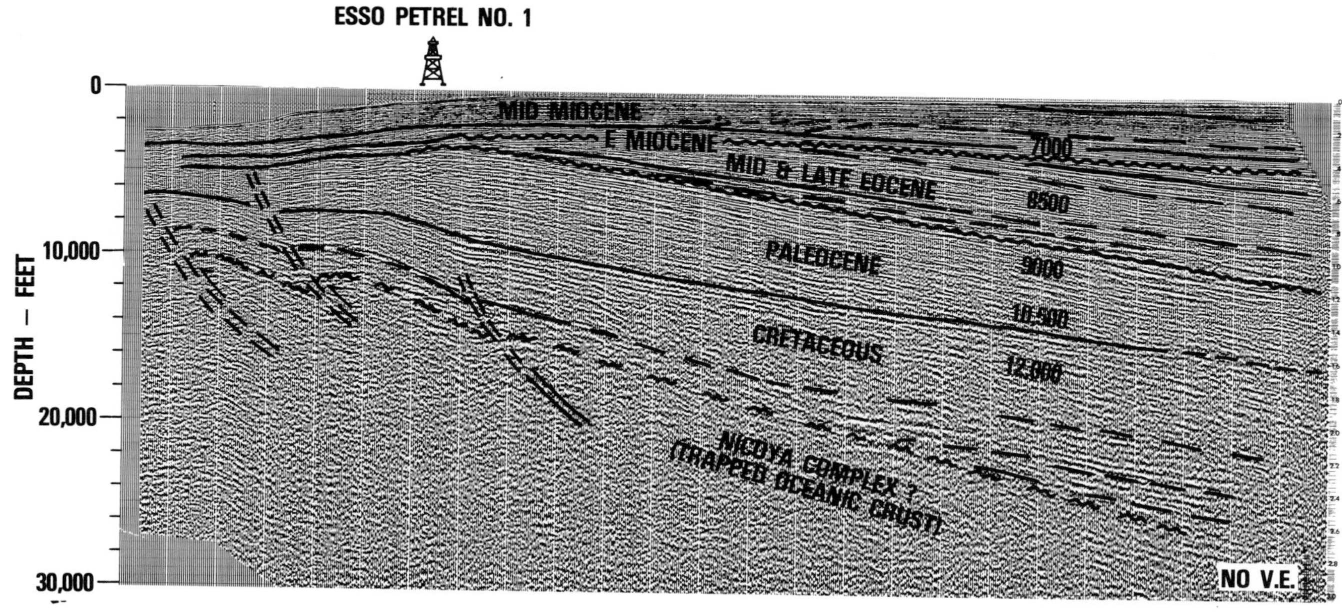

FIG. 13 Nonmigrated depth display of Figure 9. Vertical and horizontal scales are equal.

major forearc basins, therefore, appears to be a function of their locations at inception and the subsequent interplay of seaward accretion and landward understuffing, both of which result from underthrusting. Major forearc basins can develop where the highs initiate as crustal ruptures that trap large slabs of oceanic or transitional crust between the ruptures and the nascent arc. Seaward accretion broadens them. Landward understuffing elevates them and tends to move them in a landward direction, either by tilting or by compressional failure of their inner edges. Forearc basins behind them broaden or narrow depending upon the relative roles of accretion or compressional failure, respectively.

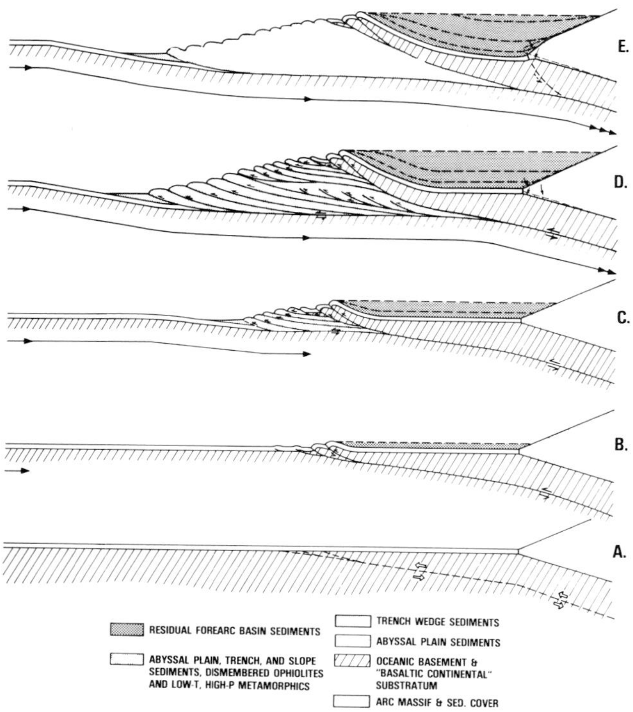

E.
D.
C.
B.
A.

RESIDUAL FOREARC BASIN SEDIMENTS
ABYSSAL PLAIN, TRENCH, AND SLOPE SEDIMENTS, DISMEMBERED OPHIOLITES AND LOW-T, HIGH-P METAMORPHICS
TRENCH WEDGE SEDIMENTS
ABYSSAL PLAIN SEDIMENTS
OCEANIC BASEMENT & "BASALTIC CONTINENTAL" SUBSTRATUM
ARC MASSIF & SED. COVER

FIG. 14 Evolution of a residual forearc basin.

REFERENCES CITED

Coulbourn, W. T., and R. Moberly, 1977, Structural evidence of the evolution of forearc basins: Canadian Jour. Earth Sci., v. 14, p. 102–116.

Curray, J. R., et al, 1977, Seismic refraction and reflection studies of crustal structure of the eastern Sunda and western Banda arcs: Jour. Geophys. Research, v. 82, p. 2479–2489.

Dengo, G., 1962, Tectonic-igneous sequence in Costa Rica, *in* A. E. Engel, H. L. James, and B. F. Leonard, eds., Petrologic studies: A volume in honor of A. F. Buddington: Geol. Soc. America, p. 133–161.

Dickinson, W. R., 1973, Widths of modern arc-trench gaps proportional to past duration of igneous activity in associated magmatic arcs: Jour. Geophys. Research, v. 78, p. 3376–3389.

——— 1974, Sedimentation within and beside ancient and modern magmatic arcs, *in* Dott, R. H., Jr., and R. H. Shaver, eds., Modern and ancient geosynclinal sedimentation: SEPM Spec. Pub. 19, p. 230–239.

——— 1976, Sedimentrary basins developed during evolution of Mesozoic-Cenozoic arc-trench system in western North America: Canadian Jour. Earth Sci., v. 13, p. 1268–1287.

——— 1977, Tectono-stratigraphic evolution of subduction-controlled sedimentary assemblages, *in* Talwani, M., and W. C. Pitman, III, eds., Island arcs, deep sea trenches and back-arc basins: Am. Geophys. Union, Maurice Ewing Ser. 1, p. 33–40.

——— and D. R. Seely, 1979, Structure and stratigraphy of forearc regions: AAPG Bull., v. 63, p. 2–31.

Grow, J. A., 1973, Crust and upper mantle structure of the central Aleutian arc: Geol. Soc. America Bull., v. 84, p. 2169–2192.

Hamilton, W., 1973, Tectonics of Indonesia region: Geol. Soc. Malaysia Bull. 6, p. 3–10.

——— 1977a, Late Mesozoic subduction tectonics of northern California (abs.): Geol. Soc. America, Cordilleran Sec., 73rd Ann. Mtg. Program, p. 430.

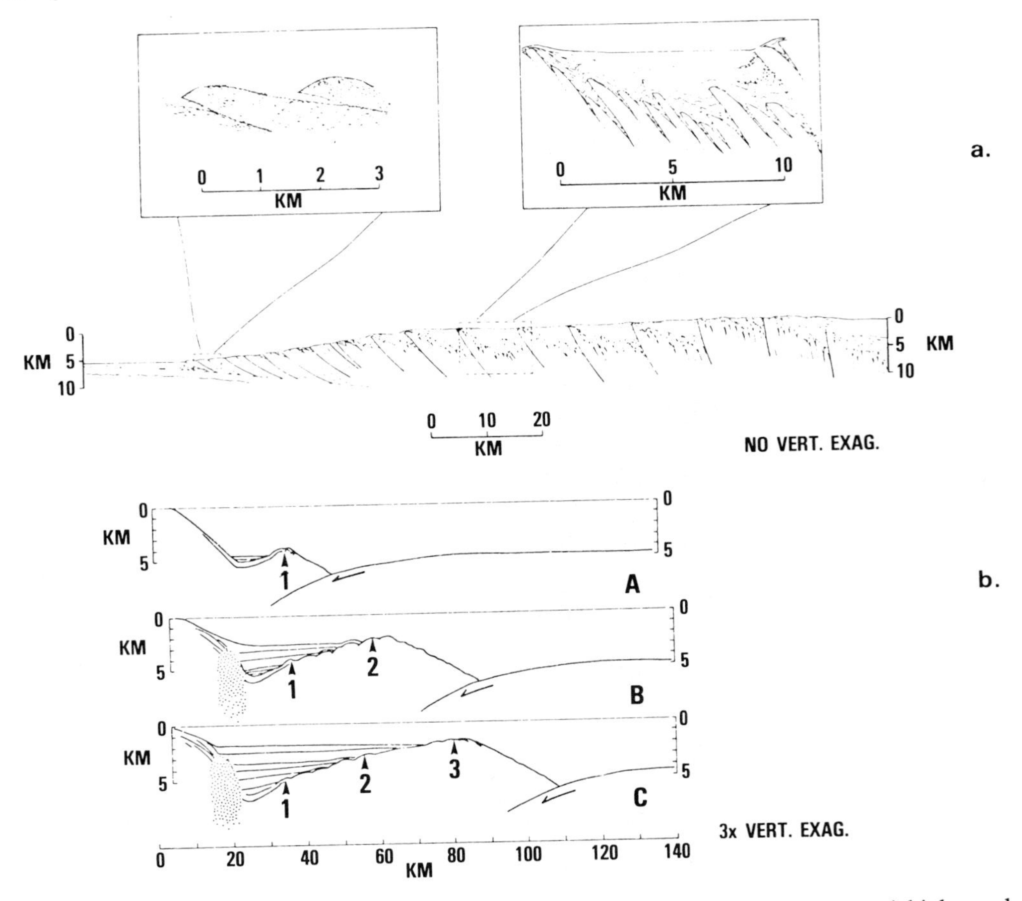

FIG. 15 Previous hypothesis for the accretionary development of structural highs and forearc basins. **a** Slope (accretionary) basins subsiding to form the floor of a forearc basin (Moore and Karig, 1976). These basins and subduction complex would underlie structural highs and forearc basins evolving in the fashion shown in 15b. **b** Progressive upward and outward growth of a structural high forming the outer edge of a (constructed) forearc basin (Karig, 1974).

——— 1977b, Subduction in the Indonesian region, *in* Talwani, M., and W. C. Pitman, III, eds., Island arcs, deep sea trenches and back-arc basins: Am. Geophys. Union, Maurice Ewing Ser. 1, p. 15–32.

Ingersoll, R. V., E. I. Rich, and W. R. Dickinson, 1977, Great Valley sequence, Sacramento Valley: Geol. Soc. America, Cordilleran Sec., Ann. Mtg., Field Trip Guide Book.

Isacks, B., and P. Molnar, 1971, Distribution of stresses in the descending lithosphere from a global survey of focal-mechanism solutions of mantle earthquakes: Rev. Geophysics and Space Physics, v. 9, p. 103–174.

Karig, D. E., 1974, Evolution of arc systems in the western Pacific: Earth and Planetary Sci. Ann. Rev., v. 2, p. 51–75.

——— 1977, Growth patterns on the upper trench slope, *in* M. Talwani, and W. C. Pitman, III, eds, Island arcs, deep sea trenches and back-arc basins: Am. Geophys. Union, Maurice Ewing Ser. 1, p. 175–185.

——— and G. F. Sharman, III, 1975, Subduction and accretion in trenches: Geol. Soc. America Bull., v. 86, p. 377–389.

King, P. B., et al, 1969, Tectonic map of North America: Washington, D.C., U.S. Geol. Survey.

Moore, G. F., and D. E. Karig, 1976, Development of sedimentary basins on the lower trench slope: Geology, v. 4, p. 693–697.

Seely, D. R., and W. R. Dickinson, 1977, Structure and stratigraphy of forearc regions, *in* Geology of continental margins: AAPG Continuing Education Course Note Ser. 5.

——— P. R. Vail, and G. G. Walton, 1974, Trench slope model, *in* C. A. Burk, and C. L. Drake, eds., The geology of continental margins: New York, Springer-Verlag, p. 249–260.

Shor, G. G., Jr., and R. L. Fisher, 1961, Middle America trench; seismic refraction studies: Geol. Soc. America Bull., v. 72, p. 621–730.

Van Bemmelen, R. W., 1949, The geology of Indonesia (vol. 1): The Hague, Staatsdrukkerij Nijhoff, 732 p.

——— 1954, Mountain building: The Hague, Martinus Nijhoff, 177 p. (p. 67).

Von Huene, R., 1972, Structure of the continental margin and tectonism at the eastern Aleutian trench: Geol. Soc. America Bull., v. 83, p. 2613–3626.

Weeks, L. A., R. N. Harbison, and G. Peter, 1967, Island arc system in Andaman Sea: AAPG Bull., v. 51, p. 1803–1815.

Structure of the Outer Convergent Margin Off Kodiak Island, Alaska, from Multichannel Seismic Records[1]

ROLAND VON HUENE[2]

Abstract In multichannel seismic reflection records off Kodiak Island, the Aleutian Trench sediment and the top of the oceanic crust can be traced 40 km landward under the slope where it is about 6 km beneath the mid-slope terrace. Under the lower slope, the section becomes tectonically consolidated, and at the mid-slope terrace it is tilted so steeply that bedding can no longer be resolved with present seismic instruments. Below the upper slope, mildly deformed beds are seen in a 0.5- to 1-km-thick downslope sediment blanket. The forearc basin directly landward of the shelf edge is even less deformed. The area most visibly affected by deformation from subduction is the lower and middle slope. This is one of the structural styles in the subduction zone along the eastern Aleutian Trench.

INTRODUCTION

Single-channel seismic reflection data across modern convergent margins do not constrain the interpretation of the geology of subduction complexes very much because a great deal of the deformation from subduction is beyond the resolving power of this technique. The multichannel seismic reflection technique enables greater penetration, and many records reveal considerably deeper information than was previously recorded. What is commonly revealed are the simple initial structures that develop at the leading edge of the upper plate. The structure of subduction zones has been largely inferred from earthquake seismicity, volcanism, and complexly deformed rocks exposed on land. The inferences are largely derived from the concepts of constant plate convergence, underthrusting of an oceanic plate, and scraping of sediment off an igneous oceanic crust. Although the general model seems sound, it is rarely specific enough to place more constraints on interpretations of the geophysical data from basins of the outer continental shelf. In this paper some new multichannel data across the continental margin off Kodiak, Alaska, are used to construct a modified conceptual model that is specific to one part of the convergent margin off Kodiak Island.

A grid of 24-channel seismic reflection lines was shot in 1976–1977 by the U.S. Geological Survey ship R/V *Lee* in an area off Kodiak Island, where earlier single-channel work (von Huene, 1972) showed some reflections continuing from the trench beneath the slope. The multichannel records were made in order to define these reflections better. Records 64 and 65 were discussed at the 1977 AAPG research symposium in Galveston, and at the AAPG continental-margin symposium during the 1977 annual meeting in Washington, D.C.

PREVIOUS STUDIES

Seely and Dickinson (1977) constructed one of the latest conceptual models across the eastern Aleutian Trench—their broad ridge model. This model is modified here (Fig. 1) to show the uplifted Kodiak insular block, the modern inner forearc basin, and the exposed ophiolitic complex along the northwest coast of the island. Seely and his colleagues (Seely, Vail and Walton, 1974, and this volume) postulate that this structure developed by imbrication of oceanic sediment and progressive rotation of the imbricate slices from a horizontal to a near-vertical position. The outer forearc basin first began to form on the continental slope and was progressively uplifted to its present position by the rising imbricate stack. This model provides a general structural framework for the discussions that follow.

The eastern Aleutian trench had only been mapped bathymetrically before 1964. The trench axis is defined by a sharp asymmetric juncture at the lowest part of a check-mark-shaped profile. In the area of this investigation (approx. lat. 57°–58°N) the axis is interrupted by Kodiak Seamount (Fig. 2). The lower continental slope (3,000 m to 5,000 m) has

[1]Manuscript received, October 24, 1977; accepted, June 14, 1978.

[2]U.S. Geological Survey, 345 Middlefield Road, Menlo Park, California 94025.

These are the first results in a regional study by myself and my colleagues, Mike Fisher, George Moore, Mark Holmes, Nardia Sasnett, Mark Sander, and Arnold Bouma. I am grateful to my colleagues for discussions that helped shape some of the ideas presented here. Mike Fisher provided a velocity analysis program and allowed me to use these data to help distinguish velocities of older rocks from velocities of sediment in the present forearc basin.

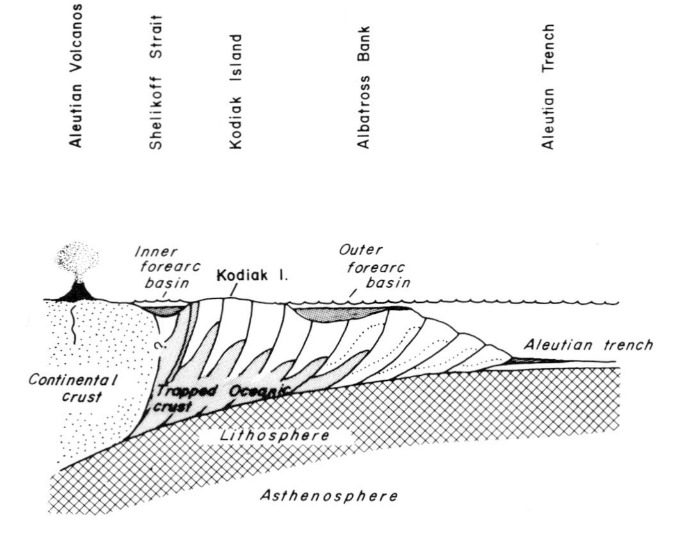

FIG. 1 General constant convergence model across the continental margin near Kodiak, after Seely and Dickinson (this volume).

the most rugged topography of the continental margin, and although the slope averages 3° to 5°, some local ridges have slopes up to 40°. The lower slope ends at a mid-slope terrace (2,500 m to 3,500 m) that has local flat areas indicative of ponded sediment. The mid-slope terrace separates the rugged lower slope from a generally smooth upper slope. Near the top of the upper slope, local over-steepened sections that cause slumping (Hampton and Bouma, 1978) are part of a bank at the seaward edge of the continental shelf. This bank is best developed off the southern part of Kodiak Island (Albatross Bank) and is absent in the northern part of the area. Glacial channels cut the bank, and they can be traced shoreward to fiords that indent the insular coast.

The first seismic reflection records across the trench were made in 1964 and 1965 using early instruments with a 20- to 40-kj sparker source (von Huene and Shor, 1969). In that study, a major emphasis was placed on the absence of observed compressional deformation in the trench fill but that absence did not preclude subduction. Three subsurface units were identified in the trench: trench fill, an underlying deep ocean basin section thought to be mainly hemipelagic and pelagic sediment, and oceanic basement. On the slope it was suggested that the downslope sediment apron might be underlain by a deformed sedimentary unit and that tectonic activity was more intense along the lower slope relative to the upper slope. The shelf-edge structure was shown to be an actively growing but discontinuous arch. Where the bathymetric bank is absent, the arch is absent also in the upper 500 m, and instead there is a prograded upper slope presumed to be underlain by a buried structural high, as suggested by gravity anomalies.

Improved records made in 1969 showed the severe limits of resolution in earlier records and resulted in interpretations more compatible with the concepts of subduction (von Huene, Shor and Malloy, 1972). It was pointed out that the Benioff zone does not emerge at the trench as the simple thrust proposed in most diagrams of the plate-tectonic model. Some graben and half-grabenlike structures on the slope were interpreted as resulting from listric faulting. On the shelf, the outer forearc basin seemed only gently deformed and was thought to be subsiding between Kodiak Island and the shelf-edge structures. The age of the shelf-edge high was inferred to be at least Pliocene from samples from a single dredging of the breached crest of the arch.

In 1970, a seismic reflection site survey for DSDP drilling made with a further improved seismic system resulted in records showing deeper features and more detail than had been seen before. Uplift of the shelf-edge high and oversteepening of the upper slope were inferred to have caused slides and slumps. Again it was emphasized that the trench-slope juncture is not a simple thrust fault. Rather, the Benioff zone must reach the surface as a complexly deformed belt with major expression along the lower slope. Three types of trench-slope juncture were identified: 1) continuation of the trench section under the slope without severe deformation, 2) simple tilted beds interspersed with folds, 3) unresolvable structure with unexplained steep slopes. The first type of juncture was crossed by the multichannel seismic traverses reported in this paper.

The result of DSDP drilling in the ocean basin, trench, and slope off Kodiak Island were reported in 1973 (Kulm, von Huene, et al, 1973). It was shown that the oldest trench fill was less than half the minimum age we had previously estimated. The deep ocean basin section was found to consist largely of turbidites of Pliocene-Pleistocene age rather than hemipelagic and pelagic sediment. Intense deformation along the lower slope was indicated by the recovery of highly dewatered and compacted sediment that had been subjected to lithostatic pressures equivalent to those found at a depth of at least 1 to 2 km. Further, the continental slope was found to have downslope sediment of Pleistocene age at least 200 m thick underlain by approximately 200 to 400 m of older downslope material.

MULTICHANNEL SEISMIC AND POTENTIAL FIELD DATA

The locations of the 24-channel seismic reflection record sections are shown in Figure 2, and the time sections are shown in Figure 3. Discussion of the record sections begins with the trench and proceeds up the slope to the shelf edge.

The first two record sections begin a short distance seaward of the axis of the Aleutian Trench and traverse the trench slope. A sequence of undeformed reflections in the trench is recorded from the sediment section that was sampled nearby at DSDP sites 178 and 180 (Kulm, von Huene et al, 1973). At the base of the section the igneous oceanic crust is represented by a series of strong diffractions and

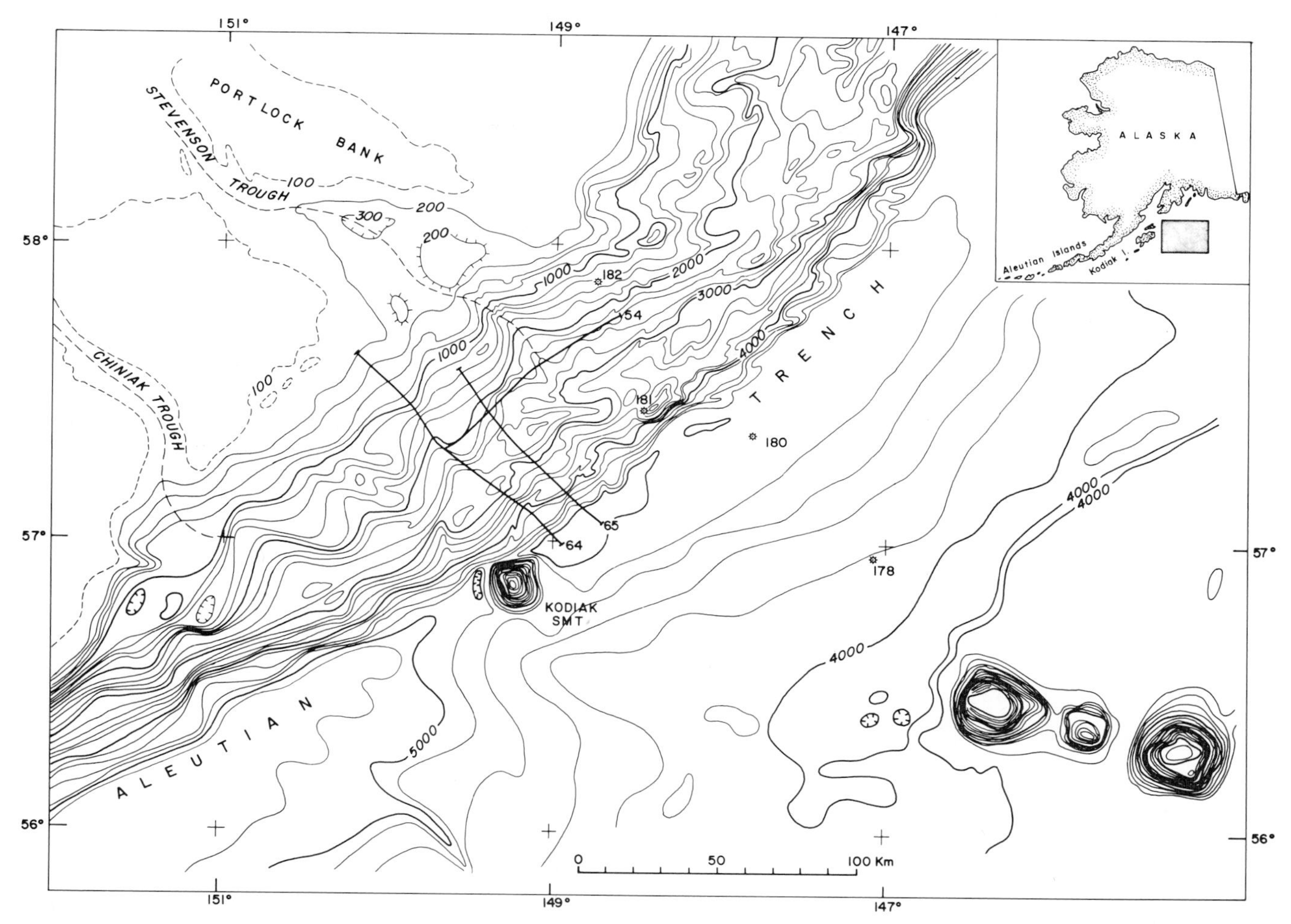

FIG. 2 Bathymetric map of outer continental margin off Kodiak Island and location of ship tracks along which multichannel seismic records were made. DSDP drill sites indicated by circles and numbers. Bathymetry south of 57° N lat. from von Huene (1972), and bathymetry north of 57° N lat. controlled by lines 5 mi apart on the continental slope (from Dunlavey, unpub.).

less commonly by smooth reflections. Above the basement the strongest reflections are from different discontinuities in each of the records. Along line 65, the upper strong reflections correspond to the base of the trench fill. Intermediate between the trench fill and the basement is another sequence of strong reflections. In line 64 the strong reflections occur about two to three reflections below the base of the trench fill, but a strong reflection from the base of the fill begins under the toe of the slope.

Along line 65 the three strong reflections can be followed for 10 km beneath the lower slope before they are broken by large faults. Steps due to faults in the basement reflector are mimicked by the sedimentary section. Beyond the faults, the three strong reflections can be followed to the midslope terrace where they then appear to plunge and are obscured by the first water-bottom multiple. Above these strong reflections are some conformable weaker reflections, but the remainder of the record is largely obscured by diffractions. The curvature of many diffractions indicates that they originated from low-velocity sources on the sea floor, out of the plane of the section. Much of the sea floor along line 65 is defined by diffractions.

A gravity meter and magnetometer were run simultaneous with shooting of the CDP records. Free-air gravity anomalies parallel the topography and show no indication of sharp vertical density changes. The magnetic anomalies have the typical oceanic ridge and trough character seaward of the mid-slope terrace. The positive anomaly is probably the west flank of anomaly 20 (von Huene, 1972). Along the upper slope the magnetic field is generally featureless.

Line 64 has about the same configuration of the strong marker reflections as line 65, except that disruption of the reflective sequence at the toe of the slope makes the correlation of the strong reflections less certain. The reflecting sequence is not as obscured by diffractions as in line 65, but the configuration of the strong reflections beneath the mid-slope area is less clear than along line 65. In the mid-slope area deep reflections roll over and become too steep to be resolved. Gravity anomalies again follow topography and show no obvious sharp den-

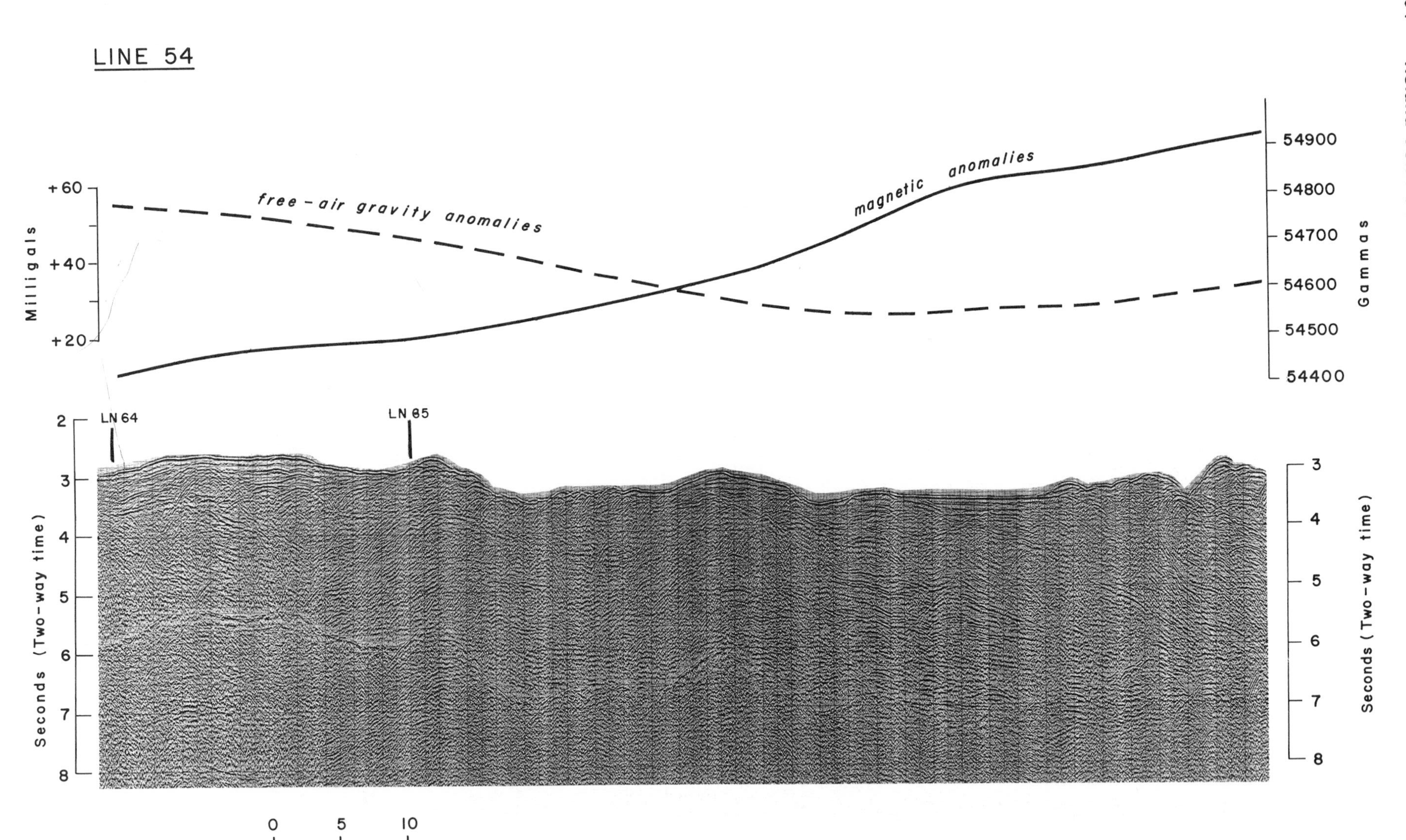

FIG. 3 (line 3-54) 24-channel seismic reflection records (54, 64, and 65). Processing by Seismic Services, Inc. includes deconvolution, 24-fold stacking, filtering, and trace equalization. Vertical exaggeration at sea floor is approximately 5×. Line 54 represents base of trench fill.

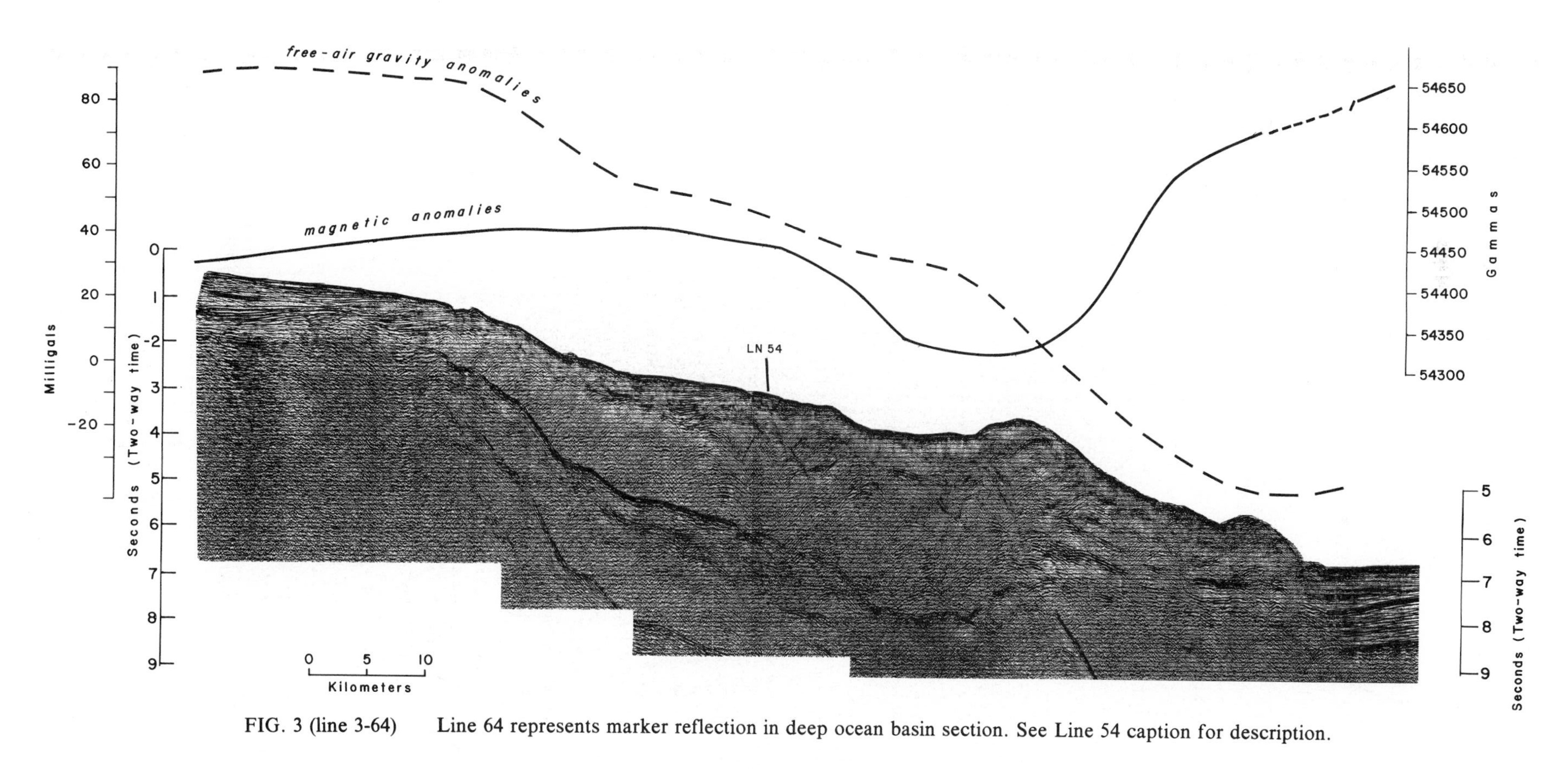

FIG. 3 (line 3-64) Line 64 represents marker reflection in deep ocean basin section. See Line 54 caption for description.

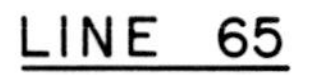

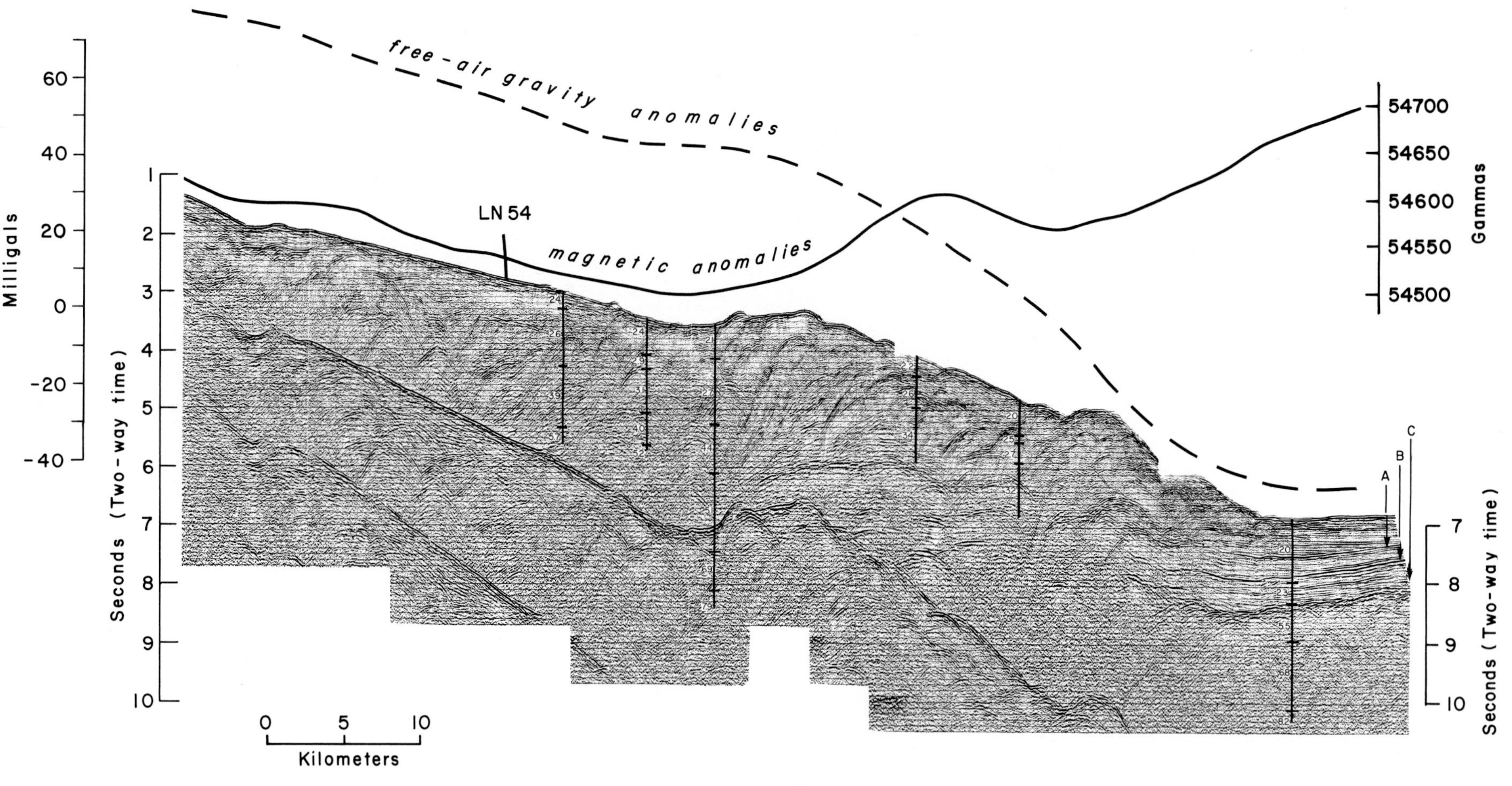

FIG. 3 (line 3-65) Line 65 represents top of second oceanic layer. See Line 54 caption for description.

sity changes. Magnetic anomalies retain an oceanic character through the mid-slope terrace and then become rather featureless.

Record 54 was made parallel to the trend of the continental slope and perpendicular to records 64 and 65. It crosses records 64 and 65 in an area that is obscured by diffractions in these latter records. In record 54, however, reflections are coherent and show a surprisingly simple structure parallel to the strike of the slope. Record 54 is presented here only to show that the structure, unresolved because of diffractions, is not chaotic. Instead it consists of reflective beds tilted more than 15° to 20°, which is the dip of diffraction legs at the depths and velocities of these records. The gravity and magnetic profiles parallel to the slopes are essentially featureless.

VELOCITY DATA

A great deal of velocity data is available around Kodiak Island from the CDP reflection work, from refraction stations (Shor and von Huene, 1972), and from 59 sonobuoy refraction measurements (Holmes, et al, 1978). The reflection velocity data are interval velocities derived from stacking velocities. The sonobuoy measurements were usually made during CDP operations and are unreversed for correction of dipping layers or topography. Sonobuoy measurements on the landward trench slope were made during runs parallel to the contours where the sea floor appeared smooth; therefore, they should be of a comparable quality to those on the shelf.

The velocity data from the slope were compared with those from the shelf to see if effects from subduction could be detected. The underlying assumption was that tectonic consolidation from subduction will result in more rapid increase in velocity with depth on the slope than on the continental shelf. Tectonic consolidation involves rapid dewatering and diagenetic changes and is expected to be much more intense in the subduction zone than along the shelf.

The reflection velocity data from the lower trench slope and mid-slope terrace areas are summarized by a least-squares curve (Fig. 4, dashed line), shown with a similar curve based upon more data on the continental shelf (Fig. 4, solid line—from Fisher, in press). The surprisingly small difference in these curves may be attributed to insensitivity of the method. The loss of discrimination in deep water is a problem with the reflection velocity technique. Thus very scattered data from the slope are probably not reliable in the range of velocity differences involved.

The more reliable sonobuoy refraction data (Fig. 4) were derived from 14 lines on the slope and 22 lines on the shelf. On the shelf the data below a pervasive velocity discontinuity (Holmes et al, 1978) were not used because these sediments may have been consolidated during an earlier tectonic episode (von Huene et al, 1976), and the discontinuity may be a hiatus as suggested to me by Fisher (1978, oral communication). These data plot in two

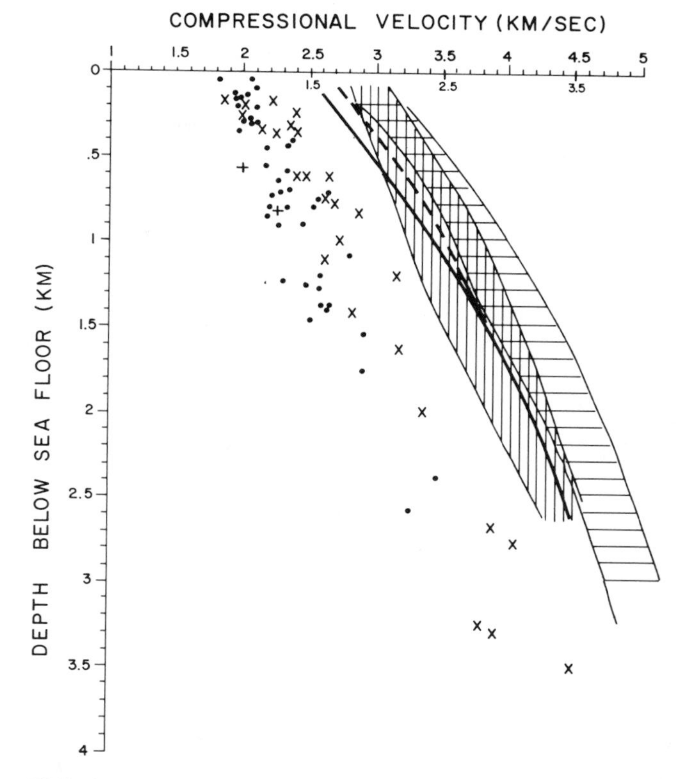

FIG. 4 Seismic velocity with depth from reflection and refraction data. Left diagram shows refraction data with dots corresponding to data points from the Kodiak shelf, x's corresponding to data points from the slope, pluses corresponding to data points from the trench. Right diagram shows data field for shelf (vertical-lined pattern) and slope (horizontal-lined pattern), as well as averaged reflection data for shelf (solid heavy line) and slope (dashed heavy line). Note that upper scale is for left diagram; lower scale for right one.

fields with considerable overlap where the data are dense and with much less overlap below 2 km, where the data are sparse (Fig. 4). In the upper 3 km of sediment there appears to be roughly a 10 percent more rapid increase of velocity with depth on the slope than on the shelf. This holds for the averaged data, whereas specific data points overlap considerably.

A more specific analysis of refraction velocity along record 65 suggests that the greatest change in velocity with depth across the subduction zone is between the trench and a sonobuoy located 18 km up the lower slope (Fig. 5). The progressive change in velocity with depth in the two lower slope measurements does not continue past the mid-slope terrace, where no progressive trend is apparent. The upper slope velocity structure is variable above the 4.0 km/sec level, which can be explained by changes in grain size and possibly in degree of deformation. Along LF-22, older single-channel reflection data show up to 2 km of ponded sediment that do not show as well in line 65. These sediments may cause the initial slow rate of velocity increase with depth at LF 22 that is very similar to the velocity

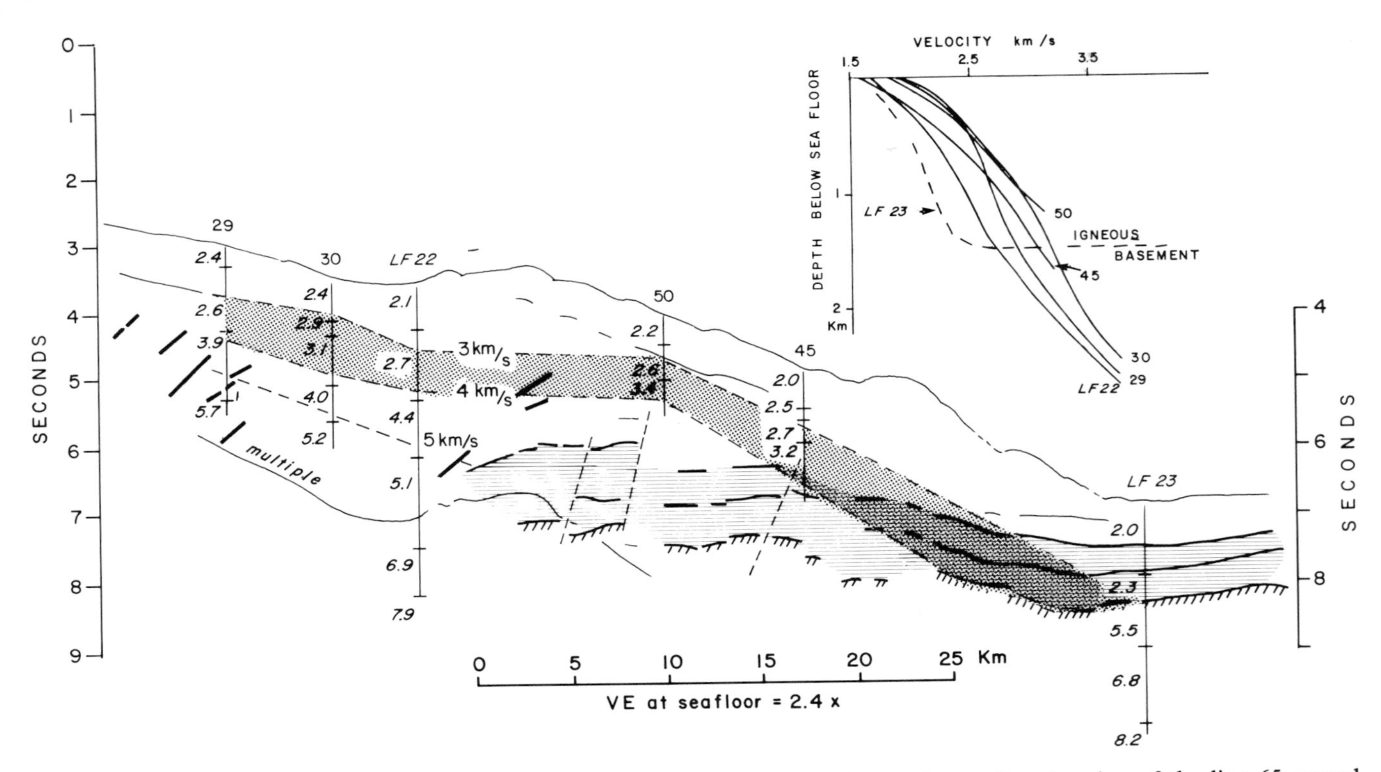

FIG. 5 Refraction velocity data along line 65. Refraction data are superimposed on a line drawing of the line 65 record section. Velocity layers indicated by dashed lines are drawn assuming each layer is of constant velocity; therefore, the value is assumed at midpoint of the layer. Inset (upper right) of velocity profiles is constructed according to these assumptions. Shaded layers, and reflection horizons, emphasize divergence between geolologic structure and velocity structure.

profile in the trench. Similarly, slope deposits containing greater amounts of sand may cause the higher near-surface velocities at stations 29 and 30 because such sand was penetrated at DSDP site 182 not far up the slope (Fig. 1). Therefore, it is possible that factors other than the degree of tectonism may influence the velocity structure of the trench slope, and the scatter in the previous diagram suggests that the relative magnitudes of the effects on velocity from nontectonic causes may be nearly as great as tectonic effects. However, those from tectonism should be more systematic from undeformed to highly deformed rock.

With a change in the rate of velocity increase with depth, there is also a divergence between geologic structure and velocity structure. This is most apparent between the trench and the first sonobuoy station on the lower continental slope (Fig. 5). It appears that along record 65, the velocity in the ocean basin section that becomes subducted increases landward. At about the mid-slope terrace (approximately 30 km) velocities approaching the upper limiting values of seismic velocity in sediment (4.5 to 5.1 km/sec) were measured. Farther landward, refraction values of 5.1 to 5.7 km/sec were measured where reflection records show bedded rocks (Figs. 3 and 5). Thus in this area, the sedimentary section achieves a velocity similar to that of the second oceanic layer and it is a problem to differentiate the second layer from consolidated sediment in the area landward of the mid-slope terrace.

INTERPRETATION OF CONTINENTAL SLOPE DATA

An averaged slope reflection velocity (Fig. 4) was used to construct the graphic depth section of part of line 65 (Fig. 6). This depth section uses two marker horizons within the deep ocean basin section as structural horizons. These acoustic boundaries are probably lithostratigraphic boundaries recognized at DSDP sites 178 and 180. The upper strong marker horizon corresponds to the discontinuity between trench fill and the underlying deep-ocean section. The present trench fill accumulated at ten times the rate of the underlying sediment of equivalent age (Piper, et al, 1973) and it is not older than 0.6 MY (Kulm, von Huene, et al, 1973). The marker horizon is time-transgressive and the underlying section is inferred to thin about 3 percent or less laterally as it passes under the slope. The intermediate marker horizon may be related to the interbedded muds and sand turbidites encountered at DSDP site 178, which are of late Miocene age. The turbidites were inferred to correspond to the onset of mountain-valley glaciation in coastal ranges around the Gulf of Alaska (Kulm, von Huene, et al, 1973). In the depth section, the units between these marker horizons thicken progressively landward under the slope (Fig. 6). This thickening is difficult to explain by sedimentary processes. One conclusion from the study of DSDP cores in this area was that rates of deep ocean basin sedimentation increased greatly during Pleistocene

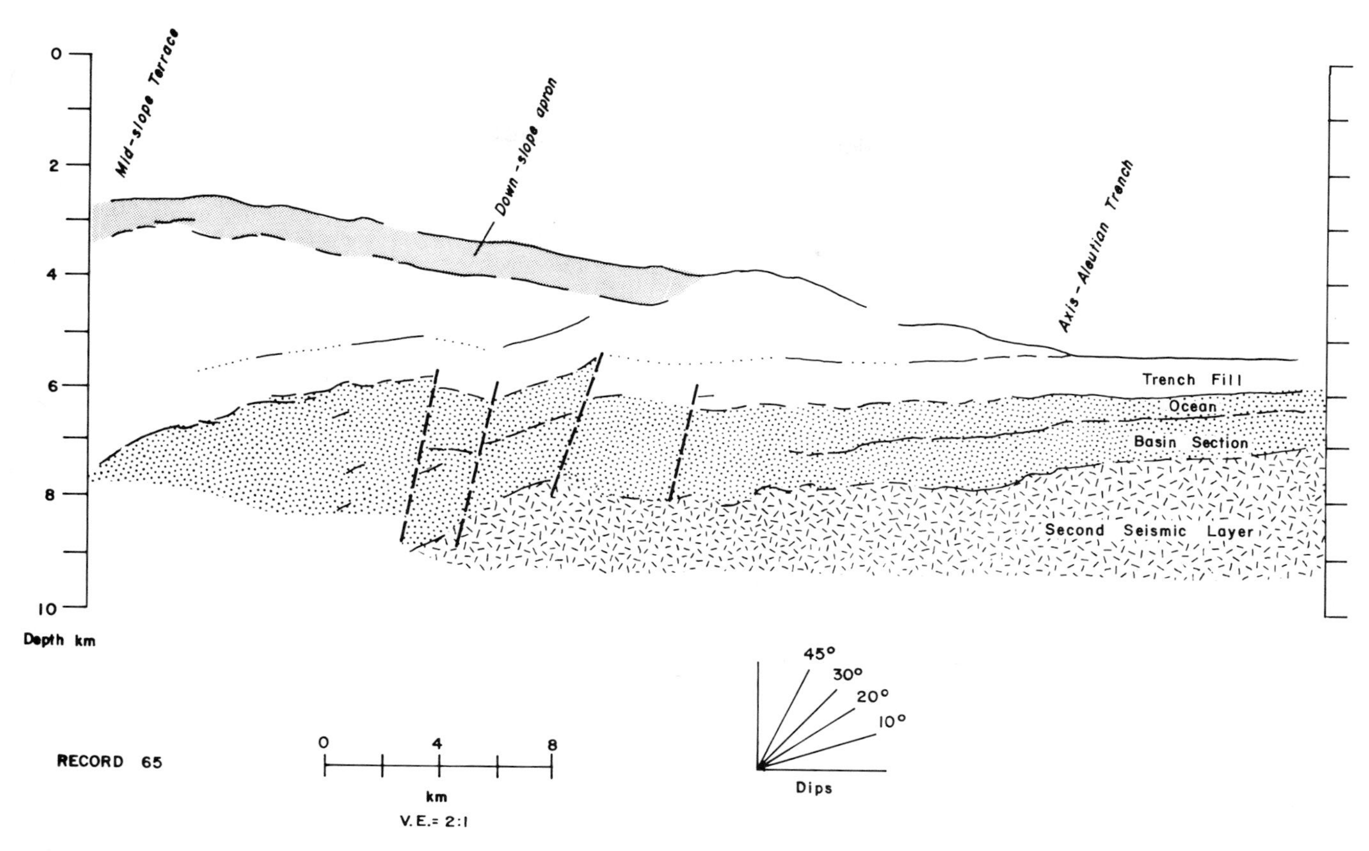

FIG. 6 Depth section of record 65 constructed graphically. The velocity-versus-time function was computed from stacking, and refraction velocities averaged across lower slope: $Z = 0.225t^2 + 0.903t$ where Z = depth and t = two-way time. Dashed lines connecting ends of beds are interpretations of faulting. Dotted line is inferred continuation of trench fill. Note landward thickening of ocean basin section.

glaciation. Therefore, the deep ocean basin section under the slope, most of which is older than Pleistocene, should be thinner than the section under the trench which includes part of the Pleistocene. Consequently, a dynamic process is sought that causes not only the thickening of section but also the tectonic consolidation observed in cores from DSDP site 181 located nearby (Kulm, von Huene, et al, 1973).

Tectonic consolidation as a process in subduction zones is not understood, and the mechanisms that cause thickening of the section other than microfolding and faulting probably involve intergranular changes. During tectonic consolidation, free fluids in the sediment are expelled, mineral grains are reoriented, and sometimes fracture cleavage develops (Moore and Geigle, 1974).

The downslope apron (Fig. 6) corresponds to a reflective sequence subparallel to the sea floor, whose base is a series of discontinuous high-amplitude reflections. Local diffractions indicate sea floor roughness and internal reflections indicate intermittent zones of more and less organized bedding. As downslope turbidites were encountered in the upper cores from DSDP sites 181 and 182 and slumps and debris flows are found on the upper slope (Hampton and Bouma, 1978), these reflections are interpreted as material from upslope terrigenous areas.

In the area between the downslope apron and the deep ocean basin section (Fig. 6), the trench fill is recorded locally in a conformable reflective sequence above the ocean basin section. Between the top of the trench fill and the bottom of the downslope apron is a wedge-shaped blank area where few reflections can be seen. This area appears thickest in line 65. The simplest explanation for the origin of material in the blank area is downslope mass movement. This explanation is consistent with earlier observations of a divergence between subsurface structure and ridges on the sea floor (von Huene, 1972). An example of such divergence is seen at the base of the slope in record 64. If this explanation is correct, the downslope sediment section is locally more than 2 km thick (Fig. 6).

The rough topography of the lower slope results from 5- to 20-km-long ridges paralleling the trench (Fig. 2). Along lines 64 and 65 the ridges may correspond with some of the thrust faults that cut the underlying marker horizons. Mud diapirism has also been suggested as an origin for some of these features (von Huene, 1972) but this idea remains untested. Whatever its origin, the ridge at site 181 was uplifted at a rate ranging from 830 to 2200 m/m.y. (Kulm, von Huene, et al, 1973); therefore, at least some of the ridges have been uplifted rapidly. It should be noted that the ridges in records 64 and 65 are not continuous between the lines, despite the fact that these lines are only 15 km apart (Fig. 1).

The mid-slope terrace is interpreted as the bathymetric feature that occurs above the roll-over and plunge of the underlying ocean basin trench fill and downslope sections. The abrupt landward steepening of dip suggests an increase in the rate of tectonism there and possibly the beginning of imbricate stacking on a large scale. The surface expression of this tectonism is the ridge that marks the outer edge of the terrace. Along line 64 and 65, the sediment ponded behind the ridge is not as thick as that seen in other records (von Huene, 1972). The sediment is probably derived from a glacial channel and canyon system that connects Cook Inlet with the Aleutian Trench (Fig. 2). Progressive tilting of deeper strata in this sediment pond documents continued uplift, as does the upper core from DSDP site 181. It should also be noted that the blank area thought to represent slope deposits is involved in the tectonism of the mid-slope area.

Tectonism appears to be much less intense on the upper slope, as seen along lines 64 and 65, than along the mid-slope and lower slope. These lines are particularly smooth in comparison to an earlier record nearby (von Huene, 1972), but the upper slope is generally the smoothest part of the whole continental slope (Fig. 2). This suggests that active faulting takes place largely in the midslope and lower slope areas.

CONTINENTAL MARGIN MODEL

My structural interpretation of records 64 and 65 is summarized in a simple diagram (Fig. 7). Below the depth of the multichannel data the interpretation is weakly constrained by the potential field, refraction, and earthquake data (Fig. 8). Extension beyond the data is based on the constant convergence model where the dominant tectonic feature is a shallow dipping thrust fault between the two lithospheric plates (Fig. 1). The diagram shows the following features not commonly seen in convergent margin models:

1. Thrust faults extend into the 2nd oceanic layer from the onset of subduction.
2. The bulk of tectonic consolidation and thrust faulting occurs on the lower and mid-slope areas. The upper slope shows much less thrusting and responds to tectonism more by the movement of large blocks.
3. The zone of tectonic consolidation prior to imbricate stacking is about 30 to 40 km wide.
4. A great amount of downslope material is involved in the deformation accompanying subduction.

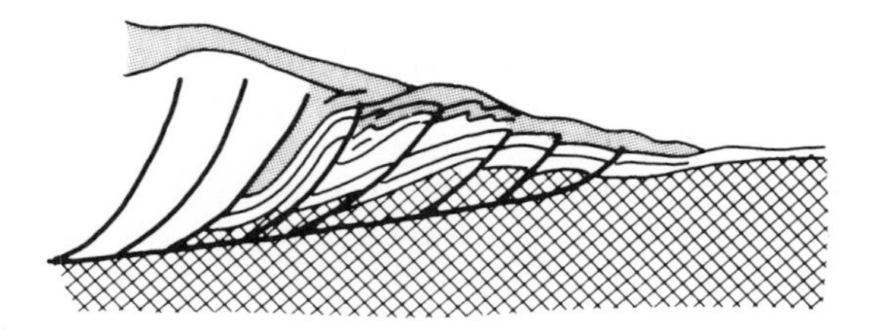

FIG. 7 Diagrammed structural interpretation of records 64 and 65.

This general structure occurs in a 150-km-long part of the margin north of Kodiak Seamount. In areas north and south of this segment the structure is different (Seely, 1977; von Huene, et al, in press).

Some observations around Kodiak have been summarized and compiled in a cross-section (von Huene, Moore, and Moore, 1976 and in press), which is shown in Figure 8 at a vertical exaggeration of 2:1. The deepest information shown is a group of earthquake hypocenters recorded between 1968 and 1976 from a 320-km-wide area including Kodiak Island and the shelf and slope directly to the southeast. Beneath Kodiak Island and the Alaska Peninsula, hypocenters located by more than 50 stations are well clustered within a 15-km-wide Benioff zone that dips landward. Only the upper end of the zone could be fit into Figure 8. In contrast, hypocenters under the shelf are diffuse and extend from the surface to below the crust-mantle boundary. Crustal layers shown on the cross-section are from seismic refraction stations (Shor and von Huene, 1972) and they give no indication of a reason for the scatter in hypocenters. The hypocenters do not define the position or existence of a master thrust fault as previously inferred, but they suggest that major faults extend deep into the igneous oceanic crustal layers.

The forearc basin contains sediment inferred to be of late Miocene and younger age (von Huene, 1972; Fisher and von Huene, in preparation). This section has been uplifted at the shelf edge to form a structural high. The structural high is cut by some steep reverse faults but farther landward faults are either normal or reverse (Figs. 8, 9). Compressional deformation from plate convergence is not strongly expressed in late Cenozoic sediment of the forearc basin.

The importance of including Kodiak Island in the cross-section is to show the striking difference in structure between the island and the geophysically defined marine section. To do this, the line of section was offset southwest of record 65 along the faulted landward flank of the forearc basin. At low vertical exaggeration, where differences in steep dips become apparent, it is evident that the complex deformation of the insular strata is not recorded in the marine geophysical data. Seismic reflection data show the beginning of subduction and the subduction complex, which is the end product, can be seen in exposures on land. The intermediate processes that change trench and slope sediment into a subduction complex are largely inferential because indirect geophysical methods cannot yet resolve such complex structure.

The observed data, extended in accord with previous discussions, and the constant convergence model are shown in Figure 9. Tectonic processes inferred to be operating are noted above the model. The master thrust fault could be drawn in a variety of ways, and the relative amounts of igneous ocean-

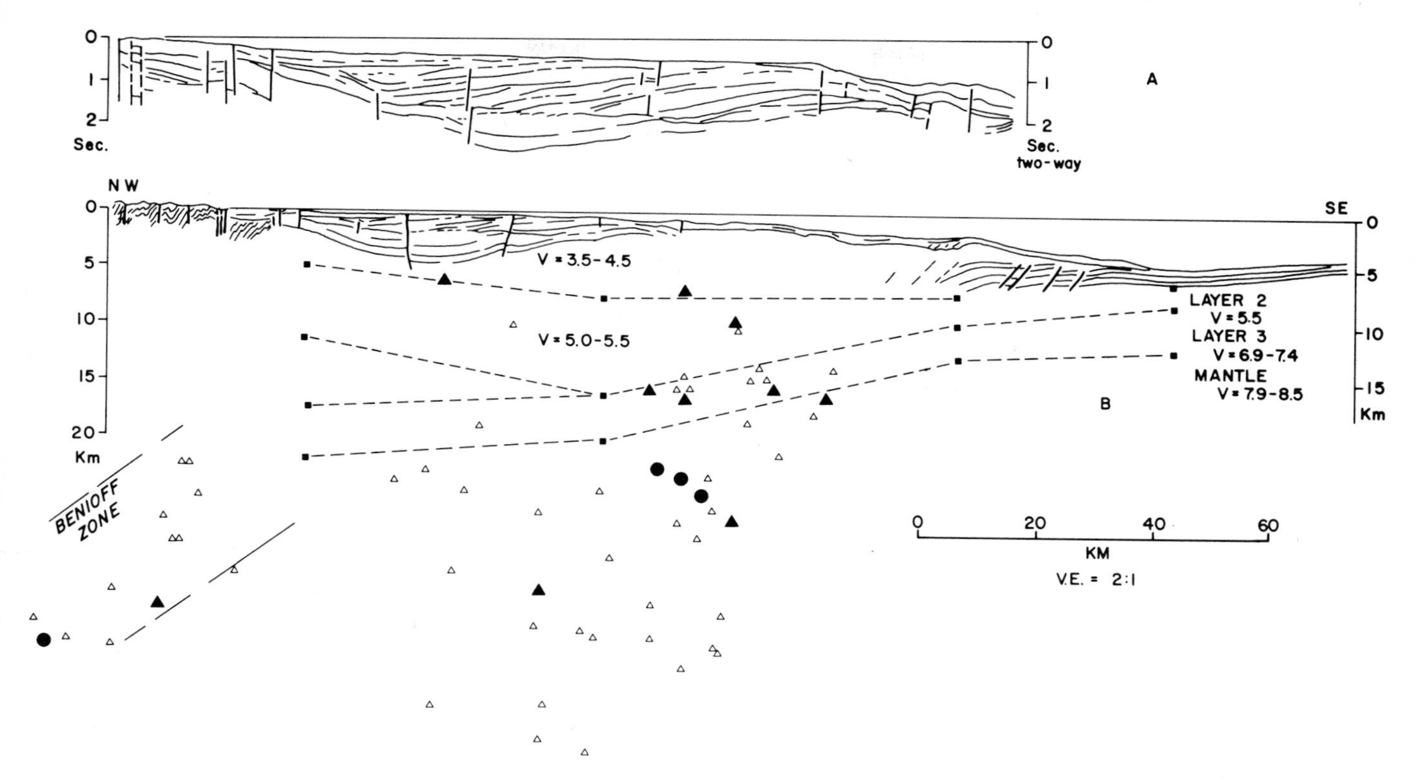

FIG. 8 Summary cross-section of geology and geophysical data from von Huene, Moore, and Moore (in press). Earthquake hypocenters represented by filled circle, if p-P phase identified and if located by more than 50 stations; solid triangle, if located by more than 50 stations; and open triangle, if located by less than 50 stations. Crustal layers determined from refraction station data shown by dashed lines, velocities in km/sec. Upper part of Benioff zone indicated at northwest end of diagram.

ic crust, deep ocean basin sediment, and slope sediment are conjectural.

The involvement of downslope sediment in the accreted deposits may be more common than was formerly recognized and it is indicated in the exposed rocks on Kodiak Island. Recently Nilsen and Moore (1977) have postulated that the upper Tertiary rocks of Kodiak Island were deposited in a slope environment. Winkler (1976) has described a similar environment of deposition for the Orca Formation on the Kenai Peninsula to the northeast.

One constraint in construction of convergent margin models, where sialic crust is formed by accretion of imbricate slices, is the length of each slice. If a slice of a given length is rotated from a nearly horizontal to a nearly vertical position, its original length becomes the maximum thickness of the imbricate stack unless more material can be thrust beneath it or be added by some sub-crustal process. Assuming that the distance between lower slope thrust faults

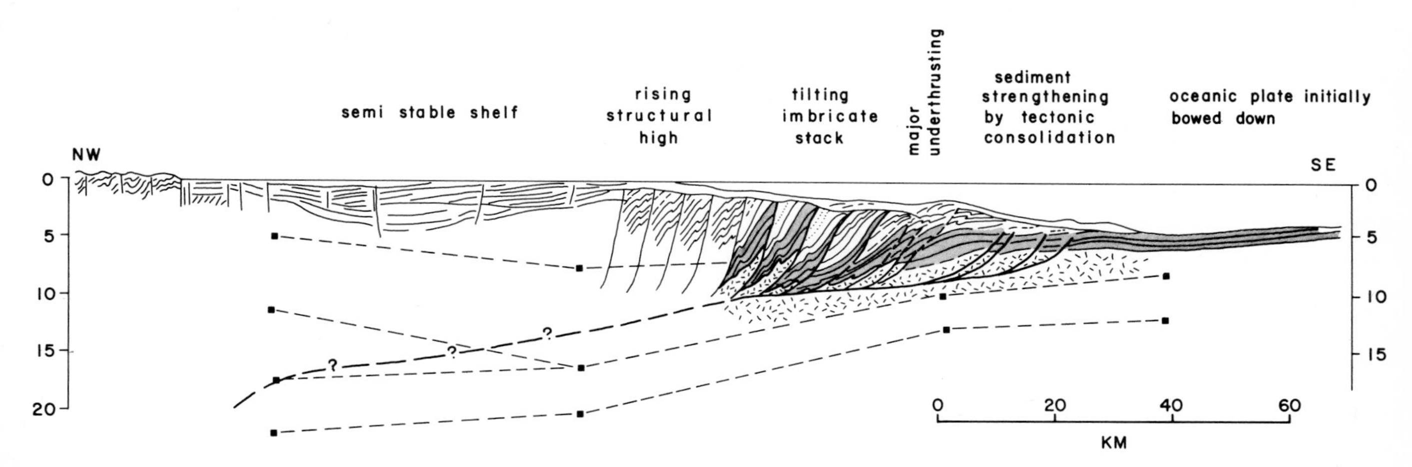

FIG. 9 Model of part of the Kodiak continental margin, extending the observational data to deeper levels in accord with general constant convergent model. Vertical exaggeration 2:1.

indicates the length of present and past imbricate slices, an average 12-km distance was established from measurements between the trench and 19 topographic ridges interpreted as related to major thrust faults (Fig. 2). An imbricate stack with a minimum thickness of 12 km projects into the seismic refraction layer with $V = > 5.5$ km/sec, a velocity common to metamorphosed sediment and igneous rock.

CONCLUSIONS

1. Subduction along a 150-km segment of the Aleutian Trench north of Kodiak Seamount, occurs on steep thrust faults that extend into the igneous oceanic crust. This interpretation is based on faulted beds under the slope where the faults are defined by the ends of disrupted beds. An inferred master thrust fault related to the Benioff zone must be deep in the second oceanic layer. The master thrust fault is inferred from the constant convergence model.

2. The upper slope seems to react to tectonism mostly by uplift and tilting of large blocks. Farther landward, in the forearc basin, there is much less indication of compressional structure than in the lower and mid-slope areas. The effects of the presumed nearly horizontal stress from plate convergence are seen mainly along the lower slope.

3. Velocity data show increase in the velocity of ocean basin sediment progressively landward, until the velocities reach values commonly found in accreted subduction complexes on land. This velocity increase occurs most rapidly beneath the lower and middle slope.

4. It is not possible to differentiate the second oceanic layer from deeply subducted or buried sediment with velocity data alone landward of the mid-slope terrace in lines 64 and 65.

5. The term *accreted deposit* often refers to sediment thought to have been scraped from the ocean floor. Along the Kodiak margin the deformation accompanying subduction involves considerable downslope material that is folded and "accreted" before it has been transported to the trench. Therefore, the term *accreted deposit* should refer to any material involved in the dynamic process of accretion.

REFERENCES CITED

Fisher, M. A., in press, Structure and tectonic setting of the continental shelf southwest of Kodiak Island, Alaska:

Hampton, M. A., and A. H. Bouma, 1978, Slope instability near the shelf break, Western Gulf of Alaska: Marine Geotechnology, v. 2, Marine Slope Stability, p. 309–331.

Holmes, M. L., C. A. Meeder, and K. C. Creager, 1978, Sonobuoy refraction data near Kodiak, Alaska: U.S. Geol. Survey Open-File Report 78-368.

Kulm, L. D., R. E. von Huene, et al, 1973, Initial reports of the Deep Sea Drilling Project—Site 178: Wash., D.C., U.S. Govt. Printing Office, v. 17, p. 287–376.

Moore, J. C., and J. E. Geigle, 1974, Slaty cleavage: Incipient occurrences in the deep sea: Science, v. 183, p. 509–510.

Nilsen, T. H., and G. W. Moore, in press, Reconnaissance study of Upper Cretaceous to Miocene sedimentary facies and subfacies, Kodiak and adjacent islands, Alaska: U.S. Geol. Survey Prof. Paper.

Piper, D. J., R. E. von Huene, and J. R. Duncan, 1973, Late Quaternary sedimentation in the active eastern Aleutian Trench: Geology, v. 1, p. 19–22.

Seely, D. R., 1977, The significance of landward vergence and oblique structural trends on trench inner slopes, *in* Maurice Ewing Series 1, Island Arcs, Deep Sea Trenches and Back Arc Basins: Amer. Geophys. Union, p. 187–198.

——— and W. R. Dickinson, 1977, Stratigraphy and structure of compressional margins: AAPG Continuing Education Short Note Series no. 5, p. C1–C23.

——— P. R. Vail, and G. G. Walton, 1974, Trench slope model, *in* C. A. Burk, and C. L. Drake, *eds.*, The geology of the continental margins: New York, Springer-Verlag, p. 249–260.

Shor, G. G., Jr., and R. E. von Huene, 1972, Marine seismic refraction studies near Kodiak, Alaska: Geophysics, v. 37, p. 697–700.

von Huene, R. E., 1972, Structure of the continental margin and tectonism at the eastern Aleutian trench: Geol. Soc. America Bull., v. 83, p. 3613–3626.

——— and G. G. Shor, Jr., 1969, The structure and tectonic history of the eastern Aleutian trench: Geol. Soc. America Bull., v. 80, p. 1899–1902.

——— J. C. Moore, and G. W. Moore, 1976, Alaska Peninsula-Kodiak Island-Aleutian Trench, *in* Maxwell, J. C., Plate margins cross-sections: Geol. Soc. America, Abs. with Programs, v. 8, p. 1002–1005.

——— G. W. Moore, and J. C. Moore, in press, Cross-section, Alaska Peninsula Kodiak Island-Aleutian Trench: Geol. Soc. America.

——— G. G. Shor, Jr., and R. J. Malloy, 1972, Offshore tectonic features in the affected region, *in* The great Alaska earthquake of 1964: Oceanography and coastal engineering: Natl. Acad. Sciences Pub. 1605, p. 266–289.

——— et al, in press, Continental margins of the Gulf of Alaska and late Cenozoic tectonic plate boundaries: Proc. Alaska Geol. Soc. Symposium, Anchorage, Alaska, 1977.

Winkler, G. R., 1976, Deep-sea fan deposition of the lower Tertiary Orca Group, eastern Prince William Sound, Alaska: U.S. Geol. Survey Open-File Report 76–83, 20 p.

Continental Margins of the Eastern Gulf of Alaska and Boundaries of Tectonic Plates[1]

ROLAND VON HUENE,[2] GEORGE G. SHOR, JR.,[3] AND JOHN WAGEMAN[4]

Abstract Between the Queen Charlotte Islands Fault and the Fairweather Fault, plate-tectonics models require a connecting transform fault. Such a zone of faulting appears to be present in seismic reflection records along most of the southeastern Alaskan continental margin. The Queen Charlotte Islands fault zone, which is most evident from structures along the continental slope, appears to continue north from the Alaska-British Columbia border to Chatham Strait. North from the strait, the fault zone is most evident along the edge of the continental shelf as the Chichagof-Baranof fault, which heads into the Fairweather Fault at a 20° angle. Their intersection is not shown in our data, but these fault zones are inferred to be related because both are main zones of active tectonism and seismicity. Along the Chichagof-Baranof Fault a glacial bank is offset hundreds of meters right-laterally. This is consistent with motion between the Pacific and North American plates in post glacial time. Models with a sharp vertical plate boundary conforming to present concepts of transform-fault configuration are permissible within the constraints of gravity, magnetic, and seismic data, but the data do not define the dip of the continental oceanic boundary. A large trough at the foot of the slope is not explained by transform-fault models.

The obliquely convergent plate boundary in the central coast of the Gulf of Alaska involves a zone of continental crust up to 300 km wide, extending from a "buried trench" at the foot of the continental slope inland to the Denali Fault. While there is some evidence of past major subduction at the seaward edge, apparently only a small part of the motion between the Pacific and the North American plates has been taken up along the seaward edge in late Cenozoic time. Evidence for this is: (1) deformation of late Cenozoic sediments at the foot of the slope is too small if this zone has taken up all of the Pacific-North American plate motion; (2) convergent deformation does not appear to have occurred between Fairweather Ground and the transform Chichagof-Baranof or Fairweather Faults in late Pleistocene time; (3) a pre-Pliocene wedge of terrigenous sediment at the foot of the continental slope has not been subducted. If late Cenozoic convergence has been small, a block bounded by the continental slope and by the Fairweather Fault and its westward splays is moving parallel to, but at a lesser rate than, the Pacific plate. The plate geometry requires that the northwestern edge of this block, in the vicinity of Kayak Island, impinge against the Alaskan continent at nearly the same rate as convergence across the Aleutian Trench and form a zone of continental convergence. The separation of oblique convergent motion into a normal subduction component and a nearby strike-slip component is similar to that proposed by Fitch (1972) to explain the major structures of Sumatra.

INTRODUCTION

Some time ago, the authors realized that their combined unpublished marine seismic data covered large areas in the central and eastern Gulf of Alaska not covered by previously published data. These data show some unreported features and confirm or modify some tectonic zones inferred from the theory of plate tectonics. In addition to our own data, other seismic and magnetic data were available from NAVOCEANO, NOAA, and DSDP. A compilation

[1]Manuscript received, November 21, 1977; accepted, March 15, 1978.

[2]U.S. Geological Survey, Menlo Park, California 94025.
[3]Marine Physical Laboratory of the Scripps Institution of Oceanography, San Diego, California 92093.
[4]1434 N. Chelton Road, Colorado Springs, Colorado 80909.

A debt of gratitude is owed to the late Fred Naugler who helped collect much of the NOAA data. We hope this paper is worthy of his memory. Thoughtful reviews by David McCulloch and Terry Bruns were very helpful in improving the paper. Jim Crouch, Mark Sander and Nardia Sasnett aided in data reduction, gravity modeling, and preparation of illustrations. Michael Loughridge made the Navy data available. The refraction data were gathered using R/V *Hugh Smith* and R/V *Stranger* on Leapfrog Expedition and the SIO reflection data from R/V *Oconostota* on Kayak Expedition of the Scripps Institution of Oceanography. The SIO work was supported by NSF grants G-19651 and OCE 76 24101 and Office of Naval Research contract N-onr-2216 (05) to the Marine Physical Laboratory of the Scripps Institution of Oceanography. We thank the captains, crews and scientific parties of the ships, especially Alan Jones, and Helen Kirk and Delpha McGowan who assisted with processing of the refraction data.
Article Identification Number:
0065–731X/78/MO29–0019/$03.00/0.
(see copyright notice, front of book)

of the seismic reflection records was used by Silver et al (1974) for a study of the Kodiak-Bowie seamount chain, but did not cover the continental margins which will be discussed here. We will relate regional late Cenozoic tectonic features to a more detailed plate-tectonic model than presented in previous publications. The simplicity of the general global plate-tectonic model diminishes rapidly as the scales of plate diagrams are expanded. The scale of Atwater's pioneering plate diagram (1970) and the Gulf of Alaska diagram of Richter and Matson (1971) differ by almost an order of magnitude. A simplified sketch (Fig. 1) based on these diagrams and one by Rogers (1977) indicates three types of plate boundaries: (1) the eastern transform boundary, (2) the western convergent boundary, and (3) the oblique convergent boundary between them. Richter and Matson speculated that the central oblique convergent boundary may involve a 300-km-wide section of the continent, taking up relative motion of the Pacific and North American plates in a series of concentric tectonic zones.

DEVELOPMENT OF THE PLATE-TECTONIC MODEL

Twenty years ago, St. Amand (1957) proposed an offshore fault along the coast of southeastern Alaska connecting the Queen Charlotte Islands Fault on the south and the Fairweather Fault on the north. Both these faults are easily recognized from their strong physiographic expression, and because both have been the sites of great earthquakes (Sykes, 1971), the continuity was appealing despite the lack of marine geologic data. St. Amand drew the fault just off the straight trace of the rugged outer coast of Baranof and the Chichagof Islands and through the single volcano along it, Mt. Edgecumbe near Sitka.

The north Pacific plate-tectonic model required a transform fault to separate the Pacific and North American plates between the spreading ridges off British Columbia and the Aleutian Trench. Concepts at the time made it more appealing to infer the extension of the Queen Charlotte Islands Fault along a morphologic boundary of greater significance, the linear continental slope, since plate boundaries were commonly thought to correspond with continental slopes. But this principle had to be violated in order to connect the northern end of the inferred fault with a line drawn across the continental shelf to the end of the Fairweather Fault. Data collected during this study indicate such a fault across the shelf near Cross Sound. Figure 2 summarizes the present tectonic knowledge of the southeastern Alaska continental margins (including data presented here) and also shows the distribution of some aftershocks accompanying great earthquakes. The zone defined by relocated epicenters of aftershocks of the 1949 Queen Charlotte Islands earthquake (Tobin and Sykes, 1968) narrows the possible location of the extension of the Queen Charlotte Islands Fault, and

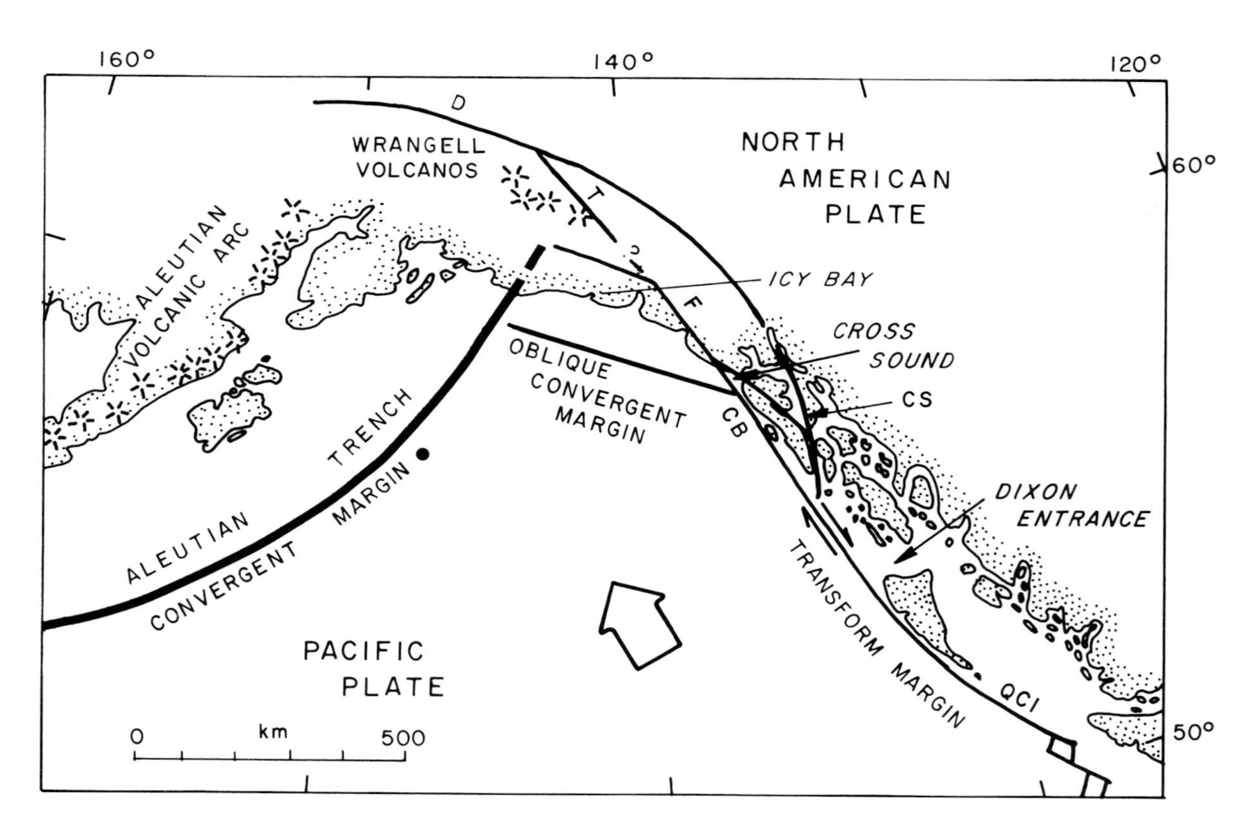

FIG. 1 Diagram of late Cenozoic tectonic plate boundaries, Gulf of Alaska, principally after Atwater (1970), Richter and Matson (1971), and Rogers (1977). The Denali, Totschund, Fairweather, Chatham Strait, Chichagof-Baranof, and Queen Charlotte Islands faults are indicated by D, T, F, CS, CB, & QCI, respectively. Location of DSDP site 178 is indicated by a dot.

first-motion studies confirm its right-lateral strike-slip nature. Additional constraint in the location of a fault off Baranof Island was provided by the aftershocks of the 1972 Sitka earthquake (Page, 1973). The northern end of this aftershock zone was shown to correspond to an offshore fault just north of Sitka (Fig. 2). This fault is traced farther north on a series of seismic reflection records reported here.

The coastal onshore geology of the central Gulf of Alaska has been discussed (Stoneley, 1967; Plafker, 1967 and 1971), but offshore geology was poorly known until the preliminary findings of a study on

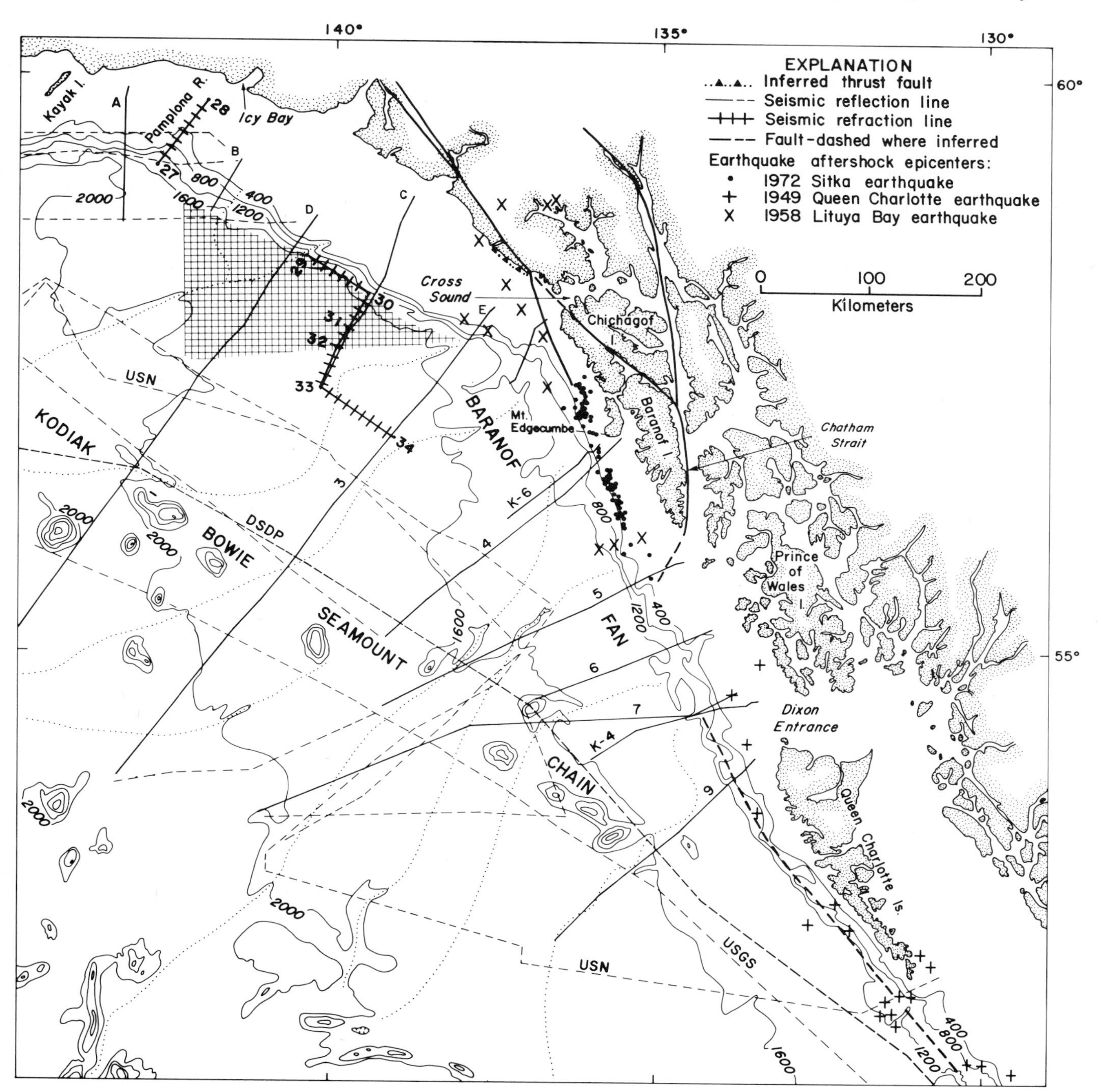

FIG. 2 Major faults and aftershocks from three major earthquakes, southeast Alaska and British Columbia, after Beikman (1975), Tobin and Sykes (1968, including only most accurately located epicenters), and Page (1973). Tracklines show location of seismic records used in this study: solid lines locate records shown in Figure 3, with K series and C from SIO, numbered series and D from NOAA, line A reinterpreted from von Huene et al (1972); the dashed track lines are from institutions as noted, and show part of the network of records used to follow the Pliocene-Pleistocene boundary from DSDP site 178 to Baranof Fan, and Miocene-Pliocene boundary to record D. Numbers along refraction lines indicate locations of stations given in Table 1. Hachured area shows location of attenuated magnetic anomalies. Bathymetry after Chase et al (1970).

the shelf by Bruns and Plafker (1975) were reported. Our data in this area are scanty but sufficient to establish the continuation of a pattern of deformed and undeformed segments on the shelf to the continental slope. Recently Rogers (1977) has published sketch maps suggesting a similar segmentation.

In this discussion zones of tectonism are proposed that may correspond to eastern plate boundaries and their tectonic style is outlined. The structure at the junction of the transform and the seaward edge of the oblique plate boundary is then dealt with and finally the structure along the offshore part of the oblique boundary is discussed. In the summary discussion, observations and the plate model for late Cenozoic time are integrated and some consequences of the model for earlier periods of geologic history are pointed out.

FAULT ZONE BETWEEN FAIRWEATHER AND QUEEN CHARLOTTE ISLANDS FAULTS

Major transform faults can look relatively unimpressive in seismic reflection records (for example, the San Andreas fault just off the Golden Gate). The records in Figure 3 are probably sufficient to identify major faults in sediments of the continental slope and rise, but these records are insufficient to show tectonic features beneath the continental slope. It is generally difficult to obtain clear subsurface information on steep rugged slopes, and these records were made with relatively low energy sources such as 90- to 160-kJ sparker or 40-in^3 air-guns. In addition, the high vertical exaggeration at which they were recorded diminishes the maximum dip that can be resolved. Therefore, the subsurface structure of the slope is poorly defined, and it becomes a place to infer a hidden fault when one cannot be found elsewhere.

The continental rise between the Queen Charlotte Islands and Cross Sound is a series of coalescing deep-sea fans collectively called the Baranof Fan. Most of the rise is of Pleistocene age (1.8 MY) based on the Pliocene-Pleistocene reflector, which can be traced through a network of seismic reflection records (partly shown, Fig. 2) from DSDP site 178 (noted in Fig. 3, lines 4, 5, 7, and 9). We have a ± 0.1 sec (about 80 m) level of confidence in the

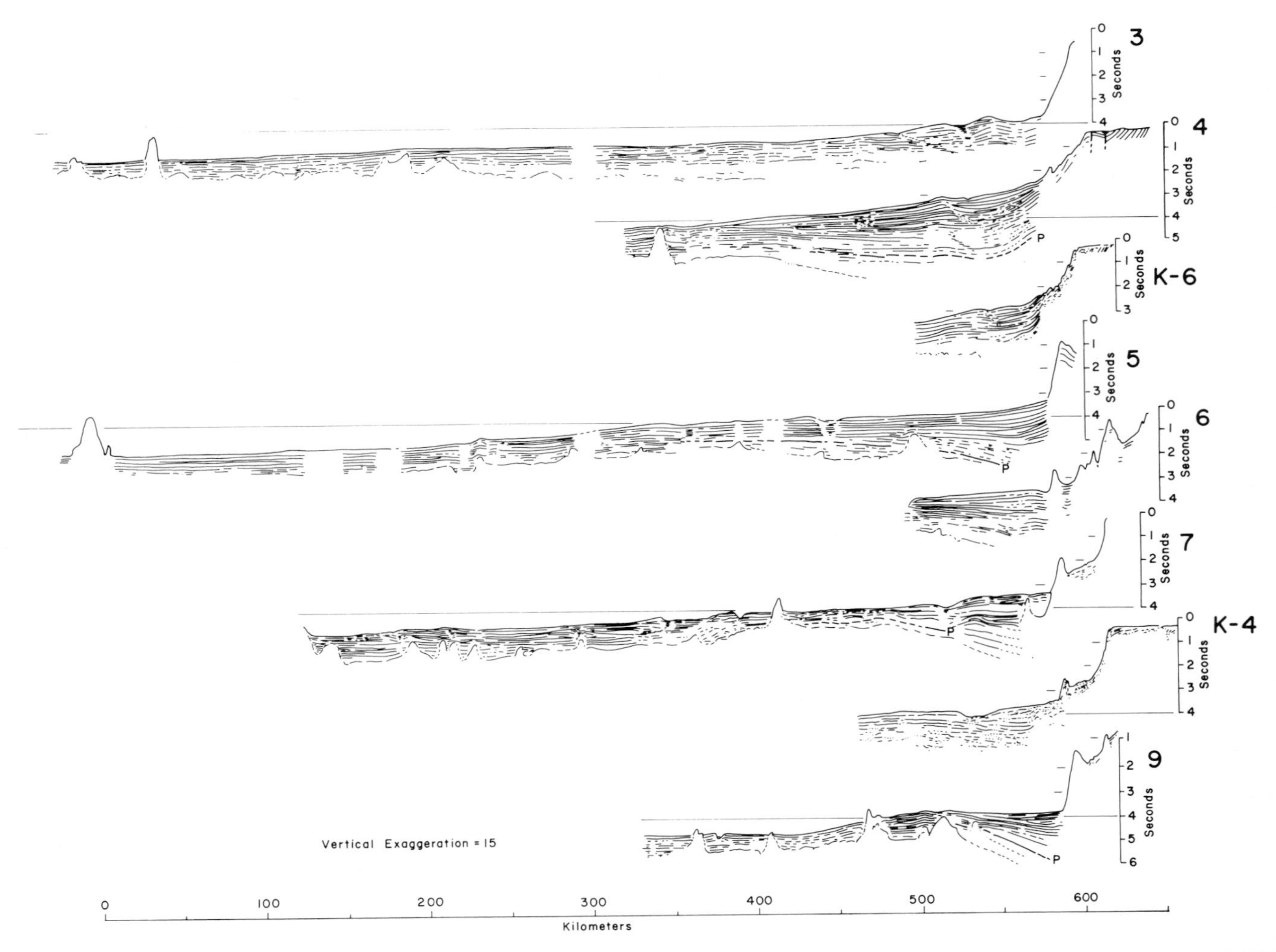

FIG. 3 Line drawings of seismic records, Cross Sound to Queen Charlotte Island. Vertical exaggeration normalized to approximately 15:1. Location of ships tracks shown in Figure 2. P is location of Pliocene-Pleistocene boundary traced from DSDP site 178.

position of the seismic reflector inferred to be equivalent to the 1.8 MY time line, which is the maximum error of closure about multiple paths in the network of seismic lines. The volume of Baranof Fan is unquestionably great; the maximum measured thickness is at least 3 km (assumed velocity = 2 km/sec), and individual turbidity current channels are 10 km or more wide and 1,000 km or more long (Figs. 2 and 3, lines K-4, 5, 7, Hamilton, 1967). The origin of this great volume of material is probably the glaciated terrain of southeast Alaska. Off the Queen Charlotte Islands, the Pleistocene fan is much thinner, because sediment was intercepted by a large trough behind the islands on the continental shelf (Shouldice, 1971). Therefore, at the base of the slope off the Queen Charlotte Islands, the glacially derived sediment is thin in contrast to the thick accumulations along the margin north of Dixon Entrance (Chase and Tiffin, 1972).

At the base of the continental slope north of Dixon Entrance, six of the eight records in Figure 3 (4, K-6, 5, 6, 7, 9) show a filled trough. Absence of the trough in two of the records may be due to insufficient penetration to reach the base of the sedimentary section and igneous basement. The trough is formed by depression of the oceanic crust, and it is not only shaped like a trench but has dimensions similar to those of the Aleutian Trench. Such a trough is not conceptually associated with transform plate boundaries. Chase and Tiffin (1972) imply local tectonism seaward of the continental slope off the Queen Charlotte Islands, possibly associated with the development of the trough. They show a fault in the rise west of Dixon Entrance (Chase and Tiffin, 1972, Profile A, Km 12, Fig. 7) but we found no faults in our nearby records (K-4 and 7, Fig. 3). To the north, between Dixon Entrance and Cross Sound, our records show no major faults in Pleistocene sediments of Baranof Fan. However, the faults generally found on the landward side of such a buried trough are not seen because our records contain little subsurface information at the head of Baranof Fan and under the continental slope. Although the origin of this trough, and the volcano, Mt. Edgecumbe, are puzzling in terms of a plate-tectonic model, this is not the first report of a trough along a proposed transform plate boundary (for example, western Aleutian Trench, Puerto Rico Trench).

Tectonism along the continental slope is largely inferred from physiography. Despite the great amount of sediment that must have passed across the slope north of Dixon Entrance in Pleistocene time, an uncovered irregular surface is seen in records across the slope (lines K-4, 6 and 7, Fig. 3), suggesting that the topography has a tectonic origin. The rate of vertical tectonic displacement may have exceeded the rate of glacial sedimentation, or most of the vertical relief may be of post-glacial age, when the slope was sediment-starved. A tectonic origin is proposed because farther south, off the Queen Charlotte Islands, such features are faults, as indicated by more complete bathymetric and seismic data (Southerland-Brown, 1968; Chase and Tiffin, 1972), whereas the seismic lines off Prince of Wales Island make this point with less certainty. It is possible that the fault-controlled topography on the continental slope off the Queen Charlotte Islands extends north of Dixon Entrance to the area off Chatham Strait, approximately where the Chatham Strait fault is projected onto the continental slope.

North of Chatham Strait, the topography of the slope is relatively smooth in records running downslope. Records parallel to the slope show numerous small canyons; however, no major ridges and troughs parallel to the slope are indicated. Therefore, if faulting occurs beneath the continental slope, displacement with major surface expression has been masked by Pleistocene sediment, or relief from postglacial deformation has not been as pronounced as to the south. Furthermore, the earthquake epicenters recorded by a local network of seismometers show tectonic activity at the edge of the shelf, but not the foot of the slope (Fig. 2).

The earlier data combined with those reported here are consistent with the fault inferred previously along southeast Alaska. A fault zone has been established with reasonable certainty by previous studies along the continental slope off the Queen Charlotte Islands (Southerland-Brown, 1968; Tobin and Sykes 1968; Chase and Tiffin, 1972) and on the outer continental shelf off Baranof Island (Page, 1973). But the continuity of faulting between southern Baranof Island and Dixon Entrance is uncertain because of insufficient data. The main evidence for faults is good physiographic expression as shown in our records. As this stretch is about 130 km of a 1,050-km-long active offshore fault system, the Queen Charlotte Islands Fault is here inferred to continue along the slope of Chatham Strait and to join with the fault outlined by the aftershocks from the 1972 Sitka earthquake. The Chatham Strait Fault, a prominent tectonic boundary, appears to separate an area to the south where faulting is now dominant along the slope from an area to the north where faulting is now dominant along the edge of the shelf.

STRUCTURE AT JUNCTURE OF TRANSFORM AND OBLIQUE CONVERGENT PLATE BOUNDARIES

Early in the study, an opportunity was recognized to define structure across the oblique convergent plate boundary at its juncture with the transform plate boundary. A projection of the oblique boundary intersects the transform boundary on the continental shelf just south of Cross Sound (Fig. 1). Thus, the zone of oblique convergence, which generally corresponds with the continental slope, might be studied on the shelf, where conditions are much more favorable for resolution with the reflection technique.

This line of investigation involved first acquiring seismic reflection lines across the shelf to define locations of faults corresponding to the transform boundary. Then the zone of convergence was examined seaward of the transform with reflection lines across its presumed tectonic fabric (see Fig. 1).

Reflection records that cross the shelf between Sitka and Cross Sound (Figs. 2–5) all show faults. In addition, the records confirm that some sharp bathymetric linear features, defined by soundings and suspected to be fault scarps, are faults with surface scarps in reflection records (Figs. 4 and 5). The fault scarps occur only off Chichagof Island and in Cross Sound, and they may not be apparent to the south because the heads of canyons have destroyed bathymetric continuity. From these data we interpret a continuous fault zone between Sitka and the Fairweather Fault (Fig. 2). The northern 75 km of the fault are well located, with the exception of a break in Cross Strath where post-glacial sedimentation partly masks it. The fault zone appears to offset the southeast bank of Cross Strath right-laterally along one or perhaps two faults (Fig. 4). The offset is estimated to be 400 ± 150 m along the well-defined offset and 300 ± 150 m along the questionable one. This bank has been eroded deeply (Fig. 4F), probably by a glacier that occupied Cross Strath and formed the moraine at its seaward end (Fig. 4E). Although the measurement of the lateral component of faulting cannot be clearly separated from vertical or erosional effects, the possible amount of offset along a glacial feature indicates rapid tectonism. The local fault scarp in the thick post-glacial sediment of Cross Strath also indicates a high rate of slip (Fig. 5). The fault has been named the Chichagof-Baranof

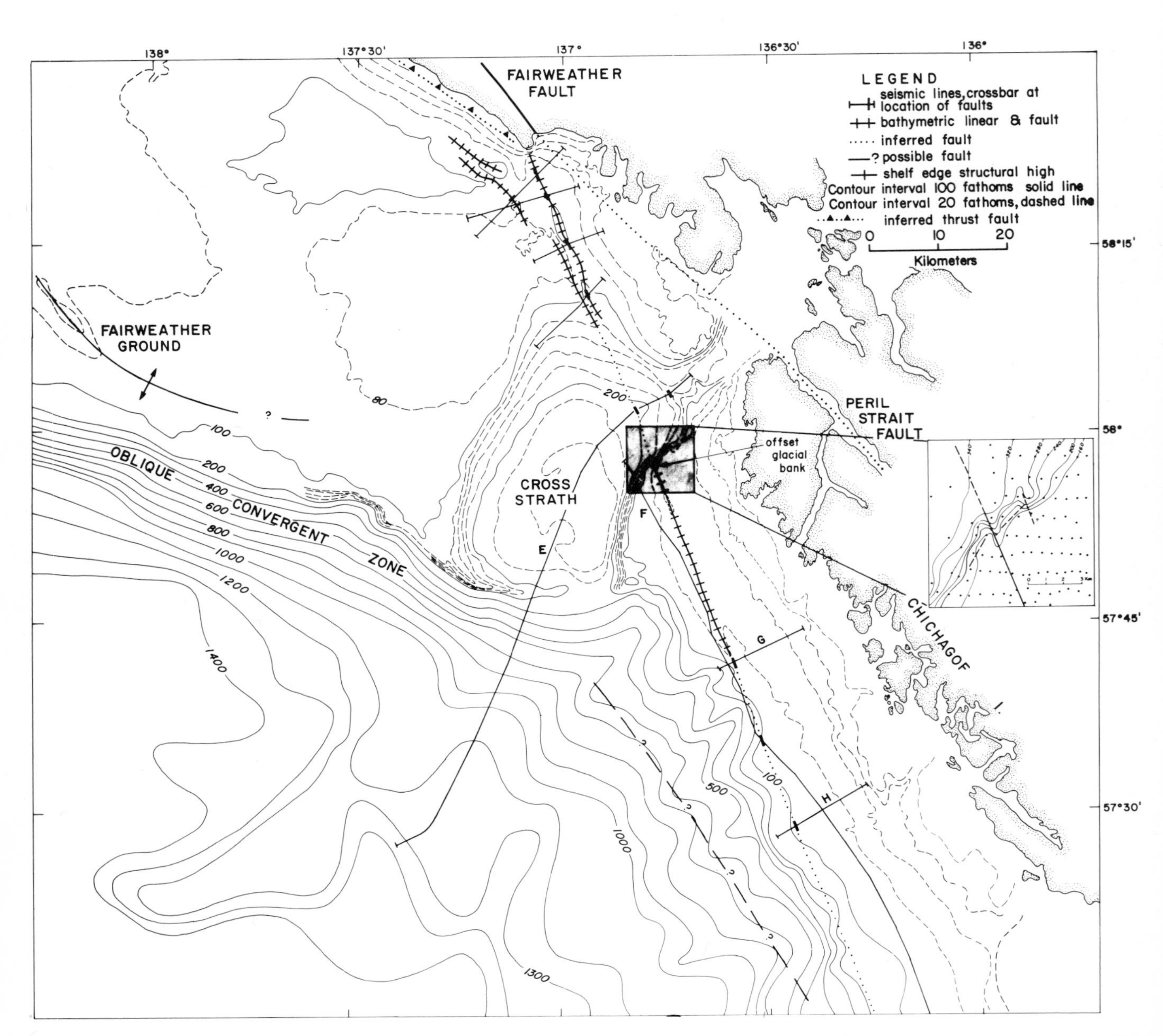

FIG. 4 Location of selected seismic records along and across the northern Chichagof-Baranof fault, Cross Sound area. Lettered lines are illustrated in Figure 5. Seismic records northwest of Cross Strath are from Molnia et al (1978). Inset shows data in area of offset glacial bank. Dots are location of soundings from USC & GS H-4529.

Fault (CB fault) (Beikman, 1975).

The most obvious trace of the CB Fault cannot be followed directly into the Fairweather Fault. A projection of the surface scarps from where the fault is last recorded, about 4 km off Icy Point, intercepts land about 2 km west of the mapped Fairweather Fault (Fig. 4). However, this mismatch may be more apparent than real, because the width of the fault zones is not shown in Fig. 4. The CB Fault is generally seen as two faults with local subsurface deformation in a zone 1.5 km to perhaps 8 km wide (Fig. 4). Similarly, the mapped traces of the Fairweather Fault and the thrust fault offshore of Icy Point (Plafker, 1967) are 2 km apart. Therefore the area of juncture is tectonically more complex and broader than indicated by the presently mapped fault traces. The strike of the Fairweather Fault is in line with the strike of the Peril Strait Fault, and since the inferred connection (Beikman, 1975) crosses a rough glaciated ocean floor, it can only be shown convincingly by a detailed survey. The juncture of the CB Fault and the Fairweather-Peril Strait Fault is also in an area of rough ocean floor and is probably difficult to follow. The CB Fault could also be related to the large offshore thrust fault subparallel to the Fairweather (Plafker, 1967). Such a relation is suggested by the physiographic lineaments just west of the CB Fault (Fig. 4).

Although the relation between the CB and Fairweather Faults is unclear, there is no reason to believe that they are not part of the same tectonic system. Both are tectonically active, with recorded seismicity and strong surface expression. The displacement of glacial features on the Fairweather (Plafker, 1976) is similar to the displacement of a glacially formed submarine bank along the CB Fault. No other faults on land are known to have Holocene displacements of this magnitude, nor have any other faults of similar continuity been traced on the continental shelf. Post-glacial displacement on the CB Fault is perhaps as great as 4 to 5 cm/yr if one assumes that the southeast bank of Cross Strath ceased being eroded by glaciers 10,000 years ago. This is about the same amount as the proposed relative motion of the Pacific and North American plates. For all these reasons we interpret the CB Fault to be part of the present transform boundary between the Pacific and North American plates.

The central Gulf of Alaska continental slope and the corresponding oblique convergent plate boundary projects across the seaward end of Cross Strath, and it should meet the CB Fault off Chichagof Island (Fig. 4). Here the continental slope changes trend through 40°, expressing the juncture of the oblique convergent and transform margins of the plate model, but no strong morphologic evidence of plate convergence is seen. Two reflection records which parallel the CB Fault where the plate bound-

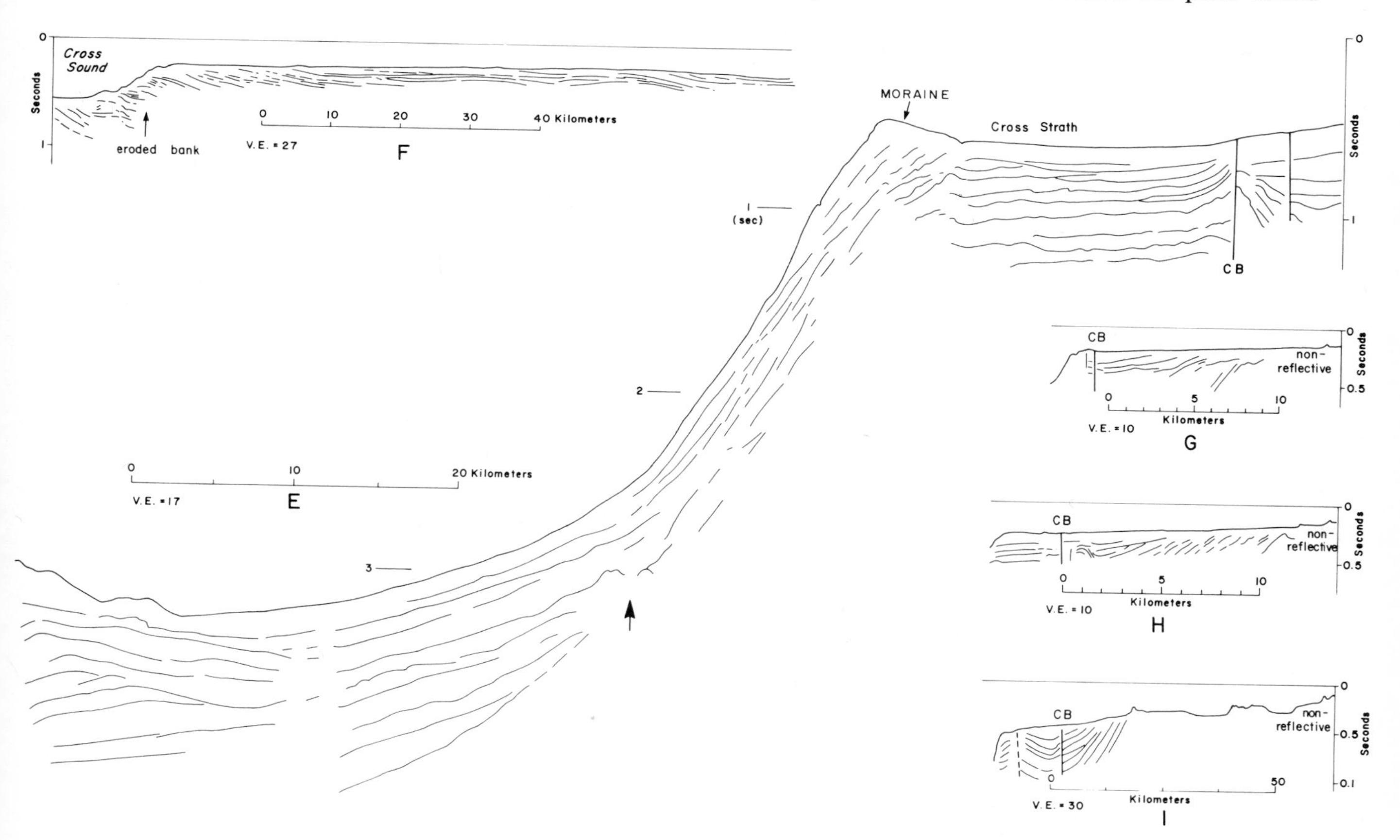

FIG. 5 Tracings of selected seismic reflection records along the Chichagof-Baranof fault and across the continental margin off Cross Sound. CB fault indicated by CB. All lines keyed to Figure 4 except I, which is off Sitka (line 4, Figure 2).

aries meet, one of which is shown in Fig. 5 (line F), also show no strong subsurface convergent deformation. The records look much the same as records elsewhere across the shelf off Chichagof Island. Therefore, if subduction is occurring across the obliquely convergent plate boundary it is either occurring at a slow rate that does not deform upper (0.5 sec) reflecting horizons, or deformation of those sediments is concentrated close to the base of the continental slope.

An alternate explanation is that an undetected transform fault paralleling the CB Fault runs beneath the rugged continental slope from the Queen Charlotte Islands Fault to the area off Cross Sound and joins with the oblique convergent zone. Such a fault zone is queried in Figure 4 and a similar fault has been shown by Plafker et al (1975). The unusual position of the CB Fault at the edge of the shelf rather than in a more conventional position under the slope lends support for the hidden fault. A reflection record across a smoother part of the continental slope shows unfaulted sediments below the sea floor and possible deformation about 430 m (0.5 sec at 1.7 km/sec) below this (arrow, line E, Fig. 5). The possible age of the unfaulted section is suggested by analogy with a similar setting off Kodiak Island at DSDP site 182, where early Pleistocene sediment of the downslope apron was recovered from 212 m depth (Kulm et al, 1973).

Because compressive deformation does not affect the upper beds across the oblique convergent plate boundary at its juncture with the transform boundary, the rate of deformation is inferred to be too slow to show through Pleistocene sedimentation. We do not propose inactivity, because some oblique convergence further along the boundary off Fairweather Ground is indicated by seismicity (Gawthrop, et al, 1973), and recent local convergent structure elsewhere along the boundary is described in a following section.

CRUSTAL MODELS ACROSS THE QUEEN CHARLOTTE-FAIRWEATHER TRANSFORM

Transform plate boundaries are generally modeled as simple vertical faults where the plates slide laterally past one another without significant accretion or consumption. Yet previously constructed crustal models based on deep geophysical data show the Queen Charlotte Islands fault zone as a wide boundary with an associated trench. The first crustal model through Dixon Entrance by Shor (1962) emphasizes the buried trough at the foot of the slope. Shor did not have sufficient data to continue velocity units across the transition between oceanic and continental crust. Couch, (in Dehlinger et al, 1970) modeled a gravity transect using the constraints from Shor's work, and in his interpretation the continental-oceanic boundary dips seaward, which suggests a wide zone of faulting across the margin. The crust-mantle boundary in Dixon Entrance was later refined by Johnson et al (1972) on the basis of new refraction information along the coastal mountain and fiord area. From these models it is uncertain if the Queen Charlotte Islands Fault is a simple vertical fault separating oceanic and continental crust.

New information has become available since the above-mentioned crustal models were constructed, and we constructed a series of models using data along line 7 (Fig. 2) where gravity, magnetic, and seismic reflection data were measured simultaneously. The line of new data crosses the slope where topography is relatively smooth, and thus the gravity data (made with a stable platform rather than gimbal instrument and with satellite navigation) are correspondingly smoother than Couch's data in Dixon Entrance. The analysis of bulk densities from DSDP Leg 18 cores (Kulm et al, 1973) has provided a general density gradient in upper Tertiary sediments of the Gulf of Alaska.

The two models in Figure 6 illustrate a vertical and a 45° landward-dipping continental-oceanic crustal boundary, thereby showing a possible latitude of interpretation within the constraints of the geophysical data. Both models indicate that the trough at the foot of the slope seen in seismic records extends 20 km or more under the slope and that it may contain a sediment section from 5 km to 8 km deep. The landward end of the trough is suggested by the inflection in observed gravity values and by a magnetic anomaly peak. The continental-oceanic boundary is not a simple break, but probably a zone of faults, as is shown by the topography off the Queen Charlotte Islands. The sediment-starved margin off the Queen Charlotte Islands has thinner sediment than must exist off the Dixon Entrance because the calculated gravity values off Dixon Entrance are difficult to match to an inflection in the observed values without including a thick block with sediment densities at the edge of the shelf (km 120–140, Fig. 6). Therefore, steps in the continental slope could be obscured by sediment north of the Queen Charlotte Islands.

From two-dimensional gravity modeling of the continental-oceanic boundary, with relatively good seismic data on either side, we conclude that a 20- to 30-km-wide tectonic zone separates the trough off Dixon Entrance from the crystalline continental crust. If a single feature in the zone is the dominant boundary, its dip is unknown. This general structure

FIG. 6 Two-dimensional gravity models through Dixon Entrance and across the continental margin along line 7, Figure 2. Landward values of gravity (dashed line) projected from Couch (in Dehlinger et al, 1970). In model A, vertical boundary between oceanic and continental crust is assumed; in model B, 45° boundary. Seismic refraction data from Shor (1962), and at km 122 from Milne (1964). Arrow indicates landward limit of seismic reflection control on oceanic basement from line 7 (Fig. 3). Refraction velocities in km/sec are indicated in italic type, assumed densities in g/cm^3 indicated in roman type. Gravity measured with a LaCoste-Romberg stable platform instrument; positions controlled by navigional satellite. Observed gravity shown by line, computed gravity by dots.

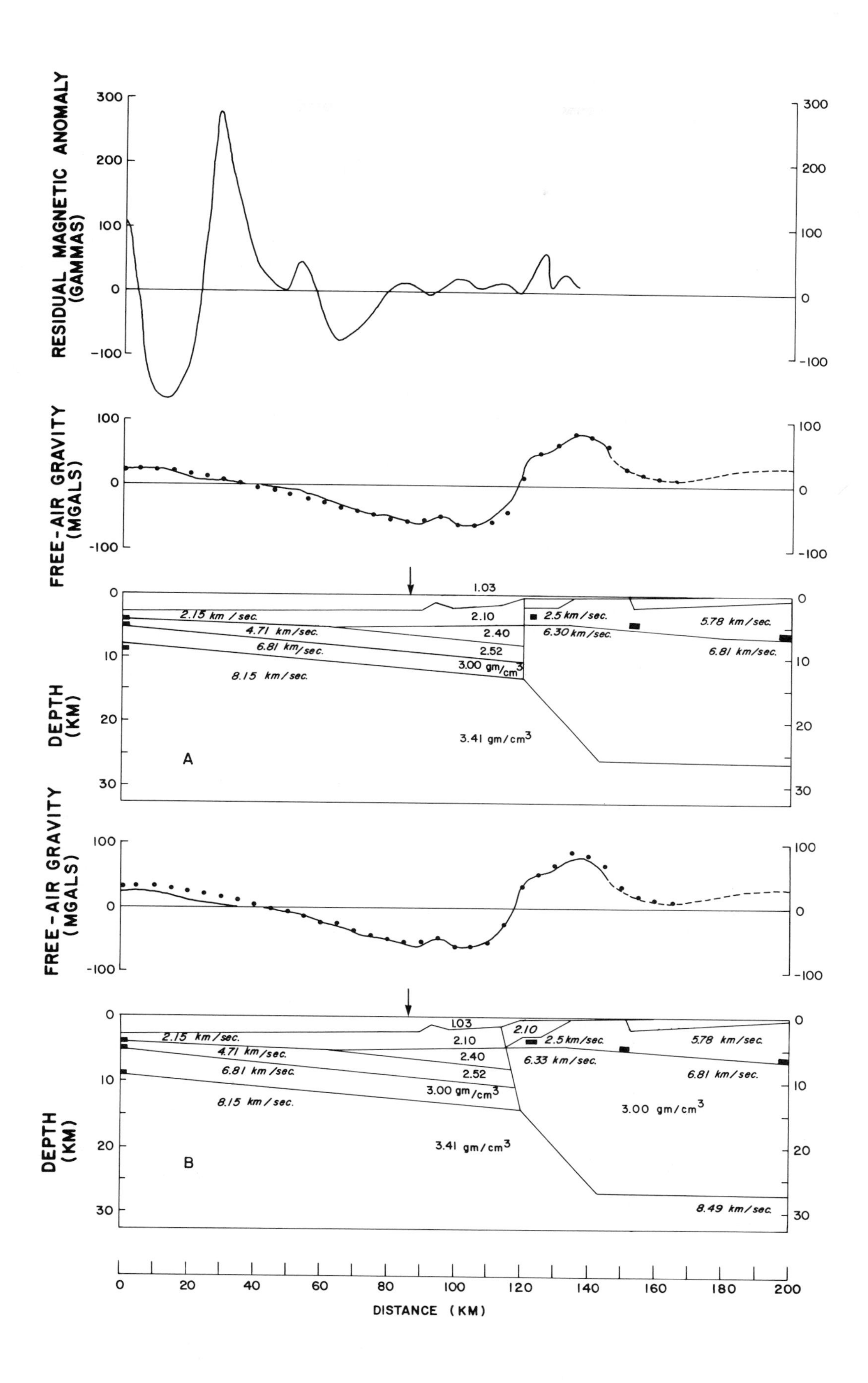
RESIDUAL MAGNETIC ANOMALY (GAMMAS)
300
200
100
0
-100
FREE-AIR GRAVITY (MGALS)
100
0
-100
DEPTH (KM)
0
10
20
30
1.03
2.15 km/sec.
2.10
2.5 km/sec.
5.78 km/sec.
4.71 km/sec.
2.40
6.30 km/sec.
6.81 km/sec.
2.52
3.00 gm/cm3
8.15 km/sec.
3.41 gm/cm3
A
2.10
6.33 km/sec.
3.00 gm/cm3
8.49 km/sec.
B
0 20 40 60 80 100 120 140 160 180 200
DISTANCE (KM)

is more like that of a trench boundary than a transform boundary. If one accepts the model of Grow and Atwater (1970), it was a trench 30 million years ago, but since that time a large amount of transform motion is thought to have occurred and thus the traces of an older trench could be found far to the north.

OBLIQUE CONVERGENT PLATE BOUNDARY

Early plate-tectonic models were not explicit about the oblique convergent plate boundary between Cross Sound and the Aleutian Trench; however, since then, studies have outlined the major structure along the continental shelf there. The structure of the continental shelf has been described by Bruns and Plafker in publications preliminary to a more complete summary discussion (Bruns and Plafker, 1975, 1976). They establish a significantly different structural style on the shelf between the oblique folded area west of and the less complex folded area east of Icy Bay. Rogers (1977) developed this point further, relying more heavily on data from the western Gulf of Alaska. He proposes that the convergent deformation of the slope and trench off Kodiak Island continues into the continent between Kayak Island and Icy Bay as the oblique-trending folds of Bruns and Plafker, which merge with onshore splays from the Fairweather Fault. Rogers also believes that in the oblique convergent zone on the shelf, folding has developed sequentially from earlier at Kayak Island to later off Icy Bay, or generally from west to east. He contrasts this with the less complexly folded shelf area between Icy Bay and Cross Sound, in which Bruns and Plafker (1975) describe a broad basin flanked by a shelf-edge structural high (Fairweather Ground). The explanation given by Rogers for this structural pattern is that the continental shelf is moving more slowly than, but in the same direction as the Pacific plate, converging with the continent along the oblique fold belt from Kayak Island to Icy Bay. The data from the continental slope in the eastern Gulf of Alaska and the floor of the Gulf of Alaska, presented here led us independently to a similar interpretation. We have both seismic refraction and reflection data bearing on this problem.

Seismic refraction stations were established off Icy Bay and Dry Bay by the Scripps Institution of Oceanography, on Leapfrog Expedition in 1961, and preliminary results were presented by Shor (1965); results presented here deviate somewhat from that presentation because of re-analysis taking into account reflection as well as refraction data. Field methods were generally as discussed by Shor (1963); corrections and structural interpretation followed the methods given by Ewing (1963), which require the assumption that layers are plane and of constant velocity. Cases where data obviously deviate from these assumptions are noted below.

Stations LF27 and LF28 (Figs. 2, 8) yield a "compound profile," with short split profiles across each receiving position and a reversed profile between the two stations. LF27 was received on the fan at the foot of Pamplona Seavalley. The reversed profile was shot up the seavalley, and across the outer part of the continental shelf. LF28 was received on the shelf, and the short run landward was shot to the entrance of Icy Bay. As there is more than 2 km difference in water depth between LF27 and LF28, and the slope is far from smooth, the plane-layer assumption is suspect here; without additional field data, however, it is the best that can be done.

Refracted arrivals from the shallowest sedimentary layer were observed on both runs for station 28. Scatter from a straight line suggests that there may be lateral variations in velocity, but that the value is at least 1.66 km/sec. A second sedimentary layer, with velocity near 3.0 km/sec, is well observed on both runs on LF28; strong second arrivals over a considerable range of distance are evidence that it is a discrete layer with a sharp interface, rather than representing a continuous increase of velocity with depth in the sediments. No arrivals from the upper of these sedimentary layers were observed at LF27 and only one arrival from the lower layer, so the same velocities were assumed here as observed in LF28. The next deeper layer, with velocity near 3.9 km/sec, and a fourth layer with velocity 4.7 km/sec may well be one continuous formation with a velocity gradient. The major volume of the continental slope and rise is comprised of this material, the lower part of which appears on reflection line B (just southeast of LF27) to be folded sediments. The plane-layer solution becomes a bit strained here, where literal application results in a solution with negative thickness of the 4.7 km/sec layer at LF27. As layer thicknesses beneath receiving stations are always extrapolations (due to ray offset), having a negative thickness at one end of a reversed profile does not reflect a physically impossible solution. The deepest observed layer has a velocity in excess of 6 km/sec; the value of the velocity is so dependent upon the large topographic corrections and upon the plane-layer assumption that no particular importance should be given to the exact value. Arrivals from this layer received at station 27 were all from shots fired on the shelf; arrivals from the same layer received at LF28 (on the shelf) were all from shots fired on the slope, and in both cases the topographic corrections were so large that the value of the calculated velocity can be changed significantly if one changes the assumption as to which interface represents the surface topography. The depth to the 6-km/sec layer at LF27 is very nearly the same as that to the oceanic crust under stations to the southwest under the Alaskan abyssal plain, which have velocities close to 6.7 km/sec. If one were to assume that this is the true velocity, the travel times could be satisfied by refractions from a layer that is horizontal beneath the continental slope, about 5.5 km below sea level, and then dips down steeply towards land beneath the shelf as shown in Figure 8 by a dashed boundary. LF27-28 crosses the shelf where structure is simple, and the results are generally consistent with sonobuoy station data taken later along strike (Bayer et al, 1977). The layers with velocity 3.9 to 4.7 km/sec are anomalously thick and of low velocity for continental crust; the velocity structure is similar to the convergent margin structure along the Aleutian Trench at Kodiak (Shor and von Huene, 1972).

A series of end-to-end ("leapfrog") profiles were shot offshore from the portion of the coast between Yakutat Bay and Cross Sound. The first pair, LF29-LF30, was shot at the foot of the continental slope from Yakutat Bay to Dry Bay; the three reversed profiles from LF30 to LF33 were perpendicular to the slope off Dry Bay; the final reversed pair, LF33-LF34 was again parallel to the slope from Dry Bay to Cross Sound.

On the reversed pair LF29-LF30, refracted arrivals were observed from sediments of relatively high velocity at the seafloor (fan deposits?) on LF29, and from a deeper sediment layer with velocity 3.6 km/sec on both runs. A deeper layer, with velocity of 5.6 km/sec was observed on both shooting runs, with a linear refraction travel-time plot. The layer with velocity of 5.6 km/sec may be either oceanic basement or high-velocity folded sediments similar to those seen on LF27-28 and on data from stations on the landward wall of the Aleutian Trench near Kodiak (Shor and von Huene, 1972). Curiously, this layer, which appears clearly on both LF29 and LF30, at the base of the continental slope, cannot be identified clearly on profiles from the contiguous station pair LF30-LF31 normal to the slope. The oceanic crust, with velocity slightly high at 7.1 km/sec, is well determined on both runs of the reversed pair. Mantle arrivals are observed on numerous shots on both runs, but an anomalously high mantle velocity of 8.8 km/sec may be an artifact created by irregularities in the thickness of the sedimentary layers where the shooting run comes close to the foot of the continental slope near LF30.

Stations LF30, LF31, LF32, and LF33 in combination yield three short reversed pairs perpendicular to the slope, and pair LF33-34 is parallel to the coast to the west; all show consistent results. High-velocity sediment arrivals from the seafloor are seen on a few shots on the first profile; thereafter no seafloor refractions are seen, and it is probable that the composition of the seafloor sediments changes sufficiently that they can be more closely approximated by a lower velocity (2.15 km/sec or less), previously determined by wide-angle reflection studies in the Alaskan abyssal plain to the southwest (Shor, 1962). Strong refracted arrivals were received on all runs from a higher velocity sedimentary layer about one km below the seafloor. The top of this layer corresponds approximately to the base of the Pliocene on reflection line D (Fig. 7). Travel-time plots on all stations from LF30 to LF34 break over directly from the sediment arrivals to refracted arrivals from the oceanic crust, with velocity near 6.8 km/sec. Strong reflections are observed from the apparent base of the sedimentary layer on profiles LF31 to LF34, and the arrivals refracted from oceanic crust are generally a little later than they should be if they were coming from this reflector. A few second arrivals with velocity 5.0 km/sec are observed on LF30-out. Basement reflections on LF30 are considerably shallower than on LF31 to LF34 and disappear at short range before the basement-refracted arrivals appear; we suspect that the basement deepens abruptly just offshore of the receiving position for LF30 and so have not tried to reconcile the solutions for the two reversed profiles that join at this point. Solutions for data from all stations from LF30 to LF34 have been calculated on the assumption that a masked basement layer is present, and that the travel-time curve is tangent to basement reflections observed on LF31 to LF34. Where no refracted arrivals were observed, a velocity of 5.55 km/sec was assumed. Results are shown in Figure 8 and Table I. This basement layer is considerably thinner than normal. Mantle refractions may be present on a few of the farthest shots on each of the lines between LF30 and LF33, but were not observed over a sufficient distance to be considered reliable for determination of velocity or depth. Good mantle arrivals are detected on the longer reversed profile parallel to the coast, LF33-34, and show mantle at nearly normal depth.

The general appearance of the section in Figure 8 is similar to ones across a trench margin, from the outer swell to the trench floor. LF29-30, at the foot of the slope, would correspond to a line in a similar position on the inner wall of a trench, except for the presence of the thick layers of fan sediments.

Tectonic structure of the continental slope is shown in seismic reflection records across the eastern, central, and western parts of the margin (Figs. 2, 4, and 7). The eastern and western records show a rather typical Pacific continental margin with a mildly deformed slope beginning at a shelf-edge structural high that merges with a downslope sediment apron and ends at a current-structured sediment wedge of the ocean basin. The continental slopes are deformed along their lower parts (at the arrows, lines A and C, Fig. 7). Uplifts on shelf-edge structural highs are recorded by erosion at the highs, and also by tilting of reflecting horizons in the shelf basins. At the foot of the slope, in record A, is a broad channel that probably developed from turbidity currents associated with a trough from the Bering Glacier. A buried turbidity-current channel is seen at the foot of the slope in record C.

Between these records is one that indicates much less tectonic deformation in the upper sediments of the slope (B, Fig. 7). The upper continental slope is a thick wedge of sediment without folds or faults that looks much like a passive continental margin. This wedge is at least 2.5 to 3 km thick along the shelf edge, thinning to 0.2 km near the foot of the slope. The Yakataga Formation of early Miocene to Pleistocene age is about 4.3 km thick directly onshore from this line, suggesting that the wedge is at least of Miocene age (Plafker et al, 1975). The wedge is underlain by a highly diffractive ridge with a rough upper surface, and, as shown on nearby refraction profiles LF27-28, has seismic velocities that

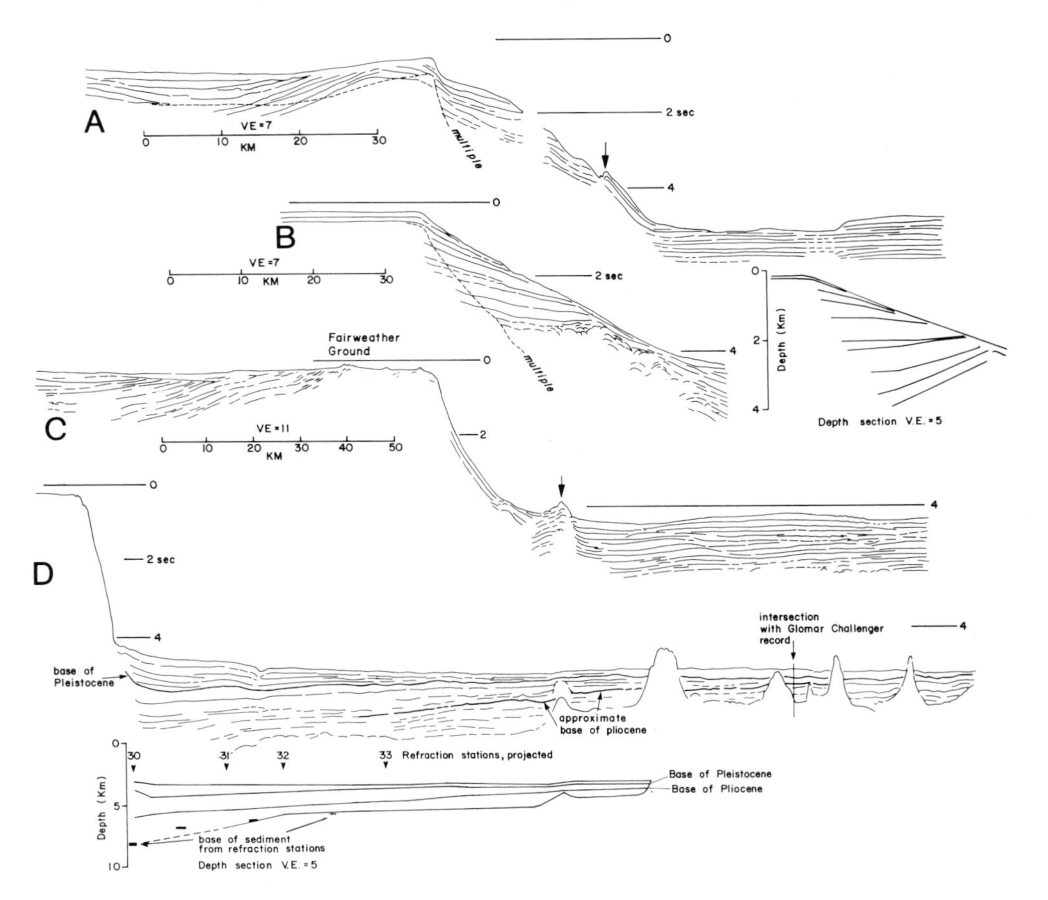

FIG. 7 Tracings of seismic records across oblique convergent margin, central Gulf of Alaska. Line A from von Huene et al (1972); Line B, von Huene et al (1975); line C, SIO; line D, NOAA. Arrow indicates deformed area at foot of slope. Depth sections constructed using velocity curve derived from refraction stations. Projection of refraction stations indicated along line D.

Table 1. Refraction Station Data

Station and run	*Type*	*Position*		*Azimuth (from N)*	*Velocity, km/sec*						*Water Depth (km)*	*Thickness, km*					*Total Depth to Mantle*
					a	*b*	*c*	*d*	*e*	*f*		*a*	*b*	*c*	*d*	*e*	
LF 27-In LF 27-Out	Split	59°14′N	142°52′W	166° 035°	1.66*		4.00				2.66	0.49					
LF 27-Out LF 28-In	Reversed	59°14′N 59°52′N	142°52′W 142°12′W	035° 216°	1.66*	2.89	3.86	4.68	6.18(?)		2.66 0.06	0.27 0.27	0.3 1.0	2.5 1.7	-0.3† 8.0		
LF 28-In LF 28-Out	Split	59°52′N	142°12′W	216° 047°	1.66	2.96	3.84				0.06	0.32	1.2				
LF 29-Out LF 30-In	Reversed	58°30′N 58°10′N	140°33′W 139°33′W	127° 302°	2.34		3.56	5.55	7.06	8.76	3.19 2.98	1.61 1.82		3.3 1.8	2.6 3.6	6.4 4.2	17.0 14.4
LF 30-Out LF 31-In	Reversed	58°10′N 57°53′N	139°33′W 139°50′W	127° 034°	(2.35)		3.16	(4.99)	6.98		2.98 3.19	1.24 1.38		3.0 2.5	2.1 0.4		
LF 31-Out LF 32-In	Reversed	57°53′N	139°50′W	219° 019°	2.15*		2.93	5.55*	6.99		3.19 3.32	0.83 1.49		2.7 1.0	0.8 1.2		
LF 32-Out LF 33-In	Reversed	57°37′N 57°21′N	140°03′W 140°18′W	212° 019°	2.15*		2.59	5.55*	6.82		3.32 3.37	0.96 1.15		1.5 1.4	0.9 1.3		
LF 33-Out LF 34	Reversed	57°21′N 56°57′N	140°18′W 139°13′W	123° 308°	2.15*		2.70	5.55*	6.73	8.15	3.37 3.35	1.53 0.81		1.0 1.2	0.8 1.4	5.1 3.4	11.8 10.2

*Assumed velocity
( )Single profile
†Due to extrapolation

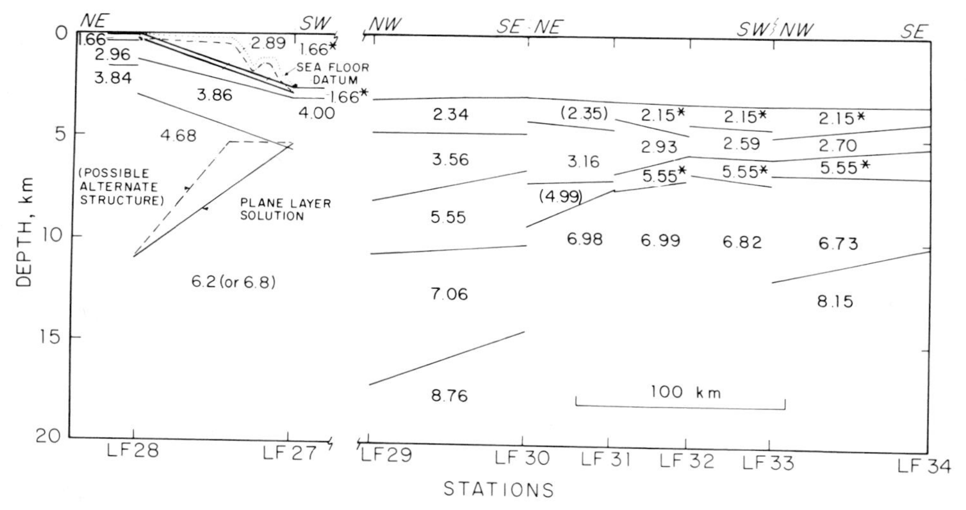

FIG. 8 Seismic refraction stations, across oblique convergent margin from Icy Bay across the shelf, and seaward of Yakutat Bay. Location of stations shown in Figure 2. On stations LF27-28, the dotted upper boundary is the true seafloor along the shooting line; the first solid boundary is the datum plane assumed for computation. The mode of topographic correction used involves the assumption that the first layer thickness varies linearly along the profile, and that all of the irregularity lies in the second layer. The second dotted boundary is, then, the base of the first layer of sediment calculated on this assumption. The deepest dashed boundary on LF27-28 is one that could produce the observed travel times if one assumes that the crust velocity is 6.8 km/sec.

are typical of consolidated (probably deformed) sediment. From this record, it appears either that the earlier period of intense tectonism has been followed by a present period of reduced compressional tectonism or that any present deformation is confined to a ridge at the foot of the slope.

Evidence from the adjacent deep ocean floor also suggests a period of reduced tectonism. In this area the deep ocean basin has magnetic anomalies that are highly attenuated (Fig. 2; Naugler and Wageman, 1973; Taylor and O'Neill, 1974). One cause of anomaly attenuation is depression of oceanic crust to greater than normal depth, as shown in refraction lines LF30-34 where the crust is depressed beneath a thick wedge of sediment. This sediment wedge is also seen in a seismic reflection line west of refraction station LF29 (D, Figs. 2 and 7) but basement is not seen in the thickest part of the wedge. In the time section the body is not as clearly wedge-shaped as in the refraction data, but a depth section indicates greater wedging than in the overlying Pliocene and Pleistocene continental-rise deposits. Perhaps a basement irregularity modifies the shape of the body as suggested in Figure 7. Part of the wedge might be an older buried deep-sea fan or a filled trough.

The age of the buried fan or trough can be estimated by tracing reflections for 275 km from the age boundaries at DSDP site 178 along a seismic record made by the *Glomar Challenger,* and then along the refraction line shown in Figure 8. The original age boundaries have recently been revised on the basis of radiometric age determinations on ash layers (Hogan et al, 1978). The Pliocene-Pleistocene boundary at site 178 can be followed easily, but the revised Miocene-Pliocene boundary can be traced with less confidence (possibly ± 200 m) because of major interruptions by four seamounts. At the thickest part of the wedge, the pre-Pliocene section may be 3 km thick. The magnetic anomalies under the wedge are between numbers 7 and 13 (Naugler and Wageman, 1973), or 27 to 38 MY old in the age chronology of Heirtzler et al (1968); hence it could be as old as Oligocene. The position of this pre-Pliocene fan suggests that there has been less convergence across the slope than the net convergence between the Pacific and North American plates in the late Cenozoic, because at that rate an old fan would have soon been subducted. The convergent vector across a plate boundary trending parallel to the slope would be about 3 cm/yr, if one assumes a constant relative motion of 5 cm/yr between the Pacific and North American plates at this latitude.

The areas of greater and lesser deformation on the shelf seem consistent with deformation along the slope in records A and B (Fig. 7) and E (Fig. 5), but in record C (Fig. 7) the deformation along the shelf edge (Fairweather Ground) and the slope seem inconsistent with the adjacent less-deformed shelf. Our records show shelf-edge structural highs that are much broader than the structures reported on the inner shelf (von Huene et al, 1971). Bruns and Plafker (1975) distinguish two types of structure in the more deformed part of the shelf, the broad shelf-edge structures paralleling the regional physiographic trend and linear folds and thrust faults that strike oblique to the regional trend. The latter are found only in the more deformed area between Kayak Island and Icy Bay. Rogers (1977) implies that it is the oblique-trending folds paralleling the Aleutian Trench that are genetically related to the regional grain of the Kodiak shelf. Therefore, the distinction between areas of different structural style on the shelf is made on the basis of the oblique rather than the shelf-edge structures. Our data are too widely spaced to define oblique structure on the slope, of which Pamplona Ridge may be an example; however, they show a discontinuity of the shelf-edge structure. The significance of this discontinuity in relation to convergence is obscure. Although it has been proposed that such structure is a result of slow convergence (Seely and Dickinson, this volume), discontinuity of the shelf-edge structure is also reported along the adjacent Kodiak margin (von Huene et al, 1972, and this volume), where the rate of convergence is presumed to be much greater. In a broader context than the records presented here, the significance of slope areas with little recent deformation and the unsubducted pre-Pliocene trough or fan deposits is that they suggest a rate of convergence considerably less than the convergent component of relative motion of the Pacific and North American plates. The shelf and slope data do not show the same division between areas of greater and lesser deformation; this is not necessarily inconsistent, because the structure of the slope is not well enough known.

It is tempting to relate the large area of seafloor with thin basement and attenuated magnetic anomalies, and the pre-Pliocene fan, to a common geologic process. The explanation for a thin layer 2 and subdued anomalies that nearly fits this situation was proposed by Larson et al (1972), and by Lawver and Hawkins (in press). The proximity of newly generated ocean crust to a large source of terrigenous sediment is thought to have prevented extrusion of magma from the ridge and formation of sills at depth rather than surface flows. This would result in a thinner layer of basement with low velocities associated with pillow lavas, and formation of larger-grained magnetic minerals that are more easily altered. This might be cited as a supporting argument that oceanic crust now at the foot of the slope was originally formed in close proximity to a continent and that underthrusting has been minimal since early Miocene time.

SUMMARY

The main results from the study of our data are first summarized and then are fitted into a regional plate-tectonic interpretation.

1. The Queen Charlotte Islands fault zone has previously been inferred to extend from Dixon Entrance to Cross Sound. Our seismic reflection records between Queen Charlotte Islands and Chatham Strait indicate that recent tectonism is mainly along the slope. The slope physiography consists of ridges and troughs similar to the fault physiography off the Queen Charlotte Islands. North of Chatham Strait the presumably contiguous Chichagof-Baranof Fault is indicated by a narrow band of earthquake epicenters (Page, 1973) located along a fault zone which can be seen in the seismic reflection records from off Sitka to Cross Sound. The dominant zone of faulting here appears to be concentrated along the outer shelf and perhaps the upper slope. The CB Fault heads into the Fairweather Fault at 20°, and their probable intersection has not been mapped. We agree with the previously inferred continuity of these three major faults, because they appear to be the most active elements of the continental margin observed in seismic data.

2. A thick wedge of continental-rise sediment (Baranof Fan) occurs along the foot of the slope between Dixon Entrance and Cross Sound. Correlation with cores from DSDP site 178 indicates a Pleistocene age. Under Baranof Fan is a buried trough which appears to have subsided as recently as Pleistocene time, as indicated by dipping beds. Corresponding buried Pliocene fans or troughs are not seen in our reflection records. These features must have been buried by Pleistocene progradation of the slope and perhaps by underthrusting.

3. The buried trough is required in modeling gravity data across the margin, and the trough may extend under most of the slope. Two-dimensional modeling indicates that the base of the trough may be as much as 8 km below sea level at Dixon Entrance. The fault zone which forms the landward wall of the trough is buried beneath the continental slope and extends as far landward as the edge of the continental shelf; however, the models fail to indicate the dip of faults in the zone. Despite the apparent simple break used in modeling, the gravity and seismic reflection data indicate a 20- to 30-km-wide zone of faults, although one fault or another could be more active than the rest at a particular time. Thus, the present position of the CB Fault at the edge of the shelf rather than under the slope may not be typical. The trough and the single volcano near Sitka suggest some underthrusting if interpreted in accord with the plate-tectonic model.

4. Along the slope of the central Gulf of Alaska, oblique convergence is predicted by the plate model. Although some parts of the slope are deformed, there are others that show little folding or faulting. At the foot of the slope there is a wedge of pre-Pliocene sediment nearly 3 km thick buried beneath a normal continental-rise deposit. This wedge may be part of a submarine fan or a trough that has not been subducted. These observations suggest that convergence across this slope is much less than the convergent vector of the motion of the Pacific and North American plates.

5. The thick pre-Pliocene wedge is associated with oceanic crust with a thinner than normal "basement" layer and with attenuated magnetic anomalies over the deep ocean basin between Cross Sound and Pamplona Ridge. The thick wedge probably originated near a continent, and the anomalous crust might also have been affected by a nearby continent during its formation.

6. Crustal structure across the continental shelf and slope just east of Pamplona Ridge consists of a normal shelf sequence underlain by a thick wedge of deformed sediments, with velocity of 3.9 to 4.7 km/sec. This type of structure is more common

along convergent continental margins than along transform or passive margins.

DISCUSSION

One possible tectonic configuration that was developed to explain the previous observations is based on the idea of Richter and Matson (1971). It consists of an arc-transform junction that involves both oceanic and continental crust. Oblique convergent splays short-cut the corner of the arc-transform system as diagrammed in the inset of Figure 9. As the corner is short-cut by the first oblique zone, slip along the transform and the normal convergent zones is decreased because of convergence along the oblique zone. Two possible short-cuts are shown, one on the continental slope in the central Gulf of Alaska and the other along the splays from the north end of the Fairweather Fault. As seen in Figure 9, not all the required faults have been identified. For instance, if Pleistocene slip along the middle of the Denali Fault (Hickman et al, 1977) is related to plate motion, a communication of stress between the Totschunda and Fairweather Faults is implied. This implication has been made by other authors, although a continuous fault has not been mapped in the intervening rugged terrain.

The limited convergence along the slope in the central Gulf of Alaska indicated by our data is ex-

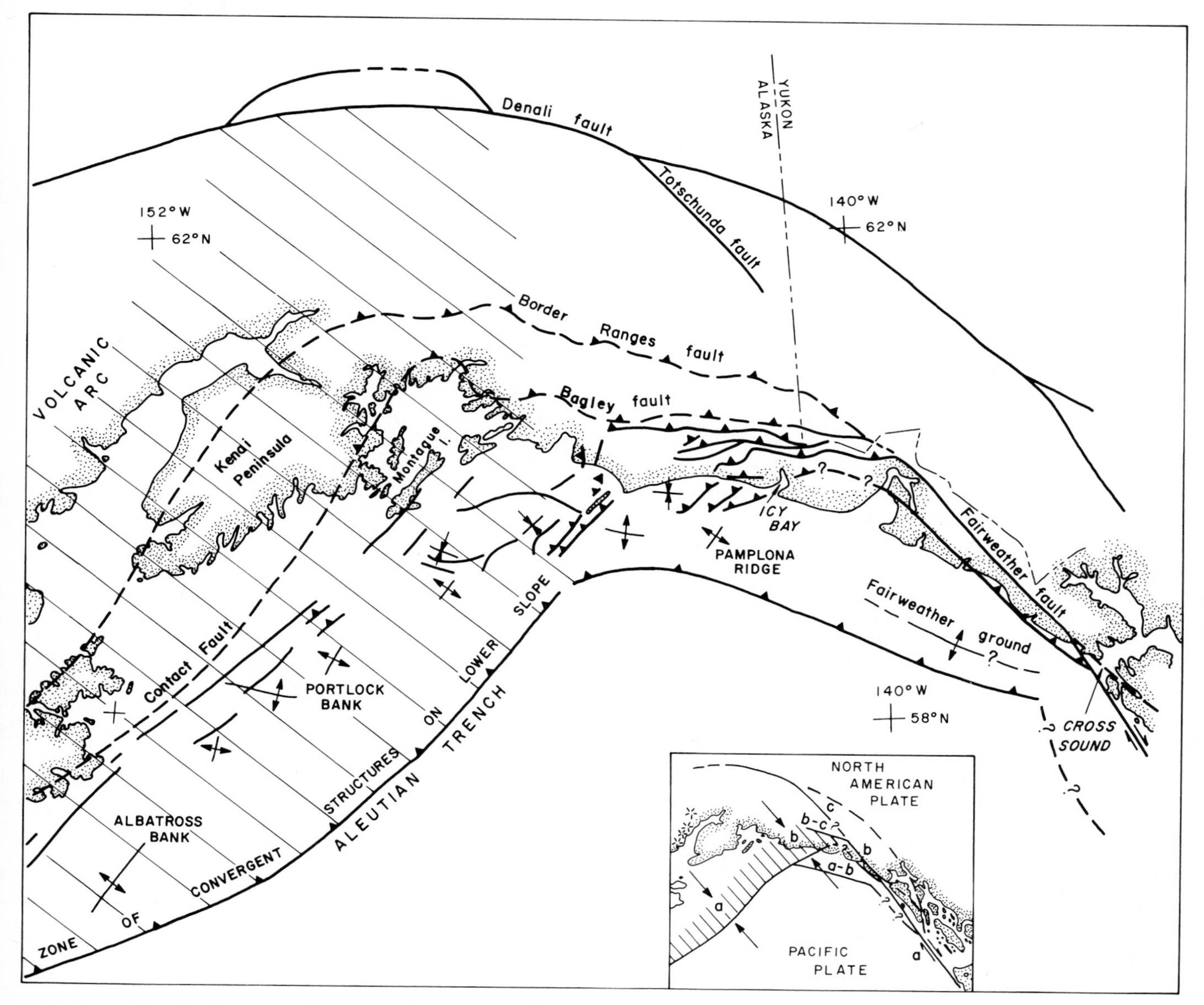

FIG. 9 Sketch of major structures, Gulf of Alaska. Structures on land from Plafker et al (1975), offshore Icy Bay to Montague Island from Bruns and Plafker (1975), Montague to Kodiak Islands, von Huene et al (in preparation). Dashed lines represent Mesozoic plate boundaries. Diagonal lines indicate extent of Benioff zone. Inset shows relative magnitude of convergent vector along plate boundaries. Letter "a" is full Pacific-North America relative plate motion and subsequent letters indicate lesser magnitudes as oblique convergent zones splay from the transform boundary. Diagonal lines in insert indicate convergent zone along landward wall of Aleutian Trench.

plained by the proposed plate configuration. The amount of convergence along the slope is the difference between a) Pacific-North American plate motion, and b) the slip along the Fairweather Fault. Pacific-North America relative motion is estimated to be 5 to 5.6 cm/yr (Silver et al, 1974, Minster et al, 1974). Along the Fairweather Fault Page (1969) estimates 4 cm/yr of late Pleistocene slip, and Plafker and others (1976) estimate 5 cm/yr. If these estimates are correct, the convergence would range from 0.6 to 1.6 cm/yr, or about 10- to 30 percent of Pacific-North American plate motion.

If most of the plate motion is transformed inland, the continental shelf must be moving with the Pacific plate but at a lesser rate. Rogers (1977) came to this conclusion and called the piece of plate southwest of the Fairweather Fault and its splays the Yakutat block. The western edge of the Yakutat block converges against the Kenai Peninsula at a rate equal to the slip on the Fairweather Fault (Fig. 9). This convergence probably results in the oblique linear folds and thrust faults on the shelf west of Icy Bay (Plafker, 1967; Plafker et al, 1975; Bruns and Plafker, 1975; Rogers, 1977). The thrusting on Kayak Island and in its vicinity (Plafker, 1974) is probably part of the same sequence of oblique folds and thrusts, which may first have developed in the northwest and progressively spread to the southeast (Rogers, 1977).

The tectonic system described above cannot be projected back for a long period of time without conflicting with other observations. Hudson and others (1977 a, b) state that the distribution of Tertiary plutons and the distribution of metamorphic facies along the Fairweather Fault during the past 20 MY do not require large strike-slip displacements. This statement suggests doubt that most Pacific-North American plate motion has occurred along the Fairweather Fault system for a prolonged period of time. Plafker et al (1976) suggest that present high displacement rates along the fault began about 100,000 years ago and that before this time plate motion was taken up on faults seaward of the Fairweather. A fault that may have been active earlier is the inferred fault seaward of the Fairweather that might join the CB Fault as suggested by the linear bathymetric features between them (Fig. 4). Rogers also found it difficult to project the present tectonic system back in time, because it would require considerable subduction of the Yakutat block beneath the Alaskan continent. Bally (1977) has proposed such subduction for the Alps (A-Subduction), but the lack of particularly high mountains on the continental shelf at the western edge of the Yakutat block makes extensive A-subduction unconvincing.

A period of tectonism in the Yakutat block that preceded the present one is suggested in the reflection and refraction data. Reflection record B (Fig. 7) shows a deformed sedimentary sequence overlain by the little-deformed sequence. The lower deformed sequence probably corresponds to the thick units with velocities of 3.9 to 4.7 km/sec (refraction profiles LF27-28, Fig. 8) overlain by the normal shelf sedimentary sequence. Similar crustal structure is seen across the margin off Kodiak (Shor and von Huene, 1972), where a 4.5-km/sec layer is interpreted as sedimentary rock deformed along a convergent margin (von Huene, this volume) and overlain by much less deformed sediments of a forearc basin.

From these observations one can suggest some aspects of pre-late Cenozoic history, as well as a mechanism. As proposed by Grow and Atwater (1970) the relative motion between the Kula plate and the North American plate in the early Tertiary would have resulted in a subduction component along the entire eastern boundary of the Gulf of Alaska, creating a trench along the coast from the Queen Charlotte Islands to the junction with the Aleutian Trench. The northward movement of the Kula Ridge continuously shortened the trench from the south, replacing it with a transform fault, as the relative motion of the Pacific and North American plates is essentially tangential to this boundary. With disappearance of the Kula Plate, all of the former Queen Charlotte margin would have been converted from a trench to a transform fault, except for the short oblique section from Cross Sound to Kayak Island. However, the present trough along southeastern Alaska and British Columbia is probably not a vestige of the pre-30-MY-old subduction event because magnetic anomalies beneath the trough (anomalies 1 through 7) are younger than 30 MY. If the pre-30-MY trench was seaward of the transform faults as indicated in seismic records and the gravity models, it must have been rafted north on the transforming Pacific plate. This is consistent with termination of Paleozoic and Mesozoic terrains seaward of the British Columbia and southeastern Alaska coasts. Also terminated is the Chatham Strait Fault, along which the latest known large offsets are of early Cenozoic age (Ovenshine and Brew, 1972). In the present trough Pliocene and Pleistocene beds tilt landward but effects of subsidence from sediment loading cannot be easily separated from those of tectonic subsidence. Thus the trough may be a late Cenozoic feature and the pre-30-MY-old trench may have been rafted north and now be buried under younger sediment.

The layers of deformed sediments seen on the reflection records and the thick accretionary wedge shown on refraction profiles LF27-28 were probably formed by convergent tectonism. Fitch (1972) has proposed that an oblique subduction system is unstable, and that frictional drag will separate the orthogonal components into pure subduction normal to the trench and into strike-slip motion along one or more faults cutting the overlying slab. We suggest that the zone of dominant deformation has shifted landward through the seaward transform faults to the Fairweather Fault and others in the Yakutat block, to reach an eventual more stable configuration with pure transform motion along a fault parallel to the vector of relative motion of the two plates.

The present orogenic episode began in Miocene time in most areas of the Gulf of Alaska. Along the eastern Gulf Monger et al (1972) postulate dominantly transform motion since Miocene time. Along the central Denali Fault, Hickman et al (1977) identify the beginning of the present strike-slip movement in mid-Miocene time. In lower Cook Inlet, Fisher and Magoon (in press) identify the beginning of the present tectonic episode in late Oligocene to early Miocene time. In the Yakutat block, Plafker et al (1975) indicated that the present episode began perhaps in mid-Miocene time and that thrusting on Kayak Island occurred after mid-Miocene time. Kayak Island is implied to be in the oldest part of the convergent zone in the western Yakutat block (Rogers, 1977). The initial age of the present tectonic episode might also be indicated by the age of the unsubducted pre-Pliocene deep-sea fan at the foot of the Yakutat block boundary. Thus, if indications of an earlier episode of tectonism in the seismic data are correct, the present tectonic pattern involving the Yakutat block might have begun between mid-Miocene and Pliocene time. However, understanding rather than speculating about the pre-late Cenozoic plate history involves knowing the timing of major movement in 300-km-wide zone of mostly rugged and glacially covered terrain.

The foregoing has made some aspects of the ideas first proposed by Richter and Matson (1971) more specific. The implications of these ideas touch on the time-frequency analysis of great north Pacific earthquakes. Sykes (1971) made a convincing case that a great earthquake was overdue in the central Gulf of Alaska, if more frequent earthquakes on either side of this area were an indication of the earthquake recurrence interval across all of the Pacific-North American plate boundary. Page (1973) considers that the 1972 Sitka earthquake filled one of the gaps identified by Sykes. The basic premise is that if great earthquakes leave between them a similar zone in which the strain is not relieved, an earthquake should follow shortly. However, since the plate boundary along the central coast of the Gulf of Alaska is dissimilar to the plate boundaries on either side, the strain may be relieved differently and the recurrence interval for large earthquakes may differ accordingly.

REFERENCES CITED

Atwater, T. M., 1970, Implications of plate tectonics for the Cenozoic tectonic evolution of western North America: Geol. Soc. America Bull., v. 81, no. 12, p. 3513–3536.

Bally, A. W., 1977, Ancient and modern continental margins, perspectives for future research: AAPG Bull., v. 61, no. 5, p. 763.

Bayer, K. C., et al, 1977, Refraction studies between Icy Bay and Kayak Island, Eastern Gulf of Alaska: U.S. Geol. Survey Open-File Report 77-550, 29 p.

Beikman, H. M., 1975, Preliminary geologic map of southeastern Alaska: U.S. Geol. Survey Misc. Field Studies Map MF-673, 2 sheets, scale 1:1,000,000.

Bruns, T. R., and G. Plafker, 1975, Preliminary structural map of part of the offshore Gulf of Alaska Tertiary province: U.S. Geol. Survey Open-File Map 75-508, scale 1:500,000.

——— ——— 1976, Structural elements of offshore Gulf of Alaska Tertiary province: AAPG Bull., v. 60, no. 12, p. 2176.

Chase, R. L., and D. L. Tiffin, 1972, Queen Charlotte fault-zone, British Columbia: 24th Internat. Geol. Conf., Montreal, Proc., section 8, p. 17–27.

Chase, T. E., H. W. Menard, and J. Mammerickx, 1970, Bathymetry of the North Pacific, sheet 4: I.M.R. Tech. Rep. Series TR-9; University of California.

Dehlinger, P., et al, 1970, Northeast Pacific structure, *in* Maxwell, A. E., ed., The Sea: New York, Wiley Interscience Publishers, v. IV, p. 133–189.

Ewing, J. I., 1963, Elementary theory of seismic refraction and reflection measurements, *in* Hill, M. N., Ed., The Sea: New York, Wiley Interscience Publishers, v. III, p. 5–19.

Fisher, M., and L. Magoon, 1978, Geologic framework of Lower Cook Inlet, Alaska: AAPG Bull., v. 62, p. 373–402.

Fitch, T. J., 1972, Plate convergence, transcurrent faults, and internal deformation adjacent to Southeast Asia and the western Pacific: Jour. Geophys. Res., v. 77, no. 23, p. 4432–4460.

Gawthrop, W. H., et al, 1973, The southeast Alaska earthquakes of July, 1973: Trans. Am. Geophys. Union, v. 54, no. 11, p. 1136.

Grow, J. A., and T. M. Atwater, 1970, Mid-Tertiary transition in the Aleutian Arc; Geol. Soc. America Bull., v. 81, p. 3715–3722.

Hamilton, E. L., 1967, Marine geology of abyssal plains in the Gulf of Alaska: Jour. Geophys. Res., v. 72, no. 16, p. 4189.

Hickman, R., C. Craddock, and K. W. Sherwood, 1977, Structural geology of the Nenan River segment of the Denali fault system, central Alaska Range: Geol. Soc. America Bull., v. 88, p. 1217–1230.

Heirtzler, J. R., et al, 1968, Marine magnetic anomalies, geomagnetic field reversals and motions of the ocean floor and continents: Jour. Geophys. Res., v. 73, no. 6, p. 2119.

Hogan, L. G., et al, 1978, Biostratigraphic and tectonic implications of ^{40}Ar- ^{39}Ar dates of ash layers from the northeast Gulf of Alaska: Geol. Soc. America Bull., v. 89, no. 8, p. 1259–1264.

Hudson, Travis, George Plafker, and M. A. Lanphere, 1977, Intrusive rocks of the Yakutat-St. Elias area, south-central Alaska: Jour. Research, U.S. Geol. Survey, v. 5, no. 2, p. 155–172.

——— ——— and D. L. Turner, 1977, Metamorphic rocks of the Yakutat-St. Elias area, south-central Alaska: Jour. Research, U.S. Geol. Survey, v. 5, no. 2, p. 173–184.

Johnson, S. H., et al, 1972, Seismic refraction measurements in southeast Alaska and western British Columbia: Can. Jour. Earth Science, v. 9, no. 12, p. 1756–1765.

Kulm, L. D., et al, 1973, Initial reports of the Deep Sea Drilling Project: Wash., D.C., U.S. Government Printing Office, v. 18.

Larson, P. A., J. D. Mudie, and R. L. Larson, 1972, Magnetic anomalies and fracture-zone trends in the Gulf of California: Geol. Soc. America Bull., v. 83, p. 3361–3368.

Lawver, L. A., and J. W. Hawkins, Diffuse magnetic anomalies in marginal basins: Their possible tectonic and petrologic significance, Proceedings of the 25th International Geological Congress, Sydney, Australia, in press.

Milne, A. R., 1964, Two seismic refraction measurements: North Pacific basin and Dixon Entrance: Bull. Seismol. Soc. America, v. 54, p. 41–50.

Minster, J. R., et al, 1974, Numerical modeling of instantaneous plate tectonics: Royal Astron. Soc. Geophys. Jour., v. 36, p. 541–576

Molnia, B. F., P. R. Carlson, L. H. Wright, 1978, Geophysical Data from the 1975 Cruise of the NOAA Ship *Surveyor:* U.S. Geol. Survey Open-File Report 78-209.

Monger, J. W., J. G. Souther, and H. Gabrielse, 1972, Evolution of the Canadian Cordillera: A plate tectonic model: Amer. Jour. Sci., v. 272, p. 577–602.

Naugler, F. P., and J. M. Wageman, 1973, Gulf of Alaska: Magnetic anomalies, fracture zones, and plate interactions: Geol. Soc. America Bull., v. 84, no. 5, p. 1575.

Ovenshine, A. T. and D. A. Brew, 1972, Separation and history of the Chatham Strait Fault, southeast Alaska, North America: 24th Internat. Geol. Cong. (Montreal), Proc., sec. 3, p. 245–254.

Page, R. A., 1969, Late Cenozoic movement on the Fairweather Fault in southeastern Alaska: Geol. Soc. America Bull., v. 80, no. 9, p. 1873–1878.

——— 1973, The Sitka, Alaska earthquake of 1972—An expected visitor: U.S. Geol. Survey Earthquake Info. Bull., v. 5, no. 5, p. 4–9.

Plafker, George, 1967, Geologic map of the Gulf of Alaska Tertiary province: U.S. Geol. Survey Misc. Geol. Map I-484, scale 1:500,000.

——— 1971, Pacific margin Tertiary basin, *in* Future petroleum provinces of North America: AAPG Mem., 15, p. 120–135.

——— 1974, Preliminary geologic map of Kayak and Wingham Island, Alaska: U.S. Geol. Survey Open-File Map 74-82, scale 1:500,000.

——— 1976, Preliminary reconnaissance geologic map of the Yakutat and Mount St. Elias quadrangles, Alaska: U.S. Geol. Survey Open-File Map, scale 1:250,000.

——— T. R. Bruns, and R. A. Page, 1975, Interim report on Petroleum resource potential and geologic hazards in the outer continental shelf of the Gulf of Alaska Tertiary province: U.S. Geol. Survey Open-File Report 75-592, 74 p.

——— Travis Hudson, and Meyer Rubin, 1976, Late Holocene offset features along the Fairweather Fault: U.S. Geol. Survey Circ. 773, p. 57–58.

Richter, D. H., and N. A. Matson, Jr., 1971, Quaternary faulting in the eastern Alaska Range: Geol. Soc. America Bull., v. 82, p. 1529–1540.

Rogers, J. F., 1977, Implications of plate tectonics for offshore Gulf of Alaska petroleum exploration: Proc., 9th Annual Offshore Technology Conf., p. 11–16.

St. Amand, P., 1957, Geological and geophysical synthesis of the tectonics of portions of British Columbia: B.C. Dept. Mines and Petroleum Res. Bull., v. 54, p. 225.

Shor, G. G. Jr., 1962, Seismic refraction studies off the coast of Alaska: 1956–1957: Seismol. Soc. America Bull., v. 52, p. 37–57.

——— 1963, Refraction and reflection techniques and procedure: *in* Maxwell, A. E., ed., The Sea: New York, Wiley Interscience, v. III, p. 20–38.

——— 1965, Continental margins and island arcs of western North America: Geol. Survey of Canada Paper 66-15, p. 216–222.

——— and R. E. von Huene, 1972, Marine seismic refraction studies near Kodiak, Alaska: Geophysics, v. 37, p. 697–700.

Shouldice, D. H., 1971, Geology of the western Canadian continental shelf: Bull. Can. Petrol. Geology, v. 19, no. 2, p. 405–436.

Silver, E. A., R. E. von Huene, and J. K. Crouch, 1974, Tectonic significance of the Kodiak-Bowie Seamount chain, northeastern Pacific: Geology, v. 2, no. 3.

Southerland-Brown, A., 1968, Geology of the Queen Charlotte Islands, British Columbia: B.C. Dept. Rivers and Petroleum Res. Bull., 54, p. 225.

Stoneley, R., 1967, The structural development of the Gulf of Alaska sedimentary province in southern Alaska: Quart. Jour. Geol. Soc. London, v. 123, p. 25–57.

Sykes, L. R., 1971, Aftershock zones of great earthquakes, seismicity gaps, and earthquake prediction for Alaska and the Aleutians: Jour. Geophys. Research, v. 76, no. 32, p. 8021–8041.

Taylor, P. T., and N. S. O'Neill, 1974, Results of an aeromagnetic survey in the Gulf of Alaska: Jour. Geophys. Research, v. 79, no. 5, p. 719–723.

Tobin, D. G., and L. R. Sykes, 1968, Seismicity and tectonics of the northeast Pacific Ocean: Jour. Geophys. Research, v. 73, p. 3821–3845.

von Huene, R. E., G. G. Shor, Jr., and R. J. Malloy, 1972, Offshore tectonic features in the region affected by the 1964 Alaska earthquake, *in* The great Alaska earthquake of 1964: Oceanography and Coastal Engineering: Natl. Acad. Science Publ. 1605, p. 266–289.

——— et al, 1975, Seismic profiles of the offshore Gulf of Alaska Tertiary province, R/V THOMPSON, September–October, 1974: U.S. Geol. Survey Open-File Report 75-664.

Small Basin Margins

Variety of Margins and Deep Basins in the Mediterranean[1]

B. BIJU-DUVAL,[2] J. LETOUZEY,[2] AND L. MONTADERT[2]

Abstract The various Mediterranean basins are similar in that all contain a Pliocene-Pleistocene clastic sequence, a late Miocene evaporite sequence, and a deep-water pre-late Miocene sequence beneath the evaporites. The basins differ, however, in age and genesis. Western, Tyrrhenian, and Aegean Basins are relatively young (Oligocene to Holocene) marginal types, whereas the Eastern Mediterranean is a remnant of an early Mesozoic ocean with its southern continental margins. The Black Sea is thought to be a back-arc basin possibly as old as Late Cretaceous.

INTRODUCTION

Our purpose is to review the geology of the various Mediterranean basins. To do this, we prepared cross-sections showing present geological structure. To justify our interpretations we evoke the genesis and the evolution of the basins and their margins through geological time.

Synthesis of a considerable amount of land data, and results of extensive marine geophysical surveys and of the Deep Sea Drilling Project (legs 13 and 42) form the basis of this review of the problems, origin, age and structure of Mediterranean basins. This paper deals with the general geological framework, emphasizing similarities and dissimilarities between the basins from the Atlantic to the Black Sea (Fig. 1). Publications by Ryan et al (1973), Mulder (1973), Finetti and Morelli (1973), Biju-Duval et al (1974), Finetti (1976), Letouzey et al (1977), Biju-Duval et al (1978), and Hsu et al (1978) contribute heavily to our synthesis.

To understand the genesis of these basins one needs to study their formation together with the evolution of the Alpine system as attempted by numerous geologists since Argand (Dewey et al, 1973; Boccaletti et al, 1974; Laubscher and Bernoulli, 1977; Biju-Duval et al, 1977).

WESTERN MEDITERRANEAN BASIN

Sedimentary Sequence

Geological cross-sections (Figs. 2, 3) show a thick sedimentary sequence (>7 km) in the whole Western Basin. The widespread extension of late Miocene evaporites and/or laterally equivalent erosional surface (Montadert et al, 1970; Alla et al, 1972; Mauffret et al, 1973; Montadert et al, 1978) permits definition of three main sequences:

Pliocene-Quaternary deposits—Thickness exceeds 500 in the deep basin; it is less on the Balearic margins and west of Sardinia. Major influxes from the Ebro and Rhone deltas considerably increase the sediment thickness in the North Balearic-Provençal Basin.

Fine-grained clastics (hemipelagic and turbiditic) sediments are dominant throughout. Extensive halokinesis disturbs the Pliocene-Quaternary layers, especially east of the basin along Corsica and Sardinia.

Pliocene-Pleistocene vertical movements are the most recent effect of the subsidence of the basin. Existence of several prograding subunits, regular tilting of the margin, and faulting give evidence of subsidence up to 2,000 m. Cooling of new lithosphere of early Miocene age and loading by active sedimentation probably caused the subsidence.

Messinian evaporites—Messinian evaporites, which are distributed throughout the whole Mediterranean (Fig. 8), can be divided into two main evaporitic sequences in the Western Basin (Montadert et al, 1970; Finetti and Morelli, 1972; Mauffret et al, 1973; Mulder, 1973; Biju-Duval et al, 1974; Montadert et al, 1978):

1. A thick (0.5 to 1.5 km) seismically homogeneous layer of salt which fills the main depression of

[1]Manuscript received, March 3, 1978; accepted, June 8, 1978.

[2]Division Geologie, Institut Francais du Petrole, Rueil Malmaison, France.

We are very indebted to numerous colleagues of IFP, Total, SNEA (P), CNEXO and of several French universities working in the Mediterranean with whom we had many helpful discussions; we also acknowledge the work of geophysicists from IFP, who made the surveys. We are very grateful to J. Watkins for reviewing and editing this paper.

Article Identification Number:
0065-731X/78/MO29-0020/$03.00/0.
(see copyright notice, front of book)

FIG. 1. Structural map

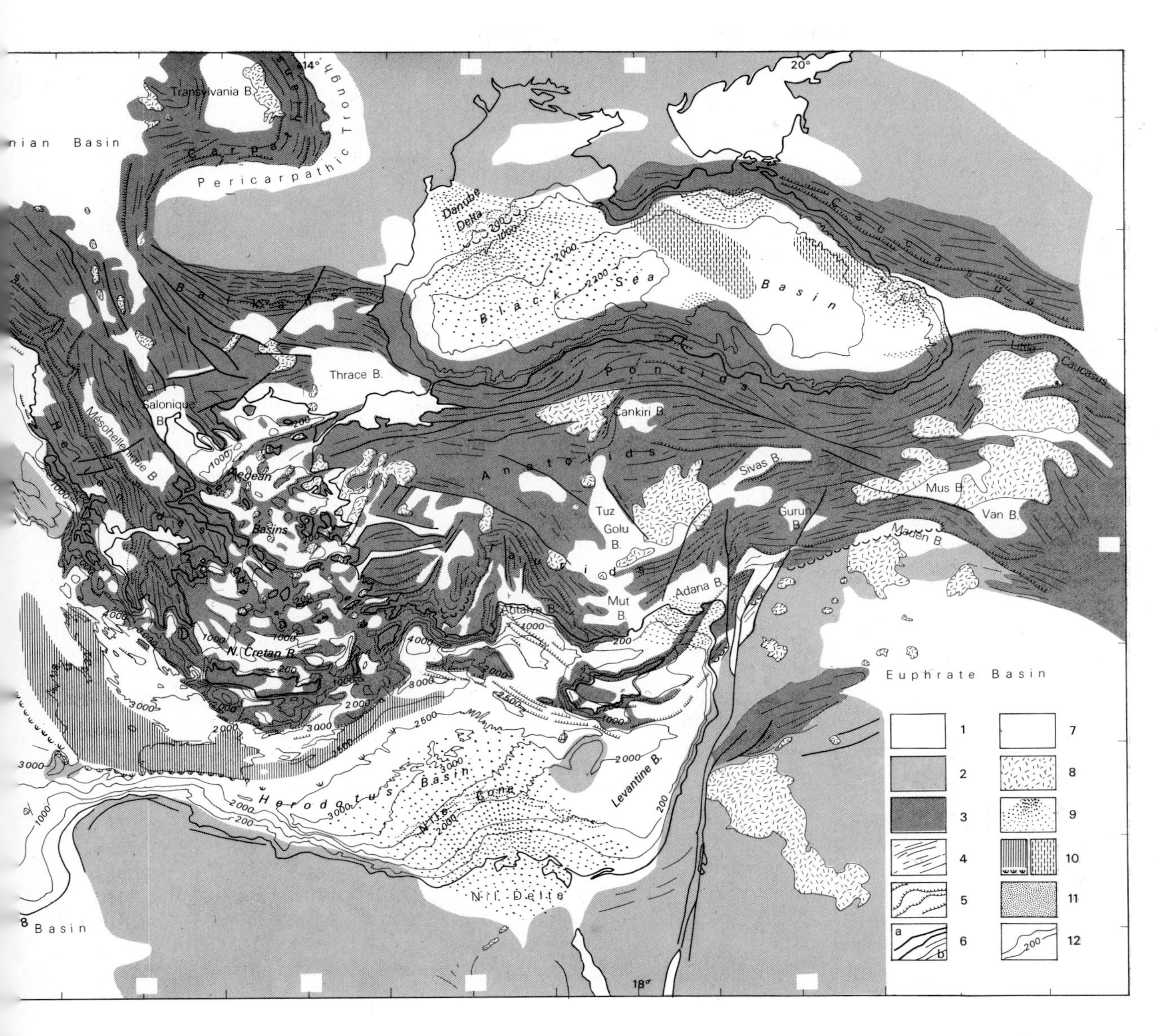

7 Front of gravity slidings 8 Volcanism 9 Major recent deltas 10 Mediterranean Ridge – Central Black Sea High 11 Tyrrhenian abyssal plain 12 Bathymetry

of the Mediterranean area.

the basin. Lateral extent of the salt more or less corresponds to the boundaries of the present abyssal plain. This layer is characterized by high velocities (4.5 km/sec) and is the source of the diapiric salt. The exact nature (halite, anhydrite, potash ?) remains unconfirmed. In some cases, intermediate reflectors appear in this member. Strong basal reflectors may indicate basal evaporites or may be limestones as known on land in Sicily.

2. An overlying upper evaporitic layer, with a maximum thickness of 0.5 to 0.6 km. Seismic profiles show that the upper evaporites overlap the salt layer and extend much farther onto the marginal areas such as Gulf of Valencia, Ligurian Sea. Drilling of this layer on legs 13 and 42A showed that it consists of interbedded evaporites (anhydrite, gypsum) and dolomitic marls (K. Hsu et al, 1973; Garrison et al, 1978). Thin salt layers are present locally.

A well-defined erosion surface can be identified on the margins landward as far as the pinching out of the salt layer. This erosion surface is unconformably overlain by the upper evaporites, which fill the topography. Sedimentary wedges in the region may be products of the erosional process (Montadert et al, 1978). This erosion surface has been recognized on land as the "Pontian" episode (Ryan et al, 1973; Clauzon, 1975) and has been a major feature in the more recent interpretation of the Messinian event (K. Hsu et al, 1978).

Pre-evaporitic (Neogene) deposits—Seismic profiles show a thick sequence (up to 3 to 4 km) of undeformed horizontal sediments below the evaporites in the deep basin (Mauffret et al, 1973). Because of the strong first multiples, acoustic basement generally is not seen in the deep basin except near margins.

Oil wells and DSDP holes near the margins give the following data on these sediments. In the North Balearic-Provençal Basin only Miocene beds have been found; Aquitanian has been found in the shelf of the Gulf of Lion (Cravatte et al, 1974) and lower Burdigalian in the eastern Menorca rise (DSDP hole 372; K. Hsu et al, 1978). Oligocene deposits may exist in margin grabens. In the South Balearic-Alboran Basin and north of Algeria the Burdigalian has been reached by drilling (Burollet et al, 1978). Because of structural conditions (as discussed later) many of the authors think it is the oldest age of this southern basin. Oligocene deposits, which appear restricted mainly to the rifted margin, could be continental clastics and evaporites. On the Provence margin, Oligocene cross-bedded sandstones have been sampled during diving (Bellaiche et al, 1977). They show an initial marine transgression in late Oligocene (Chattian). Seismic data suggest that continental deposits of DSDP borehole 134 on the Sardinia margin (Ryan et al, 1973) could be Oligocene (Mauffret et al, 1978).

Intrusions cut small and very thick sedimentary basin fill in the Alboran Sea. These intrusions may be due to Messinian salt diapirism (Auzende et al, 1975), mud diapirism from a pre-Neogene (?) layer (Mulder and Parry, 1977), or volcanism (Biju-Duval et al, 1978). DSDP hole 121 reached late Miocene rocks (Montenat et al, 1975) but older deposits probably exist at depth. DSDP hole 372 showed Miocene sediments deposited in a deep (around 1,500 m) basin before evaporite deposition (K. Hsu et al, 1978).

Structure, origin and evolution—We distinguish two different subbasins on the basis of geological history deduced from land geology; the North Balearic-Provençal Basin and the south Balearic-Algerian Basin. The North Balearic-Provençal Basin was created on a pre-existing craton (Catalonia, Provence, Sardinia; i.e., a part of the Iberia continental plate). Beneath the abyssal plain geophysical data indicate a thin crust. There also are anomalies in upper mantle velocities (Fahlquist and Hersey, 1969; Hirn et al, 1976; Morelli et al, 1977). The limited size of the oceanic basin has led to controversy about its origin but we think that it was created by rifting during Oligocene-Aquitanian, followed by spreading in Aquitanian-early Burdigalian times (Fig. 9). The geometry of the initial fit before rotation of the Corsica-Sardinia block remains uncertain (Alvarez, 1972; Auzende et al, 1973; Bayer et al, 1973; Biju-Duval et al, 1974, 1978).

In the Gulf of Lion west of Sardinia and northwest of Corsica, the continental basement extends far under the abyssal plain. The margins look like rifted margins. In some cases steep slopes may be related to strike-slip or transform motion (Provence, Ligurian Sea, east of Menorca). Magnetic anomalies and volcanism are closely related. As magmatic activity has been intermittent since the Aquitanian, it is difficult to determine linear and symmetric magnetic anomalies (Gonnard et al, 1975) but fan-shaped linear magnetic anomalies (Bayer et al, 1973) between Menorca and Tunisia correspond to the southern limit of the oceanic basin created by drifting of Corsica-Sardinia and thus to the boundary with the South Balearic-Algerian Basin.

The North Balearic-Provençal Basin could be a marginal basin related to an Apennine subduction zone to the east, active during Oligocene-early Miocene. The spreading axis jumped to the Tyrrhenian Sea during early Miocene, thereby separating Calabria and northern Sicily from Corsica-Sardinia. Alternatively, this rifted-drifted basin could be related to intraplate tensional tectonics known during Oligocene time in southwest Europe, related to collision between Africa and Europe.

The Valencia trough is probably an aborted rifted arm of the Provençal Basin (Mauffret, 1976), where attenuation of the crust and subsidence occurred (Hinz, 1972,1973). Magnetic anomalies in the area derive from volcanism active between Aquitanian and Recent times.

The South Balearic-Algerian Basin extends from Gibraltar to Sicily. The deep basin is entirely superimposed on a folded belt of Eocene to Miocene age. It is now located between the Betic range (north) and the Maghrebian range (south) that are still tectonically active today. In the Alboran area, characterized by the absence of thick evaporites and the occurrence of numerous volcanic highs, the nature

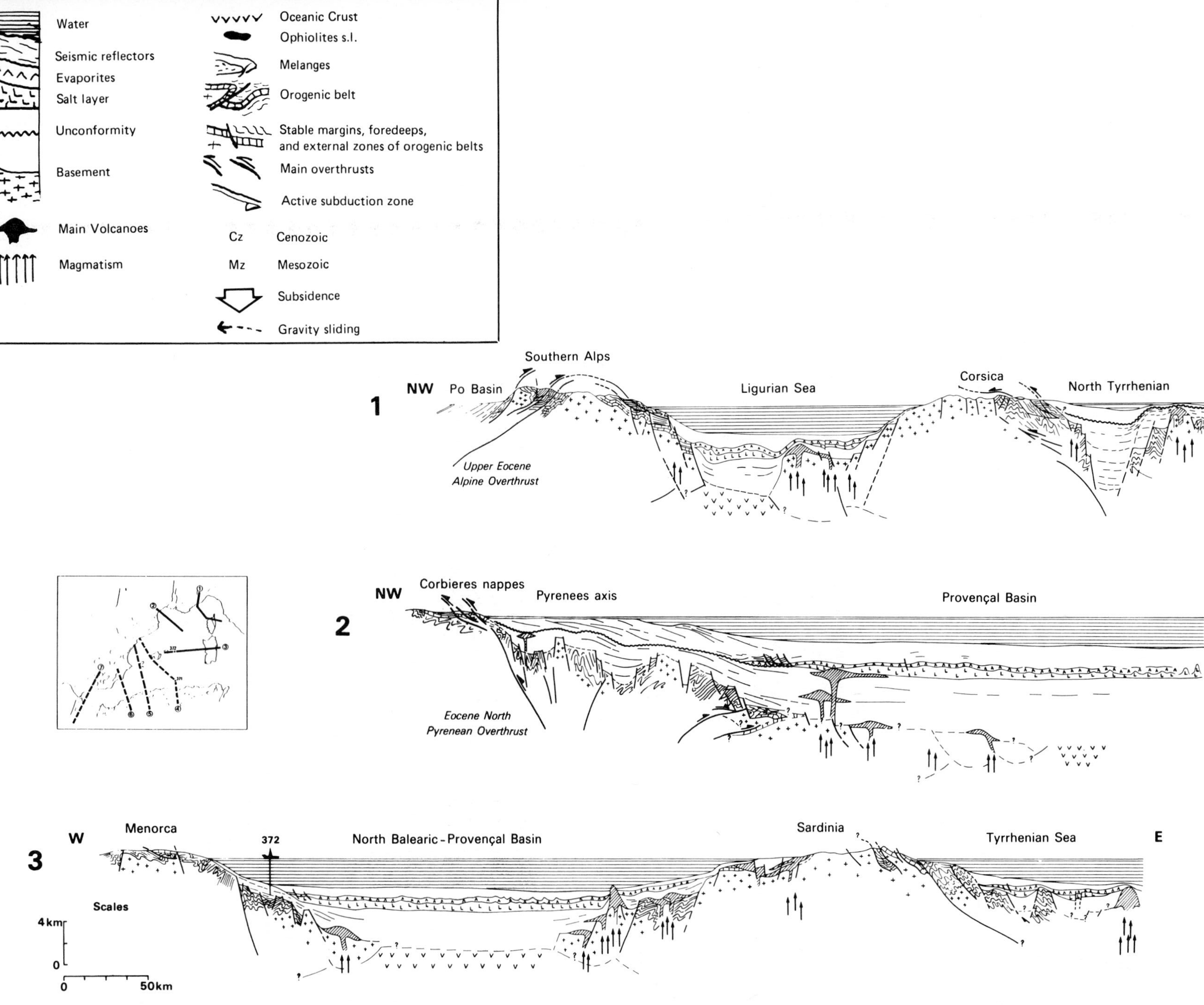

FIG. 2. Interpretive cross sections (north Balearic-Provençal basin).

of the crust is ambiguous and not clearly oceanic (Hatzfeld, 1976).

The basin corresponds to a small oceanic basin with numerous magnetic anomalies associated with volcanics emplaced from Burdigalian to Tortonian time (Gonnard et al, 1975; Bellon and Letouzey, 1977). Northward, the south Menorcan marginal plateau (Mauffret, 1976) is magnetically quiet and is separated from North Africa by a narrow basin where volcanism also was important. Northeastward, the relationship between South and North Balearic Basins is not well understood. Linear magnetic anomalies run from Menorca to southwest Sardinia but recent and relatively uniform infilling masks the deep structure between these two parts of the Western Basin.

On both sides of the South Balearic-Algerian Basin, the Betic and Maghrebian Ranges were continuously active since Eocene times (Durand-Delga, 1969; Paquet, 1974). The writers propose (Fig. 10) the following evolution: (1) A small Mesozoic arm of the Tethys running south of Rif (Morocco) to Calabria (south of Italy) closed during the late Eocene with thrusting of continental detached fragments (Rif, Kabylie, Peloritain) onto Africa; (2) An Oligocene episutural basin formed (see Bally, 1975) with deposition of Numidian flysch; (3) during early Miocene, after compression and ejection of the Numidian flysch, the basin began to take its present configuration with subsidence and volcanic activity; and (4) during late Miocene, compression on both sides of the basin resulted in thrusting and gravity sliding.

THE TYRRHENIAN BASINS

Sedimentary Sequence

The interpretive geological cross-sections of Figures 2 and 4 show significant differences between peri-Tyrrhenian basins located along the shelves and within the abyssal plain basin. The small peri-Tyrrhenian basins trap relatively thick sedimentary series. The stratigraphy is not well known because seismic profiles lack borehole calibration. Pliocene-Quaternary deposits vary enormously with local thickening and internal unconformities. The late Miocene evaporitic sequence is not well developed except in an elongated basin east of Sardinia where a local salt layer can be observed. Eocene (and Oligocene?) sediments of unknown facies fill the deepest parts of the grabens. Land data suggest that western basins along Corsica and Sardinia are older than Sicilian or Apenninic basins.

Under the abyssal plain, as observed in seismic profiles, the sedimentary sequence is limited to thin Pliocene-Pleistocene deposits (190 m in site DSDP 132) and thin evaporites (Ryan et al, 1973; Selli, 1974). Volcanism occurs in the whole area, with the well-known present activity on Eolian Island. Magnetic anomalies correspond to older volcanoes whose age can be at least late Miocene as shown by leg 42B data (Hsu et al, 1978). Volcaniclastics screen pre-late Miocene deposits in the central part of the basin. The volcaniclastics consist of tuffs and traps of late Miocene age. Underlying ancient sedimentary deposits intruded by volcanics seem likely.

Nature, Origin and Age of the Basins

Geophysical data and DSDP hole 373 show that the deep basin crust is oceanic (Hsu et al, 1978), but dredgings of sialic material (Selli and Fabbri, 1971) suggest local areas of continental crust which have been intruded and subsided. Selli and Fabbri (1971) suggested that Pliocene foundering could have been greater than 2 km. Thus, we consider two different realms: the first, to the south in the abyssal plain with typical oceanic crust, considered to be a back-arc basin related to a subduction process under Calabria (Elter et al, 1975), and the second, to the north, with a more complicated history.

The genesis of the Tyrrhenian Sea is difficult to reconstruct because we know only the more recent events of its long evolution. Tentatively, we suggest the following evolution:

1. There was a proto-Tyrrhenian marginal basin, connected with the South Balearic-Algerian Basin lying north of an Eocene subduction zone from Gibraltar to Calabria (Fig. 10) with associated volcanics (Wezel, 1974, 1977; Bellon, 1976; Bellon and Letouzey, 1977). The southeastern corner of the Iberian plate has been tectonically active since at least the Late Cretaceous (Dubois, 1970; Bousquet, 1972).
2. During Neogene, subduction stopped in front of the Apennines and Maghrebian ranges because of continental collision with Apulia and Africa, respectively. In front of the Ionian oceanic realm, subduction continued with a jump of the back-arc basin from the Western Basin to the Tyrrhenian, separating Calabria-northern Sicily from Corsica-Sardinia, in a way similar to that proposed by Alvarez et al (1974; and Fig. 11). Thus the peri-Tyrrhenian basins are thought to be of the rifted type along Corsica-Sardinia, created during the Oligocene-Miocene, and of the inter-arc type behind the Apennine-Calabria subduction zone, created during the Miocene-Pliocene. The deep basin can be divided into a new oceanic back-arc basin in the southern part, of late Miocene to Quaternary age, superimposed on an older one (late Eocene to Oligocene), and an attenuated continental zone in the northern part with aborted rifted basins of Oligocene age.

ADRIATIC BASINS

The Adriatic is a shallow area and is considered an offshore extension of the Apulian carbonate platform between Apennines and Dinarids. This area includes two main basins, the north Adriatic Basin infilled by the Po sediments, and the south Adriatic Basin where water depths do not exceed 1,250 m. The thick sedimentary section contains rocks of Triassic to Holocene age. As this area was probably an extension of Africa during Paleozoic times, Paleozoic sediments may underlie the younger section.

The sequence from Triassic to Neogene was marked by lateral variations between platform carbonate deposits and more pelagic deposits which de-

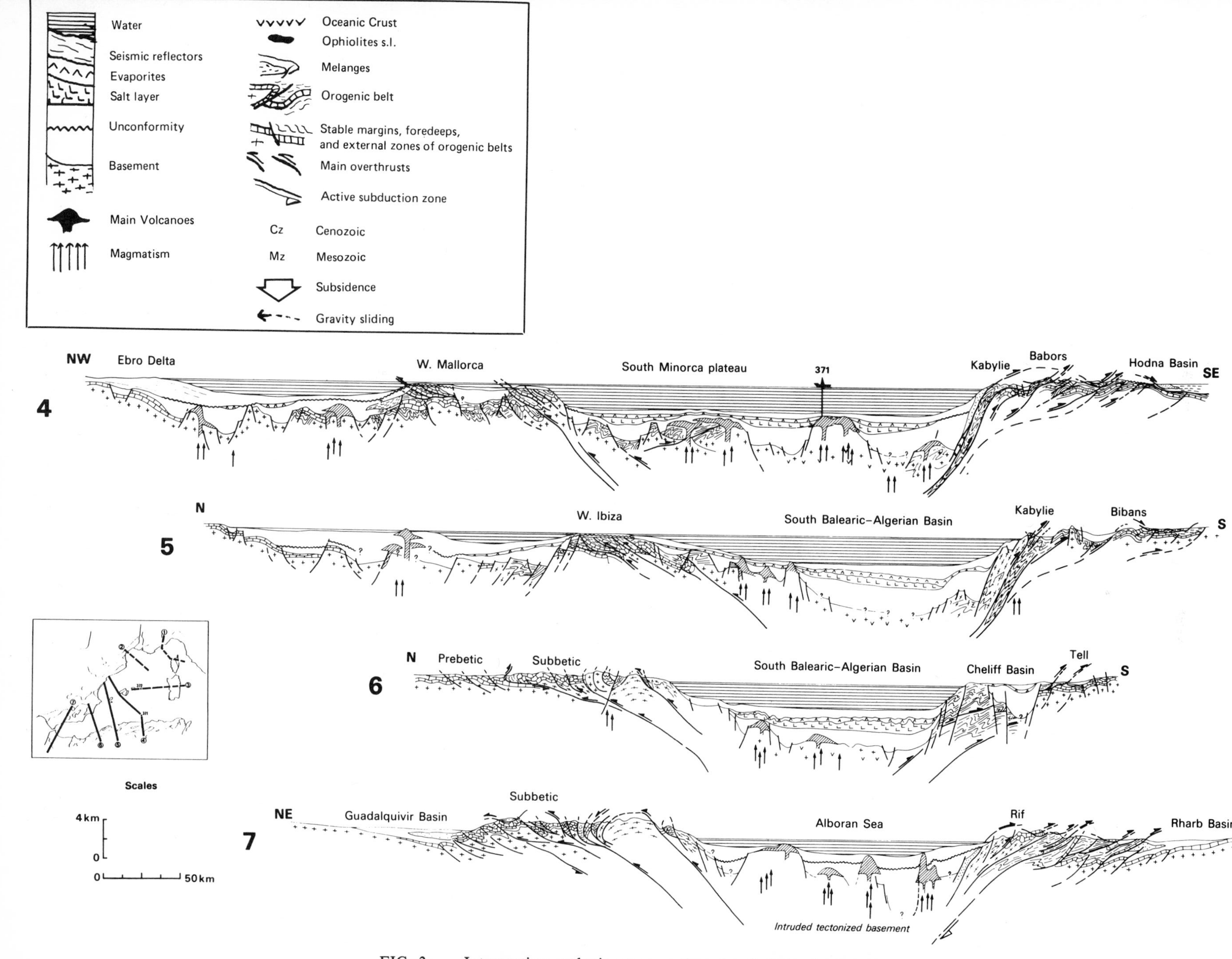

FIG. 3. Interpretive geologic cross sections (south Balearic-Algerian basin).

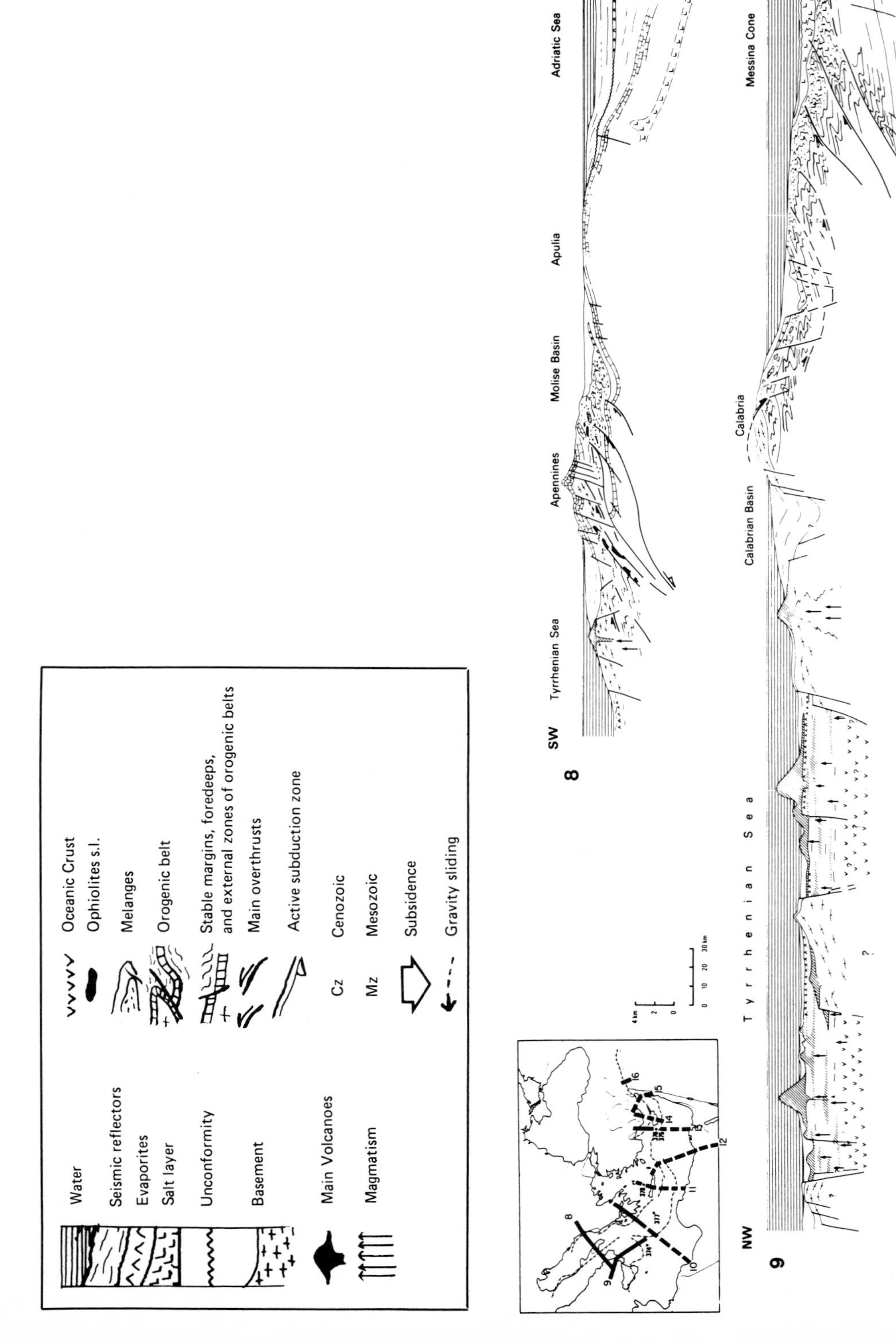

FIG. 4. Interpretive cross sections (central Mediterranean).

termine possible targets for offshore exploration (D'Argenio et al, 1971). Cretaceous reef development and diapirism from salt of Triassic or Jurassic age are evident. Paleogeography and facies distribution are well known from drilling. The Messinian event occurred here as elsewhere in the Mediterranean. Evaporites occur in Italy and can be observed on seismic lines. They are generally very thin, pinching out on an erosional surface or onlapping an erosional unconformity.

A thick sequence of Pliocene-Quaternary clastics in the Po delta infill the basin with more than 7 km of deposits. Prograding sediments infill the Albania Basin. Compression with local gravity sliding and olistostromes is evident along the Apennines up through the Quaternary.

EASTERN MEDITERRANEAN

Ionian, Herodotus, and Levantine Basins and African Margins

South of the Alpine belt (Calabrian Arc, Aegean Arc, and Cyprus Arc) and superimposed southward on the stable African margins, the eastern Mediterranean is divided into several subbasins. We discuss here only the deep basins and African margin. Basins associated with the active margin will be described later.

Before analyzing these eastern Mediterranean basins, we must briefly emphasize some facts about the Siculo-Tunisian platform (or Pelagian Sea), which is an extension of the African continent (the Ragusa platform) with a complete sedimentary sequence from Paleozoic to Holocene. Grabens running northwest to southeast separate the Siculo-Tunisian platform from the Straits of Sicily. There was extensive volcanism here from late Miocene to Quaternary times. The troughs now are filled by recent sediments. They probably are related to a rift system developed in an epicontinental sea. In the Pelagian Sea and in the Gulf of Syrte, exploration wells drilled on the carbonate platform are dry in the Sicily-Malta area but productive elsewhere, for example, in Eocene or Cretaceous limestones like in Ashtart and Isis oil fields (Byramjee et al, 1975).

Sedimentary Sequence

Thickness of sediments is very great in the whole area (Figs. 5, 6). As in the Western basin, late Miocene evaporites (Fig. 8) divide the sequence into three main segments—Pliocene-Quaternary, late Miocene evaporites, and the pre-evaporite sequence.

In the Ionian abyssal plain (as in the Pelagian Sea), low continental influx of Pliocene-Quaternary hemipelagic unconsolidated sediments (Hinz, 1974; Byramjee et al, 1975; Figs. 4, 5) limited thicknesses to 200 to 400 m. In contrast with the Ionian abyssal plain, an enormous terrigenous influx from the Nile River (Ryan et al, 1973, 1974) characterizes the Herodotus and Levantine basins. Ross and Uchupi (1977) have recently shown that a previous distinction between two offshore fans, the Rosetta and Damietta, is not valid.

Sedimentary thickness ranges from 4,000 m near the shoreline (with control by boreholes) to 2,000 m in the central part of the Herodotus abyssal plain. In the Levantine basin recent sediments (500 m) pinch out in the northwest on a former structure, the Erathosthenes plateau.

Coarse Nile sediments were found on the eastern part of the Mediterranean ridge (DSDP hole 130) where Pliocene-Quaternary series gradually thin. This shows recent uplift of the ridge relative to the abyssal plain (Ryan et al, 1973). In the upper part of the cone, growth faults and diapirism (mud diapirism?) are observed whereas in the lower part and in the Herodotus abyssal plain, Messinian salt gliding, domes, and collapses result in an irregular bottom topography.

The Nile delta development began in late Eocene-Oligocene (Salem, 1976; Biju-Duval et al, 1975); during early Miocene the Nile delta migrated to its present position at about the same time as uplift and the opening of the Red Sea.

The Messinian evaporites are widely known in the deep basins and surroundings (Biju-Duval et al, 1974; Mulder et al, 1975; geological map IFP-CNEXO, 1974). The thick salt layer is marked by high velocities, diapirism, and dissolution features. In some areas, the salt member contains internal reflectors. As in onshore basins where late Miocene evaporites have different facies (sulfates, chlorides, phosphates, carbonates, pelagic marls and diatomites, terrigenous layers, and even biostromal limestones), deep-basin evaporites probably present many local variations. Some lateral facies changes can be seen in seismic data (Montadert et al, 1978) and DSDP data. Ross and Uchupi (1977) suggested possible correlation between the evaporites of the deep basin and a middle to late Miocene prograding carbonate sequence or an erosional surface in the nearshore area.

Evaporite thickness can range from 1,000 m to more than 2,000 m, and the bottom of the salt layer can reach 8,000 m in depth under the Herodotus abyssal plain. Thickness varies widely, especially in front of the Alpine belt. On the Messina cone, thickening salt towards the northwest is probably responsible for disturbances and gliding in front of the northern active margins (Calabrian, and also Aegean and Cyprus Arcs). Thinning of evaporites on the southern edge of the Mediterranean ridge is probably due to a pre-existing high. On the ridge, thickness varies with substantial accumulations within small salt basins (see Fig. 8). The salt pinches out on the margins, where only thin interbedded evaporites are known. The formation of an extensive erosional drainage basin (Said, 1975; Girtzman and Buckbinder, 1977) should result in thick clastic accumulations. These have not been found but may be interbedded with evaporites.

Beneath the evaporites, seismic reflection and refraction data (Lort, 1972; Finetti and Morelli, 1973; Sancho, et al, 1973; Biju-Duval et al, 1974; Hinz, 1974; Wright et al, 1975) indicate a thick sedimentary pile with, in many cases, a velocity inversion just under the evaporites. There is no direct evi-

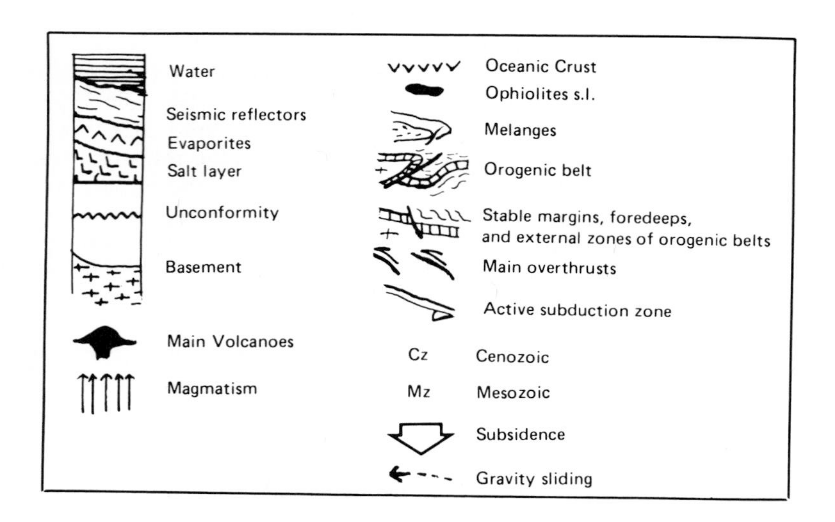

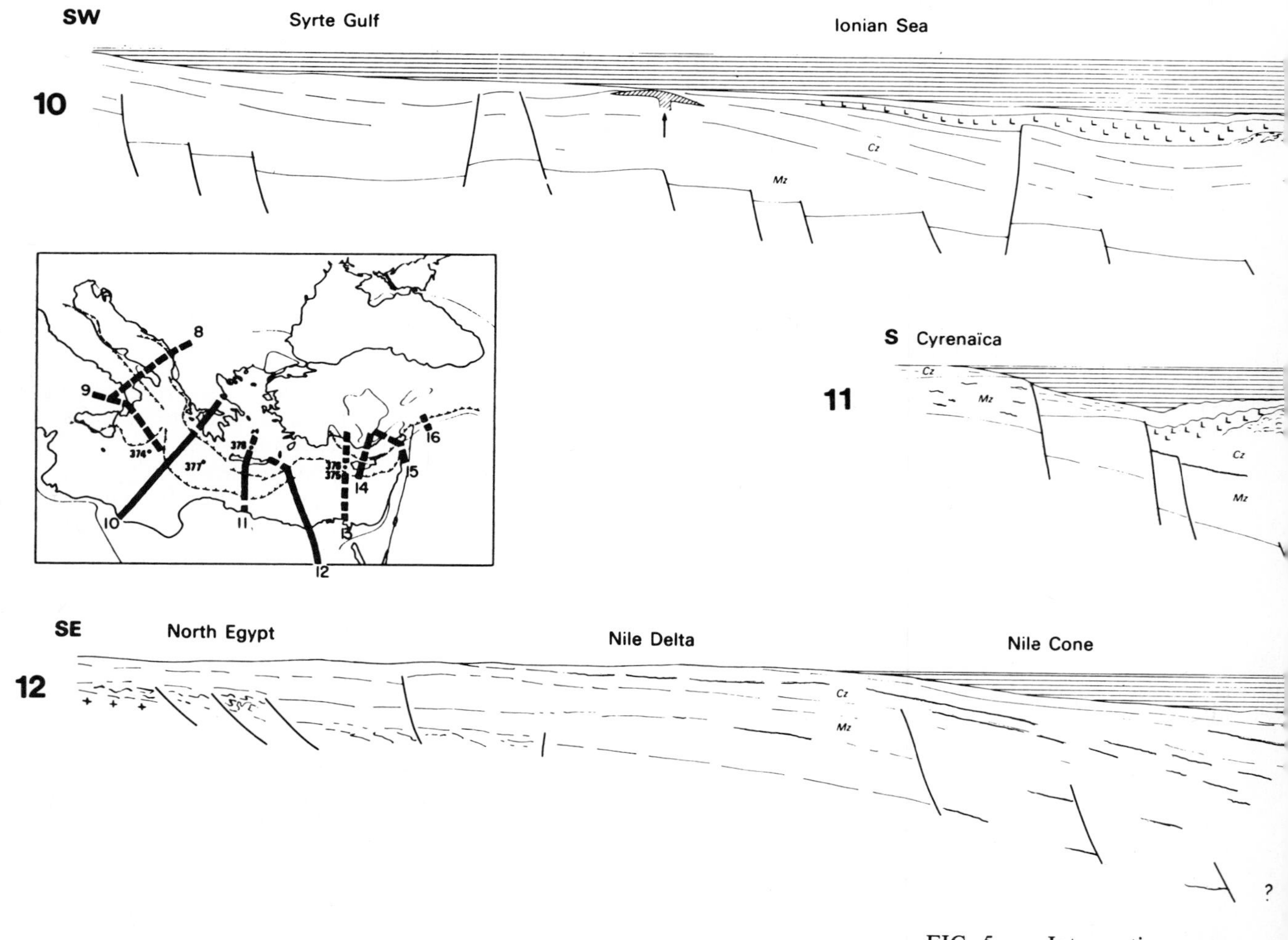

FIG. 5. Interpretive cross

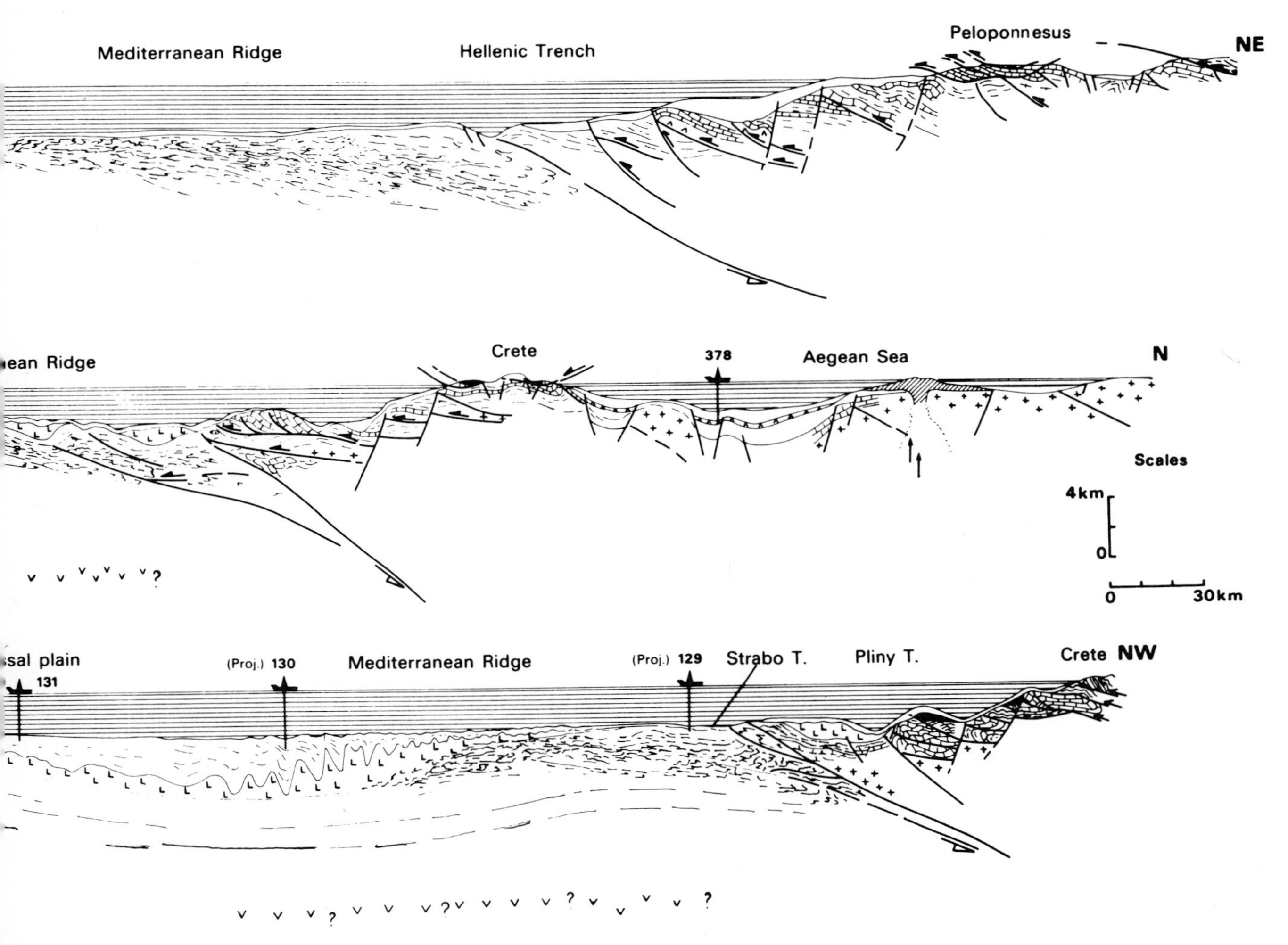

sections (eastern Mediterranean).

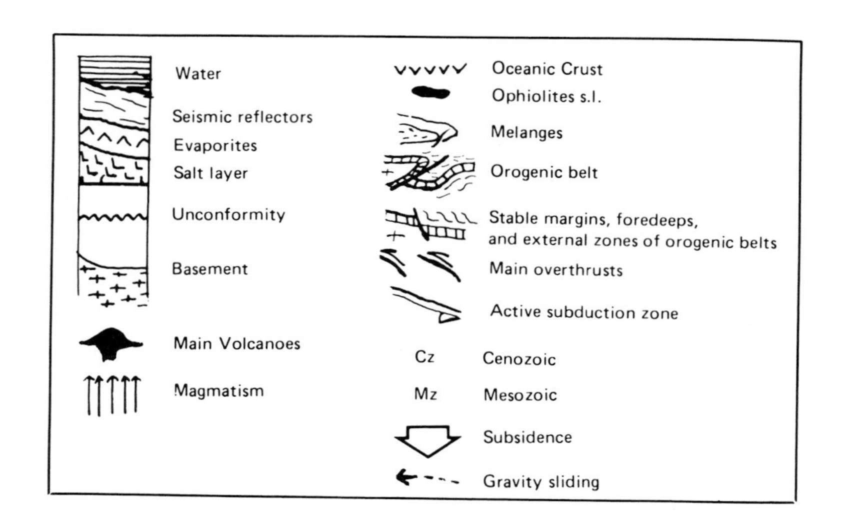

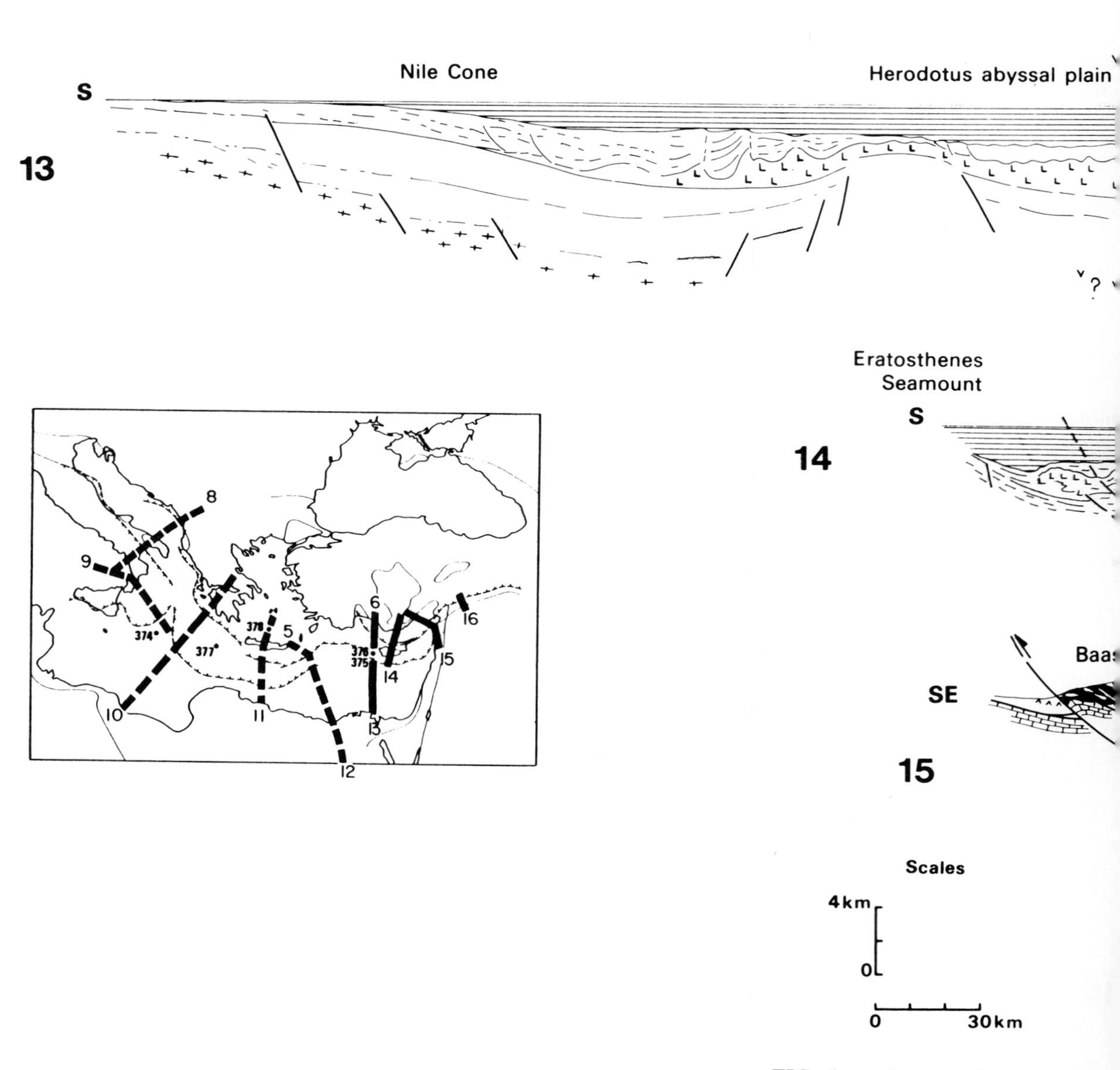

FIG. 6. Interpretive cross

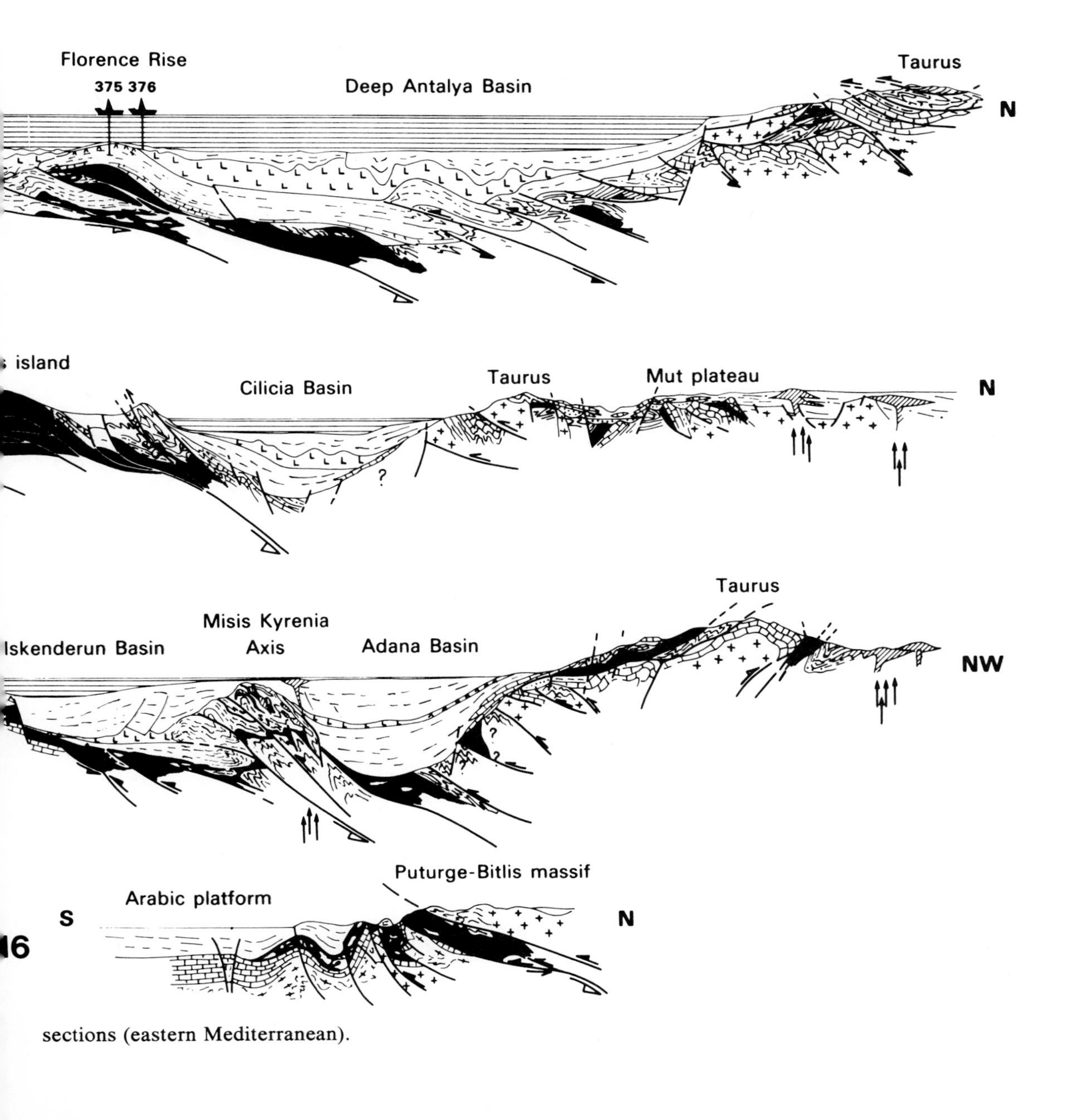

sections (eastern Mediterranean).

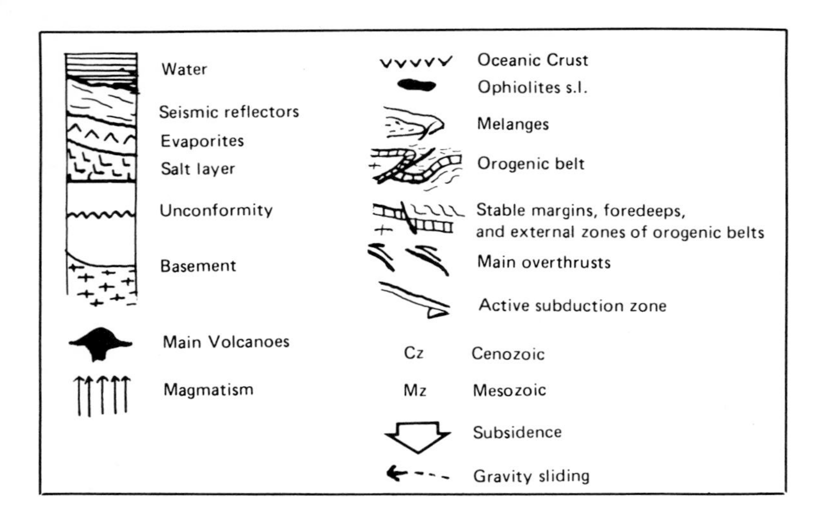

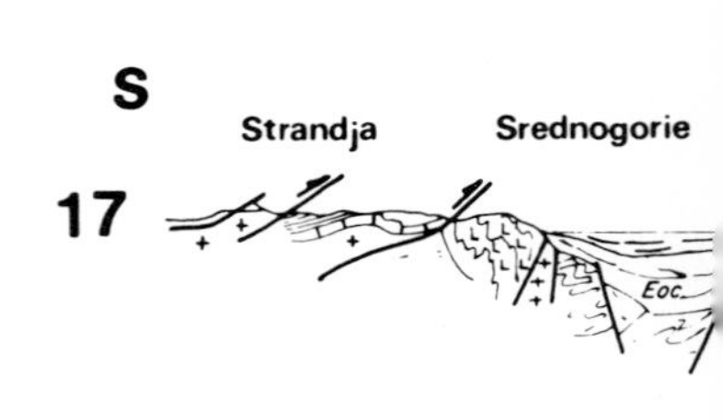

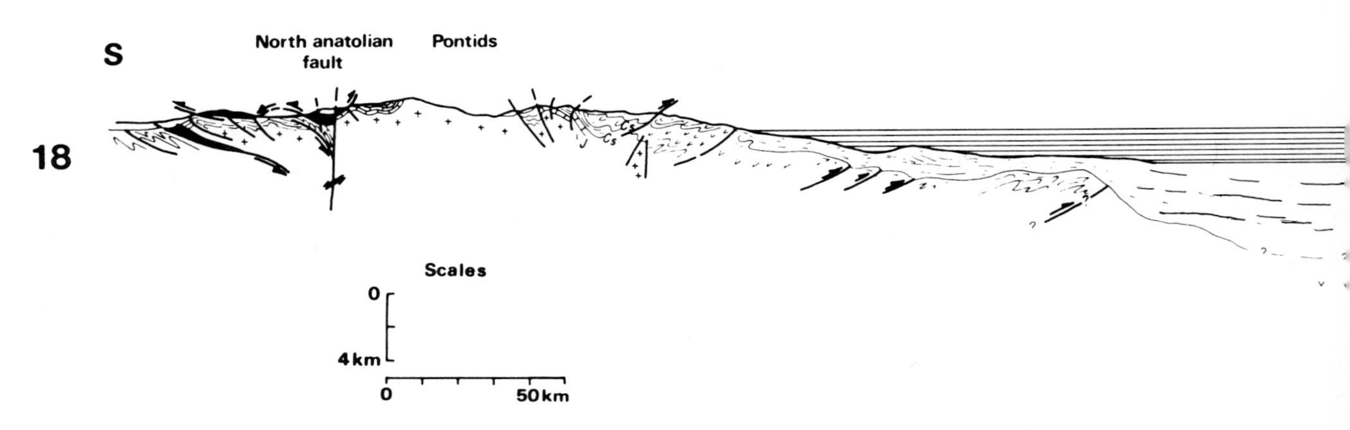

FIG. 7. Interpretive cross

dence of the age or lithology of pre-evaporitic sequences in the deep basins. However, results of DSDP boreholes 376 (Florence rise, north of the Herodotus basin), 125 and 377 (on the Mediterranean ridge), geology of nearby margins, and correlations with seismic reflection profiles suggest the following.

In contrast to the Western Basin where only a complete Neogene (+ Oligocene?) sequence has been recognized, here the series below the evaporites consists of Neogene deposits, Paleogene and Mesozoic sediments, and perhaps African Paleozoic rocks on the southern margin. On the African margin, facies changes and the progradational nature of the sequence indicate that a deep basin has existed since Jurassic time (Klitzsch, 1970; Ginzburg et al, 1975). The extent of Paleozoic Gondwanian series, and the distribution of Mesozoic-Paleogene facies (platform neritic or pelagic?) is not yet defined.

Structure, Origin and Age of the Eastern Mediterranean

Structure and genesis of the eastern Mediterranean basins remain controversial. Did they recently founder as thought by Sancho et al, (1973), Burollet and Byramjee (1974), and Byramjee et al, (1975)? If so,the Apulian carbonate platform and Tauric carbonate platform (central Turkey) would be continental extensions of the African-Arabian block (Burollet and Byramjee, 1974; Channel and Horvath, 1976; Ricou et al, 1974, 1975) and eastern Mediterranean basins would have recently collapsed.

The limited size of the deep basins has led to considerable controversy about the nature of the crust. Deep areas are complex and difficult to clearly characterize as oceanic or continental, the Moho being at around 20 km (16 to 24 km; see Lort, 1972; Weigel, 1972; Hinz, 1974; Woodside, 1975; Wright et al, 1975). As subduction is active northward beneath the Hellenic Arc and northwestward beneath the Calabrian Arc, the Ionian, Herodotus, and Levantine basis may be relics of earlier, broader oceanic basins. We wonder how to interpret the thinning of the continental crust. Is there a continental margin linked to an oceanic basin? What is the age of formation of this margin?

In contrast with Western and Tyrrhenian basins, few magnetic anomalies are recorded in the deep basins (Vogt and Higgs, 1969; Woodside, 1975; Neev et al, 1976) and the magnetism is restricted areally—e.g. Syrte Gulf, Cyrenian plateau, and Erathosth-

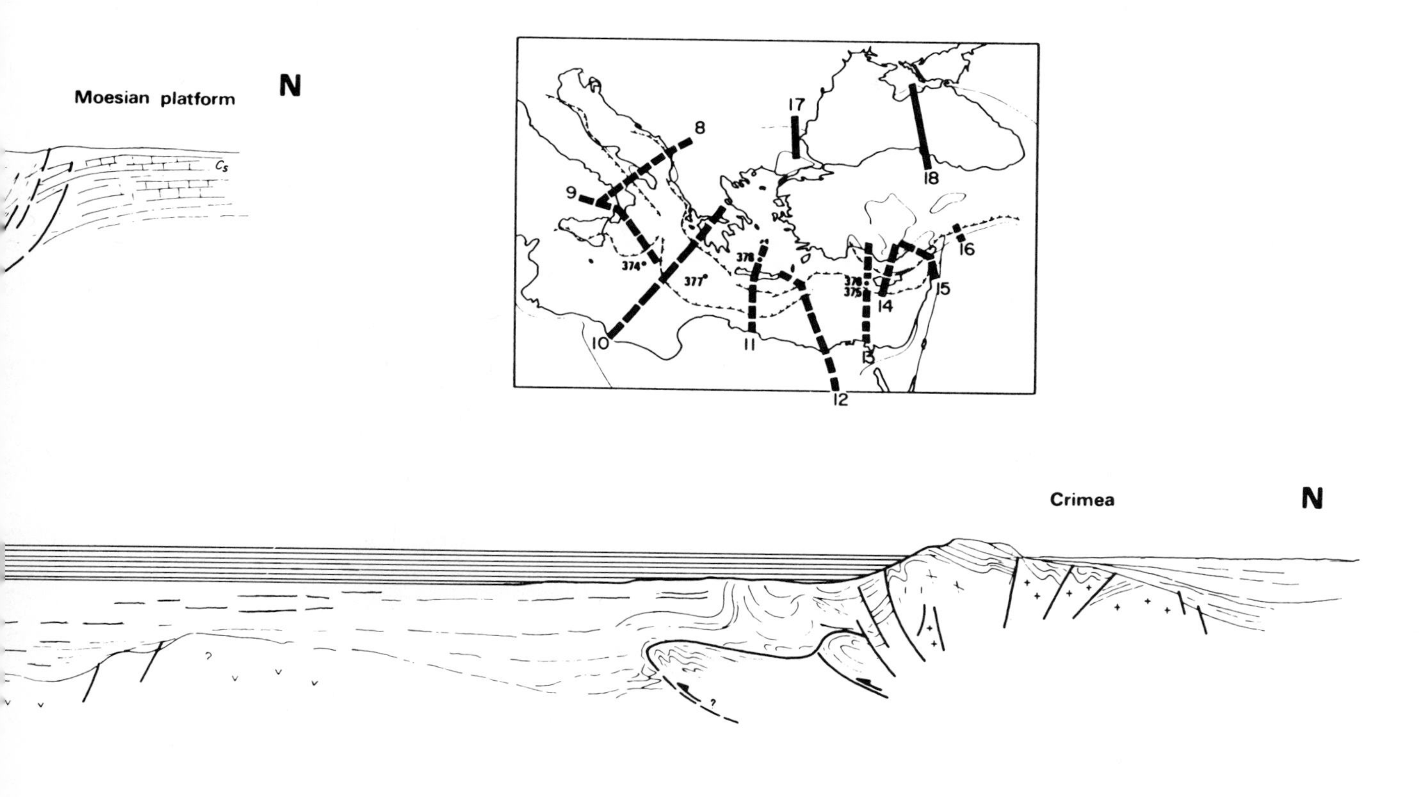

sections (Black Sea basin).

enes plateau (Gonnard et al, 1975; Bellon and Letouzey, 1977) which indicates that no new oceanic crust has been created since Mesozoic. Thus, the history of the Eastern Mediterranean is completely different from that of other Mediterranean basins. Low heat flow values (mean of 0.74 HFU) led Erickson et al (1977) to conclude similarly that the ocean basin formed as early as Mesozoic. Geological evidence of an early continental break-up north of Africa-Arabia can be interpreted as the formation of a passive continental margin during Late Triassic-Liassic. Evidence of rifted margins during this time has also been inferred from different tectonic sequences (Antalya, Mamonia, and Baassit nappes). These were previously considered margins of a proto-eastern Mediterranean, the "Pamphyllian Gulf" of Dumont et al (1972) and equivalent of South Tethys of Hortinsk (1971), but recent investigations led others to suggest that these old margins are related to a Tethys seaway north of Anatolia (Ricou et al, 1974, 1975). As discussed in Biju-Duval et al (1977, 1978), we kept the former hypothesis.

Tectonics and glidings of the Calabrian Arc (Messina cone), Aegean Arc (Mediterranean ridge and Hellenic trenches), and Cyprus Arc obscure relationships between the present deep basin and the former northern margin, and have prevented resolution of the problems of genesis.

Schematically, as shown on the interpretative geological cross-sections (Figs. 5, 6, 11, 12 and 13), eastern Mediterranean deep basins are located on remnants of an old oceanic basin and on an attenuated continental crust. Ophiolites of Antalya, Cyprus, and Kizil Dag may be pieces of obducted oceanic floor.

ACTIVE MARGINS NORTH OF EASTERN MEDITERRANEAN

Calabrian Arc

In front of the Ionian abyssal plain the deeply dissected "Messina cone" or "fan" appears to be a broad tectonic feature. Because of poor seismic penetration in the "hummocky and rolling landscape," it is difficult to determine the sedimentary sequence. Thickening of the Messinian evaporites of the abyssal plain under the cone is clear, as is the general northwestward dip of the deeper part of the sequence (Finetti and Morelli, 1973; Mulder, 1973; Letouzey et al, 1974; Finetti, 1976). Formation of an accretionary prism in the subduction zone and flowing of the salt layer accompanied by superficial glid-

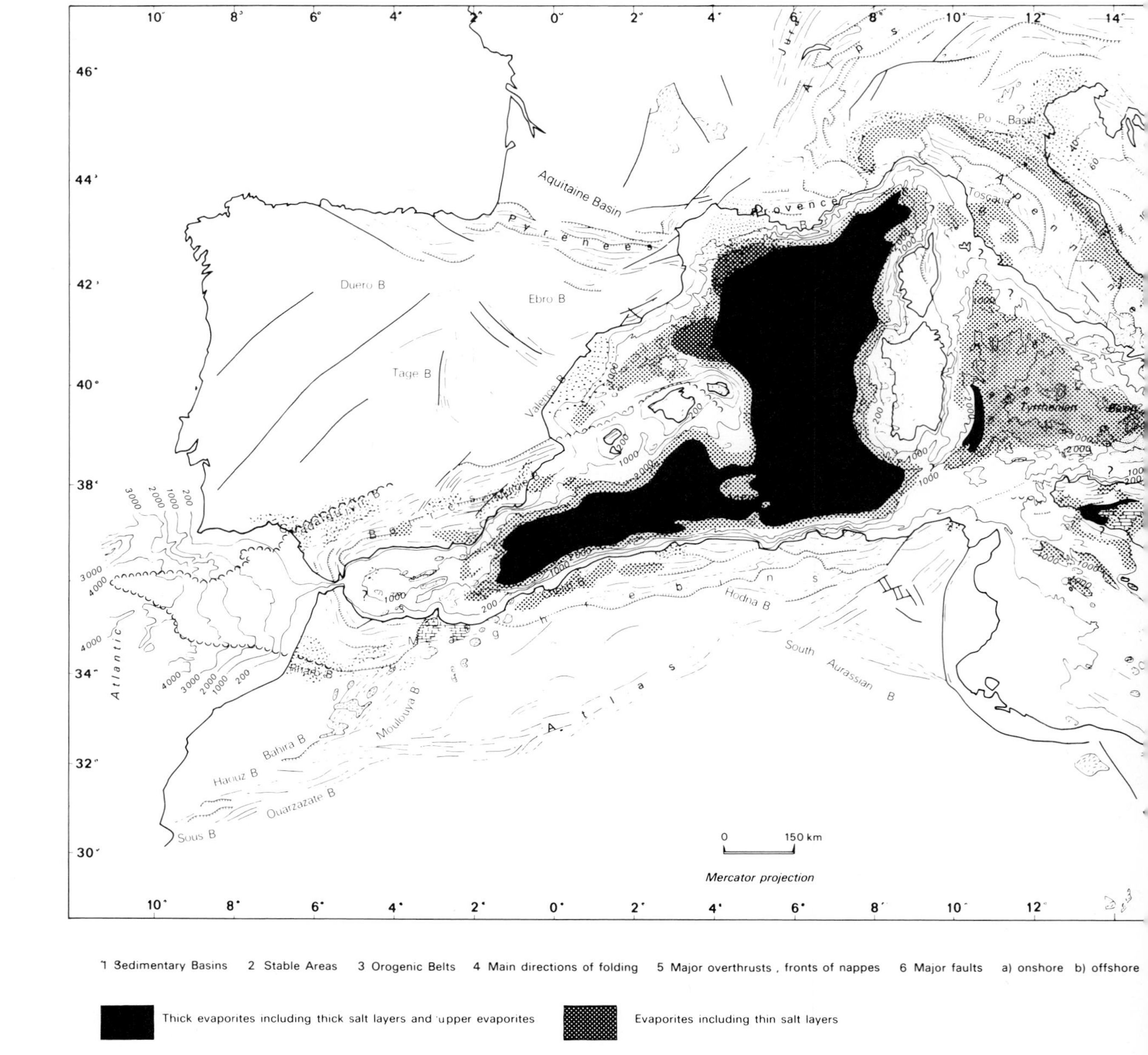

FIG. 8. Messinian evaporites

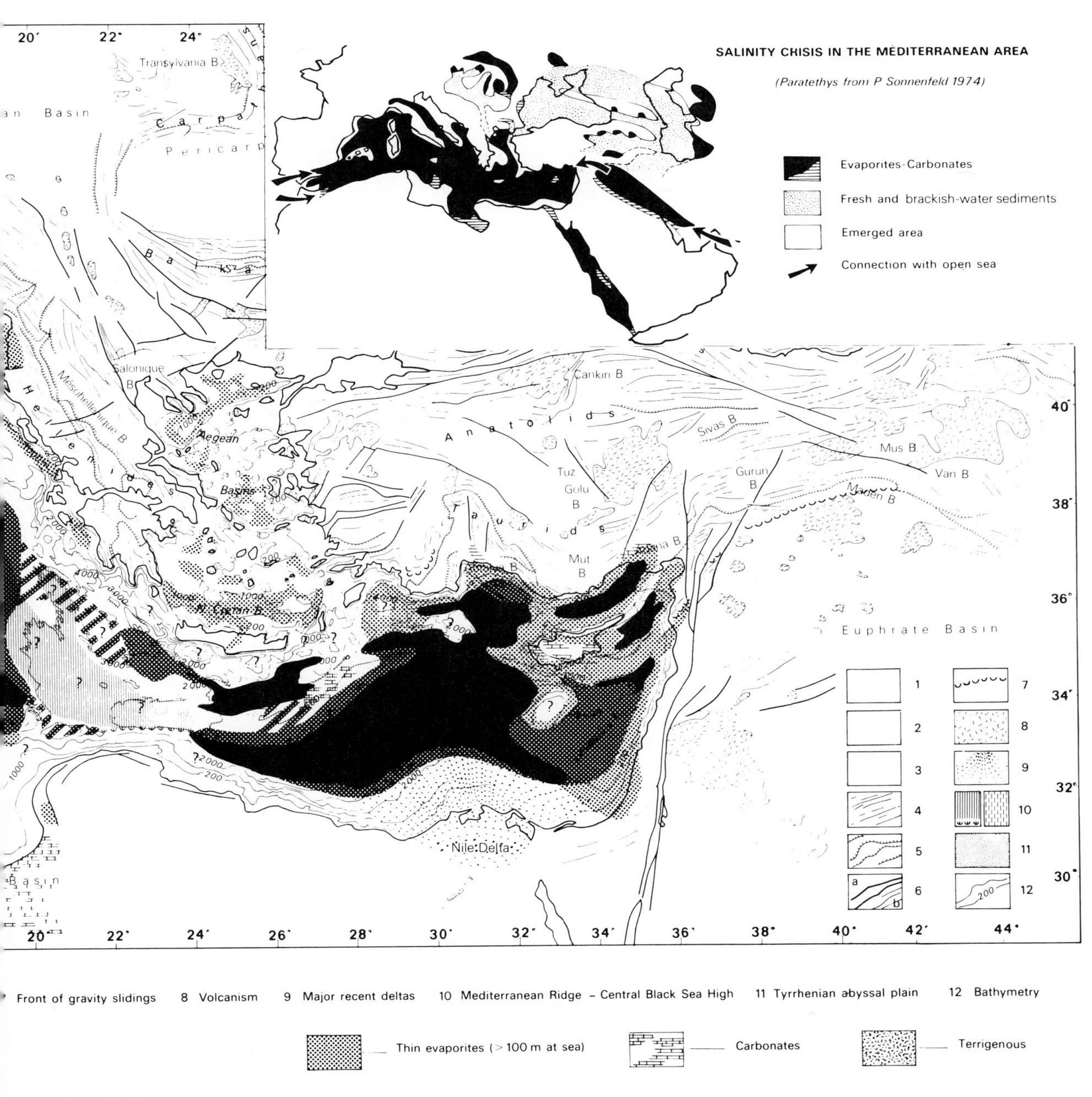

in the Mediterranean area.

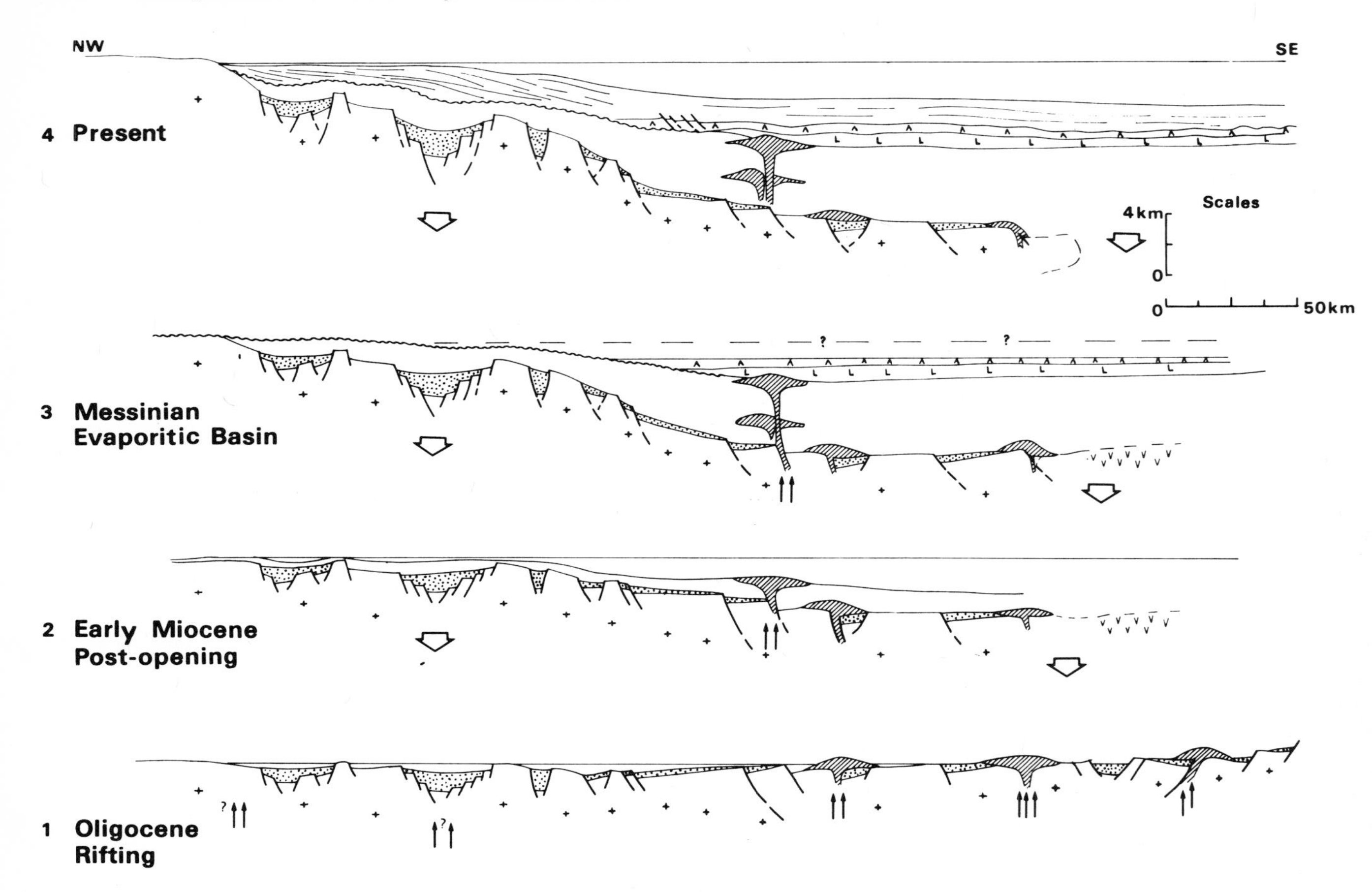

FIG. 9. Tentative geologic evolution of the north Balearic-Provençal basin.

ing explain the incoherent reflections which we interpret as gravitational allochthonous masses, slidings, slumpings, and collapses.

The arcuate system running from the southern Apennines (Taranto Basin) to Sicily is discussed by Selli and Rossi (1975), who think that the external Calabrian arc is separate from the Apenninic chain. The geology of the Calabrian arc, which has been compared with a western Pacific type subduction zone because of its general arcuate arrangement, its seismicity, and its associated volcanism (e.g., Eolian Island), is complex (Fig. 11).

The Calabrian arc has been very active during the entire Neogene. Land data in southern Italy and Sicily show that nappes and olistostromes were emplaced between Burdigalian and Tortonian times in subsiding, progressively deforming basins. The offshore data are in agreement but offshore the active zone fronted a deep marine oceanic area while in Sicily or southern Italy continental collision occurred early between the island arc, Africa, and Apulia. Seismic profiles show small forearc basins near Calabria (see Fig. 4) with an included salt layer.

Time relationships between this Neogene-Quaternary tectonic development corresponding to the formation of the Tyrrhenian Sea and previous tectonic activity known in Sicily or Calabria are poorly understood. Tectonism since the late Mesozoic may be related to uplift of the Maghrebian ranges and formation of the South Balearic-Algerian and proto-Tyrrhenian basins.

Aegean Arc

As does the Calabrian arc, the Aegean arc shows evidence of active subduction for at least 10 million years (Rabinowitz and Ryan, 1970; Le Pichon et al, 1971; MacKenzie, 1972; Berckhemer, 1977). One can distinguish different units from the trench (Hellenic trenches) to the back-arc basin (Aegean Sea) with an external feature south of the trenches, the Mediterranean ridge.

The Hellenic trenches are structurally complicated and marked by high earthquake activity. They are the surface expression of the subduction zone. Seismic data show a thick infilling of unconsolidated sediments in which turbidites are very common (Stanley et al, 1974; Nesteroff et al, 1977). Cretaceous rocks found in Quaternary sediments in DSDP hole 127 (leg 13) have been interpretated as evidence of formation of a melange in front of the accretionary prism (Hsu and Ryan, 1974).

The external arc, running from Peloponnesus to western Turkey via Crete, Karpathos, and Rhodes, consists of a complex tectonic pile including nappes with elevated ophiolites. This tectonic pile is the result of long geological evolution. Land data show that the main allochthonous units observed on the islands come far from north of the Cyclades and

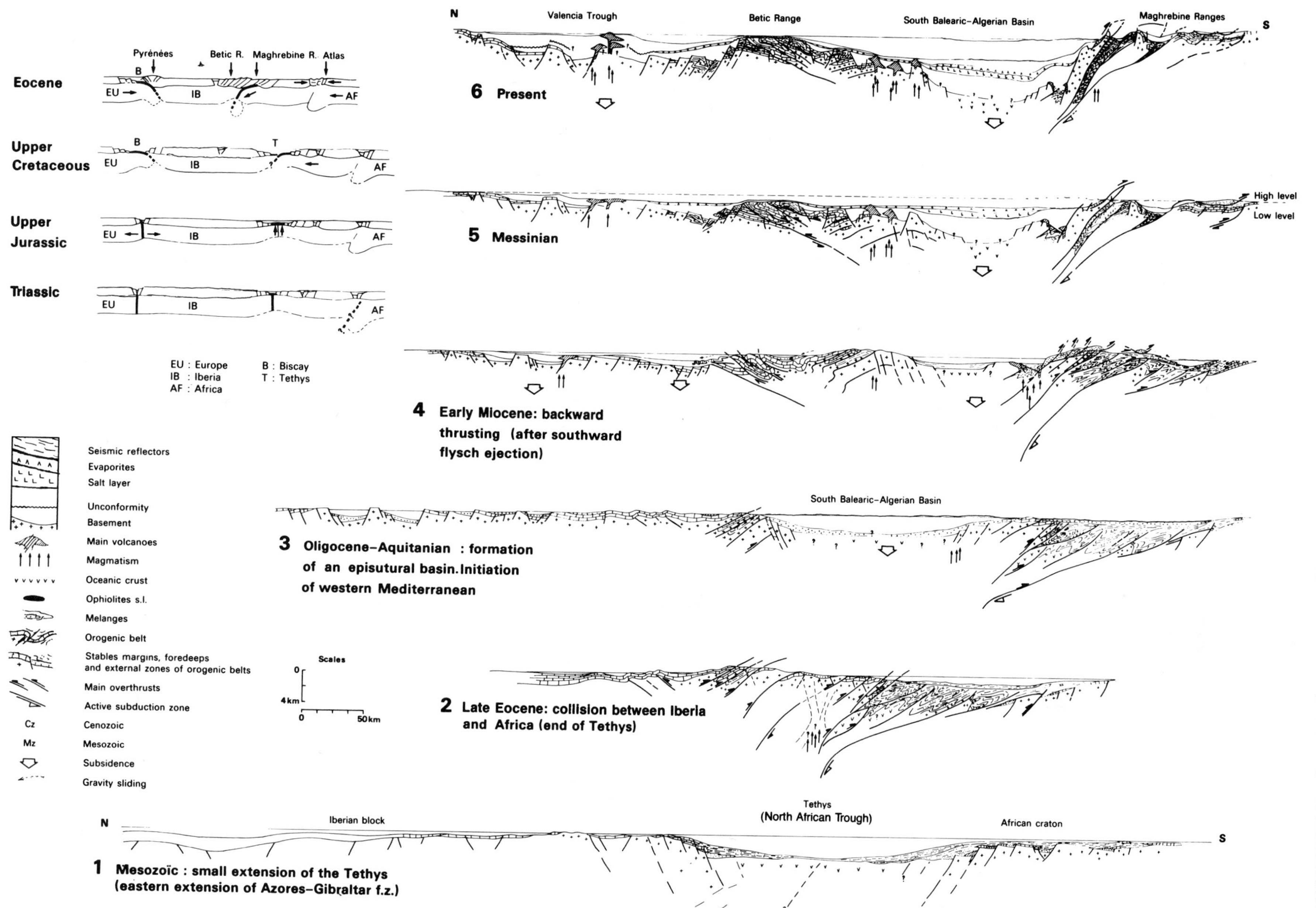

FIG. 10. Tentative geologic evolution of the south Balearic-Algerian basin.

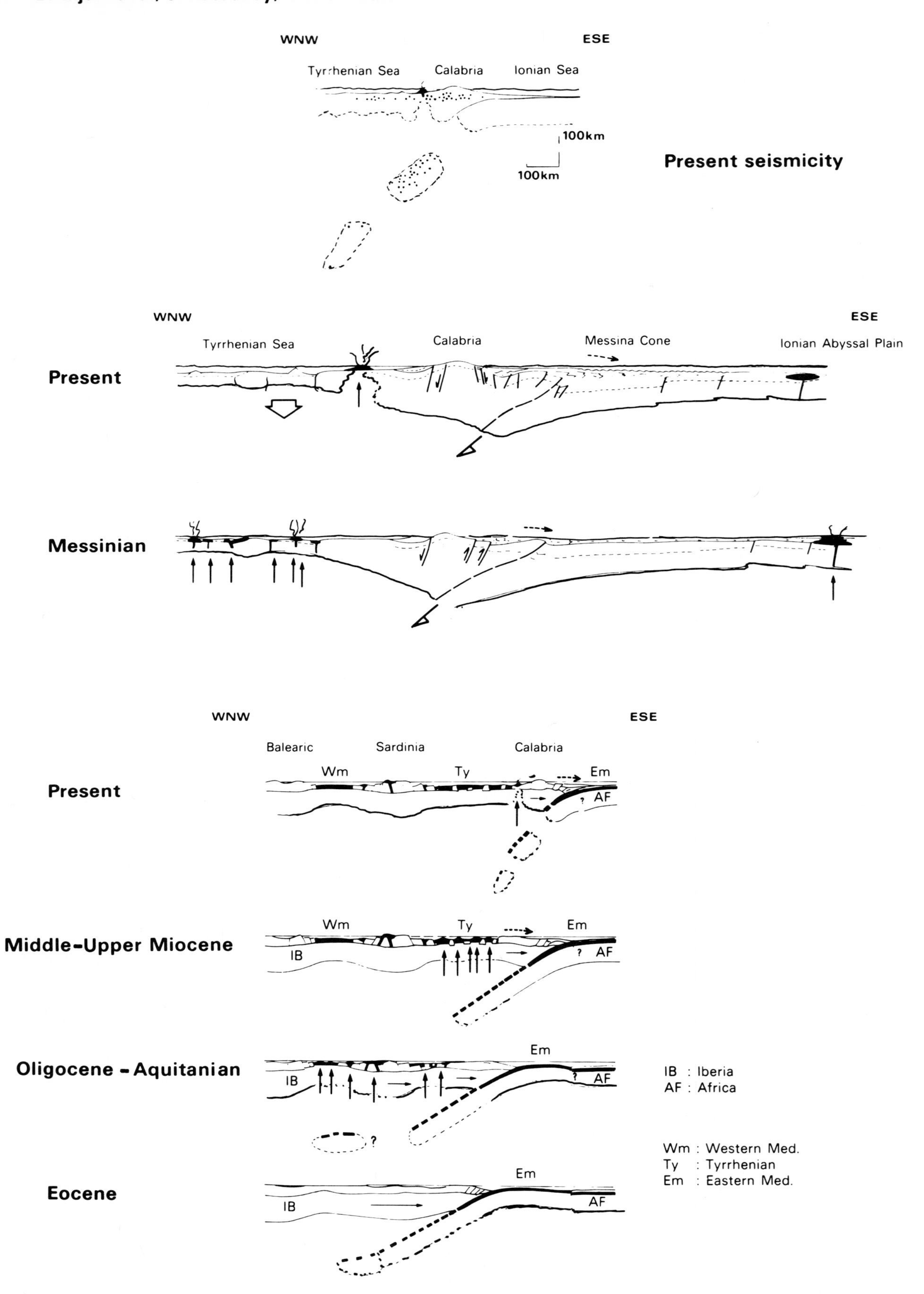

FIG. 11. Tentative geologic evolution of the Calabrian Arc.

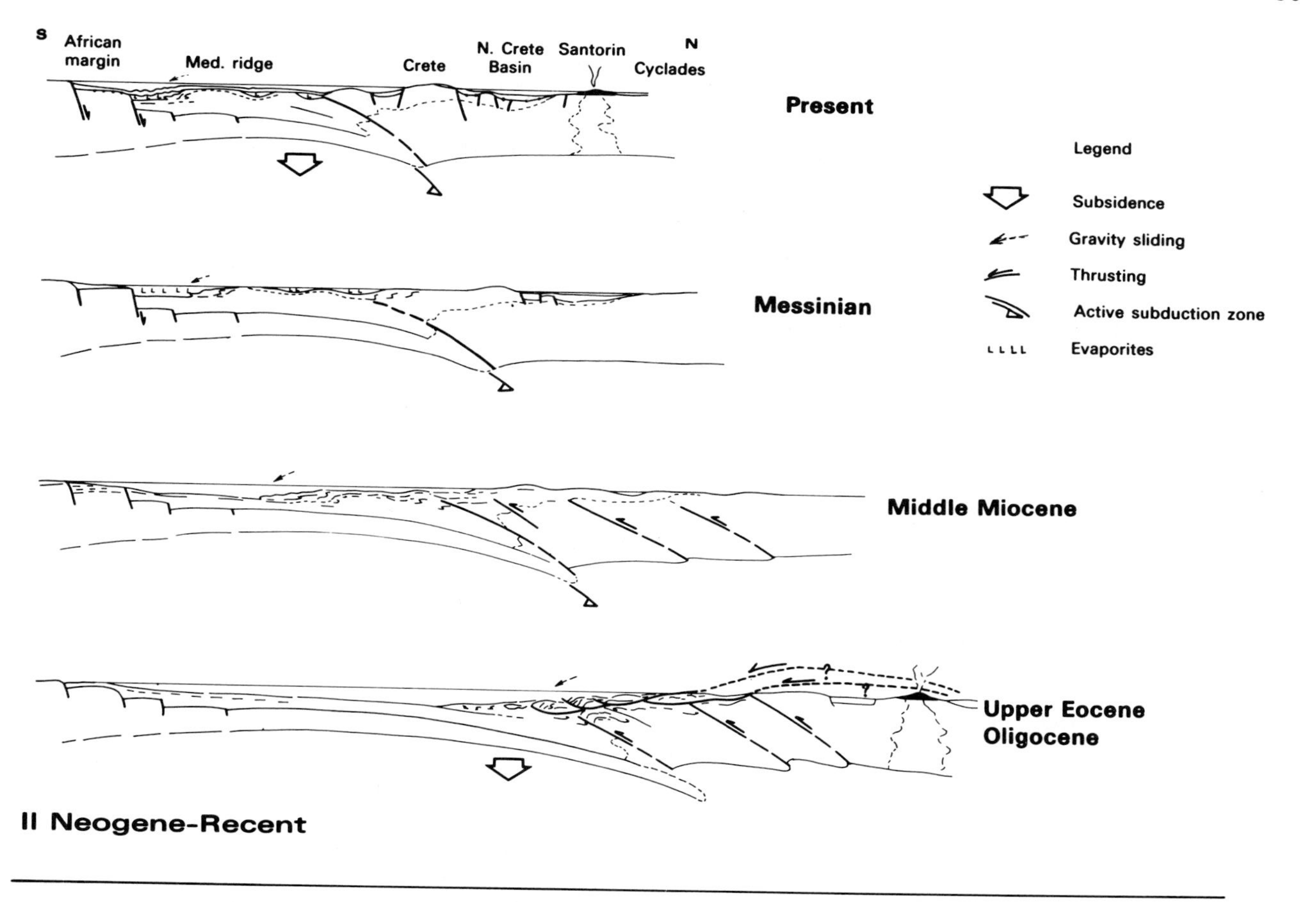

I Mesozoic-Paleogene

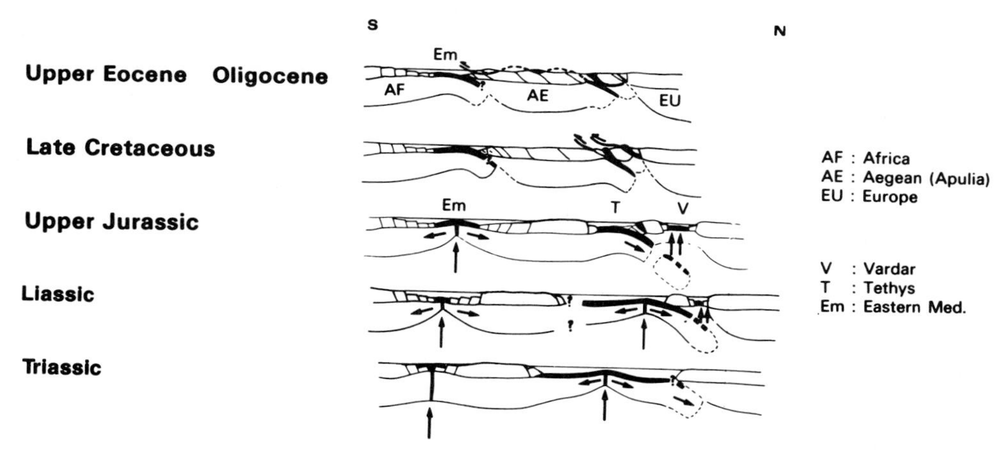

FIG. 12. Tentative geologic evolution of the Aegean Arc.

Eubee (e.g., Aubouin et al, 1976; Katsikatsos et al, 1976) between the Late Jurassic and late Eocene. Southward, Neogene and Quaternary thrusts are known (Mercier et al, 1976; Angelier, 1977). Is there a superimposition of recent thrusting on a complex tectonic pile resulting from an older subduction (Mesozoic to Paleogene) in the northern Aegean, now completely dislocated because of the effects of continental collision which results from Mesozoic consumption of a northern Tethys and from Cenozoic closure of the eastern Mediterranean? Distribution of metamorphism and volcanism (Fig. 12) remains poorly explained in terms of tectonic evolution of this area.

The interdeep basin (the north Cretan basin), the Quaternary volcanic island arc (Milos, Santorin), and a back-arc basin (the Aegean basin north of the Cyclades massif) constitute the remaining elements. These will be discussed in the section on the Aegean Sea.

To the south the Mediterranean ridge is an external arcuate swell, running from the Ionian Sea to the Anaximander Mountains, its eastern part partially drowned beneath Nile sediments. Bathymetric data show it well, especially in its central part where the Aegean arc approaches the African margin (Cyrenaica, Fig. 5, profile 11). Rabinowitz and Ryan (1969, 1970) believed the swell to be an imbricated tectonic pile dating back 10 million years. Sancho et al (1973) showed relatively quiet basins existing on the ridge. Mulder (1973) and Biju-Duval et al (1974) suggested that broad Messinian to Recent and middle Miocene gravity sliding in front of the Aegean tectonized area may be responsible. Finetti (1976) suggested that the ridge is due to shortening of a thick Mesozoic carbonate series in which two sets of overthrusts constitute the Mediterranean chain. If small suboceanic areas in the Ionian and Herodotus basins exist (the Ionian and Levantine foredeeps of Finetti), it is difficult to accept on geographical grounds that the basement of the ridge is the continental extension of the African craton. In the writers' interpretation, comparisons are possible with other broad arcuate systems of the Mediterranean or with the Barbados Ridge in the Caribbean (Biju-Duval and Letouzey, 1975; Mascle et al, 1977), where superficial gravity sliding of a thick sedimentary pile are linked to tectonics and deformation in the subduction zone (see profiles 11, 12, and 13, Fig. 5). Pinchout of the salt layer at higher elevations indicates that the topography began forming in the late Miocene and that present topography in front of the ridge is due to more recent tectonics (see the evaporite distribution, Fig. 8). We now interpret the ridge as due to large allochthonous gravity slides during the middle Miocene (as the Lycian nappes of western Turkey) with recent deformation and uplift due to the collision between the African margin and Aegean island arc. The exact deep structure is not well understood. The very poor seismic response has been interpreted as due either to intense tectonics (vertical faulting; see Hiecke, 1972), or to salt dissolution phenomena or even to karsting (Ryan, 1973). The interpretation of this outer ridge is still under discussion (Kenyon and Belderson, 1977).

Cyprus Arc

The Cyprus arc stretches from south of the Anaximander Mountains to the Levant Coast (Fig. 1). Biju-Duval et al (1976, 1978) and Baroz et al (1978) attempted to explain the formation of this broad structure formed since Late Cretaceous time, as due to closure of a proto-eastern Mediterranean. Widespread emplacement nappes of radiolarites and ophiolites during the Late Cretaceous in Oman, Iran, and southern Turkey, seem to extend into the eastern Mediterranean as evidenced by magnetic anomalies running from the Levant coast to Cyprus and thence to Antalya where ophiolites and melanges were thrusted in Late Cretaceous. These tectonized deep Mesozoic facies may represent fragments of the Mesozoic oceanic basin that joined the Tethys and proto-Mediterranean before the Late Cretaceous continental collision of Anatolia and Africa-Arabia (Biju-Duval et al, 1977; Fig. 13).

Compression was prevalent during the Cenozoic. The front of the tectonized area is marked by southward reverse faulting and can be followed south of Cyprus as far as Syria. The Florence Rise, the western continuation of Cyprus, is better known with the results of DSDP holes 375-376, which provide a complete Neogene section. The date indicates that the rise was deeper relative to the Troodos during Miocene, and that the pinchout of the salt and the thinning of evaporites mark formation of the high during the late Miocene.

The Anaximander Mountains may represent the prolongation of the autochthonous Lycian promontory with the superimposed Antalya nappes (Nesteroff et al, 1977). The Anaximander Mountains seem to be upthrust towards the south but exact relations with the Aegean arc and the Mediterranean Ridge are not well understood.

AEGEAN BASINS

Sedimentary Sequence

Aegean sedimentation is complicated, as several basins have different infillings. Lack of published data precludes description of the sedimentary sequence in each basin. However, different authors (Maley and Johnson, 1971; Morelli et al, 1975; Jongsma, 1975) have observed the following. Pliocene-Quaternary influxes from the adjacent lands are important, having been trapped in different small depressions or grabens. Thickness of unconsolidated sediments commonly reaches 600 m and reaches more than 1,000 m in some parts of the North Cretan basin. Fresh-water, brackish-water or marine sediments alternate. No concensus exists regarding links between the Black Sea and the open marine Mediterranean (Bousquet et al, 1976; Marinos et al, 1977; Guernet, 1977). Recent volcanic activity (Ninkovich and Heezen, 1965) and tectonics (see the review by Angelier, 1977) have been described for the area as a whole.

Seismic data and results from the Prinou oil field and DSDP hole 378 show that evaporites and even salt in some places are present. Evaporites are often relatively thin and it is difficult to distinguish them from an erosional unconformity also marked by a strong reflection. Interfingering of detrital material, as observed in Prinou oil field, is very common. Southward, evaporites are transgressive on the basement (Jongsma, 1975) as in Crete (Meulenkamp, 1971) where post-nappe deposition began in late Miocene. Northward, evaporites overlie an older sedimentary sequence.

Stratigraphy and lithology of the pre-evaporitic sequnce is poorly known. It may be locally very thick and affected by tectonics. In the northern part, Greek or Turkish onshore geology suggests that some interior basins with marine sedimentation may have been developed as early as late Eocene or Oligocene. In the central part (Cyclades or North Cyclades basins) and in the north Cretan basin an

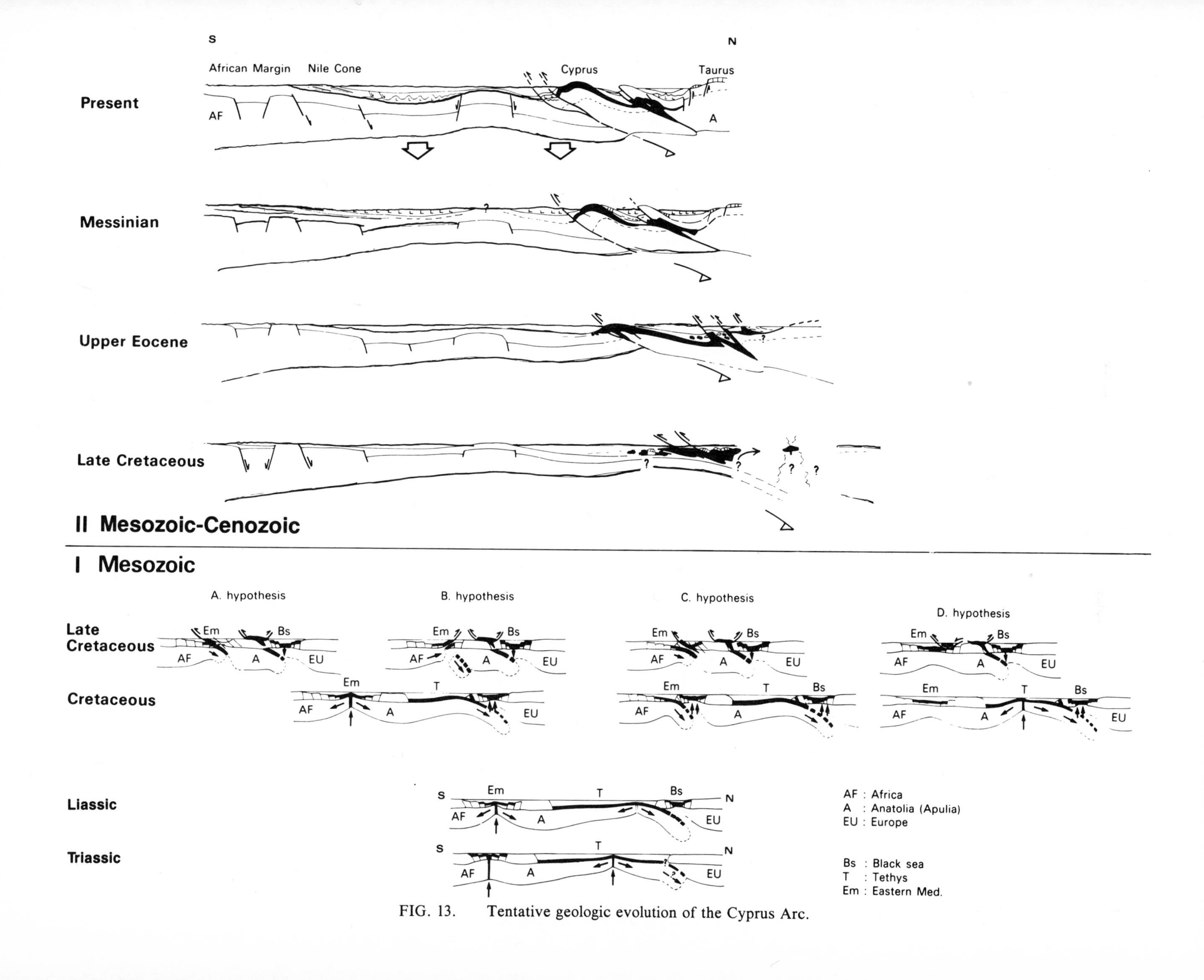

FIG. 13. Tentative geologic evolution of the Cyprus Arc.

important tectonic event has been recognized (Angelier et al, 1977; Altherr et al, 1976), probably of late Miocene age. Older deposits may therefore be tectonized.

Magnetic anomalies and numerous highs may be the expression of intrusive bodies. Magmatism is known on land (northwest Turkey) and Neogene magmatism can be distinguished from that of the recent volcanic arc of Santorin (Borsi et al, 1972; Vilminot and Robert, 1974; Innocenti et al, 1977).

Structure, Age, and Origin

The Aegean Sea comprises a set of back-arc basins (the northern Aegean), a volcanic arc (Santorin), and an inter-arc basin (the North Cretan basin). This structural framework is relatively young, and superimposed on older structures.

The structure of the Aegean is not completely explained, and several different geophysical models have been proposed (Schuilling, 1972; Makris, 1975; Morelli et al, 1975; Berckhemer, 1977; Papazachos and Comminakis, 1977). The crust is considered continental but with some thinning as (for instance) in the North Cretan basin (or in the north Aegean graben). Plate boundaries are not well defined by seismicity eastward and northward (Lort, 1971; McKenzie, 1972).

If the succession of the external arc, the interdeep basin and the volcanic arc is clear, interpretation of other basins remains difficult. The small central basins (Ikaria, Skyros, etc.) may represent the initial stages of a future back-arc basin, with a continental crust now broadly intruded and collapsed. They are superimposed on a former structure, along or on a calc-alkaline province of early Miocene age (Borsi et al, 1972). The exact significance of the north Aegean graben is also a problem. Through the deep Marmara Sea, it appears as the prolongation of the North Anatolian strike-slip fault but it has been also interpreted as the result of a new spreading center (Lort, 1971; Allan and Morelli, 1971; Le Pichon et al, 1971) or as the expression of an old subduction zone still marked by seismicity (see for instance Papazachos and Comminakis, 1977).

The age of these basins must be analyzed very carefully because of the numerous successive tectonic phases since Late Jurassic time. Reconstruction of the Mesozoic and Paleogene history based on land data is still debated in spite of recent progress.

An old subduction zone (roots of the main nappes of Hellenides, Crete, and Lycian Taurus) cuts the Aegean from the Vardar zone to the Izmir area. Cyclades Islands, south Eubee, and Attica are now considered as windows of metamorphosed Mesozoic rocks under the nappes (Aubouin et al, 1976; Katsikatsos et al, 1976). Rocks exposed in windows could represent autochthons of Aegean crust, as do the Menderes for the Lycian nappes (Durr, 1976; Poisson, 1977). Ages of nappe emplacement and of metamorphism range from Late Jurassic to late Miocene (Altherr, 1977).

Some basins, such as the interior basins of Greece or Turkey, began to form at the end of the Eocene or during the Oligocene. After several erosional episodes, they were tectonically rejuvenated in the Miocene and after. Indeed, the calc-alkaline volcanism (21 to 16 MY) of the eastern Aegean and western Turkey and granodiorite intrusions of Kos and Naxos were followed by renewed intense activity in late Miocene as demonstrated by Altherr et al (1976) and Angelier et al (1977). These events all predate recent history and recent structure of the Aegean.

ANTALYA, CILICIA-ADANA AND NORTH LEVANTINE BASINS

Sedimentary Sequence

In the deep basins, the thickness of Pliocene-Quaternary sediments often exceeds 1,000 m. Detrital influx from the Taurus Mountains, uplifted during Pliocene and Quaternary, is important, especially with the Ceyhan and Seyhan deltas which form a part of the shelf of Adana basin.

Pelagic and hemipelagic sedimentation found at sites 375-376 on the Florence Rise show that "the rise was sheltered from terrigenous influence by deeper basins" to the north (Baroz et al, 1978) in which some turbidites could be expected.

The structural trend of Misis-Kyrenia, corresponding to the offshore extension of the Misis-Andirin-Bitlis overthrust, developed continuously during Neogene and Pleistocene time. More recent vertical and reverse faulting, known onshore in the Antalya and Adana basins, also occurred. These are responsible for large escarpments in the western Antalya basin.

As in the other parts of the Mediterranean, the Messinian event is characterized by thick evaporites (Mulder, 1973; Biju-Duval et al, 1974). A thick salt layer characterized by high velocities (up to 4 km/sec) and diapirism has been confirmed by offshore drilling in the Adana basin where halite with interbedded evaporites was found. Maximum thickness reaches 2,000 m in the deep Antalya basin. Thickness is controlled by previous topography and structural evolution of the Cyprus arc. The maximum thickness corresponds to the basin axes. Salt pinches out on the main structural highs.

Land data, DSDP holes, and extensive seismic profiling show (Fig. 6) that northward, a complete Neogene sequence from at least and Burdigalian overlies a major unconformity. Southward, the sedimentary sequence may include more deposits as old as late Maestrichtian, as shown in the Kyrenia range (Baroz and Bizon, 1974). In the deep basin, thickness exceeds 5 or 6 km.

Early Miocene shallow-water facies with reef development occur onshore in Turkey but DSDP borehole 376 showed that deep-water conditions existed in the basins at that time. A deepening of the basin is clear in the onshore Adana basin during Langhian-Serravalian time. Deltaic sedimentation in Adana basin, and more generally clastic deposits, developed during the Tortonian prior to the Messinian event (Schmidt, 1961).

Southward, as known in Cyprus, continuous pelagic sedimentation has occurred since late Maestrichtian (chalk and calcareous oozes with cherty intercalations). Structural development of the Cyprus arc resulted in development of local disconformities onshore in the late Eocene (Kyrenia range) or Oligocene (south of Cyprus). The present Kyrenia-Misis Pliocene overthrusted range probably corresponded to a previous trough infilled by flysch deposits including olistolites and olistostromes before (Kyrenia) and during Miocene times (Misis; see Schietecatte, 1971; Biju-Duval et al, 1974b; Baroz et al, 1978).

Structure, Age, and Origin

Despite new data available on the evolution of the Taurides (Ozgul and Arpat, 1972; Ricou et al, 1975; Monod, 1977; Poisson, 1977) and of the Kyrenia range (Baroz, 1970; Baroz and Bizon, 1974), successive paleogeographies are very difficult to reconstruct because the basins are in an area that has been continuously tectonized since the Maestrichtian (Fig. 1 and sections of Figs. 6, 13). After the Maestrichtian tectonic event known as far as Zagros and Oman (Ricou, 1971) and in the whole of Turkey to the north (Brunn et al, 1971; Hortinsk, 1971), small basins developed within or behind the main folded zone. North of Florence rise, the Antalya deep basin constitutes a single basin but its deep structure is probably more complex. Eastward several basins (Mesaoria in Cyprus, Cilicia-Adana, north Levantine basins) can be distinguished because of the development of arches with southward overthrusts.

These basins could be interpreted as interdeep (or fore-arc) basins, but there is no well-defined volcanic arc and the nature of the crust of these basins is unknown.

BLACK SEA BASIN

Sedimentary Sequence

Geophysical surveys have determined the general framework of the area (Garkalenko et al, 1971; Goncharov, et al; 1972; Neprochnov et al, 1974, 1975). Recent multichannel seismic profiling (Letouzey et al, 1977) and DSDP holes of leg 42B drilled in the abyssal plain have also provided data.

Recent sedimentation and superficial structures observed in shallow penetration seismic reflection data were described by Ross et al (1974) but his data gave little information on the total sedimentary succession or the detailed structures of the margins. More recently Letouzey et al (1977, 1978) have shown (Fig. 7) that the margins are very narrow southward and eastward. Prograding deposits cut by sedimentary channels are affected by numerous slides. A thick section of unconsolidated sediment is present over the entire abyssal plain.

Northward, sediment from the Danube, Dniepr, and Volga prograde towards the abyssal plain. The largest of the deltas is the Danube-Dniepr. It developed mainly on the broad shelf but its prodelta covers the entire western part of the Black Sea where Pliocene-Quaternary sediment thickness reaches more than 2,500 m. In contrast with Mediterranean basins, no Messinian salt deposit has been observed in this basin.

Below the Pliocene-Quaternary sequence, well-layered acoustic reflectors indicate a continuous sedimentary section at least 3,000 m thick. Correlation with margin drilling for petroleum suggests the following interpretations (see Fig. 7). Infilling of the basin is as old as late Eocene when active subsidence took place, as known on the Bulgarian borderland and demonstrated by holes near Strandja ridge (Mannheim and Foose, 1975). Undercompaction (low velocities) and associated diapirism occur in the abyssal plain. These may correspond to the Maikopian layers (Oligocene-Miocene), as is the case in the Kerg Peninsula near Crimea. Anomalously high heat flow values may be related to the mud diapirism. On the eastern and southern margins the acoustic basement corresponds to tectonized series of Eocene age or older. Westward and northward, a Cenozoic sequence lies on an undeformed Mesozoic series (generally Cretaceous limestones but Jurassic in the Karkinit basin). Mesozoic rocks dip toward the deep basin, where they are faulted and covered by thick Danube delta sediments.

Structure, Origin, and Age

The Black Sea, as the south Caspian Sea, represents a small marine depression south of the European continent. It has evolved more or less independently of the Mediterranean since the Neogene, but previously it followed the evolution of the Alpine-Tethys system and other Mediterranean basins.

Seismic refraction surveys (Neprochnov et al, 1975) showed that granitic crust is present on the margins and that suboceanic crust is present under the abyssal plain. Surface wave studies and gravimetric data (Balavadze and Mindeli, 1975) confirmed this conclusion. They also suggested the possibility that two subbasins can be distinguished. Low heat flow values (Erickson, 1970; Lyubimova and Savostin, 1973) in the abyssal plain probably are due to the thick sedimentary sequence and to the old age of formation of the basin, as confirmed by the absence of recent volcanic activity.

Seismic reflection data show that a high separates the eastern and western subbasins. This high, buried by the thick sedimentary series, is thought to be as old as Eocene. The northwestern margin appears passive in origin while eastern and southern margins are characterized by Eocene tectonized basement well known in Caucasus and Pontids belts. Along the Caucasus, seismicity evidences continuing tectonic activity.

The preceding indicate that the Black Sea consists of one, or possibly two, small oceanic basins inserted between the European continent and the orogenic belts to the south. The orogenic belts include calc-alkaline volcanics and ophiolites upthrust on the Anatolian platform (Fig. 13). We do not know whether the Black Sea is a remnant of an old ocean or a marginal sea developed behind an island arc during consumption of the Tethys.

It remains difficult to precisely infer structural

conditions existing in Early Mesozoic because we don't know the relationships between the Cimmerian folding in Dobroudgea and in the Caucasus. But from Letouzey et al (1977) we hypothetically summarize the late Mesozoic to Holocene history as follows (Fig. 13). An active margin seems to have existed north of the present Tethyan ophiolitic suture zone along the southern edge of Europe. Intense magmatic activity has characterized this margin since Liassic times as a result of northward subduction of the Tethys. The Black Sea may have formed in Late Cretaceous as a back-arc basin. Calc-alkaline volcanism from Srednogorie (Bulgaria) to the lesser Caucasus may represent the island arc. During Eocene-Oligocene, subsidence of the basin may have resulted from cooling of oceanic crust. Loading by the thick sedimentary deposits accentuated the subsidence. The Pliocene-Pleistocene period is still marked by a very high rate of accumulation.

CONCLUSION

The Mediterranean basins exhibit marked differences. Specifically, interpretive geological cross-sections and geophysical data show that structure of the margins and of the deep basins consists of young marginal basins that evolved as part of the Alpine system, having been created behind subduction zones (e.g., western, Tyrrhenian, Aegean and Black Sea basins), and remnants of ancient oceanic crust and of old continental margins (eastern Mediterranean basins). Seismic reflection data, volcanism, and onshore data suggest the following ages of formation: Triassic-Liassic in the eastern Mediterranean, Late Cretaceous to Eocene in the Black Sea, Oligocene to early Miocene in the western basin, and late Miocene to Present in the Tyrrhenian and Aegean.

The basins also show marked similarities—specifically, a thick section of Cenozoic sediments (with the exception of the Tyrrhenian deep basin), widespread Messinian evaporites (except in the Black Sea), and deep-sea conditions in existence since at least early Miocene (except perhaps in the Aegean Sea).

The inferred oil potential of the deep basins depends strongly on the hypothesis preferred for the formation of each basin. If the basins collapsed since late Neogene time (Burollet and Byramjee, 1974) the sequence below the evaporites probably consists of shallow-water deposits, and oil potential in the deep basins is similar to that of the shelves (Byramjee et al, 1975). But if one agrees with the authors, then deep-water deposits lie below the evaporites in the central parts of the deep basins. Earliest deep deposits will be of different ages in different basins. Nevertheless, continental collision, reduced size of the oceanic areas, and width of the continental margins could result in old shallow-water deposits far into the deep basins. Thus, in the eastern Mediterranean, pre-rift early Mesozoic epicontinental facies may be found on the continental margin of Africa.

REFERENCES CITED

Adamiya, S. H., 1975, Plate tectonics and evolution of the Alpine System: Discussion: Geol. Soc. America Bull., v. 86, p. 719–720.

Alla, G., et al, 1972, Structure géologique de la marge continentale du Golfe du Lion (abstr.): 23 éme Congr. CIESM, Athénes, Rapp. Comm. Int. Mer Medit., 22, p. 38.

Allan, T. D., and C. Morelli, 1971, A geophysical study of the Mediterranean Sea: Bull. Geofis. Teor. Appl., v. 13, No. 50, p. 99–141.

Altherr, R., 1977, Abstract in VI° Aegean Colloquium, Athens, in press.

——— J. Keller, and K. Kott, 1976, Der jungtertiäre Monzonit von Kos und sein Kontakthof (Agais, Griechenland): Geol. Soc. France Bull., (7), XVIII, p. 403–412.

Alvarez, W., 1972, Rotation of the Corsica-Sardinia microplate: Nature, Phys. Sci., v. 235, p. 103–105.

——— T. Cocozza, and F. C. Wezel, 1974, Fragmentation of the Alpine orogenic belt by microplate dispersal: Nature, Phys. Sci., v. 248, p. 309–314.

Angelier, J., 1977, Sur les mouvements égéens dupuis le Miocéne supérieur: l'évolution récente de la courbure sud-hellénique (Gréce): Acad. Sci. Comptes Réndus, 284, D, 1037–1040.

——— G. Glacon and C. Muller, 1977, Sur la présence et la position tectonique du Miocéne inférieur marin dans l'archipel de Naxos (Cyclades, Gréce): Acad. Sci. Comptes Rendus, in press.

Aubouin, J., et al, 1976, Esquisse structurale de l'Arc égéen externe: des Dinarides aux Taurides: Geol. Soc. France Bull., (7), XVIII, n° 2, p. 327–336.

Auzende, J. M., J. L. Olivet, and J. Bonnin, 1972, Une structure compressive au Nord de l'Algérie: Deep-Sea Research, v. 19, p. 149–155.

——— J. Bonnin, and J. L. Olivet, 1973, The origin of the western Mediterranean basin: Jour. Geol. Soc., London, v. 129, p. 607–620.

——— et al, 1971, Upper Miocene salt layer in the western Mediterranean basin: Nature, Phys. Sci., v. 230, p. 82–84.

——— et al, 1975, Les bassins sédimentaires de la Mer d'Alboran: Geol. Soc. France Bull., 7, t. XVIII, p. 98–107.

Balavadze, B. K., and P. S. Mindeli, 1975, Methods of deep crustal studies in the Black Sea region, *in* The earth's crust and the development of the Black Sea basin: Moscow, Nauk, 274 p.

Bally, A. W., 1975, A geodynamic scenario for hydrocarbon occurrences: 9th World Petrol. Congr. (Tokyo), v. 2, p. 33–44.

Barberi, F., et al, 1978, Age and nature of basalts from the Tyrrhenian abyssal plain, *in* Initial reports of the Deep Sea Drilling Project: Washington, U.S. Government Printing Office, v. XLII, p. 509–514.

Baroz, F., 1970, Observations nouvelles sur l'Eocene de la chaine de Kyrenia (Chypre): Acad. Sci. Comptes Rendus, v. 270, p. 1205–1208.

——— and G. Bizon, 1974, Le Néogéne de la chaine de Pentadaktylos et de la partie nord de la Mesaoria (Chypre). Etude stratigraphique et micropaléontologique: Rev. IFP, v. 29, p. 327–358.

——— A. Desmet, and H. Lapierre, 1976, Les traits dominants de la géologie de Chypre: B.S.G.F., 7, T. XVIII, n° 2, p. 419–427.

——— et al, 1978, Correlations of the Neogene formations of the Florence Rise and of northern Cyprus: Paleogeographical and Structural Implications, *in* Initial reports of the Deep:Sea Drilling Project: Washington, D.C., U.S. Govt., v. XLII, p. 903–926.

Bayer, R., J. L. Le Mouel, and X. Le Pichon, 1973, Magnetic anomaly pattern in the western Mediterranean: Earth Planetary Sci. Letters, v. 19, p. 168–176.

Bein, A., and G. Gvirtzman, 1977, A Mesozoic fossil edge of the Arabian plate along the Levant coastline and its bearing on the evolution of the eastern Mediterranean, *in* B. Biju-Duval and L. Montadert, eds., Structural history of the Mediterranean basins: Paris, Technip, p. 95–110.

Bellaiche, G., et al, 1977, Etude par submersible des canyons des Stoechades et de Saint-Tropez: Acad. Sci. Comptes Rendus t. 284, p. 1631–1634.

Bellon, H., 1976, Séries magmatiques néogénes et quaternaires du pourtour de la Mediterranée occidentale, comparées dans leur cadre géochronométrique—implications géodynamiques: Thése, Paris-Sud., p. 1-367.

——— and J. Letouzey, 1977, Volcanism related to plate-tectonics in the western and eastern Mediterranean, *in* B. Biju-Duval and L. Montadert, eds., Structural history of the Mediterranean basins: Paris, Technip, p. 165–183.

Beltrandi, M. D., and P. Biro, 1975, The geology and geophysics of the Iskenderun basin, offshore southern Turkey: Rapp. Com. Int. Mer Medit., v. 23, p. 31–33.

Berckhemer, H., 1977, Some aspects of the evolution of marginal seas deduced from observations in the Aegean region, *in* B. Biju-Duval and L. Montadert, eds., Structural history of the Mediterranean basins: Paris, Technip, p. 303–314.

Bernoulli, D., 1972, North Atlantic and Mediterranean Mesozoic facies: a comparison, *in* C. D. Hollister, et al., Initial reports of the Deep Sea Drilling Project: Washington, D.C., U.S. Govt., v. 11, p. 801–871.

Biju-Duval, B., 1974, Commentaires de la carte géologique et structurale des bassins tertiaires du domaine Méditerranéan: Rev. Inst. France Pétrol., v. 29, p. 607–639.

——— and J. Letouzey, 1975, Comments about the new "Carte géologique et structurale des bassins tertiaires du domaine méditerranéen": Rap. Comm. Int. Mer Medit. 23, p. 119–120.

——— P. Courrier, and J. Letouzey, 1974, Interprétation de la structure des Monts de Misis, Turquie (Chevauchement pliocéne et masses allochtones mises en place au Miocéne) et son extension en Méditerranée orientale: Second réunion Ann. Sci. Terre (Nancy), p. 48.

——— H. Lapierre, and J. Letouzey, 1976, Is the Troodos Massif (Cyprus) allochthonous?: Geol. Soc. France Bull., v. 18, p. 1347–1356.

——— J. Dercourt, and X. Le Pichon, 1977, From the Tethys ocean to the Mediterranean Sea: a plate tectonic model of the evolution of the western Alpine system, *in* B. Biju-Duval, and L. Montadert, eds., Structural history of the Mediterranean basins: Paris, Technip, p. 143–164.

——— J. Letouzey, and L. Montadert, 1978, Structure and evolution of the Mediterranean Sea basins, *in* Initial reports of the Deep Sea Drilling Project: v. 42, Washington, D.C., U.S. Govt., p. 951–984.

——— et al, 1974, Geology of the Mediterranean Sea basins, *in* C. L. Drake and C. Burk, eds., The geology of continental margins: New York, Springer-Verlag, p. 695–721.

——— et al, 1975, Apports de l'étude des images du satellite LANDSAT-1 a la connaissance de la structure du domaine méditerranéen: Rev. Inst. Franc, du Pétrole, XXX-6, p. 841–853.

Boccaletti, M., and G. Guazzone, 1974, Remnant arcs and marginal basins in the Cainozoic development of the Mediterranean: Nature, Phys. Sci., v. 252, p. 18–21.

Borsi, S., et al, 1972, Petrology and geochronology of recent volcanics of eastern Aegean Sea (west Anatolia and Lesvos Island): Zeit. Deut. Geol. Ges., Bd. 123, Tl. 2, p. 521–522.

Bousquet, J. C., 1972, La tectonique récente de l'Apennin Calabro-Lucanien dans son cadre géologique et géophysique: Thése, Montpellier, p. 1–172.

———Bousquet, B., et al, 1976, Essai de correlations stratigraphiques entre les facies marins, lacustres et continentaux du Pliocene en Grece: Geol. Soc. France Bull., 7, t XVIII, p. 413–418.

Brunn, J. H., et al, 1971, Outline of the geology of the western Taurids, *in* A. S. Campbell, ed., Geology and history of Turkey: Tripoli, p. 225–255.

Burollet, P. F., and R. S. Byramjee, 1974, Réflexions sur la tectonique globale. Exemples africains et méditerranéens: C.F.P. Notes et Mémoires 11, p. 71–120.

——— A. Said, and Ph. Trouve, 1978, Slim holes drilled on the Algerian shelf, *in* Initial reports of the Deep Sea Drilling Project: Washington, D.C., U.S. Govt., v. 42, part 2, p. 1181–1184.

Byramjee, R. S., J. F. Mugniot, and B. Biju-Duval, 1975, Petroleum potential of deep water areas of the Mediterranean and Caribbean seas: Ninth World Petrol. Congr. (Tokyo), Proc., v. 2, p. 229–312.

Channel, J. E. T., and F. Horvath, 1976, The African-Adriatic promontory as a palaeogeographical premise for Alpine orogeny and plate movements in the Carpatho-Balkan region: Tectonophysics, v. 35, p. 71–101.

Clauzon, G., 1975, Preuves et implications de la régression endoreique messinienne au niveau des plaines abyssales: l'exemple du midi méditteranéen francais: Geo. Assoc. France Bull., no. 429–430, p. 317–333.

Commission Internationale pour l'Exploration Scientifique de la Mer Méditerranée, 1973, Rapports et Proces-Verbaux des Réunions. Symposium sur la Géodynamique de la région méditterranéenne: Athènes, 3–11, nov. 1972, 210 p.

——— 1975, Rapports et Procès-Verbaux des Réunions. Symposium Géodynamique: Monaco, 6–14 déc., 1974, 313 p.

Cravatte, J., et al, 1974, Les forages du Golfe du Lion, stratigraphie, sédimentologie Notes et Mém.: CFP, p. 209–274.

D'Argenio, B., R. Radoicic, and I. Sgrosso, 1971, A paleogeographic section through the italo-dinaric external zones during Jurassic and Cretaceous times: Nafta Zagreb, v. 22, No. 4–5, p. 195–207.

Dercourt, J., 1970, L'expansion océanique actuelle et fossile: ses implications géotectoniques: Géol. Soc. France Bull., v. 12, p. 261–317.

Dewey, J. F., 1976, Ophiolite obduction: Tectonophysics, v. 31, p. 93–120.

——— et al 1973, Plate tectonics and the evolution of the Alpine system: Geol. Soc. America Bull., v. 84, p. 3137–3180.

Dubois, R., 1970, Phases de serrage, nappes de socle et métamorphisme alpin à la jonction Calabre-Apennin: la suture calabro-apenninique: Rev. Géogr. Phys. Geol. Dyn., série 2, v. 12, fasc. 3, p. 221–254.

Dumont, J. F., et al, 1972, Le Trias des Taurides occidentales (Turquie), Définition du bassin pamphyllien: un nouveau domaine à ophiolites a la marge externe de la chaine taurique: Zeit. Dtsch. Geol. Ges., v. 123, p. 385–409.

Durand-Delga, M., 1969, Mise au point sur la structure du Nord-Est de la Berbèrie: Publi. Serv. Geol. Algérie (N.S.), p. 89–131.

Dürr, St., 1976, Über das Menderes-Kristallin und seine Aequivalente in Griechenland: Geol. Soc. France Bull., 7eme ser., XVIII, p. 429.

Elter, P., et al, 1975, Tensional and compressional areas in the recent (Tortonian to present) evolution of the Northern Apennines: Boll. Geofisica, v. 17, p. 3–18.

Erickson, A. J., 1970, The measurement and interpretation of heat flow in the Mediterranean and Black Sea: Ph.D. thesis, Mass. Inst. of Tech., 272 p.

——— G. Simmons, and W. B. F. Ryan, 1977, Review of heat-flow data from the Mediterranean and Aegean Seas, *in* B. Biju-Duval and L. Montadert, eds., Structural history of the Mediterranean basins: Paris, Technip, p. 263–280.

Fahlquist, D. A., and J. B. Hersey, 1969, Seismic refraction measurements in the western Mediterranean Sea: Bull. Inst. Océanogr. Monaco, v. 67, p. 1–52.

Finetti, I., 1976, Mediterranean Ridge: a young submerged chain associated with the Hellenic Arc.: Boll. Geofisica Teor. ed Appl., 18, n° 69, p. 31–65.

——— and C. Morelli, 1972, Wide-scale digital seismic exploration of the Mediterranean Sea: Bol. Geofisica Teor. ed Appl., v. 14, p. 291–342.

——— and ——— 1973, Geophysical exploration of the Mediterranean Sea: Boll. Geofisica Teor. ed Appl., v. 15, p. 263–341.

Foose, R. M., and F. Manheim, 1975, Geology of Bulgaria: a review: AAPG Bull., v. 59, p. 303–335.

Fourquin, C., 1975, L'Anatolie du Nord-Ouest, marge méridionale du continent européen, histoire paléogéographique, tectonique et magmatique durant le secondaire et le tertiaire: Geol. Soc. France Bull., v. 17, p. 1058–1070.

Garkalenko, I. A., Ya. P. Malovitsky, U. P. Neprochnov, 1971, Deep structure of western Black Sea (from DSS data): Acta Geol. Acad. Sc. Hungaricae, v. 15, p. 369.

Garrison, R. E., et al, 1978, Sedimentary petrology and structures of evaporitic sediments in the Mediterranean Sea, *in* Initial Reports of Deep Sea Drilling Project, leg 42A: Washington, D.C., U.S. Govt., v. 42, p. 571–612.

Gass, I. G., 1968, Is the Troodos massif of Cyprus a fragment of Mesozoic ocean floor?: Nature, v. 220, p. 39–42.

Ginzburg, A., et al, 1975, Geology of Mediterranean shelf of

Israël: AAPG Bull., v. 59-11, p. 2142–2160.
Goncharov, V. P., 1972, Geomorphology of the sea bottom, *in* Results of research on international geophysical project: Moscow, Nauka, v. 10, p. 12–17.
Gonnard, R., et al, 1975, Apports de la sismique réflection aux problèmes du volcanisme on Méditerranée et à l'intérpretation des données magnétiques (abs.): 3rd Réun. Ann. Sci. Terre Monpellier, p. 171.
Guernet, C., 1977, Le Miocene hellenique: évolution paléogéographique et tectonique: Rev. Géog. Phys. Geol. Dyn., in press.
Hatzfeld, D., 1976, Etude sismologique et gravimétrique de la structure profonde de la mer d'Alboran: mise en évidence d'un manteau anormal: Acad. Sci. Comptes Rendus, t. 283, p. 1021–1024.
Heezen, B. C., et al, 1971, Evidence of foundered continental crust beneath the central Tyrrhenian Sea: Nature, v. 229, no. 5283, p. 327–329.
Hiecke, W., 1972, Erste Ergebnisse von stratigraphisch-sedimentologischen und morphologisch-tektonischen Untersuchungen auf dem Mediterranean Rucken (Ionisches Meer): Zeit. Deutsch. Geol. Ges., Bd. 123, t. 2, p. 567–570.
Hinz, K., 1972, Results of seismic refraction investigations (Project Anna) in the western Mediterranean Sea, south and north of the island of Mallorca, *in* O. Leenhardt et al, Results of the Anna cruise; Bull. Centre Recherches. Pau, SNPA, v. 6, p. 405–426.
——— 1973, Crustal structure of the Balearic Sea: Tectonophysics, v. 20, (1-4), p. 295–302.
——— 1974, Results of seismic refraction and seismic reflection measurements in the Ionian Sea: Jour. Geol., v. 2, p. 35–65.
Hirn, A., L. Steinmetz and M. Sapin, 1976, A long-range seismic profile in the western Mediterranean basin structure of the upper mantle, in press.
Hortinsk, J., 1971, The Late Cretaceous and Tertiary geological evolution of eastern Turkey, *in* C. Keskin and F. Demirmen, eds., First Petrol. Cong. (Turkey), p. 25–41.
Hsu, K., and W. B. F. Ryan, 1974, Deep-sea drilling in the Hellenic trench: Geol. Soc. Greece Bull., t. X, p. 81–89.
——— et al, 1978, Initial reports of the Deep-Sea Drilling Project: Washington, D.C., U.S. Govt., v. 42, in press.
I.F.P.-C.N.E.X.O., 1974, Carte géologique et structurale des bassins tertiaires du Domaine méditerranéen: Paris, Technip Edit.
Innocenti, F., et al, 1977, Abstract in VI° Aegean Colloquium., Izmir (to be published).
Jongsma, D., 1974, Heat flow in the Aegean Sea: Geophys. Jour., v. 37, p. 337–346.
——— 1975, A marine geophysical study of the Hellenic Arc: Ph.D. thesis, Cambridge, p. 1–69.
Katsikatsos, G., J. L. Mercier, and Y. P. Vergely, 1976, L'Eubée méridionale: une double fenétre polyphasée dans les Hellenides internes., Acad. Sci. Comptes Rendus, 283, p. 459–462.
Kenyon, N. H., and R. H. Belderson, 1977, Young compressional structures of the Calabrian, Hellenic and Cyprus outer ridges, *in* B. Biju-Duval and L. Montadert, eds., Structural history of the Mediterranean basins: Paris, Technip, p. 233–241.
Klitzsch, E., 1970, Die Strukturgeschichte der Zentral-sahara: neue Erkenntnisse zum Dau und zur Palaogeographie eines Tafellandes: Geol. Rundschau, Bd. 59, 2, p. 459–527.
Lalechos, N. and E. Savoyat, 1977, La sédimentation néogène dans le fossé nordègéen: Abs. in VI° Aegean Colloquium, Athens, in press.
Laubscher, H., and D. Bernoulli, 1977, Mediterranean and Tethys, *in* Nairn, Kanes, and Stehli, eds., The ocean basins and margins: v. 4A, p. 1–28.
Le Borgne, E., J. L. Le Mouel, and X. Le Pichon, 1971, Aeromagnetic survey of southwestern Europe: Earth Planetary Sci. Letters, v. 12, n° 3, p. 287–299.
Le Pichon, X., et al, 1971, La Méditerranee Occidentale depuis l'Oligocène., Schéma d'évolution: Earth Planetary, Sci. Letters, v. 13, p. 145–152.
Letouzey, J., et al, 1974, Nappes de glissement actuelles au front de l'arc calabrais en mer Ionienne (d'après la sismique réflection): 2nd Annual Meeting on Earth Sciences, Nancy, p. 260.
——— et al, 1977, The Black Sea: a marginal basin, Geophysical and geological data, *in* B. Biju-Duval and L. Montadert, eds., Structural history of the Mediterranean basins: Paris, Technip, p. 363–374.
——— et al, 1978, Black Sea: Geological setting and recent deposits, distribution from seismic reflection data, *in* Initial reports of Deep-Sea Drilling Project: Washington D.C., U.S. Govt., v. 42, part 2, p. 1077–1084.
Lort, J., 1971, The tectonics of the Eastern Mediterranean, a geophysical review: Rev. Geophysics and Space Research, v. 9, No. 2, p. 189–216.
——— 1972, The crustal structure of the eastern Mediterranean: Unpub. thesis, Cambridge Univ., 117 p.
——— W. Q. Limond, and F. Gray, 1974, Preliminary seismic studies in the eastern Mediterranean: Earth and Planetary Sci. Letters, v. 21., p. 355–366.
Lyubimova, Y. A., and L. A. Savostin, 1973, Teplovoi potok v tsentralnoi i vostochnoi chasti Chernogo Morya: Doklady Akad. Nauk SSSR, v. 212, n° 2, p. 349.
Makris, J., 1975, Crustal structure of the Aegean Sea and the Hellenides obtained from geophysical surveys: Rapp. Proc. Verb. Réunions, CIESM, v. 23, p. 201–202.
Maley, T. S., and G. L. Johnson, 1971, Morphology and structure of the Aegean Sea: Deep-Sea Research, v. 18, No. 1., p. 109–122.
Marinos, G., et al, 1977, Réunion extraordinaire de la Société Géologique de Grèce en Eubée et an Afrique, Compte rendu.: Geol. Soc. France Bull., 7, t. XVX, p. 103–118.
Mascle, A., et al, 1977, Sediments and their deformations in active margins in different geological settings: Inter. Symposium Geodynamics in SW Pacific (Noumea), Paris, Technip, p. 327–343.
Mauffret, A., 1976, Etude Géodynamique de la marge des Iles Baléares: Thèses, Paris, p. 1–137.
——— et al, 1973, Northwestern Mediterranean sedimentary basin from seismic reflection profile: AAPG Bull., v. 57, p. 2245–2262.
——— et al, 1978, Geological and geophysical setting of DSDP Site 372 (Western Mediterranean), *in* Initial reports of the Deep Sea Drilling Project: Washington, D.C., U.S. Govt., v. 42, p. 889–897.
McKenzie, D., 1972, Active tectonics of the Mediterranean region: Geophys. Jour., v. 30, p. 109–185.
Mercier, J. L., et al, 1976, La néotectonique plio-quaternaire de l'Arc égéen externe et de la mer Egée et ses relations avec la séismicité: Geol. Soc. France Bull., 7., t. XVIII, p. 355–372.
Meulenkamp, J. E., 1971, The Neogene in the southern Aegean area, *in* A. Strid, ed., Evolution in the Aegean: Opera Botanica No. 30, p. 5–12.
Monod, O., 1977, Recherches géologiques dans le Taurus occidental au Sud de Baysehir (Turquie): These, Universite Paris-Sud Orsay, 442 p.
Montadert, L., A. Mauffret, and J. Letouzey, 1978, Messinian event: seismic contribution, *in* Initial reports of the Deep Sea Drilling Project: Washington, D.C., U.S. Govt., v. 42, p. 1037–1050.
——— et al, 1970, De l'age tertiaire de la série salifere responsable des structures diapiriques en Méditerranée occidentale (nord-est des Baléares): Acad. Sci. Comptes Rendus, v. 271, p. 812–815.
Montenat, C., G. Bizon, and J. J. Bizon, 1975, Remarques sur le Néogène du forage Joides 121 en Mer d'Alboran (Méditerranée Occidentale): Geol. Soc. France Bull., 7, t. XVII, p. 45–51.
Morelli, C., 1975, Geophysics of the Mediterranean: Newsl. Coop. Invest. Med. Spec. Issue 7, p. 27–111.
——— M. Pisani, and C. Gantar, 1975, Geophysical studies in the Aegean Sea and in the Eastern Mediterranean: Boll. Geofisica Teor. ed Appl., v. 18, p. 127–167.
——— et al, 1977, Seismic investigations of crustal and upper mantle structure of the Northern Apennines and Corsica, *in* B. Biju-Duval and L. Montadert, eds., Structural history of the Mediterranean basins: Paris, Technip, p. 281–286.
Moskalenko, V. N., 1966, New data on the structure of the sedimentary strata and basement in the Levant Sea: Oceanology, v. 6, p. 828–836.
Mulder, C. J., 1973, Tectonic framework and distribution of Miocene chemical sediments with emphasis on the eastern Mediterranean, *in* Drooger, ed., Messinian events in the Mediter-

ranean: Amsterdam, p. 44–59.
——— and G. R. Parry, 1977, Late Tertiary evolution of the Alboran Sea at the eastern entrance of the straits of Gibraltar, *in* B. Biju-Duval and L. Montadert, eds., Structural history of the Mediterranean basins: Paris, Technip, p. 401–410.
Mulder, J. C., P. Lehner, and D. C. K. Allen, 1975, Structural evolution of the Neogene salt basins in the eastern Mediterranean and the Red Sea: Geol. Mijnbouw, v. 54, p. 208–221.
Nairn, A. E. M., and M. Westphal, 1968, Possible implications of the paleomagnetic study of late Paleozoic igneous rocks of northwestern Corsica: Palaeogeography, Palaeoclimatology, Palaeoecology, v. 5, p. 179–204.
Needham, H. D., et al, 1973, North Aegean Sea trough: 1972 *Jean Charcot* Cruise: Geol. Soc. Greece Bull., v. 10, p. 152–153.
Neev, D., et al, 1976, The geology of the southeastern Mediterranean: Geol. Surv. Israel, v. 68, p. 1–51.
Neprochnov, U. P., A. F. Neprochnova, and Ye. G. Mirlin, 1974, Deep structure of Black Sea basin, *in* The Black Sea: geology, chemistry and biology: AAPG Memoir 20, 35 p.
——— et al, 1975, Crustal and upper mantle Structure of the Black Sea Region: Crustal cross-sections based on the deep seismic sounding, *in* The Earth's Crust and the Development of the Black Sea Basin: Moscow, Ed. Nauka, p. 284.
Nesteroff, W. D., et al, 1977, Esquisse structurale en Méditerranée orientale au front de l'Arc Egéen, *in* B. Biju-Duval and L. Montadert, eds., Structural history of the Mediterranean basins: Paris, Technip, p. 241–250.
Ninkovich, D., and B. C. Heezen, 1965, Santorini tephra, *in* W. F. Whittard and R. Bradshaw, eds., Submarine geology and geophysics: London, Butterworths, p. 143–162.
Ozgul, N., 1976, Some geological aspects of the Taurus orogenic belt (Turkey): Geol. Soc. Turkey Bull., v. 19, p. 65–78.
——— and E. Arpat, 1972, Structural units of the Taurus orogenic belt and their continuation in neighbouring regions (abs.): 23rd Cong. CIESM, Athenes, Rapp. Comm. Int. Mer Medit., 22, p. 153.
Papazachos, B. C., and P. E. Comminakis, 1977, Modes of lithospheric interaction in the Aegean area, *in* B. Biju-Duval and L. Montadert, eds., Structural history of the Mediterranean basins: Paris, Technip, p. 319–332.
Paquet, J., 1974, Tectonique Éocène dans les Cordillères Bétiques: vers une nouvelle conception de la paléogéographie en Méditerranée occidentale: Géol. Soc. France Bull., 7, XVI, p. 58–73.
Poisson, A., 1977, Recherches géologiques dans les Taurides Occidentales (Turquie): These, Paris XI.
Rabinowitz, P. D., and W. B. F. Ryan, 1970, Gravity anomalies and crustal shortening in the eastern Mediterranean: Tectonophysics, v. 10, No. 5-6, p. 585–608.
Ricou, L. E., 1971, Le croissant ophiolitique peri-arabe. Une ceinture de nappes mises en place au Cretace superieur: Rev. Geographie Phys. et Geologie Dynamique, v. 13, p. 327–350.
——— I. Argyriadis, and R. Lefevre, 1974, Proposition d'une origine interne pour les nappes d'Antalya et le massif d'Alanya (Taurides occidentales. Turquie): Géol. Soc. France Bull., v. 16, p. 107–111.
——— ——— and J. Marcoux, 1975, L'axe calcaire du Taurus, un alignement des fenetres arabo-africaines sous des nappes radiolaritiques, ophiolitiques et métamorphiques: Géol. Soc. France Bull., v. 17, p. 1024–1044.
Rigo de Righi, M., and A. Cortesini, 1964, Gravity tectonics in foothills structure belt of southeast Turkey: AAPG Bull., v. 48, p. 1911–1937.
Ross, D. A., 1974, The Black Sea, *in* C. A. Burk and C. L. Drake, eds., The geology of continental margins: New York, Springer-Verlag, p. 669–682.
——— and E. Uchupi, 1977, Structure and Sedimentary history of Southeastern Mediterranean-Nile Cone area: AAPG Bull., v. 61, p. 872–902.
——— ——— and C. O. Bowin, 1974, Shallow structure of the Black Sea, *in* The Black Sea: Geology, Chemistry, and Biology: AAPG Mem. 20, p. 11.
——— et al, 1978, Initial reports of the Deep Sea Drilling Project: v. 42, Washington, D.C., U.S. Govt., part 2.
Ryan, W. B. F. , 1969, The floor of the Mediterranean Sea, pt. 1. Structure and evolution of the sedimentary basins; pt. 2. The stratigraphy of the eastern Mediterranean: Ph.D. thesis, Columbia University, New York, p. 189–204.
——— et al, 1971, The tectonics and geology of the Mediterranean Sea, *in* A. E. Maxwell, ed., The sea: p. 387–492.
——— et al, 1973, Initial reports of the Deep Sea Drilling Project: Washington, D.C., U.S. Govt., v. 13.
Said, R., 1975, The geological evolution of the River Nile, *in* F. Wendorf and A. E, Marks, eds., Problems in prehistory: North Africa and the Levant: Dallas, Southern Methodist Univ. Press, p. 7–44.
Salem, R., 1976, Evolution of Eocene-Miocene sedimentation patterns in parts of northern Egypt: AAPG Bull., v. 60, p. 34–64.
Sancho, J., et al, 1973, New data on the structure of the eastern Mediterranean basin from seismic reflection: Earth Planet Sci. Lett., v. 18, p. 1–421.
Schietecatte, J. P., 1971, Geology in the Misis Mountains, *in* A. S. Campbell, ed., Geology and history of Turkey: PESL, p. 305–312.
Schmidt, G. C., 1961, Stratigraphic nomenclature of the Adana region petroleum district VII: Petroleum Admin. Publ. Bull. Ankara, No. 6, p. 47–62.
Schuiling, R. D., 1972, Oceanization-geothermal models: Geologie en Mijnbouw, v. 51, p. 546–547.
Selli, R., 1974, Appunti sulla geologia del Mar Tieerno in Paleogeografia del Terziaro nell'ambito del Mediterraneo occidentale: Seminario Fac. Sc. Universit. di Cagliari, Sup. 43, p. 327–351.
——— and A. Fabbri, 1971, Tyrrhenian: a Pliocene deep sea: Atti. Accad. Mazl. Lincei, v. 50, ser. 8, fasc. 5, p. 579–592.
——— and S. Rossi, 1975, The main geologic features of the Ionian Sea: Rapp. Proc. Verb. C.I.E.S.M., p. 115–116.
Stanley, D. J., 1973, The Mediterranean Sea—a natural sedimentation laboratory: Stroudsburg, Pa., Dowden, Hutchinson & Ross, 765 p.
——— et al, 1974, Subsidence of the western Mediterranean basin in Pliocene-Quaternary time: Further evidence: Geology, v. 2, p. 345–350.
——— et al, in preparation, Catalonian, Eastern Betic and Balearic Margins: definition of structural types and relation to geologically recent foundering of the western Mediterranean Basin.
Stoeckinger, W. T., 1976, Valencia Gulf offer deadline nears: Oil and Gas Jour., March, p. 197–204; April, p. 181–183.
Van Den Berg, J. and A. A. H. Wonders, 1976, Paleomagnetic evidence of large fault displacement around Po Basin: Tectonophysics, v. 33, p. 301–320.
Vilminot, J. C., and U. Robert, 1974, A propos des relations entre le volcanisme et la tectonique en Mer Egée: Acad. Sci. Comptes Rendus, t. 278, p. 2099–2102.
Vogt, P. R., and R. H. Higgs, 1969, An aeromagnetic survey of the eastern Mediterranean Sea and its interpretation: Earth and Planetary Sci. Letters, v. 5, No. 7, p. 439–448.
——— ——— and G. L. Johnson, 1971, Hypotheses on the origin of the Mediterranean basin: magnetic data: Jour. Geophys. Research, v. 76, No. 14, p. 3207–3228.
Weigel, W., 1972, Preliminary results of refractional seismic measurements in the eastern Ionian Sea: Zeit. Deut. Geol. Ges. Pd. 123, t. 2, p. 571.
Wezel, F. C., 1974, Flysch successions and the tectonic evolution of Sicily during the Oligocene and early Miocene, *in* C. H. Squyres, ed., Guidebook of the geology of Italy: Petroleum Exploration Soc. Libya, v. II, p. 105–128.
——— 1977, Widespread manifestations of Oligocene-lower Miocene volcanism around Western Mediterranean, *in* B. Biju-Duval and L. Montadert, eds., Structural history of the Mediterranean basins: Paris, Technip, p. 287–302.
Woodside, J. M., 1975, Evolution of the Eastern Mediterranean: Ph.D. thesis, Cambridge, p. 1–185.
Wright, D. W., et al, 1975, Crustal structure in the Eastern Mediterranean Sea from variable-angle reflection/refraction data: 24th Congr., C.I.E.S.M., Monaco, Rapp. Comm. Int. Mer Medit., 23, p. 278.
Zijderveld, I. D. A., and R. Van der Voo, 1973, Paleomagnetism in the Mediterranean area, *in* D. H. Tarling and S. K. Runcorn, eds., Implications of continental drift to the earth sciences: New York, Academic Press, p. 133–161.

Anatomy of the Mexican Ridges, Southwestern Gulf of Mexico[1]

RICHARD T. BUFFLER[2], F. JEANNE SHAUB[2], JOEL S. WATKINS[2,3], AND J. LAMAR WORZEL[2]

Abstract Six subparallel multichannel seismic lines across the Mexican Ridges foldbelt in the southwestern Gulf of Mexico between 21°N and 22°N latitude show details of the folds and provide new insight into their origin. The top of the seismic unit containing inferred Jurassic salt continues relatively undeformed beneath the foldbelt, suggesting that the subparallel folds are not caused by salt tectonics. Instead, the folds appear to occur in competent beds above a possible décollement. The detachment from the underlying rocks is along a deformed zone occurring within a thick Upper Cretaceous-Lower Tertiary shale section. A domal uplift observed on only one line and seemingly unrelated to the foldbelt may be cored with salt. The uniform folding of most of the upper sedimentary section without any significant diapirism in the cores of the folds, plus the numerous imbricate thrust faults associated with the folds, suggest regional compressional stresses acting in an east-west direction. Involvement of beds as young as Pliocene and Pleistocene indicates that folding and thrusting is very young and may be continuing today. Decreases in fold amplitude and the decrease in sediment ponding in the synclines in a seaward direction suggest that the zone of maximum deformation moved seaward through time. At least two separate mechanisms adequately explain the tectonic style of the Mexican Ridges folds observed on the seismic sections: 1) massive gravity sliding possibly triggered by regional uplift and supplemented by sediment loading at the head of the slide; and 2) compressional tectonic stresses originating within the deeper crust beneath Mexico and transmitted into the fold area through deep thrust zones. In both cases, detachment and deformation take place above a decollement or deformed zone located within mobile substrata (possibly geopressured slope shales).

INTRODUCTION

The Mexican Ridges comprise symmetrically folded strata that extend south from about latitude 24°N along the Mexican continental slope for almost 400 km (Fig. 1). The folds form long, linear, subparallel topographic features on the seafloor. The nature and origin of these folds has been the subject of debate since their discovery in the mid-1960's.

This paper describes the anatomy of the folds in one local area of the southwestern Gulf of Mexico (Fig. 1). The description is based on six multichannel seismic sections that cross the folds perpendicular to strike (Fig. 2). These sections show reflections below the water-bottom multiple and allow for the first time a look at the root zone of the folds, thereby providing new insights into their origin.

Seismic data presented in this paper (Fig. 3-8) were collected aboard the University of Texas Marine Science Institute (UTMSI) research vessel *Ida Green* in November–December 1975. This cruise was part of a project designated as Gulf Tectonics Phase I, a regional seismic study of the Gulf of Mexico conducted by UTMSI. The four middle lines are all 24-fold data that were processed through the stack phase by Texaco Inc. and later wave-migrated by Teledyne, Inc. (Lines GLG-18, -20, -22, -24; Figs. 4-7). The two outside lines are 12-fold data processed by Teledyne, Inc. (Line GLR 16-1-7, -8, -9 and Line GLR 16-1-10; Figs. 3 and 8).

PREVIOUS WORK

The nature and origin of the Mexican Ridges foldbelt were first discussed by Jones et al (1967) who

[1]Manuscript received, November 14, 1977; accepted, May 25, 1978.

[2]Geophysics Laboratory, University of Texas Marine Science Institute, Galveston, Texas 77550.

[3]Current address: Gulf Research and Development Company, Houston, Texas 77036.

The writers wish to acknowledge the assistance and cooperation of Cecil and Ida Green plus the captain and crew of the R/V *Ida Green*. Financial support for the project was provided by the following participants in the Gulf Tectonics Project, Phase I: Amoco Oil Company, Chevron U.S.A., Inc., Exxon Production Research Company, Mobil Oil Corporation, Phillips Petroleum Company, Texaco Inc., and U.S. Geological Survey. The writers particularly want to thank Teledyne, Inc., and Texaco Inc. for their assistance in processing the data.

This manuscript was critically reviewed by Wulf Gose, Ralph Kehle, and Kenneth McMillen.

University of Texas Marine Science Institute Contribution No. 301, Galveston Geophysics Laboratory.

Article Identification Number:
0065-731X/78/MO29-0021/$03.00/0.
(see copyright notice, front of book)

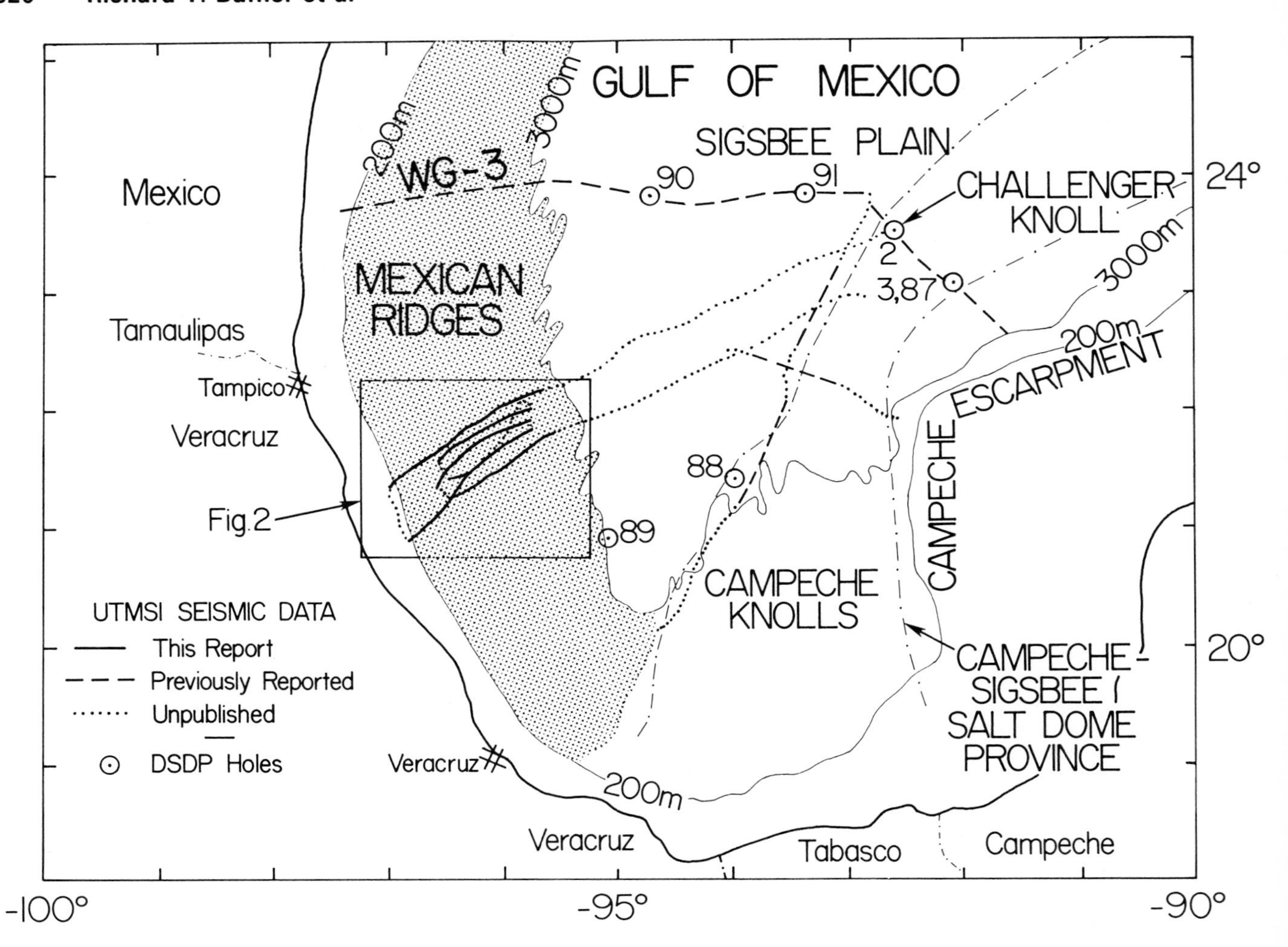

FIG. 1 Map of southwestern Gulf of Mexico showing location of study area (box) along Mexican Ridges foldbelt (stippled pattern), pertinent University of Texas Marine Science Institute (UTMSI) seismic lines, and Deep Sea Drilling Project holes.

suggested that the folds originated as salt anticlines. Bryant et al (1968) coined the term "Mexican Ridges" and described in some detail the nature and distribution of the folds, dividing them into four physiographic provinces or zones. Their Zones 2 and 4 have different trends but are both similar and are characterized by long linear folds with relief on the seafloor up to 500 m and wavelengths averaging 10 to 12 km. These zones are separated by Zone 3, a low-relief area with lower amplitude folds, supposedly characterized by a series of large east-west-trending scarps and faults. The seismic lines discussed in this paper lie in the northern part of Zone 4 near the boundary with Zone 3.

Bryant et al (1968) suggested several possible origins for the development of the folds: 1) sliding of sedimentary rocks on a decollement surface, including gravity sliding; 2) folding associated with compressional tectonic stresses; 3) vertical movement of shale or salt masses related to static loading; and 4) folds controlled by faulting. They believed that the folds probably represented the southward continuation of salt structures from the Texas-Louisiana slope, but that in the Mexican Ridges area the uniform salt features were never altered significantly by later secondary structural growth due to sedimentary overburden as on the Texas-Louisiana slope. In a later study based on single-channel seismic and other geophysical data, Massingill et al (1973) proposed that the Ridges were the result of salt tectonism or salt flowage caused by a sedimentary overburden and also influenced by movement of basement structural highs. Emery and Uchupi (1972) also favored folding of an evaporite sequence, perhaps as a result of gravity sliding on a décollement plane, but suggested that basement was not involved in the folding based upon the smoothness of the magnetic anomalies and short wavelength of the folds.

Garrison and Martin (1973) described and discussed the origin of the Mexican Ridges at some length mainly on the basis of single-channel seismic data from the 1969 *Kane* cruise. Their data showed an area of low-amplitude folds corresponding to Zone 3 but no evidence for large continuous scarps or faulting as proposed by Bryant et al (1968). Garrison and Martin (1973) favored formation of the folds by large-scale gravity sliding along a dé-

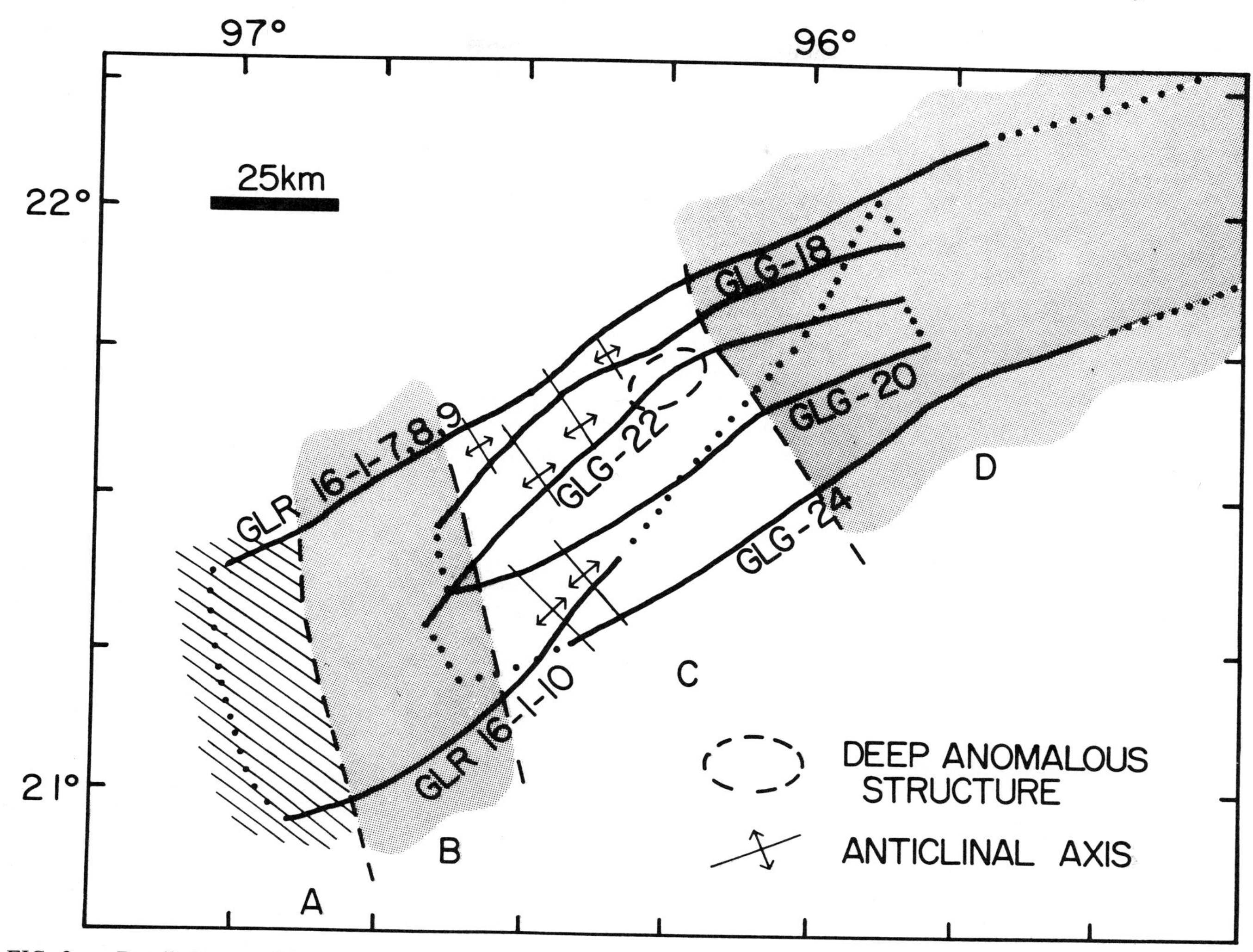

FIG. 2 Detailed map of Mexican continental slope showing designation and location of seismic data included in this paper (Figs. 3 to 8; solid lines) and associated unpublished tie-lines (dotted lines). The "Golden Lane" (GL) designation for the seismic lines is based on their proximity to the Lower Cretaceous Golden Lane carbonate bank or reef located just west of the study area beneath the continental shelf. Zone A is a system of growth faults and slumps and an associated thick sedimentary sequence along the outer shelf and upper slope. Zone B is a thick sequence of young sediments ponded behind the main Mexican Ridges foldbelt. Zone C is an area of high-to-medium-amplitude folds along the middle of the slope forming the main Mexican Ridges foldbelt. The trend of anticlinal axes is shown for only some of the more easily correlatable folds. Zone D is area of lower amplitude folds grading seaward to an area of no folds beneath Sigsbee Plain.

collement surface consisting of low-density material, either salt or shale. This conclusion was based on the lack of clear-cut diapirism or a shallow acoustic basement in the core of the folds, plus the grouping of the folds into two large lobate areas (Zones 2 and 4 of Bryant et al, 1968).

SEISMIC STRATIGRAPHY

Recent seismic studies by UTMSI outline the stratigraphy and structure of the thick sedimentary sequence underlying the deep central Gulf of Mexico basin (Fig. 1). For the convenience of discussing the stratigraphy and geologic history, Ladd et al (1976) subdivided the sedimentary sequence into six seismic units based upon seismic reflection characteristics and Deep Sea Drilling Project core hole data. The seismic units were further refined by Watkins et al (1977, 1978). They can be traced laterally throughout most of the abyssal Gulf and probably represent gross vertical changes in lithology and depositional history.

The boundaries of the units can be traced laterally into the study area from the central Gulf through unpublished seismic tie lines (Fig. 1). Table 1, modified from Ladd et al (1976) and Watkins et al (1977, 1978), summarizes the seismic characteristics, lithologies, and inferred ages of the units present in the study area. The youngest unit in the central Gulf (Sigsbee) thins, changes facies, and loses its identity in the southwestern Gulf, and thus it is included with the underlying Cinco de Mayo unit. The units are mapped on the seismic sections and provide a framework for evaluating the geologic history and origin of the folds.

The Challenger unit, previously thought to be mainly Jurassic in age (Ladd et al, 1976; Watkins et al, 1977, 1978), is now considered to include Lower Cretaceous strata (Buffler et al, 1978). Correlations using unpublished UTMSI seismic data in the central and western Gulf suggest that the strong reflections in the upper Challenger may be the deeper water equivalent of the Lower Cretaceous shallow-water carbonate banks or reef trends that rim the Gulf basin. These data also indicate that the top of the Challenger is a major unconformity and represents a significant geologic event, possibly a major sea-level change accompanied by a major shift in sedimentation pattern. We propose that this unconformity probably corresponds to the major middle Cretaceous (middle Cenomanian) global unconformity proposed by Vail et al (1977) on their cycle chart. The top of the Challenger also can be traced using unpublished UTMSI seismic lines into the eastern Gulf and Florida Straits to the vicinity of Deep Sea Drilling Project Hole 97. Here the boundary occurs just at or below the bottom of the hole, which reached Cenomanian-aged rocks. This lends further support to a Middle Cretaceous age for the top of the Challenger.

The lower part of the Challenger unit in the deep central Gulf, northwest of the Campeche Escarpment, apparently contains the inferred Jurassic salt that feeds the domes and knolls of the Campeche-Sigsbee Salt Dome Province (Ladd et al, 1976; Watkins et al, 1977, 1978). Presumably this is the only unit containing salt in the central Gulf.

The lower Mexican Ridges and Campeche units are inferred to contain mainly shales (Table 1). This is based upon the seismic characteristics of the unit and is supported by relatively low interval velocities (2 to 3 km/sec) for these units, both in the study area (preliminary unpublished data) and along seismic line WG-3 to the north (Watkins et al, 1977).

DESCRIPTION

The main Mexican Ridges foldbelt in the study area is a zone 50 to 70 km wide of relatively high-amplitude folds extending along the middle of the continental slope parallel to the shelf (Zone C, Fig. 2). The folds have wavelengths of approximately 10 km and have topographic relief on the seafloor up to 500 to 700 m. They are generally quite symmetrical, but in some places they show slight asymmetry in both a landward and a seaward direction (e.g., west half of lines GLG-22 and -24, Figs. 5 and 7). The high-amplitude nature of the folds begins abruptly midway down the slope. Landward of this point the beds are only very gently folded (Fig. 3) or not folded at all (Fig. 8). The folds decrease in amplitude seaward of Zone C and finally die out (Zone D, Figs. 2 and 3).

A sedimentary sequence equivalent to the Sigsbee-upper Cinco de Mayo (Pliocene-Holocene) fills in behind the high-amplitude folds (Figs. 3 and 8). These folds, therefore, act as barriers to sediments being transported downslope. The sediments ponded behind the first high-amplitude fold form a thick sedimentary basin approximately 20 to 40 km wide, extending along the upper slope (Zone B, Figs. 2, 3, and 8). This sequence expands and fans out updip into a large growth fault system that apparently extends along the length of the upper slope (Zone A, Figs. 2, 3, and 8). A similar feature is observed farther north along seismic line WG-3 (Watkins et al, 1977) (Fig. 1). A large rollover structure occurs just seaward of this growth fault system (Fig. 8). Small slumps with low-angle glide planes are superimposed on the growth fault system (Fig. 3). This young sedimentary sequence reaches a maximum thickness of approximately 3,000 m (almost 3 sec.) (Figs. 3 and 8).

It is difficult to correlate fold axes exactly from one line to another, even though the lines are relatively close together. The folds appear to change amplitude, relief, and overall character rapidly; some folds apparently plunge and die out along strike. Based upon a best attempt at correlating some of the fold axes, the folds appear to trend in a north-northwest direction as indicated on Figure 2. This trend forms an en echelon pattern with the generally north-south trend of the shelf and slope, adjacent growth fault system, and young sedimentary basin behind the folds (Fig. 2).

The entire foldbelt contains numerous thrust faults that form an imbricate-type structural pattern. The faults dip mainly landward and tend to flatten at depth and die out within the reflectionless zones in the lower Mexican Ridges and upper Campeche units. Most of the faults cut up through the cores of the anticlines, although some occur in the synclines. The faults are very steep in the axes of the tighter folds and are much less steep in the seaward, gentler folds. A few of the more prominent thrust faults are noted on the migrated seismic sections (Figs. 4, 5, 6, 7).

The strata in the entire upper part of the sedimentary section (Mexican Ridges and Sigsbee-Cinco de Mayo) are folded parallel to each other except for the younger ponded sediments. Very little diapirism is associated with the folds in this area except for some possible local thickening and deformation of the reflectionless zone in the lower Mexican Ridges-upper Campeche deep within the cores of the anticlines. The lack of coherent reflections in the tighter folds appears to be due simply to the steepness of the beds rather than to diapirism. The folding apparently is quite recent, as beds as young as Pliocene or possibly younger are involved in the folding.

Folding in the Mexican Ridges foldbelt appears to occur above the reflectionless zones in the lower Mexican Ridges and upper Campeche units. These zones are interpreted to consist of homogeneous, fine-grained, probably pelagic or hemipelagic sediments (Table 1) (Ladd et al, 1976; Watkins et al, 1977, 1978). Strong reflections in the lower Campeche and upper Challenger units continue relatively undeformed beneath the foldbelt.

One major exception to the observations about the nature of the folds noted in the previous paragraph occurs along line GLG-22 (Fig. 5). Here a large

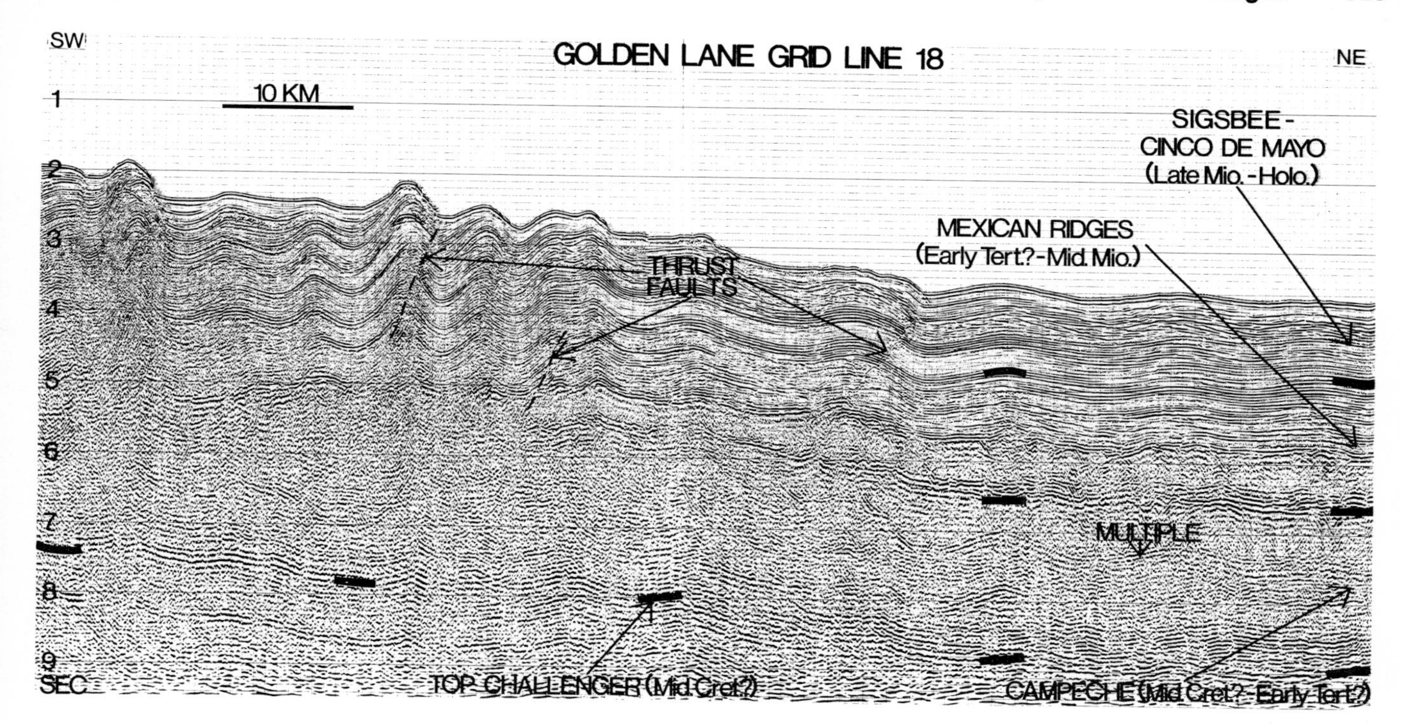

FIG. 4 Line GLG-18 (24-fold and migrated). Several prominent thrust faults are noted. Thrusts appear to flatten landward (SW) and die out within lower Mexican Ridges–upper Campeche units. Lower Campeche and upper Challenger reflections continue beneath foldbelt relatively undeformed. Migration distorts reflections in lower units. Anomalous bright reflection in first major fold at 2.7 sec (approximately 30 km from left side of section) is interpreted to be gas hydrate reflection or possible hydrocarbon accumulation.

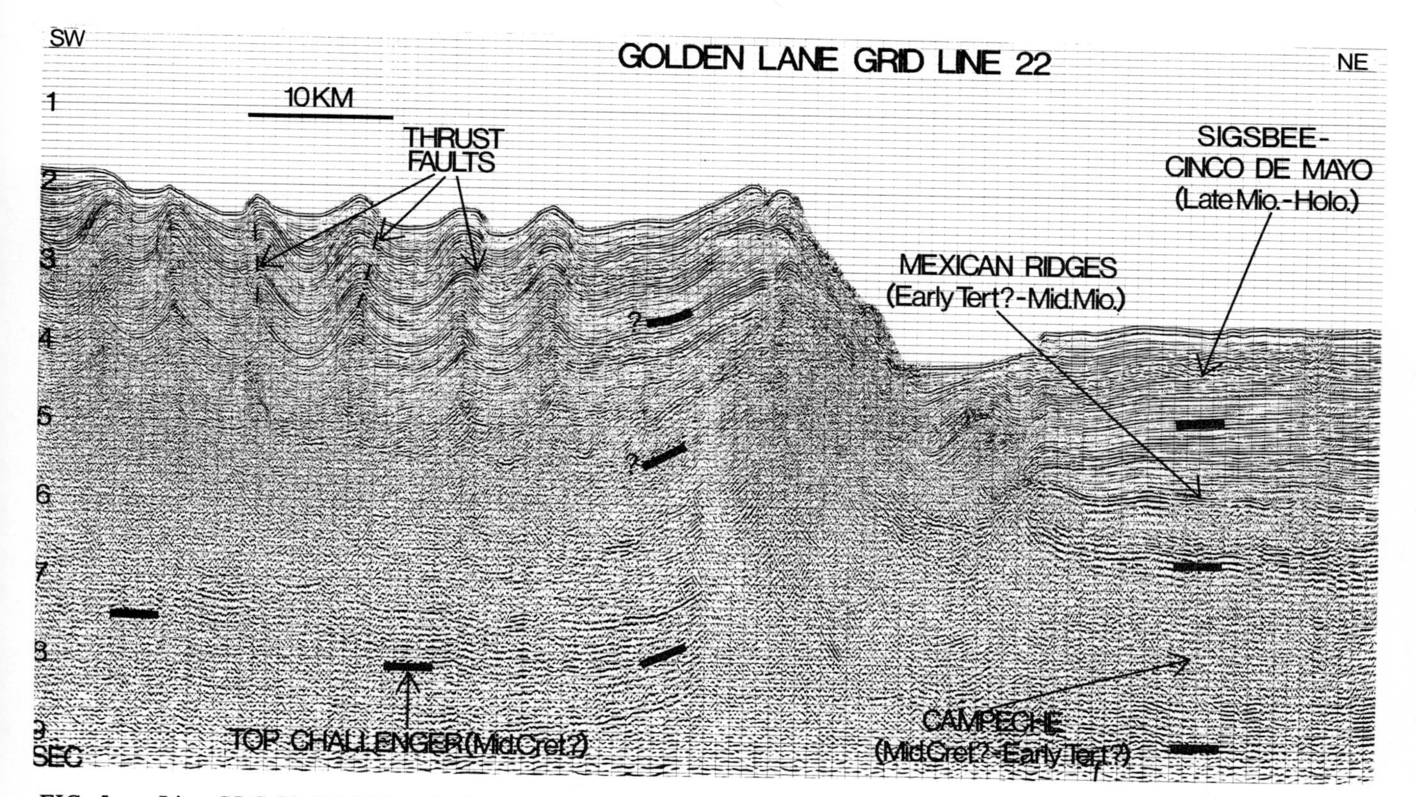

FIG. 5 Line GLG-22 (24-fold and migrated). Folds are slightly asymmetrical in both a landward and seaward direction. Note steep thrust faults. Large, anomalous, deep-seated anticlinal structure and associated syncline and normal faulting disrupt the foldbelt.

Table 1. Seismic Units, Southwestern Gulf of Mexico[1]

Unit	Seismic Characteristics	Inferred Lithology	Inferred Age
Sigsbee-Cinco de Mayo	Strong, discontinuous reflections in Mexican Ridges area. Prominent zone of large-scale cross-beds in middle of unit.	Alternating hemipelagic silty clays, fine-grained, thin turbidites and sandy turbidites.	Late Miocene to Holocene
Mexican Ridges	Strong, discontinuous reflections in upper part; reflectionless (transparent) zone in lower part; lower boundary is sequence of strong, discontinuous reflections.	Dominantly sandy turbidites in upper part; fine-grained turbidites and hemipelagic sediments in lower part; possible sandy turbidite sequence at base.	Early Tertiary? to Middle Miocene
Campeche	Generally weak reflections or reflectionless (transparent); strong, discontinuous reflections near bottom.	Mostly fine-grained, homogeneous pelagic or hemipelagic sediments (shales); possible turbidites toward base.	Middle Cretaceous? to early Tertiary?
Challenger	Upper part is sequence of strong reflections.	Upper part possible deep water carbonates equivalent to Lower Cretaceous carbonate banks rimming deep Gulf of Mexico. Lower part in central Gulf north of Campeche escarpment contains inferred Jurassic salt.	Jurassic? to middle Cretaceous?

[1]Modified from Ladd et al (1976) and Watkins et al (1977, 1978).

anomalous structure appears to originate from below the top of the Challenger unit, as the top of the Challenger is deformed along with the overlying beds. Steeply dipping beds appear to occur deep within the structure, and there is no obvious diapiric core. This structure forms a large scarp on the seafloor with an accompanying synclinal area just seaward of the scarp (Fig. 5). The outer rim of the syncline is characterized by a series of normal faults with the most seaward fault forming a small, abrupt, landward-facing scarp.

This latter deformation is also quite recent, but it appears to have an origin different from that of the folds. This scarp is apparently the same scarp described by Bryant et al (1968) and noted on two of their profiles. They used this structure as evidence for an east-west zone of scarps and faults (Zone 3). Our adjoining seismic data, however, show no evidence for any lateral extent of this feature, although conceivably it could represent a fault zone trending roughly parallel to the seismic lines. The approximate location of this feature is shown on Figure 2.

DISCUSSION

Several conclusions concerning the origin of the Mexican Ridges folds in this part of the southwestern Gulf of Mexico are inferred in light of the newer, deeper-penetration seismic data presented here.

1. The subparallel folds probably are not due to salt mobilization, as the top of the Challenger unit, presumably the only unit containing salt in the central Gulf, continues independently and relatively undeformed beneath the folds.

2. The only exception to the above conclusion is the deep, isolated structure seen on Line GLG-22 (Fig. 5). The structure apparently originates from beneath the Challenger, but the style is not typical of Gulf coast salt diapirs, and there is no strong reflection marking the top of a salt core. The presence of an adjacent withdrawal syncline, however, does support a salt origin. An alternative origin for the structure might be a major isolated zone of thrusting. The lack of any significant magnetic anomaly across the feature (unpublished data) precludes the possibility of volcanic origin.

3. The geometry of the folds fits well a two-layered model of deformation (e.g., Wiltschko and Chapple, 1977) with uniformly folded competent beds (turbidites of upper Mexican Ridges and Sigsbee-Cinco de Mayo) overlying a deformed zone in which weak or incompetent material deforms and flows from synclines into anticlines (geopressured? shales of upper Campeche-lower Mexican Ridges). The deformation may take place above a décollement or a series of sole thrust faults within the weak material. Stresses in the system are also accommodated through the many imbricate thrust faults splaying upward from the deformed zone.

4. The decrease in fold amplitude and dip of the imbricate thrusts in a seaward direction suggests that the tighter folds updip formed first and the gentler folds downdip formed later. This is supported by the ponding of younger sediments updip, while equivalent beds appear to be gently folded along with the entire section farther downdip.

5. The geometry of the Mexican Ridges foldbelt suggests that the deformation has been caused by

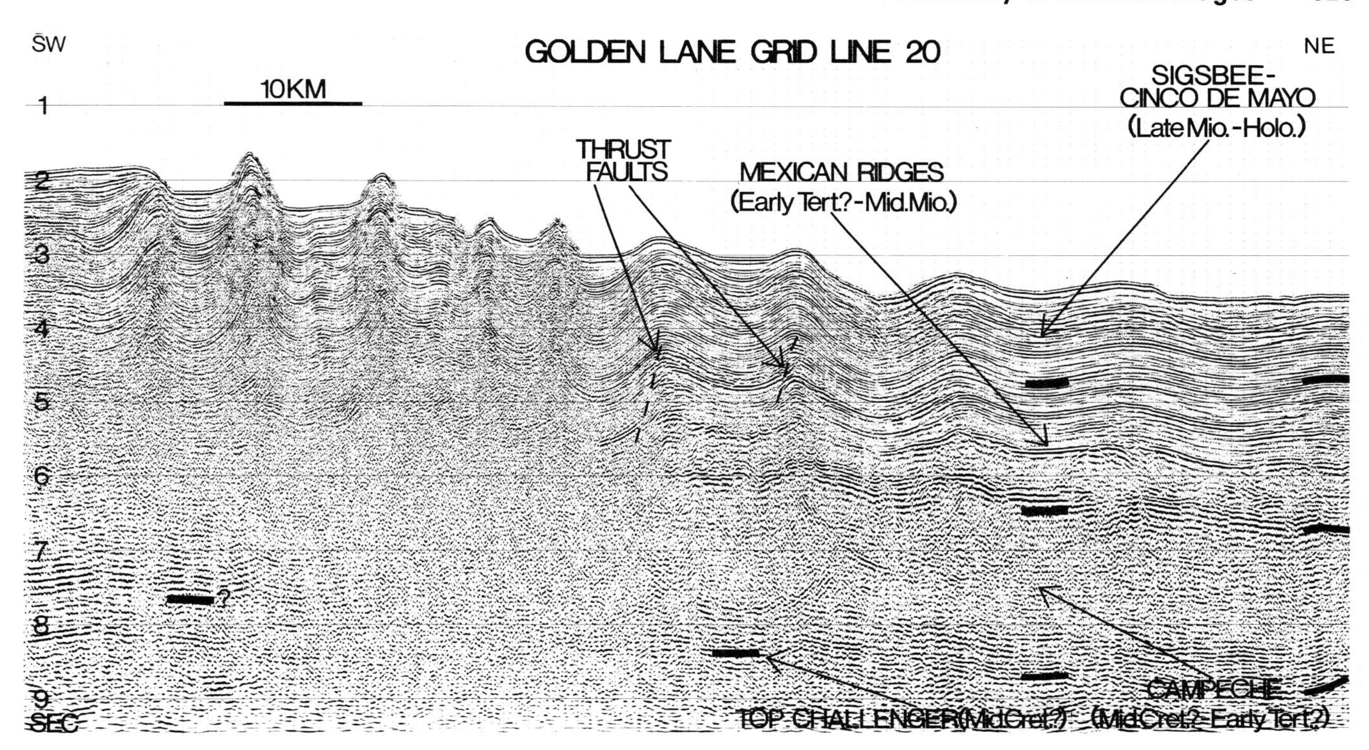

FIG. 6 Line GLG-20 (24-fold and migrated). Fold amplitudes decrease seaward. Note prominent thrust faults. Strong reflections in lower Campeche-upper Challenger appear to continue beneath folds relatively undeformed. Migration distorts deeper reflections. Young sediments are ponded behind folds.

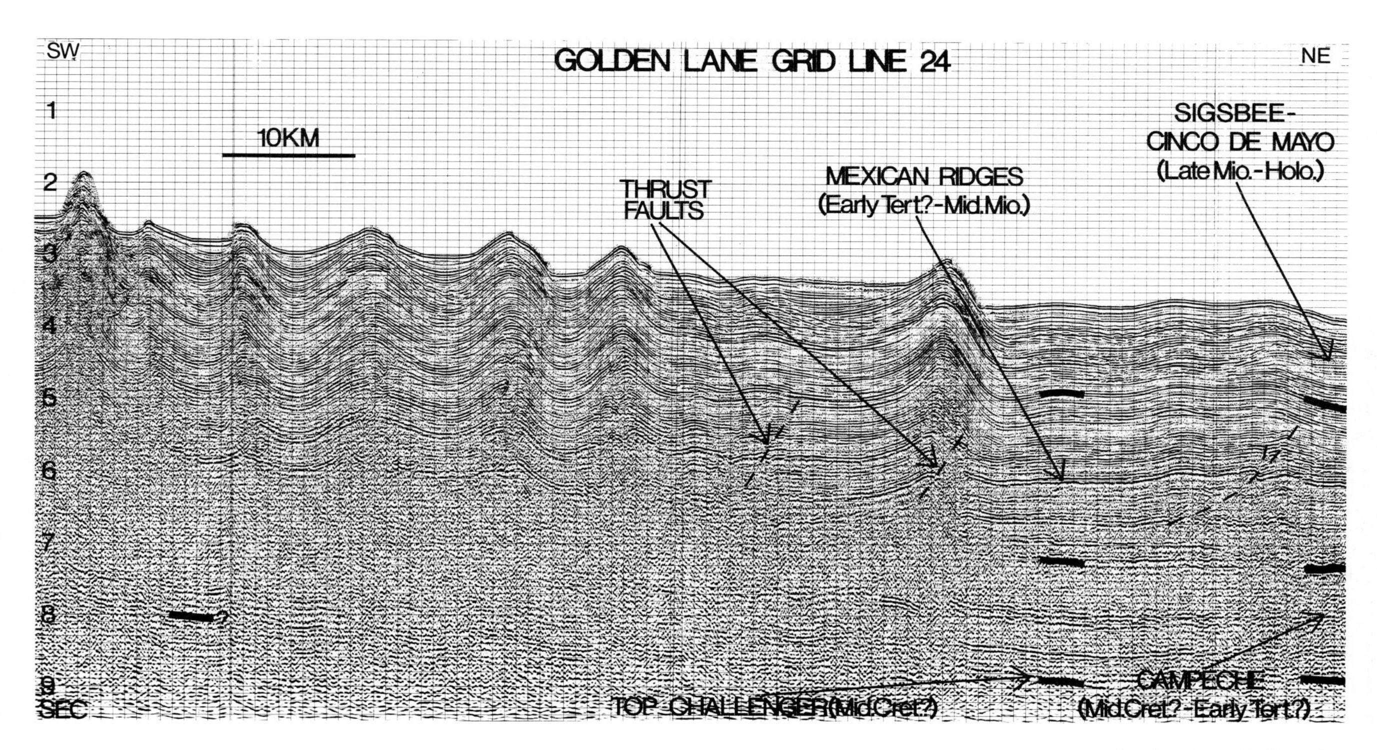

FIG. 7 Line GLG-24 (24-fold and migrated). Folds show slight asymmetry in both a seaward and a landward direction. Several prominent thrust faults are noted. Thrusts dip landward and appear to flatten out into reflectionless zones in lower Mexican Ridges-upper Campeche units. Lower Campeche-upper Challenger units continue relatively undeformed beneath folds. Migration distorts deep reflections.

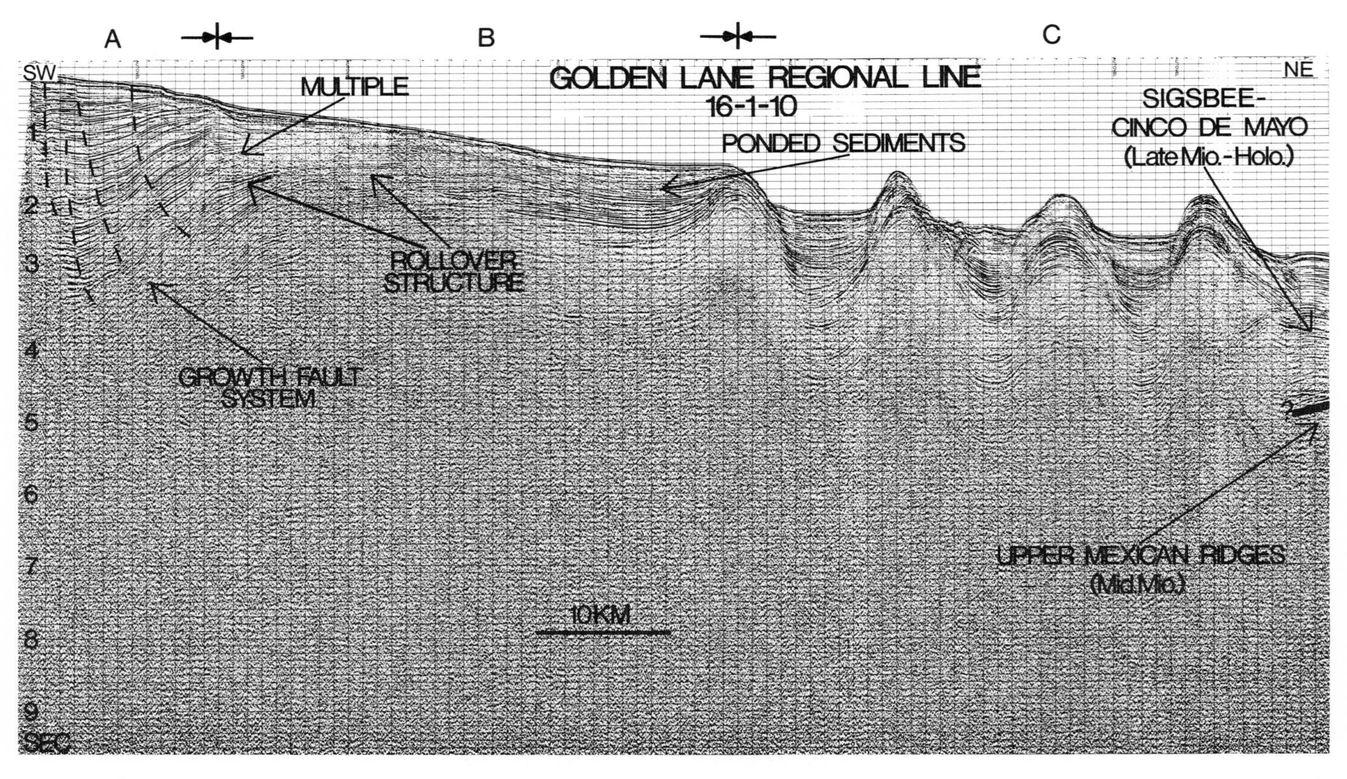

FIG. 8 Line GLR-16-1-10. Young slope sediments ponded behind high-amplitude folds of main Mexican Ridges foldbelt. Large growth-fault system and associated rollover structure prominent beneath upper slope.

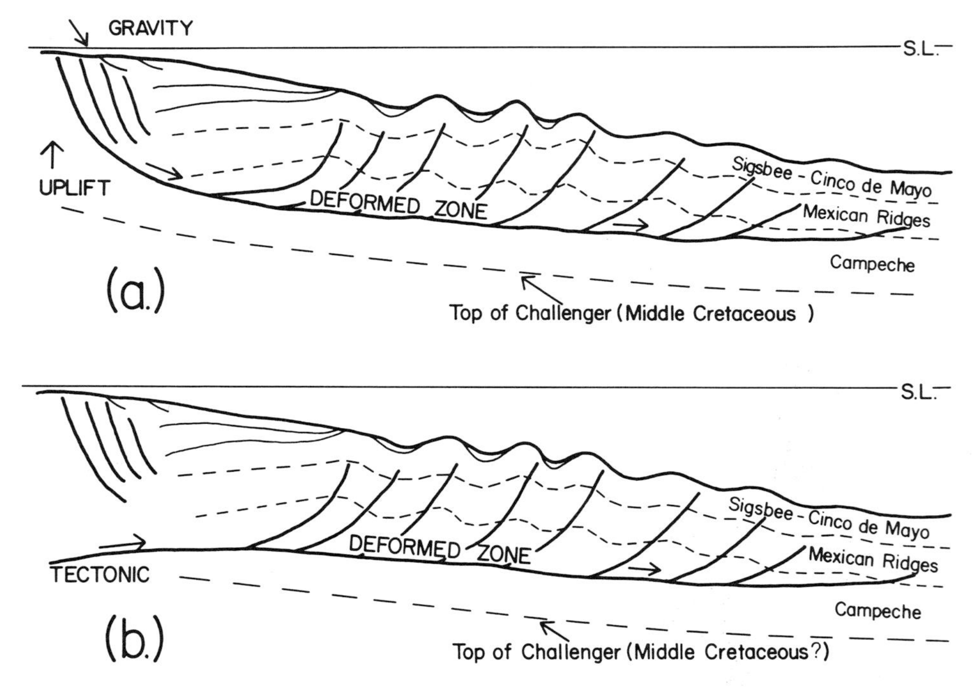

FIG. 9 Two hypothetical mechanisms to explain the structure of the Mexican Ridges foldbelt: **(a)**, gravity sliding possibly caused by uplift and supplemented by sediment loading at the head of the slide, and **(b)**, compressional tectonic stresses originating from within the deeper crust beneath Mexico and transmitted into the fold area through deep thrust zones. In both cases, deformation takes place above a décollement or deformed zone consisting of mobile substrata (possibly geopressured slope shales) in the upper Campeche unit.

compressional stresses acting in an east-west direction. Two possible mechanisms could adequately explain this structure (Fig. 9):

a) Large-scale instability of slope sediments with massive gravity sliding along weak or mobile (incompetent) substrata, possibly geopressured shales (Fig. 9a). The sliding may or may not involve actual detachment. The east-west compressional stresses created by gravity result in folding and thrusting, which is accommodated by the flow of material in the deformed zone. Space created updip allows for the formation of a large growth-fault system. The growth faults probably flatten downdip and probably merge with the deformed zone. Sediment loading in the growth-fault system and ponding behind the initial folds probably supplement the compressional stresses, intensifying the midslope folds and allowing the stress system to migrate seaward with time. The young folding possibly was triggered by recent uplift, which also could explain the influx of large volumes of fairly recent sediments ponded behind the folds and filling the growth-fault system.

b) Tectonic compressional stresses originating from within the crust beneath Mexico and transmitted to the slope along a major thrust zone (Fig. 9b). The stresses occur above a deformed zone or a zone of major sole thrust faulting and are accommodated by folding, thrusting, and the flow of material in the deformed zone. This creates space updip allowing for the formation of a large growth-fault system. This geometry is similar to that proposed for an active trench-arc system (Seely and Dickinson, 1977).

No attempt is made in this paper to analyze the mechanics of deformation and to evaluate the pros and cons of the two models.

6. The folding and thrusting are quite recent and may be continuing today, as part of a widespread tectonic event throughout the entire western Gulf of Mexico. For example, folding in the Campeche-Sigsbee Salt Dome Province to the east is also very recent, involving a similar section (Mexican Ridges and Sigsbee-Cinco de Mayo units) as well as older strata. Here the folding is associated with diapirism and appears to be related to mobilization of salt from below the Challenger unit. Although the mechanism of folding is different from that of the Mexican Ridges area, the young age of the folding suggests that it may have been triggered by the same widespread tectonic event, which Watkins et al (1977, 1978) refer to as the Maya tectonic event.

REFERENCES CITED

Bryant, W. R., J. W. Antoine, M. Ewing, and B. Jones, 1968, Structure of Mexican continental shelf and slope, Gulf of Mexico: AAPG Bull., v. 52, p. 1204–1228.

Buffler, R. T., J. S. Watkins, J. L. Worzel, and F. J. Shaub, 1978, Structure and early geologic history of the deep central Gulf of Mexico (abs.): AAPG Bull., v. 62, p. 501–502.

Emery, K. O., and E. Uchupi, 1972, Western North Atlantic Ocean: Topography, rocks, structure, water, life, and sediments: AAPG Mem. 17.

Garrison, L. E., and R. G. Martin, Jr., 1973, Geologic structures in the Gulf of Mexico basin: U.S. Geol. Survey Prof. Paper 773, 85 p.

Jones, B. R., J. W. Antoine, and W. R. Bryant, 1967, A hypothesis concerning the origin and development of salt structures in the Gulf of Mexico sedimentary basin: Gulf Coast Assoc. of Geol. Socs. Trans., v. XVII, p. 211–216.

Ladd, J. W., R. T. Buffler, J. S. Watkins, and J. L. Worzel, 1976, Deep seismic reflection results from the Gulf of Mexico: Geology, v. 4, p. 365-368.

Massingill, J. V., R. N. Bergantino, H. S. Fleming, and R. H. Feden, 1973, Geology and genesis of the Mexican Ridges: Jour. Geophys. Res., v. 78, p. 2498–2507.

Seely, D. R., and W. R. Dickinson, 1977, Structure and stratigraphy of forearc regions, *in* Geology of continental margins: AAPG Contin. Educ. Course Note Series No. 5.

Vail, P. R., et al, 1977, Seismic stratigraphy and global changes in sea level, *in* Charles E. Payton, ed., Seismic stratigraphy—applications to hydrocarbon exploration: AAPG Mem. 26, p. 49–212.

Watkins, J. S., J. W. Ladd, F. J. Shaub, R. T. Buffler, and J. L. Worzel, 1977, Seismic section WG-3, Tamaulipas shelf to Campeche Scarp, Gulf of Mexico: AAPG Seismic Section No. 1.

——— J. W. Ladd, R. T. Buffler, F. J. Shaub, M. H. Houston, and J. L. Worzel, 1978, Occurrence and evolution of salt in the deep Gulf of Mexico, *in* A. H. Bouma et al, eds., Framework, facies, and oil trapping characteristics of upper continental margin: AAPG Studies in Geology No. 7, p. 43–65.

Wiltschko, D. V., and W. M. Chapple, 1977, Flow of weak rocks in Appalachian Plateau folds: AAPG Bull., v. 61, p. 653–670.

Major Structural Elements and Evolution of Northwestern Colombia[1]

HERMAN DUQUE-CARO[2]

Abstract The northwestern coastal area of Colombia comprises two main geotectonic elements: a stable region or platform underlain by continental crust (unfolded), and an unstable region or geosyncline underlain by oceanic crust (folded).

The platform can be subdivided tectonically into four prominent structural zones: the Cicuco and El Difícil highs, the Plato geofracture and the Sucre tectonic depression, which in turn are tectonically controlled and bounded by seven major basement lineaments trending N, N 20° E, N 40° E, N 55° W, and N 20° W.

The geosynclinal region embraces the western coastal portion adjacent to the platform, and has been divided into two structural elements, the middle Eocene San Jacinto fragmented belt, trending N 20° E, and probably extending northwards into the Caribbean area, and the Pliocene-Pleistocene Sinú belt, paralleling the San Jacinto belt but turning northeastward in the Cartagena-Barranquilla area following the coastline, each with its own distinctive characteristics. These structural features are bounded by three geomorphic lineaments: the Romeral, Sinú, and Colombia lineaments, which are critical to the understanding of the tectonic and sedimentary evolution of the Caribbean northwestern coast of Colombia. These three geomorphic elements are also considered as remnant expressions of ancient trenches or paleotrenches which successively migrated westward, and their turbidite sedimentary fill was sequentially uplifted and deformed during the pre-Andean and Andean orogenies. Likewise, the sharp facies changes associated with the paleotrench margins show a progressive westward migration of the platform facies as the zone of continental accretion widened.

INTRODUCTION

The northwestern coastal area of Colombia occupies the northwestern corner of the South American continent and forms the southeastern margin of the Caribbean Colombian Basin (Fig. 1). The region is bordered by the Santa Marta massif on the east and is limited to the south by the northern extensions of the Western and Central Cordilleras of the Colombian Andes (Fig. 2). General features of the geology are shown on Figure 3, but many structural features are concealed at depth.

Three main topographic elements are recognized as outlining and characterizing the general physiography of the region (Fig. 3): (1) *A highland system* which, in addition to the Western and Central Andean Cordilleras, also includes the coastal serranías of Abibe-Las Palomas, with a maximum elevation of 2,200 m in the Alto of Quimarí, and the inland serranías of San Jacinto, San Jerónimo, and Luruaco, which increase in altitude southward to a maximum of 1,270 m at Murrucucú. The latter serranías appear to represent a low northern extension of the Western Andean Cordillera. (2) *The continental shelf and slope,* constitute a thick sedimentary wedge extending seaward from the coastline. (3) *Interhighland flatlands,* principally occupied by swampy terrain and river systems, includes the Magdalena, Cauca, San Jorge and Sinú Rivers and the Dique Canal.

To date, little information has been published on the regional stratigraphy and structure of this region, the understanding of which is critical to the interpretation of the geological history of the southwest Caribbean area and, in particular, to the interrelation

[1]Manuscript received, May 10, 1977; accepted, October 18, 1977.

[2]Instituto Nacional de Investigaciones Geológicas y Mineras, Bogotá, Colombia.

This paper represents a summary of more than 10 years of field observations and geologic thinking about northwestern Colombia, and has been prepared with the help of much unpublished data and discussions with colleagues from the Instituto de Investigaciones Geológicas y Mineras, Bogotá, and the oil industry.

Unpublished seismic data reported in this paper were kindly made available by Mario Yory, geophysicist with Ecopetrol, and by personnel of Aquitaine Colombie, to whom I am very grateful. My thanks to Geocolombia, Bogotá, for the accommodations, facilities, and geologic guidance when I had the opportunity to visit the upper Sinú River area; to Vernon Hunter and Norman Rowlinson of the oil industry for helping with the English translation of the manuscript; and to Phillips Petroleum Company, Bogotá, for drafting some of the illustrations and typing parts of the manuscript.

I feel especially indebted to J. E. Case for his valuable comments and suggestions, and to R. M. Stainforth, for critically reading the manuscript.

Article Identification Number:
0065-731X/78/MO29-0022/$03.00/0.
(see copyright notice, front of book)

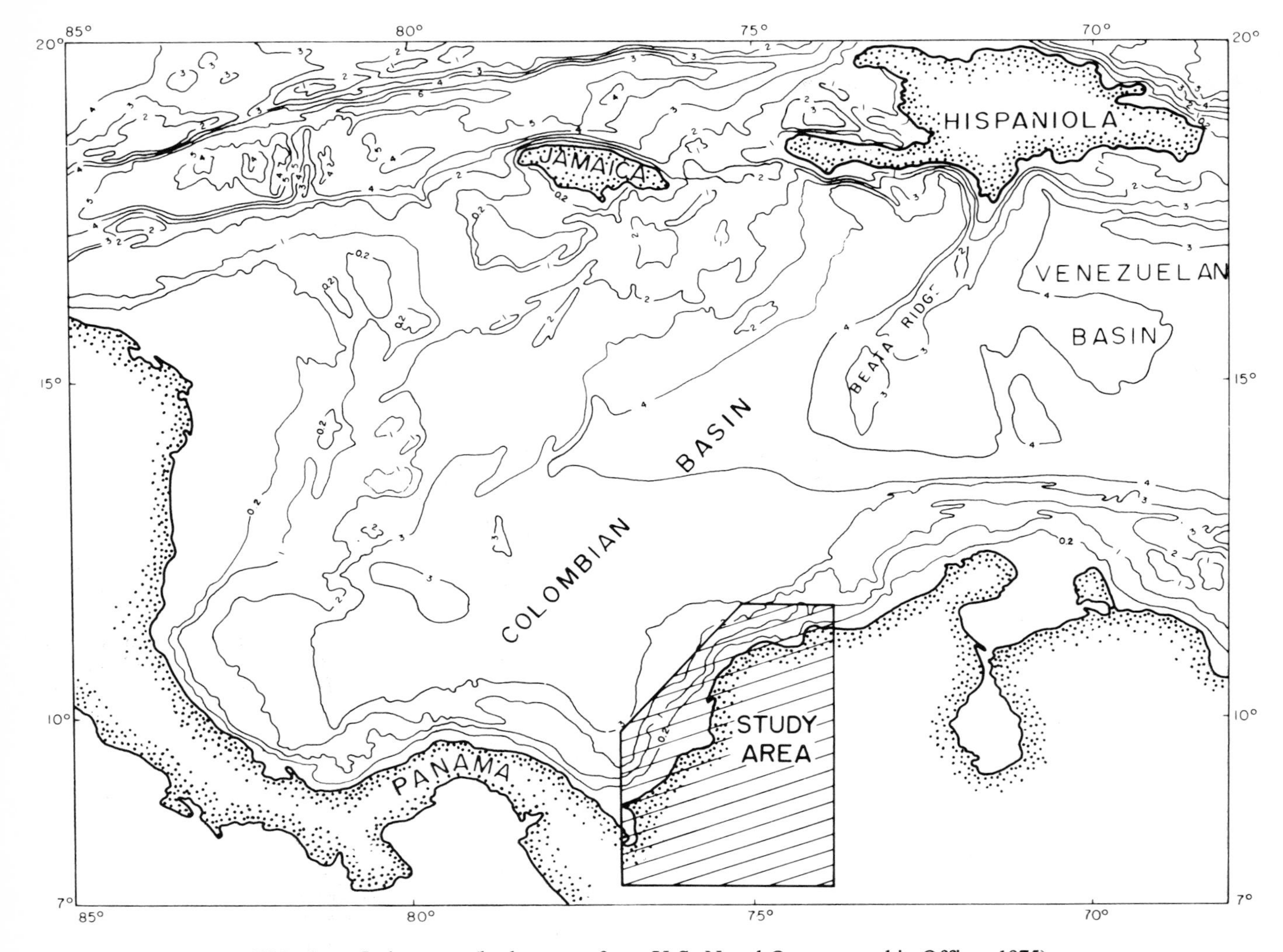

FIG. 1 Index map (bathymetry from U.S. Naval Oceanographic Office, 1975).

of the offshore and onshore geology. Publications with limited information include those of Durán (1964), Zimmerle (1968), Shepard et al (1968), Krause (1971), Shepard (1973), and Case (1974).

The principal objective of this study is to describe and interrelate the distinctive structural and sedimentary patterns of both the offshore and onshore areas in an attempt to more clearly define the geological history of this southwestern corner of the Caribbean area.

MAJOR TECTONIC ELEMENTS

For the purpose of this paper, the above-mentioned topographic features are grouped into two main geotectonic elements (Fig. 4): a stable region or platform underlain by continental crust (unfolded); and an unstable region or geosyncline underlain by oceanic crust (folded).

Stable Region or Platform

This region coincides with the so-called Lower Magdalena Valley and is bounded to the west by the San Jerónimo, San Jacinto, and Luruaco anticlinoria, and to the east and the south by the Santa Marta and San Lucas highs (Figs. 2, 4). The topography is mostly flat, swampy, and very susceptible to floods which are frequently disastrous for inhabitants of the region.

Five tectonic elements constitute the Platform, four of which are quite prominent: Cicuco high, El Difícil high, the Plato geofracture, and the Sucre tectonic depression (Fig. 4). The remaining element is the unnamed area at the southern end of the region, lying between the Sucre tectonic depression and the Central Cordillera, where insufficient information is available to clearly define it.

The Cicuco and El Difícil Highs—These two deeply buried platform highs coincide with the San Jorge Basin (Fig. 2) and El Difícil areas (Duque-Caro, 1973, 1975), and are composed of felsic igneous and metamorphic rocks of Paleozoic to Late Cretaceous age, based on absolute dating from what appears to be the geologically related Central Cordillera and Santa Marta massif (Irving, 1971). A characteristic feature of these highs is the notable steepness of their bounding flanks, specifically on

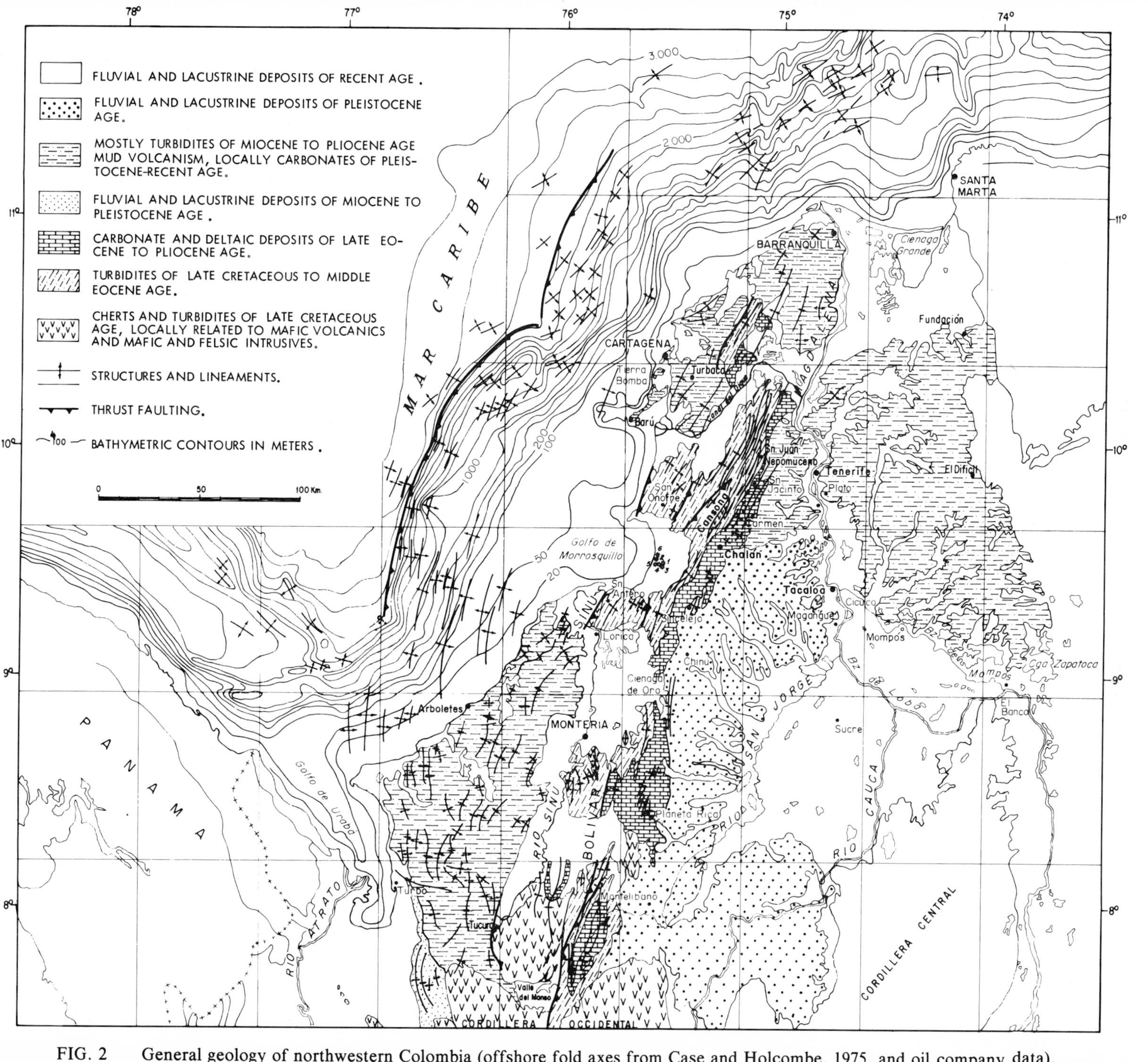

FIG. 2 General geology of northwestern Colombia (offshore fold axes from Case and Holcombe, 1975, and oil company data).

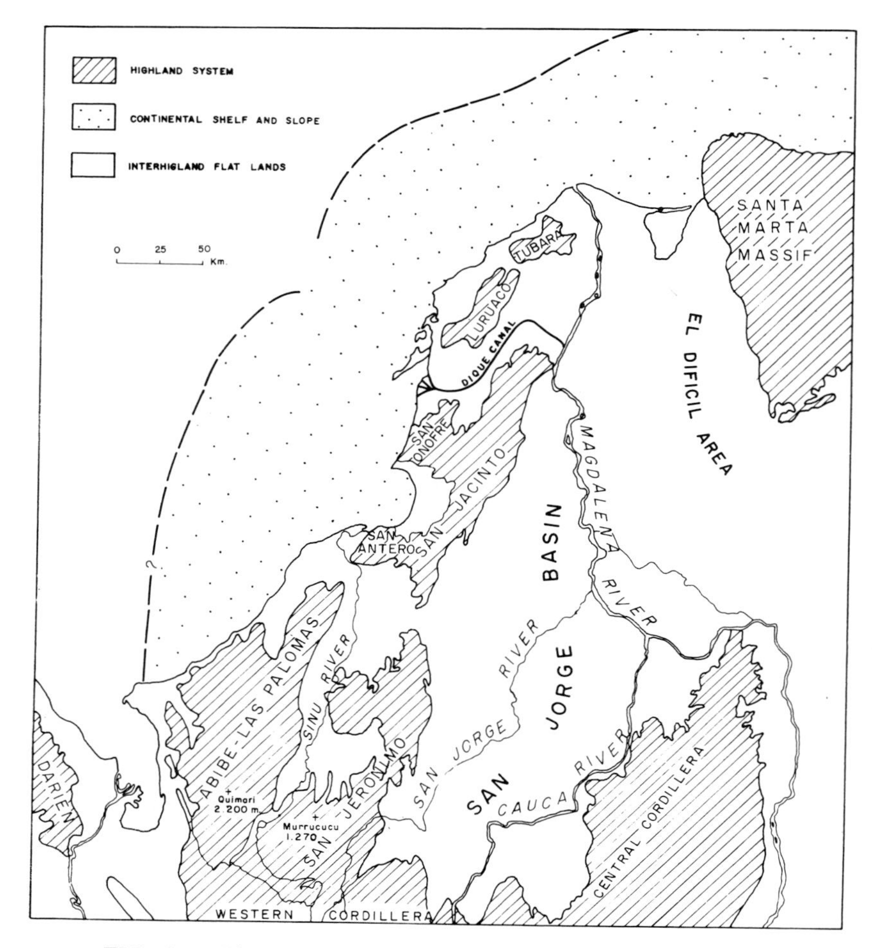

FIG. 3 Physiographic elements of northwestern Colombia.

both the northeast and southwest sides of the Cicuco high and along the west side of the El Difícil high; the east flank of the latter high is bounded by Chimichagua Fault (Fig. 4). The basement tops of these platform highs are practically horizontal and commonly controlled by block faulting, as interpreted from oil company unpublished seismic data.

No evidence of pre-Oligocene strata has been found over either the Cicuco or El Difícil highs. The stratigraphic section on both highs commences with 800 to 1,200 m of Oligocene to middle Miocene carbonate facies with occasional patches of reefal limestones resting directly on crystalline basement. These strata are overlain by up to 1,500 meters of Pleistocene to Holocene fluvial and lacustrine sediments (Fig. 5). The total thickness of this sequence does not exceed 3,000 meters in either the Cicuco or El Difícil areas. This is a relatively unfolded sequence of sediments; unpublished seismic profiles show near-horizontal reflectors with broad and gentle flexures invariably parallel to basement.

The Plato Geofracture and the Sucre Tectonic Depression—Two huge, broad, and very deep depressions have been identified within the platform area. The northern and larger of these structural lows trends north-northwestward and is here termed the Plato geofracture. This structural low coincides with what was formerly termed the Plato deep (Duque-Caro, 1973, 1975, 1976), extends northward from Cicuco between the Brazo de Mompós and the El Difícil high, and is outlined by the 5-km pre-Tertiary basement contour (Fig. 4). Based on unpublished seismic information, it has been determined that this geofracture is a very deep depression within the platform which is filled with sediments to a depth of approximately 7 km or more along the axial zone (Fig. 4). The stratigraphic section, which can only be determined from well logs, is composed of approximately 1,000 to 1,500 m of fluvial and lacustrine sediments of Pleistocene to Holocene age, and more than 3,000 m of turbidites (Duque-Caro, 1976) of Miocene-Pliocene age. The Caraballo No. 1 well, which is close to the surface projection of the 7-km basement contour (Fig. 4), penetrated about 4,000 m of Miocene to Pleistocene sediments. The presence of these turbidites suggests that this geofracture was a submarine canyon during late Tertiary time, associated with the proto-Magdalena river system, as discussed below.

I believe that the Plato geofracture was originated by the relative separation of the Santa Marta massif and the Central Cordillera during late Tertiary time. This hypothesis is supported by the following facts: (1) limestones and carbonate facies resting on base-

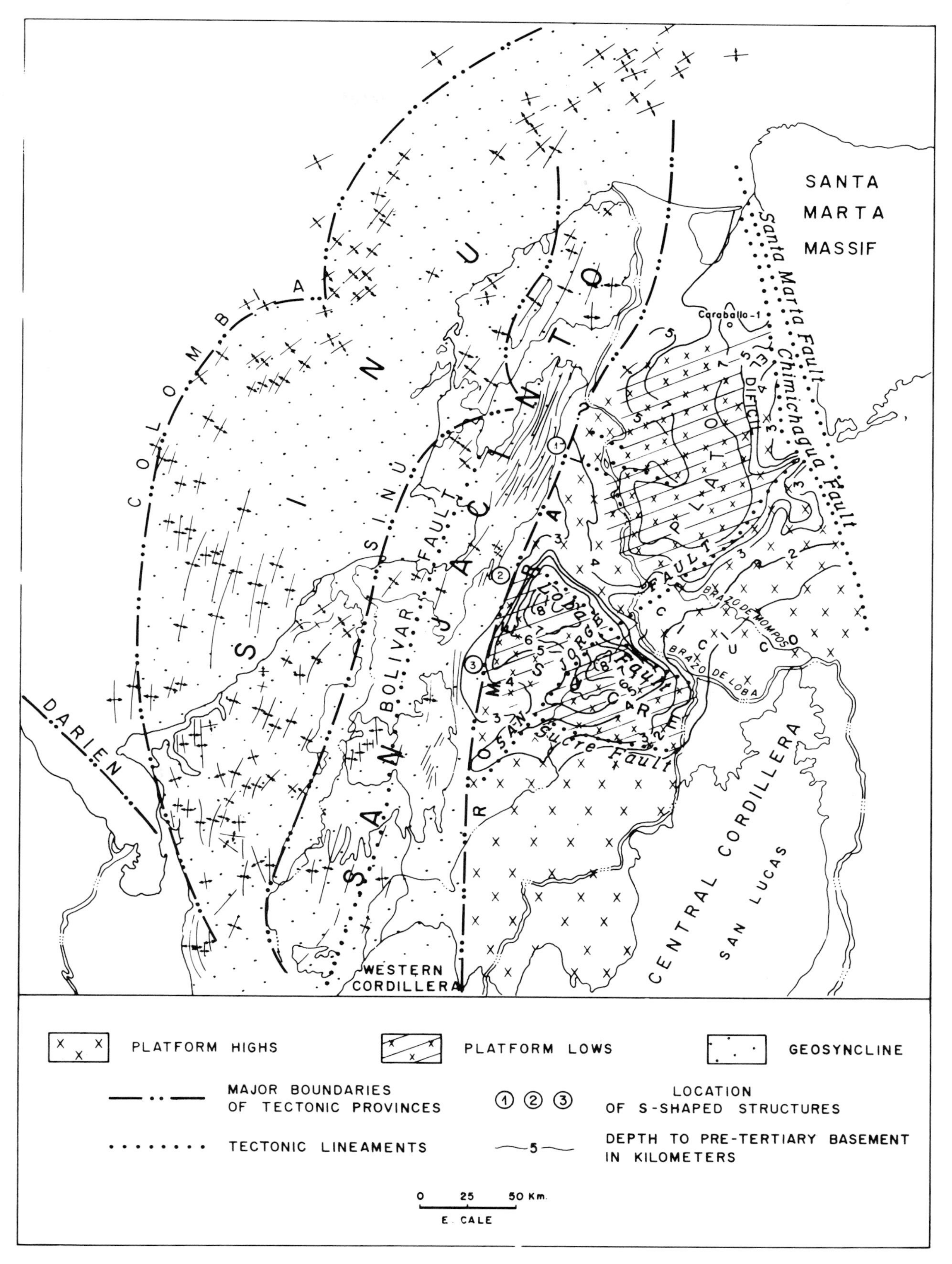

FIG. 4 Major tectonic elements of northwestern Colombia (offshore fold axes are from Case and Holcombe, 1975, and oil company data).

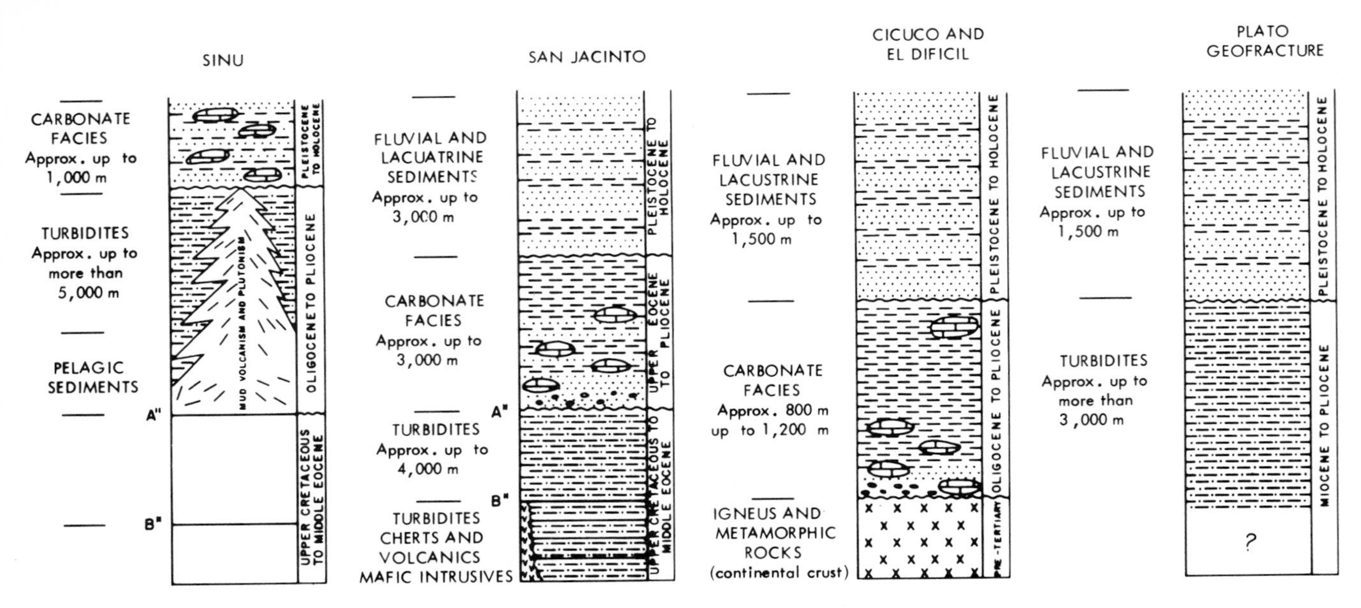

FIG. 5 Regional stratigraphy of northwestern Colombia.

ment in both the Cicuco and El Difícil highs have similar lithologic, sedimentary, environmental, and age characteristics (Fig. 5); (2) the 5-km basement contour around the El Difícil high displays a shape similar to that of the Magdalena River course northward of Cicuco between Tacaloa and Tenerife (Fig. 3), which in turn seems to be controlled between these two localities, by the same 5-km basement contour; (3) the San Jorge Fault (Fig. 4) bounds the southern end of the geofracture and coincides also with the 5-km basement contour. This suggests that this tectonic lineament could have behaved as a strike-slip fault along which the relative separation of the Central Cordillera and the Santa Marta massif took place; (4) the late Tertiary turbidites filling this depression, and related to the proto-Magdalena river system, could be a consequence of the new sedimentary episode starting during and after the opening of this tectonic feature.

The Sucre tectonic depression, which is located marginally southward of the Brazo de Loba, is a grabenlike feature, block-faulted, and controlled by the Loba and Sucre Faults (Fig. 4). The maximum thickness of the Cenozoic deposits filling this depression is estimated to reach 8 km. The general stratigraphic and sedimentary characteristics, which are similar to those of the Plato geofracture (Fig. 5), suggest that this depression could also have behaved as an ancient submarine canyon during late Tertiary time.

Basement-Controlled Lineaments—The platform area is criss-crossed by several major structural lineaments that trend N, N 20° E, N 55° W, and N 40° E (Fig. 4), which are expressions of basement faulting, and which fit the pattern of those established for the Central Cordillera (Barrero et al, 1969; Irving, 1971). The most conspicuous of these are the Romeral, which appears to represent the northernmost extension of the N to N 20° E structural lineament recognized by Barrero et al (1969) to the south of the study area, and which will be discussed in the following pages. The San Jorge, which trends N 40° E for about 250 km and which is geomorphically related to part of the course of the San Jorge River and to the southeastern boundary of the Plato geofracture, is thought to be a dextral strike-slip fault along which the relative separation of the Central Cordillera and the Santa Marta massif took place. The Loba and Sucre, trending N 55° W, bound and control the Sucre tectonic depression which is a graben-like feature. In addition to the above tectonic features, several other unnamed lineaments are shown on Figure 4, based on unpublished seismic data from oil companies.

On the eastern side of the Plato geofracture, two other structural lineaments, the Chimichagua and the Santa Marta Faults (Campbell, 1968), trend N 20° W. These tectonic directions appear to be absent from the N, N 20° E, N 40° E, N 55° W tectonic framework characteristic of the Central Cordillera and are restricted to the northeastern part of northern Colombia.

Unstable Region or Geosyncline

The geosynclinal region, with its enormous thickness of sediments, constitutes the western coastal part adjacent to the platform (Fig. 4); it has been divided into two structural elements, the San Jacinto fragmented belt and the Sinú belt, each with its own very distinctive characteristics.

The San Jacinto Fragmented Belt—This feature lies immediately adjacent to the platform and comprises three discrete structural units called (from south to north) the San Jerónimo, San Jacinto, and Luruaco anticlinoria (Figs. 2, 3), which trend N 20° E for about 360 km and have a maximum width of 60 km. Late Cretaceous pelagic rocks, such as cherts and siltstones, in some instances with interbedded diabasic and basaltic flows, commonly weathered, constitute the nuclei of these anticlinoria (Duque-Caro, 1973) and represent the upper part of a turbidite sequence more than 2,000 m thick. These turbidites locally contain substantial terrigenous components, such as quartz and mica, and shallow-water benthonic foraminifera (*Siphogenerinoides* faunal facies). The presence of this benthonic microfaunal facies contrasts with the pelagic microfauna (mostly radiolarians) of the interbedded cherts found toward the top of the sequence. Outcrops of these rocks can best be observed in the San Jerónimo anticlinorium at Tucurá in the Upper Sinú River region, at Montería, and at Planeta Rica, as well as in the San Jacinto anticlinorium at Lorica and Cerro Cansona west of El Carmen (Fig. 3). These rocks are associated with mafic and ultramafic intrusives in the vicinity of Planeta Rica and with tonalitic intrusives in the Chalán area. The sedimentary rocks are of Late Cretaceous age (Duque-Caro, 1972, 1973), and the igneous intrusive and extrusive rocks are Late Cretaceous to middle Eocene, based on field evidences and correlations with absolute dating of similar rocks of the same ages from the Central Cordillera (Irving, 1971). The cherts toward the top of the turbidite sequence are Coniacian to Campanian in age and have been correlated with the seismic reflector Horizon B″ (Duque-Caro, 1975, 1976) because of both their similar lithologic and stratigraphic characteristics.

Resting conformably on these Upper Cretaceous beds is another enormously thick sequence of turbidites (up to 4,000 m) of early Tertiary age and of a depositional environment of more than 4,000 m of water depth (based on calcium carbonate compensation depth calculations; Duque-Caro, 1972). They consist of a rhythmic succession of dirty sandstones and sandy claystones, made up of fragments of volcanic and metamorphic rocks, cherts, and detrital serpentines, which have been classified as *serpentinite graywackes* by Zimmerle (1968).

At some localities in the San Jacinto anticlinorium, such as Lorica and San Onofre (Fig. 3), chert beds and pelagic shales have been recognized interbedded with turbidites of middle Eocene age, which I also correlate with the seismic reflector Horizon A″ in the Colombian Basin (*cf.* Duque-Caro, 1975, 1976).

The youngest rocks within the belt are carbonate facies of late Eocene age and fluvial and lacustrine Pleistocene-Holocene sediments (Fig. 5). The carbonate facies are found mainly on the east flank of the belt and are of similar facies to those mentioned from the Cicuco and El Difícil highs. This carbonate facies sequence consists mainly of conglomerates, shales, sandstones, and reefal limestones, and reaches an overall thickness of 3,000 m. The sedimentary paleoenvironment which has been determined with depth foraminiferal indicators, ranges from very shallow neritic to depths of 2,000 m (Duque-Caro, 1975). The rocks on the west flank of the belt consist only of very fine sediments with minor interbedded limestones deposited at depths varying from 1,000 to 2,000 m. The fluvial and lacustrine sediments of Pleistocene-Holocene age are restricted to the east flank of the belt and have a variable thickness which ranges from a few hundred to a few thousand meters. In the Sucre tectonic depression this sediment wedge thickens abruptly to an estimated 4,000 m.

As previously stated, the San Jacinto fragmented belt is structurally composed of the San Jerónimo, San Jacinto, and Luruaco anticlinoria, three elements separated by low-lying swampy terrains, with no surface structural connection, which give it the appearance of a fragmented belt. The general structural trend is N 20° E and is characterized by very elongate tight anticlines and synclines and by normal and thrust faults parallel to the regional strike. These structural features are particularly well-developed in the San Jacinto anticlinorium, physiographically the highest of the three anticlinoria. Southward, toward the northern extension of the Western Cordillera, the general structure of the San Jacinto belt shows a westward deflection (as observed on aerial photographs and radar imagery) in the vicinity of the Manso River valley (Fig. 3), which suggests the possibility that this belt could be a separate tectonic unit from the Western Cordillera. Along the east flank of the belt, peculiar S-shaped structures have been observed associated with the Romeral lineament described from the platform area (Fig. 4). Furthermore, tectonic deformation in each of the three anticlinoria of the San Jacinto belt displays different degrees of intensity, emphasizing distinction among these three separated units. The San Jacinto anticlinorium exhibits the maximum compaction, particularly in the Upper Cretaceous and lower Tertiary sections, with very steep fold limbs, very tight structures, and thrust faulting predominating. In contrast, similar structural features in the Luruaco and northern part of the San Jerónimo anticlinoria are less intense.

In my opinion, the above general structural, tectonic and magmatic phenomena observed within the San Jacinto fragmented belt, such as (1) the very elongate tight structures, (2) the general structural trend and faulting lineations parallel to the margin (Romeral lineament) of the unfolded platform, and (3) volcanism and plutonism also parallel to the margin of the platform, have mainly been the results of the interaction between the southwestern Caribbean oceanic crust and the South American continental crust, through tensional and compressional forces along the margin of the platform especially during the pre-Andean Orogeny (middle Eocene, Van der Hammen, 1958) which uplifted, folded, and modeled this belt.

One of the most outstanding and intriguing characteristics among all of those described for the San Jacinto belt, and which has partially been discussed in an earlier paper (Duque-Caro, 1972, p. 19, fig. 3), is the islandlike structural subdivision of this feature into three elements separated by low-lying swampy terrains with no surface structural connection, giving it the appearance of a fragmented belt.

I believe that this peculiar geomorphic characteristic is the remnant expression of an ancient submarine volcanic chain which bordered the margin of the platform during Late Cretaceous to early Tertiary time, prior to the pre-Andean Orogeny. This interpretation is supported by the following field evidence:

1. Outcropping chert beds and associated volcanics, which have always been mapped as the narrow nuclei of anticlinal structures and mostly surrounded by turbidites of early Tertiary age (Fig. 3), are geomorphically expressed by very distinctive pointed hills. At some localities, such as in the Golfo de Morrosquillo area, the pointed hills of cherts and siltstones, highly folded (microfolded), emerge as isolated masses from the surrounding flat terrains. Six wildcat wells were drilled in this area (Fig. 3) to a maximum depth of about 3,000 m. None of the first five wells reached basement or penetrated sediments older than late Oligocene. However, well 6, located very close to well 2, reached basement composed of Late Cretaceous cherts and volcanics at about 300 m. A simple explanation is that the well was drilled into one of these isolated hills of volcanic origin.
2. Late Cretaceous-early Tertiary paleobathymetry along the San Jacinto fragmented belt appears to have been controlled by ancient topographic highs of this kind. In the Cerro Cansona area (Fig. 3), where the highest topographic elevations of the San Jacinto anticlinorium are located, 800 m above sea level, patches of Late Cretaceous highly fractured and somewhat metamorphosed reefal limestones crop out within a siltstone and chert sequence with volcanic interbeds. Both macro- and microfaunal evidence in this area, such as echinoid remains, ammonites, and calcareous benthonic and planktonic foraminifera, suggest much shallower water than that suggested by the exclusively siliceous planktonic microfauna (Radiolaria) found at the low topographic elevations.

Furthermore, from petrologic studies of the early Tertiary turbidites surrounding the volcanic nuclei, Crook (1974, p. 305) classifies as *quartz-poor graywackes* those *serpentinite graywackes* of Zimmerle (1968), and suggests that they are indicative of magmatic island arcs. Neither detailed petrological studies nor chemical analyses have been undertaken on the igneous rocks of the San Jacinto fragmented belt.

In view of the preceding data characterizing the belt, three questions require answers: Did the San Jacinto fragmented belt, behaving as a submarine volcanic chain, extend northward into the Caribbean area, and if so, could the Beata Ridge represent its northern extension? Or does the San Jacinto belt extend eastward to the Curaçao Ridge and Venezuelan Borderland (*cf.* Roemer et al, 1976)?

Sufficient geophysical, structural, and lithological evidence exists to suggest that both of the first two ideas are plausible (Fig. 6):

1. Structural lineaments and faulting along the San Jacinto belt onshore and faulting along the Beata Ridge under the sea (Case and Holcombe, 1975) display a similar N 20° E direction, whereas toward the eastern side of the Beata Ridge, Case and Holcombe (1975) illustrate N 20° W to N 50° W structural directions resembling those early described on the east of the San Jacinto belt on the platform (Fig. 6).
2. Ludwig et al (1975, p. 119) emphasized both the notable differences in velocity structure between the Colombian and Venezuelan Basins, and that the Horizon B″ in the Colombian basin is a *rough* surface, unlike its typically smooth counterpart in the Venezuelan Basin (Fig. 7).
3. Duque-Caro (1973, 1976, this paper) divides northwestern Colombia into two distinctive regional tectonic elements: *folded* (geosyncline) to the west and *unfolded* (platform) to the east. In my opinion, these two structural elements can be compared with the general structural style of both the Colombian and Venezuelan Basins (Fig. 7) as shown by Ludwig et al (1975, Fig. 2).
4. The change in structural style between the *rough* Colombian and *smooth* Venezuelan Basins, and that between the *folded* and *unfolded* areas of northwestern Colombia coincide, respectively, with the Beata Ridge and the San Jacinto fragmented belt (Figs. 6, 7). These first four evidences are indicative of a very distinctive structural framework and suggest that both the San Jacinto belt and the Beata Ridge represent the boundary zone between an eastern and a western tectonic province.
5. Furthermore, from analysis of magnetic anomalies, Christofferson (1973, p. 3228) suggested that a major crustal discontinuity along the trend of the median Beata Ridge may separate eastern and western regions. Similarly, Watkins and Cavanaugh (1976) postulated a northeast-southwest fault zone, parallel to fractures in the Beata Ridge and also parallel to the magnetic discontinuity in the Colombian Basin indicated by Christofferson (1973). They related it to a change in structural style in the north, which separates eastern and western Hispaniola along a line (not illustrated) more or less parallel to the hypothetical fault zone, and to the parallelism of the northwestern shoulder of the Colombian coastline to the hypothetical fault zone in the south.

In summary, all of the above statements clearly bring out the striking similarities between northwestern Colombia and the Caribbean areas immediately to the north, implying that the Beata Ridge is a northern extension of the San Jacinto fragmented belt. One of the most enigmatic and controversial questions, to date not sufficiently and convincingly resolved, concerns the crustal composition of the Caribbean and some of its physical and geophysical peculiarities which make it different from other ocean basins.

Two points appear to be fundamental in explaining

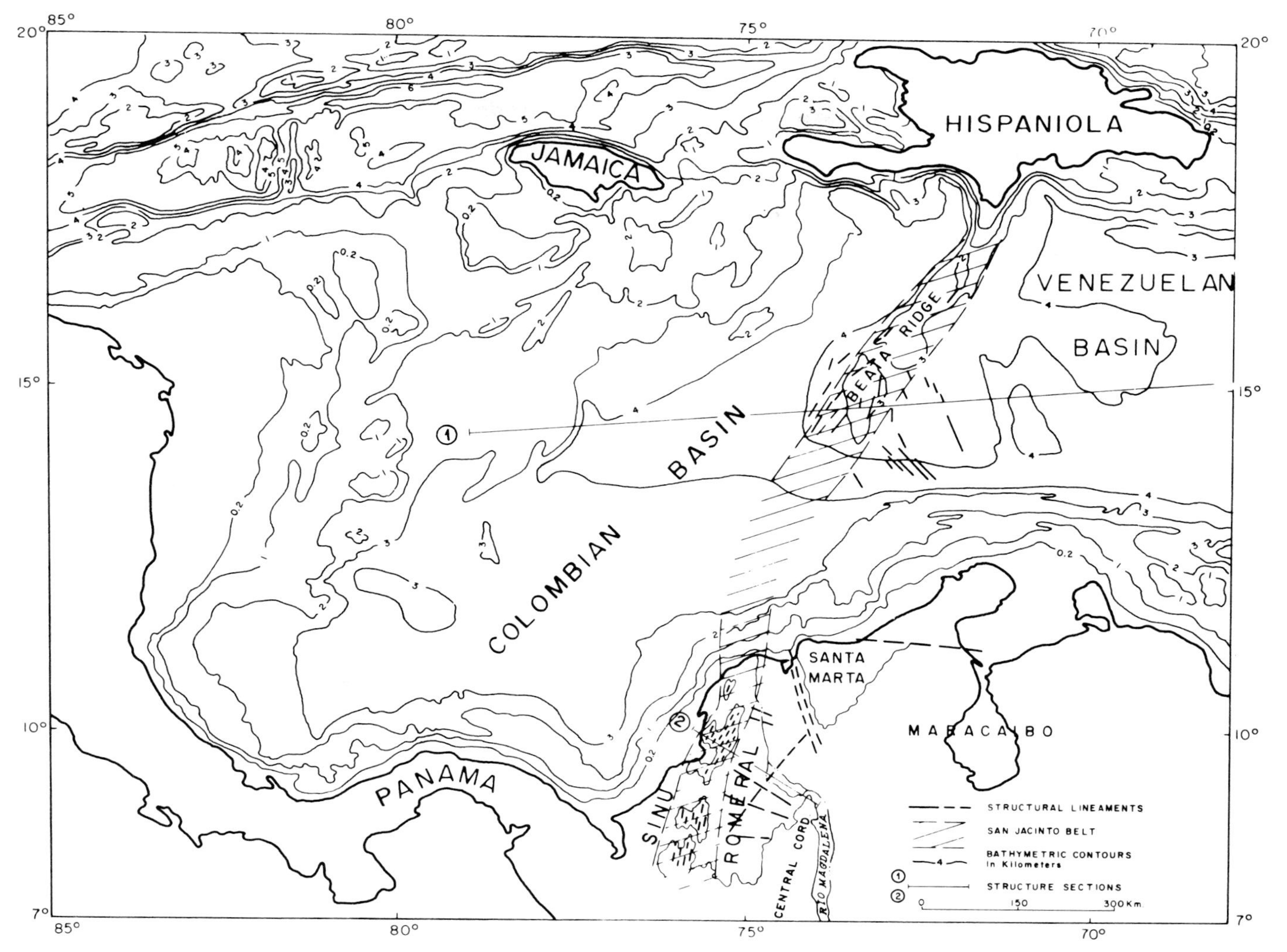

FIG. 6 Proposed structural relationship between the San Jacinto belt and the Beata Ridge (base map modified from Case and Holcombe, 1975).

the crustal composition of the Caribbean. From analyses of seismic refraction and reflection profiles, the Colombian Basin has a crustal layer that is much thicker than that of a typical ocean basin, whereas the Venezuelan Basin crustal layer is only slightly thicker than typical oceanic crust, according to Ewing et al (1971), Edgar et al (1971), and Ludwig et al (1975). Second, Horizon B″ in the Venezuelan Basin is characteristically a smooth surface, unlike the rough oceanic basement (layer 2) surface typical of most of the Atlantic and large parts of the Pacific (Ludwig et al, 1975).

The distinctive crustal and structural characteristics of northwestern Colombia and the proposed correlation of the San Jacinto fragmented belt with the Beata Ridge suggest that the smooth and rough structural character of Horizon B″ is a function of the type of crust underlying that seismic layer, and that both the San Jacinto fragmented belt and its northern extension, the Beata Ridge, represent the boundary zone between two types of crust. The crustal composition of the Venezuelan Basin would be interpreted as continental because of the smoothness of Horizon B″ and structural characteristics similar to those of the platform area to the south (Figs. 6 and 7); the crustal composition of the Colombian Basin would be interpreted as oceanic because of the roughness of Horizon B″. The Late Cretaceous basalts and dolerites drilled at the depth of Horizon B″ do not likely represent primordial crust, but reflect the last major igneous event in the history of the Caribbean, as has already been stated by Ludwig et al (1975, p. 1) and Donnelly (1975).

The Sinú Belt—Immediately west of and parallel to the San Jacinto belt lies the Sinú belt, which is more than 500 km long and as much as 125 km wide in the study area. The general structural grain parallels the San Jacinto belt trend in its southern portion, but it turns northeastward in the Cartagena-Barranquilla area following the coastline (Fig. 4) and cutting off the older middle Eocene N 20° E trend of the San Jacinto belt. It comprises the Abibe–Las Palomas and Tubará anticlinoria of the land area (Fig. 2) and the continental shelf and slope of the

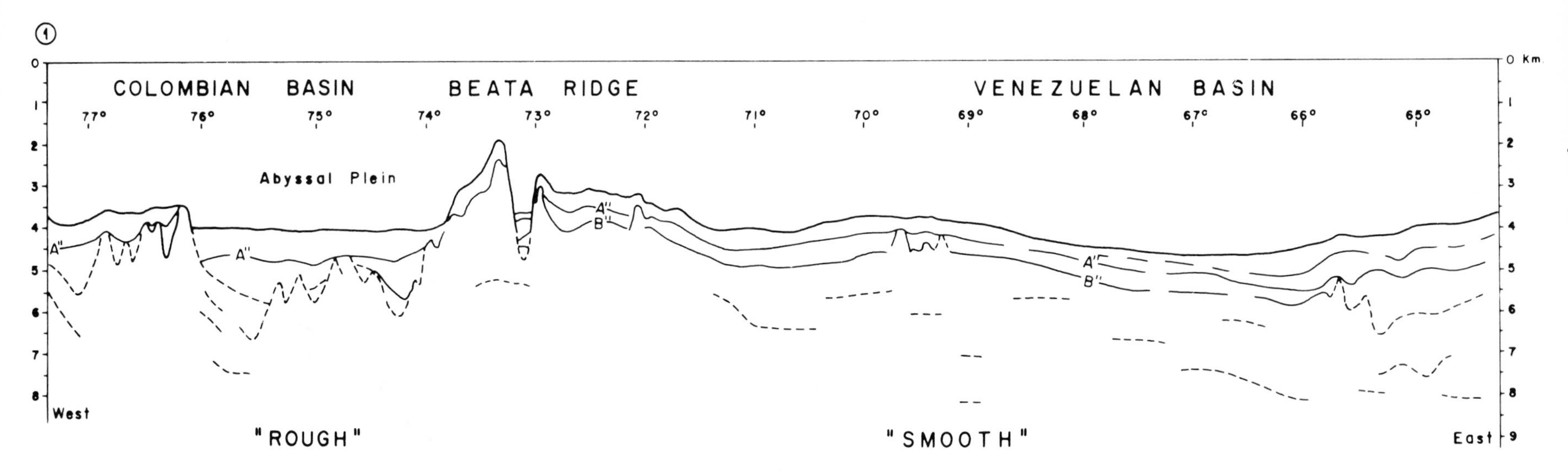

West-east seismic structure section of Colombian and Venezuelan basins showing the "rough" and "smooth" Horizon B" (from Ludwig et al. 1.975, fig. 2)

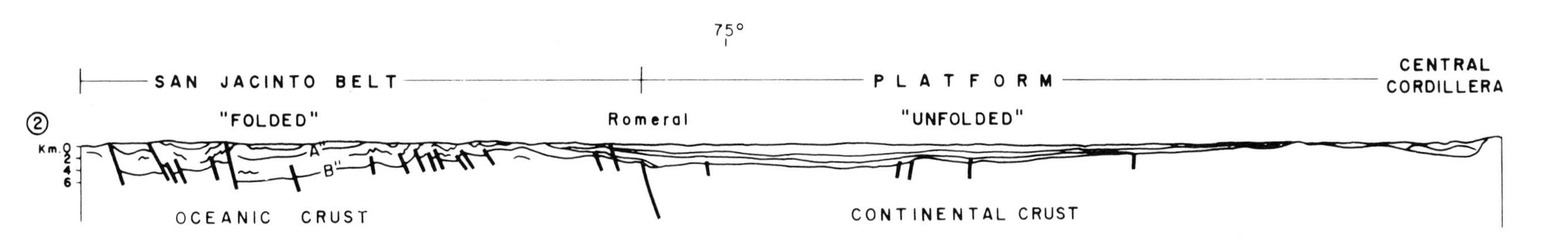

West-east structure section of the San Jacinto Belt and the platform showing "folding" on the west and "unfolding" on the east (from combination of seismic, well logs and surface information)

FIG. 7 Comparison of structural sections across the San Jacinto belt and the Beata Ridge.

offshore area. The western margin of the belt coincides with that of the deformed belt of Case (1974) and with the bathymetric change between the shelf slope and the abyssal plain (Figs. 2–4).

The outcropping rocks of mostly late Miocene–Pliocene age (Fig. 5) largely consist of a monotonous turbidite sequence of mainly fine-grained sediments up to more than 5,000 m thick, which were deposited at depths between 1,000 and 2,000 m (based upon paleobathymetrical analyses of foraminifera; Duque-Caro, 1975). Overlying this sequence is a shallow-water carbonate facies of Pleistocene-Holocene age, up to 1,000 m thick,[3] composed of shales, reef limestones, sandstones, and conglomerates. Along the eastern border of the belt, the conglomerates are brecciated and very poorly sorted toward the contact with the San Jacinto belt. The reef limestones seem to be restricted to the northern landward portion of the belt, where they are especially well developed in the Cartagena-Barranquilla area. Oldest fossil foraminiferal faunas collected to date in the outcropping rocks of the sequence indicate an age of late Oligocene–early Miocene. These fauna occur in pelagic mudstones with siliceous interbeds underlying the turbidites located immediately west of the Dique Canal (Figs. 2, 3).

Structurally, this belt is composed of narrow, steep anticlines separated by broad gentle synclines. The anticlinal trends are roughly parallel to the eastern and western margins of the belt (Fig. 4). The structure is complicated by normal, thrust, and transcurrent faults with no definite pattern and mainly related to the anticlines. This is particularly noticeable southward as the structural complexity increases. The southern end of the Sinú belt is complicated and represents a structural narrowing or squeezing near the intersection of, and possibly a result of tectonic interaction between, the northwest-southeast Darién and the north-south Western Cordillera structural elements (Fig. 4) during the Andean Orogeny (Pliocene-Pleistocene); that also is the time when the Sinú belt reached its first climax of deformation and uplifting. Another structural feature is probably represented by the bathymetric configuration northwest of Cartagena, which appears to indicate a wide inflection in the western margin of the belt. This inflection has a land equivalent in the southern geomorphic expressions of both the Dique Canal and the Luruaco anticlinorium (Fig. 4).

Perhaps the most striking feature of the Sinú belt, and considered to be the principal factor controlling the structure, is mud volcanism and plutonism.[4]

[3] Very recent information from wildcat wells in the Cartagena-Barranquilla area indicates that these carbonate facies may reach about 4,000 m in thickness.

[4] I prefer to use the term *mud plutonism,* instead of *clay diapirism,* to accompany that of *mud volcanism* (= *sedimentary volcanism* of Kugler, 1968), notwithstanding that these terms have traditionally been applied to magmatic phenomena. Both plutonism and volcanism have always been associated with the structural history of the earth, particularly with the evolution of mountain belts. Such is the case of the Sinú belt, where mud volcanism and plutonism are directly related to and restricted to the uplift and deformation of that particular belt (see also Higgins and Saunders, 1974, p. 148).

Both in the Sinú River area to the south, and in the Cartagena-Barranquilla area to the north, numerous mud volcanoes of different dimensions (most of them active) can be observed. The Tierrabomba, Barú (Fig. 3), and other islands bordering the present coastal margin of the study area are also surface expressions of the same features. In both the land and offshore subsurface sections, the same phenomena have also been recognized in seismic records by Shepard et al (1968), Edgar et al (1971), Krause (1971), Shepard (1973), and others.

The mud extruded from the land volcanoes which in general exhibits similar characteristics to that from the mud volcanoes of Trinidad (Higgins and Saunders, 1974); it is of pelagic origin with a heterogeneous mixture of fossil faunas included, a very high water content, and a low density of usually less than 2.0 g/cm^3. The clay fraction is mainly composed of chlorite and montmorillonite with the latter often representing as much as 50 percent of the clay fraction. In my opinion, this mud comes from a stratigraphic interval underlying the Miocene turbidites, called "high pressure shale" by oil industry geologists and engineers, which is recognized as a serious drilling hazard because of the characteristic high pressure. This interval on seismic profiles is a transparent or blank zone of no reflections which permits relatively easy identification; more detailed geophysical information about these mud phenomena, such as gravity and electric-log character have been described by Higgins and Saunders (1974) from the similar mud volcanoes of Trinidad. The age assigned for this mud interval is late Oligocene–early Miocene based on foraminiferal determinations from the heterogenous mixture extruded at the surface.

As to the causes giving rise to the belt, Case (1974) discussed three possible origins: (1) gravitational downslope sliding, (2) tectonic compression related to the southerly underflow of the Caribbean plate, and (3) broad, regional right-lateral shear between an eastward-moving Caribbean plate and a westward-moving South American plate. As stated earlier, the general structure is mainly controlled by mud volcanism and plutonism, which has resulted from lateral compressional stresses (Kugler, 1968; *orogenic compression*) normal to the margin of the belt, according to the axial direction of the structural trends of the Sinú belt. These stresses have been acting continuously since at least the Late Cretaceous and are the main stresses controlling the deformation of the whole northwestern folded region, as stated by Case (1974, p. 738). The first notable uplifting and important deformation of the Sinú belt took place during the Pliocene-Pleistocene Andean orogeny (Van der Hammen, 1958). This interpretation is based on the observed continuous abyssal Late Cretaceous to Miocene-Pliocene pelagic and turbidite sedimentation characteristic of this belt, in contrast with the later Pliocene-Pleistocene shallow-water carbonate facies sedimentation (Fig. 5). The carbonate facies are characteristically unfolded, and deposition appears to have been controlled by folded Miocene-Pliocene paleohighs resulting from mud volcanism and plutonism, particularly during the Andean orogeny.

Two important elements appear to be common to the deformational processes of both the San Jacinto and Sinú belts: (1) lateral compressional stresses normal to the margin of the platform, and (2) volcanism and plutonism—that of the San Jacinto belt being "igneous," and that of the Sinú belt being "sedimentary." A third common phenomenon could be added as a consequence of the volcanism and plutonism, which is control of paleobathymetry of depositional environments by the paleohighs.

Paleotrenches

The geotectonic elements described above are bounded by four roughly parallel lineaments that mark abrupt changes in the structural style of each element. These lineaments, taking into account associated structural, sedimentary, paleobathymetric, and geochronologic characteristics, have been interpreted as expressions of ancient trenches marginal to the continent.

Romeral Lineament—Barrero used the name *Romeral* to designate a fault zone dipping eastward and separating the Western and Central Cordilleras, with a length of more than 800 km extending from southern Colombia northward to Montelibano area on the south of the study area (Barrero et al, 1969; Fig. 3). One of the most important characteristics of this fault zone is that it also separates two distinct geologic provinces; continental to the east and oceanic to the west. Irving (1971) did not consider this separation in projecting the northern trend of this fault zone. Duque-Caro (1975, 1976) found it possible to extend the fault zone 140 km farther north by recognizing these two separate geologic environments. The age assigned to this fault zone by Barrero et al (1969) was Early Cretaceous.

So far the Romeral has been classified by most workers as a simple fault or a fault zone. I feel that this feature is more than that; it exhibits not only faulting, but structural, tectonic, and petrologic-sedimentary features clearly indicative of a major structural contact between oceanic and continental crusts.

The Romeral lineament in the study area is not as spectacular and physically visible as the part originally described in the type area, because the major geographic outline is covered by the Tertiary and Quaternary deposits of the San Jorge Basin. However, other characteristics aid in its identification in both surface and subsurface sections.

1. It represents the western boundary of the platform in contact with the San Jacinto belt, marked by a clear change in the structural style of these two elements: folded San Jacinto belt in contact with a nonfolded platform area.
2. Mafic volcanism and mafic, ultramafic, and tonalitic plutonism associated with cherts and turbidites of Late Cretaceous to middle Eocene age are present to the west and completely absent to the east.
3. Serpentinization of the mafic and ultramafic intrusives has occurred along the western side, which is clearly visible south of Planeta Rica (Fig. 3).
4. A low-grade metamorphic belt (greenschist facies) is found along the eastern side of the lineament, which has been recognized in wildcat wells, and which the writer correlates with the lower Tertiary Gaira Schists of the Santa Marta massif (Irving, 1971).
5. S-shaped structural closures along the lineament (Fig. 4) may possibly be related to north-south transcurrent movement.

In its initial stage of development, the Romeral feature was a deep trench with a very steep escarpment on the continental slope bordering the western margin of the platform. This conclusion is based on the presence of abyssal Upper Cretaceous and lower Tertiary pelagic sediments and turbidites lying to the west, which are completely absent to the east in the platform area (Fig. 5).

Bolivar Lineament—This name, used by Beck (1921), Zimmerle (1968), and Irving (1971) designates a very conspicuous thrust fault zone found in the lower Tertiary turbidites within the San Jacinto belt (Figs. 3, 4). Its surface trace is clearly seen in the San Jerónimo and San Jacinto anticlinoria, particularly in the latter where huge blocks of tectonic breccia can be observed. The fault dips eastward and apparently dies out to the north under Quaternary sediments in the vicinity of the Dique Canal; to the south it seems to continue into the northern part of the Western Cordillera. The fact that it parallels the Romeral lineament and is also a thrust fault, dipping eastward, suggests that the Bolivar feature might represent one of the Romeral lineament positions migrating to the west (Burk, 1972), possibly during the late Paleocene.

Sinú Lineament—This feature marks the western boundary of the San Jacinto belt and separates it from the Sinú belt (Fig. 4). Southward, in the upper Sinú River area, Hubach (1930) identified it as the Tucurá thrust fault, which dips eastward and which places Upper Cretaceous rocks in contact with Pliocene-Pleistocene beds. Northward the surface expression is masked by flat, swampy, alluvium-covered terrain, but it apparently controls the course of the Sinú River. The following distinguish this lineament.

1. It marks a change in structural style between the San Jacinto and Sinú belts.
2. Mud volcanism and mud plutonism are common to the west in the Sinú belt but completely absent to the east in the San Jacinto belt.
3. Poorly sorted and brecciated conglomerates are found along the lineament near the surface expression of the contact with the San Jacinto belt, as can be observed immediately adjacent to the west flank of the Luruaco anticlinorium; examples include the Pendales conglomerates of the Luruaco area, and conglomerates out-

cropping along the Sinú River in the Tucurá area farther south (Fig. 3) that are associated with the western side of the San Jerónimo anticlinorium.

As with the Romeral lineament, the Sinú also is interpreted to have been a trench during its initial stage of evolution, marginal to the San Jacinto belt. This interpretation is based on the occurrence of abyssal upper Oligocene to upper Miocene-Pliocene pelagic sediments and turbidites to the west and shallow-water carbonate facies of the same ages to the east (Fig. 5).

Colombia Lineament—The westernmost lineament to be described is the Colombia lineament which forms the western boundary of the Sinú belt (Fig. 4). The topographic expression is submarine and coincides with the bathymetric change at the junction of the Sinú belt slope and the abyssal plain. The Colombia lineament also marks the change from the folded structural style of the Sinú belt to the east to the simple, unfolded structure of the abyssal plain. Mud volcanism and mud plutonism are unknown and absent to the west of the lineament.

The Colombia lineament also has been interpreted as trench-like (Case, 1974, p. 738), having been marginal to the Sinú belt on the east at an even later evolutionary stage. It separates abyssal Pleistocene to Holocene pelagic and turbidite sediments in the west from mainly carbonate facies sediments of the same age to the east characteristic of both the Sinú belt shelf and coastal plain.

GEOLOGIC EVOLUTION

The geologic evolution of the northwest region of Colombia is discussed only from Late Cretaceous time onward, which represents the stratigraphic limits of this study. The sedimentary cycles and stages proposed by Duque-Caro (1972, 1975) are illustrated and correlated with the European stages (Figs. 5, 14).

Late Cretaceous–Paleocene (Early Cansonian)

The paleogeographic panorama of the region during this time interval consisted of an eastern emergent platform area, which included the Central Cordillera and Santa Marta massif, undergoing erosion, and oceanic areas to the west (Fig. 8).

The western margin of the platform was bordered by the Romeral trench, with steep margins and deep waters, extending over a length of about 1,700 km southward to the area of the Gulf of Guayaquil and, for all its length, bordering the western margin of the Central Cordillera.

The oceanic domain west of the platform was the site of pelagic sedimentation, abundant planktonic microfauna, and turbidites which included locally terrigenous material of alternating sand and conglomerate size. Spectacular conglomeratic horizons can be observed outcropping today in the Serranía de Cansona (Fig. 3). The extent and volume of turbidite sediments were very impressive in other areas also, as witnessed by the huge thicknesses in the Tucurá area of the San Jerónimo anticlinorium, where they reach more than 2,000 m in thickness. It was during this time that the cherts interpreted as seismic reflector Horizon B″ were developing toward the top of the turbidite sequence.

Within the platform region there are two puzzling areas, the Plato geofracture and the Sucre tectonic depression, which have a maximum current basement depth of 7 and 8 km, respectively, but which according to the known evidences, seem to lack any Late Cretaceous or early Tertiary sedimentary record. The absence of sediments representing this time interval would suggest that these two structural features did not exist during the Late Cretaceous and Paleocene, thereby indicating that the Santa Marta massif was coupled to the Central Cordillera at this time and was located in the vicinity of the present Brazo de Mompós (Fig. 3).

Paleocene–Middle Eocene (Middle and Late Cansonian)

During this time interval the platform area, including the Santa Marta massif and the Central Cordillera, remained emergent. However, the Romeral trench continued to deepen because of lateral compressional stresses normal to the continental margin. This sinking produced a deeper environment of deposition, with water depths in excess of 4,000 m (Duque-Caro, 1972, 1975). It is precisely at the end of this phase that the silicic sediments characterizing seismic reflector Horizon A″ were deposited west of the Bolivar Fault in the areas of San Antero and San Onofre (Fig. 3). At the same time, volcanism took place marginal to the platform, giving rise to a series of submarine volcanic cones which, according to the observations stated earlier, are interpreted as the initial phase of the uplift of the San Jacinto fragmented belt.

Pre-Andean Orogeny (Middle Eocene)

Lateral compressional stresses reached a peak during this regional diastrophic event (Van der Hammen, 1958; Irving, 1971), producing as a consequence the Chalán tonalitic plutonism along the western margin of the platform (Romeral zone) and the first important uplift and folding of the San Jacinto belt and the Western Cordillera. This uplift, estimated as on the order of 5,000 m of vertical displacement, based on calcium carbonate compensation depth calculations (Duque-Caro, 1972, 1973), caused the emergence of the San Antero and San Onofre areas and the Western Cordillera (Fig. 9). The western flank of the belt was separated from the abyssal plain by the newly formed Sinú trench, thus resulting in a "migration" to the west of the feature originally described as the Romeral trench. The platform remained emergent but underwent westward tilting, producing the initial configuration of the San Jorge Basin.

Late Eocene–Oligocene (Early Carmenian)

Following the pre-Andean orogeny the San Jorge Basin (Figs. 3, 10) lay open to progressive marine invasion from the north and west. This began from the northwest during the late Eocene and Oligocene

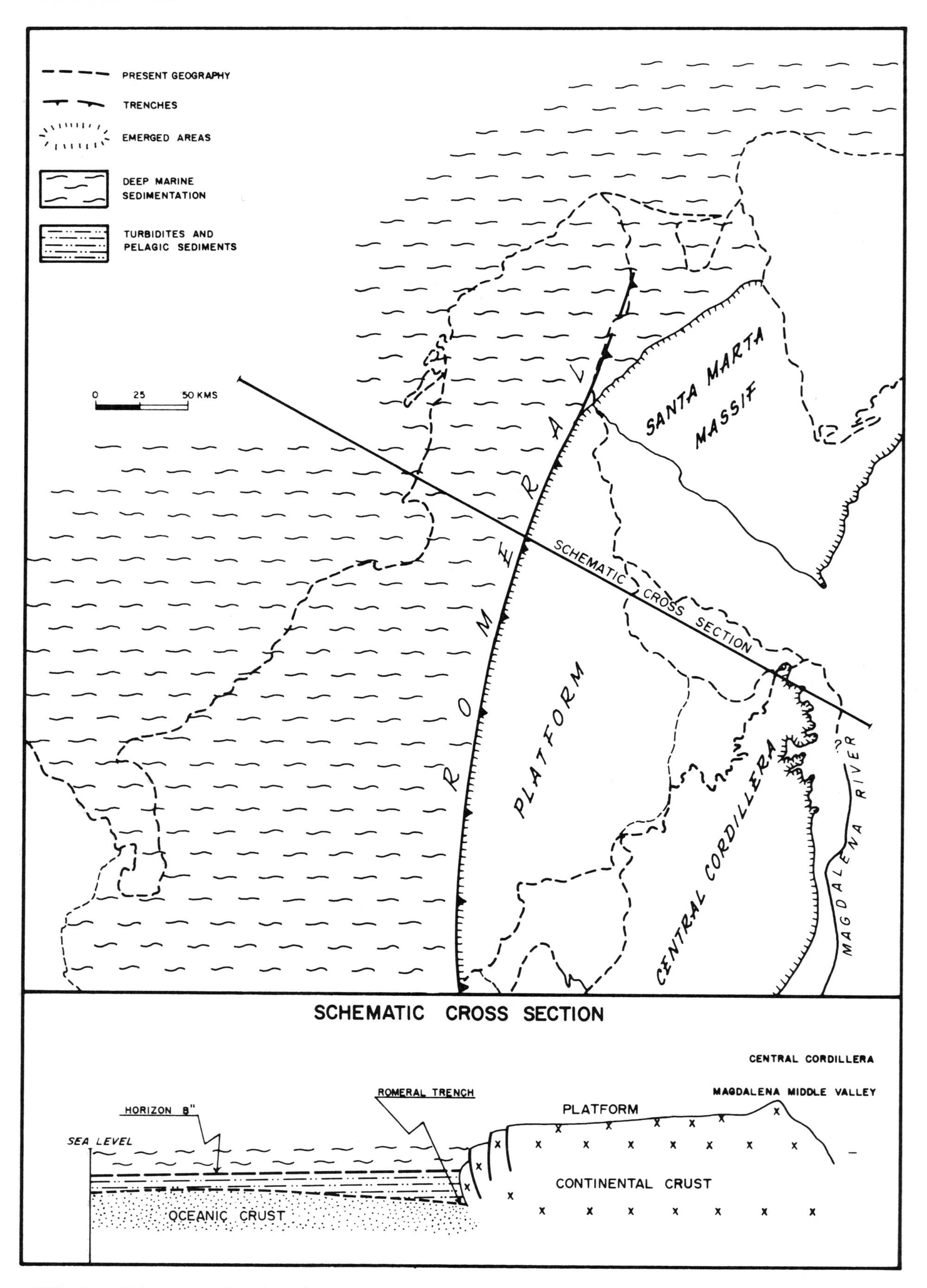

FIG. 8 Paleogeography of northwestern Colombia during Late Cretaceous to Paleocene (early Cansonian).

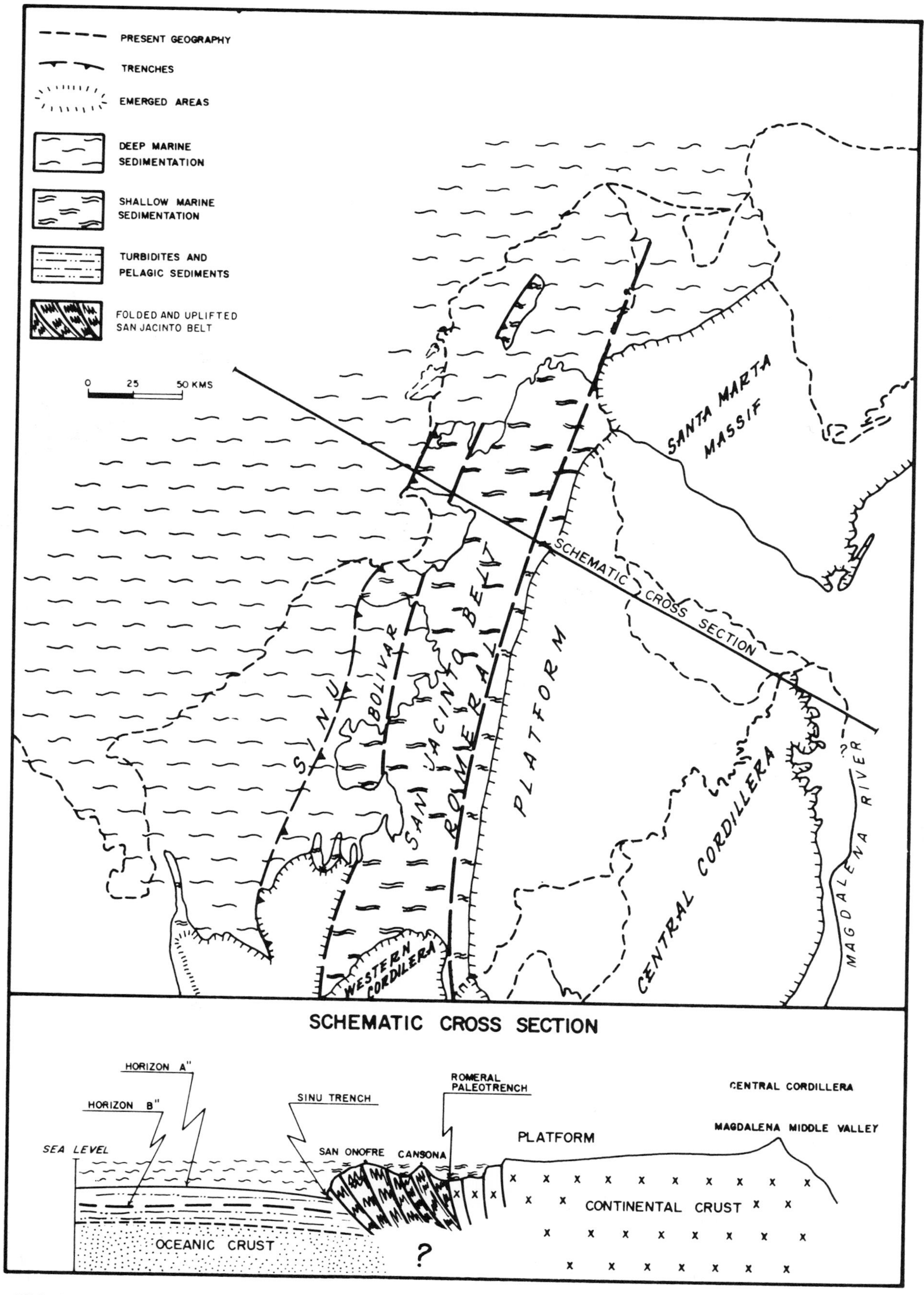

FIG. 9 Paleogeography of northwestern Colombia immediately after the pre-Andean orogeny (middle Eocene).

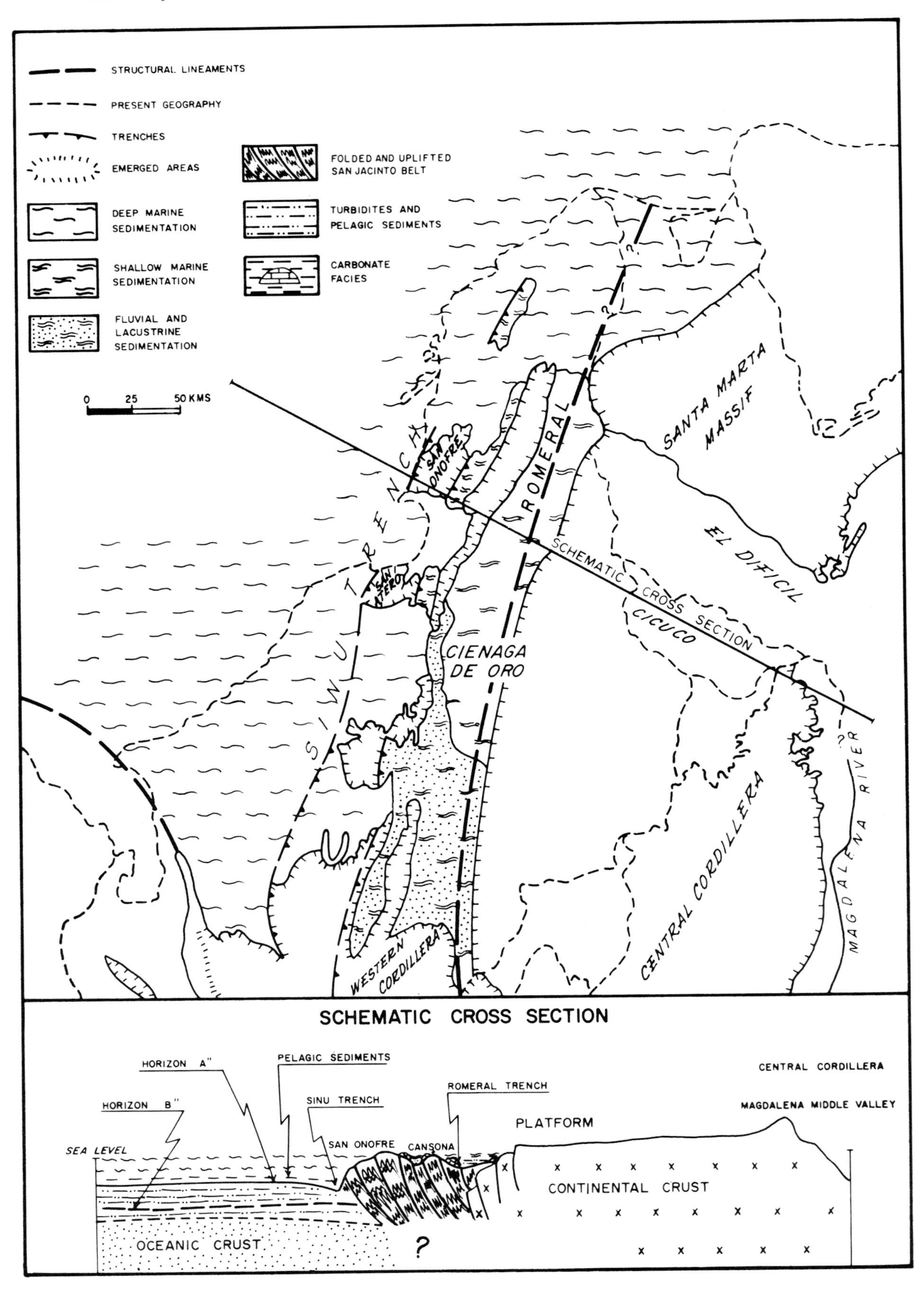

FIG. 10 Paleogeography of northwestern Colombia during late Eocene-Oligocene (early Carmenian).

with deposition of carbonate facies in the north, including conglomerates, sandstones, and shales as well as reef limestones, giving way southward to deposition of deltaic facies in the Cienaga de Oro area (Fig. 10). Eastward the sea apparently reached the margin of the platform, which still remained emergent. Thus, a new sedimentary cycle, the Carmenian, was initiated (Duque-Caro, 1972, 1975).

The east flank of the Luruaco anticlinorium was also a favorable site for carbonate facies deposition, as were depressions along the western flank of the San Jacinto anticlinorium.

At the end of the Oligocene, waters deepened accompanied by deposition of argillaceous sediments rich in planktonic foraminifera and radiolaria belonging to the *Globigerina ciperoensis* zone and lower part of the *Globorotalia kugleri* zone. Meanwhile, west of the San Jacinto belt, in the Sinú trench, pelagic sedimentation with little or no terrigenous influence was taking place, with deposition of abyssal marls and siliceous muds (containing abundant foraminifera and radiolaria).

Oligocene–Miocene Diastrophism

This tectonic phase, so named by Stainforth (1968) and corresponding to the "proto-Andean phase" (Van der Hammen, 1958), was the direct consequence of another increase in intensity of lateral compressional forces on the margins of the platform. It resulted in a tilting to the west of the platform and a renewed uplifting of the San Jacinto belt, These movements in turn produced an abrupt change in the bathymetry with a deepening of the basin, especially noticeable along the Romeral zone, associated with a marine invasion that would eventually cover the entire platform area (Duque-Caro, 1975).

Early and Middle Miocene (Middle Carmenian)

With the Oligocene-Miocene diastrophism a new sedimentary cycle, the middle Carmenian, was initiated in the San Jorge Basin (platform area). The marine invasion transgressed much farther southward to the northern extensions of the Western Cordillera, and eastward over most of the platform and the Cicuco and El Difícil highs, depositing principally carbonate facies much the same as during the preceding late Eocene phase (Fig. 11).

Little paleogeographic change, apart from continued emergence, occurred along the western flank of the San Jacinto fragmented belt. Westward of the Sinú trench pelagic sedimentation continued as during the late Eocene–Oligocene (Early Carmenian).

Late Miocene–Pliocene (Late Carmenian)

This time interval is associated with greater instability in the northwestern Colombia region, as more intense lateral compressional stresses produced the relative separation of the Santa Marta massif from the Central Cordillera along the San Jorge strike-slip fault (Fig. 4), and resulted in the formation of both the Plato geofracture and the Sucre tectonic depression. The course of the Magdalena River, which it is believed formerly flowed northeastward to the site of the present Maracaibo Basin, was diverted to the northwest toward the Plato geofracture and the Sucre tectonic depressions, which from this time began to behave as submarine canyons. Sedimentation was of high energy (turbidites) with large-scale slumping and a great supply of sediments due to uplift of the land areas to the south and east (Fig. 12).

Sedimentation in the abyssal plain, immediately west of the San Jacinto belt and along the Sinú trench, was predominantly turbiditic with strong terrigenous influence mainly from the Plato and Sucre submarine canyons, and possibly from another farther to the south; the canyons behaved as enormous discharge channels for the great volume of sediments coming from the interior of the continent. The maximum thickness of turbidites seems to occur in the upper Sinú River area where it exceeds 5,000 m. Obviously, the coarser clastic sediments would be deposited nearer the continental margin, along the Sinú trench where it intersected the Plato and Sucre submarine canyons. The Pendales and Tucurá conglomerates along the Sinú paleotrench conform to this depositional model.

Andean Orogeny (Pliocene–Pleistocene)

This important orogenic episode of the Colombian Andes evolution (Van der Hammen, 1958; Irving, 1971) extended northward into the study area. The continued lateral compression controlling the tectonic and sedimentary evolution of northwestern Colombia reached another diastrophic peak over the Pliocene-Pleistocene time interval (Fig. 13) and produced in addition to uplifting, folding and faulting normal to the principal stress—a much more varied tectonic phenomenon accompanied by transcurrent faulting and flexures parallel to the major structural trends. This variety and complexity probably was due to the fact that as continental accretion progressed along the margins bordering the Colombian Basin, the space in which the compressional stresses were acting was reduced and squeezed, thus producing more varied and complex effects than those of the previous diastrophic phases. The following are considered to have resulted from the Andean Orogeny:

1. Mud volcanism and plutonism occurred within the belt of sediments in the Sinú trench, initially uplifting and deforming the sedimentary belt, causing an abrupt change in bathymetry (Duque-Caro, 1975, Fig. 3); the Colombia trench was formed along its western margin, which represents renewed migration westward of the original Romeral trench.
2. Structural narrowing of the Sinú belt to the south took place, as well as structural inflection of the middle section, possibly as a result of convergence due to lateral compressional stresses.
3. Uplifting and folding of the San Jacinto belt, with associated S-shaped structural closures along the eastern flank, which also seem to be related to the compressional forces that pro-

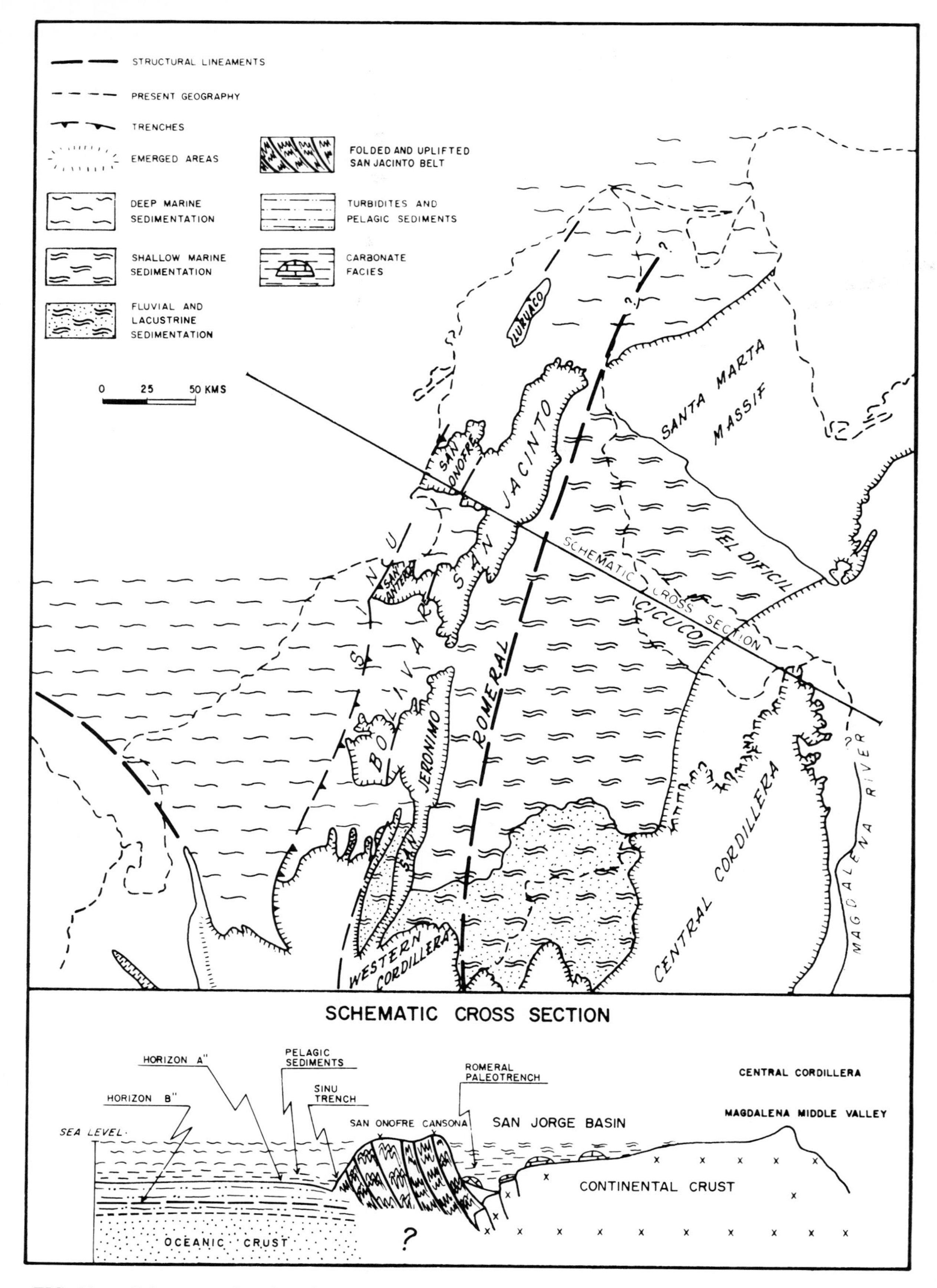

FIG. 11 Paleogeography of northwestern Colombia during early Miocene-middle Miocene (middle Carmenian).

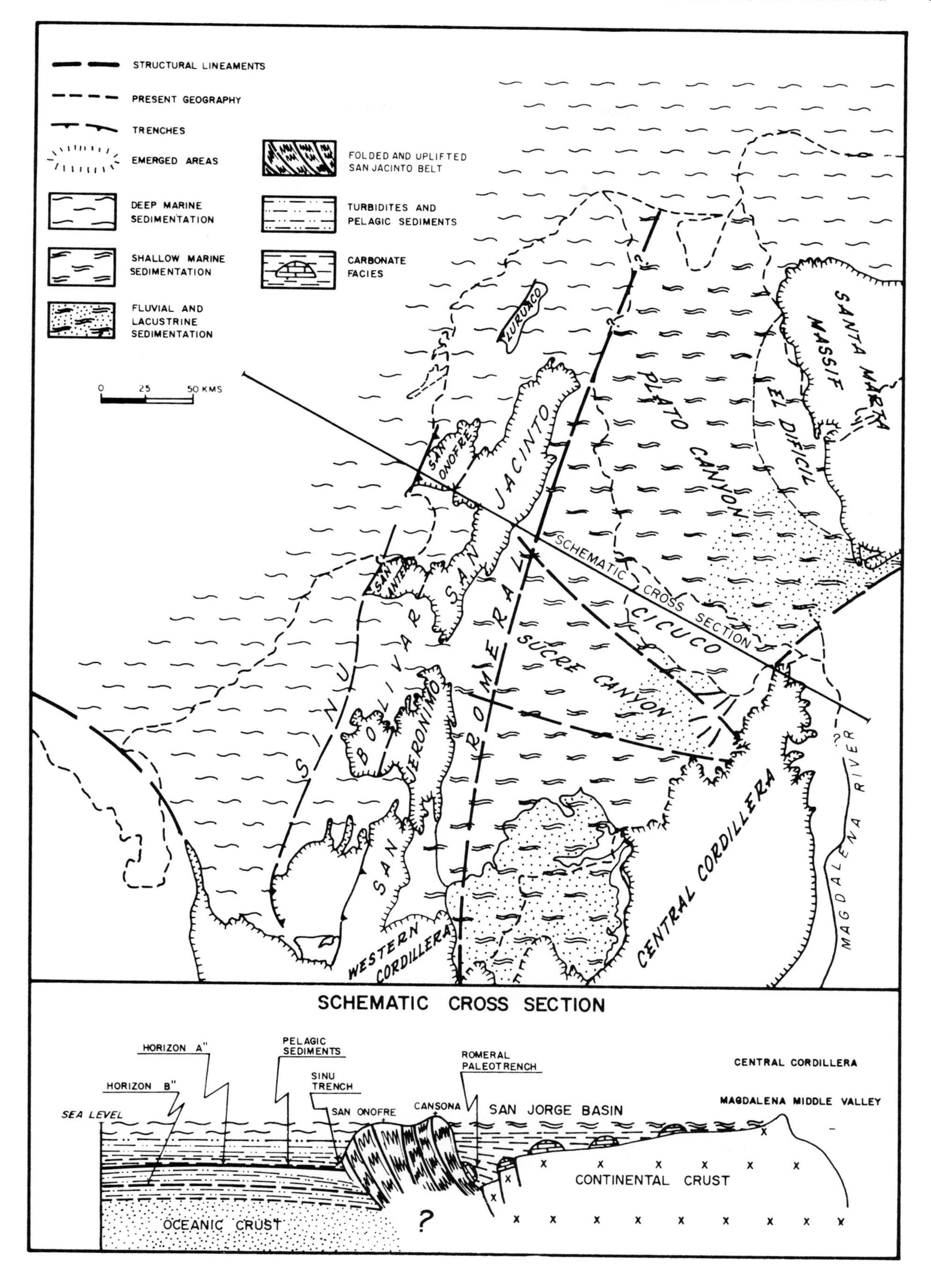

FIG. 12 Paleogeography of northwestern Colombia during late Miocene-Pliocene (late Carmenian).

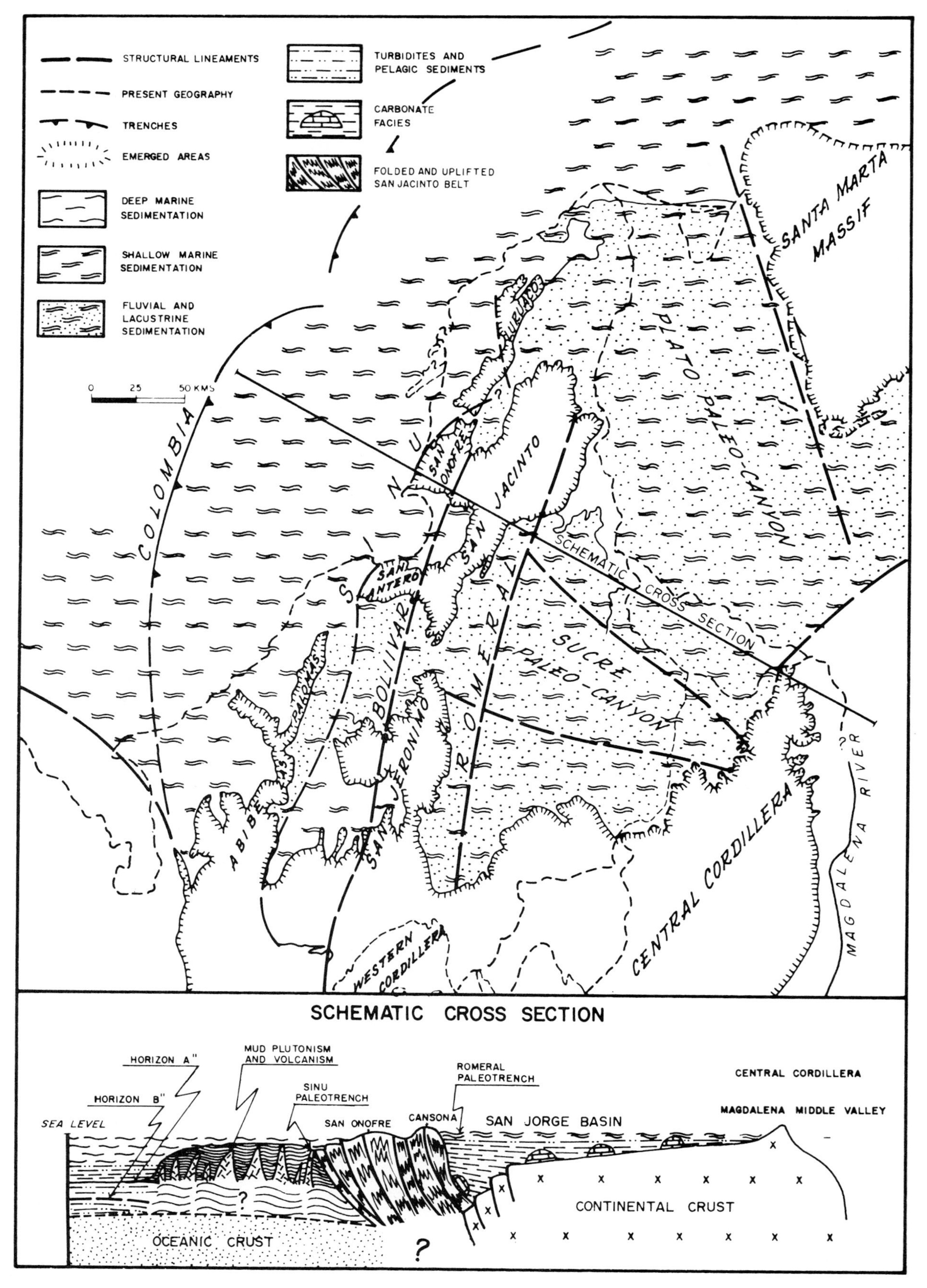

FIG. 13 Paleogeography of northwestern Colombia immediately after the Andean orogeny (Pliocene-Pleistocene).

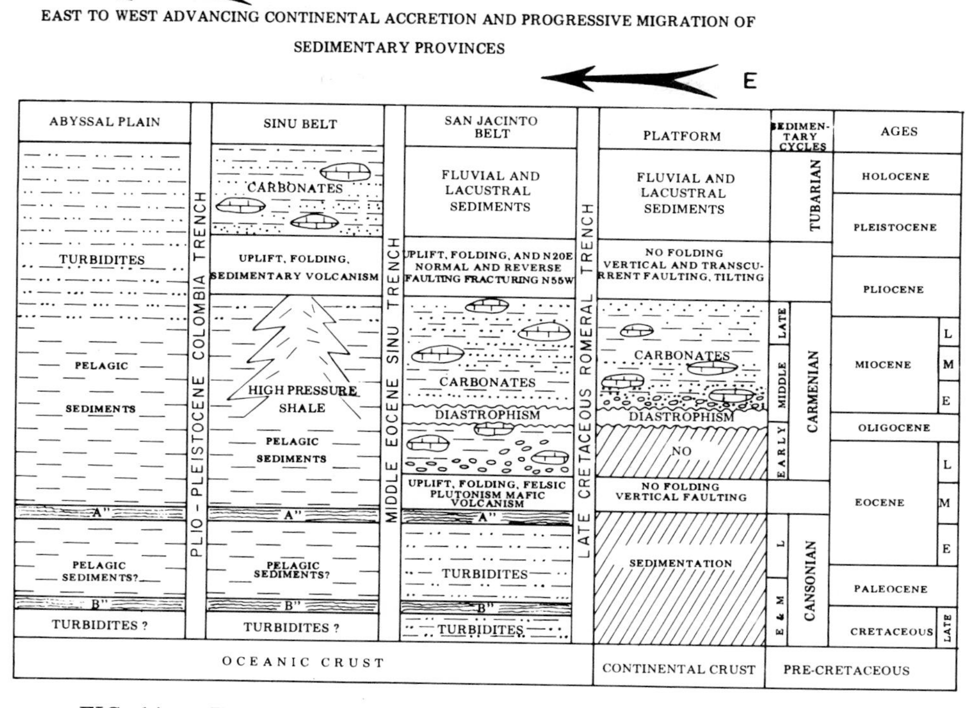

FIG. 14 Tectonic and sedimentary evolution of northwestern Colombia.

duced the inflection of the Sinú belt, reached completion. There was transcurrent movement along the Romeral lineament. In the Barranquilla area, the N 45° E-trending Sinú belt is superimposed upon, and thus hides, the original N 20° E trend of the San Jacinto belt (Fig. 4).

4. Emergence, but not folding of the sedimentary cover of the platform area, took place.
5. Left-lateral transcurrent movement occurred along the Santa Marta fault, which transferred the Santa Marta massif into its present-day position (Raasveldt, 1956; Campbell, 1968).

Pleistocene-Holocene (Tubarian)

The Tubará sedimentary cycle (Duque-Caro, 1972, 1975) began during this time interval with sedimentation in the submerged portions of the Sinú belt. The sediments were principally carbonate facies deposited in neritic paleoenvironments with water depths of as much as 200 m and composed of sandstones, shales, and reef limestones.

Meanwhile, the San Jorge Basin was dominated by fluvial and deltaic deposition along the remnant river channels left by the Plato and Sucre paleocanyons. These sediments reached a thickness of up to 4,000 m in the deepest parts of the eastern flank of the San Jacinto belt. Lacustrine sedimentation prevailed in areas adjacent to the major drainages.

On the abyssal plain, sedimentation was predominantly turbiditic (Edgar et al, 1971). The Sinú belt continued to emerge as a consequence of both the lateral compression and the great fluid pressure within the belt ("high pressure horizon"). The result was the start of mud volcanism and mud plutonism, which were the principal causes of deformation of the belt (Higgins and Saunders, 1974, p. 148), and which remain active today. This can be seen in the continuing present-day explosive volcanic eruptions of mud along the belt, particularly west of Montería in the Arboletes area (Fig. 3) and also at elevations above sea level in the islands bordering the coastal plain. All these are composed of Pleistocene to Holocene carbonate facies, mostly on the flanks of the volcanic forms. Two spectacular examples of onshore geomorphic features created by mud volcanoes are Loma de Piedras and Loma Los Volcanes, more than 200 m above sea level, respectively, and both of them are located in the Turbaco area near Cartagena (Fig. 3). Another interesting example of the uplifting force of mud plutonism is Tierrabomba Island (immediately south of Cartagena) which has a maximum elevation of 80 m above sea level. Beach-terrace deposits at an elevation of 3 m above sea level were dated as Holocene (2,850 to 150 years B.P.) by C^{14} radiometric dating (Porta et al, 1963).

CONCLUSIONS

1. Progressive westward continental accretion along the northwestern coastal margin of Colombia produced: (1) westward migration of the initial Romeral trench, occupying successively the present positions of the Bolivar, Sinú, and Colombia trenches; (2) migration in the same direction of the accompanying sedimentary province (Fig. 14); and (3) the formation and shaping of the San Jacinto belt during the pre-

Andean Orogeny, and of the Sinú belt during the Andean Orogeny.

2. Two different types of sedimentary sequences can be distinguised according to their depositional characteristics: (1) those developed immediately after each diastrophic event within each of the geotectonic elements, which are composed essentially of carbonate facies and fluvial and lacustrine sediments, and (2) those of the abyssal plain west of the trench regions, which comprise pelagic sediments and turbidites (Fig. 14).
3. Paleobathymetry along the areas currently occupied by the San Jacinto and Sinú belts was controlled by paleohighs of extrusive and intrusive origin; sedimentation along the continental margin, advancing to the west, was also strongly influenced by turbidites from the predominantly northwest-trending paleocanyons from at least Late Cretaceous time.
4. Lateral compressional stresses normal to the continental margin, the result of the continuing interaction between the Caribbean oceanic (Colombian Basin) and northern South American continental crusts, have been the main causes of all the tectonic, structural, and sedimentary features such as folding, uplifting, faulting, geofracturing, volcanism, plutonism, etc., recorded in the study area at least since Late Cretaceous time.
5. Two different structural trends originating from different orogenies converge in the northern part of the study area (Barranquilla area): the middle Eocene N 20° E trend of the San Jacinto belt (pre-Andean Orogeny), and the Pliocene-Pleistocene N 45° E trend of the Sinú belt (Andean Orogeny), paralleling the coastline.
6. Transcurrent faulting, structural flexures, and geofracturing represent the last stage of northwestern Colombian tectonic evolution and belong to the Pliocene-Pleistocene Andean Orogeny.
7. The San Jacinto fragmented belt is interpreted as an ancient submarine volcanic chain that extended northward to include the Beata Ridge, and also as the boundary zone between eastern and western tectonic provinces. I suggest that these provinces are underlain, respectively, by continental and oceanic crusts, based upon the structural and geophysical similarities of both northwestern Colombia an the Caribbean areas immediately to the north.
8. Prior to late Miocene-Pliocene time the Santa Marta massif remained part of a continuous Andean Central Cordillera system, which extended northeastward into the Caribbean area. However, during late Tertiary time, orogenic movements caused the fracturing and transcurrent faulting of this belt, bringing about the separation of the Santa Marta massif along the San Jorge strike-slip fault from what is now the northernmost geomorphic limit of the Colombian Central Cordillera.

REFERENCES CITED

Barrero, D., Alvarez, J., and T. Kassem, 1969, Actividad ígnea y tectónica en la Cordillera Central: Bol. Geol. Ingeominas, v. 17, no. 1-3, p. 145–173.

Beck, A., 1921, Geology and oil resources of Colombia. The coastal plain: Econ. Geology, v. 16, p. 457–473.

Burk, C. A., 1972, Uplifted geosynclines and continental margins: Geol. Soc. America Mem. 132, p. 75–85.

Campbell, C. J., 1968, The Santa Marta wrench fault of Colombia and its regional setting: Fourth Caribbean Geol. Conf., Trinidad (1965), Trans., p. 247–261.

Case, J. E., 1974, Major basins along the continental margin of northern South America, *in* C. A. Burk, and C. L. Drake, eds., The geology of continental margins: New York, Springer-Verlag, p. 733–741.

——— and T. L. Holcombe, 1975, Preliminary geologic-tectonic map of the Caribbean region: U.S. Geol. Survey Open-File Map, 75–146.

Christofferson, E., 1973, Linear magnetic anomalies in the Colombia Basin, central Caribbean Sea: Geol. Soc. America Bull., v. 84, no. 10, p. 3217–3230.

Crook, K. A. W., 1974, Lithogenesis and geotectonics: the significance of compositional variation in flysch arenites (graywackes), *in* R. H. Dott, Jr., and R. H. Shaver, eds., Modern and ancient geosynclinal sedimentation: SEPM Spec. Publ. 19, p. 304–310.

Donnelly, T. W., 1975, The geological evolution of the Caribbean and Gulf of Mexico—some critical problems and areas, *in* A. E. M. Nairn, and F. G. Stehli, eds., The ocean basins and margins, v. 3, The Gulf of Mexico and the Caribbean: New York, Plenum Press, p. 663–685.

Duque-Caro, H., 1972, Ciclos tectónicos y sedimentarios en el norte de Colombia y sus relaciones con la paleoecología: Bol. Geol. Ingeominas, v. 19, no. 3, p. 1–23.

——— 1973, Guidebook to the geology of the Montería area: Colombia Soc. Petrol. Geology and Geophysics, 14th Annual Field Conf., p. 1–49, Bogotá.

——— 1975, Los foraminíferos planctónicos y el terciario de Colombia: Rev. Esp. Micropal., v. 7, no. 3, p. 403–427.

——— 1976, Características estratigráficas y sedimentarias del Terciario marino de Colombia: 2nd Cong. Latin Geol., Caracas (1973), Mems. p. 945–964.

Durán, L. G., 1964, Ensayo de interpretación tectonofísica de la plataforma continental del Caribe: Bogotá, Caldasia, v. 9, no. 42, p. 138–150.

Edgar, N. T., J. I. Ewing, and J. Hennion, 1971, Seismic refraction and reflection in Caribbean sea: AAPG Bull., v. 55, no. 6, p. 833–870.

Ewing, J. I., N. T. Edgar, and J. W. Antoine, 1971, Structure of the Gulf of Mexico and Caribbean, *in* The sea, v. 4, pt. 2: New York, Wiley-Interscience, p. 321–358.

Higgins, G. E., and J. B. Saunders, 1974, Mud volcanoes—their nature and origin, *in* Contributions to the geology of the Caribbean and adjacent areas: Verhandl. Naturf. Ges. Basel, Band 84, no. 1, p. 101–152.

Hubach, E., 1930, Informe geológico de Urabá: Bol. Min. Petrol., v. 4, no. 19–20, p. 26–136.

Irving, E. M., 1971, La evolución estructural de los Andes más septentrionales de Colombia: Bol. Geol. Ingeominas, v. 19, no. 2, p. 1–89.

Krause, D. C., 1971, Bathymetry, geomagnetism, and tectonics of the Caribbean sea north of Colombia: Geol. Soc. America Mem. 130, p. 35–54.

Kugler, H. G., 1968, Sedimentary volcanism: 4th Caribbean Geol. Conf., Trinidad (1965), Trans., p. 11–13.

Ludwig, W. J., R. E. Houtz, and J. I. Ewing, 1975, Profiler-sonobuouy measurements in Colombia and Venezuela Basins, Caribbean sea: A.A.P.G., v. 59, no. 1, p. 115–123.

Porta, J. de., H. G. Richards, and E. Shapiro, 1963, Nuevas aportaciones al Holoceno de Tierrabomba: Bol. Geol. Univ. Indust. Santander, v. 12, p. 35–44.

Raasveldt, H. C., 1956, Fallas de rumbo en el nordeste de Colombia: Rev. Petrol. v. 7, no. 64, p. 19–26.

Roemer, L., W. Bryant, and D. Fahlquist, 1976, A geophysical investigation of the Beata Ridge: Seventh Caribbean Geol. Conf., Guadaloupe (1974), Trans., p. 115–125.

Shepard, F. P., 1973, Seafloor off Magdalena delta and Santa Marta area, Colombia: Geol. Soc. America Bull., v. 84, no. 6, p. 1955–1972.

——— Dill, R. F., and B. C. Heezen, 1968, Diapiric intrusions in foreset slope sediments off Magdalena delta, Colombia: AAPG Bull., v. 52, no. 11, p. 2197–2207.

Stainforth, R. M., 1968, Mid-Tertiary diastrophism in northern South America: 4th Caribbean Geol. Conf., Trinidad (1965), Trans., p. 159–177.

U.S. Naval Oceanographic Office, 1975, Bathymetric map of the Caribbean region: U.S. Geol. Surv. Open-File Map, 75–146.

Van der Hammen, T., 1958, Estratigrafía del Terciario y Maestrichtiano continentales y tectogénesis de los Andes Colombianos: Bol. Geol. Ingeominas, v. 6, no. 1–3, p. 67–128.

Watkins, J. S., and T. Cavanaugh, 1976, Implications of magnetic anomalies in the Venezuela Basin: 7th Caribbean Geol. Conf., Guadaloupe (1974), Trans., p. 127–138.

Zimmerle, W., 1968, Serpentine graywackes from the north coast basin of Colombia and their geotectonic significance: Neues Jahrb. Mineralogie Abh. 109, no. 1/2, p. 156–182.

Geomorphology and Subsurface Geology West of St. Croix, U.S. Virgin Islands[1]

TROY L. HOLCOMBE[2]

Abstract West of St. Croix, U.S. Virgin Islands, a small plateau (Fredericksted Plateau) sits atop the St. Croix Ridge at a depth of 800–1200 m. It is separated from St. Croix by an island slope which is probably a fault scarp, and is bounded on the north by a steep escarpment which leads down to the floor of the Virgin Islands Trough. Numerous small submarine channels indent the island slope and coalesce downslope into three canyons (Shepard Canyon, Sprat Hall Canyon, and Fredericksted Canyon) which interrupt the otherwise gently sloping surface of the Plateau. Fredericksted Canyon is flanked by natural levees and a submarine "floodplain," suggesting that it is associated with an aggrading sedimentation regime. Shepard and Sprat Hall Canyons, on the other hand, are erosional features cut through hard limestone strata of varying resistance to erosion. Fredericksted Plateau is underlain by an unknown thickness of sedimentary strata; a northeast-southwest-trending syncline underlies Fredericksted Canyon. Whereas the older strata were folded with or blanketed on the synclinal structure, the youngest strata include horizontally bedded or gently inclined strata, filling the synclinal valley. The north escarpment coincides with the northwest limb of the syncline. The aggradation associated with Fredericksted Canyon probably proceeded until the level of the valley floor reached the elevation of the rim of the north escarpment in the vicinity of Shepard and Sprat Hall Canyons. Then, with sediments available as cutting tools for erosion, Shepard and Sprat Hall Canyons eroded headward across the Fredericksted Canyon depositional plain, capturing part of the island slope drainage.

INTRODUCTION

West of St. Croix, U.S. Virgin Islands, the study of the geomorphology and history of sediment deposition and erosion is enhanced by the availability of bathymetric and seismic reflection data of unusually high resolution and detail. In 1976, 12-kHz and 3.5-kHz bathymetric data and 30-kilojoule sparker seismic reflection data (single-channel, unprocessed) were collected by USNS *Lynch* at sufficient line spacing (300 m, with high navigation accuracy), relative to the natural dimensions of major topographic features, to establish the morphology of sea floor and subsurface reflecting horizons with minimum ambiguity. With coincident bathymetric and seismic reflection data, it has been possible to do what could not have been accomplished to the same degree with bathymetric data alone: tentatively establish a sequence of structural, depositional, and erosional events which account for the present observed bottom form, and which are based primarily on the spatial distribution of acoustic reflectors. Absolute chronology of events can only be speculated upon due to lack of stratigraphic control.

The area of interest and the island of St. Croix are crestal portions of the St. Croix Ridge, a predominantly submarine feature capped by plateaulike, flat or gently tilted segments probably separated by fault scarps, and underlain by an unknown thickness of stratified, moderately deformed sedimentary and probably metasedimentary rocks. The St. Croix Ridge extends between the northern end of the Aves Ridge and the southeastern corner of Puerto Rico in the northeastern Caribbean (Fig. 1); north of the

[1]Manuscript received, August 5, 1977; accepted, December 22, 1977.

[2]Sea Floor Division, Naval Ocean Research and Development Activity, NSTL Station, Mississippi 39529.

Bathymetric and seismic reflection data collected by crews of USNS *Lynch*, R/V *Sheldrake*, and R/V *Oceanic;* bottom photographs collected by crews of R/V *Oceanic* and the submersible *Alvin;* dive narratives obtained by crews of *Alvin* and *Aluminaut;* and cores collected by the crew of USNS *San Pablo* were made available to the author by the Chesapeake Division, Naval Facilities Engineering Command (CHESNAVFACENGCOM). Acknowledgment is due the scientific parties and crews of these vessels, and to E. C. Escowitz and S. Ling of CHESNAVFACENGCOM. Rock samples were made available by R. F. Dill of the West Indies Laboratory, Fairleigh Dickinson University, and by CHESNAVFACENGCOM. W. C. Ward, J. A. Klasik, and W. W. Craig of the University of New Orleans made petrographic descriptions of the rock samples from thin sections. R. F. Dill, J. E. Matthews, and F. A. Bowles made valuable criticisms and suggestions. A. Einwich studied the bottom photographs from which examples and notes are drawn. J. Egloff contoured the bathymetric data. Notes on surficial sediments are drawn from a synthesis by F. A. Bowles. J. Green assisted with the illustrations. Editorial assistance was provided by L. McRaney.

This study was supported by CHESNAVFACENGCOM under contract N62477-PO-6-0002 and the Chief of Naval Research through the Naval Ocean Research and Development Activity.

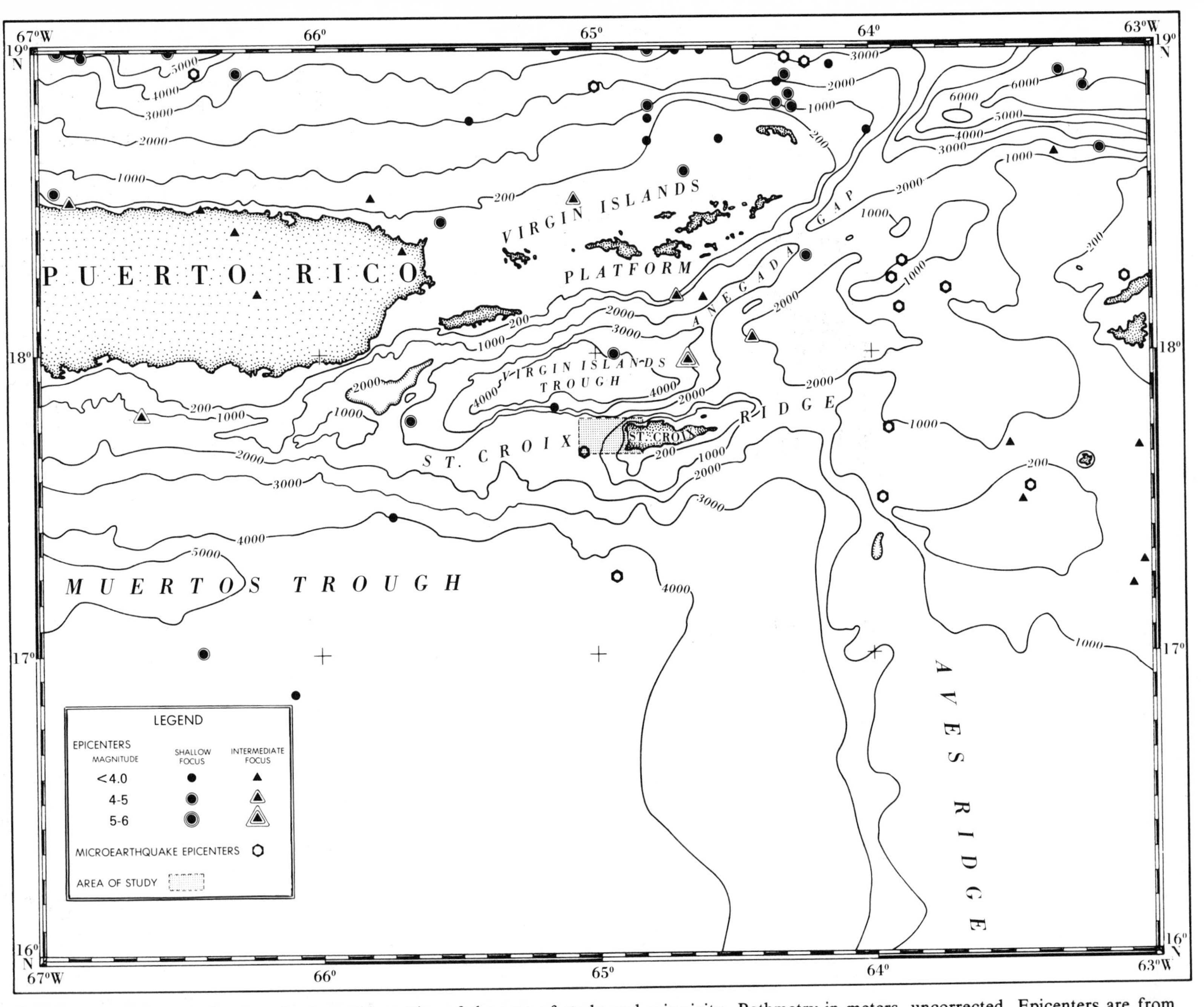

FIG. 1 Index map showing physiographic setting of the area of study and seismicity. Bathmetry in meters, uncorrected. Epicenters are from Sykes and Ewing (1965) for the period 1950 to 1964, and from the Environmental Data Service worldwide earthquake file for the period 1964 to 1975. Microearthquake epicenters are from Murphy et al, 1970.

ridge, a steep escarpment leads down to the deep (~4500 m) floor of the Virgin Islands Trough. South of the plateau, a less steep, probably structurally deformed, slope leads down to the Venezuelan Basin at about 4,500 m depth. The area of interest consists of a 12-by-18-km portion of the flattened ridge crest (Fredericksted Plateau) which lies immediately west of the island of St. Croix.

A northeast-trending fault zone (Anegada Fault), which coincides with the Virgin Islands Trough and the Anegada Gap, separates the St. Croix Ridge from the Virgin Islands Platform. Present seismic activity suggests tectonic movements along the fault zone (Donnelly, 1964), but the sense and direction of motion has not been resolved. The linearity of the associated seismic belt, and the fact that the epicenters do not define a sloping seismic zone, suggest shear, tensional, or transtensional movement (Fig. 1). Therefore, the St. Croix Ridge may once have been part of the Virgin Islands Platform (Matthews, 1970). Initial separation probably occurred in the Late Cretaceous to pre-Oligocene time interval, a time of structural activity and stratigraphic breaks on St. Croix (Cederstrom, 1950; Whetten, 1966). Even though the Late Cretaceous spilites and keratophyres of the Virgin Islands Platform contrast with the Late Cretaceous epiclastic volcanic and tuffaceous sedimentary strata of St. Croix, the two suites of rocks were probably formed in an adjacent or juxtaposed marine environment (Whetten, 1966). A series of northeast-trending structures along the St. Croix Ridge extend through the subject area; these might represent secondary structural effects of relative plate motion along the Anegada fault zone (Fig. 1). West of the Anegada fault zone, the southern margin of Puerto Rico and Hispaniola exhibits all the morphologic and structural features of a plate-convergent continental margin (Case, 1975; Garrison et al, 1972; Matthews and Holcombe, 1974).

This paper presents highlights of the acoustic stratigraphy and morphology of the subject area which establish the proposed geological history. For systematic treatment of the regional geology, geomorphology, surficial and subsurface geology, and seismicity of the area, please see Holcombe et al, 1977.

SEAFLOOR MORPHOLOGY

The seafloor morphology of the area is illustrated by bathymetry and track control (Fig. 2). Along the eastern margin of Fredericksted Plateau, the island slope (presumed to be a fault scarp) has a relief of ~750 m and slopes of 5 to 30°, increasing northward. The plateau itself lies at a depth of 800 to 1200 m; it is flat to gently sloping, and partly dissected by canyons. Slopes are westward to southwestward at 2 to 5° in the east and 1 to 3° in the west, except for the tilted northern edge which slopes south-southeastward at 2 to 4°. Abruptly terminating the plateau at its upturned northern edge is the steep (10 to 30°) escarpment facing the Virgin Islands Trough. Although the escarpment trends generally east–west, it is indented in the eastern part of the area near the island slope, where canyons extend over the scarp; to the west, the scarp rim elevation increases as it juts northward. Many small submarine channels of 20 m or less relief originate on the St. Croix Island slope and coalesce downslope to form three canyons: Fredericksted Canyon, Shepard Canyon, and Sprat Hall Canyon (Fig. 2).

Fredericksted Canyon extends westward, then curves southwestward, across the plateau. It drains the southern half of the island slope seen within the area of study. It varies in relief from 20 to 100 m, increasing in size and depth downslope. Canyon slopes average 5 to 15°. Downslope, the floor of Fredericksted Canyon is flat in cross-profile. Adjacent to Fredericksted Canyon, the plateau surface resembles that of a subaerial river floodplain. The plateau surface slopes gently westward and southwestward in conformity to the curvature and gradient of the canyon.

Natural levees border Fredericksted Canyon itself, with the right (north) levee being 50 to 100 m higher than the left (Fig. 2). The width of the canyon varies from 0.75 to 1.2 km. Beyond the limits of the study area, Fredericksted Canyon continues to curve southward, and gradients increase as it reaches the southward slope of the St. Croix Ridge.

Asymmetry of natural levees and preferential curvature of channels have been widely observed in association with mid-ocean canyons and with distributary channels on deep-sea fans (Menard, 1955; Heezen, 1963; Menard et al, 1965, and Heezen et al, 1969). These phenomena are thought to be manifestations of the Coriolis effect (Menard, 1955). Higher right-bank levees and leftward curvature, both characteristic of Fredericksted Canyon, are in agreement with what is observed and ascribed to Coriolis effect in the northern hemisphere. The leftward curvature of Fredericksted Canyon, however, may be due to structural control.

Shepard and Sprat Hall Canyons curve northwestward, then northward, extending beyond the edge of the north escarpment (Fig. 2). Shepard Canyon increases from 20 to 180 m relief downslope, reaching maximum relief at the scarp rim. Its relief decreases beyond the rim of the escarpment. Sprat Hall Canyon reaches a maximum relief of 60 m at the plateau edge. Most of the submersible dives were made in or near Shepard and Sprat Hall Canyons. Observations made from the submersibles reveal that the canyon walls comprise alternating cliffs and benches cut in horizontal or moderately dipping sedimentary rock strata.

Consolidated rock crops out in the walls of Shepard and Sprat Hall Canyons and in smaller canyons on and near the base of the island slope, including the tributaries of Fredericksted Canyon. Undoubtedly, extensive outcrops also occur on the precipitous northern island slope and on the north escarpment. The total surface area encompassed by outcrops, however, is quite small; the outcrops are probably largely confined to the aforementioned areas. Distinct stratification is seen in Shepard and

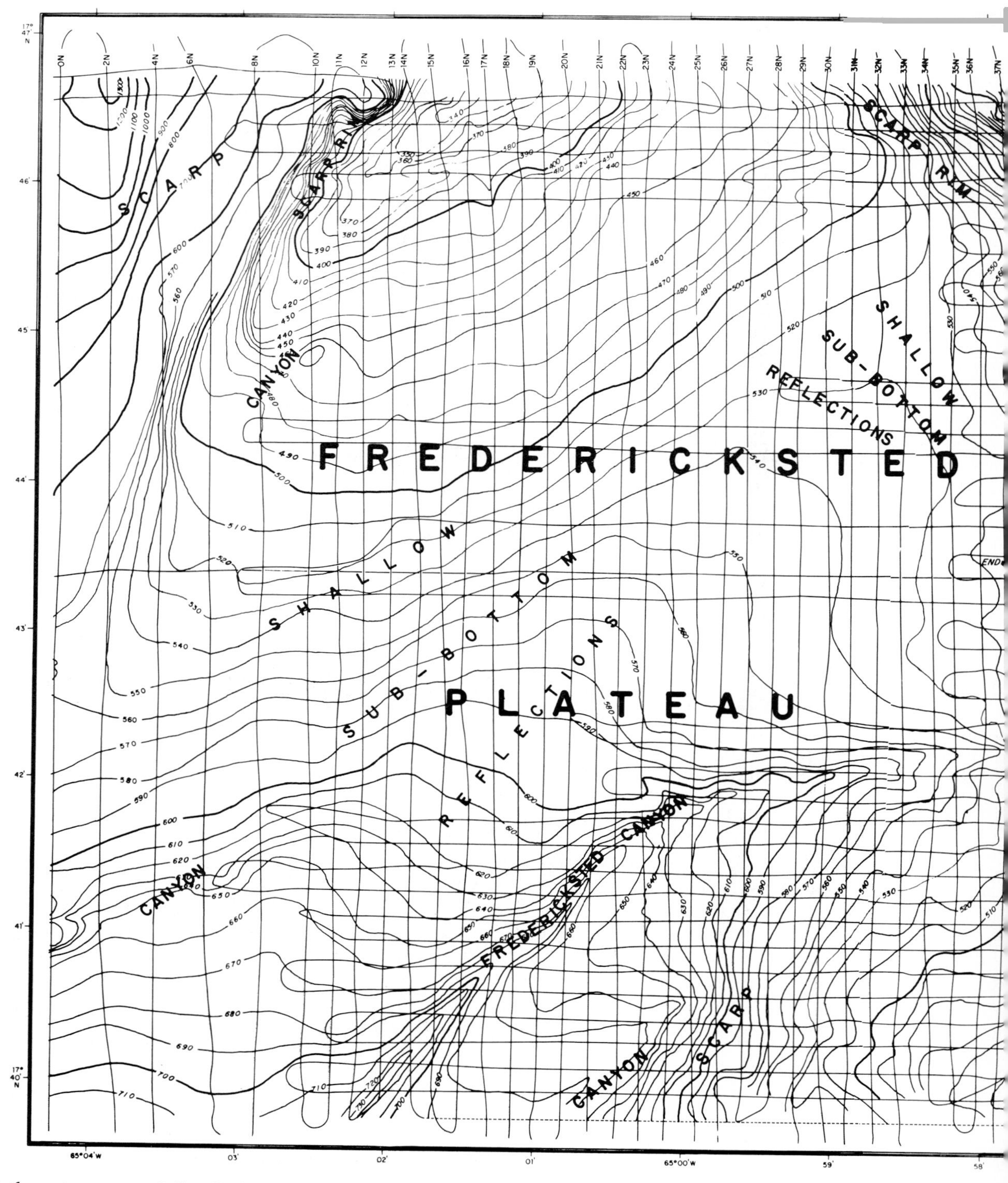

FIG. 2 Bathymetry west of St. Croix. Depths are in units of 1/400 sec two-way travel time (1 sec = 750 nominal meters, based on a sound speed in water of 1,500 m/sec, or 400 nominal fathoms, based on a sound speed in water

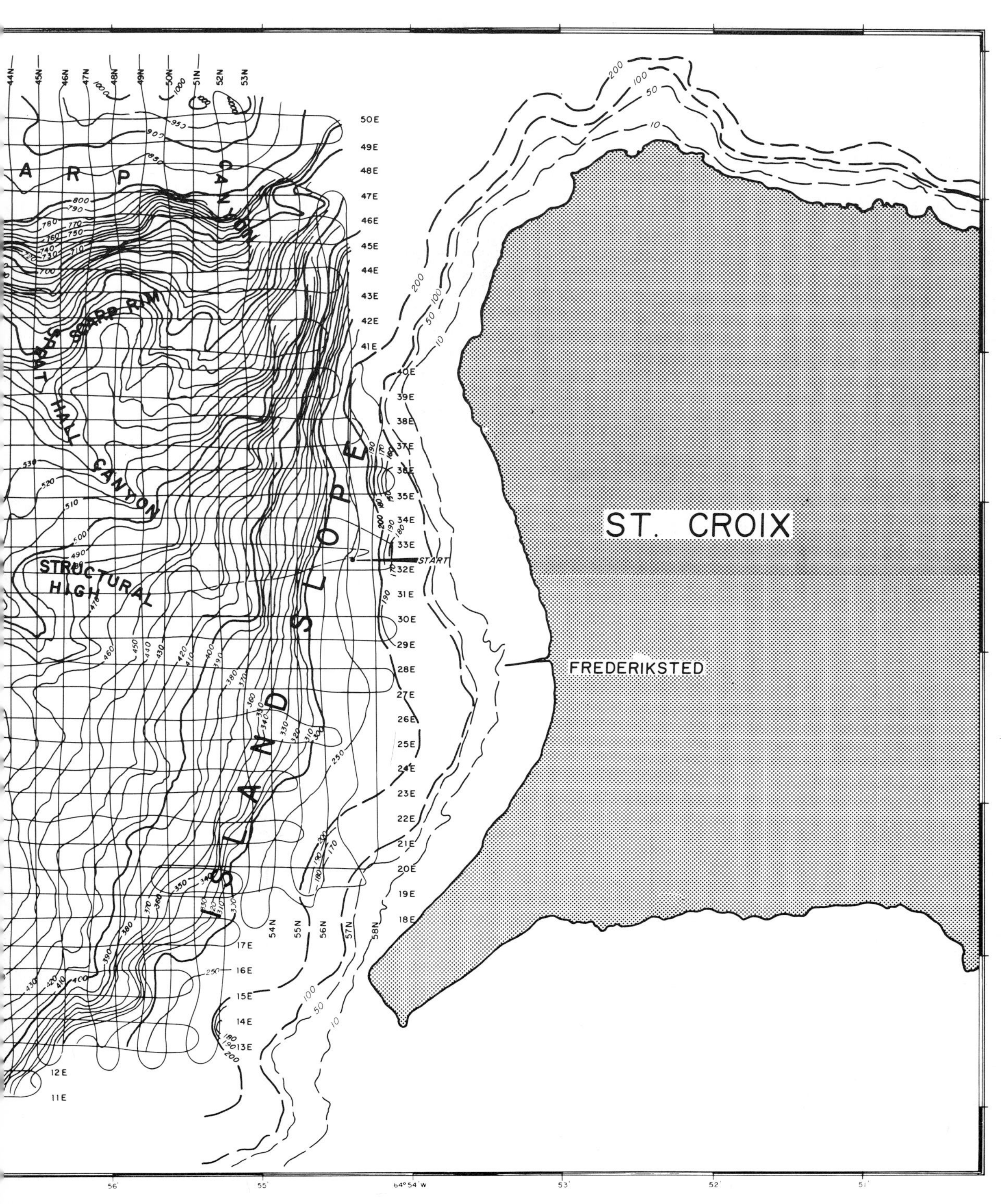

of 800 fm/sec). Contour interval is 10/400 sec. Tracklines are numbered as a means of locating section lines.

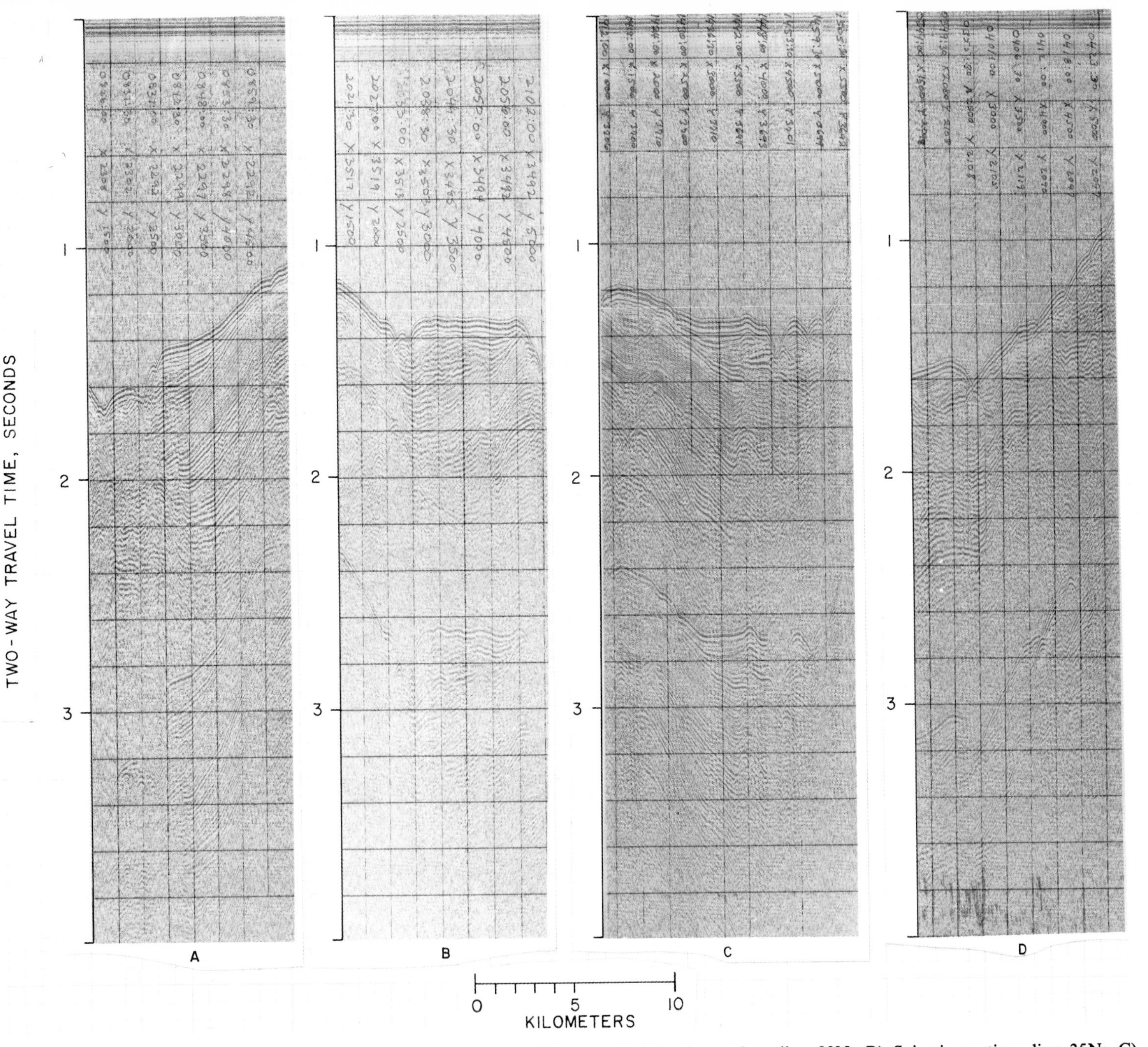

FIG. 3 Representative seismic sections across Fredericksted Plateau. A) Seismic section, line 23N; B) Seismic section, line 35N; C) Seismic section, line 37E; D) Seismic section, line 21E.

Sprat Hall Canyons, with resistant strata forming cliffs or steep slopes, and less resistant strata forming benches. Bed thicknesses are typically on the order of a few centimeters, with beds in many cases displaying varying resistance to erosion, resulting in a corrugated surface expression similar to that observed in the eroded walls of Scripps Canyon (Dill, 1964).

SUBSURFACE GEOLOGY

Three stratigraphic units, termed units A, B, and C in descending order, are identifiable in the seismic reflection sections (Figs. 3–5). An underlying unit, termed unit D, consists of whatever rock sequence lies beneath the deepest consistently observed reflectors. Delineation of units is based upon character of reflective sequences rather than upon intervals between specific reflectors, because of discontinuities and discordances in individual reflectors. The reflective sequences defined, however, are consistently correlatable beneath the Fredericksted Plateau. All stratigraphic units apparently crop out on the steep north escarpment where stratigraphic control could probably be established by dredging.

The principal structure of the area consists of a major northeast-southwest-trending syncline (Fig. 6). Its north limb forms the north rim of the escarpment; its south limb coincides with the northeast-southwest-trending escarpment occurring in the south-central portion of the area, which is also the north limb of an anticline extending across the southeastern part of the area. Subsidiary folds having the same northeast-southwest trend occur within the syncline.

Unit C is defined by a series of strong internal reflectors displaying moderate structural deformation. It ranges in thickness from 0.3 to 0.8 seconds (two-way travel time), reaching its greatest thickness in the axis of the syncline and thinning on the syncline flanks. The attitude of unit C strata is concordant with the syncline, and its beds dip southeast and northwest, respectively, on the northwest and southeast limbs. The strata seem to be least structurally disturbed on the northwest limb of the syncline.

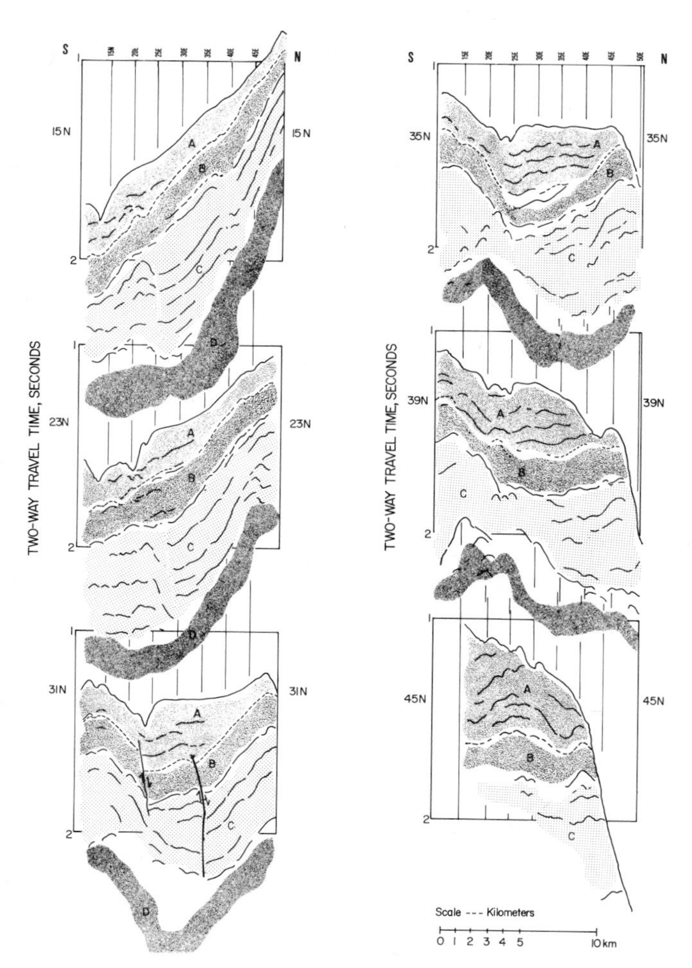

FIG. 4 North-south stratigraphic cross-sections made from seismic sections. Crossing points of east-west tracklines shown by ruled vertical lines and labeled at top. Stratigraphic units A, B, C, and D are clearly shown.

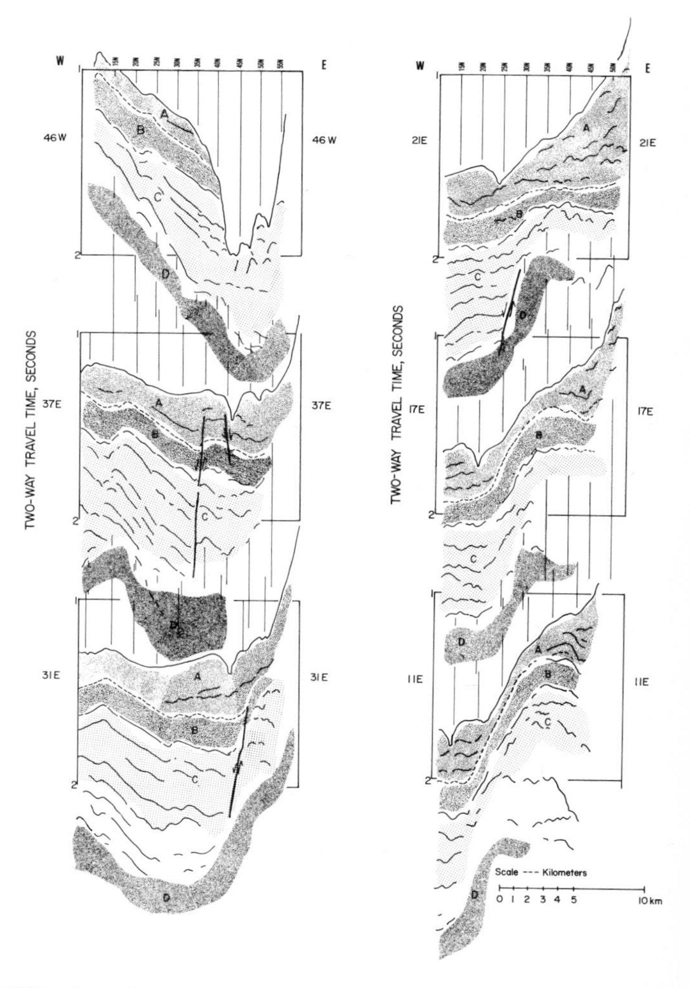

FIG. 5 East-west stratigraphic cross-sections made from seismic sections. Crossing points of north-south tracklines are shown by ruled vertical lines and labeled at top. Stratigraphic units A, B, C, and D are clearly shown.

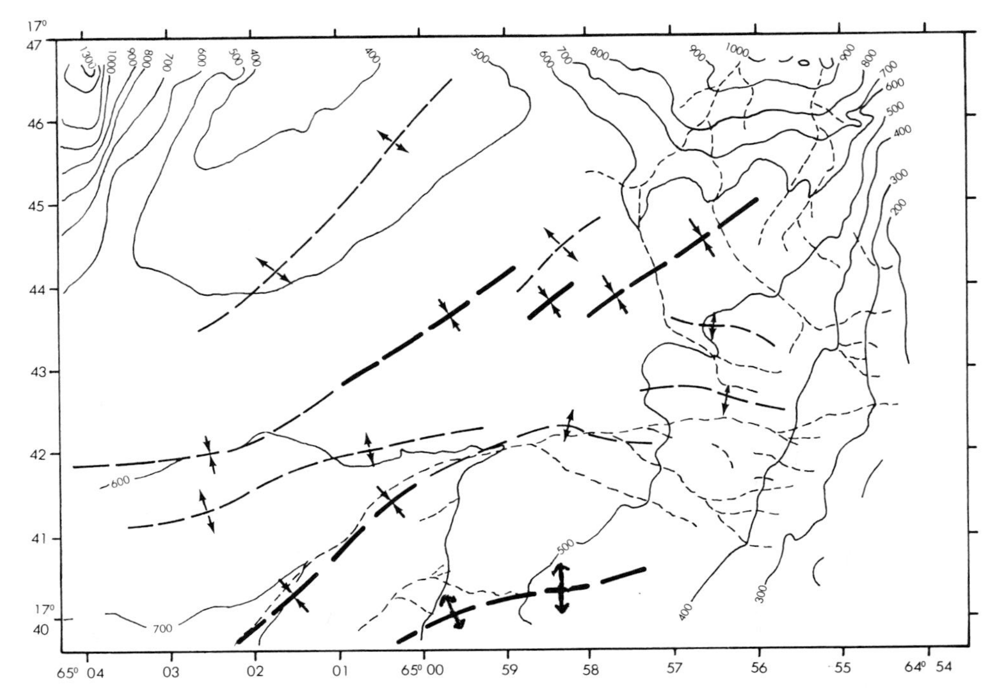

FIG. 6 Bathymetric map showing main features of Fredericksted Plateau structure. Syncline extends broadly across area from northeast to southwest. Its principal axis shifts between small subsidiary folds. Canyon axes are shown by dashed lines.

They do not thicken near the island slope.

Unit B consists of a moderately transparent interval which does not have prominent internal reflectors. Its thickness ranges from 0.1 to 0.25 seconds. It does not thicken appreciably in the axis of the syncline or near the base of the island slope.

Unit A averages 0.1 to 0.3 seconds in thickness (thickness extremes are 0.05 and 0.6 seconds). It is thin and relatively transparent in the limbs of the syncline. Strong reflectors, moderately to slightly structurally deformed, occur within the thicker part of unit A, coinciding with the axis of the syncline. Unlike unit C, these reflectors are not concordant with the synclinal structure; instead, they are of horizontal or subhorizontal attitude resembling formerly flat-lying turbidites or other deposits which form a valley-fill sequence. A particularly strong and persistent reflector occurs near the base of the unit. Unit A is the only unit which thickens beneath the island slope.

Unit A is also the only unit for which lithologic control is available. Three rock samples collected using the submersible *Alvin* are from about the middle levels of the valley-fill sequence of unit A, which has been exposed in the walls of Shepard and Sprat Hall Canyons. All samples are well-cemented limestones.

One rock sample is a pelagic calcilutite composed predominantly of planktonic foraminifera cemented by carbonate mud. A few quartz grains occur that are probably eolian. The two other samples are bioclastic calcarenites composed predominantly of coral and algal reef fragments and other clasts, including carbonate rock fragments of shallow-water origin. Minor constituents of the samples include pelagic carbonate mud and planktonic foraminifera, and terrigenous silicate minerals and rock fragments. Both samples are interpreted as reflecting a two-stage history, first of shallow-water deposition, followed by removal of the sediments to a deeper water environment. The neritic and pelagic components are mixed throughout one of the samples, good evidence that shallow-water sediments were reworked, then transported to a deeper water environment, probably by turbidity currents. The occurrence of two probable turbidite horizons and one pelagic horizon within the subhorizontal valley-fill sequence of unit A demonstrates that these strata probably constitute a turbidite sequence of interstratified turbidite and pelagic layers.

DISCUSSION

The history of Fredericksted Canyon (excluding the upper portion and its tributaries, on and near the island slope) apparently differs from that of Shepard and Sprat Hall Canyons. Fredericksted Canyon shows evidence of having been part of an aggrading sedimentation regime in which there has built up a valley-fill sequence of alternating pelagic and turbidite strata. The natural levee, "floodplain," and occurrence of turbidites in the underlying strata all suggest that shallow-water, reef-environment sediments have been transported down the island slope and deposited out on the gently sloping valley floor. Occurrence of the channel suggests, as it does on deep-sea fans, that some of the turbidity-current flows were partially or completely confined to the channel. Slight southward migration of the channel of Fredericksted Canyon is suggested by a southward shift in the deep axis of subsurface sedimentary strata, but the confining walls of the syncline have no doubt prevented substantial horizontal migration of the channel, as occurs on deep-sea fans.

Shepard and Sprat Hall Canyons, on the other hand, are clearly erosional features. Both canyons are cut through flat-lying to moderately dipping, well-cemented limestone strata which display the effects of differential resistance to erosion. The fact that Shepard and Sprat Hall Canyons slice through turbidites that probably were deposited in the Fredericksted Canyon aggrading regime, and the fact that both canyons interrupt the gentle east-to-west downslope gradient of the Fredericksted Canyon "floodplain," which farther south is continuous with the island slope, suggest that formation of these canyons represents a late event in the geomorphologic development of the region, compared to the history of aggradation of the Fredericksted Canyon valley-fill sequence.

In view of the poor sorting and the pelagic biota observed in the surficial sediments of Fredericksted Plateau, one can conclude that no present-day turbidites are being deposited over any part of the Fredericksted Canyon "floodplain." Whether any significant channel-confined turbidity-current events are occurring now, or have occurred in Holocene time, is a matter for speculation. Observers in submersibles have noted the sediment-free appearance of Shepard and Sprat Hall Canyon bottoms, but these canyons may be kept free of sediment by sand flow, by bottom currents, or both.

Occurrence of the well-cemented, cliff-forming strata on the upper slopes of the erosional canyons suggests that cementation is occurring soon after deposition, and without significant burial. Ample evidence has accumulated that recent submarine lithification of carbonates can and does occur in widely scattered localities at depths ranging from shallow water to 3,500 m, and without significant burial (Gevirtz and Friedman, 1966; Fischer and Garrison, 1967; Land and Goreau, 1969). Such lithification observed in Holocene shallow-water reef deposits has been referred to as "syndepositional" (Ginsberg and James, 1976).

Certain relative time limits can be placed on the two major structural events which affected the history of the area as revealed by the subsurface geology: (1) formation of the northeast-southwest syncline, and (2) faulting which formed the island slope. Both events occurred prior to deposition of the A stratigraphic unit. Unit A thickens in wedge-shaped fashion beneath the island slope, whereas older units do not. Therefore, it seems likely that faulting producing the north–south fault scarp which is now the island slope occurred in the post-B, pre-A time period. This was probably the structural event which resulted in the lithologic change associated with the difference in acoustic reflectivity between the B and A units, and with the initiation of the valley-fill mode of deposition. On the other hand, it seems probable that formation of the syncline was an earlier event. Unit C deposits show thickening related to the synclinal axis. However, the thickening might also be related to downslope gravity movement of previously deposited unit C strata over synclinal limbs which were formed later. Structural deformation may have been a slow process continuing over a long period of time. It seems likely that the major syncline-forming deformation was near completion by the end of "C" time, in view of the less disturbed appearance of unit B relative to that of unit C. Certainly, some structural deformation has continued as late as the time of deposition of unit A, as evidenced by apparent disturbance of reflectors which define the valley-fill sequence, and by faults on which displacement is propagated into the lower part of the unit A strata.

Due to absence of stratigraphic control, discussion of the absolute age of the stratigraphic sequence is purely speculative. However, one suggestion may be based on structural correlation with the island of St. Croix. One notes that the northeast-southwest-trending syncline underlying the Fredericksted Plateau has the same trend and similar dimensions as the syncline or graben which forms the central structure of St. Croix (Cederstrom, 1950; Whetten, 1966). Structural deformation which produced the St. Croix syncline (or graben) may be loosely dated as Maestrictian to Oligocene (Whetten, 1966). If the syncline beneath the Fredericksted Plateau were formed in the same time interval, as seems likely, then we may say that either units A, B, and C are post-Oligocene, or that only units A and B are post-Oligocene, depending on whether we interpret that formation of the syncline was completed before or after deposition of unit C. The corollary to this is that unit D only, or units C and D, respectively, may correlate with the Late Cretaceous Caledonia and Judith Fancy Formations of St. Croix.

Simple block diagrams serve to illustrate the probable geomorphologic evolution of the area subsequent to deposition of acoustic-stratigraphic unit B (Fig. 7). It appears that unit A time was a time of fan-type sediment accumulation which resulted in sedimentary filling of the synclinal valley which underlies Fredericksted Canyon. Once aggrading sediments reached the level of the northern rim of the plateau in the eastern part of the area where the rim occurs at lower elevation, sediments then spilled over the north escarpment. With sediments available as cutting tools for erosion, Shepard and Sprat Hall Canyons eroded headward, slicing across the sediment sequence associated with Fredericksted Canyon and capturing part of the island slope drainage.

Submarine stream piracy due to headward erosion is probably an unusual, but not unique, feature of continental margins. Another such occurrence is noted by F. P. Shepard (1963). On the Southern California borderland, Coronado Canyon has apparently eroded through Coronado Bank and captured drainage that formerly followed Loma Sea

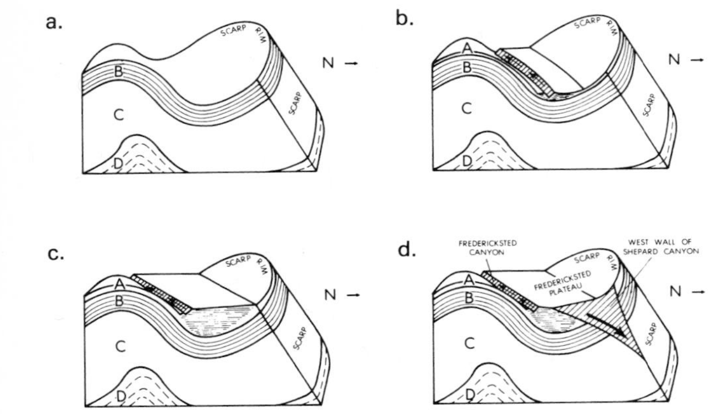

FIG. 7 Block diagrams illustrating geomorphologic development of Fredericksted Plateau. Front surface of block represents schematic north-south section through the northern part of Fredericksted Plateau at about the longitude of line 40N (see Fig. 2). Line 40N approximately coincides with the axis of Shepard Canyon where it encounters the north rim of the escarpment. Stratigraphic units A, B, C, and D are labeled on the front surface of the block. The view is to the west, across the plateau. The north escarpment is represented by the sloping surface of each block on the right. Block a represents the plateau and syncline as it must have appeared prior to deposition of stratigraphic unit A. Block b represents an early stage of deposition of stratigraphic unit A, with pelagic deposition on the side slopes of the valley and valley-fill deposition in the valley floor. Fredericksted Canyon and natural levees are shown. Block c represents a later stage when level of valley-fill reached level of the north rim of the escarpment at the longitude of Shepard Canyon. Block d shows representation of west wall of Shepard Canyon where it has dissected the northern rim of the plateau.

Valley, which extends northwestward around the end of Coronado Bank.

REFERENCES CITED

Case, J. E., 1975, Geophysical studies in the Caribbean Sea, *in* A. E. M. Nairn and F. G. Stehli, eds., The ocean basins and margins: Plenum Press, v. 3, p. 107–173.

Cederstrom, D. J., 1950, Geology and groundwater resources of St. Croix, Virgin Islands: U.S. Geol. Survey Water Supply Pap. 1067, 117 p.

Dill, R. F., 1964, Sedimentation and erosion in Scripps submarine canyon head, *in* R. L. Miller, ed., Papers in Marine Geology, Shepard Commemorative Volume: New York, MacMillan, p. 23–41.

Donnelly, T. W., 1964, Evolution of Eastern Antillean island arc: AAPG Bull., v. 48, p. 680–696.

Fischer, A. G. and R. E. Garrison, 1967, Carbonate lithification on the sea floor: Jour. Geology, v. 75, p. 488–496.

Garrison, L. E. et al, 1972, USGS-IDOE, Leg 3: Geotimes, v. 17, no. 3, p. 14–15.

Gevirtz, J. L. and G. M. Friedman, 1966, Deep-sea carbonate sediments of the Red Sea and their implications on marine lithification: Jour. Sed. Petrology, v. 36, p. 143–151.

Ginsburg, R. N. and N. P. James, 1976, Submarine botryoidal aragonite in Holocene reef limestones, Belize: Geology, v. 4, p. 431–436.

Heezen, B. C., 1963, The Tonga-Kermadec and Hikurangi trenches (Abs.): Pap. Int. Assoc. Phys. Oceanography, XIII Assembly Int. Union of Geodesy and Geophysics, v. 6, p. 70.

——— G. L. Johnson, and C. D. Hollister, 1969, The northwest Atlantic mid-ocean canyon: Can. Jour. Earth Science, v. 6, p. 1441–1453.

Holcombe, T. L., et al, 1977, The geological environment west of St. Croix: Naval Ocean Research and Development Activity Publ. No. 5, 80 p.

Land, L. S. and T. F. Goreau, 1969, Submarine lithification of Jamaican reefs: Jour. Sed. Petrology, v. 40, p. 457–462.

Matthews, J. E., 1970, Geology of the northeastern Caribbean: Salt Lake City, University of Utah, M.S. thesis, unpublished.

——— and T. L. Holcombe, 1974, Possible Caribbean underthrusting of the Greater Antilles along the Muertos Trough: Trans. Seventh Caribbean Geol. Conf., p. 235–242.

Menard, H. W., Jr., 1955, Deep-sea channels, topography, and sedimentation: AAPG Bull., v. 39, p. 236–255.

——— S. M. Smith, and R. M. Pratt, 1965, The Rhone deep sea fan, *in* W. F. Hittard and R. Bradshaw, eds., Submarine geology and geophysics: London, Butterworth, p. 271–284.

Murphy, A. J., L. R. Sykes, and T. W. Donnelly, 1970, Preliminary survey of the microseismicity of the northeastern Caribbean: Geol. Soc. America Bull., v. 81, p. 2459–2464.

Shepard, F. P., 1963, Submarine geology: New York, Harper and Row, 2nd edition, 557 p.

Sykes, L. R. and M. Ewing, 1965, The seismicity of the Caribbean region: Jour. Geophys. Research, v. 70, no. 20, p. 5065–5074.

Whetten, J. T. , 1966, Geology of St. Croix, United States Virgin Islands: Geol. Soc. Amer. Mem. 98, p. 177–239.

Tectonic Development of Trench-Arc Complexes on the Northern and Southern Margins of the Venezuela Basin[1]

JOHN W. LADD[2] AND JOEL S. WATKINS[2,3]

Abstract Multichannel seismic reflection records from the northern and southern margins of the Venezuela Basin reveal compressional structures analogous to structures in active Pacific trench-arc systems and indicate that these Caribbean margins have probably evolved during a Tertiary period of north–south convergence and consequent compression across the Venezuela Basin. Seismic records show evidence for underthrusting beneath and sediment accretion against the margins of the basin. A trench-slope break anticline has formed on the landward slope of both the Muertos Trench and the Venezuela Trench. This anticline has ponded thick sediment accumulations behind it in sediment basins homologous to forearc basins in other active trench-arc complexes. Progressively steeper landward dips deeper in the section on midslope terraces and in the forearc basin of the Muertos Trench suggest syndepositional tectonic rotations of the landward slopes.

Caribbean convergence rates calculated from published plate tectonic models are consistent with convergence rates calculated from considerations of sediment volume within the landward slopes of Venezuela Basin trenches. These sediment calculations suggest that over half of the accreted sediment beneath the landward slopes may be terrigenous material slumped down the landward slope or trapped as turbidite fill in the trench axis and later incorporated into the inner slope.

INTRODUCTION

During most of the Tertiary the northern and southern margins of the Venezuela Basin have been characterized by oblique slip motion. Bucher (1947) suggested that the symmetry of folds along the southern and northern margins of the Caribbean is the same symmetry of structures that would develop along the sides of a valley glacier moving from west to east. He concluded that "the great plate of the Caribbean Sea basin under compression from all sides, yielded throughout its length, elongating in the east-west direction while shortening at right angles, i.e., from north to south." More recently marine magnetic studies in the North and South Atlantic have emphasized Tertiary convergence between North and South America in the region of the Caribbean (Ladd, 1976), while studies of the Tertiary development of the Lesser Antilles (Nagle, 1971), interpreted as a subduction zone (Malfait and Dinkleman, 1972), have emphasized the eastward motion of the Caribbean basins with respect to the Americas during the Tertiary. The instantaneous poles of rotation determined by Jordan (1975) primarily from the fault plane solutions of Molnar and Sykes (1969) and the bathymetric studies of Holcombe et al (1973) and Uchupi (1973) indicate that the Tertiary pattern of plate motion is continuing.

Though the strike-slip component of motion along the northern and southern margins of the Caribbean has probably been much larger than the convergent component of motion throughout the later Tertiary, local mapping onshore and offshore has been more successful at documenting compressional features related to convergence than at finding evidence for extensive strike-slip faulting. The Caribbean mountain system of Venezuela was gently folded by north-south compressive forces from early Eocene

[1]Manuscript received, March 25, 1977; accepted, February 21, 1978.

[2]Geophysics Laboratory, University of Texas Marine Science Institute, Galveston, TX 77550.

[3]Present address: Gulf Research and Development Corp., Houston, Texas 77036.

The writers acknowledge the help given them by the crew of the R/V *Ida Green*, Otis Murray commanding, and the scientific staff. We are grateful to Cecil and Ida Green who made the ship and the basic equipment available for our work. Much geophysical equipment and valuable advice were provided by Chevron Oil Company, Continental Oil Company, Exxon Production Research Company, Mobil Oil Corporation, Shell Oil Company, and Texaco Inc.

Thomas H. Shipley and Mark H. Houston provided internal review of this manuscript.

This work was supported principally by NSF Grant DES 75-06249.

University of Texas Marine Science Institute Contribution No. 147, Galveston Geophysics Laboratory.

Article Identification Number:
0065–731X/78/MO29–0024/$03.00/0.
(see copyright notice, front of book)

to Miocene time (Menendez, 1967) as was the Paria Peninsula farther east.

Similar regional north-south compression led to gentle folding in Puerto Rico (Berryhill, 1965) and Hispaniola (Bowin, 1966). The marine seismic reflection work of Edgar et al (1971), Garrison (1972), and Silver (1972) indicates that more intense deformation has been occurring offshore where large sediment wedges have been accumulating beneath the north slope of the Muertos Trench and beneath the South Caribbean Basin offshore Venezuela (Case, 1974). Edgar et al (1971) indicate that the central Venezuela Basin is only mildly deformed with a gentle upbowing of the central basin from north to south.

Because of the deformation of Venezuela Basin margin sediment wedges, earlier single-channel seismic work did not clearly resolve some of the structure of these areas. With the intent of more clearly defining the internal structure and evolutionary history of the margins of the Venezuela Basin, the writers ran several multichannel seismic reflection lines in the region. These data, some of which were reviewed in an earlier paper (Ladd et al, 1977), suggest an underthrusting mode of deformation beneath the Curaçao Ridge and the southern margin of Hispaniola and Puerto Rico and accretion of sediments onto these basin margins. These two zones may have absorbed much of the continuous Tertiary north-south compression across the Caribbean.

DATA COLLECTION AND PROCESSING

University of Texas multichannel track in the Venezuela Basin is shown in Figure 1. Our sound source consisted of three 1,500 cu. in. Bolt air guns fired at 500 psi. Data were recorded digitally, using a Texas Instruments DFS 10,000. Demultiplexing and final data display were done commercially; the rest of the data processing, including editing, sort, velocity analysis, filter, and stack, were done in-house, using an IBM 370/155. After 24-fold stacking, data were filtered with a pass band of 15 to 50 Hz. During display a time-varying AGC was applied with a short window length of a few tens of milliseconds at the water bottom and with window length increasing with time in the section. This mode of AGC reduced dynamic range of the data without losing all information on relative amplitudes and without losing weak reflectors just below water bottom.

INTERPRETATION

Curaçao Ridge and Venezuela Trench

Our ship track crossed the northern margin of the deformed sediments of the South Caribbean Basin near 69°W long. where the main ridges of the basin are the Curaçao Ridge and the Dutch Leeward Antilles separated by the sediment-filled Los Roques Trench (Figs. 1, 2, and 3). At the foot of the north flank of the Curaçao Ridge which joins the southward-sloping Venezuela Basin floor, a linear topographic low of ponded sediments, presumably turbidites, is the westward extension of the Venezuela Basin abyssal plain. This linear trough, which runs from Aruba Gap to the east end of the Curaçao

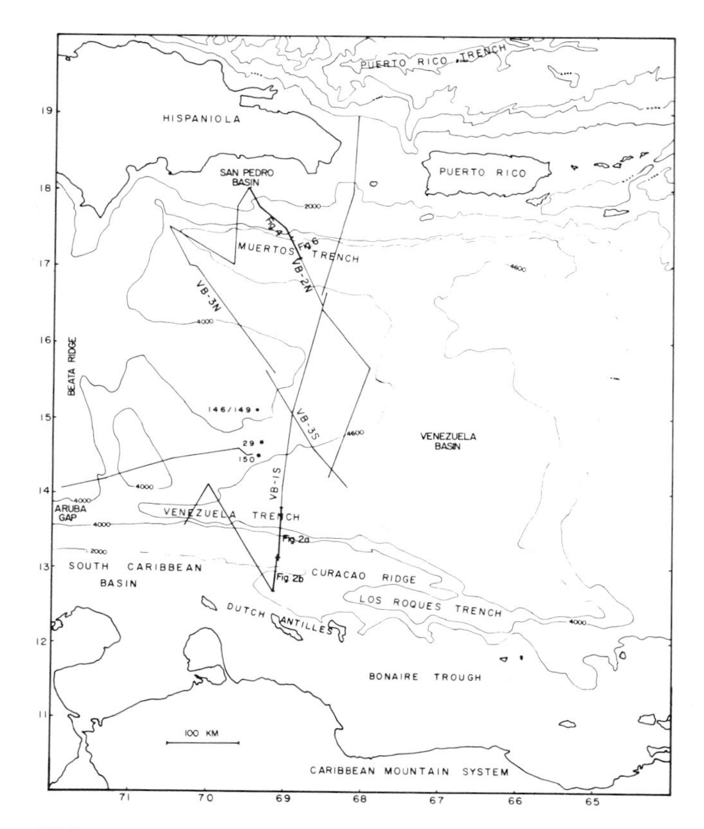

FIG. 1 Track chart showing UTMSI multichannel seismic lines in the Venezuela Basin. DSDP sites 146/149, 29, and 150 are indicated. Bathymetry is from Matthews and Holcombe (1976).

Ridge at about 67°W, we have called the Venezuela Trench.

As with the Muertos Trench, the Venezuela Trench marks the boundary between mildly deformed sediments of the Venezuela Basin floor and strongly deformed sediments of the Venezuela Basin margin. The Curaçao Ridge is an anticlinorium of intensely deformed sediments that is homologous to the trench-slope break of active Pacific trench-arc systems (Karig and Sharman, 1975). The mildly deformed sediments that fill the Los Roques Trench south of the Curaçao Ridge are homologous to the forearc basin sediments found between island arcs or continental landmasses and the trench-slope break of most active Pacific trench systems. The Dutch Leeward Antilles are the expression of a once volcanically active island arc.

These regional structural similarities between the South Caribbean Basin and active trench-arc complexes have suggested to many that the South Caribbean Basin is a zone of compression and underthrusting of Venezuela Basin and Colombia Basin crust (Krause, 1971; Case, 1974; Jordan, 1975; Silver et al, 1975); however, as with the Muertos Trench to the north, there is a paucity of late Tertiary volcanism, and there is no documented landward-dipping seismic zone along the southern margin of the Venezuela Basin.

The Curaçao Ridge is a locus of intense deforma-

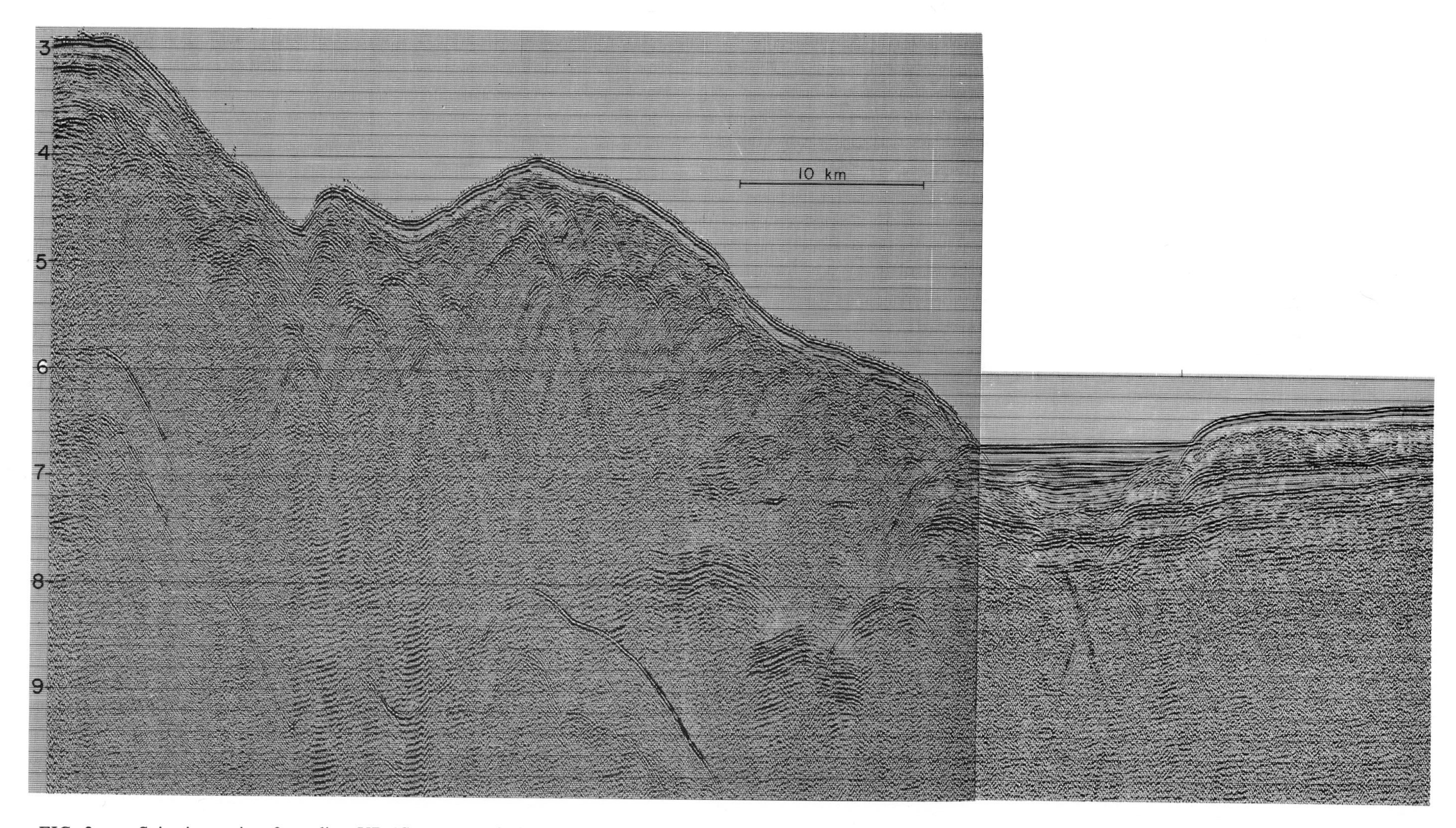

FIG. 2a Seismic section from line VB-1S over north flank of the Curaçao Ridge and the southern margin of the Venezuela Basin. Figure location is indicated in Figure 1. Vertical scale is seconds of two-way reflection time.

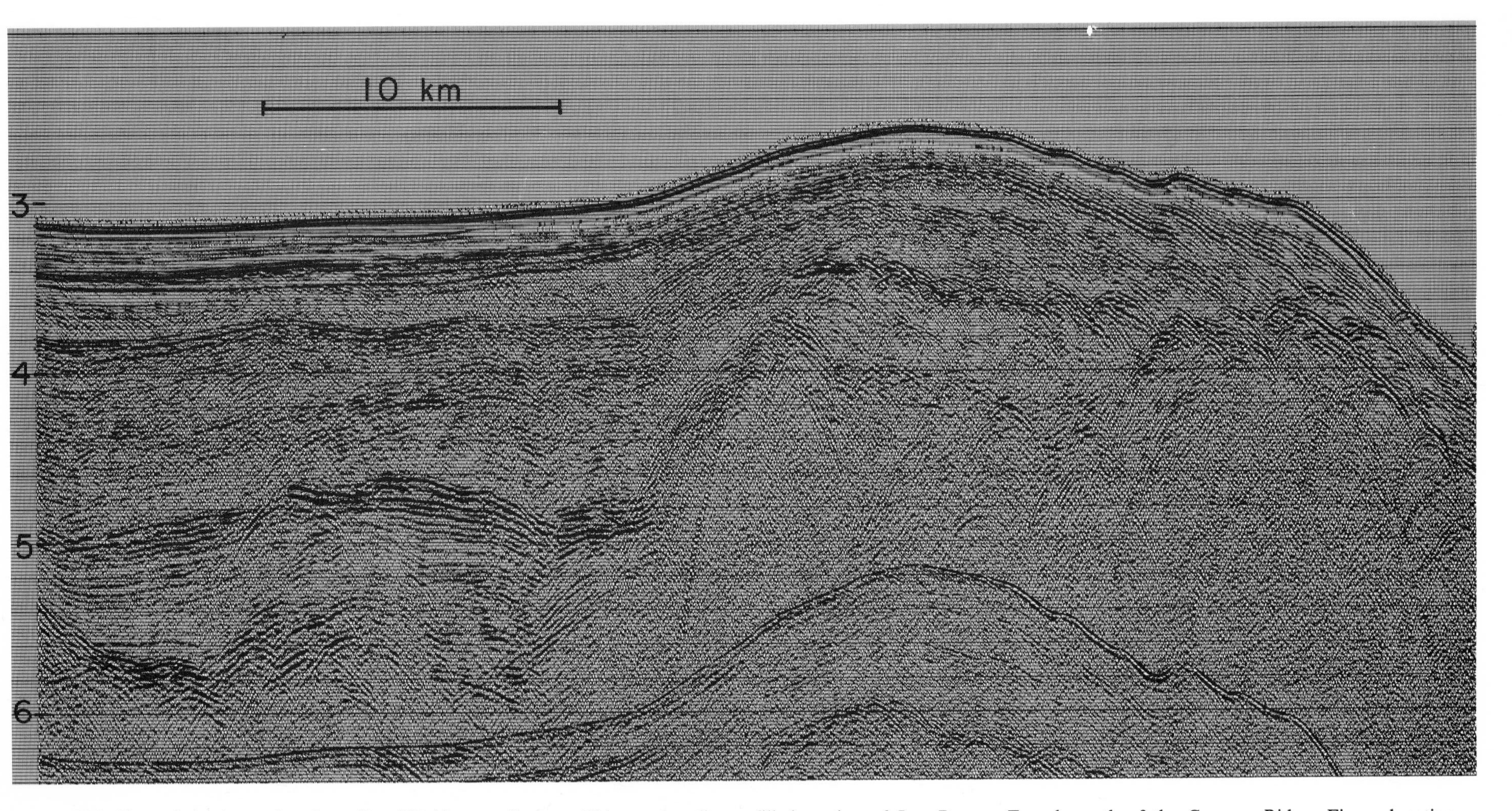

FIG. 2b Seismic section from line VB-1S over Curaçao Ridge and sediment-filled portion of Los Roques Trench south of the Curaçao Ridge. Figure location is indicated in Figure 1. Vertical scale is seconds of two-way reflection time.

tion between the less-deformed Los Roques Trench and the Venezuela Trench. The Curaçao Ridge, for the most part, lacks coherent reflections. It is characterized by many diffraction hyperbolas which in some cases coalesce, suggesting a deformed sedimentary section folded parallel to the topography. Farther south in the Los Roques Trench the thick (3 to 4 km) sediment section seen in Figures 2 and 3 has a strongly deformed basal reflector at about 5.6 sec, but shallower reflectors are only mildly folded and faulted. Midslope basins on the north flank of the Curaçao Ridge show no ponded sediment, possibly indicating recent tectonic development of these basins or nearly perfect isolation from detrital sediment supply. The contrast between mildly-deformed trench fill and the adjacent strongly-deformed Curaçao Ridge is striking, especially since the turbidite fill has probably been in existence during the time of deformation of some of the Curaçao Ridge. A similar sharp horizontal gradient of deformation occurs from the Curaçao Ridge to the Los Roques Trench. However, as in the Venezuela Trench, older units in the Los Roques Trench are more strongly deformed than younger units, probably the result of mild, continual deformation in the marginal trenches.

The Venezuela Basin reflector sequence north of the Venezuela Trench includes the sedimentary layers named the Carib Beds by J. Ewing and others (1968; see also J. Ewing et al, 1967) and which were drilled during DSDP legs 4 and 15 (Bader et al, 1970; Edgar, Saunders et al, 1973). The strong reflector at the base of the Carib Beds has been called B″ and corresponds to the top of the Coniacian basalt drilled by DSDP (unit 7 at DSDP site 146/149). A″ is a strong reflector about midway down in the Carib Beds and corresponds to an Eocene disconformity between younger oozes above and older, more lithified limestones and chalks below. Our seismic data define another reflector which is continuous across the Venezuela Basin between the water bottom and A″. For purposes of this discussion, we call this reflector a″ and suggest that it may correspond to an interface between units 1 and 2 drilled at DSDP site 146/149. Unit 1 is a chalk and marl ooze of early Miocene age and younger. Unit 2 is radiolarian chalk and indurated radiolarian ooze of middle Eocene to early Miocene age. Figures 2 and 3 also show that there is approximately 0.8 sec of section beneath B″. This high-velocity section, which has not been drilled, presumably is a series of basalt sills and flows similar to B″.

The seismic line across the Venezuela Trench and Curaçao Ridge shows part of the Venezuela Basin reflector sequence continuing landward beneath the trench fill and possible 20 km landward beneath the Curaçao Ridge (Figs. 2 and 3). Starting at the seaward edge of the trench fill, the reflectors that

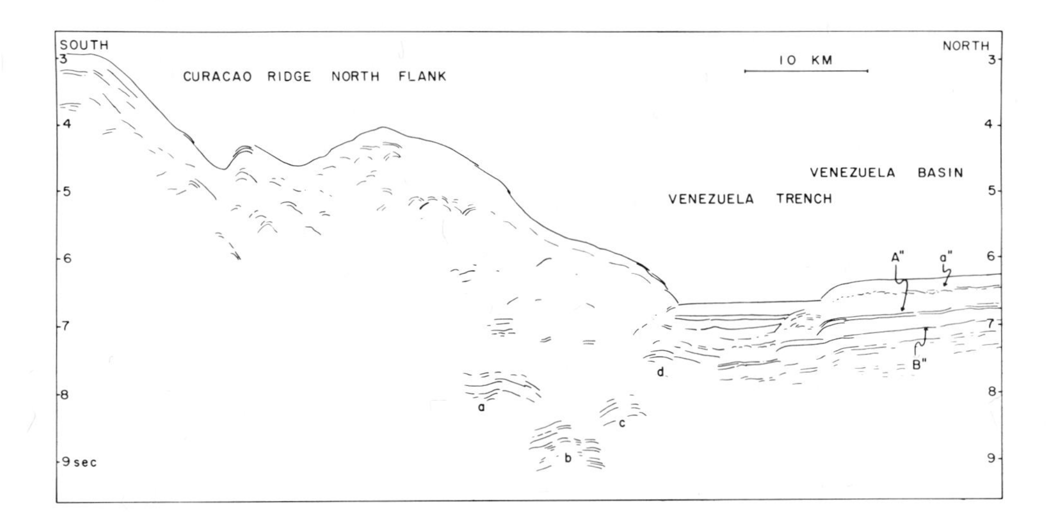

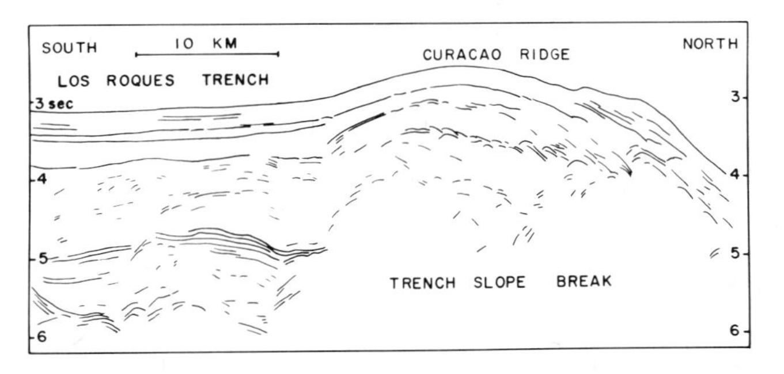

FIG. 3 Tracing of Figures 2a and 2b showing reflecting horizons within the Curaçao Ridge–Venezuela Trench complex.

can be traced southward are broken and offset by a series of faults which downdrop the reflector sequence by almost 2 sec beneath the north flank of the Curaçao Ridge.

The section within the Venezuela Trench appears to contain the entire section between B″ and a″ overlain by about 0.6 sec of turbidite fill that becomes more deformed with depth, indicating syntectonic sedimentation. The dip of most major fault planes offsetting the Carib Beds beneath the trench fill is indeterminate from these data, but gentle folds within the turbidite fill suggest compression. Thrust faulting may be the primary fault mode. A compressional regime is also indicated by the coalesced hyperbolas in the upper second of sediments beneath the Curaçao Ridge that suggest bedding folded parallel to the surface topography of the ridge. If a″ is indeed present beneath the trench fill and if a″ is the boundary between the two ooze units found above A″ at DSDP site 146/149, then the oldest turbidite in the Venezuela Trench is no older than early Miocene.

The style of deformation of the older Carib Beds beneath the toe of the Curaçao Ridge (Figures 2 and 3) may indicate one possible mode of underthrusting of oceanic crust beneath a margin. It is possible that surficial layers of crust are being stacked beneath the Curaçao Ridge in multiple slices like cards in a deck in much the same way that Hsu and Ryan (1973) envisioned underthrusting in the eastern Mediterranean or as Seely et al (1974) postulated for the Middle America Trench off Guatemala. Beneath the Curaçao Ridge the most landward piece of probable Venezuela Basin crust (designated "a" in Fig. 3) is at about 7.8 sec. Seaward from this is a piece at 8.4 sec ("b" in Fig. 3) and then other shallower pieces ("c" and "d"). It appears that the slice at 8.4 sec (b) is being thrust under the piece at 7.8 sec (a). There are also some pieces of reflector at 7.2 sec on the section above the faulted reflectors discussed above. These shallower reflectors may also be pieces of Venezuela Basin crust that have been isolated higher in the Curaçao Ridge. The Curaçao Lava Formation on the island of Curaçao includes a kilometer-thick section of Cretaceous tholeiitic pillow basalts (Beets, 1977) which may also be Venezuela Basin crust that was uplifted during underthrusting in Late Cretaceous time.

Muertos Trench

The same major structural elements found in the Curaçao Ridge complex are found in the Muertos Trench and its associated north slope (Figs. 4, 5, 6): the turbidite wedge in the trench floor, the contorted sediments of the inner slope and trench-slope break lying above an apparently underthrust Venezuela Basin sequence, and the sediment pond or forearc basin landward of the trench-slope break. The Greater Antilles island arc, which for the most part has been volcanically inactive since Eocene time, is homologous to the Dutch Leeward Antilles.

Muertos Trench sections which have been discussed more fully elsewhere (Ladd et al, 1977) show several similarities in detail as well as gross structure to the Curaçao Ridge sections. As in the Venezuela Trench, most of the Carib Beds section can be followed beneath the trench fill in the Muertos Trench, indicating a probable late Tertiary age for the base of the turbidite fill. The work of Edgar et al (1971) and Saunders et al (1973) demonstrated the paucity of turbidites in the central Venezuela Basin since Late Cretaceous time, implying the existence of sediment traps including the Muertos Trench around the margins since that time. The seismic data of Figure 6 suggest that the trench fill is of late Tertiary age, indicating that some tectonic process must have deformed or otherwise altered older trench fill. The warping of all but the uppermost reflections within the Muertos Trench fill, as well as the lack of continuous reflectors within the north slope of the trench above 8 sec, suggest that a compressional process that incorporates the trench fill and underlying Carib Beds into the north slope has been active quite recently. As in the Curaçao Ridge complex there is an abrupt change from the relatively undeformed trench fill and reflectors beneath the trench fill into the incoherent hyperbolic returns of the trench-inner slope and trench-slope break. As on the southern margin landward of the anticlinal ridge the intense deformation abruptly ends, and a thick, mildly deformed sediment basin occupies the region landward to the island arc. Within the north slope of the Muertos Trench near the trench-slope anticline, diffraction hyperbolas coalesce to define reflecting horizons folded parallel to the topography of the trench anticline. These distorted reflecting horizons, which have their counterpart within the Curaçao Ridge, suggest that the trench-slope break is an anticlinal feature created under compression.

Despite many similarities between the Curaçao Ridge complex and the Muertos Trench complex there are certain differences which may reflect local variations in the style of deformation. Unlike the Venezuela Basin sequence beneath the toe of the Curaçao Ridge, the continuation of Venezuela Basin reflectors beneath the north slope of the Muertos Trench is relatively undeformed. Near the trench axis diffraction hyperbolas attest to some disturbance, but this is minor compared to the large offsets seen within the Curaçao Ridge. Perhaps the intense deformation within the Curaçao Ridge is causally related to the sharp offset in Bowin's (1976) negative free-air anomaly at 69°W long. Both the gravity and the seismic data may be indicating a north-south-trending fault zone connecting the Venezuela Trench and the Los Roques Trench, though the bathymetric contours of the Curaçao Ridge are not noticeably offset. The progressive landward tilting of sediments in the forearc basin north of the Muertos Trench trench-slope break contrasts with the generally horizontal nature of deposits within the sediment-filled portion of the Los Roques Trench south of Curaçao Ridge. On the Muertos north slope there are some shallow

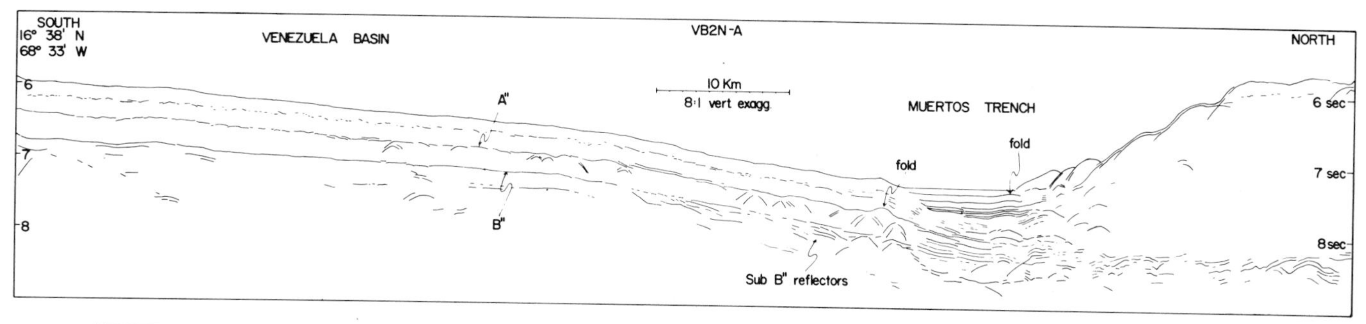

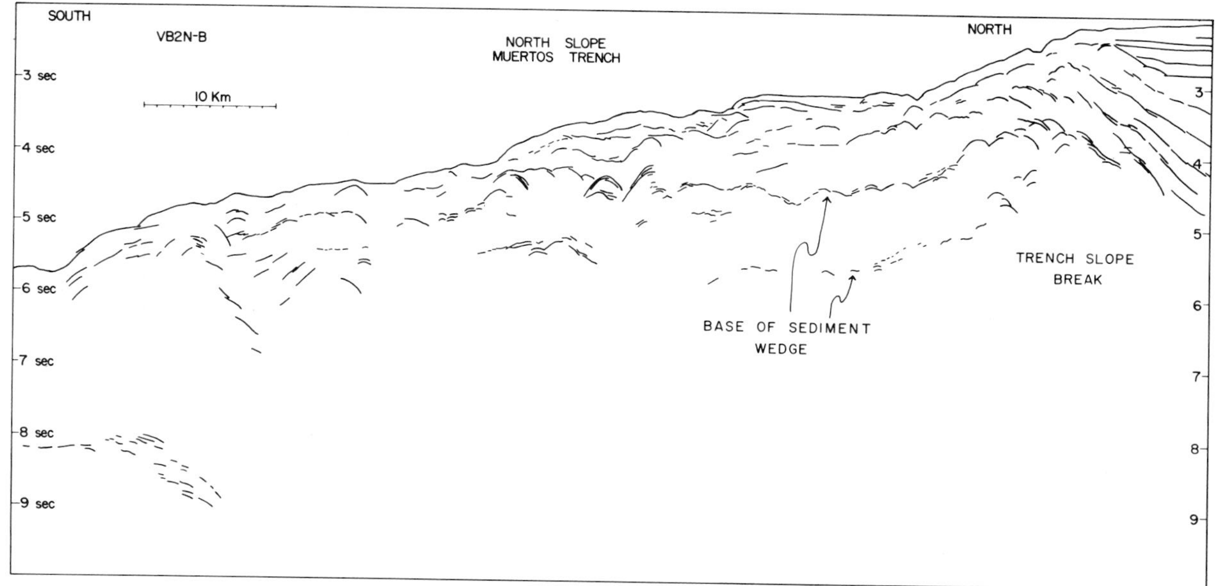

FIG. 5 Tracings of Figure 4. This figure is from Ladd et al, 1977.

sediment ponds with gentle northward dips of the entrained strata. Such terraces are absent from the section across the Curaçao Ridge.

Convergence Rates

Given the thickness of the Carib Beds in the Venezuela Basin as well as the approximate cross-sectional area of the deformed sediment wedges of the Curaçao Ridge and the north slope of the Muertos Trench, one should be able to calculate convergence rates across these trench systems, assuming that the deformed sediment wedges are composed entirely of Carib Bed sediments. It is of interest to compare convergence rates calculated in this manner with the rates determined by Ladd (1976) and Minster et al (1974) for convergence of North and South America across the Caribbean.

Minster et al gave an instantaneous rotation of South America with respect to North America of 0.18°/m.y. about a pole at 3°S lat., 53°W long. Ladd gave a finite difference rotation for the period of 38 mybp to 9 mybp of 0.21°/m.y. about a pole at 17°N lat., 53°W long. Using an equation by Morgan (1968) and considering a point near the Venezuela Trench, both poles give a convergence rate of 0.7 cm/yr of North America with respect to South America.

If the Curaçao Ridge with an approximate cross-sectional area of 200 sq km has been built entirely of Carib Bed sediments scraped off an underthrust slab over the last 84 m.y. since the Coniacian Venezuela Basin sills were formed, a convergence rate of 1 cm/yr is obtained. Similarly, if the Muertos Trench inner slope with cross-sectional area of approximately 500 sq km was formed entirely of Carib Bed sediments since 84 m.y. BP, then this requires a convergence rate in that zone of 2 cm/yr. This would indicate a convergence rate of North America with respect to South America of 3 cm/yr, not counting any convergence across the Puerto Rico Trench. Allowance for compaction would require a higher convergence rate.

Perhaps we should be satisfied with an agreement within an order of magnitude between convergence rates based upon crude sediment volume and calculations and convergence rates based upon global plate models. However, it is encouraging to find that the sediment calculations give rates higher than the plate rotation calculations, because the sediment calculations assume no source of sediments other than the pelagics of the Carib Beds. If one assumes that the deformed sediment wedges are one-third pelagic and two-thirds terrigenous debris from nearby island

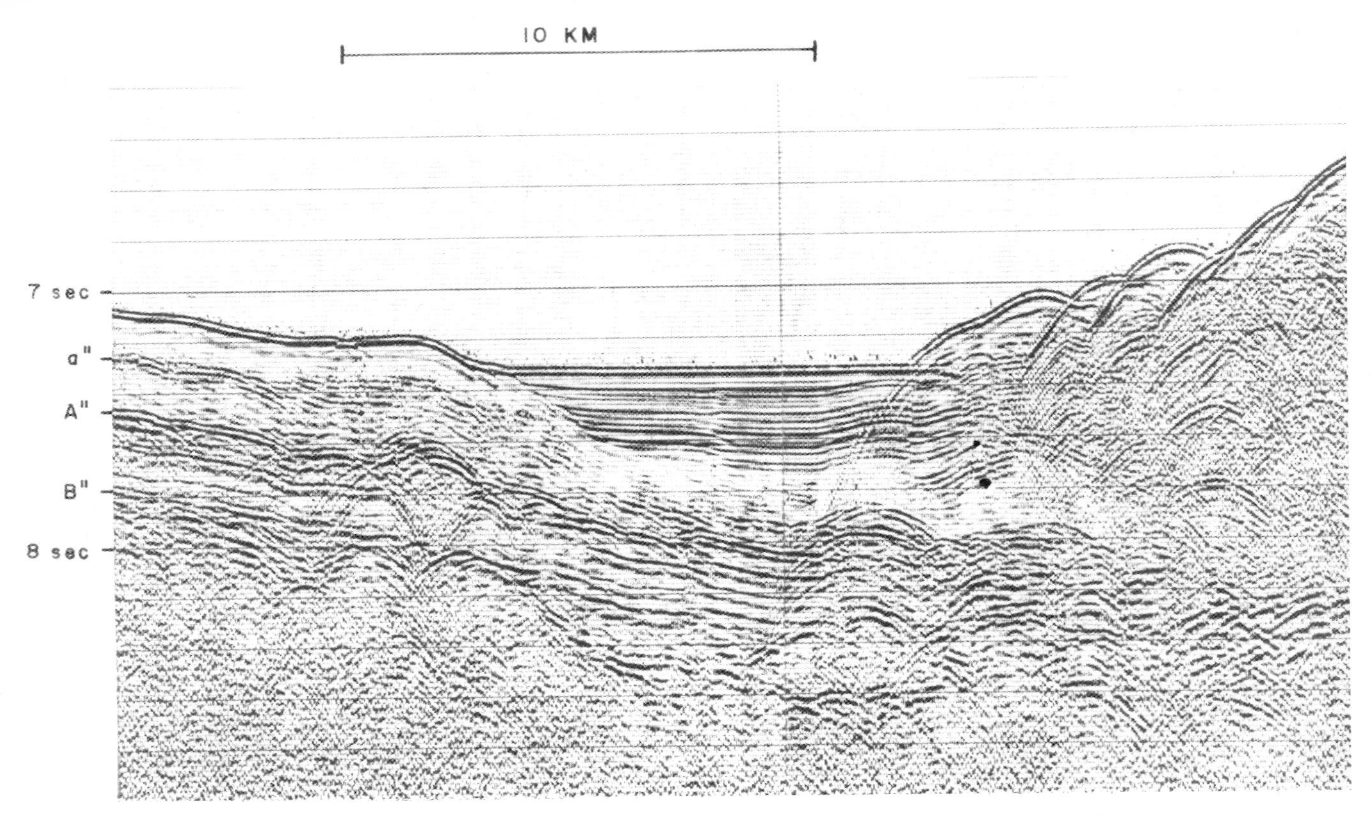

FIG. 6 Enlarged portion of Figure 4 showing details of the Muertos Trench.

chains, then the different methods of calculating convergence rates give much closer agreement. Certainly it is reasonable to assume that various forms of debris flow have added sediment from the islands directly to the landward slopes and indirectly via the turbidite traps of the trench floors.

CONCLUSIONS

The multichannel seismic technique has allowed us to improve the resolution of structures within the deformed margins of the Venezuela Basin and to show that these structures are probably due to convergence of lithospheric plates across these margins. The structural elements of the deformed margins have their counterparts in active Pacific trench-arc systems, although the Venezuela Basin trenches lack the associated Benioff zone of earthquakes and active volcanism that is common to most Pacific systems. A southward-dipping Benioff zone may be associated with the Puerto Rico Trench (Carver and Tarr, 1976), however.

The free-air gravity map of Bowin (1976) also suggests similarities between the Venezuela Trench, the Muertos Trench, and other deep-sea trenches where underthrusting is occurring. Negative anomalies are common over active trench systems (Watts and Talwani, 1974), and Bowin's map shows a negative anomaly over the Muertos Trench and the Venezuela Trench. Near 69°W the negative anomaly of the Venezuela Trench is offset to the south to the Los Roques Trench. East of 69°W the negative follows the Los Roques Trench.

The seismic sections across the Venezuela Basin margin show in detail some of the local zones of deformation that are judged to have existed throughout the Late Cretaceous and Tertiary from plate tectonic models based upon relative rotations of South America with respect to North America, as well as upon known geology of the circum-Caribbean region. The seismic sections demonstrate that much of the convergence has probably been accommodated by thrusting along the margins of the Venezuela Basin with concomitant off-scraping of surficial sediments.

During the middle and late Tertiary the convergence across the Venezuela Basin was probably at rates of an order of magnitude less than rates of strike-slip motion along these same borders. This strike-slip motion, which has been a consequence of eastward motion of the Caribbean with respect to the Americas, may have been accommodated by faulting landward of the seismic sections under consideration or possibly by shear across a zone which included the margin sediment wedges. Perhaps the distortion of the underthrust section beneath the Curaçao Ridge is an expression of this shear.

Alternatively the distortion of the underthrust slab seen on our line across the Curaçao Ridge may be a local phenomenon associated with a possible offset in a zone of crustal convergence from the Venezuela Trench to the Los Roques Trench. Judging from the offset of the gravity anomaly on Bowin's map, our line crosses the Curaçao Ridge near the location of a tectonic offset. This offset, which may be viewed as a tectonic segment boundary (Carr et al, 1974), may actually be a small transform fault zone with a north-south strike-slip component of motion.

Caribbean convergence rates calculated from studies of fracture zone geometry, marine magnetics, and earthquake focal mechanisms related to a worldwide plate pattern agree with convergence rates calculated within the Caribbean from sediment volume considerations. Comparison of these two methods of calculating convergence rates allows some inference about relative amounts of pelagic sediment and terrigenous sediment within the accreting landward slopes of the Muertos Trench and the Venezuela Trench. However, the large sources of error in these calculations prevent more than the general statement that about one-third of the accretionary wedge is of pelagic origin with the balance being derived from terrigenous sources, possibly through incorporation of turbidite fill accumulating in the trench axis.

REFERENCES CITED

Bader, R. G., et al, eds., 1970, Initial report of the Deep-Sea Drilling Project: Washington, D.C., U.S. Govt., v. 4, 753 p.

Beets, D. J., 1977, Cretaceous and Early Tertiary in Curaçao, *in* Guide to the field excursions on Curaçao, Bonaire, and Aruba, Netherlands Antilles: 8th Caribbean Geol. Conf. (Curaçao) p. 7–17.

Berryhill, H. L., 1965, Geology of the Ciales quadrangle Puerto Rico: U.S. Geol. Survey Bull., v. 1184, p. 116.

Bowin, C. O., 1966, Geology of the central Dominican Republic, *in* H. H. Hess, ed., Caribbean geological investigations: Geol. Soc. America Mem. 98, p. 11–84.

——— 1976, The Caribbean gravity field and plate tectonics: Geol. Soc. America Special Paper 169, 79 p.

Bucher, W. H., 1947, Problems of earth deformation illustrated by the Caribbean Sea basin: Trans. N.Y. Acad. Sci. Ser. 2, v. 9, p. 98–116.

Carr, M. J., R. E. Stoiber, and C. L. Drake, 1974, The segmented nature of some continental margins, *in* C. A. Burk and C. L. Drake, eds., The geology of continental margins: New York, Springer-Verlag, p. 105–114.

Carver, D., and A. C. Tarr, 1976, Recent seismicity of the Puerto Rico region: Abstracts with Programs, 1976 Annual Meeting of the Geol. Soc. America, v. 8, p. 805.

Case, J. E., 1974, Major basins along the continental margin of northern South America, *in* C. A. Burk and C. L. Drake, eds., The geology of continental margins: New York, Springer-Verlag, p. 733–742.

Edgar, N. T., Saunders, J. B., et al, 1973, Initial reports of the Deep Sea Drilling Project: Washington, D.C., U.S. Govt., v. 15, 1137 p.

——— J. I. Ewing, and J. Hennion, 1971, Seismic refraction and reflection in Caribbean Sea: AAPG Bull., v. 55, p. 833–870.

Ewing, J., M. Talwani, and M. Ewing, 1968, Sediment distribution in the Caribbean Sea: 4th Caribbean Geol. Conf., Proc. (Trinidad and Tobago, 1965), p. 317–323.

——— et al, 1967, Sediments of the Caribbean, *in* Studies in tropical oceanography: Miami, Univ. of Miami, v. 5, p. 88–102.

Garrison, L. E., 1972, Acoustic-reflection profiles, eastern Greater Antilles: U.S. Geological Survey Pub. No. USGS-GD-72-004.

Holcombe, T. L., et al, 1973, Evidence for seafloor spreading in the Cayman Trough: Earth and Planetary Sci. Letters, v. 20, p. 357–371.

Hsu, K. J., and W. B. F. Ryan, 1973, Summary of the evidence for extensional and compressional tectonics in the Mediterranean *in* W. B. F. Ryan, K. J. Hsu, et al, eds., Initial reports of the Deep Sea Drilling Project: Washington, D.C., U.S. Govt., v. 13, p. 1011–1020.

Jordan, T. H., 1975, The present-day motions of the Caribbean plate: Jour. Geophys. Research, v. 80, p. 4433–4439.

Karig, D. E., and G. F. Sharman III, 1975, Subduction and accretion in trenches: Geol. Soc. America Bull., v. 86, p. 377–389.

Krause, D. C., 1971, Bathymetry, geomagnetism, and tectonics of the Caribbean Sea north of Colombia, *in* T. W. Donelly, ed., Caribbean geophysical, tectonic, and petrologic studies: Geol. Soc. America Mem. 130, p. 35–54.

Ladd, J. W., 1976, Relative motion of South America with respect to North America and Caribbean tectonics: Geol. Soc. America Bull., v. 87, p. 969–976.

——— J. L. Worzel, and J. S. Watkins, 1977, Multifold seismic reflection records from the northern Venezuela Basin and the north slope of the Muertos Trench, *in* M. Talwani and W. C. Pitman III, eds., Island arcs, deep sea trenches, and back arc basins: Washington, D.C., Am. Geophys. Union, p. 41–56.

Malfait, B. T., and M. G. Dinkleman, 1972, Circum-Caribbean tectonic and igneous activity and the evolution of the Caribbean plate: Geol. Soc. America Bull.,v. 83, p. 251–272.

Matthews, J. E., and T. L. Holcombe, 1976, Regional geological/geophysical study of the Caribbean Sea (Navy Ocean Area NA-9): 1. Geophysical maps of the Eastern Caribbean: Washington, D.C., Naval Oceanogr. Office, 43 p.

Menendez, V. de V., 1967, Tectonics of the central part of the western Caribbean Mountains, Venezuela, *in* Studies in tropical oceanography: Miami, Univ. Miami Press, v. 5, p. 103–130.

Minster, J. G., et al, 1974, Numerical modelling of instantaneous plate tectonics: Royal Astron. Soc. Geophys. Jour., v. 36, p. 541–576.

Molnar, P., and L. R. Sykes, 1969, Tectonics of the Caribbean and middle America regions from focal mechanisms and seismicity: Geol. Soc. America Bull., v. 80, p. 1639–1684.

Morgan, W. J., 1968, Rises, trenches, great faults, and crustal blocks: Jour. Geophys. Research, v. 73, p. 1959–1982.

Nagle, F., 1971, Caribbean geology, 1970: Bull. Marine Sci., v. 21, p. 375–439.

Saunders, J. B., et al, 1973, Cruise synthesis, *in* N. T. Edgar, J. B. Saunders, et al, Initial reports of the Deep Sea Drilling Project: Washington, D.C., U.S. Govt., v. 15, p. 1077–1111.

Seely, D. R., P. R. Vail, and G. G. Walton, 1974, Trench slope model, *in* C. A. Burk and C. L. Drake, eds., The geology of continental margins: New York, Springer-Verlag, p. 249–260.

Silver, E. A., 1972, Acoustic-reflection profiles Venezuela continental borderland: U.S. Geological Survey Pub. No. USGS-GD-72-005.

——— J. E. Case, and H. J. MacGillavry, 1975, Geophysical study of the Venezuelan borderland: Geol. Soc. America Bull., v. 86, p. 213–226.

Uchupi, E., 1973, Eastern Yucatan continental margin and western Caribbean tectonics: AAPG Bull., v. 57, p. 1075–1085.

The Sulu Sea: A Marginal Basin in Southeast Asia[1]

A. MASCLE[2] AND P. A. BISCARRAT[3]

Abstract Southeast Asia is located at the junction of four plates: the Eurasian, Indian Ocean/Australian, Pacific, and Philippine Sea plates. Their interaction during Tertiary times resulted in a complex active margin composed of a mosaic of small geotectonic units such as microcontinental blocks, island arcs, and marginal seas. The Sulu Sea is an example of a marginal sea with two distinct basins of different types: (1) the Outer Sulu Sea basin, formed inside an old island arc, the Palawan arc; (2) the Inner Sulu Sea, a basin with an oceanic crust. It is fringed to the southeast and the east by an active margin, supposedly the remains of a larger active margin which during Tertiary times extended along the western side of the Philippines from Luzon to Negros and perhaps from the Sulu Archipelago to the northeastern part of Sabah.

INTRODUCTION

The Sulu Sea (Fig. 1) is a small marginal sea located between the islands of the Philippine Archipelago (Palawan, Mindoro, Panay, Negros, Mindanao, and the Sulu Archipelago) and Sabah. It covers approximately 260,000 square kilometers and comprises two basins separated by the Cagayan Ridge, marked by the small islands of San Miguel, Cavili, and Cagayan. The two basins, according to Irwing's terminology (1951), are (Fig. 2) the Outer Sulu Sea, where the water depth increases from 1,500 m in the northeast to 2,000 m in the southwest and the Inner Sulu Sea with an abyssal plain at depths ranging from 3,800 m in the northwest to more than 5,000 m in the southeast (Sulu Trough). This abyssal plain is separated from the Cagayan Ridge by a shallower area of more complex topography.

During 1974, the Comité d'Etudes Pétrolières Marines (CEPM), which includes two French oil companies, Compagnie Française des Pétroles (CFP) and Société Nationale Elf-Aquitaine (SNEA-P), and the Institut Français du Pétrole (IFP), conducted a multichannel seismic survey across the Sulu Sea. These lines improve our knowledge of the Sulu Sea itself, and also help to increase our understanding of its position within the tectonic framework of Southeast Asia.

STRUCTURAL FRAMEWORK OF SOUTHEAST ASIA

Southeast Asia is located at the junction of four major plates (Fitch, 1970): the Eurasian, Indian Ocean-Australian, Pacific and Philippine Sea plates (Fig. 3); Figure 1 gives the present arrangement of those plates. The configuration of their margins is irregular and probably unstable, particularly around the Molucca Sea and around the Philippine Archipelago. In this area, for instance, active margins resulting from the convergence of the Eurasian and Philippine Sea plates seem to exist both on the eastern and on the western side of the Archipelago in such a way that the boundary between the plates is not obvious and may involve some intermediate microplates.

The convergence of these plates in Cenozoic times has built a cluster of small geotectonic units resulting from their constant interaction (Murphy, 1975). Among them is the Sulu Sea; we intend to study its general structure and relationship with the surrounding continental microblocks and Tertiary island arcs.

RESULTS OF SEISMIC SURVEYS

A previous geophysical survey by Murauchi et al (1973) has already shown that the two deep basins of the Sulu Sea (Outer and Inner Sulu Sea) correspond to two distinct geotectonic areas (Fig. 4). The Outer

[1]Manuscript received, July 5, 1977; accepted, November 16, 1977.

[2]Institut Français du Pétrole, 1 et 4 avenue de Bois Préau, 92500 RUEILL MALMAISON, France.

[3]Compagnie Française des Pétroles, 39/43 quai André Citroën, 75739 PARIS CEDEX 15, France.

The writers would like to thank B. Biju-Duval and L. Montadert of IFP, and J. A. D. Johnson of CFP for their critical reading of the manuscript. They would also like to thank the Comité d'Etudes Pétrolières Marines and the French petroleum companies, CFP and SNEA-P, for authorizing the publication of seismic lines.

Article Identification Number:
0065–731X/78/MO29-0025/$03.00/0.
(see copyright notice, front of book)

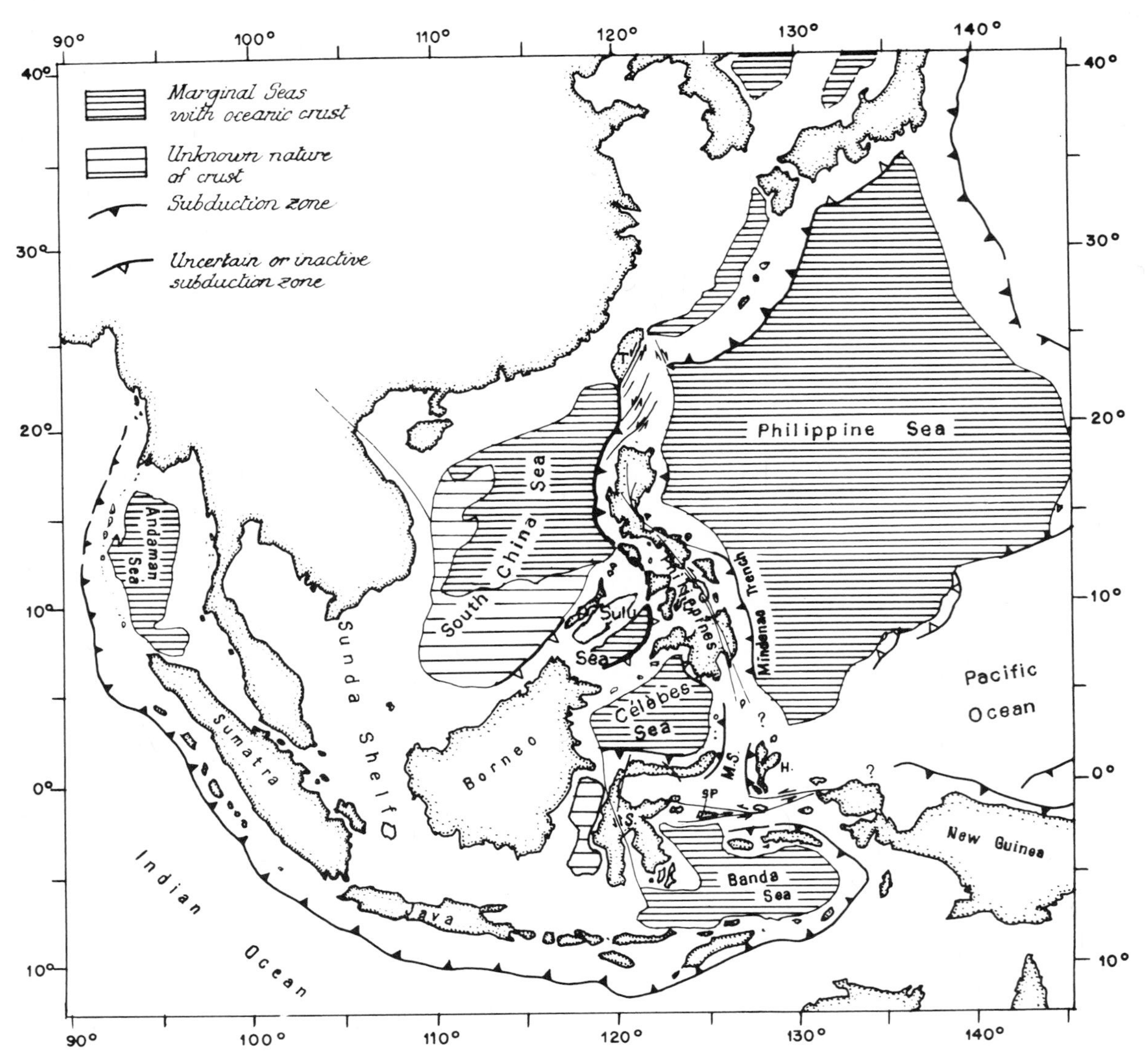

FIG. 1 Structural framework of Southeast Asia: T, Taiwan; MS, Molucca Sea; H, Halmahera; S, Sulawesi (or Celebes); SP, Sulu Spur.

Sulu Sea has a thick crust. The velocities of the two lower layers (5.7 and 6.7 km/s) are close to the velocities encountered below island arcs. The 4.8-km/s layer could represent either the volcanic basement of the Cagayan Ridge or metasediments. It is overlain by a great thickness of sediments, especially near Palawan.

The Inner Sulu Sea has an oceanic crust with a layer 3 a little thinner than expected. The associated heat flow (2.4 HFU as an average) suggests recent origin (Nagasaka et al, 1970).

Outer Sulu Sea

The deeper part of the Outer Sulu Sea (terraces at 1,500 and 2,000 m depth) represents a deep sedimentary basin, orientated roughly northeast–southwest, slightly off the axis of Palawan (Fig. 5). This basin is limited to the northeast by the Calamian-Cuyo Shelf (the submerged part of the Palawan-Busuanga microcontinent) and the Cagayan Ridge and is linked to the southwest to the Balabac subbasin (Bell and Jessop, 1974).

As shown on Figure 4 (modified from Murauchi et al, 1973), this basin has two asymmetrical flanks. The northwestern flank is very steep. It corresponds to the basement of Palawan and the Calamian-Cuyo Shelf. The steepness of this flank is illustrated by the CEPM line (Fig. 6), which shows that the basement plunges 6,000 m over a horizontal distance of 20,000 m. On the bathymetric map (Fig. 2) the abrupt change in slope of the seabed suggests fault systems, some of them extending across Palawan (Ulugan fault zone, for instance; Fig. 5).

The southeastern flank is smoother. The upper part represents a platform with some local relief (volcanoes?). The greatest elevations are located on

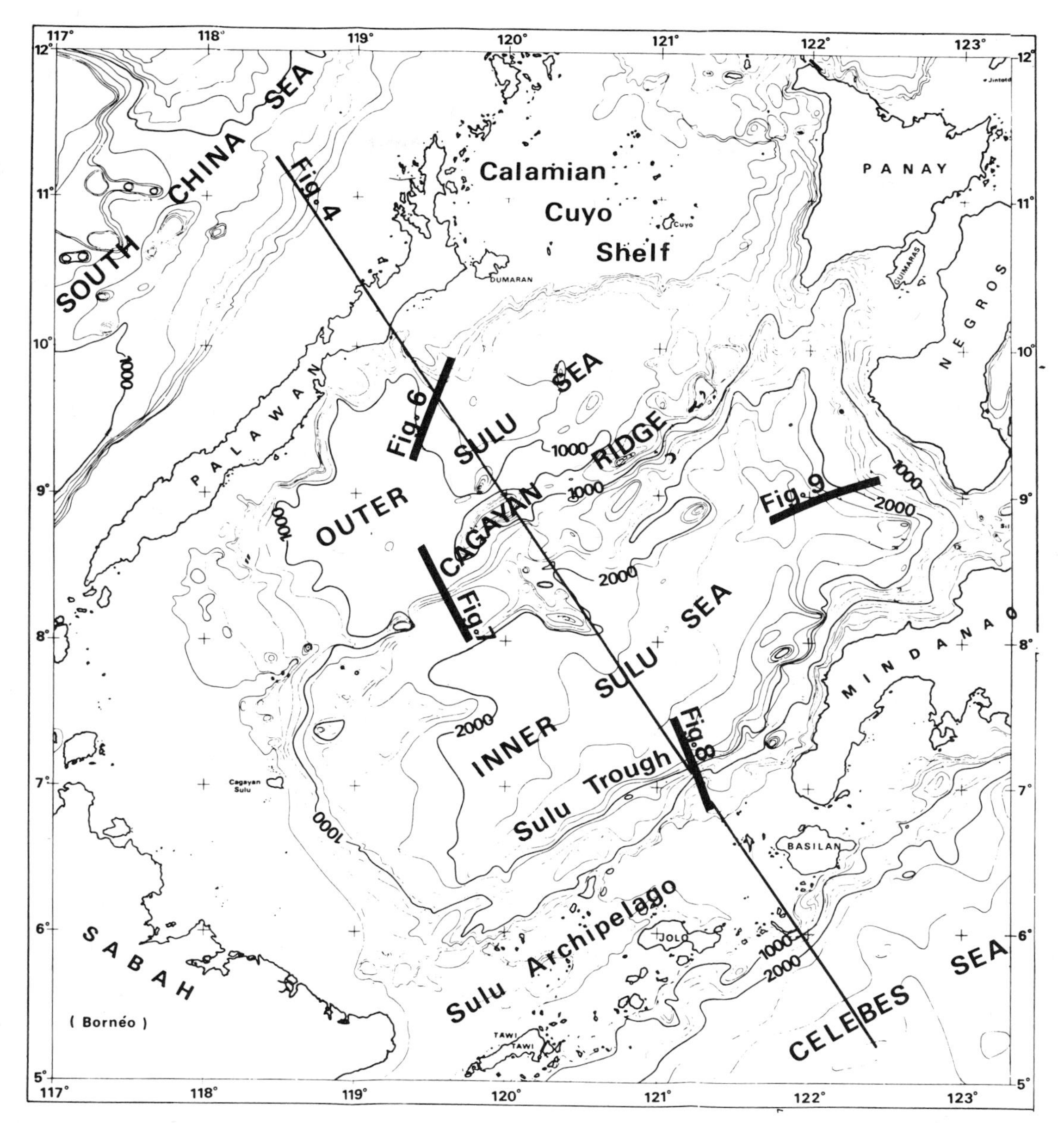

FIG. 2 Sulu Sea. Bathymetric map and location of the sections. Water depth is in fathoms.

the far edge of the basin and form the Cagayan Ridge. This ridge is topographically rough, has associated high-amplitude magnetic anomalies (Murauchi et al, 1973), and is considered to be a volcanic ridge of Tertiary age.

The CEPM survey confirms the division into two sedimentary sequences previously defined by Murauchi et al (1973) and reveals additional characteristics. For example, the lower sequence is restricted to the deeper part of the basin (Unit 2, Fig. 6). The velocities at its top are close to 3,000 m/sec and increase with depth. This sequence seems to be tectonically disturbed but the tectonic style is not clearly apparent on our lines.

The upper sequence (unit 1, Fig. 6) is very well defined throughout the basin. Velocities range from 1,700 m/sec at the top to 3,000 m/sec lower in the unit. The northern part of the CEPM line shows that the recent series has built a prograding shelf in the vicinity of Palawan. This upper unit is probably of Neogene age.

Both units are affected by deformation interpreted as mud diapirism. Similar events have already been described westward in the Bancauan subbasin (Bell and Jessop, 1974). Diapirism seems to be still active, as associated fault systems affect the sea bottom. It leads us to wonder whether the deformation observed within the lower sedimentary sequence (unit 2) is a consequence of this diapirism, or is older and due to compression or shear. Additional deep seismic investigations are needed to resolve this problem.

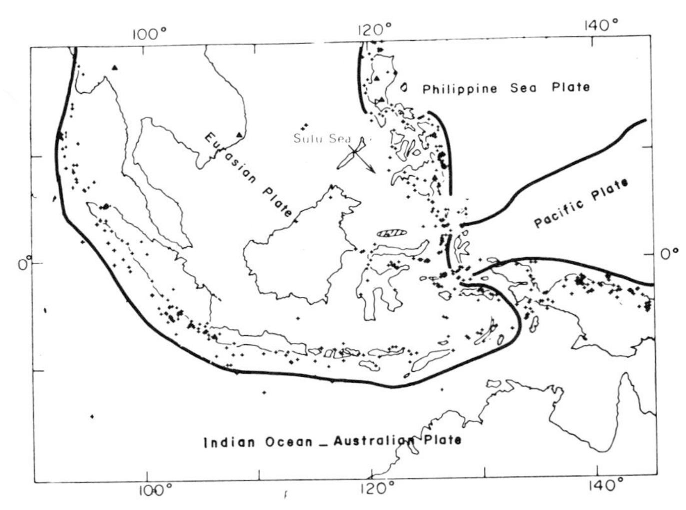

Epicenters of recent (1963_1967) shallow focus earthquakes (depths <70km) and limits of the four major plates.

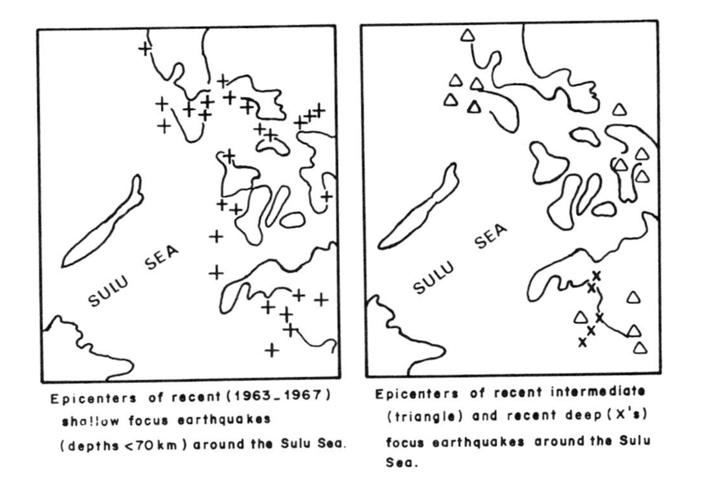

Epicenters of recent (1963_1967) shallow focus earthquakes (depths <70 km) around the Sulu Sea.

Epicenters of recent intermediate (triangle) and recent deep (X's) focus earthquakes around the Sulu Sea.

FIG. 3 Limits of the four major plates and earthquake epicenters distribution (from Fitch, 1970).

Inner Sulu Sea

The Inner Sulu Sea is the oceanic basin. To the northwest, the southeastern flank of the Cagayan Ridge constitutes an escarpment probably generated by a system of normal faults. Within a distanceof 10 km, the oceanic basement of the Inner Sulu Sea appears 3,000 m deeper than the basement of the Outer Sulu Sea and Cagayan Ridge (Fig. 7). The top of this oceanic basement shows rough morphology along an 80-kilometer-wide strip parallel to the Cagayan Ridge (Fig. 5). Its effect on the sea bottom is smoothed by sedimentary fill in the depression. The sediment thickness seldom exceeds 2,000 m (Fig. 7). Farther eastward, the top of the basement is more regular, horizontal at first, then getting deeper toward Negros and the Sulu Archipelago. This deepening is reflected in bathymetry by the Sulu Trough. Another local deepening exists in the southern part of the basin at the foot of the Sandakan subbasin. The thickness of sediments, generally less than 2,000 m in the central part of the basin, in-

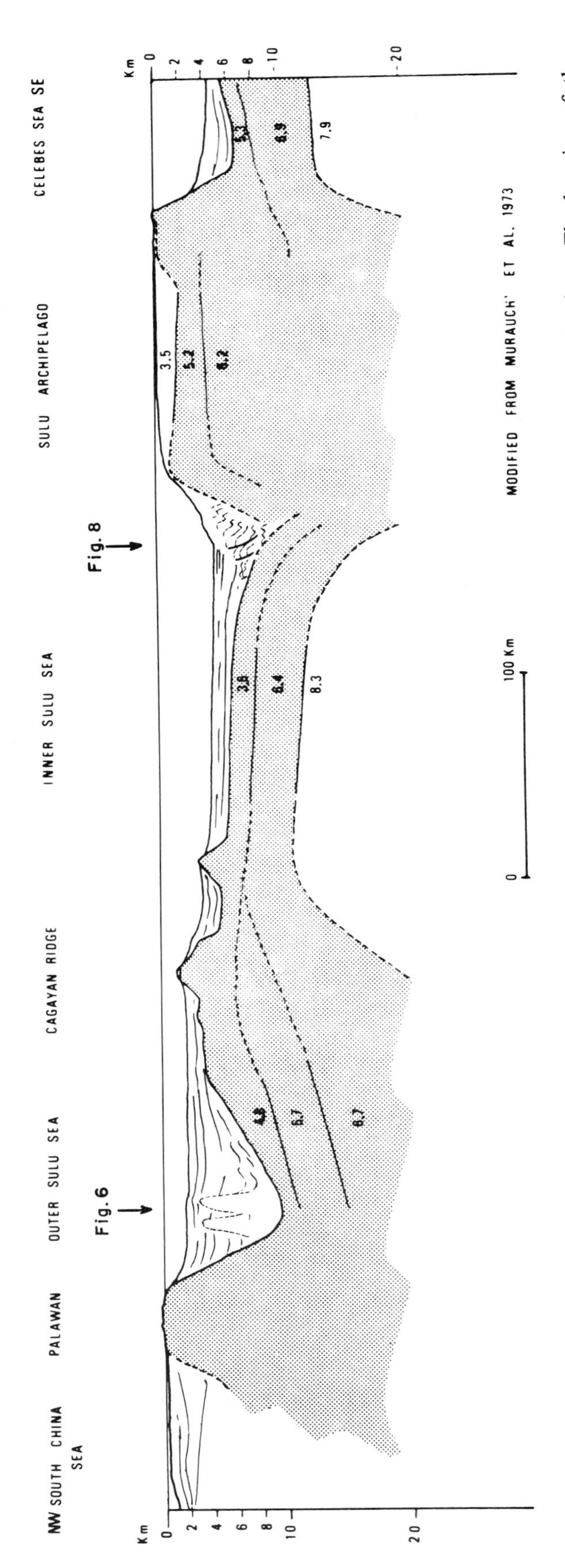

FIG. 4 Schematic section across the Sulu Sea, modified from Murauchi et al, 1973. Numbers are seismic velocities in km/sec. The location of the section is shown on Figure 2.

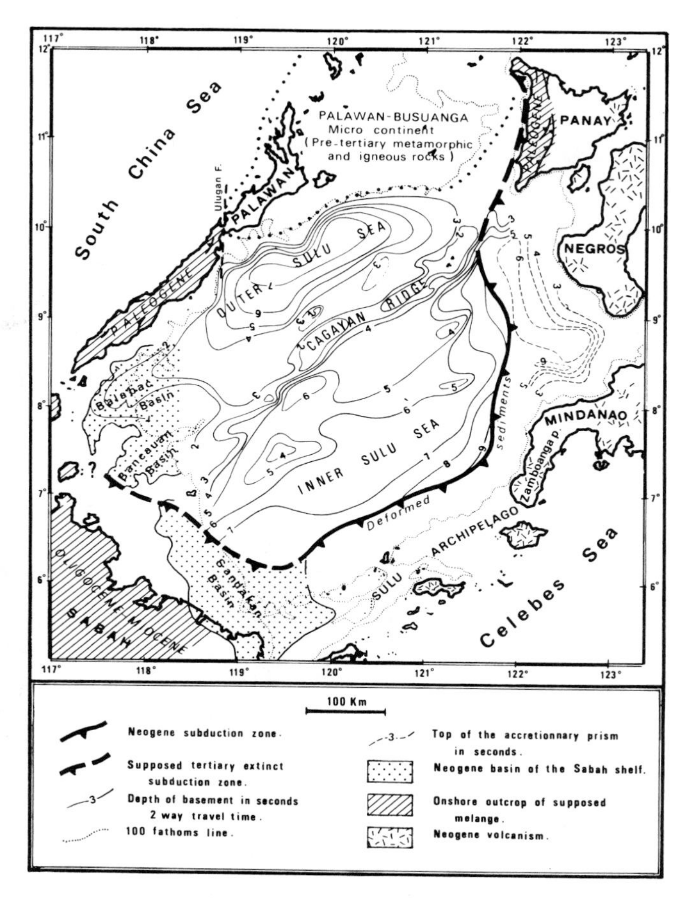

Fig. 5 Sulu Sea. Structural map (modified from Hamilton, 1974; and Beddoes, 1976) and time-structure map of the basement.

creases progressively toward the edges. Their nature and distribution is linked to the nature of these margins.

Three types of margins can be identified around the Inner Sulu Sea:

1. The southwest margin of the Inner Sulu Sea corresponds to the edge of the Sabah Shelf, in front of the Sandakan subbasin (Bell and Jessop, 1974). The deepening of the basement at the foot of this margin (Fig. 5) is associated with a thickening of the sediments above, as the water depths decrease. This local thickening of the sediments in the Inner Sulu Sea is probably linked to the deep distal part of the Sandakan subbasin of Neogene age.

2. The southeast margin is associated with the Sulu Archipelago where it is possible to find structural features similar to those of typical subduction zones in front of volcanic arcs in intra-oceanic domains (Mascle et al, 1977). From NW to SE (Figs. 4, 8) the following structural elements are observed:

A trough, the Sulu Trough, resulting from the deepening of the oceanic basement of the Inner Sulu Sea toward the Sulu Archipelago. This trough is partially filled by a general thickening of the sediments (Fig. 8) and in places has undergone recent filling by a horizontally layered sequence (turbidites?; see Figs. 20–21, Murphy, 1976). In the inside half of the trough, only the older sediments show deformation; the lack of important deformation in the younger sequence is probably related to the lack of present subduction.

The inner wall of the trough is very steep. It should, at least partially, correspond to an accretionary prism. Its upper edge is, in fact, a frontal arc forming the northwest part of the Sulu Archipelago. Behind this frontal arc a small mid-slope sedimentary basin is shown on the extremity of the CEPM line of Figure 8.

The volcanic arc is represented by the southern islands of the Sulu Archipelago (Philippine Bureau of Mines, 1963). The age of this volcanism (late Miocene to Pleistocene, if we include the volcanism of the Zamboanga Peninsula), the absence of seismicity along the arc (Fig. 3, Fitch, 1970), and the lack of deformation in the younger sequence of the Sulu

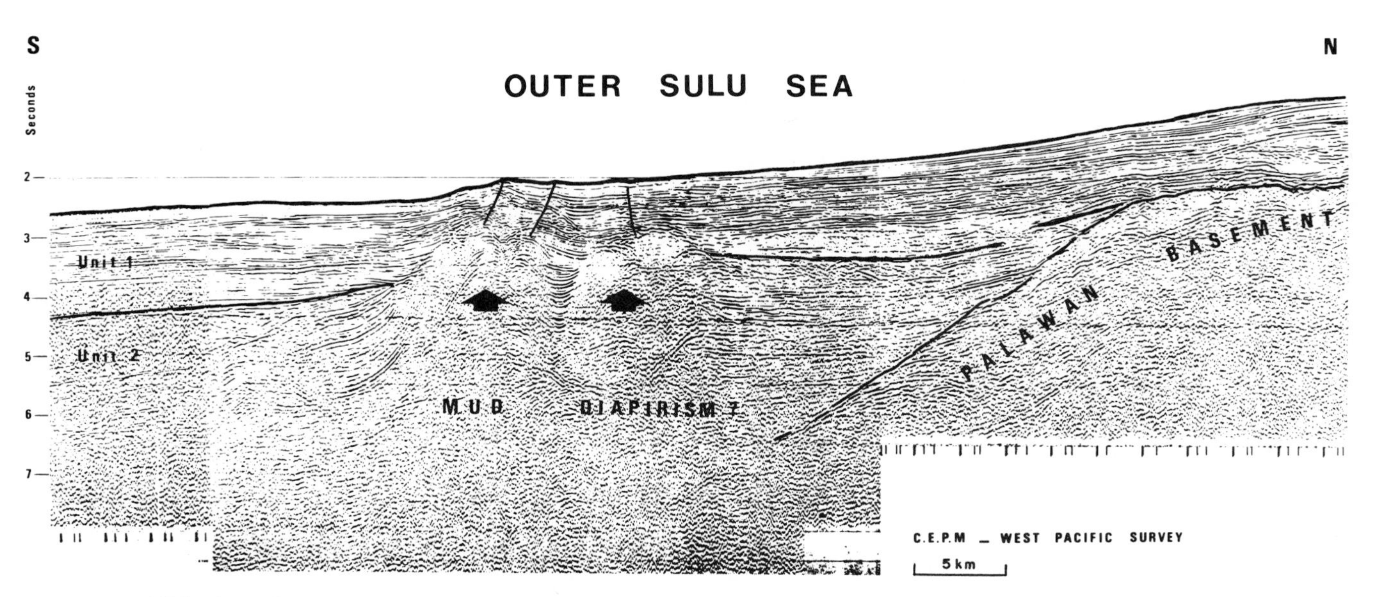

FIG. 6 CEPM multichannel seismic line south of Palawan. The location is shown on Figure 2.

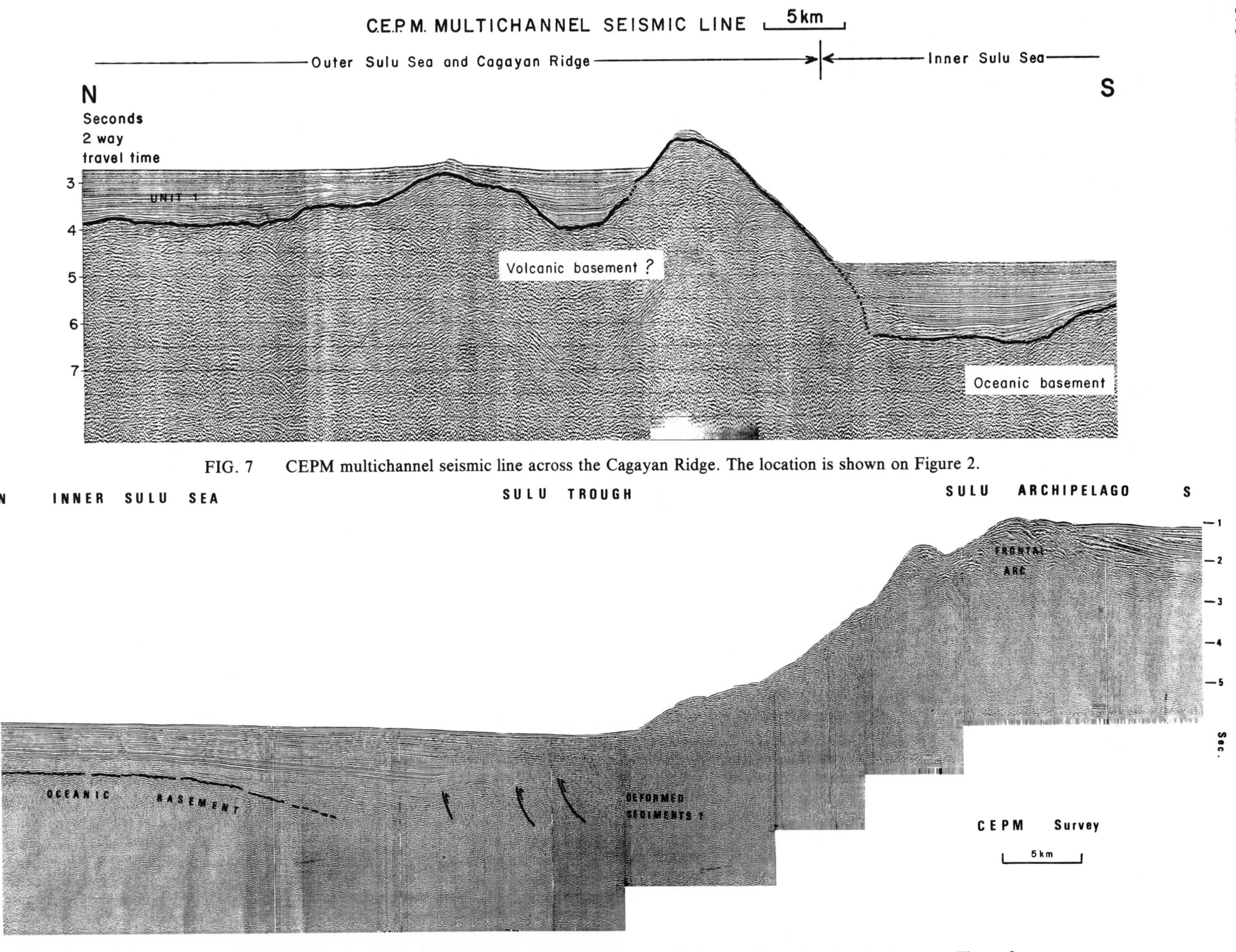

FIG. 7 CEPM multichannel seismic line across the Cagayan Ridge. The location is shown on Figure 2.

FIG. 8 CEPM multichannel seismic line across the Sulu Trough west of Mindanao. The location is shown on Figure 2.

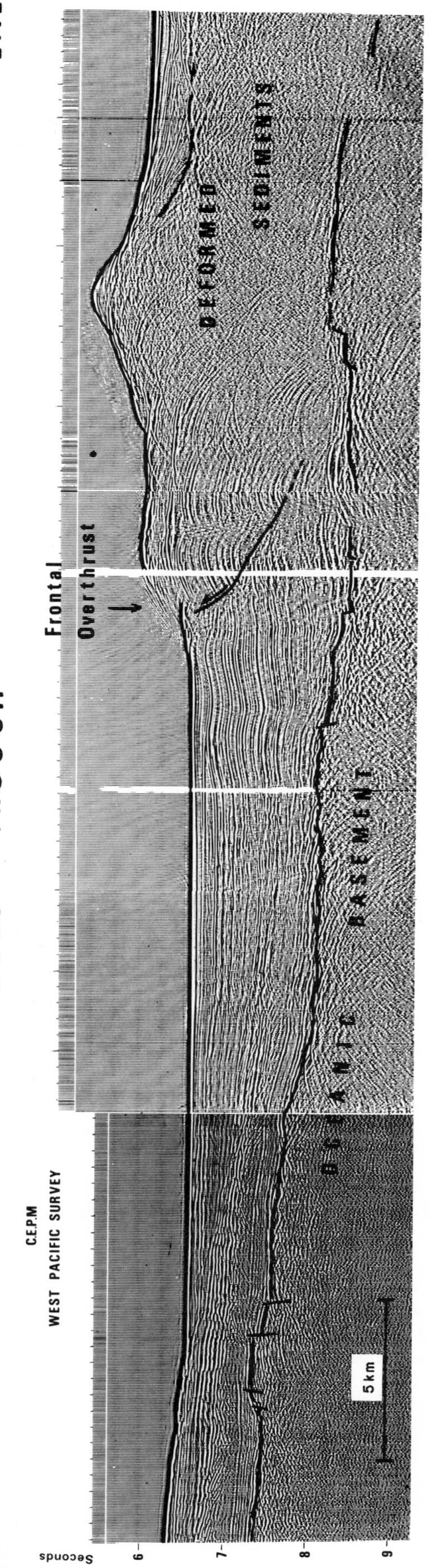

FIG. 9 CEPM multichannel seismic line across the Sulu Trough southwest of Negros. The location is shown on Figure 2.

Trough suggest that this active margin was a subduction zone during part of the Tertiary but became inactive later on.

3. The eastern margin of the basin is continuous with the margin already described. It is contiguous to the Zamboanga Peninsula and Negros, and is essentially the same margin as that of the Sulu Archipelago but is probably still active. Seismic lines recorded across the Sulu Trough (line CEPM, Fig. 9) show deformation of recent sediments very clearly. The downwarping of the oceanic basement has produced a topographic depression in which recent sediments (turbidites?) accumulated. In the inner half of the trough, the early compressional movements are shown by smooth and gentle folding of 100-m amplitude and a period close to 3,000 m.

A major thrust at the foot of the inner wall leads to progressively more tectonically disrupted zones (highly deformed sediments in the accretionary prism), with the exception of the oceanic basement, which can be seen at depth for at least 20 km below the deformed sediments. A line recorded by Gulf (Montecchi, 1976) across the Sulu Trough and the accretionary prism north of the CEPM line shows a forearc basin (or mid-slope basin) on this prism. This basin, part of which can be seen at the end of the CEPM line of Figure 8, is of wide extent inside the area limited by the Negros and Mindanao Islands behind the Sulu Trough.

This very recent and perhaps currently active deformation suggests that subduction is still in progress. However, the seismicity in this area (Fig. 3) is limited to a few shallow-focus earthquakes (depths less than 70 km), which do not suggest the presence of an active subduction zone. Further, the present volcanism of Negros is probably due to the activity of the Mindanao Trench to the east (Fig. 10). This lack of organization of the earthquakes and the lack of volcanic activity has been already described along larger active margins such as the North Venezuelan margin (Silver et al, 1975) or in the Gulf of Oman (White et al, 1976). Such margins have tentatively been explained by recent, or oblique, or very slow subduction.

The Inner Sulu Sea is thus limited to the southeast and to the east by a subduction zone of Neogene age, probably still active in its northern half, which demands shortening during Neogene times behind the Philippines Archipelago. This shortening must be taken into account when making paleogeographic or strain reconstitutions. Compressional components oriented NW to SE and E to W, i.e. perpendicular to the active margins should have appeared during this period of time. The first attempt at such a structural analysis based on the study of the Balabac subbasin to the northeast of the Sulu Sea (Beddoes, 1976) is in agreement with this hypothesis, as two primary compressional stress directions are defined: one roughly oriented N 80° W the other N 30° W.

DISCUSSION

The nature of the crust and its position between Palawan and the Cagayan Ridge indicate that the

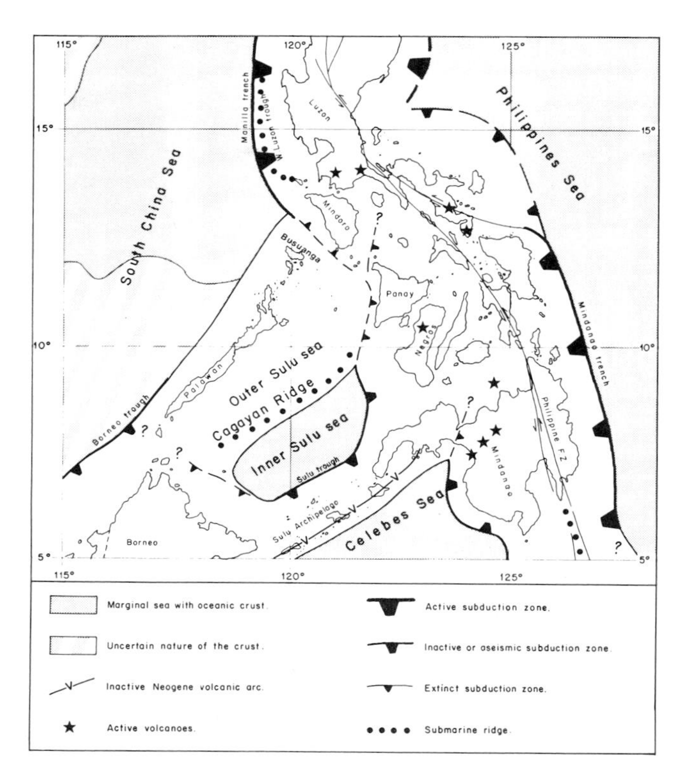

FIG. 10 Structural map of central Southeast Asia showing the complex arrangement of present and past subduction zones.

Outer Sulu Sea is a basin located in the middle of an old island arc. The southern half of Palawan is generally interpreted as a nonvolcanic outer arc of Paleogene age (Hamilton, 1974).

The geological map of the Philippines shows Upper Cretaceous to lower Miocene sediments, and Cretaceous to Paleogene ultramafic and mafic rocks with metamorphosed submarine flows. The tectonic style is not clear but thrust faults have been drawn. The Cagayan Ridge could be the associated volcanic arc (Fig. 6); however, the contemporaneity of these two units has not been proved. As to the origins of the Outer Sulu Sea basin two possibilities must be considered: either it was formed during the period of activity of the island arc, i.e., during the Paleogene, and must therefore be considered as a forearc basin; or it was formed later in the Neogene as a result of the extension or shear which caused part of the island arc to collapse. The lack of age-dating on the deeper sequence of the basin does not allow a choice to be made between these two possible origins.

The Inner Sulu Sea represents a marginal basin with an oceanic crust. The origin of such oceanic basins, associated with active margins, is a subject of controversy. Three types of models have been proposed: 1) collapse of a continental domain (or insular), with oceanization; 2) opening by distension and extension behind an island arc, and upwelling of mantle material to form new crust; and 3) cutting off of oceanic crust, formerly part of a large plate, by island arc development. Because of the lack of geological data about the nature and the age of the crust of the Inner Sulu Sea, it is not easy to choose among these three hypotheses.

The southern and eastern margins of the Inner Sulu Sea correspond to an active margin of Neogene age, which is probably still active in the northern half. Some authors (Hamilton, 1974; Beddoes, 1976) have postulated an extension to the northwest of this active margin along the Sabah Shelf in Oligocene and Miocene times. The deformed sediments and the ophiolites (Brondjik, 1963) of the land basin of Sabah should then correspond to the accretionary prism (or outer arc) of this margin.

Unfortunately, recent deposits of the Sandakan Basin (Fig. 5) have blanketed this extinct active margin so that its structure is no longer obvious. Farther west however, deformations of sediments at the inner slope of the Borneo Trough suggest that subduction is still active along the northwest side of Borneo (Fig. 10).

At the other end of the eastern margin of the Sulu Sea we can suggest a similar explanation (Fig. 5 and 10). The geological map of the Philippines shows overthrusts affecting even the Miocene, east of Panay and Mindoro. An old section by L. Santo Ynigo (1945) published by Irwing (1950) shows strong tangential deformation west of Panay. The rocks involved were pre-Tertiary(?) cherts, volcanics, serpentines and diorite-gabbro, and Miocene limestones, shales, schists. Beddoes (1976) also sees west of Panay an area of "melange" of Paleogene age. The data are incomplete but suggest the presence of a folded (and thrusted ?) complex west of Panay and Mindoro.

Farther north, subduction is still active along the Manila Trench, where the oceanic abyssal plain of the South China Sea has been subducted westward below Luzon since the middle or late Miocene (Ludwig, 1970; Karig, 1973). At the southern extremity of the Philippine Archipelago, some authors (Hamilton, 1974; Roeder, 1977) have mapped a subduction zone west of Mindanao.

All the data suggest that during Tertiary time several (probably short-lived) active margins existed west of the Philippine Archipelago, and that some of them are still active in places. They were perhaps connected from time to time to make a single active margin west of the Philippines, but the present framework suggests several unstable active margins. Activity along them led to the closure of the small oceanic areas west of the Philippines (South China Sea, Inner Sulu Sea, Celebes Sea) and the collapse of the island arcs to build the complex Philippine Archipelago.

The Inner Sulu Sea is thus one of these remaining oceanic areas, subducted in Neogene times along its southern and eastern margins. Prior to that time—particularly if the deformed belts of Panay and Mindoro are true melanges—an oceanic area existed between those islands and the Palawan-Busuanga microblock. It would have connected the South China Sea and the Inner Sulu Sea, but subduction northeastward under Panay and Mindoro led to the closure of this oceanic area and to the collision with the Palawan-Busuanga microblock in middle or late Tertiary time.

Such a model assumes an early Tertiary origin for the Inner Sulu Sea (contemporaneous with, or earlier than the start of the active margins, and recent enough to justify high heat flows). A similar age for the opening of the South China Sea (connected with Sulu Sea?) has already been put forth by Karig (1973). On the other hand, the pre-Tertiary cherts, volcanics, serpentines and diorite-gabbro already described from west of Panay (Irwing, 1950) argue for an older age.

CONCLUSIONS

The Sulu Sea contains two distinct basins. The Outer Sulu Sea basin represents a collapsed basin inside an old island arc. It is not possible to say whether the arc and the basin originated simultaneously or if one succeeded the other, but at least sedimentation took place in Neogene time when the island arc was already extinct. A thick sedimentary sequence with mud diapirs fills this basin towards Palawan. Inner Sulu Sea is a marginal sea with an oceanic crust. The thickness of the sediments is not as great there with the exception of along the southwestern margin (subbasin of Sandakan) and on the active margin southeast of the Sulu Sea (deformed sediments in the accretionary prism).

The active margin to the south and to the east of the Sulu Sea developed in Tertiary time. The best evidence for this assumption is the Sulu Archipelago (Miocene to Pleistocene island arc) and the West Negros margin (still active today). The deformed belts observed onshore in northeast Sabah and on the western side of Panay and Mindoro tend to prove the large extent of this active margin. Consumption of oceanic crust under the Philippines must have led to collision of the arc with the microcontinent of Palawan-Busuanga and the consolidation of the system during the Neogene.

The structure of the Sulu Sea compares favorably with that of other marginal seas of Southeast Asia. A more detailed study of its evolution may help to improve our understanding of island arcs and marginal seas in Southeast Asia. The deep-sea well data that are needed for a study of this kind, could be augmented with outcrop geological data from active margins around the Sulu Sea.

REFERENCES CITED

Beddoes, L. R., 1976, The Balabac subbasin, southwestern Sulu Sea, Philippines: SEAPEX Program, Offshore Southeast Asia Conf., paper 15.

Bell, R. M., and R. G. C. Jessop, 1974, Exploration and geology of the West Sulu Basin, Philippines: Austral. Petrol. Explor. Assoc. Jour., v. 1, p. 21.

Brondjick, J. F., 1963, The Danau Formation in northwest Borneo: Borneo region, Malaysia Geol. Survey, Annual Rept., p. 167–178.

Fitch, T., 1970, Earthquake mechanism and island arc tectonics in the Indonesian-Philippine region: Seismol. Soc. America Bull., v. 60, No. 2, p. 565–591.

——— 1972, Plate convergence, transcurrent faults and internal deformation adjacent to Southeast Asia and the Western Pacific: Jour. Geophys. Research, v. 77, no. 23.

Hamilton W., 1974, Map of sedimentary basins of the Indonesian region: U.S. Geol. Survey.

——— 1975, Subduction in the Indonesian region: SEAPEX Proceedings v. II, p. 37–40.

Hatherton, T., and W. R. Dickinson, 1969, The relationship between andesitic volcanism and seismicity in Indonesia, the Lesser Antilles, and other island arcs: Jour. Geophys. Research, v. 74, no. 22.

Irwing, E., 1950, Review of Philippines basement geology and its problems: Philippine Jour. Science, v. 79.

——— 1951, Submarine morphology of the Philippines Archipelago and its geological significance: Philippine Jour. Science, v. 80.

Karig, D., 1973, Plate convergence between the Philippines and the Ryukyu Islands: Marine Geology, v. 14, p. 153–168.

Ludwig, W. J., 1970, The Manila Trench and West Luzon Trough, III. Seismic refraction measurements: Deep Sea Research, v. 17, p. 553–571.

Mascle, A., et al, 1977, Sediment deformations in active margins of different geological settings: Geodynamics in Southwest Pacific, Technip Edition.

Montecchi, P. A., 1976, Some shallow tectonic consequences of subduction and their meaning to the hydrocarbon explorationist, *in* M. T. Halbouty, J. C. Maher, and H. M. Lian, eds., Circum-Pacific Energy and Mineral Resources: AAPG Memoir 25, p. 189.

Murauchi, S., et al, 1973, Structure of the Sulu Sea and the Celebes Sea: Jour. Geophys. Research, v. 78, no. 17, p. 3437.

Murphy, R. W., 1975, Tertiary basins of Southeast Asia: SEAPEX Proceedings, v. II, p. 1–36.

——— 1976, Pre-Tertiary framework of Southeast Asia: SEAPEX Program, Offshore Southeast Asia Conf., Paper 3.

Nagasaka, K., et al, 1970, Terrestrial heat flow in the Celebes and Sulu Seas: Marine Geophys. Research, v. I, p. 99–103.

Philippine Bureau of Mines, 1963, Geological map of the Philippines: Manila, Philippines, Bureau of Mines.

Roeder, D., 1977, Philippine arc system, collision or flipped subduction zones?: Geology, v. 5, p. 203–206.

Silver, E. I., J. E. Case, and H. J. MacGillavry, 1975, Geophysical study of the Venezuelan Borderland: Geol. Soc. America Bull., v. 86, p. 213–226.

White, et al, 1976, Sediment deformation and plate tectonics in the Gulf of Oman: Earth and Planetary Sci. Letters, v. 32, p. 199–209.

Investigation of Mississippi Fan, Gulf of Mexico[1]

G. T. MOORE,[2] H. O. WOODBURY,[3] J. L. WORZEL,[4]
J. S. WATKINS,[5] G. W. STARKE[2]

Abstract The Mississippi Fan, in the northeast part of the Gulf of Mexico, is an arcuate pile of clastic sediments derived primarily from the ancestral Mississippi River. The radius of the fan from its apex, at a water depth of 1,200 m, to its margin is about 350 km. The fan merges to the southeast with the Florida Plain (3,300 m) and to the southwest with the Sigsbee Plain (3,500 m).

The upper fan contains a partly leveed channel cut into older sediments, the upper portion of which is filled with late Pleistocene (Wisconsin) fine, clastic sediments. The middle part of the fan, typical of many, is a massive complex of channels forming a crown up to 500 m above the surrounding fan surface. The lower fan has smooth, gentle slopes and depositional-type distributary channels.

Sufficient continuity exists on seismic profiles to permit subdividing the upper sedimentary section into three seismic stratigraphic units. We term the youngest Unit A. It consists of a proximal facies of disrupted seismic zones and a distal facies of generally parallel reflections, separated by transparent zones. The former is interpreted to be channel, slump, and debris flow deposits; the latter, turbidite flows interbedded with hemipelagic sediments.

Correlation to DSDP sites shows that Unit A represents a thick Pleistocene sequence. The fan formed during the time represented by Unit A. Two older intervals spanning the period from Pliocene to middle Miocene were mapped to show the relationships in this region prior to fan development.

Computer-generated isopach maps of each unit permit study of the Neogene through Quaternary regional depositional patterns. Unit A is over 3 km thick on the mid-fan.

When the reservoir-type rocks, source beds, stratostructural traps, and the age/time/depth-of-burial relations of the fan are placed in perspective, the petroleum potential of this province remains an open question.

INTRODUCTION

The Mississippi Fan, located in the northeast Gulf of Mexico, is a broad, arcuate pile of sediments marked by a gentle outward bowing of the bathymetric contours (Fig. 1). The fan lies between the west Florida carbonate platform on the east and the Texas-Louisiana-Florida continental slope on the north and west. The lower portion of the fan bifurcates around the north-jutting salient of the Yucatan carbonate platform. The lower portions merge with the Florida Plain to the southeast and the Sigsbee Plain to the southwest.

The fan, originally recognized by Bates (1953), has since been studied by numerous investigators (Bergantino, 1971; Ewing et al, 1958; Ewing et al, 1960; Ewing et al, 1962; Fisk and McFarlan, 1955; Garrison and Martin, 1973; Huang and Goodell, 1970; Martin, 1976; Newman et al, 1971; Stuart and Caughey, 1976; Uchupi 1967; Walker and Massingill, 1970; Wilhelm and Ewing, 1972). Various workers have alternatively called this feature a cone or a fan;

[1]Manuscript received, November 8, 1977; accepted, July 14, 1978.

[2]Chevron Oil Field Research Company, La Habra, California 90631

[3]Chevron U.S.A. Inc., Eastern Region, New Orleans, Louisiana 70112

[4]Geophysics Laboratory, University of Texas, Marine Science Institute, Galveston, Texas 77550

[5]Gulf Research and Development Company, Houston, Texas 77036

We are grateful to Exxon Production Research Company and Chevron (New Orleans), Mobil, and Gulf Oil Companies for release of core data and seismic profiles collected during the 1966 Gulf of Mexico Upper Continental Slope Project. We wish to thank Chevron Oil Field Research Company and Chevron U.S.A. Inc., Eastern Region for permission to publish this paper.

Many of the results reported here were taken from analytical data developed by others. In particular we acknowledge the work of W. H. Akers, I. B. Murray, Jr., P. J. Pickford, C. C. Humphris, Jr., J. H. Spotts, and S. R. Silverman. Fay Dennis, Edward Compise, and D. T. Teaford drafted the illustrations; W. F. Dreyer developed the computer programs; and D. A. Ivey typed the manuscript. We appreciate their efforts.

Microfilm copies of the USNS *Kane* seismic profiles and high-resolution fathometer records were obtained from the National Geophysical and Solar-Terrestrial Data Center, Boulder, Colorado.

The interpretations, results, and conclusions expressed in this paper are those of the authors. We accept full responsibility for their accuracy.

Article Identification Number:
0065-731X/78/MO29-0026/$03.00/0.
(see copyright notice, front of book)

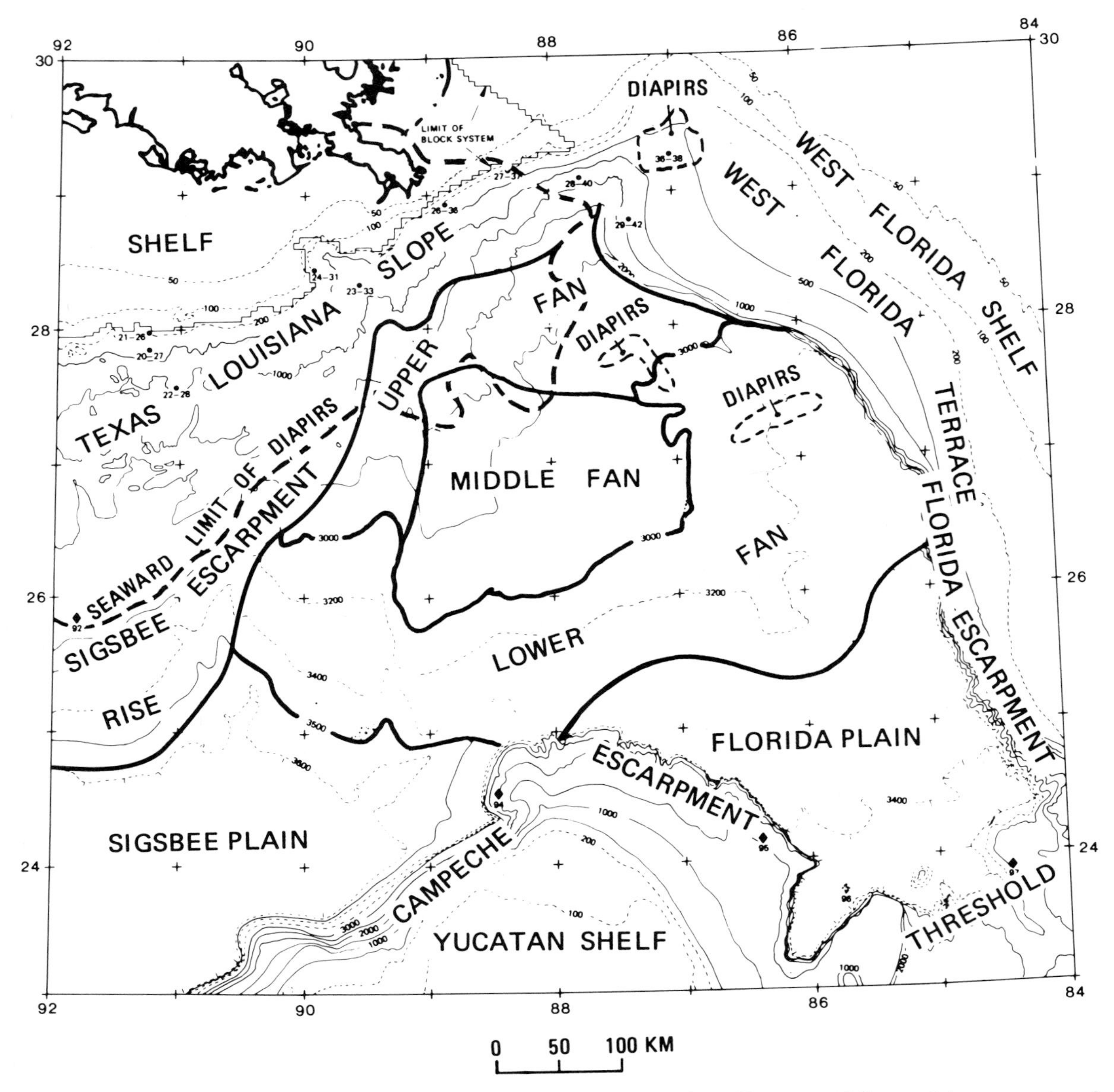

FIG. 1 Bathymetry and physiography of eastern Gulf of Mexico. Upper, middle, and lower parts of Mississippi Fan outlined. Seaward limits of diapirs from Martin (1976) and Watkins et al (1976). Contours from Uchupi (1967); solid circles are core holes; diamonds indicate DSDP sites. C.I. = 500 m, except on shelf where interval is smaller.

however in this paper we use the USGS terminology and call it the Mississippi Fan following Martin and Bouma (1978).

The pattern of deposition on the fan has changed with the fluctuation of sea level. During high stands the ancestral Mississippi River constructed a delta at one or another of its numerous mouths, and relatively little sediment has reached the fan. However, in the very late Pleistocene (18,000 Y.B.P.), sea level was at least 85 m lower than at present (CLIMAP Project Members, 1976). During such low sea-level stands, associated with glacial maxima, the ancestral Mississippi flowed across the continental shelf, cut canyons, and disgorged its sediment load directly onto the slope (Carsey, 1950; Fisk and McFarland, 1955; Osterhoudt, 1946; and Woodbury et al, 1976).

Many factors contribute to sedimentation rates, sediment types, volume, and deformation on any one fan. Moore et al (1978) compared these parameters for major fans. Each fan may be affected by a unique set of conditions. By this study we hope to place the Mississippi Fan in proper perspective.

The bathymetric map of the eastern Gulf of Mexico, (Uchupi, 1967), shows the locations of single-channel seismic profiles studied and those used in this paper (Fig. 2). The fan morphology, sedimentational history, and deformation were studied on 3.5-kHz high-resolution records and seismic profiles. These were acquired during the joint U.S. Naval Oceanographic Office and U.S. Geological Survey investigation by the USNS *Elisha Kane* cruise in the Gulf (solid lines). These regional lines were supple-

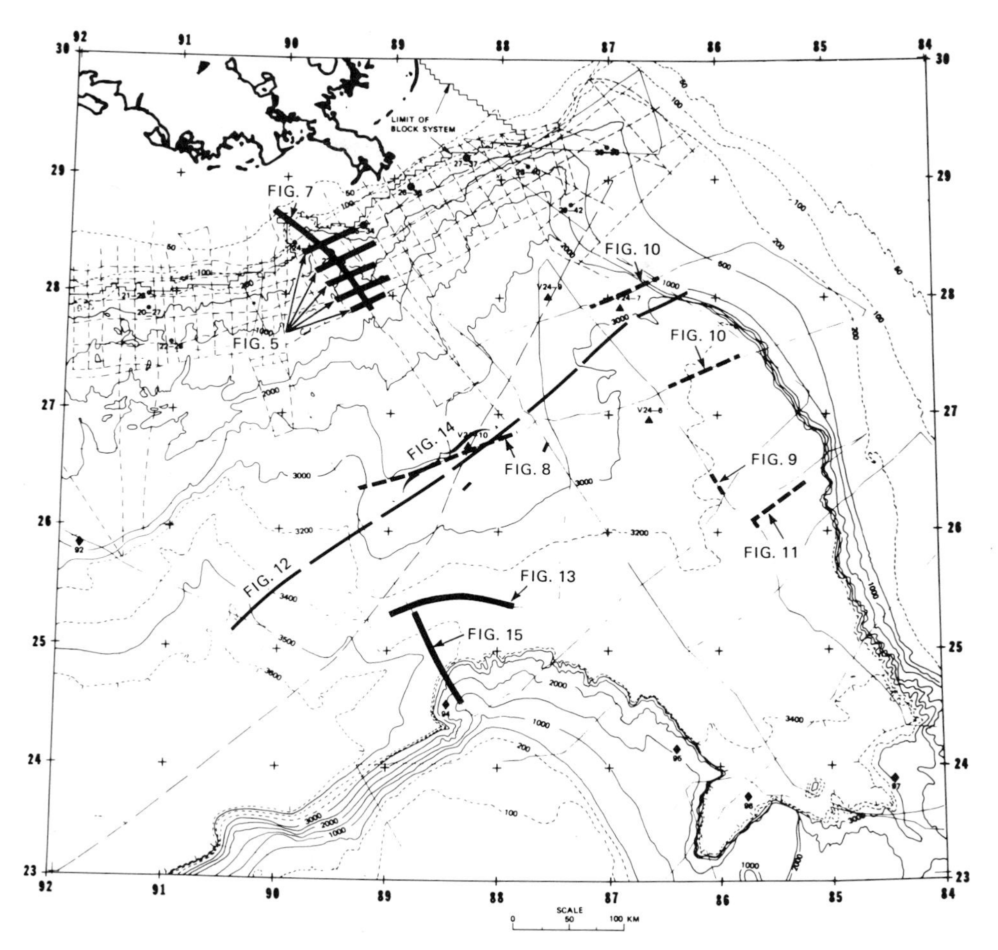

FIG. 2 Index map to tracklines and profiles, used in this paper; oil company core holes (solid circles); and DSDP sites (diamonds). Tracklines: USNS Kane (solid); UTMSI R/V Ida Green (long dash); Exxon-Chevron-Mobil-Gulf (short dash). Heavy solid lines indicate deep-penetration seismic profiles; heavy dashed lines, 3.5-kHz high-resolution profiles used in paper.

mented by a detailed rectangular network (small dashes) on the upper continental slope, made using data from a joint Exxon-Chevron-Mobil-Gulf (E-C-(ECMG) survey in 1966. While both surveys used single-channel equipment, the energy sources (Garrison and Martin, 1973 and Sangree et al, 1976) were large enough to obtain deep penetration.

A northeast-southwest multichannel seismic profile was recorded across the lower and middle fan in 1976 by the University of Texas Marine Science Institute's vessel, R/V *Ida Green* (Worzel and Burk, 1978). The high quality of the processed section permits study of this part of the fan.

Core holes as much as 305 m (1,000 ft) deep were drilled on the upper continental slope by Superior (Caughey and Stuart, 1976), Shell (Lehner, 1969), and Exxon, Chevron, Mobil and Gulf oil companies (ECMG), from 1965–1968. Figure 2 shows the ECMG sites on the Texas-Louisiana-Florida Slope.

PHYSIOGRAPHY

On a bathymetric map (Fig. 1) the Mississippi Fan is shown as a gentle outward bowing of the contours. Its lower part is divided by the northward projection of the Yucatan Shelf. The length of the fan, from apex to merger with the Florida and Sigsbee Plains, ranges from 330 to about 380 km. In area the fan is approximately 170,000 sq km, somewhat smaller than New England or the state of Missouri.

The fan was deposited on an abyssal plain and rise, against the original continental slope. It represents now a much expanded version of the original rise. As the lower fan merges with two abyssal plains, the base is not at a single level. Based on gradient and morphology, we consider the lower limits of the fan to be 3,300 m on the southeast and 3,500 m on the southwest. The fan abuts three escarpments: two carbonate (the Florida and Campeche) and one of deformed sediments intruded by diapirs (the Sigsbee).

The fan is divisible into three geomorphic units, upper, middle, and lower. The 3.5-kHz records and bathymetric data were utilized for this subdivision. Each part is characterized by a different morphology and system of channels. Figure 3 shows the distribution of channels (arrows) found on the USNS *Kane*

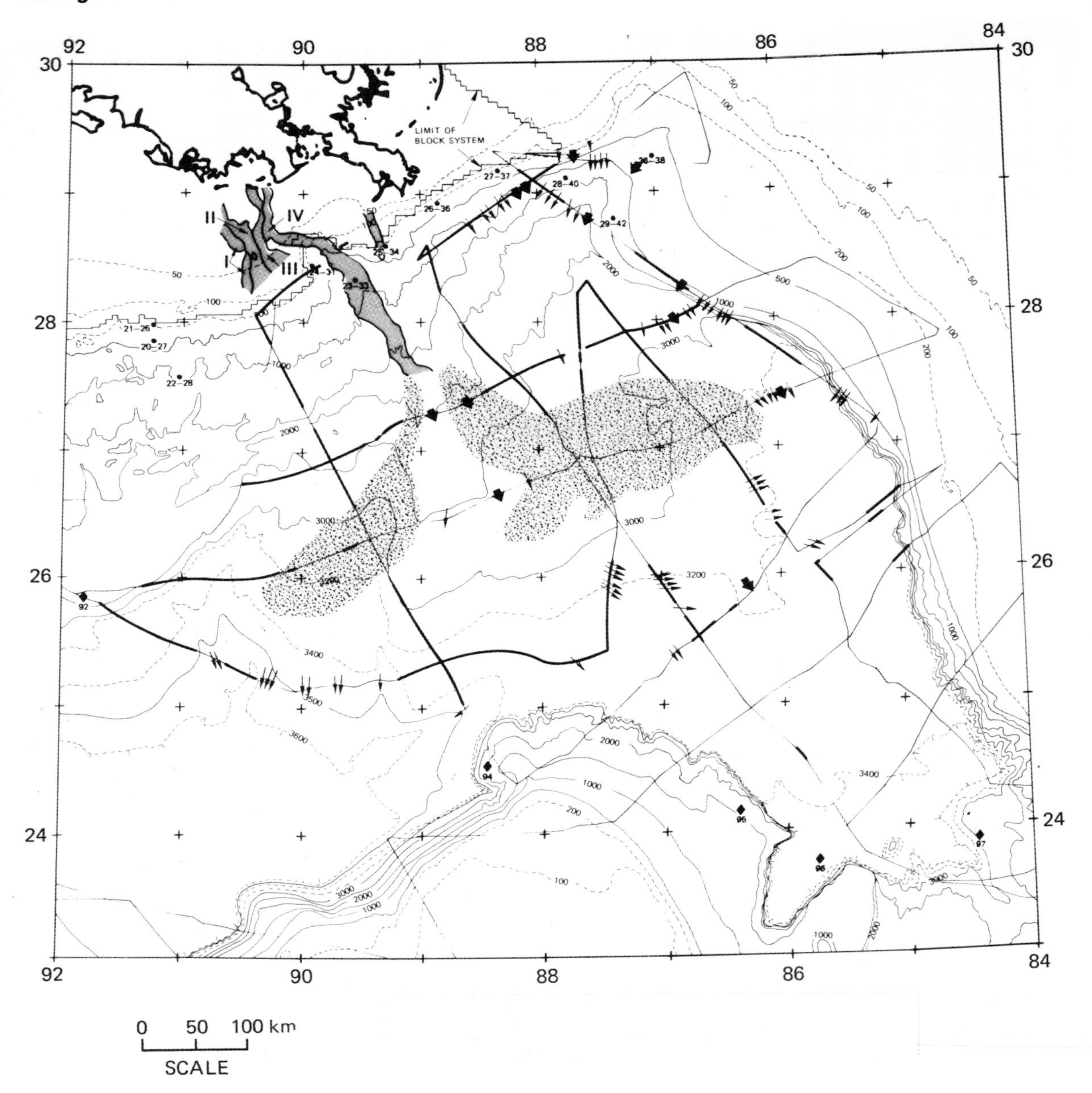

FIG. 3 Channel crossings (large and small arrows) and subbottom reflectivity of near-surface sediments (wide parts of track lines) on Mississippi Fan and plains. Two large slumps (shown stippled) are from Walker and Massingill (1970). Late Pleistocene channels I to IV from Woodbury et al, 1976.

3.5-kHz profiles and mapped by Woodbury et al (1976). Surprisingly few distributary channels are found on the lower fan.

Subbottom penetration to 100 m was obtained on these records where the sedimentary sequence was reflective. Figure 3 shows the areas of subbottom reflectivity (heavy lines) on these records. These areas are interpreted to contain bedded turbidites alternating with hemipelagic sediments. Much of the upper and lower geomorphic portions of the fan and the plains have an acoustically transparent record character. On the upper part this is interpreted to result from the nature of the sediments, which include characterless slumps and homogeneous channel fill. Most of the larger channels on the middle and lower fan are filled. The lack of subbottom reflections on the lower fan and abyssal plains is interpreted to result from homogeneous, hemipelagic sediments containing marly silt and fine sand beds too uniform to reflect energy.

The basinward extent of what is termed the Mississippi Trough is a matter of debate or conjecture. Uchupi (1967) places the mouth at the 2,200-m isobath, whereas Garrison and Martin (1973) place it at 1,650 m. More recently, Shih et al (1975) proposed that a topographic break at 1,200 to 1,350 m may be caused by the buried extension of the Sigsbee Escarpment and marks the limit of the trough.

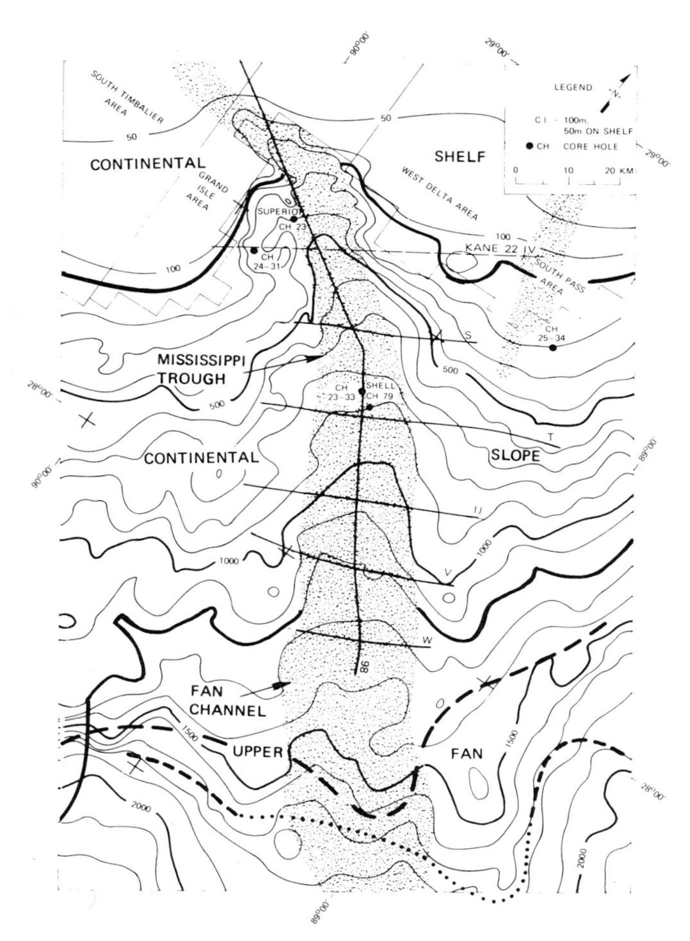

FIG. 4 Detailed bathymetry showing character of Mississippi Trough, fan channel, and smaller channel to east (stipple). Modified from Woodbury et al, (1976). Salt ridge forming eastern continuation of Sigsbee Escarpment shown by heavy dashes and second ridge to south, partly buried (heavy dots) across upper fan from Shih et al (1978). See Figure 2 for location.

Because the apex of the fan reaches the upper part of the continental slope (Martin and Bouma, 1978), we suggest that the basinward limit of the trough be restricted to that part incised into the upper continental slope. Thus, we agree with Shih et al (1975) and place the downslope termination of the trough at the 1,200-m isobath (Fig. 4).

Upper Fan

The upper fan is characterized by a central erosional channel. Its flanks, some of which may be levees, can exceed 100 m in height. The channel passes shoreward into the broad Mississippi Trough, where the relief on one profile across it exceeds 400 m (Moore et al, 1978). The channel is relatively flat-bottomed throughout its length; but, as Figure 4 shows, it changes character with depth. Its gradient is approximately 1:100, versus 1:30 to 1:60 for the continental slope where it abuts the fan.

The trough and channel on the upper fan are crossed by five profiles (Fig. 5). The average gradient between Line S, the upper, and Line W, the lower, is 1:117, or 8.6 m/km. The outline of this late Wisconsin channel, probably the youngest of at least four mapped in the subsurface on the shelf and upper slope in this region (I, II, III and IV, Fig. 3), is shown on five sections. Core Hole 23-33, just above line T, was drilled on the continental slope in the channel to a depth of 192 m. It penetrated the upper part of the channel fill, consisting of very fine-grained upper Wisconsin sediments (Fig. 6).

An axial profile along the channel passes 3 km northeast of Superior core hole 23, about 10 km south of an erosional remnant (Fig. 7). Core hole 23 bottoms in late Illinoisan sediments. Because there is not a direct tie to the hole, we are not certain whether it passed through the channel and into the underlying beds or whether the channel is the site of an earlier excavated and refilled one.

Diapiric intrusions consisting of individual domes and ridges, presumed largely to be salt, have both modified the physiography and deformed the sedimentary sequence on the shallower part of the upper fan. The area so affected lies between the upper boundary of the fan and the line showing the seaward limit of diapirs (Fig. 1). Shih et al, (1978) mapped these diapirs in detail; showed that the Sigsbee Scarp, the shallowest and most extensive of four ridges, can be traced eastward across the upper fan and adjoining continental slope; and presented a scenario of scarp development and emplacement. The distribution of these scarps and buried ridges is shown on the central part of the upper fan in Figure 4.

Middle Fan

The middle part of the fan beyond the diapirs (Fig. 1), has a generally smooth, slightly convex-upward profile representing the suprafan (of Normark, 1970; Fig. 8). The crossings show that most channels are filled with sediments. The high-resolution record shows at least three channels, filled almost to the tops of their natural levees. The suprafan locally rises over 400 m above the general contour of the fan and on the upper middle fan, divides into southeast and southwest systems. These supplied sediments to the respective parts of the fan, shown diagramatically by Stuart and Caughey (1976). The gradient in this region of the suprafan ranges from 1:120 to 1:250.

Lower Fan

The lower fan is a very smooth, gently sloping surface that merges imperceptibly with the plains. The bases of separation are the low gradient (1:250 to 1:550), a decrease in bedded turbidites, and a commensurate increase in acoustically transparent hemipelagic sediment. This region contains numerous small distributary channels (Fig. 3). Most channels show clear evidence of natural levee construction from overspilling during periods of high-

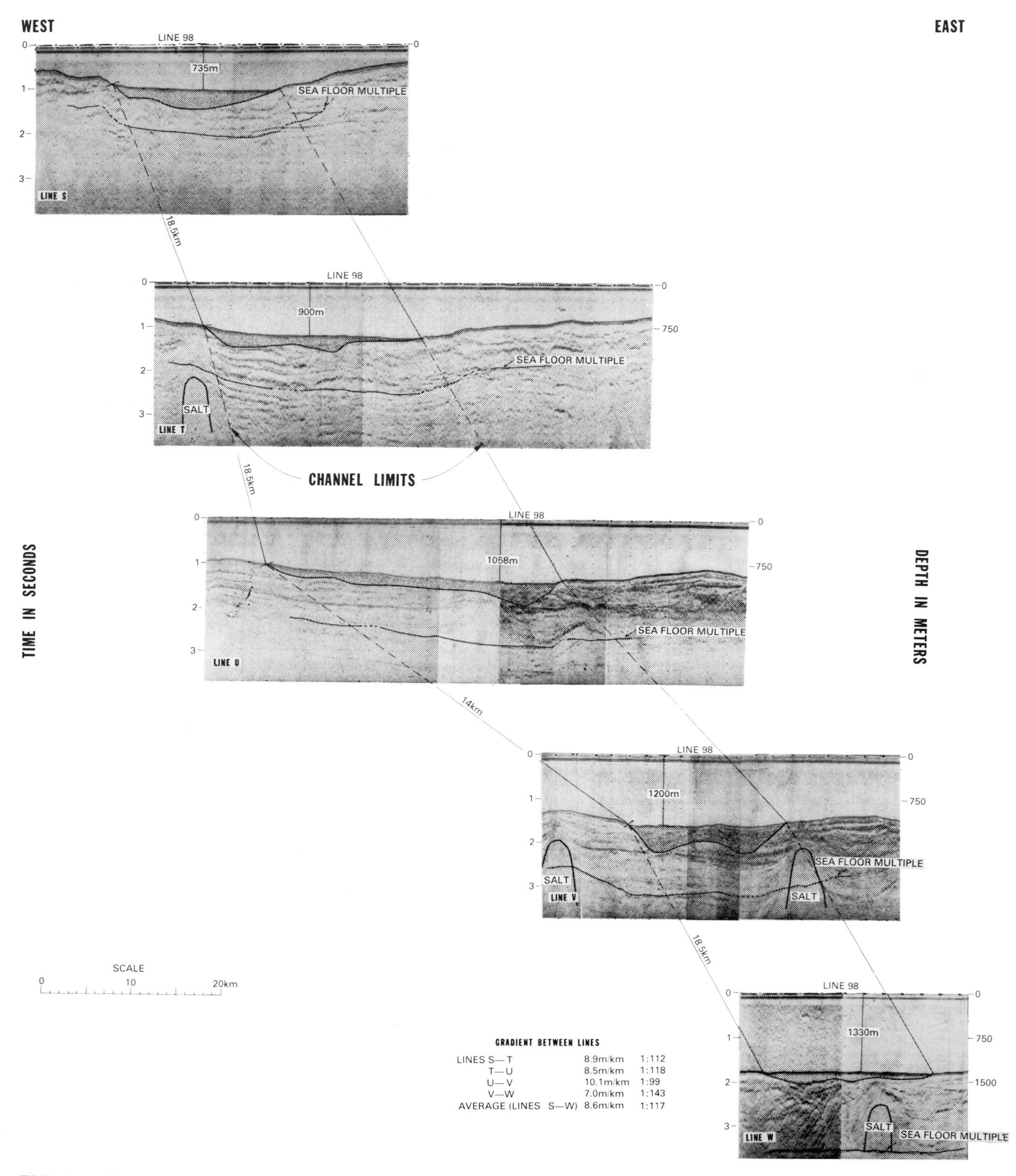

FIG. 5 Sequence of seismic profiles across late Pleistocene channel, showing boundaries, channel bottom, and late Pleistocene fill. Profiles are perpendicular to channel axis and aligned parallel to its trend. Average gradient decreases from approximately 1:110 between lines S and V to less then 1:140 between lines V and W. See Figures 2 and 4 for location.

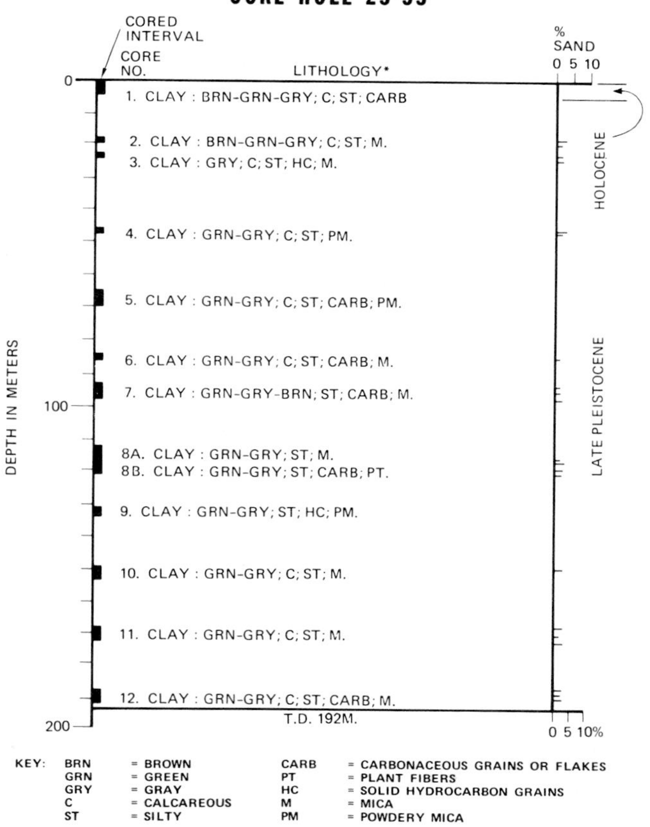

FIG. 6 Lithologic log from ECMG Core Hole 23-33. Description by J. W. Low. Extent of cored intervals shown by bars (left), percent sand (right). Located on Figure 4.

volume flow (Fig. 9). These sediments are deposited on the older fan surface and can be traced for 15 km beyond the channel axis.

Most of the distributary channels on the eastern part of the lower fan are in clusters (Fig. 3). This suggests sedimentation by turbidity currents moving through discrete channels. During periods of high-volume flow, the turbidity currents topped the natural levees, depositing the sediments on them and in interchannel areas. In contrast, few channels are evident on the western part of the lower fan.

DeSoto Deep-Sea Channel

An extension of the DeSoto Canyon, here termed the DeSoto Deep-Sea Channel, was observed on several profiles on the eastern flank of the fan. This extension subparallels the Florida Escarpment. Figure 10 shows two cross-sections of this channel. The northern profile shows natural levees up to 15 m high, projecting above the seafloor. The channel itself is 3.3 km wide. About 89 km southeast of this line the channel is 328 m deeper, but is filled with debris to near the top of the natural levees.

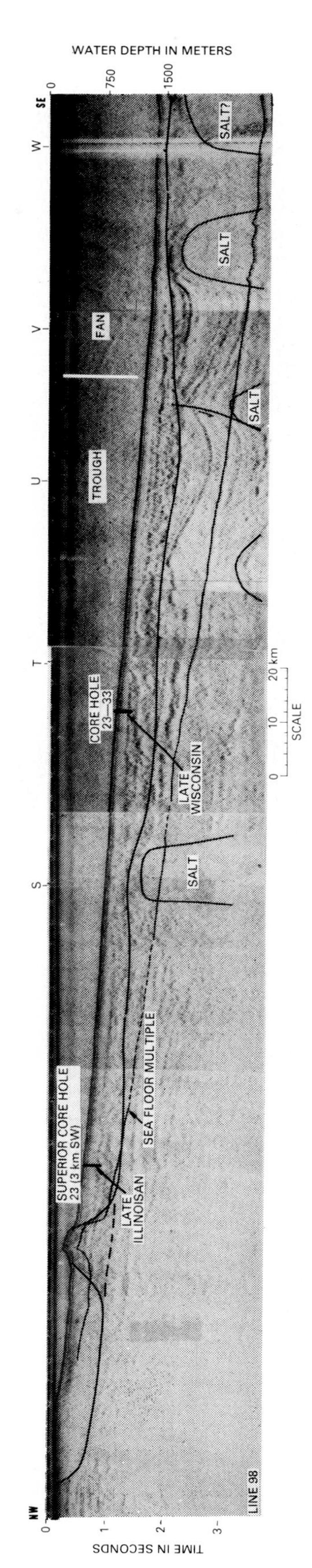

FIG. 7 Longitudinal seismic profile along Mississippi Trough and upper Mississippi Fan channel. Line follows axis of late Pleistocene channel. Profile also shows ECMG core hole 23-33 and Superior core hole 23 (3 km southwest of line). See Figure 4 for location.

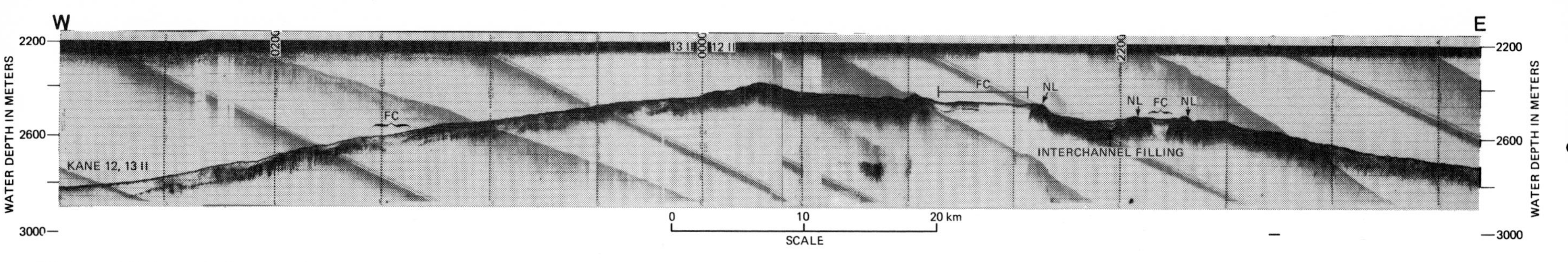

FIG. 8 High-resolution (3.5 kHz) profile of middle part of Mississippi Fan (*Kane* 12, 13 II)—FC: Filled channel; NL: Natural levee. See Figure 2 for location. Diagonal pairs of lines on this and succeeding *Kane* high-resolution profiles are a secondary outgoing signal and seafloor reflection picked up by recorder. They could result from a second source—e.g., 12 kHz—operating simultaneously; from reflection of signal off another piece of equipment; or from lack of syncronization of outgoing signal at 3.5 kHz.

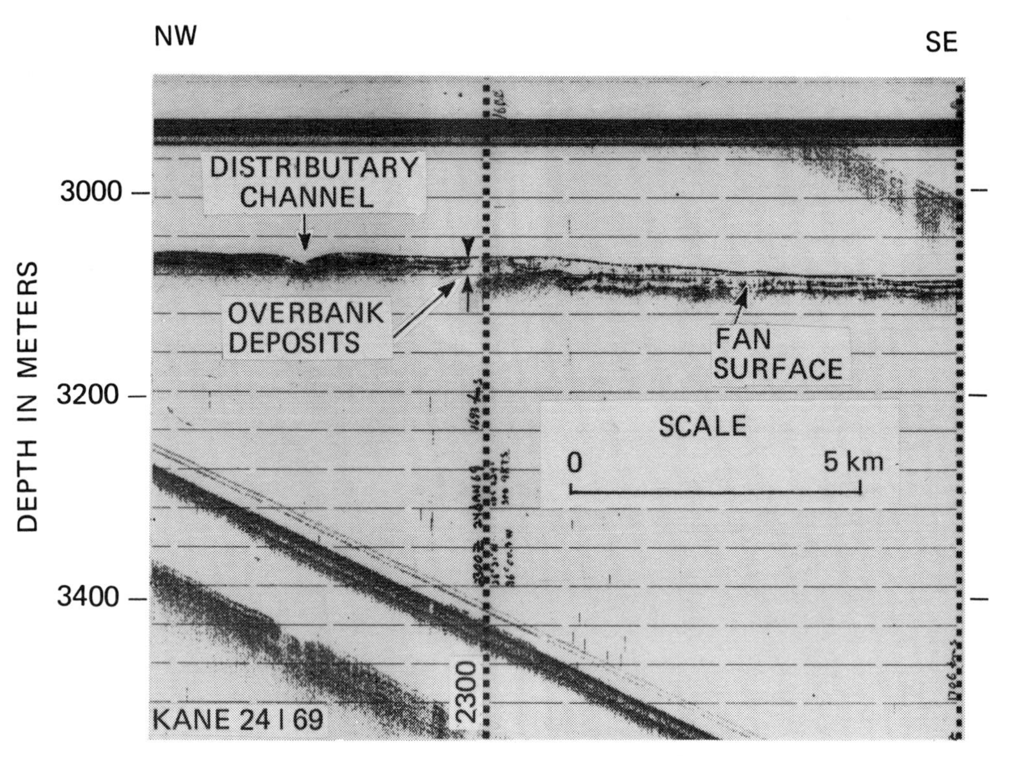

FIG. 9 High-resolution (3.5 kHz) profile showing overbank deposits (25 m minimum thickness) laid down by turbidity currents overflowing distributary channel on eastern flank of Mississippi Fan. Channel about 500 m wide and 15 m deep. See Figure 2 for location. For discussion of diagonal lines see Figure 8.

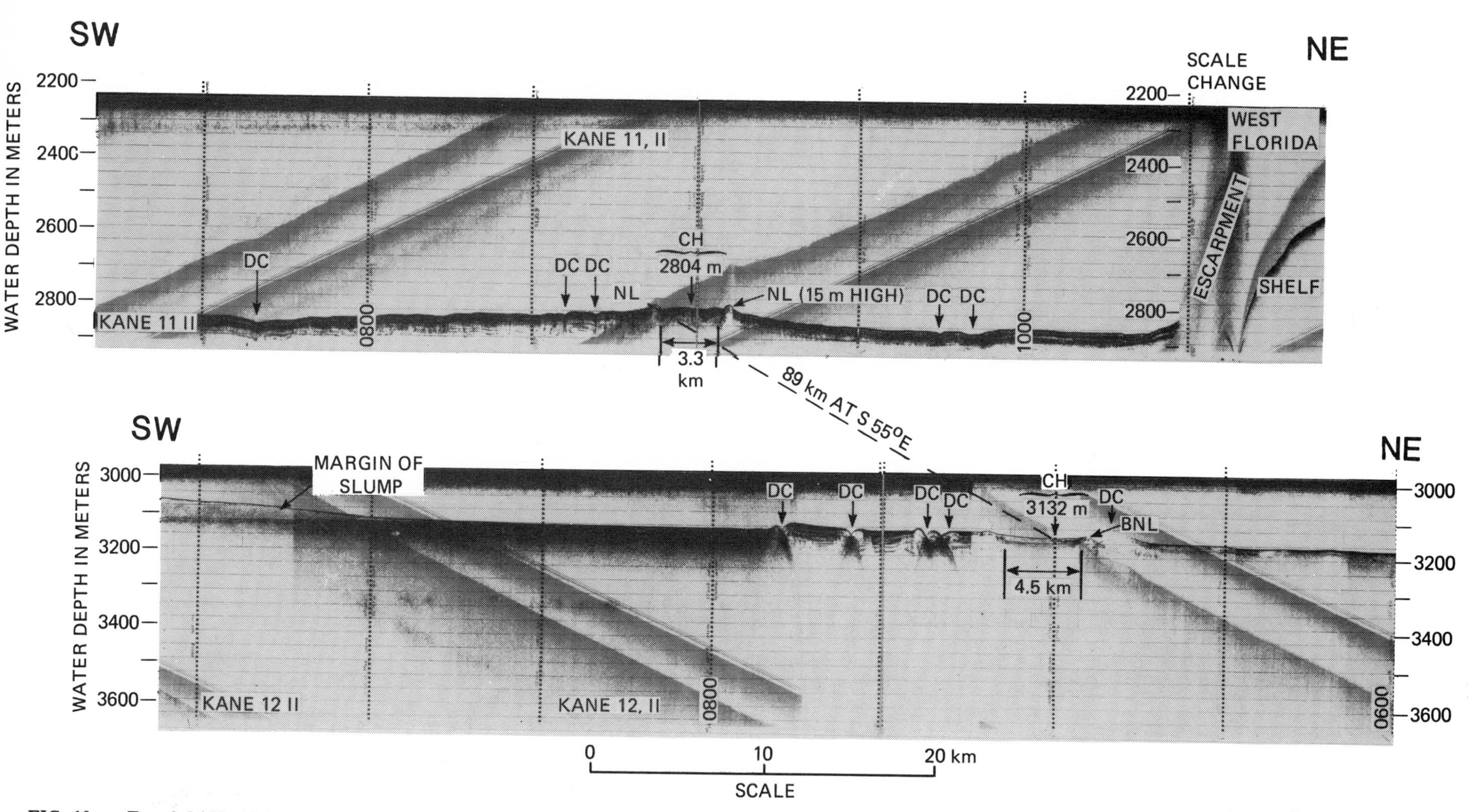

FIG. 10 Two 3.5-kHz high-resolution profiles of DeSoto Deep-Sea Channel on east flank of Mississippi Fan. Upper profile shows channel developed on a mound of sediments constructed by vertical and lateral accretion during periods of active channel flow. Channel subparallels Florida Escarpment. For location see Figure 2. For discussion of diagonal lines see Figure 8.

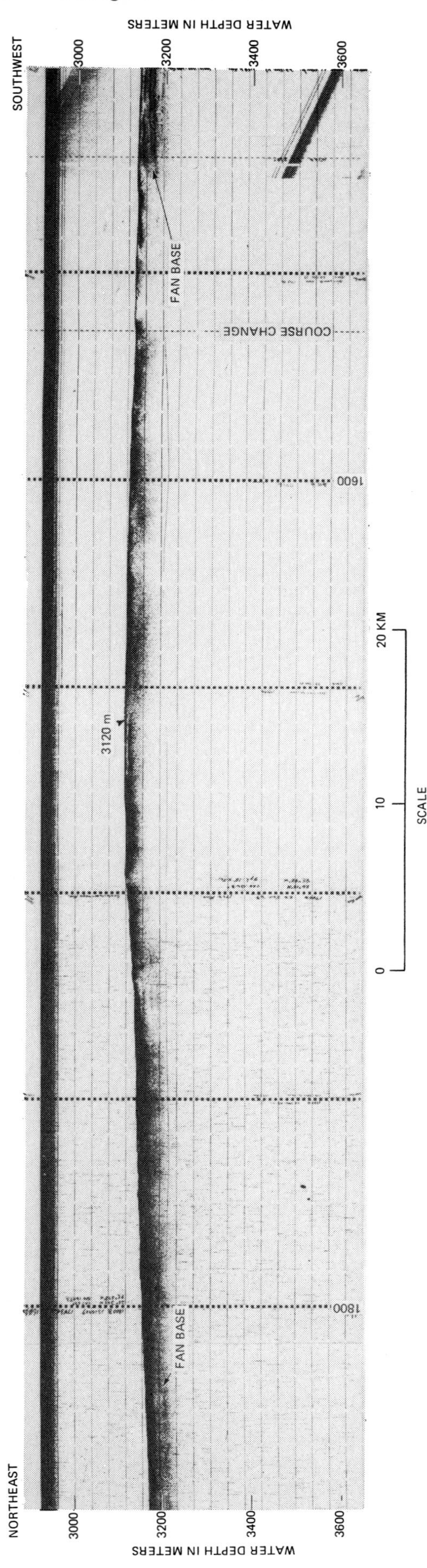

FIG. 11 High-resolution 3.5-kHz profile across small fan on east part of Mississippi Fan. For location see Figure 2.

Plainly a change in the capacity of the channel to transport the volumes of sediment occurred between these two lines. Originally the channel, with its natural levees, was constructed on the fan. Somewhere between the two profiles the channel became choked with debris. Turbidity currents moving down the channel overflowed the banks, deposited much of their load on top of the original natural levees, particularly to the southwest, and raised the level of the seafloor.

A 3.5-kHz profile, approximately 140 km south of the lower profile in Figure 10 shows a broad arching of the surface about 70 km in width (Fig. 11). The surface is relatively uniform. The subbottom is acoustically opaque. There are no indications of cut-and-fill channels, natural levees, or distributary channels. This lack suggests that the feature may have been constructed by uncontained turbidity currents flowing across a broad front rather than in channels. On the margins, traces of the original seafloor on which the fan was constructed can be observed (Fig. 11). The regional gradient on this part of the lower fan is very gentle. The crest of the fan (3,120 m, Fig. 11) is 12 m shallower than at the last cross-section of the DeSoto channel (Fig. 10 lower profile); however, continuity with it cannot be established. In all probability this is the lobe of a depositional feature related to sedimentation on the middle fan.

Methods of aggradation on the lower fan are overspilling of distributaries (Fig. 9), sequential channel development and abandonment or debris flows (Moore et al, 1978), and small subsidiary fans (Fig. 11).

AGE OF SEDIMENTS

Because no wells have been drilled on the fan, the methods of establishing its age are all indirect. However, various authors deduce that it is a youthful feature (Garrison and Martin, 1973; Moore et al, 1978; Stuart and Caughey, 1976). A Pleistocene age is indicated by correlations from DSDP sites (Moore et al, 1978; Stuart and Caughey, 1976; Worzel and Burk, 1978), maps of late Cenozoic depocenters (Woodbury et al, 1973), and volumetric calculations (Moore, 1969).

The sediment mass constituting the fan has been neither drilled nor cored to any significant depth, leaving us with a frustratingly small amount of stratigraphic information. The data can be classed into three categories:

1. Core holes ranging to 305 m (1,000 ft) depth on the upper continental slope.
2. Piston cores with penetration to 20 m (65 ft) on the upper, middle, and lower parts of the fan.
3. DSDP penetrations to 800 m (2,630 ft) on the Sigsbee Plain, beyond the margins of the lower fan.

Because of distance and changing facies, correlation of stratigraphic intervals from DSDP sites to the fan is tenuous. However, Watkins et al (1976) established a stratigraphic sequence based on the character of seismic intervals on a series of multichannel profiles in the western Gulf. The lines tie DSDP Sites 2, 3, 87, 90, and 91. Preliminary correlation between the two areas indicates that the mapped interval termed Unit A in this paper correlates with the Pleistocene Sigsbee seismic unit of Watkins et al (1976) and Worzel and Burk (1978).

FAN STRATIGRAPHY

Seismic Interpretation

The USNS *Kane* survey covered the fan and adjoining provinces with a gridded network of seismic profile lines (Garrison and Martin, 1973). The profiles, even though single-channel, indicate that geological and recording conditions were optimum throughout much of the fan for continuous and deep reflections (to 4 sec, two-way time). The record quality over the fan and seaward of the diapirs is excellent to fair; correlations are reasonably accurate; but deterioration occurs over the crown of the fan and near the diapirs. As no core holes or DSDP sites have been drilled on the fan and only shallow penetrating piston or gravity cores exist, discussion of the sediments beneath the immediate surface must be considered speculative.

Loss of deep energy returns over the suprafan is attributed to: the thicker section; the heterogeneous composition of major slumps (Walker and Massingill, 1970); channel cut and fill; and local slides, slumps, and debris flow deposits. Correlations are lost near shallow diapirs. Along several lines the shallower reflectors can be carried only until a diapir is crossed that penetrates to the seafloor.

The University of Texas Marine Science Institute (UTMSI) recently recorded a southwest-northeast multichannel seismic profile that crossed the lower and middle parts of the fan. Figure 12 is a 610-km-long portion of that profile, over the fan. The profile was recorded to 12.0 sec, but only the upper 7.0 sec, containing Unit A, are reproduced here.

Seismic-Stratigraphic Units

With the available data, three units can be readily traced over most of the middle and lower fan, adjoining abyssal plain, and rise (Fig. 13). The nature and character of these units on regional seismic lines are shown by Moore et al (1978). Stuart and Caughey (1976) show similar relations on the middle and upper fan. However, this discussion is limited to the upper interval, Unit A, that constitutes the Mississippi Fan. This is basically equivalent to the Pleistocene-aged Sigsbee seismic unit of Watkins et al (1976).

Unit A regionally is divisible into upper and lower intervals. On Figure 12 the overall character of Unit A can be traced across most of the fan; however, it changes character and is difficult to follow on the eastern part of the lower fan. This we attribute primarily to a change in the character of the sediments with greater distance from the source area. However, we cannot totally discount an influx of carbonate material, caused by turbidity currents flowing out beyond the escarpment.

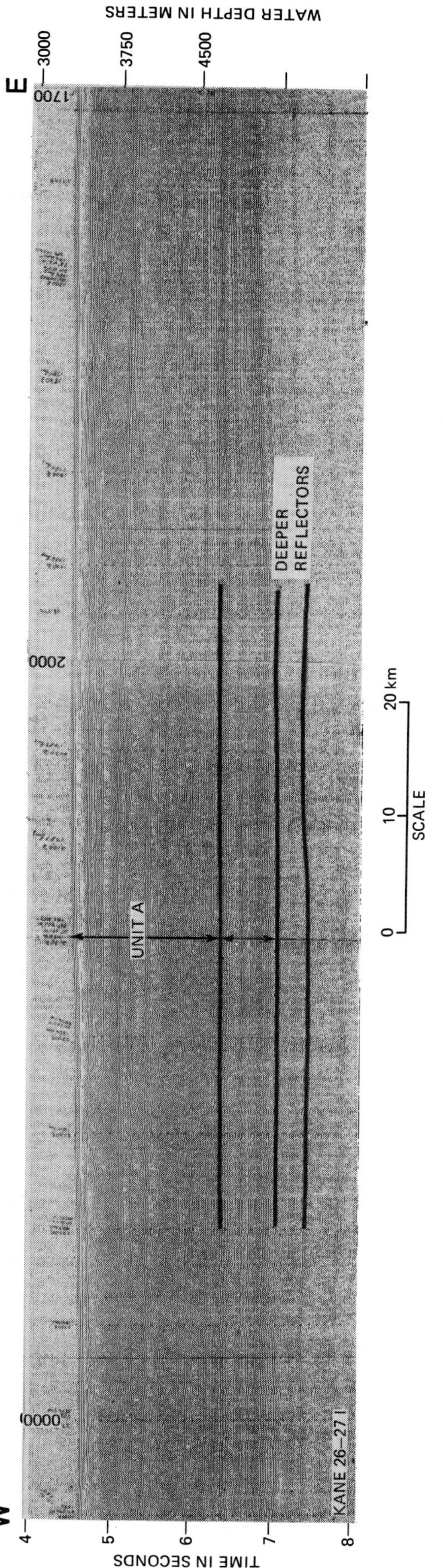

FIG. 13 Typical seismic pattern for Unit A. See Figure 2 for location.

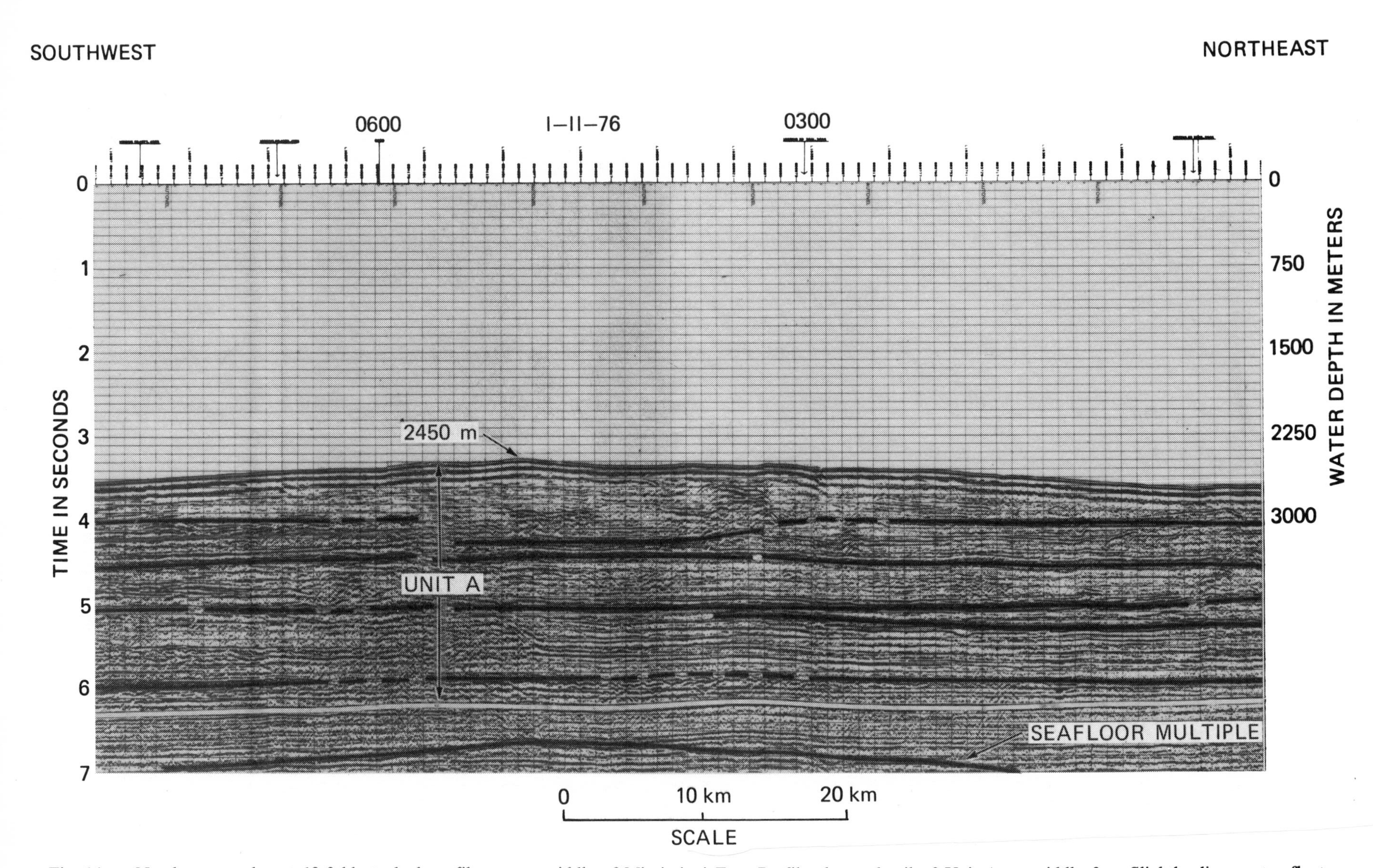

Fig. 14 Northeast-southwest 12-fold stacked profile across middle of Mississippi Fan. Profile shows detail of Unit A on middle fan. Slightly divergent reflectors on southwest end of profile between 5 and 6 sec suggests older middle fan may lie to west. Continuity of reflectors deteriorates from lower to upper portion of Unit A. For uninterpreted section, see Figure 12, upper profile. For location, see Figure 2.

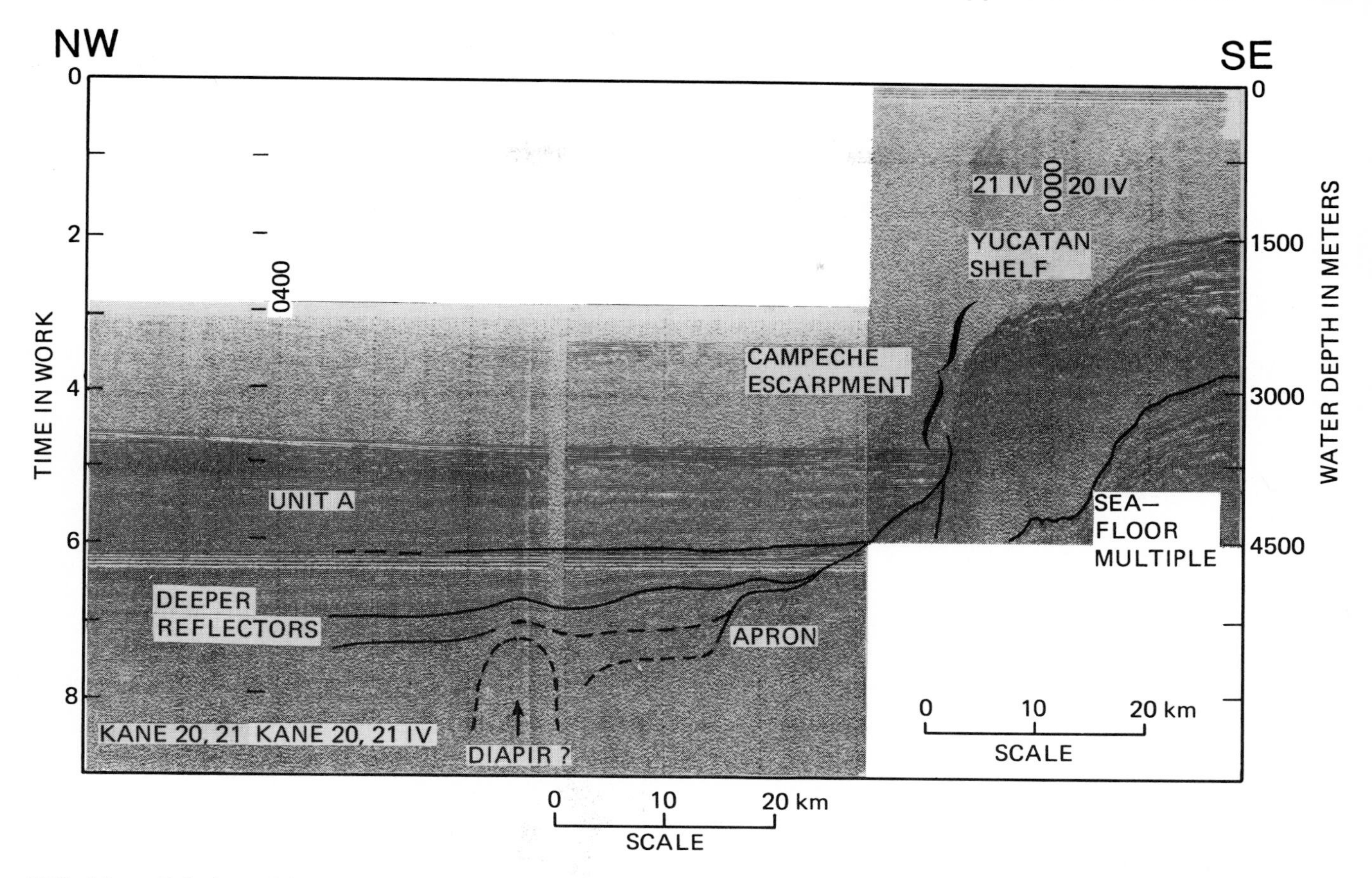

FIG. 15 Relation of lower part of Mississippi Fan to Campeche Escarpment. Airgun signature at 6.2 to 6.4 sec. For location see Figure 2.

The basal part of Unit A is characterized by a group of parallel reflectors (0.2 to 0.35 seconds record time), which we interpret to be turbidite sandstones or siltstones interbedded with hemipelagic sediments (Figs. 12, 13). This section is distinctive on much of the lower fan and in the Sigsbee and Florida Plains. Northeast of the suprafan, it appears to thin by onlap of older units and lose its typically strong reflection character. The interval is barely recognizable northeast of the middle fan.

Farther northeast this deterioration of record character extends upward into the unit. The lower part of the upper interval is gradually lost by regional onlap of the older section, influenced in part at least by sedimentation off the Florida Escarpment, and an undetermined amount of diapiric uplift. These intrusions are believed to be salt.

Areally, the upper part of Unit A is divisible into two distinct facies. One is proximal to the sources of sediment and, by our definition, is more than 10 percent disrupted seismic zones (Figs. 14, 15 [near escarpment], 16). The other consists of generally parallel (though discontinuous) reflections, separated by disturbed reflections or transparent zones (Figs. 12, 13, 16).

The distal facies of Unit A (less than 10% disrupted zones) becomes dominant beyond the suprafan and the immediate vicinity of the escarpments (Figs. 12, 15, 16). Overall, the entire unit is characterized by parallel to slightly divergent, discontinuous single or multiple reflectors.

The Florida and Campeche Escarpments are gradually being buried by Unit A (Figs. 12, 15). Where still in contact with the escarpments' debris apron, the lower part of A normally thins by onlapping, with a progressive updip loss of the deeper units. Where basin filling has buried the apron, the contact between an escarpment and Unit A is an abrupt (emphasized by vertical exaggeration) unconformity.

Whereas the contact with the Sigsbee Escarpment appears relatively sharp on many profiles, the deeper section is known to continue under the often strongly deformed shallow section (Humphris, 1978; Watkins et al, 1976). As suggested by several sections the uppermost interval may have continuity, particularly where the escarpment is not too precipitous and where diapirs are not too close to the seafloor (Moore et al, 1978).

Sediment Distribution

Over 200 time values for Unit A were measured from the USNS *Elisha Kane* seismic profiles. The network of lines was suitable for mapping the unit on all but the uppermost part of the fan.

Computer-generated time interval (one-way, Fig. 17) and isopach maps (Fig. 18) were constructed for

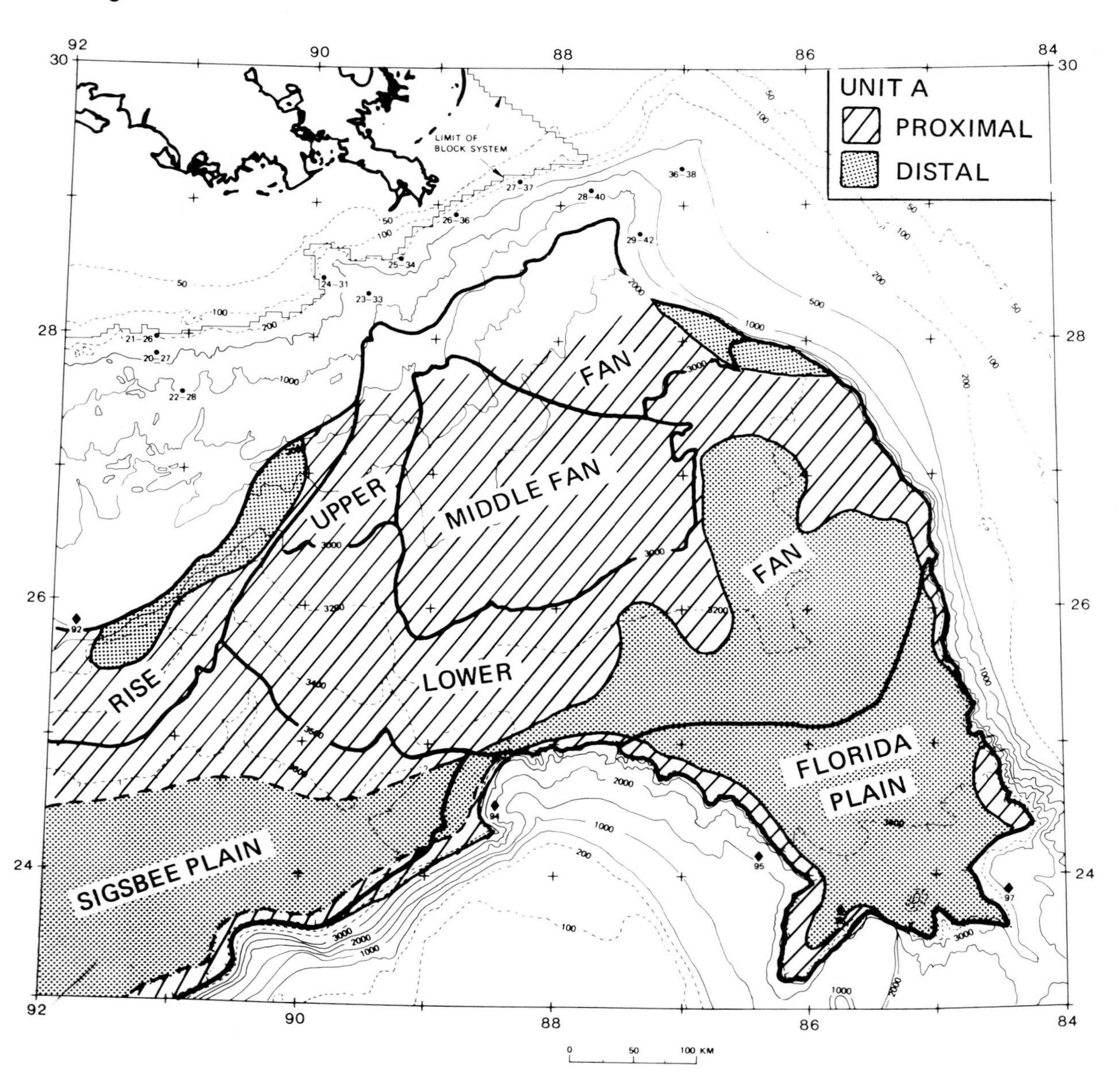

FIG. 16 Map showing proximal and distal facies of Unit A. Fan physiography shown for reference.

Unit A. Using refraction sonobuoy data from Houtz et al (1968) a velocity of 2.15 km/s was assigned to Unit A for computing the unit thickness (Moore et al, 1978). Because we used an average velocity in the central part of the fan, Unit A may be slightly thicker there than mapped (and thinner on the margins). Stuart and Caughey (1976, 1977) mapped the Pleistocene using an average velocity determined from stacking velocities in the northern Gulf. The differences in the maps are more a function of the control than of the velocity function used.

The outbowing of contours characterized by Unit A are not repeated in deeper, geologically older intervals (Moore et al, 1978, Worzel and Burk, 1978). Thus, we conclude that, while this region was receiving a moderate amount of sediment from the north prior to this time, the vast bulk of sediments constituting the middle and lower fan are represented by Unit A, particularly the upper part.

SEDIMENTATION RATES

The rates of sedimentation have varied significantly in the deep Gulf. This has been documented particularly well for the upper Miocene and younger strata cored by the DSDP (Table 1). However, the sites listed in the table are all on the periphery of the Mississippi Fan.

Pleistocene

On Figure 12 the section crosses the lower and middle fan obliquely from the Sigsbee Plain to the West Florida Terrace. This section shows that the upper part of Unit A thickens by a factor of two toward the middle fan. As the lower part of the sec-

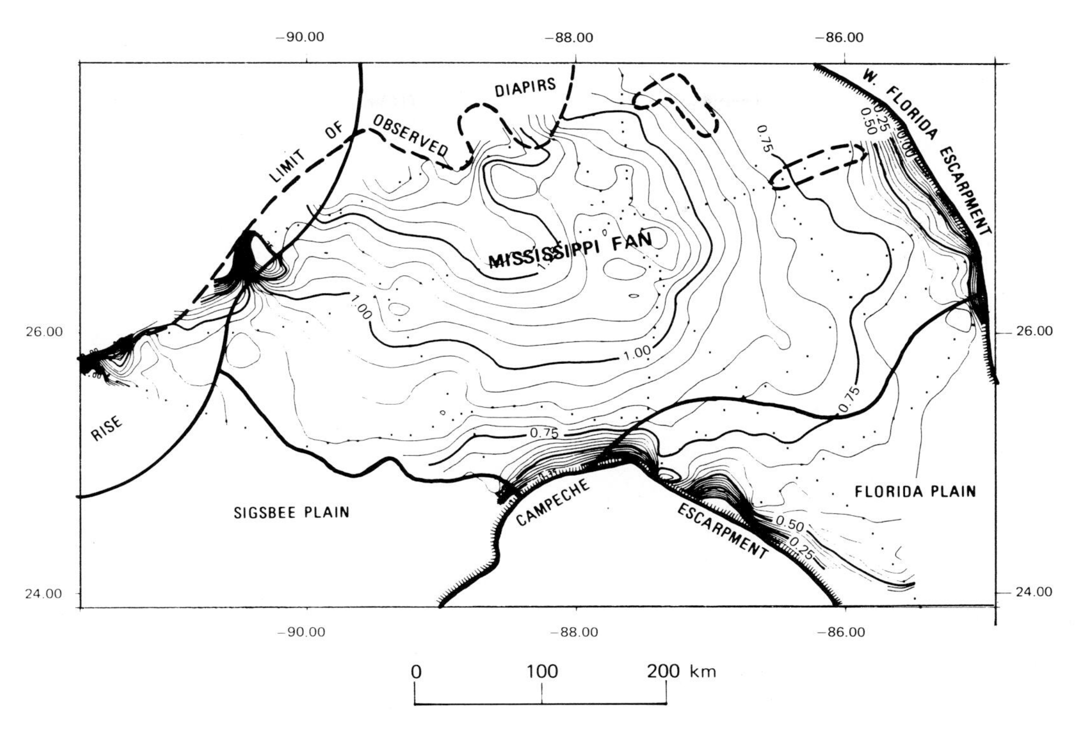

FIG. 17 Time-interval map of Unit A. One-way time in seconds. Heavy lines outline physiographic subdivisions. Data points shown by dots.

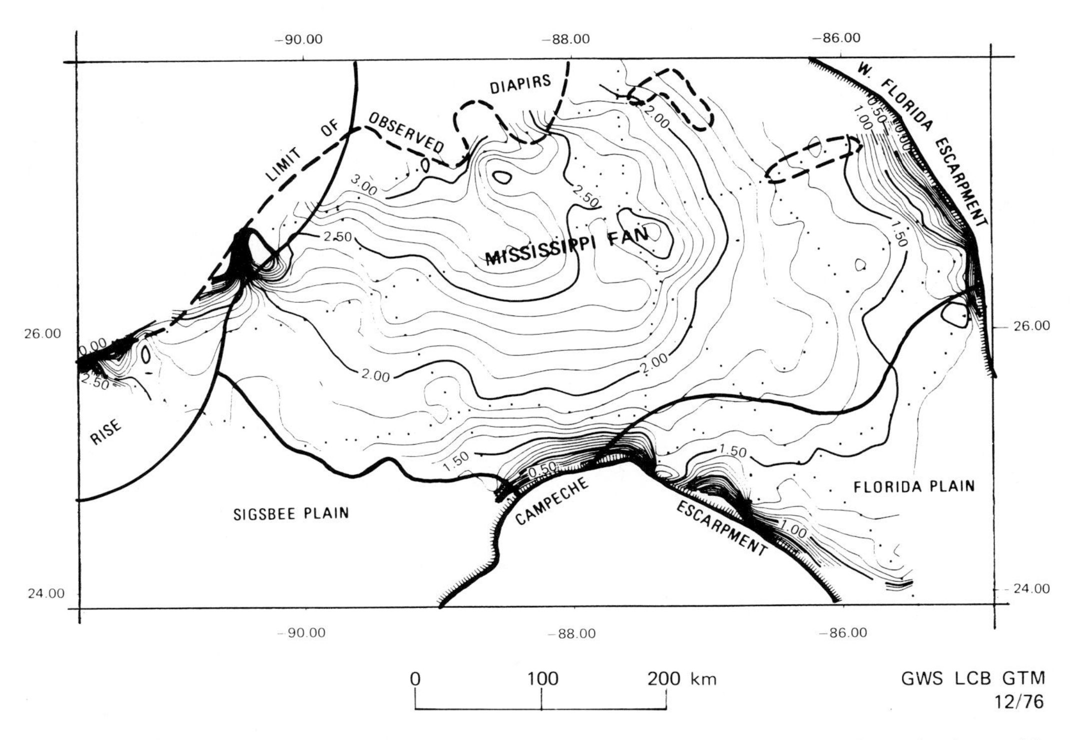

FIG. 18 Isopach of Unit A. Thickness in kilometers. Heavy lines outline physiographic units. Data points shown by dots.

Table 1. Sedimentation Rates in cm/10^3y at DSDP Sites Marginal to Mississippi Fan†

Province	Sigsbee Escarpment			Sigsbee Plain	
	South of Escarp.	On Escarp.	Southeast	Southeast	Center
Age of Sites	1	92	3/87	85	91
Late Pleistocene	38	32	28/32	31	*
Early Pleistocene			6		10
Pliocene			3.5		4
Upper Miocene			4.6		3.7

†From Ewing et al (1969) and Worzel et al (1973).
*Middle and late Pleistocene at Site 91.
Holocene - 40,000 B.P., 150 cm/10^3y; 40,000 - 60,000 B.P., 320 cm/10^3y; 60,000 - 600,000 B.P., 55 cm/10^3y (middle and early late Pleistocene).

tion is relatively uniform in thickness, the section does not represent uniform deposition throughout the Pleistocene. With this thickness variability in mind, we can compute an average thickness. Thus, the sedimentation rate over the last two million years for the thicker part of the middle fan, using a 2.15 km/s constant velocity, averaged 160 cm/10^3 years. This compares favorably with a 124 cm/10^3 years value suggested for the entire Pleistocene on the continental shelf from a number of petroleum company wells (Poag, 1971).

On a cross-section constructed from data at six core hole sites (Fig. 19), we show a correlation of the Pleistocene on the slope in this region, based upon both planktonic foraminifera (as temperature indicators) and the coiling directions of *Globorotalia truncatulinoides*. The coiling directions, however, do not appear to be diagnostic. Numbers indicate the environment of deposition, based on an interpretation of the benthonic foraminifera. The age of the sediments below the unconformity in Core Hole 24-31 is Pleistocene. This section shows that during late Wisconsin time a thick lobe of sediments was deposited beyond the present birdfoot delta of the Mississippi. The thickest section of the lobe is at Core Hole 25-34, which penetrated about 300 m of upper Wisconsin sediments, a minimum thickness for this interval. We do not have a radiometric date for the Wisconsin interstadial, so we cannot determine a rate of sedimentation here.

Holocene

Phleger (1960) determined the thickness of sediments containing a Holocene planktonic foraminiferal fauna in the northern Gulf of Mexico. Figure 20 shows the locations and thicknesses of Holocene sediments in 23 piston cores taken on the Mississippi Fan and others nearby. The plot shows Holocene sediments more than 200 cm thick on the outer shelf and upper continental slope in the vicinity of the Mississippi Delta. Thus, even though the Mississippi River has prograded across the continental shelf and is currently debouching its load of sediment near the shelf break, most of the 600 × 10^6 tons of sediment annually brought to the Gulf (NEDECO, 1959) is initially deposited on the shelf or slope, largely landward of the 1,000 m isobath. Holocene sediments on the upper and middle fan are thicker on the west side (nearer the effects of the Mississippi River) than on the east. On the lower fan these sediments are up to 167 cm thick; however, most are less than the 100 cm typical of deep Gulf sedimentation.

Berggren and van Couvering (1974) place the beginning of the Holocene at 10,000 Y.B.P. in the Netherlands. This date fits well with Curray's (1960) postglacial sea level curve for the Gulf, determined from radiocarbon dating of nearshore shells and a paleotemperature curve from an ice core in the Greenland Ice Cap (Poag, 1973). On the basis of this date, the Holocene sedimentation rate on the lower fan and adjoining plain was about 10 cm/10^3 years, and less than 20 cm/10^3 years on the middle and western upper fan. However, most of the thicknesses on the continental slope and upper fan are minimum values; the maximum sedimentation rate is certainly greater on the slope.

Holocene sedimentation rates on the lower and eastern upper fan and abyssal plain are about double those determined for the pre-Pleistocene at DSDP Sites 3/87 and 91 (Table 1). Rates for the remaining fan are somewhat less than for the late Pleistocene (Table 1). However, both rates are an order of magnitude less than those for the middle and late Pleistocene at DSDP Site 91 (on the Sigsbee Plain) and those from the seismic profile crossing the middle fan.

This discussion supports the results and observation by Milliman (1976) regarding Atlantic-type continental margins: that during periods of high sea level stands little sediment is introduced into the deep sea. The exceptions are where a river has prograded across the shelf or where the continental shelf is very narrow, as at the mouth of the Congo River. Even though the Mississippi River has nearly prograded across the broad shelf, these effects are only reflected in depositional rates on the upper con-

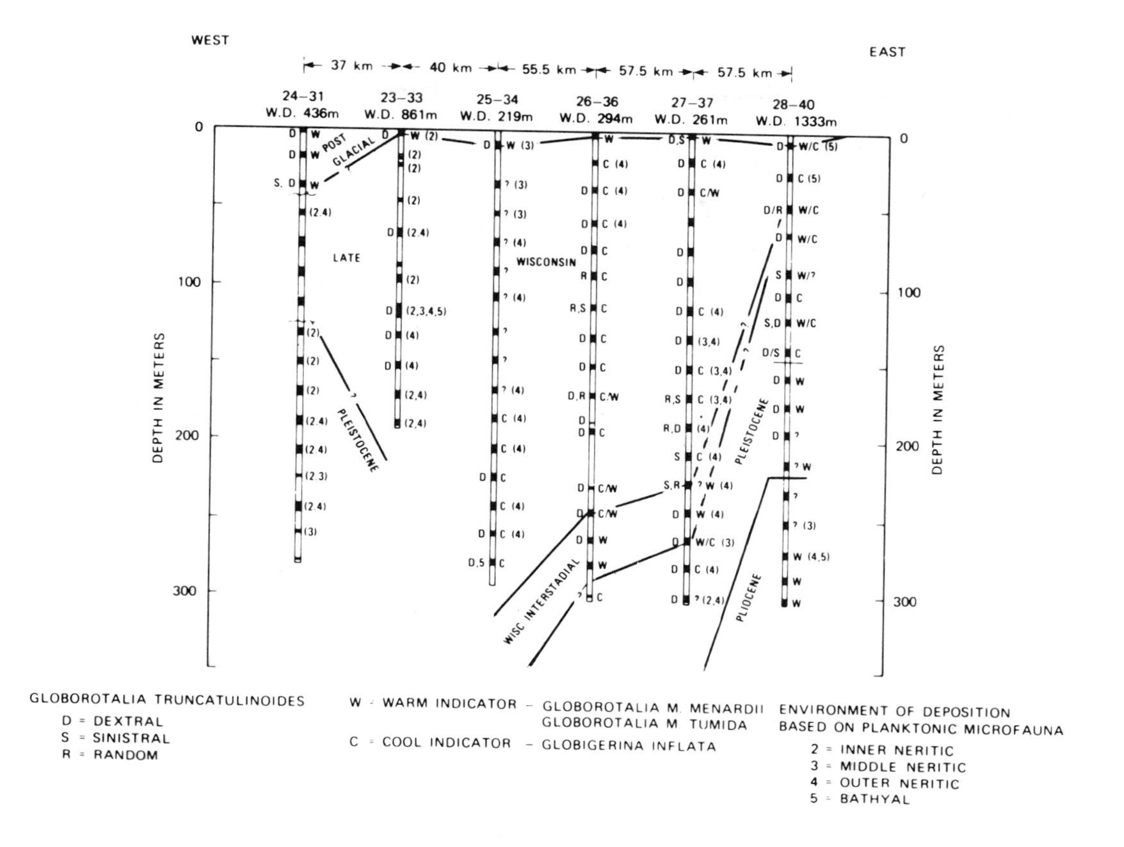

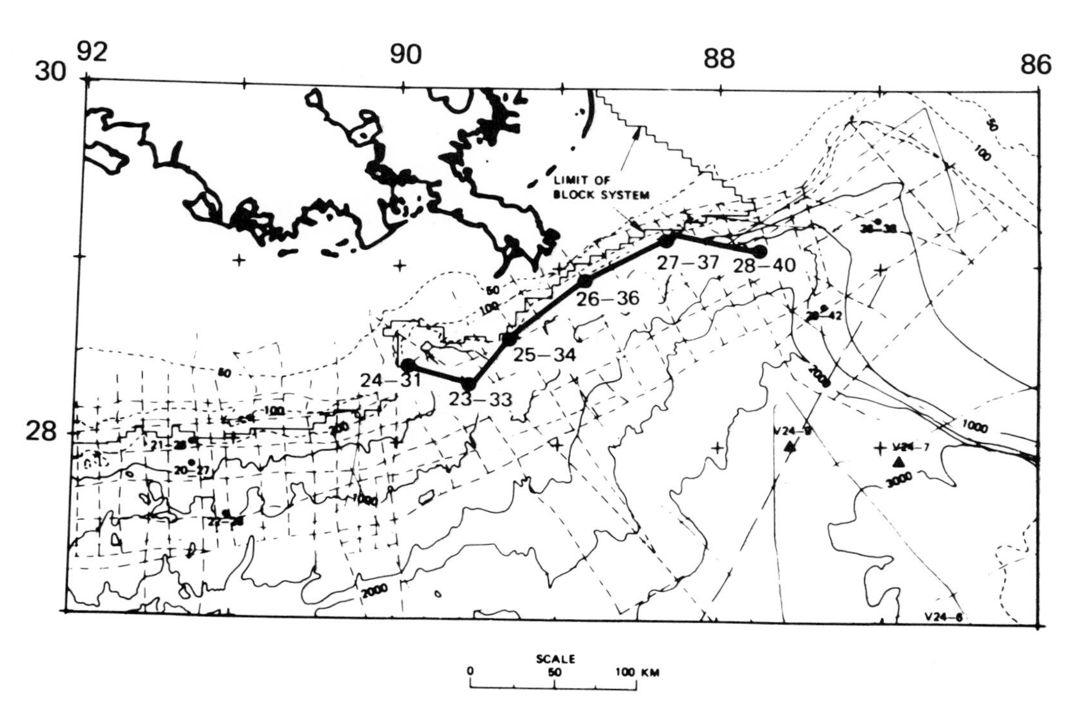

FIG. 19 Correlation of drill holes across Louisiana continental slope and in Mississippi Trough.

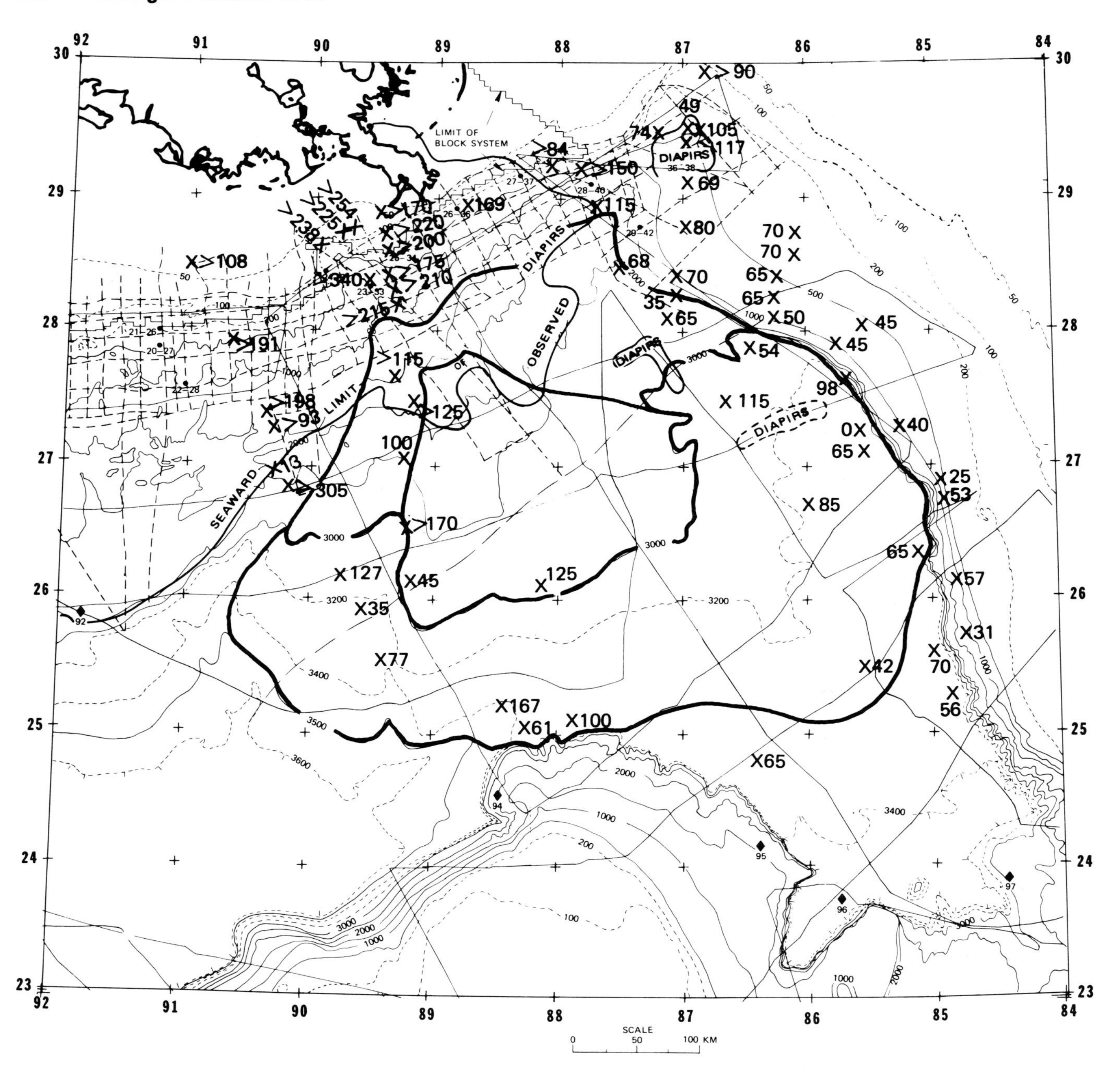

FIG. 20 Map showing location (X) and thickness of Holocene sediments (in cm) recovered in piston cores on and adjacent to Mississippi Fan. Data from Phleger (1960). In Core Hole 23-33, first interval cored was totally in Holocene (Fig. 6); thus, value is minimum. Heavy lines are fan subdivisions and boundaries.

tinental slope, upper, and possibly middle fan, not in the deep sea.

PETROLEUM POTENTIAL

Within this framework of data on fan stratigraphy, sedimentational patterns, and deformation, but keeping in mind our degree of ignorance, we can consider the petroleum potential of the fan. Caughey and Stuart (1976), Moore et al (1978), and Stuart and Caughey (1976) address this question. Table 2 presents facts regarding not only the stratigraphy, but the many other parameters, and evaluates each in its effects on the petroleum potential of the Pleistocene Mississippi Fan. Details of the material summarized here, plus references, are given in Moore et al (1978).

FAN HISTORY

The imprint of upper Wisconsin sedimentational patterns is very strong, and our knowledge of the great volumes of rocks constituting the Mississippi Fan is quite limited. These factors prevent making up a total chronological scenario of its evolution. However, using what data we have, we can achieve

Table 2. Petroleum Potential of Pleistocene Mississippi Fan

Parameter	Fan Position	Data	Interpretation
Reservoir type; Sedimentary environment	Upper Middle	High-energy. Channels, coarser parts of overflow deposits, slumps, and debris flow deposits.	Quality varies with grain size, particle/ matrix ratio, thickness, number of beds, areal extent, etc.
	Lower	Low-energy. Thin, fine-grained distal turbidites.	
	Upper	Folds, faults, sand pinchouts or terminations, channel sands on fan and in interdiapiric lows.	Familiar types of structures, with stratigraphic variations.
Traps	Middle	Disruption of channels and overbank deposits by later channeling, slumping, etc.	Stratigraphic, effective, but traps difficult to explore for.
	Lower	Sand pinchouts on diapirs or escarpments. Fining out of individual beds up- or downdip.	Stratigraphic, but poor economic risk.
Source Rocks (%organic carbon)		<1.0	Low, but normal for the Cenozoic Gulf Coast.
Age-Time		Pleistocene ≤2 m.y.	Younger sediments would require high geothermal gradient.
Depth of Burial		>3.0 km on Middle Fan.	Significant depocenter.
H. F. and Geothermal Gradient		Normal: 0.6 to 1.19 H.F.U.	Poor for Pleistocene sediments.

some perspective. This requires making assumptions as to (1) the complex interplay in an uplifted region of the continent; (2) a lowered sea level, with attendant increased river gradient; and (3) the enormous flows of glacial meltwaters that must have accompanied each deglaciation.

Regional uplift of the western part of the continent in the Pliocene and Pleistocene (Dunbar 1960) rejuvenated streams and rivers, causing them to begin transporting to the Gulf of Mexico the vast blanket of largely nonmarine sediments eroded off the Rocky Mountains through the Paleogene. Much of this sediment was transported to the northwestern and north-central Gulf by the ancestral Mississippi and other rivers.

Woodbury et al (1973) mapped the upper Pliocene and Pleistocene depocenters for the last 3.5 M.Y. on the continental shelf and upper slope. They showed that maximum deposition during this interval shifted over 320 km westward, from the present mouth of the Mississippi to near the present shelf edge south of the Texas-Louisiana border. Thus, while sedimentation in the west was prograding the continental shelf and upper slope 80 km southward, to the east most of the sediment was being carried into deep water during periods of lowered sea level. In this way the ancestral Mississippi constructed a fan on the abyssal plain more than 3 km thick (Fig. 18), largely over the last two million years.

REFERENCES CITED

Bates, C. C., 1953, Rational theory of delta formation: AAPG Bull., v. 37, no. 9, p. 2119–2162.

Bergantino, R. N., 1971, Submarine regional geomorphology of the Gulf of Mexico: Geol. Soc. America Bull., v. 82, no. 3, p. 741–752.

Berggren, W. A., and J. A. van Couvering, 1974, The late Neogene, biostratigraphy, geochronology, and paleoclimatology of the last 15 million years in marine and continental sequences: Amsterdam, Elsevier, 216 p.

Carsey, J. B., 1950, Geology of Gulf coastal area and continental shelf: AAPG Bull., v. 34, no. 3, p. 361–385.

Caughey, C. A., and C. J. Stuart, 1976, Where the potential is in the deep Gulf of Mexico: World Oil, v. 183, no. 1, July, p. 67–72.

CLIMAP Project Members, 1976, The surface of the ice-age Earth: Science, v. 191, no. 4232, p. 1131–1137.

Curray, J. R., 1960, Sediments and history of Holocene transgression, continental shelf, northwest Gulf of Mexico, *in* F. P. Shepard, F. B. Phleger, and T. H. van Andel, eds., Recent sediments, northwest Gulf of Mexico: Tulsa, AAPG, p. 221–266.

Dunbar, C. O. 1960, Historical geology: New York, John Wiley and Sons 500 p.

Ewing, J., J. Antoine, and M. Ewing, 1960, Geophysical measurements in the western Caribbean Sea and in the Gulf of Mexico: Jour. Geophys. Research, v. 65, no. 12, p. 4087–4126.

——— J. L. Worzel, and M. Ewing, 1962, Sediments and oceanic structural history of the Gulf of Mexico: Jour. Geophys. Research, v. 67, no. 6, p. 2509–2527.

Ewing, M., D. B. Ericson, and B. C. Heezen, 1958, Sediments and topography of the Gulf of Mexico, *in* L. G. Weeks, ed., Habitat of oil: Tulsa, AAPG, p. 995–1053.

——— et al, 1969, Initial reports of the Deep Sea Drilling Project: Washington, D.C., U.S. Govt., vol. 1, 672 p.

Fisk, H. N., and E. McFarlan, Jr., 1955; Late Quaternary deltaic deposits of the Mississippi River, *in* A. Poldervaart, ed., Crust of the earth: Geol. Soc. America Spec. Paper 62, p. 279–302.

Garrison, L. E., and R. G. Martin, Jr., 1973, Geologic structures in the Gulf of Mexico basin: U.S. Geol. Survey Prof. Paper 773, 85 p.

Houtz, R., J. Ewing, and X. Le Pichon, 1968, Velocity of deep-sea sediments from sonobuoy data: Jour. Geophys. Research, v. 73, no. 8, p. 2615–2641.

Huang, T. C., and H. G. Goodell, 1970, Sediments and sedimentary processes of eastern Mississippi Cone, Gulf of Mexico: AAPG Bull., v. 54, no. 11, p. 2070–2100.
Humphris, C. C., Jr., 1978, Salt movement on the continental slope, northern Gulf of Mexico; *in* A. H. Bouma, G. T. Moore, and J. M. Coleman, eds., Framework, facies and oil trapping mechanisms on upper continental margin: AAPG Studies in Geology, No. 7.
Lehner, P., 1969, Salt tectonics and Pleistocene stratigraphy on continental slope of northern Gulf of Mexico: AAPG Bull., v. 53, no. 12, p. 2431–2479.
Martin, R. G., Jr., 1976, Geologic framework of northern and eastern continental margins, Gulf of Mexico, *in* A. H. Bouma, G. T. Moore, and J. M. Coleman, eds., Beyond the shelf break: AAPG Marine Geology Committee Short Course, p. A-1—A-28.
——— and A. H. Bouma, 1978, Physiography of Gulf of Mexico, *in* A. H. Bouma, G. T. Moore, J. M. Coleman, eds., Framework, facies, and oil trapping characteristics on upper continental margin: AAPG Studies in Geology, No. 7.
Milliman, J. D., 1976, Late Quaternary sedimentation on Atlantic continental margins and the deep sea, *in* Continental margins of Atlantic type: Anais da Acad. Brasileira de Ciências (Sao Paulo), v. 48, Suplemento, p. 199–206.
Moore, G. T., 1969, Interaction of rivers and oceans—Pleistocene petroleum potential: AAPG Bull., v. 53, no. 12, p. 2421–2430.
——— et al, 1978, Mississippi fan, Gulf of Mexico—Physiography, stratigraphy, and sedimentational patterns, *in* A. H. Bouma, G. T. Moore, and J. M. Coleman, eds., Framework, facies, and oil trapping characteristics on upper continental margin: AAPG Studies in Geology, No. 7.
NEDECO, 1959, River studies and recommendations on improvement of Niger and Benue: The Hague, Netherlands Engineering Consultants, 1000 p.
Newman, J. W., P. L. Parker, and E. W. Behrens, 1971, Quaternary sediments from the Gulf of Mexico: characterization by organic carbon isotope ratios (abs.): Geol. Soc. America & Assoc. Soc. Mtg., Abs. with Program, v. 3, no. 7, p. 658–659.
Normark, W. R., 1970, Growth patterns of deep-sea fans: AAPG Bull., v. 54, no. 11, p. 2170–2195.
Osterhoudt, W. J., 1946, The seismograph discovery of an ancient Mississippi River channel (abs.): Geophysics, v. XI, no. 3, p. 417.
Phleger, F. B., 1960, Sedimentary patterns of microfaunas in northern Gulf of Mexico, *in* F. P. Shepard, F. B. Phleger, and T. H. van Andel, Recent sediments, northwest Gulf of Mexico: Tulsa, AAPG, p. 267–301.
Poag, C. W., 1971, A reevaluation of the Gulf Coast Pliocene-Pleistocene boundary: Gulf Coast Assoc. Geol. Socs. Trans, v. 21, p. 291–308.
——— 1973, Late Quaternary sea levels in the Gulf of Mexico: Gulf Coast Assoc. Geol. Socs. Trans., v. 23, p. 394–400.
Sangree, J. B., et al, 1976, Recognition of continental slope seismic facies offshore Texas-Louisiana, *in* A. H. Bouma, G. T. Moore, and J. M. Coleman, eds., Beyond the shelf break: AAPG Marine Geology Committee Short Course, p. F-1–F-54.
Shih, T., J. L. Worzel, and J. S. Watkins, 1978, Northeast extension of Sigsbee scarp, Gulf of Mexico: AAPG Bull., in press.
——— et al, 1975, Origin of giant slump features on the Mississippi Fan, Gulf of Mexico (abs.): EOS, v. 56, no. 6, p. 372.
Stuart, C. J., and C. A. Caughey, 1976, Form and composition of the Mississippi fan: Gulf Coast Assoc. Geol. Socs. Trans., v. 26, p. 333–343.
——— ——— 1977, Seismic facies and sedimentology of terrigenous Pleistocene deposits in the northwest and central Gulf of Mexico: AAPG Memoir 26, p. 249–275.
Uchupi, E., 1967, Bathymetry of the Gulf of Mexico: Gulf Coast Assoc. Geol. Socs. Trans., v. 17, p. 161–172.
Walker, J. R., and J. V. Massingill, 1970, Slump features on the Mississippi fan, northeastern Gulf of Mexico: Geol. Soc. America Bull., v. 81, no. 10, p. 3101–3108.
Watkins, J. S., J. L. Worzel, and J. W. Ladd, 1976, Deep seismic reflection investigation of occurrence of salt in Gulf of Mexico, *in* A. H. Bouma, G. T. Moore, and J. M. Coleman, eds., Beyond the shelf break: AAPG Marine Geology Committee Short Course, p. G-1–G-34.
Wilhelm, O., and M. Ewing, 1972, Geology and history of the Gulf of Mexico: Geol. Soc. America Bull., v. 83, no. 3, p. 575–600.
Woodbury, H. O., et al, 1973, Pliocene and Pleistocene depocenters, outer continental shelf, Louisiana and Texas: AAPG Bull., v. 57, no. 12, p. 2428–2439.
——— J. H. Spotts, and W. H. Akers, 1976, Gulf of Mexico continental slope sediments and sedimentation, *in* A. H. Bouma, G. T. Moore, and J. M. Coleman, eds., Beyond the shelf break: AAPG Marine Geology Committee Short Course, p. C-1–C-28.
Worzel, J. L., et al, 1973, Initial reports of the Deep Sea Drilling Project: Washington, D.C., U.S. Govt., v. 10, 748 p.
——— and C. A. Burk, 1978, The Margins of the Gulf of Mexico: this volume.

The Margins of the Gulf of Mexico[1]

J. LAMAR WORZEL[2] AND C. A. BURK[2]

Abstract Preliminary results of a long-term program to investigate the tectonic history of the Gulf of Mexico are summarized principally through multichannel seismic reflection methods. Knowledge of the deep part of the Gulf contributes significantly to understanding the evolution of its continental margins. Reasonable and usual stratigraphic principles, based on marine reflection data, can be used successfully to interpret the geological history of such regions.

The stratigraphic unit which contains the Late Jurassic salt responsible for the diapirs of the deep Gulf of Mexico can now be recognized throughout this region, and diapirs have now been located as far east as the base of the West Florida Escarpment. Basement structural arches which seem responsible for the reef growth controlling the Florida and Campeche scarps are also present in the Yucatan Straits. All post-Jurassic units of the deep Gulf pinch out by depositional overlap against the Florida Escarpment to the east, against the Campeche Escarpment to the south, and against the newly discovered basement feature in the Strait of Florida. There is no evidence of faulting associated with these scarps.

The post-Jurassic through Miocene sediments of the deep Gulf originally continued north of the Sigsbee scarp, and completely across the Mexican Ridges to the west. Deformation by detachment sliding and diapirism progressed southward on the United States margin throughout the Cenozoic and occurred suddenly in the folded Mexican Ridges in late Pliocene or Pleistocene time. There is no evidence in the deep Gulf of Mexico of pre-Pleistocene deep-sea cones. The large deep-sea cone of the present Mississippi River is a prominent and unique feature of the Gulf.

INTRODUCTION

In September of 1975, the Galveston Geophysics Laboratory of the University of Texas Marine Science Institute began a long-term study of the tectonics of the Gulf of Mexico. The successful application of multichannel seismic reflection techniques made such a study possible to depths of 6 to 8 km below the sea floor (Watkins, et al, 1975; Ladd, et al, 1976a, 1976b, 1976c; Watkins, et al, 1976).

To date, we have collected 4,000 nautical miles of 12-fold reflection data and have surveyed 7 detailed grids consisting of 4,000 nm of 24-fold data (Fig. 1). The regional traverses have been arranged to intersect at the Challenger Knoll, which was drilled (Hole #2) during the first leg of the Deep Sea Drilling Project (Burk, et al, 1969; Ewing, et al, 1969). The most recent data are still being processed and interpreted, but we will attempt here to summarize some of the more obvious conclusions which already can be drawn.

SEISMIC STRATIGRAPHY

It became obvious in the early phases of study that the multichannel seismic reflection technique permitted recognition of stratigraphic units throughout very large oceanic regions. Not only could these units be followed continuously, but their internal acoustic characteristics were also consistent and recognizable over large regions. Consequently, these units were named and defined to facilitate discus-

[1]Manuscript received, October 13, 1977; accepted, March 28, 1978.

[2]University of Texas Marine Science Institute, 700 The Strand, Galveston Texas, 77550.

We gratefully acknowledge financial and material contribution to this work by the following: Cecil and Ida Green; Amoco Production Company; Chevron, U.S.A.; Continental Oil Company; Exxon Production Research Company; General Crude Oil Company; Gulf Energy and Minerals Company, U.S.; Mobil Oil Corporation; Phillips Petroleum Corporation; Shell Oil Company; Texaco Incorporated; Teledyne Exploration; Texas Instruments Company; U.S. Geological Survey, and Western Geophysical Company.

The assistance of the officers and crew of R/V *Ida Green*, Otis Murray commanding, is sincerely appreciated.

Critical reading of the manuscript by F. Jeanne Shaub and Richard T. Buffler was very helpful.

Many staff members of the Galveston Geophysics Laboratory, Marine Science Institute participated in various phases of the work for which we are grateful; particularly noteworthy were the contributions of Buffler, Houston, Ladd, Shipley, Watkins, Shaub, and Kunselman.

University of Texas Marine Science Institute contribution no. 289, Geophysical Laboratory.

Article Identification Number:
0065-731X/78/MO29-0027/$03.00/0.
(see copyright notice, front of book)

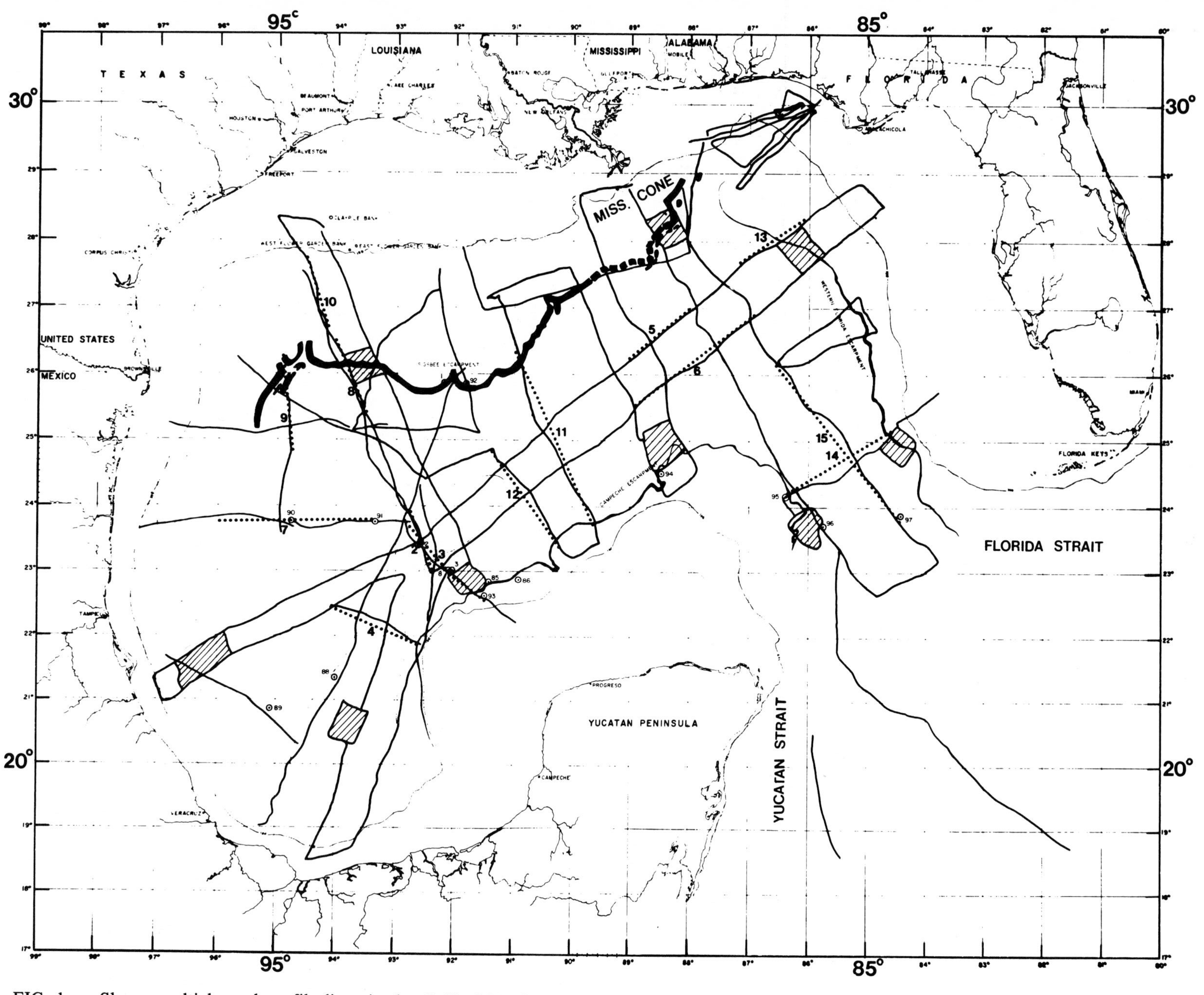

FIG. 1 Shows multichannel profile lines in the Gulf of Mexico. Circles with numbers show locations of the drilled holes. The cross-hatched areas are where 10-mile-spaced grids were observed. Some of the multichannel data are still being processed. Dotted lines show profile sections reproduced in figures, numbers of which are adjacent.

Table 1. Seismic Units, Southwestern Gulf of Mexico[1]

Unit	Seismic Characteristics	Inferred Lithology	Inferred Age
Sigsbee	Strong persistant multiple reflections interrupted within the Mississippi Cone.	Multiple turbidites with fractures resembling slumps within the Mississippi Cone. The most extensive turbidites have their distal edges at the western termination of the Abyssal Plain and in the approaches to Florida Strait.	Pleistocene and Holocene.
Cinco de Mayo	Weak, laterally persistent reflections beneath Sigsbee Plain; contains strong, discontinuous reflections to west in Mexican Ridges area. Prominent zone of large-scale cross-beds in middle of unit.	Hemipelagic silty clays and fine-grained, thin turbidites beneath Sigsbee Plain. Grades west into dominantly sandy turbidites in Mexican Ridges area.	Late Miocene and Pliocene
Mexican Ridges	Strong, discontinuous reflections in upper part; reflectionless (transparent) zone in lower part; lower boundary is sequence of strong, discontinuous reflections.	Dominantly sandy turbidites in upper part; fine-grained turbidites and hemipelagic sediments in lower part; possible sandy turbidite sequence at base.	Early Tertiary? to middle Miocene
Campeche	Generally weak reflections or reflectionless (transparent); strong, discontinuous reflections near bottom.	Mostly fine-grained, homogeneous pelagic or hemipelagic sediments (shales); possible turbidites toward base.	Middle Cretaceous? to early Tertiary?
Challenger	Top marked by strong reflection sequence.	Upper part possible carbonate turbidites equivalent to Early Cretaceous carbonate banks rimming Gulf. Lower part contains Jurassic salt in Central Gulf.	Jurassic? to middle Cretaceous?

[1]Modified from Ladd et al (1976a)and Watkins et al (1975, 1976).

sion—just as in onshore stratigraphy (Ladd, et al, 1976b). In many cases, especially where these units were sampled during the Deep-Sea Drilling Project, their ages and lithologies could be estimated.

The stratigraphic units which the writers recognize in the Gulf of Mexico (Table 1) are the Challenger Unit at the base, which includes the salt layers of the Gulf and is believed to be Jurassic in age. The overlying Campeche Unit is believed to be entirely Cretaceous in age. The Mexican Ridges Unit consists of Paleogene turbidites and is as young as mid-Miocene. The Cinco de Mayo Unit is late Miocene and Pliocene in age and consists of pelagic muds and turbidites. The Sigsbee Unit consists of Pleistocene and Holocene turbidites.

CENTRAL GULF OF MEXICO

Much of the Sigsbee Abyssal Plain is characterized by abundant salt diapirs which form knolls and domes (Fig. 2). These features extend from the Tabasco-Campeche Salt Province of Mexico, through the Sigsbee Abyssal Plain to the foot of the Florida Escarpment. The Jurassic Challenger Unit, which contains the salt, can now be mapped throughout the entire deep part of the Gulf of Mexico.

Reflectors beneath the Challenger Unit are observed in a number of localities and are absent in others. Locally they are essentially parallel to the surface of the Challenger Unit and elsewhere they are independently dipping. This suggests that the Challenger Unit rests on an unconformable surface and that pre-Jurassic sedimentary rocks may be present locally.

The Challenger Knoll was drilled and named during the first leg of the Deep Sea Drilling Project. Cores contained typical caprock minerals, sulphur, oil, spores, and pollen which indicate a probable Late Jurassic age (Burk, et al, 1969; Ewing, et al, 1969). The Challenger Knoll shows a prominent rim syncline, filled largely with the Mexican Ridges Unit (Fig. 3). This indicates that salt movement was taking place at least from latest Cretaceous to mid-Miocene times. There also is an erosional discontinuity at the base of the Sigsbee Unit, indicating that salt movement was continuing at least in the Pleistocene. It is likely that salt movement may still be active at Challenger and other knolls and domes (Ewing and Ewing, 1962). To the south, the overlap of all younger units onto the base of the Campeche Escarpment (Fig. 3) is apparent.

The writers obtained several sections which seem to show the roots of these diapirs. One in particular shows flow from the Challenger Unit into a dome, whereas the bottom of this unit appears to continue beneath the domes substantially undisturbed (Fig. 4). This is one of the best indications that the Challenger Unit contains the salt which forms the abundant diapirs of the deep Gulf of Mexico.

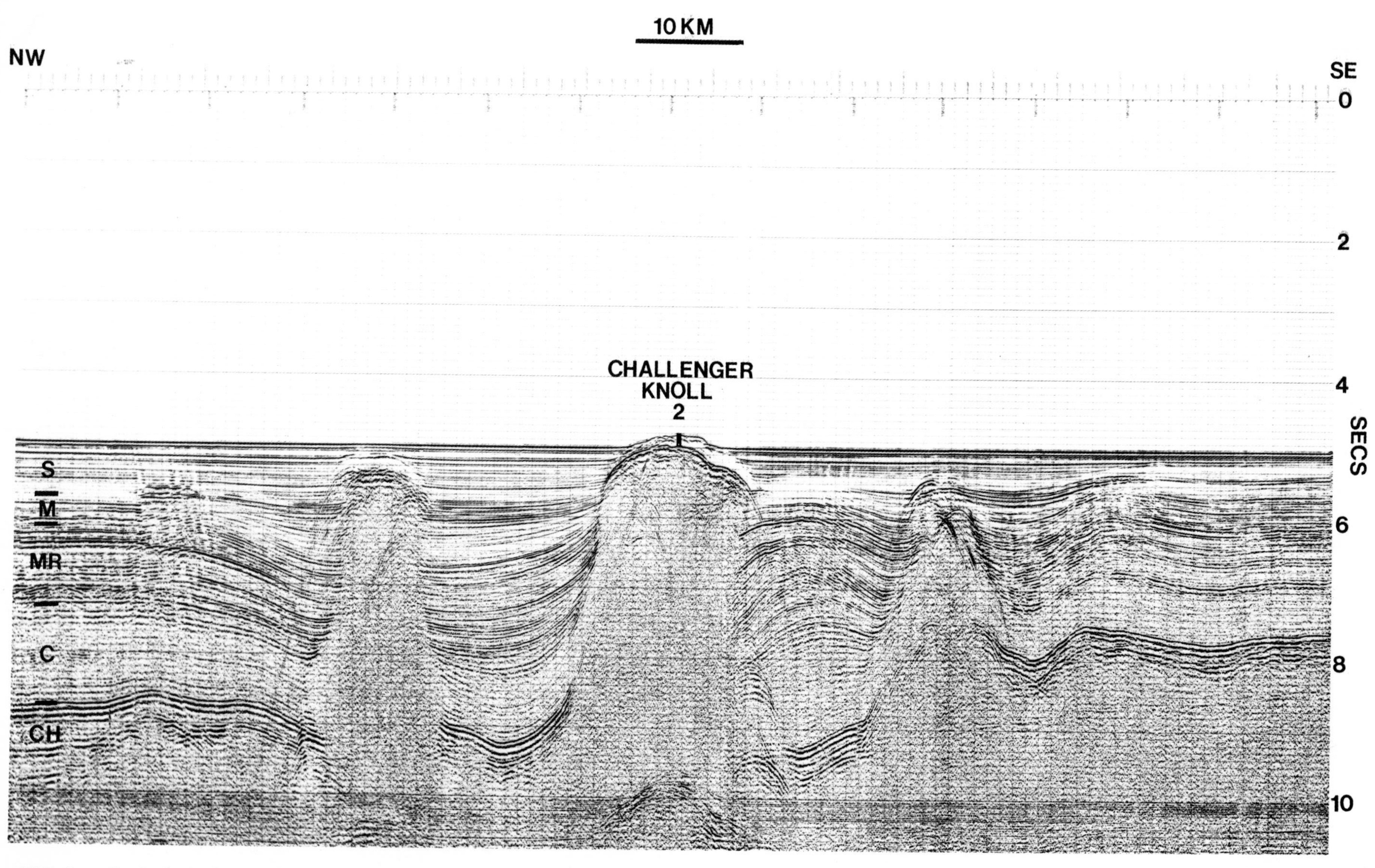

FIG. 2 Stacked 12-fold section in the Sigsbee Knolls region. Challenger Knoll was drilled in Hole 2 of Leg 1 of the Deep Sea Drilling Project. Time scale is two-way reflection time. Seismic stratigraphic units are as follows: S, Sigsbee; M, Cinco de Mayo; MR, Mexican Ridges; C, Campeche; CH, Challenger; and B, Seismic Basement.

The Campeche Unit is largely acoustically transparent, although locally there are prominent reflectors. It is considered to be mostly pelagic deposits of Cretaceous age. It is thickest in the northwestern Sigsbee Abyssal Plain, thinning sharply toward the Campeche (south) and Florida (east) scarps, gently toward the Sigsbee Escarpment (north), and thickens slightly toward the Mexican slope (west). This suggests possible sedimentary sources for the Campeche Unit from the east coast of Mexico, as well as ubiquitous Cretaceous pelagic sedimentation.

The Mexican Ridges Unit is thickest in rim synclines associated with the Sigsbee diapirs, suggesting that this diapir province was well established in the early Tertiary. Regionally, however, this unit has the most uniform thickness throughout the deep Gulf of Mexico. The best present interpretation is that the Mexican Ridges Unit is composed largely of turbidites originating mostly from the west and north, related to Laramide orogenic and volcanic influences onshore.

The Cinco de Mayo Unit is relatively thin and largely transparent acoustically. In Deep Sea Drilling Project holes, it consists predominantly of pelagic muds and low-energy turbidites of late Miocene to Pliocene age. It thickens slightly from the central Gulf both to the north and west, and thins to the east and south, suggesting that the pelagic sediments have been added to by turbidite sources from the north and west during the waning phase of sedimentation in the Mexican Ridges Unit.

The Sigsbee Unit shows the greatest variation in thickness in the deep Gulf of Mexico. It fills the bathymetric lows, smoothing the topography to form the Sigsbee Abyssal Plain and the Mississippi Deep Sea Fan. In drill holes (Fig. 1), it is made up of many closely spaced seismic reflectors identified as Pleistocene turbidites. In the northern part of the abyssal Gulf of Mexico the Sigsbee Unit has a thickness of about 3 km, adjacent to the Mississippi Submarine Canyon (Fig. 5). The individual layers, as well as the total interval, appear to thin to the east, west, and south from this region. Note the uniform thickness of the underlying Tertiary and Cretaceous units. Farther south, the Mississippi Cone is greatly suppressed and tapers gently to both the east and the west, as well as to the south (Fig. 6).

The most surprising and important aspect of this is that the Pleistocene to Recent Mississippi Cone is the only such feature that can be recognized in all of the sediments of the deep Gulf of Mexico. At present, it must be considered that previous cones must have been trapped entirely on the continental slope or shelf.

WESTERN GULF OF MEXICO

The eastern continental slope of Mexico is marked by elongate topographic ridges, parallel to the strike of the coastline (Ewing, et al, 1960; Bryant, et al, 1968; Ensminger and Matthews, 1972; Wilhelm and Ewing, 1972). These are known to be the sea-floor expression of anticlines which involve great thicknesses of sediments. Our multifold seismic surveys

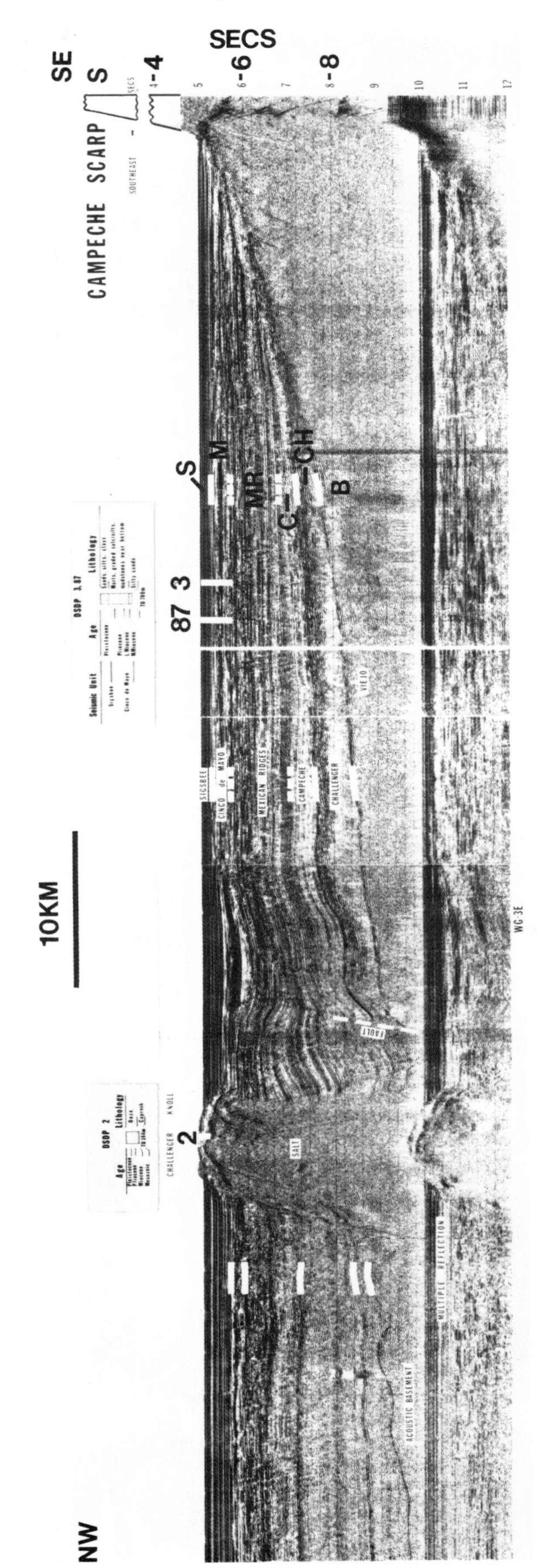

FIG. 3 12-fold reflection section across Challenger Knoll to the Campeche Scarp. Note thickening of Mexican Ridges Unit in the Rim Syncline discussed in text. Seismic stratigraphic units as in Figure 2.

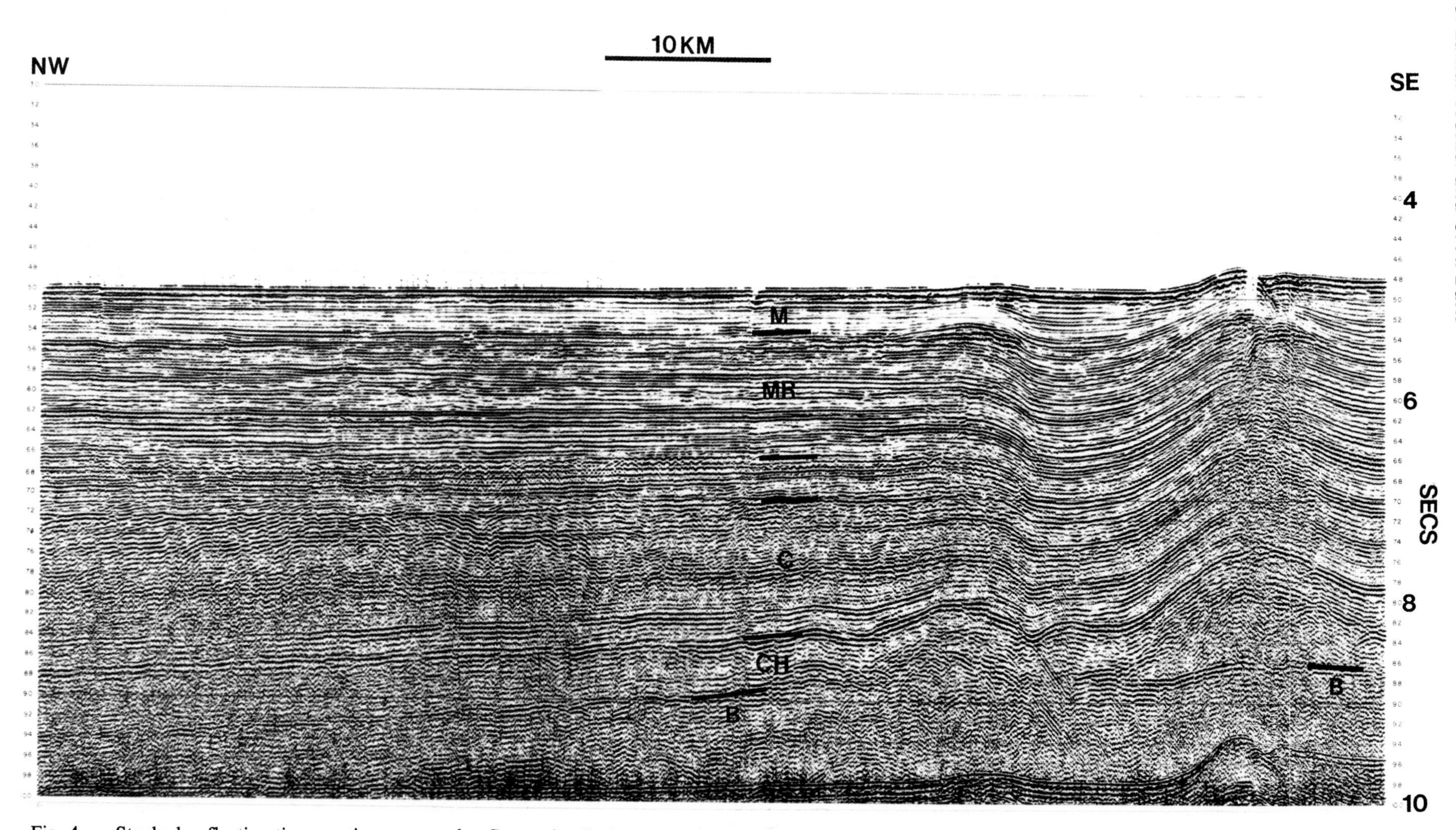

Fig. 4 Stacked reflection-time section across the Campeche-Sigsbee Salt Dome Province south of the Sigsbee Knolls. The upper boundary of the Challenger Unit shows salt tectonics while the lower boundary shows none. This is the best evidence that the salt flow is from a part of the Challenger Unit. Seismic stratigraphic units as in Figure 2.

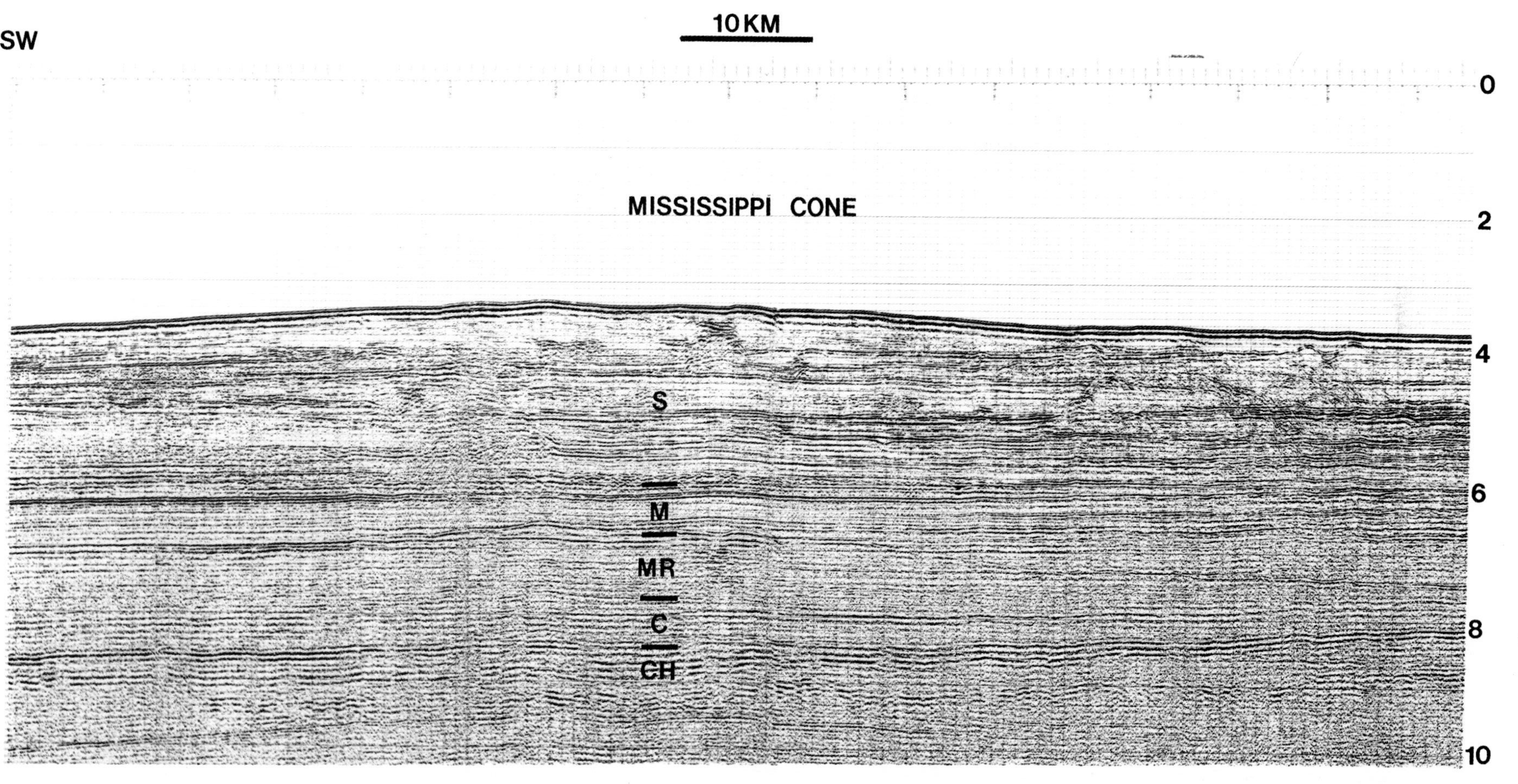

FIG. 5 Northerly reflection-time section across the Mississippi Cone. Note that there is no evidence of a cone earlier than the Sigsbee Unit, which is Pleistocene and Holocene in age. Seismic stratigraphic units as in Figure 2.

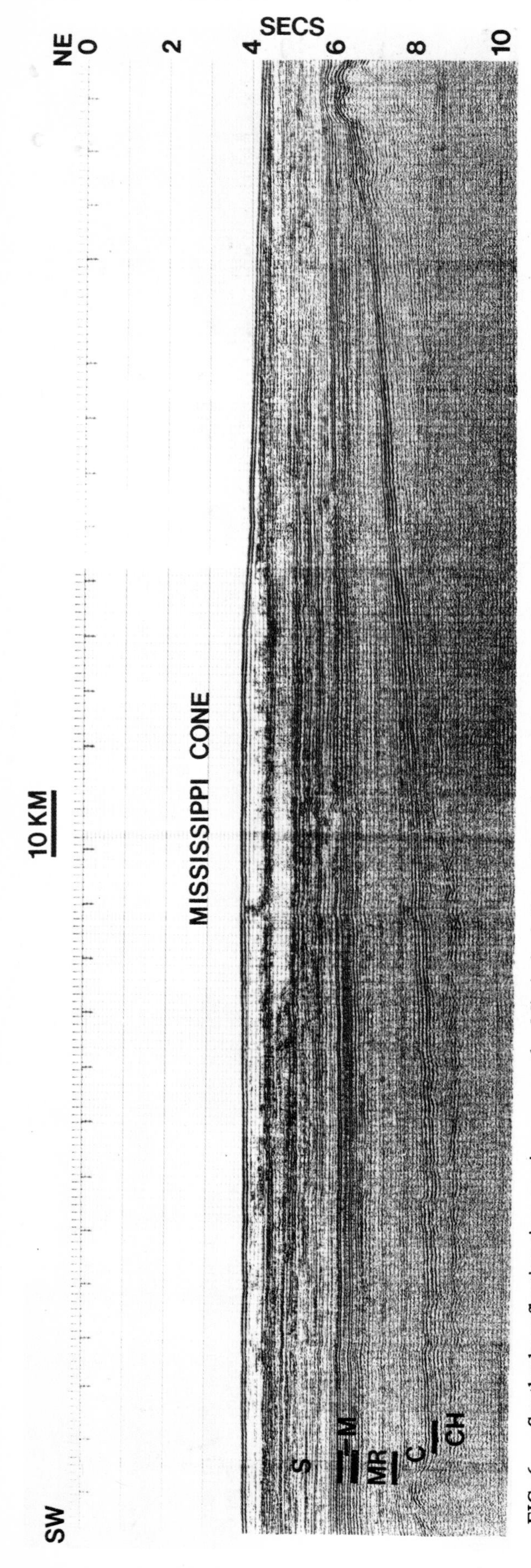

FIG. 6 Southerly reflection-time section across the Mississippi Cone. Here the cone has a broad, nearly level crest. It thins appreciably to the east, west, and south of this section. Note affects of salt tectonics at the top of the Challenger Unit and the Campeche Unit in the bottom right corner. Seismic stratigraphic units as in Figure 2.

have shown that the Mexican Ridges Unit (Paleogene) and the Cinco de Mayo Unit (Miocene-Pliocene) are clearly involved in the folding (Fig. 7). This relationship is even clearer in other traverses, where these two units can be followed without change in thickness or acoustic character into the Mexican Ridges fold belt.

These observations suggest that the Mexican Ridges fold belt is the result of deformation which occurred in late or post-Pliocene times. The symmetry and alignment of the folds suggests that they are décollement features, detached along Cretaceous clays or Jurassic evaporites, or both. Whether these anticlines also include Cretaceous as well as Tertiary sediments, they must be considered as potentially major petroleum producers, and their very large dimensions suggest that they may contain enormous quantities of hydrocarbons.

NORTHERN GULF OF MEXICO

The Sigsbee Escarpment is one of the most prominent features in the Gulf of Mexico, separating the northern continental slope from the deep continental rise and abyssal plain. The characteristic sedimentary relationships are shown in Figure 8. However, in nearby sections as first pointed out by Amery (1969) and confirmed by several of our sections, deep stratigraphic units can be followed landward of the scarp for distances up to 18 km.

Thus, it appears that these sediments recently continued across what is now the Sigsbee Escarpment. The scarp appears to be associated with a series of parallel ridges, which seem to grow outward. Early single-channel data have already indicated such an evolution of the Sigsbee scarp (Shih, et al, 1977). Figure 9 shows a newly formed ridge, with only minor surface expression, but which as elevation continues, would be responsible for the outward growth of the Sigsbee Escarpment.

On Figure 10, northwest along a nearby line of traverse, a large filled valley occurs at a water depth of about 1,500 km. The writers believe that these sediments represent a normal deep-Gulf sedimentary sequence, plus deposits from local slumping, and more recent sedimentation.

A typical seismic section from the Sigsbee Escarpment to the Mexican-Campeche scarp shows involvement of the deep-Gulf stratigraphic units in the Sigsbee scarp, and depositional overlap of these same units at the base of the Campeche scarp, with thinning of all layers to the south (Fig. 11).

SOUTHERN GULF OF MEXICO

All sedimentary units of the deep Gulf appear to pinch out by overlap against the Campeche Escarpment (Fig. 12). It is possible that the Jurassic Challenger Unit also pinches out, but the high acoustic velocity of this unit would make it difficult to distinguish from the volcanic-evaporite bearing Jurassic units known to exist in the Yucatan Peninsula of Mexico.

All Cretaceous to Recent sediments of the deep Gulf of Mexico thin, overlap, and disappear at the

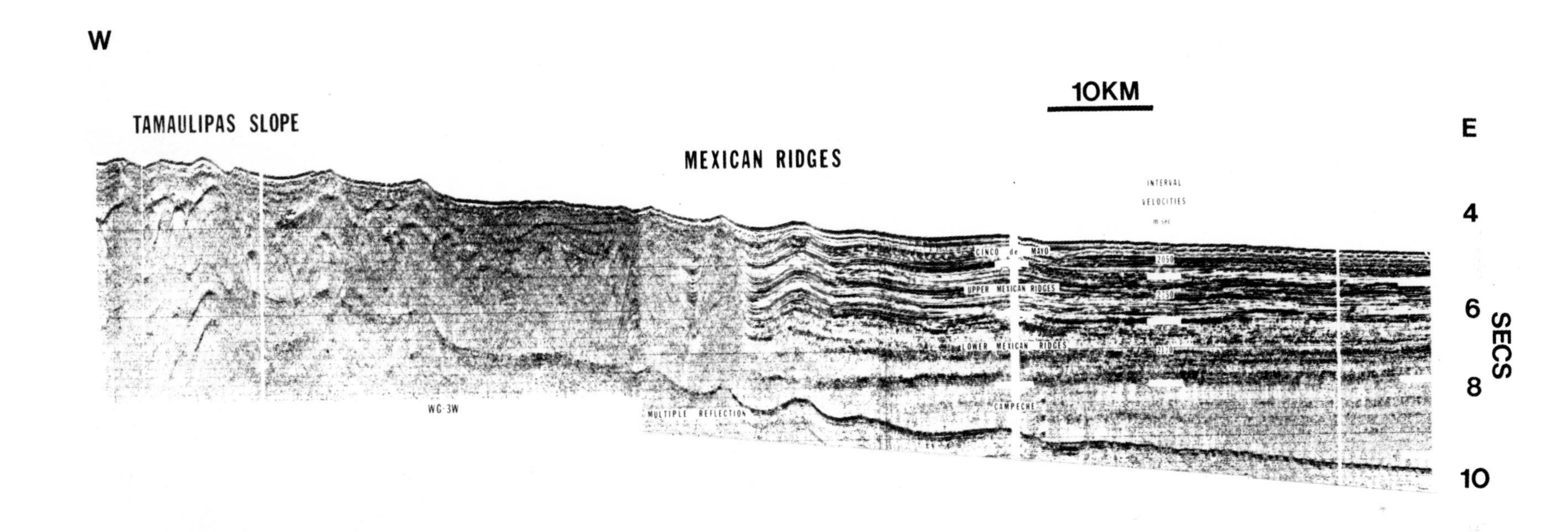

FIG. 7 Reflection-time section near the front of the Mexican Ridges off the coast of Mexico. The layers in the Mexican Ridges Unit and the Cinco de Mayo Unit are clearly involved in the ridge tectonics, leading to the conclusion that ridge deformation must have started no earlier than the late Pliocene. Seismic stratigraphic units as in Figure 2.

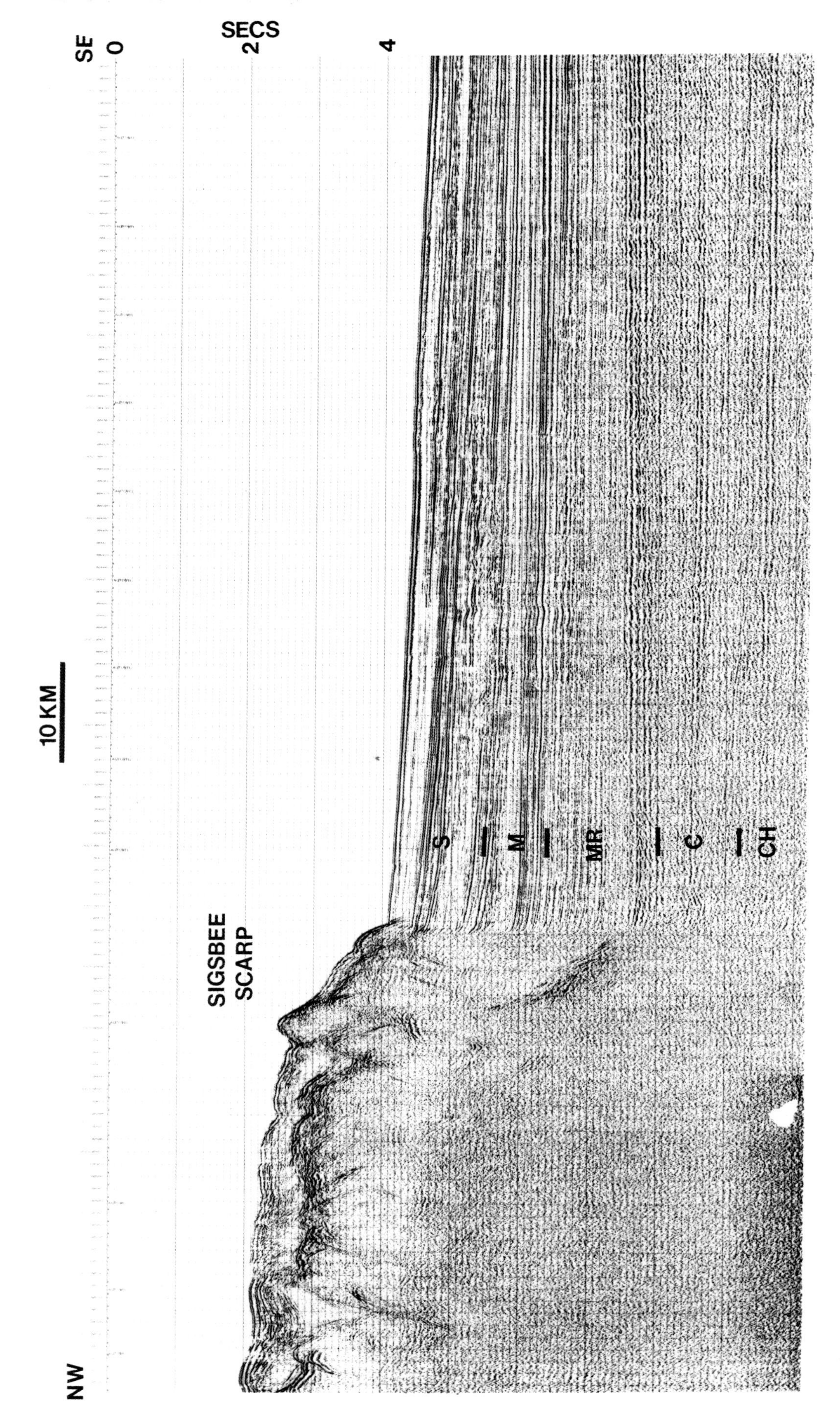

FIG. 8 Reflection-time section across the Sigsbee Escarpment. Note that the layering above the Mexican Ridges Unit can be followed well landward of the scarp. In other sections, deeper layering can be similarly traced and in some cases much farther landward. Seismic stratigraphic units as in Figure 2.

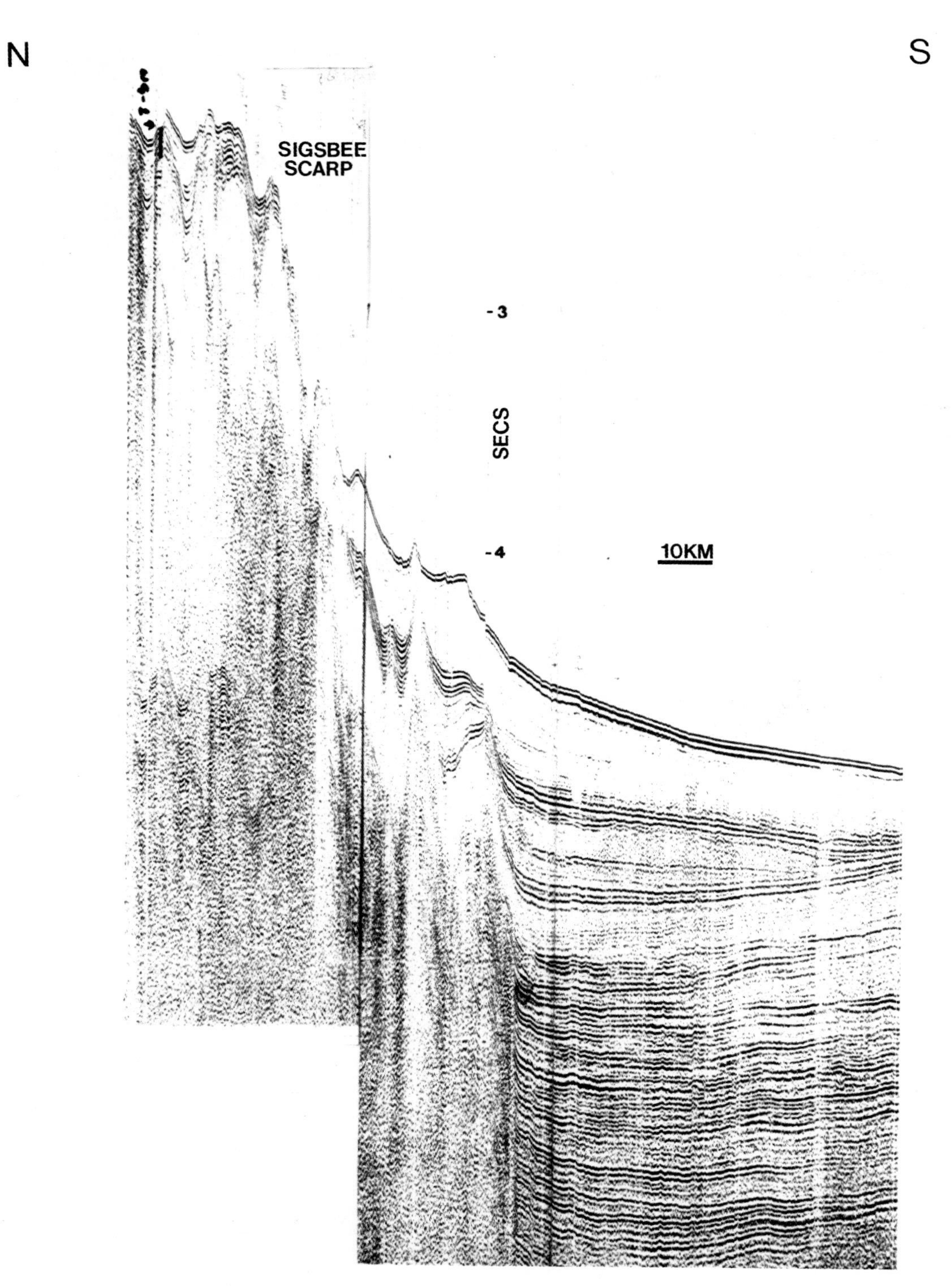

FIG. 9 Single-channel monitor record across the Sigsbee Escarpment a little west of Figure 8. This shows a buried ridge in front of one with topographic expression, suggesting the successive encroachment of the scarp into the deep basin.

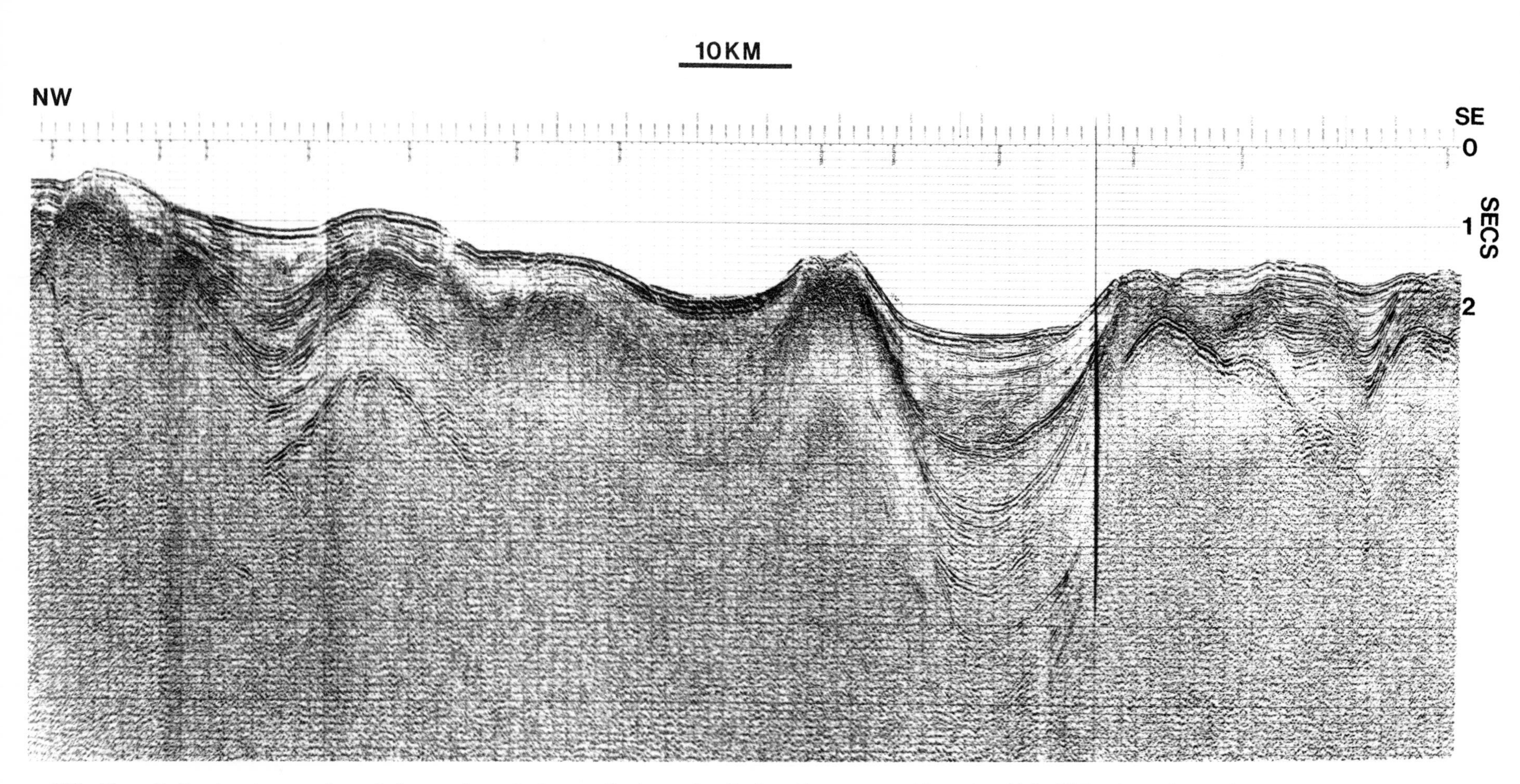

FIG. 10 Reflection-time section of the continental slope well above the Sigsbee Escarpment. Note the thick fill in the valley. These sediments are believed to be former basin sediments augmented by slumping and slope sedimentation after "capture" by the scarp.

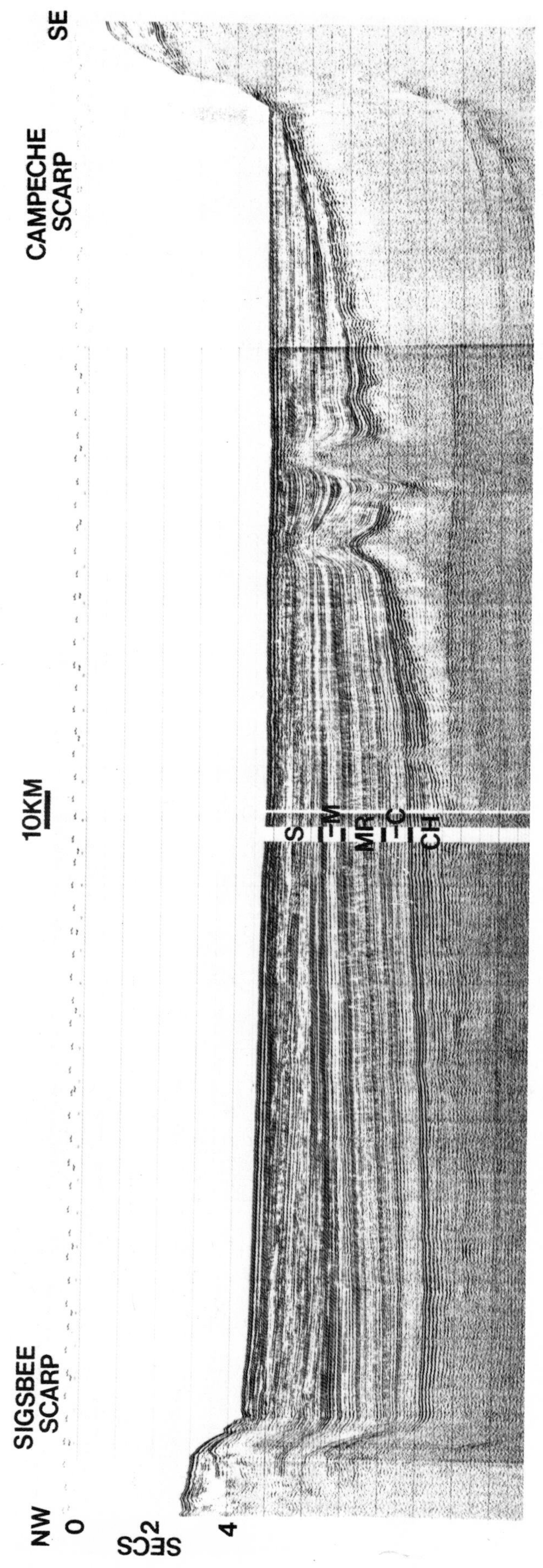

FIG. 11 Reflection-time section from the Sigsbee Escarpment on the northwest (left) to the Campeche Escarpment on the southeast (right). Note that the units thin to the southeast and overlap against the Campeche Escarpment; most units continue across the Sigsbee Escarpment. Seismic stratigraphic units as in Figure 2.

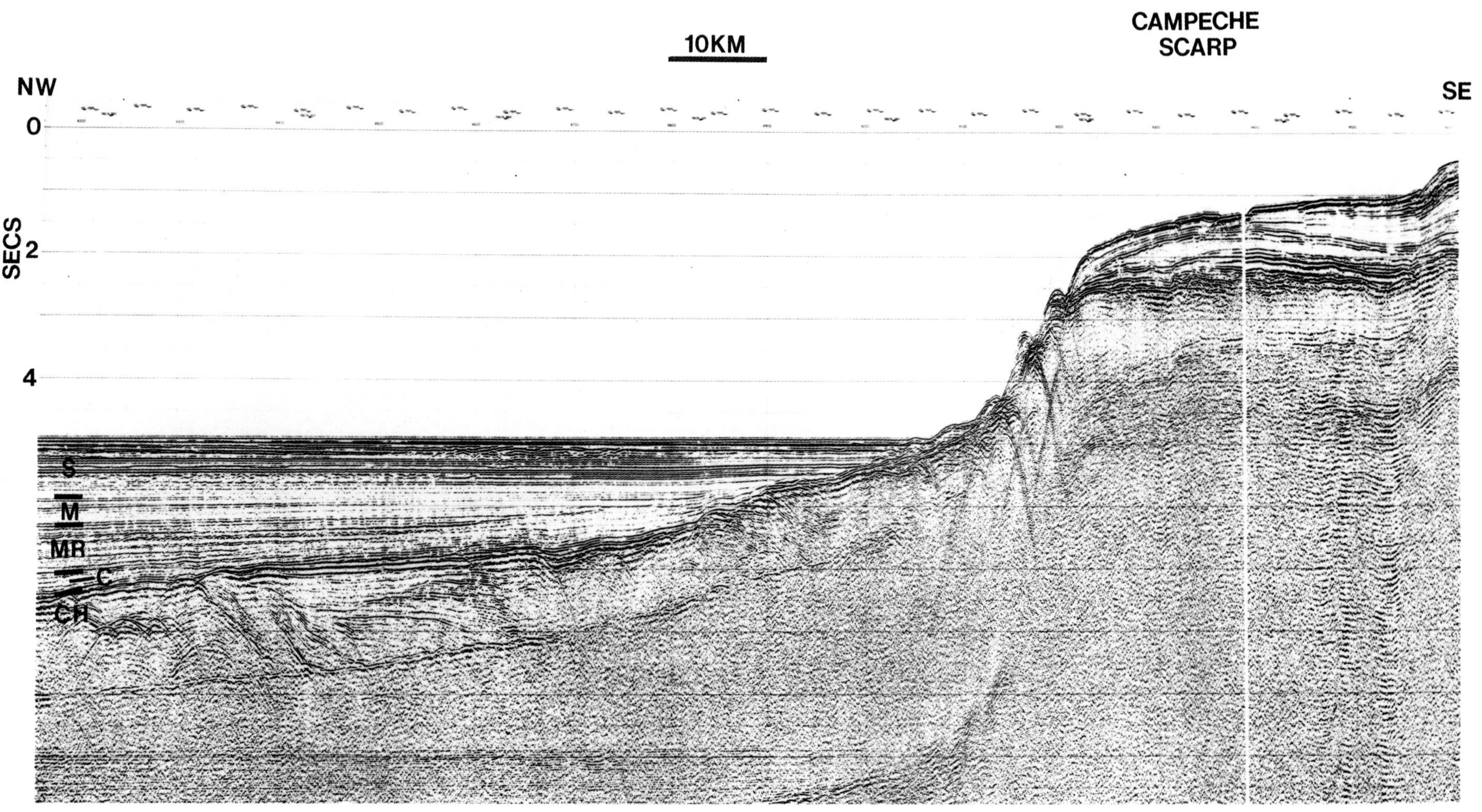

FIG. 12 Multichannel reflection-time section at the Campeche Escarpment. All post-Challenger units pinch out by overlap against the scarp face. Seismic stratigraphic units as in Figure 2.

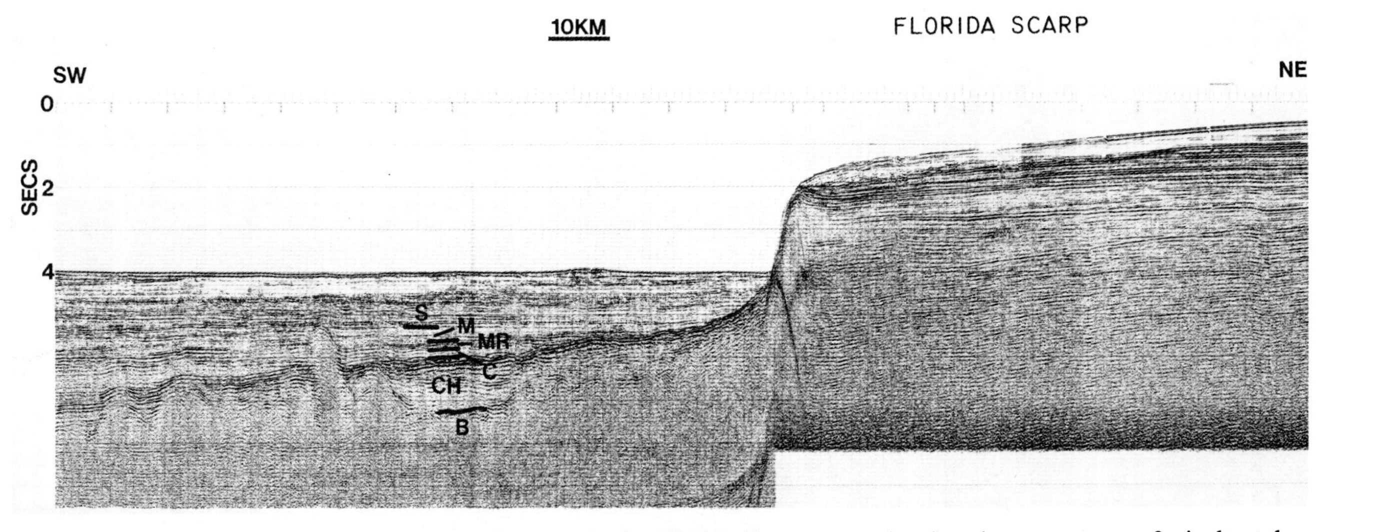

FIG. 13 Multichannel reflection-time section across the Florida Escarpment showing the same type of pinchout by overlap against the scarp face as shown in Figure 12. Seismic stratigraphic units as in Figure 2.

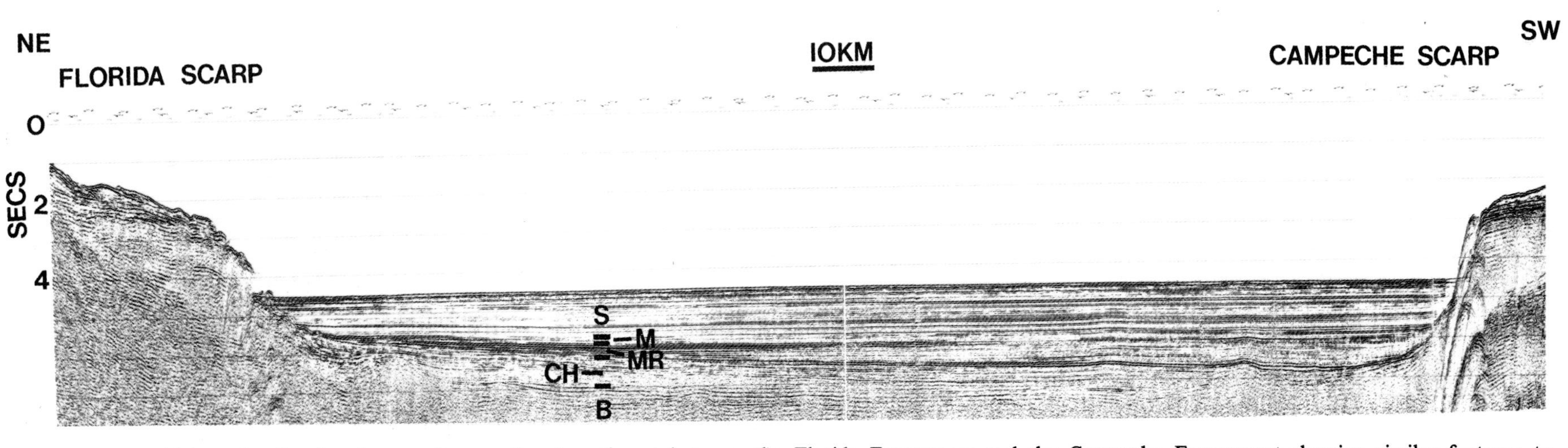

FIG. 14 Multichannel reflection-time section northeast-southwest between the Florida Escarpment and the Campeche Escarpment showing similar features to Figures 12 and 13. Note the southwestward-thickening section just north of the Florida Straits–Yucatan Straits. Note also that the Cinco de Mayo Unit is missing and the Campeche Unit is very thin. Seismic stratigraphic units as in Figure 2.

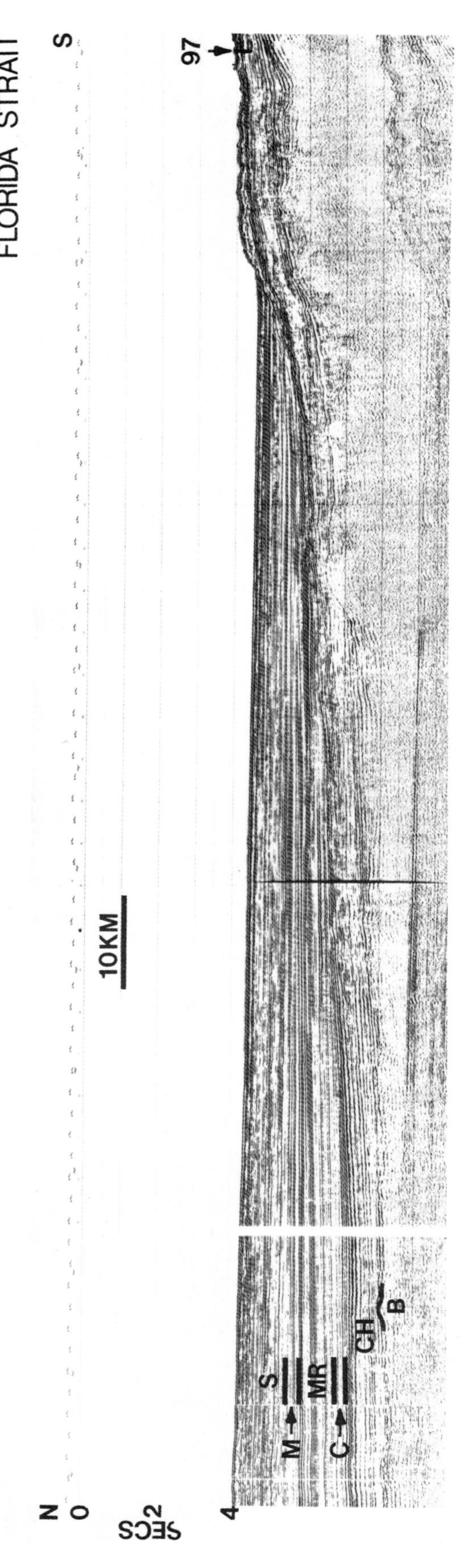

FIG. 15 Multichannel reflection-time section on a north-to-south approach to the Florida Straits and Cuba. The "basement scarp" is similar to those of the West Florida Banks and the Campeche Bank. Pinchout of Cretaceous and younger layers by overlap against this surface is evident. DSDP cores from hole 97 included sediments from mid-Eocene to Late Cretaceous. Seismic stratigraphic units as in Figure 2.

Campeche and West Florida escarpments; no faulting is evident at or beneath these scarps. This must reflect major isostatic subsidence within the Gulf of Mexico, at least since the Jurassic.

EASTERN GULF OF MEXICO

The West Florida Escarpment generally is similar in all respects to the Campeche Escarpment. Jurassic evaporites are known in the subsurface of Florida, and at least all younger units overlap and disappear from the deep Gulf against the West Florida Escarpment (Fig. 13). Thus, we would expect that deep-Gulf sediments typical of the Cretaceous and younger would not be present within the Florida shelf.

An east-west section from the Florida Escarpment to the Campeche Escarpment shows these similarities clearly (Fig. 14). Even though this is a single-channel monitor record, it is obvious that at least the Cretaceous and younger sediments pinch out against both scarps by depositional overlap. The Jurassic Challenger Unit also appears to pinch out, but as noted earlier this may be due to problems of acoustic contrast.

SOUTHEASTERN APPROACH TO THE GULF

A particular effort was made to obtain a seismic traverse from the Gulf of Mexico across the Florida Strait toward Cuba. This traverse shows a basement feature similar to the Campeche and West Florida Escarpments and a corresponding overlap of Cretaceous and younger deposits of the deep Gulf of Mexico (Fig. 15). There is good evidence, following the Challenger Unit in the record section, that it extends at least into these straits, and possibly into Cuba.

DSDP site 97 was drilled into the region just beyond this buried scarp, and the post-Challenger section included middle Eocene to Late Cretaceous sediments (Chapter 14, Worzel, et al, 1973), which were overlapped by deep basin units.

We consider this buried ridge between the Campeche (Yucatan) and West Florida Escarpment to represent a possible continuation between the two, having its origin probably in the Early Cretaceous. It appears to represent part of a single tectonic flexure, which upon subsidence did not maintain the shallow-water reef growth typical of the adjacent areas. Had such reef growth been maintained, the Gulf of Mexico might now be a completely isolated deep-ocean basin.

REFERENCES CITED

Amery, G. B., 1969, Structure of Sigsbee scarp, Gulf of Mexico: AAPG Bull., v. 53, p. 2480–2482.

Bryant, W. R., et al, 1968, Structure of Mexican continental shelf and slope, Gulf of Mexico: AAPG Bull., v. 52, n. 7, p. 1204–1228.

Burk, C. A., et al, 1969, Deep-sea drilling into the Challenger Knoll, central Gulf of Mexico: AAPG Bull., v. 53, n. 7, p. 1338–1347.

Ensminger, H. R., and J. E. Matthews, 1972, Origin of salt domes in Bay of Campeche, Gulf of Mexico: AAPG Bull., v. 56, n. 4, p. 802–807.

Ewing, J., J. Antoine, and M. Ewing, 1960, Geophysical measurements in the Gulf of Mexico: Jour. Geophy. Research, v. 65, p. 4087–4126.
Ewing, M., and J. Ewing, 1962, Rate of salt-dome growth: AAPG Bull., v. 46, n. 5, p. 708–709.
——— J. L. Worzel, et al, 1969, Initial reports of the Deep Sea Drilling Project: Washington, D.C., U.S. Govt., v. 1. p. 672.
Ladd, J. W., et al, 1976a, Interpretation of multi-channel seismic reflection records from the Gulf of Mexico: Phys. Earth Planet. Inter., v. 12, p. 241–247.
——— et al, 1976b, Deep seismic reflection results from the Gulf of Mexico: Geology, v. 4, p. 365–368.
——— et al, 1976c, Multichannel seismic reflection results from the western Gulf of Mexico: An. Acad. Bras. Cienc., v. 48 (Supplemento), p. 145–151.
Shih, T. C., J. L. Worzel, and J. S. Watkins, 1977, Northeastern extension of the Sigsbee Scarp, Gulf of Mexico: AAPG Bull., v. 61, no. 11, p. 1962.
Watkins, J. S., et al, 1975, Deep seismic reflection results from the Gulf of Mexico, Part I: Science, v. 186, p. 834–836.
———et al, 1976, Seismic section WG-3, Tamaulipas shelf to Campeche scarp, Gulf of Mexico: AAPG Seismic Section No. 1.
Wilhelm, O., and M. Ewing, 1972, Geology and history of the Gulf of Mexico: Geol. Soc. America Bull., v. 83, p. 575–600.
Worzel, J. L., W. Bryant, et al, 1973, Initial reports of the Deep Sea Drilling Project: Washington, D.C., U.S. Govt., v. 10, p. 311–334.

Resources, Comparative Structure, and Eustatic Changes of Sea Level

Petroleum Source Beds on Continental Slopes and Rises[1]

WALLACE G. DOW[2]

Abstract Continental slopes commonly are sites of high marine organic productivity and frequently contain reducing bottom conditions, quiet water, and intermediate sedimentation rates, all of which favor deposition of organic-rich sediments. These deposits typically have high percentages of aquatic organic matter with high petroleum yields as contrasted to relatively organic-lean shelf deposits which contain primarily low-yield terrestrial organic matter.

Conversion of organic matter in potential source beds to oil and gas requires a combination of temperature and time. These variables are controlled primarily by the geothermal gradient, the rate of burial, and the age of the source interval. Most divergent margins need between 2 and 4 km of overburden for oil generation and from 3 to 7 km for gas generation. Typically cooler and younger convergent margins and deltaic margins must have even greater burial depth to achieve the same results.

Continental margins, including present slopes and rises, can contain oil and gas source beds when minimum requirements of organic content, kerogen type, and thermal maturity are met. Migration and accumulation are most efficient, however, where reservoir sequences prograde over source beds in areas of structural complexity. Preservation of trapped petroleum requires effective seals and minimal structural readjustment after accumulation. All these conditions can be found on present slopes and rises, although they are not common, and must be considered as part of any economic evaluation of these largely untested deep-water realms.

INTRODUCTION

Continental slopes and rises are physiographic features of continental margins (Fig. 1), and there is no a priori reason why petroleum source beds should not exist there as they do on continental shelves. Deep sediments on slopes and rises have not been drilled and much of their geology, including reservoir and source potential, must be inferred. Recent data, especially from Deep Sea Drilling Project (DSDP) cores, have revealed certain features of slopes and rises which favor the preservation of organic matter in sediments. Some data defining the rate of maturation with depth also have been developed. Other aspects such as structural style and sediment thickness and approximate age can be postulated from regional geophysical profiles, especially when used in conjunction with DSDP core data. Analogy with relatively abundant data from test drilling on continental shelves is also useful, but present slopes and rises typically contain younger sediments than do shelves (Fig. 1), some of which may not have been deposited in deep-water environments. Some slopes and rises are underlain by older sediments that are not related to slope and rise sedimentation, and some of these might contain good petroleum source beds.

DEFINITION OF SOURCE BEDS

The question of what constitutes oil or gas source beds is still debated, and it is useful to define criteria for their recognition. In this paper, we apply the modern geochemical concept that petroleum and gas are formed from disseminated sedimentary organic matter (kerogen) by a series of predominantly first-order chemical reactions, the rates of which are dependent primarily on temperature and the duration of heating. The quantity and variety of petroleum generated are related to the concentration of organic matter in the source bed and the type of kerogen present. Whether petroleum-generating reactions have occurred is determined by the thermal maturity of the primary organic matter. Although geochemical criteria for the recognition of source beds are fairly well agreed on by most geochemists, factors governing expulsion and migration are less understood.

[1]Manuscript received, October 7, 1977; Accepted, February 14, 1978.

[2]Robertson Research (U.S.) Inc., Houston, Texas 77060.

Sincere appreciation is expressed to Don Mathews, formerly of Superior Oil Company, for releasing material on which the original version of this paper (Dow, 1977c) was based. Appreciation also is due Getty Oil Company for permission to publish the revised manuscript, and to Linda Samuels for her editorial assistance.

Article Identification Number:
0065-731X/78/MO29-0028/$03.00/0.
(See copyright notice, front of book)

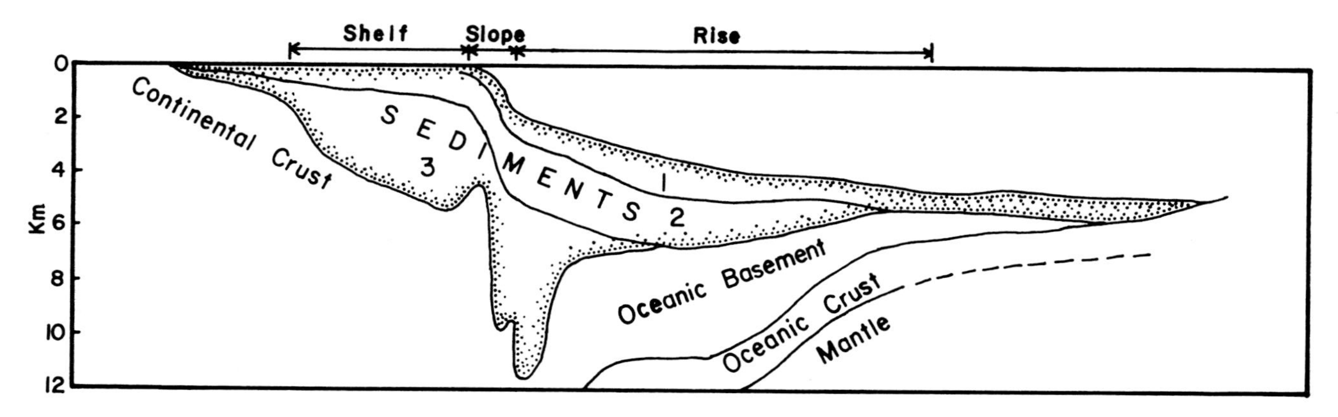

FIG. 1 Schematic section across typical divergent continental margin showing classification of shelf, slope, and rise physiographic provinces. Slopes and rises typically contain younger sediments (1 and 2) than shelves (3). Modified from Roberts and Caston (1975).

Some favorable and unfavorable parameters can, however, be discussed.

Kerogen Concentration

Recent papers on kerogen (Dow, 1977b; Harwood, 1977) review criteria for the amount, type, and maturity of organic matter in source beds, and the types of hydrocarbons they are capable of yielding. Under the most favorable conditions, an arbitrary empirical lower limit of 0.4 weight % organic carbon, (about 0.5 weight % organic matter) is generally accepted as the minimum kerogen concentration required before significant expulsion can be effected. Most acknowledged source beds contain between 0.8 and 2.0 weight % organic carbon, and some of the best contain as much as 10 weight %.

Kerogen Type

Most sedimentary kerogens are mixtures of two basic types of organic matter: terrestrial or humic organic matter derived from higher land plants, and aquatic or liptinitic organic matter derived from lower marine or lacustrine plants. Aquatic organic matter is generally deposited under reducing conditions and will yield both oil and gas. Terrestrial organic matter commonly is incorporated into sediments under oxidizing conditions and will yield primarily gas. When deposited under reducing conditions, some forms of terrestrial organic matter will yield wet gas and paraffinic crude oil (Harwood, 1977). Most sedimentary kerogens are mixtures of these two basic types, and many also contain a third type of recycled or oxidized material termed "inertinite." Inertinite is essentially "dead carbon," because it is practically unaffected by heat and yields no oil or gas (Erdman, 1975).

Kerogen Maturity

The effect of temperature and burial in increasing the rank of coal and sedimentary kerogen is well known and can be measured by several techniques including pyrolysis, elemental composition, color in transmitted light (TAI), and reflectivity in incident light (vitrinite reflectance). The random reflectance of vitrinite particles (R_o) is perhaps the best rank indicator for sedimentary kerogens because it is quantitative, discriminatory, and accounts for both time and temperature; is useful over the entire maturity range; and can be applied to most sedimentary rock types (Dow, 1977b).

Kerogen maturation studies are used primarily to determine whether petroleum has been generated in source beds and preserved in reservoirs. The zones of oil and gas generation and their correlation with the coal-rank scale and various maturation indices are shown on Figure 2. The peak generation zones for oil, wet gas, and dry gas occur at different maturity levels, and the type of petroleum generated depends on the kerogen composition as well as its maturity. The transition from oil to dry gas source beds is continuous and the "bubble" size (Fig. 2) indicates the approximate relative yield for three basic kerogen types. Very little oil or gas, except biogenic methane, is generated in rocks with maturities less than 0.6 R_o, and gas is formed at higher maturity levels than oil (reviewed at length by Dow, 1977b, and by Harwood, 1977).

Even if the criteria used to define source beds are accepted, the words employed are often ambiguous and confusing. For example, the *economic potential* of an unexplored area depends in part on the *source potential* of the area, and whether or not *potential source beds* are present. *Potential* is defined as something "that can, but has not yet, come into being; unrealized; undeveloped." Thus, the *economic potential* in the previous sentence is a correct usage, but potential source beds are those which are capable of yielding oil or gas, but which have not yet done so.

To alleviate this confusion, the terms "source bed" and "potential source bed" should be more rigorously defined, and care should be employed in their usage. These terms are defined as follows:

Source Bed—A unit of rock that has generated and expelled oil or gas in sufficient quantity to form commercial accumulations. Must meet minimum cri-

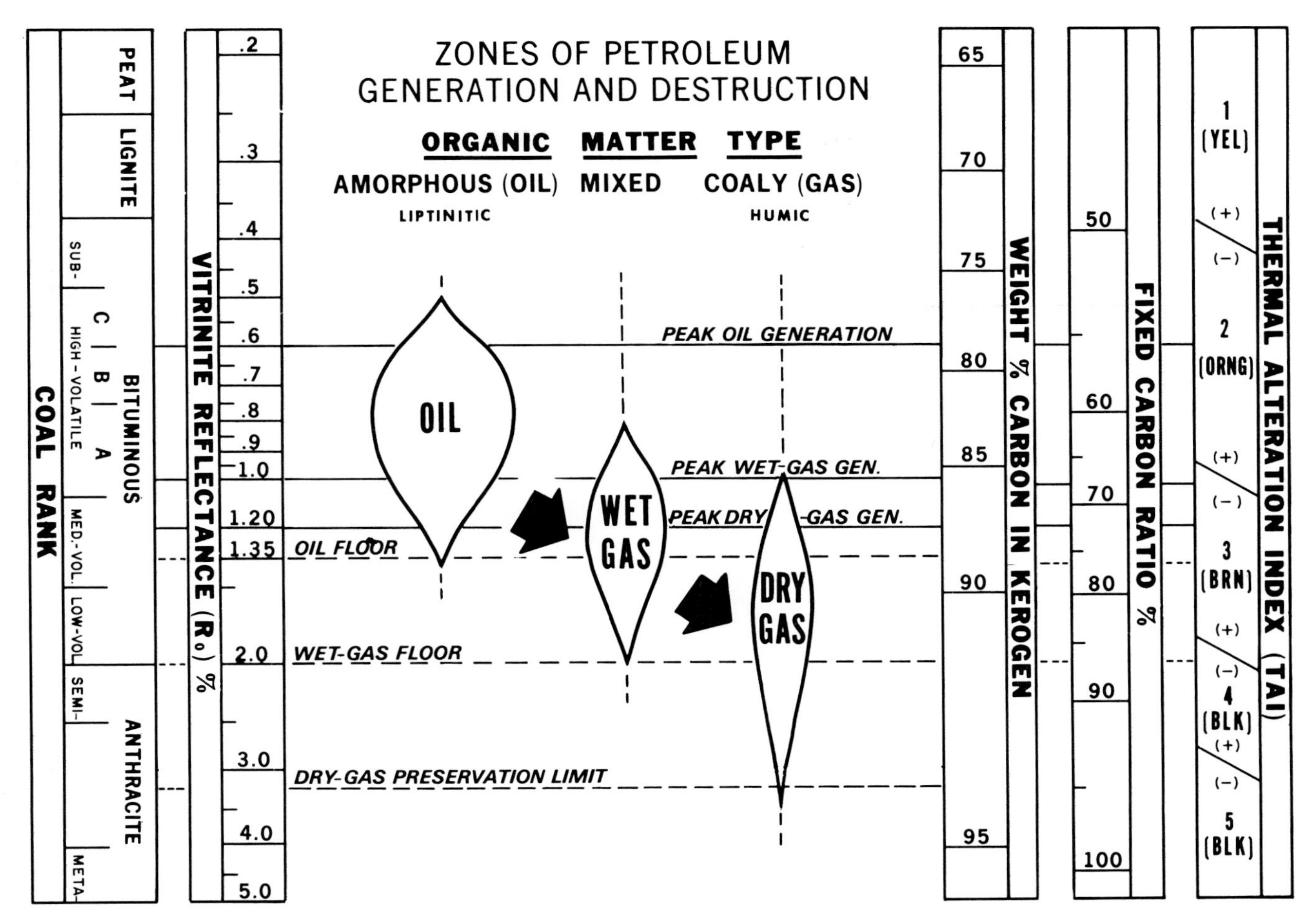

FIG. 2 Correlation of coal-rank scale with various maturation indices and zones of petroleum generation and destruction. Relative importance of each petroleum-generation zone depends on composition of original kerogen. From Dow (1977b).

teria of organic richness, kerogen type, and thermal maturity.

Potential Souce Bed—A unit of rock that has the capacity to generate oil or gas in sufficient quantities to form commercial accumulations but has not yet done so because of insufficient thermal maturation.

PRODUCTION AND ACCUMULATION OF ORGANIC MATTER

Organic Productivity

Production and accumulation of organic matter are the first stages in the formation of petroleum source beds. The basis for the production of primary organic matter is photosynthesis which occurs on land or in the upper 200 m of the sea. At present, terrestrial plants and marine phytoplankton produce about equal quantities of organic carbon. Terrestrial organic matter is most common along continental margins, especially in areas of major river runoff. Marine organic-matter production is controlled by light, temperature, and the nutrient content of seawater. Upwelling of deep ocean water can introduce large quantities of nutrients such as phosphate and nitrate into the euphotic zone, resulting in areas of high organic production. Upwelling is controlled primarily by trade winds and Coriolis forces and is most common along the west coasts of continents (Fig. 3). Areas of upwelling correlate closely with areas of high, marine organic productivity (Fig. 4). These areas contain some of the richest known source beds.

Sediments deposited along the east coasts of continents typically contain a greater percentage of detrital terrestrial-organic matter as a result of lower productivity of marine organic matter. Marine productivity is important because marine phytoplankton contain abundant lipids and lipid-related compounds which are the most important precursors of petroleum. Terrestrial plants, however, are rich in hydrogen-deficient skeletal material which primarily yields gas rather than oil. Some components of terrestrial kerogen such as spores, cutin, resins, and waxes are capable of yielding wet gas and paraffinic oils, if they are sufficiently concentrated. Nearshore reducing depositional environments, either marine or more commonly lacustrine, typically contain this kerogen type.

Organic-Matter Accumulation

Incorporation of organic matter into sediments ultimately depends on processes which not only con-

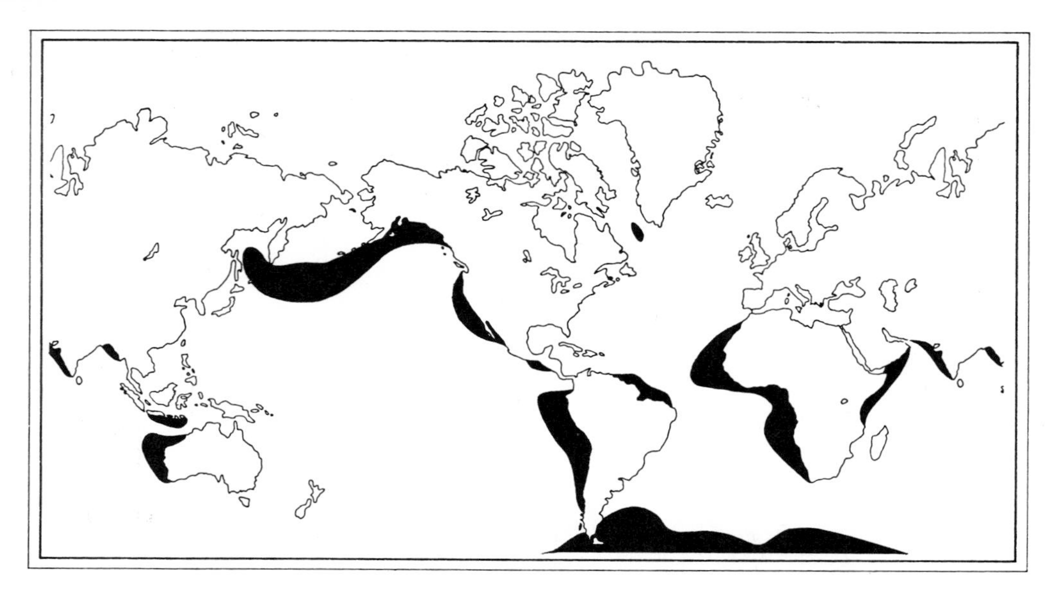

FIG. 3 General world areas of deep-water upwelling. Compare with areas of high, primary organic productivity (Fig. 4). Redrawn from Fairbridge (1966).

serve and concentrate, but also destroy and dilute. It is estimated that less than 1% of the organic matter produced ultimately is incorporated into sediments. The main processes that destroy organic matter are chemical oxidation and consumption by grazing heterotrophic organisms. Organic-matter deposition is favorable where bottom conditions are not highly oxidizing. Such conditions may occur in closed anoxic basins, in the oxygen minimum zone of the sea along certain upper continental slopes (Fig. 5), and in areas where organic productivity exceeds the ability of available free oxygen to oxidize the organic matter.

Organic matter in aquatic environments is either dissolved or suspended as fine particles and is easily transported by water currents. Some mineral particles, especially clays, adsorb certain polar organic compounds and therefore, convert dissolved organic matter to particulate form and reduce its residence time in the water column. Because clays have a greater adsorption capacity than carbonate rocks, shales are typically more organic-rich than carbonate muds deposited under similar conditions.

Clay-bearing sediments (muds) accumulate in areas of quiet water, such as deep-water or restricted basins, where wave and current activity is at a minimum. If the supply of organic matter is constant, its concentration in sediments should be inversely related to the depositional rate of mineral particles. Therefore, areas of high sedimentation

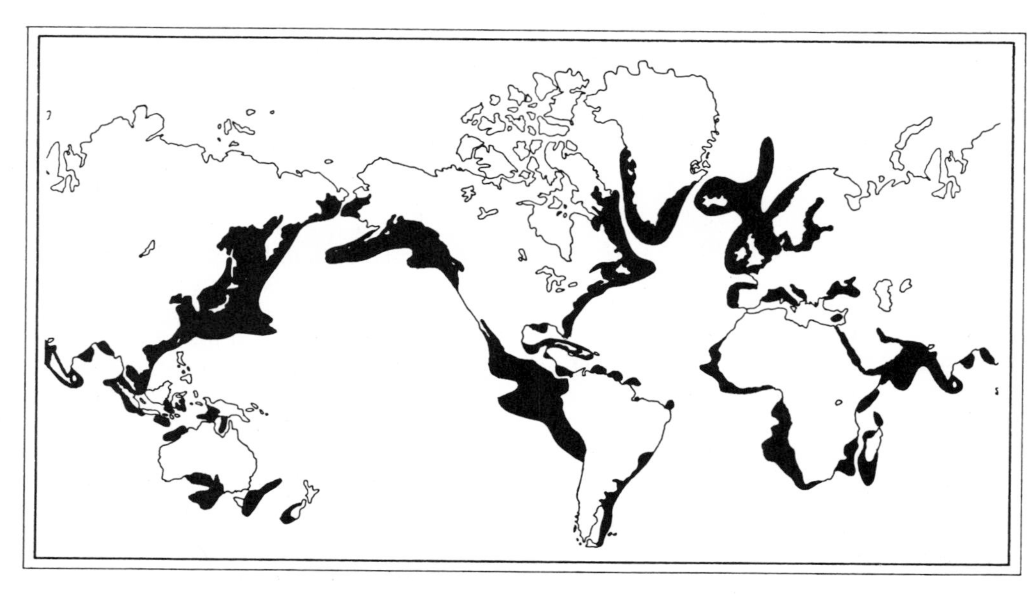

FIG. 4 General world areas of high, primary marine, organic productivity. Compare with Figure 3. Redrawn from Degens and Mopper (1976).

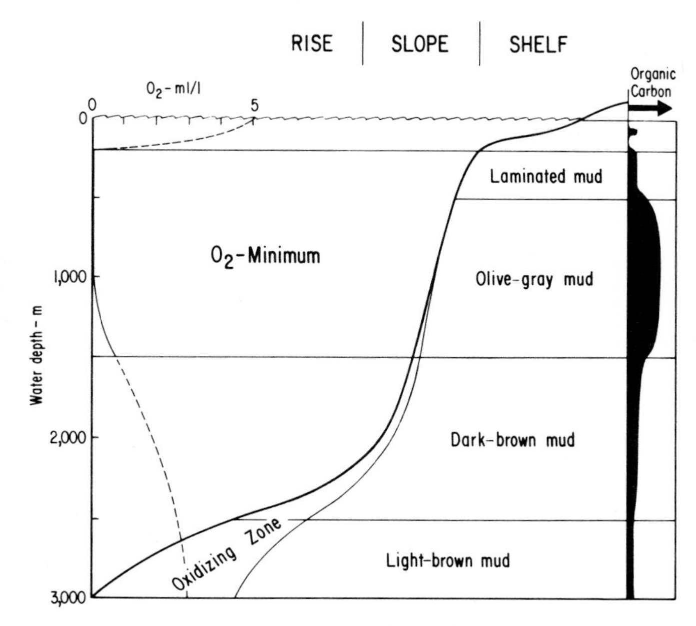

FIG. 5 Schematic section across India-Pakistan continental margin showing oxygen contents in seawater and qualitative contents of selected components of surface sediments. Organic carbon is most abundant in upper slope oxygen-minimum zone. From Closs et al (1974).

rates, such as deltas, should contain sediments with relatively low organic-carbon concentrations. If the sedimentation rate is too low, however, much of the organic matter reaching the bottom may be consumed by heterotrophic organisms before it can be protected by burial. Intermediate sedimentation rates which minimize the effects of both dilution and consumption commonly result in the most organic-rich sediments.

In summary, the most organic-rich sediments are deposited in areas of high organic productivity, where the supply of bottom oxygen is minimal, the water is reasonably quiet, and the sedimentation rate of mineral particles is intermediate. Restricted basins such as the Black Sea and the upper parts of some continental slopes are the most favorable sites for the deposition of organic-rich sediments. Continental shelves and rises commonly are oxidizing environments (Fig. 5), contain high-energy coarse clastic deposits (sands), or have high sedimentation rates (turbidites). Organic-carbon content in most shelf and rise deposits consequently is reduced.

ORGANIC CONTENT OF CONTINENTAL-MARGIN SEDIMENTS

General

A brief review of the distribution of organic carbon in marine sediments as determined by analysis of DSDP cores (McIver, 1975) concluded that the overall distribution of organic carbon is skewed toward the lower values. The average value of 7,300 measurements in DSDP Legs 1 to 23 is 0.3 weight % and the median value is 0.1 weight %. Only 17.3% of the samples analyzed exceeded 0.5 weight % organic carbon and 4.3% exceeded 1.0 weight % organic carbon. A few core holes, however, have surprisingly high organic-carbon content, suggesting that "no rock should be condemned as a poor potential source on the basis of present paleobathymetry or position with respect to continents" (McIver, 1975, p. 270). The previous discussion suggests, however, that certain areas and depositional environments should be more favorable than others in terms of the factors which control organic-matter accumulation. The following review considers some typical continental margins in more detail.

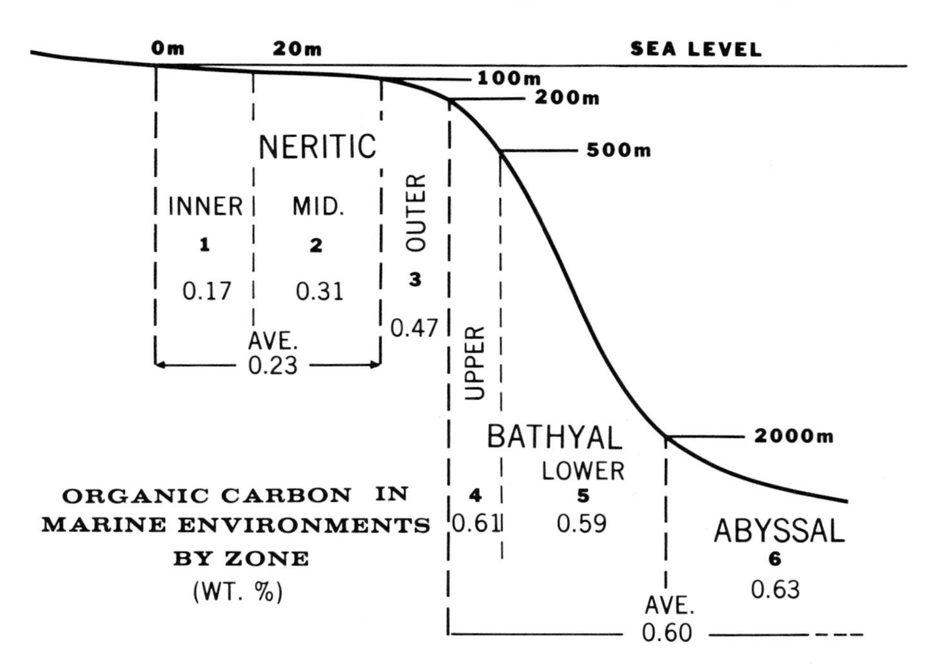

FIG. 6 Mean organic-carbon content by environmental depth zones in Louisiana Gulf Coast Tertiary. From Dow and Pearson (1975).

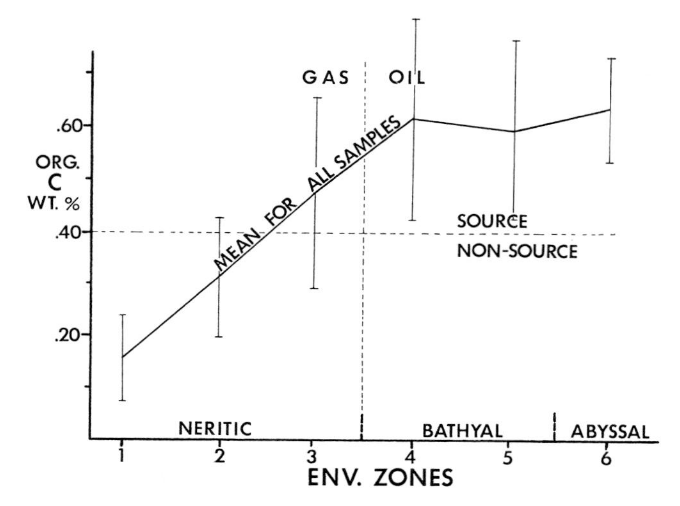

FIG. 7 Mean organic-carbon values in various Gulf Coast environmental zones. Bars represent standard deviation of mean and approximate range in values. From Dow and Pearson (1975).

Louisiana Gulf Coast—Divergent Deltaic Margin

Organic-carbon content and organic-matter type were determined in 264 cuttings samples from 12 deep wells and 62 core samples from eight shallow Caldrill holes in the Louisiana Gulf Coast area (Dow and Pearson, 1975). The mean organic-carbon content in each of six depositional zones, defined by paleontology, revealed a systematic increase from a low of 0.17 weight % in inner shelf shales to a high of 0.63 weight % in deep-water abyssal shales (Fig. 6). The range in values (standard deviation) is greatest in the outer shelf and slope zones and indicates a greater diversity in depositional conditions than in either shallow- or deep-water environments (Fig. 7). Aquatic organic matter is found almost exclusively in slope and rise environments, although many samples from these zones contain predominantly terrestrial organic matter similar to that of the three continental-shelf zones.

A bathymetric chart of the area (Fig. 8) shows considerable irregularity in bottom topography on the continental slope, including the presence of numerous small, closed basins. The complex bathymetry of the open continental slope is the result of salt diapirism and resulting slump features. One of these intraslope depressions, tentatively termed the Orca basin (Fig. 8), contains a pool of anoxic and hypersaline water over an area of about 400 sq km (Shokes et al, 1977). The brine composition indicates it originated by dissolution of a near-surface salt deposit, and the stable density gradient it creates depletes oxygen in the first 200 m above the bottom. Such conditions favor the accumulation of organic matter which is winnowed and oxidized in adjacent, open-slope environments (Rogers et al, 1973). Similar conditions probably exist in other slope basins (Fig. 9). Because the basins are caused by salt movement, and may have persisted for a considerable period of geologic time, vertical columns of organic-rich, potential oil source beds probably would have been created, surrounded by areas of relatively

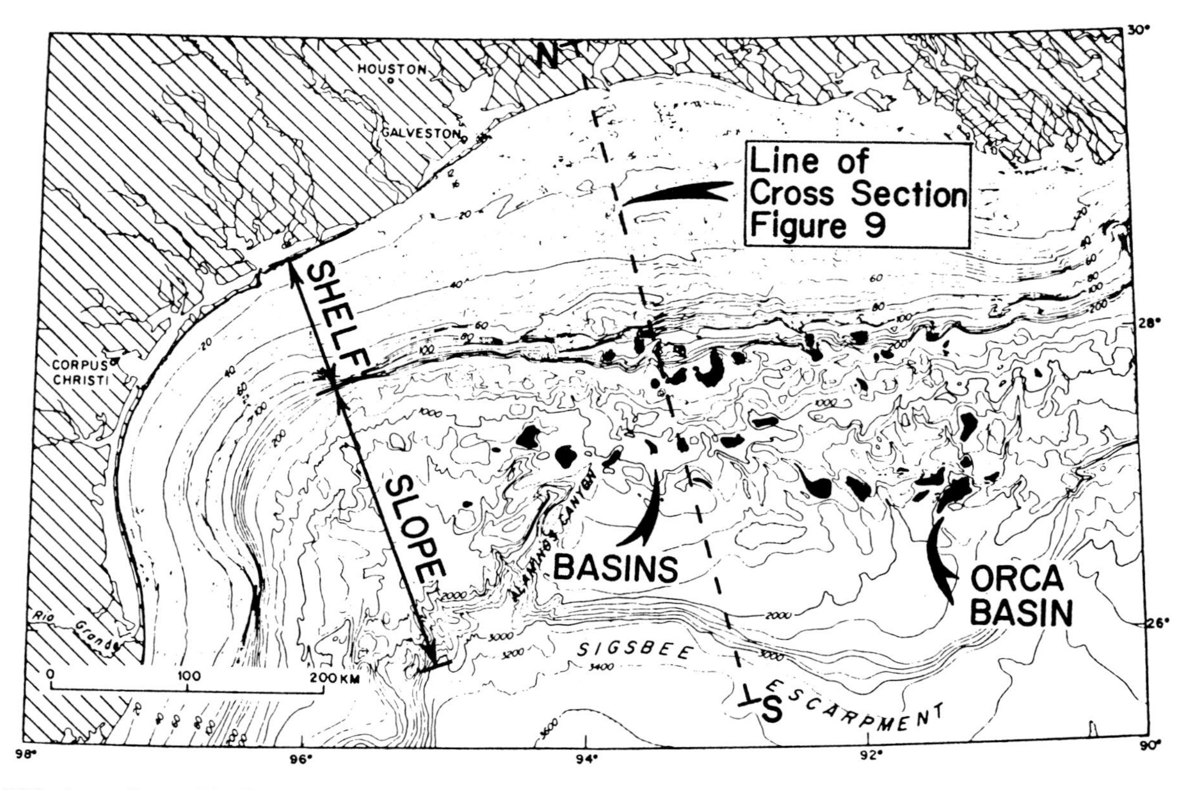

FIG. 8 General bathymetric chart of Gulf of Mexico showing continental slope with closed basins (black) caused by salt movement. From Emery and Uchupi (1972).

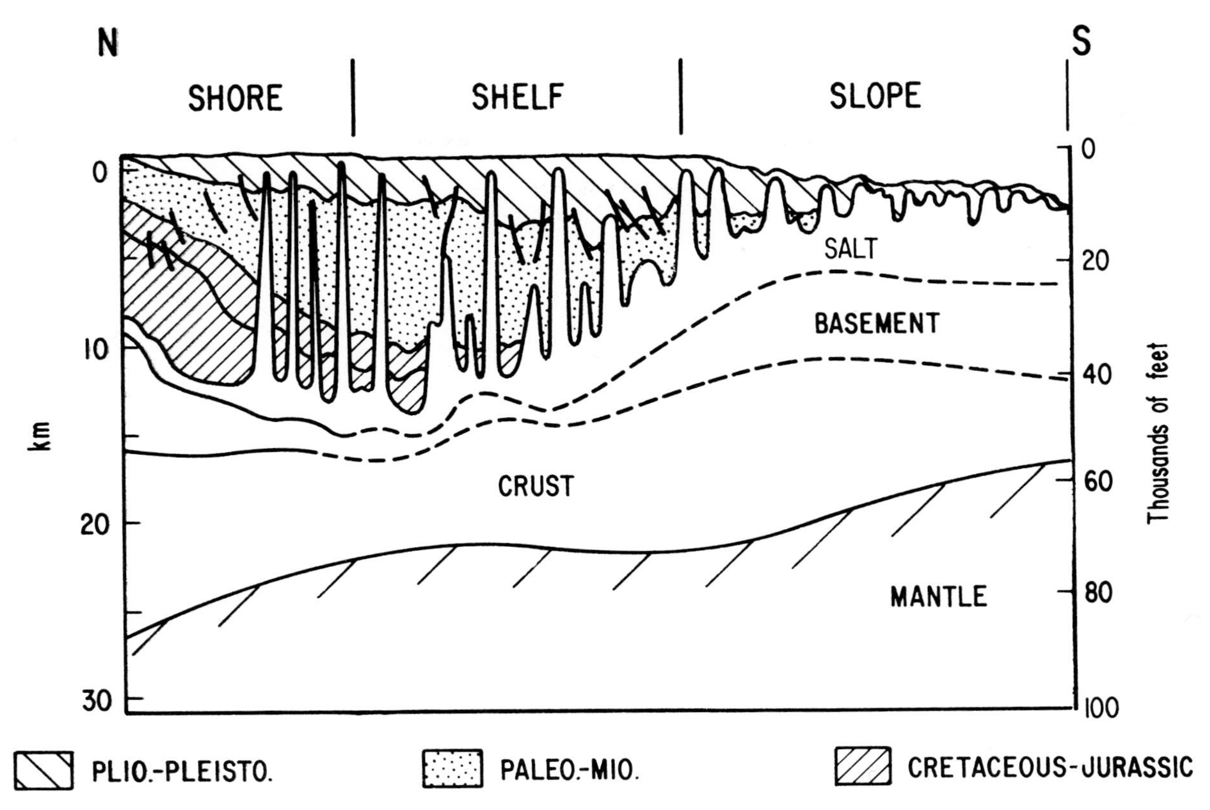

FIG. 9 Cross section showing irregular slope topography caused by salt movement. For location see Figure 8. From Antoine et al (1974).

organic-lean shales containing primarily more resistant terrestrial and recycled organic matter. The proximity of organic-rich shales to salt piercement and their attendant fault networks which serve as migration pathways is largely responsible for the productivity of the Gulf Coast Tertiary province (Frey and Grimes, 1970).

Oregon and Alaska—Convergent Margins

Organic carbon in sediments from convergent continental margins is illustrated by samples collected on DSDP Leg 18 (Kulm et al, 1973). The range and average organic-carbon concentration from 11 core holes are shown on Table 1. The highest organic-carbon concentrations are in slope deposits, especially in local slope basins caused by tectonic folding (Fig. 10) or faulting (Fig. 11). Shelf, rise, and abyssal sediments are distinctly leaner in organic carbon, but sediments from the Aleutian Trench contain about the same organic-carbon concentration as the adjacent slope deposits (Fig. 11).

The average organic-carbon contents are surpris-

Table 1. Average Organic Carbon Contents for DSDP Sites 172-182, Leg 18, Oregon and Alaska Continental Margins*

Site	Location	Ave. Organic Carbon	Range	Number of Samples
172	Abyssal plain	0.12	0.0-0.9	15
173	Lower slope	0.69	0.0-1.6	126
174	Rise	0.30	0.1-1.0	95
175	Lower slope basin	0.69	0.2-1.1	80
176	Outer shelf	0.36	0.2-0.6	27
177	Rise(?)	0.35	0.2-0.7	94
178	Abyssal plain	0.38	0.1-1.2	157
179	Abyssal plain	0.30	0.1-0.6	41
180	Trench	0.58	0.4-1.1	58
181	Lower slope	0.55	0.2-0.9	70
182	Upper slope	0.66	0.5-0.8	5

* Data from Musich and Weser (1971).

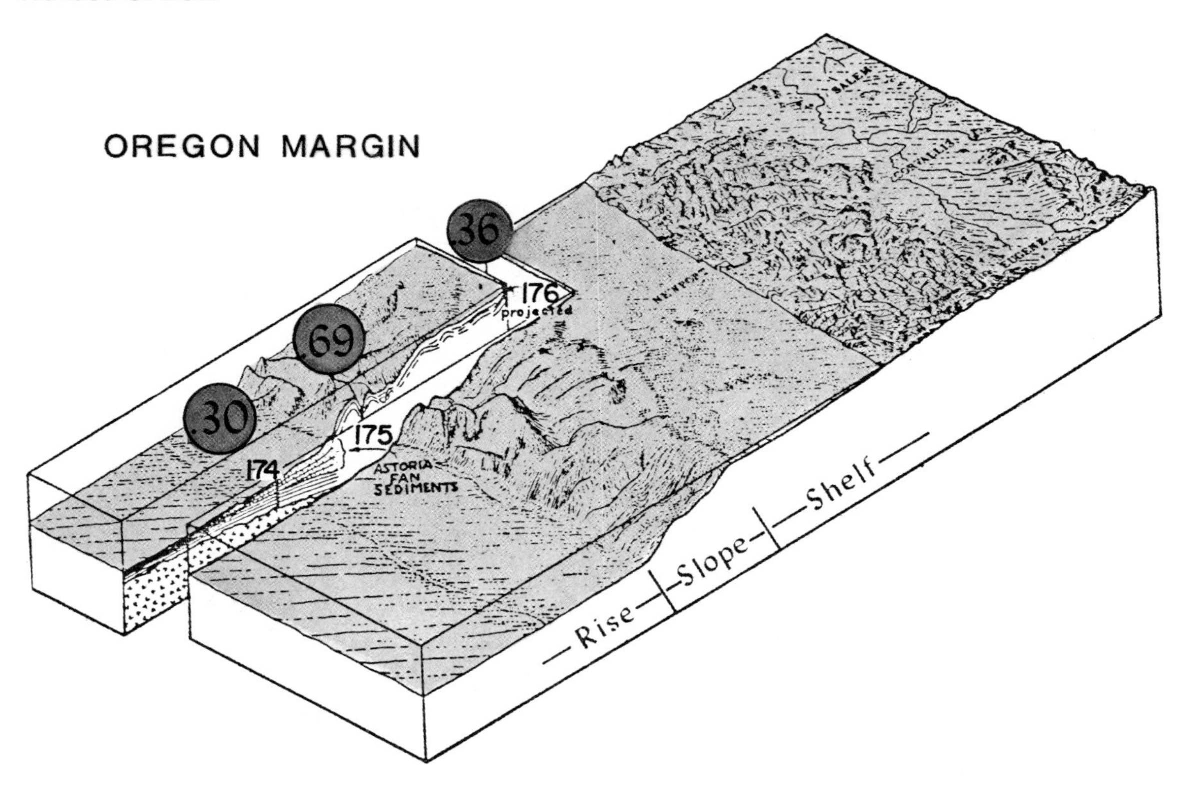

FIG. 10 Orthographic diagram of Oregon continental margin showing selected DSDP site locations and average organic-carbon contents. See Table 1. From Musich and Weser (1971).

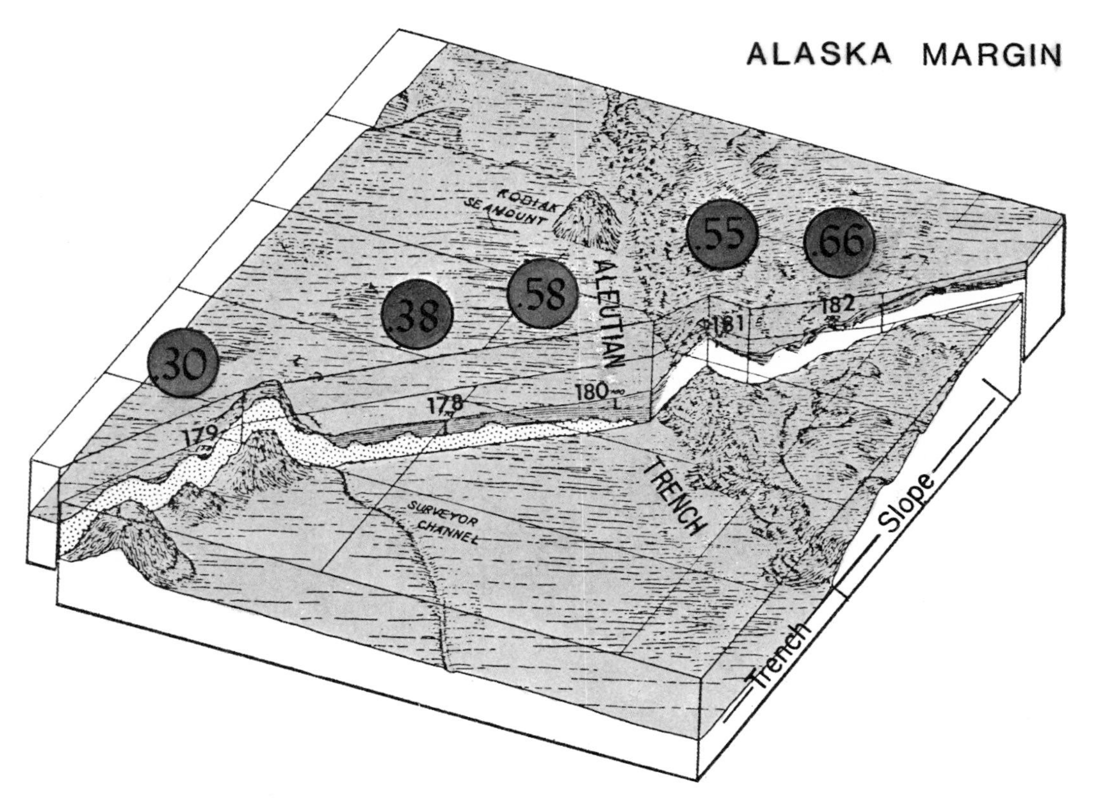

FIG. 11 Orthographic diagram of Alaska continental margin showing selected DSDP site locations and average organic-carbon contents. See Table 1. From Musich and Weser (1971).

ingly similar to those observed in Gulf Coast sediments, even though the Alaska and Oregon continental margins are located in areas of high organic productivity caused by deep-water upwelling (Figs. 3, 4). The increased organic productivity may be compensated partly by the dilution effect of rapid sedimentation combined with less anoxic depositional conditions in the slope basins. Data on the type of organic matter present are sketchy, but paleontologic reports indicate a mixture of aquatic, terrestrial, and recycled material in most samples. The overall capacity of this type of kerogen mixture to yield hydrocarbons, especially crude oil, is limited. The greatest concentration of organic matter in the Alaskan and Oregon convergent margins appears to be in slope basins, although the absence of abundant aquatic kerogen indicates that quiet sedimentation, rather than highly anoxic bottom conditions, may be the primary factor in its accumulation.

Small Ocean Basins

Small ocean basins such as the present Black Sea and Mediterranean Sea may become periodically anoxic and allow the deposition of organic-rich, sapropelic sediments. Similar anoxic conditions also existed in some faulted marine grabens formed during initial continental rifting or aborted rifting, for example, the black Jurassic shales of the North Sea. Relatively large tectonic basins, such as those in the southern California area, periodically were anoxic, and the high organic productivity (partly from upwelling) resulted in thick sections of organic-rich source beds. Slumping of organic-rich slope sediments deposited in oxygen-minimum zones also may be responsible for some organic-rich deep-water turbidites.

Poorly oxygenated bottom water is also probably responsible for the preservation of organic matter in the Cretaceous black shales in the proto-Atlantic Ocean. Bottom morphology, periodic nonreducing conditions, and variable amounts of terrestrial organic contribution also appear to affect the kerogen composition of these "black shales." The configuration of the northern Atlantic at the end of the Early Cretaceous and DSDP sites where Cretaceous "black shale" was recovered are shown on Figure 12. Organic-carbon content ranges from only traces to 10% or more. Some locations contain abundant oil-generating kerogen (Dow, 1977a), and others contain primarily the gas-generating type (Dow, 1978).

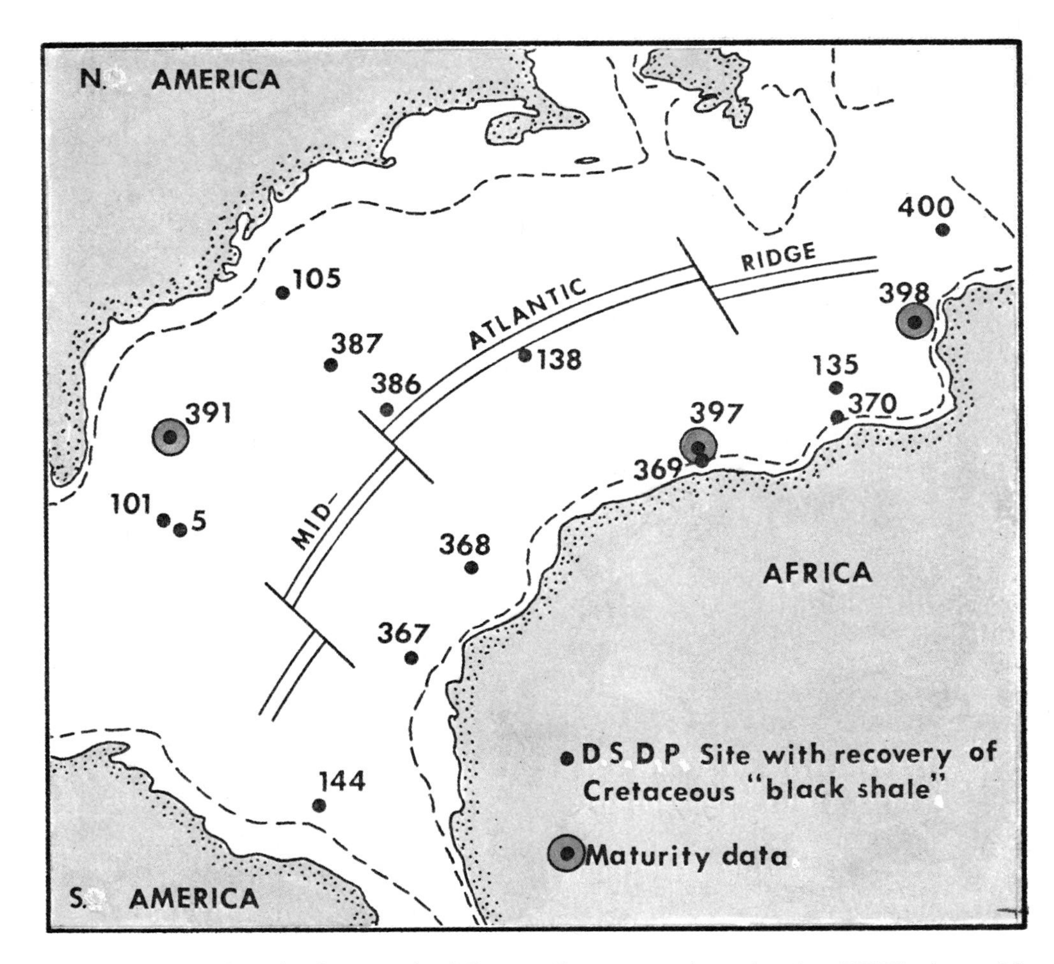

FIG. 12 North Atlantic at end of Lower Cretaceous time showing DSDP sites with recovery of Cretaceous "black shale" and location of three sites with vitrinite-reflectance maturity data. From Dow (1977c).

The significance of the Cretaceous black shales is that they may be oil or gas source beds if buried deeply enough below continental-margin deposits.

Summary

Three basic types of organic matter are available for incorporation into sediments: (1) terrestrial material derived from higher land plants; (2) amorphous material from lower aquatic plants; and (3) recycled material from erosion of uplifted sedimentary rocks. The first type will yield primarily gas and some condensate; the second type oil; and the third type very little gas and no oil. The relative importance of each type depends on the amount that enters the depositional site and the conditions which control its preservation in sediments. In general, nearshore facies have higher percentages of terrestrial organic material, especially near deltas of rivers with large drainage areas containing high land-plant productivity. Rivers draining areas with arid or arctic climates, and those with limited drainage basins, typically contain less terrestrial material. The supply of recycled organic matter is erratic and is controlled by the nature of the rocks being eroded in the drainage areas. The abundance of marine organic matter depends on factors such as nutrient supply from upwelling and river runoff, water temperature, and light, which affect primary phytoplankton productivity.

Incorporation of organic matter into sediments depends on a supply greater than the ability of dissolved oxygen and heterotrophic organisms to destroy it, reasonably quiet water with minimum current activity, and an intermediate sedimentation rate. Therefore, favorable sites for the deposition of organic-rich sediments rich in oil-generating, aquatic organic matter include: (1) irregular bottom topography and closed bathymetric basins on some continental slopes (caused by folding, faulting, or salt diapirism and associated slumping); (2) the oxygen-minimum zone on some continental slopes where organic productivity, usually the result of upwelling, is so high that anoxic conditions develop in unrestricted, open-marine environments; and (3) on continental rises which receive organic-rich turbidites from unstable continental-slope deposits.

It is apparent that continental slopes are favorable sites for the deposition of potential petroleum source beds and that the formation of organic-rich sediments requires special conditions which are the exception. Most continental-margin deposits are relatively low in organic matter and contain primarily gas-generating, terrestrial, and recycled kerogens. Numerous exceptions do occur, however, especially along margins with high organic productivity, usually caused primarily by deep-water upwelling, and in large, restricted basins associated with some margins. Local structure commonly is a significant factor in creating bottom conditions which result in the deposition and preservation of organic-rich sediments in otherwise oxidizing margins. These conditions should be used to help judge the odds of petroleum source beds existing in untested deep-water realms.

THERMAL MATURITY OF CONTINENTAL MARGIN SEDIMENTS

General

It is well documented in the geochemical literature that oil and gas are formed from disseminated organic matter in sedimentary rocks by a complex of reactions of overall first-order aspect, the rates of which are dependent primarily on temperature and the duration of heating. Catalysts and pressure play ancillary roles and affect the types of products formed more than causing the reactions themselves. Because the reactions are basically first order, there is no reaction threshold and the effects of temperature begin early as the sediments are exposed to the earth's geothermal gradient. The reactions begin very slowly, however, and the rate generally will double as the exposure time doubles or the temperature is increased by about 18°F or 10°C (Philippi, 1965). This exponential increase in products with depth of burial ultimately results in a period of very rapid generation which may be termed "peak generation" or the "principal phase of oil formation." At this time, organic extracts in shales containing oil-generating kerogen increase markedly and become more oil-like in composition (Tissot et al, 1974); the depth of the occurrence depends on the geothermal gradient and burial history (Fig. 13). Thermal destruction of oil with increasing time and temperature results in the generation of gas at higher maturity levels. Even though generation begins early in the thermal history of sediments, the small quantities of oil and gas formed are readily adsorbed on clay minerals or the kerogen itself or become dissolved in intrastratal waters. The formation of "commercial

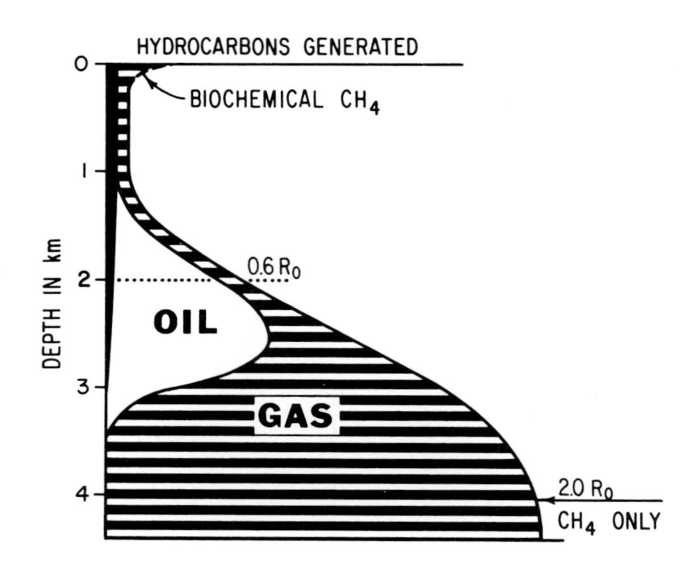

FIG. 13 General scheme of oil and gas generation in pre-Tertiary basins. Depth is indicative only and depends on burial history and geothermal gradient. Thermal cracking of crude oil yields gas and the oil phase may not exist unless oil source beds are present. From Tissot et al (1974).

quantities'' of hydrocarbons and expulsion from the source bed do not occur until the principal phase of generation is attained.

The principal phase of oil and gas generation is correlated with the random vitrinite reflectance scale on Figure 2. The advantage of using this scale is that reflectance is the product of both time and temperature and does not depend on variations in geothermal gradient or burial history. Therefore, it can be used in complex situations where burial history is difficult to determine (Dow, 1977b).

Louisiana Gulf Coast

The relative effects of time and temperature on the reflectance of disseminated vitrinite (and on the generation of oil) are illustrated in a series of wells between the Cretaceous and Pleistocene producing trends (Fig. 14). A sequence of R_o values in wells from each trend gives a ''kerogen-maturation profile'' which is an expression of increasing rank with depth, or temperature and age. The wells selected have very similar geothermal gradients, and the variation in maturation-gradient slopes among trends is due primarily to exposure time or average sediment age (Fig. 15). The depths and present temperatures at the top of the peak oil-generation zone (0.6 R_o) and the oil floor (1.35 R_o) in each trend are different. For example, the 0.6 R_o maturity level occurs at 18,300 ft and 327° F (5,578 m and 164°C) in the Plio-Pleistocene trend and only 8,100 ft and 183°F (2,469 m and 84°C) in the Cretaceous trend (Table 2). The difference is due to exposure time.

However, if the exposure time is constant, the depth to the top of the oil-generation zone will change as the geothermal gradient (temperature) changes. If the geothermal gradient were doubled, the 0.6 R_o maturity level would still occur at 327°F (164°C) in the Pleistocene trend, but the depth would be only 9,150 ft (2,789 m) instead of 18,300 ft (5,578 m). With the same geothermal gradient in the Cretaceous trend, oil could be generated below 4,050 ft and 183°F (1,234 m and 84°C). It is evident that the depth of burial and temperature required to reach the principal zone of oil generation (0.6 R_o to 1.35 R_o) vary according to both exposure time and the geothermal gradient. The numbers for the Louisiana Gulf Coast will not apply everywhere because of differences in geologic histories. Changing sedimentation rates, periods of erosion or nondeposition, and varying geothermal gradients can all affect the actual maturities in a given basin (Dow, 1977b). The Louisiana Gulf Coast has a minimum of complications except for a continuously increasing sedimentation rate, and it can be used as a model for most continental-margin provinces.

A highly schematic, regional stratigraphic section of the Louisiana Gulf Coast basin (Fig. 16) shows the oil-generation maturity zone (0.6 R_o to 1.35 R_o) is increasingly thicker and deeper in younger rocks. The position of the sandstone-shale magnafacies (shelf deposits) in which most of the production is found, with relation to the top of the oil-generation

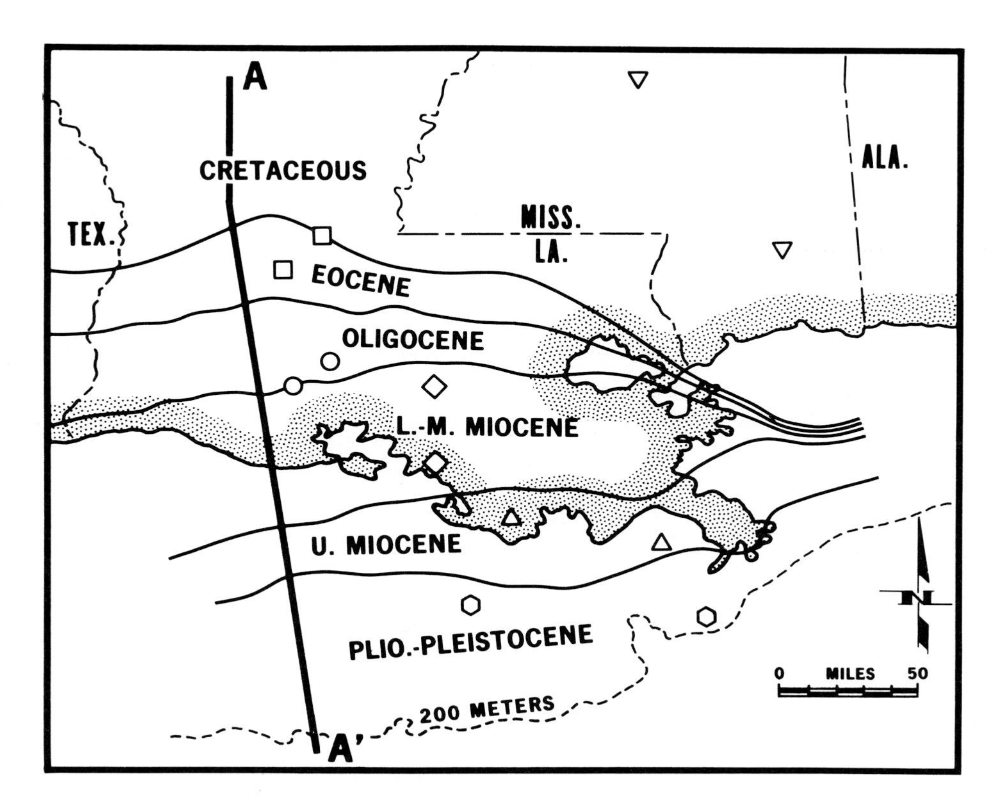

FIG. 14 Index map showing locations of 12 wells, two in each producing trend, from the Louisiana Gulf Coast, on which maturity data was obtained (Fig. 15). Figures 14–17 from Dow (1977c).

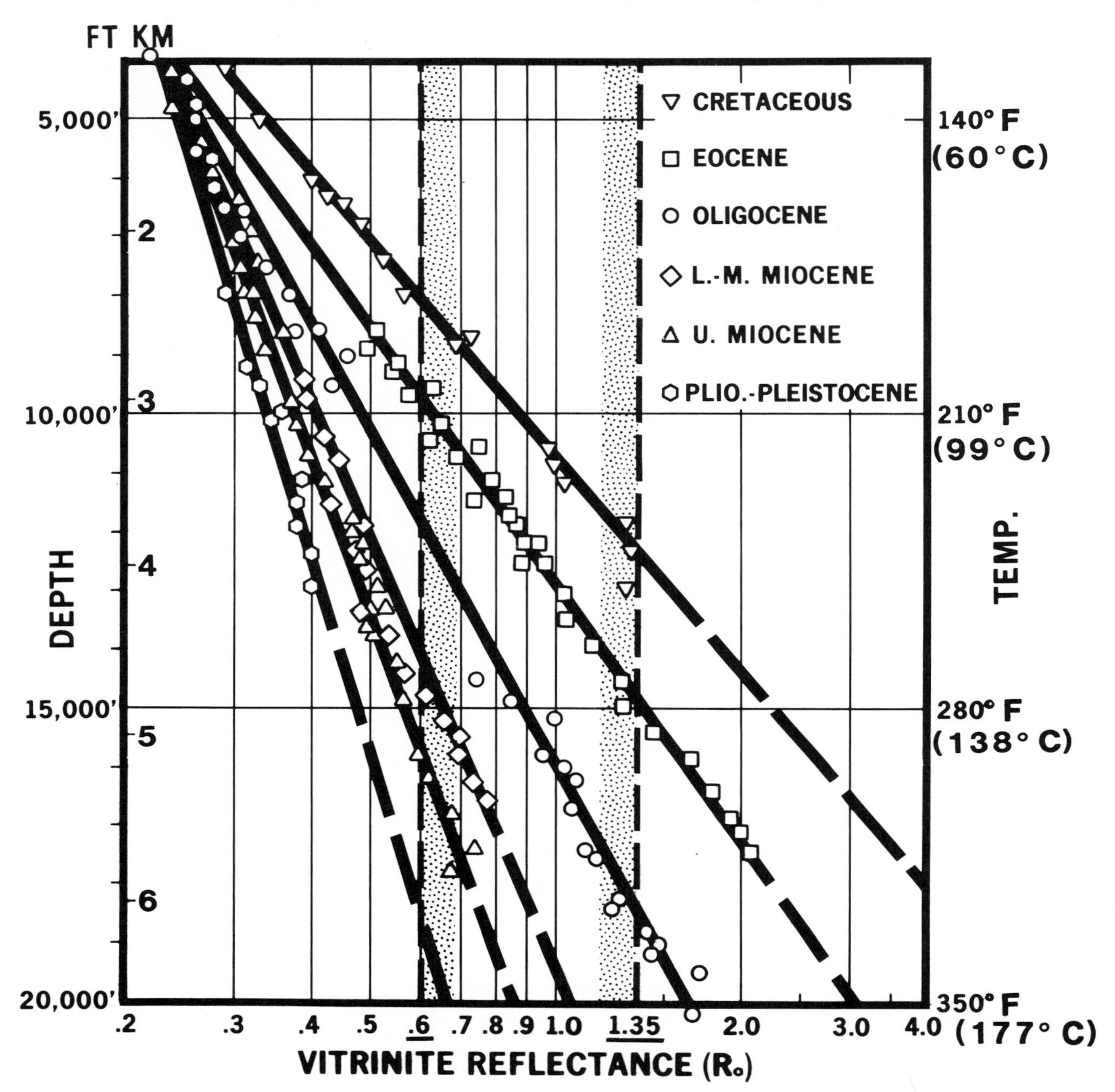

FIG. 15 Composite maturation profiles for two wells in each Gulf Coast producing trend. All wells have uniform geothermal gradients close to 1.4° F/1,000 ft. Note that higher temperatures are required to achieve equivalent maturities in younger rocks than in older ones. Well locations shown on Figure 14.

maturity zone, is also shown in Figure 16. Most of the Louisiana Gulf Coast production is from thermally immature rocks (Fig. 17), and emplacement of oil must have occurred by vertical migration from more mature source beds in the underlying slope and rise deposits of the shale magnafacies. Deep-seated faults and piercements with their attendant fracture systems serve as pathways for oil and gas migration (Frey and Grimes, 1970). Periodic fracturing also permits expulsion from high-pressure shales, and lack of fractures in shallow, less indurated rock, provides seals. Oil age calculations (Young et al, 1977) showed that the ages of offshore Gulf Coast oils average 8.7 m.y. older than their reservoirs and that this time difference indicates an average vertical migration of 11,000 ft (3,353 m).

The Gulf Coast Tertiary province is especially productive because mature slope shales are overlain by prograding shelf sands which are connected by structural "plumbing." The effects of source beds,

Table 2. Depth, Temperature, and Age of Rocks in Which 0.6 R_o Maturity Level is Attained, Louisiana Gulf Coast.

Producing Trend	M.Y. Ago	Oil Generation (0.6R_o) Depth (Ft)	Oil Generation (0.6R_o) Temp. (Deg.)	Oil Destruction (1.35R_o) Depth (Ft)	Oil Destruction (1.35R_o) Temp. (Deg.)
Cretaceous	100	8,100	183	12,400	244
Eocene	58	9,800	207	14,800	278
Oligocene	34	11,800	235	18,400	328
Lower to Middle Miocene	20	14,100	268	22,600	388
Upper Miocene	12	15,600	289	25,400	430
Pliocene	5	18,300	327	30,100	495

* Data from Louisiana Gulf Coast wells with geothermal gradients close to 1.4°F/100 ft. Underlined are projected values. Data are indicative only because wells analyzed do not necessarily represent midpoints of each producing trend.

migration pathways, reservoirs, traps, and seals combine in uncommon harmony in this geologic province. Such tectonically active, thick, prograding deltaic sequences are not common along continental margins but the geochemical principles illustrated here are universal.

Gulf Coast continental slopes almost certainly contain source beds, at least for oil. Many slope deposits contain relatively abundant oil-generating organic matter which is apparently mature enough for oil generation in inter-piercement areas, especially on the upper part of the slope. Shows of oil have been reported in the cap rocks of salt piercements on the upper continental slope itself. Higher heat flow in proximity to thick salt deposits (Rashid and McAlary, 1977) and within high-pressure shale zones (Steiner, 1976) might also reduce the depth requirements for maturation of young sediments. The primary problem seems to be lack of prograded shelf sandstone reservoirs in the present slope areas.

Convergent Margins

Very little published material is available on maturity data from convergent margins, and the question of petroleum generation is more difficult to assess there than in well-studied areas like the Gulf Coast. Active thermal generation has been reported from depths of only 250 m in the Aleutian Trench

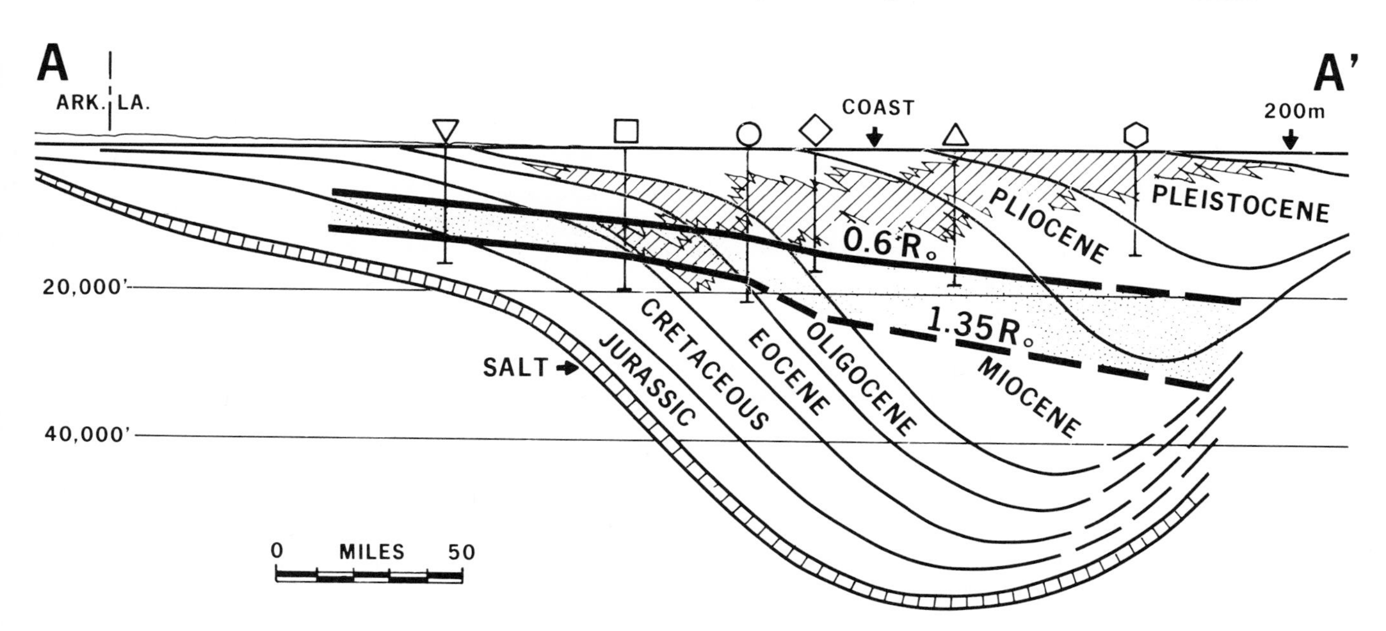

FIG. 16 Schematic section of Gulf Coast showing location of wells analyzed (Fig. 15), depths to 0.6 and 1.35 R_o maturity levels, and of sand-shale magnafacies or shelf deposits (shaded) in each producing trend. Rocks above shelf facies are continental and those below are deep-water slope and rise deposits (shale magnafacies). Location shown on Figure 14.

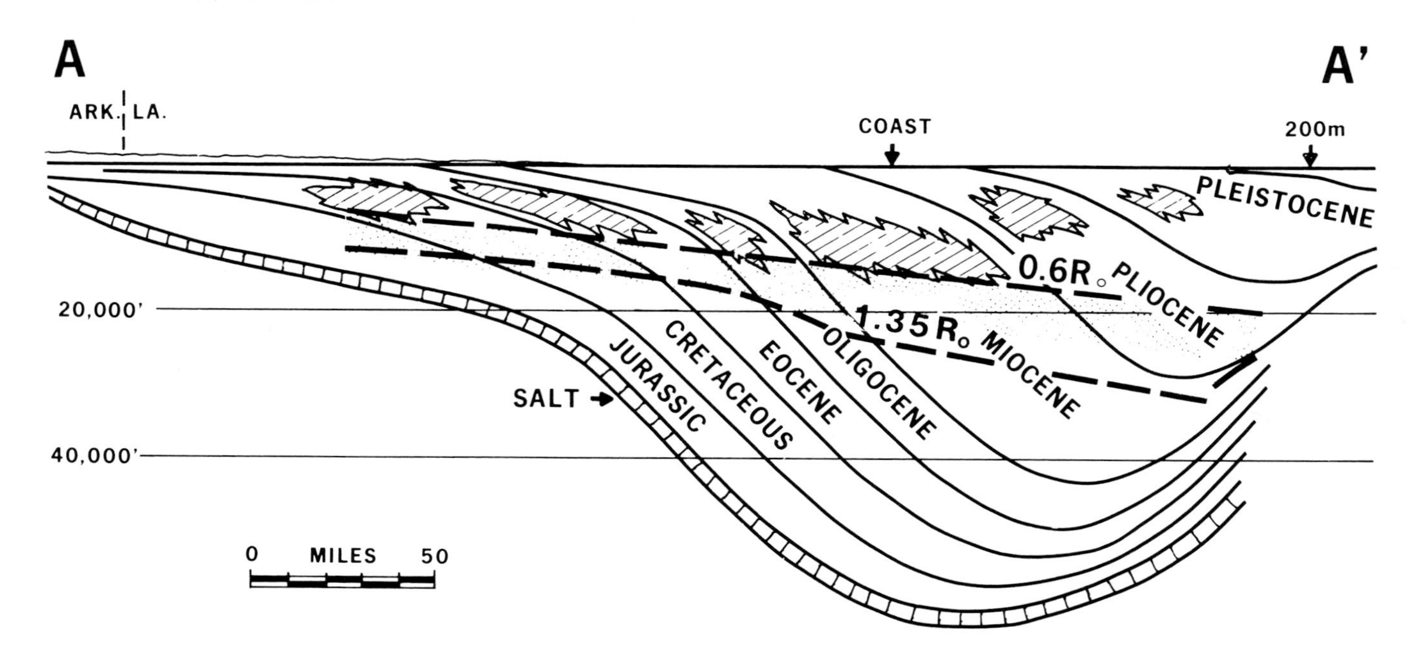

FIG. 17 Distribution of most productive intervals in each Gulf Coast producing trend. Note relation of production to 0.6 R_o horizon and age of reservoir rocks. Location shown on Figure 14.

(Claypool et al, 1973) although geothermal gradients in the vicinity of trenches are typically below average (Roberts and Caston, 1975). However, the relative quantities of light hydrocarbon gases described from the Aleutian Trench are very low and probably represent early thermal generation rather than peak generation. On the basis of spore-coloration data, Grayson and LaPlante (1973) described a level of maturation in indurated Pleistocene sediments at DSDP Site 181 on the Alaskan continental slope (Fig. 11) which is equivalent to about 8,500 ft (2,591 m) of overburden in the Gulf Coast Pleistocene. Assuming a geothermal gradient similar to that of the Gulf Coast, the R_o value would be about 0.30 (Fig. 15) which is far above the top of the oil-generation maturity zone.

Thompson (1976) used these data to construct a thrust-fault hypothesis for petroleum generation during continental-margin subduction. According to his hypothesis, thrust faulting may transport organic-rich slope sediments to anomalously great depths (±5 km) where the increased temperatures accelerate petroleum-generating reactions. A trend of petroleum source beds thus may exist along the outer shelf and upper slope of certain subduction margins. Thompson's hypothesis seems plausible if the relatively young sediments typical of subduction margins could be heated to the temperatures required for oil generation (300 to 350°F or 149 to 177°C). With the apparently low geothermal gradient in such areas, a burial of between 5 and 6 km, thrust-induced or otherwise, would be required to effect the generation of oil, and even more would be required for gas generation.

Hedberg (1970) facetiously offered a cartoon illustrating how organic-rich slope and rise sediments could be carried downward along subduction zones until the juice (water and oil) was cooked and squeezed out of them. The problem is that great concentrations of oil do not seem to be found near the downthrust, sediment-filled, sinks or hoppers, which may be due, in part, to the buoyancy of the slope and rise sediments and their inability to be carried beneath continents and, in part, to the below-average geothermal gradients typical of such areas. Early expulsion of interstitial fluids and over-compaction by tectonic squeezing prior to peak hydrocarbon generation should reduce expulsion efficiencies within the subduction complex itself.

The best petroleum potential of convergent continental margins is most likely to be in several types of basins associated with these margins, rather than in the areas of present slopes and rises (Seely and Dickinson, 1977). The Cook Inlet basin of Alaska, the Piura basin of Peru, and the Tertiary basins of Indonesia are typical examples. In such areas, the sedimentary section is thicker, the heat flow greater, and the geochemical prerequisites for production are usually present.

In a series of excellent papers, Philippi (1965, 1974, 1975, 1977) offers compelling evidence for the thermal generation of oil and gas especially in the Los Angeles and Ventura basins of California. Philippi found that oil was generated in upper Miocene shales below 11,000 ft (3,353 m) in the hot Los Angeles basin, and below 15,000 ft (4,572 m) in the cooler Ventura basin. The temperatures at which oil is generated in each basin (302° and 298°F, or 150° and 148°C) is about the same and very similar to the temperature at the top of the oil-generation zone (0.6 R_o) in the upper Miocene trend of the Gulf Coast (Table 2). Therefore, conditions which cause oil generation in different basins are very much the same, and the Gulf Coast model is a valid approximation for any Tertiary basin, regardless of tectonic setting, if it is adapted for varying geothermal gradients and sedimentation rates. Philippi (1975) also argued convincingly that wet-gas generation requires greater maturity than oil generation, which is consistent with vitrinite-reflectance data (Fig. 2).

Divergent Continental Margins

Sediments beneath continental slopes and rises of stable, Atlantic-type margins are typically older than those beneath slopes and rises of deltaic areas like the Gulf Coast or beneath active subduction margins of the Pacific type. The longer exposure times available reduce the burial depth (temperature) required to achieve peak oil generation. Geothermal gradient information is scant in deep-water areas in the Atlantic but gradients are probably very similar to, or slightly higher than, those in the Gulf Coast (1.4°F/100 ft or 2.6°C/100 m). Gradients of this magnitude have been reported from the passive East African margin (Roberts and Caston, 1975). If this is true, then the predominantly Mesozoic sequences along the North Atlantic margins should attain oil-generation maturities (0.6 R_o) at about 2.4 km, similar to the Gulf Coast Cretaceous trend (Fig. 15). The average top of the oil-generation zone would occur at about 2 km (Fig. 13) in Mesozoic and Paleozoic source beds (Tissot et al, 1974). Oil generation therefore should occur below 2 to 2.5 km in stable, Atlantic-type continental margins where the geothermal gradient is about average and the Tertiary cover is minimal.

In basins with uniform geothermal gradients, maturation profiles based on vitrinite-reflectance data will invariably be linear if the reflectance values are plotted on a log scale and depth on a linear scale (Dow, 1977b). The slope is dependent on the temperature gradient and the exposure time and will project to about 0.2 R_o at the present surface in continuously subsiding basins such as the Gulf Coast (Fig. 15). These principles have many applications, such as estimating former depth of burial in uplifted areas and the amount of section lost at unconformities, as well as determining the depth to the oil- and gas-generation zones (Dow, 1977b).

A maturation profile for DSDP Site 391 (Fig. 12) is shown on Figure 18. The rate of maturation increase (slope) in the Mesozoic section is only slightly greater than in the Gulf Coast Cretaceous trend (Fig. 15) where the geothermal gradient is about 0.3°F/100 ft below average for a clastic sequence. The profile at Site 391 is complicated by a maturation anomaly, or "offset," at the Cretaceous-Cenozoic unconformity. The amount of offset can be used to estimate the thickness of section lost at the unconformity. For example, if the Cretaceous profile at Site 391 (Fig. 18) is projected to the rank of the organic matter at the base of the Miocene, it is estimated that about 800 m of section has been lost to erosion. The rationale for making such an estimate is described by Dow (1977a,1977b). Similar estimates have been made for DSDP Sites 397 and 398, in the eastern North Atlantic (Fig. 12), both of which have similar maturation rates but less erosion at the Mesozoic-Cenozoic unconformity (Dow, 1977c; Pearson and Dow, 1978).

The present depth to the top of the oil-generation maturity zone (0.6 R_o) at DSDP Site 391 is only 1.2

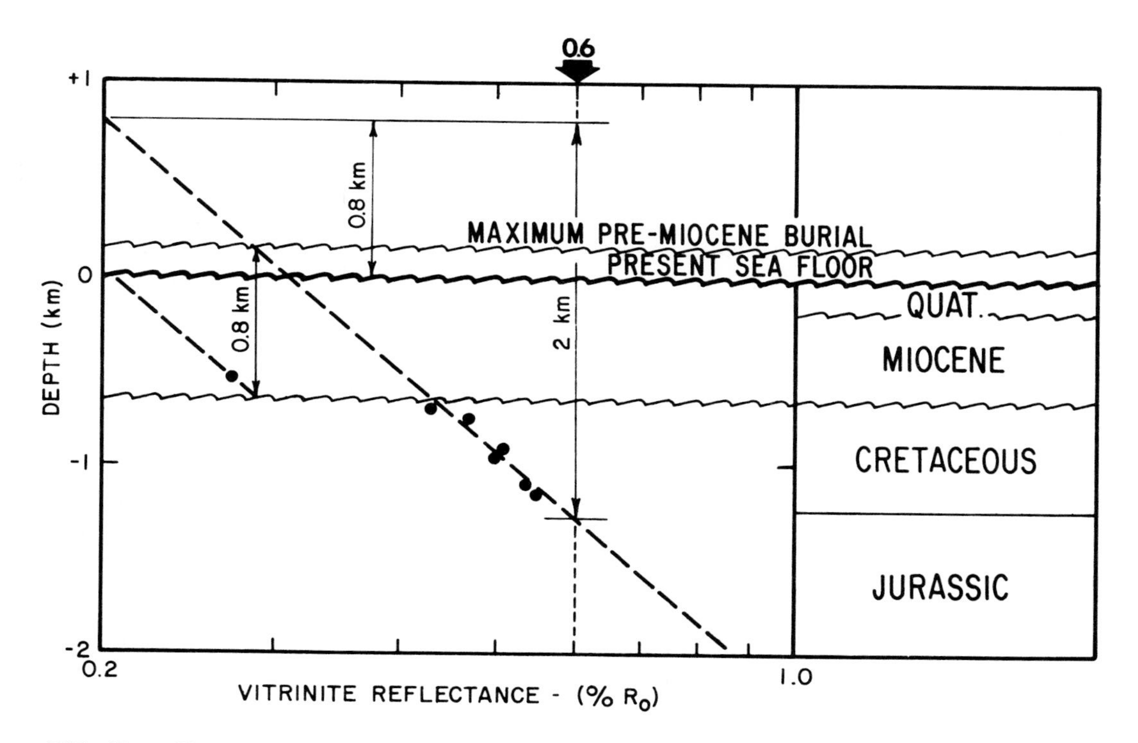

FIG. 18 Kerogen-maturation profile for DSDP Site 391, continental rise, western North Atlantic. For location see Figures 12, 20, and 21. From Dow (1977c).

km (Fig. 18) but without the major erosional event between the Mesozoic and Cenozoic sections, the 0.6 R_o maturity level would be very close to 2 km. Less erosion has occurred at DSDP Sites 397 and 398, but the reconstructed depths to the same maturity horizon are, at both sites, close to 2 km (Dow, 1977c). The 2-km burial required for oil generation on these continental-rise sites is about average for pre-Cenozoic basins of the world (Fig. 14) and only slightly less than the 2.4 km required in the Gulf Coast Cretaceous trend (Fig. 15). The maturity of the Cretaceous black shale is not sufficient for oil generation and even if the organic-rich black-shale sequence contains oil-generating kerogen, it must be considered to be a potential source bed in the areas of the three sites studied. Confirmation of thermal immaturity is provided by a distinct odd-carbon predominance (high C.P.I.) in the organic extracts from cores below the black shale interval (Dow, 1977c). High C.P.I. values indicate the presence of terrestrial organic matter as well as thermal immaturity.

The three DSDP sites studied are on continental rises where erosional events have resulted in complex thermal histories and maturation profiles. Less erosion probably has occurred between the Mesozoic and Cenozoic sediments in the upper rises, slopes, and outer shelves, and the 2-km burial depth for oil generation can be considered a minimum in these parts of the Atlantic continental margin.

Thick Cenozoic sediments would be expected to increase the burial depth to the 0.6 R_o maturity level, the amount being proportional to the age of the Tertiary cover (Fig. 15). The depth to the top of the oil-generation zone in the Baltimore Canyon B-2 COST well (Fig. 19) is about 3.4 km, but the well penetrated about 1.5 km of Cenozoic sediments and has a low geothermal gradient of about 1.3°F/100 ft (Scholle, 1977). Wells in areas with higher geothermal gradients and less Cenozoic cover should encounter the top of the oil-generation maturity zone somewhere between 2 and 3.4 km.

The maximum areal extent of sediments mature enough to generate oil along the eastern continental margin of North America can be defined by the area containing at least 2 km of total sedimentary rocks (Fig. 20). The thickness of sediments required to reach the 0.6 R_o maturity level will be greatest in areas with thick Cenozoic cover because of shorter exposure times available.

Figure 20 shows the distribution of thick Cenozoic sediments along the western North Atlantic continental margin. The depth to the 0.6 R_o maturity level must be increased from 2 km to about 3 km in areas with 1 to 2 km of Cenozoic cover and to 4 km in areas with more than 2 km of Cenozoic cover, assuming there are no appreciable geothermal gradient variations. The maximum extent of oil source beds is reduced to about half the area containing greater than 2 km of total sediments (Fig. 21). The outer shelves, slopes, and upper rises beneath the western North Atlantic continental margin have the best chance of containing oil source beds because of their thermal maturity. Gas source beds require

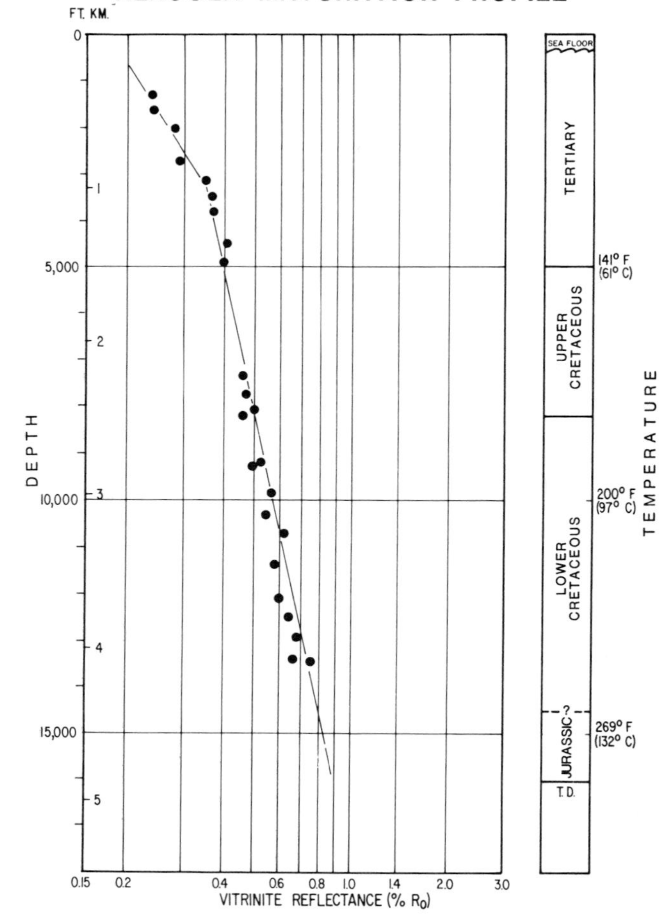

FIG. 19 Kerogen-maturation profile for Baltimore Canyon, B-2 COST well. For location see Figures 20 and 21. Replotted from Scholle (1977).

higher maturation levels equivalent to about 1 km of additional burial in this region. Although source beds could exist in areas so defined, the key is the presence of organic-rich rocks within the thermally mature portion of the sedimentary section.

Figure 22, a diagrammatic cross section of the continental margin in the vicinity of the Baltimore Canyon B-2 COST well (Fig. 21), indicates the part of the sedimentary section which could contain oil source beds on the basis of thermal maturity alone. Rocks in this section in the B-2 well, however, consist primarily of evaporite, carbonate, and terrigenous rocks, and are either organically lean or contain primarily gas-generating kerogen types (Scholle, 1977). Some Lower Cretaceous coaly deposits might be capable of generating gas, but an additional kilometer or so of burial would be required (Figs. 2, 19). It is not likely that the Cretaceous "black shales" which occur between reflecting horizons A and B under the continental rise (Fig. 22) are mature enough to generate oil even if they are present beneath the upper rise and slope. The geochemical characteristics of much of the Lower Cretaceous and Jurassic section along the western North Atlan-

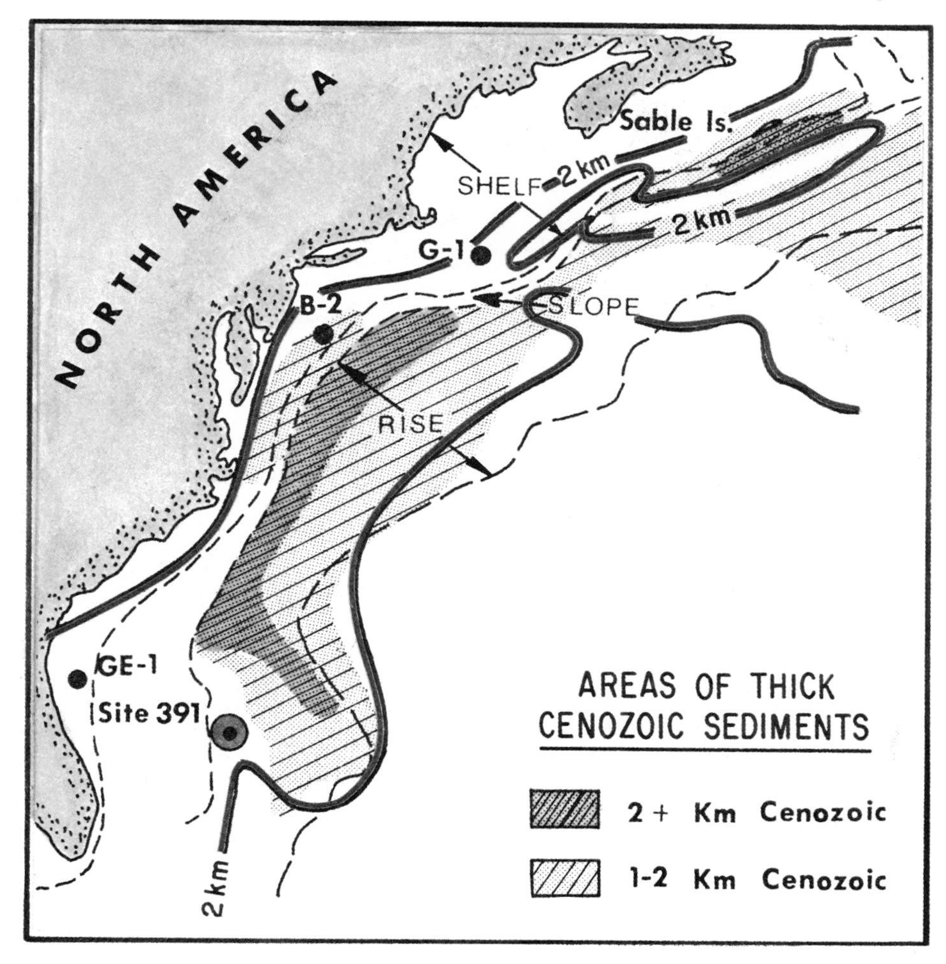

Fig. 20 Map of eastern North America continental margin showing maximum extent of area with mature sediments (>2 km) and areas of thick Cenozoic sediments. Thickest Cenozoic accumulations are located on the continental slope or upper continental rise. Isopach data from Emery et al (1970).

tic continental margin have yet to be defined, and the future of economic development in this region depends on whether source-quality rocks exist in the thermally mature part of the section. It does not seem likely, however, that these rocks were deposited under conditions typical of slope and rise sedimentation. Slope and rise sediments in this region are thermally immature and cannot be oil or gas source beds even if they have all the other geochemical prerequisites.

CONCLUSIONS

The purpose of this brief review has been to present some factors which control the formation of petroleum source beds, especially on outer continental margins. Depositional environments favorable for the production of aquatic, petroleum-generating organic matter and for its preservation in sediments are most commonly found on continental slopes. Generation of oil and gas requires thermal maturation of organic-rich sediments under appropriate time and temperature conditions. The amount of burial required to achieve oil and gas generation at rates sufficient for expulsion of commercial quantities varies according to the geothermal gradient and the exposure time. It is not enough to say that oil and gas are generated at certain temperatures unless the burial rates and exposure times are fairly constant. Before petroleum can accumulate, migration of expelled oil and gas between source beds and reservoirs must occur, and certain structural and sedimentologic features are most favorable for migration.

Other variables not described in this report may ultimately affect source-bed performance. Increasing sedimentation rates during peak oil or gas generation accelerate the generation rate, increase expulsion efficiency, and favor concentration rather than dispersion of expelled hydrocarbons. High-pressure shales typically contain above-average geothermal gradients which accelerate maturation rates and retard dewatering until oil and gas are formed. Tectonic activity associated with faulting and salt or shale piercement can fracture high-pressure shales and release generated oil and gas to tectonically controlled migration pathways. Many petroleum occurrences on continental margins, especially those associated

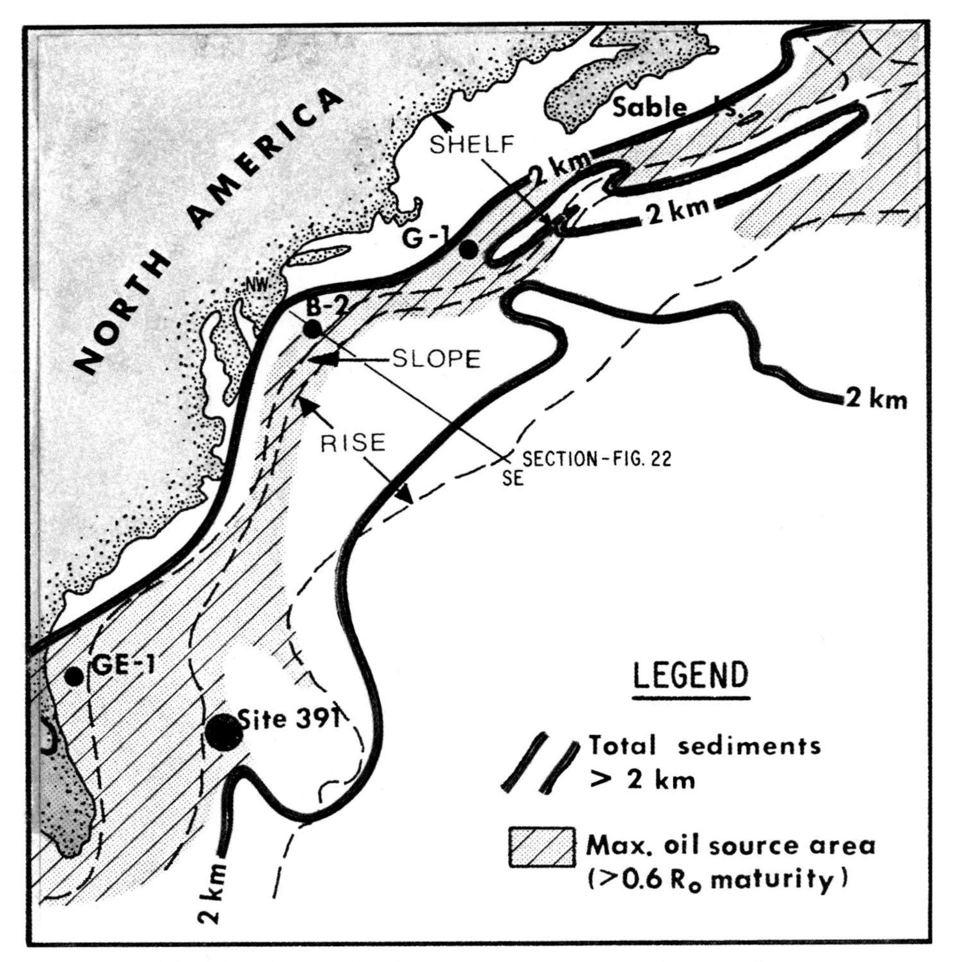

Fig. 21 Map showing probable extent of area containing sediments mature enough (>0.6 R_o) to generate oil. A part of area with more than 2 km of sediments has been eliminated because of thick Cenozoic cover. Oil source beds could be present in the shaded areas if organic-rich sediments occur below the 0.6 maturation horizon.

with thick Cenozoic sedimentation, are closely related to thermally mature, high-pressure shales. Some examples include the MacKenzie basin and Scotian Shelf of Canada, the Louisiana Gulf Coast, offshore Trinidad, and the Niger delta.

Paramount to the formation of commercial petroleum accumulations are the geometric relations among source beds, migration pathways, reservoirs, traps, seals, and the timing of geologic events. For example, traps cannot form after migration has occurred. These variables cannot be considered as separate, detached entities. Each exploration province is unique and must be evaluated on the basis of its own merit, although the basic principles used to study it are universal. Intelligent forecasts on the economic potential of unexplored regions can be made from principles developed in analogous, better known areas. Specifically, the source capacity of slope and rise sediments can be predicted with some reliability from data collected from wells on the better explored continental shelves and from core holes of the Deep Sea Drilling Project.

The physiographic constraints which define present continental slopes and rises should not impose arbitrary limits to exploration. Features favoring production on shelves may extend beyond the shelf-slope boundary, but because the rocks seaward of this boundary are generally younger, the thickness of the sedimentary section within the oil and gas maturity zones is less. Reserve estimates should not be based solely on the volume of sediments (Krueger, 1978) but on the volume of sediments mature enough to generate oil and gas. Also, if possible, it should be determined whether these mature sediments are likely to have the capacity to yield commercial quantities of petroleum. Estimating reserves in the early stages of exploration is difficult but, if geochemical principles are applied, far more rational and realistic predictions can be developed. Attention must be focused not only on traps and reservoirs but on whether there is a reasonable expectation that they contain oil and gas before committing to expensive ventures in the largely unknown realm of continental slopes and rises.

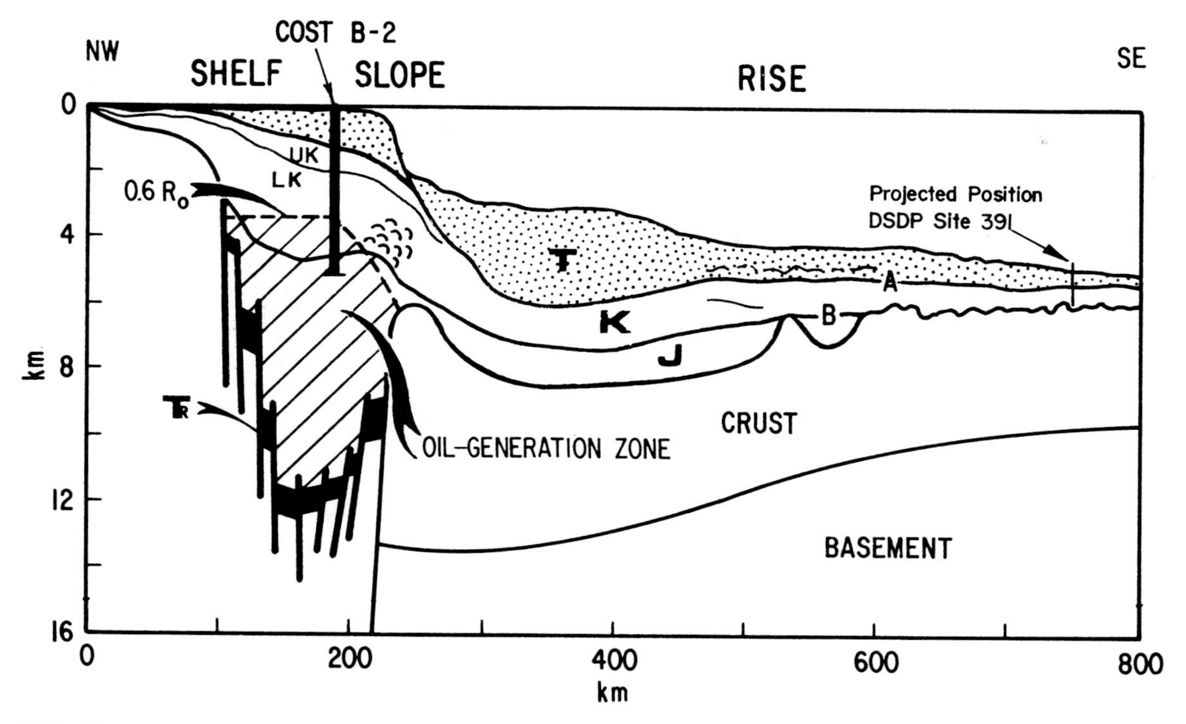

FIG. 22 Diagrammatic cross section of continental margin off New Jersey showing distribution of sedimentary rocks mature enough to generate crude oil. Position of DSDP Site 391 is projected and is approximate. From Sheridan (1974).

SELECTED REFERENCES

Antoine, J. W., et al, 1974, Continental margins of the Gulf of Mexico, *in* The geology of continental margins: New York, Springer-Verlag, p. 683–693.

Claypool, G. E., B. J. Presley, and I. R. Kaplan, 1973, Gas analyses in sediment samples from Legs 10, 11, 13, 14, 15, 18, and 19, *in* Initial reports of the Deep Sea Drilling Project, v. 19: Washington, D.C., U.S. Govt. Printing Office, p. 879–884.

Closs, H., H. Narain, and S. C. Garde, 1974, Continental margins of India, *in* The geology of continental margins: New York, Springer-Verlag, p. 629–639.

Degens, E. T., and K. Mopper, 1976, Factors controlling the distribution and early diagenesis of organic material in marine sediments: Chem. Oceanography, v. 6, p. 59–113.

Dow, W. G., 1977a, Contact metamorphism in sediments from Leg 41: Cape Verde Rise and Basin, *in* Initial reports of the Deep Sea Drilling Project, v. 41: Washington, D.C., U.S. Govt. Printing Office, p. 821–824.

——— 1977b, Kerogen studies and geological interpretations: Jour. Geochem. Exploration, v. 7, p. 79–99.

——— 1977c, Petroleum source beds on continental slopes and rises, *in* Geology of continental margins: AAPG Cont. Education Course Note Ser. 5, p. D1-D37.

——— 1978, Geochemical analysis of samples from Sites 391A and 391C, *in* Initial reports of the Deep Sea Drilling Project, v. 44; Washington, D.C., U.S. Govt. Printing Office (in press).

——— and D. B. Pearson, 1975, Organic matter in Gulf Coast sediments: 7th Offshore Tech. Conf. Preprints, paper OTC 2343.

Emery, K. O., et al, 1970, Continental rise off eastern North America: AAPG Bull., v. 54, p. 44–108.

——— and E. Uchupi, 1972, Western North Atlantic Oceans-topography, rocks, structure, water, life, and sediments: AAPG Mem. 17, 532 p.

Erdman, J. G., 1975, Time and temperature relations affecting the origin, expulsion and preservation of oil and gas: 9th World Petroleum Cong. Proc., v. 2, p. 139–148.

Ewing, M., et al, 1973, Sediment distribution in the oceans—the Atlantic: Geol. Soc. America Bull., v. 84, p. 71–87.

Fairbridge, R. W., ed., 1966, The encyclopedia of oceanography, v. 1: Encyclopedia of Earth Science Ser., New York, Reinhold Pub. Corp.

Frey, M. G., and W. H. Grimes, 1970, Bay Marchand-Timbalier Bay-Caillou Island salt complex, Louisiana, *in* Geology of giant petroleum fields: AAPG Mem. 14, p. 277–291.

Grayson, J., and R. E. LaPlante, 1973, Estimated temperature history in the lower part of hole 181 from carbonization measurements, *in* Initial reports of the Deep Sea Drilling Project, v. 18: Washington, D.C., U.S. Govt. Printing Office, p. 1077.

Harwood, R. J., 1977, Oil and gas generation by laboratory pyrolysis of kerogen: AAPG Bull., v. 61, p. 2082–2102.

Hedberg, H. D., 1970, Continental margins from viewpoint of the petroleum geologist: AAPG Bull., v. 54, p. 3–43.

Heezen, B. C., 1974, Atlantic-type continental margins, *in* The geology of continental margins: New York, Springer-Verlag, p. 13–24.

Krueger, W. C., 1978, Sediment thickness and percentage estimate of offshore petroleum reserves: Oil and Gas Jour., v. 76, no. 3, p. 88–90.

Kulm, L. D., et al, 1973, Initial reports of the Deep Sea Drilling Project, v. 18: Washington, D.C., U.S. Govt. Printing Office, 1077 p.

Lijmbach, G. W. M., 1975, On the origin of petroleum: 9th World Petroleum Cong. Proc., v. 2, p. 357–369.

McIver, R. D., 1975, Hydrocarbon occurrences from JOIDES Deep Sea Drilling Project: 9th World Petroleum Cong. Proc., v. 2, p. 269–280.

Pearson, D. D., and W. G. Dow, 1978, Geochemical analysis of samples from Sites 397 and 398, *in* Initial reports of the Deep Sea Drilling Project, v. 47: Washington, D.C., U.S. Govt. Printing Office (in press).

Philippi, G. T., 1965, On the depth, time, and mechanism of petroleum generation: Geochim. et Cosmochim. Acta, v. 29, p. 1021–1049.

——— 1974, The influence of marine and terrestrial source material on the composition of petroleum: Geochim. et Cosmochim. Acta, v. 38, p. 947–966.

——— 1975, The deep subsurface temperature controlled origin of the gaseous and gasoline-range hydrocarbons of petroleum: Geochim. et Cosmochim. Acta, v. 39, p. 1353–1373.

——— 1977, On the depth, time, and mechanism of origin of the heavy to medium gravity naphthenic crude oils: Geochim. et Cosmochim. Acta, v. 41, p. 33–52.

Rashid, M. A., and J. D. McAlary, 1977, Early maturation of organic matter and genesis of hydrocarbons as results of heat from a shallow piercement salt dome: Jour. Geochem. Exploration, v. 8, p. 549–569.

Roberts, D. G., and V. D. Caston, 1975, Petroleum potential of the deep Atlantic Ocean: 9th World Petroleum Cong. Proc., v. 2, p. 281–298.

Rogers, M. A., et al, 1973, Geologic controls on the hydrocarbon source potential of young sediments (abs): Gulf Coast Assoc. Geol. Socs. Trans., v. 23, p. 194.

Ross, D. A., 1974, The Black Sea, *in* The geology of continental margins: New York, Springer-Verlag, p. 669-682.

Scholle, P. A., 1977, Geologic studies on the C.O.S.T. No. B-2 well, United States Mid-Atlantic outer continental shelf area: U.S. Geol. Survey Circ. 750, 71 p.

Seely, D. R., and W. R. Dickinson, 1977, Structure and stratigraphy of forearc basins, *in* Geology of continental margins: AAPG Cont. Education Course Note Ser. 5, p. C1–C23.

Sheridan, R. E., 1974, Atlantic continental margin of North America, *in* The geology of continental margins: New York, Springer-Verlag, p. 391–407.

Shokes, R. F., et al, 1977, Anoxic, hypersaline basin in the northern Gulf of Mexico: Science, v. 196, p. 1443–1446.

Steiner, R. J., 1976, Grand Isle Block 16 field, offshore Louisiana, *in* North American oil and gas fields: AAPG Mem. 24, p. 229–238.

Thompson, T. L., 1976, Plate tectonics in oil and gas exploration of continental margins: AAPG Bull., v. 60, p. 1463–1501.

Tissot, B., et al, 1974, Influence of nature and diagenesis of organic matter in formation of petroleum: AAPG Bull., v. 58, p. 499–506.

Uchupi, E., and K. O. Emery, 1968, Structure of continental margin off Gulf Coast of United States: AAPG Bull., v. 52, p. 1162–1193.

Weeks, L. G., 1974, Petroleum resources, potential of continental margins, *in* The geology of continental margins: New York, Springer-Verlag, p. 953–964.

Young, A., P. H. Monaghan, and R. T. Schweisberger, 1977, Calculation of ages of hydrocarbons in oils—physical chemistry applied to petroleum geochemistry: AAPG Bull., v. 61, p. 573–600.

Explosion Seismology Studies of Active and Passive Continental Margins[1]

B. KELLER,[2] B. T. R. LEWIS,[2] C. MEEDER,[2]
C. HELSLEY,[3] AND R. P. MEYER[4]

Abstract Data from seismic refraction experiments across the trenches of Peru, Colombia, and Mexico, and across the U.S. East Coast show distinctive differences between active and passive margins, especially in the mantle paths. Active margins show clear mantle arrivals across the trench whose velocities are consistent with the subduction of oceanic lithosphere at relatively shallow angles of 6° to 14° near the trench axis. Data across the U.S. East Coast have discontinuous weak mantle arrivals that can be partly explained by a relatively sharp change in depth between the continental and oceanic Moho.

INTRODUCTION

Several onshore/offshore seismic experiments have been conducted in the past 12 years to examine the structure of continental margins. They have been conducted across passive margins: the U.S. East Coast ECOOE experiment (Hales et al, 1968) and the Gulf of Mexico (Hales et al, 1970); and active margins: the Nariño experiment off Colombia, the Mexico experiment across the Middle America Trench, and the Peru Trench experiment across the Southern Peru Trench.

In the ECOOE experiment, arrays of seismometers were deployed on land to record shots fired offshore. In the trench experiments, both land and sea detectors were used to record shot lines. Results from the Nariño experiment were reported by Meyer et al (1976) and Wade et al (1974).

In this paper the writers present new data from the Peru and Mexico experiments and compare those results with the other active and passive margin results to see whether diagnostic differences between active and passive margins can be seen. Locations of experiments are shown in Figure 1.

THE PERU EXPERIMENT

Seismic refraction and reflection data were recorded off the coast of southern Peru in an experiment involving the Universities of Washington, Wisconsin, and Texas, Carnegie Institute, and Instituto Geofisico del Peru. Data were recorded on ocean bottom seismometers (OBS), land seismometers, and towed hydrophones. Large explosive sources allowed refraction data to be recorded at distances greater than 350 km. These data, in conjunction with previous shallow reflection and refraction data, provide constraints on the structure of the subduction of the Nazca oceanic plate under South America and the structure of the continental slope and shelf.

Previous seismic studies several hundred kilometers north of the present study have been reported (Kulm et al, 1973; Prince et al, 1974; Prince and Kulm, 1975; Hussong et al, 1976; Hussong et al, 1977). All of the previous studies were north of the Nazca Ridge, a major aseismic bathymetric feature. These earlier airgun and sonobuoy reflection and refraction investigations suggest a wedge-shaped toe of low velocity sediment on the continental slope and a ridge of oceanic basalt in the bottom of the trench. The last feature might have been caused by imbricate thrusting due to compressional forces at the zone of plate convergence.

The Nazca–South American plate boundary has an exceptionally high convergence rate of about 10 cm/yr (Minster et al, 1974). A plot of earthquake hypocenters transverse to the trend of the trench and South American structural features in the area of this study show a steeply dipping (≈30°) seismic zone about 30 km thick (Barazangi and Isacks, 1976). The line of active Andean volcanoes is about 280 km inland from the trench. These features are shown in Figure 2. The width of the seismic zone and depth of hypocenters near the trench suggest that failure may be occurring below the oceanic crust.

[1]Manuscript received, September 28, 1977; accepted, July 17, 1978.

[2]Department of Oceanography and Geophysics, University of Washington, Seattle, Washington 98195.
[3]Hawaii Institute of Geophysics, Honolulu, Hawaii.
[4]Department of Geology and Geophysics, University of Wisconsin, Madison, Wisconsin.

Article Identification Number:
0065–731X/78/MO29–0029/$03.00/0.
(see copyright notice, front of book)

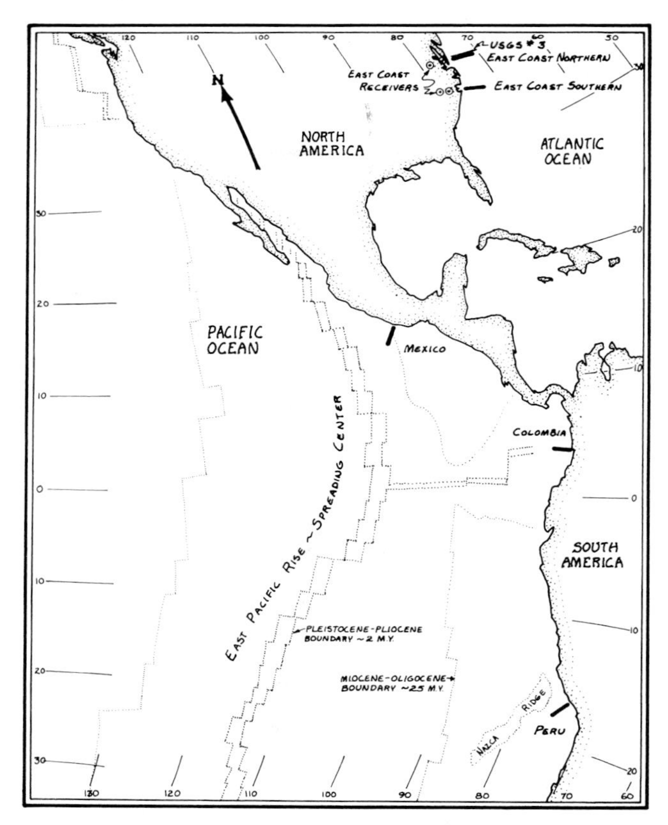

FIG. 1 Map showing locations of explosion seismology experiments across continental margins.

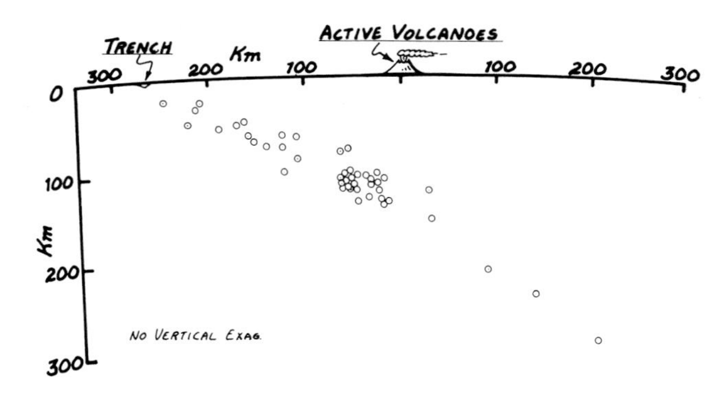

FIG. 2 Cross section of earthquake hypocenters on a line crossing the Peru Trench at 19°S lat., perpendicular to the trench (after Barazangi and Isacks, 1976). This is about 100 km south of the refraction lines of this study, but the distribution is typical of the subduction zone from 15°S to 25°S. 0 km is at the line of active volcanoes. Dip of the Benioff zone is about 30°.

Exposed on the coast in this area are old crystalline rocks (Bellido, 1969). Precambrian gneiss, schist, phyllites and migmatites are intruded by Paleozoic and Mesozoic granites and diorites. Geophysical evidence of this study indicates that rocks with similar properties may extend under a sediment basin to the edge of the continental shelf.

Figures 3, 4, and 5 show the seismic refraction record sections as recorded by two OBSs and two land seismometers. The OBS records form a reversed profile across the Peru Trench. Arrival times have been corrected for water depth, thus the records are as if the shots were on the bottom. Prominent features of all the record sections represent delays of several tenths of a second over the lower continental slope, high apparent velocities across the lower slope and oceanic crust (as seen from the shelf OBS and land receivers), and low apparent velocities across the slope (as seen from the seaward OBS).

The relatively large time delays over the lower slope are suggestive of low velocity material in this area. To better define the zone a reflection profiling experiment was run along line 5. Because no large repetitive sound source was available on this cruise, explosive charges were used as sources and a 100 m hydrophone streamer as the detector.

Two interesting features are apparent on the reflection profile (Fig. 6). One is the strong reflector which continues from the mid-trench ridge up the continental slope at about 1 to 2 sec below the sea floor. The material above this reflector probably causes the delays seen in the refraction data. The other is the lack of coherent reflectors in the oceanic crustal material on the seaward side of the trench.

The reflection records were filtered at 3 to 10 Hz to enhance deep crustal reflections. Over the ocean floor, reflections are coherent only over distances of several kilometers and further processing is essential for proper interpretation. Reflectors are clearly seen from the section of the profile over the trench axis and continental slope.

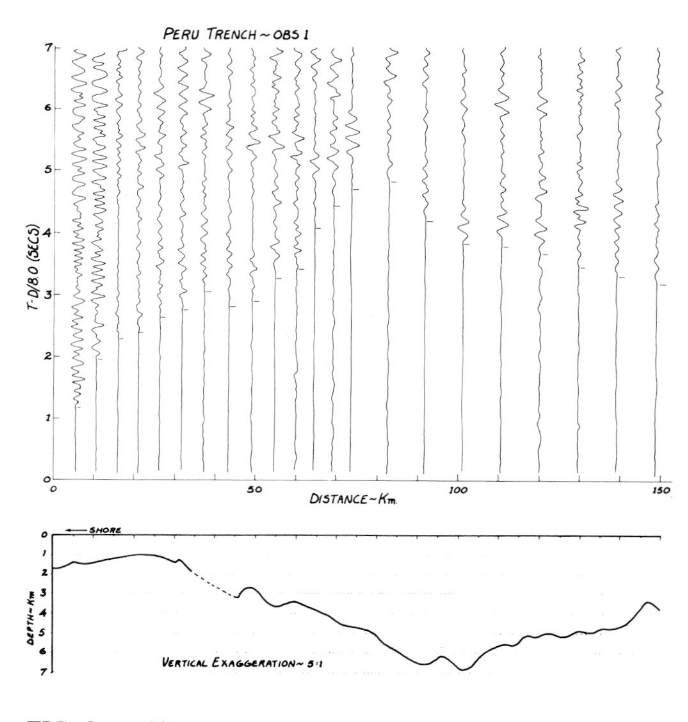

FIG. 3 Shoreward OBS record section and bathymetry of Peru Trench. OBS 1 is located in a sediment basin on the continental shelf (0 km in Figure).

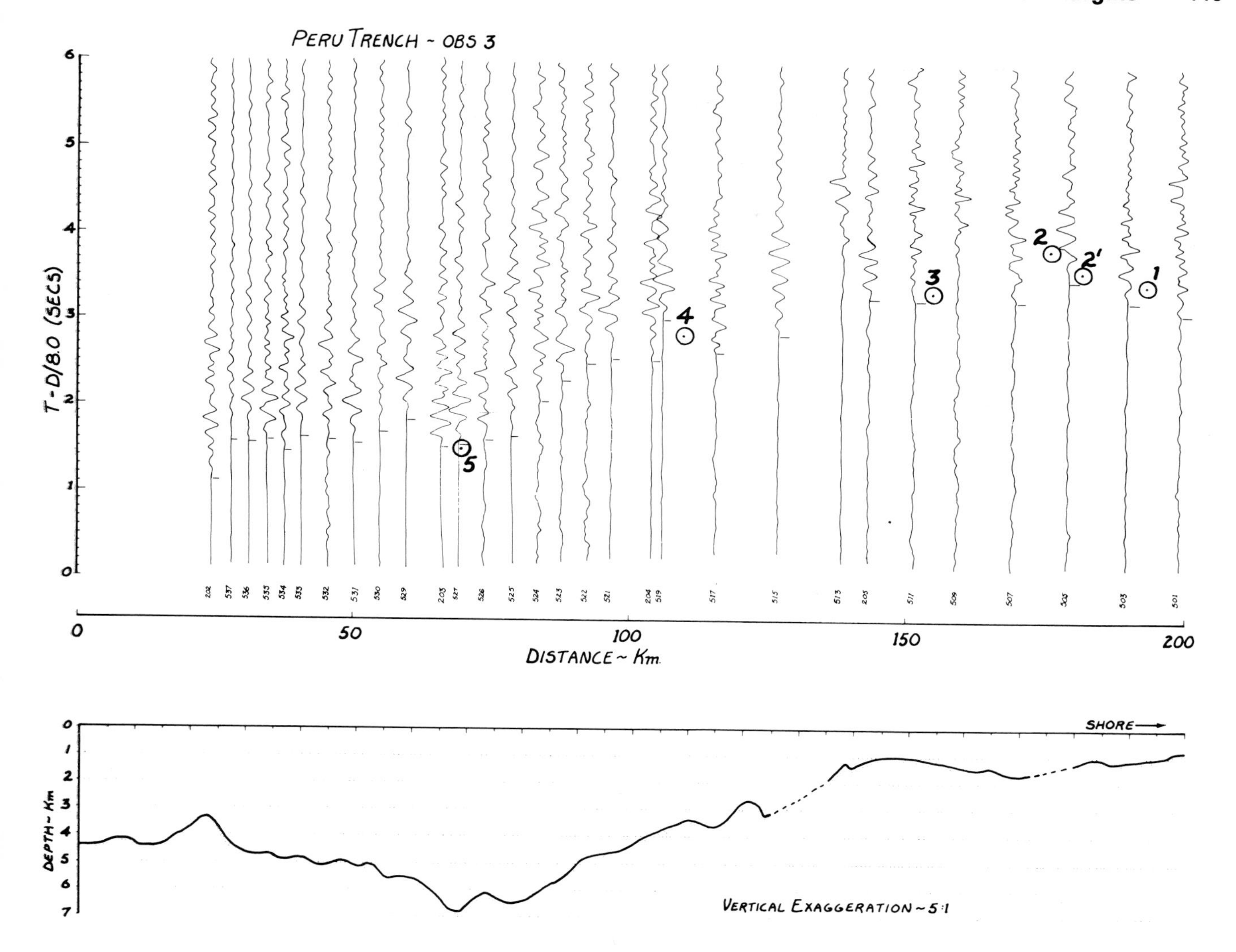

FIG. 4 Seaward OBS record section and bathymetry of Peru Trench. OBS 3 is located in a sediment pond on the Nazca Plate ocean floor (0 km in Fig.). Numbered points are model arrival times from Figure 7.

Bottom profile records over the trench often show a basalt ridge (Kulm et al, 1973) at the bottom of the trench which extends at least 10 km along the axis of the trench. The maximum relief of the ridge is 500 m above the trench floor and the reflection record shows ponds of sediment on either side of the ridge. On the seaward side, beneath the sediment pond, are reflectors with the same apparent dips as the sides of the pond, that might be interpreted as thrust fault surfaces. On the landward side, the reflector surface of the ridge continues under the toe of the continental slope where it is covered with turbidites (Prince et al, 1974). This sediment layer is continuous up the continental slope with thicknesses of about 2.0 km, based on velocities of about 2.0 km/sec. The morphology of the trench is consistent with imbricate thrusting, as suggested by Prince and Kulm (1975).

To model the deep structure of the shelf and slope, we used the reversed travel times between the shelf and ocean crust OBS, an unreversed OBS line on the shelf, gravity data published by Whitsett (1975), and some constraints provided by the reflection line. In the modeling procedure, a velocity model was constructed which satisfied the reversed travel times and this model was then converted into a density model using a linear relation between P velocity and density. Gravity anomalies computed from this model were compared to the data, then the velocity and density models were modified until a satisfactory fit to both gravity and travel times was obtained. A more detailed description of this analysis is in preparation. On the ocean crust side, the crust and upper mantle models were constrained by OBS refraction results published by Meeder et al (1977). The resulting velocity and density models are shown in Figures 7 and 8. Important features are the low velocity material of the continental slope, high velocity material underlying a sediment basin on the continental shelf, and the dip of $\approx 6°$ of the oceanic Moho as it is subducted under the slope and shelf. This dip requires some bending of the oceanic plate near the trench, but is much shallower than the 30° dip of Benioff zone earthquakes.

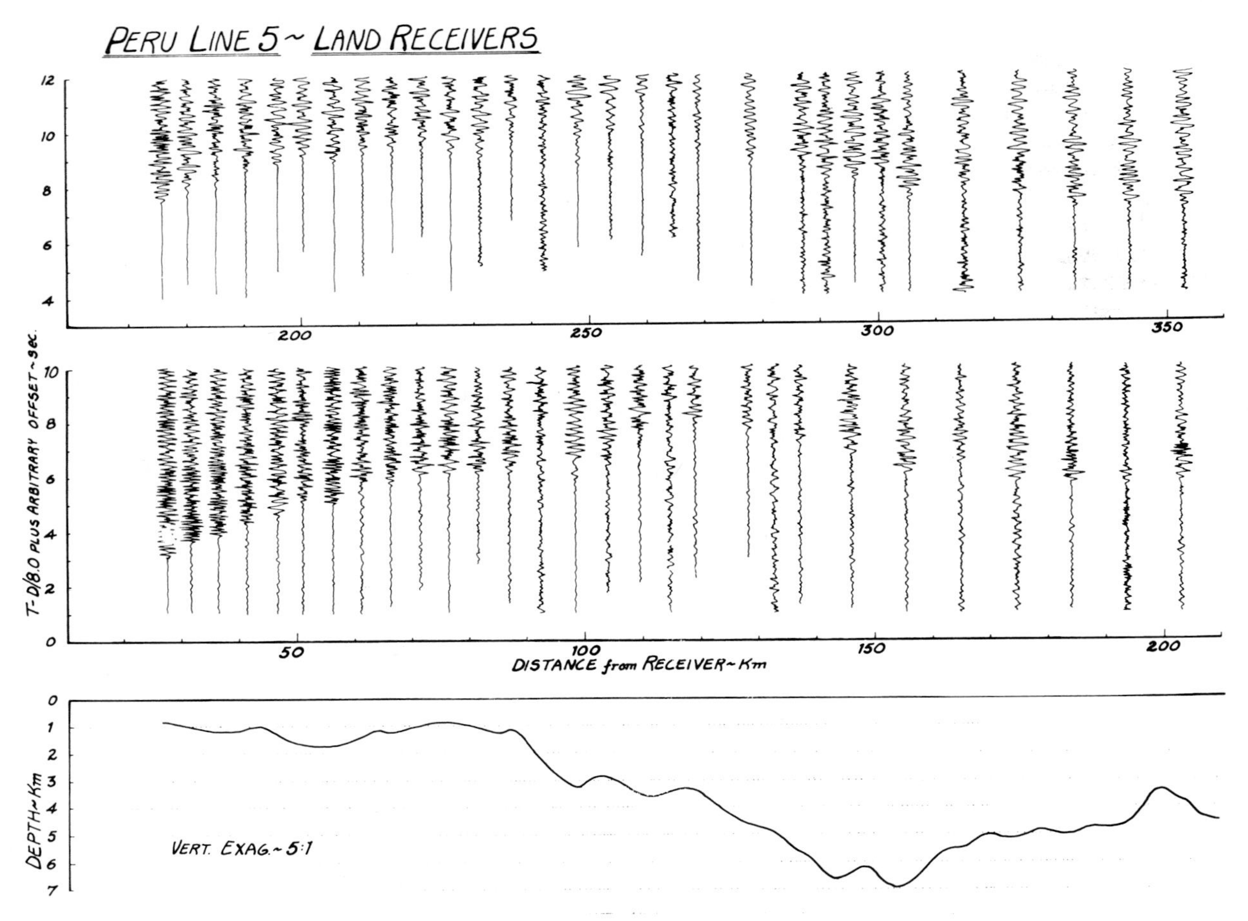

FIG. 5 Land seismometer record sections and bathymetry of Peru Trench.

THE MEXICO EXPERIMENT

A refraction experiment was run across the Middle America Trench at about the latitude of Acapulco (Fig. 1) by the University of Washington in conjunction with the Universities of Texas, Wisconsin, and Mexico. Presented here are data recorded on land by the University of Wisconsin and on telemetering buoys by the University of Washington (Figs. 9, 10).

The depths used are from a 12-kHz sonar bathymetry profile recorded along the ship's course as the shots were fired. An example of quirky results which can come from very irregular topography can be seen in the buoy line (Fig. 10). The third and fourth records appear to be about 0.5 sec late. This is due to a smaller water time correction over the 1-km hill in the bathymetry profile. If the feature was in fact a thin spire or small seamount, this correction may be inappropriate and a more appropriate correction would have used the depths of the surrounding valleys. Such a correction would bring the two records more in line with the rest of the record section. Fortunately, most of the topography was not so irregular.

The travel times from the land and sea stations in this experiment show patterns similar to the Peru results. A low apparent velocity is seen by the buoy for shots crossing the slope and a corresponding high apparent velocity is seen by the land station. This is probably caused by the dip in the Moho refractor coupled with the increasing thickness of continental slope material as one proceeds landward from the trench. No obvious time delays in shots over the lower slope are seen which correspond to those observed in Peru. This suggests that no large body of low velocity material exists under the lower slope as was modeled for the Peru trench. However, there are anomalous time delays from shots seaward of the trench that are not yet explained (Fig. 9).

A model which fits the major features of the records over the trench and slope, shown in Figure 11, is a schematic representation of the subduction features of a model presented by Mooney et al (1975 and personal commun.). The entire continental slope is composed of fairly high-velocity material. The oceanic subduction dip is 13°, which is somewhat steeper than that indicated for southern Peru in a comparable position.

Seismicity data from this part of the Middle America Trench do not define a clear Benioff zone (Molnar and Sykes, 1969) and the dip of this zone could be anywhere between 0° and 30°. Therefore, it is unclear how the 13° dip suggested by these data relates to the earthquake hypocenters.

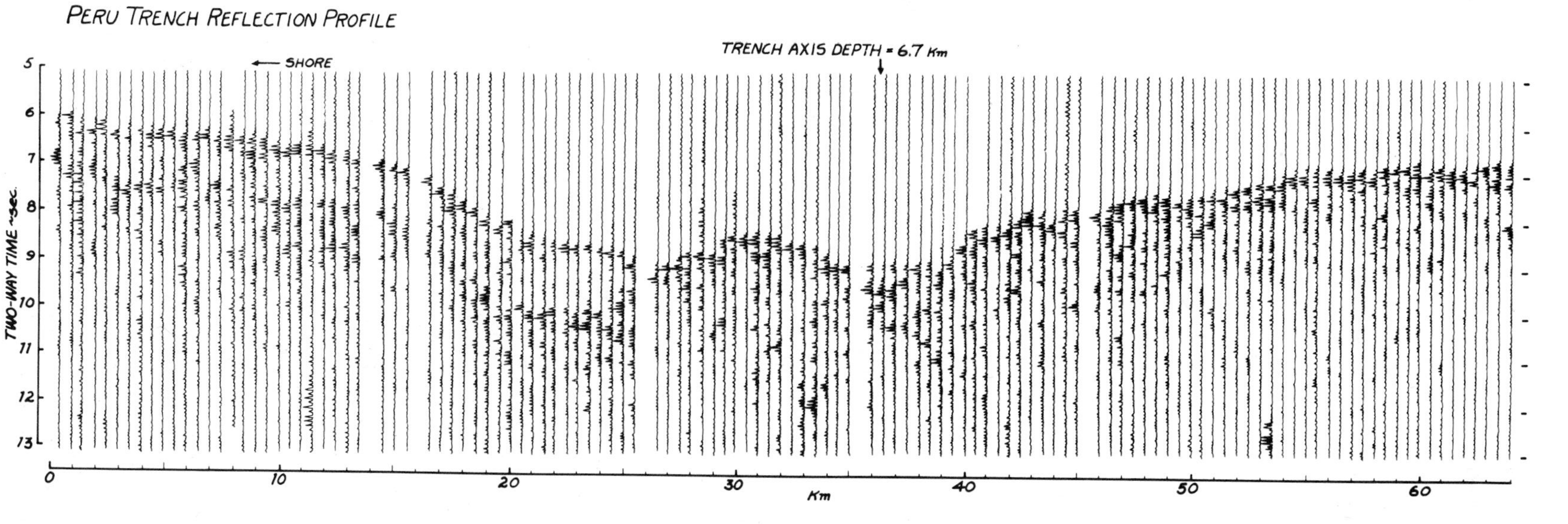

FIG. 6 Low frequency explosion reflection profile across the Peru Trench. Note the continental slope reflector which continues into the basalt ridge in the bottom of the trench.

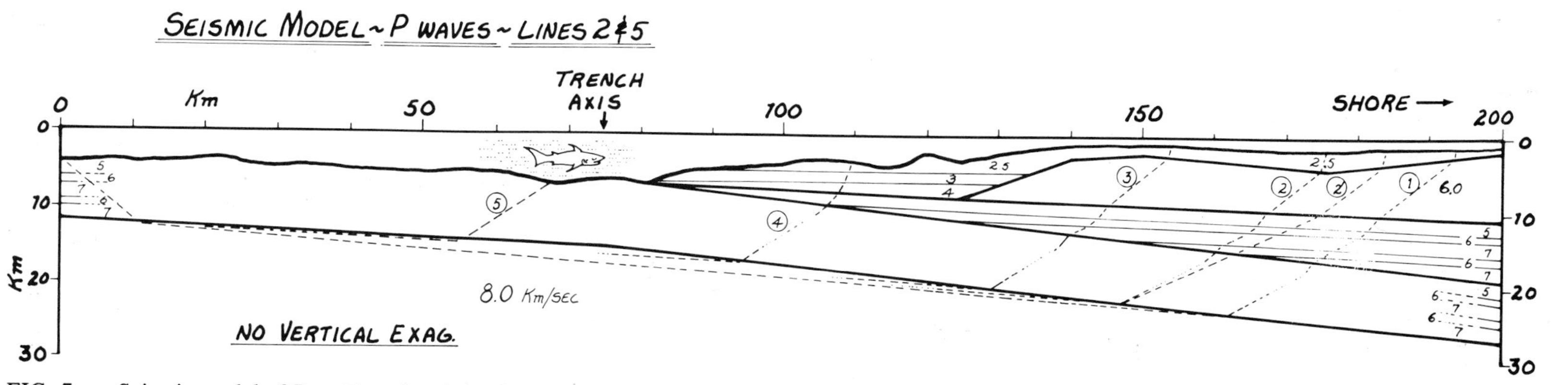

FIG. 7 Seismic model of Peru Trench subduction zone. Outlined by heavy lines are oceanic crust and upper mantle, sediment wedge of the continental slope along with sediment basin on the continental shelf, higher velocity material underlying the continental shelf, and a triangular region with crustal crystalline rock velocities. This last region is shown schematically as a thrust slice of oceanic crust, but could also be continental material. Unprimed rays are for velocities as shown. Ray 2′ is for the same structure with all the material of the triangular region and subducted ocean crust having $V_p = 7.0$, which is reasonable for oceanic crust rocks at 5 to 10 kbar pressure.

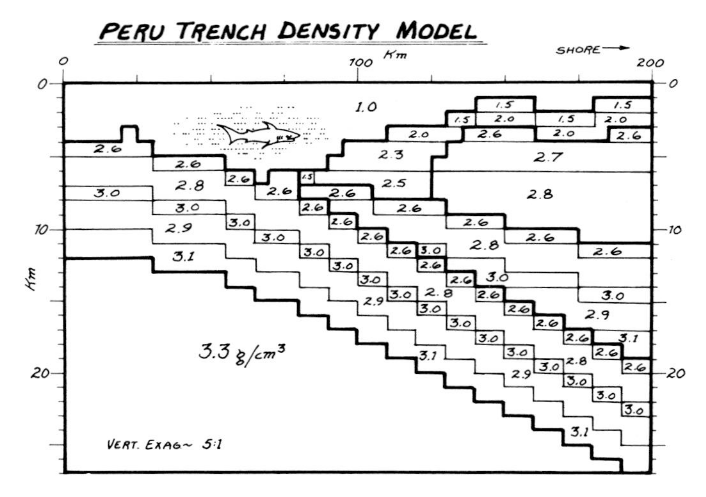

FIG. 8 Density model of Peru Trench subduction zone. This model has the same structure as the seismic model, with appropriate densities.

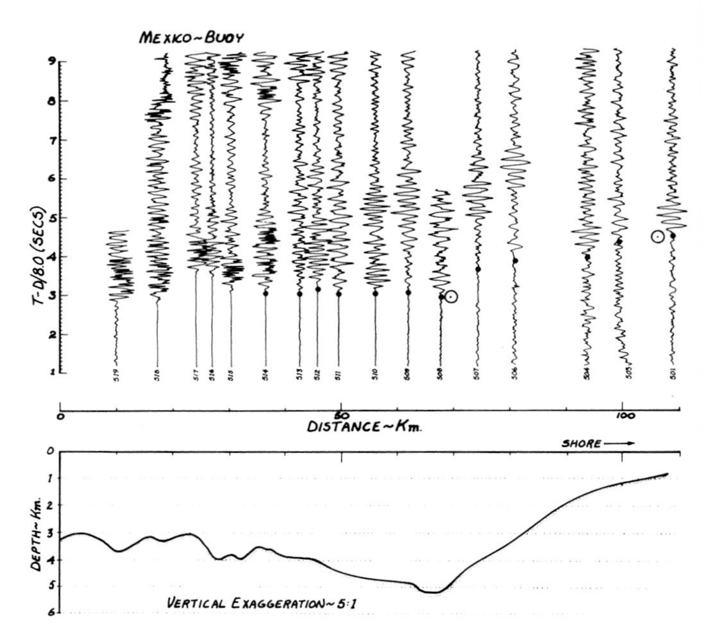

FIG. 10 Buoy record section of the Mexico experiment. Circled points are arrival times for model rays shown in Figure 11.

THE COLOMBIA EXPERIMENT

A refraction experiment across an active continental margin in Colombia was performed as part of Project Nariño and has been previously reported by Meyer et al (1976). The shots were at sea, the recorders on land, so the geometry is similar to the Mexico land record section and Figure 5 for Peru. The record section, bathymetry, and structure inferred by Meyer et al (1976) are shown in Figure 12. As in the other trench experiments, the arrivals across the slope and trench are clearly readable and have a high apparent velocity, as seen from land. The interpretation of these data by Meyer et al (1976) is consistent with subduction models. The model shown has an oceanic Moho dip equal to 14°.

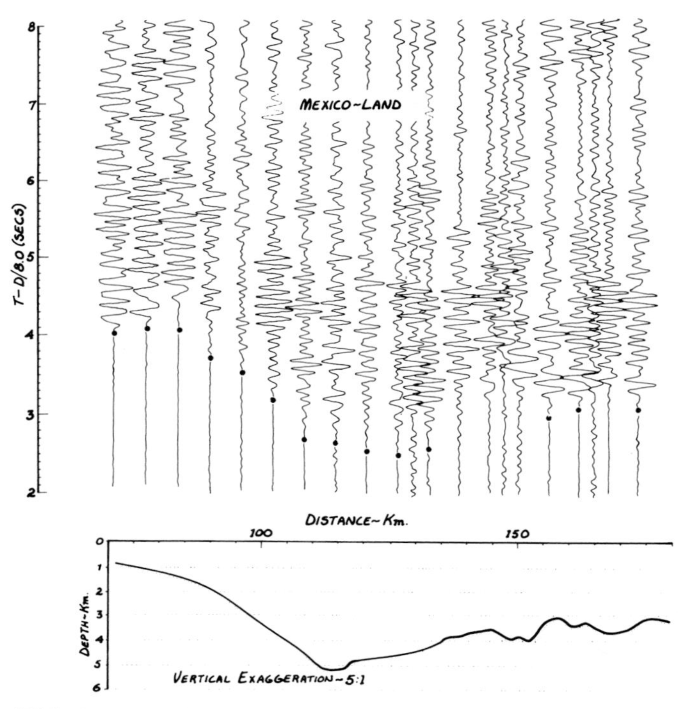

FIG. 9 Mexico experiment record section of shots at sea as recorded by a land seismometer. Low point on bathymetry is the Middle Americas Trench.

U.S. EAST COAST EXPERIMENT

In contrast to the experiments described above, all on the active margin of the eastern Pacific Ocean, one record section is presented here showing the results from a passive margin experiment on the East Coast of the United States (Fig. 13). A more complete description of these results is given in Hales et al (1968) and Lewis and Meyer (1977). The most striking feature of the record section is that the mantle arrivals are discontinuous. They can be separated into weak continental mantle arrivals which die out on the continental shelf, about 220 km from the receiver, and stronger but greatly delayed oceanic mantle arrivals past 350 km. There are no clear mantle arrivals at intermediate distances, where a very thick consolidated sediment accumulation is indicated by multichannel reflection profiles obtained by the U.S. Geological Survey and IPOD (Grow and Schlee, 1976). Selected rays through the two-dimensional model used by Lewis and Meyer (1977) to explain these data are shown in Figure 14.

Low amplitude Pn arrivals across the margin are predicted by this model for two reasons. First, the constant velocity, or even slightly decreasing velocity, from about 40 to 70 km depth will cause low amplitude. Second, the step in the depth to Moho

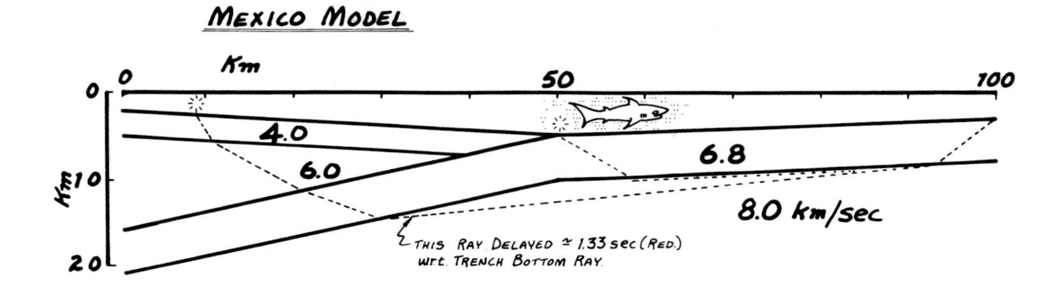

FIG. 11 A model with the subduction angle and continental slope structure derived for Mexico land data by Mooney et al (1975, and personal commun.). The rays shown here indicate that the delay times for shots over the continental slope relative to the trench bottom are in agreement with the buoy data (See Fig. 10). The subduction dip is 13°.

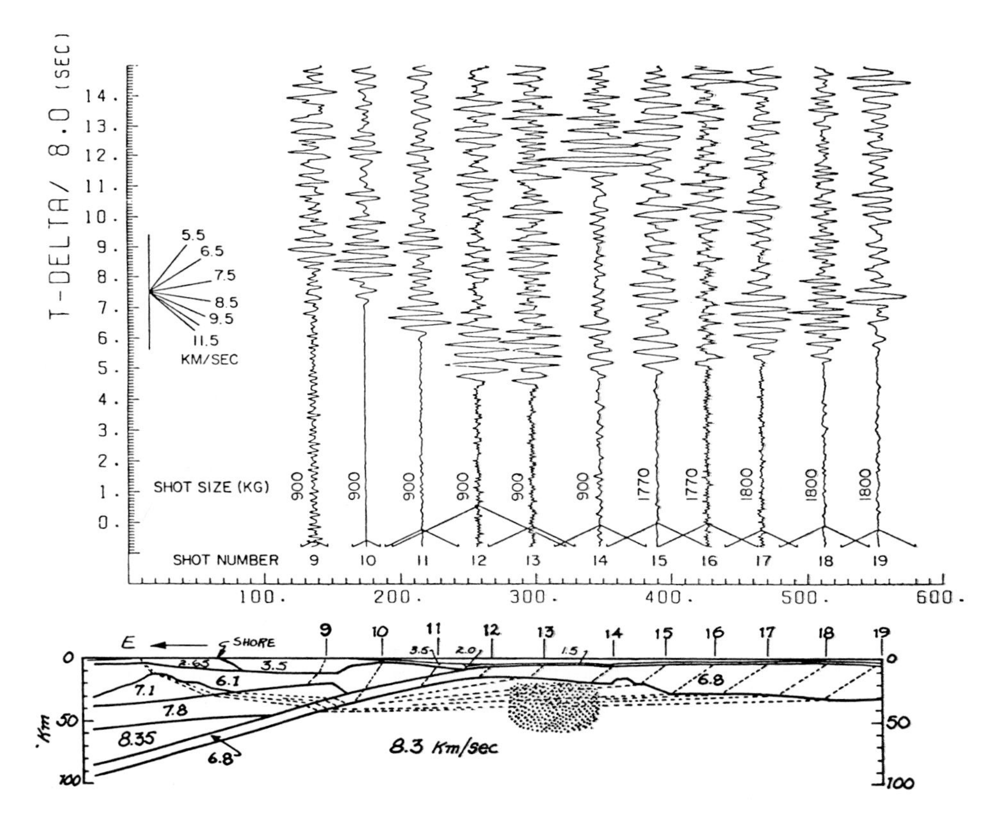

FIG. 12 Colombia experiment record section, bathymetry and structural model after Meyer et al (1976). The subduction dip in the model is 14°. The grey spot in the oceanic mantle is a possible zone of partial melting.

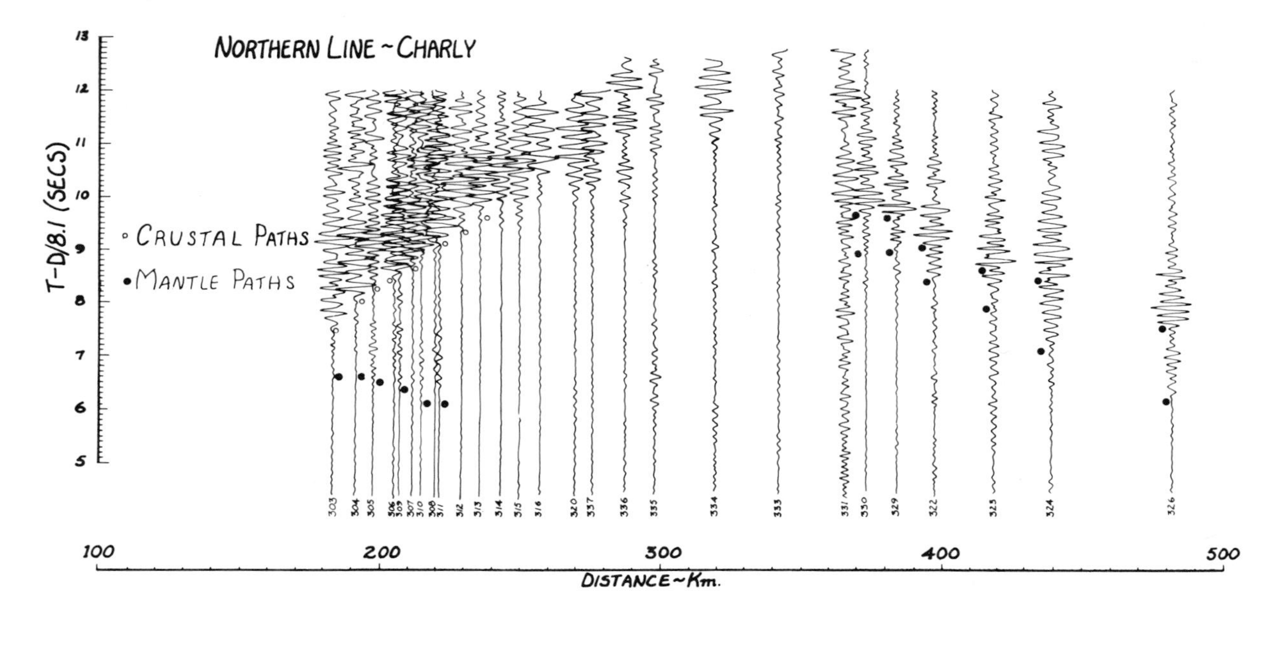

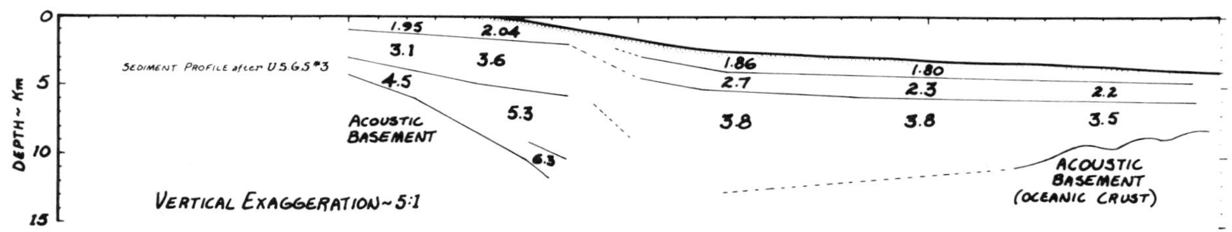

FIG. 13 East Coast experiment record section, bathymetry and shallow crustal structure off Chesapeake Bay. Travel times have been corrected to the sea floor. Note discontinuous mantle arrivals. Basement topography from Grow and Schlee (1976).

associated with the change from continental to oceanic crust will cause a shadow zone for some distance beyond the shelf for a land-based receiver. An increase in mantle velocity at about 70 km depth is used by Lewis and Meyer (1977) to explain the high apparent velocity of the oceanic mantle arrivals. One might expect to see conversions at the continent-ocean boundary from oceanic mantle paths to continental crustal paths, but these will appear on the travel time curve as extensions of the continental crustal travel time curve (Bott et al, 1976; Lewis and Meyer, 1977; Fig. 15). A comparison of these passive margin data with trench data shows obvious differences in the character of mantle energy propagation across these margins.

CONCLUSION

A comparison of experiments using refraction explosion seismology techniques with both land and sea receivers shows that for the passive and active margins considered here there appear to be characteristic differences in the propagation of seismic waves across these margins. From the seismic record sections presented here, it appears that there are distinctive forms for the two classes of margins. The three active margins considered all show record sections consistent with downgoing oceanic slabs being subducted at shallow angles under overlying continental slope topography. The details of the shape of the record section may be modified by sedimentary prisms of accretionary or turbidite material, but the general shape is consistent with a dipping interface and continuous mantle paths.

Figure 15 is a schematic representation of the ray paths for both cases. The "shadow zone" causing the discontinuous arrivals on the passive margin may be the result of a step in the Moho. The details of the step are not resolvable by the refraction line, but it is apparent that the abrupt change from oceanic to continental crustal thickness can cause the discontinuous mantle arrivals. Another possible cause may be the nature of the mantle velocity depth function, as in Figure 14.

One curious feature of our results for active margins is the very shallow dips of 6° to 14° indicated for the subducting oceanic slabs landward of the trench. It seems that the topographic dip seen on the oceanic side continues under the continental slope with only slight bending. This is in contrast with the earthquake Benioff zones which may dip as much as 30°. If the Benioff zone is taken to indicate the position of the interface between the colliding litho-

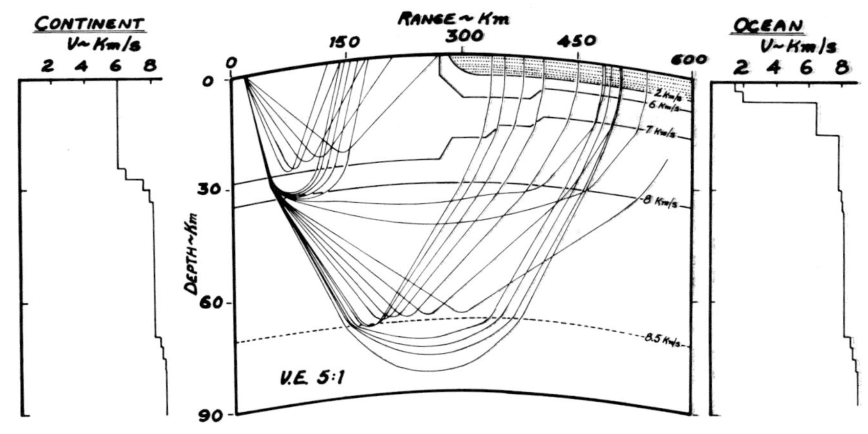

FIG. 14 East Coast seismic model showing ray paths. Continental and oceanic mantle arrivals form distinct groups of rays.

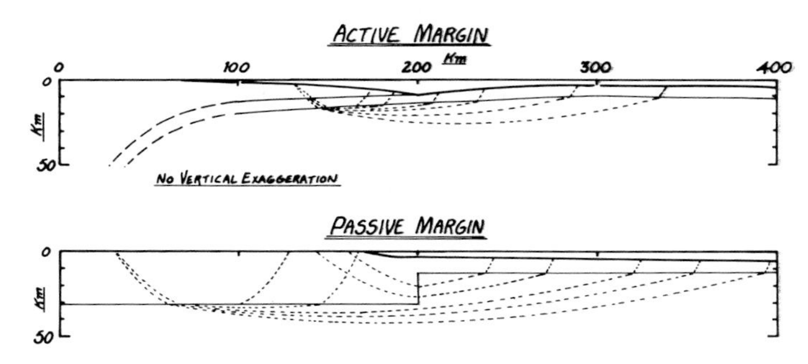

FIG. 15 Schematic models of active and passive margin ray paths.

spheric plates, then most of the requisite bending of the oceanic plate must begin landward of the trench in the areas studied.

REFERENCES CITED

Barazangi, M., and B. L. Isacks, 1976, Spatial distribution of earthquakes and subduction of the Nazca plate beneath South America: Geology, v. 4, p. 686–692.

Bellido, E., 1969, Sinopsis de la geologia del Peru: Servicio de Geologia y Mineria., No. 22.

Bott, M. H. P., P. H. Nielsen, and J. Sunderland, 1976, Converted P waves originating at the continental margin between the Iceland-Faeroe ridge and the Faeroe block: Royal Astron. Soc. Geophys. Jour., v. 44, p. 229–238.

Grow, J. A., and J. Schlee, 1976, Interpretation and velocity analysis of U.S. Geological Survey multichannel reflection profiles 4, 5, and 6, Atlantic continental margin: U.S. Geol. Survey Misc. Field Studies Map, MF-808.

Hales, A. L., C. E. Helsley, and J. B. Nation, 1970, P travel times for an oceanic path: Jour. Geophys. Research. v. 75, p. 7362–7381.

——— et al, 1968, The East Coast onshore/offshore experiment I. The first arrival phases: Seismol. Soc. America Bull., v. 58, p. 757–820.

Hussong, D. M., M. E. Odegard, and L. K. Wipperman. 1975. Compressional faulting of the oceanic crust prior to subduction in the Peru-Chile trench: Geology, v. 3, p. 601–604.

——— et al, 1976, Crustal structure of the Peru-Chile trench: 8°–12°S latitude: Am. Geophys. Union Geophys. Mon. Ser. 19, p. 71–85.

Kulm, L. D., et al, 1973, Tholeiitic basalt ridge in the Peru Trench: Geology, v. 1, p. 11–14.

Lewis, B. T. R., and R. P. Meyer, 1977, Upper mantle velocities under the East Coast margin of the U.S.: Geophys. Research Letters. v. 4, p. 341–344.

Meeder, C. A., B. T. R. Lewis, and J. McLain, 1977, The structure of the ocean crust off southern Peru determined from an ocean bottom seismometer: Earth and Planetary Sci. Letters, v. 37, p. 13–28.

Meyer, R. P., et al, 1976, Project Nariño III: Refraction and observation across a leading edge, Malpelo Island to the Colombian Cordillera Occidental: Am. Geophys. Union Geophys. Mon. Ser. 19, p. 105–132.

Minster, J. B., et al, 1974, Numerical modeling of instantaneous plate tectonics: Royal Astron. Soc. Geophys. Jour., v. 36, p. 541–576.

Molnar, P., and L. Sykes, 1969, Tectonics of the Caribbean and Middle America regions from focal mechanisms and seismicity. Geol. Soc. America Bull., v. 86, p. 1639–1653.

Mooney, W. M., et al, 1975, Refracted waves across a leading edge: Observations of Pacific shots in southern Mexico (abs.): EOS, v. 56, p. 452.

Prince, R. A., and L. D. Kulm, 1975, Crustal rupture and the initiation of imbricate thrusting in the Peru-Chile trench: Geol. Soc. America Bull.: v. 86, p. 1639–1653.

———et al, 1974, Uplifted turbidite basins on the seaward wall of the Peru Trench: Geology, v. 2, no. 12, p. 607–611.

Wade, U. S., B. T. R. Lewis, and C. R. B. Lister, 1974, Structure near Malpelo ridge from seismic refraction results: Historia y Resultados del Proyecto Nariño, Instituto Geophysico, Bogotá, Colombia.

Whitsett, R. M., 1975, Gravity measurements and their structural implications for the continental margin of southern Peru: Ph.D. thesis, Oregon State Univ.

The Effect of Eustatic Sea Level Changes on Stratigraphic Sequences at Atlantic Margins[1]

WALTER C. PITMAN, III[2]

Abstract It commonly is thought that transgressive or regressive events which may have occurred simultaneously at geographically separated continental margins have been caused by worldwide sea-level rise or fall, respectively. Instead, it will be shown here that these events commonly may be caused by changes in the rates of sea-level rise or fall. The subsidence of an Atlantic-type (passive) margin is modeled here as a platform subsiding about a landward hingeline. The rate of subsidence is greatest at the seaward side of the platform and decreases landward to zero at the hingeline. It appears that the rate of subsidence at the seaward edge of the platform (shelf edge) is greater than the rate at which sea level may possibly rise or fall (excepting sea-level changes due to glaciation, dessication, flooding of small ocean basins, and other sudden events). Thus, if sea level is falling, the shoreline will move toward that point on the subsiding platform at which the rate of sea-level fall is equal to the rate of subsidence minus the sedimentation rate. If the rate of sea-level fall decreases the shoreline will move landward, if the rate increases the shoreline will migrate seaward. If sea level is rising the shoreline will move to that point where the rate of sea-level rise is equal to the sedimentation rate minus the subsidence rate. Thus, if the rate of sea-level rise decreases the shoreline will move seaward; if the rate increases the shoreline will move landward.

The position of the shoreline also is a function of the sedimentation rate. These relationships have been quantified so that the position of the shoreline and the thickness of the sediments deposited during discrete time intervals may be computed as a function of the rate of sea-level change and the sedimentation rate.

INTRODUCTION

The writer will first examine various models that explain eustatic sea level in terms of their quantitative significance. A hypothetical sea level curve will be developed for the past 86 m.y. based on changes in the volume of the mid-ocean ridge system. The dependence of the position of the shoreline upon changes in the rate of sea-level change will be quantified. This will be used with the calculated sea level curve to compare a hypothetical transgressive and regressive sequence for the upper Mesozoic and Cenozoic and to calculate a hypothetical stratigraphic section.

CONSIDERATION OF MECHANISMS

Eustatic sea-level changes may be caused by changes in the volume of the ocean waters or change in the volume of the ocean basins. A number of specific causes, which have been considered quantitatively and in detail by Pitman (1978), will be summarized here.

Differentiation of Mantle Materials

The plate tectonic process involves persistent convective turnover of the oceanic lithosphere including incorporated elements of the continents. As a consequence, differentiation of the mantle may occur: water may be added to the oceans at volcanic regions, or removed by hydrothermal alteration or at subduction zones. Continents may accrete at their margins (Hurley, 1968) or gain vertically by underplating (Hallam, 1971). Because the plate tectonic process may have been operating continuously for at least 2 b.y. (J. F. Dewey, personal commun.) a near steady state may have been reached. From the Phanerozoic sea-level curves presented by Hallam (1971) it may be inferred that the effect on eustacy of continued differentiation has been to increase the freeboard of the continents at a rate of about 0.02 cm/1,000 years.

[1]Manuscript received, April 11, 1978; accepted, August 1, 1978.
[2]Lamont-Doherty Geological Observatory of Columbia University, Palisades, New York 10964.
Discussions with J. Hays, W. B. F. Ryan, A. Watts, B. C. Schreiber, and J. Conolly were particularly helpful in formulating many of the ideas that are in this paper. The author is grateful to R. L. Larson and J. L. LaBrecque for their helpful reviews.

This work was supported by Contract N00014-75-C-210 from the Office of Naval Research, and by Grant DES 75 15141 from the Department of Earth Sciences of the National Science Foundation.

Lamont-Doherty Geological Observatory contribution 2738.

A more detailed version of this paper appears in The Geological Society of America Bulletin (September 1978).

Article Identification number:
0065-731X/78/MO29-0030/$03.00/0.
(see copyright notice, front of book)

Sediment Influx and Removal in the Oceans

Eroded material from the continents is continuously transported to the sea where it occupies some volume in the ocean basins. On the other hand, sediment is removed from the oceans at subduction zones. If continental accretion occurs by the lateral "plastering" of these sedimentary wedges onto the continents the result is still to reduce the volumetric capacity of the oceans. However, much of this sedimentary material is lifted to a subaerial position, thus offsetting the effect of its original emplacement. Sea level fluctuations that might be caused by the imbalance between the rates of sediment deposition and removal have been estimated to be a maximum of about 0.2 cm/1,000 yrs (Pitman, 1978).

Crustal Shortening and Thermal Welts

Continental collision phenomena (such as the Himalayan orogeny) caused an increase in the volume of the ocean basins. The great arc length of the Himalayan front is 3,000 km, the depth of the ocean basin thus created is 5.5 km; the rate of convergence has been 5 cm/year since the beginning of the Eocene. The rate at which continental freeboard would be increased is 0.16 cm/1,000 years.

Where ocean plates traverse hot spots, volcanic welts (such as the Hawaiian island chain) may be created. Assume that the welt thus created has a trapezoidal cross-section: 550 km wide at the base, 100 km wide at the top, 5.5 km high and that the rate of motion of the plate to the hot spot is 6 cm/year. The net effect would be to reduce continental freeboard about 0.02 cm/1,000 years.

Volume Changes in the Mid-Oceanic Ridge System

Menard (1963), Hallam (1963), Russell (1968), and Valentine and Moores (1970) suggested that major sea level changes may have been caused by volume changes in the mid-ocean ridge system. Hays and Pitman (1973) showed that the magnitude of the Late Cretaceous transgression could be explained in this way.

All parts of the mid-oceanic ridge system generally follow the same age-versus-depth curve. This curve gives subsidence as a function of age and may be attributed to time-dependent exponential cooling. As a consequence the cross-sectional areas of ridges are a function of the spreading rate history. A ridge that has been spreading at 6 cm/year for 70 m.y. will have three times the cross-sectional area (and hence three times the volume per unit length of a ridge that has been spreading at 2 cm/year ; Fig. 1, top). If the spreading rate of the 2 cm/year ridge is increased to 6 cm/year the cross-sectional area (and hence the volume) of the ridge will gradually be increased to that of the 6 cm/year ridge. This transition will take 70 m.y. Conversely, if the spreading rate at the 6 cm/year ridge is reduced to 2 cm/year the cross-sectional area of this ridge will gradually be reduced (Fig. 1).

Other ways of expanding or contracting the volume of the ridges are: (1) creation of a new ridge system by rifting, or (2) destruction of an old ridge system by the cessation of spreading or by subduction. The ridge volume and hence sea level is a function of spreading rate and ridge lengths at any given time.

To convert changes in ridge volume to changes in freeboard of the continents two corrections must be made. An isostatic correction because increase of the depth of the ocean waters (h) will cause subsidence of the ocean basins (d). The change in freeboard of the continents $(h-d) = 0.7h$. The second correction is because of the shape of the vessel that contains the ocean waters. This is described by the hypsometric curve (Sverdrup et al, 1942). At or near sea level, the hypsometric curve has a constant slope. Approximately as freeboard decreases (increases), the area covered by the seas increases (decreases) linearly ($0.01 \times 10^6 \text{km}^2$ for each 0.001 km change in freeboard). Thus, the change in freeboard due to a change in ridge volume may be calculated from the equation

$$\Delta V = A_o \cdot h + (0.7h)^2 \cdot \frac{170}{2}$$

ΔV = the difference in volume between the present ridges and the volume of time t.
h = the change in depth of the oceans (above or below present).
$0.7h$ = the change in freeboard of the continents with respect to the present.

Sea Level Changes from Late Mesozoic to Present (as a function of ridge volume only)

Calculated sea-level changes for the interval from −85 m.y. to −15 m.y. are shown in Figure 2. The ridge volume (and change in volume) was calculated for each 10 m.y. from −85 m.y. to −15 m.y. To compute the ridge volume back to −85 m.y., the spreading history of each of the ridge segments must be known back to −155 m.y. However, many segments of the older segments have been subducted. The spreading data for these ridge segments have been obtained by extrapolation and thus, are a likely source of significant error. The computed spreading rates also are likely to be a source of significant error as they depend on the reliability of the magnetic polarity time scale (Heirtzler et al, 1968; Larson and Pitman, 1972; and Larson and Hilde, 1975).

Key parts of the time scale have been tested paleontologically. LaBrecque et al, (1977) proposed modifications. Calibration with absolute time depends on the correlation of paleontologic age with absolute age. The time scale used for this latter purpose by Larson and Pitman (1972) is the Phanerozoic (Anonymous, 1964) as modified by Berggren (1969) for the Cenozoic. Sea-level is at a maximum in the latest Cretaceous (Maestrichtian).

Sea-level dropped rapidly during the Paleocene; the rate of fall increased slightly in the early Eocene, decreased in the late Eocene, and increased once more in the Oligocene. The rate of sea level fall is seen to be a maximum of nearly 0.67 cm/1,000 years, a rate far below that caused by the fluctuations in continental glaciation but greater by

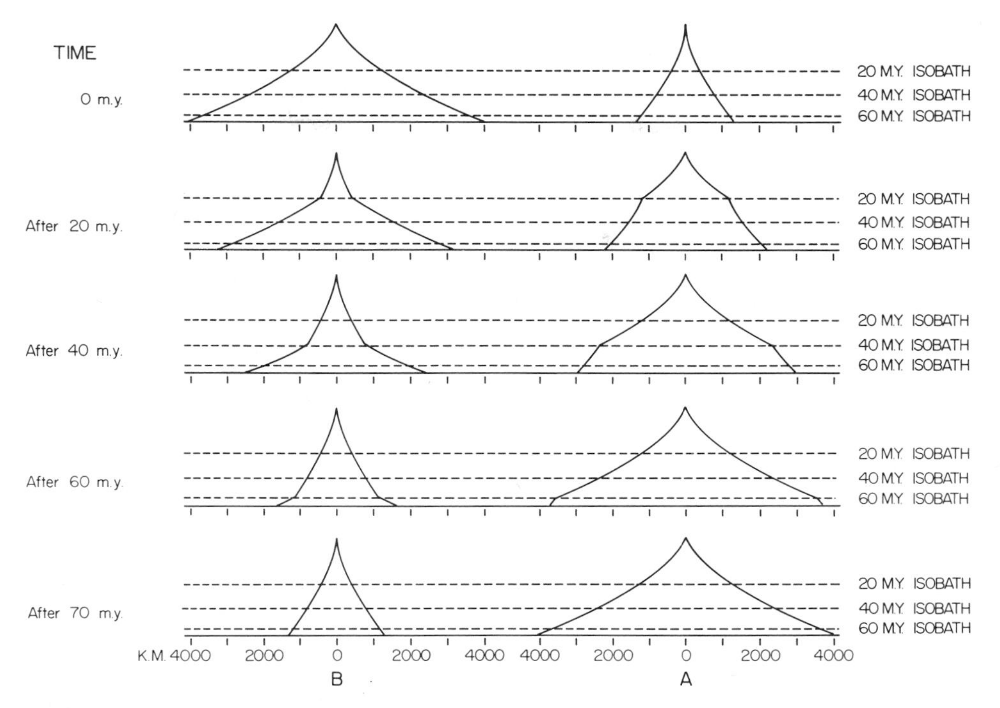

FIG. 1 **A,** at the top is the profile of a ridge that has been spreading at 2 cm/yr for 70 m.y. At time 0 m.y. the spreading rate is increased to 6 cm/year. Sequential stages in the consequent expansion of the ridge profile are shown; first at 20 m.y. after the spreading rate change; then at 40 m.y., at 60 m.y., and finally at 70 m.y., after the spreading rate change. At 70 m.y. the ridge will be at a new steady state profile; the cross sectional area at this time will be three times what it was at 0 m.y. (Modified from Pitman, 1978).

B, at the top is the profile of a ridge that has spread at 6 cm/yr for 70 m.y. At 0 m.y. the spreading rate is reduced to 2 cm/yr. The sequential stages in the subsequent contractions of the ridge are shown. At 70 m.y. after the change in spreading rate the ridge will be at a new steady state profile; the cross sectional area will be one-third of what it was at 0 m.y. (Modified from Pitman, 1978).

at least a factor of 3 than that caused by any other process.

SEDIMENTARY HISTORY OF ATLANTIC-TYPE MARGINS

Position of Shoreline on a Subsiding Shelf-Coastal Plain Platform as a Function of the Rate of Eustatic Sea-Level Change

Sloss (1962) pointed out that the occurrence of transgression and regression is in part controlled by the relative magnitude of the rate of sea-level change and the rate of subsidence. The problem is that of the interaction of a slowly subsiding continental margin platform with a more slowly changing sea level.

The general structure of Atlantic-type margins is that of a seaward thickening mass of systematically stratified sediment overlying a deeply subsided, faulted, basement platform (see Rona, 1973; Jansa and Wade, 1975; Brown et al, 1972; and Schlee et al, 1976). The sedimentary strata consist of seaward thickening wedges separated at least in the shallower sections by remarkably undisturbed planar horizons. The deepest strata are often disturbed by basement horsts and grabens, by reefal structures, and diapirs.

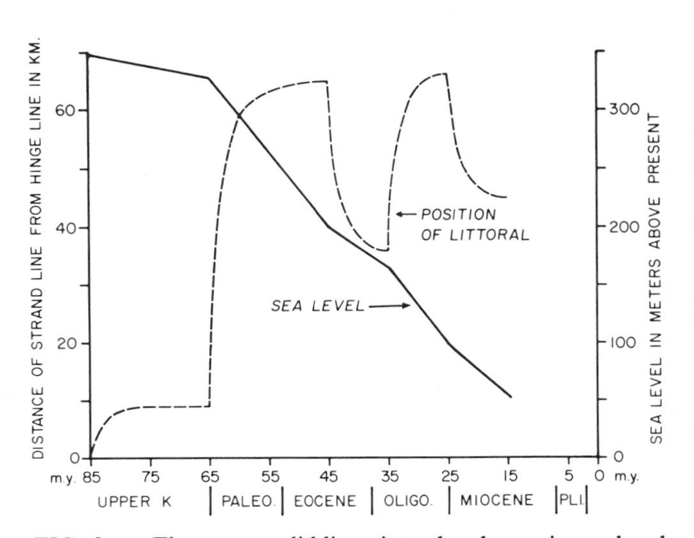

FIG. 2 The upper solid line gives the change in sea level due to the change in ridge volume for the period from −85 m.y. to −15 m.y. It is assumed that by −15 m.y., glacial buildup in Antarctica may have been sufficient that fluctuations in the glacial mass became the dominant mechanism causing sea-level changes. The dashed line gives the position of the shoreline with respect to the hingeline as a function of the rate of sea-level change (Modified from Pitman, 1978).

Borehole and other data (e.g. Fox et al, 1970; Brown et al, 1972; and Jansa and Wade, 1975) show that the entire sedimentary section that underlies the shelf (from the surface to the basement) was deposited in water depths rarely exceeding several hundred meters. This means that the basement that floors the section has subsided slowly but persistently through time from a near-sea-level position at the time of the rift-drift transition to its present depth.

Subsidence appears to be caused by two simultaneously occurring mechanisms (Sleep, 1971; Kinsman, 1975; and Watts and Ryan, 1976): the first of these is a driving subsidence which is attributed to increased density of the rifted basement caused by thermal contraction and/or phase changes; secondly, the weight of the sediment that fills the basin causes further subsidence. The total rate of subsidence is generally so slow that sediment influx is sufficient to keep pace. The rate of subsidence at the shelf edge of all margins (young or old) where such measurements have been made commonly is greater than 2 cm/1,000 years old and commonly is greater than the rate at which sea level may normally fall (or rise).

At a margin such as the east coast of the United States where the rift-drift transition probably occurred in the Early Jurassic, the post-Cretaceous subsidence rate at the shelf edge (east of Cape Hatteras) has averaged 2.5 cm/1,000 years (Rona, 1973). This means that even if sea level falls persistently for millions of years, it cannot migrate over the shelf edge. (Geologic periods of significant glacial fluctuation will not be considered here).

The rate of subsidence decreases with time, however the rate of decrease is quite slow for older margins. For simplicity, assume that the subsidence of a mature Atlantic-type margin may be modeled as a marginal platform subsiding about a fixed landward hingeline. Also assume that the rate of subsidence at the seaward edge of the platform is constant and that there is always sufficient sediment to keep pace with any combination of subsidence and sea-level rise, and that the sedimentation rates across the coastal plain and continental shelf are adjusted so as to maintain a quasiequilibrium profile (constant slope) relative to the sea surface.

Computations of the Position of the Shoreline

The sedimentation rate at any point, X, on the shelf or coastal plain needed to maintain a constant slope S_L, is $dSed/dt$

$$\frac{dSed}{dt} = \frac{X \cdot R}{D} SS - \frac{X}{D} L \cdot R_{SS} + S \tag{1}$$

To include the possibility that sediment buildup may occur on the coastal plain an additional uniform sedimentation term "s" has been added. (For definitions of the other terms see Fig. 3). In this equation the sedimentation rate is made to vary spatially such that the slope of the coastal plain–shelf surface is constant in spite of the subsidence. When $S = 0$ the sedimentation rate is such that erosion $(-dSed/dt)$ occurs landward of the shoreline and deposition $(+dSed/dt)$ occurs seaward. The rate of vertical movement (dy_{SS}/dt) of the shelf surface with respect to a horizontal plane through the hingeline equals the subsidence rate minus the sedimentation rate. The effects of compaction have not been included here.

$$\frac{dy_{SS}}{dt} = \frac{X_L \cdot R_{SS}}{D} - S$$

The rate of movement of the sea surface with respect to the shelf surface is

$$\frac{dy}{dt} WS = R_{SL} - \frac{X_L \cdot R_{SS}}{D} + S;$$

and the rate of change of the position of the shoreline along the coastal plain shelf surface is:

$$\frac{dx_L}{dt} = \frac{dy_{WS}}{dt} \div S_L; \text{ and}$$

$$\frac{dx_L}{dt} = \frac{R_{SL}}{S_L} - \frac{X_L \cdot R_{SS}}{D \cdot S_L} + \frac{S}{S_L}$$

then integrating we get

$$X_L = R_{SL}\frac{D}{R_S S} + S \cdot \frac{D}{R_{SS}} - e^{T \cdot R_{SS}/D \cdot S_L}\left(R_{SL}\frac{D}{R_{SS}} + S \cdot \frac{D}{R_{SS}} - X_{LI}\right) \tag{2}$$

X_L = the position of the shoreline after time T (T is in thousands of years). During the time interval T the rate of sea level change R_{SL} is constant and X_{LI} is the position of the shoreline at the beginning of the time interval.

If T is large, say 10^7 years (10^4 thousands of years) and selecting values of D, S_L and R_{SS} ($D = 250 \times 10^5$ cm, $S_L = 1/5{,}000$ and $R_{SS} = 2.5$ cm/1,000 yrs) that might have been typical of the East Coast of the United States in the early Tertiary and setting $S = 0$, then:

$$e - {}^{T \cdot R_{SS}/D \cdot S_L} = \frac{1}{148}$$

$$\text{thus } X_L \cong \frac{R_{SL}}{R_{SS}} \cdot D \tag{3}$$

In this case the equation is useful only when sea level is falling (when R_{SL} is positive). It means that when sea level is lowering, the shoreline will move to that point on the subsiding shelf where the rate of sea level fall is equal to the rate of shelf subsidence. If $S \neq 0$; that is, if there is significant uniform sedimentation over the shelf and coastal plain (and T is again large):

$$X_L = (R_{SL} + S)\frac{D}{R_{SS}}. \tag{4}$$

Then if sea level is falling the shoreline will be shifted seaward by a constant distance $(S/R_{SS} \times D)$ as

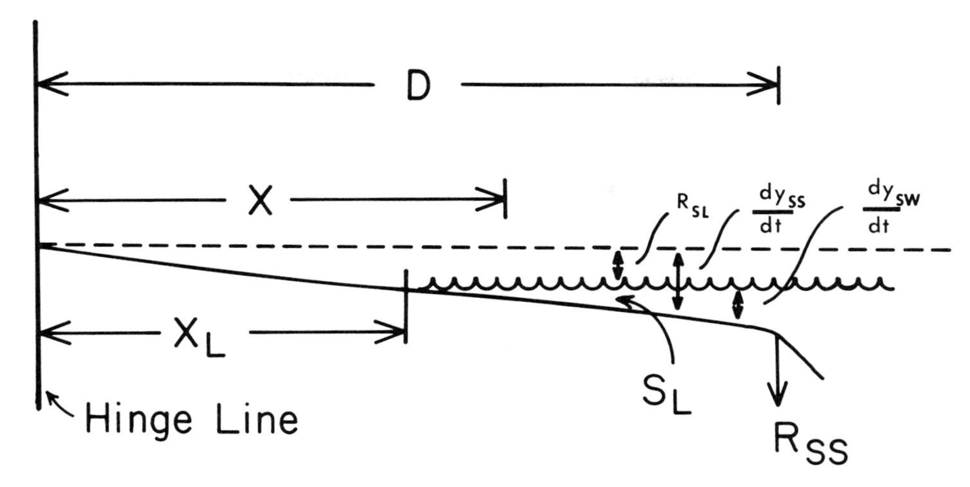

FIG. 3 A schematic and simplified model of an Atlantic-type margin is shown. It is modeled as a platform subsiding at a constant rate about a fixed hingeline. Thus, the subsidence rate decreases linearly from a maximum at the shelf edge to zero at the hingeline. It is assumed that sedimentation (and erosion) rates are distributed so as to maintain a constant slope.

D = the distance from the hingeline to the shelf edge; X_L = the distance from the hingeline to the shoreline; X = the distance from the hingeline to any point on the shelf or coastal plain; S_L = the shelf and coastal plain slope; R_{SS} = the rate of subsidence at the shelf edge of the basement platform relative to a horizontal plane that extends through the hingeline; R_{SL} = the rate of sea-level change (positive downward) relative to the same horizontal plane; $\frac{dy_{ss}}{dt}$ = the rate of vertical movement of the shelf surface with respect to the same horizontal plane; $\frac{dy_{ws}}{dt}$ = the rate of vertical movement of the sea-level surface with respect to the shelf surface; $\frac{dx}{dt}$ = the rate of movement of the shoreline with respect to the hingeline. (Modified from Pitman, 1978.)

compared with the position given by equation (3). A rise in sea level (R_{SL} is negative) may be offset by sedimentation "S" in which case the shoreline will stabilize at an equilibrium point where the rate of subsidence plus the sedimentation rate is equal to the rate of sea-level rise.

Using equation (2) and the calculated sea-level curve given in Figure 3, the position of the shoreline X_L has been calculated. The values of D, S_L, and R_{SS} used, are as given above. For the first set of computations $S = 0$ cm/1,000 years. R_{SL} has been calculated for each 10 m.y. time interval from the sea level curve of Figure 3. It is assumed that Late Cretaceous sea level was at a maximum height; that $X_{LI} = 0$ km at −85 m.y. From −85 m.y. to −65 m.y., the rate of sea level fall is 0.09 cm/1,000 years so the shoreline exponentially approaches a stable point 9 km from the hingeline. At −65 m.y., the rate of sea-level fall increases to 0.63 cm/1,000 years and so the strandline moves rapidly seaward approaching an equilibrium point 63 km from the hingeline. At −45 m.y. (mid-Eocene), the rate of sea-level fall decreased to 0.37 cm/1,000 years causing a large transgression. In the early Oligocene (at −35 m.y.) the rate of sea-level fall increased to 0.66 cm/1,000 years with a consequent regression, and in early Miocene the rate decreased again to 0.40 cm/1,000 years causing a further transgression. Since middle Miocene time sea-level rise and fall probably has been dominated by the waxing and waning of continental glaciers.

Although sea level fell continuously from −85 m.y. to −15 m.y., the rate of sea level change varied. Transgressions occur during both the late Eocene and the Miocene, because the rate of sea level fall is at a minimum during these intervals. Intervening periods are regressive because the rate of sea level fall is greater. These calculations are not accurate enough to synthesize the history of worldwide Tertiary transgressive and regressive sequences. The model used is probably too simplistic. The computed sea level curve is inaccurate because it is based on an imprecise knowledge of the ocean ridge geometry in the past; the geomagnetic reversal time scale used is inaccurate and ridge volumes have been averaged over 10 m.y. time periods. However, the calculations do illustrate the dependence of the position of the shoreline on the *rate* of sea level change and the sediment flux.

Note that the magnitude of the change in the position of the shoreline (X_L) depends on the magnitude of the change of the rate of sea-level change but it is also a function of the ratio D/R_{SS}. The larger the ratio of D/R_{SS}, the greater will be the movement of the strand line. Minor variations in spreading rates and other factors such as orogeny and sediment flux may affect the rate of sea-level change. For example, the rate of sea-level fall due to ridge contraction was 0.65 cm/1,000 years from −55 to −45 m.y. If other causes combine to change the rate of sea level fall by 0.1 cm/1,000 years every 2 m.y., such that the rate changes from 0.7 to 0.6 to 0.7 every 4 m.y., the position of the strand line will oscillate between 67.3 km and 62.7 km with the same periodicity. And in the case where $D/R_{SS} = 10^{11}$ years, the oscillation will be between 673 and 627 km. The effect of these kinds of oscillations would be to give a sawtooth shape to the strand line curves of Figure 3, which is a situation very much more in keeping with geologic reality. Changes in the rate of uniform sedimentation, "S," will have the same effect as variations in the rate of sea-level change. Taking the differential form of equation (3), but holding R_{SL} as well as D and R_{SS} constant, we get

$$X_L = S \cdot \frac{D}{R_{SS}} \qquad (5)$$

which is the same form as equation (4) and thus the same argument and conclusions are applicable.

Computation of a Theoretical Stratigraphic Cross-Section

By substituting equation (2) into equation (1) for X_L and integrating, an equation is obtained that gives the total thickness of the sedimentary section deposited and/or eroded during the time interval T at point X:

$$Sed = \frac{X \cdot R_{SS}}{D} T - R_{SL}T - \frac{D \cdot S_L}{R_{SS}} \left(R_{SL} + S - X_{LI} \frac{R_{SS}}{D} \right) (e^{TR_{SS}/-D_{SL}} - 1) \qquad (6)$$

R_{SL} must be constant during the time interval T; X_{LI} is the position of the shoreline at the beginning of the time interval, and all other parameters are as defined previously. Using this equation and the rates of sea-level change given in Figure 3, hypothetical sedimentary sections have been computed. In the first set of computations shown in Figure 4, $S = 0$ cm/1,000 years in which case the moving shoreline

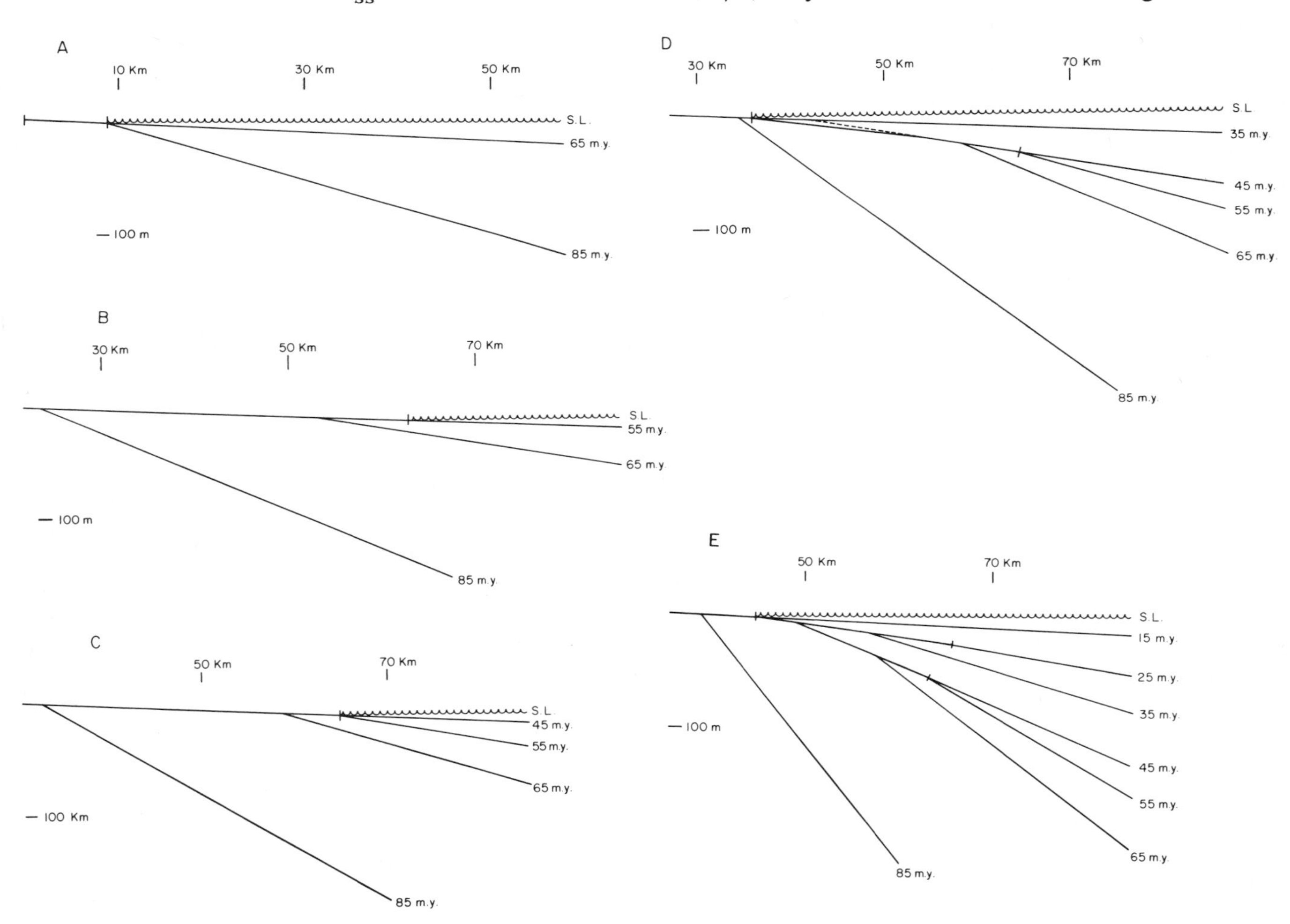

FIG. 4 Hypothetical stratigraphic sections are shown for a sequence of stages from −85 m.y. to −15 m.y. The distance from the hingeline is given in km. The only variable is the rate of sea-level fall given by the slope of the sea-level curve in Figure 2. (Modified from Pitman, 1978).

is always the boundary between erosion and deposition.

At −65 m.y., the point of pinchout is at 10 km where the net sedimentation is zero for the interval. Landward the net sedimentation is negative; seaward it is positive (Fig. 4).

The Paleocene/mid-Eocene regression (−65 m.y. to −45 m.y.) moved the shoreline rapidly seaward. Note that the point of pinchout of Paleocene deposits lies landward of the final position of the shoreline. This is because as the shoreline moved from its position at −65 m.y. to its more seaward position at −45 m.y. the shelf over which it passed experienced first deposition and then erosion. From the point of pinchout seaward to the shoreline and beyond there was net deposition; all sediments deposited within this time interval should be found in this wedge (excepting those that were subsequently reworked).

During the interval from −45 m.y. to −35 m.y., the shoreline moved rapidly landward. Again because $S = 0$ the moving shoreline acted as a boundary between a zone of erosion and a zone of deposition. The wedge of net positive sedimentation for the interval −45 to −35 m.y. (shown by the dashed line) is seen to pinch out seaward of the −35 m.y. shoreline. However, the lens of sediments of −45 m.y. to −35 m.y. age extends all the way to the shoreline. This is because during this transgressive phase the shoreline moved landward, hence the coastal plain–shelf surface experienced first erosion and finally deposition. This transgressive event was followed by a large Oligocene regression and a Miocene transgression. The comments made previously with respect to the Paleocene regression and the Eocene transgression are applicable here.

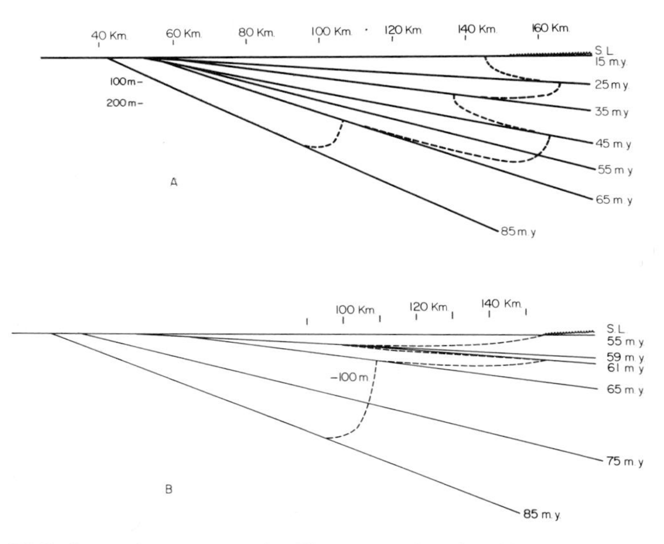

FIG. 5 **A,** same as in Figure 4E but in this case S = 1 cm/1,000 yrs. Note that the pattern of pinchout and truncation of the various sedimentary wedges is the same as in Figure 4E, but that the position of the shoreline marked by the heavy dashed line looping up through the section is pushed seaward.

B, as in Figure 5A S = 1 cm/1,000 yrs but in this case S is reduced to zero for the interval from −61 m.y. to −59 m.y. and then increased back to 1 cm/1,000 yrs. (Modified from Pitman, 1978).

The stratigraphic section has been computed for the case where $S = 1$ to simulate sedimentation occurring on a coastal plain. The geometric configuration of the resultant succession of sediment is the same as when $S = 0$ but in this case the shoreline is always 100 km farther seaward (Fig. 5A). The position of the shoreline zone up through the section is traced by the dashed line. If, as discussed (see equation 6), S is allowed to range from 1 to 0 to 1 during a 2 m.y. interval, a brief but extensive transgression occurs (Fig. 5B). In this case, where S was reduced to zero for the interval from −61 m.y. to −59 m.y., the shoreline moved over 50 km landward and (at the same time) there was over 50 km of offlap as the point of pinchout moved seaward. At −59 m.y., "S" was increased to 1 cm/1,000 years. This caused the shoreline to move 50 km seaward but simultaneously 60 km of onlap occurred. During this entire interval, sea level was falling at 0.67 cm/1,000 years.

GENERAL RESULTS AND DISCUSSION

The most important conclusion of this work is that transgressive or regressive events recorded as being synchronous at various margins may not be indicative of eustatic sea level rise or fall, respectively, but may be caused by changes in the rate of sea level change. The effect of a decrease in the rate of sea-level rise will be regressive as will that of an increase in the rate of sea-level fall. And conversely, the effect of an increase in the rate of sea-level rise or a decrease in the rate of fall will be transgressive. A key element in this model is the hypothesis that rates of sea-level change (except those due to glacial fluctuations and catastrophes) are almost always less than 2 cm/1,000 years and are thus less than the rate of subsidence at the shelf edge of Atlantic-type margins (see Fox et al, 1974). It is also assumed that sedimentation rates on subsiding margins commonly vary spatially and temporally such that a constant shelf/coastal plain slope is maintained. In the model the sedimentation rate seaward of the strand line is always positive; landward, it may be negative or positive (but varies spatially away from the shoreline so as to maintain a constant slope). If the rate of sedimentation landward of the strand line is sufficient, then where sea level is rising the strand line will stabilize at that point where the rate of sedimentation, minus the rate of subsidence, is equal to the rate of sea-level fall. In this case "S" in equation 4 must be large enough to offset a negative R_{SL}. Thus, if the rate of sea-level change varies, the strand will move to a new equilibrium point. Calculations (Fig. 2) indicate that since the Late Cretaceous time sea level may have been falling continuously. However, sea level fell slowly during the latest Cretaceous, rapidly during the Paleocene, less rapidly during the late Eocene, more rapidly during the Oligocene, and less rapidly during the early Miocene. Thus, there was a minor regression during the Late Cretaceous, a large Paleocene regression, a large Eocene transgression, an Oligocene regression, and an early Miocene transgression. Vail et al (1977) interpreted seismic stratigraphic data from a number of passive margins and basins to obtain a relative sea level curve. The Late Cretaceous–Tertiary section of their curve bears general resemblance to the sea level curve shown here in Figure 2, however their sea level curve shows considerably more detail.

The most significant source of error is the inaccuracy of the polarity-reversal time scale which has been used to calculate spreading rates and from which age-versus-depth is determined. Results of the Deep Sea Drilling Project indicate several modifications that might be made to the magnetic time scale. Also, in this model (for the sake of simplicity) the subsidence rate was assumed to be constant—in reality, the subsidence due to thermal cooling decreases exponentially with time (Sleep, 1971; Watts and Ryan, 1976). The space created by this subsidence is filled with sediment and this causes further subsidence. The subsidence due to the sediment influx may be synthesized by flexural loading of a rigid beam or by an Airy model (Watts and Ryan, 1976). Thus, even under idealized conditions of constant sediment flux, the rate of subsidence will decrease slowly with time. Also, it was assumed above that sedimentation rates are always sufficient to keep pace with subsidence; excess sediments are deposited on the slope and rise. If the sediment flux is just sufficient, there will be little or no deposition on the slope and rise, but when it is excessive a great deal of sediment may be deposited on the slope and rise. Because the flux of sediment coming into the system may vary with time the loading of the slope and rise will also vary. All of these factors must eventually be taken into consideration in attempting to construct predictive stratigraphic models such as those shown in Figures 4 and 5.

REFERENCES CITED

Anonymous, 1964, Geological society Phanerozoic time scale: Geol. Soc. London, Quart. Jour., v. 120, p. 260–262.

Berggren, W. A., 1969, Cenozoic chronostratigraphy, planktonic foraminiferal zonation and the radiometric time scale: Nature, v. 224, p. 1073 (table).

Brown, P. M., J. A. Miller, and F. M. Swain, 1972, Structural and stratigraphic framework and spatial distribution of permeability of the Atlantic coastal plain, North Carolina to New York: U.S. Geol. Survey Prof. Paper 796, U.S. Gov't Printing Office.

Fox, P. J., B. C. Heezen, and G. L. Johnson, 1970, Jurassic sandstone from the tropical Atlantic: Science, v. 170, p. 1402–1404.

Hallam, A., 1963, Major epeirogenic and eustatic changes since the Cretaceous, and their possible relationship to crustal structure: Am. Jour. Science, v. 261, p. 397–423.

——— 1971, Re-evaluation of the paleogeographic argument for an expanding earth: Nature, v. 232, p. 180–182.

Hays, J. D., and W. C. Pitman, III, 1973, Lithospheric plate motion, sea-level changes and climatic and ecological consequences: Nature, v. 246, p. 18–22.

Heirtzler, J. R., et al, 1968, Marine magnetic anomalies, geomagnetic field reversals and motions of the ocean floor and continents: Jour. Geophys. Research, v. 73, p. 2119–2136.

Holmes, A., 1965, Principles of physical geology: New York, Ronald Press, 1288 p.

Hurley, P. M., 1968, Absolute abundance and distribution of Rb, K and Sr in the earth: Geochim. et Cosmochim. Acta, v. 32, p. 273–283.

Jansa, L. F., and J. A. Wade, 1975, Geology of the continental margins off Nova Scotia and Newfoundland, *in* Offshore geology of eastern Canada: Geol. Survey Canada, Paper 74-3, v. 2, p. 51–105.

Kinsman, D. J. J., 1975, Rift valley basins and sedimentary history of trailing continental margins, *in* A. G. Fischer and S. Judson, eds., Petroleum and global tectonics: Princeton Univ. Press, 322 p.

LaBrecque, J. L., et al, 1977, Revised magnetic polarity time scale for Late Cretaceous and Cenozoic time: Geology, v. 5, no. 6, p. 330–335.

Larson, R. L., and W. C. Pitman, III, 1972, World-wide correlation of Mesozoic magnetic anomalies and its implications: Geol. Soc. America Bull., v. 83, p. 3645–3662.

——— and T. W. C. Hilde, 1975, A revised time scale of magnetic reversals for the Early Cretaceous and Late Jurassic: Jour. Geophys. Research, v. 80, p. 2586–2594.

Le Pichon, X., 1968, Sea-floor spreading and continental drift: Jour. Geophys. Research, v. 73, p. 3661–3697.

McIver, N. L., 1972, Cenozoic and Mesozoic stratigraphy of the Nova Scotia shelf: Canadian Jour. Earth Sci, v. 9, p. 54–70.

Menard, H. W., 1964, Marine geology of the Pacific: New York, McGraw Hill, 271 p.

——— 1969, Elevation and subsidence of oceanic crust: Earth and Planetary Sci. Letters, v. 6, p. 275–284.

Pitman, W. C., 1978, The relationship between eustacy and stratigraphic sequences of passive margins: Geol. Soc. America Bull., v. 89, p. 1389–1403.

Rona, P. A., 1973, Relations between rates of sediment accumulation on continental shelves, sea-floor spreading and eustacy inferred from the Central North Atlantic: Geol. Soc. America Bull., v. 84, p. 2851–2872.

Russell, K. L., 1968, Oceanic ridges and eustatic changes in sea level: Nature, v. 218, p. 861–862.

Schlee, J., et al, 1976, Regional geologic framework of northeastern U.S.: AAPG Bull., v. 60, p. 926–951.

Sclater, J. G., R. N. Anderson, and M. L. Bell, 1971, The elevation of ridges and the evolution of the central eastern Pacific: Jour. Geophys. Research, v. 76, p. 7888–7915.

Sleep, N. H., 1971, Thermal effects of the formation of Atlantic continental margins by continental breakup: Royal Astron. Soc. Geophys. Jour., v. 24, p. 325–350.

Sloss, L. L., 1962, Stratigraphic models in exploration: Jour. Sed. Petrology, v. 32, p. 415–422.

Sverdrup, H. U., M. W. Johnson, and R. H. Fleming, 1942, The oceans: Englewood Cliffs, N. J., Prentice Hall, 1087 p.

Vallentine, J. W., and E. M. Moores, 1970, Plate tectonic regulation of faunal diversity and sea level—A model: Nature, v. 228, p. 657–659.

Vail, P. R., R. M., Mitchum, Jr., and S. Thompson, 1977, Seismic stratigraphy and global changes of sea level, part 4, *in* Seismic stratigraphy—applications to hydrocarbon exploration: AAPG Memoir 26, 516 p.

Watts, A. B., and W. B. F. Ryan, 1976, Flexure of lithosphere and continental margin basins; Tectonophysics, v. 36, p. 25–44

Global Sea Level Change: A View from the Craton[2]

L. L. SLOSS[2]

Abstract Measurement of the volumes of sediment preserved in sedimentary basins permits consideration of subsidence history. Analysis of six Mesozoic-Cenozoic basins, including passive and active margins and cratonic interiors, indicates a significant agreement from basin to basin in terms of episodes of nondeposition or erosion. Further, the episodes of rapid deposition match the times of elevated sea levels established from other data. This observation suggests that subsidence of craton-interior basins and of continental margins are equally responsive to some underlying global-tectonic force and that sea-level change is a second-order concomitant. If so, the subsidence history of Paleozoic basins should be applicable to an understanding of global-tectonic modes and events in the absence of a preserved oceanic history.

Three widely separated mid-Paleozoic basins are analyzed to show the same degree of inter-basin synchrony displayed by Mesozoic-Cenozoic basins and shelves. It is natural to speculate that times of rapid subsidence of Paleozoic basins equate with times of accelerated Paleozoic plate motions, inflated mid-ocean ridge systems, and marine highstands.

INTRODUCTION

Analysis of geophysical data on continental slopes and rises is incomplete without consideration of the well-documented geologic history of adjacent continental shelves and interiors. Further, in the absence of undeformed autochthonous slope/rise rocks significantly older than Late Jurassic, it is necessary to turn to the record preserved on the continents for interpretation of much of the plate-tectonic history that is the common heritage of both oceans and continents. This paper has a three-fold purpose: (1) to review the possible relationship between events at ocean-continent margins and those of continental interiors; (2) to consider the proposition that the sedimentary-tectonic history of cratonic basins may be used as a guide to the history of pre-Jurassic margins; and (3) to point out the need for more than a single working hypothesis to explain alternations of submergence and emergence or of deposition and erosion at continental margins and within continental interiors.

MARGINS VS. INTERIORS

Visualization and conceptualization of the interplay between sea-level change and sedimentation on currently passive continental margins is relatively easy. The stratigraphic record is well preserved and commonly represents the complete spectrum of depositional environments, up to and including intertidal and supratidal regimes; chronostratigraphy and paleobathymetry are reasonably precise; intuitively, geophysical calculations of responses to loading and to subsidence of the adjacent cooling oceanic lithosphere appear to be relatively straightforward; the eustatic effect on freeboard is apparent; and processes operative over the past 150 m.y. continue to operate today.

Convergent margins are more troublesome. We have no unequivocal and currently active examples of colliding margins for study and present-day subduction tends to be concentrated along arc-trench systems at a distance from continental blocks. Where plates now converge at continental margins, the corresponding foredeep basins, (if identifiable) lie well above sea level and have not been responsive to eustatic changes for tens of millions of years. Nevertheless, analysis of ancient foredeeps (the Alpine molasse, the Cretaceous of the North American West, the Paleozoic clastic wedges of the Appalachians) leads to the rational view that the foredeeps are dragged down by the loads imposed

[1]Manuscript received, June 27, 1977; accepted, November 29, 1977.

[2]Department of Geological Sciences, Northwestern University, Evanston, Illinois 60201.

The writer is deeply indebted to James H. Fisher for supplying isopach maps of Ordovician and certain Silurian units in the Michigan basin; an equal acknowledgement is due a number of Northwestern students whose anonymous labors are represented by part of the Michigan and Elk Point Devonian data. Much of the research reported here was supported by NSF grants DES 74-22337 and EAR 76-22499.

Article Identification Number:
0065-731X/78/MO29-0031/$03.00/0.
(see copyright notice, front of book)

by orogenesis at the continental margin and that subsidence is accelerated by the mass of resultant detrital sediment. Eustatic control is considered a second-order factor in this model, modulating the more dramatic concomitants of plate interaction but remaining effective in determining the distribution of depositional environments.

Continental interiors pose more challenging problems. Here the observations, although available in indigestible abundance and subject to analysis in near-stupefying detail, do not suggest clearly readable patterns. Much of the most interesting part of cratonic history took place in Paleozoic and early Mesozoic time, a time represented by biostratigraphy of low chronologic resolution and by paleobathymetric measures that are, at best, relative rather than absolute. Worse, the stratigraphic record of cratonic interiors is interrupted by major unconformities beneath which hundreds of meters to kilometers of stratal thickness and depositional history have been stripped. Destruction of the record is particularly severe from the margins of sedimentary basins to the interiors of arches and shields and tends to have the most pervasive effect on the rocks deposited near the close of a transgressive cycle. Strandlines which existed during the early phases of a major cycle (supersequence of Mitchum, Vail, and Thompson, 1977) commonly are preserved in the sediments through protection under the cover of successive onlapping deposits; the record of later stages of a cycle is subject to increasing destruction as the sequence-bounding unconformity is approached. As a result, the evolution of cratonic interiors during the initial and medial phases of a major cycle (e.g., Cambrian, Middle Ordovician, Middle Devonian to early Viséan) can be delineated with a degree of confidence unjustified in interpretation of the latter half or third of the same cycle. Indeed, the discovery of an isolated fault block, or diatreme xenolith, or glacial erratic, can shift the purported extent of late-cycle seas by hundreds of kilometers and alter the supposed maximum elevation of sea level by tens to hundreds of meters. The net effect of inherent uncertainties is to make analysis of the changing position of sea level or base level with reference to the interiors of cratons a highly hazardous occupation capable of producing wildly inaccurate interpretations. Obviously, investigation of the course of emergence and submergence of continental interiors merits an approach that does not require tracking the migration of strandlines and nearshore environments nor detailing of onlap and offlap relationships among strata.

ANALYSIS OF MESOZOIC-CENOZOIC BASINS

Sedimentary basins of cratons are the loci of accumulation and preservation of the most complete depositional history of continental interiors. Basins are the first interior positions to subside below base level and receive sediment at the beginning of a cycle, and the last to emerge above base level and suffer erosion (i.e., the lacunas represented by regional and interregional unconformities have minimum time values in basins). Cratonic-interior sediments, with rare exceptions in starved basins, give every evidence of deposition at or close to base level—that is, in a narrow bathymetric range within meters (or at most a few tens of meters) above or below sea level. Therefore, the rate of accumulation of sediment in a basin is a measure of the rate of depression of the basin floor with respect to the position of base level at the initiation of deposition, whether such depression relates to sea-level rise or to tectonic subsidence. From isopach maps of chronostratigraphic units, the volumes of mapped units can be calculated. To the extent that the absolute time spans of such units can be measured, the rate of accumulation of each unit is determinable; plots of the accumulation rates of successive units reveal much about subsidence history without recourse to the commonly subjective interpretation of ancient strandlines or other paleogeographic reconstructions. Inasmuch as isopach maps represent *preserved* sediments only, the derived rates are minima that exclude the influence of erosion; however, in sedimentary basins erosion plays a relatively minor role in reduction of volumes and calculated subsidence rates are probably close to those actually represented, decreasing in accuracy with approach to superjacent sequence-bounding unconformities.

Figure 1 shows subsidence rates for Mesozoic-Cenozoic time derived by analysis of isopach maps of six sedimentary basins, including both cratonic-interior and cratonic-margin examples. The method and the time scale are those described in an earlier paper (Sloss, 1976). A broad spectrum of tectonic settings and degrees of completeness of data sets is represented; no data from submerged continental shelves or other submarine areas are included, nor are data from allochthonous blocks. The Aquitaine Basin was subject to severe intra-Mesozoic vertical movements, presumably in response to plate-tectonic events in the North Atlantic and Bay of Biscay. Information on the Gulf Coast Basin is constrained to a narrow belt between the outcrop edge of a particular unit and the line along which the unit passes below depths reached by the drill. Integration of thickness data on the Sahara Basin is impeded by a relative lack of exploratory drilling in critical areas. Much of the Volga-Ural (Paracaspian) Basin is covered by the Caspian Sea, whereas almost the entire extent of the West Siberian Basin is available for analysis. The Western Interior of North America is treated as a single region but, in reality, the plot displays Neocomian and older data from the Rocky Mountains and Great Plains of the U.S., and younger data from the Western Canada Basin. The split is introduced to take advantage of optimum resolution of chronostratigraphic units on either side of the border.

Considering that most of the differences among the areas illustrated have no dependence upon depositional tectonics or the position of sea level at the time of accumulation and considering further the random errors introduced by miscorrelation, inaccuracies in geochronology, and the biases of tens of

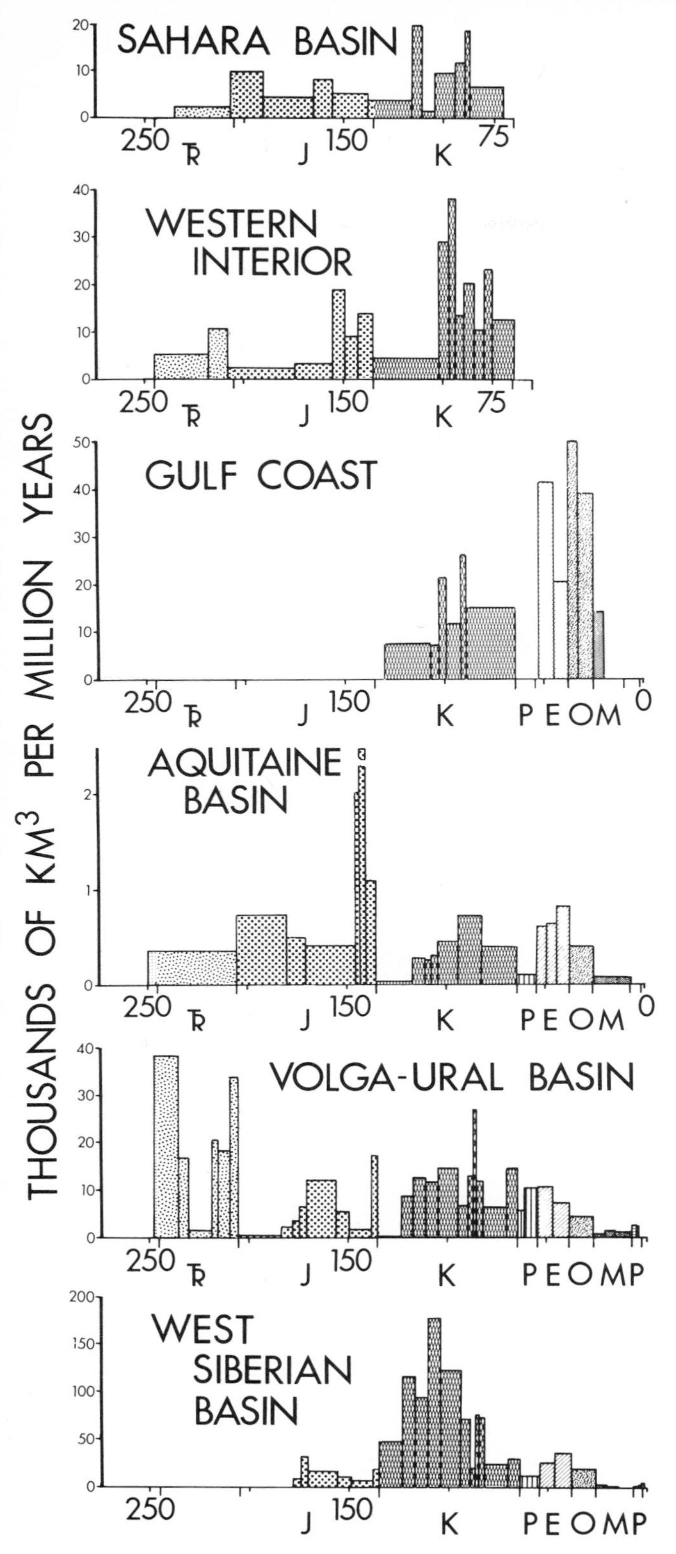

FIG. 1 Volume of sediment preserved per unit time in 6 Mesozoic-Cenozoic basins. Time scale from several sources as in Sloss (1976). Data sources: Aquitaine Basin—Bureau de Récherche Géologique et al (1974); Gulf Coast Basin—Shinn (1971), Tipsword et al (1971), Lofton and Adams (1971), Holcomb (1971), and Rainwater (1971); Sahara Basin—Busson (1972); Volga-Ural and West Siberian Basins—Kiparisova and Khabakov (1968), Krymholz, Sachs, and Tasikhin (1968), Vereshchagin and Ronov (1968), Ronov and Vereshchagin (1968), Grossheim (1967), and Khain and Eberzin (1967); Western Interior Basin—MacLachlan (1972), Peterson (1972), McGookey et al (1972), and Williams and Burk (1964).

individual map-makers, it would be difficult to predict the emergence of anything resembling a pattern. Yet, the four basins (Aquitaine, Sahara, Western Interior, and Volga-Ural) that yield data on the Triassic-Early Jurassic period indicate a marked acceleration of subsidence in Late Triassic or Early Jurassic (220 to 190 m.y.). The West Siberian, Western Interior, Volga-Ural, and Sahara Basins display a Middle Jurassic (175 to 150 m.y.) pulse of high subsidence rates. Among these basins the Western Interior and Volga-Ural are marked by a secondary and apparently short-lived latest Jurassic episode of rapid subsidence that is shared by the Aquitaine Basin and is evident in Gulf Coast history although unsupported by mapped data. All areas studied reveal a general mid-Cretaceous increase in subsidence rates, commonly comprising subsidiary peaks in Aptian-Albian time (111 to 94 m.y.) and in the Cenomanian-Santonian interval (94 to 83 m.y.). Finally, the four map sets covering Tertiary data reveal high Eocene (54 to 38 m.y.) subsidence rates; with one exception, these rates decline exponentially through the younger Tertiary. The exception is that part of the Gulf Coast discussed in this report; here, an apparent second peak in the Oligocene (38 to 26 m.y.) is exhibited.

Figure 2 is an attempt to reduce the noise level inherent in volumetric data based on stratigraphic units of irregular time values while removing the effects of scale. The plot summarizes the findings in the six basins by: (1) normalizing the volume/rate figure for each unit as a percent of the median rate for each basin; (2) viewing the data through successive ten-million-year windows; (3) assigning weighted percent-of-median values for each unit in terms of the fraction of the unit "seen" in each window; and (4) deriving the mean percent-of-median for each 10-million-year span. The resulting smoothed plot preserves many of the features discerned with greater difficulty by eye-ball analysis of the six separate plots while suppressing short-term perturbations and certain interesting idiosyncrasies of individual basins.

Thus far, my admittedly small sample suggests that a common trend exists among the six basins analyzed even though a mix of interior and pericratonic basins is represented. The suggestion is strongly reinforced by comparison with the Cretaceous subsidence data on the U.S. Atlantic Shelf by Whitten (1976) and with the plot by Vail, Mitchum, and Thompson (1977, Fig. 2) of quite different derivation. If it can be assumed that subsidence rates determined from volumes in six separated basins, plus similar rates measured by thickness on the Atlantic Shelf, display a common time trend that mimics global patterns read from the seismic stratigraphy of continental shelves, what is the significance of this shared trend? It is useful in this regard to consider the areas analyzed in terms of their positions with reference to craton interiors and continental margins.

Three study areas (Gulf Coast, Atlantic Shelf, Aquitaine Basin) are clearly astride passive margins,

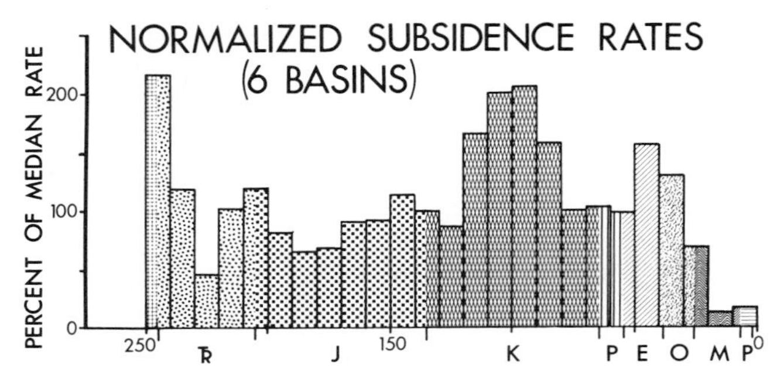

FIG. 2 Average Mesozoic-Cenozoic subsidence rates of 6 basins shown on Figure 1 expressed in 10-m.y. increments as described in text.

although the history and nature of the Gulf margin remains obscure and the Aquitaine Basin may well reflect the complex tectonics of both the Bay of Biscay and the Pyrenees. The Western Interior lies adjacent to a convergent margin, at least from Late Jurassic to Paleocene, but the westernmost deposits included in the data set were formed above cratonic crust at least 200 km from the Mesozoic margin.

The position of the Sahara Basin is obscured by lack of resolution of Tethyan and Mediterranean events but it is difficult, in the light of its northern closure to call it anything but cratonic.

Data on the Volga-Ural Basin are taken from the north end of the Caspian Sea and beyond, at a minimum distance of 550 km from the Caucasus trend and well removed from active participation in continental-margin events. It is noteworthy that, although Mesozoic volumes measured in the Volga-Ural Basin amount to only 20 percent of the total Mesozoic volume on the Russian Platform, the volume/rate plot for the Volga-Ural Basin (Fig. 1) differs in no important manner from the plot covering the entire platform (Sloss, 1976, Fig. 2). Therefore, if the subsidence history of the Volga-Ural Basin is related to the orogenic history of the Caucasus, then the other 80% of sediments deposited in subsiding areas thousands of kilometers to the north are equally related to the same plate-margin mechanism.

Last, but far from least, the West Siberian Basin is unequivocally intracratonic.

Thus, we are presented with three basinal areas directly or indirectly related to Atlantic margins, one immediately adjacent to a convergent margin, and three surrounded by craton. The Atlantic-margin examples are not united by similarities or differences not equally shared with other sites; nor is there any special kinship exhibited among craton-interior basins that sets them apart from basins at continental margins; and the convergent-margin case evinces about the same degrees of similarity and uniqueness that are documented in the other areas.

INTERPRETATION OF MESOZOIC-CENOZOIC BASINS

Several points emerge from the foregoing analysis of the subsidence history of Mesozoic and Cenozoic sedimentary basins:

1. Subsidence episodes of sedimentary basins are divisible into short-term pulses (<20 m.y.) which may be confined to individual basins or regions, and long term episodes (20 to 40 m.y.) which are evident from basin to basin and from craton to craton.
2. Long-term components of the subsidence history of sedimentary basins are independent of the positions of basins with respect to continental margins.
3. Sedimentary basins display long-term subsidence episodes in phase with continental shelf/slope records.
4. Mutually synchronous intracratonic, pericratonic, and shelf/slope subsidence episodes are grossly in phase with episodes of oceanic history reflecting eustatic highstands, accelerated spreading rates, and elevated seawater temperatures.

If a substantial part of the above represents fact, it follows that the sedimentary basins of cratonic interiors may be viewed as guides to oceanic tectonism. Thus, for that part of Phanerozoic time (pre-Late Jurassic) that is unrepresented by oceanic sediments, volcanics, or magnetic anomalies, it should be possible to reconstruct spreading and convergence rates, volumes of mid-ocean ridges, eustatic levels, and other concomitants of plate motions, by analysis of the sedimentary record preserved on cratons.

ANALYSIS AND INTERPRETATION OF MID-PALEOZOIC BASINS

Subsidence data covering three intracratonic basins from Middle Ordovician through Devonian time are presented on Figure 3. Note that mean thickness per unit time is the dependent variable of the plots rather than volume per unit time; no essential change in relative rates is introduced by using mean thickness instead of volume, and certain problems of scale are avoided. All cratonic interiors suffered one or more episodes of severe erosion, marked by the sub-Kaskaskia unconformity in North America between Early Devonian (post-Gedinnian, ~425 m.y.) and the onset of renewed subsidence in Middle or Late Devonian. The result is evident in the degree to which the underlying strata, particularly those representing Late Silurian and earliest Devonian time, have been stripped from broad areas and drastically reduced in volume in two of the study areas of Figure 3 (Williston-Elk Point and Moscow Basins). Plots of the preserved volumes show misleading orders-of-magnitude differences that are partially obviated by consideration of mean thicknesses.

The Ordovician-Silurian data of Figure 3 cover the Tippecanoe Sequence of North America and its equivalents in eastern Europe. Both the Michigan and Moscow Basins, although differently resolved by the time spans of the chronostratigraphic units mapped, display mid-Ordovician (Llanvirnian-Llandelian) pulses of subsidence not represented in the Williston-Elk Point area. All three basins indicate declining subsidence rates in the Late Ordovician or Early Silurian and marked acceleration in the Late Silurian. The timing of the extreme subsidence

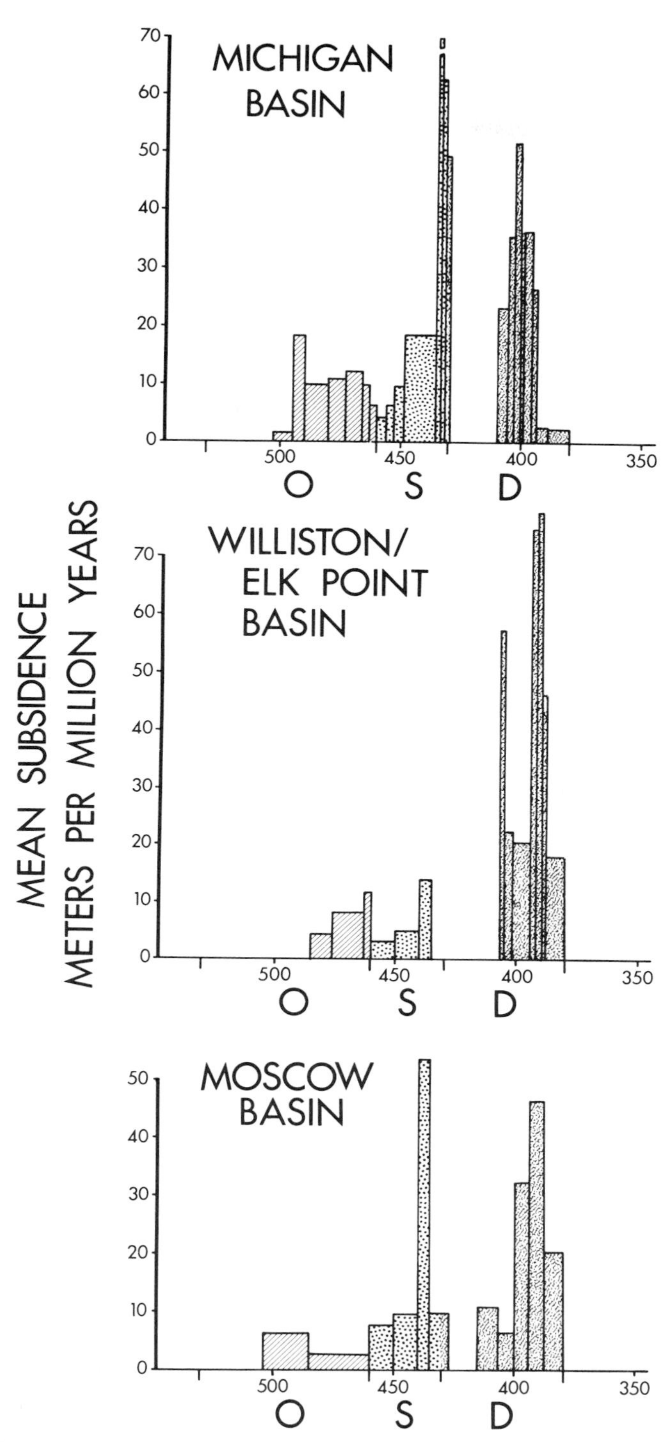

FIG. 3 Mean thickness per unit time of stratigraphic units preserved in three mid-Paleozoic basins. Time scale as in Sloss (1976). Data sources: Michigan Basin—Fisher (personal commun., 1976), Mesolella et al (1974), Allen (1974), and Gardner (1974); Williston-Elk Point Basin—Porter and Fuller (1964), Foster (1972), Gibbs (1972), Grayston et al (1964), Belyea et al (1964), and Baars (1972); Moscow Basin—Sokolov and Krylov (1968), Nikiforova and Predtechensky (1968), and Tikhy (1969).

rates registered in the Michigan Basin involves strong differences in opinion as to the correlation of thick evaporite units; nevertheless, it is apparent that high rates prevailed in late Ludlovian time (~440 to 435 m.y.) in the Michigan as in the Moscow and Williston-Elk Point Basins. Both the Michigan and Moscow Basins show evidence of sharp slowing of subsidence approaching the end of the Silurian, while an equivalent phase in the Williston-Elk Point region can only be surmised in the absence of latest Silurian strata below the sub-Kaskaskia unconformity.

The remainder of Figure 3 describes Devonian subsidence history in the three study areas. All three basins show markedly accelerated subsidence through Middle Devonian time, but continued rapid subsidence into early Late Devonian (Frasnian, ~385 m.y.), expressed in the Moscow and Williston-Elk Point Basins, is not recorded in the Michigan Basin. All three basinal areas appear to have entered an episode of sharply reduced subsidence rates before the end of Devonian time. Parenthetically, the Williston-Elk Point and Moscow Basins, which share strikingly similar Devonian subsidence histories, also exhibit radical shifts of 1,000 km in the centers of maximum subsidence between Late Silurian or earliest Devonian and Middle Devonian time (between post-Tippecanoe emergence and the beginning of Kaskaskia deposition.).

In sum, the three mid-Paleozoic intracratonic basins analyzed display degrees of similarity and idiosyncrasy not radically different from that expressed by Mesozoic-Cenozoic basins despite the fact that the state of preservation of the mid-Paleozoic record is significantly poorer and the reliability of inter-basin correlation is more questionable. Obvious similarities include synchronous erosional episodes marked by long stratigraphic lacunas and coeval pulses of rapid subsidence in Late Silurian and Middle Devonian time. An episode of slackened subsidence near the Ordovician-Silurian boundary is suggested but poorly resolved by the data presented (the occurrence of such an episode of global proportions is supported by the frequency of local unconformities at this stratigraphic position and by widespread evidence of continental glaciation). The most significant difference is in the Frasnian (early Late Devonian) cessation of rapid subsidence in the Michigan Basin while the other basinal areas continued to accomodate sediment at a high rate. The Michigan Basin deceleration is not a purely local phenomenon, as it is matched in the Illinois and Anadarko Basin areas and elsewhere in the Eastern Interior and Midcontinent.

There are, of course, real differences among basins in the timing of the initiation of renewed sedimentation following episodes of uplift and erosion. Such differences can be viewed as an indication of nonsynchronous subsidence from area to area; however, the same observations can equally well be explained, if subsidence began more or less simultaneously and some areas (e.g., Williston-Elk Point in Ordovician and Devonian) were not carried below depositional base level as soon as others.

The purely cratonic basins here analyzed for mid-Paleozoic time display periodicities and relative amplitudes of variations in rates of subsidence that are similar to those expressed by Mesozoic-Cenozoic basins. The times of most rapid subsidence of the latter basins can be shown to correlate with episodes of rapid seafloor spreading, as these are documented by the preserved chrono- and magnetostratigraphy of the oceans. It follows, therefore, that the subsidence histories of Paleozoic basins can be used as indices to the histories of long-departed oceans for which a directly decipherable record no longer exists.

DISCUSSION

The coincidence of continental-margin orogenesis and plutonism with widespread submergence of continental interiors has been noted by many, and the causal relationships between rapid spreading rates, broadly elevated mid-ocean ridges, eustatic rise of sea level, and marine transgression of the continents has been invoked repeatedly since the principle was first enunciated by Hallam (1963). Indeed, with quantification of the displacement volumes of ridges under varying spreading rates by Hays and Pitman (1973) and with the further refinement of the dependence of strandline positions on the rate of sea-level change as elegantly developed by Pitman (1978), there is wide and almost unquestioning acceptance of eustatic change as *the* control on the submergence of continental margins and interiors and on the subsidence of sedimentary basins.

I have long maintained that basin subsidence is incompletely accounted for by eustatic rise of sea level and concomitant imposition of sedimentary load. This largely intuitive view now finds more rigorous support in the work of Watts and Ryan (1976) who find it necessary to call upon unidentified "driving forces" to account for amplitudes of subsidence not attributable to sedimentary loading. Thermal contraction following a sub-basinal heating event (e.g., Sleep and Snell, 1976) has many attractions as a possible mechanism, but synchrony among widely separated basins, the repeated episodic nature of basin subsidence, and the continuation or reinitiation of subsidence over time spans too long for thermal decay reduce the credibility of the mechanism. A number of workers have invoked compressional stresses to produce the swell-and-swale geometry of alternating interior arches and basins; others note the possibility of localized tensional thinning of continental crust and consequent basinal subsidence. These models encounter difficulties related to the transmissivity of crustal stresses over sufficiently large areas to account for the synchronous behavior of distant basins. An hypothesis that offers possible acceptability in terms of the stratigraphic and historical data emphasized in this report is one that combines buckling under gravitational forces with mass transfer from the continental asthenosphere (Sloss and Speed, 1974); however, supporting geophysical data are largely unavailable.

CONCLUSIONS

1. Cratonic-interior basins and pericratonic basins adjacent to continental margins have roughly synchronous subsidence histories. Therefore, a common mechanism must exist to control both cratonic and pericratonic subsidence.

2. Mid-Paleozoic cratonic basins exhibit no marked distinctions to separate them from younger basins in rates and amplitudes of subsidence or in degrees of synchrony over wide areas. In view of conclusion 1, it is reasonable to assume that mid-Paleozoic pericratonic and continent-margin basins subsided at the same times as interior basins in response to similar processes.

3. A correlation has been established between episodes of rapid seafloor spreading and phases of rapid subsidence of Mesozoic-Cenozoic basins. It is suggested that an equivalent interdependence of oceanic and continental events can be logically extended to Paleozoic time.

4. If eustatic rise of sea level is an ineffective mechanism to explain global acceleration of basin subsidence rates, then a more satisfactory globally effective mechanism requires identification. Such a mechanism cannot be identified and made acceptable without intensive cooperation by geologists and geophysicists with specialized knowledge of oceans, continental margins, and continental interiors.

REFERENCES CITED

Allen, R. F., 1974, Stratigraphy of the Sanilac Group (Silurian) of the Michigan basin: Northwestern University, unpublished M.S. thesis.

Baars, D. J., 1972, Devonian system, *in* W. W. Mallory, ed., Geologic atlas of the Rocky Mountain region: Denver, Rocky Mtn. Assoc. Geologists, p. 90–99.

Belyea, H., et al., 1964, Upper Devonian series, *in* R. G. McCrossan, and R. P. Glaister, eds., Geological history of western Canada: Calgary, Alberta Soc. Petroleum Geologists, p. 60–88.

Bureau de Récherches Géologiques et Minières, et al., 1974, Géologie du bassin d'Aquitaine: Paris, Bur. Récherches Geol. Minieres, 26 maps.

Busson, G., 1972, Principes, méthodes et résultats d'une étude stratigraphique du Mésozoique saharien: Mus. National d'Histoire Naturelle, Mem. ser. C, v. 26, 441 pp.

Foster, N. H., 1972, Ordovician system, *in* W. W. Mallory, ed., Geologic atlas of the Rocky Mountain region: Denver, Rocky Mtn. Assoc. Geologists, p. 76–85.

Gardner, W. C., 1974, Middle Devonian stratigraphy and depositional environments in the Michigan basin: Michigan Basin Geol. Soc. Spec. Paper 1, 133 pp.

Gibbs, F. K., 1972, Silurian system, *in* W. W. Mallory, ed., Geologic atlas of the Rocky Mountain region: Denver, Rocky Mtn. Assoc. Geologists, pp. 86–89.

Grayston, L. D., D. F. Sherwin, and J. F. Allan, 1964, Middle Devonian Series, *in* R. G. McCrossan, and R. P. Glaister, eds., Geological history of western Canada: Calgary, Alberta Soc. of Petroleum Geologists, p. 49–59.

Grossheim, V. A., 1967, Paleogene series, *in* A. P. Vinogradov, ed., Atlas of the lithological-paleogeographical maps of the USSR: Moscow, USSR Ministry Geol. and USSR Acad. Sci. v. IV, maps 2–9.

Hallam, A., 1963, Major epeirogenic and eustatic changes since the Cretaceous and their possible relationship to crustal structure: Am. Jour. Science, v. 261, p. 397–423.

Hays, J. D., and W. C. Pitman, III, 1973, Lithospheric plate motion, sea-level changes and climatic and ecological consequences: Nature, v. 246, p. 18–21.

Holcomb, C. W., 1971, Hydrocarbon potential of Gulf series of western Gulf basin, *in* I. H. Cram, ed., Future petroleum provinces of the United States: AAPG Memoir 15, v. 2, p. 887–900.

Johnson, J. G., 1971, Timing and coordination of orogenic, epeirogenic, and eustatic events: Geol. Soc. America Bull.,

v. 82, p. 3263–3298.

Khain, V. E., and A. G. Eberzin, 1967, Neogene Series, *in* A. P. Vinogradov, ed., Atlas of the lithological-paleogeographical maps of the USSR: Moscow, USSR Ministry Geol. and USSR Acad Sci., v. IV, maps 10–21.

Kiparisova, L. D., and A. V. Khabakov, 1968, Triassic System, *in* A. P. Vinogradov, ed., Atlas of the lithological-paleogeographical maps of the USSR: Moscow, USSR Ministry Geol., and USSR Acad. Sci., v. III, maps 7–11.

Krymholz, G. Y., V. N. Sachs, and N. N. Tasikhin, 1968, Jurassic systems, *in* A. P. Vinogradov, ed., Atlas of the lithological-paleogeographical maps of the USSR: Moscow, USSR Ministry Geol., and USSR Acad. Sci., v. III, maps 12–23.

Lofton, C. L., and W. M. Adams, 1971, Possible future petroleum provinces of Eocene and Paleocene, western Gulf basin, *in* I. H. Cram, ed., Future petroleum provinces of the United States: AAPG Memoir 15, v. 2, p. 855–886.

MacLachlan, M. M., 1972, Triassic system, *in* W. W. Mallory, ed., Geologic atlas of the Rocky Mountain region: Denver, Rocky Mtn. Assoc. Geologists, p. 166–176.

McGookey, D. P., et al, 1972, Cretaceous system, *in* W. W. Mallory, ed., Geologic atlas of the Rocky Mountain region: Denver, Rocky Mtn. Assoc. Geologists, p. 190–199.

Mesolella, K. J., et al, 1974, Cyclic deposition of Silurian carbonates and evaporites in the Michigan basin: AAPG Bull., v. 58, p. 34–62.

Mitchum, R. M., Jr., P. R. Vail, and S. Thompson, III, 1977, The depositional sequence as a basic unit for stratigraphic analysis, *in* Seismic stratigraphy—applications to hydrocarbon exploration: AAPG Memoir 26, p. 53–62.

Nikiforova, O. I., and N. N. Predtechensky, 1968, Silurian system, *in* A. P. Vinogradov, ed., Atlas of the lithological-paleogeographical maps of the USSR: Moscow, USSR Ministry Geol., and USSR Acad. Sci., v. I, maps 26–33.

Peterson, J. A., 1972, Jurassic system, *in* W. W. Mallory, ed., Geologic atlas of the Rocky Mountain region: Denver, Rocky Mtn. Assoc. Geologists, p. 177–189.

Pitman, W. C., III, 1978, The relationship between eustacy and stratigraphic sequences of passive margins: Geol. Soc. America Bull., v. 89, no. 9, p. 1389–1403.

Porter, J. W., J. G. C. M. Fuller, and B. S. Norford, 1964, Ordovician and Silurian systems, *in* R. G. McCrossan, and R. P. Glaister, eds., Geological history of western Canada: Calgary, Alberta Soc. of Petroleum Geologists, p. 34–48.

Rainwater, E. H., 1971, Possible future petroleum potential of Lower Cretaceous, western Gulf basin, *in* I. H. Cram, ed., Future petroleum provinces of the United States: AAPG Memoir 15, v. 2, p. 901–926.

Ronov, A. B., and V. N. Vereshchagin, 1968, Upper Cretaceous series, *in* A. P. Vinogradov, ed., Atlas of the lithological-paleogeographical maps of the USSR: Moscow, USSR Ministry of Geol. USSR Acad. Sci., v. III, maps 38–50.

Shinn, A. D., 1971, Possible future petroleum potential of upper Miocene and Pliocene, western Gulf basin, *in* I. H. Cram, ed., Future petroleum provinces of the United States: AAPG Memoir 15, v. 2, p. 824–835.

Sleep, N. H., and N. S. Snell, 1976, Thermal contraction and flexure of Midcontinent and Atlantic marginal basins: Royal Astron. Soc. Geophys. Jour., v. 45, p. 125–154.

Sloss, L. L., 1976, Areas and volumes of cratonic sediments, western North America and eastern Europe: Geology, v. 4, p. 272–276.

——— and R. C. Speed, 1974, Relationships of cratonic and continental margin tectonic episodes, *in* W. R. Dickinson, ed., Tectonics and sedimentation: SEPM Spec. Pub. 22, p. 98–119.

Sokolov, B. S. and N. S. Krylov, 1968, Ordovician system, *in* A. P. Vinogradov, ed., Atlas of the lithological-paleogeographical maps of the USSR: Moscow, USSR Ministry Geol., and USSR Acad. Sci., v. I, maps 20–25.

Tikhy, V. N., 1969, Devonian system, *in* A. P. Vinogradov, ed., Atlas of the lithological-paleogeographical maps of the USSR: Moscow, USSR Ministry Geol., and USSR Acad. Sci., v. II, maps 2–11.

Tipsword, H. L., W. A. Fowler, Jr., and B. J. Sorrell, 1971, Possible future petroleum potential of lower Miocene-Oligocene, western Gulf basin, *in* I. H. Cram, ed., Future petroleum provinces of the United States: AAPG Memoir 15, v. 2, p. 836–854.

Vail, P. R., R. M. Mitchum, Jr., and S. Thompson, III, 1977, Global cycles of relative changes of sea level, *in* Seismic stratigraphy—applications to hydrocarbon exploration: AAPG Memoir 26, p. 83–97.

Vereshchagin, V. N., and A. B. Ronov, 1968, Lower Cretaceous series, *in* A. P. Vinogradov, ed., Atlas of the lithological-paleogeographical maps of the USSR: Moscow, USSR Ministry Geol., and USSR Acad. Sci., v. III, maps 28–37.

Watts, A. B., and W. B. F. Ryan, 1976, Flexure of lithosphere and continental margin basins: Tectonophysics, v. 36, p. 25–44.

Whitten, E. H. T., 1976, Cretaceous phases of rapid sediment accumulation, continental shelf, eastern U.S.A.: Geology, v. 4, p. 237–240.

Williams, G. D., and C. F. Burk, Jr., 1964, Upper Cretaceous series, *in* R. G. McCrossan, and R. P. Glaister, eds., Geological history of Western Canada: Calgary, Alberta Soc. of Petroleum Geologists, p. 169–190.

Global Cycles of Relative Changes of Sea Level from Seismic Stratigraphy[1]

P. R. VAIL[2] AND R. M. MITCHUM, JR.[2]

DISCUSSION

Cycles of relative change of sea level on a global scale are evident throughout Phanerozoic time (Vail et al, 1977). The evidence is based on the facts that many regional cycles determined on different continents are simultaneous, and that the relative magnitudes of the changes generally are similar. Because global cycles are records of geotectonic, glacial, and other large-scale processes, they reflect major events of Phanerozoic history.

A global cycle of relative change of sea level is an interval of geologic time during which a relative rise and fall of mean sea level takes place on a global scale. A global cycle may be determined from a modal average of correlative regional cycles derived from seismic stratigraphic studies.

On a global cycle curve for Phanerozoic time, three major orders of cycles are superimposed on the sea-level curve. Cycles of first, second, and third order have durations of 200 to 300 million, 10 to 80 million, and 1 to 10 million years, respectively (Figs. 1,2, and 3). Two cycles of the first order, over 14 of the second order and approximately 80 of the third order, are present in the Phanerozoic (not counting late Paleozoic cyclothems). Third-order cycles for the pre-Jurassic and Cretaceous are not shown. Sea-level changes from Cambrian through Early Triassic are not as well documented globally as are those from Late Triassic through Holocene.

Relative changes of sea level from Late Triassic to the present are reasonably well documented with respect to the ages, durations, and relative magnitudes of the second- and third-order cycles. Magnitudes of eustatic changes of sea level are only approximations. Our best estimate is that sea level reached a high point near the end of the Campanian (Late Cretaceous) about 350 m above present sea level, and had low points during the Early Jurassic, middle Oligocene, and late Miocene about 150, 250, and 200 m, respectively, below present sea level.

Facies and general patterns of distribution of many depositional sequences are related to cycles of global highstands and lowstands of sea level. Interregional unconformities occur at times of lowstand. Geotectonic and glacial phenomena are the most likely causes of the sea-level cycles, although there may be other as yet unrecognized factors.

Major applications of the global cycle chart include (1) improved stratigraphic and structural analysis within a basin, (2) estimation of the geological age of strata prior to drilling, and (3) development of a global system of geochronology.

REFERENCES CITED

Foreman, H. P., 1973, Radiolaria of leg 10 with systematics and ranges for the families *Amphipyndacidae, Artostrobiidae,* and *Theoperidae*: Deep Sea Drilling Project, leg 10, *in* Initial reports of the Deep Sea Drilling Project: Washington, D.C., U.S. Govt., v. 10, p. 407–474.

Hardenbol, J., and W. A. Berggren, 1978, A new Paleogene numerical time scale, *in* Contributions to the geologic time scale: AAPG Studies in Geology 6, p. 213–234.

Moore, T., 1971, Radiolaria, Deep Sea Drilling Project, leg 8, *in* Initial reports of the Deep Sea Drilling Project; Washington, D.C., U.S. Govt., v. 8, p. 727–748.

Riedel, W. R., and A. Sanfilippo, 1970, Radiolaria, Deep Sea Drilling Project, leg 4, *in* Initial reports of the Deep Sea Drilling Project: Washington, D.C., U.S. Govt., v. 4, p. 503–575.

——— ——— 1971, Cenozoic radiolaria from the western tropical Pacific, Deep Sea Drilling Project, leg 7, *in* Initial reports of the Deep Sea Drilling Project: Washington, D.C., U.S. Govt., p. 1529–1672.

Ryan, W. F. B., et al, 1974, A paleomagnetic assignment of Neogene stage boundaries and the development of isochronous datum planes between the Mediterranean, the Pacific, and Indian oceans in order to investigate the response of the world ocean to the Mediterranean salinity crisis: Riv. Italiana Paleontologia, v. 80, no. 4, p. 631–688.

Theyer, F., and S. R. Hammond, 1974, Cenozoic magnetic time scale in deep-sea cores—completion of the Neogene: Geology, v. 2, no. 10, p. 487–492.

Vail, P. R., et al, 1977, Seismic stratigraphy and global changes of sea level, *in* Seismic stratigraphy—applications to hydrocarbon exploration: AAPG Memoir 26, p. 49–212.

Van Hinte, J. E., 1976a, A Jurassic time scale: AAPG Bull., v. 60, p. 489–497.

——— 1976b, A Cretaceous time scale: AAPG Bull., v. 60, p. 498–516.

[1]Manuscript received, January 5, 1978; accepted, January 21, 1978.

[2]Exxon Production Research Co., Houston, Texas 77001.

Article Identification Number:
0065-731X/78/MO29-0032/$03.00/0.
(see copyright notice, front of book)

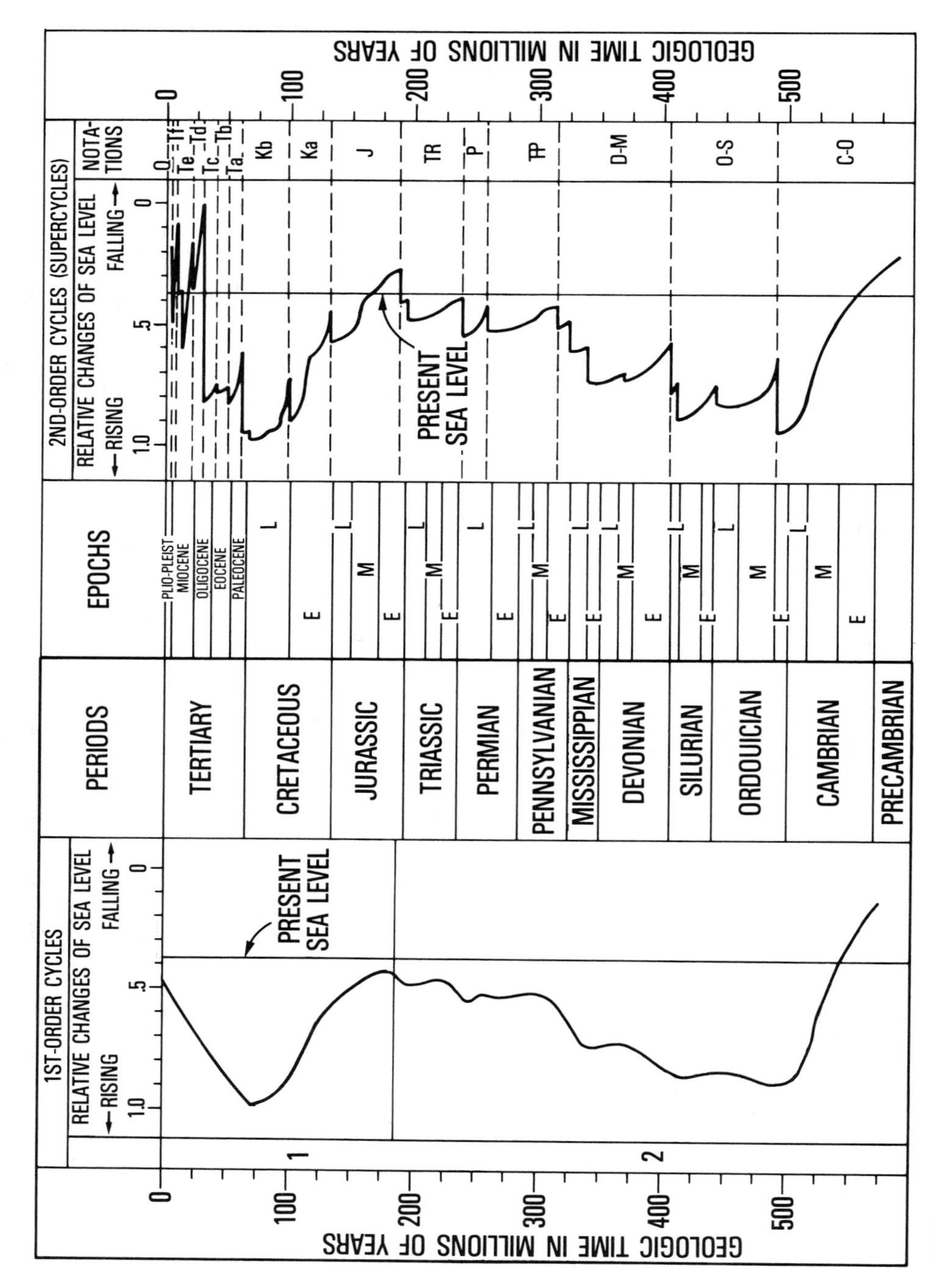

FIG. 1 First- and second-order global cycles of relative change of sea level during Phanerozoic time (Modified from Vail et al, 1977).

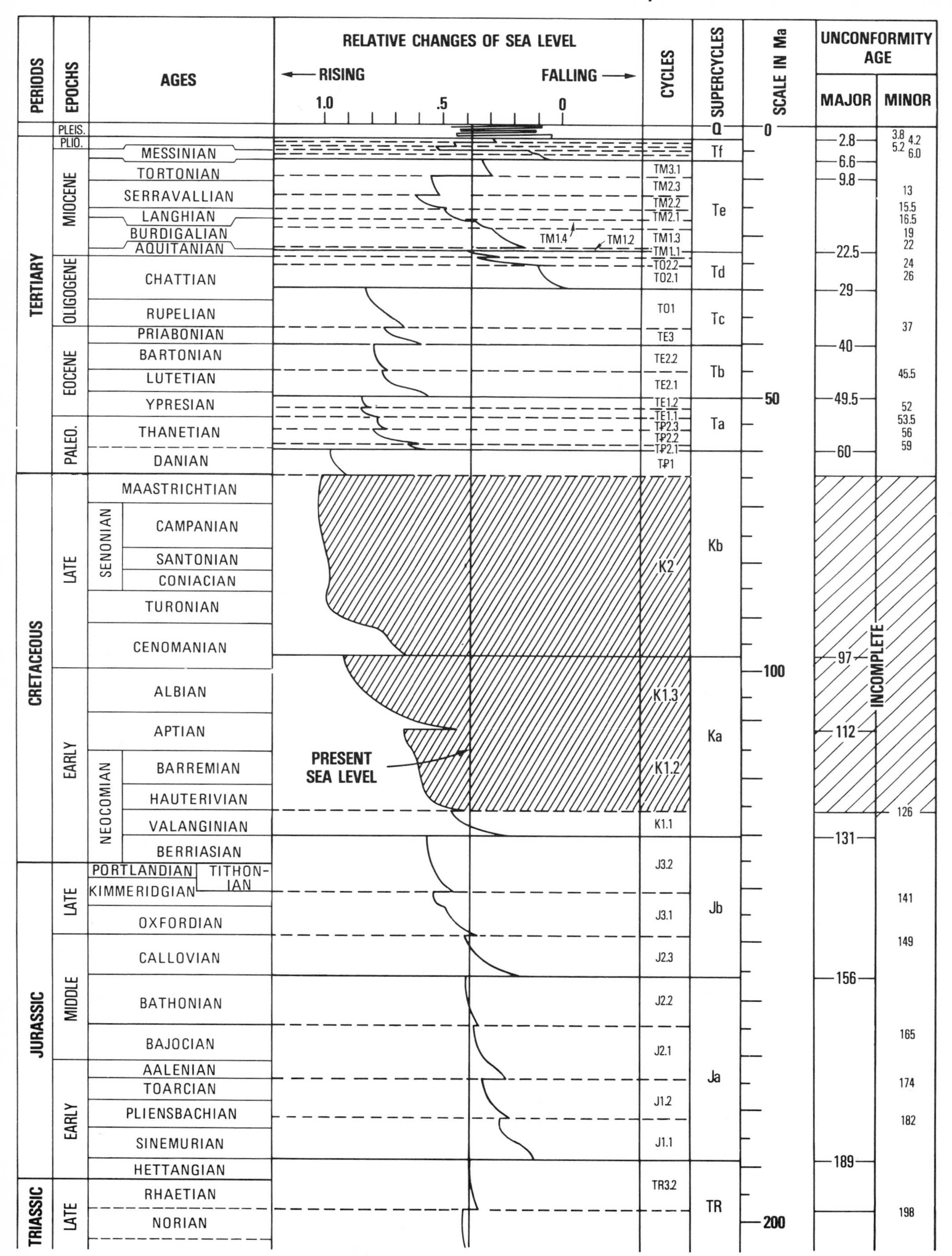

FIG. 2 Global cycles of relative change of sea level during Jurassic-Tertiary time. Cretaceous cycles (hatchured area) have not been released for publication (Modified from Vail et al, 1977).

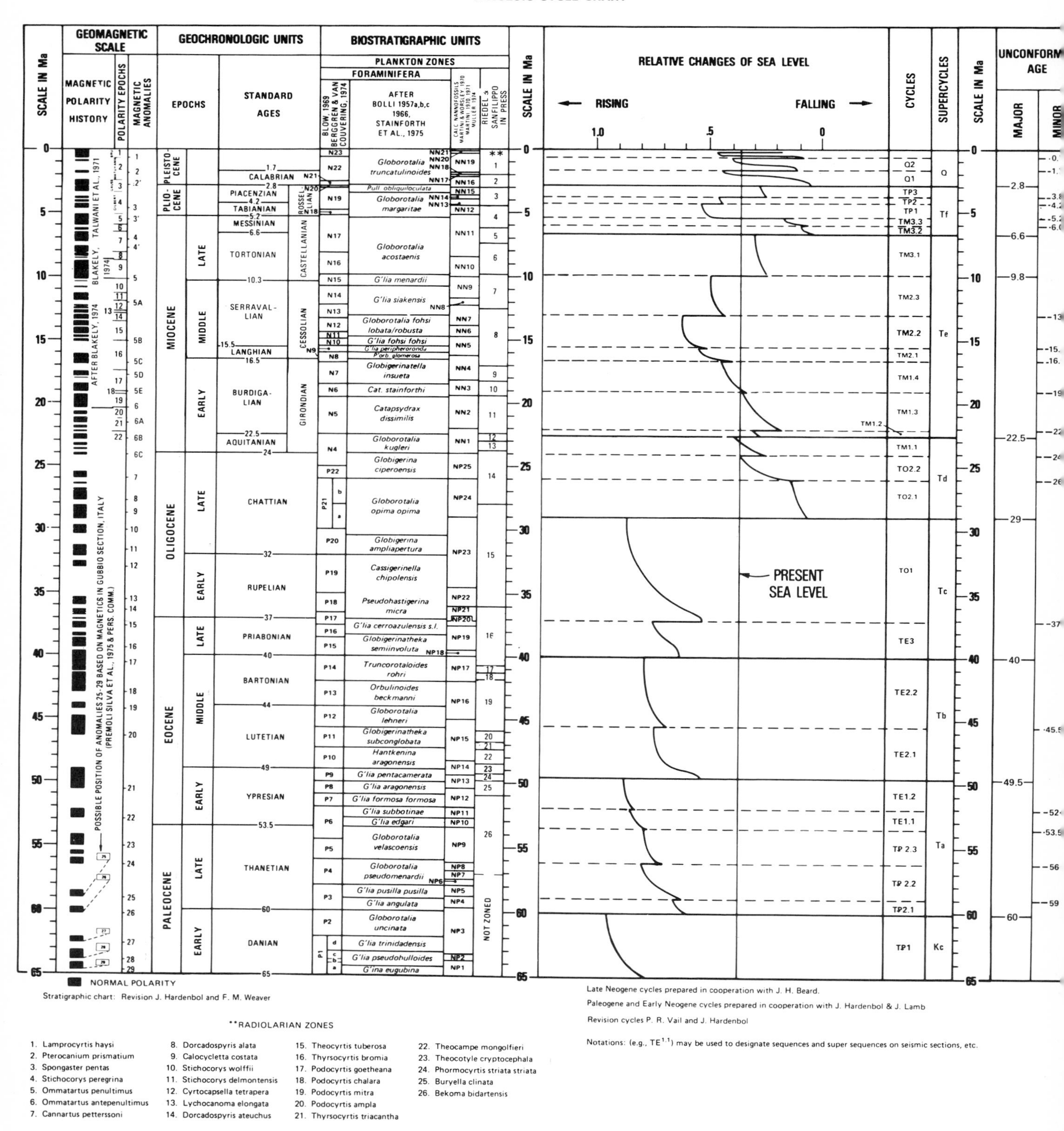

FIG. 3 Global cycles of relative change of sea level during Cenozoic time. Basic references for the stratigraphic part of the chart are Hardenbol (unpublished after Ryan et al, 1974) and Hardenbol and Berggren (1977) (Modified from Vail, et al, 1977).

Index

Explanation of Indexing

A reference is indexed according to its important, or "key" words. Authors and titles are also represented here; where more than one author has contributed to a paper, each person is cited, alphabetically, according to his last name.

Three columns are to the left of the keyword entries. The first column, a letter entry, represents the AAPG book series from which the reference originated. In this case, M stands for Memoir Series. Every five years, AAPG merges all its indexes together, and the letter M will differentiate this reference from those of the AAPG Studies in Geology Series (S) or from the AAPG Bulletin (B).

The following number is the series number. In this case, 29 represents a reference from Memoir 29.

The last column entry is the page number in this volume where the reference will be found.

A small dagger symbol (†) is used to highlight a manuscript title entry.

Note: This index is set up for single-line entry. Where entries exceed one line of type, the line is terminated. (This is especially evident with manuscript titles, which tend to be long and descriptive.) The reader sometimes must be able to realize key-words, although commonly taken out of context.